Dick Aufmann

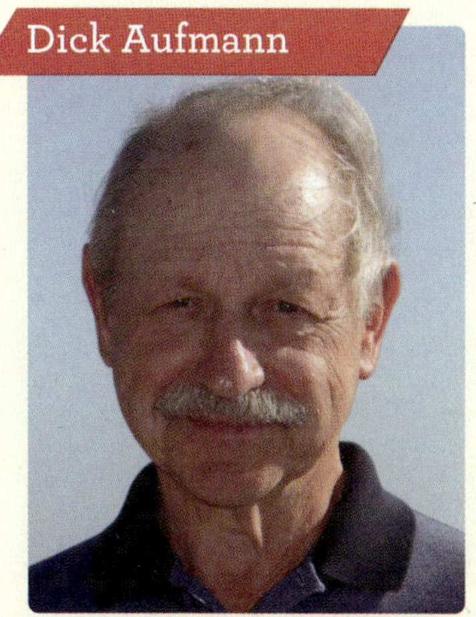

Joanne Lockwood

We have taught math for many years. During that time, we have had students ask us a number of questions about mathematics and this course. Here you find some of the questions we have been asked most often, starting with the big one.

Why do I have to take this course? You may have heard that *"Math is everywhere."* That is probably a slight exaggeration but math does find its way into many disciplines. There are obvious places like engineering, science, and medicine. There are other disciplines such as business, social science, and political science where math may be less obvious but still essential. If you are going to be an artist, writer, or musician, the direct connection to math may be even less obvious. Even so, as art historians who have studied the Mona Lisa have shown, there is a connection to math. But, suppose you find these reasons not all that compelling. **There is still a reason to learn basic math skills: You will be a better consumer and able to make better financial choices for you and your family.** For instance, is it better to buy a car or lease a car? Math can provide an answer.

I find math difficult. Why is that? It is true that some people, even very smart people, find math difficult. Some of this can be traced to previous math experiences. If your basic skills are lacking, it is more difficult to understand the math in a new math course. Some of the difficulty can be attributed to the ideas and concepts in math. They can be quite challenging to learn. Nonetheless, most of us can learn and understand the ideas in the math courses that are required for graduation. **If you want math to be less difficult, practice. When you have finished practicing, practice some more.** Ask an athlete, actor, singer, dancer, artist, doctor, skateboarder, or (name a profession) what it takes to become successful and the one common characteristic they all share is that they practiced—a lot.

Why is math important? As we mentioned earlier, math is found in many fields of study. There are, however, other reasons to take a math course. Primary among these reasons is to become a better problem solver. Math can help you learn critical thinking skills. It can help you develop a logical plan to solve a problem. Math can help you see relationships between ideas and to identify patterns. **When employers are asked what they look for in a new employee, being a problem solver is one of the highest ranked criteria.**

What do I need to do to pass this course? The most important thing you must do is to know and understand the requirements outlined by your instructor. These requirements are usually given to you in a syllabus. Once you know what is required, you can chart a course of action. Set time aside to study and do homework. If possible, choose your classes so that you have a free hour after your math class. Use this time to review your lecture notes, rework examples given by the instructor, and to begin your homework. All of us eventually need help, so know where you can get assistance with this class. This means knowing your instructor's office hours, the hours of the math help center, and how to access available online resources. And finally, do not get behind. **Try to do some math EVERY day, even if it is for only 20 minutes.**

Prealgebra and Introductory Algebra

An Applied Approach

SECOND EDITION

Richard N. Aufmann
Palomar College

Joanne S. Lockwood
Nashua Community College

Australia • Brazil • Japan • Korea • Mexico • Singapore • Spain • United Kingdom • United States

***Prealgebra and Introductory Algebra: An Applied Approach*, Second Edition**
Richard N. Aufmann and Joanne S. Lockwood

Acquisitions Editor: Marc Bove

Developmental Editor: Erin Brown

Assistant Editor: Shaun Williams

Editorial Assistant: Kyle O'Loughlin

Media Editor: Heleny Wong

Marketing Manager: Gordon Lee

Marketing Assistant: Angela Kim

Marketing Communications Manager: Katy Malatesta

Content Project Manager: Cheryll Linthicum

Creative Director: Rob Hugel

Art Director: Vernon Boes

Print Buyer: Becky Cross

Rights Acquisitions Account Manager, Text: Roberta Broyer

Rights Acquisitions Account Manager, Image: Don Schlotman

Production Service: Graphic World Inc.

Text Designer: The Davis Group

Photo Researcher: Scott Rosen

Copy Editor: Jean Bermingham

Illustrator: Graphic World Inc.

Cover Designer: Irene Morris

Cover Image: © iStockphoto exclusive/Corbis

Compositor: Graphic World Inc.

Library of Congress Control Number: 2010922076

ISBN-13: 978-0-8400-4808-0

ISBN-10: 0-8400-4808-4

Brooks/Cole
20 Davis Drive
Belmont, CA 94002-3098
USA

Cengage Learning is a leading provider of customized learning solutions with office locations around the globe, including Singapore, the United Kingdom, Australia, Mexico, Brazil, and Japan. Locate your local office at **www.cengage.com/global**

Cengage Learning products are represented in Canada by Nelson Education, Ltd.

To learn more about Brooks/Cole, visit **www.cengage.com/brookscole**
Purchase any of our products at your local college store or at our preferred online store **www.CengageBrain.com**

Printed in the United States of America
4 5 6 7 14 13 12

Contents

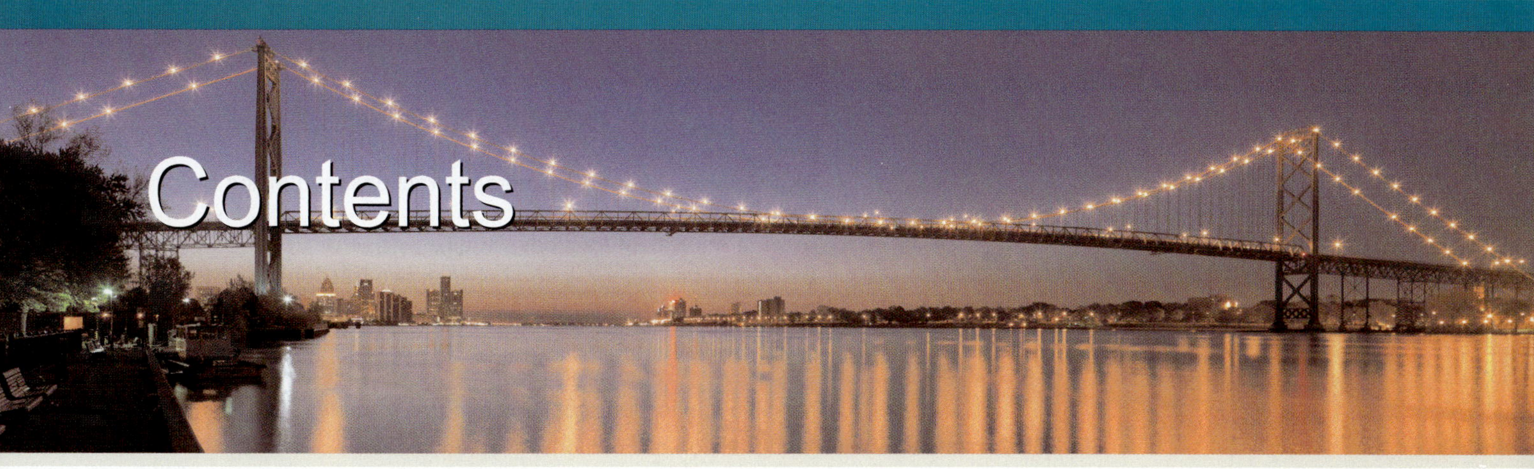

© iStockphoto exclusive/Corbis

CHAPTER 16 Quadratic Equations 789

CHAPTER R Review Topics 831

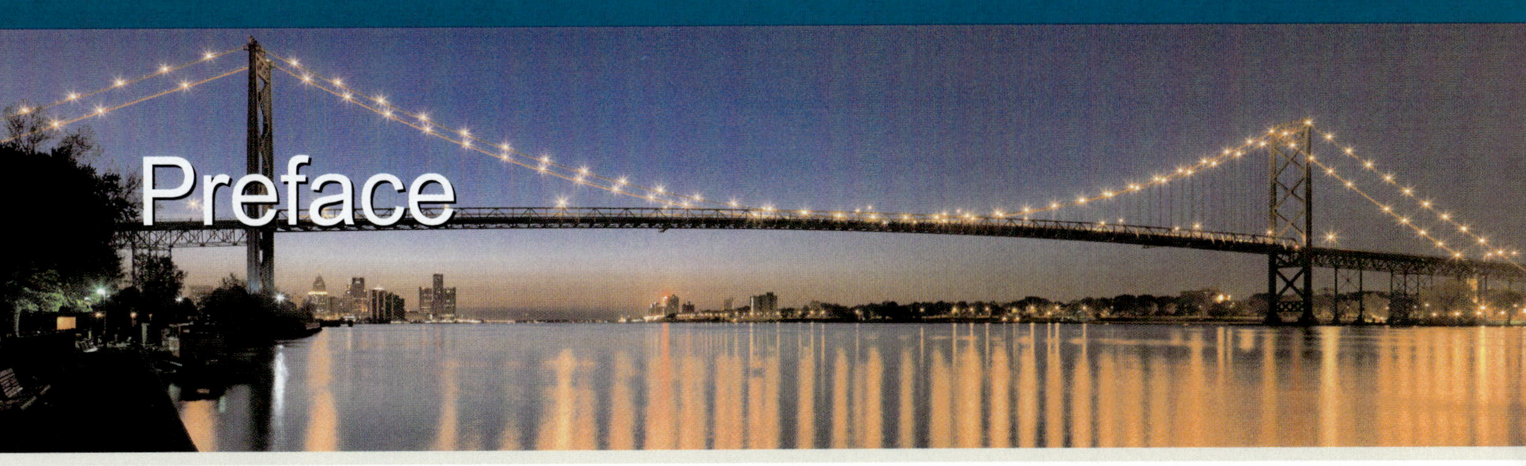

The goal in any textbook revision is to improve upon the previous edition, taking advantage of new information and new technologies, where applicable, in order to make the book more current and appealing to students and instructors. While change goes hand-in-hand with revision, a revision must be handled carefully, without compromise to valued features and pedagogy. In the second edition of *Prealgebra and Introductory Algebra,* we endeavored to meet these goals.

As in previous editions, the focus remains on the **Aufmann Interactive Method (AIM).** Students are encouraged to be active participants in the classroom and in their own studies as they work through the How To examples and the paired Examples and You Try It problems. The role of "active participant" is crucial to success. Providing students with worked examples, and then affording them the opportunity to immediately work similar problems, helps them build their confidence and eventually master the concepts.

To this point, simplicity plays a key factor in the organization of this edition, as in all other editions. All lessons, exercise sets, tests, and supplements are organized around a carefully constructed hierarchy of objectives. This "objective-based" approach not only serves the needs of students, in terms of helping them to clearly organize their thoughts around the content, but instructors as well, as they work to design syllabi, lesson plans, and other administrative documents.

In order to enhance the AIM and the organization of the text around objectives, we have introduced a new design. We believe students and instructors will find the page even easier to follow. Along with this change, we have introduced several new features and modifications that we believe will increase student interest and renew the appeal of presenting the content to students in the classroom, be it live or virtual.

Changes to the Second Edition

With the second edition, previous users will recognize many of the features that they have come to trust. Yet, they will notice some new additions and changes:

- Enhanced WebAssign® now accompanies the text
- Revised exercise sets with new applications
- New **In the News** applications
- New **Think About It** exercises
- Revised Chapter Review Exercises and Chapter Tests
- End-of-chapter materials now include Concept Reviews
- Revised Chapter Openers, now with Prep Tests

Take AIM and Succeed!

Prealgebra and Introductory Algebra is organized around a carefully constructed hierarchy of **OBJECTIVES**. This "objective-based" approach provides an integrated learning environment that allows students and professors to find resources such as assessment (both within the text and online), videos, tutorials, and additional exercises.

Chapter Openers are set up to help you organize your study plan for the chapter. Each opener includes: Objectives, Are You Ready? and a Prep Test.

Each Chapter Opener outlines the **OBJECTIVES** that appear in each section. The list of objectives serves as a resource to guide you in your study and review of the topics.

ARE YOU READY? outlines what you need to know to be successful in the coming chapter.

Complete each **PREP TEST** to determine which topics you may need to study more carefully, versus those you may only need to skim over to review.

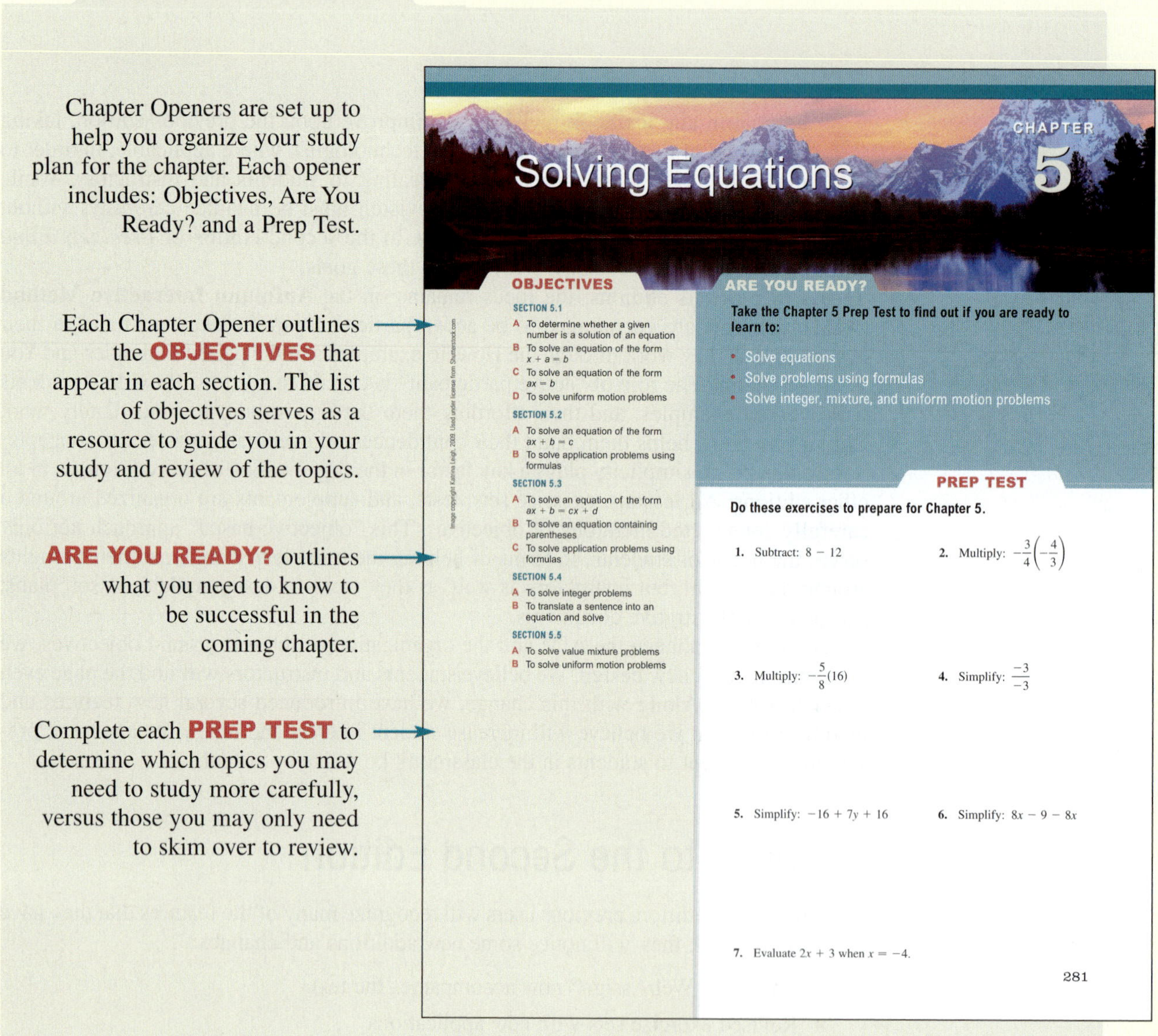

Solving Equations
CHAPTER 5
OBJECTIVES
SECTION 5.1
A To determine whether a given number is a solution of an equation
B To solve an equation of the form x + a = b
C To solve an equation of the form ax = b
D To solve uniform motion problems
SECTION 5.2
A To solve an equation of the form ax + b = c
B To solve application problems using formulas
SECTION 5.3
A To solve an equation of the form ax + b = cx + d
B To solve an equation containing parentheses
C To solve application problems using formulas
SECTION 5.4
A To solve integer problems
B To translate a sentence into an equation and solve
SECTION 5.5
A To solve value mixture problems
B To solve uniform motion problems
ARE YOU READY?
Take the Chapter 5 Prep Test to find out if you are ready to learn to:
Solve equations
Solve problems using formulas
Solve integer, mixture, and uniform motion problems
PREP TEST
Do these exercises to prepare for Chapter 5.
1. Subtract: 8 − 12
2. Multiply: −3/4(−4/3)
3. Multiply: −5/8(16)
4. Simplify: −3/−3
5. Simplify: −16 + 7y + 16
6. Simplify: 8x − 9 − 8x
7. Evaluate 2x + 3 when x = −4.
281

302 CHAPTER 5 • Solving Equations

SECTION

5.3 General Equations—Part II

OBJECTIVE A To solve an equation of the form $ax + b = cx + d$

Tips for Success
Have you considered joining a study group? Getting together regularly with other students in the class to go over material and quiz each other can be very beneficial. See *AIM for Success* in the front of the book.

In solving an equation of the form $ax + b = cx + d$, the goal is to rewrite the equation in the form *variable* = *constant*. Begin by rewriting the equation so that there is only one variable term in the equation. Then rewrite the equation so that there is only one constant term.

HOW TO 1 Solve: $2x + 3 = 5x - 9$

$2x + 3 = 5x - 9$

$2x - 5x + 3 = 5x - 5x - 9$ • Subtract $5x$ from each side of the equation.

$-3x + 3 = -9$ • Simplify. There is only one variable term.

$-3x + 3 - 3 = -9 - 3$ • Subtract 3 from each side of the equation.

$-3x = -12$ • Simplify. There is only one constant term.

$\frac{-3x}{-3} = \frac{-12}{-3}$ • Divide each side of the equation by -3.

$x = 4$ • The equation is in the form *variable* = *constant*.

The solution is 4. You should verify this by checking this solution.

EXAMPLE 1
Solve: $4x - 5 = 8x - 7$

Solution

$4x - 5 = 8x - 7$

$4x - 8x - 5 = 8x - 8x - 7$ • Subtract $8x$ from each side.

$-4x - 5 = -7$

$-4x - 5 + 5 = -7 + 5$ • Add 5 to each side.

$-4x = -2$

$\frac{-4x}{-4} = \frac{-2}{-4}$ • Divide each side by -4.

YOU TRY IT 1
Solve: $5x + 4 = 6 + 10x$

Your solution

In each section, **OBJECTIVE STATEMENTS** introduce each new topic of discussion.

In each section, the **HOW TO'S** provide detailed explanations of problems related to the corresponding objectives.

The **EXAMPLE/YOU TRY IT** matched pairs are designed to actively involve you in learning the techniques presented. The You Try Its are based on the Examples. They appear side-by-side so you can easily refer to the steps in the Examples as you work through the You Try Its.

Complete, **WORKED-OUT SOLUTIONS** to the You Try It problems are found in an appendix at the back of the text. Compare your solutions to the solutions in the appendix to obtain immediate feedback and reinforcement of the concept(s) you are studying.

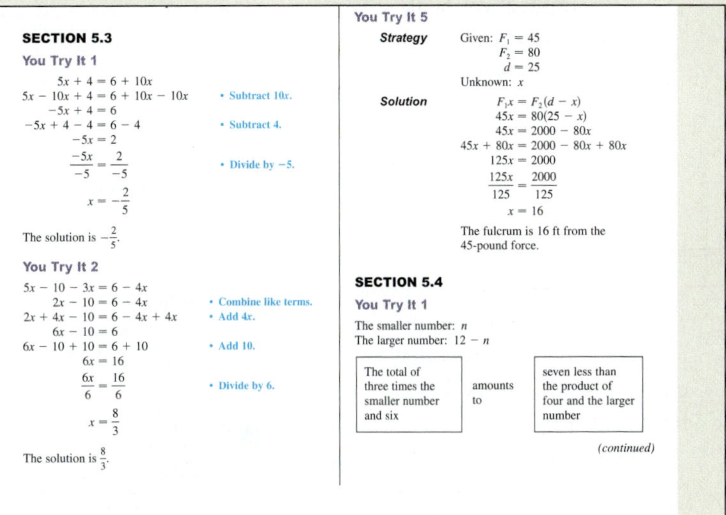

SECTION 5.3

You Try It 1

$5x + 4 = 6 + 10x$

$5x - 10x + 4 = 6 + 10x - 10x$ • Subtract $10x$.

$-5x + 4 = 6$

$-5x + 4 - 4 = 6 - 4$ • Subtract 4.

$-5x = 2$

$\frac{-5x}{-5} = \frac{2}{-5}$ • Divide by -5.

$x = -\frac{2}{5}$

The solution is $-\frac{2}{5}$.

You Try It 2

$5x - 10 - 3x = 6 - 4x$

$2x - 10 = 6 - 4x$ • Combine like terms.

$2x + 4x - 10 = 6 - 4x + 4x$ • Add $4x$.

$6x - 10 = 6$

$6x - 10 + 10 = 6 + 10$ • Add 10.

$6x = 16$

$\frac{6x}{6} = \frac{16}{6}$ • Divide by 6.

$x = \frac{8}{3}$

The solution is $\frac{8}{3}$.

You Try It 5

Strategy Given: $F_1 = 45$, $F_2 = 80$, $d = 25$
Unknown: x

Solution

$F_1x = F_2(d - x)$

$45x = 80(25 - x)$

$45x = 2000 - 80x$

$45x + 80x = 2000 - 80x + 80x$

$125x = 2000$

$\frac{125x}{125} = \frac{2000}{125}$

$x = 16$

The fulcrum is 16 ft from the 45-pound force.

SECTION 5.4

You Try It 1

The smaller number: n
The larger number: $12 - n$

The total of three times the smaller number and six	amounts to	seven less than the product of four and the larger number

(continued)

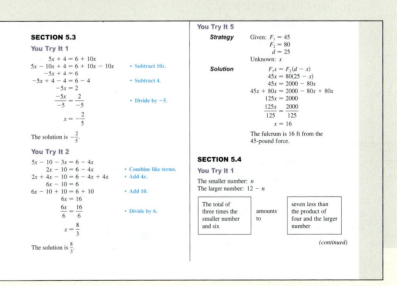

Prealgebra and Introductory Algebra: An Applied Approach contains **A WIDE VARIETY OF EXERCISES** that promote skill building, skill maintenance, concept development, critical thinking, and problem solving.

THINK ABOUT IT exercises promote conceptual understanding. Completing these exercises will deepen your understanding of the concepts being addressed.

Completing the **WRITING** exercises will help you to improve your communication skills, while increasing your understanding of mathematical concepts.

Working through the application exercises that contain **REAL DATA** will help prepare you to answer questions and/or solve problems based on your own experiences, using facts or information you gather.

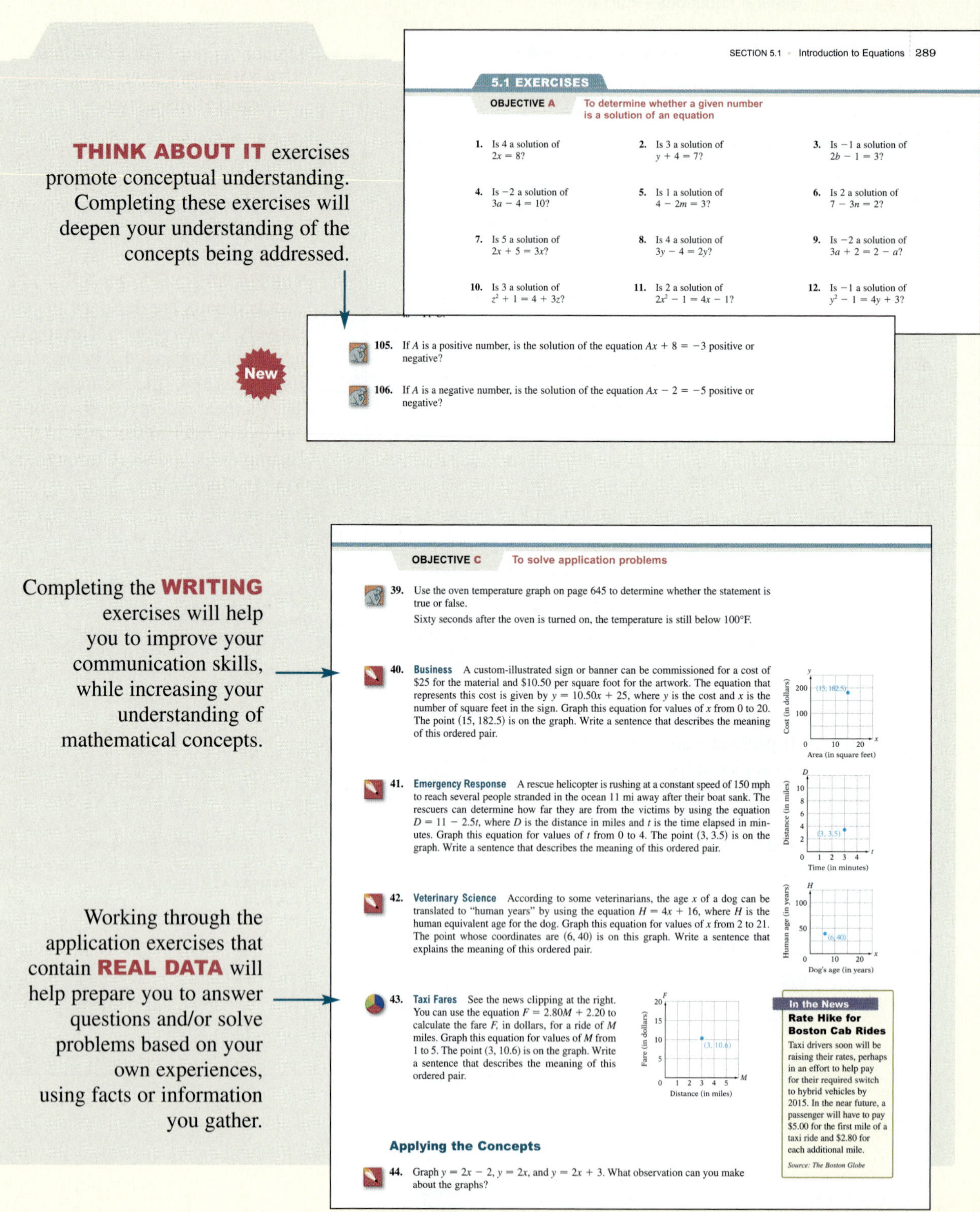

SECTION 5.1 Introduction to Equations 289
5.1 EXERCISES
OBJECTIVE A To determine whether a given number is a solution of an equation
1. Is 4 a solution of 2x = 8?
2. Is 3 a solution of y + 4 = 7?
3. Is −1 a solution of 2b − 1 = 3?
4. Is −2 a solution of 3a − 4 = 10?
5. Is 1 a solution of 4 − 2m = 3?
6. Is 2 a solution of 7 − 3n = 2?
7. Is 5 a solution of 2x + 5 = 3x?
8. Is 4 a solution of 3y − 4 = 2y?
9. Is −2 a solution of 3a + 2 = 2 − a?
10. Is 3 a solution of z² + 1 = 4 + 3z?
11. Is 2 a solution of 2x² − 1 = 4x − 1?
12. Is −1 a solution of y² − 1 = 4y + 3?
New
105. If A is a positive number, is the solution of the equation Ax + 8 = −3 positive or negative?
106. If A is a negative number, is the solution of the equation Ax − 2 = −5 positive or negative?
OBJECTIVE C To solve application problems
39. Use the oven temperature graph on page 645 to determine whether the statement is true or false.
Sixty seconds after the oven is turned on, the temperature is still below 100°F.
40. Business A custom-illustrated sign or banner can be commissioned for a cost of $25 for the material and $10.50 per square foot for the artwork. The equation that represents this cost is given by y = 10.50x + 25, where y is the cost and x is the number of square feet in the sign. Graph this equation for values of x from 0 to 20. The point (15, 182.5) is on the graph. Write a sentence that describes the meaning of this ordered pair.
Cost (in dollars)
Area (in square feet)
41. Emergency Response A rescue helicopter is rushing at a constant speed of 150 mph to reach several people stranded in the ocean 11 mi away after their boat sank. The rescuers can determine how far they are from the victims by using the equation D = 11 − 2.5t, where D is the distance in miles and t is the time elapsed in minutes. Graph this equation for values of t from 0 to 4. The point (3, 3.5) is on the graph. Write a sentence that describes the meaning of this ordered pair.
Distance (in miles)
Time (in minutes)
42. Veterinary Science According to some veterinarians, the age x of a dog can be translated to "human years" by using the equation H = 4x + 16, where H is the human equivalent age for the dog. Graph this equation for values of x from 2 to 21. The point whose coordinates are (6, 40) is on this graph. Write a sentence that explains the meaning of this ordered pair.
Human age (in years)
Dog's age (in years)
43. Taxi Fares See the news clipping at the right. You can use the equation F = 2.80M + 2.20 to calculate the fare F, in dollars, for a ride of M miles. Graph this equation for values of M from 1 to 5. The point (3, 10.6) is on the graph. Write a sentence that describes the meaning of this ordered pair.
Fare (in dollars)
Distance (in miles)
In the News
Rate Hike for Boston Cab Rides
Taxi drivers soon will be raising their rates, perhaps in an effort to help pay for their required switch to hybrid vehicles by 2015. In the near future, a passenger will have to pay $5.00 for the first mile of a taxi ride and $2.80 for each additional mile.
Source: The Boston Globe
Applying the Concepts
44. Graph y = 2x − 2, y = 2x, and y = 2x + 3. What observation can you make about the graphs?

19. 387.8 mi in 7 h

20. 364.8 mi on 9.5 gal of gas

21. $19.08 for 4.5 lb

22. $20.16 for 15 oz

23. **Sports** NCAA statistics show that for every 2800 college seniors playing college basketball, only 50 will play as rookies in the National Basketball Association. Write the ratio of the number of National Basketball Association rookies to the number of college seniors playing basketball.

24. **Energy** A transformer has 40 turns in the primary coil and 480 turns in the secondary coil. State the ratio of the number of turns in the primary coil to the number of turns in the secondary coil.

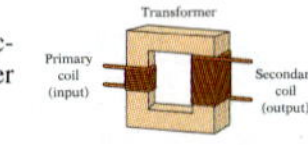

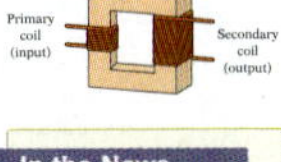

25. **Travel** An airplane flew 1155 mi in 2.5 h. Find the rate of travel.

26. **Facial Hair** Using the data in the news clipping at the right and the figure 50 million for the number of adult males in the United States, write the ratio of the number of men who participated in Movember to the number of adult males in the U.S. Write the ratio as a fraction in simplest form.

In the News

Grow a Mustache, Save a Life

Last fall, in an effort to raise money for the Prostate Cancer Foundation, approximately 2000 men participated in a month-long mustache-growing competition. The event was dubbed Movember.

Source: Time, February 18, 2008

27. **Investments** An investor purchased 100 shares of stock for $2500. One year later the investor sold the stock for $3200. What was the investor's profit per share?

For Exercises 28 to 30, complete the unit rate.

28. 5 miles in ___ hour

29. 15 feet in ___ second

30. 5 grams of fat in ___ serving

IN THE NEWS application exercises help you master the utility of mathematics in our everyday world. They are based on information found in popular media sources, including newspapers and magazines, and the Web.

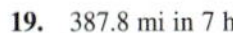

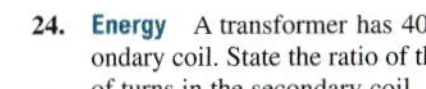

APPLYING THE CONCEPTS exercises may involve further exploration of topics, or they may involve analysis. They may also integrate concepts introduced earlier in the text. **Optional** calculator exercises are included, denoted by .

Applying the Concepts

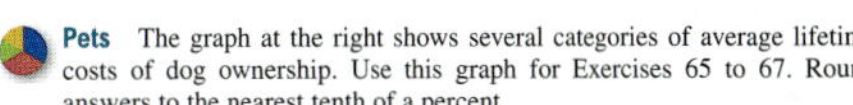

Pets The graph at the right shows several categories of average lifetime costs of dog ownership. Use this graph for Exercises 65 to 67. Round answers to the nearest tenth of a percent.

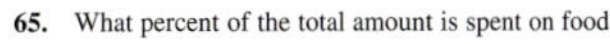

65. What percent of the total amount is spent on food?

66. What percent of the total is spent on veterinary care?

67. What percent of the total is spent on all categories except training?

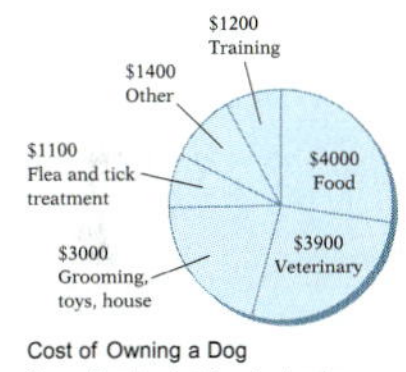

Cost of Owning a Dog

Source: Based on data from the American Kennel Club, *USA Today* research

PROJECTS AND GROUP ACTIVITIES

Consumer Price Index The consumer price index (CPI) is a percent that is written without the percent sign. For instance, a CPI of 160.1 means 160.1%. This number means that an item that cost $100 between 1982 and 1984 (the base years) would cost $160.10 today. Determining the cost is an application of the basic percent equation.

$$\text{Percent} \times \text{base} = \text{amount}$$
$$\text{CPI} \times \text{cost in base year} = \text{cost today}$$
$$1.601 \times 100 = 160.1 \quad \bullet\ 160.1\% = 1.601$$

The table below gives the CPI for various products in March of 2008. You can obtain current data for the items below, as well as other items not on this list, by visiting the website of the Bureau of Labor Statistics.

Product	*CPI*
All items	213.5
Food and beverages	209.7
Housing	214.4
Clothes	120.9
Transportation	195.2
Medical care	363.0
Entertainment[1]	112.7
Education[1]	121.8

[1]Indexes on December 1997 = 100

PROJECTS AND GROUP ACTIVITIES appear at the end of each chapter. Your instructor may assign these to you individually, or you may be asked to work through the activity in groups.

Prealgebra and Introductory Algebra: An Applied Approach addresses students' broad range of study styles by offering **A WIDE VARIETY OF TOOLS FOR REVIEW**.

At the end of each chapter you will find a **SUMMARY** with **KEY WORDS** and **ESSENTIAL RULES AND PROCEDURES**. Each entry includes an example of the summarized concept, an objective reference, and a page reference to show where each concept was introduced.

CHAPTER 6

SUMMARY

KEY WORDS	EXAMPLES
A *ratio* is the comparison of two quantities with the same unit. A ratio can be written in three ways: as a fraction, as two numbers separated by a colon, or as two numbers separated by the word *to*. A ratio is in simplest form when the two quantities do not have a common factor. [6.1A, p. 336]	The comparison 16 oz to 24 oz can be written as a ratio in simplest form: $\frac{2}{3}$, 2:3, or 2 to 3
A *rate* is the comparison of two quantities with different units. A rate is in simplest form when the two quantities do not have a common factor. [6.1A, p. 336]	You earned \$63 for working 6 h. The rate is written $\frac{\$21}{2\text{ h}}$.
A *unit rate* is a rate in which the denominator is 1. [6.1A, p. 336]	You traveled 144 mi in 3 h. The unit rate is 48 mph.
A *proportion* is the equality of two ratios or rates. Each of the four members in a proportion is called a *term*.	In the proportion $\frac{3}{5} = \frac{12}{20}$, 5 and 12 are the means; 3 and 20 are the extremes.

CHAPTER 6

CONCEPT REVIEW

Test your knowledge of the concepts presented in this chapter. Answer each question. Then check your answers against the ones provided in the Answer Section.

1. If the units in a comparison are different, is it a ratio or a rate?
2. How do you find a unit rate?
3. How do you write the ratio 12:15 in simplest form?
4. How do you write the ratio 19:6 as a fraction?

CONCEPT REVIEWS actively engage you as you study and review the contents of a chapter. The **ANSWERS** to the questions are found in an appendix at the back of the text. After each answer, look for an objective reference that indicates where the concept was introduced.

By completing the chapter **REVIEW EXERCISES**, you can practice working problems that appear in an order that is different from the order they were presented in the chapter. The **ANSWERS** to these exercises include references to the section objectives upon which they are based. This will help you to quickly identify where to go to review the concepts if needed.

CHAPTER 6

REVIEW EXERCISES

1. Write the comparison 100 lb to 100 lb as a ratio in simplest form using a fraction, a colon, and the word *to*.
2. Write 18 roof supports for every 9 ft as a rate in simplest form.
3. Write \$628 earned in 40 h as a unit rate.
4. Write 8 h to 15 h as a ratio in simplest form using a fraction.
5. Solve: $\frac{n}{3} = \frac{8}{15}$
6. Write 15 lb of fertilizer for 12 trees as a rate in simplest form.
7. Write 171 mi driven in 3 h as a unit rate.
8. Solve $\frac{2}{3.5} = \frac{n}{12}$. Round to the nearest hundredth.
9. Write 32% as a fraction.
10. Write 22% as a decimal.
11. Write 25% as a fraction and as a decimal.
12. Write $3\frac{2}{5}\%$ as a fraction.

Each chapter **TEST** is designed to simulate a possible test of the concepts covered in the chapter. The **ANSWERS** include references to section objectives. References to How Tos, worked Examples, and You Try Its, that provide solutions to similar problems, are also included.

CHAPTER 6

TEST

1. Write the comparison 3 yd to 24 yd as a ratio in simplest form using a fraction, a colon, and the word *to*.
2. Write 16 oz of sugar for 64 cookies as a rate in simplest form.
3. Write 120 mi driven in 200 min as a unit rate.
4. Write 200 ft to 100 ft as a ratio in simplest form using a fraction.
5. Solve: $\frac{n}{5} = \frac{3}{20}$
6. Write 8 ft walked in 4 s as a unit rate.
7. Write 2860 ft^2 mowed in 6 h as a unit rate. Round to the nearest hundredth.
8. Solve: $\frac{n}{4} = \frac{8}{9}$. Round to the nearest hundredth.

CUMULATIVE REVIEW EXERCISES, which appear at the end of each chapter (beginning with Chapter 2), help you maintain skills you previously learned. The **ANSWERS** include references to the section objectives upon which the exercises are based.

CUMULATIVE REVIEW EXERCISES

1. Simplify: $18 \div \frac{6-3}{9} - (-3)$
2. Evaluate 5^4.
3. Subtract: $7\frac{5}{12} - 3\frac{5}{9}$
4. Simplify: $\frac{4}{5} \div \frac{4}{5} + \frac{2}{3}$
5. Find the quotient of 342 and −3.
6. Evaluate $2a - 3ab$ when $a = 2$ and $b = -3$.
7. Solve: $5x - 20 = 0$
8. Solve: $3(x - 4) + 2x = 3$
9. Simplify: $-\frac{5}{8} - \left(-\frac{3}{4}\right) + \frac{5}{6}$
10. Find the product of 1.005 and 10^5.

A **FINAL EXAM** appears after the last chapter in the text. It is designed to simulate a possible examination of all the concepts covered in the text. The **ANSWERS** to the exam questions are provided in the answer appendix at the back of the text and include references to the section objectives upon which the questions are based.

FINAL EXAM

1. Evaluate $-|-3|$.
2. Subtract: $-15 - (-12) - 3$
3. Simplify: $-\frac{4}{5} - \left(-\frac{3}{10}\right)$
4. Simplify: $-7 - \frac{12 - 15}{2 - (-1)} \cdot (-4)$
5. Evaluate $\frac{a^2 - 3b}{2a - 2b^2}$ when $a = 3$ and $b = -2$.
6. Simplify: $6x - (-4y) - (-3x) + 2y$
7. Simplify: $(-15z)\left(-\frac{2}{5}\right)$
8. Simplify: $-2[5 - 3(2x - 7) - 2x]$
9. Solve: $20 = -\frac{2}{5}x$
10. Solve: $4 - 2(3x + 1) = 3(2 - x) + 5$
11. Write $\frac{1}{8}$ as a percent.
12. Find 19% of 80.

Other Key Features

MARGINS Within the margins, students can find the following.

Take Note boxes alert students to concepts that require special attention.

Point of Interest boxes, which may be historical in nature or be of general interest, relate to topics under discussion.

Integrating Technology boxes, which are offered as optional instruction in the proper use of the scientific calculator, appear for selected topics under discussion.

Tips for Success boxes outline good study habits.

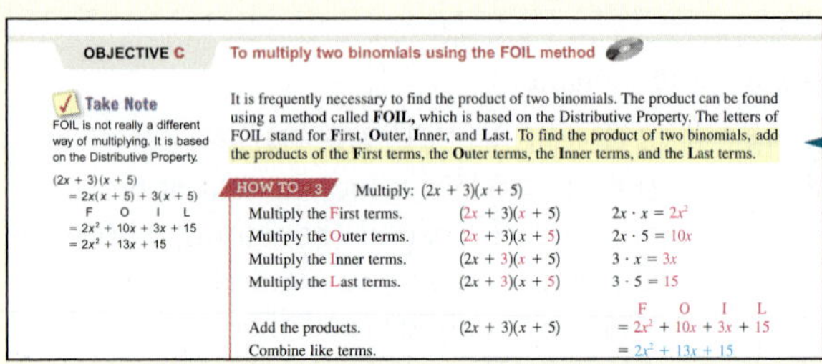

OBJECTIVE C To multiply two binomials using the FOIL method

Take Note
FOIL is not really a different way of multiplying. It is based on the Distributive Property.

$(2x + 3)(x + 5)$
$= 2x(x + 5) + 3(x + 5)$
F O I L
$= 2x^2 + 10x + 3x + 15$
$= 2x^2 + 13x + 15$

It is frequently necessary to find the product of two binomials. The product can be found using a method called **FOIL,** which is based on the Distributive Property. The letters of FOIL stand for **F**irst, **O**uter, **I**nner, and **L**ast. To find the product of two binomials, add the products of the **F**irst terms, the **O**uter terms, the **I**nner terms, and the **L**ast terms.

HOW TO 3 Multiply: $(2x + 3)(x + 5)$

Multiply the First terms.	$(2x + 3)(x + 5)$	$2x \cdot x = 2x^2$
Multiply the Outer terms.	$(2x + 3)(x + 5)$	$2x \cdot 5 = 10x$
Multiply the Inner terms.	$(2x + 3)(x + 5)$	$3 \cdot x = 3x$
Multiply the Last terms.	$(2x + 3)(x + 5)$	$3 \cdot 5 = 15$
		F O I L
Add the products.	$(2x + 3)(x + 5)$	$= 2x^2 + 10x + 3x + 15$
Combine like terms.		$= 2x^2 + 13x + 15$

IMPORTANT POINTS Passages of text are now highlighted to help students recognize what is most important and to help them study more effectively.

PROBLEM-SOLVING STRATEGIES The text features a carefully developed approach to problem solving that encourages students to develop a Strategy for a problem and then to create a Solution based on the Strategy.

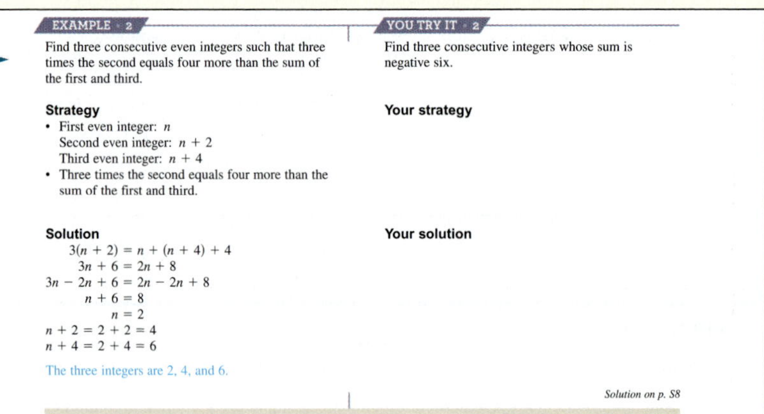

EXAMPLE 2

Find three consecutive even integers such that three times the second equals four more than the sum of the first and third.

Strategy
- First even integer: n
  Second even integer: $n + 2$
  Third even integer: $n + 4$
- Three times the second equals four more than the sum of the first and third.

Solution

$$3(n + 2) = n + (n + 4) + 4$$
$$3n + 6 = 2n + 8$$
$$3n - 2n + 6 = 2n - 2n + 8$$
$$n + 6 = 8$$
$$n = 2$$
$$n + 2 = 2 + 2 = 4$$
$$n + 4 = 2 + 4 = 6$$

The three integers are 2, 4, and 6.

YOU TRY IT 2

Find three consecutive integers whose sum is negative six.

Your strategy

Your solution

Solution on p. S8

FOCUS ON PROBLEM SOLVING At the end of each chapter, the Focus on Problem Solving fosters further discovery of new problem-solving strategies, such as applying solutions to other problems, working backwards, inductive reasoning, and trial and error.

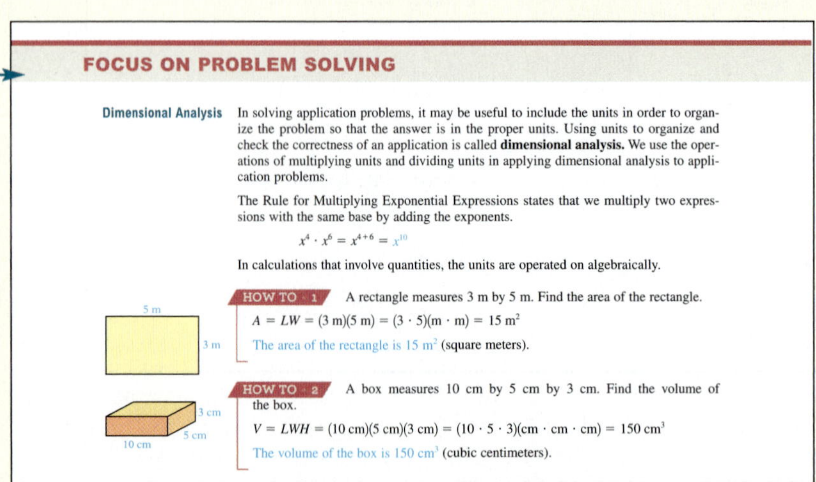

FOCUS ON PROBLEM SOLVING

Dimensional Analysis In solving application problems, it may be useful to include the units in order to organize the problem so that the answer is in the proper units. Using units to organize and check the correctness of an application is called **dimensional analysis.** We use the operations of multiplying units and dividing units in applying dimensional analysis to application problems.

The Rule for Multiplying Exponential Expressions states that we multiply two expressions with the same base by adding the exponents.

$$x^4 \cdot x^6 = x^{4+6} = x^{10}$$

In calculations that involve quantities, the units are operated on algebraically.

HOW TO 1 A rectangle measures 3 m by 5 m. Find the area of the rectangle.

$A = LW = (3\text{ m})(5\text{ m}) = (3 \cdot 5)(\text{m} \cdot \text{m}) = 15\text{ m}^2$

The area of the rectangle is 15 m^2 (square meters).

HOW TO 2 A box measures 10 cm by 5 cm by 3 cm. Find the volume of the box.

$V = LWH = (10\text{ cm})(5\text{ cm})(3\text{ cm}) = (10 \cdot 5 \cdot 3)(\text{cm} \cdot \text{cm} \cdot \text{cm}) = 150\text{ cm}^3$

The volume of the box is 150 cm^3 (cubic centimeters).

General Revisions

- Section 14.1 was revised to include an introduction to interval notation. In the remainder of the chapter, the solution sets to inequalities are written either in set-builder notation or in interval notation.
- Chapter Openers now include Prep Tests for students to test their knowledge of prerequisite skills for the new chapter.
- Each exercise set has been thoroughly reviewed to ensure that the pace and scope of the exercises adequately cover the concepts introduced in the section.
- The variety of word problems has increased. This will appeal to instructors who teach to a range of student abilities and want to address different learning styles.
- Think About It exercises, which are conceptual in nature, have been added. They are meant to assess and strengthen a student's understanding of the material presented in an objective.
- In the News exercises have been added and are based on a media source such as a newspaper, a magazine, or the Web. The exercises demonstrate the pervasiveness and utility of mathematics in a contemporary setting.
- Concept Reviews now appear in the end-of-chapter materials to help students more actively study and review the contents of the chapter.
- The Chapter Review Exercises and Chapter Tests have been adjusted to ensure that there are questions that assess the key ideas in the chapter.
- The design has been significantly modified to make the text even easier to follow.

Acknowledgments

The authors would like to thank the people who have reviewed this manuscript and provided many valuable suggestions.

Chris Bendixen, *Lake Michigan College*
Dorothy Fujimura, *CSU East Bay*
Oxana Grinevich, *Lourdes College*
Joseph Phillips, *Warren County Community College*
Melissa Rossi, *Southwestern Illinois College*
Daryl Schrader, *St. Petersburg College*
Yan Tian, *Palomar College*

The authors would also like to thank the people who reviewed the first edition.

Dorothy A. Brown, *Camden County College*
Kim Doyle, *Monroe Community College*
Said Fariabi, *San Antonio College*
Kimberly A. Gregor, *Delaware Technical and Community College*
Allen Grommet, *East Arkansas Community College*
Anne Haney
Rose M. Kaniper, *Burlington County College*
Mary Ann Klicka, *Bucks County Community College*
Helen Medley, *Kent State University*
Steve Meidinger, *Merced College*
James R. Perry, *Owens Community College*
Gowribalan Vamadeva, *University of Cincinnati*
Susan Wessner, *Tallahassee Community College*

Special thanks go to Jean Bermingham for her work copyediting and proofreading, to Pat Foard for preparing the solutions manuals, and to Cindy Trimble for her work in ensuring the accuracy of the text. We would also like to thank the many people at Cengage Learning who worked to guide the manuscript from development through production.

Instructor Resources

Print Ancillaries

Complete Solutions Manual (0-840-05350-9)
Pat Foard, *South Plains College*

The Complete Solutions Manual provides worked-out solutions to all of the problems in the text.

Instructor's Resource Binder (0-840-05349-5)
Maria H. Andersen, *Muskegon Community College*

New The Instructor's Resource Binder contains uniquely designed Teaching Guides, which include instruction tips, examples, activities, worksheets, overheads, and assessments, with answers to accompany them.

Appendix to accompany Instructor's Resource Binder (0-840-05349-5)
Richard N. Aufmann, *Palomar College*
Joanne S. Lockwood, *Nashua Community College*

New! The Appendix to accompany the Instructor's Resource Binder contains teacher resources that are tied directly to *Prealgebra and Introductory Algebra: An Applied Approach,* 2e. Organized by objective, the Appendix contains additional questions and short, in-class activities. The Appendix also includes answers to Writing Exercises, Focus on Problem Solving, and Projects and Group Activities found in the text.

Electronic Ancillaries

Enhanced WebAssign ENHANCED WebAssign

New Used by over one million students at more than 1,100 institutions, WebAssign allows you to assign, collect, grade, and record homework assignments via the Web. This proven and reliable homework system includes thousands of algorithmically generated homework problems, links to relevant textbook sections, video examples, problem-specific tutorials, and more.

Solution Builder (0-495-91390-1)

This online solutions manual allows instructors to create customizable solutions that they can print out to distribute or post as needed. This is a convenient and expedient way to deliver solutions to specific homework sets.

PowerLecture with Diploma® (0-840-05353-3)

New This CD-ROM provides the instructor with dynamic media tools for teaching. Create, deliver, and customize tests (both print and online) in minutes with Diploma's Computerized Testing featuring algorithmic equations. Easily build solution sets for homework or exams using Solution Builder's online solutions manual. Quickly and easily update your syllabus with the new Syllabus Creator, which was created by the authors and contains the new edition's table of contents. Practice Sheets, First Day of Class PowerPoint® lecture slides, art and figures from the book, and a test bank in electronic format are also included on this CD-ROM.

Text Specific DVDs (0-840-05802-0)

Hosted by Dana Mosely and captioned for the hearing-impaired, these DVDs cover all sections in the text. Ideal for promoting individual study and review, these comprehensive DVDs also support students in online courses or those who may have missed a lecture.

Student Resources

Print Ancillaries

Student Solutions Manual (0-840-04930-7)
Pat Foard, *South Plains College*

The Student Solutions Manual provides worked-out solutions to the odd-numbered problems in the textbook.

Student Workbook (0-840-05351-7)
Maria H. Andersen, *Muskegon Community College*

Get a head-start! The Student Workbook contains assessments, activities, and worksheets from the Instructor's Resource Binder. Use them for additional practice to help you master the content.

Electronic Ancillaries

Enhanced WebAssign ENHANCED WebAssign

New If you are looking for extra practice or additional support, Enhanced WebAssign offers practice problems, videos, and tutorials that are tied directly to the problems found in the textbook.

Text Specific DVDs (0-840-05802-0)

Hosted by Dana Mosley, an experienced mathematics instructor, the DVDs will help you to get a better handle on topics found in the textbook. A comprehensive set of DVDs for the entire course is available to order.

AIM for Success: Getting Started

Welcome to *Prealgebra and Introductory Algebra: An Applied Approach!* Students come to this course with varied backgrounds and different experiences in learning math. We are committed to your success in learning mathematics and have developed many tools and resources to support you along the way. Want to excel in this course? Read on to learn the skills you'll need and how best to use this book to get the results you want.

Motivate Yourself

You'll find many real-life problems in this book, relating to sports, money, cars, music, and more. We hope that these topics will help you understand how you will use mathematics in your real life. However, to learn all of the necessary skills and how you can apply them to your life outside this course, you need to stay motivated.

Take Note

Motivation alone won't lead to success. For example, suppose a person who cannot swim is rowed out to the middle of a lake and thrown overboard. That person has a lot of motivation to swim, but will most likely drown without some help. You'll need motivation and learning in order to succeed.

THINK ABOUT WHY YOU WANT TO SUCCEED IN THIS COURSE. LIST THE REASONS HERE (NOT IN YOUR HEAD . . . ON THE PAPER!):

____

____

We also know that this course may be a requirement for you to graduate or complete your major. That's OK. If you have a goal for the future, such as becoming a nurse or a teacher, you will need to succeed in mathematics first. Picture yourself where you want to be, and use this image to stay on track.

Make the Commitment

Stay committed to success! With practice, you will improve your math skills. Skeptical? Think about when you first learned to ride a bike or drive a car. You probably felt self-conscious and worried that you might fail. But with time and practice, it became second nature to you.

You will also need to put in the time and practice to do well in mathematics. Think of us as your "driving" instructors. We'll lead you along the path to success, but we need you to stay focused and energized along the way.

LIST A SITUATION IN WHICH YOU ACCOMPLISHED YOUR GOAL BY SPENDING TIME PRACTICING AND PERFECTING YOUR SKILLS (SUCH AS LEARNING TO PLAY THE PIANO OR PLAYING BASKETBALL):

____

____

____

____

If you spend time learning and practicing the skills in this book, you will also succeed in math.

© Cengage Learning/Photodisc

Think You Can't Do Math? Think Again!

You can do math! When you first learned the skills you just listed, you may have not done them well. With practice, you got better. With practice, you will be better at math. Stay focused, motivated, and committed to success.

It is difficult for us to emphasize how important it is to overcome the "I Can't Do Math Syndrome." If you listen to interviews of very successful athletes after a particularly bad performance, you will note that they focus on the positive aspect of what they did, not the negative. Sports psychologists encourage athletes to always be positive—to have a "Can Do" attitude. Develop this attitude toward math and you will succeed.

Skills for Success

GET THE BIG PICTURE If this were an English class, we wouldn't encourage you to look ahead in the book. But this is mathematics—go right ahead! Take a few minutes to read the table of contents. Then, look through the entire book. Move quickly: scan titles, look at pictures, notice diagrams.

Getting this big picture view will help you see where this course is going. To reach your goal, it's important to get an idea of the steps you will need to take along the way.

As you look through the book, find topics that interest you. What's your preference? Horse racing? Sailing? TV? Amusement parks? Find the Index of Applications at the back of the book and pull out three subjects that interest you. Then, flip to the pages in the book where the topics are featured and read the exercises or problems where they appear.

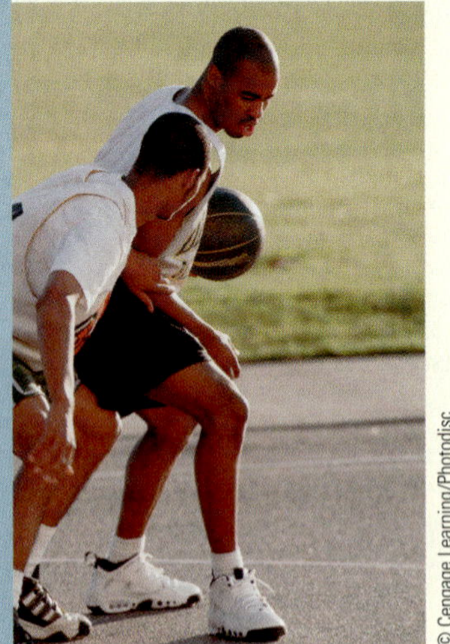
© Cengage Learning/Photodisc

WRITE THE TOPIC HERE:	WRITE THE CORRESPONDING EXERCISE/PROBLEM HERE:
____________	____________
____________	____________
____________	____________

You'll find it's easier to work at learning the material if you are interested in how it can be used in your everyday life.

Use the following activities to think about more ways you might use mathematics in your daily life. Flip open your book to the following exercises to answer the questions.

- (see p. 316, #33) I'm trying to figure out how many text messages I can afford to send in a month. I need to use algebra to . . .

____

____

____

____

- (see p. 594, #81) I'm comparing the gas mileage between two cars. I need algebra to . . .

____

____

- (see p. 155, #144) I want to rent a car, but I have to find the company that offers the best overall price. I need algebra to . . .

____

____

____

You know that the activities you just completed are from daily life, but do you notice anything else they have in common? That's right—they are **word problems.** Try not to be intimidated by word problems. You just need a strategy. It's true that word problems can be challenging because we need to use multiple steps to solve them:

- Read the problem.
- Determine the quantity we must find.
- Think of a method to find it.
- Solve the problem.
- Check the answer.

In short, we must come up with a **strategy** and then use that strategy to find the **solution.**

We'll teach you about strategies for tackling word problems that will make you feel more confident in branching out to these problems from daily life. After all, even though no one will ever come up to you on the street and ask you to solve a multiplication problem, you will need to use math every day to balance your checkbook, evaluate credit card offers, etc.

Take a look at the following example. You'll see that solving a word problem includes finding a *strategy* and using that strategy to find a *solution.* If you find yourself struggling with a word problem, try writing down the information you know about the problem. Be as specific as you can. Write out a phrase or a sentence that states what you are trying to find. Ask yourself whether there is a formula that expresses the known and unknown quantities. Then, try again!

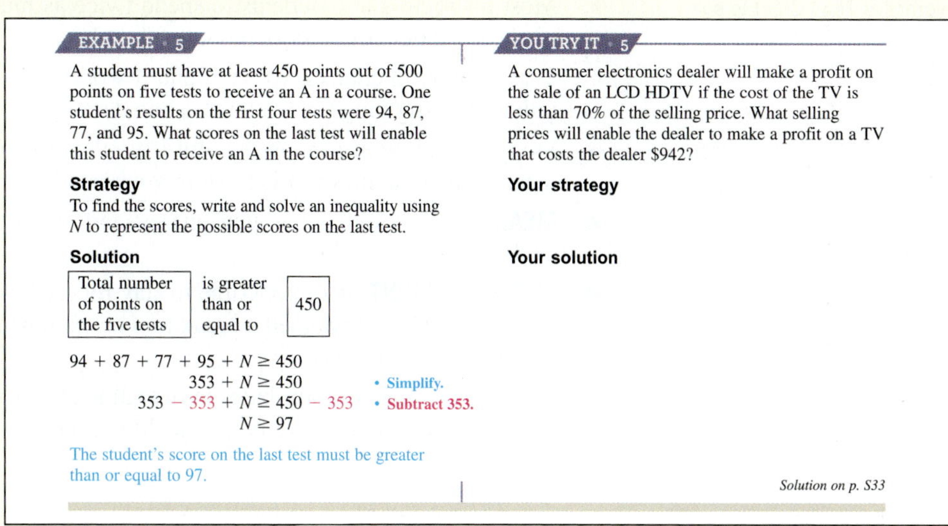

EXAMPLE 5

A student must have at least 450 points out of 500 points on five tests to receive an A in a course. One student's results on the first four tests were 94, 87, 77, and 95. What scores on the last test will enable this student to receive an A in the course?

Strategy

To find the scores, write and solve an inequality using N to represent the possible scores on the last test.

Solution

Total number of points on the five tests	is greater than or equal to	450

$$94 + 87 + 77 + 95 + N \geq 450$$

$$353 + N \geq 450 \quad \text{• Simplify.}$$

$$353 - 353 + N \geq 450 - 353 \quad \text{• Subtract 353.}$$

$$N \geq 97$$

The student's score on the last test must be greater than or equal to 97.

YOU TRY IT 5

A consumer electronics dealer will make a profit on the sale of an LCD HDTV if the cost of the TV is less than 70% of the selling price. What selling prices will enable the dealer to make a profit on a TV that costs the dealer $942?

Your strategy

Your solution

Solution on p. S33

Page 732

Take Note

Take a look at your syllabus to see if your instructor has an **attendance policy** that is part of your overall grade in the course.

The attendance policy will tell you:

- How many classes you can miss without a penalty
- What to do if you miss an exam or quiz
- If you can get the lecture notes from the professor if you miss a class

GET THE BASICS On the first day of class, your instructor will hand out a **syllabus** listing the requirements of your course. Think of this syllabus as your personal roadmap to success. It shows you the destinations (topics you need to learn) and the dates you need to arrive at those destinations (by when you need to learn the topics). Learning mathematics is a journey. But, to get the most out of this course, you'll need to know what the important stops are and what skills you'll need to learn for your arrival at those stops.

You've quickly scanned the table of contents, but now we want you to take a closer look. Flip open to the table of contents and look at it next to your syllabus. Identify when your major exams are and what material you'll need to learn by those dates. For example, if you know you have an exam in the second month of the semester, how many chapters of this text will you need to learn by then? What homework do you have to do during this time? Managing this important information will help keep you on track for success.

Take Note

When planning your schedule, give some thought to how much time you realistically have available each week. For example, if you work 40 hours a week, take 15 units, spend the recommended study time given at the right, and sleep 8 hours a day, you will use over 80% of the available hours in a week. That leaves less than 20% of the hours in a week for family, friends, eating, recreation, and other activities.

Visit http://college.cengage.com/masterstudent/shared/content/time_chart/chart.html and use the Interactive Time Chart to see how you're spending your time—you may be surprised.

MANAGE YOUR TIME We know how busy you are outside of school. Do you have a full-time or a part-time job? Do you have children? Visit your family often? Play basketball or write for the school newspaper? It can be stressful to balance all of the important activities and responsibilities in your life. Making a **time management plan** will help you create a schedule that gives you enough time for everything you need to do.

© Cengage Learning/Photodisc

Let's get started! Create a weekly schedule.

First, list all of your responsibilities that take up certain set hours during the week. Be sure to include:

- each class you are taking
- time you spend at work
- any other commitments (child care, tutoring, volunteering, etc.)

Then, list all of your responsibilities that are more flexible. Remember to make time for:

- **STUDYING** You'll need to study to succeed, but luckily you get to choose what times work best for you. Keep in mind:
  - Most instructors ask students to spend twice as much time studying as they do in class (3 hours of class = 6 hours of study).
  - Try studying in chunks. We've found it works better to study an hour each day, rather than studying for 6 hours on one day.
  - Studying can be even more helpful if you're able to do it right after your class meets, when the material is fresh in your mind.
- **MEALS** Eating well gives you energy and stamina for attending classes and studying.
- **ENTERTAINMENT** It's impossible to stay focused on your responsibilities 100% of the time. Giving yourself a break for entertainment will reduce your stress and help keep you on track.
- **EXERCISE** Exercise contributes to overall health. You'll find you're at your most productive when you have both a healthy mind and a healthy body.

Here is a sample of what part of your schedule might look like:

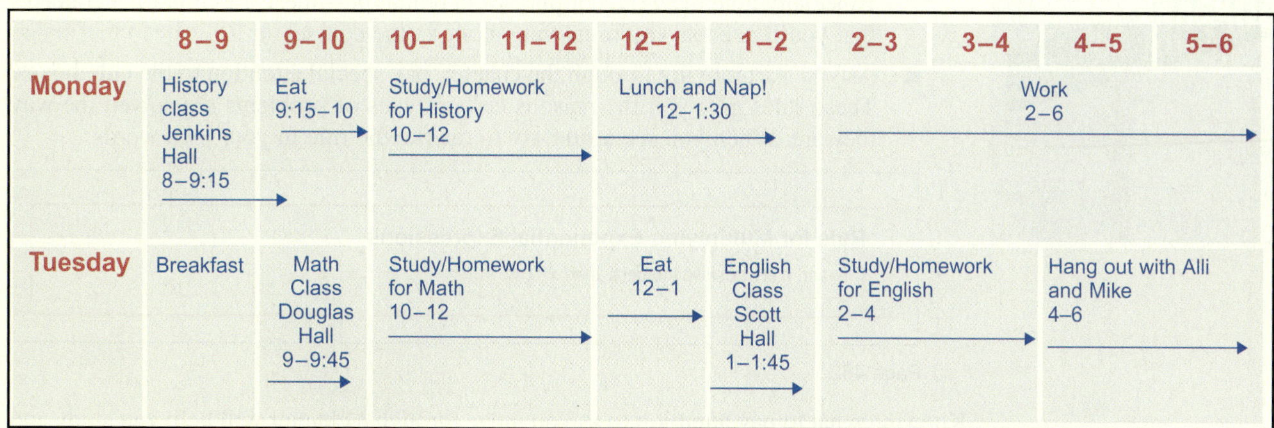

	8–9	9–10	10–11	11–12	12–1	1–2	2–3	3–4	4–5	5–6
Monday	History class Jenkins Hall 8–9:15	Eat 9:15–10	Study/Homework for History 10–12		Lunch and Nap! 12–1:30		Work 2–6			
Tuesday	Breakfast	Math Class Douglas Hall 9–9:45	Study/Homework for Math 10–12		Eat 12–1	English Class Scott Hall 1–1:45	Study/Homework for English 2–4		Hang out with Alli and Mike 4–6	

Features for Success in This Text

ORGANIZATION Let's look again at the table of contents. There are 16 chapters in this book. You'll see that every chapter is divided into **sections,** and each section contains a number of **learning objectives.** Each learning objective is labeled with a letter from A to E. Knowing how this book is organized will help you locate important topics and concepts as you're studying.

PREPARATION Ready to start a new chapter? Take a few minutes to be sure you're ready, using some of the tools in this book.

- **CUMULATIVE REVIEW EXERCISES:** You'll find these exercises after every chapter, starting with Chapter 2. The questions in the Cumulative Review Exercises are taken from the previous chapters. For example, the Cumulative Review for Chapter 3 will test all of the skills you have learned in Chapters 1, 2, and 3. Use this to refresh yourself before moving on to the next chapter, or to test what you know before a big exam.

Here's an example of how to use the Cumulative Review:

- Turn to page 243 and look at the questions for the Chapter 3 Cumulative Review, which are taken from the current chapter and the previous chapters.
- We have the answers to all of the Cumulative Review Exercises in the back of the book. Flip to page A9 to see the answers for this chapter.
- Got the answer wrong? We can tell you where to go in the book for help! For example, scroll down page A9 to find the answer for the first exercise, which is 5. You'll see that after this answer, there is an **objective reference** [3.2B]. This means that the question was taken from Chapter 3, Section 2, Objective B. Go here to restudy the objective.

- **PREP TESTS:** These tests are found at the beginning of every chapter and will help you see if you've mastered all of the skills needed for the new chapter.

Here's an example of how to use the Prep Test:

- Turn to page 245 and look at the Prep Test for Chapter 4.
- All of the answers to the Prep Tests are in the back of the book. You'll find them in the first set of answers in each answer section for a chapter. Turn to page A10 to see the answers for this Prep Test.
- Restudy the objectives if you need some extra help.

© Cengage Learning/Photodisc

- Before you start a new section, take a few minutes to read the **Objective Statement** for that section. Then, browse through the objective material. Especially note the words or phrases in bold type—these are important concepts that you'll need as you're moving along in the course.
- As you start moving through the chapter, pay special attention to the **rule boxes.** These rules give you the reasons certain types of problems are solved the way they are. When you see a rule, try to rewrite the rule in your own words.

Rule for Multiplying Exponential Expressions

If m and n are positive integers, then $x^m \cdot x^n = x^{m+n}$.

Page 482

Knowing what to pay attention to as you move through a chapter will help you study and prepare.

INTERACTION We want you to be actively involved in learning mathematics and have given you many ways to get hands-on with this book.

- **HOW TO EXAMPLES** Take a look at page 285 shown here. See the HOW TO example? This contains an explanation by each step of the solution to a sample problem.

HOW TO 5 Solve: $6x = 14$

$6x = 14$ • The goal is to rewrite the equation in the form *variable = constant.*

$\frac{6x}{6} = \frac{14}{6}$ • Divide each side of the equation by 6.

$x = \frac{7}{3}$ • Simplify. The equation is in the form *variable = constant.*

The solution is $\frac{7}{3}$.

Page 285

Grab a paper and pencil and work along as you're reading through each example. When you're done, get a clean sheet of paper. Write down the problem and try to complete the solution without looking at your notes or at the book. When you're done, check your answer. If you got it right, you're ready to move on.

- **EXAMPLE/YOU TRY IT PAIRS** You'll need hands-on practice to succeed in mathematics. When we show you an example, work it out beside our solution. Use the Example/You Try It pairs to get the practice you need.

  Take a look at page 285, Example 5 and You Try It 5 shown here:

EXAMPLE 5

Solve: $5x - 9x = 12$

Solution

$5x - 9x = 12$

$-4x = 12$ • Combine like terms.

$\frac{-4x}{-4} = \frac{12}{-4}$

$x = -3$

The solution is -3.

YOU TRY IT 5

Solve: $4x - 8x = 16$

Your solution

Solutions on p. S12

Page 285

You'll see that each Example is fully worked-out. Study this Example carefully by working through each step. Then, try your hand at it by completing the You Try It. If you get stuck, the solutions to the You Try Its are provided in the back of the book. There is a page number following the You Try It, which shows you where you can find the completely worked-out solution. Use the solution to get a hint for the step on which you are stuck. Then, try again!

When you've finished the solution, check your work against the solution in the back of the book. Turn to page S12 to see the solution for You Try It 5.

Remember that sometimes there can be more than one way to solve a problem. But, your answer should always match the answers we've given in the back of the book. If you have any questions about whether your method will always work, check with your instructor.

REVIEW We have provided many opportunities for you to practice and review the skills you have learned in each chapter.

- **SECTION EXERCISES** After you're done studying a section, flip to the end of the section and complete the exercises. If you immediately practice what you've learned, you'll find it easier to master the core skills. Want to know if you answered the questions correctly? The answers to the odd-numbered exercises are given in the back of the book.
- **CHAPTER SUMMARY** Once you've completed a chapter, look at the Chapter Summary. This is divided into two sections: *Key Words* and *Essential Rules and Procedures.* Flip to page 271 to see the Chapter Summary for Chapter 4. This summary shows all of the important topics covered in the chapter. See the reference following each topic? This shows you the objective reference and the page in the text where you can find more information on the concept.
- **CONCEPT REVIEW** Following the Chapter Summary for each chapter is the Concept Review. Flip to page 273 to see the Concept Review for Chapter 4. When you read each question, jot down a reminder note on the right about whatever you feel will be most helpful to remember if you need to apply that concept during an exam. You can also use the space on the right to mark what concepts your instructor expects you to know for the next test. If you are unsure of the answer to a concept review question, flip to the answers appendix at the back of the book.
- **CHAPTER REVIEW EXERCISES** You'll find the Chapter Review Exercises after the Concept Review. Flip to page 371 to see the Chapter Review Exercises for Chapter 6. When you do the review exercises, you're giving yourself an important opportunity to test your understanding of the chapter. The answer to each review exercise is given at the back of the book, along with the objective the question relates to. When you're done with the Chapter Review Exercises, check your answers. If you had trouble with any of the questions, you can restudy the objectives and retry some of the exercises in those objectives for extra help.

- **CHAPTER TESTS** The Chapter Tests can be found after the Chapter Review Exercises and can be used to prepare for your exams. The answer to each test question is given at the back of the book, along with a reference to a How To, Example, or You Try It that the question relates to. Think of these tests as "practice runs" for your in-class tests. Take the test in a quiet place and try to work through it in the same amount of time you will be allowed for your exam.

Here are some strategies for success when you're taking your exams:

- Scan the entire test to get a feel for the questions (get the big picture).
- Read the directions carefully.
- Work the problems that are easiest for you first.
- Stay calm, and remember that you will have lots of opportunities for success in this class!

EXCEL Visit **www.cengage.com/math/aufmann** to learn about additional study tools!

- *Enhanced* ***WebAssign***® online practice exercises and homework problems match the textbook exercises.
- **DVDs** Hosted by Dana Mosley, an experienced mathematics instructor, the DVDs will help you to get a better handle on topics that may be giving you trouble. A comprehensive set of DVDs for the entire course is available to order.

Get Involved Have a question? Ask! Your professor and your classmates are there to help. Here are some tips to help you jump in to the action:

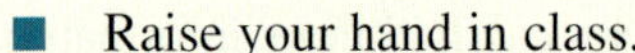

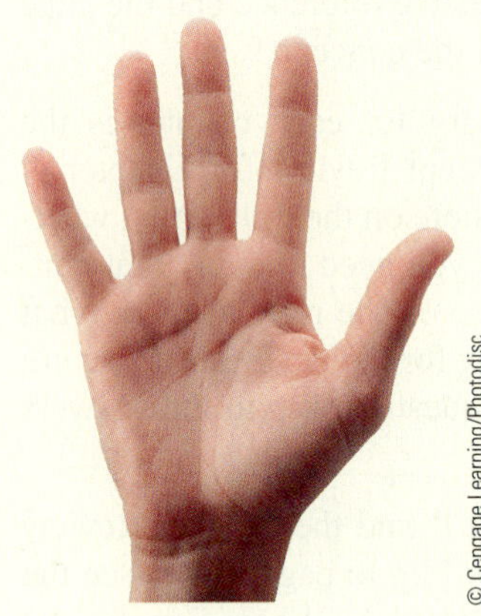

© Cengage Learning/Photodisc

- Raise your hand in class.
- If your instructor prefers, email or call your instructor with your question. If your professor has a website where you can post your question, also look there for answers to previous questions from other students. Take advantage of these ways to get your questions answered.
- Visit a **math center.** Ask your instructor for more information about the math center services available on your campus.
- Your instructor will have **office hours** where he or she will be available to help you. Take note of where and when your instructor holds office hours. Use this time for one-on-one help, if you need it.
- Form a **study group** with students from your class. This is a great way to prepare for tests, catch up on topics you may have missed, or get extra help on problems you're struggling with. Here are a few suggestions to make the most of your study group:

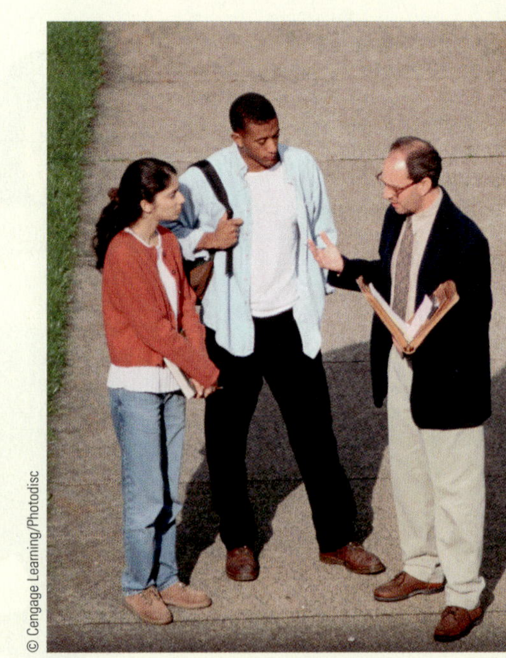

© Cengage Learning/Photodisc

- **Test each other by asking questions.** Have each person bring a few sample questions when you get together.

- **Practice teaching each other.** We've found that you can learn a lot about what you know when you have to explain it to someone else.
- **Compare class notes.** Couldn't understand the last five minutes of class? Missed class because you were sick? Chances are someone in your group has the notes for the topics you missed.
- **Brainstorm test questions.**
- **Make a plan for your meeting.** Agree on what topics you'll talk about and how long you'll be meeting. When you make a plan, you'll be sure that you make the most of your meeting.

© Cengage Learning/Photodisc

Ready, Set, Succeed! It takes hard work and commitment to succeed, but we know you can do it! Doing well in mathematics is just one step you'll take along the path to success.

I succeeded in Prealgebra and Introductory Algebra!

We are confident that if you follow our suggestions, you will succeed. Good luck!

Rubberball

CHAPTER 1

Whole Numbers

OBJECTIVES

SECTION 1.1

- **A** To identify the order relation between two whole numbers
- **B** To write whole numbers in words, in standard form, and in expanded form
- **C** To round a whole number to a given place value
- **D** To solve application problems and read statistical graphs

SECTION 1.2

- **A** To add whole numbers
- **B** To subtract whole numbers
- **C** To solve application problems and use formulas

SECTION 1.3

- **A** To multiply whole numbers
- **B** To evaluate expressions that contain exponents
- **C** To divide whole numbers
- **D** To factor numbers and find the prime factorization of numbers
- **E** To solve application problems and use formulas

SECTION 1.4

- **A** To use the Order of Operations Agreement to simplify expressions

ARE YOU READY?

Take the Chapter 1 Prep Test to find out if you are ready to learn to:

- Order whole numbers
- Round whole numbers
- Add, subtract, multiply, and divide whole numbers
- Factor numbers and find their prime factorization
- Simplify numerical expressions

PREP TEST

Do these exercises to prepare for Chapter 1.

1. Name the number of ♦s shown below.

♦♦♦♦♦♦♦♦

2. Write the numbers from 1 to 10.

1 ___ ___ ___ ___ ___ ___ ___ ___ 10

3. Match the number with its word form.

a. 4	**A.** five
b. 2	**B.** one
c. 5	**C.** zero
d. 1	**D.** four
e. 3	**E.** two
f. 0	**F.** three

4. How many American flags contain the color green?

5. Write the number of states in the United States of America as a word, not a number.

SECTION

1.1 Introduction to Whole Numbers

OBJECTIVE A To identify the order relation between two whole numbers 

The **natural numbers** are 1, 2, 3, 4, 5, 6, 7, 8, 9, 10, 11, ...

The three dots mean that the list continues on and on and there is no largest natural number. The natural numbers are also called the **counting numbers.**

Point of Interest

Among the slang words for zero are *zilch, zip,* and *goose egg.* The word *love* for zero in scoring a tennis game comes from the French for "the egg": *l'oeuf.*

The **whole numbers** are 0, 1, 2, 3, 4, 5, 6, 7, 8, 9, 10, 11, ... Note that the whole numbers include the natural numbers and zero.

Just as distances are associated with markings on the edge of a ruler, the whole numbers can be associated with points on a line. This line is called the **number line** and is shown below.

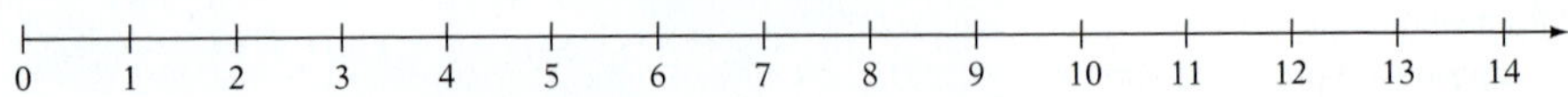

The arrowhead at the right indicates that the number line continues to the right.

The **graph of a whole number** is shown by placing a heavy dot on the number line directly above the number. Shown below is the graph of 6 on the number line.

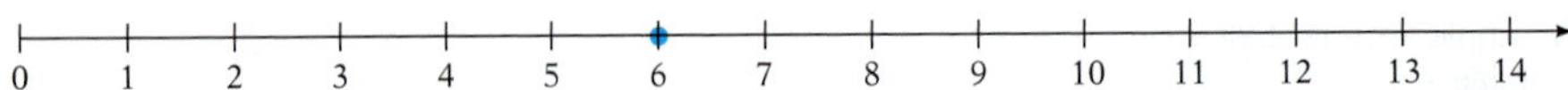

On the number line, the numbers get larger as we move from left to right. The numbers get smaller as we move from right to left. Therefore, the number line can be used to visualize the order relation between two whole numbers.

A number that appears to the right of a given number is **greater than** the given number. The symbol for *is greater than* is $>$.

8 is to the right of 3.
8 is greater than 3.
$8 > 3$

0 1 2 3 4 5 6 7 8 9 10 11 12 13 14

A number that appears to the left of a given number is **less than** the given number. The symbol for *is less than* is $<$.

5 is to the left of 12.
5 is less than 12.
$5 < 12$

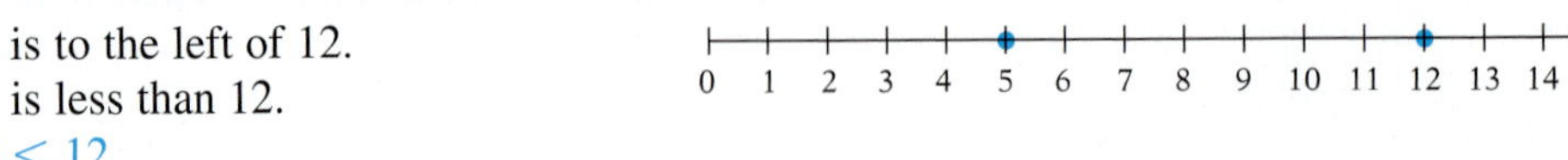

Take Note

An inequality symbol, $<$ or $>$, points to the smaller number. The symbol opens toward the larger number.

An **inequality** expresses the relative order of two mathematical expressions. $8 > 3$ and $5 < 12$ are inequalities.

EXAMPLE • 1

Graph 4 on the number line.

Solution

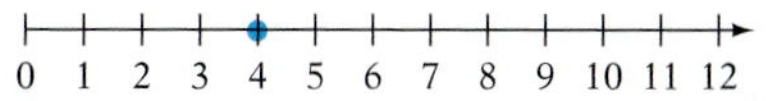

YOU TRY IT • 1

Graph 9 on the number line.

Your solution

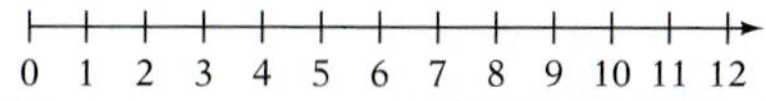

EXAMPLE • 2

On the number line, what number is 3 units to the right of 4?

Solution

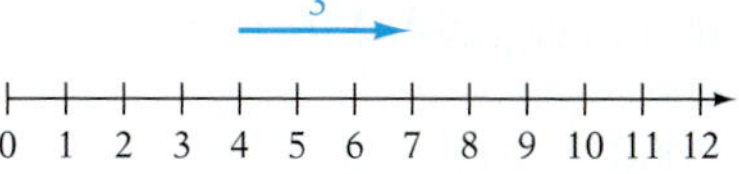

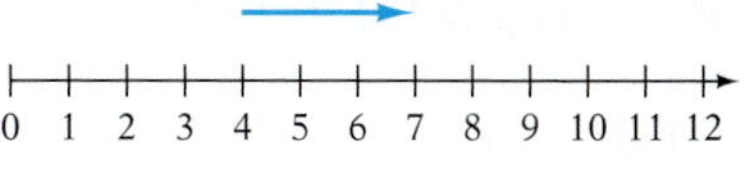

7 is 3 units to the right of 4.

YOU TRY IT • 2

On the number line, what number is 4 units to the left of 11?

Your solution

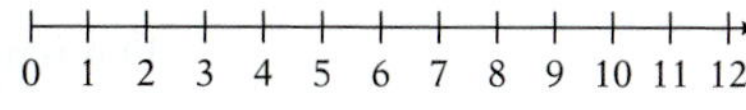

EXAMPLE • 3

Place the correct symbol, < or >, between the two numbers.

a. 38 23 **b.** 0 54

Solution **a.** 38 > 23 **b.** 0 < 54

YOU TRY IT • 3

Place the correct symbol, < or >, between the two numbers.

a. 47 19 **b.** 26 0

Your solution

EXAMPLE • 4

Write the given numbers in order from smallest to largest.

16, 5, 47, 0, 83, 29

Solution 0, 5, 16, 29, 47, 83

YOU TRY IT • 4

Write the given numbers in order from smallest to largest.

52, 17, 68, 0, 94, 3

Your solution

Solutions on p. S1

OBJECTIVE B To write whole numbers in words, in standard form, and in expanded form

Point of Interest

The Romans represented numbers using M for 1000, D for 500, C for 100, L for 50, X for 10, V for 5, and I for 1. For example, MMDCCCLXXVI represented 2876. The Romans could represent any number up to the largest they would need for their everyday life, except zero.

When a whole number is written using the digits 0, 1, 2, 3, 4, 5, 6, 7, 8, and 9, it is said to be in **standard form.** The position of each digit in the number determines the digit's **place value.** The diagram below shows a **place-value chart** naming the first twelve place values. The number 64,273 is in standard form and has been entered in the chart.

In the number 64,273, the position of the digit 6 determines that its place value is ten-thousands.

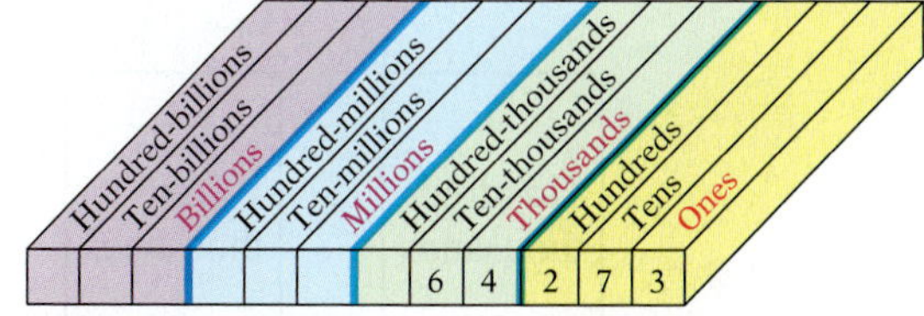

When a number is written in standard form, each group of digits separated by a comma is called a **period.** The number 5,316,709,842 has four periods. The period names are shown in red in the place-value chart above.

Point of Interest

George Washington used a code to communicate with his men. He had a book in which each word or phrase was represented by a three-digit number. The numbers were arbitrarily assigned to each entry. Messages appeared as a string of numbers and thus could not be decoded by the enemy.

To write a number in words, start from the left. Name the number in each period. Then write the period name in place of the comma.

5,316,709,842 is read "five billion three hundred sixteen million seven hundred nine thousand eight hundred forty-two."

To write a whole number in standard form, write the number named in each period, and replace each period name with a comma.

Six million fifty-one thousand eight hundred seventy-four is written 6,051,874. The zero is used as a place holder for the hundred-thousands place.

The whole number 37,286 can be written in **expanded form** as

$$30{,}000 + 7000 + 200 + 80 + 6$$

The place-value chart can be used to find the expanded form of a number.

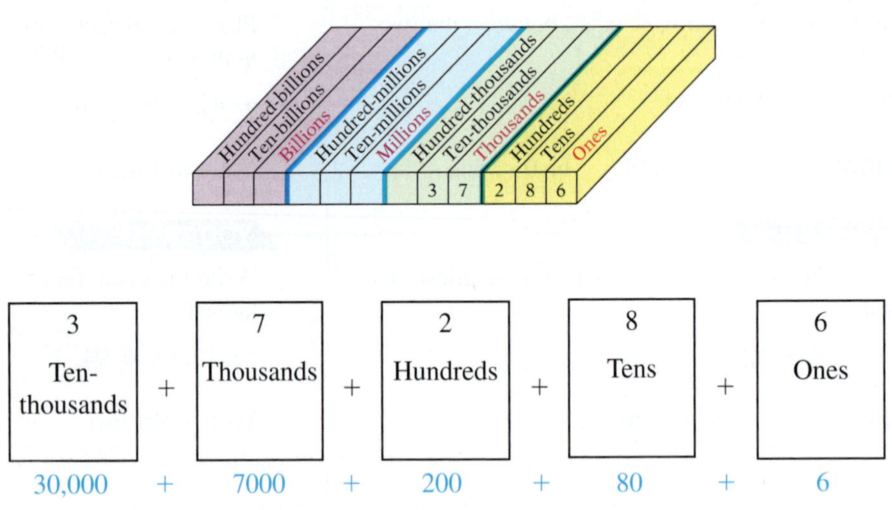

Write the number 510,409 in expanded form.

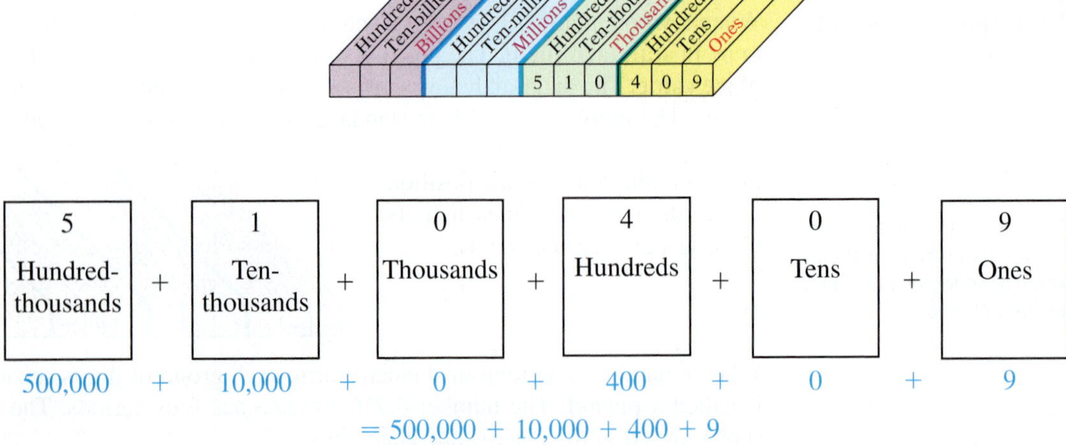

EXAMPLE • 5

Write 82,593,071 in words.

Solution

Eighty-two million five hundred ninety-three thousand seventy-one

YOU TRY IT • 5

Write 46,032,715 in words.

Your solution

EXAMPLE • 6

Write four hundred six thousand nine in standard form.

Solution

406,009

YOU TRY IT • 6

Write nine hundred twenty thousand eight in standard form.

Your solution

EXAMPLE • 7

Write 32,598 in expanded form.

Solution

30,000 + 2000 + 500 + 90 + 8

YOU TRY IT • 7

Write 76,245 in expanded form.

Your solution

Solutions on p. S1

OBJECTIVE C To round a whole number to a given place value

When the distance to the sun is given as 93,000,000 mi, the number represents an approximation to the true distance. Giving an approximate value for an exact number is called **rounding.** A number is rounded to a given place value.

48 is closer to 50 than it is to 40. 48 rounded to the nearest ten is 50.

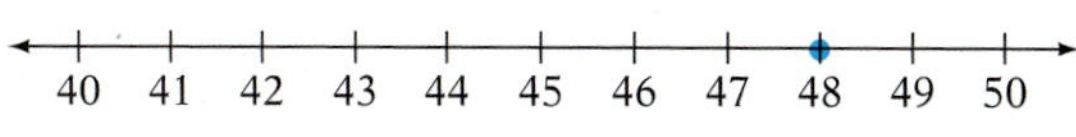

4872 rounded to the nearest ten is 4870.

4872 rounded to the nearest hundred is 4900.

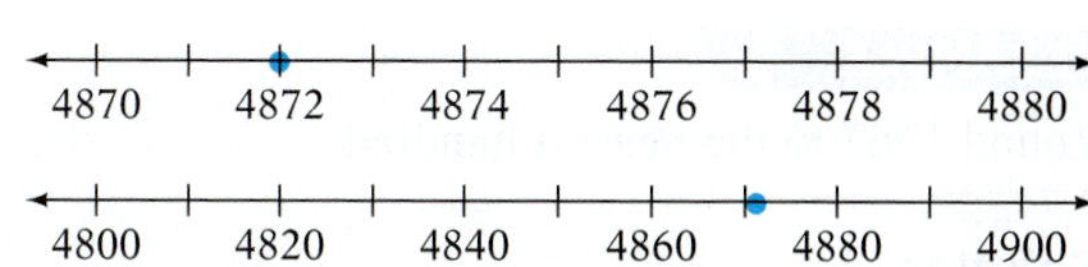

A number is rounded to a given place value without using the number line by looking at the first digit to the right of the given place value.

If the digit to the right of the given place value is less than 5, replace that digit and all digits to the right of it by zeros.

Round 12,743 to the nearest hundred.

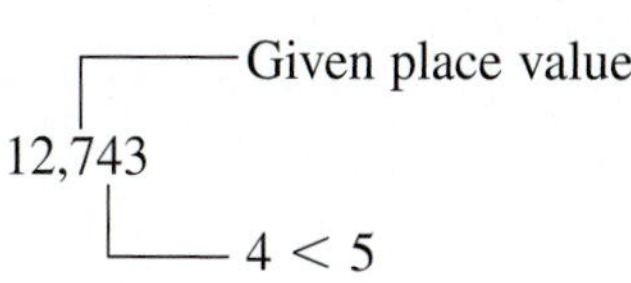

12,743 rounded to the nearest hundred is 12,700.

If the digit to the right of the given place value is greater than or equal to 5, increase the digit in the given place value by 1, and replace all other digits to the right by zeros.

Round 46,738 to the nearest thousand.

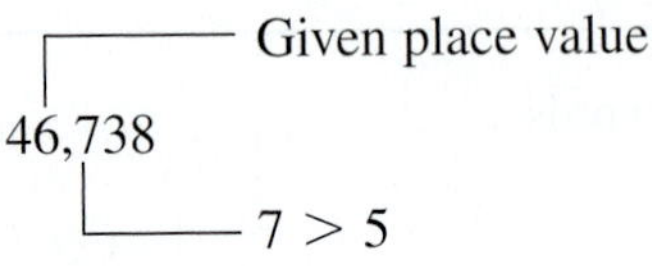

46,738 rounded to the nearest thousand is 47,000.

HOW TO • 1 Round 29,873 to the nearest thousand.

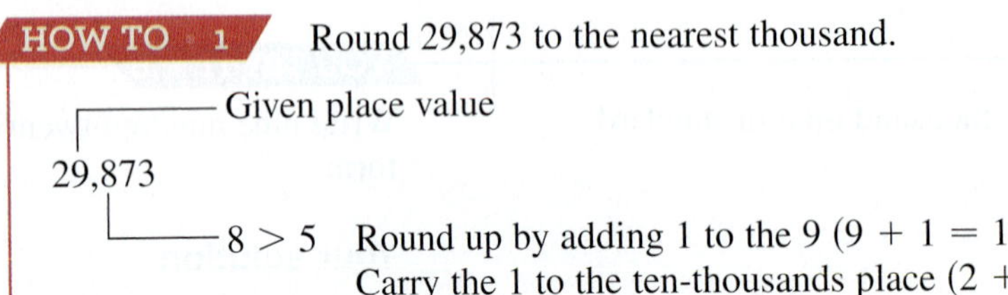

Given place value
29,873
8 > 5 Round up by adding 1 to the 9 (9 + 1 = 10).
Carry the 1 to the ten-thousands place (2 + 1 = 3).

29,873 rounded to the nearest thousand is 30,000.

EXAMPLE • 8

Round 435,278 to the nearest ten-thousand.

Solution

Given place value

435,278

5 = 5

435,278 rounded to the nearest ten-thousand is 440,000.

YOU TRY IT • 8

Round 529,374 to the nearest ten-thousand.

Your solution

EXAMPLE • 9

Round 1967 to the nearest hundred.

Solution

Given place value

1967

6 > 5

1967 rounded to the nearest hundred is 2000.

YOU TRY IT • 9

Round 7985 to the nearest hundred.

Your solution

Solutions on p. S1

OBJECTIVE D To solve application problems and read statistical graphs

Graphs are displays that provide a pictorial representation of data. The advantage of graphs is that they present information in a way that is easily read.

A **pictograph** uses symbols to represent information. The symbol chosen usually has a connection to the data it represents.

Figure 1.1 represents the net worth of America's richest billionaires. Each symbol represents ten billion dollars.

AP Photo/Joe Cavaretta

Bill Gates

	Net Worth (in tens of billions of dollars)
Bill Gates	$ $ $ $ $
Warren Buffett	$ $ $ $
Paul Allen	$ $ $
Larry Ellison	$ $
Jim C. Walton	$ $

FIGURE 1.1 Net Worth of America's Richest Billionaires
Source: www.Forbes.com

From the pictograph, we can see that Bill Gates has the greatest net worth. Warren Buffett's net worth is $10 billion more than Paul Allen's net worth.

A typical household in the United States has an average after-tax income of $40,550. The **circle graph** in Figure 1.2 represents how this annual income is spent. The complete circle represents the total amount, $40,550. Each sector of the circle represents the amount spent on a particular expense.

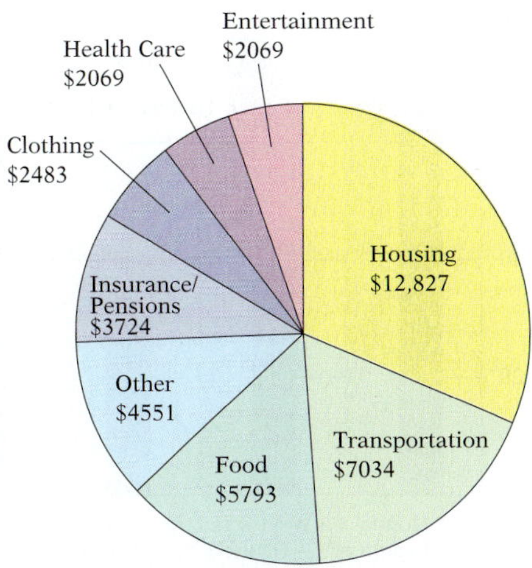

FIGURE 1.2 Average Annual Expenses in a U.S. Household
Source: American Demographics

From the circle graph, we can see that the largest amount is spent on housing. We can see that the amount spent on food ($5793) is less than the amount spent on transportation ($7034).

The **bar graph** in Figure 1.3 shows the expected U.S. population aged 100 and over.

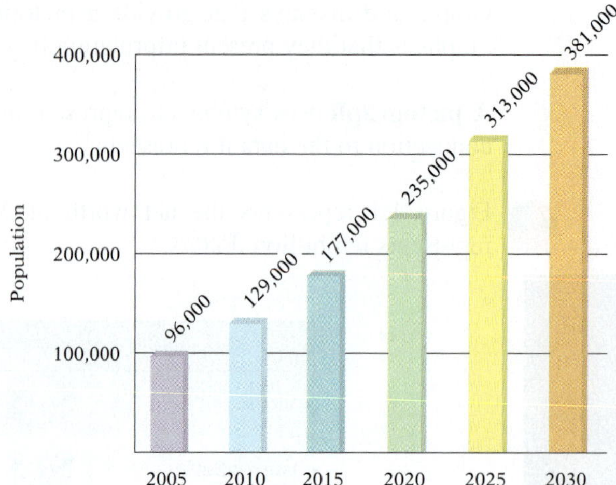

FIGURE 1.3 Expected U.S. Population Aged 100 and Over
Source: Census Bureau

In this bar graph, the horizontal axis is labeled with the years (2005, 2010, 2015, etc.) and the vertical axis is labeled with the numbers for the population. For each year, the height of the bar indicates the population for that year. For example, we can see that the expected population of those aged 100 and over in the year 2015 is 177,000. The graph indicates that the population of people aged 100 and over is projected to increase.

A **double-bar graph** is used to display data for purposes of comparison.

The double-bar graph in Figure 1.4 shows the fuel efficiency of four vehicles, as rated by the Environmental Protection Agency. These were among the most fuel-efficent cars for city and highway mileage in a recent year.

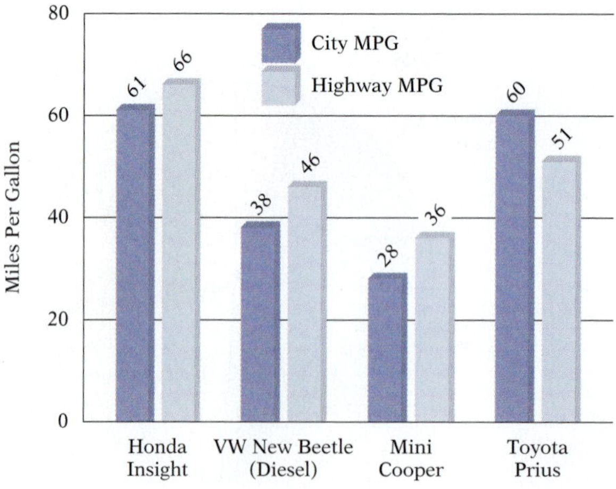

FIGURE 1.4

From the graph, we can see that the fuel efficiency of the Honda Insight is greater on the highway (66 mpg) than it is for city driving (61 mpg).

The **broken-line graph** in Figure 1.5 shows the effect of inflation on the value of a \$100,000 life insurance policy. (An inflation rate of 5 percent is used here.)

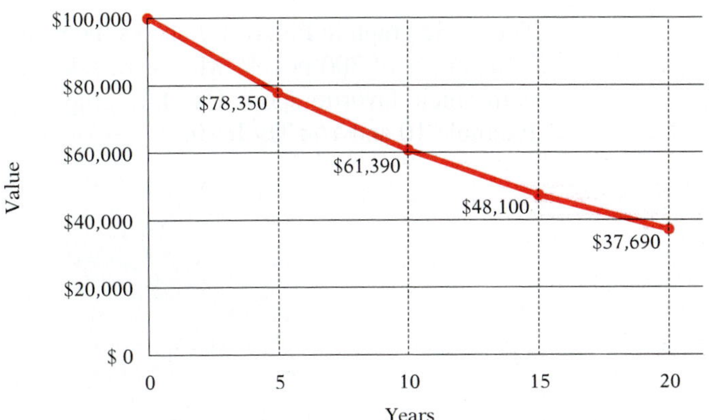

FIGURE 1.5 Effect of Inflation on the Value of a \$100,000 Life Insurance Policy

According to the line graph, after 5 years the purchasing power of the \$100,000 has decreased to \$78,350. We can see that the value of the \$100,000 keeps decreasing over the 20-year period.

Two broken-line graphs are used so that data can be compared. Figure 1.6 shows the populations of California and Texas. The figures are those of the U.S. Census for the years 1900, 1925, 1950, 1975, and 2000. The numbers are rounded to the nearest thousand.

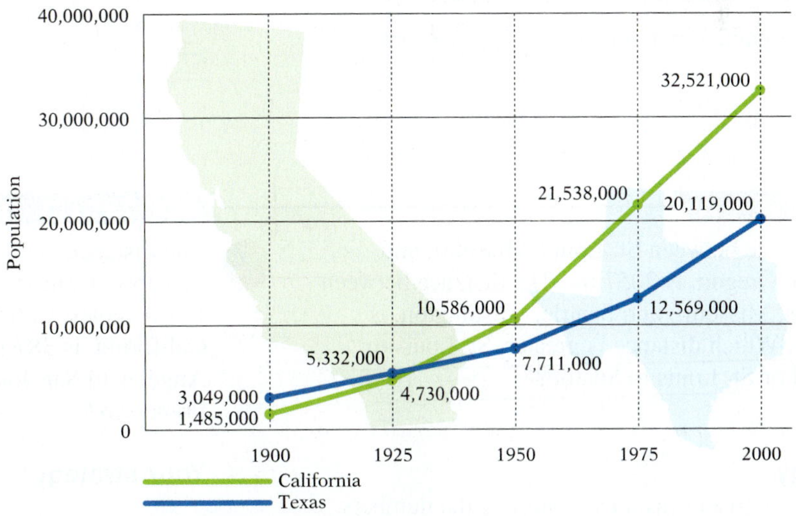

FIGURE 1.6 Populations of California and Texas

From the graph, we can see that the population was greater in Texas in 1900 and 1925, while the population was greater in California in 1950, 1975, and 2000.

To solve an application problem, first read the problem carefully. The **Strategy** involves identifying the quantity to be found and planning the steps that are necessary to find that quantity. The **Solution** involves performing each operation stated in the Strategy and writing the answer.

The circle graph in Figure 1.7 shows the result of a survey of 300 people who were asked to name their favorite sport. Use this graph for Example 10 and You Try It 10.

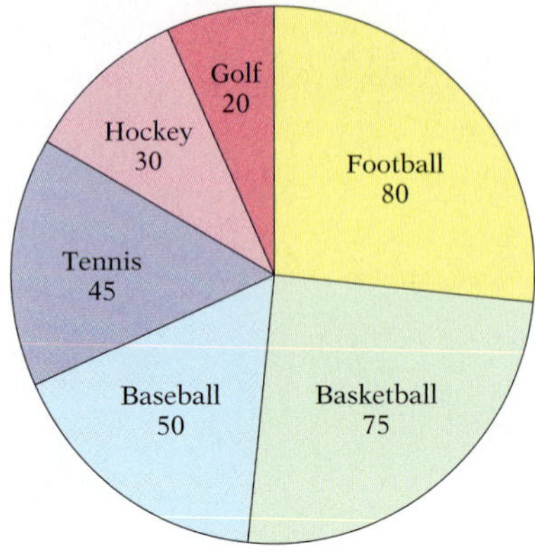

FIGURE 1.7 Distribution of Responses in a Survey

EXAMPLE • 10

According to Figure 1.7, which sport was named by the least number of people?

Strategy
To find the sport named by the least number of people, find the smallest number given in the circle graph.

Solution
The smallest number given in the graph is 20.

The sport named by the least number of people was golf.

YOU TRY IT • 10

According to Figure 1.7, which sport was named by the greatest number of people?

Your strategy

Your solution

EXAMPLE • 11

The distance between St. Louis, Missouri, and Portland, Oregon, is 2057 mi. The distance between St. Louis, Missouri, and Seattle, Washington, is 2135 mi. Which distance is greater, St. Louis to Portland or St. Louis to Seattle?

Strategy
To find the greater distance, compare the numbers 2057 and 2135.

Solution
$2135 > 2057$

The greater distance is from St. Louis to Seattle.

YOU TRY IT • 11

The distance between Los Angeles, California, and San Jose, California, is 347 mi. The distance between Los Angeles, California, and San Francisco, California, is 387 mi. Which distance is shorter, Los Angeles to San Jose or Los Angeles to San Francisco?

Your strategy

Your solution

Solutions on p. S1

The bar graph in Figure 1.8 shows the states with the most sanctioned league bowlers. Use this graph for Example 12 and You Try It 12.

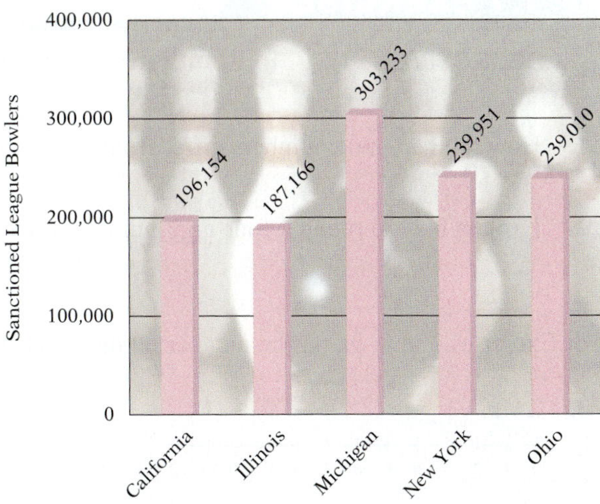

FIGURE 1.8 States with the Most Sanctioned League Bowlers
Sources: American Bowling Congress, Women's International Bowling Congress, Young American Bowling Alliance

EXAMPLE • 12

According to Figure 1.8, which state has the most sanctioned league bowlers?

Strategy
To determine which state has the most sanctioned league bowlers, locate the state that corresponds to the highest bar.

Solution
The highest bar corresponds to Michigan.

Michigan is the state with the most sanctioned league bowlers.

YOU TRY IT • 12

According to Figure 1.8, which state has fewer sanctioned league bowlers, New York or Ohio?

Your strategy

Your solution

EXAMPLE • 13

The land area of the United States is 3,539,341 mi^2. What is the land area of the United States to the nearest ten-thousand square miles?

Strategy
To find the land area to the nearest ten-thousand square miles, round 3,539,341 to the nearest ten-thousand.

Solution
3,539,341 rounded to the nearest ten-thousand is 3,540,000.

To the nearest ten-thousand square miles, the land area of the United States is 3,540,000 mi^2.

YOU TRY IT • 13

The land area of Canada is 3,851,809 mi^2. What is the land area of Canada to the nearest thousand square miles?

Your strategy

Your solution

Solutions on p. S1

1.1 EXERCISES

OBJECTIVE A **To identify the order relation between two whole numbers**

1. How do the whole numbers differ from the natural numbers?

2. Explain how to round a four-digit number to the nearest hundred.

For Exercises 3 to 8, graph the number on the number line.

3. 2

0 1 2 3 4 5 6 7 8 9 10 11 12

4. 7

0 1 2 3 4 5 6 7 8 9 10 11 12

5. 10

0 1 2 3 4 5 6 7 8 9 10 11 12

6. 1

0 1 2 3 4 5 6 7 8 9 10 11 12

7. 5

0 1 2 3 4 5 6 7 8 9 10 11 12

8. 11

0 1 2 3 4 5 6 7 8 9 10 11 12

On the number line, which number is:

9. 4 units to the left of 9

10. 5 units to the left of 8

11. 3 units to the right of 2

12. 4 units to the right of 6

13. 7 units to the left of 7

14. 8 units to the left of 11

For Exercises 15 to 26, place the correct symbol, $<$ or $>$, between the two numbers.

15. 27 39

16. 68 41

17. 0 52

18. 61 0

19. 273 194

20. 419 502

21. 2761 3857

22. 3827 6915

23. 4610 4061

24. 5600 56,000

25. 8005 8050

26. 92,010 92,001

27. Do the inequalities $21 < 30$ and $30 > 21$ express the same order relation?

For Exercises 28 to 36, write the given numbers in order from smallest to largest.

28. 21, 14, 32, 16, 11

29. 18, 60, 35, 71, 27

30. 72, 48, 84, 93, 13

31. 54, 45, 63, 28, 109

32. 26, 49, 106, 90, 77

33. 505, 496, 155, 358, 271

34. 736, 662, 204, 981, 399

35. 440, 404, 400, 444, 4000

36. 377, 370, 307, 3700, 3077

OBJECTIVE B To write whole numbers in words, in standard form, and in expanded form

For Exercises 37 to 48, write the number in words.

37. 704

38. 508

39. 374

40. 635

41. 2861

42. 4790

43. 48,297

44. 53,614

45. 563,078

46. 246,053

47. 6,379,482

48. 3,842,905

For Exercises 49 to 60, write the number in standard form.

49. Seventy-five

50. Four hundred ninety-six

51. Two thousand eight hundred fifty-one

52. Fifty-three thousand three hundred forty

53. One hundred thirty thousand two hundred twelve

54. Five hundred two thousand one hundred forty

55. Eight thousand seventy-three

56. Nine thousand seven hundred six

57. Six hundred three thousand one hundred thirty-two

58. Five million twelve thousand nine hundred seven

59. Three million four thousand eight

60. Eight million five thousand ten

61. What is the place value of the first number on the left in a seven-digit whole number?

For Exercises 62 to 73, write the number in expanded form.

62. 6398

63. 7245

64. 46,182

65. 532,791

66. 328,476

67. 5064

68. 90,834

69. 20,397

70. 400,635

71. 402,708

72. 504,603

73. 8,000,316

74. The expanded form of a number consists of four numbers added together. Must the number be a four-digit number?

OBJECTIVE C To round a whole number to a given place value

For Exercises 75 to 90, round the number to the given place value.

75. 3049; tens

76. 7108; tens

77. 1638; hundreds

78. 4962; hundreds

79. 17,639; hundreds

80. 28,551; hundreds

81. 5326; thousands

82. 6809; thousands

83. 84,608; thousands

84. 93,825; thousands

85. 389,702; thousands

86. 629,513; thousands

87. 746,898; ten-thousands

88. 352,876; ten-thousands

89. 36,702,599; millions

90. 71,834,250; millions

91. True or false? If a number rounded to the nearest ten is less than the original number, then the ones digit of the original number is greater than 5.

OBJECTIVE D To solve application problems and read statistical graphs

92. **Baseball** During his baseball career, Eddie Collins had a record of 743 stolen bases. Max Carey had a record of 738 stolen bases during his baseball career. Who had more stolen bases, Eddie Collins or Max Carey?

93. **Baseball** During his baseball career, Ty Cobb had a record of 892 stolen bases. Billy Hamilton had a record of 937 stolen bases during his baseball career. Who had more stolen bases, Ty Cobb or Billy Hamilton?

94. **Nutrition** The figure at the right shows the annual per capita turkey consumption in different countries.
a. What is the annual per capita turkey consumption in the United States?
b. In which country is the annual per capita turkey consumption the highest?

Britain	🦃🦃🦃🦃
Canada	🦃🦃🦃🦃🦃
France	🦃🦃🦃🦃🦃🦃
Ireland	🦃🦃🦃🦃
Israel	🦃🦃🦃🦃🦃🦃🦃🦃🦃🦃🦃
Italy	🦃🦃🦃🦃🦃
U.S.	🦃🦃🦃🦃🦃🦃🦃🦃🦃
	Each 🦃 represents 2 lb.

Per Capita Turkey Consumption
Source: National Turkey Federation

95. **The Arts** The play *Hello Dolly* was performed 2844 times on Broadway. The play *Fiddler on the Roof* was performed 3242 times on Broadway. Which play had the greater number of performances, *Hello Dolly* or *Fiddler on the Roof*?

96. **The Arts** The play *Annie* was performed 2377 times on Broadway. The play *My Fair Lady* was performed 2717 times on Broadway. Which play had the greater number of performances, *Annie* or *My Fair Lady*?

97. **Nutrition** Two tablespoons of peanut butter contain 190 calories. Two tablespoons of grape jelly contain 114 calories. Which contains more calories, two tablespoons of peanut butter or two tablespoons of grape jelly?

98. History In 1892, the diesel engine was patented. In 1844, Samuel F. B. Morse patented the telegraph. Which was patented first, the diesel engine or the telegraph?

99. Geography The distance between St. Louis, Missouri, and Reno, Nevada, is 1892 mi. The distance between St. Louis, Missouri, and San Diego, California, is 1833 mi. Which is the shorter distance, St. Louis to Reno or St. Louis to San Diego?

100. Movie Theaters The circle graph at the right shows the result of a survey of 150 people who were asked, "What bothers you most about movie theaters?"

a. Among the respondents, what was the most often mentioned complaint?

b. What was the least often mentioned complaint?

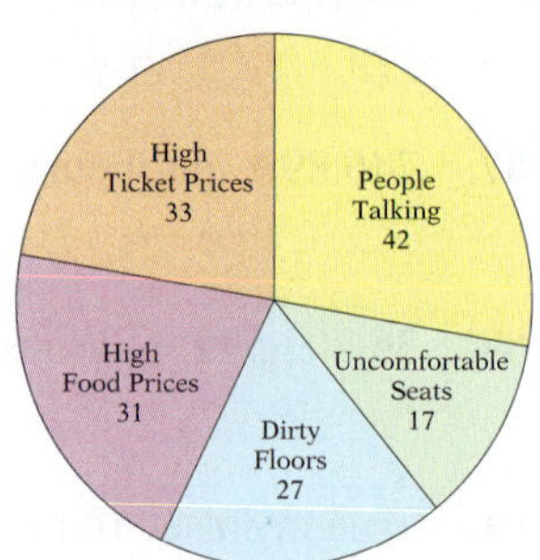

Distribution of Responses in a Survey

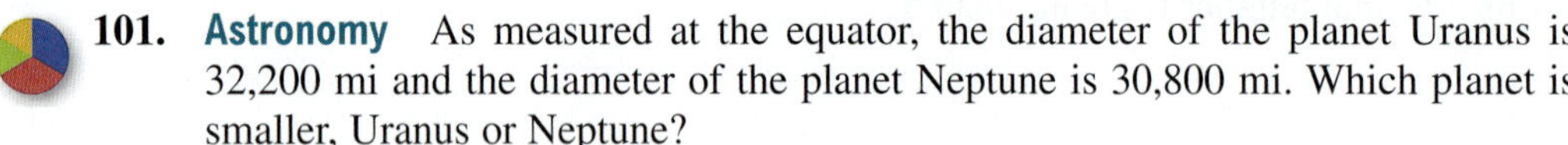

101. Astronomy As measured at the equator, the diameter of the planet Uranus is 32,200 mi and the diameter of the planet Neptune is 30,800 mi. Which planet is smaller, Uranus or Neptune?

102. Astronomy The diameter of Callisto, one of the moons orbiting Jupiter, is 4890 mi. The diameter of Ganymede, another of Jupiter's moons, is 5216 mi. Which is the larger moon, Callisto or Ganymede?

103. Geography The land area of Alaska is 570,833 mi^2. What is the land area of Alaska to the nearest thousand square miles?

104. Geography The acreage of the Appalachian Trail is 161,546 acres. What is the acreage of the Appalachian Trail to the nearest ten thousand acres?

105. Automobile Accidents The figure below shows the number of crashes on U.S. roadways during each of the last six months of a recent year. Also shown is the number of vehicles involved in those crashes.

a. Which was greater, the number of crashes in July or in October?

b. Were there fewer vehicles involved in these crashes in July or in December?

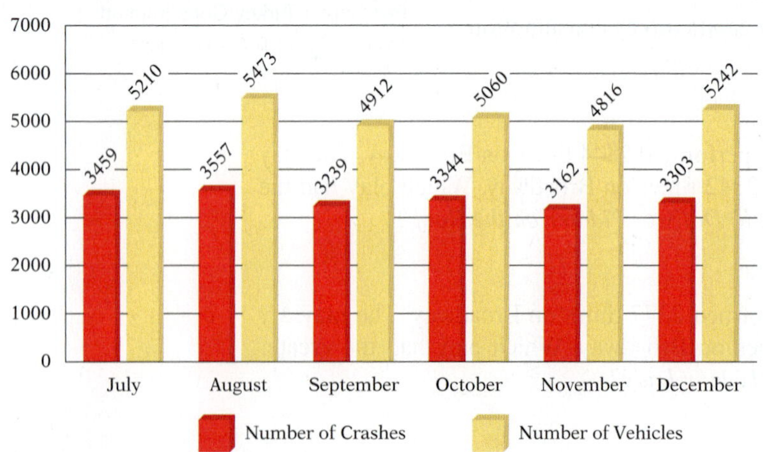

Accidents on U.S. Roadways

Source: National Highway Traffic Safety Administration

106. **a.** See Figure 1.3. Is the U.S. population aged 100 and over increasing or decreasing?
b. See Figure 1.5. Does the value of the life insurance policy increase or decrease?
c. See Figure 1.6. Has the population of Texas increased or decreased?

107. **School Enrollment** Actual and projected student enrollment in elementary and secondary schools in the United States is shown in the figure at the right. Enrollment figures are for the fall of each year. The jagged line at the bottom of the vertical axis indicates that this scale is missing the tens of millions from 0 to 30,000,000.
a. During which year was enrollment the lowest?
b. Did enrollment increase or decrease between 1975 and 1980?

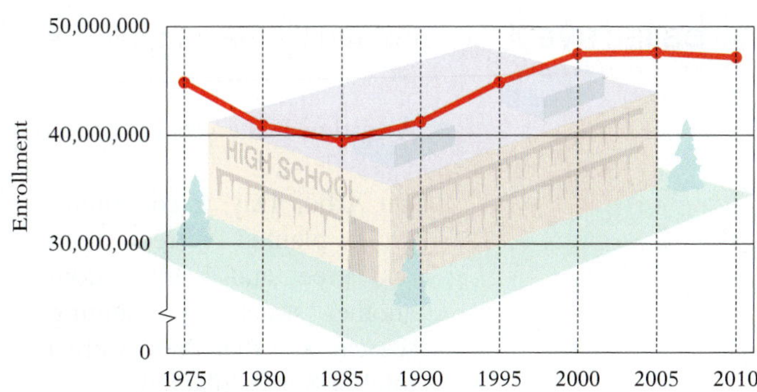

Enrollment in Elementary and Secondary Schools
Source: National Center for Education Statistics

108. **Aviation** The cruising speed of a Boeing 747 is 589 mph. What is the cruising speed of a Boeing 747 to the nearest ten miles per hour?

109. **Physics** Light travels at a speed of 299,800 km/s. What is the speed of light to the nearest thousand kilometers per second?

Applying the Concepts

110. **Geography** Find the land areas of the seven continents. List the continents in order from largest to smallest. List the oceans on Earth from largest to smallest.

111. **Mathematics** What is the largest three-digit number? What is the smallest five-digit number?

112. What is the total enrollment of your school? To what place value would it be reasonable to round this number? Why? To what place value is the population of your town or city rounded? Why? To what place value is the population of your state rounded? To what place value is the population of the United States rounded?

For Exercise 113, answer true or false. If the answer is false, give an example to show that it is false.

113. **a.** If you are given two different whole numbers, then one of the numbers is always greater than the other number.
b. A rounded number is always less than its exact value.

SECTION

1.2 Addition and Subtraction of Whole Numbers

OBJECTIVE A To add whole numbers 

Addition is the process of finding the total of two or more numbers.

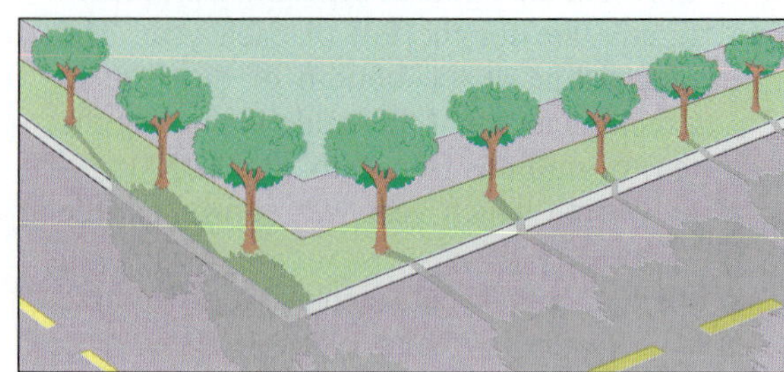

On Arbor Day, a community group planted 3 trees along one street and 5 trees along another street. By counting, we can see that there were a total of 8 trees planted.

$3 + 5 = 8$

The 3 and 5 are called **addends.** The **sum** is 8.

The basic addition facts for adding one digit to one digit should be memorized. Addition of larger numbers requires the repeated use of the basic addition facts.

To add large numbers, begin by arranging the numbers vertically, keeping the digits of the same place value in the same column.

HOW TO • 1 Add: 321 + 6472

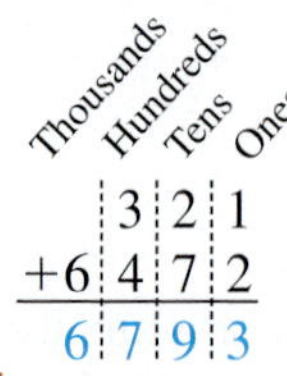

$$\begin{array}{r} 321 \\ +6472 \\ \hline 6793 \end{array}$$

• **Add the digits in each column.**

Tips for Success

The HOW TO feature indicates an example with explanatory remarks. Using paper and pencil, you should work through the example. See *AIM for Success* in the Preface.

HOW TO • 2 Find the sum of 211, 45, 23, and 410.

$$\begin{array}{r} 211 \\ 45 \\ 23 \\ +\ 410 \\ \hline 689 \end{array}$$

• **Remember that a sum is the answer to an addition problem.**
• **Arrange the numbers vertically, keeping digits of the same place value in the same column.**
• **Add the numbers in each column.**

The phrase *the sum of* was used in the example above to indicate the operation of addition. All of the phrases listed below indicate addition. An example of each is shown at the right of each phrase.

added to	6 added to 9	9 + 6
more than	3 more than 8	8 + 3
the sum of	the sum of 7 and 4	7 + 4
increased by	2 increased by 5	2 + 5
the total of	the total of 1 and 6	1 + 6
plus	8 plus 10	8 + 10

Integrating Technology

Most scientific calculators use algebraic logic: the add (**+**), subtract (**−**), multiply (**x**), and divide (**÷**) keys perform the indicated operation on the number in the display and the next number keyed in. For instance, for the example at the right, enter 359 **+** 478 **=** . The display reads 837.

When the sum of the numbers in a column exceeds 9, addition involves **carrying**.

HOW TO • 3 Add: 359 + 478

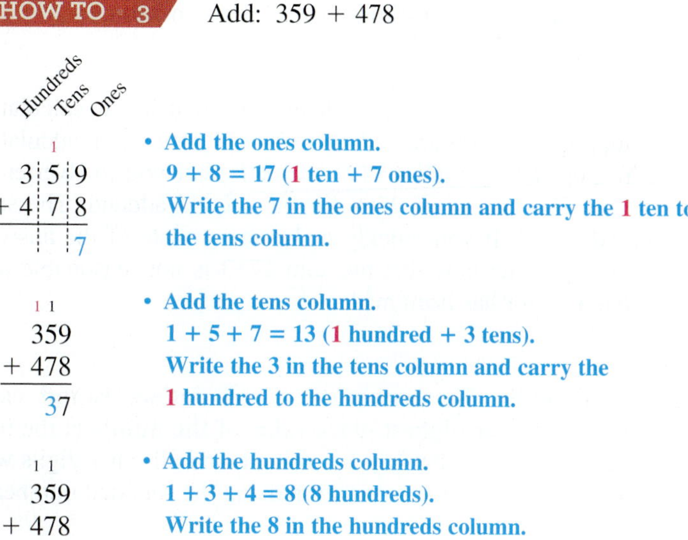

- **Add the ones column.**
  9 + 8 = 17 (1 ten + 7 ones).
  Write the 7 in the ones column and carry the 1 ten to the tens column.

- **Add the tens column.**
  1 + 5 + 7 = 13 (1 hundred + 3 tens).
  Write the 3 in the tens column and carry the 1 hundred to the hundreds column.

- **Add the hundreds column.**
  1 + 3 + 4 = 8 (8 hundreds).
  Write the 8 in the hundreds column.

HOW TO • 4  The bar graph in Figure 1.9 shows the population of each of the six New England states at the 2000 Census. What was the population of New England at the 2000 Census?

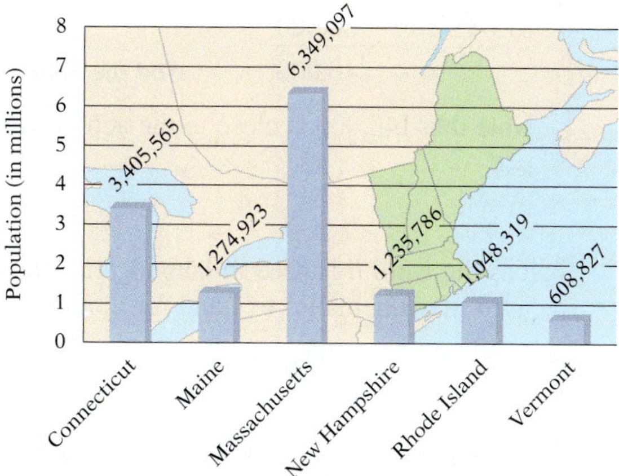

FIGURE 1.9 Population of the Six New England States

$$\begin{array}{r} 3{,}405{,}565 \\ 1{,}274{,}923 \\ 6{,}349{,}097 \\ 1{,}235{,}786 \\ 1{,}048{,}319 \\ +\quad 608{,}827 \\ \hline 13{,}922{,}517 \end{array}$$

The population of New England at the 2000 Census was 13,922,517 people.

An important skill in mathematics is the ability to determine whether an answer to a problem is reasonable. One method of determining whether an answer is reasonable is to use estimation. An **estimate** is an approximation.

Integrating Technology

Here is an example of why estimation is important when using a calculator.

Estimation is especially valuable when using a calculator. Suppose that you are adding 1497 and 2568 on a calculator. You enter the number 1497 correctly, but you inadvertently enter 256 instead of 2568 for the second addend. The sum reads 1753. If you quickly make an estimate of the answer, you can determine that the sum 1753 is not reasonable and that an error has been made.

$$\begin{array}{r} 1497 \\ 2568 \\ \hline 4065 \end{array} \qquad \begin{array}{r} 1497 \\ +\ 256 \\ \hline 1753 \end{array}$$

To estimate the answer to a calculation, round each number to the highest place value of the number; the first digit of each number will be nonzero and all other digits will be zero. Perform the calculation using the rounded numbers.

$$\begin{array}{rr} 1497 \rightarrow & 1000 \\ 2568 \rightarrow & +3000 \\ \hline & 4000 \end{array}$$

As shown above, the sum 4000 is an estimate of the sum of 1497 and 2568; it is very close to the actual sum, 4065. 4000 is not close to the incorrectly calculated sum, 1753.

HOW TO • 5 Estimate the sum of 35,498, 17,264, and 81,093.

$$\begin{array}{rr} 35{,}498 \rightarrow & 40{,}000 \\ 17{,}264 \rightarrow & 20{,}000 \\ 81{,}093 \rightarrow & +\ 80{,}000 \\ \hline & 140{,}000 \end{array}$$

- Round each number to the nearest ten-thousand.
- Add the rounded numbers.

Note that 140,000 is close to the actual sum, 133,855.

Just as the word *it* is used in language to stand for an object, a letter of the alphabet can be used in mathematics to stand for a number. Such a letter is called a **variable.**

A mathematical expression that contains one or more variables is a **variable expression.** Replacing the variables in a variable expression with numbers and then simplifying the numerical expression is called **evaluating the variable expression.**

HOW TO • 6 Evaluate $a + b$ when $a = 678$ and $b = 294$.

$a + b$

$678 + 294$ • Replace a with 678 and b with 294.

$$\begin{array}{r} {\scriptstyle 1\,1} \\ 678 \\ +294 \\ \hline 972 \end{array}$$

• Arrange the numbers vertically.

• Add.

Variables are often used in algebra to describe mathematical relationships. Variables are used below to describe three properties, or rules, of addition. An example of each property is shown at the right.

The Addition Property of Zero

$a + 0 = a$ or $0 + a = a$

$$5 + 0 = 5$$

The Addition Property of Zero states that the sum of a number and zero is the number. The variable a is used here to represent any whole number. It can even represent the number zero because $0 + 0 = 0$.

The Commutative Property of Addition

$a + b = b + a$

$$5 + 7 = 7 + 5$$
$$12 = 12$$

The Commutative Property of Addition states that two numbers can be added in either order; the sum will be the same. Here the variables a and b represent any whole numbers. Therefore, if you know that the sum of 5 and 7 is 12, then you also know that the sum of 7 and 5 is 12, because $5 + 7 = 7 + 5$.

The Associative Property of Addition

$(a + b) + c = a + (b + c)$

$$(2 + 3) + 4 = 2 + (3 + 4)$$
$$5 + 4 = 2 + 7$$
$$9 = 9$$

The Associative Property of Addition states that when adding three or more numbers, we can group the numbers in any order; the sum will be the same. Note in the example at the right above that we can add the sum of 2 and 3 to 4, or we can add 2 to the sum of 3 and 4. In either case, the sum of the three numbers is 9.

HOW TO • 7 Rewrite the expression by using the Associative Property of Addition.

$(3 + x) + y$

$(3 + x) + y = 3 + (x + y)$ • **The Associative Property of Addition states that addends can be grouped in any order.**

Point of Interest

The equals sign (=) is generally credited to Robert Recorde. In his 1557 treatise on algebra, *The Whetstone of Whit,* he wrote, "No two things could be more equal (than two parallel lines)." His equals sign gained popularity, even though continental mathematicians preferred a dash.

An **equation** expresses the equality of two numerical or variable expressions. In the preceding example, $(3 + x) + y$ is an expression; it does not contain an equals sign. $(3 + x) + y = 3 + (x + y)$ is an equation; it contains an equals sign.

Here is another example of an equation. The **left side** of the equation is the variable expression $n + 4$. The **right side** of the equation is the number 9.

$n + 4 = 9$

Just as a statement in English can be true or false, an equation may be true or false. The equation shown above is *true* if the variable is replaced by 5.

$n + 4 = 9$
$5 + 4 = 9$ True

The equation is *false* if the variable is replaced by 8.

$8 + 4 = 9$ False

A **solution** of an equation is a number that, when substituted for the variable, results in a true equation. The solution of the equation $n + 4 = 9$ is 5 because replacing n by 5 results in a true equation. When 8 is substituted for n, the result is a false equation; therefore, 8 is not a solution of the equation.

Tips for Success

One of the key instructional features of this text is the Example/You Try It pairs. Each Example is completely worked. You are to solve the You Try It problems. When you are ready, check your solution against the one in the Solutions section. The solutions for You Try Its 1 and 2 below are on page S1 (see the reference at the bottom right of the You Try It box). *See AIM for Success* in the Preface.

10 is a solution of $x + 5 = 15$ because $10 + 5 = 15$ is a true equation.

20 is not a solution of $x + 5 = 15$ because $20 + 5 = 15$ is a false equation.

HOW TO • 8 Is 9 a solution of the equation $11 = 2 + x$?

$$\begin{array}{c|c} 11 = & 2 + x \\ \hline 11 & 2 + 9 \\ 11 = & 11 \end{array}$$

- **Replace x by 9.**
- **Simplify the right side of the equation. Compare the results. If the results are equal, the given number is a solution of the equation. If the results are not equal, the given number is not a solution.**

Yes, 9 is a solution of the equation.

EXAMPLE • 1

Estimate the sum of 379, 842, 693, and 518.

Solution

$$\begin{array}{lr} 379 \rightarrow & 400 \\ 842 \rightarrow & 800 \\ 693 \rightarrow & 700 \\ 518 \rightarrow & +\ 500 \\ \hline & 2400 \end{array}$$

YOU TRY IT • 1

Estimate the total of 6285, 3972, and 5140.

Your solution

EXAMPLE • 2

Identify the property that justifies the statement.
$7 + 2 = 2 + 7$

Solution

The Commutative Property of Addition

YOU TRY IT • 2

Identify the property that justifies the statement.
$33 + 0 = 33$

Your solution

Solutions on p. S1

The topic of the circle graph in Figure 1.10 is the eggs produced in the United States in a recent year. It shows where the eggs that were produced went or how they were used. Use this graph for Example 3 and You Try It 3.

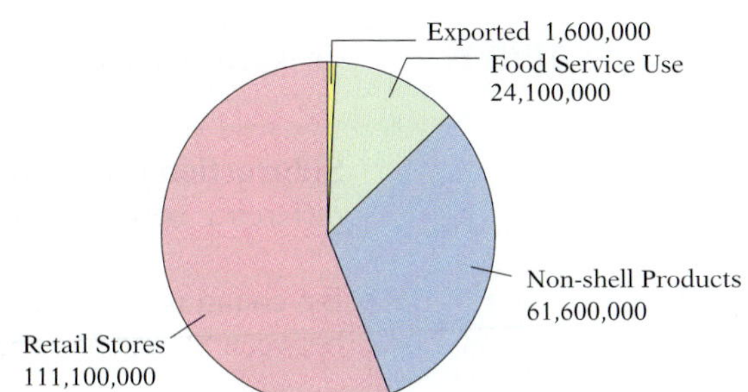

FIGURE 1.10 Distribution of Eggs Produced in the United States (in cases)
Source: American Egg Board. *USA Today.* Copyright © November 27, 2001.

EXAMPLE • 3

Use Figure 1.10 to determine the sum of the number of cases of eggs sold by retail stores and the number used for non-shell products.

Solution

111,100,000 cases of eggs were sold by retail stores. 61,600,000 cases of eggs were used for non-shell products.

$$\begin{array}{r} 111{,}100{,}000 \\ +\ 61{,}600{,}000 \\ \hline 172{,}700{,}000 \end{array}$$

172,700,000 cases of eggs were sold by retail stores and used for non-shell products.

YOU TRY IT • 3

Use Figure 1.10 to determine the total number of cases of eggs produced during the year.

Your solution

EXAMPLE • 4

Evaluate $x + y + z$ when $x = 8427$, $y = 3659$, and $z = 6281$.

Solution

$x + y + z$

$8427 + 3659 + 6281$

$$\begin{array}{r} {\scriptstyle 1\,1\,1} \\ 8427 \\ 3659 \\ +\ 6281 \\ \hline 18{,}367 \end{array}$$

YOU TRY IT • 4

Evaluate $x + y + z$ when $x = 1692$, $y = 4783$, and $z = 5046$.

Your solution

EXAMPLE • 5

Is 6 a solution of the equation $9 + y = 14$?

Solution

$$\begin{array}{c|c} 9 + y = 14 & \\ \hline 9 + 6 & 14 \\ 15 \neq 14 & \end{array}$$

• **The symbol $\neq$ is read "is not equal to."**

No, 6 is not a solution of the equation $9 + y = 14$.

YOU TRY IT • 5

Is 7 a solution of the equation $13 = b + 6$?

Your solution

Solutions on pp. S1–S2

OBJECTIVE B To subtract whole numbers 

Subtraction is the process of finding the difference between two numbers.

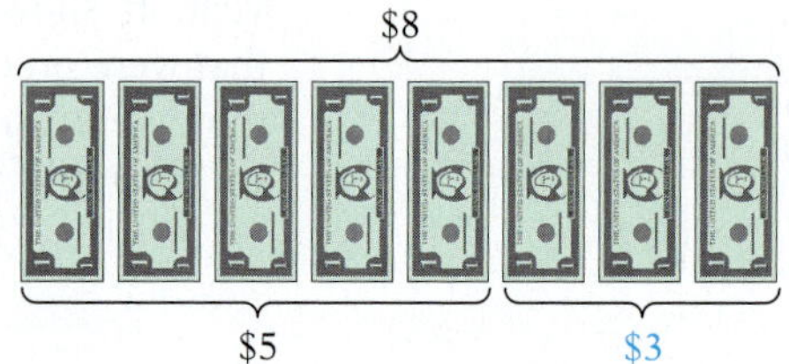

By counting, we see that the difference between \$8 and \$5 is \$3.

\$8 − \$5 = \$3

Minuend − Subtrahend = Difference

Note that addition and subtraction are related.

Subtrahend	5
+ Difference	+3
= Minuend	8

The fact that the sum of the subtrahend and the difference equals the minuend can be used to check subtraction.

To subtract large numbers, begin by arranging the numbers vertically, keeping the digits of the same place value in the same column. Then subtract the numbers in each column.

HOW TO • 9 Find the difference between 8955 and 2432.

A *difference* is the answer to a subtraction problem.

Thousands	Hundreds	Tens	Ones
8	9	5	5
− 2	4	3	2
6	5	2	3

Check:

Subtrahend	2432
+ Difference	+6523
= Minuend	8955

In the subtraction example above, the lower digit in each place value is smaller than the upper digit. When the lower digit is larger than the upper digit, subtraction involves **borrowing.**

HOW TO • 10 Subtract: 692 − 378

Hundreds	Tens	Ones
	8 + 1	
6	9	2
− 3	7	8

8 > 2
Borrowing is necessary.
9 tens = 8 tens + 1 ten

Hundreds	Tens	Ones
	8 + ①	10
6	9	2
− 3	7	8

Borrow 1 ten from the tens column and write 10 in the ones column.

Hundreds	Tens	Ones
	8	12
6	9	2
− 3	7	8

Add the borrowed 10 to 2.

Hundreds	Tens	Ones
	8	12
6	9	2
− 3	7	8
3	1	4

Subtract the numbers in each column.

Subtraction may involve repeated borrowing.

HOW TO • 11 Subtract: 7325 − 4698

Step 1	Step 2	Step 3
$\begin{array}{r} \scriptstyle 1\ 15 \\ 7\,3\,\not2\,\not5 \\ -\,4\,6\,9\,8 \\ \hline 7 \end{array}$	$\begin{array}{r} \scriptstyle 11 \\ \scriptstyle 2\ \not1\ 15 \\ 7\,\not3\,\not2\,\not5 \\ -\,4\,6\,9\,8 \\ \hline 2\,7 \end{array}$	$\begin{array}{r} \scriptstyle 12\ 11 \\ \scriptstyle 6\ 2\ \not1\ 15 \\ \not7\,\not3\,\not2\,\not5 \\ -\,4\,6\,9\,8 \\ \hline 2\,6\,2\,7 \end{array}$
Borrow 1 ten (10 ones) from the tens column and add 10 to the 5 in the ones column. Subtract 15 − 8.	Borrow 1 hundred (10 tens) from the hundreds column and add 10 to the 1 in the tens column. Subtract 11 − 9.	Borrow 1 thousand (10 hundreds) from the thousands column and add 10 to the 2 in the hundreds column. Subtract 12 − 6 and 6 − 4.

When there is a zero in the minuend, subtraction involves repeated borrowing.

HOW TO • 12 Subtract: 3904 − 1775

Step 1	Step 2	Step 3
$\begin{array}{r} \scriptstyle 8\ 10 \\ 3\,\not9\,\not0\,4 \\ -\,1\,7\,7\,5 \\ \hline \end{array}$	$\begin{array}{r} \scriptstyle 9 \\ \scriptstyle 8\ \not{10}\ 14 \\ 3\,\not9\,\not0\,\not4 \\ -\,1\,7\,7\,5 \\ \hline \end{array}$	$\begin{array}{r} \scriptstyle 9 \\ \scriptstyle 8\ \not{10}\ 14 \\ 3\,\not9\,\not0\,\not4 \\ -\,1\,7\,7\,5 \\ \hline 2\,1\,2\,9 \end{array}$
There is a 0 in the tens column. Borrow 1 hundred (10 tens) from the hundreds column and write 10 in the tens column.	Borrow 1 ten from the tens column and add 10 to the 4 in the ones column.	Subtract the numbers in each column.

Note that, for the preceding example, the borrowing could be performed as shown below.

Borrow 1 from 90. (90 − 1 = 89. The 8 is in the hundreds column. The 9 is in the tens column.) Add 10 to the 4 in the ones column. Then subtract the numbers in each column.

$$\begin{array}{r} \scriptstyle 8\ 9\ 14 \\ 3\,\not9\,\not0\,\not4 \\ -\,1\,7\,7\,5 \\ \hline 2\,1\,2\,9 \end{array}$$

HOW TO • 13 Estimate the difference between 49,601 and 35,872.

$$\begin{array}{rr} 49{,}601 \rightarrow & 50{,}000 \\ 35{,}872 \rightarrow & -40{,}000 \\ \hline & 10{,}000 \end{array}$$

• **Round each number to the nearest ten-thousand.**

• **Subtract the rounded numbers.**

Note that 10,000 is close to the actual difference, 13,729.

Take Note

Note the order in which the numbers are subtracted when the phrase *less than* is used. Suppose that you have \$10 and I have \$6 *less than* you do; then I have \$6 *less than* \$10, or \$10 − \$6 = \$4.

The phrase *the difference between* was used in the example above to indicate the operation of subtraction. All of the phrases listed below indicate subtraction. An example of each is shown at the right of each phrase.

minus	10 minus 3	$10 - 3$
less	8 less 4	$8 - 4$
less than	2 less than 9	$9 - 2$
the difference between	the difference between 6 and 1	$6 - 1$
decreased by	7 decreased by 5	$7 - 5$
subtract . . . from	subtract 11 from 20	$20 - 11$

HOW TO • 14 Evaluate $c - d$ when $c = 6183$ and $d = 2759$.

$c - d$

$6183 - 2759$ • **Replace *c* with 6183 and *d* with 2759.**

$$\begin{array}{r} \overset{5\ 11\ 7\ 13}{\cancel{6}\cancel{1}\cancel{8}\cancel{3}} \\ -\ 2759 \\ \hline 3424 \end{array}$$

• **Arrange the numbers vertically and then subtract.**

Point of Interest

Someone who is our equal is our peer. Two make a pair. Both of the words *peer* and *pair* come from the Latin *par, paris,* meaning "equal."

HOW TO • 15 Is 23 a solution of the equation $41 - n = 17$?

$$\begin{array}{r|l} 41 - n = & 17 \\ \hline 41 - 23 & 17 \\ 18 \neq & 17 \end{array}$$

• **Replace *n* by 23.**

• **Simplify the left side of the equation. The results are not equal.**

No, 23 is not a solution of the equation.

EXAMPLE • 6

Subtract and check: 57,004 − 26,189

Solution

$$\begin{array}{r} \overset{6\ \ 9\ 9\ 14}{5\cancel{7},\cancel{0}\cancel{0}\cancel{4}} \\ -\ 26,189 \\ \hline 30,815 \end{array}$$

Check:

$$\begin{array}{r} 26,189 \\ +\ 30,815 \\ \hline 57,004 \end{array}$$

YOU TRY IT • 6

Subtract and check: 49,002 − 31,865

Your solution

EXAMPLE • 7

Estimate the difference between 7261 and 4315. Then find the exact answer.

Solution

$$\begin{array}{lrr} 7261 \longrightarrow & 7000 & 7261 \\ 4315 \longrightarrow & -4000 & -4315 \\ \hline & 3000 & 2946 \end{array}$$

YOU TRY IT • 7

Estimate the difference between 8544 and 3621. Then find the exact answer.

Your solution

Solutions on p. S2

The graph in Figure 1.11 shows the actual and projected world energy consumption in quadrillion British thermal units (Btu). Use this graph for Example 8 and You Try It 8.

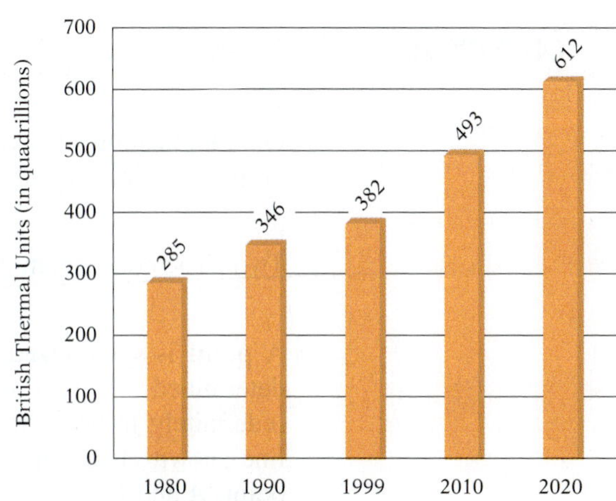

FIGURE 1.11 World Energy Consumption (in quadrillion British thermal units)
Sources: Energy Information Administration; Office of Energy Markets and End Use; *International Statistics Database and International Energy Annual;* World Energy Projection System

EXAMPLE • 8

Use Figure 1.11 to find the difference between the world energy consumption in 1980 and that projected for 2010.

Solution

2010: 493 quadrillion Btu
1980: 285 quadrillion Btu

$$\begin{array}{r} 493 \\ -285 \\ \hline 208 \end{array}$$

The difference between the world energy consumption in 1980 and that projected for 2010 is 208 quadrillion Btu.

YOU TRY IT • 8

Use Figure 1.11 to find the difference between the world energy consumption in 1990 and that projected for 2020.

Your solution

EXAMPLE • 9

Evaluate $x - y$ when $x = 3506$ and $y = 2477$.

Solution

$x - y$
$3506 - 2477$

$$\begin{array}{r} \scriptstyle 4\ 9\ 16 \\ 3\,\not{5}\,\not{0}\,\not{6} \\ -\ 2\,4\,7\,7 \\ \hline 1\,0\,2\,9 \end{array}$$

YOU TRY IT • 9

Evaluate $x - y$ when $x = 7061$ and $y = 3229$.

Your solution

EXAMPLE • 10

Is 39 a solution of the equation $24 = m - 15$?

Solution

$$\begin{array}{l|l} 24 & m - 15 \\ \hline 24 & 39 - 15 \\ 24 & = 24 \end{array}$$

• **Replace *m* by 39.**

Yes, 39 is a solution of the equation.

YOU TRY IT • 10

Is 11 a solution of the equation $46 = 58 - p$?

Your solution

Solutions on p. S2

OBJECTIVE C To solve application problems and use formulas 

One application of addition is calculating the perimeter of a figure. However, before defining perimeter, we will introduce some terms from geometry.

Two basic concepts in the study of geometry are point and line.

A **point** is symbolized by drawing a dot. A **line** is determined by two distinct points and extends indefinitely in both directions, as the arrows on the line shown at the right indicate. This line contains points A and B.

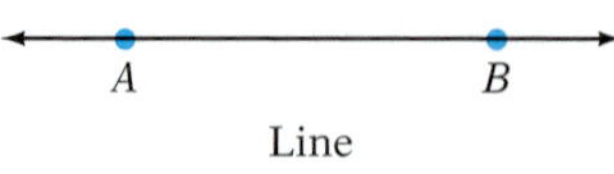

Line

A **ray** starts at a point and extends indefinitely in *one* direction. The point at which a ray starts is called the **endpoint** of the ray. Point A is the endpoint of the ray shown at the right.

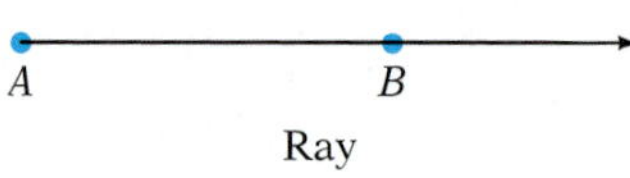

Ray

A **line segment** is part of a line and has two endpoints. The line segment shown at the right has endpoints A and B.

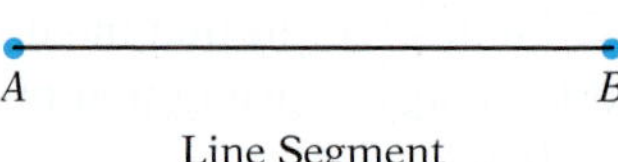

Line Segment

The corner of a page of this book is a good model of a right angle.

An **angle** is formed by two rays with the same endpoint. An angle is measured in **degrees.** The symbol for degrees is a small raised circle, °. A **right angle** is an angle whose measure is 90°.

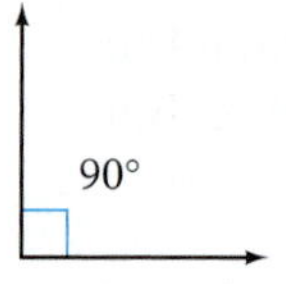

Right Angle

A **plane** is a flat surface and can be pictured as a floor or a wall. Figures that lie in a plane are called **plane figures.**

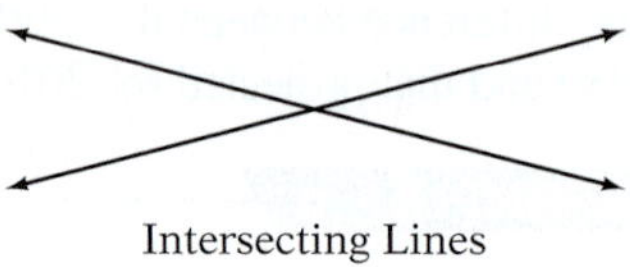
Intersecting Lines

Lines in a plane can be intersecting or parallel. **Intersecting lines** cross at a point in the plane. **Parallel lines** never meet. The distance between them is always the same.

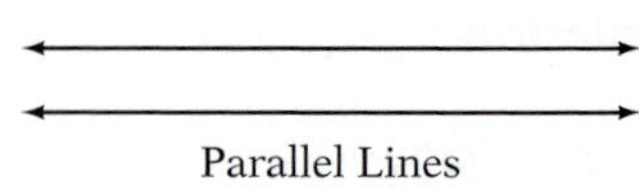
Parallel Lines

A **polygon** is a closed figure determined by three or more line segments that lie in a plane. The line segments that form the polygon are called its **sides.** The figures below are examples of polygons.

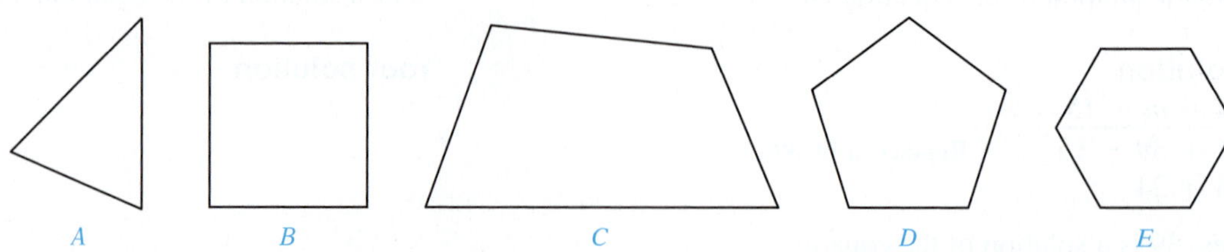

The name of a polygon is based on the number of its sides. A polygon with three sides is a **triangle.** Figure *A* on the previous page is a triangle. A polygon with four sides is a **quadrilateral.** Figures *B* and *C* are quadrilaterals.

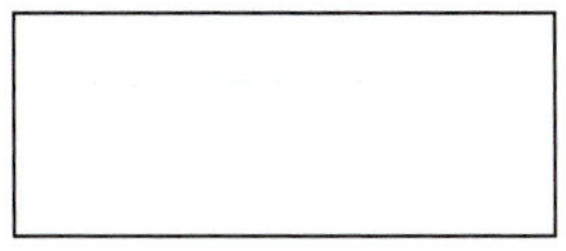

Rectangle

Quadrilaterals are one of the most common types of polygons. Quadrilaterals are distinguished by their sides and angles. For example, a **rectangle** is a quadrilateral in which opposite sides are parallel, opposite sides are equal in length, and all four angles measure 90°.

The **perimeter** of a plane geometric figure is a measure of the distance around the figure.

The perimeter of a triangle is the sum of the lengths of the three sides.

Perimeter of a Triangle

The formula for the perimeter of a triangle is $P = a + b + c$, where P is the perimeter of the triangle and a, b, and c are the lengths of the sides of the triangle.

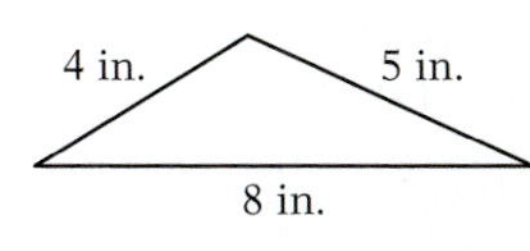

HOW TO • 16 Find the perimeter of the triangle shown at the left.

$P = a + b + c$ • **Use the formula for the perimeter of a triangle.**
$P = 4 + 5 + 8$ • **It does not matter which side you label *a, b,* or *c*.**
$P = 17$ • **Add.**

The perimeter of the triangle is 17 in.

The perimeter of a quadrilateral is the sum of the lengths of its four sides.

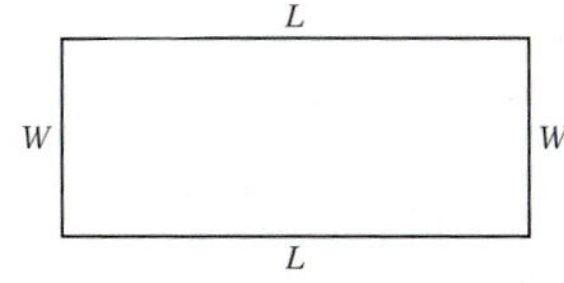

In a rectangle, opposite sides are equal in length. Usually the length, L, of a rectangle refers to the length of one of the longer sides of the rectangle, and the width, W, refers to the length of one of the shorter sides. The perimeter can then be represented as $P = L + W + L + W$.

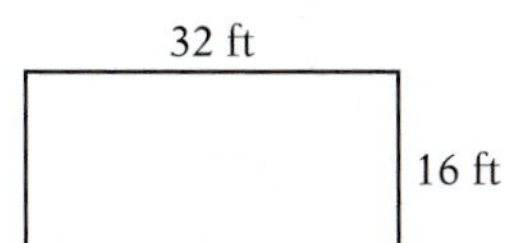

HOW TO • 17 Use the formula $P = L + W + L + W$ to find the perimeter of the rectangle shown at the left.

$P = L + W + L + W$ • **Write the given formula for the perimeter of a rectangle.**
$P = 32 + 16 + 32 + 16$ • **Substitute 32 for *L* and 16 for *W*.**
$P = 96$ • **Add.**

The perimeter of the rectangle is 96 ft.

In this section, some of the phrases used to indicate the operations of addition and subtraction were presented. In solving application problems, you might also look for the types of questions listed below.

Addition	**Subtraction**
How many . . . altogether?	How many more (or fewer) . . . ?
How many . . . in all?	How much is left?
How many . . . and . . . ?	How much larger (or smaller) . . . ?

The bar graph in Figure 1.12 shows the number of fatal accidents on amusement rides in the United States each year during the 1990s. Use this graph for Example 11 and You Try It 11.

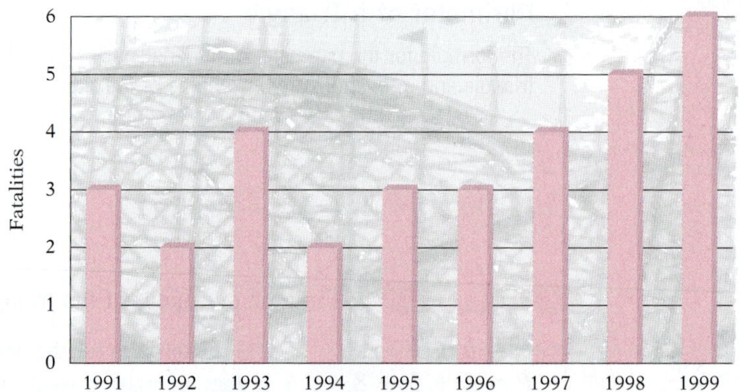

FIGURE 1.12 Number of Fatal Accidents on Amusement Rides
Source: USA Today, April 7, 2000

EXAMPLE • 11

Use Figure 1.12 to determine how many more fatal accidents occurred during the years 1995 through 1998 than occurred during the years 1991 through 1994.

Strategy

To find how many more fatalities occurred in 1995 through 1998 than occurred in 1991 through 1994:

- Find the total number of fatalities that occurred from 1995 through 1998 and the total number that occurred from 1991 through 1994.
- Subtract the smaller number from the larger.

Solution

Fatalities during 1995–1998: 15
Fatalities during 1991–1994: 11

$15 - 11 = 4$

4 more fatalities occurred from 1995 to 1998 than occurred from 1991 to 1994.

YOU TRY IT • 11

Use Figure 1.12 to find the total number of fatal accidents on amusement rides during 1991 through 1999.

Your strategy

Your solution

Solution on p. S2

EXAMPLE • 12

What is the price of a pair of skates that cost a business \$109 and has a markup of \$49? Use the formula $P = C + M$, where P is the price of a product paid by the consumer, C is the cost paid by the store for the product, and M is the markup.

Strategy

To find the price, replace C by 109 and M by 49 in the given formula and solve for P.

Solution

$P = C + M$

$P = 109 + 49$ • **Replace C by 109 and M by 49.**

$P = 158$

The price of the skates is \$158.

YOU TRY IT • 12

What is the price of a leather jacket that cost a business \$148 and has a markup of \$74? Use the formula $P = C + M$, where P is the price of a product paid by the consumer, C is the cost paid by the store for the product, and M is the markup.

Your strategy

Your solution

EXAMPLE • 13

Find the length of decorative molding needed to edge the tops of the walls in a rectangular room that is 12 ft long and 8 ft wide.

Strategy

Draw a diagram.

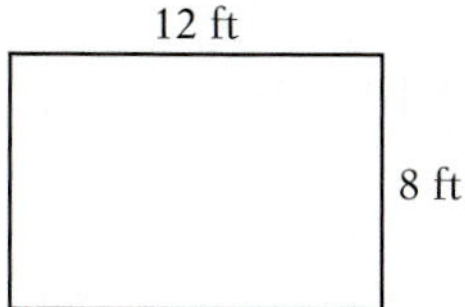

To find the length of molding needed, use the formula for the perimeter of a rectangle, $P = L + W + L + W$. $L = 12$ and $W = 8$.

Solution

$P = L + W + L + W$

$P = 12 + 8 + 12 + 8$ • **Replace L by 12 and W by 8.**

$P = 40$

40 ft of decorative molding is needed.

YOU TRY IT • 13

Find the length of fencing needed to surround a rectangular corral that measures 60 ft on each side.

Your strategy

Your solution

Solutions on p. S2

1.2 EXERCISES

OBJECTIVE A To add whole numbers

1. Provide at least three examples of situations in which we add numbers.

2. Explain how to estimate the sum of 3287 and 4916.

For Exercises 3 to 14, add.

3. $\begin{array}{r} 732{,}453 \\ +\ 651{,}206 \\ \hline \end{array}$

4. $\begin{array}{r} 563{,}841 \\ +\ 726{,}053 \\ \hline \end{array}$

5. $\begin{array}{r} 2879 \\ +\ 3164 \\ \hline \end{array}$

6. $\begin{array}{r} 9857 \\ +\ 1264 \\ \hline \end{array}$

7. $\begin{array}{r} 45{,}825 \\ +\ 66{,}327 \\ \hline \end{array}$

8. $\begin{array}{r} 56{,}442 \\ +\ 71{,}289 \\ \hline \end{array}$

9. $\begin{array}{r} 4037 \\ 3342 \\ +\ 5169 \\ \hline \end{array}$

10. $\begin{array}{r} 5242 \\ 7883 \\ +\ 4165 \\ \hline \end{array}$

11. $\begin{array}{r} 67{,}390 \\ 42{,}761 \\ +\ 89{,}405 \\ \hline \end{array}$

12. $\begin{array}{r} 34{,}801 \\ 97{,}302 \\ +\ 68{,}945 \\ \hline \end{array}$

13. $\begin{array}{r} 54{,}097 \\ 33{,}432 \\ 97{,}126 \\ 64{,}508 \\ +\ 78{,}310 \\ \hline \end{array}$

14. $\begin{array}{r} 23{,}086 \\ 44{,}697 \\ 67{,}302 \\ 83{,}441 \\ +\ 19{,}843 \\ \hline \end{array}$

15. What is 88,123 increased by 80,451?

16. What is 44,765 more than 82,003?

17. What is 654 added to 7293?

18. Find the sum of 658, 2709, and 10,935.

19. Find the total of 216, 8707, and 90,714.

20. Write the sum of x and y.

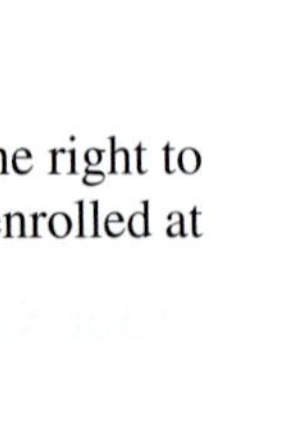

21. **College Enrollment** Use the figure at the right to find the total number of undergraduates enrolled at the college in 2011.

22. **College Enrollment** Use the figure at the right to find the total number of undergraduates enrolled at the college in 2012.

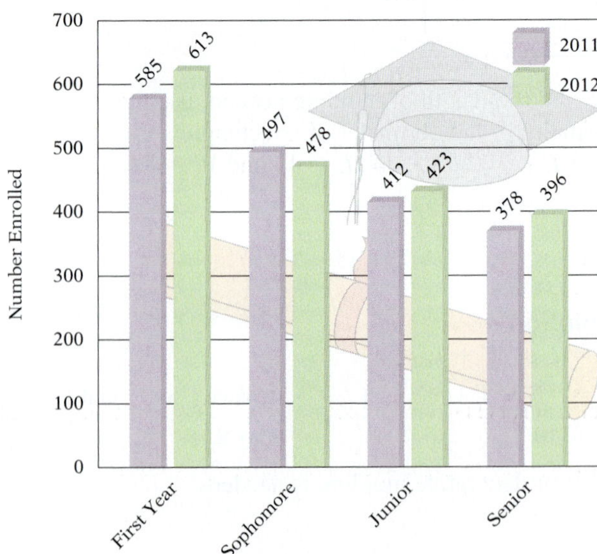

Undergraduates Enrolled in a Private College

For Exercises 23 to 30, estimate by rounding. Then find the exact answer.

23. 6742 + 8298

24. 5426 + 1732

25. 972,085 + 416,832

26. 23,774 + 38,026

27. $\begin{array}{r} 387 \\ 295 \\ 614 \\ +\ 702 \\ \hline \end{array}$

28. $\begin{array}{r} 528 \\ 163 \\ 947 \\ +\ 275 \\ \hline \end{array}$

29. $\begin{array}{r} 224{,}196 \\ 7{,}074 \\ +\ 98{,}531 \\ \hline \end{array}$

30. $\begin{array}{r} 1{,}607 \\ 873{,}925 \\ +\ 28{,}744 \\ \hline \end{array}$

For Exercises 31 to 36, evaluate the variable expression $x + y$ for the given values of x and y.

31. $x = 574$; $y = 698$

32. $x = 359$; $y = 884$

33. $x = 4752$; $y = 7398$

34. $x = 6047$; $y = 9283$

35. $x = 38{,}229$; $y = 51{,}671$

36. $x = 74{,}376$; $y = 19{,}528$

For Exercises 37 to 42, evaluate the variable expression $a + b + c$ for the given values of a, b, and c.

37. $a = 693$; $b = 508$; $c = 371$

38. $a = 177$; $b = 892$; $c = 405$

39. $a = 4938$; $b = 2615$; $c = 7038$

40. $a = 6059$; $b = 3774$; $c = 5136$

41. $a = 12{,}897$; $b = 36{,}075$; $c = 7038$

42. $a = 52{,}847$; $b = 3774$; $c = 5136$

For Exercises 43 to 48, identify the property that justifies the statement.

43. $9 + 12 = 12 + 9$

44. $8 + 0 = 8$

45. $11 + (13 + 5) = (11 + 13) + 5$

46. $0 + 16 = 16 + 0$

47. $0 + 47 = 47$

48. $(7 + 8) + 10 = 7 + (8 + 10)$

For Exercises 49 to 54, use the given property of addition to complete the statement.

49. The Addition Property of Zero
$28 + 0 = ?$

50. The Commutative Property of Addition
$16 + ? = 7 + 16$

51. The Associative Property of Addition
$9 + (? + 17) = (9 + 4) + 17$

52. The Addition Property of Zero
$0 + ? = 51$

53. The Commutative Property of Addition
$? + 34 = 34 + 15$

54. The Associative Property of Addition
$(6 + 18) + ? = 6 + (18 + 4)$

55. Which property of addition (see page 21) allows you to use either arrangement shown at the right to find the sum of 691 and 452?

$$\begin{array}{r} 691 \\ +\,452 \\ \hline \end{array} \qquad \begin{array}{r} 452 \\ +\,691 \\ \hline \end{array}$$

56. Is 38 a solution of the equation
$42 = n + 4$?

57. Is 17 a solution of the equation
$m + 6 = 13$?

58. Is 13 a solution of the equation
$2 + h = 16$?

59. Is 41 a solution of the equation
$n = 17 + 24$?

60. Is 30 a solution of the equation
$32 = x + 2$?

61. Is 29 a solution of the equation
$38 = 11 + z$?

62. What is the difference between an expression and an equation?

OBJECTIVE B To subtract whole numbers

63. Provide at least three examples of situations in which we subtract numbers.

For Exercises 64 to 79, subtract.

64. $\begin{array}{r} 883 \\ -\,467 \\ \hline \end{array}$

65. $\begin{array}{r} 591 \\ -\,238 \\ \hline \end{array}$

66. $\begin{array}{r} 360 \\ -\,172 \\ \hline \end{array}$

67. $\begin{array}{r} 950 \\ -\,483 \\ \hline \end{array}$

68. $\begin{array}{r} 657 \\ -\ 193 \\ \hline \end{array}$

69. $\begin{array}{r} 762 \\ -\ 659 \\ \hline \end{array}$

70. $\begin{array}{r} 407 \\ -\ 199 \\ \hline \end{array}$

71. $\begin{array}{r} 805 \\ -\ 147 \\ \hline \end{array}$

72. $\begin{array}{r} 6814 \\ -\ 3257 \\ \hline \end{array}$

73. $\begin{array}{r} 7361 \\ -\ 4575 \\ \hline \end{array}$

74. $\begin{array}{r} 5000 \\ -\ 2164 \\ \hline \end{array}$

75. $\begin{array}{r} 4000 \\ -\ 1873 \\ \hline \end{array}$

76. $\begin{array}{r} 3400 \\ -\ 1963 \\ \hline \end{array}$

77. $\begin{array}{r} 7300 \\ -\ 2562 \\ \hline \end{array}$

78. $\begin{array}{r} 30{,}004 \\ -\ 9{,}856 \\ \hline \end{array}$

79. $\begin{array}{r} 70{,}003 \\ -\ 8{,}246 \\ \hline \end{array}$

80. Suppose three whole numbers, called *minuend, subtrahend,* and *difference,* are related by the subtraction statement *minuend* − *subtrahend* = *difference*. State whether the given relationship *must be true, might be true,* or *cannot be true.*
a. minuend > difference **b.** subtrahend < difference

81. Find the difference between 2536 and 918.

82. What is 1623 minus 287?

83. What is 5426 less than 12,804?

84. Find 14,801 less 3522.

85. Find 85,423 decreased by 67,875.

86. Write the difference between x and y.

87. Geysers Use the figure at the right to find the difference between the maximum height to which Great Fountain erupts and the maximum height to which Valentine erupts.

88. Geysers According to the figure at the right, how much higher is the eruption of the Giant than that of Old Faithful?

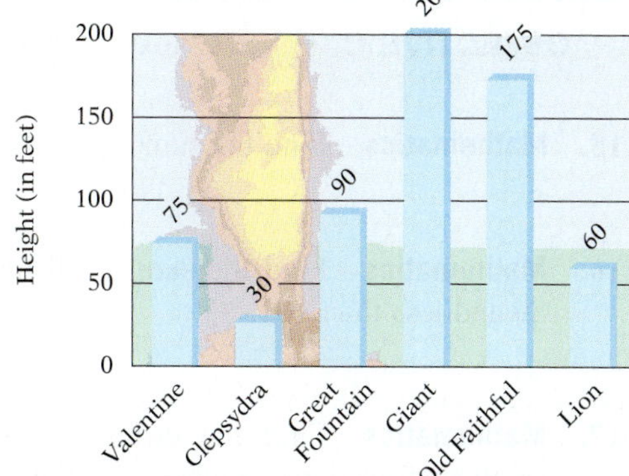

The Maximum Heights of the Eruptions of Six Geysers at Yellowstone National Park

For Exercises 89 to 96, estimate by rounding. Then find the exact answer.

89. 7355 − 5219

90. 8953 − 2217

91. 59,126 − 20,843

92. 63,051 − 29,478

93. $\begin{array}{r} 36{,}287 \\ -\ 5{,}092 \\ \hline \end{array}$

94. $\begin{array}{r} 58{,}316 \\ -\ 19{,}072 \\ \hline \end{array}$

95. $\begin{array}{r} 224{,}196 \\ -\ 98{,}531 \\ \hline \end{array}$

96. $\begin{array}{r} 873{,}925 \\ -\ 28{,}744 \\ \hline \end{array}$

For Exercises 97 to 108, evaluate the variable expression $x - y$ for the given values of x and y.

97. $x = 50; y = 37$

98. $x = 80; y = 33$

99. $x = 914; y = 271$

100. $x = 623; y = 197$

101. $x = 740; y = 385$

102. $x = 870; y = 243$

103. $x = 8672; y = 3461$

104. $x = 7814; y = 3512$

105. $x = 1605; y = 839$

106. $x = 1406; y = 968$

107. $x = 23{,}409; y = 5178$

108. $x = 56{,}397; y = 8249$

109. Is 24 a solution of the equation $29 = 53 - y$?

110. Is 31 a solution of the equation $48 - p = 17$?

111. Is 44 a solution of the equation $t - 16 = 60$?

112. Is 25 a solution of the equation $34 = x - 9$?

113. Is 27 a solution of the equation $82 - z = 55$?

114. Is 28 a solution of the equation $72 = 100 - d$?

OBJECTIVE C To solve application problems and use formulas

115. **Mathematics** Find the sum of all the whole numbers less than 21.

116. **Mathematics** Find the sum of all the natural numbers greater than 89 and less than 101.

117. **Mathematics** Find the difference between the smallest four-digit number and the largest two-digit number.

118. **Demography** The figure at the right shows the expected U.S. population aged 100 and over every two years from 2010 to 2020.

a. Which two-year period has the smallest increase in the number of people aged 100 and over?

b. Which two-year period has the greatest increase?

119. In the figure at the right, what does the difference $208{,}000 - 166{,}000$ represent?

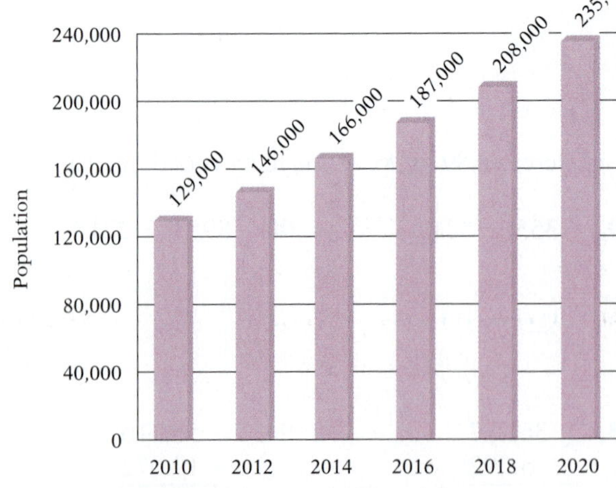

Expected U.S. Population Aged 100 and Over

Source: Census Bureau

120. Nutrition You eat an apple and one cup of cornflakes with one tablespoon of sugar and one cup of milk for breakfast. Find the total number of calories consumed if one apple contains 80 calories, one cup of cornflakes has 95 calories, one tablespoon of sugar has 45 calories, and one cup of milk has 150 calories.

121. Health You are on a diet to lose weight and are limited to 1500 calories per day. If your breakfast and lunch contained 950 calories, how many more calories can you consume during the rest of the day?

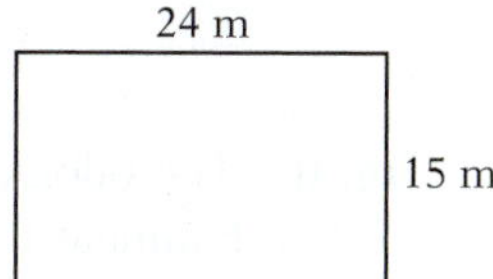

122. Geometry A rectangle has a length of 24 m and a width of 15 m. Find the perimeter of the rectangle.

123. Geometry Find the perimeter of a rectangle that has a length of 18 ft and a width of 12 ft.

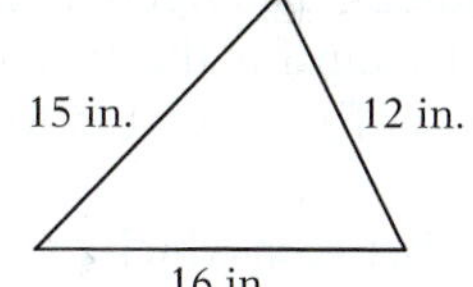

124. Geometry Find the perimeter of a triangle that has sides that measure 16 in., 12 in., and 15 in.

125. Geometry A triangle has sides of lengths 36 cm, 48 cm, and 60 cm. Find the perimeter of the triangle.

126. Geometry A rectangular playground has a length of 160 ft and a width of 120 ft. Find the length of hedge that surrounds the playground.

127. Geometry A rectangular vegetable garden has a length of 20 ft and a width of 14 ft. How many feet of wire fence should be purchased to surround the garden?

128. Space Flights The Gemini-Titan 7 space flight made 206 orbits of Earth. The Apollo-Saturn 7 space flight made 163 orbits of Earth. How many more orbits did the Gemini-Titan 7 flight make than the Apollo-Saturn 7 flight?

129. Finances You had $1054 in your checking account before making a deposit of $870. Find the amount in your checking account after you made the deposit.

130. Baseball Fields The seating capacity of SAFECO Field in Seattle is 47,116. The seating capacity of Fenway Park in Boston is 36,298. Find the difference between the seating capacities of SAFECO Field and Fenway Park.

Dave Bartruff/Corbis

Fenway Park

131. Repair Bills The repair bill on your car includes $358 for parts, $156 for labor, and a sales tax of $30. What is the total amount owed?

132. **Purchasing** The computer system you would like to purchase includes an operating system priced at \$830, a monitor that costs \$245, an extended keyboard priced at \$175, and a printer that sells for \$395. What is the total cost of the computer system?

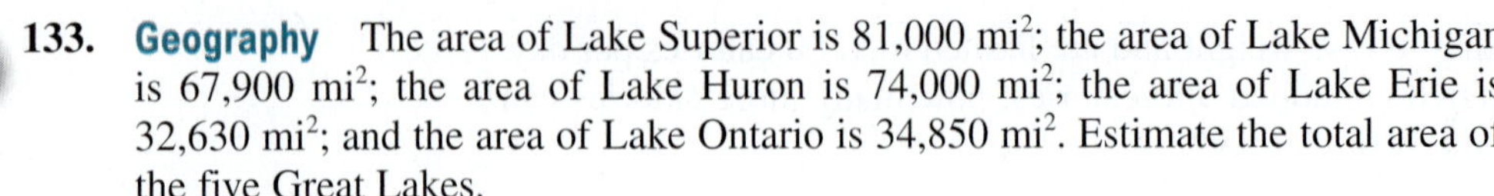

133. **Geography** The area of Lake Superior is 81,000 mi^2; the area of Lake Michigan is 67,900 mi^2; the area of Lake Huron is 74,000 mi^2; the area of Lake Erie is 32,630 mi^2; and the area of Lake Ontario is 34,850 mi^2. Estimate the total area of the five Great Lakes.

The Great Lakes

134. **Travel** The odometer on your car read 58,376 this time last year. It now reads 77,912. Estimate the number of miles your car has been driven during the past year.

Car Sales The figure at the right shows the number of cars sold by a dealership for the first four months of 2011 and 2012. Use this graph for Exercises 135 to 137.

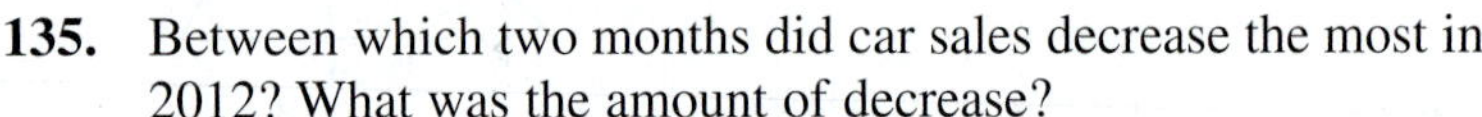

135. Between which two months did car sales decrease the most in 2012? What was the amount of decrease?

136. Between which two months did car sales increase the most in 2011? What was the amount of increase?

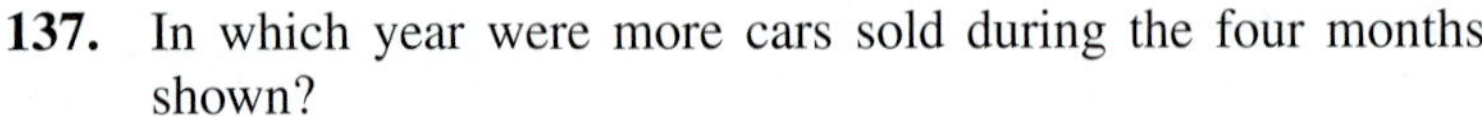

137. In which year were more cars sold during the four months shown?

Car Sales at a Dealership

138. **Investments** Use the formula $A = P + I$, where A is the value of an investment, P is the original investment, and I is the interest earned, to find the value of an investment that earned \$775 in interest on an original investment of \$12,500.

139. **Investments** Use the formula $A = P + I$, where A is the value of an investment, P is the original investment, and I is the interest earned, to find the value of an investment that earned \$484 in interest on an original investment of \$8800.

140. **Mortgages** What is the mortgage loan amount on a home that sells for \$290,000 with a down payment of \$29,000? Use the formula $M = S - D$, where M is the mortgage loan amount, S is the selling price, and D is the down payment.

141. **Mortgages** What is the mortgage loan amount on a home that sells for \$236,000 with a down payment of \$47,200? Use the formula $M = S - D$, where M is the mortgage loan amount, S is the selling price, and D is the down payment.

142. Air Travel What is the ground speed of an airplane traveling into a 25-mph head wind with an air speed of 375 mph? Use the formula $g = a - h$, where g is the ground speed, a is the air speed, and h is the speed of the head wind.

143. Air Travel Find the ground speed of an airplane traveling into a 15-mph head wind with an air speed of 425 mph. Use the formula $g = a - h$, where g is the ground speed, a is the air speed, and h is the speed of the head wind.

Speeds In some states, the speed limit on certain sections of highway is 70 mph. To test drivers' compliance with the speed limit, the highway patrol conducted a one-week study during which it recorded the speeds of motorists on one of these sections of highway. The results are recorded in the table at the right. Use this table for Exercises 144 to 147.

Speed	*Number of Cars*
> 80	1708
76–80	2503
71–75	3651
66–70	3717
61–65	2984
< 61	2870

144. **a.** How many drivers were traveling at 70 mph or less?
b. How many drivers were traveling at 76 mph or more?

145. Looking at the data in the table, is it possible to tell how many motorists were driving at 70 mph? Explain your answer.

146. Looking at the data in the table, is it possible to tell how many motorists were driving at less than 70 mph? Explain your answer.

147. Are more people driving at or below the posted speed limit, or are more people driving above the posted speed limit?

Applying the Concepts

148. Dice If you roll two ordinary six-sided dice and add the two numbers that appear on top, how many different sums are possible?

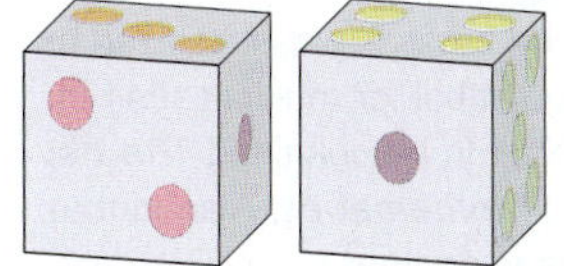

149. Mathematics How many two-digit numbers are there? How many three-digit numbers are there?

150. Determine whether the statement is always true, sometimes true, or never true.
a. If a is any whole number, then $a - 0 = a$.
b. If a is any whole number, then $a - a = 0$.

151. Find the circulation of your local newspaper and the population of the area served by that paper. What is the difference between the area's population and the newspaper's circulation? Why would this figure be of concern to the owner of the newspaper?

152. What estimate is given for the size of the population of your state by the year 2050? What is the estimate of the size of the population of the United States by the year 2050? Estimates differ. On what basis was the estimate you recorded derived?

SECTION 1.3 Multiplication and Division of Whole Numbers

OBJECTIVE A To multiply whole numbers

A store manager orders six boxes of telephone answering machines. Each box contains eight answering machines. How many answering machines are ordered?

The answer can be calculated by adding six 8's.

$8 + 8 + 8 + 8 + 8 + 8 = 48$

This problem involves repeated addition of the same number. The answer can be calculated by a shorter process called multiplication. **Multiplication** is the repeated addition of the same number.

There is a total of 48 dots on the 6 dominoes.

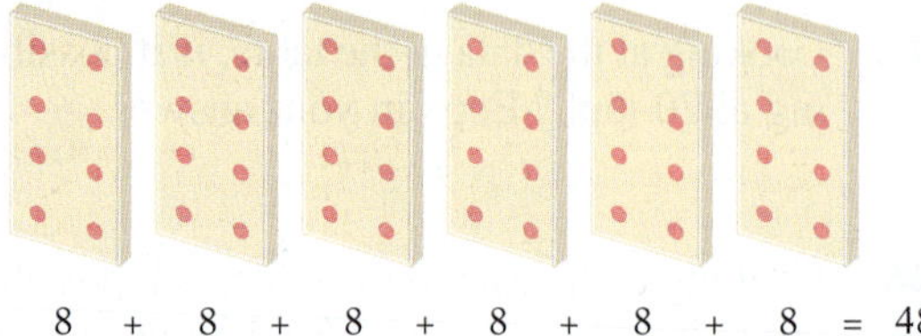

$8 + 8 + 8 + 8 + 8 + 8 = 48$

The numbers that are multiplied are called **factors.** The answer is called the **product.**

or

$6 \times 8 = 48$

Factor Factor Product

The times sign "×" is one symbol that is used to mean multiplication. Each of the expressions below also represents multiplication.

$6 \cdot 8 \qquad 6(8) \qquad (6)(8) \qquad 6a \qquad 6(a) \qquad ab$

The expression $6a$ means "6 times a." The expression ab means "a times b."

Point of Interest

The cross X was first used as a symbol for multiplication in 1631 in a book titled *The Key to Mathematics.* Also in that year, another book, *Practice of the Analytical Art,* advocated the use of a dot to indicate multiplication.

The basic facts for multiplying one-digit numbers should be memorized. Multiplication of larger numbers requires the repeated use of the basic multiplication facts.

HOW TO • 1 Multiply: 37(4)

$$\begin{array}{r} \scriptstyle 2 \\ 37 \\ \times \; 4 \\ \hline 8 \end{array}$$

- **Multiply 4 · 7.**
  4 · 7 = 28 (2 tens + 8 ones).
  Write the 8 in the ones column and carry the 2 to the tens column.

$$\begin{array}{r} \scriptstyle 2 \\ 37 \\ \times \; 4 \\ \hline 148 \end{array}$$

- **The 3 in 37 is 3 tens.**
  Multiply 4 · 3 tens = 12 tens.
  Add the carry digit: 12 tens + 2 tens = 14 tens.
  Write the 14.

In the preceding example, a number was multiplied by a one-digit number. The examples that follow illustrate multiplication by larger numbers.

HOW TO • 2 Multiply: (47)(23)

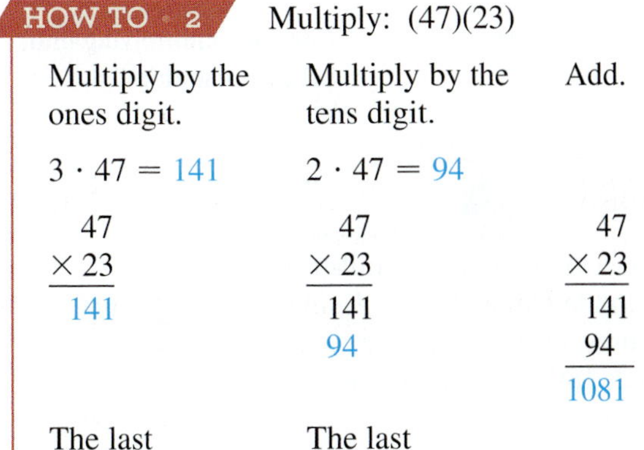

Multiply by the ones digit.

$3 \cdot 47 = 141$

$$\begin{array}{r} 47 \\ \times 23 \\ \hline 141 \end{array}$$

The last digit is written in the ones column.

Multiply by the tens digit.

$2 \cdot 47 = 94$

$$\begin{array}{r} 47 \\ \times 23 \\ \hline 141 \\ 94 \end{array}$$

The last digit is written in the tens column.

Add.

$$\begin{array}{r} 47 \\ \times 23 \\ \hline 141 \\ 94 \\ \hline 1081 \end{array}$$

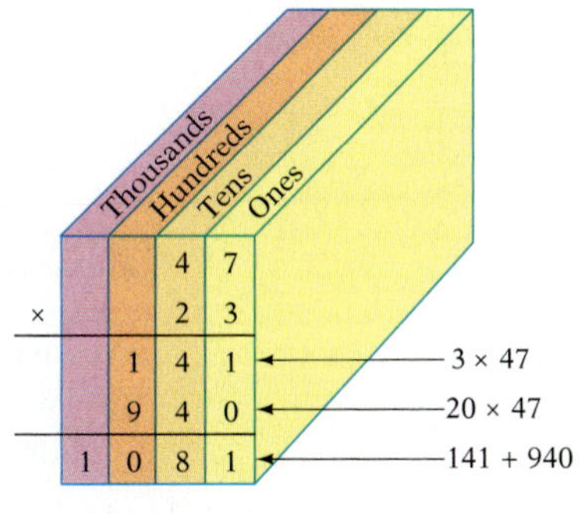

The place-value chart illustrates the placement of the products.

Note the placement of the products when we are multiplying by a factor that contains a zero.

HOW TO • 3 Multiply: 439(206)

$$\begin{array}{r} 439 \\ \times\ 206 \\ \hline 2\ 634 \\ 0\ 00 \\ 87\ 8 \\ \hline 90{,}434 \end{array}$$

When working the problem, we usually write only one zero, as shown at the right. Writing this zero ensures the proper placement of the products.

$$\begin{array}{r} 439 \\ \times\ 206 \\ \hline 2\ 634 \\ 87\ 80 \\ \hline 90{,}434 \end{array}$$

Note the pattern when the following numbers are multiplied.

Multiply the nonzero parts of the factors.

Now attach the same number of zeros in the product as the total number of zeros in the factors.

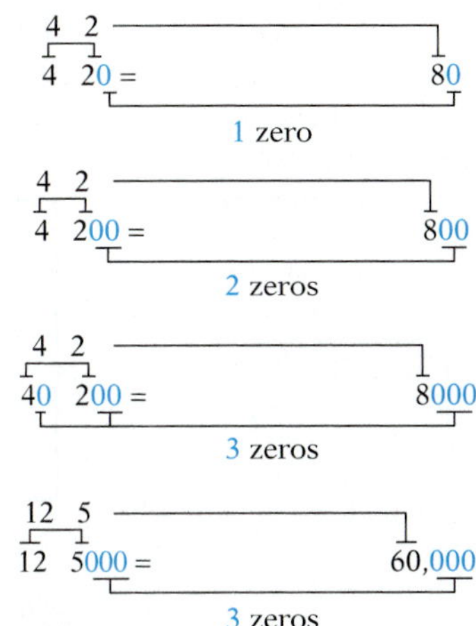

HOW TO • 4 Find the product of 600 and 70.

$600 \cdot 70 = 42{,}000$ • **Remember that a *product* is the answer to a multiplication problem.**

HOW TO • 5 Multiply: 3(20)(10)(4)

3(20)(10)4 = 60(10)(4) • **Multiply the first two numbers.**

= (600)(4) • **Multiply the product by the third number.**

= 2400 • **Continue multiplying until all the numbers have been multiplied.**

HOW TO • 6  Figure 1.13 shows the average weekly earnings of full-time workers in the United States. Using these figures, calculate the earnings of a female full-time worker, age 22, for working for 4 weeks.

Multiply the number of weeks (4) times the amount earned for one week ($354).

4(354) = 1416

The average earnings of a 22-year-old, female, full-time worker for working for 4 weeks are $1416.

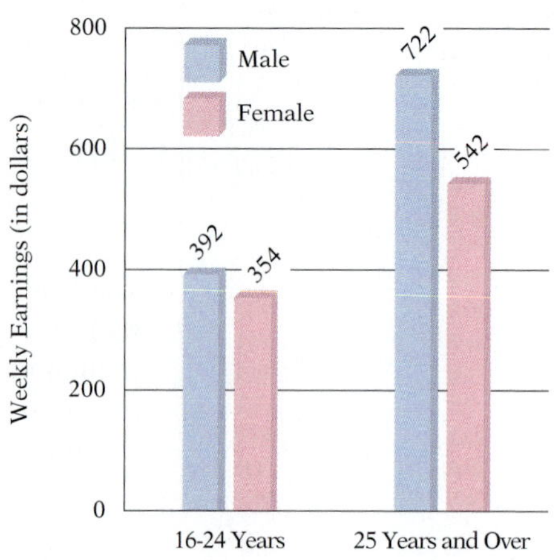

FIGURE 1.13 Average Weekly Earnings of Full-Time Workers
Source: Bureau of Labor Statistics

HOW TO • 7 Estimate the product of 345 and 92.

345 → 300 • **Round each number to its highest place value.**
92 → 90

300 · 90 = 27,000 • **Multiply the rounded numbers.**

27,000 is an estimate of the product of 345 and 92.

The phrase *the product of* was used in the example above to indicate the operation of multiplication. All of the phrases below indicate multiplication. An example of each is shown at the right of each phrase.

times	8 times 4	8 · 4
the product of	the product of 9 and 5	9 · 5
multiplied by	7 multiplied by 3	3 · 7
twice	twice 6	2 · 6

HOW TO • 8 Evaluate xyz when $x = 50$, $y = 2$, and $z = 7$.

xyz • **xyz means $x \cdot y \cdot z$.**

50 · 2 · 7 • **Replace each variable by its value.**

= 100 · 7 • **Multiply the first two numbers.**

= 700 • **Multiply the product by the next number.**

Tips for Success

Some students think they can "coast" at the beginning of this course because the topic of Chapter 1 is whole numbers. However, this chapter lays the foundation for the entire course. Be sure you know and understand all the concepts presented. For example, study the properties of multiplication presented in this lesson.

As for addition, there are properties of multiplication.

The Multiplication Property of Zero

$a \cdot 0 = 0$ or $0 \cdot a = 0$

$8 \cdot 0 = 0$

The Multiplication Property of Zero states that the product of a number and zero is zero. The variable a is used here to represent any whole number. It can even represent the number zero because $0 \cdot 0 = 0$.

The Multiplication Property of One

$a \cdot 1 = a$ or $1 \cdot a = a$

$1 \cdot 9 = 9$

The Multiplication Property of One states that the product of a number and 1 is the number. Multiplying a number by 1 does not change the number.

The Commutative Property of Multiplication

$a \cdot b = b \cdot a$

$$4 \cdot 9 = 9 \cdot 4$$
$$36 = 36$$

The Commutative Property of Multiplication states that two numbers can be multiplied in either order; the product will be the same. Here the variables a and b represent any whole numbers. Therefore, for example, if you know that the product of 4 and 9 is 36, then you also know that the product of 9 and 4 is 36 because $4 \cdot 9 = 9 \cdot 4$.

The Associative Property of Multiplication

$(a \cdot b) \cdot c = a \cdot (b \cdot c)$

$$(2 \cdot 3) \cdot 4 = 2 \cdot (3 \cdot 4)$$
$$6 \cdot 4 = 2 \cdot 12$$
$$24 = 24$$

The Associative Property of Multiplication states that when multiplying three numbers, the numbers can be grouped in any order; the product will be the same. Note in the example at the right above that we can multiply the product of 2 and 3 by 4, or we can multiply 2 by the product of 3 and 4. In either case, the product of the three numbers is 24.

HOW TO • 9 What is the solution of the equation $5x = 5$?

By the Multiplication Property of One, the product of a number and 1 is the number.

$$\begin{array}{c|c} \multicolumn{2}{c}{5x = 5} \\ \hline 5(1) & 5 \\ \multicolumn{2}{c}{5 = 5} \end{array}$$

The solution is 1.

The check is shown at the right.

HOW TO • 10 Is 7 a solution of the equation $3m = 21$?

$$\begin{array}{c|c} \multicolumn{2}{c}{3m = 21} \\ \hline 3(7) & 21 \\ \multicolumn{2}{c}{21 = 21} \end{array}$$

- **Replace *m* by 7.**
- **Simplify the left side of the equation. The results are equal.**

Yes, 7 is a solution of the equation.

Figure 1.14 shows the average monthly savings of individuals in seven different countries. Use this graph for Example 1 and You Try It 1.

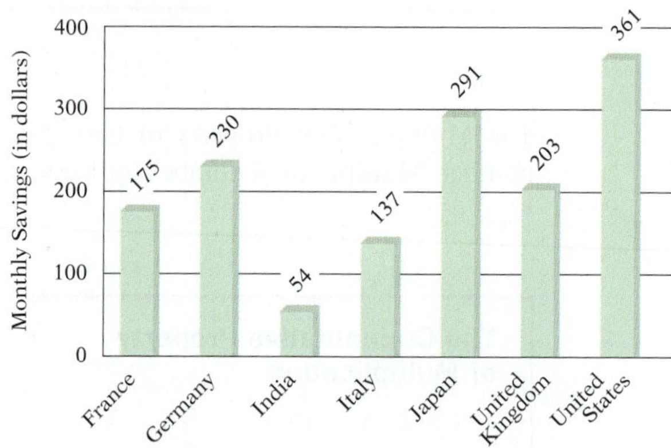

FIGURE 1.14 Average Monthly Savings

Source: Taylor Nelson - Sofres for American Express

EXAMPLE • 1

Use Figure 1.14 to determine the average annual savings of individuals in Japan.

Solution

The average monthly savings in Japan is \$291. The number of months in one year is 12.

$$\begin{array}{r} 291 \\ \times\ 12 \\ \hline 582 \\ 291 \\ \hline 3492 \end{array}$$

The average annual savings of individuals in Japan is \$3492.

YOU TRY IT • 1

According to Figure 1.14, what is the average annual savings of individuals in France?

Your solution

Solution on p. S2

EXAMPLE • 2

Estimate the product of 2871and 49.

Solution

$$2871 \rightarrow 3000$$
$$49 \rightarrow 50$$

$3000 \cdot 50 = 150{,}000$

YOU TRY IT • 2

Estimate the product of 8704 and 93.

Your solution

EXAMPLE • 3

Evaluate $3ab$ when $a = 10$ and $b = 40$.

Solution

$3ab$

$$\begin{aligned} 3(10)(40) &= 30(40) \\ &= 1200 \end{aligned}$$

YOU TRY IT • 3

Evaluate $5xy$ when $x = 20$ and $y = 60$.

Your solution

EXAMPLE • 4

What is 800 times 300?

Solution

$800 \cdot 300 = 240{,}000$

YOU TRY IT • 4

What is 90 multiplied by 7000?

Your solution

EXAMPLE • 5

Complete the statement by using the Associative Property of Multiplication.

$(7 \cdot 8) \cdot 5 = 7 \cdot (? \cdot 5)$

Solution

$(7 \cdot 8) \cdot 5 = 7 \cdot (8 \cdot 5)$

YOU TRY IT • 5

Complete the statement by using the Multiplication Property of Zero.

$? \cdot 10 = 0$

Your solution

EXAMPLE • 6

Is 9 a solution of the equation $82 = 9q$?

Solution

$$\begin{array}{c|c} \multicolumn{2}{c}{82 = 9q} \\ \hline 82 & 9(9) \end{array}$$

$82 \neq 81$

No, 9 is not a solution of the equation.

YOU TRY IT • 6

Is 11 a solution of the equation $7a = 77$?

Your solution

Solutions on p. S2

OBJECTIVE B To evaluate expressions that contain exponents

Point of Interest

Lao-tzu, founder of Taoism, wrote: Counting gave birth to Addition, Addition gave birth to Multiplication, Multiplication gave birth to Exponentiation, Exponentiation gave birth to all the myriad operations.

Repeated multiplication of the same factor can be written in two ways:

$$4 \cdot 4 \cdot 4 \cdot 4 \cdot 4 \text{ or } \underset{\uparrow \text{ base}}{4}^{5} \longleftarrow \text{exponent}$$

The expression 4^5 is in **exponential form.** The **exponent,** 5, indicates how many times the **base,** 4, occurs as a factor in the multiplication.

Point of Interest

One billion is too large a number for most of us to comprehend. If a computer were to start counting from 1 to 1 billion, writing to the screen one number every second of every day, it would take over 31 years for the computer to complete the task.

And if a billion is a large number, consider a googol. A googol is 1 with 100 zeros after it, or 10^{100}. Edward Kasner is the mathematician credited with thinking up this number, and his nine-year-old nephew is said to have thought up the name. The two then coined the word googolplex, which is 10^{googol}.

It is important to be able to read numbers written in exponential form.

$2 = 2^1$	Read "two to the first power" or just "two." Usually the 1 is not written.
$2 \cdot 2 = 2^2$	Read "two squared" or "two to the second power."
$2 \cdot 2 \cdot 2 = 2^3$	Read "two cubed" or "two to the third power."
$2 \cdot 2 \cdot 2 \cdot 2 = 2^4$	Read "two to the fourth power."
$2 \cdot 2 \cdot 2 \cdot 2 \cdot 2 = 2^5$	Read "two to the fifth power."

Variable expressions can contain exponents.

$x^1 = x$	x to the first power is usually written simply as x.
$x^2 = x \cdot x$	x^2 means x times x.
$x^3 = x \cdot x \cdot x$	x^3 means x occurs as a factor 3 times.
$x^4 = x \cdot x \cdot x \cdot x$	x^4 means x occurs as a factor 4 times.

Each place value in the place-value chart can be expressed as a power of 10.

Ten =	10	= 10	$= 10^1$
Hundred =	100	= $10 \cdot 10$	$= 10^2$
Thousand =	1000	= $10 \cdot 10 \cdot 10$	$= 10^3$
Ten-thousand =	10,000	= $10 \cdot 10 \cdot 10 \cdot 10$	$= 10^4$
Hundred-thousand =	100,000	= $10 \cdot 10 \cdot 10 \cdot 10 \cdot 10$	$= 10^5$
Million =	1,000,000	= $10 \cdot 10 \cdot 10 \cdot 10 \cdot 10 \cdot 10$	$= 10^6$

Note that the exponent on 10 when the number is written in exponential form is the same as the number of zeros in the number written in standard form. For example, $10^5 = 100{,}000$; the exponent on 10 is 5, and the number 100,000 has 5 zeros.

To evaluate a numerical expression containing exponents, write each factor as many times as indicated by the exponent and then multiply.

$$5^3 = 5 \cdot 5 \cdot 5 = 25 \cdot 5 = 125$$

$$2^3 \cdot 6^2 = (2 \cdot 2 \cdot 2) \cdot (6 \cdot 6) = 8 \cdot 36 = 288$$

Integrating Technology

A calculator can be used to evaluate an exponential expression. The y^x key (or on some calculators an x^y key or ^ key) is used to enter the exponent. For instance, for the example at the right, enter 4 y^x 3 =. The display reads 64.

HOW TO • 11 Evaluate the variable expression c^3 when $c = 4$.

$c^3 = c \cdot c \cdot c$

$4^3 = 4 \cdot 4 \cdot 4$ • **Replace c with 4 and then evaluate the exponential expression.**

$= 16 \cdot 4 = 64$

EXAMPLE • 7

Write $7 \cdot 7 \cdot 7 \cdot 4 \cdot 4$ in exponential form.

Solution

$7 \cdot 7 \cdot 7 \cdot 4 \cdot 4 = 7^3 \cdot 4^2$

YOU TRY IT • 7

Write $2 \cdot 2 \cdot 2 \cdot 3 \cdot 3 \cdot 3 \cdot 3$ in exponential form.

Your solution

Solution on p. S2

EXAMPLE • 8

Evaluate 8^3.

Solution

$8^3 = 8 \cdot 8 \cdot 8 = 64 \cdot 8 = 512$

YOU TRY IT • 8

Evaluate 6^4.

Your solution

EXAMPLE • 9

Evaluate 10^7.

Solution

$10^7 = 10{,}000{,}000$

(The exponent on 10 is 7. There are 7 zeros in 10,000,000.)

YOU TRY IT • 9

Evaluate 10^8.

Your solution

EXAMPLE • 10

Evaluate $3^3 \cdot 5^2$.

Solution

$$3^3 \cdot 5^2 = (3 \cdot 3 \cdot 3) \cdot (5 \cdot 5)$$
$$= 27 \cdot 25 = 675$$

YOU TRY IT • 10

Evaluate $2^4 \cdot 3^2$.

Your solution

EXAMPLE • 11

Evaluate x^2y^3 when $x = 4$ and $y = 2$.

Solution

x^2y^3 (x^2y^3 means x^2 times y^3.)

$$4^2 \cdot 2^3 = (4 \cdot 4) \cdot (2 \cdot 2 \cdot 2)$$
$$= 16 \cdot 8$$
$$= 128$$

YOU TRY IT • 11

Evaluate x^4y^2 when $x = 1$ and $y = 3$.

Your solution

Solutions on pp. S2–S3

OBJECTIVE C To divide whole numbers

Division is used to separate objects into equal groups.

A store manager wants to display 24 new objects equally on 4 shelves. From the diagram, we see that the manager would place 6 objects on each shelf.

The manager's division problem can be written as follows:

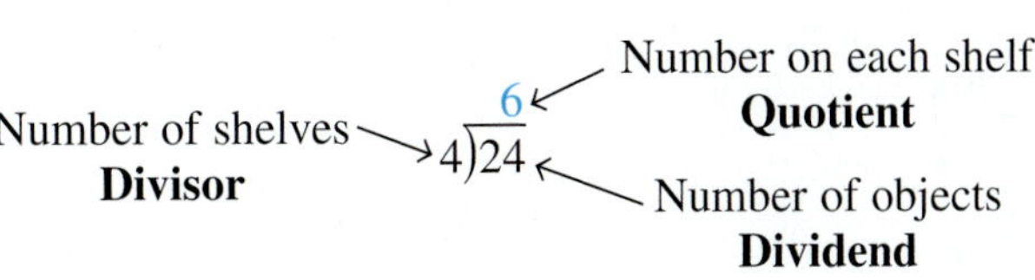

Point of Interest

The Chinese divided a day into 100 k'o, which was a unit equal to a little less than 15 min. Sundials were used to measure time during the daylight hours, and by A.D. 500, candles, water clocks, and incense sticks were used to measure time at night.

Note that the quotient multiplied by the divisor equals the dividend.

$4\overline{)24}$ with quotient 6 because 6 Quotient × 4 Divisor = 24 Dividend

Division is also represented by the symbol ÷ or by a fraction bar. Both are read "divided by."

$$9\overline{)54}^{\,6} \qquad 54 \div 9 = 6 \qquad \frac{54}{9} = 6$$

The fact that the quotient times the divisor equals the dividend can be used to illustrate properties of division.

$0 \div 4 = 0$ because $0 \cdot 4 = 0$.

Integrating Technology

Enter 4 ÷ 0 =. An error message is displayed because division by zero is undefined.

$4 \div 4 = 1$ because $1 \cdot 4 = 4$.

$4 \div 1 = 4$ because $4 \cdot 1 = 4$.

$4 \div 0 = ?$ What number can be multiplied by 0 to get 4? There is no number whose product with 0 is 4 because the product of a number and zero is 0. **Division by zero is undefined.** $? \cdot 0 = 4$

The properties of division are stated below. In these statements, the symbol ≠ is read "is not equal to."

Take Note

Recall that the variable a represents any whole number. Therefore, for the first two properties, we must state that $a \neq 0$ in order to ensure that we are not dividing by zero.

Division Properties of Zero and One

If $a \neq 0$, $0 \div a = 0$.	Zero divided by any number other than zero is zero.
If $a \neq 0$, $a \div a = 1$.	Any number other than zero divided by itself is one.
$a \div 1 = a$	A number divided by one is the number.
$a \div 0$ is undefined.	Division by zero is undefined.

The example below illustrates division of a larger whole number by a one-digit number.

HOW TO • 12 Divide and check: $3192 \div 4$

```
   7
4)3192      • Think 31 ÷ 4.
 -28        • Subtract 7 × 4.
  39        • Bring down the 9.

   79
4)3192
 -28
  39        • Think 39 ÷ 4.
 -36        • Subtract 9 × 4.
   32       • Bring down the 2.

   798
4)3192
 -28
  39
 -36
   32       • Think 32 ÷ 4.
  -32       • Subtract 8 × 4.
    0
```

Check:

```
   798
 ×   4
  3192
```

The place-value chart can be used to show why this method works.

```
          Hundreds Tens Ones
            7    9    8
     4) 3   1    9    2
       -2   8    0    0    7 hundreds × 4
            3    9    2
           -3    6    0    9 tens × 4
                 3    2
                -3    2    8 ones × 4
                      0
```

Sometimes it is not possible to separate objects into a whole number of equal groups.

A packer at a bakery has 14 muffins to pack into 3 boxes. Each box holds 4 muffins. From the diagram, we see that after the baker places 4 muffins in each box, there are 2 muffins left over. The 2 is called the **remainder.**

The packer's division problem can be written as follows:

```
                                  Number in each box
                          4 ←     Quotient                          4 r2
Number of boxes ⟶ 3)  14 ←Total number of muffins         or      3)14
   Divisor             −12        Dividend
                          2 ←—— Number left over
                                  Remainder
```

For any division problem, **(quotient · divisor) + remainder = dividend.** This result can be used to check a division problem.

HOW TO • 13 Find the quotient of 389 and 24.

```
    16 r5
24)389
  −24
   149
  −144
     5
```

Check: $(16 \cdot 24) + 5 = 384 + 5 = 389$

The phrase *the quotient of* was used in the example above to indicate the operation of division. The phrase *divided by* also indicates division.

the quotient of	the quotient of 8 and 4	$8 \div 4$
divided by	9 divided by 3	$9 \div 3$

HOW TO • 14 Estimate the result when 56,497 is divided by 28.

56,497 ⟶ 60,000 • **Round each number to its highest place value.**

28 ⟶ 30

60,000 ÷ 30 = 2000 • **Divide the rounded numbers. 2000 is an estimate of 56,497 ÷ 28.**

HOW TO • 15 Evaluate $\frac{x}{y}$ when $x = 4284$ and $y = 18$.

$\frac{x}{y}$ • **Replace x with 4284 and y with 18.**

$\frac{4284}{18} = 238$ • **$\frac{4284}{18}$ means 4284 ÷ 18.**

HOW TO • 16 Is 42 a solution of the equation $\frac{x}{6} = 7$?

$\frac{x}{6} = 7$

$\frac{42}{6} \mid 7$ • **Replace x by 42.**

$7 = 7$ • **Simplify the left side of the equation. The results are equal.**

42 is a solution of the equation.

EXAMPLE • 12

What is the quotient of 8856 and 42?

Solution

```
     210 r36
 42)8856
   - 84
      45
    - 42
       36    • Think 42)36.
      - 0    • Subtract 0 · 42.
       36
```

Check: $(210 \cdot 42) + 36$
$= 8820 + 36 = 8856$

YOU TRY IT • 12

What is 7694 divided by 24?

Your solution

Solution on p. S3

Figure 1.15 shows a household's annual expenses of $44,000. Use this graph for Example 13 and You Try It 13.

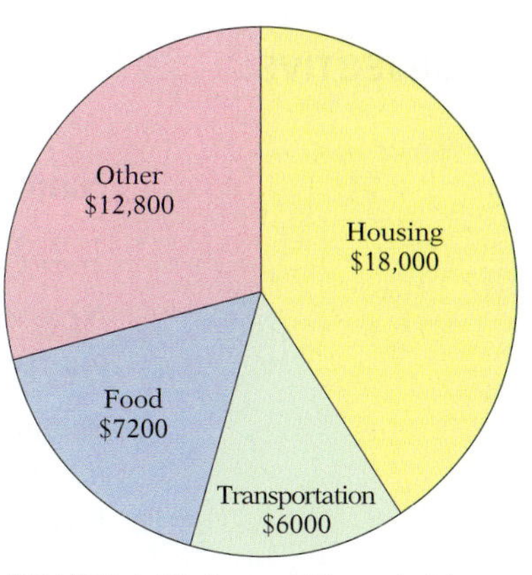

FIGURE 1.15 Annual Household Expenses

EXAMPLE 13

Use Figure 1.15 to find the household's monthly expense for housing.

Solution

The annual expense for housing is $18,000.

$18{,}000 \div 12 = 1500$

The monthly expense is $1500.

YOU TRY IT 13

Use Figure 1.15 to find the household's monthly expense for food.

Your solution

EXAMPLE 14

Estimate the quotient of 55,272 and 392.

Solution

$55{,}272 \longrightarrow 60{,}000$
$392 \longrightarrow 400$

$60{,}000 \div 400 = 150$

YOU TRY IT 14

Estimate the quotient of 216,936 and 207.

Your solution

EXAMPLE 15

Evaluate $\frac{x}{y}$ when $x = 342$ and $y = 9$.

Solution

$$\frac{x}{y}$$

$$\frac{342}{9} = 38$$

YOU TRY IT 15

Evaluate $\frac{x}{y}$ when $x = 672$ and $y = 8$.

Your solution

EXAMPLE 16

Is 28 a solution of the equation $\frac{x}{7} = 4$?

Solution

$$\frac{x}{7} = 4$$

$$\begin{array}{c|c} \frac{28}{7} & 4 \end{array}$$

$$4 = 4$$

Yes, 28 is a solution of the equation.

YOU TRY IT 16

Is 12 a solution of the equation $\frac{60}{y} = 2$?

Your solution

Solutions on p. S3

OBJECTIVE D To factor numbers and find the prime factorization of numbers 

Natural number factors of a number divide that number evenly (there is no remainder).

1, 2, 3, and 6 are natural number factors of 6 because they divide 6 evenly.

Note that both the divisor and the quotient are factors of the dividend.

$$1\overline{)6}^{\,6} \quad 2\overline{)6}^{\,3} \quad 3\overline{)6}^{\,2} \quad 6\overline{)6}^{\,1}$$

To find the factors of a number, try dividing the number by 1, 2, 3, 4, 5, ... Those numbers that divide the number evenly are its factors. Continue this process until the factors start to repeat.

Point of Interest

Twelve is the smallest abundant number, or number whose proper divisors add up to more than the number itself. The proper divisors of a number are all of its factors except the number itself. The proper divisors of 12 are 1, 2, 3, 4, and 6, which add up to 16, which is greater than 12. There are 246 abundant numbers between 1 and 1000.

A perfect number is one whose proper divisors add up to exactly that number. For example, the proper divisors of 6 are 1, 2, and 3, which add up to 6. There are only three perfect numbers less than 1000: 6, 28, and 496.

HOW TO • 17 Find all the factors of 42.

$42 \div 1 = 42$	1 and 42 are factors.
$42 \div 2 = 21$	2 and 21 are factors.
$42 \div 3 = 14$	3 and 14 are factors.
$42 \div 4$	Will not divide evenly
$42 \div 5$	Will not divide evenly
$42 \div 6 = 7$	6 and 7 are factors.
$42 \div 7 = 6$	7 and 6 are factors.

The factors are repeating.
All the factors of 42 have been found.

The factors of 42 are 1, 2, 3, 6, 7, 14, 21, and 42.

The following rules are helpful in finding the factors of a number.

2 is a factor of a number if the digit in the ones place of the number is 0, 2, 4, 6, or 8.	436 ends in 6. Therefore, 2 is a factor of 436 $(436 \div 2 = 218)$.
3 is a factor of a number if the sum of the digits of the number is divisible by 3.	The sum of the digits of 489 is $4 + 8 + 9 = 21$. 21 is divisible by 3. Therefore, 3 is a factor of 489 $(489 \div 3 = 163)$.
4 is a factor of a number if the last two digits of the number are divisible by 4.	556 ends in 56. 56 is divisible by 4 $(56 \div 4 = 14)$. Therefore, 4 is a factor of 556 $(556 \div 4 = 139)$.
5 is a factor of a number if the ones digit of the number is 0 or 5.	520 ends in 0. Therefore, 5 is a factor of 520 $(520 \div 5 = 104)$.

A **prime number** is a natural number greater than 1 that has exactly two natural number factors, 1 and the number itself. 7 is prime because its only factors are 1 and 7. If a number is not prime, it is a **composite** number. Because 6 has factors of 2 and 3, 6 is a composite number. The prime numbers less than 50 are

2, 3, 5, 7, 11, 13, 17, 19, 23, 29, 31, 37, 41, 43, 47

The **prime factorization** of a number is the expression of the number as a product of its prime factors. To find the prime factorization of 90, begin with the smallest prime number as a trial divisor and continue with prime numbers as trial divisors until the final quotient is prime.

HOW TO • 18 Find the prime factorization of 90.

$$2\overline{)90}\ \text{quotient } 45$$
Divide 90 by 2.

$$2\overline{)90},\ 3\overline{)45}\ \text{quotient } 15$$
45 is not divisible by 2. Divide 45 by 3.

$$2\overline{)90},\ 3\overline{)45},\ 3\overline{)15}\ \text{quotient } 5$$
Divide 15 by 3. 5 is prime.

The prime factorization of 90 is $2 \cdot 3 \cdot 3 \cdot 5$, or $2 \cdot 3^2 \cdot 5$.

Finding the prime factorization of larger numbers can be more difficult. Try each prime number as a trial divisor. Stop when the square of the trial divisor is greater than the number being factored.

HOW TO • 19 Find the prime factorization of 201.

$$3\overline{)201}\ \text{quotient } 67$$

• **67 cannot be divided evenly by 2, 3, 5, 7, or 11. Prime numbers greater than 11 need not be tried because $11^2 = 121$ and $121 > 67$.**

The prime factorization of 201 is $3 \cdot 67$.

EXAMPLE • 17

Find all the factors of 40.

Solution

$40 \div 1 = 40$
$40 \div 2 = 20$
$40 \div 3$ • **Does not divide evenly.**
$40 \div 4 = 10$
$40 \div 5 = 8$
$40 \div 6$ • **Does not divide evenly.**
$40 \div 7$ • **Does not divide evenly.**
$40 \div 8 = 5$ • **The factors are repeating.**

The factors of 40 are 1, 2, 4, 5, 8, 10, 20, and 40.

YOU TRY IT • 17

Find all the factors of 30.

Your solution

EXAMPLE • 18

Find the prime factorization of 84.

Solution

$$2\overline{)84},\ 2\overline{)42},\ 3\overline{)21}\ \text{quotient } 7$$

$84 = 2 \cdot 2 \cdot 3 \cdot 7 = 2^2 \cdot 3 \cdot 7$

YOU TRY IT • 18

Find the prime factorization of 88.

Your solution

Solutions on p. S3

EXAMPLE • 19

Find the prime factorization of 141.

Solution

$$3\overline{)141}^{\;47}$$

• Try only 2, 3, 5, and 7 because $7^2 = 49$ and $49 > 47$.

$141 = 3 \cdot 47$

YOU TRY IT • 19

Find the prime factorization of 295.

Your solution

Solution on p. S3

OBJECTIVE E To solve application problems and use formulas

Take Note

Remember that $2L$ means 2 times L, and $2W$ means 2 times W.

In Section 1.2, we defined perimeter as the distance around a plane figure. The perimeter of a rectangle was given as $P = L + W + L + W$. This formula is commonly written as $P = 2L + 2W$.

Perimeter of a Rectangle

The formula for the perimeter of a rectangle is $P = 2L + 2W$, where P is the perimeter of the rectangle, L is the length, and W is the width.

32 ft

16 ft

HOW TO • 20 Find the perimeter of the rectangle shown at the left.

$P = 2L + 2W$ • Use the formula for the perimeter of a rectangle.

$P = 2(32) + 2(16)$ • Substitute 32 for L and 16 for W.

$P = 64 + 32$ • Find the product of 2 and 32 and the product of 2 and 16.

$P = 96$ • Add.

The perimeter of the rectangle is 96 ft.

s

s s

s

$P = s + s + s + s$
$P = 4s$

A **square** is a rectangle in which each side has the same length. Letting s represent the length of each side of a square, the perimeter of the square can be represented $P = s + s + s + s$. Note that we are adding *four* s's. We can write the addition as multiplication: $P = 4s$.

Perimeter of a Square

The formula for the perimeter of a square is $P = 4s$, where P is the perimeter and s is the length of a side of the square.

28 km

Find the perimeter of the square shown at the left.

$P = 4s$ • **Use the formula for the perimeter of a square.**

$P = 4(28)$ • **Substitute 28 for *s*.**

$P = 112$ • **Multiply.**

The perimeter of the square is 112 km.

Area is the amount of surface in a region. Area can be used to describe the size of a skating rink, the floor of a room, or a playground. Area is measured in square units.

1 in²

1 cm²

A square that measures 1 inch on each side has an area of 1 square inch, which is written 1 in^2. A square that measures 1 centimeter on each side has an area of 1 square centimeter, which is written 1 cm^2.

Larger areas can be measured in square feet (ft^2), square meters (m^2), acres (43,560 ft^2), square miles (mi^2), or any other square unit.

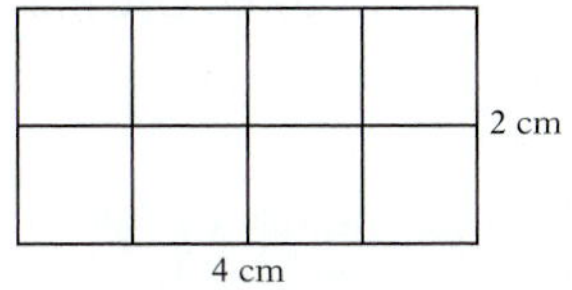

The area of the rectangle is 8 cm^2.

The area of a geometric figure is the number of squares that are necessary to cover the figure. In the figure at the left, a rectangle has been drawn and covered with squares. Eight squares, each of area 1 cm^2, were used to cover the rectangle. The area of the rectangle is 8 cm^2. Note from this figure that the area of a rectangle can be found by multiplying the length of the rectangle by its width.

Area of a Rectangle

The formula for the area of a rectangle is $A = LW$, where A is the area, L is the length, and W is the width of the rectangle.

HOW TO 22

Find the area of the rectangle shown at the left.

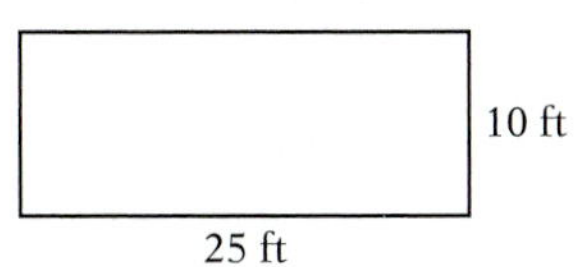

$A = LW$ • **Use the formula for the area of a rectangle.**

$A = 25(10)$ • **Substitute 25 for *L* and 10 for *W*.**

$A = 250$ • **Multiply.**

The area of the rectangle is 250 ft^2.

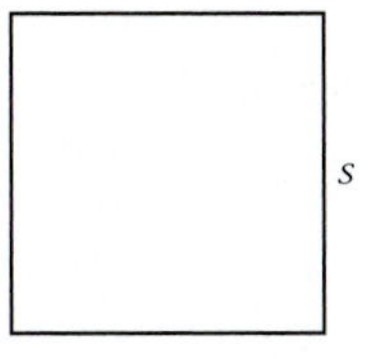

$A = s \cdot s = s^2$

A square is a rectangle in which all sides are the same length. Therefore, both the length and the width of a square can be represented by s, and $A = LW = s \cdot s = s^2$.

Area of a Square

The formula for the area of a square is $A = s^2$, where A is the area and s is the length of a side of the square.

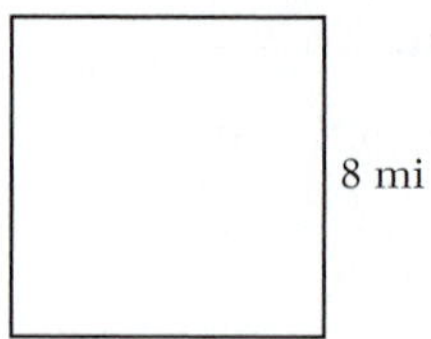

HOW TO • 23 Find the area of the square shown at the left.

$A = s^2$ • **Use the formula for the area of a square.**

$A = 8^2$ • **Substitute 8 for s.**

$A = 64$ • **Multiply.**

The area of the square is 64 mi^2.

Integrating Technology

Many scientific calculators have an x^2 key. This key is used to square the displayed number. For example, after pressing 8 x^2 =, the display reads 64.

In this section, some of the phrases used to indicate the operations of multiplication and division were presented. In solving application problems, you might also look for the following types of questions:

Multiplication	**Division**
per . . . How many altogether?	What is the hourly rate?
each . . . What is the total number of . . ?	Find the amount per . . .
every . . . Find the total . . .	How many does each . . . ?

Take Note

Each of the following indicates multiplication:

"You purchased 6 boxes of doughnuts with 12 doughnuts *per* box. *How many* doughnuts did you purchase *altogether*?"

"If *each* bottle of apple juice contains 32 oz, *what is the total number of* ounces in 8 bottles of the juice?"

"You purchased 5 bags of oranges. *Every* bag contained 10 oranges. *Find the total* number of oranges purchased."

Figure 1.16 shows the cost of a first-class postage stamp from the 1950s to 2009. Use this graph for Example 20 and You Try It 20.

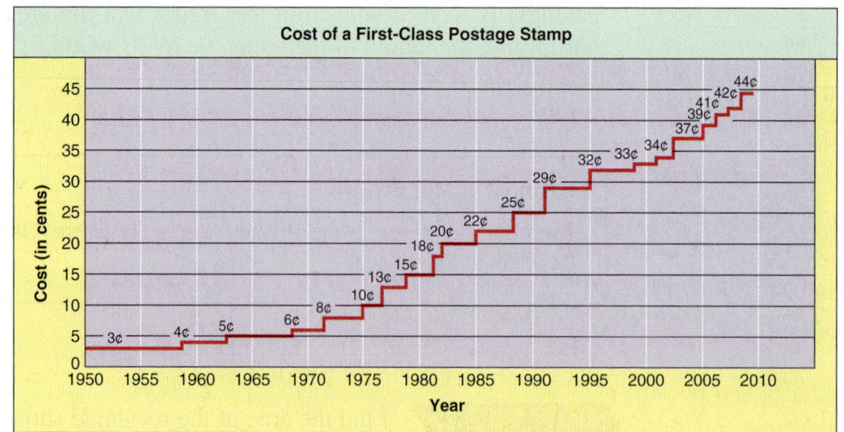

FIGURE 1.16 Cost of a First-Class Postage Stamp

EXAMPLE • 20

How many times more expensive was a stamp in 1980 than in 1950? Use Figure 1.16.

Strategy

To find how many times more expensive a stamp was, divide the cost in 1980 (15) by the cost in 1950 (3).

Solution

$15 \div 3 = 5$

A stamp was 5 times more expensive in 1980.

YOU TRY IT • 20

How many times more expensive was a stamp in 1997 than in 1960? Use Figure 1.16.

Your strategy

Your solution

Solution on p. S3

EXAMPLE • 21

Find the amount of sod needed to cover a football field. A football field measures 120 yd by 50 yd.

Strategy
Draw a diagram.

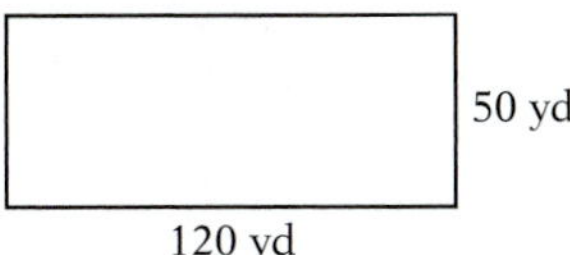

To find the amount of sod needed, use the formula for the area of a rectangle, $A = LW$. $L = 120$ and $W = 50$.

Solution
$A = LW$
$A = 120(50)$
$A = 6000$

6000 ft^2 of sod are needed.

YOU TRY IT • 21

A homeowner wants to carpet the family room. The floor is square and measures 6 m on each side. How much carpet should be purchased?

Your strategy

Your solution

EXAMPLE • 22

At what rate of speed would you need to travel in order to drive a distance of 294 mi in 6 h? Use the formula $r = \frac{d}{t}$, where r is the average rate of speed, d is the distance, and t is the time.

Strategy
To find the rate of speed, replace d by 294 and t by 6 in the given formula and solve for r.

Solution

$$r = \frac{d}{t}$$

$$r = \frac{294}{6} = 49$$

You would need to travel at a speed of 49 mph.

YOU TRY IT • 22

At what rate of speed would you need to travel in order to drive a distance of 486 mi in 9 h? Use the formula $r = \frac{d}{t}$, where r is the average rate of speed, d is the distance, and t is the time.

Your strategy

Your solution

Solutions on p. S3

1.3 EXERCISES

OBJECTIVE A To multiply whole numbers

1. Explain how to rewrite the addition 6 + 6 + 6 + 6 + 6 as multiplication.

2. Provide at least three examples of situations in which we multiply numbers.

For Exercises 3 to 14, multiply.

3. (9)(127)

4. (4)(623)

5. (6709)(7)

6. (3608)(5)

7. 8 · 58,769

8. 7 · 60,047

9. $\begin{array}{r} 683 \\ \times\ 71 \\ \hline \end{array}$

10. $\begin{array}{r} 591 \\ \times\ 92 \\ \hline \end{array}$

11. $\begin{array}{r} 7053 \\ \times\ \ 46 \\ \hline \end{array}$

12. $\begin{array}{r} 6704 \\ \times\ \ 58 \\ \hline \end{array}$

13. $\begin{array}{r} 3285 \\ \times\ 976 \\ \hline \end{array}$

14. $\begin{array}{r} 5327 \\ \times\ 624 \\ \hline \end{array}$

15. Find the product of 500 and 3.

16. Find 30 multiplied by 80.

17. What is 40 times 50?

18. What is twice 700?

19. What is the product of 400, 3, 20, and 0?

20. Write the product of *f* and *g*.

21. Write the product of *q*, *r*, and *s*.

22. **Physical Exercise** The figure at the right shows the number of calories burned on three different exercise machines during 1 h of a light, moderate, or vigorous workout. How many calories would you burn by **a.** working out vigorously on a stair climber for a total of 6 h? **b.** working out moderately on a treadmill for a total of 12 h?

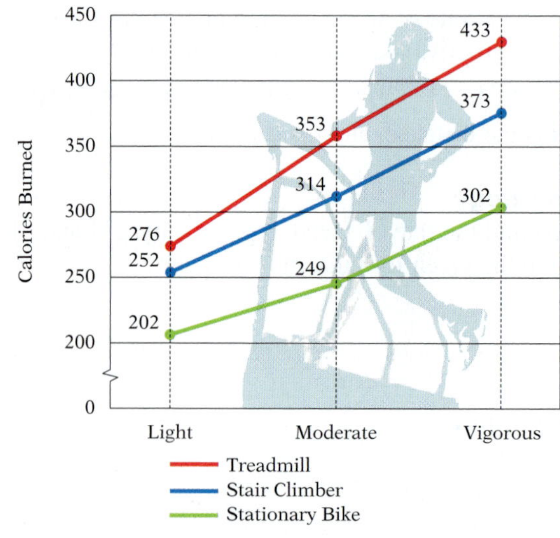

Calories Burned on Exercise Machines

Source: Journal of American Medical Association

For Exercises 23 to 30, estimate by rounding. Then find the exact answer.

23. 3467 · 359

24. 8745(63)

25. (39,246)(29)

26. 64,409 · 67

27. 745(63)

28. $432 \cdot 91$

29. (8941)(726)

30. 2837(216)

For Exercises 31 to 38, evaluate the expression for the given values of the variables.

31. ab, when $a = 465$ and $b = 32$

32. cd, when $c = 381$ and $d = 25$

33. $7a$, when $a = 465$

34. $6n$, when $n = 382$

35. xyz, when $x = 5$, $y = 12$, and $z = 30$

36. abc, when $a = 4$, $b = 20$, and $c = 50$

37. $2xy$, when $x = 67$ and $y = 23$

38. $4ab$, when $a = 95$ and $b = 33$

39. Find a one-digit number and a two-digit number whose product is a number that ends in two zeros.

For Exercises 40 to 43, identify the property that justifies the statement.

40. $1 \cdot 29 = 29$

41. $(10 \cdot 5) \cdot 8 = 10 \cdot (5 \cdot 8)$

42. $43 \cdot 1 = 1 \cdot 43$

43. $0(76) = 0$

For Exercises 44 to 47, use the given property of multiplication to complete the statement.

44. The Commutative Property of Multiplication
$19 \cdot ? = 30 \cdot 19$

45. The Associative Property of Multiplication
$(? \cdot 6)100 = 5(6 \cdot 100)$

46. The Multiplication Property of Zero
$45 \cdot 0 = ?$

47. The Multiplication Property of One
$? \cdot 77 = 77$

48. Is 6 a solution of the equation $4x = 24$?

49. Is 0 a solution of the equation $4 = 4n$?

50. Is 23 a solution of the equation $96 = 3z$?

51. Is 14 a solution of the equation $56 = 4c$?

52. Is 19 a solution of the equation $2y = 38$?

53. Is 11 a solution of the equation $44 = 3a$?

OBJECTIVE B To evaluate expressions that contain exponents

For Exercises 54 to 61, write in exponential form.

54. $2 \cdot 2 \cdot 2 \cdot 7 \cdot 7 \cdot 7 \cdot 7 \cdot 7$

55. $3 \cdot 3 \cdot 3 \cdot 3 \cdot 3 \cdot 3 \cdot 5 \cdot 5 \cdot 5$

56. $2 \cdot 2 \cdot 3 \cdot 3 \cdot 3 \cdot 5 \cdot 5 \cdot 5 \cdot 5$

57. $7 \cdot 7 \cdot 11 \cdot 11 \cdot 11 \cdot 19 \cdot 19 \cdot 19 \cdot 19$

58. $c \cdot c$

59. $d \cdot d \cdot d$

60. $x \cdot x \cdot x \cdot y \cdot y \cdot y$

61. $a \cdot a \cdot b \cdot b \cdot b \cdot b$

For Exercises 62 to 73, evaluate.

62. 2^5

63. 2^6

64. 10^6

65. 10^9

66. $2^3 \cdot 5^2$

67. $2^5 \cdot 3^2$

68. $3^2 \cdot 10^3$

69. $2^4 \cdot 10^2$

70. $0^2 \cdot 6^2$

71. $4^3 \cdot 0^3$

72. $2^2 \cdot 5 \cdot 3^3$

73. $5^2 \cdot 2 \cdot 3^4$

74. Rewrite the expression using the numbers 3 and 5 exactly once. Then simplify the expression.
a. $3 + 3 + 3 + 3 + 3$
b. $3 \cdot 3 \cdot 3 \cdot 3 \cdot 3$

75. Find the square of 12.

76. What is the cube of 6?

77. Find the cube of 8.

78. What is the square of 11?

79. Write the fourth power of a.

80. Write the fifth power of t.

For Exercises 81 to 86, evaluate the expression for the given values of the variables.

81. x^3y, when $x = 2$ and $y = 3$

82. x^2y, when $x = 3$ and $y = 4$

83. ab^6, when $a = 5$ and $b = 2$

84. ab^3, when $a = 7$ and $b = 4$

85. c^2d^2, when $c = 3$ and $d = 5$

86. m^3n^3, when $m = 5$ and $n = 10$

OBJECTIVE C To divide whole numbers

87. Provide at least three examples of situations in which we divide numbers.

88. In what situation does a division problem have a remainder?

For Exercises 89 to 104, divide.

89. $9\overline{)2763}$

90. $4\overline{)2160}$

91. $5\overline{)1549}$

92. $8\overline{)1636}$

93. $15{,}300 \div 6$

94. $43{,}500 \div 5$

95. $681 \div 32$

96. $879 \div 41$

97. $9152 \div 62$

98. $4161 \div 23$

99. $7408 \div 37$

100. $5207 \div 26$

101. $31{,}546 \div 78$

102. $38{,}976 \div 64$

103. $7713 \div 476$

104. $8947 \div 223$

105. Find the quotient of 7256 and 8.

106. What is the quotient of 8172 and 9?

107. What is 6168 divided by 7?

108. Find 4153 divided by 9.

109. Write the quotient of c and d.

110. Insurance Claims The table at the right shows the sources of laptop insurance claims in a recent year. Claims have been rounded to the nearest ten-thousand dollars.

a. What was the average monthly claim for theft?

b. For all sources combined, find the average claims per month.

Source	*Claims (in dollars)*
Accidents	560,000
Theft	300,000
Power Surge	80,000
Lightning	50,000
Transit	20,000
Water/flood	20,000
Other	110,000

Source: Safeware, The Insurance Company

For Exercises 111 to 118, estimate by rounding. Then find the exact answer.

111. $36{,}472 \div 47$ **112.** $62{,}176 \div 58$ **113.** $389{,}804 \div 76$ **114.** $637{,}072 \div 29$

115. $79\overline{)38{,}984}$ **116.** $53\overline{)11{,}792}$ **117.** $219\overline{)332{,}004}$ **118.** $324\overline{)632{,}124}$

For Exercises 119 to 124, evaluate the variable expression $\frac{x}{y}$ for the given values of x and y.

119. $x = 48; y = 1$ **120.** $x = 56; y = 56$ **121.** $x = 79; y = 0$

122. $x = 0; y = 23$ **123.** $x = 39{,}200; y = 4$ **124.** $x = 16{,}200; y = 3$

125. True or false? When a three-digit number is divided by a one-digit number, the quotient can be a one-digit number.

126. Is 9 a solution of the equation $\frac{36}{z} = 4$? **127.** Is 60 a solution of the equation $\frac{n}{12} = 5$?

128. Is 49 a solution of the equation $56 = \frac{x}{7}$? **129.** Is 16 a solution of the equation $6 = \frac{48}{y}$?

OBJECTIVE D To factor numbers and find the prime factorization of numbers

For Exercises 130 to 149, find all the factors of the number.

130. 10 **131.** 20 **132.** 12 **133.** 9 **134.** 8

135. 16 **136.** 13 **137.** 17 **138.** 18 **139.** 24

140. 25 **141.** 36 **142.** 56 **143.** 45 **144.** 28

145. 32 **146.** 48 **147.** 64 **148.** 54 **149.** 75

150. True or false? If a number has exactly four factors, then the product of those four factors must be the number.

For Exercises 151 to 170, find the prime factorization of the number.

151. 16 **152.** 24 **153.** 12 **154.** 27 **155.** 15

156. 36 **157.** 40 **158.** 50 **159.** 37 **160.** 83

161. 65 **162.** 80 **163.** 28 **164.** 49 **165.** 42

166. 81 **167.** 51 **168.** 89 **169.** 46 **170.** 120

OBJECTIVE E To solve application problems and use formulas

171. Nutrition One ounce of cheddar cheese contains 115 calories. Find the number of calories in 4 oz of cheddar cheese.

Nutrition Facts	Amount/Serving	% DV*	Amount/Serving	% DV*
Serv. Size 1 oz.	Total Fat 9g	14%	Total Carb. 1g	0%
Servings Per Package 12	Sat Fat 5g	25%	Fiber 0g	0%
Calories 115	Cholest. 30mg	10%	Sugars 0g	
Fat Cal. 80	Sodium 170mg	7%	Protein 7g	
*Percent Daily Values (DV) are based on a 2,000 calorie diet	Vitamin A 6% • Vitamin C 0% • Calcium 20% • Iron 0%			

172. Matchmaking Services See the news clipping at the right. **a.** How many marriages occur between eHarmony members each week? **b.** How many marriages occur each year? Use a 365-day year.

In the News

Find Your Match Online

eHarmony, the online matchmaking service, boasts marriages among its members at the rate of 90 a day.

Source: Time, January 17, 2008

173. Aviation A plane flying from Los Angeles to Boston uses 865 gal of jet fuel each hour. How many gallons of jet fuel are used on a 5-hour flight?

174. Geometry Find **a.** the perimeter and **b.** the area of a square that measures 16 mi on each side.

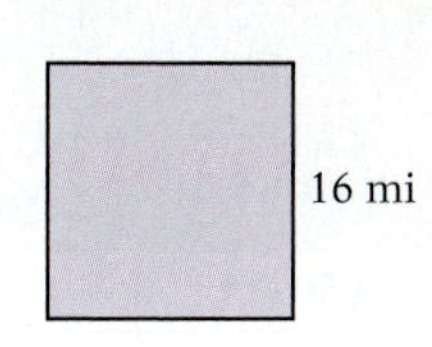

175. Geometry Find **a.** the perimeter and **b.** the area of a rectangle with a length of 24 m and a width of 15 m.

176. Geometry Find the length of fencing needed to surround a square corral that measures 55 ft on each side.

177. Geometry A homeowner plans to fence in the area around a swimming pool in the backyard. The area to be fenced in is a square measuring 24 ft on each side. How many feet of fencing should the homeowner purchase?

178. Geometry A solar panel is in the shape of a rectangle that has a width of 2 ft and a length of 3 ft. Find the area of the solar panel.

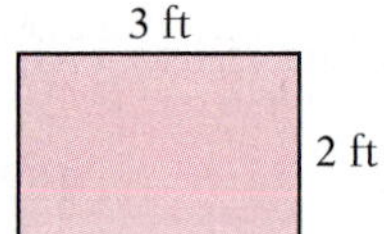

179. Geometry What is the area of the floor of a two-car garage that is in the shape of a square that measures 24 ft on a side?

180. Geometry A fieldstone patio is in the shape of a square that measures 9 ft on each side. What is the area of the patio?

181. Geometry Find the amount of fabric needed for a rectangular flag that measures 308 cm by 192 cm.

182. U.S. Postal Service There are 114 million households in the United States. Use the information in the news clipping at the right to determine, on average, how many pieces of mail each household will receive between Thanksgiving and Christmas this year. Round to the nearest whole number.

In the News

Holiday Mail Delivery

The U.S. Postal Service expects to deliver 20 billion pieces of mail between Thanksgiving and Christmas this year.

Source: www.usps.com

183. Purchasing A buyer for a department store purchased 215 suits at $83 each. Estimate the total cost of the order.

184. Finances Financial advisors may predict how much money we should have saved for retirement by the ages of 35, 45, 55, and 65. One such prediction is included in the table below.

a. A couple has earnings of $100,000 per year. According to the table, by how much should their savings grow per year from age 45 to 55?

b. A couple has earnings of $50,000 per year. According to the table, by how much should their savings grow per year from age 55 to 65?

Minimum Levels of Savings Required for Married Couples to Be Prepared for Retirement

Earnings	Savings Accumulation by Age			
	35	45	55	65
$50,000	8,000	23,000	90,000	170,000
$75,000	17,000	60,000	170,000	310,000
$100,000	34,000	110,000	280,000	480,000
$150,000	67,000	210,000	490,000	840,000

185. Loan Payments Find the total amount paid on a loan when the monthly payment is $285 and the loan is paid off in 24 months. Use the formula $A = MN$, where A is the total amount paid, M is the monthly payment, and N is the number of payments.

186. Loan Payments Find the total amount paid on a loan when the monthly payment is $187 and the loan is paid off in 36 months. Use the formula $A = MN$, where A is the total amount paid, M is the monthly payment, and N is the number of payments.

187. Travel Use the formula $t = \frac{d}{r}$, where t is the time, d is the distance, and r is the average rate of speed, to find the time it would take to drive 513 mi at an average speed of 57 mph.

188. Travel Use the formula $t = \frac{d}{r}$, where t is the time, d is the distance, and r is the average rate of speed, to find the time it would take to drive 432 mi at an average speed of 54 mph.

Henry Ray Abrams/AFP/Getty Images

New York Stock Exchange

189. Mutual Funds The current value of the stocks in a mutual fund is $10,500,000. The number of shares outstanding is 500,000. Find the value per share of the fund. Use the formula $V = \frac{C}{S}$, where V is the value per share, C is the current value of the stocks in the fund, and S is the number of shares outstanding.

190. Mutual Funds The current value of the stocks in a mutual fund is $4,500,000. The number of shares outstanding is 250,000. Find the value per share of the fund. Use the formula $V = \frac{C}{S}$, where V is the value per share, C is the current value of the stocks in the fund, and S is the number of shares outstanding.

191. The price of Braeburn apples is $1.29 per pound, and the price of Cameo apples is $1.79 per pound. Which of the following represents the price of 3 lb of Braeburn apples and 2 lb of Cameo apples?

(i) $(3 \times 1.29) + (3 \times 1.79)$ **(ii)** $(2 \times 1.29) + (3 \times 1.79)$
(iii) $5 \times (1.29 + 1.79)$ **(iv)** $(3 \times 1.29) + (2 \times 1.79)$

Applying the Concepts

192. Time There are 52 weeks in a year. Is this an exact figure or an approximation?

193. Mathematics 13,827 is not divisible by 4. By rearranging the digits, find the largest possible number that is divisible by 4.

194. According to the National Safety Council, in a recent year, a death resulting from an accident occurred at the rate of one death every 5 min. At this rate, how many accidental deaths occurred each hour? each day? throughout the year? Explain how you arrived at your answers.

SECTION

1.4 The Order of Operations Agreement

OBJECTIVE A To use the Order of Operations Agreement to simplify expressions 

More than one operation may occur in a numerical expression. For example, the expression

$$4 + 3(5)$$

includes two arithmetic operations, addition and multiplication. The operations could be performed in different orders.

If we multiply first and then add, we have:	$4 + 3(5)$ $4 + 15$ 19	If we add first and then multiply, we have:	$4 + 3(5)$ $7(5)$ 35

To prevent more than one answer to the same problem, an Order of Operations Agreement is followed. By this agreement, 19 is the only correct answer.

Integrating Technology

Many calculators use the Order of Operations Agreement shown at the right.

Enter 4 + 3 x 5 = into your calculator. If the answer is 19, your calculator uses the Order of Operations Agreement.

The Order of Operations Agreement

Step 1 Do all operations inside parentheses.

Step 2 Simplify any numerical expressions containing exponents.

Step 3 Do multiplication and division as they occur from left to right.

Step 4 Do addition and subtraction as they occur from left to right.

Integrating Technology

Here is an example of using the parentheses keys on a calculator. To evaluate 28(103 − 78), enter:

28 x (103 − 78) =. Note that x is required on most calculators.

HOW TO 1 Simplify: $2(4 + 1) - 2^3 + 6 \div 2$

$2(4 + 1) - 2^3 + 6 \div 2$

$= 2(5) - 2^3 + 6 \div 2$ • Perform operations in parentheses.

$= 2(5) - 8 + 6 \div 2$ • Simplify expressions with exponents.

$= 10 - 8 + 6 \div 2$ • Do multiplication and division as they occur from left to right.

$= 10 - 8 + 3$

$= 2 + 3$ • Do addition and subtraction as they occur from left to right.

$= 5$

One or more of the foregoing steps may not be needed to simplify an expression. In that case, proceed to the next step in the Order of Operations Agreement.

HOW TO 2 Simplify: $8 + 9 \div 3$

$8 + 9 \div 3$ • There are no parentheses (Step 1). There are no exponents (Step 2).

$= 8 + 3$ • Do the division (Step 3).

$= 11$ • Do the addition (Step 4).

Point of Interest

Try this: Use the same one-digit number three times to write an expression that is equal to 30.

HOW TO • 3 Evaluate $5a - (b + c)^2$ when $a = 6$, $b = 1$, and $c = 3$.

$5a - (b + c)^2$

$5(6) - (1 + 3)^2$ • Replace a with 6, b with 1, and c with 3.

$= 5(6) - (4)^2$ • Use the Order of Operations Agreement to simplify the resulting numerical expression. Perform operations inside parentheses.

$= 5(6) - 16$ • Simplify expressions with exponents.

$= 30 - 16$ • Do the multiplication.

$= 14$ • Do the subtraction.

EXAMPLE • 1

Simplify: $18 \div (6 + 3) \cdot 9 - 4^2$

Solution

$$\begin{aligned} 18 \div (6 + 3) \cdot 9 - 4^2 &= 18 \div 9 \cdot 9 - 4^2 \\ &= 18 \div 9 \cdot 9 - 16 \\ &= 2 \cdot 9 - 16 \\ &= 18 - 16 \\ &= 2 \end{aligned}$$

YOU TRY IT • 1

Simplify: $4 \cdot (8 - 3) \div 5 - 2$

Your solution

EXAMPLE • 2

Simplify: $20 + 24(8 - 5) \div 2^2$

Solution

$$\begin{aligned} 20 + 24(8 - 5) \div 2^2 &= 20 + 24(3) \div 2^2 \\ &= 20 + 24(3) \div 4 \\ &= 20 + 72 \div 4 \\ &= 20 + 18 \\ &= 38 \end{aligned}$$

YOU TRY IT • 2

Simplify: $16 + 3(6 - 1)^2 \div 5$

Your solution

EXAMPLE • 3

Evaluate $(a - b)^2 + 3c$ when $a = 6$, $b = 4$, and $c = 1$.

Solution

$(a - b)^2 + 3c$

$$\begin{aligned} (6 - 4)^2 + 3(1) &= (2)^2 + 3(1) \\ &= 4 + 3(1) \\ &= 4 + 3 \\ &= 7 \end{aligned}$$

YOU TRY IT • 3

Evaluate $(a - b)^2 + 5c$ when $a = 7$, $b = 2$, and $c = 4$.

Your solution

Solutions on p. S3

1.4 EXERCISES

OBJECTIVE A To use the Order of Operations Agreement to simplify expressions

1. Why do we need an Order of Operations Agreement?

2. What are the steps in the Order of Operations Agreement?

For Exercises 3 to 32, simplify.

3. $8 \div 4 + 2$
4. $12 - 9 \div 3$
5. $6 \cdot 4 + 5$
6. $5 \cdot 7 + 3$
7. $4^2 - 3$
8. $6^2 - 14$
9. $5 \cdot (6 - 3) + 4$
10. $8 + (6 + 2) \div 4$
11. $9 + (7 + 5) \div 6$
12. $14 \cdot (3 + 2) \div 10$
13. $13 \cdot (1 + 5) \div 13$
14. $14 - 2^3 + 9$
15. $6 \cdot 3^2 + 7$
16. $18 + 5 \cdot 3^2$
17. $14 + 5 \cdot 2^3$
18. $20 + (9 - 4) \cdot 2$
19. $10 + (8 - 5) \cdot 3$
20. $3^2 + 5 \cdot (6 - 2)$
21. $2^3 + 4(10 - 6)$
22. $3^2 \cdot 2^2 + 3 \cdot 2$
23. $6(7) + 4^2 \cdot 3^2$
24. $14 - 2(6)$
25. $18 + 3(7)$
26. $2(9 - 2) + 5$
27. $6(8 - 3) - 12$
28. $15 - (7 - 1) \div 3$
29. $16 - (13 - 5) \div 4$
30. $11 + 2 - 3 \cdot 4 \div 3$
31. $17 + 1 - 8 \cdot 2 \div 4$
32. $3(5 + 3) \div 8$

For Exercises 33 to 42, evaluate the expression for the given values of the variables.

33. $x - 2y$, where $x = 8$ and $y = 3$

34. $x + 6y$, where $x = 5$ and $y = 4$

35. $x^2 + 3y$, where $x = 6$ and $y = 7$

36. $3x^2 + y$, where $x = 2$ and $y = 9$

37. $x^2 + y \div x$, where $x = 2$ and $y = 8$

38. $x + y^2 \div x$, where $x = 4$ and $y = 8$

39. $4x + (x - y)^2$, where $x = 8$ and $y = 2$

40. $(x + y)^2 - 2y$, where $x = 3$ and $y = 6$

41. $x^2 + 3(x - y) + z^2$, where $x = 2$, $y = 1$, and $z = 3$

42. $x^2 + 4(x - y) \div z^2$, where $x = 8$, $y = 6$, and $z = 2$

43. Use the inequality symbol $>$ to compare the expressions $11 + (8 + 4) \div 6$ and $12 + (9 - 5) \cdot 3$.

44. Use the inequality symbol $<$ to compare the expressions $3^2 + 7(4 - 2)$ and $14 - 2^3 + 20$.

For Exercises 45 to 47, insert parentheses as needed in the expression $8 - 2 \cdot 3 + 1$ in order to make the statement true.

45. $8 - 2 \cdot 3 + 1 = 3$

46. $8 - 2 \cdot 3 + 1 = 0$

47. $8 - 2 \cdot 3 + 1 = 24$

Applying the Concepts

48. Arrange the expressions in order from greatest value to least value.

$27 \div 9 + 8$ $\quad$ $4 + 3 \cdot 12$

$81 - 8^2$ $\quad$ $50 - 6(8)$

$5(10 - 2) \div 4$ $\quad$ $2(1 + 4)^2 \div 10$

49. What is the smallest prime number greater than $15 + (8 - 3)(2^4)$?

FOCUS ON PROBLEM SOLVING

Questions to Ask

© Brownie Harris/Corbis

You encounter problem-solving situations every day. Some problems are easy to solve, and you may mentally solve these problems without considering the steps you are taking in order to draw a conclusion. Others may be more challenging and may require more thought and consideration.

Suppose a friend suggests that you both take a trip over spring break. You'd like to go. What questions go through your mind? You might ask yourself some of the following questions:

How much will the trip cost? What will be the cost for travel, hotel rooms, meals, and so on?

Are some costs going to be shared by both my friend and me?

Can I afford it?

How much money do I have in the bank?

How much more money than I have now do I need?

How much time is there to earn that much money?

How much can I earn in that amount of time?

How much money must I keep in the bank in order to pay the next tuition bill (or some other expense)?

These questions require different mathematical skills. Determining the cost of the trip requires **estimation;** for example, you must use your knowledge of air fares or the cost of gasoline to arrive at an estimate of these costs. If some of the costs are going to be shared, you need to **divide** those costs by 2 in order to determine your share of the expense. The question regarding how much more money you need requires **subtraction:** the amount needed minus the amount currently in the bank. To determine how much money you can earn in the given amount of time requires **multiplication**—for example, the amount you earn per week times the number of weeks to be worked. To determine if the amount you can earn in the given amount of time is sufficient, you need to use your knowledge of **order relations** to compare the amount you can earn with the amount needed.

Facing the problem-solving situation described above may not seem difficult to you. The reason may be that you have faced similar situations before and, therefore, know how to work through this one. You may feel better prepared to deal with a circumstance such as this one because you know what questions to ask. An important aspect of learning to solve problems is learning what questions to ask. As you work through the application problems in this text, try to become more conscious of the mental process you are going through. You might begin the process by asking yourself the following questions whenever you are solving an application problem.

1. Have I read the problem enough times to be able to understand the situation being described?
2. Will restating the problem in different words help me to understand the problem situation better?
3. What facts are given? (You might make a list of the information contained in the problem.)
4. What information is being asked for?

5. What relationships exist among the given facts? What relationships exist among the given facts and the solution?

6. What mathematical operations are needed in order to solve the problem?

Try to focus on the problem-solving situation, not on the computation or on getting the answer quickly. And remember, the more problems you solve, the better able you will be to solve other problems in the future, partly because you are learning what questions to ask.

PROJECTS AND GROUP ACTIVITIES

Order of Operations

Does your calculator use the Order of Operations Agreement? To find out, try this problem:

$$2 + 4 \cdot 7$$

If your answer is 30, then the calculator uses the Order of Operations Agreement. If your answer is 42, it does not use the agreement.

Even if your calculator does not use the Order of Operations Agreement, you can still correctly evaluate numerical expressions. The parentheses keys, (and), are used for this purpose.

Remember that $2 + 4 \cdot 7$ means $2 + (4 \cdot 7)$ because the multiplication must be completed before the addition. To evaluate this expression, enter the following:

Enter:	2	+	(	4	×	7	)	=
Display:	2	2	(	4	4	7	28	30

When using your calculator to evaluate numerical expressions, insert parentheses around multiplications and around divisions. This has the effect of forcing the calculator to do the operations in the order you want.

For Exercises 1 to 10, evaluate.

1. $3 \cdot 8 - 5$

2. $6 + 8 \div 2$

3. $3 \cdot (8 - 2)^2$

4. $24 - (4 - 2)^2 \div 4$

5. $3 + (6 \div 2 + 4)^2 - 2$

6. $16 \div 2 + 4 \cdot (8 - 12 \div 4)^2 - 50$

7. $3 \cdot (15 - 2 \cdot 3) - 36 \div 3$

8. $4 \cdot 2^2 - (12 + 24 \div 6) + 5$

9. $16 \div 4 \cdot 3 + (3 \cdot 4 - 5) + 2$

10. $15 \cdot 3 \div 9 + (2 \cdot 6 - 3) + 4$

Surveys On page 16 there is a circle graph showing the results of a survey of 150 people who were asked, "What bothers you most about movie theaters?" Note that the responses included (1) people talking in the theater, (2) high ticket prices, (3) high prices for food purchased in the theater, (4) dirty floors, and (5) uncomfortable seats.

Conduct a similar survey in your class. Ask each classmate which of the five conditions stated above is most irritating. Record the number of students who choose each one of the five possible responses. Prepare a bar graph to display the results of the survey. A model is provided below to help you get started.

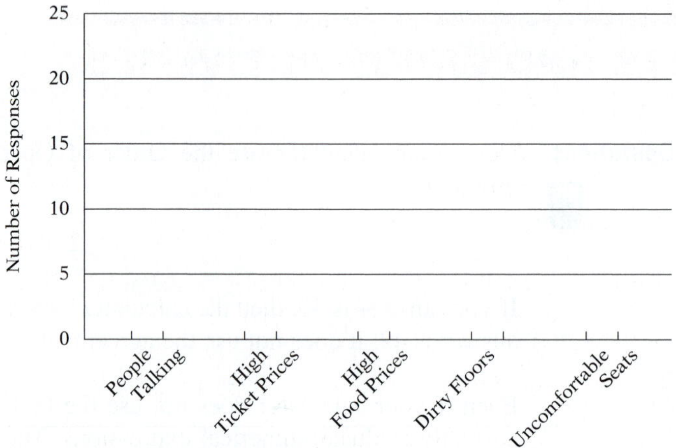

Responses to Theater-Goers Survey

Patterns in Mathematics For the circle at the left, use a straight line to connect each dot on the circle with every other dot on the circle. How many different straight lines are there?

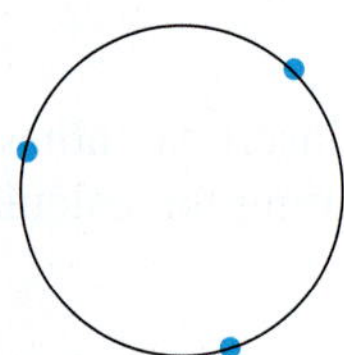

Follow the same procedure for each of the circles shown below. How many different straight lines are there in each?

Find a pattern to describe the number of dots on a circle and the corresponding number of different lines drawn. Use the pattern to determine the number of different lines that would be drawn in a circle with 7 dots and in a circle with 8 dots.

Now use the pattern to answer the following question. You are arranging a tennis tournament with 9 players. How many singles matches will be played among the 9 players if each player plays each of the other players only once?

CHAPTER 1

SUMMARY

KEY WORDS	EXAMPLES
The *natural numbers* or *counting numbers* are 1, 2, 3, 4, 5, 6, 7, 8, 9, 10, ... [1.1A, p. 2]	
The *whole numbers* are 0, 1, 2, 3, 4, 5, 6, 7, 8, 9, 10, ... [1.1A, p. 2]	
The symbol for "is less than" is $<$. The symbol for "is greater than" is $>$. A statement that uses the symbol $<$ or $>$ is an *inequality.* [1.1A, p. 2]	$3 < 7$ $9 > 2$
When a whole number is written using the digits 0, 1, 2, 3, 4, 5, 6, 7, 8, and 9, it is said to be in *standard form.* The position of each digit in the number determines the digit's *place value.* [1.1B, p. 3]	The number 598,317 is in standard form. The digit 8 is in the thousands place.
A *pictograph* represents data by using a symbol that is characteristic of the data. A *circle graph* represents data by the size of the sectors. A *bar graph* represents data by the height of the bars. A *broken-line graph* represents data by the position of the lines and shows trends or comparisons. [1.1D, pp. 7–9]	
Addition is the process of finding the total of two or more numbers. The numbers being added are called *addends.* The answer is the *sum.* [1.2A, pp. 18–19]	1 1 1 8762 + 1359 10,121
Subtraction is the process of finding the difference between two numbers. The *minuend* minus the *subtrahend* equals the *difference.* [1.2B, pp. 24–25]	4 11 11 6 13 52,173 −34,968 17,205
Multiplication is the repeated addition of the same number. The numbers that are multiplied are called *factors.* The answer is the *product.* [1.3A, p. 40]	4 5 358 × 7 2506
The expression 3^5 is in *exponential form.* The *exponent,* 5, indicates how many times the *base,* 3, occurs as a factor in the multiplication. [1.3B, p. 45]	$5^4 = 5 \cdot 5 \cdot 5 \cdot 5 = 625$
Division is used to separate objects into equal groups. The *dividend* divided by the *divisor* equals the *quotient.* For any division problem, (*quotient* · *divisor*) + *remainder* = *dividend.* [1.3C, pp. 47–49]	93 r3 7)654 −63 24 −21 3 *Check:* $(93 \cdot 7) + 3 = 651 + 3 = 654$

Natural number *factors* of a number divide that number evenly (there is no remainder). [1.3D, p. 52]

$18 \div 1 = 18$
$18 \div 2 = 9$
$18 \div 3 = 6$
$18 \div 4$ 4 does not divide 18 evenly.
$18 \div 5$ 5 does not divide 18 evenly.
$18 \div 6 = 3$ The factors are repeating.
The factors of 18 are 1, 2, 3, 6, 9, and 18.

A number greater than 1 is a *prime number* if its only whole number factors are 1 and itself. If a number is not prime, it is a *composite number*. [1.3D, p. 52]

The prime numbers less than 20 are 2, 3, 5, 7, 11, 13, 17, and 19.

The composite numbers less than 20 are 4, 6, 8, 9, 10, 12, 14, 15, 16, and 18.

The *prime factorization* of a number is the expression of the number as a product of its prime factors. [1.3D, p. 53]

$$\begin{array}{r} 7 \\ 3\overline{)21} \\ 2\overline{)42} \end{array}$$

The prime factorization of 42 is $2 \cdot 3 \cdot 7$.

A *variable* is a letter that is used to stand for a number. A mathematical expression that contains one or more variables is a *variable expression*. Replacing the variables in a variable expression with numbers and then simplifying the numerical expression is called *evaluating the variable expression.* [1.2A, p. 20]

To evaluate the variable expression $4ab$ when $a = 3$ and $b = 2$, replace a with 3 and b with 2. Simplify the resulting expression.

$4ab$
$4(3)(2) = 12(2) = 24$

An *equation* expresses the equality of two numerical or variable expressions. An equation contains an equals sign. A *solution* of an equation is a number that, when substituted for the variable, results in a true equation. [1.2A, p. 22]

6 is a solution of the equation $5 + x = 11$ because $5 + 6 = 11$ is a true equation.

Parallel lines never meet; the distance between them is always the same. [1.2C, p. 28]

Parallel Lines

An angle is measured in *degrees*. A 90° angle is a *right angle*. [1.2C, p. 28]

90°

Right Angle

A *polygon* is a closed figure determined by three or more line segments. The line segments that form the polygon are its *sides*. A *triangle* is a three-sided polygon. A *quadrilateral* is a four-sided polygon. A *rectangle* is a quadrilateral in which opposite sides are parallel, opposite sides are equal in length, and all four angles are right angles. A *square* is a rectangle in which all sides have the same length. The *perimeter* of a plane figure is a measure of the distance around the figure, and its *area* is the amount of surface in the region. [1.2C, pp. 28–29; 1.3E, pp. 54–55]

Triangle

Rectangle

Square

ESSENTIAL RULES AND PROCEDURES	EXAMPLES
To round a whole number to a given place value: If the digit to the right of the given place value is less than 5, replace that digit and all digits to the right by zeros. If the digit to the right of the given place value is greater than or equal to 5, increase the digit in the given place value by 1, and replace all other digits to the right by zeros. [1.1C, pp. 5–6]	36,178 rounded to the nearest thousand is 36,000. 4592 rounded to the nearest thousand is 5000.
To estimate the answer to a calculation: Round each number to the highest place value of that number. Perform the calculation using the rounded numbers. [1.2A, p. 20]	39,471 ⟶ 40,000 12,586 ⟶ +10,000 50,000 50,000 is an estimate of the sum of 39,471 and 12,586.
Properties of Addition [1.2A, p. 21]	
Addition Property of Zero $a + 0 = a$ or $0 + a = a$	$7 + 0 = 7$
Commutative Property of Addition $a + b = b + a$	$8 + 3 = 3 + 8$
Associative Property of Addition $(a + b) + c = a + (b + c)$	$(2 + 4) + 6 = 2 + (4 + 6)$
Properties of Multiplication [1.3A, p. 43]	
Multiplication Property of Zero $a \cdot 0 = 0$ or $0 \cdot a = 0$	$3 \cdot 0 = 0$
Multiplication Property of One $a \cdot 1 = a$ or $1 \cdot a = a$	$6 \cdot 1 = 6$
Commutative Property of Multiplication $a \cdot b = b \cdot a$	$2 \cdot 8 = 8 \cdot 2$
Associative Property of Multiplication $(a \cdot b) \cdot c = a \cdot (b \cdot c)$	$(2 \cdot 4) \cdot 6 = 2 \cdot (4 \cdot 6)$
Division Properties of Zero and One [1.3C, p. 48]	
If $a \neq 0$, $0 \div a = 0$.	$0 \div 3 = 0$
If $a \neq 0$, $a \div a = 1$.	$3 \div 3 = 1$
$a \div 1 = a$	$3 \div 1 = 3$
$a \div 0$ is undefined.	$3 \div 0$ is undefined.
Order of Operations Agreement [1.4A, p. 66]	
Step 1 Do all operations inside parentheses.	$5^2 - 3(2 + 4) = 5^2 - 3(6)$
Step 2 Simplify any numerical expressions containing exponents.	$= 25 - 3(6)$
Step 3 Do multiplication and division as they occur from left to right.	$= 25 - 18$
Step 4 Do addition and subtraction as they occur from left to right.	$= 7$
Geometric Formulas [1.2C, p. 29; 1.3E, pp. 54–55]	
Perimeter of a Triangle $P = a + b + c$ Perimeter of a Rectangle $P = 2L + 2W$ Perimeter of a Square $P = 4s$ Area of a Rectangle $A = LW$ Area of a Square $A = s^2$	Find the perimeter of a triangle with sides that measure 9 m, 6 m, and 5 m. $P = a + b + c$ $P = 9 + 6 + 5$ $P = 20$ The perimeter of the triangle is 20 m.

CHAPTER 1

CONCEPT REVIEW

Test your knowledge of the concepts presented in this chapter. Answer each question. Then check your answers against the ones provided in the Answer Section.

1. What is the difference between the symbols $<$ and $>$?

2. How do you round a four-digit whole number to the nearest hundred?

3. What is the difference between the Commutative Property of Addition and the Associative Property of Addition?

4. How do you estimate the sum of two numbers?

5. When is it necessary to borrow when performing subtraction?

6. What is the difference between the Multiplication Property of Zero and the Multiplication Property of One?

7. How do you multiply a whole number by 100?

8. How do you estimate the product of two numbers?

9. What is the difference between $0 \div 9$ and $9 \div 0$?

10. How do you check the answer to a division problem that has a remainder?

11. What are the steps in the Order of Operations Agreement?

12. How do you know if a number is a factor of another number?

13. What is a quick way to determine if 3 is a factor of a number?

CHAPTER 1

REVIEW EXERCISES

1. Graph 8 on the number line.

   0 1 2 3 4 5 6 7 8 9 10 11 12

2. Evaluate 10^4.

3. Find the difference between 4207 and 1624.

4. Write $3 \cdot 3 \cdot 5 \cdot 5 \cdot 5 \cdot 5$ in exponential notation.

5. Add: $319 + 358 + 712$

6. Round 38,729 to the nearest hundred.

7. Place the correct symbol, < or >, between the two numbers.
   247   163

8. Write thirty-two thousand five hundred nine in standard form.

9. Evaluate $2xy$ when $x = 50$ and $y = 7$.

10. Find the quotient of 15,642 and 6.

11. Subtract: $6407 - 2359$

12. Estimate the sum of 482, 319, 570, and 146.

13. Find all the factors of 50.

14. Is 7 a solution of the equation $24 - y = 17$?

15. Simplify: $16 + 4(7 - 5)^2 \div 8$

16. Identify the property that justifies the statement.
    $10 + 33 = 33 + 10$

17. Write 4,927,036 in words.

18. Evaluate x^3y^2 when $x = 3$ and $y = 5$.

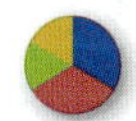

19. **Film Ratings** The circle graph at the right categorizes the 655 films released during a recent year by their ratings.
    a. How many times more PG-13 films were released than NC-17 films?
    b. How many times more R-rated films were released than NC-17 films?

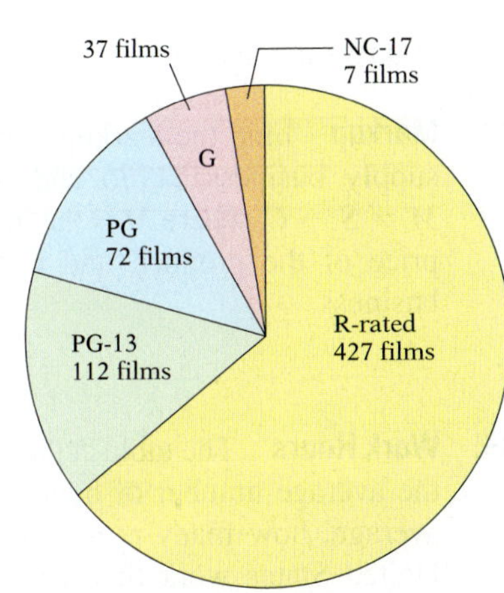

Ratings of Films Released
Source: MPA Worldwide Market Research

20. Divide: $6234 \div 92$

21. Find the product of 4 and 659.

22. Evaluate $x - y$ when $x = 270$ and $y = 133$.

23. Find the prime factorization of 90.

24. Evaluate $\frac{x}{y}$ when $x = 480$ and $y = 6$.

25. Complete the statement by using the Multiplication Property of One.
$? \cdot 82 = 82$

26. Simplify: $58 - 3 \cdot 4^2$

27. Evaluate $x + y$ when $x = 683$ and $y = 249$.

28. Multiply: $18 \cdot 24$

29. Evaluate $(a + b)^2 - 2c$ when $a = 5$, $b = 3$, and $c = 4$.

30. Basketball During his professional basketball career, Kareem Abdul-Jabbar had 17,440 rebounds. Elvin Hayes had 16,279 rebounds during his professional basketball career. Who had more rebounds, Abdul-Jabbar or Hayes?

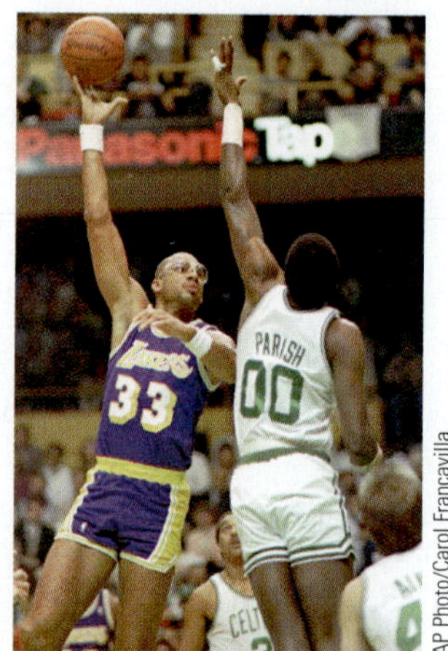

Kareem Abdul-Jabbar

31. Construction A contractor quotes the cost of work on a new house, which is to have 2800 ft^2 of floor space, at \$65 per square foot. Find the total cost of the contractor's work on the house.

32. Geometry A rectangle has a length of 25 m and a width of 12 m. Find **a.** the perimeter and **b.** the area of the rectangle.

33. Travel Use the formula $d = rt$, where d is distance, r is rate of speed, and t is time, to find the distance traveled in 3 h by a cyclist traveling at a speed of 14 mph.

34. Markup Find the markup on a copy machine that cost an office supply business \$1775 and sold for \$2224. Use the formula $M = S - C$, where M is the markup on a product, S is the selling price of the product, and C is the cost of the product to the business.

35. Work Hours The table at the right shows, for different countries, the average number of hours per year that employees work. On average, how many more hours per week do employees in the United States work than employees in France? Use a 50-week year. Round answers to the nearest whole number.

Country	*Annual Number of Hours Worked*
Britain	1731
France	1656
Japan	1889
Norway	1399
United States	1966

Source: International Labor Organization

CHAPTER 1

TEST

1. Multiply: 3297×100

2. Evaluate $2^4 \cdot 10^3$.

3. Find the difference between 4902 and 873.

4. Write $x \cdot x \cdot x \cdot x \cdot y \cdot y \cdot y$ in exponential notation.

5. Is 7 a solution of the equation $23 = p + 16$?

6. Round 2961 to the nearest hundred.

7. Place the correct symbol, $<$ or $>$, between the two numbers.
7177 7717

8. Write eight thousand four hundred ninety in standard form.

9. Write 382,904 in words.

10. Estimate the sum of 392, 477, 519, and 648.

11. Find the product of 8 and 1376.

12. Estimate the product of 36,479 and 58.

13. Find all the factors of 92.

14. Find the prime factorization of 240.

15. Evaluate $x - y$ when $x = 39{,}241$ and $y = 8375$.

16. Identify the property that justifies the statement.
$14 + y = y + 14$

17. Evaluate $\frac{x}{y}$ when $x = 3588$ and $y = 4$.

18. Simplify: $27 - (12 - 3) \div 9$

19. **Education** The table at the right shows the average annual earnings, based on level of education, for people aged 25 and older. What is the difference between average annual earnings for an individual with some college, but no degree, and for an individual with a bachelor's degree?

Education Level	*Average Annual Earnings*
No high school diploma	\$21,400
High school diploma	\$28,800
Some college, no degree	\$32,400
Associate degree	\$35,400
Bachelor's degree	\$46,300
Master's degree	\$55,300

Source: Census Bureau; Bureau of Labor Statistics

20. Simplify: $5 + 2(4 - 3)^6$

21. Write 3972 in expanded form.

22. Evaluate $5x + (x - y)^2$ when $x = 8$ and $y = 4$.

23. Complete the statement by using the Associative Property of Addition.
$(3 + 7) + x = 3 + (? + x)$

24. **Mathematics** What is the product of all the natural numbers less than 7?

25. **Purchasing** You purchase a computer system that includes an operating system priced at \$850, a monitor that cost \$270, an extended keyboard priced at \$175, and a printer for \$425. You pay for the purchase by check. You had \$2276 in your checking account before making the purchase. What was the balance in your account after making the purchase?

26. **Geometry** The length of each side of a square is 24 cm. Find **a.** the perimeter and **b.** the area of the square.

27. **Pay Deductions** A data processor receives a total salary of \$5690 per month. Deductions from the paycheck include \$854 for taxes, \$272 for retirement, and \$108 for insurance. Find the data processor's monthly take-home pay.

28. **College Education** See the news clipping at the right. **a.** Find the average cost of tuition, room, and board for 4 years at a public college. **b.** Find the average cost of tuition, room, and board for 4 years at a private college. **c.** Find the difference in cost for tuition, room, and board between 4 years at a private college and 4 years at a public college.

In the News

Comparing Tuition Costs

The average annual cost of tuition, room, and board at a four-year public college is \$12,796. At a four-year private college, the average cost is \$30,367.

Source: Kiplinger.com, January 24, 2007

29. **Commissions** Use the formula $C = U \cdot R$, where C is the commission earned, U is the number of units sold, and R is the commission rate per unit, to find the commission earned from selling 480 boxes of greeting cards when the commission rate per box is \$2.

30. **Mutual Funds** The current value of the stocks in a mutual fund is \$5,500,000. The number of shares outstanding is 500,000. Find the value per share of the fund. Use the formula $V = \frac{C}{S}$, where V is the value per share, C is the current value of the stocks in the fund, and S is the number of shares outstanding.

CHAPTER 2

Fractions and Decimals

OBJECTIVES

SECTION 2.1

- A To find the least common multiple (LCM)
- B To find the greatest common factor (GCF)

SECTION 2.2

- A To write proper fractions, improper fractions, and mixed numbers
- B To write equivalent fractions
- C To identify the order relation between two fractions

SECTION 2.3

- A To add fractions
- B To subtract fractions
- C To solve application problems and use formulas

SECTION 2.4

- A To multiply fractions
- B To divide fractions
- C To simplify a complex fraction
- D To solve application problems and use formulas

SECTION 2.5

- A To read and write decimals
- B To identify the order relation between two decimals
- C To round a decimal to a given place value
- D To solve application problems

SECTION 2.6

- A To add and subtract decimals
- B To multiply decimals
- C To divide decimals
- D To convert between decimals and fractions and identify the order relation between a decimal and a fraction
- E To solve application problems and use formulas

SECTION 2.7

- A To use the Order of Operations Agreement

ARE YOU READY?

Take the Chapter 2 Prep Test to find out if you are ready to learn to:

- Find the LCM or GCF of two or more numbers
- Write equivalent fractions
- Compare fractions
- Add, subtract, multiply, and divide fractions
- Compare decimals
- Add, subtract, multiply, and divide decimals
- Convert between decimals and fractions

PREP TEST

Do these exercises to prepare for Chapter 2.

For Exercises 1 to 6, add, subtract, multiply, or divide.

1. 4×5

2. $2 \cdot 2 \cdot 2 \cdot 3 \cdot 5$

3. 9×1

4. $6 + 4$

5. $10 - 3$

6. $63 \div 30$

7. Which of the following numbers divide evenly into 12?
1 2 3 4 5 6 7 8 9 10 11 12

8. Simplify: $8 \times 7 + 3$

9. Complete: $8 = ? + 1$

10. Place the correct symbol, $<$ or $>$, between the two numbers.
44 48

11. Round 36,852 to the nearest hundred.

SECTION

2.1 The Least Common Multiple and Greatest Common Factor

OBJECTIVE A To find the least common multiple (LCM) 

Tips for Success

Before you begin a new chapter, you should take some time to review previously learned skills. One way to do this is to complete the Prep Test. See page 81. This test focuses on the particular skills that will be required for the new chapter.

The **multiples of a number** are the products of that number and the numbers 1, 2, 3, 4, 5, ...

$3 \times 1 = 3$
$3 \times 2 = 6$
$3 \times 3 = 9$
$3 \times 4 = 12$ The multiples of 3 are 3, 6, 9, 12, 15, ...
$3 \times 5 = 15$
.
.
.

A number that is a multiple of two or more numbers is a **common multiple** of those numbers.

The multiples of 4 are 4, 8, 12, 16, 20, 24, 28, 32, 36, ...
The multiples of 6 are 6, 12, 18, 24, 30, 36, 42, ...
Some common multiples of 4 and 6 are 12, 24, and 36.

The **least common multiple (LCM)** is the smallest common multiple of two or more numbers.

The least common multiple of 4 and 6 is 12.

Listing the multiples of each number is one way to find the LCM. Another way to find the LCM uses the prime factorization of each number.

To find the LCM of 450 and 600, find the prime factorization of each number and write the factorization of each number in a table. Circle the greatest product in each column. The LCM is the product of the circled numbers.

	2	3	5
450 =	2	(3 · 3)	(5 · 5)
600 =	(2 · 2 · 2)	3	5 · 5

• In the column headed by 5, the products are equal. Circle just one product.

The LCM is the product of the circled numbers.
The LCM = $2 \cdot 2 \cdot 2 \cdot 3 \cdot 3 \cdot 5 \cdot 5 = 1800$.

EXAMPLE • 1

Find the LCM of 24, 36, and 50.

Solution

	2	3	5
24 =	(2 · 2 · 2)	3	
36 =	2 · 2	(3 · 3)	
50 =	2		(5 · 5)

The LCM = $2 \cdot 2 \cdot 2 \cdot 3 \cdot 3 \cdot 5 \cdot 5 = 1800$.

YOU TRY IT • 1

Find the LCM of 12, 27, and 50.

Your solution

Solution on p. S4

OBJECTIVE B To find the greatest common factor (GCF) 

Recall that a number that divides another number evenly is a factor of that number. The number 64 can be evenly divided by 1, 2, 4, 8, 16, 32, and 64, so the numbers 1, 2, 4, 8, 16, 32, and 64 are factors of 64.

A number that is a factor of two or more numbers is a **common factor** of those numbers.

The factors of 30 are 1, 2, 3, 5, 6, 10, 15, and 30.
The factors of 105 are 1, 3, 5, 7, 15, 21, 35, and 105.
The common factors of 30 and 105 are 1, 3, 5, and 15.

The **greatest common factor (GCF)** is the largest common factor of two or more numbers.

The greatest common factor of 30 and 105 is 15.

Listing the factors of each number is one way of finding the GCF. Another way to find the GCF is to use the prime factorization of each number.

To find the GCF of 126 and 180, find the prime factorization of each number and write the factorization of each number in a table. Circle the least product in each column that does not have a blank. The GCF is the product of the circled numbers.

	2	3	5	7
126 =	(2)	(3 · 3)		7
180 =	2 · 2	3 · 3	5	

- In the column headed by 3, the products are equal. Circle just one product.
- Columns 5 and 7 have a blank, so 5 and 7 are not common factors of 126 and 180. Do not circle any number in these columns.

The GCF is the product of the circled numbers.
The GCF = $2 \cdot 3 \cdot 3 = 18$.

EXAMPLE 2

Find the GCF of 90, 168, and 420.

Solution

	2	3	5	7
90 =	(2)	3 · 3	5	
168 =	2 · 2 · 2	(3)		7
420 =	2 · 2	3	5	7

The GCF = $2 \cdot 3 = 6$.

YOU TRY IT 2

Find the GCF of 36, 60, and 72.

Your solution

EXAMPLE 3

Find the GCF of 7, 12, and 20.

Solution

	2	3	5	7
7 =				7
12 =	2 · 2	3		
20 =	2 · 2		5	

Because no numbers are circled, the GCF = 1.

YOU TRY IT 3

Find the GCF of 11, 24, and 30.

Your solution

Solutions on p. S4

2.1 EXERCISES

OBJECTIVE A To find the least common multiple (LCM)

For Exercises 1 to 34, find the LCM.

1. 5, 8 **2.** 3, 6 **3.** 3, 8 **4.** 2, 5 **5.** 5, 6

6. 5, 7 **7.** 4, 6 **8.** 6, 8 **9.** 8, 12 **10.** 12, 16

11. 5, 12 **12.** 3, 16 **13.** 8, 14 **14.** 6, 18 **15.** 3, 9

16. 4, 10 **17.** 8, 32 **18.** 7, 21 **19.** 9, 36 **20.** 14, 42

21. 44, 60 **22.** 120, 160 **23.** 102, 184 **24.** 123, 234 **25.** 4, 8, 12

26. 5, 10, 15 **27.** 3, 5, 10 **28.** 2, 5, 8 **29.** 3, 8, 12 **30.** 5, 12, 18

31. 9, 36, 64 **32.** 18, 54, 63 **33.** 16, 30, 84 **34.** 9, 12, 15

35. True or false? If two numbers have no common factors, then the LCM of the two numbers is their product.

36. True or false? If one number is a multiple of a second number, then the LCM of the two numbers is the second number.

OBJECTIVE B To find the greatest common factor (GCF)

For Exercises 37 to 70, find the GCF.

37. 3, 5 **38.** 5, 7 **39.** 6, 9 **40.** 18, 24 **41.** 15, 25

42. 14, 49 **43.** 25, 100 **44.** 16, 80 **45.** 32, 51 **46.** 21, 44

47. 12, 80

48. 8, 36

49. 16, 140

50. 12, 76

51. 24, 30

52. 48, 144

53. 44, 96

54. 18, 32

55. 3, 5, 11

56. 6, 8, 10

57. 7, 14, 49

58. 6, 15, 36

59. 10, 15, 20

60. 12, 18, 20

61. 24, 40, 72

62. 3, 17, 51

63. 17, 31, 81

64. 14, 42, 84

65. 25, 125, 625

66. 12, 68, 92

67. 28, 35, 70

68. 1, 49, 153

69. 32, 56, 72

70. 24, 36, 48

71. True or false? If two numbers have a GCF of 1, then the LCM of the two numbers is their product.

72. True or false? If the LCM of two numbers is one of the two numbers, then the GCF of the numbers is the other of the two numbers.

Applying the Concepts

73. **Work Schedules** Joe Salvo, a lifeguard, works 3 days and then has a day off. Joe's friend works 5 days and then has a day off. How many days after Joe and his friend have a day off together will they have another day off together?

© Johnny Buzzerio/Corbis

74. Find the LCM of each of the following pairs of numbers: 2 and 3, 5 and 7, and 11 and 19. Can you draw a conclusion about the LCM of two prime numbers? Suggest a way of finding the LCM of three distinct prime numbers.

75. Find the GCF of each of the following pairs of numbers: 3 and 5, 7 and 11, and 29 and 43. Can you draw a conclusion about the GCF of two prime numbers? What is the GCF of three distinct prime numbers?

76. Using the pattern for the first two triangles at the right, determine the center number of the last triangle.

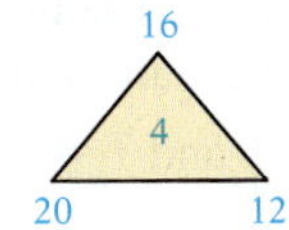

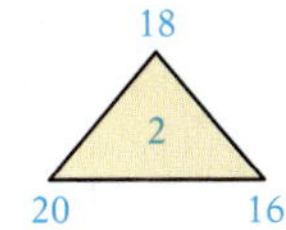

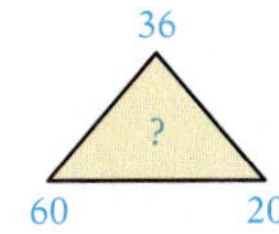

SECTION 2.2 Introduction to Fractions

OBJECTIVE A To write proper fractions, improper fractions, and mixed numbers

A recipe calls for $\frac{1}{2}$ cup of butter; a carpenter uses a $\frac{3}{8}$-inch screw; and a tailor might need $\frac{3}{4}$ yd of fabric. The numbers $\frac{1}{2}$, $\frac{3}{8}$, and $\frac{3}{4}$ are fractions.

A **fraction** can represent the number of equal parts of a whole. The circle at the right is divided into 8 equal parts. 3 of the 8 parts are shaded. The shaded portion of the circle is represented by the fraction $\frac{3}{8}$.

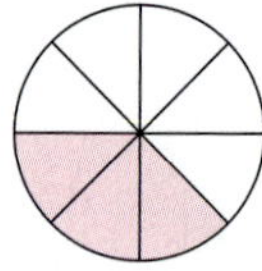

Point of Interest

The fraction bar was first used in 1050 by al-Hassar. It is also called a vinculum.

Each part of a fraction has a name.

Fraction bar ⟶ $\frac{3}{8}$ ← **Numerator** / ← **Denominator**

In a **proper fraction,** the numerator is smaller than the denominator. A proper fraction is less than 1.

$\frac{1}{2}$ $\frac{3}{8}$ $\frac{3}{4}$

Proper fractions

In an **improper fraction,** the numerator is greater than or equal to the denominator. An improper fraction is a number greater than or equal to 1.

$\frac{7}{3}$ $\frac{4}{4}$

Improper fractions

The shaded portion of the circles at the right is represented by the improper fraction $\frac{7}{3}$.

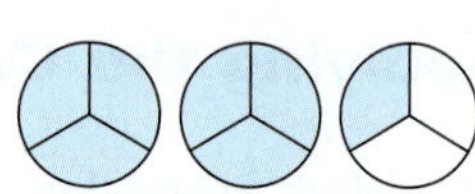

The shaded portion of the square at the right is represented by the improper fraction $\frac{4}{4}$.

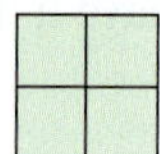

A fraction bar can be read "divided by." Therefore, the fraction $\frac{4}{4}$ can be read "$4 \div 4$."

Because a number divided by itself is equal to 1, $4 \div 4 = 1$ and $\frac{4}{4} = 1$.

The shaded portion of the square above can be represented as $\frac{4}{4}$ or 1.

Since the fraction bar can be read as "divided by" and any number divided by 1 is the number, any whole number can be represented as an improper fraction. For example, $5 = \frac{5}{1}$ and $7 = \frac{7}{1}$.

Because zero divided by any number other than zero is zero, **the numerator of a fraction can be zero.**

For example, $\frac{0}{6} = 0$ because $0 \div 6 = 0$.

Recall that division by zero is not defined. Therefore, **the denominator of a fraction cannot be zero.**

For example, $\frac{9}{0}$ is not defined because $\frac{9}{0} = 9 \div 0$, and division by zero is not defined.

A **mixed number** is a number greater than 1 with a whole number part and a fractional part.

The shaded portion of the circles at the right is represented by the mixed number $2\frac{1}{2}$.

Note from the diagram at the right that the improper fraction $\frac{5}{2}$ is equal to the mixed number $2\frac{1}{2}$.

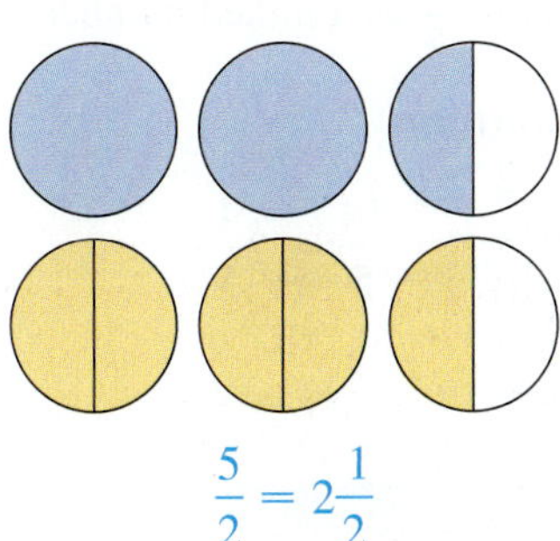

$$\frac{5}{2} = 2\frac{1}{2}$$

An improper fraction can be written as a mixed number.

To write $\frac{5}{2}$ as a mixed number, read the fraction bar as "divided by."

$\frac{5}{2}$ means $5 \div 2$.

Divide the numerator by the denominator.	To write the fractional part of the mixed number, write the remainder over the divisor.	Write the answer.
$\begin{array}{r} 2 \\ 2\overline{)5} \\ -4 \\ \hline 1 \end{array}$	$\begin{array}{r} 2\frac{1}{2} \\ 2\overline{)5} \\ -4 \\ \hline 1 \end{array}$	$\frac{5}{2} = 2\frac{1}{2}$

To write a mixed number as an improper fraction, multiply the denominator of the fractional part of the mixed number by the whole number part. The sum of this product and the numerator of the fractional part is the numerator of the improper fraction. The denominator remains the same.

HOW TO • 1 Write $4\frac{5}{6}$ as an improper fraction.

$$4\frac{5}{6} = \frac{(6 \cdot 4) + 5}{6} = \frac{24 + 5}{6} = \frac{29}{6}$$

EXAMPLE • 1

Express the shaded portion of the circles as an improper fraction and as a mixed number.

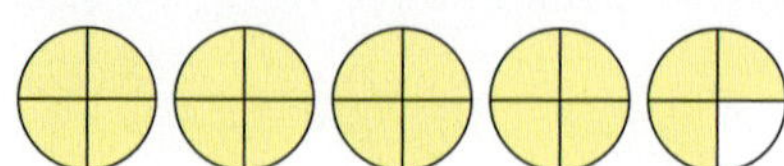

Solution

$\frac{19}{4}; 4\frac{3}{4}$

YOU TRY IT • 1

Express the shaded portion of the circles as an improper fraction and as a mixed number.

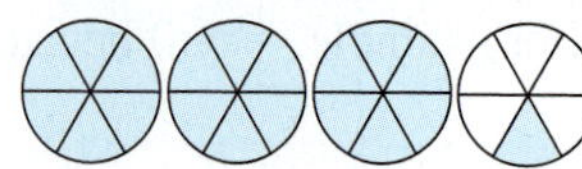

Your solution

EXAMPLE • 2

Write $\frac{14}{5}$ as a mixed number.

Solution

$$\begin{array}{r} 2 \\ 5\overline{)14} \\ -10 \\ \hline 4 \end{array} \qquad \frac{14}{5} = 2\frac{4}{5}$$

YOU TRY IT • 2

Write $\frac{26}{3}$ as a mixed number.

Your solution

EXAMPLE • 3

Write $\frac{35}{7}$ as a whole number.

Solution

$$\begin{array}{r} 5 \\ 7\overline{)35} \\ -35 \\ \hline 0 \end{array} \qquad \frac{35}{7} = 5$$

• Note: The remainder is zero.

YOU TRY IT • 3

Write $\frac{36}{4}$ as a whole number.

Your solution

EXAMPLE • 4

Write $12\frac{5}{8}$ as an improper fraction.

Solution

$$12\frac{5}{8} = \frac{(8 \cdot 12) + 5}{8} = \frac{96 + 5}{8}$$
$$= \frac{101}{8}$$

YOU TRY IT • 4

Write $9\frac{4}{7}$ as an improper fraction.

Your solution

EXAMPLE • 5

Write 9 as an improper fraction.

Solution

$9 = \frac{9}{1}$

YOU TRY IT • 5

Write 3 as an improper fraction.

Your solution

Solutions on p. S4

OBJECTIVE B To write equivalent fractions

Point of Interest

Leonardo of Pisa, who was also called Fibonacci (c. 1175–1250), is credited with bringing the Hindu–Arabic number system to the Western world and promoting its use instead of the cumbersome Roman numeral system. He was also influential in promoting the idea of the fraction bar. His notation, however, was very different from what we use today. For instance, he wrote $\frac{3\ \ 5}{4\ \ 7}$ to mean $\frac{5}{7} + \frac{3}{7 \cdot 4}$.

Fractions can be graphed as points on a number line. The number lines at the right show thirds, sixths, and ninths graphed from 0 to 1.

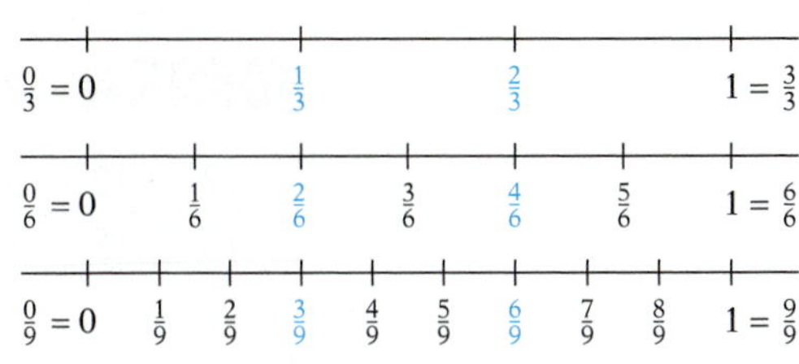

A particular point on the number line may be represented by different fractions, all of which are equal.

For example, $\frac{0}{3} = \frac{0}{6} = \frac{0}{9}$, $\frac{1}{3} = \frac{2}{6} = \frac{3}{9}$, $\frac{2}{3} = \frac{4}{6} = \frac{6}{9}$, and $\frac{3}{3} = \frac{6}{6} = \frac{9}{9}$.

Equal fractions with different denominators are called **equivalent fractions.** $\frac{1}{3}$, $\frac{2}{6}$, and $\frac{3}{9}$ are equivalent fractions. $\frac{2}{3}$, $\frac{4}{6}$, and $\frac{6}{9}$ are equivalent fractions.

Note that we can rewrite $\frac{2}{3}$ as $\frac{4}{6}$ by multiplying both the numerator and denominator of $\frac{2}{3}$ by 2.

$$\frac{2}{3} = \frac{2 \cdot 2}{3 \cdot 2} = \frac{4}{6}$$

Also, we can rewrite $\frac{4}{6}$ as $\frac{2}{3}$ by dividing both the numerator and denominator of $\frac{4}{6}$ by 2.

$$\frac{4}{6} = \frac{4 \div 2}{6 \div 2} = \frac{2}{3}$$

This suggests the following property of fractions.

Equivalent Fractions

The numerator and denominator of a fraction can be multiplied or divided by the same nonzero number. The resulting fraction is equivalent to the original fraction.

$$\frac{a}{b} = \frac{a \cdot c}{b \cdot c}, \quad \frac{a}{b} = \frac{a \div c}{b \div c}, \qquad \text{where } b \neq 0 \text{ and } c \neq 0$$

HOW TO • 2 Write an equivalent fraction with the given denominator.

$$\frac{3}{8} = \frac{}{40}$$

$40 \div 8 = 5$ • Divide the larger denominator by the smaller one.

$\frac{3}{8} = \frac{3 \cdot 5}{8 \cdot 5} = \frac{15}{40}$ • Multiply the numerator and denominator of the given fraction by the quotient (5).

A fraction is in **simplest form** when the numerator and denominator have no common factors other than 1. The fraction $\frac{3}{8}$ is in simplest form because 3 and 8 have no common factors other than 1. The fraction $\frac{15}{50}$ is not in simplest form because the numerator and denominator have a common factor of 5.

To write a fraction in simplest form, divide the numerator and denominator of the fraction by their common factors.

HOW TO • 3 Write $\frac{12}{15}$ in simplest form.

$$\frac{12}{15} = \frac{12 \div 3}{15 \div 3} = \frac{4}{5}$$

- **12 and 15 have a common factor of 3. Divide the numerator and denominator by 3.**

Simplifying a fraction requires that you recognize the common factors of the numerator and denominator. One way to do this is to write the prime factorizations of the numerator and denominator and then divide by the common prime factors.

HOW TO • 4 Write $\frac{30}{42}$ in simplest form.

$$\frac{30}{42} = \frac{\overset{1}{\cancel{2}} \cdot \overset{1}{\cancel{3}} \cdot 5}{\underset{1}{\cancel{2}} \cdot \underset{1}{\cancel{3}} \cdot 7} = \frac{5}{7}$$

- **Write the prime factorization of the numerator and denominator. Divide by the common factors.**

HOW TO • 5 Write $\frac{2x}{6}$ in simplest form.

$$\frac{2x}{6} = \frac{\overset{1}{\cancel{2}} \cdot x}{\underset{1}{\cancel{2}} \cdot 3} = \frac{x}{3}$$

- **Factor the numerator and denominator. Then divide by the common factors.**

EXAMPLE • 6

Write an equivalent fraction with the given denominator.

$$\frac{2}{5} = \frac{}{30}.$$

Solution

$30 \div 5 = 6$

$$\frac{2}{5} = \frac{2 \cdot 6}{5 \cdot 6} = \frac{12}{30}$$

$\frac{12}{30}$ is equivalent to $\frac{2}{5}$.

YOU TRY IT • 6

Write an equivalent fraction with the given denominator.

$$\frac{5}{8} = \frac{}{48}.$$

Your solution

EXAMPLE • 7

Write an equivalent fraction with the given denominator.

$$3 = \frac{}{15}.$$

Solution

$3 = \frac{3}{1}$ $\quad 15 \div 1 = 15$

$$3 = \frac{3}{1} = \frac{3 \cdot 15}{1 \cdot 15} = \frac{45}{15}$$

$\frac{45}{15}$ is equivalent to 3.

YOU TRY IT • 7

Write an equivalent fraction with the given denominator.

$$8 = \frac{}{12}.$$

Your solution

Solutions on p. S4

EXAMPLE 8

Write $\frac{18}{54}$ in simplest form.

Solution

$$\frac{18}{54} = \frac{\overset{1}{\cancel{2}} \cdot \overset{1}{\cancel{3}} \cdot \overset{1}{\cancel{3}}}{\underset{1}{\cancel{2}} \cdot \underset{1}{\cancel{3}} \cdot \underset{1}{\cancel{3}} \cdot 3} = \frac{1}{3}$$

YOU TRY IT 8

Write $\frac{21}{84}$ in simplest form.

Your solution

EXAMPLE 9

Write $\frac{36}{20}$ in simplest form.

Solution

$$\frac{36}{20} = \frac{\overset{1}{\cancel{2}} \cdot \overset{1}{\cancel{2}} \cdot 3 \cdot 3}{\underset{1}{\cancel{2}} \cdot \underset{1}{\cancel{2}} \cdot 5} = \frac{9}{5}$$

YOU TRY IT 9

Write $\frac{32}{12}$ in simplest form.

Your solution

EXAMPLE 10

Write $\frac{10m}{12}$ in simplest form.

Solution

$$\frac{10m}{12} = \frac{\overset{1}{\cancel{2}} \cdot 5 \cdot m}{\underset{1}{\cancel{2}} \cdot 2 \cdot 3} = \frac{5m}{6}$$

YOU TRY IT 10

Write $\frac{11t}{11}$ in simplest form.

Your solution

Solutions on p. S4

OBJECTIVE C To identify the order relation between two fractions

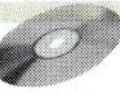

The number line can be used to determine the order relation between two fractions.

A fraction that appears to the left of a given fraction on the number line is less than the given fraction.

$\frac{3}{8}$ is to the left of $\frac{5}{8}$.

$$\frac{3}{8} < \frac{5}{8}$$

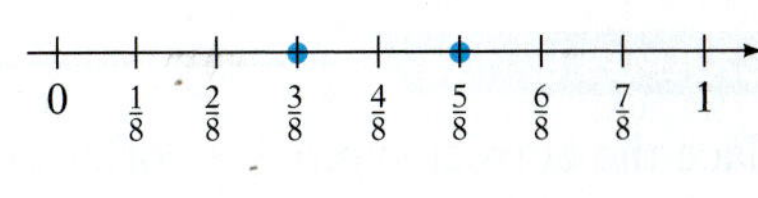

A fraction that appears to the right of a given fraction on the number line is greater than the given fraction.

$\frac{7}{8}$ is to the right of $\frac{3}{8}$.

$$\frac{7}{8} > \frac{3}{8}$$

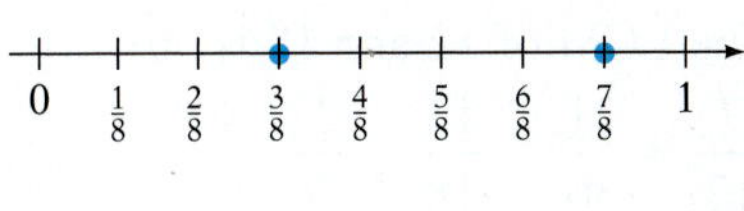

To find the order relation between two fractions with the *same* denominator, compare the numerators. The fraction with the smaller numerator is the smaller fraction. The larger fraction is the fraction with the larger numerator.

$\frac{3}{8}$ and $\frac{5}{8}$ have the same denominator. $\frac{3}{8} < \frac{5}{8}$ because $3 < 5$.

$\frac{7}{8}$ and $\frac{3}{8}$ have the same denominator. $\frac{7}{8} > \frac{3}{8}$ because $7 > 3$.

To compare two fractions with *different* denominators, rewrite the fractions with a common denominator. The common denominator is the least common multiple (LCM) of the denominators of the fractions. The LCM of the denominators is sometimes called the **least common denominator** or **LCD.**

HOW TO • 6 Find the order relation between $\frac{5}{12}$ and $\frac{7}{18}$.

The LCM of 12 and 18 is 36. • **Find the LCM of the denominators.**

$\frac{5}{12} = \frac{5 \cdot 3}{12 \cdot 3} = \frac{15}{36}$ ⟵ Larger numerator • **Write each fraction as an equivalent fraction with the LCM as the denominator.**

$\frac{7}{18} = \frac{7 \cdot 2}{18 \cdot 2} = \frac{14}{36}$ ⟵ Smaller numerator

$\frac{15}{36} > \frac{14}{36}$ • **Compare the fractions.**

$\frac{5}{12} > \frac{7}{18}$

EXAMPLE • 11

Place the correct symbol, < or >, between the two numbers.

$\frac{2}{3} \quad \frac{4}{7}$

Solution

The LCM of 3 and 7 is 21.

$\frac{2}{3} = \frac{14}{21} \quad \frac{4}{7} = \frac{12}{21}$

$\frac{14}{21} > \frac{12}{21}$

$\frac{2}{3} > \frac{4}{7}$

YOU TRY IT • 11

Place the correct symbol, < or >, between the two numbers.

$\frac{4}{9} \quad \frac{8}{21}$

Your solution

EXAMPLE • 12

Place the correct symbol, < or >, between the two numbers.

$\frac{7}{12} \quad \frac{11}{18}$

Solution

The LCM of 12 and 18 is 36.

$\frac{7}{12} = \frac{21}{36} \quad \frac{11}{18} = \frac{22}{36}$

$\frac{21}{36} < \frac{22}{36}$

$\frac{7}{12} < \frac{11}{18}$

YOU TRY IT • 12

Place the correct symbol, < or >, between the two numbers.

$\frac{17}{24} \quad \frac{7}{9}$

Your solution

Solutions on p. S4

2.2 EXERCISES

OBJECTIVE A **To write proper fractions, improper fractions, and mixed numbers**

For Exercises 1 to 4, express the shaded portion of the circle as a fraction.

1. 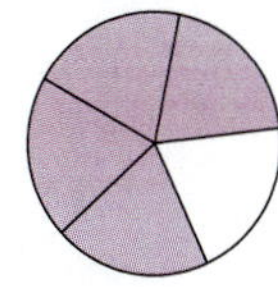

2. 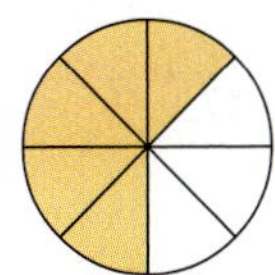

3. 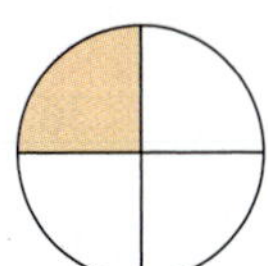

4. 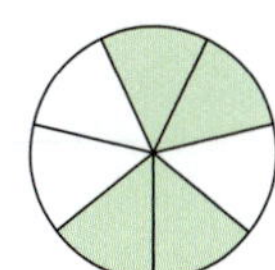

For Exercises 5 to 8, express the shaded portion of the circles as an improper fraction and as a mixed number.

5. 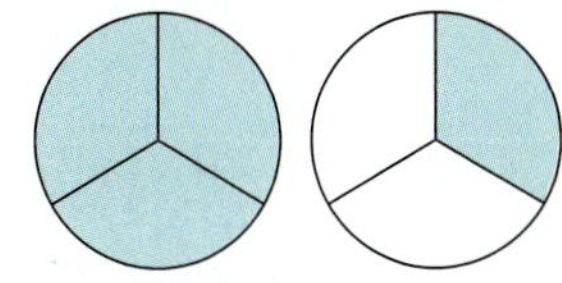

6. 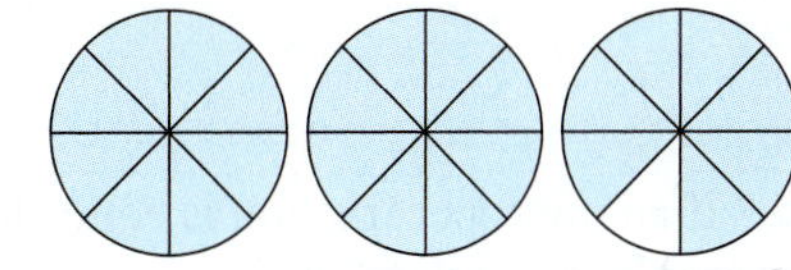

7. 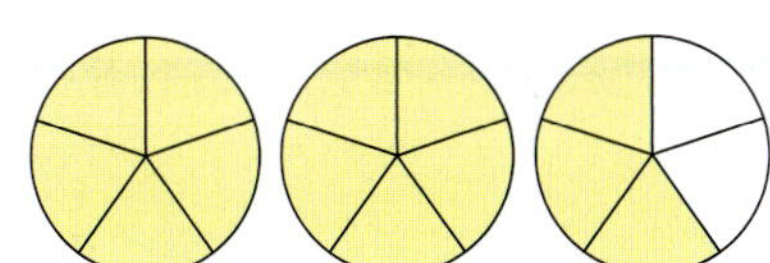

8. 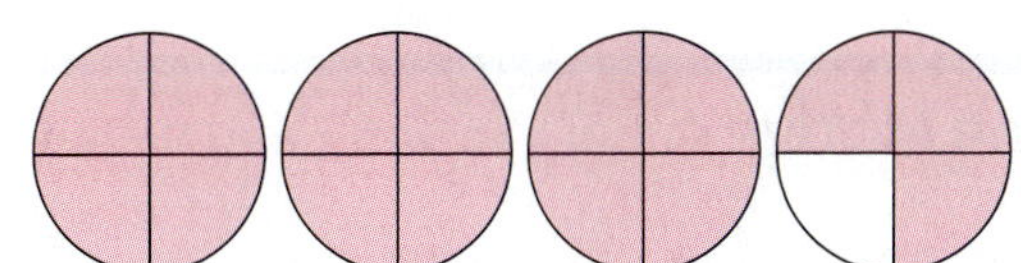

For Exercises 9 to 28, write the improper fraction as a mixed number or a whole number.

9. $\frac{13}{4}$ **10.** $\frac{14}{3}$ **11.** $\frac{20}{5}$ **12.** $\frac{18}{6}$ **13.** $\frac{27}{10}$

14. $\frac{31}{3}$ **15.** $\frac{56}{8}$ **16.** $\frac{27}{9}$ **17.** $\frac{17}{9}$ **18.** $\frac{8}{3}$

19. $\frac{12}{5}$ **20.** $\frac{19}{8}$ **21.** $\frac{18}{1}$ **22.** $\frac{21}{1}$ **23.** $\frac{32}{15}$

24. $\frac{39}{14}$ **25.** $\frac{8}{8}$ **26.** $\frac{12}{12}$ **27.** $\frac{28}{3}$ **28.** $\frac{43}{5}$

For Exercises 29 to 48, write the mixed number or whole number as an improper fraction.

29. $2\frac{1}{4}$ **30.** $4\frac{2}{5}$ **31.** $5\frac{1}{2}$ **32.** $3\frac{2}{3}$ **33.** $2\frac{4}{5}$

34. $6\frac{3}{8}$ **35.** $7\frac{5}{6}$ **36.** $9\frac{1}{5}$ **37.** 7 **38.** 4

39. $8\frac{1}{4}$ **40.** $1\frac{7}{9}$ **41.** $10\frac{1}{3}$ **42.** $6\frac{3}{7}$ **43.** $4\frac{7}{12}$

44. $5\frac{4}{9}$ **45.** 8 **46.** 6 **47.** $12\frac{4}{5}$ **48.** $11\frac{5}{8}$

49. True or false? If an improper fraction is equivalent to 1, then the numerator and the denominator are the same number.

OBJECTIVE B To write equivalent fractions

50. When you multiply the numerator and denominator of a fraction by the same number, you are actually multiplying the fraction by the number _____.

For Exercises 51 to 70, write an equivalent fraction with the given denominator.

51. $\frac{1}{2} = \frac{}{12}$ **52.** $\frac{1}{4} = \frac{}{20}$ **53.** $\frac{3}{8} = \frac{}{24}$ **54.** $\frac{9}{11} = \frac{}{44}$ **55.** $\frac{2}{17} = \frac{}{51}$

56. $\frac{9}{10} = \frac{}{80}$ **57.** $\frac{3}{4} = \frac{}{32}$ **58.** $\frac{5}{8} = \frac{}{32}$ **59.** $6 = \frac{}{18}$ **60.** $5 = \frac{}{35}$

61. $\frac{1}{3} = \frac{}{90}$ **62.** $\frac{3}{16} = \frac{}{48}$ **63.** $\frac{2}{3} = \frac{}{21}$ **64.** $\frac{4}{9} = \frac{}{36}$ **65.** $\frac{6}{7} = \frac{}{49}$

66. $\frac{7}{8} = \frac{}{40}$ **67.** $\frac{4}{9} = \frac{}{18}$ **68.** $\frac{11}{12} = \frac{}{48}$ **69.** $7 = \frac{}{4}$ **70.** $9 = \frac{}{6}$

For Exercises 71 to 100, write the fraction in simplest form.

71. $\frac{3}{12}$ **72.** $\frac{10}{22}$ **73.** $\frac{33}{44}$ **74.** $\frac{6}{14}$ **75.** $\frac{4}{24}$

76. $\frac{25}{75}$ **77.** $\frac{8}{33}$ **78.** $\frac{9}{25}$ **79.** $\frac{0}{8}$ **80.** $\frac{0}{11}$

81. $\frac{42}{36}$ **82.** $\frac{30}{18}$ **83.** $\frac{16}{16}$ **84.** $\frac{24}{24}$ **85.** $\frac{21}{35}$

86. $\frac{11}{55}$ **87.** $\frac{16}{60}$ **88.** $\frac{8}{84}$ **89.** $\frac{12}{20}$ **90.** $\frac{24}{36}$

91. $\frac{12m}{18}$ **92.** $\frac{20x}{25}$ **93.** $\frac{4y}{8}$ **94.** $\frac{14z}{28}$ **95.** $\frac{24a}{36}$

96. $\frac{28z}{21}$ **97.** $\frac{8c}{8}$ **98.** $\frac{9w}{9}$ **99.** $\frac{18k}{3}$ **100.** $\frac{24t}{4}$

OBJECTIVE C To identify the order relation between two fractions

For Exercises 101 to 120, place the correct symbol, $<$ or $>$, between the two numbers.

101. $\frac{3}{8}\quad\frac{2}{5}$ **102.** $\frac{5}{7}\quad\frac{2}{3}$ **103.** $\frac{3}{4}\quad\frac{7}{9}$ **104.** $\frac{7}{12}\quad\frac{5}{8}$

105. $\frac{2}{3}\quad\frac{7}{11}$ **106.** $\frac{11}{14}\quad\frac{3}{4}$ **107.** $\frac{17}{24}\quad\frac{11}{16}$ **108.** $\frac{11}{12}\quad\frac{7}{9}$

109. $\frac{7}{15}\quad\frac{5}{12}$ **110.** $\frac{5}{8}\quad\frac{4}{7}$ **111.** $\frac{5}{9}\quad\frac{11}{21}$ **112.** $\frac{11}{30}\quad\frac{7}{24}$

113. $\frac{7}{12}$ $\frac{13}{18}$

114. $\frac{9}{11}$ $\frac{7}{8}$

115. $\frac{4}{5}$ $\frac{7}{9}$

116. $\frac{3}{4}$ $\frac{11}{13}$

117. $\frac{9}{16}$ $\frac{5}{9}$

118. $\frac{2}{3}$ $\frac{7}{10}$

119. $\frac{5}{8}$ $\frac{13}{20}$

120. $\frac{3}{10}$ $\frac{7}{25}$

121. Without writing the fractions $\frac{4}{5}$ and $\frac{1}{7}$ with a common denominator, decide which fraction is larger.

Applying the Concepts

122. **Weight** A ton is equal to 2000 lb. What fractional part of a ton is 250 lb?

123. **Time** If a history class lasts 50 min, what fractional part of an hour is the history class?

124. **Jewelry** Gold is designated by karats. Pure gold is 24 karats. What fractional part of an 18-karat gold bracelet is pure gold?

125. **The Food Industry** The table at the right shows the results of a survey that asked fast-food patrons their criteria for choosing where to go for fast food. For example, 3 out of every 25 people surveyed said that the speed of the service was most important.

a. According to the survey, do more people choose a fast-food restaurant on the basis of its location or the quality of the food?

b. Which criterion was cited by the most people?

Fast-Food Patrons' Top Criteria for Fast-Food Restaurants	
Food quality	$\frac{1}{4}$
Location	$\frac{13}{50}$
Menu	$\frac{4}{25}$
Price	$\frac{2}{25}$
Speed	$\frac{3}{25}$
Other	$\frac{13}{100}$

Source: Maritz Marketing Research, Inc.

126. **Card Games** A standard deck of playing cards consists of 52 cards.
a. What fractional part of a standard deck of cards is spades?
b. What fractional part of a standard deck of cards is aces?

127. **Geography** What fraction of the states in the United States begin with the letter A?

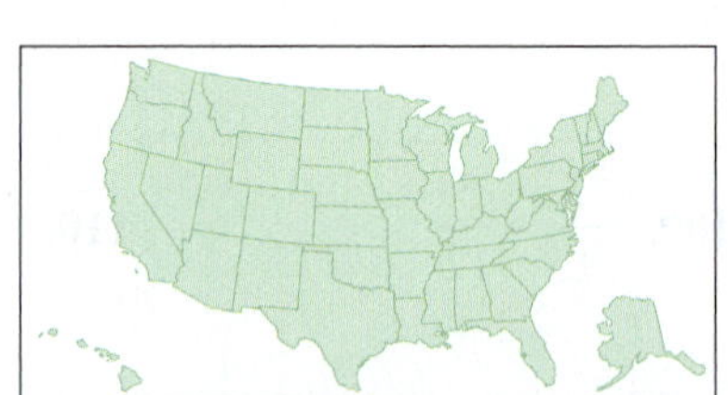

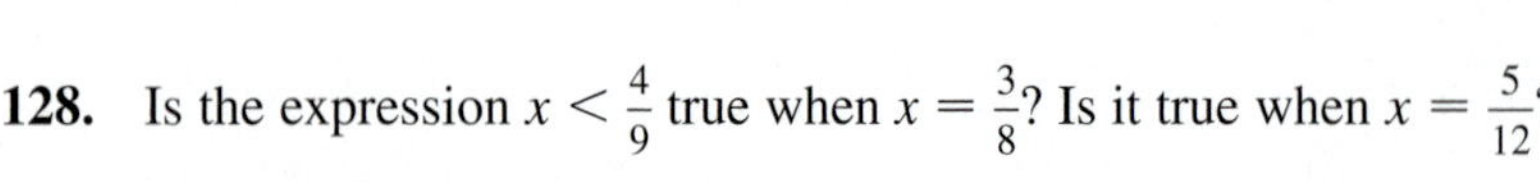

128. Is the expression $x < \frac{4}{9}$ true when $x = \frac{3}{8}$? Is it true when $x = \frac{5}{12}$?

SECTION

2.3 Addition and Subtraction of Fractions

OBJECTIVE A To add fractions

Tips for Success

Before the class meeting in which your professor begins a new section, you should read each objective statement for that section. Next, browse through the objective material. The purpose of browsing through the material is so that your brain will be prepared to accept and organize the new information when it is presented to you. See *AIM for Success* in the front of the book.

Suppose you and a friend order a pizza. The pizza has been cut into 8 equal pieces. If you eat 3 pieces of the pizza and your friend eats 2 pieces, then together you have eaten $\frac{5}{8}$ of the pizza.

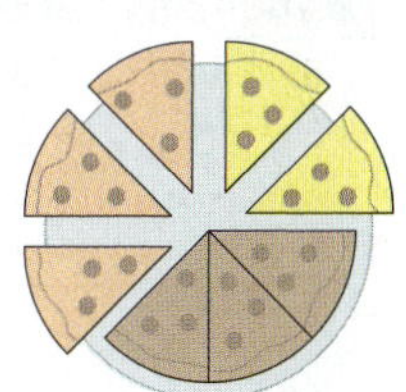

Note that in adding the fractions $\frac{3}{8}$ and $\frac{2}{8}$, the numerators are added and the denominator remains the same.

$$\frac{3}{8} + \frac{2}{8} = \frac{3+2}{8} = \frac{5}{8}$$

Addition of Fractions

To add fractions with the same denominator, add the numerators and place the sum over the common denominator.

$$\frac{a}{b} + \frac{c}{b} = \frac{a+c}{b}, \quad \text{where } b \neq 0$$

HOW TO • 1 Add: $\frac{5}{16} + \frac{7}{16}$

$$\frac{5}{16} + \frac{7}{16} = \frac{5+7}{16}$$

• **The denominators are the same. Add the numerators and place the sum over the common denominator.**

$$= \frac{12}{16} = \frac{3}{4}$$

• **Write the answer in simplest form.**

HOW TO • 2 Add: $\frac{4}{x} + \frac{8}{x}$

$$\frac{4}{x} + \frac{8}{x} = \frac{4+8}{x}$$

• **The denominators are the same. Add the numerators and place the sum over the common denominator.**

$$= \frac{12}{x}$$

Before two fractions can be added, the fractions must have the same denominator. To add fractions with different denominators, first rewrite the fractions as equivalent fractions with a common denominator. The common denominator is the least common multiple (LCM) of the denominators of the fractions. Recall that the LCM of denominators is sometimes called the least common denominator (LCD).

Integrating Technology

Some scientific calculators have a fraction key, **a^b/c**. It is used to perform operations on fractions. To use this key to simplify the expression at the right, enter

5 **a^b/c** 6 **+** 3 **a^b/c** 8 **=**

($\frac{5}{6}$) ($\frac{3}{8}$)

HOW TO • 3 Find the sum of $\frac{5}{6}$ and $\frac{3}{8}$.

The LCM of 6 and 8 is 24.
- The common denominator is the LCM of 6 and 8.

$$\frac{5}{6} + \frac{3}{8} = \frac{20}{24} + \frac{9}{24}$$
- Write the fractions as equivalent fractions with the common denominator.

$$= \frac{20 + 9}{24}$$
- Add the fractions.

$$= \frac{29}{24} = 1\frac{5}{24}$$

HOW TO • 4 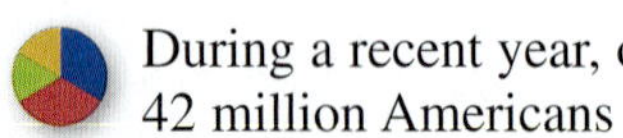 During a recent year, over 42 million Americans changed homes. Figure 2.1 shows what fractions of the people moved within the same county, moved to a different county in the same state, and moved to a different state. What fractional part of those who changed homes moved outside the county they had been living in?

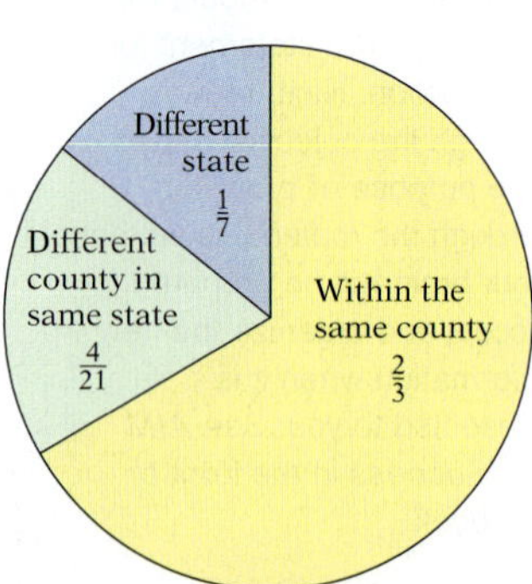

FIGURE 2.1 Where Americans Moved
Source: Census Bureau; *Geographical Mobility*

Add the fraction of the people who moved to a different county in the same state and the fraction who moved to a different state.

$$\frac{4}{21} + \frac{1}{7} = \frac{4}{21} + \frac{3}{21} = \frac{7}{21} = \frac{1}{3}$$

$\frac{1}{3}$ of the Americans who changed homes moved outside of the county they had been living in.

The mixed number $2\frac{1}{2}$ is the sum of 2 and $\frac{1}{2}$.

$$2\frac{1}{2} = 2 + \frac{1}{2}$$

Therefore, the sum of a whole number and a fraction is a mixed number.

$$2 + \frac{1}{2} = 2\frac{1}{2}$$
$$3 + \frac{4}{5} = 3\frac{4}{5}$$
$$8 + \frac{7}{9} = 8\frac{7}{9}$$

The sum of a whole number and a mixed number is a mixed number.

Take Note

$$5 + 4\frac{2}{7} = 5 + \left(4 + \frac{2}{7}\right)$$
$$= (5 + 4) + \frac{2}{7}$$
$$= 9 + \frac{2}{7} = 9\frac{2}{7}$$

HOW TO • 5 Add: $5 + 4\frac{2}{7}$

$$5 + 4\frac{2}{7} = 9\frac{2}{7}$$
- Add the whole numbers (5 and 4). Write the fraction.

Integrating Technology

Use the fraction key on a calculator to enter mixed numbers. For the example at the right, enter

3 **a b/c** 5 **a b/c** 8 **+** → $3\frac{5}{8}$

4 **a b/c** 7 **a b/c** 12 **=** → $4\frac{7}{12}$

To add two mixed numbers, first write the fractional parts as equivalent fractions with a common denominator. Then add the fractional parts and add the whole numbers.

HOW TO • 6 Add: $3\frac{5}{8} + 4\frac{7}{12}$

$3\frac{5}{8} + 4\frac{7}{12} = 3\frac{15}{24} + 4\frac{14}{24}$ • Write the fractions as equivalent fractions with a common denominator. The common denominator is the LCM of 8 and 12 (24).

$= 7\frac{29}{24}$ • Add the fractional parts and add the whole numbers.

$= 7 + \frac{29}{24}$ • Write the sum in simplest form.

$= 7 + 1\frac{5}{24}$

$= 8\frac{5}{24}$

HOW TO • 7 Evaluate $x + y$ when $x = 2\frac{3}{4}$ and $y = 7\frac{5}{6}$.

$x + y$

$2\frac{3}{4} + 7\frac{5}{6}$ • Replace x with $2\frac{3}{4}$ and y with $7\frac{5}{6}$.

$= 2\frac{9}{12} + 7\frac{10}{12}$ • Write the fractions as equivalent fractions with a common denominator.

$= 9\frac{19}{12}$ • Add the fractional parts and add the whole numbers.

$= 10\frac{7}{12}$ • Write the sum in simplest form.

EXAMPLE • 1

Add: $\frac{9}{16} + \frac{5}{12}$

Solution

$\frac{9}{16} + \frac{5}{12} = \frac{27}{48} + \frac{20}{48}$

$= \frac{27 + 20}{48} = \frac{47}{48}$

YOU TRY IT • 1

Add: $\frac{7}{12} + \frac{3}{8}$

Your solution

Solution on p. S4

EXAMPLE 2

Add: $\frac{4}{5} + \frac{3}{4} + \frac{5}{8}$

Solution

$\frac{4}{5} + \frac{3}{4} + \frac{5}{8} = \frac{32}{40} + \frac{30}{40} + \frac{25}{40} = \frac{87}{40} = 2\frac{7}{40}$

YOU TRY IT 2

Add: $\frac{3}{5} + \frac{2}{3} + \frac{5}{6}$

Your solution

EXAMPLE 3

Find the sum of $12\frac{4}{7}$ and 19.

Solution

$12\frac{4}{7} + 19 = 31\frac{4}{7}$

YOU TRY IT 3

What is the sum of 16 and $8\frac{5}{9}$?

Your solution

EXAMPLE 4

Is $\frac{2}{3}$ a solution of $\frac{1}{4} + y = \frac{11}{12}$?

Solution

$$\begin{array}{c|c} \frac{1}{4} + y & \frac{11}{12} \\ \hline \frac{1}{4} + \frac{2}{3} & \frac{11}{12} \\ \frac{3}{12} + \frac{8}{12} & \frac{11}{12} \\ \frac{11}{12} = & \frac{11}{12} \end{array}$$

Yes, $\frac{2}{3}$ is a solution of $\frac{1}{4} + y = \frac{11}{12}$.

YOU TRY IT 4

Is $\frac{3}{8}$ a solution of $\frac{2}{3} + z = \frac{23}{24}$?

Your solution

EXAMPLE 5

Evaluate $x + y + z$ when $x = 2\frac{1}{6}$, $y = 4\frac{3}{8}$, and $z = 7\frac{5}{9}$.

Solution

$x + y + z$

$$2\frac{1}{6} + 4\frac{3}{8} + 7\frac{5}{9} = 2\frac{12}{72} + 4\frac{27}{72} + 7\frac{40}{72}$$

$$= 13\frac{79}{72} = 14\frac{7}{72}$$

YOU TRY IT 5

Evaluate $x + y + z$ when $x = 3\frac{5}{6}$, $y = 2\frac{1}{9}$, and $z = 5\frac{5}{12}$.

Your solution

Solutions on p. S4

OBJECTIVE B To subtract fractions 

Point of Interest

The first woman mathematician for whom documented evidence exists is Hypatia (370–415). She lived in Alexandria, Egypt, and lectured at the Museum, the forerunner of our modern university. She made important contributions in mathematics, astronomy, and philosophy.

In the last objective, it was stated that in order for fractions to be added, the fractions must have the same denominator. The same is true for subtracting fractions: The two fractions must have the same denominator.

Subtraction of Fractions

To subtract fractions with the same denominator, subtract the numerators and place the difference over the common denominator.

$\frac{a}{b} - \frac{c}{b} = \frac{a-c}{b}$, where $b \neq 0$

HOW TO • 8 Subtract: $\frac{5}{8} - \frac{3}{8}$

$\frac{5}{8} - \frac{3}{8} = \frac{5-3}{8}$
- The denominators are the same. Subtract the numerators and place the difference over the common denominator.

$= \frac{2}{8} = \frac{1}{4}$
- Write the answer in simplest form.

To subtract fractions with different denominators, first rewrite the fractions as equivalent fractions with a common denominator. The common denominator is the least common multiple (LCM) of the denominators of the fractions.

HOW TO • 9 Subtract: $\frac{5}{12} - \frac{3}{8}$

The LCM of 12 and 8 is 24.
- The common denominator is the LCM of 12 and 8.

$\frac{5}{12} - \frac{3}{8} = \frac{10}{24} - \frac{9}{24}$
- Write the fractions as equivalent fractions with the common denominator.

$= \frac{10-9}{24} = \frac{1}{24}$
- Subtract the fractions.

To subtract mixed numbers when borrowing is not necessary, subtract the fractional parts and then subtract the whole numbers.

HOW TO • 10 Find the difference between $5\frac{8}{9}$ and $2\frac{5}{6}$.

The LCM of 9 and 6 is 18.

$5\frac{8}{9} - 2\frac{5}{6} = 5\frac{16}{18} - 2\frac{15}{18}$
- Write the fractions as equivalent fractions with the LCM as the common denominator.

$= 3\frac{1}{18}$
- Subtract the fractional parts and subtract the whole numbers.

As in subtraction with whole numbers, subtraction of mixed numbers may involve borrowing.

HOW TO • 11 Subtract: $7 - 4\frac{2}{3}$

$7 - 4\frac{2}{3} = 6\frac{3}{3} - 4\frac{2}{3}$

- Borrow 1 from 7. Write the 1 as a fraction with the same denominator as the fractional part of the mixed number (3).

Note: $7 = 6 + 1 = 6 + \frac{3}{3} = 6\frac{3}{3}$

$= 2\frac{1}{3}$

- Subtract the fractional parts and subtract the whole numbers.

HOW TO • 12 Subtract: $9\frac{1}{8} - 2\frac{5}{6}$

$9\frac{1}{8} - 2\frac{5}{6} = 9\frac{3}{24} - 2\frac{20}{24}$

- Write the fractions as equivalent fractions with a common denominator.

$= 8\frac{27}{24} - 2\frac{20}{24}$

- $3 < 20$. Borrow 1 from 9. Add the 1 to $\frac{3}{24}$.

Note: $9\frac{3}{24} = 9 + \frac{3}{24} = 8 + 1 + \frac{3}{24}$

$= 8 + \frac{24}{24} + \frac{3}{24} = 8 + \frac{27}{24} = 8\frac{27}{24}$

$= 6\frac{7}{24}$

- Subtract.

HOW TO • 13 Evaluate $x - y$ when $x = 7\frac{2}{9}$ and $y = 3\frac{5}{12}$.

$x - y$

$7\frac{2}{9} - 3\frac{5}{12}$

- Replace x with $7\frac{2}{9}$ and y with $3\frac{5}{12}$.

$= 7\frac{8}{36} - 3\frac{15}{36}$

- Write the fractions as equivalent fractions with a common denominator.

$= 6\frac{44}{36} - 3\frac{15}{36}$

- $8 < 15$. Borrow 1 from 7. Add the 1 to $\frac{8}{36}$.

Note: $7\frac{8}{36} = 6 + \frac{36}{36} + \frac{8}{36} = 6\frac{44}{36}$

$= 3\frac{29}{36}$

- Subtract.

EXAMPLE • 6

Subtract: $\frac{5}{6} - \frac{3}{8}$

Solution

$\frac{5}{6} - \frac{3}{8} = \frac{20}{24} - \frac{9}{24} = \frac{11}{24}$

YOU TRY IT • 6

Subtract: $\frac{5}{6} - \frac{7}{9}$

Your solution

Solution on p. S5

EXAMPLE • 7

Find the difference between $8\frac{5}{6}$ and $2\frac{3}{4}$.

Solution

$$8\frac{5}{6} - 2\frac{3}{4} = 8\frac{10}{12} - 2\frac{9}{12} = 6\frac{1}{12}$$

YOU TRY IT • 7

Find the difference between $9\frac{7}{8}$ and $5\frac{2}{3}$.

Your solution

EXAMPLE • 8

Subtract: $7 - 3\frac{5}{13}$

Solution

$$7 - 3\frac{5}{13} = 6\frac{13}{13} - 3\frac{5}{13} = 3\frac{8}{13}$$

YOU TRY IT • 8

Subtract: $6 - 4\frac{2}{11}$

Your solution

Solutions on p. S5

OBJECTIVE C **To solve application problems and use formulas** 

EXAMPLE • 9

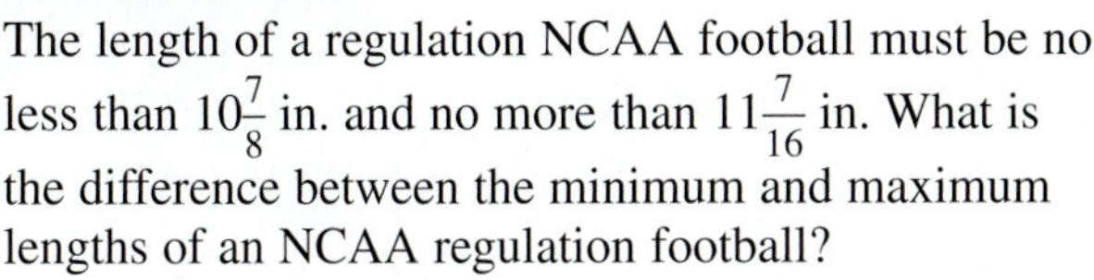

The length of a regulation NCAA football must be no less than $10\frac{7}{8}$ in. and no more than $11\frac{7}{16}$ in. What is the difference between the minimum and maximum lengths of an NCAA regulation football?

Strategy

To find the difference, subtract the minimum length $\left(10\frac{7}{8}\right)$ from the maximum length $\left(11\frac{7}{16}\right)$.

Solution

$$11\frac{7}{16} - 10\frac{7}{8} = 11\frac{7}{16} - 10\frac{14}{16}$$
$$= 10\frac{23}{16} - 10\frac{14}{16} = \frac{9}{16}$$

The difference is $\frac{9}{16}$ in.

YOU TRY IT • 9

The Heller Research Group conducted a survey to determine favorite doughnut flavors. $\frac{2}{5}$ of the respondents named glazed doughnuts, $\frac{8}{25}$ named filled doughnuts, and $\frac{3}{20}$ named frosted doughnuts. What fraction of the respondents did not name glazed, filled, or frosted as their favorite type of doughnut?

Your strategy

Your solution

Solution on p. S5

2.3 EXERCISES

OBJECTIVE A To add fractions

For Exercises 1 to 32, add.

1. $\frac{4}{11} + \frac{5}{11}$

2. $\frac{3}{7} + \frac{2}{7}$

3. $\frac{2}{3} + \frac{1}{3}$

4. $\frac{1}{2} + \frac{1}{2}$

5. $\frac{5}{6} + \frac{5}{6}$

6. $\frac{3}{8} + \frac{7}{8}$

7. $\frac{7}{18} + \frac{13}{18} + \frac{1}{18}$

8. $\frac{8}{15} + \frac{2}{15} + \frac{11}{15}$

9. $\frac{7}{b} + \frac{9}{b}$

10. $\frac{3}{y} + \frac{6}{y}$

11. $\frac{5}{c} + \frac{4}{c}$

12. $\frac{2}{a} + \frac{8}{a}$

13. $\frac{1}{x} + \frac{4}{x} + \frac{6}{x}$

14. $\frac{8}{n} + \frac{5}{n} + \frac{3}{n}$

15. $\frac{1}{4} + \frac{2}{3}$

16. $\frac{2}{3} + \frac{1}{2}$

17. $\frac{7}{15} + \frac{9}{20}$

18. $\frac{4}{9} + \frac{1}{6}$

19. $\frac{2}{3} + \frac{1}{12} + \frac{5}{6}$

20. $\frac{3}{8} + \frac{1}{2} + \frac{5}{12}$

21. $\frac{7}{12} + \frac{3}{4} + \frac{4}{5}$

22. $\frac{7}{11} + \frac{1}{2} + \frac{5}{6}$

23. $8 + 7\frac{2}{3}$

24. $6 + 9\frac{3}{5}$

25. $2\frac{1}{6} + 3\frac{1}{2}$

26. $1\frac{3}{10} + 4\frac{3}{5}$

27. $8\frac{3}{5} + 6\frac{9}{20}$

28. $7\frac{5}{12} + 3\frac{7}{9}$

29. $5\frac{5}{12} + 4\frac{7}{9}$

30. $2\frac{11}{12} + 3\frac{7}{15}$

31. $2\frac{1}{4} + 3\frac{1}{2} + 1\frac{2}{3}$

32. $1\frac{2}{3} + 2\frac{5}{6} + 4\frac{7}{9}$

For Exercises 33 to 36, each statement concerns a pair of fractions that have the same denominator. State whether the sum of the fractions is a proper fraction, the number 1, a mixed number, or a whole number other than 1.

33. The sum of the numerators is a multiple of the denominator.

34. The sum of the numerators is one more than the denominator.

35. The sum of the numerators is the denominator.

36. The sum of the numerators is smaller than the denominator.

37. Find the total of $\frac{2}{7}$, $\frac{3}{14}$, and $\frac{1}{4}$.

38. Find the total of $\frac{1}{3}$, $\frac{5}{18}$, and $\frac{2}{9}$.

39. Find $3\frac{7}{12}$ plus $2\frac{5}{8}$.

40. Find $5\frac{4}{9}$ plus $6\frac{5}{6}$.

41. Find $\frac{7}{8}$ increased by $1\frac{1}{3}$.

42. Find the sum of $7\frac{11}{15}$, $2\frac{7}{10}$, and $5\frac{2}{5}$.

For Exercises 43 to 46, evaluate the variable expression $x + y$ for the given values of x and y.

43. $x = \frac{3}{5}, y = \frac{4}{5}$

44. $x = \frac{5}{8}, y = \frac{3}{8}$

45. $x = \frac{5}{6}, y = \frac{8}{9}$

46. $x = \frac{5}{8}, y = \frac{1}{6}$

For Exercises 47 to 52, evaluate the variable expression $x + y + z$ for the given values of x, y, and z.

47. $x = \frac{3}{8}, y = \frac{1}{4}, z = \frac{7}{12}$

48. $x = \frac{5}{6}, y = \frac{2}{3}, z = \frac{7}{24}$

49. $x = 1\frac{1}{2}, y = 3\frac{3}{4}, z = 6\frac{5}{12}$

50. $x = 7\frac{2}{3}, y = 2\frac{5}{6}, z = 5\frac{4}{9}$

51. $x = 4\frac{3}{5}, y = 8\frac{7}{10}, z = 1\frac{9}{20}$

52. $x = 2\frac{3}{14}, y = 5\frac{5}{7}, z = 3\frac{1}{2}$

53. Is $\frac{3}{5}$ a solution of the equation $z + \frac{1}{4} = \frac{17}{20}$?

54. Is $\frac{3}{8}$ a solution of the equation $\frac{3}{4} = t + \frac{3}{8}$?

OBJECTIVE B To subtract fractions

For Exercises 55 to 86, subtract.

55. $\frac{7}{12} - \frac{5}{12}$

56. $\frac{17}{20} - \frac{9}{20}$

57. $\frac{11}{24} - \frac{7}{24}$

58. $\frac{39}{48} - \frac{23}{48}$

59. $\frac{8}{d} - \frac{3}{d}$

60. $\frac{12}{y} - \frac{7}{y}$

61. $\frac{10}{n} - \frac{5}{n}$

62. $\frac{13}{c} - \frac{6}{c}$

63. $\frac{3}{7} - \frac{5}{14}$

64. $\frac{7}{8} - \frac{5}{16}$

65. $\frac{2}{3} - \frac{1}{6}$

66. $\frac{5}{21} - \frac{1}{6}$

67. $\frac{11}{12} - \frac{2}{3}$

68. $\frac{9}{20} - \frac{1}{30}$

69. $4\frac{11}{18} - 2\frac{5}{18}$

70. $3\frac{7}{12} - 1\frac{1}{12}$

71. $8\frac{3}{4} - 2$

72. $6\frac{5}{9} - 4$

73. $8\frac{5}{6} - 7\frac{3}{4}$

74. $5\frac{7}{8} - 3\frac{2}{3}$

75. $7 - 3\frac{5}{8}$

76. $6 - 2\frac{4}{5}$

77. $10 - 4\frac{8}{9}$

78. $5 - 2\frac{7}{18}$

79. $7\frac{3}{8} - 4\frac{5}{8}$

80. $11\frac{1}{6} - 8\frac{5}{6}$

81. $12\frac{5}{12} - 10\frac{17}{24}$

82. $16\frac{1}{3} - 11\frac{5}{12}$

83. $6\frac{2}{3} - 1\frac{7}{8}$

84. $7\frac{7}{12} - 2\frac{5}{6}$

85. $10\frac{2}{5} - 8\frac{7}{10}$

86. $5\frac{5}{6} - 4\frac{7}{8}$

For Exercises 87 and 88, each statement describes the difference between a pair of fractions that have the same denominator. State whether the difference of the fractions will need to be rewritten in order to be in simplest form. Answer *yes* or *no*.

87. The difference between the numerators is a factor of the denominator.

88. The difference between the numerators is 1.

89. What is $\frac{2}{3}$ less than $\frac{7}{8}$?

90. Find the difference between $\frac{8}{9}$ and $\frac{1}{6}$.

91. Find 8 less $1\frac{7}{12}$.

92. Find 9 minus $5\frac{3}{20}$.

For Exercises 93 to 100, evaluate the variable expression $x - y$ for the given values of x and y.

93. $x = \frac{8}{9}, y = \frac{5}{9}$

94. $x = \frac{5}{6}, y = \frac{1}{6}$

95. $x = \frac{7}{15}, y = \frac{3}{10}$

96. $x = \frac{5}{6}, y = \frac{2}{15}$

97. $x = 5\frac{7}{9}, y = 4\frac{2}{3}$

98. $x = 9\frac{5}{8}, y = 2\frac{3}{16}$

99. $x = 5, y = 2\frac{7}{9}$

100. $x = 8, y = 4\frac{5}{6}$

101. Is $\frac{3}{4}$ a solution of the equation $\frac{4}{5} = \frac{31}{20} - y$?

102. Is $\frac{2}{3}$ a solution of the equation $\frac{2}{3} - x = 0$?

OBJECTIVE C To solve application problems and use formulas

103. Carpentry A $2\frac{3}{4}$-foot piece is cut from a 6-foot board. Find the length of the remaining piece of board.

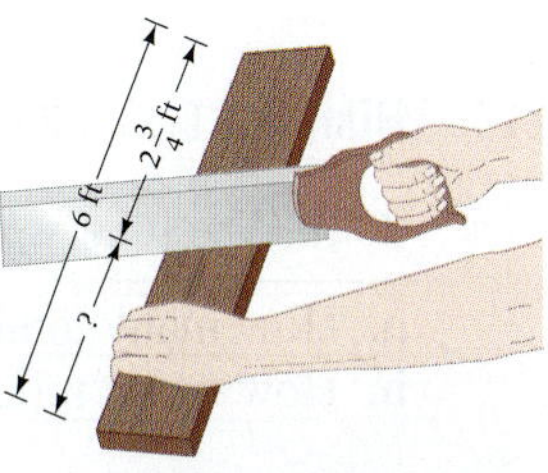

104. Boxing A boxer is put on a diet to gain 15 lb in 4 weeks. The boxer gains $4\frac{1}{2}$ lb the first week and $3\frac{3}{4}$ lb the second week. How much weight must the boxer gain during the third and fourth weeks in order to gain a total of 15 lb?

AP Photo/Roberto Borea

105. Horseracing The 3-year-olds in the Kentucky Derby run $1\frac{1}{4}$ mi. The horses in the Belmont Stakes run $1\frac{1}{2}$ mi, and they run $1\frac{3}{16}$ mi in the Preakness Stakes. How much farther do the horses run in the Kentucky Derby than in the Preakness Stakes? How much farther do they run in the Belmont Stakes than in the Preakness Stakes?

106. Sociology The table at the right shows the results of a survey in which adults in the United States were asked how many evening meals they cook at home during an average week.

a. Which response was given most frequently?

b. What fraction of the adult population cooks two or fewer dinners at home per week?

c. What fraction of the adult population cooks five or more dinners at home per week? Is this less than half or more than half of the people?

Responses to the question, "How many evening meals do you cook at home each week?"	
0	$\frac{2}{25}$
1	$\frac{1}{20}$
2	$\frac{1}{10}$
3	$\frac{13}{100}$
4	$\frac{3}{20}$
5	$\frac{21}{100}$
6	$\frac{9}{100}$
7	$\frac{19}{100}$

Source: Millward Brown for Whirlpool

Golf During the second half of the 1900s, greenskeepers mowed the grass on golf putting surfaces progressively lower. The table at the right below shows the average grass height by decade. Use this table for Exercises 107 and 108.

107. What was the difference between the average height of the grass in the 1980s and the 1950s?

108. Calculate the difference between the average grass height in the 1970s and the 1960s.

Average Height of Grass on Golf Putting Surfaces	
Decade	***Height (in inches)***
1950s	$\frac{1}{4}$
1960s	$\frac{7}{32}$
1970s	$\frac{3}{16}$
1980s	$\frac{5}{32}$
1990s	$\frac{1}{8}$

Source: Golf Course Superintendents Association of America

109. Geometry You want to fence in the triangular plot of land shown below. How many feet of fencing do you need? Use the formula $P = a + b + c$.

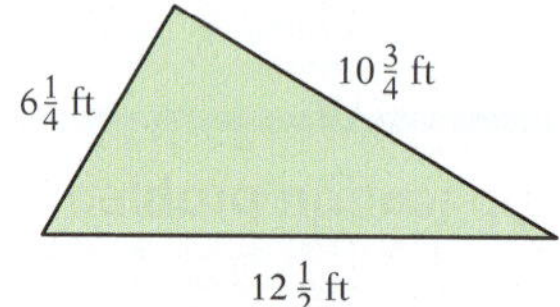

110. Hiking Two hikers plan a 3-day, $27\frac{1}{2}$-mile backpack trip carrying a total of 80 lb. The hikers plan to travel $7\frac{3}{8}$ mi the first day and $10\frac{1}{3}$ mi the second day.

a. How many total miles do the hikers plan to travel the first two days?

b. How many miles will be left to travel on the third day?

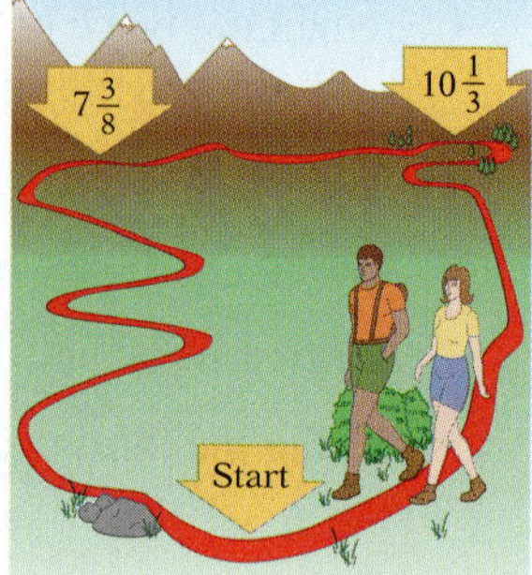

For Exercises 111 and 112, refer to Exercise 110. Describe what each difference represents.

111. $27\frac{1}{2} - 7\frac{3}{8}$

112. $10\frac{1}{3} - 7\frac{3}{8}$

The Olympics The table at the right shows the jump heights of three U.S. Olympic gold medalists in the high jump during the 1900s. Use this table for Exercises 113 and 114.

Olympic Gold Medalists in the High Jump		
Year	***Athlete***	***Height of Jump (in feet)***
1920	Richard Landon	$6\frac{1}{3}$
1924	Harold Osborn	$6\frac{1}{2}$
1996	Charles Austin	$7\frac{5}{6}$

Source: The World Almanac and Book of Facts

113. Find the difference between the height of Charles Austin's jump and the height of Richard Landon's jump.

114. How much higher was Charles Austin's jump than Harold Osborn's jump?

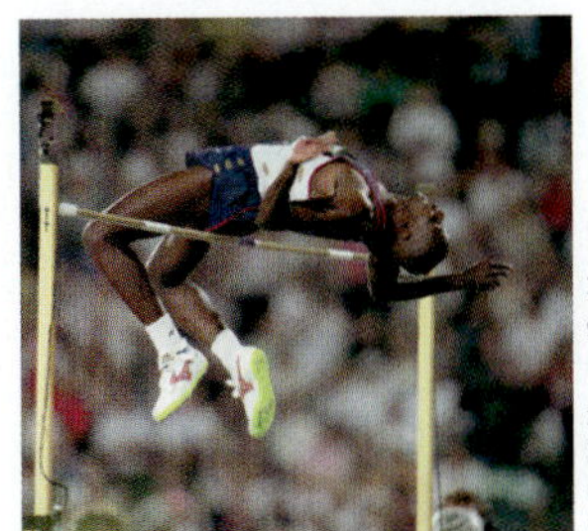

Wally McNamee/Corbis

Charles Austin

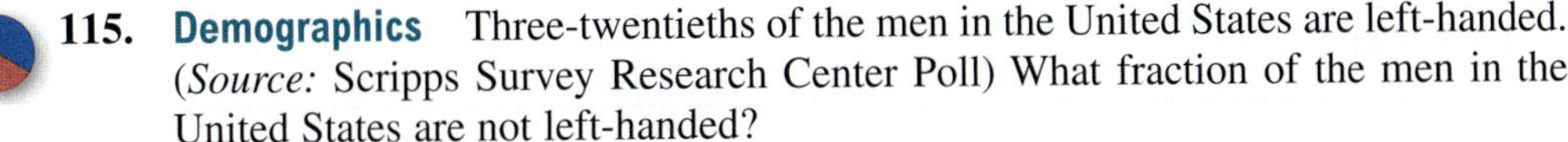

115. **Demographics** Three-twentieths of the men in the United States are left-handed. (*Source:* Scripps Survey Research Center Poll) What fraction of the men in the United States are not left-handed?

Construction The size of an interior door frame is determined by the width of the wall into which it is installed. The width of the wall is determined by the width of the stud in the wall and the thickness of the sheets of drywall installed on each side of the wall. A 2×4 stud is $3\frac{5}{8}$ in. thick. A 2×6 stud is $5\frac{5}{8}$ in. thick. Use this information for Exercises 116 and 117.

116. Find the thickness of a wall constructed with 2×4 studs and drywall that is $\frac{1}{2}$ in. thick.

Ryan McVay/Photodisc/Getty Images

117. Find the thickness of a wall constructed with 2×6 studs and drywall that is $\frac{1}{2}$ in. thick.

Applying the Concepts

118. The figure at the right is divided into five parts. Is each part of the figure $\frac{1}{5}$ of the figure? Why or why not?

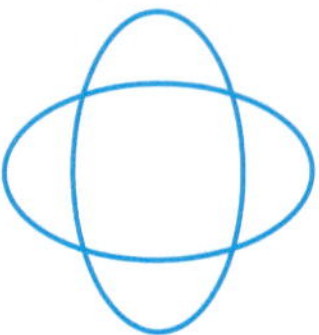

119. Use the diagram at the right to illustrate the sum of $\frac{1}{8}$ and $\frac{5}{6}$. Why does the figure contain 24 squares? Would it be possible to illustrate the sum of $\frac{1}{8}$ and $\frac{5}{6}$ if there were 48 squares in the figure? What if there were 16 squares? Make a list of the possible numbers of squares that could be used to illustrate the sum of $\frac{1}{8}$ and $\frac{5}{6}$.

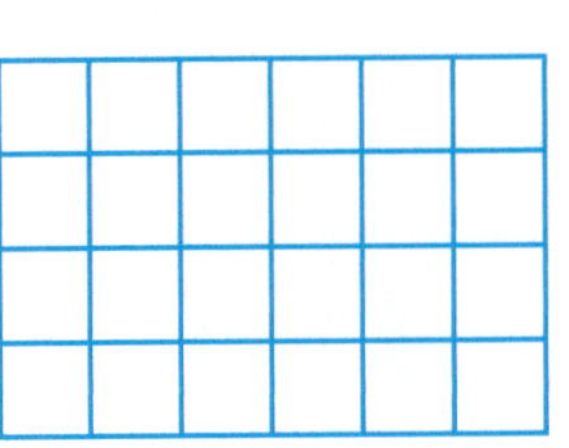

SECTION

2.4 Multiplication and Division of Fractions

OBJECTIVE A To multiply fractions 

To multiply two fractions, multiply the numerators and multiply the denominators.

Multiplication of Fractions

The product of two fractions is the product of the numerators over the product of the denominators.

$\frac{a}{b} \cdot \frac{c}{d} = \frac{ac}{bd}$, where $b \neq 0$ and $d \neq 0$

Note that **fractions do not need to have the same denominator in order to be multiplied.**

HOW TO • 1 Multiply: $\frac{2}{5} \cdot \frac{1}{3}$

$\frac{2}{5} \cdot \frac{1}{3} = \frac{2 \cdot 1}{5 \cdot 3} = \frac{2}{15}$ • **Multiply the numerators. Multiply the denominators.**

The product $\frac{2}{5} \cdot \frac{1}{3}$ can be read "$\frac{2}{5}$ times $\frac{1}{3}$" or "$\frac{2}{5}$ of $\frac{1}{3}$."

Reading the times sign as "of" is useful in diagramming the product of two fractions.

$\frac{1}{3}$ of the bar at the right is shaded.

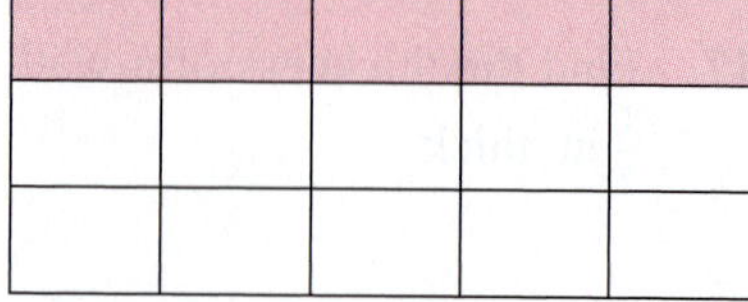

Shade $\frac{2}{5}$ of the $\frac{1}{3}$ already shaded.

$\frac{2}{15}$ of the bar is now shaded.

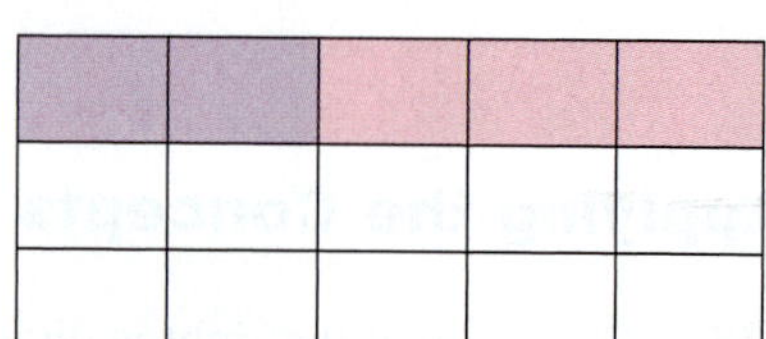

$$\frac{2}{5} \text{ of } \frac{1}{3} = \frac{2}{5} \cdot \frac{1}{3} = \frac{2}{15}$$

If a is a natural number, then $\frac{1}{a}$ is called the **reciprocal** or **multiplicative inverse** of a. Note that $a \cdot \frac{1}{a} = \frac{a}{1} \cdot \frac{1}{a} = \frac{a}{a} = 1$.

The product of a number and its multiplicative inverse is 1.

$$\frac{1}{8} \cdot 8 = 8 \cdot \frac{1}{8} = 1$$

After multiplying two fractions, write the product in simplest form.

HOW TO • 2 Multiply: $\frac{3}{8} \cdot \frac{4}{9}$

$$\frac{3}{8} \cdot \frac{4}{9} = \frac{3 \cdot 4}{8 \cdot 9}$$
- Multiply the numerators. Multiply the denominators.

$$= \frac{3 \cdot 2 \cdot 2}{2 \cdot 2 \cdot 2 \cdot 3 \cdot 3}$$
- Express the fraction in simplest form by first writing the prime factorization of each number.

$$= \frac{1}{6}$$
- Divide by the common factors and write the product in simplest form.

Point of Interest

Try this: What is the result if you take one-third of a half-dozen and add to it one-fourth of the product of the result and 8?

To multiply a whole number by a fraction or a mixed number, first write the whole number as a fraction with a denominator of 1.

HOW TO • 3 Multiply: $3 \cdot \frac{5}{8}$

$$3 \cdot \frac{5}{8} = \frac{3}{1} \cdot \frac{5}{8}$$
- Write the whole number 3 as the fraction $\frac{3}{1}$.

$$= \frac{3 \cdot 5}{1 \cdot 8}$$
- Multiply the fractions. There are no common factors in the numerator and denominator.

$$= \frac{15}{8} = 1\frac{7}{8}$$
- Write the improper fraction as a mixed number.

HOW TO • 4 Multiply: $\frac{x}{7} \cdot \frac{y}{5}$

$$\frac{x}{7} \cdot \frac{y}{5} = \frac{x \cdot y}{7 \cdot 5}$$
- Multiply the numerators. Multiply the denominators.

$$= \frac{xy}{35}$$
- Write the product in simplest form.

When a factor is a mixed number, first write the mixed number as an improper fraction. Then multiply.

HOW TO • 5 Find the product of $4\frac{1}{6}$ and $2\frac{7}{10}$.

$$4\frac{1}{6} \cdot 2\frac{7}{10} = \frac{25}{6} \cdot \frac{27}{10}$$
- Write each mixed number as an improper fraction.

$$= \frac{25 \cdot 27}{6 \cdot 10}$$
- Multiply the fractions.

$$= \frac{5 \cdot 5 \cdot 3 \cdot 3 \cdot 3}{2 \cdot 3 \cdot 2 \cdot 5}$$

$$= \frac{45}{4} = 11\frac{1}{4}$$
- Write the product in simplest form.

HOW TO • 6 Is $\frac{2}{3}$ a solution of the equation $\frac{3}{4}x = \frac{1}{2}$?

$$\frac{3}{4}x = \frac{1}{2}$$

$$\frac{3}{4}\left(\frac{2}{3}\right) \;\Big|\; \frac{1}{2}$$ • Replace x by $\frac{2}{3}$ and then simplify.

$$\frac{3 \cdot 2}{4 \cdot 3} \;\Big|\; \frac{1}{2}$$

$$\frac{3 \cdot 2}{2 \cdot 2 \cdot 3} \;\Big|\; \frac{1}{2}$$

$$\frac{1}{2} = \frac{1}{2}$$ • The results are equal.

Yes, $\frac{2}{3}$ is a solution of the equation.

Point of Interest

René Descartes (1596–1650) was the first mathematician to extensively use exponential notation as it is used today. However, for some unknown reason, he always used xx for x^2.

Recall that an exponent indicates the repeated multiplication of the same factor. For example,

$$3^5 = 3 \cdot 3 \cdot 3 \cdot 3 \cdot 3$$

The exponent, 5, indicates how many times the base, 3, occurs as a factor in the multiplication.

The base of an exponential expression can be a fraction; for example, $\left(\frac{2}{3}\right)^4$. To evaluate this expression, write the factor as many times as indicated by the exponent and then multiply.

$$\left(\frac{2}{3}\right)^4 = \frac{2}{3} \cdot \frac{2}{3} \cdot \frac{2}{3} \cdot \frac{2}{3} = \frac{2 \cdot 2 \cdot 2 \cdot 2}{3 \cdot 3 \cdot 3 \cdot 3} = \frac{16}{81}$$

HOW TO • 7 Evaluate $\left(\frac{3}{5}\right)^2 \cdot \left(\frac{5}{6}\right)^3$.

$$\left(\frac{3}{5}\right)^2 \cdot \left(\frac{5}{6}\right)^3$$

$$= \frac{3}{5} \cdot \frac{3}{5} \cdot \frac{5}{6} \cdot \frac{5}{6} \cdot \frac{5}{6}$$ • Write each factor as many times as indicated by the exponent.

$$= \frac{3 \cdot 3 \cdot 5 \cdot 5 \cdot 5}{5 \cdot 5 \cdot 6 \cdot 6 \cdot 6}$$ • Multiply.

$$= \frac{5}{24}$$ • Write the product in simplest form.

EXAMPLE • 1

Multiply: $\frac{6}{x} \cdot \frac{8}{y}$

Solution $\frac{6}{x} \cdot \frac{8}{y} = \frac{6 \cdot 8}{x \cdot y} = \frac{48}{xy}$

YOU TRY IT • 1

Multiply: $\frac{y}{10} \cdot \frac{z}{7}$

Your solution

Solution on p. S5

EXAMPLE • 2

Multiply: $\frac{7}{9} \cdot \frac{3}{14} \cdot \frac{2}{5}$

Solution

$$\frac{7}{9} \cdot \frac{3}{14} \cdot \frac{2}{5} = \frac{7 \cdot 3 \cdot 2}{9 \cdot 14 \cdot 5}$$
$$= \frac{7 \cdot 3 \cdot 2}{3 \cdot 3 \cdot 2 \cdot 7 \cdot 5} = \frac{1}{15}$$

YOU TRY IT • 2

Multiply: $\frac{5}{12} \cdot \frac{9}{35} \cdot \frac{7}{8}$

Your solution

EXAMPLE • 3

What is the product of $\frac{7}{12}$ and 4?

Solution

$$\frac{7}{12} \cdot 4 = \frac{7}{12} \cdot \frac{4}{1}$$
$$= \frac{7 \cdot 4}{12 \cdot 1}$$
$$= \frac{7 \cdot 2 \cdot 2}{2 \cdot 2 \cdot 3 \cdot 1}$$
$$= \frac{7}{3} = 2\frac{1}{3}$$

YOU TRY IT • 3

Find the product of $\frac{8}{9}$ and 6.

Your solution

EXAMPLE • 4

Multiply: $7\frac{1}{2} \cdot 4\frac{2}{5}$

Solution

$$7\frac{1}{2} \cdot 4\frac{2}{5} = \frac{15}{2} \cdot \frac{22}{5} = \frac{15 \cdot 22}{2 \cdot 5}$$
$$= \frac{3 \cdot 5 \cdot 2 \cdot 11}{2 \cdot 5}$$
$$= \frac{33}{1} = 33$$

YOU TRY IT • 4

Multiply: $3\frac{6}{7} \cdot 2\frac{4}{9}$

Your solution

EXAMPLE • 5

Evaluate x^2y^2 when $x = 1\frac{1}{2}$ and $y = \frac{2}{3}$.

Solution

x^2y^2

$$\left(1\frac{1}{2}\right)^2 \cdot \left(\frac{2}{3}\right)^2 = \left(\frac{3}{2}\right)^2 \cdot \left(\frac{2}{3}\right)^2$$
$$= \frac{3}{2} \cdot \frac{3}{2} \cdot \frac{2}{3} \cdot \frac{2}{3}$$
$$= \frac{3 \cdot 3 \cdot 2 \cdot 2}{2 \cdot 2 \cdot 3 \cdot 3} = 1$$

YOU TRY IT • 5

Evaluate x^4y^3 when $x = 2\frac{1}{3}$ and $y = \frac{3}{7}$.

Your solution

Solutions on p. S5

OBJECTIVE B To divide fractions 

Recall that the **reciprocal** of a fraction is that fraction with the numerator and denominator interchanged.

The reciprocal of $\frac{3}{4}$ is $\frac{4}{3}$.

The reciprocal of $\frac{a}{b}$ is $\frac{b}{a}$.

Point of Interest

Try this: What number when multiplied by its reciprocal is equal to 1?

The process of interchanging the numerator and denominator of a fraction is called **inverting** the fraction.

To find the reciprocal of a whole number, first rewrite the whole number as a fraction with a denominator of 1. Then invert the fraction.

$6 = \frac{6}{1}$

The reciprocal of 6 is $\frac{1}{6}$.

Reciprocals are used to rewrite division problems as related multiplication problems. Look at the following two problems:

$6 \div 2 = 3$

6 divided by 2 equals 3.

$6 \cdot \frac{1}{2} = 3$

6 times the reciprocal of 2 equals 3.

Division is defined as multiplication by the reciprocal. Therefore, "divided by 2" is the same as "times $\frac{1}{2}$." Fractions are divided by making this substitution.

Division of Fractions

To divide two fractions, multiply by the reciprocal of the divisor.

$\frac{a}{b} \div \frac{c}{d} = \frac{a}{b} \cdot \frac{d}{c}$, where $b \neq 0$, $c \neq 0$, and $d \neq 0$

HOW TO • 8 Divide: $\frac{2}{5} \div \frac{3}{4}$

$\frac{2}{5} \div \frac{3}{4} = \frac{2}{5} \cdot \frac{4}{3}$ • Rewrite the division as multiplication by the reciprocal.

$= \frac{2 \cdot 4}{5 \cdot 3}$ • Multiply the fractions.

$= \frac{2 \cdot 2 \cdot 2}{5 \cdot 3} = \frac{8}{15}$

To divide a fraction and a whole number, first write the whole number as a fraction with a denominator of 1.

Take Note

$\frac{3}{4} \div 6 = \frac{1}{8}$ means that if $\frac{3}{4}$ is divided into 6 equal parts, each equal part is $\frac{1}{8}$ of the whole. For example, if 6 people share $\frac{3}{4}$ of a pizza, each person eats $\frac{1}{8}$ of the pizza.

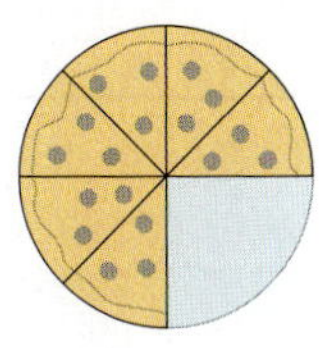

HOW TO • 9 Find the quotient of $\frac{3}{4}$ and 6.

$$\frac{3}{4} \div 6 = \frac{3}{4} \div \frac{6}{1}$$
• Write the whole number 6 as the fraction $\frac{6}{1}$.

$$= \frac{3}{4} \cdot \frac{1}{6}$$
• Rewrite the division as multiplication by the reciprocal.

$$= \frac{3 \cdot 1}{4 \cdot 6}$$
• Multiply the fractions.

$$= \frac{3 \cdot 1}{2 \cdot 2 \cdot 2 \cdot 3}$$

$$= \frac{1}{8}$$

When a number in a quotient is a mixed number, first write the mixed number as an improper fraction. Then divide the fractions.

HOW TO • 10 Divide: $\frac{2}{3} \div 1\frac{1}{4}$

$$\frac{2}{3} \div 1\frac{1}{4} = \frac{2}{3} \div \frac{5}{4}$$
• Write the mixed number $1\frac{1}{4}$ as an improper fraction.

$$= \frac{2}{3} \cdot \frac{4}{5}$$
• Rewrite the division as multiplication by the reciprocal.

$$= \frac{2 \cdot 4}{3 \cdot 5} = \frac{8}{15}$$
• Multiply the fractions.

EXAMPLE • 6

Divide: $\frac{4}{5} \div \frac{8}{15}$

Solution

$$\frac{4}{5} \div \frac{8}{15} = \frac{4}{5} \cdot \frac{15}{8}$$
$$= \frac{4 \cdot 15}{5 \cdot 8}$$
$$= \frac{2 \cdot 2 \cdot 3 \cdot 5}{5 \cdot 2 \cdot 2 \cdot 2}$$
$$= \frac{3}{2} = 1\frac{1}{2}$$

YOU TRY IT • 6

Divide: $\frac{5}{6} \div \frac{10}{27}$

Your solution

Solution on p. S5

EXAMPLE • 7

Divide: $\frac{x}{2} \div \frac{y}{4}$

Solution

$$\frac{x}{2} \div \frac{y}{4} = \frac{x}{2} \cdot \frac{4}{y} = \frac{x \cdot 4}{2 \cdot y} = \frac{x \cdot 2 \cdot 2}{2 \cdot y} = \frac{2x}{y}$$

YOU TRY IT • 7

Divide: $\frac{x}{8} \div \frac{y}{6}$

Your solution

EXAMPLE • 8

Divide: $3\frac{4}{15} \div 2\frac{1}{10}$

Solution

$$3\frac{4}{15} \div 2\frac{1}{10} = \frac{49}{15} \div \frac{21}{10} = \frac{49}{15} \cdot \frac{10}{21} = \frac{49 \cdot 10}{15 \cdot 21} = \frac{7 \cdot 7 \cdot 2 \cdot 5}{3 \cdot 5 \cdot 3 \cdot 7} = \frac{14}{9} = 1\frac{5}{9}$$

YOU TRY IT • 8

Divide: $4\frac{3}{8} \div 3\frac{1}{2}$

Your solution

EXAMPLE • 9

Evaluate $x \div y$ when $x = 3\frac{1}{8}$ and $y = 5$.

Solution

$x \div y$

$$3\frac{1}{8} \div 5 = \frac{25}{8} \div \frac{5}{1} = \frac{25}{8} \cdot \frac{1}{5} = \frac{25 \cdot 1}{8 \cdot 5} = \frac{5 \cdot 5 \cdot 1}{2 \cdot 2 \cdot 2 \cdot 5} = \frac{5}{8}$$

YOU TRY IT • 9

Evaluate $x \div y$ when $x = 2\frac{1}{4}$ and $y = 9$.

Your solution

Solutions on p. S5

OBJECTIVE C To simplify a complex fraction 

A **complex fraction** is a fraction whose numerator or denominator contains one or more fractions. Examples of complex fractions are shown below.

Main fraction bar ⟶ $\dfrac{\frac{3}{4}}{\frac{7}{8}}$ $\qquad \dfrac{4}{3-\frac{1}{2}}$ $\qquad \dfrac{\frac{9}{10}+\frac{3}{5}}{\frac{5}{6}}$ $\qquad \dfrac{3\frac{1}{2}\cdot 2\frac{5}{8}}{\left(4\frac{2}{3}\right)\div\left(3\frac{1}{5}\right)}$

Look at the first example given above and recall that the fraction bar can be read "divided by."

Therefore, $\dfrac{\frac{3}{4}}{\frac{7}{8}}$ can be read "$\frac{3}{4}$ divided by $\frac{7}{8}$" and can be written $\frac{3}{4}\div\frac{7}{8}$. This is the division of two fractions and can be simplified by multiplying by the reciprocal, as shown.

$$\frac{\frac{3}{4}}{\frac{7}{8}}=\frac{3}{4}\div\frac{7}{8}=\frac{3}{4}\cdot\frac{8}{7}=\frac{3\cdot 8}{4\cdot 7}=\frac{6}{7}$$

To simplify a complex fraction, first simplify the expression above the main fraction bar and the expression below the main fraction bar; the result is one number in the numerator and one number in the denominator. Then rewrite the complex fraction as a division problem by reading the main fraction bar as "divided by."

HOW TO • 11 Simplify: $\dfrac{4}{3-\frac{1}{2}}$

$\dfrac{4}{3-\frac{1}{2}}=\dfrac{4}{\frac{5}{2}}$

• **The numerator (4) is already simplified. Simplify the expression in the denominator.**

Note: $3-\frac{1}{2}=\frac{6}{2}-\frac{1}{2}=\frac{5}{2}$

$=4\div\frac{5}{2}$ • **Rewrite the complex fraction as division.**

$=\frac{4}{1}\div\frac{5}{2}$ • **Divide.**

$=\frac{4}{1}\cdot\frac{2}{5}$

$=\frac{8}{5}=1\frac{3}{5}$ • **Write the answer in simplest form.**

HOW TO • 12 Evaluate $\frac{wx}{yz}$ when $w = 1\frac{1}{3}$, $x = 2\frac{5}{8}$, $y = 4\frac{1}{2}$, and $z = 3\frac{1}{3}$.

$$\frac{wx}{yz}$$

$$\frac{1\frac{1}{3} \cdot 2\frac{5}{8}}{4\frac{1}{2} \cdot 3\frac{1}{3}}$$

• Replace each variable with its given value.

$$= \frac{\frac{7}{2}}{15}$$

• Simplify the numerator.

Note: $1\frac{1}{3} \cdot 2\frac{5}{8} = \frac{4}{3} \cdot \frac{21}{8} = \frac{7}{2}$

Simplify the denominator.

Note: $4\frac{1}{2} \cdot 3\frac{1}{3} = \frac{9}{2} \cdot \frac{10}{3} = 15$

$$= \frac{7}{2} \div 15$$

• Rewrite the complex fraction as division.

$$= \frac{7}{2} \cdot \frac{1}{15} = \frac{7}{30}$$

• Divide by multiplying by the reciprocal.

Note: $15 = \frac{15}{1}$; the reciprocal of $\frac{15}{1}$ is $\frac{1}{15}$.

EXAMPLE • 10

Is $\frac{2}{3}$ a solution of $\frac{x + \frac{1}{2}}{x} = \frac{7}{4}$?

Solution

$$\frac{x + \frac{1}{2}}{x} = \frac{7}{4}$$

$$\frac{\frac{2}{3} + \frac{1}{2}}{\frac{2}{3}} \quad \Big| \quad \frac{7}{4}$$

$$\frac{\frac{7}{6}}{\frac{2}{3}} \quad \Big| \quad \frac{7}{4}$$

$$\frac{7}{6} \div \frac{2}{3} \quad \Big| \quad \frac{7}{4}$$

$$\frac{7}{6} \cdot \frac{3}{2} \quad \Big| \quad \frac{7}{4}$$

$$\frac{7}{4} = \frac{7}{4}$$

Yes, $\frac{2}{3}$ is a solution of the equation.

YOU TRY IT • 10

Is $\frac{1}{2}$ a solution of $\frac{2y + 3}{y} = 2$?

Your solution

Solution on p. S5

EXAMPLE • 11

Evaluate the variable expression $\frac{x-y}{z}$ when $x = 4\frac{1}{8}$, $y = 2\frac{5}{8}$, and $z = \frac{3}{4}$.

Solution

$$\frac{x-y}{z}$$

$$\frac{4\frac{1}{8} - 2\frac{5}{8}}{\frac{3}{4}} = \frac{\frac{3}{2}}{\frac{3}{4}}$$

$$= \frac{3}{2} \div \frac{3}{4}$$

$$= \frac{3}{2} \cdot \frac{4}{3} = 2$$

YOU TRY IT • 11

Evaluate the variable expression $\frac{x}{y-z}$ when $x = 2\frac{4}{9}$, $y = 3$, and $z = 1\frac{1}{3}$.

Your solution

Solution on p. S6

OBJECTIVE D To solve application problems and use formulas

Figure ABC is a triangle. AB is the **base,** b, of the triangle. The line segment from C that forms a right angle with the base is the **height,** h, of the triangle. The formula for the area of a triangle is given below. Use this formula for Example 12 and You Try It 12.

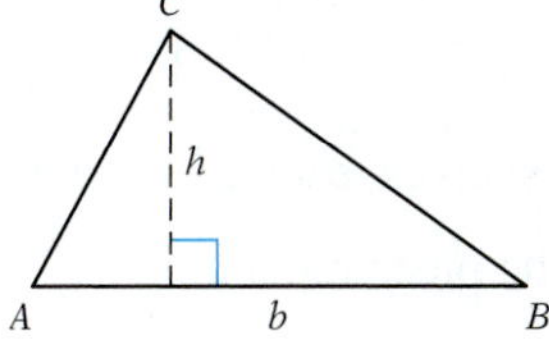

Area of a Triangle

The formula for the area of a triangle is $A = \frac{1}{2}bh$, where A is the area of the triangle, b is the base, and h is the height.

EXAMPLE • 12

A riveter uses metal plates that are in the shape of a triangle and have a base of 12 cm and a height of 6 cm. Find the area of one metal plate.

Strategy

To find the area, use the formula for the area of a triangle, $A = \frac{1}{2}bh$. $b = 12$ and $h = 6$.

Solution

$$A = \frac{1}{2}bh$$

$$A = \frac{1}{2}(12)(6)$$

$$A = 36$$

6 cm

12 cm

The area is 36 cm^2.

YOU TRY IT • 12

Find the amount of felt needed to make a banner that is in the shape of a triangle with a base of 18 in. and a height of 9 in.

Your strategy

Your solution

EXAMPLE • 13

A 12-foot board is cut into pieces $2\frac{1}{2}$ ft long for use as bookshelves. What is the length of the remaining piece after as many shelves as possible are cut?

Strategy

To find the length of the remaining piece:

- Divide the total length (12) by the length of each shelf $\left(2\frac{1}{2}\right)$. The quotient is the number of shelves cut, with a certain fraction of a shelf left over.
- Multiply the fraction left over by the length of a shelf.

Solution

$$12 \div 2\frac{1}{2} = \frac{12}{1} \div \frac{5}{2} = \frac{12}{1} \cdot \frac{2}{5} = \frac{12 \cdot 2}{1 \cdot 5} = \frac{24}{5} = 4\frac{4}{5}$$

4 shelves, each $2\frac{1}{2}$ ft long, can be cut from the board.

The piece remaining is $\frac{4}{5}$ of $2\frac{1}{2}$ ft long.

$$\frac{4}{5} \cdot 2\frac{1}{2} = \frac{4}{5} \cdot \frac{5}{2} = \frac{4 \cdot 5}{5 \cdot 2} = 2$$

The length of the remaining piece is 2 ft.

YOU TRY IT • 13

The Booster Club is making 22 sashes for the high school band members. Each sash requires $1\frac{3}{8}$ yd of material at a cost of \$12 per yard. Find the total cost of the material.

Your strategy

Your solution

Solutions on p. S6

2.4 EXERCISES

OBJECTIVE A To multiply fractions

1. Explain why you need a common denominator when adding or subtracting two fractions and why you don't need a common denominator when multiplying or dividing two fractions.

2. The product of 1 and a number is $\frac{3}{8}$. Find the number. Explain how you arrived at the answer.

For Exercises 3 to 26, multiply.

3. $\frac{2}{3} \cdot \frac{9}{10}$

4. $\frac{3}{8} \cdot \frac{4}{5}$

5. $\frac{14}{15} \cdot \frac{6}{7}$

6. $\frac{15}{16} \cdot \frac{4}{9}$

7. $\frac{6}{7} \cdot \frac{0}{10}$

8. $\frac{5}{12} \cdot \frac{3}{0}$

9. $\frac{9}{x} \cdot \frac{7}{y}$

10. $\frac{4}{c} \cdot \frac{8}{d}$

11. $\frac{2}{3} \cdot \frac{3}{8} \cdot \frac{4}{9}$

12. $\frac{5}{7} \cdot \frac{1}{6} \cdot \frac{14}{15}$

13. $6 \cdot \frac{1}{6}$

14. $\frac{1}{10} \cdot 10$

15. $\frac{3}{4} \cdot 8$

16. $\frac{5}{7} \cdot 14$

17. $\frac{6}{7} \cdot 0$

18. $0 \cdot \frac{9}{11}$

19. $\frac{5}{22} \cdot 2\frac{1}{5}$

20. $\frac{4}{15} \cdot 1\frac{7}{8}$

21. $3\frac{1}{2} \cdot 5\frac{3}{7}$

22. $2\frac{1}{4} \cdot 1\frac{1}{3}$

23. $8 \cdot 5\frac{1}{4}$

24. $3 \cdot 2\frac{1}{9}$

25. $3\frac{1}{2} \cdot 1\frac{5}{7} \cdot \frac{11}{12}$

26. $2\frac{2}{3} \cdot \frac{8}{9} \cdot 1\frac{5}{16}$

27. Give an example of a proper and an improper fraction whose product is 1.

28. True or false? If the product of a whole number and a fraction is a whole number, then the denominator of the fraction is a factor of the original whole number.

29. Find the product of $\frac{3}{4}$ and $\frac{14}{15}$.

30. Find the product of $\frac{12}{25}$ and $\frac{5}{16}$.

31. What is $4\frac{4}{5}$ times $\frac{3}{8}$?

32. What is $5\frac{1}{3}$ times $\frac{3}{16}$?

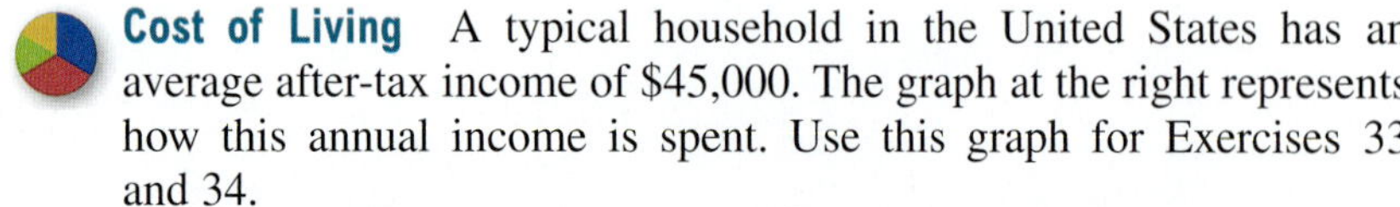

Cost of Living A typical household in the United States has an average after-tax income of \$45,000. The graph at the right represents how this annual income is spent. Use this graph for Exercises 33 and 34.

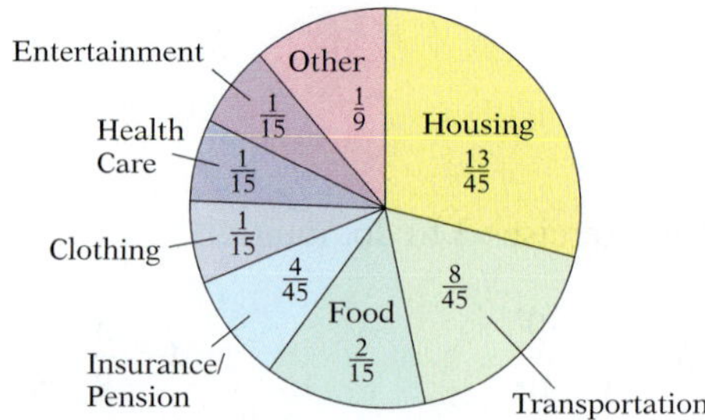

How a Typical U.S. Household Spends Its Annual Income

Source: Based on data from American Demographics

33. Find the amount of money a typical household in the United States spends on housing per year.

34. How much money does a typical household in the United States spend annually on food?

For Exercises 35 to 38, evaluate the variable expression xy for the given values of x and y.

35. $x = \frac{5}{16}, y = \frac{7}{15}$

36. $x = \frac{2}{5}, y = \frac{5}{6}$

37. $x = \frac{4}{7}, y = 6\frac{1}{8}$

38. $x = 6\frac{3}{5}, y = 3\frac{1}{3}$

For Exercises 39 to 44, evaluate the variable expression xyz for the given values of x, y, and z.

39. $x = \frac{3}{8}, y = \frac{2}{3}, z = \frac{4}{5}$

40. $x = 4, y = \frac{0}{8}, z = 1\frac{5}{9}$

41. $x = 2\frac{3}{8}, y = \frac{3}{19}, z = \frac{4}{9}$

42. $x = \frac{4}{5}, y = 15, z = \frac{7}{8}$

43. $x = \frac{5}{6}, y = 3, z = 1\frac{7}{15}$

44. $x = 4\frac{1}{2}, y = 3\frac{5}{9}, z = 1\frac{7}{8}$

45. Is $\frac{3}{4}$ a solution of the equation $\frac{4}{5}x = \frac{5}{3}$?

46. Is $\frac{1}{2}$ a solution of the equation $\frac{3}{4}p = \frac{3}{2}$?

For Exercises 47 to 54, evaluate.

47. $\left(\frac{3}{4}\right)^2$

48. $\left(\frac{5}{8}\right)^2$

49. $\left(\frac{5}{8}\right)^3 \cdot \left(\frac{2}{5}\right)^2$

50. $\left(\frac{3}{5}\right)^3 \cdot \left(\frac{1}{3}\right)^2$

51. $\left(\frac{18}{25}\right)^2 \cdot \left(\frac{5}{9}\right)^3$

52. $\left(\frac{2}{3}\right)^3 \cdot \left(\frac{5}{6}\right)^2$

53. $7^2 \cdot \left(\frac{2}{7}\right)^3$

54. $4^3 \cdot \left(\frac{5}{12}\right)^2$

For Exercises 55 to 58, evaluate the variable expression for the given values of *x* and *y*.

55. x^4, when $x = \frac{2}{3}$

56. y^3, when $y = \frac{3}{4}$

57. x^3y^2, when $x = \frac{2}{3}$ and $y = 1\frac{1}{2}$

58. x^2y^4, when $x = 2\frac{1}{3}$ and $y = \frac{3}{7}$

OBJECTIVE B To divide fractions

For Exercises 59 to 78, divide.

59. $\frac{5}{7} \div \frac{2}{5}$

60. $\frac{3}{8} \div \frac{2}{3}$

61. $0 \div \frac{7}{9}$

62. $0 \div \frac{4}{5}$

63. $6 \div \frac{3}{4}$

64. $8 \div \frac{2}{3}$

65. $\frac{3}{4} \div 6$

66. $\frac{2}{3} \div 8$

67. $\frac{9}{10} \div 0$

68. $\frac{2}{11} \div 0$

69. $\frac{b}{6} \div \frac{5}{d}$

70. $\frac{y}{10} \div \frac{4}{z}$

71. $3\frac{1}{3} \div \frac{5}{8}$

72. $5\frac{1}{2} \div \frac{1}{4}$

73. $5\frac{1}{2} \div 11$

74. $4\frac{2}{3} \div 7$

75. $5\frac{2}{7} \div 1$

76. $9\frac{5}{6} \div 1$

77. $2\frac{4}{13} \div 1\frac{5}{26}$

78. $3\frac{3}{8} \div 2\frac{7}{16}$

79. True or false? If a fraction has a numerator of 1, then the reciprocal of the fraction is a whole number.

80. True or false? The reciprocal of an improper fraction that is not equal to 1 is a proper fraction.

81. Find the quotient of $\frac{9}{10}$ and $\frac{3}{4}$.

82. Find the quotient of $\frac{3}{5}$ and $\frac{12}{25}$.

83. Find $\frac{7}{8}$ divided by $3\frac{1}{4}$.

84. Find $\frac{3}{8}$ divided by $2\frac{1}{4}$.

For Exercises 85 to 88, evaluate the variable expression $x \div y$ for the given values of x and y.

85. $x = \frac{5}{8}, y = \frac{15}{2}$

86. $x = \frac{14}{3}, y = \frac{7}{9}$

87. $x = 18, y = \frac{3}{8}$

88. $x = 20, y = \frac{5}{6}$

The Food Industry The table at the right shows the net weight of four different boxes of cereal. Use this table for Exercises 89 and 90.

Cereal	*Net Weight*
Kellogg Honey Crunch Corn Flakes	24 oz
Nabisco Instant Cream of Wheat	28 oz
Post Shredded Wheat	18 oz
Quaker Oats	41 oz

89. Find the number of $\frac{3}{4}$-ounce servings in a box of Kellogg Honey Crunch Corn Flakes.

90. Find the number of $1\frac{1}{4}$-ounce servings in a box of Post Shredded Wheat.

OBJECTIVE C To simplify a complex fraction

Simplify.

91. $\dfrac{\frac{9}{16}}{\frac{3}{4}}$

92. $\dfrac{\frac{7}{24}}{\frac{3}{8}}$

93. $\dfrac{\frac{2}{3} + \frac{1}{2}}{7}$

94. $\dfrac{5}{\frac{3}{8} - \frac{1}{4}}$

95. $\dfrac{2 + \frac{1}{4}}{\frac{3}{8}}$

96. $\dfrac{1 - \frac{3}{4}}{\frac{5}{12}}$

97. $\dfrac{\frac{9}{25}}{\frac{4}{5} - \frac{1}{10}}$

98. $\dfrac{\frac{9}{14} - \frac{1}{7}}{\frac{9}{14} + \frac{1}{7}}$

99. $\dfrac{3 + 2\frac{1}{3}}{5\frac{1}{6} - 1}$

100. $\dfrac{4 - 3\frac{5}{8}}{2\frac{1}{2} - \frac{3}{4}}$

101. $\dfrac{5\frac{2}{3} - 1\frac{1}{6}}{3\frac{5}{8} - 2\frac{1}{4}}$

102. $\dfrac{3\frac{1}{4} - 2\frac{1}{2}}{4\frac{3}{4} + 1\frac{1}{2}}$

103. True or false? If the denominator of a complex fraction is the same as its numerator, then the complex fraction is equal to 1.

For Exercises 104 to 107, evaluate the expression for the given values of the variables.

104. $\frac{x + y}{z}$, when $x = \frac{2}{3}$, $y = \frac{3}{4}$, and $z = \frac{1}{12}$

105. $\frac{x}{y + z}$, when $x = \frac{8}{15}$, $y = \frac{3}{5}$, and $z = \frac{2}{3}$

106. $\frac{x - y}{z}$, when $x = 2\frac{5}{8}$, $y = 1\frac{1}{4}$, and $z = 1\frac{3}{8}$

107. $\frac{x}{y - z}$, when $x = 2\frac{3}{10}$, $y = 3\frac{2}{5}$, and $z = 1\frac{4}{5}$

108. Is $\frac{3}{4}$ a solution of the equation $\frac{4x}{x + 5} = \frac{4}{3}$?

109. Is $\frac{4}{5}$ a solution of the equation $\dfrac{15y}{\frac{3}{10} + y} = 24$?

OBJECTIVE D To solve application problems and use formulas

Dennis MacDonald/PhotoEdit, Inc.

110. Polo A chukker is one period of play in a polo match. A chukker lasts $7\frac{1}{2}$ min. Find the length of time in four chukkers.

111. Calendars The Assyrian calendar was based on the phases of the moon. One lunation was $29\frac{1}{2}$ days long. There were 12 lunations in 1 year. Find the number of days in 1 year in the Assyrian calendar.

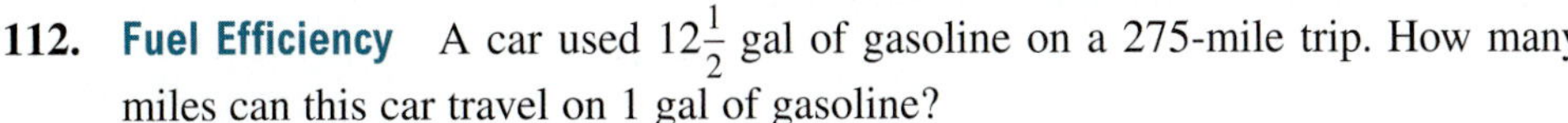

112. Fuel Efficiency A car used $12\frac{1}{2}$ gal of gasoline on a 275-mile trip. How many miles can this car travel on 1 gal of gasoline?

For Exercises 113 and 114, give your answer without actually doing a calculation.

113. Read Exercise 115. Will the requested cost be greater than or less than $12?

114. Read Exercise 116. Will the requested distance be greater than or less than 1 mi?

115. Consumerism Salmon costs $4 per pound. Find the cost of $2\frac{3}{4}$ lb of salmon.

116. Exercise Maria Rivera can walk $3\frac{1}{2}$ mi in 1 h. At this rate, how far can Maria walk in $\frac{1}{3}$ h?

117. Manufacturing A factory worker can assemble a product in $7\frac{1}{2}$ min. How many products can the worker assemble in 1 h?

118. Real Estate A developer purchases $25\frac{1}{2}$ acres of land and plans to set aside 3 acres for an entranceway to a housing development to be built on the property. Each house will be built on a $\frac{3}{4}$-acre plot of land. How many houses does the developer plan to build on the property?

119. Party Planning You are planning a barbecue for 25 people. You want to serve $\frac{1}{4}$-pound hamburger patties to your guests and you estimate each person will eat two hamburgers. How much hamburger meat should you buy for the barbecue?

120. Nutrition According to the Center for Science in the Public Interest, the average teenage boy drinks $3\frac{1}{3}$ cans of soda per day. The average teenage girl drinks $2\frac{1}{3}$ cans of soda per day.

a. The average teenage boy drinks how many cans of soda per week?

b. If a can of soda contains 150 calories, how many calories does the average teenage boy consume each week in soda?

c. How many more cans of soda per week does the average teenage boy drink than the average teenage girl?

121. Wages Find the total wages of an employee who worked $26\frac{1}{2}$ h this week and who earns an hourly wage of $12.

122. Geometry Find the area of a rectangle that has a length of $8\frac{1}{2}$ yd and a width of 5 yd.

123. **Biofuels** See the news clipping at the right. How many bushels of corn produced each year are turned into ethanol?

In the News

A New Source of Energy

Of the 11 billion bushels of corn produced each year, half is converted into ethanol. The majority of new cars are capable of running on E10, a fuel consisting of 10% ethanol and 90% gas.

Source: Time, April 9, 2007

124. **Geometry** What is the area of a rectangular recreational area that has a length of $3\frac{1}{4}$ mi and a width of $1\frac{1}{2}$ mi?

125. **Geometry** A city plans to plant grass seed in a public playground that has the shape of a triangle with a height of 24 m and a base of 20 m. Each bag of grass seed will seed 120 m^2. How many bags of seed should be purchased?

126. **Oceanography** The pressure on a submerged object is given by $P = 15 + \frac{1}{2}D$, where D is the depth in feet and P is the pressure measured in pounds per square inch. Find the pressure on a diver who is at a depth of $12\frac{1}{2}$ ft.

Stephen Frink/Corbis

127. **Hiking** Find the rate of a hiker who walked $4\frac{2}{3}$ mi in $1\frac{1}{3}$ h. Use the equation $r = \frac{d}{t}$, where r is the rate in miles per hour, d is the distance, and t is the time.

128. **Physics** Find the amount of force necessary to push a 75-pound crate across a floor, where the coefficient of friction is $\frac{3}{8}$. Use the equation $F = \mu N$, where F is the force, μ is the coefficient of friction, and N is the weight of the crate. Force is measured in pounds.

Applying the Concepts

129. **Cartography** On a map, two cities are $3\frac{1}{8}$ in. apart. If $\frac{1}{8}$ in. on the map represents 50 mi, what is the number of miles between the two cities?

Amsterdam
London
Frankfurt
Paris
Rome
Athens

130. Determine whether the statement is always true, sometimes true, or never true.

a. Let n be an even natural number. Then $\frac{1}{2}n$ is a whole number.

b. Let n be an odd number. Then $\frac{1}{2}n$ is an improper fraction.

SECTION

2.5 Introduction to Decimals

OBJECTIVE A To read and write decimals 

The price tag on a sweater reads $61.88. The number 61.88 is in **decimal notation.** A number written in decimal notation is often called simply a **decimal.**

A number written in decimal notation has three parts.

61	.	88
Whole-number part	**Decimal point**	**Decimal part**

The decimal part of the number represents a number less than one. For example, $.88 is less than one dollar. The decimal point (.) separates the whole-number part from the decimal part.

The position of a digit in a decimal determines the digit's place value. The place-value chart is extended to the right to show the place values of digits to the right of a decimal point.

In the decimal 458.302719, the position of the digit 7 determines that its place value is ten-thousandths.

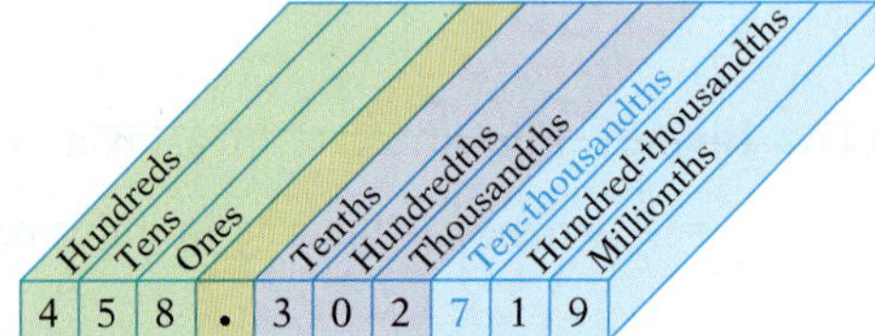

Note the relationship between fractions and numbers written in decimal notation.

seven tenths	seven hundredths	seven thousandths
$\frac{7}{10} = 0.7$	$\frac{7}{100} = 0.07$	$\frac{7}{1000} = 0.007$
1 zero in 10	2 zeros in 100	3 zeros in 1000
1 decimal place in 0.7	2 decimal places in 0.07	3 decimal places in 0.007

To write a decimal in words, write the decimal part of the number as though it were a whole number, and then name the place value of the last digit.

0.9684 nine thousand six hundred eighty-four ten-thousandths

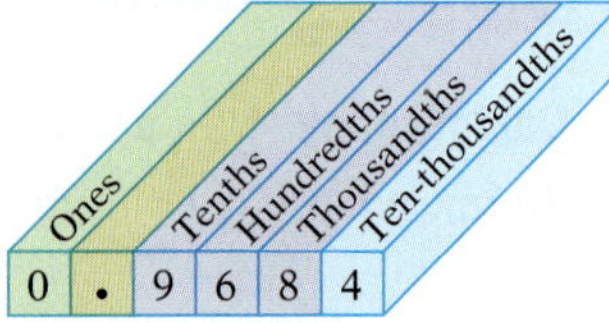

The decimal point in a decimal is read as "and."

372.516 three hundred seventy-two and five hundred sixteen thousandths

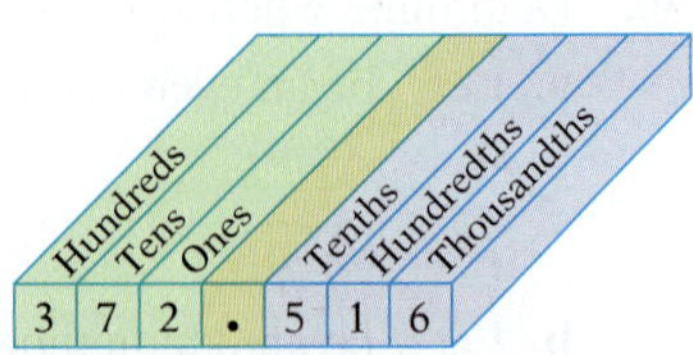

Point of Interest

The idea that all fractions should be represented in tenths, hundredths, and thousandths was presented in 1585 in Simon Stevin's publication *De Thiende* and its French translation, *La Disme,* which was widely read and accepted by the French. This may help to explain why the French accepted the metric system so easily 200 years later.

In *De Thiende,* Stevin argued in favor of his notation by including examples for astronomers, tapestry makers, surveyors, tailors, and the like. He stated that using decimals would enable calculations to be "performed . . . with as much ease as counterreckoning."

To write a decimal in standard form when it is written in words, write the whole number part, replace the word *and* with a decimal point, and write the decimal part so that the last digit is in the given place-value position.

four and twenty-three hundredths
3 is in the hundredths place. 4.23

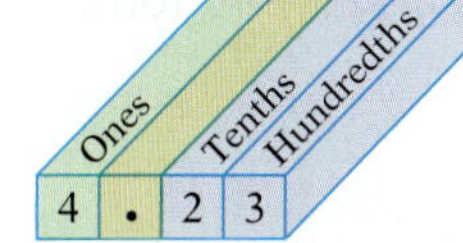

When writing a decimal in standard form, you may need to insert zeros after the decimal point so that the last digit is in the given place-value position.

ninety-one and eight thousandths
8 is in the thousandths place. 91.008

Insert two zeros so that the 8 is in the thousandths place.

sixty-five ten-thousandths
5 is in the ten-thousandths place. 0.0065

Insert two zeros so that the 5 is in the ten-thousandths place.

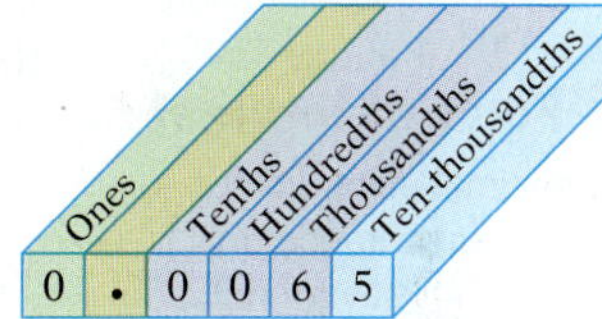

EXAMPLE • 1

Name the place value of the digit 8 in the number 45.687.

Solution

The digit 8 is in the hundredths place.

YOU TRY IT • 1

Name the place value of the digit 4 in the number 907.1342.

Your solution

EXAMPLE • 2

Write $\frac{43}{100}$ as a decimal.

Solution

$\frac{43}{100} = 0.43$ • **Forty-three hundredths**

YOU TRY IT • 2

Write $\frac{501}{1000}$ as a decimal.

Your solution

EXAMPLE • 3

Write 0.589 as a fraction.

Solution

$0.589 = \frac{589}{1000}$ • **589 thousandths**

YOU TRY IT • 3

Write 0.67 as a fraction.

Your solution

EXAMPLE • 4

Write 293.50816 in words.

Solution

Two hundred ninety-three and fifty thousand eight hundred sixteen hundred-thousandths

YOU TRY IT • 4

Write 55.6083 in words.

Your solution

Solutions on p. S6

EXAMPLE • 5

Write twenty-three and two hundred forty-seven millionths in standard form.

Solution

23.000247

YOU TRY IT • 5

Write eight hundred six and four hundred ninety-one hundred-thousandths in standard form.

Your solution

Solution on p. S6

OBJECTIVE B To identify the order relation between two decimals

A whole number can be written as a decimal by writing a decimal point to the right of the last digit. For example,

$$62 = 62. \qquad 497 = 497.$$

You know that \$62 and \$62.00 both represent 62 dollars. Any number of zeros may be written to the right of the decimal point in a whole number without changing the value of the number.

$$62 = 62.00 = 62.0000 \qquad 497 = 497.0 = 497.000$$

Also, any number of zeros may be written to the right of the last digit in a decimal without changing the value of the number.

$$0.8 = 0.80 = 0.800 \qquad 1.35 = 1.350 = 1.3500 = 1.35000 = 1.350000$$

This fact is used to find the order relation between two decimals.

To compare two decimals, write the decimal part of each number so that each has the same number of decimal places. Then compare the two numbers.

Point of Interest

The decimal point did not make its appearance until the early 1600s. Stevin's notation used subscripts with circles around them after each digit: 0 for ones, 1 for tenths (which he called "primes"), 2 for hundredths (called "seconds"), 3 for thousandths ("thirds"), and so on. For example, 1.375 would have been written

1 3 7 5
⓪ ① ② ③

HOW TO • 1 Place the correct symbol, < or >, between the two numbers 0.693 and 0.71.

$0.71 = 0.710$ • **0.693 has 3 decimal places. 0.71 has 2 decimal places. Write 0.71 with 3 decimal places.**

$0.693 < 0.710$ • **Compare 0.693 and 0.710. 693 thousandths < 710 thousandths**

$0.693 < 0.71$ • **Remove the zero written in 0.710.**

HOW TO • 2 Place the correct symbol, < or >, between the two numbers 5.8 and 5.493.

$5.8 = 5.800$ • **Write 5.8 with 3 decimal places.**

$5.800 > 5.493$ • **Compare 5.800 and 5.493. The whole number part (5) is the same. 800 thousandths > 493 thousandths**

$5.8 > 5.493$ • **Remove the extra zeros written in 5.800.**

EXAMPLE • 6

Place the correct symbol, < or >, between the two numbers.

0.039   0.1001

Solution

0.039 = 0.0390

0.0390 < 0.1001

0.039 < 0.1001

YOU TRY IT • 6

Place the correct symbol, < or >, between the two numbers.

0.065   0.0802

Your solution

EXAMPLE • 7

Write the given numbers in order from smallest to largest.

1.01, 1.2, 1.002, 1.1, 1.12

Solution

1.010, 1.200, 1.002, 1.100, 1.120

1.002, 1.010, 1.100, 1.120, 1.200

1.002, 1.01, 1.1, 1.12, 1.2

YOU TRY IT • 7

Write the given numbers in order from smallest to largest.

3.03, 0.33, 0.3, 3.3, 0.03

Your solution

Solutions on p. S6

OBJECTIVE C   To round a decimal to a given place value

In general, rounding decimals is similar to rounding whole numbers except that the digits to the right of the given place value are dropped instead of being replaced by zeros.

If the digit to the right of the given place value is less than 5, that digit and all digits to the right are dropped.

Round 6.9237 to the nearest hundredth.

Given place value (hundredths)

6.9237

3 < 5   Drop the digits 3 and 7.

• 6.9237 rounded to the nearest hundredth is 6.92.

If the digit to the right of the given place value is greater than or equal to 5, increase the digit in the given place value by 1, and drop all digits to its right.

Round 12.385 to the nearest tenth.

Given place value (tenths)

12.385

8 > 5   Increase 3 by 1 and drop all digits to the right of 3.

12.385 rounded to the nearest tenth is 12.4.

HOW TO • 3 Round 0.46972 to the nearest thousandth.

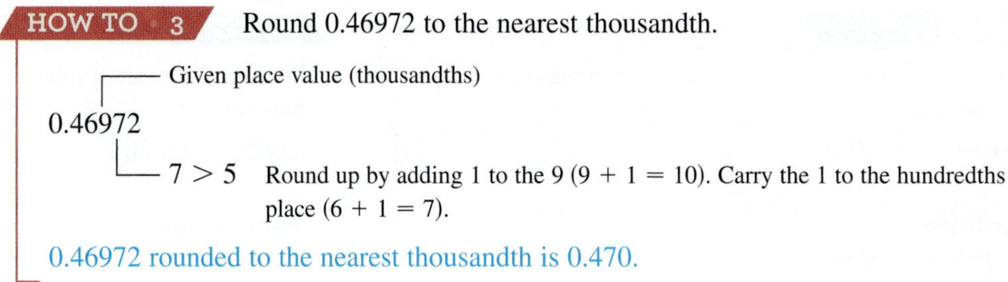

Given place value (thousandths)
0.46972
7 > 5 Round up by adding 1 to the 9 (9 + 1 = 10). Carry the 1 to the hundredths place (6 + 1 = 7).
0.46972 rounded to the nearest thousandth is 0.470.

Note that in this example, the zero in the given place value is not dropped. This indicates that the number is rounded to the nearest thousandth. If we dropped the zero and wrote 0.47, it would indicate that the number was rounded to the nearest hundredth.

EXAMPLE • 8

Round 0.9375 to the nearest thousandth.

Solution

Given place value

0.9375

$5 = 5$

0.9375 rounded to the nearest thousandth is 0.938.

YOU TRY IT • 8

Round 3.675849 to the nearest ten-thousandth.

Your solution

EXAMPLE • 9

Round 2.5963 to the nearest hundredth.

Solution

Given place value

2.5963

$6 > 5$

2.5963 rounded to the nearest hundredth is 2.60.

YOU TRY IT • 9

Round 48.907 to the nearest tenth.

Your solution

EXAMPLE • 10

Round 72.416 to the nearest whole number.

Solution

Given place value

72.416

$4 < 5$

72.416 rounded to the nearest whole number is 72.

YOU TRY IT • 10

Round 31.8652 to the nearest whole number.

Your solution

Solutions on p. S6

OBJECTIVE D To solve application problems

Bettmann/Corbis

Babe Ruth

The table below shows the number of home runs hit, for every 100 times at bat, by four Major League baseball players. Use this table for Example 11 and You Try It 11.

Home Runs Hit for Every 100 At-Bats	
Harmon Killebrew	7.03
Ralph Kiner	7.09
Babe Ruth	8.05
Ted Williams	6.76

Source: Major League Baseball

EXAMPLE • 11

According to the table above, who had more home runs for every 100 times at bat, Ted Williams or Babe Ruth?

Strategy

To determine who had more home runs for every 100 times at bat, compare the numbers 6.76 and 8.05.

Solution

$8.05 > 6.76$

Babe Ruth had more home runs for every 100 at-bats.

YOU TRY IT • 11

According to the table above, who had more home runs for every 100 times at bat, Harmon Killebrew or Ralph Kiner?

Your strategy

Your solution

EXAMPLE • 12

On average, an American goes to the movies 4.56 times per year. To the nearest whole number, how many times per year does an American go to the movies?

Strategy

To find the number, round 4.56 to the nearest whole number.

Solution

4.56 rounded to the nearest whole number is 5.

An American goes to the movies about 5 times per year.

YOU TRY IT • 12

One of the driest cities in the Southwest is Yuma, Arizona, with an average annual precipitation of 2.65 in. To the nearest inch, what is the average annual precipitation in Yuma?

Your strategy

Your solution

Solutions on p. S6

2.5 EXERCISES

OBJECTIVE A To read and write decimals

For Exercises 1 to 6, name the place value of the digit 5.

1. 76.31587 **2.** 291.508 **3.** 432.09157

4. 0.0006512 **5.** 38.2591 **6.** 0.0000853

For Exercises 7 to 14, write the fraction as a decimal.

7. $\frac{3}{10}$ **8.** $\frac{9}{10}$ **9.** $\frac{21}{100}$ **10.** $\frac{87}{100}$

11. $\frac{461}{1000}$ **12.** $\frac{853}{1000}$ **13.** $\frac{93}{1000}$ **14.** $\frac{61}{1000}$

For Exercises 15 to 22, write the decimal as a fraction.

15. 0.1 **16.** 0.3 **17.** 0.47 **18.** 0.59

19. 0.289 **20.** 0.601 **21.** 0.09 **22.** 0.013

For Exercises 23 to 31, write the number in words.

23. 0.37 **24.** 25.6 **25.** 9.4

26. 1.004 **27.** 0.0053 **28.** 41.108

29. 0.045 **30.** 3.157 **31.** 26.04

For Exercises 32 to 39, write the number in standard form.

32. Six hundred seventy-two thousandths

33. Three and eight hundred six ten-thousandths

34. Nine and four hundred seven ten-thousandths

35. Four hundred seven and three hundredths

36. Six hundred twelve and seven hundred four thousandths

37. Two hundred forty-six and twenty-four thousandths

38. Two thousand sixty-seven and nine thousand two ten-thousandths

39. Seventy-three and two thousand six hundred eighty-four hundred-thousandths

40. Suppose the first nonzero digit to the right of the decimal point in a decimal number is in the hundredths place. If the number has three consecutive nonzero digits to the right of the decimal point, and all other digits are zero, what place value names the number?

OBJECTIVE B To identify the order relation between two decimals

For Exercises 41 to 52, place the correct symbol, $<$ or $>$, between the two numbers.

41. 0.16 0.6

42. 0.7 0.56

43. 5.54 5.45

44. 3.605 3.065

45. 0.047 0.407

46. 9.004 9.04

47. 1.0008 1.008

48. 9.31 9.031

49. 7.6005 7.605

50. 4.6 40.6

51. 0.31502 0.3152

52. 0.07046 0.07036

53. Use the inequality symbol $<$ to rewrite the order relation expressed by the inequality $17.2 > 0.172$.

54. Use the inequality symbol $>$ to rewrite the order relation expressed by the inequality $0.0098 < 0.98$.

For Exercises 55 to 60, write the given numbers in order from smallest to largest.

55. 0.39, 0.309, 0.399

56. 0.66, 0.699, 0.696, 0.609

57. 0.24, 0.024, 0.204, 0.0024

58. 1.327, 1.237, 1.732, 1.372

59. 0.06, 0.059, 0.061, 0.0061

60. 21.87, 21.875, 21.805, 21.78

OBJECTIVE C To round a decimal to a given place value

For Exercises 61 to 75, round the number to the given place value.

61. 6.249; tenths

62. 5.398; tenths

63. 21.007; tenths

64. 30.0092; tenths

65. 18.40937; hundredths

66. 413.5972; hundredths

67. 72.4983; hundredths

68. 6.061745; thousandths

69. 936.2905; thousandths

70. 96.8027; whole number

71. 47.3192; whole number

72. 5439.83; whole number

73. 7014.96; whole number

74. 0.023591; ten-thousandths

75. 2.975268; hundred-thousandths

For Exercises 76 and 77, give an example of a decimal number that satisfies the given condition.

76. The number rounded to the nearest tenth is greater than the number rounded to the nearest hundredth.

77. The number rounded to the nearest hundredth is equal to the number rounded to the nearest thousandth.

OBJECTIVE D To solve application problems

78. Boston Marathon Runners in the Boston Marathon run a distance of 26.21875 mi. To the nearest tenth of a mile, find the distance an entrant who completes the Boston Marathon runs.

Frank Siteman/PhotoEdit, Inc.

The table at the right lists National Football League leading lifetime rushers. Use the table for Exercises 79 and 80.

Football Player	*Average Number of Yards per Carry*
Eric Dickerson	4.43
Tony Dorsett	4.34
Walter Payton	4.36
Barry Sanders	4.99
Emmitt Smith	4.24

Source: Pro Football Hall of Fame

79. Football Who had the greater average number of yards per carry, Tony Dorsett or Emmitt Smith?

80. Football Of all the players listed in the table, who has the greatest average number of yards per carry?

81. **Life Expectancy** The average life expectancy in Great Britain is 75.3 years. The average life expectancy in Italy is 75.5 years (*Source:* U.S. Centers for Disease Control). In which country is the average life expectancy higher, Great Britain or Italy?

82. **Measurement** Can a piece of rope 4 ft long be wrapped around the box shown at the right?

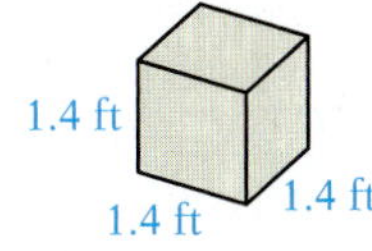

83. **The Olympics** The length of the marathon footrace in the Olympics is 42.195 km. What is the length of this race to the nearest tenth of a kilometer?

84. **Credit Cards** Credit card companies generally require a minimum payment on the balance of the account each month. Use the minimum payment schedule shown at the right to determine the minimum payment due on the given account balances.
a. \$187.93
b. \$342.55
c. \$261.48
d. \$16.99
e. \$310.00
f. \$158.32
g. \$200.10

If the New Balance Is:	*The Minimum Required Payment Is:*
Up to \$20.00	The new balance
\$20.01 to \$200.00	\$20.00
\$200.01 to \$250.00	\$25.00
\$250.01 to \$300.00	\$30.00
\$300.01 to \$350.00	\$35.00
\$350.01 to \$400.00	\$40.00

85. **Shipping and Handling** Shipping and handling charges when ordering online generally are based on the dollar amount of the order. Use the table shown at the right to determine the cost of shipping each order.
a. \$12.42
b. \$23.56
c. \$47.80
d. \$66.91
e. \$35.75
f. \$20.00
g. \$18.25

If the Amount Ordered Is:	*The Shipping and Handling Charge Is:*
\$10.00 and under	\$1.60
\$10.01 to \$20.00	\$2.40
\$20.01 to \$30.00	\$3.60
\$30.01 to \$40.00	\$4.70
\$40.01 to \$50.00	\$6.00
\$50.01 and up	\$7.00

Applying the Concepts

86. Indicate which digits of the number, if any, need not be entered on a calculator.
a. 1.500 **b.** 0.908 **c.** 60.07 **d.** 0.0032

87. Find a number between **a.** 0.1 and 0.2, **b.** 1 and 1.1, and **c.** 0 and 0.005.

88. Provide an example of a situation in which a decimal is always rounded up, even if the digit to the right is less than 5. Provide an example of a situation in which a decimal is always rounded down, even if the digit to the right is 5 or greater than 5. (*Hint:* Think about situations in which money changes hands.)

SECTION

2.6 Operations on Decimals

OBJECTIVE A To add and subtract decimals 

To add decimals, write the numbers so that the decimal points are on a vertical line. Add as you would with whole numbers. Then write the decimal point in the sum directly below the decimal points in the addends.

HOW TO 1 Add: 0.326 + 4.8 + 57.23

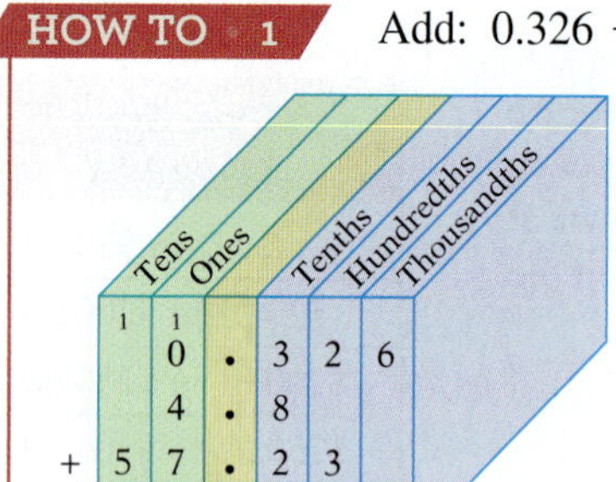

• Note that placing the decimal points on a vertical line ensures that digits of the same place value are added.

The sum is 62.356.

Point of Interest

Try this: Six different numbers are added together and their sum is 11. Four of the six numbers are 4, 3, 2, and 1. Find the other two numbers.

HOW TO 2 Find the sum of 0.64, 8.731, 12, and 5.9.

```
 1 2
  0.64
  8.731
 12.
+ 5.9
 27.271
```

• Arrange the numbers vertically, placing the decimal points on a vertical line.

• Add the numbers in each column. Write the decimal point in the sum directly below the decimal points in the addends.

The sum is 27.271.

To subtract decimals, write the numbers so that the decimal points are on a vertical line. Subtract as you would with whole numbers. Then write the decimal point in the difference directly below the decimal point in the subtrahend.

HOW TO 3 Subtract and check: 31.642 − 8.759

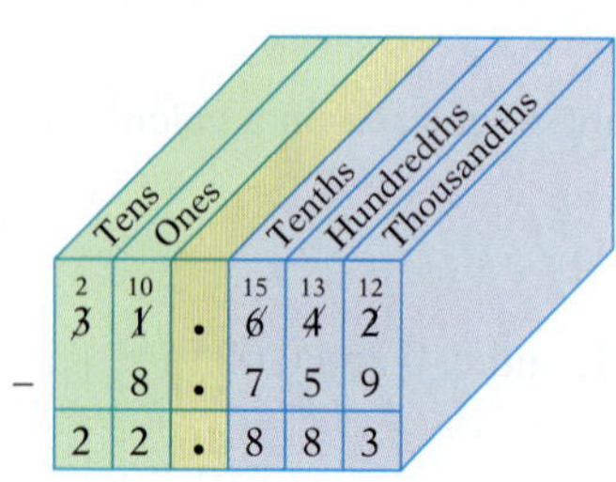

• Note that placing the decimal points on a vertical line ensures that digits of the same place value are subtracted.

The difference is 22.883.

Check:

Subtrahend	8.759
+ Difference	+22.883
= Minuend	31.642

HOW TO • 4 Subtract and check: 5.4 − 1.6832

$$\begin{array}{r} 5.4000 \\ -1.6832 \\ \hline \end{array}$$

• Insert zeros in the minuend so that it has the same number of decimal places as the subtrahend.

$$\begin{array}{r} \overset{4\ 13\ 9\ 9\ 10}{\cancel{5}.\cancel{4}\cancel{0}\cancel{0}\cancel{0}} \\ -1.6832 \\ \hline 3.7168 \end{array} \qquad \textit{Check: } \begin{array}{r} 1.6832 \\ +3.7168 \\ \hline 5.4000 \end{array}$$

• Subtract and then check.

Point of Interest

Try this brain teaser. You have two U.S. coins that add up to $.55. One is not a nickel. What are these two coins?

Recall that to estimate the answer to a calculation, round each number to the highest place value of the number; the first digit of each number will be nonzero and all other digits will be zero. Perform the calculation using the rounded numbers.

HOW TO • 5 Estimate the sum of 23.037 and 16.7892.

$$\begin{array}{rcr} 23.037 & \longrightarrow & 20 \\ 16.7892 & \longrightarrow & +20 \\ \hline & & 40 \end{array}$$

• Round each number to the nearest ten.

• Add the rounded numbers.

$$\begin{array}{r} 23.037 \\ +\ 16.7892 \\ \hline 39.8262 \end{array}$$

40 is an estimate of the sum of 23.037 and 16.7892. Note that 40 is very close to the actual sum of 39.8262.

When a number in an estimation is a decimal less than 1, round the decimal so that there is one nonzero digit.

HOW TO • 6 Estimate the difference between 4.895 and 0.6193.

$$\begin{array}{rcr} 4.895 & \longrightarrow & 5.0 \\ 0.6193 & \longrightarrow & -\ 0.6 \\ \hline & & 4.4 \end{array}$$

• Round 4.895 to the nearest one. Round 0.6193 to the nearest tenth.

• Subtract the rounded numbers.

$$\begin{array}{r} 4.8950 \\ -0.6193 \\ \hline 4.2757 \end{array}$$

4.4 is an estimate of the difference between 4.895 and 0.6193. It is close to the actual difference of 4.2757.

EXAMPLE • 1

Add: 35.8 + 182.406 + 71.0934

Solution

$$\begin{array}{r} \overset{1\ 1}{35.8} \\ 182.406 \\ +\ 71.0934 \\ \hline 289.2994 \end{array}$$

YOU TRY IT • 1

Add: 8.64 + 52.7 + 0.39105

Your solution

Solution on p. S6

EXAMPLE • 2

Subtract and check: $73 - 8.16$

Solution

$$\begin{array}{r} \overset{6\ 12}{\not7\not3}.\overset{9\ 10}{\not0\not0} \\ -\ 8.16 \\ \hline 64.84 \end{array} \qquad \begin{array}{r} \textit{Check:} \\ 8.16 \\ +64.84 \\ \hline 73.00 \end{array}$$

YOU TRY IT • 2

Subtract and check: $25 - 4.91$

Your solution

EXAMPLE • 3

Estimate the sum of 0.3927, 0.4856, and 0.2104.

Solution

$$\begin{array}{lr} 0.3927 \longrightarrow & 0.4 \\ 0.4856 \longrightarrow & 0.5 \\ 0.2104 \longrightarrow & +0.2 \\ \hline & 1.1 \end{array}$$

YOU TRY IT • 3

Estimate the sum of 6.514, 8.903, and 2.275.

Your solution

EXAMPLE • 4

Evaluate $x + y + z$ when $x = 1.6$, $y = 7.9$, and $z = 4.8$.

Solution

$x + y + z$

$$\begin{aligned} 1.6 + 7.9 + 4.8 &= 9.5 + 4.8 \\ &= 14.3 \end{aligned}$$

YOU TRY IT • 4

Evaluate $x + y + z$ when $x = 7.84$, $y = 3.05$, and $z = 2.19$.

Your solution

Solutions on pp. S6–S7

OBJECTIVE B To multiply decimals 

Decimals are multiplied as though they were whole numbers; then the decimal point is placed in the product. Writing the decimals as fractions shows where to write the decimal point in the product.

$$0.4 \cdot 2 = \frac{4}{10} \cdot \frac{2}{1} = \frac{8}{10} = 0.8$$

1 decimal place in 0.4   1 decimal place in 0.8

$$0.4 \cdot 0.2 = \frac{4}{10} \cdot \frac{2}{10} = \frac{8}{100} = 0.08$$

1 decimal place in 0.4
1 decimal place in 0.2   2 decimal places in 0.08

$$0.4 \cdot 0.02 = \frac{4}{10} \cdot \frac{2}{100} = \frac{8}{1000} = 0.008$$

1 decimal place in 0.4
2 decimal places in 0.02
3 decimal places in 0.008

Integrating Technology

Scientific calculators have a floating decimal point. This means that the decimal point is automatically placed in the answer. For example, for the product at the right, enter

32 . 41 X 7 . 6 =

The display reads 246.316, with the decimal point in the correct position.

To multiply decimals, multiply the numbers as you would whole numbers. Then write the decimal point in the product so that the number of decimal places in the product is the sum of the numbers of decimal places in the factors.

HOW TO • 7 Multiply: (32.41)(7.6)

```
   32.41      2 decimal places
 ×   7.6      1 decimal place
   19446
  22687
 246.316      3 decimal places
```

Estimating the product of 32.41 and 7.6 shows that the decimal point has been correctly placed in the example above.

32.41 ⟶ 30 • **Round 32.41 to the nearest ten.**
7.6 ⟶ × 8 • **Round 7.6 to the nearest one.**
240 • **Multiply the two numbers.**

240 is an estimate of (32.41)(7.6). It is close to the actual product 246.316.

HOW TO • 8 Multiply: 0.061(0.08)

```
   0.061      3 decimal places
 ×  0.08      2 decimal places
 0.00488      5 decimal places
```

• **Insert two zeros between the 4 and the decimal point so that there are 5 decimal places in the product.**

To multiply a decimal by a power of 10 (10, 100, 1000, . . .), move the decimal point to the right the same number of places as there are zeros in the power of 10.

2.7935 · 10 = 27.935
1 zero — 1 decimal place

2.7935 · 100 = 279.35
2 zeros — 2 decimal places

2.7935 · 1000 = 2793.5
3 zeros — 3 decimal places

2.7935 · 10,000 = 27,935.
4 zeros — 4 decimal places

2.7935 · 100,000 = 279,350.
5 zeros — 5 decimal places

• **A zero must be inserted before the decimal point.**

Note that if the power of 10 is written in exponential notation, the exponent indicates how many places to move the decimal point.

$2.7935 \cdot 10^1 = 27.935$

1 decimal place

$2.7935 \cdot 10^2 = 279.35$

2 decimal places

$2.7935 \cdot 10^3 = 2793.5$

3 decimal places

$2.7935 \cdot 10^4 = 27{,}935.$

4 decimal places

$2.7935 \cdot 10^5 = 279{,}350.$

5 decimal places

HOW TO • 9 Find the product of 64.18 and 10^3.

$64.18 \cdot 10^3 = 64{,}180$

- **The exponent on 10 is 3. Move the decimal point in 64.18 three places to the right.**

EXAMPLE • 5

Multiply: 0.00073(0.052)

Solution

$$\begin{array}{r} 0.00073 \\ \times \quad 0.052 \\ \hline 146 \\ 365 \\ \hline 0.00003796 \end{array}$$

YOU TRY IT • 5

Multiply: 0.000081(0.025)

Your solution

EXAMPLE • 6

Estimate the product of 0.7639 and 0.2188.

Solution

$0.7639 \longrightarrow 0.8$

$0.2188 \longrightarrow \times\, 0.2$

0.16

YOU TRY IT • 6

Estimate the product of 6.407 and 0.959.

Your solution

EXAMPLE • 7

What is 835.294 multiplied by 1000?

Solution

$835.294 \cdot 1000 = 835{,}294$

YOU TRY IT • 7

Find the product of 1.756 and 10^4.

Your solution

Solutions on p. S7

EXAMPLE • 8

Evaluate $50ab$ when $a = 0.9$ and $b = 0.2$.

Solution

$50ab$

$50(0.9)(0.2) = 45(0.2)$
$= 9$

YOU TRY IT • 8

Evaluate $25xy$ when $x = 0.8$ and $y = 0.6$.

Your solution

Solution on p. S7

OBJECTIVE C To divide decimals 

Point of Interest

Benjamin Banneker (1731–1806) was the first African American to earn distinction as a mathematician and a scientist. He was on the survey team that determined the boundaries of Washington, D.C. The mathematics of surveying requires extensive use of decimals.

To divide decimals, move the decimal point in the divisor to the right so that the divisor is a whole number. Move the decimal point in the dividend the same number of places to the right. Place the decimal point in the quotient directly above the decimal point in the dividend. Then divide as you would with whole numbers.

HOW TO • 10 Divide: $29.585 \div 4.85$

$4.85.\overline{)29.58.5}$

Move the decimal point 2 places to the right in the divisor. Move the decimal point 2 places to the right in the dividend. Place the decimal point in the quotient. Then divide as shown at the right.

$$\begin{array}{r} 6.1 \\ 485.\overline{)2958.5} \\ -2910 \\ \hline 48\,5 \\ -48\,5 \\ \hline 0 \end{array}$$

Moving the decimal point the same number of places in the divisor and the dividend does not change the quotient because the process is the same as multiplying the numerator and denominator of a fraction by the same number. For the last example,

$$4.85\overline{)29.585} = \frac{29.585}{4.85} = \frac{29.585 \cdot 100}{4.85 \cdot 100} = \frac{2958.5}{485} = 485\overline{)2958.5}$$

In division of decimals, rather than writing the quotient with a remainder, we usually round the quotient to a specified place value. **The symbol** $\approx$, which is read **"is approximately equal to,"** is used to indicate that the quotient is an approximate value after being rounded.

HOW TO • 11 Divide and round to the nearest tenth: $0.86 \div 0.7$

$$\begin{array}{r} 1.22 \approx 1.2 \\ 0.7.\overline{)0.8.60} \\ -7 \\ \hline 1\,6 \\ -1\,4 \\ \hline 20 \\ -14 \\ \hline 6 \end{array}$$

← To round the quotient to the nearest tenth, the division must be carried to the hundredths place. Therefore, zeros must be inserted in the dividend so that the quotient has a digit in the hundredths place.

To divide a decimal by a power of 10 (10, 100, 1000, 10,000, . . .), move the decimal point to the left the same number of places as there are zeros in the power of 10.

462.81 ÷ 10 = 46.281
1 zero — 1 decimal place

462.81 ÷ 100 = 4.6281
2 zeros — 2 decimal places

462.81 ÷ 1000 = 0.46281
3 zeros — 3 decimal places

462.81 ÷ 10,000 = 0.046281 • A zero must be inserted between the decimal point and the 4.
4 zeros — 4 decimal places

462.81 ÷ 100,000 = 0.0046281 • Two zeros must be inserted between the decimal point and the 4.
5 zeros — 5 decimal places

If the power of 10 is written in exponential notation, the exponent indicates how many places to move the decimal point.

$462.81 \div 10^1 = 46.281$
1 decimal place

$462.81 \div 10^2 = 4.6281$
2 decimal places

$462.81 \div 10^3 = 0.46281$
3 decimal places

$462.81 \div 10^4 = 0.046281$
4 decimal places

$462.81 \div 10^5 = 0.0046281$
5 decimal places

HOW TO • 12 Find the quotient of 3.59 and 100.

3.59 ÷ 100 = 0.0359 • **There are two zeros in 100. Move the decimal point in 3.59 two places to the left.**

HOW TO • 13 What is the quotient of 64.79 and 10^4?

$64.79 \div 10^4 = 0.006479$ • **The exponent on 10 is 4. Move the decimal point in 64.79 four places to the left.**

EXAMPLE 9

Divide: 431.97 ÷ 7.26

Solution

$$\begin{array}{r} 59.5 \\ 7.26.\overline{)431.97.0} \\ -363\,0 \\ \hline 68\,97 \\ -65\,34 \\ \hline 3\,630 \\ -3\,630 \\ \hline 0 \end{array}$$

YOU TRY IT 9

Divide: 314.746 ÷ 6.53

Your solution

EXAMPLE 10

Estimate the quotient of 8.37 and 0.219.

Solution

8.37 ⟶ 8

0.219 ⟶ 0.2

$8 \div 0.2 = 40$

YOU TRY IT 10

Estimate the quotient of 62.7 and 3.45.

Your solution

EXAMPLE 11

Divide and round to the nearest hundredth:
448.2 ÷ 53

Solution

$$\begin{array}{r} 8.456 \approx 8.46 \\ 53\overline{)448.200} \\ -424 \\ \hline 24\,2 \\ -21\,2 \\ \hline 3\,00 \\ -2\,65 \\ \hline 350 \\ -318 \\ \hline 32 \end{array}$$

YOU TRY IT 11

Divide and round to the nearest thousandth:
519.37 ÷ 86

Your solution

EXAMPLE 12

Find the quotient of 592.4 and 10^4.

Solution

$592.4 \div 10^4 = 0.05924$

YOU TRY IT 12

What is 63.7 divided by 100?

Your solution

EXAMPLE 13

Evaluate $\frac{x}{y}$ when $x = 76.8$ and $y = 0.8$.

Solution

$\frac{x}{y}$

$\frac{76.8}{0.8} = 96$

YOU TRY IT 13

Evaluate $\frac{x}{y}$ when $x = 40.6$ and $y = 0.7$.

Your solution

Solutions on p. S7

OBJECTIVE D To convert between decimals and fractions and identify the order relation between a decimal and a fraction

Because the fraction bar can be read "divided by," any fraction can be written as a decimal. To write a fraction as a decimal, divide the numerator of the fraction by the denominator.

Take Note

The fraction bar can be read "divided by."

$\frac{3}{4} = 3 \div 4$

Dividing the numerator by the denominator results in a remainder of 0. The decimal 0.75 is a terminating decimal.

HOW TO 14 Convert $\frac{3}{4}$ to a decimal.

$$\begin{array}{r} 0.75 \\ 4\overline{)3.00} \\ -2\,8 \\ \hline 20 \\ -20 \\ \hline 0 \end{array}$$

0.75 ← This is a **terminating decimal.**

0 ← The remainder is zero.

$\frac{3}{4} = 0.75$

HOW TO 15 Convert $\frac{5}{11}$ to a decimal.

$$\begin{array}{r} 0.4545 \\ 11\overline{)5.0000} \\ -4\,4 \\ \hline 60 \\ -55 \\ \hline 50 \\ -44 \\ \hline 60 \\ -55 \\ \hline 5 \end{array}$$

0.4545 ← This is a **repeating decimal.**

5 ← The remainder is never zero.

$\frac{5}{11} = 0.\overline{45}$ The bar over the digits 45 is used to show that these digits repeat.

Take Note

No matter how far we carry out the division, the remainder is never zero. The decimal $0.\overline{45}$ is a repeating decimal.

HOW TO 16 Convert $2\frac{4}{9}$ to a decimal.

$9\overline{)4.000}$ gives $0.444 = 0.\overline{4}$

- **Write the fractional part of the mixed number as a decimal. Divide the numerator by the denominator.**

$2\frac{4}{9} = 2.\overline{4}$

- **The whole number part of the mixed number is the whole number part of the decimal.**

To convert a decimal to a fraction, remove the decimal point and place the decimal part over a denominator equal to the place value of the last digit in the decimal.

hundredths: $0.57 = \frac{57}{100}$

hundredths: $7.65 = 7\frac{65}{100} = 7\frac{13}{20}$

tenths: $8.6 = 8\frac{6}{10} = 8\frac{3}{5}$

Integrating Technology

Some calculators *truncate* a decimal number that exceeds the calculator display. This means that the digits beyond the calculator's display are not shown. For this type of calculator, $\frac{2}{3}$ would be shown as 0.66666666. Other calculators *round* a decimal number when the calculator display is exceeded. For this type of calculator, $\frac{2}{3}$ would be shown as 0.66666667.

HOW TO • 17 Convert 4.375 to a fraction.

$4.375 = 4\frac{375}{1000}$ • The 5 in 4.375 is in the thousandths place. Write 0.375 as a fraction with a denominator of 1000.

$= 4\frac{3}{8}$ • Simplify the fraction.

To find the order relation between a fraction and a decimal, first rewrite the fraction as a decimal. Then compare the two decimals.

HOW TO • 18 Find the order relation between $\frac{6}{7}$ and 0.855.

$\frac{6}{7} \approx 0.8571$ • Write the fraction as a decimal. Round to one more place value than the given decimal. (0.855 has 3 decimal places; round to 4 decimal places.)

$0.8571 > 0.8550$ • Compare the two decimals.

$\frac{6}{7} > 0.855$ • Replace the decimal approximation of $\frac{6}{7}$ with $\frac{6}{7}$.

EXAMPLE • 14

Convert $\frac{5}{8}$ to a decimal.

Solution

$$8\overline{)5.000} = 0.625 \qquad \frac{5}{8} = 0.625$$

YOU TRY IT • 14

Convert $\frac{4}{5}$ to a decimal.

Your solution

EXAMPLE • 15

Convert $3\frac{1}{3}$ to a decimal.

Solution

Write $\frac{1}{3}$ as a decimal.

$$3\overline{)1.000} = 0.333 = 0.\overline{3}$$

$$3\frac{1}{3} = 3.\overline{3}$$

YOU TRY IT • 15

Convert $1\frac{5}{6}$ to a decimal.

Your solution

EXAMPLE • 16

Convert 7.25 to a fraction.

Solution

$$7.25 = 7\frac{25}{100} = 7\frac{1}{4}$$

YOU TRY IT • 16

Convert 6.2 to a fraction.

Your solution

Solutions on p. S7

EXAMPLE • 17

Place the correct symbol, < or >, between the two numbers.

$0.845 \quad \frac{5}{6}$

Solution

$\frac{5}{6} \approx 0.8333$

$0.8450 > 0.8333$

$0.845 > \frac{5}{6}$

YOU TRY IT • 17

Place the correct symbol, < or >, between the two numbers.

$0.588 \quad \frac{7}{12}$

Your solution

Solution on p. S7

OBJECTIVE E To solve application problems and use formulas

EXAMPLE • 18

A 1-year subscription to a monthly magazine costs \$93. The price of each issue at the newsstand is \$9.80. How much would you save per issue by buying a year's subscription rather than buying each issue at the newsstand?

Strategy

To find the amount saved:

- Find the subscription price per issue by dividing the cost of the subscription (93) by the number of issues (12).
- Subtract the subscription price per issue from the newsstand price (9.80).

Solution

$$\begin{array}{r} 7.75 \\ 12\overline{)93.00} \\ -84 \\ \hline 9\,0 \\ -8\,4 \\ \hline 60 \\ -60 \\ \hline 0 \end{array} \qquad \begin{array}{r} 9.80 \\ -7.75 \\ \hline 2.05 \end{array}$$

The savings would be \$2.05 per issue.

YOU TRY IT • 18

You hand a postal clerk a ten-dollar bill to pay for the purchase of twelve 44¢ stamps. How much change do you receive?

Your strategy

Your solution

Solution on p. S7

EXAMPLE • 19

An overseas flight charges \$12.80 for each kilogram or part of a kilogram over 50 kg of luggage weight. How much extra must be paid for three pieces of luggage weighing 21.4 kg, 19.3 kg, and 16.8 kg?

Strategy

To find the extra charge:

- Add the three weights (21.4, 19.3, and 16.8) to find the total weight of the luggage.
- Subtract 50 kg from the total weight of the luggage to find the excess weight.
- Round the difference up to the nearest whole number.
- Multiply the charge per kilogram of excess weight (12.80) by the excess weight.

Solution

$21.4 + 19.3 + 16.8 = 57.5$

$57.5 - 50 = 7.5$

7.5 rounded up to the nearest whole number is 8.

$12.80(8) = 102.40$

The extra charge for the luggage is \$102.40.

YOU TRY IT • 19

A health food store buys nuts in 100-pound containers and repackages the nuts in cellophane bags for resale. Each cellophane bag costs \$.06, 2 lb of nuts are placed in each bag, and each bag of nuts is then sold for \$12.50. Find the profit on a 100-pound container of nuts costing \$475.

Your strategy

Your solution

EXAMPLE • 20

Use the formula $P = BF$, where P is the insurance premium, B is the base rate, and F is the rating factor, to find the insurance premium due on an insurance policy with a base rate of \$342.50 and a rating factor of 2.2.

Strategy

To find the insurance premium due, replace B by 342.50 and F by 2.2 in the given formula and solve for P.

Solution

$P = BF$
$P = 342.50(2.2)$
$P = 753.50$

The insurance premium due is \$753.50.

YOU TRY IT • 20

Use the formula $P = BF$, where P is the insurance premium, B is the base rate, and F is the rating factor, to find the insurance premium due on an insurance policy with a base rate of \$276.25 and a rating factor of 1.8.

Your strategy

Your solution

Solutions on p. S7

2.6 EXERCISES

OBJECTIVE A **To add and subtract decimals**

For Exercises 1 to 9, add or subtract.

1. $1.864 + 39 + 25.0781$

2. $2.04 + 35.6 + 4.918$

3. $35.9 + 8.217 + 146.74$

4. $12 + 73.59 + 6.482$

5. $36.47 - 15.21$

6. $85.69 - 2.13$

7. $28 - 6.74$

8. $5 - 1.386$

9. $6.02 - 3.252$

10. Find the sum of 2.536, 14.97, 8.014, and 21.67.

11. Find the total of 6.24, 8.573, 19.06, and 22.488.

12. What is 6.9217 decreased by 3.4501?

13. What is 8.9 less than 62.57?

For Exercises 14 to 22, estimate by rounding. Then find the exact answer.

14. $45.06 + 80.71$

15. $6.408 + 5.917$

16. $0.24 + 0.38 + 0.96$

17. $56.87 - 23.24$

18. $6.272 - 1.848$

19. $0.931 - 0.628$

20. $5.37 + 26.49$

21. $87.65 - 49.032$

22. $387.6 - 54.92$

For Exercises 23 and 24, evaluate the variable expression $x + y + z$ for the given values of x, y, and z.

23. $x = 41.33$; $y = 26.095$; $z = 70.08$

24. $x = 6.059$; $y = 3.884$; $z = 15.71$

For Exercises 25 to 27, evaluate the variable expression $x - y$ for the given values of x and y.

25. $x = 43.29$; $y = 18.76$

26. $x = 6.029$; $y = 4.708$

27. $x = 16.329$; $y = 4.54$

For Exercises 28 to 30, use the relationship between addition and subtraction to write the subtraction problem you would use to find the missing addend.

28. ______ + 2.325 = 7.01 **29.** 5.392 + ______ = 8.07 **30.** ______ + 8.967 = 19.35

OBJECTIVE B To multiply decimals

For Exercises 31 to 34, multiply.

31. 0.9(0.3) **32.** (3.4)(0.5) **33.** (0.72)(3.7) **34.** 8.29(0.004)

35. What is the product of 5.92 and 100? **36.** What is 1000 times 4.25?

37. Find 0.82 times 10^2. **38.** Find the product of 6.71 and 10^4.

39. A number is rounded to the nearest thousandth. What is the smallest power of 10 the number must be multiplied by to give a product that is a whole number?

For Exercises 40 to 45, estimate by rounding. Then find the exact answer.

40. 86.4(4.2) **41.** (9.81)(0.77) **42.** 0.238(8.2)

43. (6.88)(9.97) **44.** (8.432)(0.043) **45.** 28.45(1.13)

For Exercises 46 to 50, evaluate the expression for the given values of the variables.

46. xy, when $x = 5.68$ and $y = 0.2$ **47.** ab, when $a = 6.27$ and $b = 8$ **48.** $40c$, when $c = 2.5$

49. $10t$, when $t = 4.8$ **50.** xy, when $x = 3.71$ and $y = 2.9$

OBJECTIVE C To divide decimals

For Exercises 51 to 54, divide.

51. 16.15 ÷ 0.5 **52.** 7.02 ÷ 3.6 **53.** 27.08 ÷ 0.4 **54.** 8.919 ÷ 0.9

For Exercises 55 to 58, divide. Round to the nearest tenth.

55. $55.63 \div 8.8$ **56.** $1.873 \div 1.4$ **57.** $52.8 \div 9.1$ **58.** $6.824 \div 0.053$

For Exercises 59 to 62, divide. Round to the nearest hundredth.

59. $6.457 \div 8$ **60.** $19.07 \div 0.54$ **61.** $0.0416 \div 0.53$ **62.** $31.792 \div 0.86$

63. Find the quotient of 52.78 and 10.

64. What is 37,942 divided by 1000?

65. What is the quotient of 48.05 and 10^2?

66. Find 9.407 divided by 10^3.

67. A four-digit whole number is divided by 1000. Is the quotient less than 1 or greater than 1?

For Exercises 68 to 75, estimate by rounding. Then divide and round to the nearest hundredth.

68. $42.43 \div 3.8$ **69.** $678 \div 0.71$ **70.** $6.398 \div 5.5$ **71.** $0.994 \div 0.456$

72. $1.237 \div 0.021$ **73.** $421.093 \div 4.087$ **74.** $33.14 \div 4.6$ **75.** $129.38 \div 4.47$

For Exercises 76 to 81, evaluate the variable expression $\frac{x}{y}$ for the given values of x and y.

76. $x = 52.8; y = 0.4$ **77.** $x = 3.542; y = 0.7$ **78.** $x = 2.436; y = 0.6$

79. $x = 0.648; y = 2.7$ **80.** $x = 26.22; y = 6.9$ **81.** $x = 8.034; y = 3.9$

OBJECTIVE D To convert between decimals and fractions and identify the order relation between a decimal and a fraction

For Exercises 82 to 96, convert the fraction to a decimal. Place a bar over repeating digits of a repeating decimal.

82. $\frac{3}{8}$ **83.** $\frac{7}{15}$ **84.** $\frac{8}{11}$ **85.** $\frac{9}{16}$ **86.** $\frac{7}{12}$

87. $\frac{5}{3}$ **88.** $\frac{7}{4}$ **89.** $2\frac{3}{4}$ **90.** $1\frac{1}{2}$ **91.** $3\frac{2}{9}$

92. $4\frac{1}{6}$ **93.** $\frac{3}{25}$ **94.** $2\frac{1}{4}$ **95.** $6\frac{3}{5}$ **96.** $3\frac{8}{9}$

For Exercises 97 to 100, without actually doing any division, state whether the decimal equivalent of the given fraction is greater than 1 or less than 1.

97. $\frac{54}{57}$ **98.** $\frac{176}{129}$ **99.** $\frac{88}{80}$ **100.** $\frac{2007}{2008}$

For Exercises 101 to 115, convert the decimal to a fraction.

101. 0.6 **102.** 0.2 **103.** 0.25 **104.** 0.75 **105.** 0.48

106. 0.125 **107.** 0.325 **108.** 2.5 **109.** 3.4 **110.** 4.55

111. 9.95 **112.** 1.72 **113.** 5.68 **114.** 0.045 **115.** 0.085

For Exercises 116 to 127, place the correct symbol, $<$ or $>$, between the two numbers.

116. $\frac{9}{10}$ 0.89 **117.** $\frac{7}{20}$ 0.34 **118.** $\frac{4}{5}$ 0.803 **119.** $\frac{3}{4}$ 0.706

120. 0.444 $\frac{4}{9}$ **121.** 0.72 $\frac{5}{7}$ **122.** 0.13 $\frac{3}{25}$ **123.** 0.25 $\frac{13}{50}$

124. $\frac{5}{16}$ 0.312 **125.** $\frac{7}{18}$ 0.39 **126.** $\frac{10}{11}$ 0.909 **127.** $\frac{8}{15}$ 0.543

OBJECTIVE E To solve application problems and use formulas

128. You have \$30 to spend, and you make purchases that cost \$6.74 and \$13.68. Which expressions correctly represent the amount of money you have left?
(i) 30 − 6.74 + 13.68 (ii) (6.74 + 13.68) − 30
(iii) 30 − (6.74 + 13.68) (iv) 30 − 6.74 − 13.68

129. Salaries If you earn an annual salary of \$47,619, what is your monthly salary?

130. Education The graph at the right shows where U.S. children in grades K–12 are being educated. Figures are in millions of children.

a. Find the total number of children in grades K–12.

b. How many more children are being educated in public school than in private school?

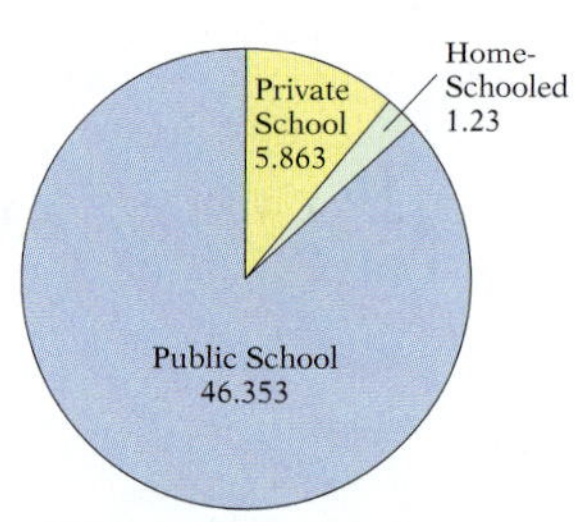

Where Children in Grades K–12 are Being Educated in the United States

Source: U.S. Department of Education; Home School Legal Defense Association

131. Banking You had a balance of \$347.08 in your checking account. You then made a deposit of \$189.53 and wrote a check for \$62.89. Find the new balance in your checking account.

132. Income A bookkeeper earns a salary of \$660 for a 40-hour week. This week the bookkeeper worked 6 h of overtime at a rate of \$24.75 for each hour of overtime worked. Find the bookkeeper's total income for the week.

133. Computers The table below shows the average number of hours per week that students use a computer. On average, how many more hours per year does a second-grade student use a computer than a fifth-grade student?

Grade Level	*Average Number of Hours of Computer Use Per Week*
Pre Kindergarten – Kindergarten	3.9
1st – 3rd	4.9
4th – 6th	4.2
7th – 8th	6.9
9th – 12th	6.7

Source: Find/SVP American Learning Household Survey

134. Business For \$135, a druggist purchases 5 L of cough syrup and repackages it in 250-milliliter bottles. Each bottle costs the druggist \$.55. Each bottle of cough syrup is sold for \$11.89. Find the profit on the 5 L of cough syrup. (*Hint:* There are 1000 milliliters in 1 liter.)

135. Geometry The length of each side of a square is 3.5 ft. Find the perimeter of the square. Use the formula $P = 4s$.

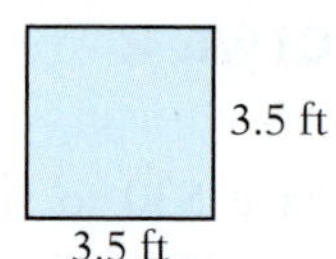

136. Geometry Find the perimeter of a rectangle that measures 4.5 in. by 3.25 in. Use the formula $P = 2L + 2W$.

137. Geometry Find the perimeter of a rectangle that measures 2.8 m by 6.4 m. Use the formula $P = 2L + 2W$.

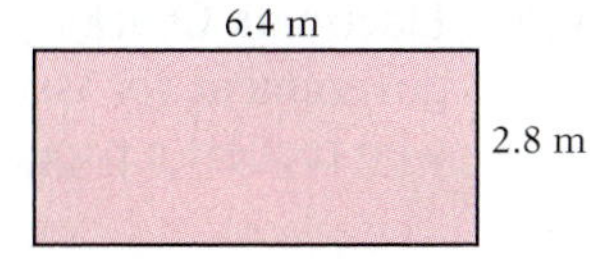

138. Geometry Find the area of a rectangle that measures 4.5 in. by 3.25 in. Use the formula $A = LW$.

139. Geometry Find the area of a rectangle that has a length of 7.8 cm and a width of 4.6 cm. Use the formula $A = LW$.

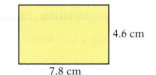

140. Geometry Find the perimeter of a triangle with sides that measure 2.8 m, 4.75 m, and 6.4 m. Use the formula $P = a + b + c$.

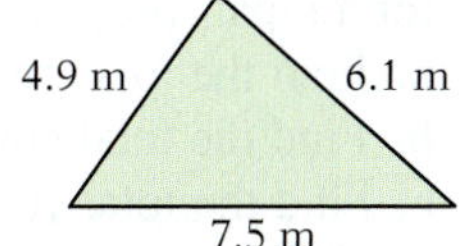

141. Geometry The lengths of three sides of a triangle are 7.5 m, 6.1 m, and 4.9 m. Find the perimeter of the triangle. Use the formula $P = a + b + c$.

142. Markup Use the formula $M = S - C$, where M is the markup on a consumer product, S is the selling price, and C is the cost of the product to the business, to find the markup on a product that cost a business \$1653.19 and has a selling price of \$2231.81.

143. Accounting The amount of an employee's earnings that is subject to federal withholding is called federal earnings. Find the federal earnings for an employee who earns \$694.89 and has a withholding allowance of \$132.69. Use the formula $F = E - W$, where F is the federal earnings, E is the employee's earnings, and W is the withholding allowance.

144. Consumerism Use the formula $M = \frac{C}{N}$, where M is the cost per mile for a rental car, C is the total cost, and N is the number of miles driven, to find the cost per mile when the total cost of renting a car is \$260.16 and you drive the car 542 mi.

145. Physics Find the force exerted on a falling object that has a mass of 4.25 kg. Use the formula $F = ma$, where F is the force exerted by gravity on a falling object, m is the mass of the object, and a is the acceleration due to gravity. The acceleration due to gravity is 9.8 m/s^2 (meters per second squared). The force is measured in newtons.

146. Utilities Find the cost of operating an 1800-watt TV set for 5 h at a cost of \$.06 per kilowatt-hour. Use the formula $c = 0.001wtk$, where c is the cost of operating an appliance, w is the number of watts, t is the time in hours, and k is the cost per kilowatt-hour.

147. Home Equity Find the equity on a home that is valued at \$225,000 when the homeowner has \$167,853.25 in loans on the property. Use the formula $E = V - L$, where E is the equity, V is the value of the home, and L is the loan amount on the property.

148. Electronic Checks See the news clipping at the right. Find the added cost to the government for issuing paper checks to the 4 million Social Security recipients who request a paper check because they do not have a bank account.

In the News

Paper Checks Cost Government Millions

The federal government's cost to issue a paper check is \$.89, while the cost for an electronic check is \$.09.

Source: finance.yahoo.com

149. Transportation A taxi costs \$2.50 and \$.20 for each $\frac{1}{8}$ mi driven. Find the cost of hiring a taxi to get from the airport to the hotel, a distance of $5\frac{1}{2}$ mi.

150. Business The table at the right lists three pieces of steel required for a repair project.

a. Find the total cost of grade 1.
b. Find the total cost of grade 2.
c. Find the total cost of grade 3.
d. Find the total cost of the three pieces of steel.

Grade of Steel	*Weight (pounds per foot)*	*Required Number of Feet*	*Cost per Pound*
1	2.2	8	\$3.20
2	3.4	6.5	\$3.35
3	6.75	15.4	\$3.94

151. Business A confectioner ships holiday packs of candy and nuts anywhere in the United States. At the right is a price list for nuts and candy, and below is a table of shipping charges to zones in the United States. For any fraction of a pound, use the next higher weight. Sixteen ounces (16 oz) is equal to 1 lb.

Code	*Description*	*Price*
112	Almonds 16 oz	\$6.75
116	Cashews 8 oz	\$5.90
117	Cashews 16 oz	\$8.50
130	Macadamias 7 oz	\$7.25
131	Macadamias 16 oz	\$11.95
149	Pecan halves 8 oz	\$8.25
155	Mixed nuts 8 oz	\$6.80
160	Cashew brittle 8 oz	\$5.95
182	Pecan roll 8 oz	\$6.70
199	Chocolate peanuts 8 oz	\$5.90

Pounds	*Zone 1*	*Zone 2*	*Zone 3*	*Zone 4*
1–3	\$7.55	\$7.85	\$8.25	\$8.75
4–6	\$8.10	\$8.40	\$8.80	\$9.30
7–9	\$8.50	\$8.80	\$9.20	\$9.70
10–12	\$8.90	\$9.20	\$9.60	\$10.10

Find the cost of sending the following orders to the given mail zone.

a. *Code*	*Quantity*	**b.** *Code*	*Quantity*	**c.** *Code*	*Quantity*
116	2	112	1	117	3
130	1	117	4	131	1
149	3	131	2	155	2
182	4	160	3	160	4
Mail to zone 4.		182	5	182	1
		Mail to zone 3.		199	3
				Mail to zone 2.	

© Steve Lupton/Corbis

Applying the Concepts

152. Show how the decimal is placed in the product of 1.3 × 2.31 by first writing each number as a fraction and then multiplying. Then change the product back to decimal notation.

153. **Automotive Repair** Chris works at B & W Garage as an auto mechanic and has just completed an engine overhaul for a customer. To determine the cost of the repair job, Chris keeps a list of times worked and parts used. A parts list and a list of the times worked are shown below.

Parts Used		*Time Spent*	
Item	***Quantity***	***Day***	***Hours***
Gasket set	1	Monday	7.0
Ring set	1	Tuesday	7.5
Valves	8	Wednesday	6.5
Wrist pins	8	Thursday	8.5
Valve springs	16	Friday	9.0
Rod bearings	8		
Main bearings	5		
Valve seals	16		
Timing chain	1		

Price List		
Item Number	***Description***	***Unit Price***
27345	Valve spring	$9.25
41257	Main bearing	$17.49
54678	Valve	$16.99
29753	Ring set	$169.99
45837	Gasket set	$174.90
23751	Timing chain	$50.49
23765	Fuel pump	$229.99
28632	Wrist pin	$23.55
34922	Rod bearing	$13.69
2871	Valve seal	$1.69

a. Organize a table of data showing the parts used, the unit price for each part, and the price of the quantity used. *Hint:* Use the following headings for the table.

Quantity Item Number Description Unit Price Total

b. Add up the numbers in the "Total" column to find the total cost of the parts.

c. If the charge for labor is $46.75 per hour, compute the cost of labor.

d. What is the total cost for parts and labor?

154. Explain how the decimal point is placed when a number is multiplied by 10, 100, 1000, 10,000, etc.

155. Explain how the decimal point is placed in the product of two decimals.

SECTION

2.7 The Order of Operations Agreement

OBJECTIVE A To use the Order of Operations Agreement 

The Order of Operations Agreement applies when simplifying expressions containing fractions and decimals.

The Order of Operations Agreement

Step 1. Do all operations inside parentheses.

Step 2. Simplify any numerical expressions containing exponents.

Step 3. Do multiplication and division as they occur from left to right.

Step 4. Do addition and subtraction as they occur from left to right.

HOW TO 1 Simplify: $\left(\frac{1}{2}\right)^2 + \left(\frac{2}{3} \div \frac{5}{9}\right) \cdot \frac{5}{6}$

$\left(\frac{1}{2}\right)^2 + \left(\frac{2}{3} \div \frac{5}{9}\right) \cdot \frac{5}{6}$

$= \left(\frac{1}{2}\right)^2 + \left(\frac{6}{5}\right) \cdot \frac{5}{6}$ • Do the operation inside the parentheses (Step 1).

$= \frac{1}{4} + \left(\frac{6}{5}\right) \cdot \frac{5}{6}$ • Simplify the exponential expression (Step 2).

$= \frac{1}{4} + 1$ • Do the multiplication (Step 3).

$= 1\frac{1}{4}$ • Do the addition (Step 4).

A fraction bar acts like parentheses. Therefore, simplify the numerator and denominator of a fraction as part of Step 1 in the Order of Operations Agreement.

HOW TO 2 Simplify: $6 - \frac{2+1}{15-8} \div \frac{3}{14}$

$6 - \frac{2+1}{15-8} \div \frac{3}{14}$

$= 6 - \frac{3}{7} \div \frac{3}{14}$ • Perform operations above and below the fraction bar.

$= 6 - \left(\frac{3}{7} \cdot \frac{14}{3}\right)$ • Do the division.

$= 6 - 2$

$= 4$ • Do the subtraction.

HOW TO • 3 Evaluate $\frac{w+x}{y} - z$ when $w = \frac{3}{4}$, $x = \frac{1}{4}$, $y = 2$, and $z = \frac{1}{3}$.

$$\frac{w+x}{y} - z$$

$$\frac{\frac{3}{4}+\frac{1}{4}}{2} - \frac{1}{3}$$ • Replace each variable with its given value.

$$= \frac{1}{2} - \frac{1}{3}$$ • Simplify the numerator of the complex fraction.

$$= \frac{3}{6} - \frac{2}{6} = \frac{1}{6}$$ • Do the subtraction.

EXAMPLE • 1

Simplify: $0.2(5.6 - 2.5) + (1.4)^2$

Solution

$0.2(5.6 - 2.5) + (1.4)^2$

$= 0.2(3.1) + (1.4)^2$ • Parentheses

$= 0.2(3.1) + 1.96$ • Exponents

$= 0.62 + 1.96$ • Multiply.

$= 2.58$ • Add.

YOU TRY IT • 1

Simplify: $(1.2 - 0.8)^2 + (1.5)(6)$

Your solution

EXAMPLE • 2

Simplify: $\left(\frac{2}{3}\right)^2 \div \frac{7-2}{13-4} - \frac{1}{3}$

Solution

$$\left(\frac{2}{3}\right)^2 \div \frac{7-2}{13-4} - \frac{1}{3}$$

$$= \left(\frac{2}{3}\right)^2 \div \frac{5}{9} - \frac{1}{3}$$

$$= \frac{4}{9} \div \frac{5}{9} - \frac{1}{3}$$

$$= \frac{4}{9} \cdot \frac{9}{5} - \frac{1}{3}$$

$$= \frac{4}{5} - \frac{1}{3} = \frac{12}{15} - \frac{5}{15} = \frac{7}{15}$$

YOU TRY IT • 2

Simplify: $\left(\frac{1}{2}\right)^3 \cdot \frac{7-3}{9-4} + \frac{4}{5}$

Your solution

Solutions on p. S8

2.7 EXERCISES

OBJECTIVE A To use the Order of Operations Agreement

For Exercises 1 to 21, simplify.

1. $\frac{3}{7} \cdot \frac{14}{15} + \frac{4}{5}$

2. $\frac{3}{5} \div \frac{6}{7} + \frac{4}{5}$

3. $\left(\frac{5}{6}\right)^2 - \frac{5}{9}$

4. $\left(\frac{3}{5}\right)^2 - \frac{3}{10}$

5. $\frac{3}{4} \cdot \left(\frac{11}{12} - \frac{7}{8}\right) + \frac{5}{16}$

6. $\frac{7}{18} + \frac{5}{6} \cdot \left(\frac{2}{3} - \frac{1}{6}\right)$

7. $\frac{11}{16} - \left(\frac{3}{4}\right)^2 + \frac{7}{8}$

8. $\left(\frac{2}{3}\right)^2 - \frac{7}{18} + \frac{5}{6}$

9. $\left(1\frac{1}{3} - \frac{5}{6}\right) + \frac{7}{8} \div \left(\frac{1}{2}\right)^2$

10. $\left(\frac{1}{4}\right)^2 \div \left(2\frac{1}{2} - \frac{3}{4}\right) + \frac{5}{7}$

11. $\left(\frac{2}{3}\right)^2 + \frac{8 - 7}{9 - 3} \div \frac{3}{8}$

12. $\left(\frac{1}{3}\right)^2 \cdot \frac{14 - 5}{10 - 6} + \frac{3}{4}$

13. $(0.5)(0.2)^2 + 1.7$

14. $0.3(4.8 - 1.7) + (1.2)^2$

15. $(1.8)^2 - 2.52 \div 1.8$

16. $(1.65 - 1.05)^2 \div 0.4 + 0.9$

17. $0.4(3 - 1.5) + (1.2)^2$

18. $(5 - 3.5)^2 + (0.75)(8)$

19. $\frac{1}{2} + \frac{\frac{13}{25}}{4 - \frac{3}{4}} \div \frac{1}{5}$

20. $\frac{4}{5} + \frac{3 - \frac{7}{9}}{\frac{5}{6}} \cdot \frac{3}{8}$

21. $\left(\frac{2}{3}\right)^2 + \frac{\frac{5}{8} - \frac{1}{4}}{\frac{2}{3} - \frac{1}{6}} \cdot \frac{8}{9}$

For Exercises 22 to 31, evaluate the expression for the given values of the variables.

22. $x^2 + \frac{y}{z}$, when $x = \frac{2}{3}$, $y = \frac{5}{8}$, and $z = \frac{3}{4}$

23. $\frac{x}{y} - z^2$, when $x = \frac{5}{6}$, $y = \frac{1}{3}$, and $z = \frac{3}{4}$

24. $x - y^3z$, when $x = \frac{5}{6}$, $y = \frac{1}{2}$, and $z = \frac{8}{9}$

25. $xy^3 + z$, when $x = \frac{9}{10}$, $y = \frac{1}{3}$, and $z = \frac{7}{15}$

26. $\frac{wx}{y} + z$, when $w = \frac{4}{5}$, $x = \frac{5}{8}$, $y = \frac{3}{4}$, and $z = \frac{2}{3}$

27. $\frac{w}{xy} - z$, when $w = 2\frac{1}{2}$, $x = 4$, $y = \frac{3}{8}$, and $z = \frac{2}{3}$

28. $c^2 - ab$, when $a = 1.7$, $b = 0.6$, and $c = 2.8$

29. $(a + b)^2 - c$, when $a = 2.5$, $b = 1.8$, and $c = 0.4$

30. $\frac{b^2}{c} + 4a$, when $a = 1.5$, $b = 0.2$, and $c = 0.4$

31. $\frac{x}{y^2} + 3z$, when $x = 7.2$, $y = 0.6$, and $z = 3.5$

32. Insert parentheses into the expression $\frac{2}{9} \cdot \frac{5}{6} + \frac{3}{4} \div \frac{3}{5}$ so that **a.** the first operation to be performed is addition and **b.** the first operation to be performed is division.

Applying the Concepts

33. Simplify: $\dfrac{\frac{3}{x} + \frac{2}{x}}{\frac{5}{6}}$

34. Given that x is a whole number, for what value of x will the expression $\left(\frac{3}{4}\right)^2 + x^5 \div \frac{7}{8}$ have a minimum value? What is the minimum value?

35. A farmer died and left 17 horses to be divided among 3 children. The first child was to receive one-half of the horses, the second child one-third of the horses, and the third child one-ninth of the horses. The executor for the family's estate realized that 17 horses could not be divided by halves, thirds, or ninths and so added a neighbor's horse to the farmer's. With 18 horses, the executor gave 9 horses to the first child, 6 horses to the second child, and 2 horses to the third child. This accounted for the 17 horses, so the executor returned the borrowed horse to the neighbor. Explain why this worked.

FOCUS ON PROBLEM SOLVING

Common Knowledge An application problem may not provide all the information that is needed to solve the problem. Sometimes, however, the necessary information is common knowledge.

HOW TO • 1 You are traveling by bus from Boston to New York. The trip is 4 h long. If the bus leaves Boston at 10 A.M., what time should you arrive in New York?

What other information do you need to solve this problem?

You need to know that, using a 12-hour clock, the hours run

10 A.M.
11 A.M.
12 P.M.
1 P.M.
2 P.M.

Four hours after 10 A.M. is 2 P.M.

You should arrive in New York at 2 P.M.

HOW TO • 2 You purchase a 44¢ stamp at the Post Office and hand the clerk a one-dollar bill. How much change do you receive?

What information do you need to solve this problem?

You need to know that there are 100¢ in one dollar.

Your change is 100¢ − 44¢.

$100 - 44 = 56$

You receive 56¢ in change.

What information do you need to know to solve each of the following problems?

1. You sell a dozen tickets to a fundraiser. Each ticket costs $10. How much money do you collect?
2. The weekly lab period for your science course is 1 h and 20 min long. Find the length of the science lab period in minutes.
3. An employee's monthly salary is $3750. Find the employee's annual salary.
4. A survey revealed that eighth graders spend an average of 3 h each day watching television. Find the total time an eighth grader spends watching TV each week.
5. You want to buy a carpet for a room that is 15 ft wide and 18 ft long. Find the amount of carpet that you need.

PROJECTS AND GROUP ACTIVITIES

Music

In musical notation, notes are printed on a **staff,** which is a set of five horizontal lines and the spaces between them. The notes of a musical composition are grouped into **measures,** or **bars.** Vertical lines separate measures on a staff. The shape of a note indicates how long it should be held. The whole note has the longest time value of any note. Each time value is divided by 2 in order to find the next smallest time value.

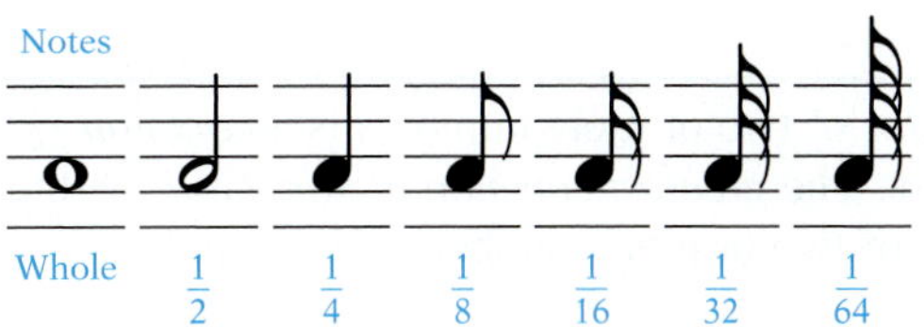

The **time signature** is a fraction that appears at the beginning of a piece of music. The numerator of the fraction indicates the number of beats in a measure. The denominator indicates what kind of note receives 1 beat. For example, music written in $\frac{2}{4}$ time has 2 beats to a measure, and a quarter note receives 1 beat. One measure in $\frac{2}{4}$ time may have 1 half note, 2 quarter notes, 4 eighth notes, or any other combination of notes totaling 2 beats. Other common time signatures are $\frac{4}{4}$, $\frac{3}{4}$, and $\frac{6}{8}$.

1. Explain the meaning of the 6 and the 8 in the time signature $\frac{6}{8}$.
2. Give some possible combinations of notes in one measure of a piece written in $\frac{4}{4}$ time.
3. What does a dot at the right of a note indicate? What is the effect of a dot at the right of a half note? At the right of a quarter note? At the right of an eighth note?
4. Symbols called rests are used to indicate periods of silence in a piece of music. What symbols are used to indicate the different time values of rests?
5. Find some examples of musical compositions written in different time signatures. Use a few measures from each to show that the sum of the time values of the notes and rests in each measure equals the numerator of the time signature.

Construction

Suppose you are involved in building your own home. Design a stairway from the first floor of the house to the second floor. Some of the questions you will need to answer follow.

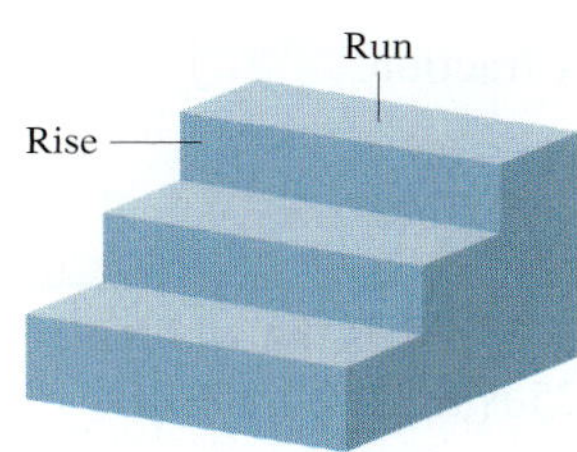

What is the distance from the floor of the first story to the floor of the second story?

Typically, what is the number of steps in a stairway?

What is a reasonable length for the run of each step?

What is the width of the wood being used to build the staircase?

In designing the stairway, remember that each riser should be the same height, that each run should be the same length, and that the width of the wood used for the steps will have to be incorporated into the calculation.

CHAPTER 2

SUMMARY

KEY WORDS	EXAMPLES
A number that is a multiple of two or more numbers is a *common multiple* of those numbers. The *least common multiple (LCM)* is the smallest common multiple of two or more numbers. [2.1A, p. 82]	12, 24, 36, 48, . . . are common multiples of 4 and 6. The LCM of 4 and 6 is 12.
A number that is a factor of two or more numbers is a *common factor* of those numbers. The *greatest common factor (GCF)* is the largest common factor of two or more numbers. [2.1B, p. 83]	The common factors of 12 and 16 are 1, 2, and 4. The GCF of 12 and 16 is 4.
A *fraction* can represent the number of equal parts of a whole. In a fraction, the *fraction bar* separates the *numerator* and the *denominator.* [2.2A, p. 86]	In the fraction $\frac{3}{4}$, the numerator is 3 and the denominator is 4.
In a *proper fraction*, the numerator is smaller than the denominator; a proper fraction is a number less than 1. In an *improper fraction*, the numerator is greater than or equal to the denominator; an improper fraction is a number greater than or equal to 1. A *mixed number* is a number greater than 1 with a whole-number part and a fractional part. [2.2A, pp. 86–87]	$\frac{2}{5}$ is a proper fraction. $\frac{7}{6}$ is an improper fraction. $4\frac{1}{10}$ is a mixed number; 4 is the whole-number part and $\frac{1}{10}$ is the fractional part.
Equal fractions with different denominators are called *equivalent fractions*. [2.2B, p. 89]	$\frac{3}{4}$ and $\frac{6}{8}$ are equivalent fractions.
A fraction is in *simplest form* when the numerator and denominator have no common factors other than 1. [2.2B, p. 89]	The fraction $\frac{11}{12}$ is in simplest form.
The *reciprocal* of a fraction is the fraction with the numerator and denominator interchanged. [2.4B, p. 114]	The reciprocal of $\frac{3}{8}$ is $\frac{8}{3}$. The reciprocal of 5 is $\frac{1}{5}$.
A *complex fraction* is a fraction whose numerator or denominator contains one or more fractions. [2.4C, p. 117]	$\dfrac{\frac{2}{3} - \frac{5}{8}}{\frac{1}{9}}$ is a complex fraction.
A number written in *decimal notation* has three parts: a whole number part, a decimal point, and a decimal part. The *decimal part* of a number represents a number less than 1. A number written in decimal notation is often called simply a *decimal*. [2.5A, p. 128]	For the decimal 31.25, 31 is the whole-number part and 25 is the decimal part.

ESSENTIAL RULES AND PROCEDURES	EXAMPLES
To find the LCM of two or more numbers, find the prime factorization of each number and write the factorization of each number in a table. Circle the greatest product in each column. The LCM is the product of the circled numbers. [2.1A, p. 82]	Columns 2, 3: $12 =$ (2 · 2) circled, 3; $18 =$ 2, (3 · 3) circled. The LCM of 12 and 18 is $2 \cdot 2 \cdot 3 \cdot 3 = 36$.
To find the GCF of two or more numbers, find the prime factorization of each number and write the factorization of each number in a table. Circle the least product in each column that does not have a blank. The GCF is the product of the circled numbers. [2.1B, p. 83]	Columns 2, 3: $12 =$ 2 · 2, (3) circled; $18 =$ (2) circled, 3 · 3. The GCF of 12 and 18 is $2 \cdot 3 = 6$.
To write an improper fraction as a mixed number, divide the numerator by the denominator. [2.2A, p. 87]	$\frac{29}{6} = 29 \div 6 = 4\frac{5}{6}$
To write a mixed number as an improper fraction, multiply the denominator of the fractional part of the mixed number by the whole number part. Add this product and the numerator of the fractional part. The sum is the numerator of the improper fraction. The denominator remains the same. [2.2A, p. 87]	$3\frac{2}{5} = \frac{5 \times 3 + 2}{5} = \frac{17}{5}$
To write a fraction in simplest form, divide the numerator and denominator of the fraction by their common factors. [2.2B, p. 90]	$\frac{30}{45} = \frac{2 \cdot \overset{1}{\cancel{3}} \cdot \overset{1}{\cancel{5}}}{\underset{1}{\cancel{3}} \cdot 3 \cdot \underset{1}{\cancel{5}}} = \frac{2}{3}$
To add fractions with the same denominators, add the numerators and place the sum over the common denominator. $\frac{a}{b} + \frac{c}{b} = \frac{a+c}{b}$, where $b \neq 0$ [2.3A, p. 97]	$\frac{5}{12} + \frac{11}{12} = \frac{16}{12} = 1\frac{1}{3}$
To subtract fractions with the same denominators, subtract the numerators and place the difference over the common denominator. $\frac{a}{b} - \frac{c}{b} = \frac{a-c}{b}$, where $b \neq 0$ [2.3B, p. 101]	$\frac{9}{16} - \frac{5}{16} = \frac{4}{16} = \frac{1}{4}$
To add or subtract fractions with different denominators, first rewrite the fractions as equivalent fractions with a common denominator. The common denominator is the least common multiple (LCM) of the denominators of the fractions. Then add or subtract the fractions. [2.3A/2.3B, pp. 97, 101]	$\frac{7}{8} + \frac{5}{6} = \frac{21}{24} + \frac{20}{24} = \frac{41}{24} = 1\frac{17}{24}$ $\frac{2}{3} - \frac{7}{16} = \frac{32}{48} - \frac{21}{48} = \frac{11}{48}$
To multiply two fractions, multiply the numerators; this is the numerator of the product. Multiply the denominators; this is the denominator of the product. $\frac{a}{b} \cdot \frac{c}{d} = \frac{ac}{bd}$, where $b \neq 0$ and $d \neq 0$ [2.4A, p. 110]	$\frac{3}{4} \cdot \frac{2}{9} = \frac{3 \cdot 2}{4 \cdot 9} = \frac{3 \cdot 2}{2 \cdot 2 \cdot 3 \cdot 3} = \frac{1}{6}$

To divide two fractions, multiply the first fraction by the reciprocal of the second fraction.
$\frac{a}{b} \div \frac{c}{d} = \frac{a}{b} \cdot \frac{d}{c}$, where $b \neq 0$, $c \neq 0$, and $d \neq 0$ [2.4B, p. 114]

$$\frac{8}{15} \div \frac{4}{5} = \frac{8}{15} \cdot \frac{5}{4} = \frac{8 \cdot 5}{15 \cdot 4}$$
$$= \frac{2 \cdot 2 \cdot 2 \cdot 5}{3 \cdot 5 \cdot 2 \cdot 2} = \frac{2}{3}$$

To simplify a complex fraction, simplify the expression above the main fraction bar and simplify the expression below the main fraction bar. Then rewrite the complex fraction as a division problem by reading the main fraction bar as "divided by." [2.4C, p. 117]

$$\frac{\frac{8}{9} - \frac{2}{3}}{1\frac{1}{5}} = \frac{\frac{8}{9} - \frac{6}{9}}{\frac{6}{5}} = \frac{\frac{2}{9}}{\frac{6}{5}}$$
$$= \frac{2}{9} \div \frac{6}{5} = \frac{2}{9} \cdot \frac{5}{6} = \frac{5}{27}$$

The formula for the area of a triangle is $A = \frac{1}{2}bh$, where A is the area of the triangle, b is the base, and h is the height. [2.4D, p. 119]

Find the area of a triangle with a base measuring 6 ft and a height of 3 ft.
$A = \frac{1}{2}bh = \frac{1}{2}(6)(3) = 9$
The area is 9 ft^2.

To write a decimal in words, write the decimal part as though it were a whole number. Then name the place value of the last digit. The decimal point is read as "and." [2.5A, p. 128]

The decimal 12.875 is written in words as twelve and eight hundred seventy-five thousandths.

To write a decimal in standard form when it is written in words, write the whole number part, replace the word *and* with a decimal point, and write the decimal part so that the last digit is in the given place-value position. [2.5A, p. 129]

The decimal forty-nine and sixty-three thousandths is written in standard form as 49.063.

To compare two decimals, write the decimal part of each number so that each has the same number of decimal places. Then compare the two numbers. [2.5B, p. 130]

$1.790 > 1.789$
$0.8130 < 0.8315$

To round a decimal, use the same rules used with whole numbers, except drop the digits to the right of the given place value instead of replacing them with zeros. [2.5C, p. 131]

2.7134 rounded to the nearest tenth is 2.7.

0.4687 rounded to the nearest hundredth is 0.47.

To add or subtract decimals, write the decimals so that the decimal points are on a vertical line. Add or subtract as you would with whole numbers. Then write the decimal point in the answer directly below the decimal points in the given numbers. [2.6A, p. 138]

$$\begin{array}{r} {}^{1\ 1} \\ 1.35 \\ 20.8 \\ +\ 0.76 \\ \hline 22.91 \end{array} \qquad \begin{array}{r} {}^{2\ 15\ \ \ 6\ 10} \\ \not{3}\not{5}.8\not{7}\not{0} \\ -\ \ 9.641 \\ \hline 26.229 \end{array}$$

To estimate the answer to a calculation, round each number to the highest place value of the number; the first digit of each number will be nonzero, and all other digits will be zero. If a number is a decimal less than 1, round the decimal so that there is one nonzero digit. Perform the calculation using the rounded numbers. [2.6A, p. 139]	35.87 ⟶ 40 61.09 ⟶ +60 100 0.3876 ⟶ 0.4 0.5472 ⟶ +0.5 0.9
To multiply decimals, multiply the numbers as you would whole numbers. Then write the decimal point in the product so that the number of decimal places in the product is the sum of the numbers of decimal places in the factors. [2.6B, p. 141]	26.83 2 decimal places × 0.45 2 decimal places 13415 10732 12.0735 4 decimal places
To multiply a decimal by a power of 10, move the decimal point to the right the same number of places as there are zeros in the power of 10. If the power of 10 is written in exponential notation, the exponent indicates how many places to move the decimal point. [2.6B, pp. 141–142]	$3.97 \cdot 10{,}000 = 39{,}700$ $0.641 \cdot 10^5 = 64{,}100$
To divide decimals, move the decimal point in the divisor to the right so that the divisor is a whole number. Move the decimal point in the dividend the same number of places to the right. Place the decimal point in the quotient directly above the decimal point in the dividend. Then divide as you would with whole numbers. [2.6C, p. 143]	6.2 0.39.)2.41.8 −2 34 7 8 −7 8 0
To divide a decimal by a power of 10, move the decimal point to the left the same number of places as there are zeros in the power of 10. If the power of 10 is written in exponential notation, the exponent indicates how many places to move the decimal point. [2.6C, p. 144]	$972.8 \div 1000 = 0.9728$ $61.305 \div 10^4 = 0.0061305$
To write a fraction as a decimal, divide the numerator of the fraction by the denominator. [2.6D, p. 146]	$\frac{7}{8} = 7 \div 8 = 0.875$
To convert a decimal to a fraction, remove the decimal point and place the decimal part over a denominator equal to the place value of the last digit in the decimal. [2.6D, p. 146]	0.85 is eighty-five <u>hundredths</u>. $0.85 = \frac{85}{100} = \frac{17}{20}$
To find the order relation between a decimal and a fraction, first rewrite the fraction as a decimal. Then compare the two decimals. [2.6D, p. 147]	Because $\frac{3}{11} \approx 0.273$, and $0.273 > 0.26$, $\frac{3}{11} > 0.26$.

Order of Operations Agreement [2.7A, p. 158]

Step 1 Do all operations inside parentheses.

Step 2 Simplify any numerical expressions containing exponents.

Step 3 Do multiplication and division as they occur from left to right.

Step 4 Do addition and subtraction as they occur from left to right.

$$\left(\frac{1}{3}\right)^2 + \left(\frac{11}{12} - \frac{5}{6}\right) \cdot 4$$

$$= \left(\frac{1}{3}\right)^2 + \frac{1}{12} \cdot 4$$

$$= \frac{1}{9} + \frac{1}{12} \cdot 4 = \frac{1}{9} + \frac{1}{3} = \frac{4}{9}$$

CHAPTER 2

CONCEPT REVIEW

Test your knowledge of the concepts presented in this chapter. Answer each question. Then check your answers against the ones provided in the Answer Section.

1. How do you find the LCM of 75, 30, and 50?

2. How do you find the GCF of 42, 14, and 21?

3. How do you write an improper fraction as a mixed number?

4. When is a fraction in simplest form?

5. When adding fractions, why do you have to convert to equivalent fractions with a common denominator?

6. How do you add mixed numbers?

7. If you are subtracting a mixed number from a whole number, why do you need to borrow?

8. Where do you put the decimal point in the product of two decimals?

9. How do you estimate the product of two decimals?

10. What do you do with the decimal point when dividing decimals?

11. How many zeros must be inserted when dividing 0.763 by 0.6 and rounding to the nearest hundredth?

12. How do you subtract a decimal from a whole number that has no decimal point?

CHAPTER 2

REVIEW EXERCISES

1. Write $\frac{19}{2}$ as a mixed number.

2. Subtract: $6\frac{2}{9} - 3\frac{7}{18}$

3. Evaluate $x \div y$ when $x = 2\frac{5}{8}$ and $y = 1\frac{3}{4}$.

4. Write five and thirty-four thousandths in standard form.

5. Convert 0.28 to a fraction.

6. Find the product of 3 and $\frac{8}{9}$.

7. Place the correct symbol, < or >, between the two numbers.

   8.039 8.31

8. Place the correct symbol, < or >, between the two numbers.

   $\frac{3}{5}$ $\frac{7}{15}$

9. Find the LCM of 50 and 75.

10. Find the product of 0.918 and 10^5.

11. Evaluate xy when $x = 8$ and $y = \frac{5}{12}$.

12. Express the shaded portion of the circles as an improper fraction and as a mixed number.

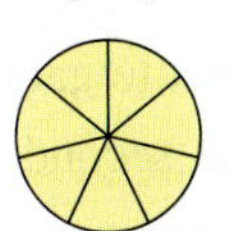 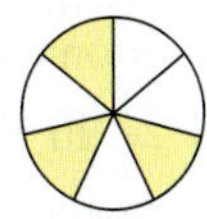

13. Place the correct symbol, < or >, between the two numbers.

    $\frac{3}{7}$ 0.429

14. Simplify: $\dfrac{\frac{5}{8} - \frac{1}{4}}{\frac{1}{2} + \frac{1}{8}}$

15. Write a fraction that is equivalent to $\frac{4}{9}$ and has a denominator of 72.

16. Evaluate x^2y^3 when $x = \frac{8}{9}$ and $y = \frac{3}{4}$.

17. Evaluate $ab^2 - c$ when $a = 4$, $b = \frac{1}{2}$, and $c = \frac{5}{7}$.

18. Find the GCF of 42 and 63.

19. Find the quotient of 14.2 and 10^3.

20. Divide and round to the nearest tenth: $6.8 \div 47.92$

21. Find the quotient of $\frac{5}{9}$ and $\frac{2}{3}$.

22. Evaluate $\frac{x}{y}$ when $x = 0.396$ and $y = 3.6$.

23. Estimate the difference between 506.81 and 64.1.

24. Multiply: (9.47)(0.26)

25. Evaluate $a - b$ when $a = 80.32$ and $b = 29.577$.

26. Evaluate $\left(\frac{3}{8}\right)^2 \cdot 4^2$.

27. Find the sum of $3\frac{7}{12}$ and $5\frac{1}{2}$.

28. Write $\frac{30}{105}$ in simplest form.

29. Evaluate $a - b$ when $a = 7$ and $b = 2\frac{3}{10}$.

30. **Tourism** See the news clipping at the right. Find the average amount spent by each visitor to the United States. Round to the nearest cent.

In the News

Tourists Boost the Economy

This past summer, 14.3 million visitors came to the United States, spending a record $30.7 billion.

Source: Commerce Department

31. **Wrestling** A wrestler is put on a diet to gain 12 lb in 4 weeks. The wrestler gains $3\frac{1}{2}$ lb the first week and $2\frac{1}{4}$ lb the second week. How much weight must the wrestler gain during the third and fourth weeks in order to gain a total of 12 lb?

32. **Manufacturing** An employee hired for piecework can assemble a unit in $2\frac{1}{2}$ min. How many units can this employee assemble during an 8-hour day?

33. **Wages** Find the overtime pay due an employee who worked $6\frac{1}{4}$ h of overtime this week. The employee's overtime rate is $24 an hour.

34. **Physics** What is the final velocity, in feet per second, of an object dropped from a plane with a starting velocity of 0 ft/s and a fall of $15\frac{1}{2}$ s? Use the formula $V = S + 32t$, where V is the final velocity of a falling object, S is its starting velocity, and t is the time of the fall.

Sandor Szabo/EPA/Landov

CHAPTER 2

TEST

1. Write $\frac{18}{7}$ as a mixed number.

2. Subtract: $7\frac{3}{4} - 3\frac{5}{6}$

3. Evaluate xy when $x = 6\frac{3}{7}$ and $y = 3\frac{1}{2}$.

4. Find the product of $\frac{2}{3}$ and $\frac{7}{8}$.

5. Find the LCM of 30 and 45.

6. Write nine and thirty-three thousandths in standard form.

7. Evaluate x^3y^2 when $x = 1\frac{1}{2}$ and $y = \frac{5}{6}$.

8. Write $3\frac{4}{5}$ as an improper fraction.

9. What is $\frac{7}{12}$ divided by $\frac{3}{4}$?

10. Place the correct symbol, < or >, between the two numbers.
    4.003 4.009

11. Evaluate $\frac{x}{yz}$ when $x = \frac{7}{20}$, $y = \frac{2}{15}$, and $z = \frac{3}{8}$.

12. Find the GCF of 18 and 54.

13. How much larger is $\frac{13}{14}$ than $\frac{16}{21}$?

14. Write $\frac{60}{75}$ in simplest form.

15. Evaluate $x + y + z$ when $x = 1\frac{3}{8}$, $y = \frac{1}{2}$, and $z = \frac{5}{6}$.

16. Place the correct symbol, < or >, between the two numbers.
    $\frac{5}{6}$ $\frac{11}{15}$

17. Evaluate $a^2b - c^2$ when $a = \frac{2}{3}$, $b = 9$, and $c = \frac{3}{5}$.

18. Place the correct symbol, < or >, between the two numbers.
    0.22 $\frac{2}{9}$

19. Round 6.051367 to the nearest thousandth.

20. Evaluate $x \div y$ when $x = \frac{8}{9}$ and $y = \frac{16}{27}$.

21. Find the difference between 30 and 7.247.

22. Estimate the difference between 92.34 and 17.95.

23. Find the total of 4.58, 3.9, and 6.017.

24. Evaluate $20cd$ when $c = 0.5$ and $d = 6.4$.

25. Find the quotient of 84.96 and 100.

26. Write a fraction that is equivalent to $\frac{3}{7}$ and has a denominator of 28.

27. **The Film Industry** The table at the right shows six James Bond films released between 1960 and 1970 and their gross box office income, in millions of dollars, in the United States. How much greater was the gross from *Thunderball* than the gross from *On Her Majesty's Secret Service*?

Film	*U.S. Box Office Gross*
Dr. No	\$16.1
On Her Majesty's Secret Service	\$22.8
From Russia with Love	\$24.8
You Only Live Twice	\$43.1
Goldfinger	\$51.1
Thunderball	\$63.6

Source: www.worldwideboxoffice.com

28. **Community Service** You are required to contribute 20 h of community service to the town in which your college is located. On one occasion you work $7\frac{1}{4}$ h, and on another occasion you work $2\frac{3}{4}$ h. How many more hours of community service are still required of you?

29. **Manufacturing** An employee hired for piecework can assemble a unit in $4\frac{1}{2}$ min. How many units can this employee assemble in 6 h?

30. **Accounting** The fundamental accounting equation is $A = L + S$, where A is the assets of the company, L is the liabilities of the company, and S is the stockholders' equity. Find the stockholders' equity in a company whose assets are \$48.2 million and whose liabilities are \$27.6 million.

31. **Geometry** The lengths of the three sides of a triangle are 8.75 m, 5.25 m, and 4.5 m. Find the perimeter of the triangle. Use the formula $P = a + b + c$.

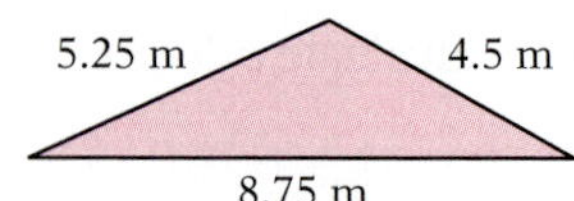

CUMULATIVE REVIEW EXERCISES

1. Find the quotient of 387.9 and 10^4.

2. Evaluate $(x + y)^2 - 2z$ when $x = 3$, $y = 2$, and $z = 5$.

3. Find the prime factorization of 140.

4. Write eight million seventy-two thousand ninety-two in standard form.

5. Place the correct symbol, < or >, between the two numbers.
   $\frac{7}{11} \quad \frac{4}{5}$

6. Find the GCF of 72 and 108.

7. Find the difference between $\frac{5}{14}$ and $\frac{9}{42}$.

8. Estimate the sum of 372, 541, 608, and 429.

9. Add: $6847 + 3501 + 924$

10. Evaluate $x \div y$ when $x = 3\frac{2}{3}$ and $y = 2\frac{4}{9}$.

11. What is 36.92 increased by 18.5?

12. Simplify: $\left(\frac{5}{9}\right)\left(\frac{3}{10}\right)\left(\frac{6}{7}\right)$

13. Evaluate x^4y^2 when $x = 2$ and $y = 10$.

14. Find the prime factorization of 260.

15. Convert $\frac{19}{25}$ to a decimal.

16. Estimate the difference between 89,357 and 66,042.

17. **Vacation Days** The figure at the right shows the number of vacation days per year that are legally mandated in several countries.
    a. Which country mandates more vacation days, Ireland or Sweden?
    b. How many times more vacation days does Austria mandate than Switzerland?

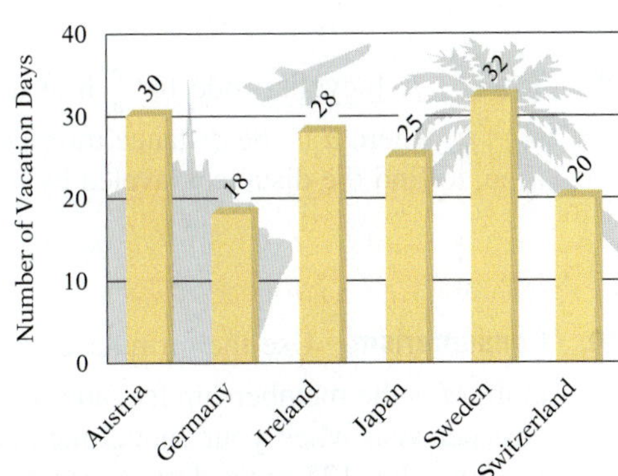

Number of Legally Mandated Vacation Days
Source: Economic Policy Institute; *World Almanac*

18. Divide: $\frac{8}{0}$

19. Simplify: $\frac{5}{7} + \frac{4}{21}$

20. Subtract: $8\frac{3}{4} - 1\frac{5}{7}$

21. Evaluate $3a + (a - b)^3$ when $a = 4$ and $b = 1$.

22. Simplify: $5(7 - 3) \div (4) + 6(2)$

23. Evaluate $\frac{a}{b + c}$ when $a = \frac{3}{8}$, $b = \frac{1}{2}$, and $c = \frac{3}{4}$.

24. Evaluate x^3y^4 when $x = \frac{7}{12}$ and $y = \frac{6}{7}$.

25. Divide and round to the nearest tenth:
$2.617 \div 0.93$

26. Physical Fitness The chart at the right shows the calories burned per hour as a result of different aerobic activities. Suppose you weigh 150 lb. According to the chart, how many more calories would you burn by bicycling at 12 mph for 4 h than by walking at a rate of 3 mph for 5 h?

Activity	*100 lb*	*150 lb*
Bicycling, 6 mph	160	240
Bicycling, 12 mph	270	410
Jogging, $5\frac{1}{2}$ mph	440	660
Jogging, 7 mph	610	920
Jumping rope	500	750
Tennis, singles	265	400
Walking, 2 mph	160	240
Walking, 3 mph	210	320
Walking, $4\frac{1}{2}$ mph	295	440

27. Demographics The Census Bureau projects that the population of New England will increase to 15,321,000 in 2020 from 13,581,000 in 2000. Find the projected increase in the population of New England during the 20-year period.

28. Sales The figure at the right shows how the average salesperson spends the workweek.

a. On average, how many hours per week does a salesperson work?

b. Does the average salesperson spend more time face-to-face selling or doing both administrative work and placing service calls?

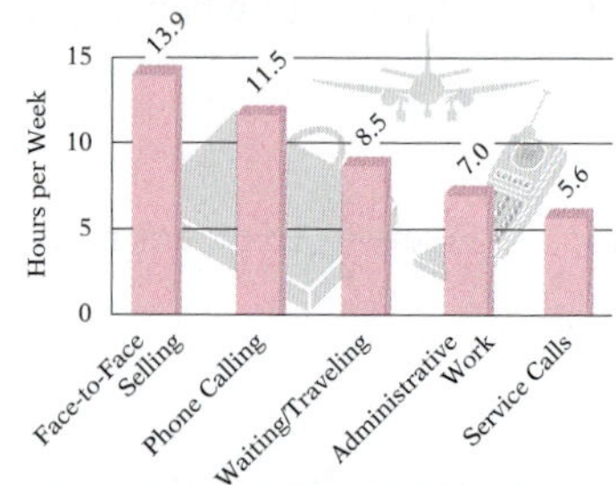

Average Salesperson's Workweek

Source: Dartnell's 28th Survey of Sales Force Compensation

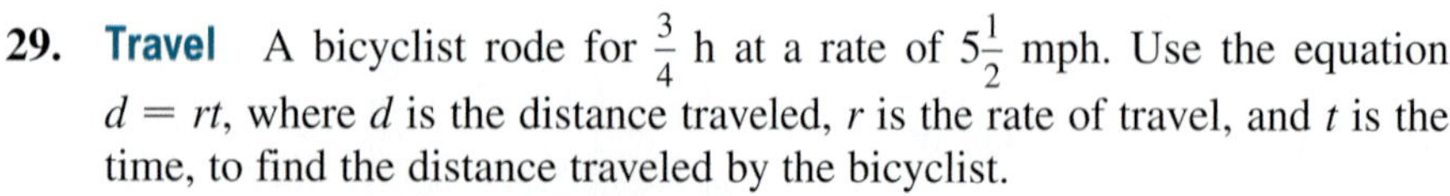

29. Travel A bicyclist rode for $\frac{3}{4}$ h at a rate of $5\frac{1}{2}$ mph. Use the equation $d = rt$, where d is the distance traveled, r is the rate of travel, and t is the time, to find the distance traveled by the bicyclist.

30. Consumerism Use the formula $C = \frac{M}{N}$, where C is the cost per visit at a health club, M is the membership fee, and N is the number of visits to the club, to find the cost per visit when your annual membership fee at a health club is \$390 and you visit the club 125 times during the year.

CHAPTER 3

Rational Numbers

OBJECTIVES

SECTION 3.1

- A To identify order relations between integers
- B To find the opposite of a number
- C To evaluate expressions that contain the absolute value symbol
- D To solve application problems

SECTION 3.2

- A To add integers
- B To subtract integers
- C To solve application problems

SECTION 3.3

- A To multiply integers
- B To divide integers
- C To solve application problems

SECTION 3.4

- A To add or subtract rational numbers
- B To multiply or divide rational numbers
- C To solve application problems

SECTION 3.5

- A To use the Order of Operations Agreement to simplify expressions

ARE YOU READY?

Take the Chapter 3 Prep Test to find out if you are ready to learn to:

- Order integers
- Evaluate expressions that contain the absolute value symbol
- Add, subtract, multiply, and divide integers and rational numbers
- Simplify numerical expressions

PREP TEST

Do these exercises to prepare for Chapter 3.

1. Place the correct symbol, $<$ or $>$, between the two numbers.
54 45

2. What is the distance from 4 to 8 on the number line?

For Exercises 3 to 14, add, subtract, multiply, or divide.

3. $7654 + 8193$ **4.** $6097 - 2318$ **5.** 472×56

6. $\dfrac{144}{24}$ **7.** $\dfrac{2}{3} + \dfrac{3}{5}$ **8.** $\dfrac{3}{4} - \dfrac{5}{16}$

9. $0.75 + 3.9 + 6.408$ **10.** $5.4 - 1.619$

11. $\dfrac{3}{4} \times \dfrac{8}{15}$ **12.** $\dfrac{5}{12} \div \dfrac{3}{4}$

13. 23.5×0.4 **14.** $0.96 \div 2.4$

15. Simplify: $(8 - 6)^2 + 12 \div 4 \cdot 3^2$

SECTION

3.1 Introduction to Integers

OBJECTIVE A To identify order relations between integers 

In Chapters 1 and 2, only zero and numbers greater than zero were discussed. In this chapter, numbers less than zero are introduced. Phrases such as "7 degrees below zero," "$50 in debt," and "20 feet below sea level" refer to numbers less than zero.

Numbers greater than zero are called **positive numbers.** Numbers less than zero are called **negative numbers.**

Point of Interest

Chinese manuscripts dating from about 250 B.C.E. contain the first recorded use of negative numbers. However, it was not until late in the 14th century that mathematicians generally accepted these numbers.

Positive and Negative Numbers

A number n is positive if $n > 0$.

A number n is negative if $n < 0$.

A positive number can be indicated by placing the sign $+$ in front of the number. For example, we can write $+4$ instead of 4. Both $+4$ and 4 represent "positive 4." Usually, however, the plus sign is omitted and it is understood that the number is a positive number.

A negative number is indicated by placing a negative sign $(-)$ in front of the number. The number -1 is read "negative one," -2 is read "negative two," and so on.

The number line can be extended to the left of zero to show negative numbers.

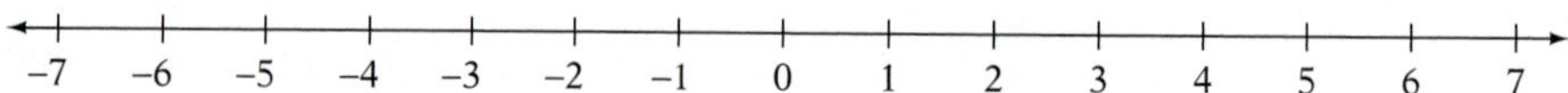

The **integers** are . . . , $-4, -3, -2, -1, 0, 1, 2, 3, 4,$ . . . . The integers to the right of zero are the **positive integers.** The integers to the left of zero are the **negative integers.** Zero is an integer, but it is neither positive nor negative. The point corresponding to 0 on the number line is called the **origin.**

On a number line, the numbers get larger as we move from left to right. The numbers get smaller as we move from right to left. Therefore, a number line can be used to visualize the order relation between two integers.

A number that appears to the right of a given number on the number line is greater than $(>)$ the given number. A number that appears to the left of a given number on the number line is less than $(<)$ the given number.

2 is to the right of -3 on the number line.
2 is greater than -3.
$2 > -3$

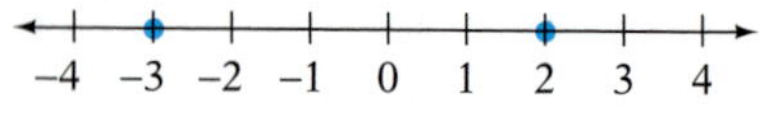

-4 is to the left of 1 on the number line.
-4 is less than 1.

$-4 < 1$

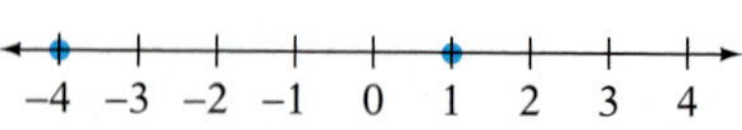

Order Relations

$a > b$ if a is to the right of b on the number line.

$a < b$ if a is to the left of b on the number line.

EXAMPLE • 1

On the number line, what number is 5 units to the right of -2?

Solution

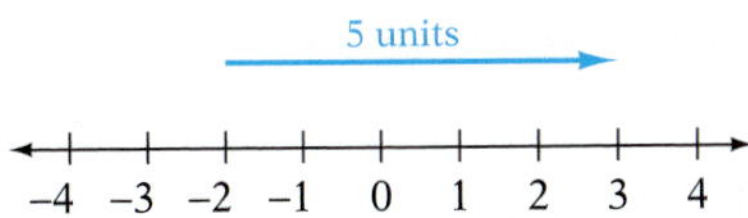

3 is 5 units to the right of -2.

YOU TRY IT • 1

On the number line, what number is 4 units to the left of 1?

Your solution

EXAMPLE • 2

If G is 2 and I is 4, what numbers are B and D?

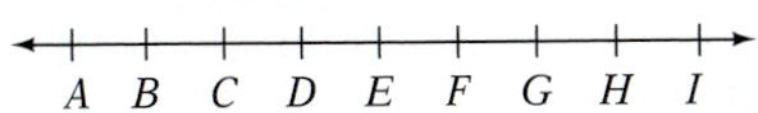

Solution

–4 –3 –2 –1 0 1 2 3 4

B is -3, and D is -1.

YOU TRY IT • 2

If G is 1 and H is 2, what numbers are A and C?

A B C D E F G H I

Your solution

EXAMPLE • 3

Place the correct symbol, $<$ or $>$, between the two numbers.

a. $-3 \quad -1$ **b.** $1 \quad -2$

Solution

a. -3 is to the left of -1 on the number line.
$-3 < -1$

b. 1 is to the right of -2 on the number line.
$1 > -2$

YOU TRY IT • 3

Place the correct symbol, $<$ or $>$, between the two numbers.

a. $2 \quad -5$ **b.** $-4 \quad 3$

Your solution

EXAMPLE • 4

Write the given numbers in order from smallest to largest.

$5, -2, 3, 0, -6$

Solution

$-6, -2, 0, 3, 5$

YOU TRY IT • 4

Write the given numbers in order from smallest to largest.

$-7, 4, -1, 0, 8$

Your solution

Solutions on p. S8

OBJECTIVE B To find the opposite of a number 

The distance from 0 to 3 on the number line is 3 units. The distance from 0 to -3 on the number line is 3 units. 3 and -3 are the same distance from 0 on the number line, but 3 is to the right of 0 and -3 is to the left of 0.

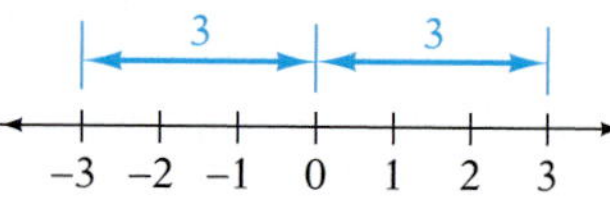

Integrating Technology

The **+/–** key on your calculator is used to find the opposite of a number. The **–** key is used to perform the operation of subtraction.

Two numbers that are the same distance from zero on the number line but are on opposite sides of zero are called **opposites.**

-3 is the opposite of 3 and 3 is the opposite of -3.

For any number n, the opposite of n is $-n$ and the opposite of $-n$ is n.

We can now define the **integers** as the whole numbers and their opposites.

A negative sign can be read as "the opposite of."

$-(3) = -3$ The opposite of positive 3 is negative 3.

$-(-3) = 3$ The opposite of negative 3 is positive 3.

Therefore, $-(a) = -a$ and $-(-a) = a$.

Note that with the introduction of negative integers and opposites, the symbols + and − can be read in different ways.

$6 + 2$	"six plus two"	+ is read "plus"
$+2$	"positive two"	+ is read "positive"
$6 - 2$	"six minus two"	− is read "minus"
-2	"negative two"	− is read "negative"
$-(-6)$	"the opposite of negative six"	− is read first as "the opposite of" and then as "negative"

When the symbols + and − indicate the operations of addition and subtraction, spaces are inserted before and after the symbol. When the symbols + and − indicate the sign of a number (positive or negative), there is no space between the symbol and the number.

EXAMPLE 5

Find the opposite number.

a. -8 **b.** 15 **c.** a

Solution

a. 8 **b.** -15 **c.** $-a$

YOU TRY IT 5

Find the opposite number.

a. 24 **b.** -13 **c.** $-b$

Your solution

Solution on p. S8

EXAMPLE • 6

Write the expression in words.

a. $7 - (-9)$ **b.** $-4 + 10$

Solution

a. Seven minus negative nine

b. Negative four plus ten

YOU TRY IT • 6

Write the expression in words.

a. $-3 - 12$ **b.** $8 + (-5)$

Your solution

EXAMPLE • 7

Simplify.

a. $-(-27)$ **b.** $-(-c)$

Solution

a. $-(-27) = 27$

b. $-(-c) = c$

YOU TRY IT • 7

Simplify.

a. $-(-59)$ **b.** $-(y)$

Your solution

Solutions on p. S8

OBJECTIVE C To evaluate expressions that contain the absolute value symbol

The **absolute value** of a number is the distance from zero to the number on the number line. Distance is never a negative number. Therefore, the absolute value of a number is a positive number or zero. The symbol for absolute value is "| |."

The distance from 0 to 3 is 3 units. Thus $|3| = 3$ (the absolute value of 3 is 3).

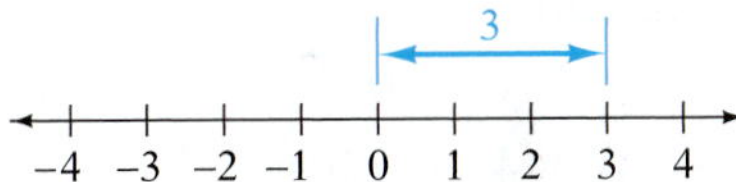

The distance from 0 to −3 is 3 units. Thus $|-3| = 3$ (the absolute value of −3 is 3).

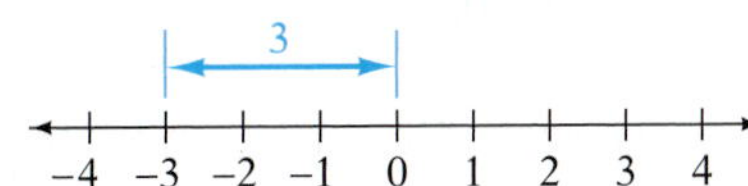

Because the distance from 0 to 3 and the distance from 0 to −3 are the same,

$$|3| = |-3| = 3$$

Absolute Value

The absolute value of a positive number is positive.	$|5| = 5$
The absolute value of a negative number is positive.	$|-5| = 5$
The absolute value of zero is zero.	$|0| = 0$

Take Note

It is important to be aware that the negative sign is *in front of the absolute value symbol.* This means $-|7| = -7$, but $|-7| = 7$.

HOW TO • 1 Evaluate $-|7|$.

The negative sign is *in front of* the absolute value symbol.

Recall that a negative sign can be read as "the opposite of."

Therefore, $-|7|$ can be read "the opposite of the absolute value of 7."

$-|7| = -7$

EXAMPLE • 8

Find the absolute value of **a.** 6 and **b.** -9.

Solution

a. $|6| = 6$

b. $|-9| = 9$

YOU TRY IT • 8

Find the absolute value of **a.** -8 and **b.** 12.

Your solution

EXAMPLE • 9

Evaluate **a.** $|-27|$ and **b.** $-|-14|$.

Solution

a. $|-27| = 27$

b. $-|-14| = -14$

YOU TRY IT • 9

Evaluate **a.** $|0|$ and **b.** $-|35|$.

Your solution

EXAMPLE • 10

Evaluate $|-x|$, where $x = -4$.

Solution

$|-x| = |-(-4)| = |4| = 4$

YOU TRY IT • 10

Evaluate $|-y|$, where $y = 2$.

Your solution

EXAMPLE • 11

Write the given numbers in order from smallest to largest.

$|-7|, -5, |0|, -(-4), -|-3|$

Solution

$|-7| = 7, |0| = 0,$

$-(-4) = 4, -|-3| = -3$

$-5, -|-3|, |0|, -(-4), |-7|$

YOU TRY IT • 11

Write the given numbers in order from smallest to largest.

$|6|, |-2|, -(-1), -4, -|-8|$

Your solution

Solutions on p. S8

OBJECTIVE D To solve application problems

EXAMPLE • 12

Which is the colder temperature, -18°F or -15°F?

Strategy

To determine which is the colder temperature, compare the numbers -18 and -15. The lower number corresponds to the colder temperature.

Solution

$-18 < -15$

The colder temperature is -18°F.

YOU TRY IT • 12

Which is closer to blastoff, -9 s and counting or -7 s and counting?

Your strategy

Your solution

Solution on p. S8

3.1 EXERCISES

OBJECTIVE A **To identify order relations between integers**

For Exercises 1 to 8, graph the number on the number line.

1. -5

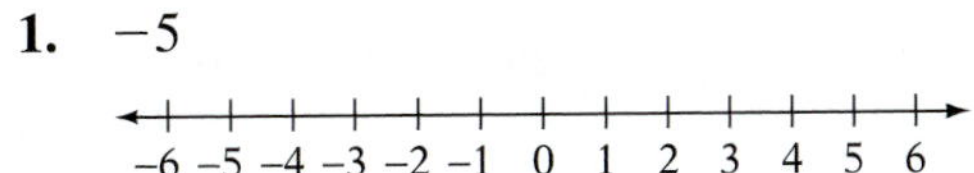

2. -1

−6 −5 −4 −3 −2 −1 0 1 2 3 4 5 6

3. -6

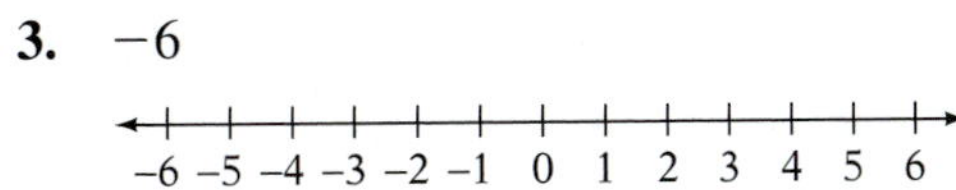

4. -2

−6 −5 −4 −3 −2 −1 0 1 2 3 4 5 6

5. x, where $x = 5$

−6 −5 −4 −3 −2 −1 0 1 2 3 4 5 6

6. x, where $x = 0$

−6 −5 −4 −3 −2 −1 0 1 2 3 4 5 6

7. x, where $x = -4$

−6 −5 −4 −3 −2 −1 0 1 2 3 4 5 6

8. x, where $x = -3$

−6 −5 −4 −3 −2 −1 0 1 2 3 4 5 6

On the number line, which number is:

9. 3 units to the right of -2?

10. 5 units to the right of -3?

11. 4 units to the left of 3?

12. 2 units to the left of -1?

13. 6 units to the right of -3?

14. 4 units to the right of -4?

For Exercises 15 to 18, use the following number line.

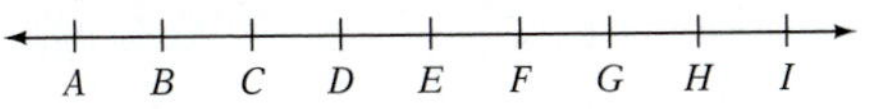

15. If F is 1 and G is 2, what numbers are A and C?

16. If G is 1 and H is 2, what numbers are B and D?

17. If H is 0 and I is 1, what numbers are A and D?

18. If G is 2 and I is 4, what numbers are B and E?

For Exercises 19 to 30, place the correct symbol, $<$ or $>$, between the two numbers.

19. $-2 \quad -5$ **20.** $-6 \quad -1$ **21.** $3 \quad -7$ **22.** $-11 \quad -8$

23. $-42 \quad -7$ **24.** $-21 \quad -34$ **25.** $53 \quad -46$ **26.** $-27 \quad -39$

27. $-51 \quad -20$ **28.** $-136 \quad 0$ **29.** $-131 \quad 101$ **30.** $127 \quad -150$

For Exercises 31 to 39, write the given numbers in order from smallest to largest.

31. $3, -7, 0, -2$ **32.** $-4, 8, 6, -1$ **33.** $-3, 1, -5, 4$

34. $-6, 2, -8, 7$ **35.** $9, -4, 5, 0$ **36.** $6, -9, -12, 8$

37. $-10, 4, 12, -5, -7$ **38.** $11, -8, -1, 7, -6$ **39.** $10, -11, -2, 5, -7$

For Exercises 40 to 43, determine whether the statement is *always true, never true,* or *sometimes true.*

40. A number that is to the right of -6 on the number line is a negative number.

41. A number that is to the left of -2 on the number line is a negative number.

42. A number that is to the right of 7 on the number line is a negative number.

43. A number that is to the left of 4 on the number line is a negative number.

OBJECTIVE B To find the opposite of a number

For Exercises 44 to 51, find the opposite of the number.

44. 22 **45.** 45 **46.** -31 **47.** -88

48. c **49.** n **50.** $-w$ **51.** $-d$

For Exercises 52 to 63, write the expression in words.

52. $-(-11)$ **53.** $-(-13)$ **54.** $-(-d)$ **55.** $-(-p)$

56. $-2 + (-5)$ **57.** $5 + (-10)$ **58.** $6 - (-7)$ **59.** $-14 - (-3)$

60. $9 - 12$ **61.** $-13 - 8$ **62.** $-a - b$ **63.** $m + (-n)$

For Exercises 64 to 75, simplify.

64. $-(-5)$ **65.** $-(-7)$ **66.** $-(-38)$ **67.** $-(-61)$

68. $-(29)$ **69.** $-(46)$ **70.** $-(-52)$ **71.** $-(-73)$

72. $-(-m)$ **73.** $-(-z)$ **74.** $-(b)$ **75.** $-(p)$

For Exercises 76 and 77, determine whether the statement is true for *positive integers, negative integers,* or *all integers.*

76. The opposite of an integer is less than the integer.

77. The opposite of an integer is negative.

OBJECTIVE C To evaluate expressions that contain the absolute value symbol

For Exercises 78 to 85, find the absolute value of the number.

78. 4 **79.** -4 **80.** -7 **81.** 9

82. -1 **83.** -11 **84.** 10 **85.** -12

For Exercises 86 to 101, evaluate.

86. $|-15|$ **87.** $|-23|$ **88.** $-|33|$ **89.** $-|27|$

90. $|32|$ **91.** $|25|$ **92.** $-|-36|$ **93.** $-|-41|$

94. $-|-81|$ **95.** $-|-93|$

96. $|x|$, where $x = 7$ **97.** $|x|$, where $x = -10$

98. $|-x|$, where $x = 2$

99. $|-x|$, where $x = 8$

100. $|-y|$, where $y = -3$

101. $|-y|$, where $y = -6$

For Exercises 102 to 109, place the correct symbol, $<$, $=$, or $>$, between the two numbers.

102. $|7| \quad |-9|$

103. $|-12| \quad |8|$

104. $|-5| \quad |-2|$

105. $|6| \quad |13|$

106. $|-8| \quad |3|$

107. $|-1| \quad |-17|$

108. $|-14| \quad |14|$

109. $|x| \quad |-x|$

For Exercises 110 to 115, write the given numbers in order from smallest to largest.

110. $|-8|, -(-3), |2|, -|-5|$

111. $-|6|, -(4), |-7|, -(-9)$

112. $-(-1), |-6|, |0|, -|3|$

113. $-|-7|, -9, -(5), |4|$

114. $-|2|, -(-8), 6, |1|, -7$

115. $-(-3), -|-8|, |5|, -|10|, -(-2)$

For Exercises 116 and 117, determine whether the statement is true for *positive integers, negative integers,* or *all integers.*

116. The absolute value of an integer is the opposite of the integer.

117. The absolute value of an integer is greater than the integer.

OBJECTIVE D To solve application problems

Environmental Science The table below gives equivalent temperatures for combinations of temperature and wind speed. For example, the combination of a temperature of 15°F and a wind blowing at 10 mph has a cooling power equal to 3°F. Use this table for Exercises 118 to 123.

Wind Chill Factors

Wind Speed (mph)	*Thermometer Reading (degrees Fahrenheit)*														
	25	20	15	10	5	0	−5	−10	−15	−20	−25	−30	−35	−40	−45
5	19	13	7	1	−5	−11	−16	−22	−28	−34	−40	−46	−52	−57	−63
10	15	9	3	−4	−10	−16	−22	−28	−35	−41	−47	−53	−59	−66	−72
15	13	6	0	−7	−13	−19	−26	−32	−39	−45	−51	−58	−64	−71	−77
20	11	4	−2	−9	−15	−22	−29	−35	−42	−48	−55	−61	−68	−74	−81
25	9	3	−4	−11	−17	−24	−31	−37	−44	−51	−58	−64	−71	−78	−84
30	8	1	−5	−12	−19	−26	−33	−39	−46	−53	−60	−67	−73	−80	−87
35	7	0	−7	−14	−21	−27	−34	−41	−48	−55	−62	−69	−76	−82	−89
40	6	−1	−8	−15	−22	−29	−36	−43	−50	−57	−64	−71	−78	−84	−91
45	5	−2	−9	−16	−23	−30	−37	−44	−51	−58	−65	−72	−79	−86	−93

118. Find the wind chill factor when the temperature is 5°F and the wind speed is 15 mph.

119. Find the wind chill factor when the temperature is 10°F and the wind speed is 20 mph.

120. Find the cooling power of a temperature of −10°F and a 5-mph wind.

121. Find the cooling power of a temperature of −15°F and a 10-mph wind.

122. Which feels colder, a temperature of 0°F with a 15-mph wind or a temperature of 10°F with a 25-mph wind?

123. Which would feel colder, a temperature of −30°F with a 5-mph wind or a temperature of −20°F with a 10-mph wind?

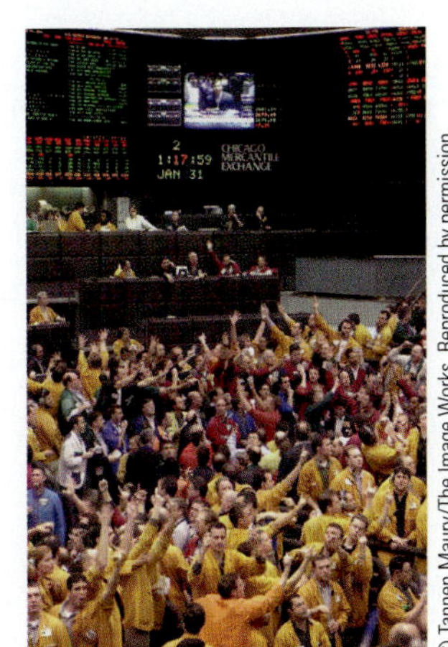

124. **Rocketry** Which is closer to blastoff, −12 min and counting or −17 min and counting?

125. **Stocks** In the stock market, the net change in the price of a share of stock is recorded as a positive or a negative number. If the price rises, the net change is positive. If the price falls, the net change is negative. If the net change for a share of Stock A is −2 and the net change for a share of Stock B is −1, which stock showed the least net change?

126. **Business** Some businesses show a profit as a positive number and a loss as a negative number. During the first quarter of this year, the loss experienced by a company was recorded as −12,575. During the second quarter of this year, the loss experienced by the company was −11,350. During which quarter was the loss greater?

127. **Mathematics** *A* is a point on the number line halfway between −9 and 3. *B* is a point halfway between *A* and the graph of 1 on the number line. *B* is the graph of what number?

Applying the Concepts

128. Find the values of a for which $|a| = 7$.

129. Find the values of y for which $|y| = 11$.

130. **a.** Name two numbers that are 4 units from 2 on the number line.
b. Name two numbers that are 5 units from 3 on the number line.

SECTION 3.2 Addition and Subtraction of Integers

OBJECTIVE A To add integers

Not only can an integer be graphed on a number line, an integer can be represented anywhere along a number line by an arrow. A positive number is represented by an arrow pointing to the right. A negative number is represented by an arrow pointing to the left. The absolute value of the number is represented by the length of the arrow. The integers 5 and -4 are shown on the number line in the figure below.

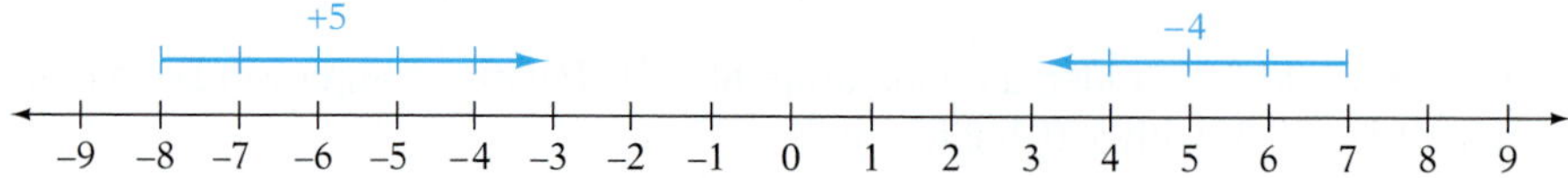

The sum of two integers can be shown on a number line. To add two integers, find the point on the number line corresponding to the first addend. At that point, draw an arrow representing the second addend. The sum is the number directly below the tip of the arrow.

$4 + 2 = 6$

$-4 + (-2) = -6$

$-4 + 2 = -2$

$4 + (-2) = 2$

The sums shown above can be categorized by the signs of the addends.

The addends have the same sign.

$4 + 2$	positive 4 plus positive 2
$-4 + (-2)$	negative 4 plus negative 2

The addends have different signs.

$-4 + 2$	negative 4 plus positive 2
$4 + (-2)$	positive 4 plus negative 2

The rule for adding two integers depends on whether the signs of the addends are the same or different.

Tips for Success

Be sure to do all you need to do in order to be successful at adding and subtracting integers: Read through the introductory material, work through the HOW TO examples, study the paired Examples, do the You Try Its and check your solutions against those in the back of the book, and do the exercises in the 3.2 Exercise set. See *AIM for Success* in the front of the book.

Rule for Adding Two Integers

To add two integers with the same sign, add the absolute values of the numbers. Then attach the sign of the addends.

To add two integers with different signs, find the absolute values of the numbers. Subtract the smaller absolute value from the larger absolute value. Then attach the sign of the addend with the larger absolute value.

HOW TO 1 Add: $(-4) + (-9)$

$|-4| = 4, |-9| = 9$ • The signs of the addends are the same. Find the absolute values of the numbers.

$4 + 9 = 13$ • Add the absolute values of the numbers.

$(-4) + (-9) = -13$ • Attach the sign of the addends. (Both addends are negative. The sum is negative.)

Integrating Technology

To add $-14 + (-47)$ with your calculator, enter the following:

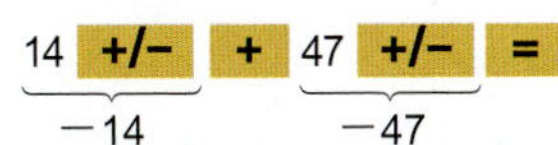

HOW TO 2 Add: $-14 + (-47)$

$-14 + (-47) = -61$ • The signs are the same. Add the absolute values of the numbers. Attach the sign of the addends.

HOW TO 3 Add: $6 + (-13)$

$|6| = 6, |-13| = 13$ • The signs of the addends are different. Find the absolute values of the numbers.

$13 - 6 = 7$ • Subtract the smaller absolute value from the larger absolute value.

$|-13| > |6|$ • Attach the sign of the number with the larger absolute value. Attach the negative sign.

$6 + (-13) = -7$

HOW TO 4 Add: $162 + (-247)$

$247 - 162 = 85$ • The signs are different. Find the difference between the absolute values of the numbers.

$162 + (-247) = -85$ • Attach the sign of the number with the larger absolute value.

HOW TO 5 Add: $-8 + 8$

$8 - 8 = 0$ • The signs are different. Find the difference between the absolute values of the numbers.

$-8 + 8 = 0$

Note in this last example that we are adding a number and its opposite (−8 and 8), and the sum is 0. The opposite of a number is called its **additive inverse.** The opposite or additive inverse of −8 is 8, and the opposite or additive inverse of 8 is −8. **The sum of a number and its additive inverse is always zero.** This is known as the Inverse Property of Addition.

The properties of addition presented in Chapter 1 hold true for integers as well as for whole numbers. These properties are repeated below, along with the Inverse Property of Addition.

Take Note
With the Commutative Properties, the order in which the numbers appear changes. With the Associative Properties, the order in which the numbers appear remains the same.

The Addition Property of Zero	$a + 0 = a$ or $0 + a = a$
The Commutative Property of Addition	$a + b = b + a$
The Associative Property of Addition	$(a + b) + c = a + (b + c)$
The Inverse Property of Addition	$a + (-a) = 0$ or $-a + a = 0$

Take Note
For the example at the right, check that the sum is the same if the numbers are added in a different order.

HOW TO • 6 Add: $(-4) + (-6) + (-8) + 9$

$(-4) + (-6) + (-8) + 9$

$= (-10) + (-8) + 9$ • Add the first two numbers.

$= (-18) + 9$ • Add the sum to the third number.

$= -9$ • Continue until all the numbers have been added.

HOW TO • 7 The price of Byplex Corporation's stock fell each trading day of the first week of June 2005. Use Figure 3.1 to find the change in the price of Byplex stock over the week's time.

FIGURE 3.1 Change in Price of Byplex Corporation Stock

Add the five changes in price.

$-2 + (-3) + (-1) + (-2) + (-1)$

$= (-5) + (-1) + (-2) + (-1)$

$= -6 + (-2) + (-1)$

$= -8 + (-1) = -9$

The change in the price was −9.

This means that the price of the stock fell \$9 per share.

HOW TO • 8 Evaluate $-x + y$ when $x = -15$ and $y = -5$.

$-x + y$
$-(-15) + (-5)$ • Replace x with -15 and y with -5.
$= 15 + (-5)$ • Simplify $-(-15)$.
$= 10$ • Add.

Take Note
Recall that a solution of an equation is a number that, when substituted for the variable, results in a true equation.

HOW TO • 9 Is -7 a solution of the equation $x + 4 = -3$?

$x + 4 = -3$
$-7 + 4 \mid -3$ • Replace x by -7 and then simplify.
$-3 = -3$ • The results are equal.

-7 is a solution of the equation.

EXAMPLE • 1

Add: $-97 + (-45)$

Solution
$-97 + (-45)$
$= -142$ • The signs of the addends are the same.

YOU TRY IT • 1

Add: $-38 + (-62)$

Your solution

EXAMPLE • 2

Add: $81 + (-79)$

Solution
$81 + (-79)$
$= 2$ • The signs of the addends are different.

YOU TRY IT • 2

Add: $47 + (-53)$

Your solution

EXAMPLE • 3

Add: $42 + (-12) + (-30)$

Solution
$42 + (-12) + (-30)$
$= 30 + (-30)$
$= 0$

YOU TRY IT • 3

Add: $-36 + 17 + (-21)$

Your solution

EXAMPLE • 4

What is -162 increased by 98?

Solution
$-162 + 98 = -64$

YOU TRY IT • 4

Find the sum of -154 and -37.

Your solution

EXAMPLE • 5

Evaluate $-x + y$ when $x = -11$ and $y = -2$.

Solution
$-x + y$
$-(-11) + (-2) = 11 + (-2)$
$= 9$

YOU TRY IT • 5

Evaluate $-x + y$ when $x = -3$ and $y = -10$.

Your solution

Solutions on p. S8

EXAMPLE • 6

Is -6 a solution of the equation $3 + y = -2$?

Solution

$$\begin{array}{r|l} 3 + y & = -2 \\ \hline 3 + (-6) & -2 \\ -3 & \neq -2 \end{array}$$

• Replace y by -6.

No, -6 is not a solution of the equation.

YOU TRY IT • 6

Is -9 a solution of the equation $2 = 11 + a$?

Your solution

Solution on p. S8

OBJECTIVE B **To subtract integers** 

Before the rules for subtracting two integers are explained, look at the translation into words of expressions that represent the difference of two integers.

$9 - 3$	positive 9 minus positive 3
$-9 - 3$	negative 9 minus positive 3
$9 - (-3)$	positive 9 minus negative 3
$-9 - (-3)$	negative 9 minus negative 3

Note that the sign $-$ is used in two different ways. One way is as a negative sign, as in -9 (negative 9). The second way is to indicate the operation of subtraction, as in $9 - 3$ (9 minus 3).

Look at the next four expressions and decide whether the second number in each expression is a positive number or a negative number.

1. $(-10) - 8$
2. $(-10) - (-8)$
3. $10 - (-8)$
4. $10 - 8$

In expressions 1 and 4, the second number is positive 8. In expressions 2 and 3, the second number is negative 8.

Opposites are used to rewrite subtraction problems as related addition problems. Notice below that the subtraction of whole numbers is the same as the addition of the opposite number.

Subtraction		*Addition of the Opposite*	
$8 - 4$	$=$	$8 + (-4)$	$= 4$
$7 - 5$	$=$	$7 + (-5)$	$= 2$
$9 - 2$	$=$	$9 + (-2)$	$= 7$

Subtraction of integers can be written as the addition of the opposite number. To subtract two integers, rewrite the subtraction expression as the first number plus the opposite of the second number. Some examples are shown below.

First number	−	second number	=	First number	+	opposite of the second number
8	−	15	=	8	+	$(-15) = -7$
8	−	(-15)	=	8	+	$15 = 23$
−8	−	15	=	−8	+	$(-15) = -23$
−8	−	(-15)	=	−8	+	$15 = 7$

Rule for Subtracting Two Integers

To subtract two integers, add the opposite of the second integer to the first integer.

HOW TO • 10 Subtract: $(-15) - 75$

$(-15) - 75$

$= (-15) + (-75)$

$= -90$

- **Rewrite the subtraction operation as the sum of the first number and the opposite of the second number. The opposite of 75 is −75.**
- **Add.**

HOW TO • 11 Subtract: $6 - (-20)$

$6 - (-20)$

$= 6 + 20$

$= 26$

- **Rewrite the subtraction operation as the sum of the first number and the opposite of the second number. The opposite of −20 is 20.**

HOW TO • 12 Subtract: $11 - 42$

$11 - 42$

$= 11 + (-42)$

$= -31$

- **Rewrite the subtraction operation as the sum of the first number and the opposite of the second number. The opposite of 42 is −42.**

Take Note

$42 - 11 = 31$

$11 - 42 = -31$

$42 - 11 \neq 11 - 42$

By the Commutative Property of Addition, the order in which two numbers are added does not affect the sum; $a + b = b + a$. However, note that the order in which two numbers are subtracted *does* affect the difference. The operation of subtraction is not commutative.

Integrating Technology

To subtract $-13 - 5 - (-8)$ with your calculator, enter the following:

13 **+/−** **−** 5 **−** 8 **+/−** **=**

(13 +/− gives −13; 8 +/− gives −8)

When subtraction occurs several times in an expression, rewrite each subtraction as addition of the opposite and then add.

HOW TO • 13 Subtract: $-13 - 5 - (-8)$

$-13 - 5 - (-8)$
$= -13 + (-5) + 8$ • Rewrite each subtraction as addition of the opposite.
$= -18 + 8$ • Add.
$= -10$

HOW TO • 14 Simplify: $-14 + 6 - (-7)$

$-14 + 6 - (-7)$ • This problem involves both addition and subtraction.
$= -14 + 6 + 7$ Rewrite the subtraction as addition of the opposite.
$= -8 + 7$ • Add.
$= -1$

HOW TO • 15 Evaluate $a - b$ when $a = -2$ and $b = -9$.

$a - b$
$-2 - (-9)$ • Replace a with -2 and b with -9.
$= -2 + 9$ • Rewrite the subtraction as addition of the opposite.
$= 7$ • Add.

HOW TO • 16 Is -4 a solution of the equation $3 - a = 11 + a$?

$3 - a$	$= 11 + a$
$3 - (-4)$	$11 + (-4)$
$3 + 4$	7

• Replace a by -4 and then simplify.

$7 = 7$ • The results are equal.

Yes, -4 is a solution of the equation.

EXAMPLE • 7

Subtract: $-12 - (-17)$

Solution

$-12 - (-17)$
$= -12 + 17$ • Rewrite "−" as "+". The opposite of -17 is 17.
$= 5$

YOU TRY IT • 7

Subtract: $-35 - (-34)$

Your solution

EXAMPLE • 8

Subtract: $66 - (-90)$

Solution

$66 - (-90)$
$= 66 + 90$ • Rewrite "−" as "+". The opposite of -90 is 90.
$= 156$

YOU TRY IT • 8

Subtract: $83 - (-29)$

Your solution

Solutions on p. S8

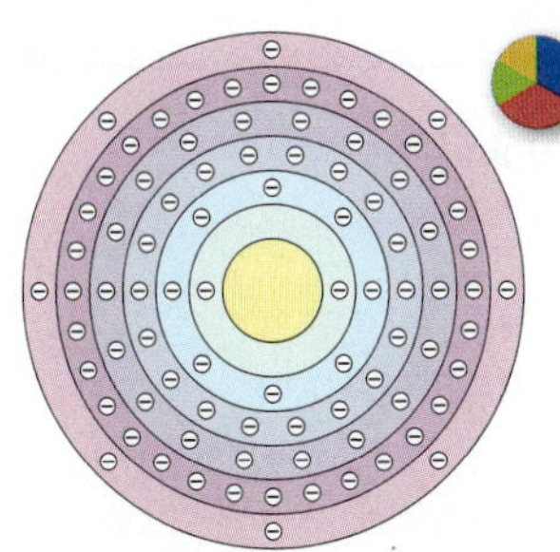
Radon

The table below shows the boiling point and the melting point in degrees Celsius of three chemical elements. Use this table for Example 9 and You Try It 9.

Chemical Element	*Boiling Point*	*Melting Point*
Mercury	357	−39
Radon	−62	−71
Xenon	−108	−112

EXAMPLE • 9

Use the table above to find the difference between the boiling point and the melting point of mercury.

Solution

The boiling point of mercury is 357.

The melting point of mercury is −39.

$$357 - (-39) = 357 + 39$$
$$= 396$$

The difference is 396°C.

YOU TRY IT • 9

Use the table above to find the difference between the boiling point and the melting point of xenon.

Your solution

EXAMPLE • 10

What is −12 minus 8?

Solution

$-12 - 8$

$= -12 + (-8)$ • **Rewrite "−" as "+". The opposite of 8 is −8.**

$= -20$

YOU TRY IT • 10

What is 14 less than −8?

Your solution

EXAMPLE • 11

Subtract 91 from 43.

Solution

$43 - 91$

$= 43 + (-91)$ • **Rewrite "−" as "+". The opposite of 91 is −91.**

$= -48$

YOU TRY IT • 11

What is 25 decreased by 68?

Your solution

EXAMPLE • 12

Simplify: $-8 - 30 - (-12) - 7 - (-14)$

Solution

$-8 - 30 - (-12) - 7 - (-14)$

$= -8 + (-30) + 12 + (-7) + 14$

$= -38 + 12 + (-7) + 14$

$= -26 + (-7) + 14$

$= -33 + 14$

$= -19$

YOU TRY IT • 12

Simplify: $-4 - (-3) + 12 - (-7) - 20$

Your solution

Solutions on p. S8

EXAMPLE • 13

Evaluate $-x - y$ when $x = -4$ and $y = -3$.

Solution

$-x - y$

$$-(-4) - (-3) = 4 - (-3)$$
$$= 4 + 3$$
$$= 7$$

YOU TRY IT • 13

Evaluate $x - y$ when $x = -9$ and $y = 7$.

Your solution

EXAMPLE • 14

Is 8 a solution of the equation $-2 = 6 - x$?

Solution

$$-2 = 6 - x$$
$$-2 \mid 6 - 8 \qquad \text{• Replace } x \text{ by } 8.$$
$$-2 \mid 6 + (-8)$$
$$-2 = -2$$

Yes, 8 is a solution of the equation.

YOU TRY IT • 14

Is -3 a solution of the equation $a - 5 = -8$?

Your solution

Solutions on p. S8

OBJECTIVE C

To solve application problems

Figure 3.2 shows the melting points in degrees Celsius of six chemical elements. The abbreviations of the elements are:

F—Fluorine	H—Hydrogen
S—Sulfur	N—Nitrogen
O—Oxygen	Li—Lithium

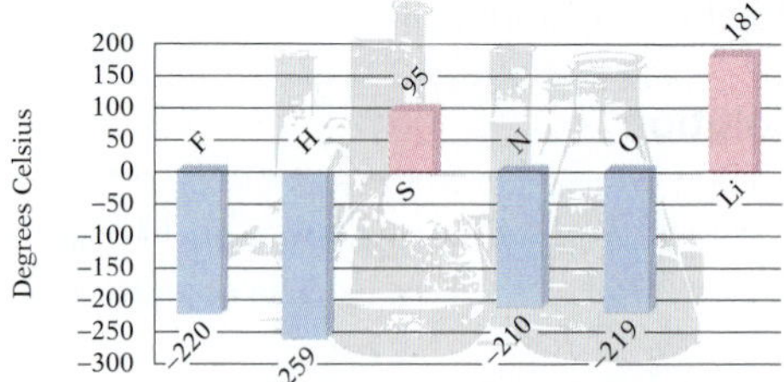

FIGURE 3.2 Melting Points of Chemical Elements

Use this graph for Example 15 and You Try It 15.

EXAMPLE • 15

Find the difference between the two lowest melting points shown in Figure 3.2.

Strategy

To find the difference, subtract the lowest melting point shown (-259) from the second lowest melting point shown (-220).

Solution

$$-220 - (-259) = -220 + 259 = 39$$

The difference is 39°C.

YOU TRY IT • 15

Find the difference between the highest and lowest melting points shown in Figure 3.2.

Your strategy

Your solution

Solution on p. S9

EXAMPLE • 16

Find the temperature after an increase of 8°C from -5°C.

Strategy
To find the temperature, add the increase (8) to the previous temperature (-5).

Solution
$-5 + 8 = 3$

The temperature is 3°C.

YOU TRY IT • 16

Find the temperature after an increase of 10°C from -3°C.

Your strategy

Your solution

EXAMPLE • 17

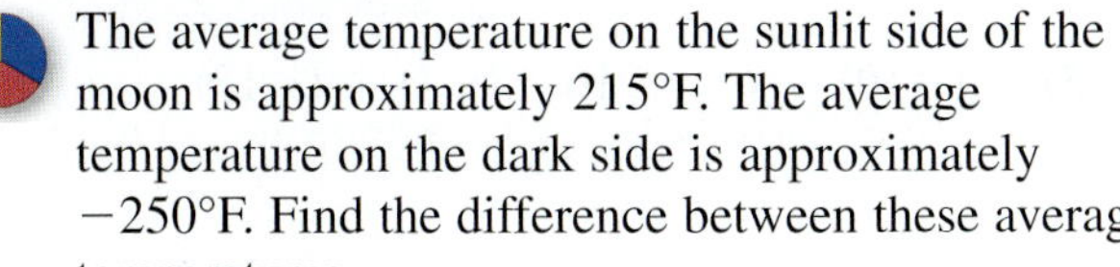

The average temperature on the sunlit side of the moon is approximately 215°F. The average temperature on the dark side is approximately -250°F. Find the difference between these average temperatures.

Strategy
To find the difference, subtract the average temperature on the dark side of the moon (-250) from the average temperature on the sunlit side (215).

Solution
$$215 - (-250) = 215 + 250$$
$$= 465$$

The difference is 465°F.

YOU TRY IT • 17

The average temperature on Earth's surface is 57°F. The average temperature throughout Earth's stratosphere is -70°F. Find the difference between these average temperatures.

Your strategy

Your solution

EXAMPLE • 18

The distance, d, between point a and point b on the number line is given by the formula $d = |a - b|$. Use the formula to find d when $a = 7$ and $b = -8$.

Strategy
To find d, replace a by 7 and b by -8 in the given formula and solve for d.

Solution
$d = |a - b|$
$d = |7 - (-8)|$
$d = |7 + 8|$
$d = |15|$
$d = 15$

The distance between the two points is 15 units.

YOU TRY IT • 18

The distance, d, between point a and point b on the number line is given by the formula $d = |a - b|$. Use the formula to find d when $a = -6$ and $b = 5$.

Your strategy

Your solution

Solutions on p. S9

3.2 EXERCISES

OBJECTIVE A To add integers

For Exercises 1 to 36, add.

1. $-3 + (-8)$
2. $-6 + (-9)$
3. $-8 + 3$
4. $-7 + 2$
5. $-5 + 13$
6. $-4 + 11$
7. $6 + (-10)$
8. $8 + (-12)$
9. $3 + (-5)$
10. $6 + (-7)$
11. $-4 + (-5)$
12. $-12 + (-12)$
13. $-6 + 7$
14. $-9 + 8$
15. $(-5) + (-10)$
16. $(-3) + (-17)$
17. $-7 + 7$
18. $-11 + 11$
19. $(-15) + (-6)$
20. $(-18) + (-3)$
21. $0 + (-14)$
22. $-19 + 0$
23. $73 + (-54)$
24. $-89 + 62$
25. $2 + (-3) + (-4)$
26. $7 + (-2) + (-8)$
27. $-3 + (-12) + (-15)$
28. $9 + (-6) + (-16)$
29. $-17 + (-3) + 29$
30. $13 + 62 + (-38)$
31. $11 + (-22) + 4 + (-5)$
32. $-14 + (-3) + 7 + (-6)$
33. $-22 + 10 + 2 + (-18)$
34. $-6 + (-8) + 13 + (-4)$
35. $-25 + (-31) + 24 + 19$
36. $10 + (-14) + (-21) + 8$
37. What is 3 increased by -21?
38. Find 12 plus -9.
39. What is 16 more than -5?
40. What is 17 added to -7?
41. Find the total of -3, -8, and 12.
42. Find the sum of 5, -16, and -13.
43. Write the sum of x and -7.
44. Write the total of $-a$ and b.

For Exercises 45 to 52, evaluate the expression for the given values of the variables.

45. $x + y$, where $x = -5$ and $y = -7$

46. $-a + b$, where $a = -8$ and $b = -3$

47. $a + b$, where $a = -8$ and $b = -3$

48. $-x + y$, where $x = -5$ and $y = -7$

49. $a + b + c$, where $a = -4$, $b = 6$, and $c = -9$

50. $a + b + c$, where $a = -10$, $b = -6$, and $c = 5$

51. $x + y + (-z)$, where $x = -3$, $y = 6$, and $z = -17$

52. $-x + (-y) + z$, where $x = -2$, $y = 8$, and $z = -11$

For Exercises 53 to 56, identify the property that justifies the statement.

53. $-12 + 5 = 5 + (-12)$

54. $-33 + 0 = -33$

55. $-46 + 46 = 0$

56. $-7 + (3 + 2) = (-7 + 3) + 2$

For Exercises 57 to 60, use the given property of addition to complete the statement.

57. The Associative Property of Addition
$-11 + (6 + 9) = (? + 6) + 9$

58. The Addition Property of Zero
$-13 + ? = -13$

59. The Commutative Property of Addition
$-2 + ? = -4 + (-2)$

60. The Inverse Property of Addition
$? + (-18) = 0$

61. Is -3 a solution of the equation $x + 4 = 1$?

62. Is -8 a solution of the equation $6 = -3 + z$?

63. Is -2 a solution of the equation $3 + y = y + 3$?

64. Is -4 a solution of the equation $1 + z = z + 2$?

For Exercises 65 to 68, determine whether the statement is *always true, never true,* or *sometimes true.*

65. The sum of an integer and its opposite is zero.

66. The sum of two negative integers is a positive integer.

67. The sum of two negative integers and one positive integer is a negative integer.

68. If the absolute value of a negative integer is greater than the absolute value of a positive integer, then the sum of the integers is negative.

OBJECTIVE B To subtract integers

69. **a.** Explain how to rewrite the subtraction $8 - (-6)$ as addition of the opposite.

b. Explain the meanings of the words *minus* and *negative*.

For Exercises 70 to 89, subtract.

70. $7 - 14$ **71.** $6 - 9$ **72.** $-7 - 2$ **73.** $-9 - 4$

74. $7 - (-2)$ **75.** $3 - (-4)$ **76.** $-6 - (-6)$ **77.** $-4 - (-4)$

78. $-12 - 16$ **79.** $-10 - 7$ **80.** $(-9) - (-3)$ **81.** $(-7) - (-4)$

82. $4 - (-14)$ **83.** $-4 - (-16)$ **84.** $(-14) - (-7)$ **85.** $3 - (-24)$

86. $9 - (-9)$ **87.** $(-41) - 65$ **88.** $57 - 86$ **89.** $-95 - (-28)$

90. How much larger is 5 than -11?

91. What is -10 decreased by -4?

92. Find -13 minus -8.

93. What is 6 less than -9?

94. Write the difference between $-y$ and 5.

95. Write $-t$ decreased by r.

Temperature The figure at the right shows the highest and lowest temperatures ever recorded for selected regions of the world. Use this graph for Exercises 96 to 98.

96. What is the difference between the highest and lowest temperatures ever recorded in Africa?

97. What is the difference between the highest and lowest temperatures ever recorded in South America?

98. What is the difference between the lowest temperature recorded in Europe and the lowest temperature recorded in Asia?

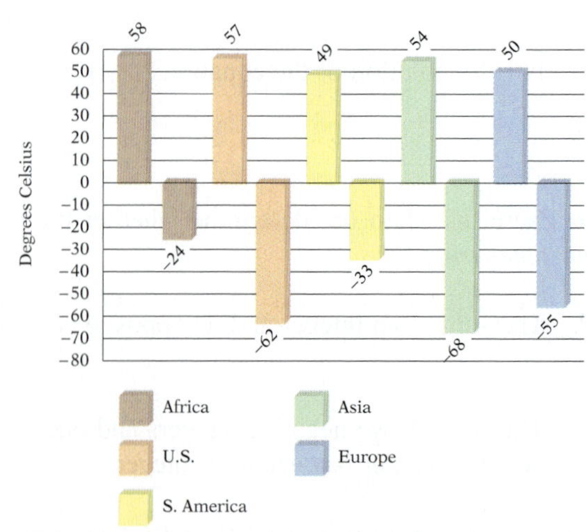

Highest and Lowest Temperatures Recorded (in degrees Celsius)

For Exercises 99 to 116, simplify.

99. $-4 - 3 - 2$

100. $4 - 5 - 12$

101. $12 - (-7) - 8$

102. $-12 - (-3) - (-15)$

103. $4 - 12 - (-8)$

104. $-30 - (-65) - 29 - 4$

105. $-16 - 47 - 63 - 12$

106. $42 - (-30) - 65 - (-11)$

107. $12 - (-6) + 8$

108. $-7 + 9 - (-3)$

109. $-8 - (-14) + 7$

110. $-4 + 6 - 8 - 2$

111. $9 - 12 + 0 - 5$

112. $11 - (-2) - 6 + 10$

113. $5 + 4 - (-3) - 7$

114. $-1 - 8 + 6 - (-2)$

115. $-13 + 9 - (-10) - 4$

116. $6 - (-13) - 14 + 7$

For Exercises 117 to 124, evaluate the expression for the given values of the variables.

117. $-x - y$, where $x = -3$ and $y = 9$

118. $x - (-y)$, where $x = -3$ and $y = 9$

119. $-x - (-y)$, where $x = -3$ and $y = 9$

120. $a - (-b)$, where $a = -6$ and $b = 10$

121. $a - b - c$, where $a = 4$, $b = -2$, and $c = 9$

122. $a - b - c$, where $a = -1$, $b = 7$, and $c = -15$

123. $x - y - (-z)$, where $x = -9$, $y = 3$, and $z = 30$

124. $-x - (-y) - z$, where $x = 8$, $y = 1$, and $z = -14$

125. Is -3 a solution of the equation $x - 7 = -10$?

126. Is -4 a solution of the equation $1 = 3 - y$?

127. Is -6 a solution of the equation $-t - 5 = 7 + t$?

128. Is -7 a solution of the equation $5 + a = -9 - a$?

For Exercises 129 to 132, determine whether the statement is *always true, never true,* or *sometimes true.*

129. The difference between a positive integer and a negative integer is zero.

130. A negative integer subtracted from a positive integer is a positive integer.

131. The difference between two negative integers is a positive integer.

132. The difference between an integer and its absolute value is zero.

OBJECTIVE C To solve application problems

Geography The elevation, or height, of places on Earth is measured in relation to sea level, or the average level of the ocean's surface. The table below shows height above sea level as a positive number and depth below sea level as a negative number. Use the table below for Exercises 133 to 135.

Mt. Everest

Continent	*Highest Elevation (in meters)*		*Lowest Elevation (in meters)*	
Africa	Mt. Kilimanjaro	5895	Lake Assal	−156
Asia	Mt. Everest	8850	Dead Sea	−411
North America	Denali	6194	Death Valley	−28
South America	Mt. Aconcagua	6960	Valdes Peninsula	−86

133. What is the difference in elevation (a) between Mt. Aconcagua and Death Valley and (b) between Mt. Kilimanjaro and Lake Assal?

134. For which continent shown is the difference between the highest and lowest elevations greatest?

135. For which continent shown is the difference between the highest and lowest elevations smallest?

136. Temperature The news clipping at the right was written on February 11, 2008. The record low temperature for Minnesota is −51°C. Find the difference between the low temperature in International Falls on February 11, 2008, and the record low temperature for Minnesota.

> **In the News**
>
> **Minnesota Town Named "Icebox of the Nation"**
>
> In International Falls, Minnesota, the temperature fell to −40°C just days after the citizens received word that the town had won a federal trademark naming it the "Icebox of the Nation."
>
> *Source:* news.yahoo.com

137. Temperature Find the temperature after a rise of 9°C from −6°C.

Aviation The table at the right shows the average temperatures at different cruising altitudes for airplanes. Use the table for Exercises 138 and 139.

Cruising Altitude	*Average Temperature*
12,000 ft	16°F
20,000 ft	−12°F
30,000 ft	−48°F
40,000 ft	−70°F
50,000 ft	−70°F

138. What is the difference between the average temperatures at 12,000 ft and at 40,000 ft?

139. What is the difference between the average temperatures at 40,000 ft and at 50,000 ft?

140. If the temperature begins at $-54°C$ and rises more than $60°C$, is the new temperature above or below $0°C$?

141. If the temperature begins at $-37°C$ and falls more than $40°C$, is the new temperature above or below $0°C$?

142. **Golf** Use the equation $S = N - P$, where S is a golfer's score relative to par in a tournament, N is the number of strokes made by the golfer, and P is par, to find a golfer's score relative to par when the golfer made 196 strokes and par is 208.

143. **Golf** Use the equation $S = N - P$, where S is a golfer's score relative to par in a tournament, N is the number of strokes made by the golfer, and P is par, to find a golfer's score relative to par when the golfer made 49 strokes and par is 52.

144. **Mathematics** The distance, d, between point a and point b on the number line is given by the formula $d = |a - b|$. Find d when $a = 6$ and $b = -15$.

145. **Mathematics** The distance, d, between point a and point b on the number line is given by the formula $d = |a - b|$. Find d when $a = 7$ and $b = -12$.

Applying the Concepts

146. **Mathematics** Given the list of numbers at the right, find the largest difference that can be obtained by subtracting one number in the list from a different number in the list.

$5, -2, -9, 11, 14$

147. Determine whether the statement is always true, sometimes true, or never true.
a. The difference between a number and its additive inverse is zero.
b. The sum of a negative number and a negative number is a negative number.

148. The sum of two negative integers is -7. Find the two integers.

149. Describe the steps involved in using a calculator to simplify $-17 - (-18) + (-5)$.

SECTION

3.3 Multiplication and Division of Integers

OBJECTIVE A To multiply integers 

When 5 is multiplied by a sequence of decreasing integers, each product decreases by 5.

$5(3) = 15$
$5(2) = 10$
$5(1) = 5$
$5(0) = 0$

The pattern developed can be continued so that 5 is multiplied by a sequence of negative numbers. To maintain the pattern of decreasing by 5, the resulting products must be negative.

$5(-1) = -5$
$5(-2) = -10$
$5(-3) = -15$
$5(-4) = -20$

This example illustrates that the product of a positive number and a negative number is negative.

Point of Interest

Operations with negative numbers were not accepted until the late 14th century. One of the first attempts to prove that the product of two negative numbers is positive was made in the book *Ars Magna,* by Girolamo Cardan, in 1545.

When -5 is multiplied by a sequence of decreasing integers, each product increases by 5.

$-5(3) = -15$
$-5(2) = -10$
$-5(1) = -5$
$-5(0) = 0$

The pattern developed can be continued so that -5 is multiplied by a sequence of negative numbers. To maintain the pattern of increasing by 5, the resulting products must be positive.

$-5(-1) = 5$
$-5(-2) = 10$
$-5(-3) = 15$
$-5(-4) = 20$

This example illustrates that the product of two negative numbers is positive.

The pattern for multiplication shown above is summarized in the following rule for multiplying integers.

Rule for Multiplying Two Integers

To multiply two integers with the same sign, multiply the absolute values of the factors. The product is **positive.**

To multiply two integers with different signs, multiply the absolute values of the factors. The product is **negative.**

Integrating Technology

To multiply $(-6)(-15)$ with your calculator, enter the following:

6 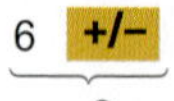  15 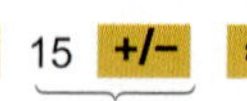 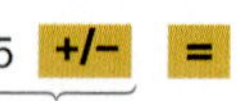

-6 -15

HOW TO 1 Multiply: $-9(12)$

$-9(12) = -108$ • The signs are different. The product is negative.

HOW TO 2 Multiply: $(-6)(-15)$

$(-6)(-15) = 90$ • The signs are the same. The product is positive.

HOW TO • 3  Figure 3.3 shows the melting points of bromine and mercury. The melting point of helium is seven times the melting point of mercury. Find the melting point of helium.

Multiply the melting point of mercury (−39°C) by 7.

$$-39(7) = -273$$

The melting point of helium is −273°C.

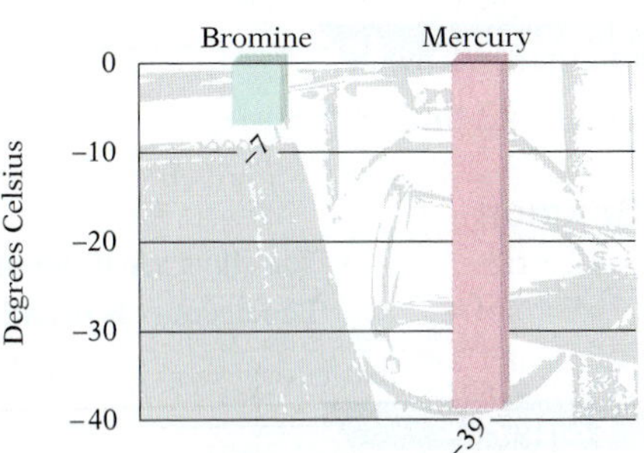

FIGURE 3.3 Melting Point of Chemical Elements (in degrees Celsius)

The properties of multiplication presented in Chapter 1 hold true for integers as well as whole numbers. These properties are repeated below.

The Multiplication Property of Zero	$a \cdot 0 = 0$ or $0 \cdot a = 0$
The Multiplication Property of One	$a \cdot 1 = a$ or $1 \cdot a = a$
The Commutative Property of Multiplication	$a \cdot b = b \cdot a$
The Associative Property of Multiplication	$(a \cdot b) \cdot c = a \cdot (b \cdot c)$

Take Note

For the example at the right, the product is the same if the numbers are multiplied in a different order. For instance,

$$2(-3)(-5)(-7)$$
$$= 2(-3)(35)$$
$$= 2(-105)$$
$$= -210$$

HOW TO • 4 Multiply: $2(-3)(-5)(-7)$

$2(-3)(-5)(-7)$

$= -6(-5)(-7)$ • **Multiply the first two numbers.**

$= 30(-7)$ • **Then multiply the product by the third number.**

$= -210$ • **Continue until all the numbers have been multiplied.**

By the Multiplication Property of One, $1 \cdot 6 = 6$ and $\mathbf{1 \cdot x = x}$. Applying the rules for multiplication, we can extend this to $-1 \cdot 6 = -6$ and $\mathbf{-1 \cdot x = -x}$.

Take Note

When variables are placed next to each other, it is understood that the operation is multiplication. $-ab$ means "the opposite of a times b."

HOW TO • 5 Evaluate $-ab$ when $a = -2$ and $b = -9$.

$-ab$

$-(-2)(-9)$ • **Replace a with −2 and b with −9.**

$= 2(-9)$ • **Simplify −(−2).**

$= -18$ • **Multiply.**

HOW TO • 6 Is −4 a solution of the equation $5x = -20$?

$5x$	$= -20$
$5(-4)$	-20
-20	$= -20$

• **Replace x by −4 and then simplify.**
• **The results are equal.**

Yes, −4 is a solution of the equation.

EXAMPLE • 1

Find -42 times 62.

Solution

$-42 \cdot 62$ • The signs are different. The product is negative.

$= -2604$

YOU TRY IT • 1

What is -38 multiplied by 51?

Your solution

EXAMPLE • 2

Multiply: $-5(-4)(6)(-3)$

Solution

$$\begin{aligned} -5(-4)(6)(-3) &= 20(6)(-3) \\ &= 120(-3) \\ &= -360 \end{aligned}$$

YOU TRY IT • 2

Multiply: $-7(-8)(9)(-2)$

Your solution

EXAMPLE • 3

Evaluate $-5x$ when $x = -11$.

Solution

$-5x$

$-5(-11) = 55$

YOU TRY IT • 3

Evaluate $-9y$ when $y = 20$.

Your solution

EXAMPLE • 4

Is 5 a solution of the equation $30 = -6z$?

Solution

$30 = -6z$

$30 \mid -6(5)$

$30 \neq -30$

No, 5 is not a solution of the equation.

YOU TRY IT • 4

Is -3 a solution of the equation $12 = -4a$?

Your solution

Solutions on p. S9

OBJECTIVE B To divide integers

Take Note

Recall that the fraction bar can be read "divided by." Therefore, $\frac{8}{2}$ can be read "8 divided by 2."

For every division problem, there is a related multiplication problem.

$$\text{Division: } \frac{8}{2} = 4 \qquad \text{Related multiplication: } 4(2) = 8$$

This fact can be used to illustrate a rule for dividing integers.

$$\frac{12}{3} = 4 \quad \text{because} \quad 4(3) = 12 \quad \text{and} \quad \frac{-12}{-3} = 4 \quad \text{because} \quad 4(-3) = -12.$$

These two division examples suggest that the quotient of two numbers with the same sign is positive. Now consider these two examples.

$$\frac{12}{-3} = -4 \qquad \text{because} \qquad -4(-3) = 12$$

$$\frac{-12}{3} = -4 \qquad \text{because} \qquad -4(3) = -12$$

These two division examples suggest that the quotient of two numbers with different signs is negative. This property is summarized next.

Rule for Dividing Two Integers

To divide two numbers with the same sign, divide the absolute values of the numbers. The quotient is **positive**.

To divide two numbers with different signs, divide the absolute values of the numbers. The quotient is **negative**.

Note from this rule that $\frac{12}{-3}$, $\frac{-12}{3}$, and $-\frac{12}{3}$ are all equal to -4.

If a and b are integers ($b \neq 0$), then $\frac{a}{-b} = \frac{-a}{b} = -\frac{a}{b}$.

HOW TO • 7 Divide: $-36 \div 9$

$-36 \div 9 = -4$ • The signs are different. The quotient is negative.

Integrating Technology

To divide (-105) by (-5) with your calculator, enter the following:

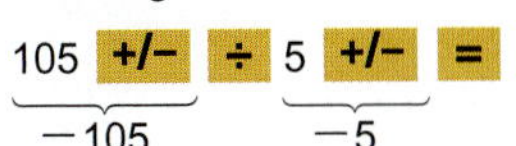

HOW TO • 8 Divide: $(-105) \div (-5)$

$(-105) \div (-5) = 21$ • The signs are the same. The quotient is positive.

HOW TO • 9 Figure 3.4 shows the record high and low temperatures in the United States for the first four months of the year. We can read from the graph that the record low temperature for April is -36°F. This is four times the record low temperature for September. What is the record low temperature for September?

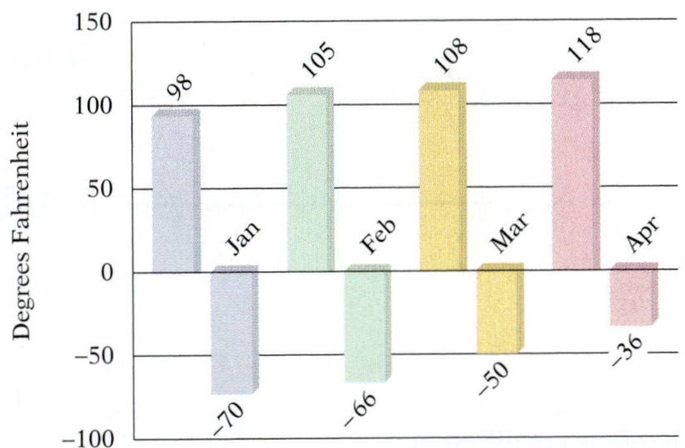

FIGURE 3.4 Record High and Low Temperatures, in Degrees Fahrenheit, in the United States for January, February, March, and April
Source: National Climatic Data Center, Asheville, NC, and Storm Phillips, STORMFAX, Inc.

To find the record low temperature for September, divide the record low for April (-36) by 4. $-36 \div 4 = -9$

The record low temperature in the United States for the month of September is -9°F.

The division properties of zero and one, which were presented in Chapter 1, hold true for integers as well as whole numbers. These properties are repeated here.

Point of Interest

Historical manuscripts indicate that mathematics is at least 4000 years old. Yet it was only 400 years ago that mathematicians started using variables to stand for numbers. Before that time, mathematics was written in words.

Division Properties of Zero and One

If $a \neq 0, \frac{0}{a} = 0.$ If $a \neq 0, \frac{a}{a} = 1.$

$\frac{a}{1} = a$ $\frac{a}{0}$ is undefined.

HOW TO • 10 Evaluate $a \div (-b)$ when $a = -28$ and $b = -4$.

$a \div (-b)$

$-28 \div [-(-4)]$ • Replace a with -28 and b with -4.

$= -28 \div (4)$ • Simplify $-(-4)$.

$= -7$ • Divide.

HOW TO • 11 Is -4 a solution of the equation $\frac{-20}{x} = 5$?

$\frac{-20}{x} = 5$

$\frac{-20}{-4} \mid 5$ • Replace x by -4 and then simplify.

$5 = 5$ • The results are equal.

Yes, -4 is a solution of the equation.

EXAMPLE • 5

Find the quotient of -23 and -23.

Solution

$-23 \div (-23) = 1$ • If $a \neq 0, \frac{a}{a} = 1.$

YOU TRY IT • 5

What is 0 divided by -17?

Your solution

EXAMPLE • 6

Divide: $\frac{95}{-5}$

Solution

$\frac{95}{-5} = -19$ • The signs are different. The quotient is negative.

YOU TRY IT • 6

Divide: $\frac{84}{-6}$

Your solution

EXAMPLE • 7

Divide: $x \div 0$

Solution

Division by zero is not defined.

$x \div 0$ is undefined.

YOU TRY IT • 7

Divide: $x \div 1$

Your solution

Solutions on p. S9

EXAMPLE • 8

Evaluate $\frac{-a}{b}$ when $a = -6$ and $b = -3$.

Solution

$$\frac{-a}{b}$$

$$\frac{-(-6)}{-3} = \frac{6}{-3} = -2$$

YOU TRY IT • 8

Evaluate $\frac{a}{-b}$ when $a = -14$ and $b = -7$.

Your solution

EXAMPLE • 9

Is -9 a solution of the equation $-3 = \frac{x}{3}$?

Solution

$$-3 = \frac{x}{3}$$

$$-3 \;\Big|\; \frac{-9}{3} \quad \text{• Replace } x \text{ by } -9.$$

$$-3 = -3$$

Yes, -9 is a solution of the equation.

YOU TRY IT • 9

Is -3 a solution of the equation $\frac{-6}{y} = -2$?

Your solution

Solutions on p. S9

OBJECTIVE C To solve application problems 

EXAMPLE • 10

The daily low temperatures during one week were recorded as follows: −10°, 2°, −1°, −9°, 1°, 0°, 3°. Find the average daily low temperature for the week.

Strategy

To find the average daily low temperature:

- Add the seven temperature readings.
- Divide by 7.

Solution

$-10 + 2 + (-1) + (-9) + 1 + 0 + 3 = -14$

$-14 \div 7 = -2$

The average daily low temperature was −2°.

YOU TRY IT • 10

The daily high temperatures during one week were recorded as follows: −7°, −8°, 0°, −1°, −6°, −11°, −2°. Find the average daily high temperature for the week.

Your strategy

Your solution

Solution on p. S9

3.3 EXERCISES

OBJECTIVE A To multiply integers

1. Name the operation in each expression and explain how you determined that it was that operation.
a. $8(-7)$ **b.** $8 - 7$ **c.** $8 - (-7)$ **d.** $-xy$ **e.** $x(-y)$ **f.** $-x - y$

For Exercises 2 to 33, multiply.

2. $-4 \cdot 6$
3. $-7 \cdot 3$
4. $-2(-3)$
5. $-5(-1)$
6. $(9)(2)$
7. $(3)(8)$
8. $5(-4)$
9. $4(-7)$
10. $-8(2)$
11. $-9(3)$
12. $(-5)(-5)$
13. $(-3)(-6)$
14. $(-7)(0)$
15. $-11(1)$
16. $14(3)$
17. $62(9)$
18. $-32(4)$
19. $-24(3)$
20. $(-8)(-26)$
21. $(-4)(-35)$
22. $9(-27)$
23. $8(-40)$
24. $-5 \cdot (23)$
25. $-6 \cdot (38)$
26. $-7(-34)$
27. $-4(-51)$
28. $4 \cdot (-8) \cdot 3$
29. $5 \cdot 7 \cdot (-2)$
30. $(-6)(5)(7)$
31. $(-9)(-9)(2)$
32. $-8(-7)(-4)$
33. $-1(4)(-9)$
34. What is twice -20?
35. Find the product of 100 and -7.
36. What is -30 multiplied by -6?
37. What is -9 times -40?
38. Write the product of $-q$ and r.
39. Write the product of $-f$, g, and h.

For Exercises 40 to 43, state whether the given product will be *positive, negative,* or *zero*.

40. The product of three negative integers

41. The product of two negative integers and one positive integer

42. The product of one negative integer, one positive integer, and zero

43. The product of five positive integers and one negative integer

For Exercises 44 to 47, identify the property that justifies the statement.

44. $0(-7) = 0$

45. $1p = p$

46. $-8(-5) = -5(-8)$

47. $-3(9 \cdot 4) = (-3 \cdot 9)4$

For Exercises 48 to 51, use the given property of multiplication to complete the statement.

48. The Commutative Property of Multiplication
$-3(-9) = -9(?)$

49. The Associative Property of Multiplication
$?(5 \cdot 10) = (-6 \cdot 5)10$

50. The Multiplication Property of Zero
$-81 \cdot ? = 0$

51. The Multiplication Property of One
$?(-14) = -14$

For Exercises 52 to 61, evaluate the expression for the given values of the variables.

52. xy, when $x = -3$ and $y = -8$

53. $-xy$, when $x = -3$ and $y = -8$

54. $x(-y)$, when $x = -3$ and $y = -8$

55. $-xyz$, when $x = -6$, $y = 2$, and $z = -5$

56. $-8a$, when $a = -24$

57. $-7n$, when $n = -51$

58. $5xy$, when $x = -9$ and $y = -2$

59. $8ab$, when $a = 7$ and $b = -1$

60. $-4cd$, when $c = 25$ and $d = -8$

61. $-5st$, when $s = -40$ and $t = -8$

62. Is -4 a solution of the equation $6m = -24$?

63. Is 0 a solution of the equation $-8 = -8a$?

64. Is 7 a solution of the equation $-3c = 21$?

65. Is 9 a solution of the equation $-27 = -3c$?

OBJECTIVE B To divide integers

For Exercises 66 to 85, divide.

66. $12 \div (-6)$

67. $18 \div (-3)$

68. $(-72) \div (-9)$

69. $(-64) \div (-8)$

70. $0 \div (-6)$

71. $-49 \div 1$

72. $81 \div (-9)$

73. $-40 \div (-5)$

74. $\frac{72}{-3}$

75. $\frac{44}{-4}$

76. $\frac{-93}{-3}$

77. $\frac{-98}{-7}$

78. $-114 \div (-6)$

79. $-91 \div (-7)$

80. $-53 \div 0$

81. $(-162) \div (-162)$

82. $-128 \div 4$

83. $-130 \div (-5)$

84. $(-200) \div 8$

85. $(-92) \div (-4)$

For Exercises 86 to 89, determine whether the statement is *always true, never true,* or *sometimes true.*

86. The quotient of a negative integer and its absolute value is –1.

87. The quotient of zero and a positive integer is a positive integer.

88. A negative integer divided by zero is zero.

89. The quotient of two negative numbers is the same as the quotient of the absolute values of the two numbers.

90. Find the quotient of -700 and 70.

91. Find 550 divided by -5.

92. What is -670 divided by -10?

93. What is the quotient of -333 and -3?

94. Write the quotient of $-a$ and b.

95. Write -9 divided by x.

For Exercises 96 to 103, evaluate the expression for the given values of the variables.

96. $a \div b$, where $a = -36$ and $b = -4$

97. $-a \div b$, where $a = -36$ and $b = -4$

98. $a \div (-b)$, where $a = -36$ and $b = -4$

99. $(-a) \div (-b)$, where $a = -36$ and $b = -4$

100. $\frac{x}{y}$, where $x = -42$ and $y = -7$

101. $\frac{-x}{y}$, where $x = -42$ and $y = -7$

102. $\frac{x}{-y}$, where $x = -42$ and $y = -7$

103. $\frac{-x}{-y}$, where $x = -42$ and $y = -7$

104. Is 20 a solution of the equation $\frac{m}{-2} = -10$?

105. Is -3 a solution of the equation $\frac{21}{n} = 7$?

106. Is -6 a solution of the equation $\frac{x}{2} = \frac{-18}{x}$?

107. Is 8 a solution of the equation $\frac{m}{-4} = \frac{-16}{m}$?

OBJECTIVE C To solve application problems

108. **Golf** The combined scores of the top five golfers in a tournament equaled -10 (10 under par). What was the average score of the five golfers?

AP Photo/Phil Coale

109. **Golf** The combined scores of the top four golfers in a tournament equaled -12 (12 under par). What was the average score of the four golfers?

Temperature The following figure shows the record low temperatures, in degrees Fahrenheit, in the United States for each month. Use this figure for Exercises 110 to 112.

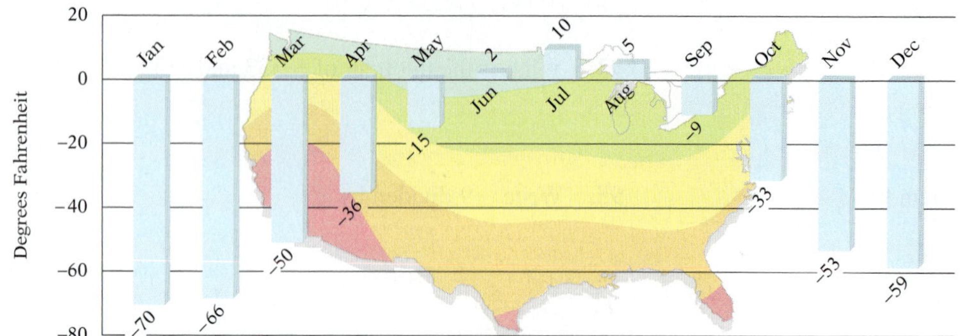

Record Low Temperatures, in Degrees Fahrenheit, in the United States

Source: National Climatic Data Center, Asheville, NC, and Storm Phillips, STORMFAX, Inc.

110. What is the average record low temperature for July, August, and September?

111. What is the average record low temperature for the first three months of the year?

112. What is the average record low temperature for the four months with the lowest record low temperatures?

113. **Meteorology** The high temperatures for a 6-day period in Barrow, Alaska, were $-23°F$, $-29°F$, $-21°F$, $-28°F$, $-28°F$, and $-27°F$. Calculate the average daily high temperature.

Luciana Whitaker/Getty Images

Barrow, Alaska

114. **Meteorology** The low temperatures for a 10-day period in a midwestern city were $-4°F$, $-9°F$, $-5°F$, $-2°F$, $4°F$, $-1°F$, $-1°F$, $-2°F$, $-2°F$, and $2°F$. Calculate the average daily low temperature for this city.

115. **Meteorology** The average low temperature for five consecutive days was $-12°C$. If the average low temperature after the sixth day was $-13°C$, was the low temperature on the sixth day higher or lower than $-12°C$?

116. **Testing** To discourage random guessing on a multiple-choice exam, a professor assigns 5 points for a correct answer, -2 points for an incorrect answer, and 0 points for leaving the question blank. What is the score for a student who had 20 correct answers, had 13 incorrect answers, and left 7 questions blank?

117. **Testing** To discourage random guessing on a multiple-choice exam, a professor assigns 7 points for a correct answer, -3 points for an incorrect answer, and -1 point for leaving the question blank. What is the score for a student who had 17 correct answers, had 8 incorrect answers, and left 2 questions blank?

118. **Newspapers** See the news clipping at the right. The table below shows the declining number of evening newspapers published in the United States. (*Source:* Newspaper Association of America) Find the average annual change in the number of evening newspapers published.

Year	*01–02*	*02–03*	*03–04*	*04–05*	*05–06*
Change in number of evening newspapers	−12	−12	−27	−8	−31

In the News

Evening Newspapers Face Extinction

The *Daily Mail,* Hagertown, Maryland's evening newspaper, first went to press on July 4, 1828. It ceased publication on September 28, 2007. This newspaper is another casualty amid the ever-declining interest in afternoon editions of newspapers in the United States.

Source: Newspaper Association of America

119. **Environmental Science** The wind chill factor when the temperature is -20°F and the wind is blowing at 15 mph is five times the wind chill factor when the temperature is 10°F and the wind is blowing at 20 mph. If the wind chill factor at 10°F with a 20-mph wind is -9°F, what is the wind chill factor at -20°F with a 15-mph wind?

Mathematics A geometric sequence is a list of numbers in which each number after the first is found by multiplying the preceding number in the list by the same number. For example, in the sequence 1, 3, 9, 27, 81, . . . , each number after the first is found by multiplying the preceding number in the list by 3. To find the multiplier in a geometric sequence, divide the second number in the sequence by the first number; for the example above, $3 \div 1 = 3$.

120. Find the next three numbers in the geometric sequence $-5, 15, -45, \ldots$.

121. Find the next three numbers in the geometric sequence $2, -4, 8, \ldots$.

122. Find the next three numbers in the geometric sequence $-3, -12, -48, \ldots$.

123. Find the next three numbers in the geometric sequence $-1, -5, -25, \ldots$.

Applying the Concepts

124. Use repeated addition to show that the product of two integers with different signs is a negative number.

125. **Mathematics**

a. Find the largest possible product of two negative integers whose sum is -18.

b. Find the smallest possible sum of two negative integers whose product is 16.

126. Determine whether the statement is always true, sometimes true, or never true.

a. The product of a number and its additive inverse is a negative number.

b. The product of an odd number of negative numbers is a negative number.

c. The square of a negative number is a positive number.

SECTION

3.4 Operations with Rational Numbers

OBJECTIVE A To add or subtract rational numbers

In this section, operations with rational numbers are discussed. A **rational number** is the quotient of two integers.

Rational Numbers

A rational number is a number that can be written in the form $\frac{a}{b}$, where a and b are integers and $b \neq 0$.

Each of the three numbers shown at the right is a rational number.

$$\frac{3}{4} \qquad \frac{-2}{9} \qquad \frac{13}{-5}$$

An integer can be written as the quotient of the integer and 1. Therefore, **every integer is a rational number.**

$$6 = \frac{6}{1} \qquad -8 = \frac{-8}{1}$$

A mixed number can be written as the quotient of two integers. Therefore, **every mixed number is a rational number.**

$$1\frac{4}{7} = \frac{11}{7} \qquad 3\frac{2}{5} = \frac{17}{5}$$

Take Note

Rational numbers are fractions such as $-\frac{6}{7}$ and $\frac{10}{3}$, in which the numerator and denominator are integers. Rational numbers are also represented by repeating decimals such as 0.25767676 . . . or terminating decimals such as 1.73. An irrational number is neither a terminating decimal nor a repeating decimal. For example, 2.45445444544445 . . . is an irrational number.

Take Note

The three dots mean that the number continues without end.

Recall that every fraction can be written as a decimal by dividing the numerator of the fraction by the denominator. The result is either a terminating decimal or a repeating decimal.

To write $\frac{5}{8}$ as a decimal, divide 5 by 8.

$$8\overline{)5.000} = 0.625$$ ← This is a **terminating decimal.**

To write $\frac{4}{11}$ as a decimal, divide 4 by 11.

$$11\overline{)4.0000} = 0.3636\ldots = 0.\overline{36}$$ ← This is a **repeating decimal.**

Every rational number can be written as a terminating or a repeating decimal. Some numbers, for example, $\sqrt{7}$ and π, have decimal representations that never terminate or repeat. These numbers are called **irrational numbers.**

$$\sqrt{7} \approx 2.6457513\ldots \qquad \pi \approx 3.1415926\ldots$$

The rational numbers and the irrational numbers taken together are called the **real numbers.**

We begin the presentation of operations on rational numbers by adding rational numbers in fractional form. If an addend is a fraction containing a negative sign, rewrite the fraction with the negative sign in the numerator. Then add the numerators and place the sum over the common denominator.

Tips for Success

Have you considered joining a study group? Getting together regularly with other students in the class to go over material and quiz each other can be very beneficial. See *AIM for Success* in the front of the book.

HOW TO • 1 Add: $-\frac{5}{6}+\frac{3}{4}$

The LCM of 4 and 6 is 12.
• The common denominator is the LCM of 4 and 6.

$$-\frac{5}{6}+\frac{3}{4}=\frac{-5}{6}+\frac{3}{4}$$
• Rewrite with the negative sign in the numerator.

$$=\frac{-10}{12}+\frac{9}{12}$$
• Rewrite each fraction in terms of the common denominator.

$$=\frac{-10+9}{12}$$
• Add the fractions.

$$=\frac{-1}{12}=-\frac{1}{12}$$
• Simplify the numerator and write the negative sign in front of the fraction.

Although the sum in the last example could have been left as $\frac{-1}{12}$, **all answers in this text are written with the negative sign in front of the fraction.**

HOW TO • 2 Add: $-\frac{2}{3}+\left(-\frac{4}{5}\right)$

$$-\frac{2}{3}+\left(-\frac{4}{5}\right)=\frac{-2}{3}+\frac{-4}{5}$$
• Rewrite each negative fraction with the negative sign in the numerator.

$$=\frac{-10}{15}+\frac{-12}{15}$$
• Rewrite each fraction as an equivalent fraction using the LCM as the denominator.

$$=\frac{-10+(-12)}{15}$$
• Add the fractions.

$$=\frac{-22}{15}=-1\frac{7}{15}$$

HOW TO • 3 Is $-\frac{2}{3}$ a solution of the equation $\frac{3}{4}+y=-\frac{1}{12}$?

$$\frac{3}{4}+y=-\frac{1}{12}$$

$$\frac{3}{4}+\left(-\frac{2}{3}\right) \quad \Big| \quad -\frac{1}{12}$$
• Replace y by $-\frac{2}{3}$. Then simplify.

$$\frac{9}{12}+\left(\frac{-8}{12}\right) \quad \Big| \quad -\frac{1}{12}$$
• The common denominator is 12.

$$\frac{9+(-8)}{12} \quad \Big| \quad -\frac{1}{12}$$

$$\frac{1}{12}\neq-\frac{1}{12}$$
• The results are not equal.

No, $-\frac{2}{3}$ is not a solution of the equation.

To subtract fractions with negative signs, first rewrite the fractions with the negative signs in the numerators.

HOW TO • 4 Simplify: $-\frac{2}{9} - \frac{5}{12}$

$$-\frac{2}{9} - \frac{5}{12} = \frac{-2}{9} - \frac{5}{12}$$

- Rewrite the negative fraction with the negative sign in the numerator.

$$= \frac{-8}{36} - \frac{15}{36}$$

- Write the fractions as equivalent fractions with a common denominator.

$$= \frac{-8 - 15}{36} = \frac{-23}{36}$$

- Subtract the numerators and place the difference over the common denominator.

$$= -\frac{23}{36}$$

- Write the negative sign in front of the fraction.

HOW TO • 5 Subtract: $\frac{2}{3} - \left(-\frac{4}{5}\right)$

$$\frac{2}{3} - \left(-\frac{4}{5}\right) = \frac{2}{3} + \frac{4}{5}$$

- Rewrite subtraction as addition of the opposite.

$$= \frac{10}{15} + \frac{12}{15}$$

- Write the fractions as equivalent fractions with a common denominator.

$$= \frac{10 + 12}{15}$$

- Add the fractions.

$$= \frac{22}{15} = 1\frac{7}{15}$$

HOW TO • 6 Evaluate $x - y$ when $x = -\frac{2}{5}$ and $y = -\frac{3}{10}$.

$$x - y$$

$$-\frac{2}{5} - \left(-\frac{3}{10}\right)$$

- Replace x with $-\frac{2}{5}$ and y with $-\frac{3}{10}$.

$$= -\frac{2}{5} + \frac{3}{10}$$

- Rewrite subtraction as addition of the opposite.

$$= \frac{-2}{5} + \frac{3}{10}$$

- Rewrite the negative fraction with the negative sign in the numerator.

$$= \frac{-4}{10} + \frac{3}{10}$$

- Write the fractions as equivalent fractions with a common denominator. *Note:* The LCM of 5 and 10 is 10.

$$= \frac{-4 + 3}{10} = \frac{-1}{10}$$

- Add the numerators and place the sum over the common denominator.

$$= -\frac{1}{10}$$

- Write the negative sign in front of the fraction.

The sign rules for adding and subtracting decimals are the same rules used to add and subtract integers.

Take Note

Recall that the absolute value of a number is the distance from zero to the number on the number line. The absolute value of a number is a positive number or zero.
$|54.29| = 54.29$
$|-36.087| = 36.087$

HOW TO • 7 Simplify: $-36.087 + 54.29$

$54.29 - 36.087 = 18.203$ • The signs of the addends are different. Subtract the smaller absolute value from the larger absolute value.

$|54.29| > |-36.087|$ • Attach the sign of the number with the larger absolute value.

$-36.087 + 54.29 = 18.203$ • The sum is positive.

Recall that the opposite of n is $-n$ and the opposite of $-n$ is n. To find the opposite of a number, change the sign of the number.

HOW TO • 8 Simplify: $-2.86 - 10.3$

$-2.86 - 10.3$
$= -2.86 + (-10.3)$ • Rewrite subtraction as addition of the opposite. The opposite of 10.3 is -10.3.
$= -13.16$ • The signs of the addends are the same. Add the absolute values of the numbers. Attach the sign of the addends.

HOW TO • 9 Evaluate $c - d$ when $c = 9.34$ and $d = -8.7$.

$c - d$
$9.34 - (-8.7)$ • Replace c with 9.34 and d with -8.7.
$= 9.34 + 8.7$ • Rewrite subtraction as addition of the opposite.
$= 18.04$ • Add.

EXAMPLE • 1

Add: $-\frac{3}{8} + \frac{3}{4} + \left(-\frac{5}{6}\right)$

Solution

$-\frac{3}{8} + \frac{3}{4} + \left(-\frac{5}{6}\right)$

$= \frac{-3}{8} + \frac{3}{4} + \frac{-5}{6}$ • Rewrite with negative signs in the numerators.

$= \frac{-9}{24} + \frac{18}{24} + \frac{-20}{24}$ • The LCM of the denominators is 24.

$= \frac{-9 + 18 + (-20)}{24}$

$= \frac{-11}{24} = -\frac{11}{24}$ • Add the numerators.

YOU TRY IT • 1

Add: $-\frac{5}{12} + \frac{5}{8} + \left(-\frac{1}{6}\right)$

Your solution

Solution on p. S9

EXAMPLE • 2

Subtract: $-\frac{5}{6} - \left(-\frac{3}{8}\right)$

Solution

$$-\frac{5}{6} - \left(-\frac{3}{8}\right) = -\frac{5}{6} + \frac{3}{8}$$

• Rewrite "−" as "+". The opposite of $-\frac{3}{8}$ is $\frac{3}{8}$.

$$= \frac{-20}{24} + \frac{9}{24}$$
$$= \frac{-20 + 9}{24}$$
$$= \frac{-11}{24} = -\frac{11}{24}$$

YOU TRY IT • 2

Subtract: $-\frac{5}{6} - \frac{7}{9}$

Your solution

EXAMPLE • 3

What is −251.49 more than −638.7?

Solution

$-638.7 + (-251.49) = -890.19$

YOU TRY IT • 3

What is 4.002 minus 9.378?

Your solution

EXAMPLE • 4

Evaluate $x + y + z$ when $x = -1.6$, $y = 7.9$, and $z = -4.8$.

Solution

$x + y + z$

$-1.6 + 7.9 + (-4.8) = 6.3 + (-4.8)$
$= 1.5$

YOU TRY IT • 4

Evaluate $x + y + z$ when $x = -7.84$, $y = -3.05$, and $z = 2.19$.

Your solution

EXAMPLE • 5

Is $\frac{3}{8}$ a solution of the equation $\frac{2}{3} = w - \frac{5}{6}$?

Solution

$$\frac{2}{3} = w - \frac{5}{6}$$

$$\frac{2}{3} \;\Big|\; \frac{3}{8} - \frac{5}{6}$$

• Replace w by $\frac{3}{8}$.

$$\frac{2}{3} \;\Big|\; \frac{9}{24} - \frac{20}{24}$$
$$\frac{2}{3} \;\Big|\; \frac{-11}{24}$$
$$\frac{2}{3} \neq -\frac{11}{24}$$

No, $\frac{3}{8}$ is not a solution of the equation.

YOU TRY IT • 5

Is $-\frac{1}{4}$ a solution of the equation $\frac{2}{3} - v = \frac{11}{12}$?

Your solution

Solutions on p. S10

OBJECTIVE B To multiply or divide rational numbers 

The product of two rational numbers written in fractional form is the product of the numerators over the product of the denominators. The sign rules are the same rules used to multiply integers.

The product of two numbers with the same sign is positive.

The product of two numbers with different signs is negative.

HOW TO • 10 Multiply: $-\frac{3}{4} \cdot \frac{8}{15}$

$$-\frac{3}{4} \cdot \frac{8}{15} = -\left(\frac{3}{4} \cdot \frac{8}{15}\right)$$ • The signs are different. The product is negative.

$$= -\frac{3 \cdot 8}{4 \cdot 15}$$ • Multiply the numerators. Multiply the denominators.

$$= -\frac{3 \cdot 2 \cdot 2 \cdot 2}{2 \cdot 2 \cdot 3 \cdot 5}$$ • Write the product in simplest form.

$$= -\frac{2}{5}$$

HOW TO • 11 Multiply: $-\frac{3}{8}\left(-\frac{2}{5}\right)\left(-\frac{10}{21}\right)$

$$-\frac{3}{8}\left(-\frac{2}{5}\right)\left(-\frac{10}{21}\right)$$

$$= \left(\frac{3}{8} \cdot \frac{2}{5}\right)\left(-\frac{10}{21}\right)$$ • Multiply the first two fractions. The product is positive.

$$= -\left(\frac{3}{8} \cdot \frac{2}{5} \cdot \frac{10}{21}\right)$$ • The product of the first two fractions and the third fraction is negative.

$$= -\frac{3 \cdot 2 \cdot 10}{8 \cdot 5 \cdot 21}$$ • Multiply the numerators. Multiply the denominators.

$$= -\frac{3 \cdot 2 \cdot 2 \cdot 5}{2 \cdot 2 \cdot 2 \cdot 5 \cdot 3 \cdot 7}$$ • Write the product in simplest form.

$$= -\frac{1}{14}$$

Thus, the product of three negative fractions is negative. We can modify the rule for multiplying positive and negative fractions to say that **the product of an odd number of negative fractions is negative and the product of an even number of negative fractions is positive.**

The sign rules for dividing positive and negative fractions are the same rules used to divide integers.

The quotient of two numbers with the same sign is positive.

The quotient of two numbers with different signs is negative.

HOW TO • 12 Simplify: $-\frac{7}{10} \div \left(-\frac{14}{15}\right)$

$$-\frac{7}{10} \div \left(-\frac{14}{15}\right) = \frac{7}{10} \div \frac{14}{15}$$
• The signs are the same. The quotient is positive.

$$= \frac{7}{10} \cdot \frac{15}{14}$$
• Rewrite the division as multiplication by the reciprocal.

$$= \frac{7 \cdot 15}{10 \cdot 14}$$
• Multiply the fractions.

$$= \frac{7 \cdot 3 \cdot 5}{2 \cdot 5 \cdot 2 \cdot 7}$$

$$= \frac{3}{4}$$

The sign rules for multiplying decimals are the same rules used to multiply integers.

The product of two numbers with the same sign is positive.
The product of two numbers with different signs is negative.

HOW TO • 13 Multiply: $(-3.2)(-0.008)$

$(-3.2)(-0.008) = 0.0256$ • The signs are the same. The product is positive. Multiply the absolute values of the numbers.

HOW TO • 14 Is -0.6 a solution of the equation $4.3a = -2.58$?

$4.3a$	$=$	-2.58	
$4.3(-0.6)$		-2.58	• Replace a by -0.6 and then simplify.
-2.58	$=$	-2.58	• The results are equal.

Yes, -0.6 is a solution of the equation.

The sign rules for dividing decimals are the same rules used to divide integers.

The quotient of two numbers with the same sign is positive.
The quotient of two numbers with different signs is negative.

HOW TO • 15 Divide: $-1.16 \div 2.9$

$-1.16 \div 2.9 = -0.4$ • The signs are different. The quotient is negative. Divide the absolute values of the numbers.

HOW TO • 16 Evaluate $c \div d$ when $c = -8.64$ and $d = -0.4$.

$c \div d$

$(-8.64) \div (-0.4)$ • Replace c with -8.64 and d with -0.4.

$= 21.6$ • The signs are the same. The quotient is positive. Divide the absolute values of the numbers.

EXAMPLE • 6

Multiply: $-\frac{3}{4}\left(\frac{1}{2}\right)\left(-\frac{8}{9}\right)$

Solution

$$-\frac{3}{4}\left(\frac{1}{2}\right)\left(-\frac{8}{9}\right)$$

$$= \frac{3}{4} \cdot \frac{1}{2} \cdot \frac{8}{9}$$

• The product of two negative fractions is positive.

$$= \frac{3 \cdot 1 \cdot 8}{4 \cdot 2 \cdot 9}$$

$$= \frac{3 \cdot 1 \cdot 2 \cdot 2 \cdot 2}{2 \cdot 2 \cdot 2 \cdot 3 \cdot 3} = \frac{1}{3}$$

YOU TRY IT • 6

Multiply: $-\frac{1}{3}\left(-\frac{5}{12}\right)\left(\frac{8}{15}\right)$

Your solution

EXAMPLE • 7

What is the product of $-\frac{1}{2}$ and $\frac{2}{5}$?

Solution

$$-\frac{1}{2} \cdot \frac{2}{5} = -\left(\frac{1}{2} \cdot \frac{2}{5}\right)$$

• The signs are different. The product is negative.

$$= -\frac{1 \cdot 2}{2 \cdot 5}$$

$$= -\frac{1}{5}$$

YOU TRY IT • 7

Multiply $3\frac{6}{7}$ by $-\frac{4}{9}$.

Your solution

EXAMPLE • 8

What is the quotient of 6 and $-\frac{3}{5}$?

Solution

$$6 \div \left(-\frac{3}{5}\right)$$

$$= -\left(\frac{6}{1} \div \frac{3}{5}\right)$$

• The signs are different. The quotient is negative.

$$= -\left(\frac{6}{1} \cdot \frac{5}{3}\right)$$

$$= -\frac{6 \cdot 5}{1 \cdot 3}$$

$$= -\frac{2 \cdot 3 \cdot 5}{1 \cdot 3}$$

$$= -\frac{10}{1} = -10$$

YOU TRY IT • 8

Find the quotient of 4 and $-\frac{6}{7}$.

Your solution

Solutions on p. S10

EXAMPLE • 9

Multiply: $-3.42(6.1)$

Solution

$-3.42(6.1)$
$= -20.862$

• The signs are different. The product is negative.

YOU TRY IT • 9

Multiply: $(-0.7)(-5.8)$

Your solution

EXAMPLE • 10

Divide and round to the nearest tenth: $-6.94 \div -1.5$

Solution

$-6.94 \div (-1.5)$
≈ 4.6

• The signs are the same. The quotient is positive.

YOU TRY IT • 10

Divide and round to the nearest tenth: $-25.7 \div 0.31$

Your solution

EXAMPLE • 11

Evaluate the variable expression xy when $x = 1\frac{4}{5}$ and $y = -\frac{5}{6}$.

Solution

xy

$$1\frac{4}{5}\left(-\frac{5}{6}\right) = -\left(\frac{9}{5} \cdot \frac{5}{6}\right)$$
$$= -\frac{9 \cdot 5}{5 \cdot 6}$$
$$= -\frac{3 \cdot 3 \cdot 5}{5 \cdot 2 \cdot 3}$$
$$= -\frac{3}{2} = -1\frac{1}{2}$$

YOU TRY IT • 11

Evaluate the variable expression xy when $x = -5\frac{1}{8}$ and $y = -\frac{2}{3}$.

Your solution

EXAMPLE • 12

Evaluate $\frac{x}{y}$ when $x = -76.8$ and $y = 0.8$.

Solution

$\frac{x}{y}$

$$\frac{-76.8}{0.8} = -96$$

YOU TRY IT • 12

Evaluate $\frac{x}{y}$ when $x = -40.6$ and $y = -0.7$.

Your solution

Solutions on p. S10

EXAMPLE • 13

Evaluate $50ab$ when $a = -0.9$ and $b = -0.2$.

Solution

$50ab$

$50(-0.9)(-0.2) = -45(-0.2)$

$= 9$

YOU TRY IT • 13

Evaluate $25xy$ when $x = -0.8$ and $y = 0.6$.

Your solution

EXAMPLE • 14

Is -0.4 a solution of the equation $\frac{8}{x} = -20$?

Solution

$$\frac{8}{x} = -20$$

$$\frac{8}{-0.4} \;\Big|\; -20$$ • Replace x by -0.4.

$$-20 = -20$$

Yes, -0.4 is a solution of the equation.

YOU TRY IT • 14

Is -1.2 a solution of the equation $-2 = \frac{d}{-0.6}$?

Your solution

Solutions on p. S10

OBJECTIVE C To solve application problems 

EXAMPLE • 15

In Fairbanks, Alaska, the average temperature during the month of July is 61.5°F. During the month of January, the average temperature in Fairbanks is −12.7°F. What is the difference between the average temperature in Fairbanks during July and the average temperature during January?

Strategy

To find the difference, subtract the average temperature in January (−12.7°F) from the average temperature in July (61.5°F).

Solution

$61.5 - (-12.7) = 61.5 + 12.7 = 74.2$

The difference between the average temperature during July and the average temperature during January in Fairbanks is 74.2°F.

YOU TRY IT • 15

On January 10, 1911, in Rapid City, South Dakota, the temperature fell from 12.78°C at 7:00 A.M. to −13.33°C at 7:15 A.M. How many degrees did the temperature fall during the 15-minute period?

Your strategy

Your solution

Solution on p. S10

3.4 EXERCISES

OBJECTIVE A To add or subtract rational numbers

For Exercises 1 to 39, simplify.

1. $\frac{5}{8} - \frac{5}{6}$

2. $\frac{1}{9} - \frac{5}{27}$

3. $-\frac{5}{12} - \frac{3}{8}$

4. $-\frac{5}{6} - \frac{5}{9}$

5. $-\frac{6}{13} + \frac{17}{26}$

6. $-\frac{7}{12} + \frac{5}{8}$

7. $-\frac{5}{8} - \left(-\frac{11}{12}\right)$

8. $-\frac{7}{12} - \left(-\frac{7}{8}\right)$

9. $\frac{5}{12} - \frac{11}{15}$

10. $\frac{2}{5} - \frac{14}{15}$

11. $-\frac{3}{4} - \frac{5}{8}$

12. $-\frac{2}{3} - \frac{5}{8}$

13. $-\frac{5}{2} - \left(-\frac{13}{4}\right)$

14. $-\frac{7}{3} - \left(-\frac{3}{2}\right)$

15. $-\frac{3}{8} - \frac{5}{12} - \frac{3}{16}$

16. $-\frac{5}{16} + \frac{3}{4} - \frac{7}{8}$

17. $\frac{1}{2} - \frac{3}{8} - \left(-\frac{1}{4}\right)$

18. $\frac{3}{4} - \left(-\frac{7}{12}\right) - \frac{7}{8}$

19. $\frac{1}{3} - \frac{1}{4} - \frac{1}{5}$

20. $\frac{5}{16} + \frac{1}{8} - \frac{1}{2}$

21. $\frac{1}{2} + \left(-\frac{3}{8}\right) + \frac{5}{12}$

22. $-\frac{3}{8} + \frac{3}{4} - \left(-\frac{3}{16}\right)$

23. $3.4 + (-6.8)$

24. $-4.9 + 3.27$

25. $-8.32 + (-0.57)$

26. $-3.5 + 7$

27. $-4.8 + (-3.2)$

28. $6.2 + (-4.29)$

29. $-4.6 + 3.92$

30. $7.2 + (-8.42)$

31. $-45.71 + (-135.8)$

32. $-35.274 + 12.47$

33. $4.2 + (-6.8) + 5.3$

34. $6.7 + 3.2 + (-10.5)$

35. $-4.5 + 3.2 + (-19.4)$

36. $2.09 - 6.72 - 5.4$

37. $-18.39 + 4.9 - 23.7$

38. $19 - (-3.72) - 82.75$

39. $-3.09 - 4.6 - 27.3$

40. What is $-\frac{5}{6}$ added to $\frac{4}{9}$?

41. What is $\frac{7}{12}$ added to $-\frac{11}{16}$?

42. What is $-\frac{2}{3}$ more than $-\frac{5}{6}$?

43. What is $-\frac{7}{12}$ more than $-\frac{5}{9}$?

44. What is $-\frac{7}{12}$ minus $\frac{7}{9}$?

45. What is $\frac{3}{5}$ decreased by $-\frac{7}{10}$?

46. What is the sum of -65.47 and -32.91?

47. Find -138.72 minus 510.64.

For Exercises 48 to 53, evaluate the variable expression $x + y$ for the given values of x and y.

48. $x = -\frac{3}{8}, y = \frac{2}{9}$

49. $x = \frac{3}{10}, y = -\frac{7}{15}$

50. $x = -\frac{5}{8}, y = -\frac{1}{6}$

51. $x = -\frac{3}{8}, y = -\frac{5}{6}$

52. $x = 62.97, y = -43.85$

53. $x = -6.175, y = -19.49$

For Exercises 54 to 60, evaluate the variable expression $x - y$ for the given values of x and y.

54. $x = -\frac{11}{12}, y = \frac{5}{12}$

55. $x = -\frac{15}{16}, y = \frac{5}{16}$

56. $x = -\frac{2}{3}, y = -\frac{3}{4}$

57. $x = -\frac{5}{12}, y = -\frac{5}{9}$

58. $x = -21.073, y = 6.48$

59. $x = -3.69, y = -1.527$

60. $x = -8.21, y = -6.798$

61. Is $-\frac{5}{6}$ a solution of the equation $\frac{1}{4} + x = -\frac{7}{12}$?

62. Is $\frac{5}{8}$ a solution of the equation $-\frac{1}{4} = x - \frac{7}{8}$?

63. Is -1.2 a solution of the equation $6.4 = 5.2 + a$?

64. Is -2.8 a solution of the equation $0.8 - p = 3.6$?

For Exercises 65 to 68, estimate each sum to the nearest integer. Do not find the exact sum.

65. $\frac{7}{8} + \frac{4}{5}$

66. $\frac{1}{3} + \left(-\frac{1}{2}\right)$

67. $-0.125 + 1.25$

68. $-1.3 + 0.2$

OBJECTIVE B To multiply or divide rational numbers

For Exercises 69 to 104, simplify.

69. $\frac{1}{2}\left(-\frac{3}{4}\right)$

70. $-\frac{2}{9}\left(-\frac{3}{14}\right)$

71. $\left(-\frac{3}{8}\right)\left(-\frac{4}{15}\right)$

72. $\left(-\frac{3}{4}\right)\left(-\frac{8}{27}\right)$

73. $-\frac{1}{2}\left(\frac{8}{9}\right)$

74. $\frac{5}{12}\left(-\frac{8}{15}\right)$

75. $\left(-\frac{5}{12}\right)\left(\frac{42}{65}\right)$

76. $\left(\frac{3}{8}\right)\left(-\frac{15}{41}\right)$

77. $\left(-\frac{15}{8}\right)\left(-\frac{16}{3}\right)$

78. $\left(-\frac{5}{7}\right)\left(-\frac{14}{15}\right)$

79. $\frac{5}{8}\left(-\frac{7}{12}\right)\left(\frac{16}{25}\right)$

80. $\left(\frac{1}{2}\right)\left(-\frac{3}{4}\right)\left(-\frac{5}{8}\right)$

81. $\frac{1}{3} \div \left(-\frac{1}{2}\right)$

82. $-\frac{3}{8} \div \frac{7}{8}$

83. $\left(-\frac{3}{4}\right) \div \left(-\frac{7}{40}\right)$

84. $\frac{5}{6} \div \left(-\frac{3}{4}\right)$

85. $-\frac{5}{12} \div \frac{15}{32}$

86. $-\frac{5}{16} \div \left(-\frac{3}{8}\right)$

87. $\left(-\frac{3}{8}\right) \div \left(-\frac{5}{12}\right)$

88. $\left(-\frac{8}{19}\right) \div \frac{7}{38}$

89. $\left(-\frac{2}{3}\right) \div 4$

90. $-6 \div \frac{4}{9}$

91. $-6.7(-4.2)$

92. $-8.9(-3.5)$

93. $-1.6(4.9)$

94. $-14.3(7.9)$

95. $(-0.78)(-0.15)$

96. $(-1.21)(-0.03)$

97. $(-8.919) \div (-0.9)$

98. $-77.6 \div (-0.8)$

99. $59.01 \div (-0.7)$

100. $(-7.04) \div (-3.2)$

101. $(-84.66) \div 1.7$

102. $-3.312 \div (0.8)$

103. $1.003 \div (-0.59)$

104. $26.22 \div (-6.9)$

105. Find $-\frac{9}{16}$ multiplied by $\frac{4}{27}$.

106. Find $\frac{3}{7}$ multiplied by $-\frac{14}{15}$.

107. What is the product of $-\frac{7}{24}$, $\frac{8}{21}$, and $\frac{3}{7}$?

108. What is the product of $-\frac{5}{13}$, $-\frac{26}{75}$, and $\frac{5}{8}$?

109. What is $-\frac{15}{24}$ divided by $\frac{3}{5}$?

110. What is $\frac{5}{6}$ divided by $-\frac{10}{21}$?

111. Find the product of 2.7, -16, and 3.04.

112. What is the product of 0.06, -0.4, and -1.5?

In Exercises 113 and 114, round answers to the nearest tenth.

113. Find the quotient of -19.04 and 0.75.

114. What is -13.97 divided by 28.4?

For Exercises 115 and 116, evaluate the variable expression xy for the given values of x and y.

115. $x = -49, y = \frac{5}{14}$

116. $x = -\frac{3}{10}, y = -35$

For Exercises 117 and 118, evaluate the variable expression xyz for the given values of x, y, and z.

117. $x = 2\frac{3}{8}, y = -\frac{3}{19}, z = -\frac{4}{9}$

118. $x = \frac{4}{5}, y = -15, z = \frac{7}{8}$

For Exercises 119 and 120, evaluate the expression for the given values of the variables.

119. $10t$, when $t = -4.8$

120. ab, when $a = 452$ and $b = -0.86$

For Exercises 121 to 127, evaluate the variable expression $x \div y$ for the given values of x and y.

121. $x = -\frac{5}{8}, y = -\frac{15}{2}$

122. $x = -\frac{14}{3}, y = -\frac{7}{9}$

123. $x = -18, y = \frac{3}{8}$

124. $x = 20, y = -\frac{5}{6}$

125. $x = -64.05, y = -6.1$

126. $x = -2.501, y = 0.41$

127. $x = 1.173, y = -0.69$

128. Is $-\frac{1}{6}$ a solution of the equation $6x = 1$?

129. Is $-\frac{4}{5}$ a solution of the equation $\frac{5}{4}n = -1$?

130. Is -8 a solution of the equation $1.6 = -0.2z$?

131. Is -1 a solution of the equation $-7.9c = -7.9$?

132. Without finding the product, determine whether $\frac{11}{13} \cdot \frac{50}{51}$ is greater than 1 or less than 1.

133. Without finding the quotient, determine whether $8.713 \div 7.2$ is greater than 1 or less than 1.

OBJECTIVE C To solve application problems

134. **Meteorology** On January 23, 1916, the temperature in Browing, Montana, was 6.67°C. On January 24, 1916, the temperature in Browing was −48.9°C. Find the difference between the temperatures in Browing on these two days.

135. **Meteorology** On January 22, 1943, in Spearfish, South Dakota, the temperature fell from 12.22°C at 9 A.M. to −20°C at 9:27 A.M. How many degrees did the temperature fall during the 27-minute period?

136. **Temperature** The date of the news clipping at the right is July 20, 2007. Find the difference between the record high and low temperatures for Slovakia.

> **In the News**
>
> **Slovakia Hits Record High**
>
> Slovakia, which became an independent country in 1993 with the peaceful division of Czechoslovakia, hit a record high temperature today of 104.5°F. The record low temperature, set on February 11, 1929, was −41.8°F.
>
> *Source:* wikipedia.org

Slovakia

DEA/W. Buss/De Agostini Picture Library/Getty Images

137. If the temperature begins at 4.8°C and ends up below 0°C, is the difference between the starting and ending temperatures less than or greater than 4.8?

138. If the temperature rose 20.3°F during one day and ended up at a high temperature of 15.7°F, did the temperature begin above or below 0°F?

139. **Chemistry** The boiling point of nitrogen is −195.8°C, and the melting point is −209.86°C. Find the difference between the boiling point and the melting point of nitrogen.

140. **Chemistry** The boiling point of oxygen is −182.962°C. Oxygen's melting point is −218.4°C. What is the difference between the boiling point and the melting point of oxygen?

Applying the Concepts

141. Determine whether the statement is true or false.
- **a.** Every integer is a rational number.
- **b.** Every whole number is an integer.
- **c.** Every integer is a positive number.
- **d.** Every rational number is an integer.

142. **Number Problems**
- **a.** Find a rational number between 0.1 and 0.2.
- **b.** Find a rational number between 1 and 1.1.
- **c.** Find a rational number between 0 and 0.005.

SECTION 3.5 The Order of Operations Agreement

OBJECTIVE A To use the Order of Operations Agreement to simplify expressions

The Order of Operations Agreement, introduced in Chapter 1, is repeated here for your reference.

The Order of Operations Agreement

Step 1 Do all operations inside parentheses.

Step 2 Simplify any numerical expressions containing exponents.

Step 3 Do multiplication and division as they occur from left to right.

Step 4 Do addition and subtraction as they occur from left to right.

Take Note

The -3 is squared only when the negative sign is *inside* the parentheses. In $(-3)^2$, we are squaring -3; in -3^2, we are finding the opposite of 3^2.

Note how the following expressions containing exponents are simplified.

$(-3)^2 = (-3)(-3) = 9$ The (-3) is squared. Multiply -3 by -3.

$-(3^2) = -(3 \cdot 3) = -9$ Read $-(3^2)$ as "the opposite of three squared." 3^2 is 9. The opposite of 9 is -9.

$-3^2 = -(3^2) = -9$ The expression -3^2 is the same as $-(3^2)$.

HOW TO • 1 Simplify: $8 - 4 \div (-2)$

$8 - 4 \div (-2) = 8 - (-2)$
- There are no operations inside parentheses (Step 1). There are no exponents (Step 2). Do the division (Step 3).

$= 8 + 2 = 10$
- Do the subtraction (Step 4).

Integrating Technology

As shown above and at the right, the value of -3^2 is different from the value of $(-3)^2$. The keystrokes to evaluate each of these on your calculator are different. To evaluate -3^2, enter

3 **x²** **+/−**

To evaluate $(-3)^2$, enter

3 **+/−** **x²**

HOW TO • 2 Simplify: $(-3)^2 - 2(8 - 3) + (-5)$

$(-3)^2 - 2(8 - 3) + (-5)$

$= (-3)^2 - 2(5) + (-5)$
- Perform operations inside parentheses.

$= 9 - 2(5) + (-5)$
- Simplify expressions with exponents.

$= 9 - 10 + (-5)$
- Do multiplication and division as they occur from left to right.

$= 9 + (-10) + (-5)$
- Do addition and subtraction as they occur from left to right.

$= -1 + (-5)$

$= -6$

HOW TO • 3 Evaluate $ab - b^2$ when $a = 2$ and $b = -6$.

$ab - b^2$

$2(-6) - (-6)^2$ • **Replace a with 2 and each b with −6.**

$= 2(-6) - 36$ • **Use the Order of Operations Agreement to simplify the resulting numerical expression. Simplify the exponential expression.**

$= -12 - 36$ • **Do the multiplication.**

$= -12 + (-36)$ • **Do the subtraction.**

$= -48$

EXAMPLE • 1

Simplify: $12 \div (-2)^2 - 5$

Solution

$12 \div (-2)^2 - 5$

$= 12 \div 4 - 5$ • **Exponents**

$= 3 - 5$ • **Division**

$= 3 + (-5)$ • **Subtraction**

$= -2$

YOU TRY IT • 1

Simplify: $8 \div 4 \cdot 4 - (-2)^2$

Your solution

EXAMPLE • 2

Simplify: $\left(-\frac{2}{3}\right)^2 \div \frac{7-2}{13-4} - \frac{1}{3}$

Solution

$\left(-\frac{2}{3}\right)^2 \div \frac{7-2}{13-4} - \frac{1}{3}$

$= \left(-\frac{2}{3}\right)^2 \div \frac{5}{9} - \frac{1}{3}$ • **Simplify above and below the fraction bar.**

$= \frac{4}{9} \div \frac{5}{9} - \frac{1}{3}$ • **Exponents**

$= \frac{4}{9} \cdot \frac{9}{5} - \frac{1}{3}$ • **Division**

$= \frac{4}{5} - \frac{1}{3} = \frac{7}{15}$ • **Subtraction**

YOU TRY IT • 2

Simplify: $\left(-\frac{1}{2}\right)^3 \cdot \frac{7-3}{4-9} + \frac{4}{5}$

Your solution

EXAMPLE • 3

Evaluate $6a \div (-b)$ when $a = -2$ and $b = -3$.

Solution

$6a \div (-b)$

$6(-2) \div (-(-3))$

$= 6(-2) \div (3)$

$= -12 \div 3$

$= -4$

YOU TRY IT • 3

Evaluate $3a - 4b$ when $a = -2$ and $b = 5$.

Your solution

Solutions on p. S11

3.5 EXERCISES

OBJECTIVE A To use the Order of Operations Agreement to simplify expressions

For Exercises 1 to 38, simplify.

1. $3 - 12 \div 2$

2. $-16 \div 2 + 8$

3. $2(3 - 5) - 2$

4. $2 - (8 - 10) \div 2$

5. $4 - (-3)^2$

6. $(-2)^2 - 6$

7. $4 \cdot (2 - 4) - 4$

8. $6 - 2 \cdot (1 - 3)$

9. $4 - (-2)^2 + (-3)$

10. $-3 + (-6)^2 - 1$

11. $3^3 - 4(2)$

12. $9 \div 3 - (-3)^2$

13. $3 \cdot (6 - 2) \div 6$

14. $4 \cdot (2 - 7) \div 5$

15. $2^3 - (-3)^2 + 2$

16. $6(8 - 2) \div 4$

17. $6 - 2(1 - 5)$

18. $(-2)^2 - (-3)^2 + 1$

19. $6 - (-4)(-3)^2$

20. $4 - (-5)(-2)^2$

21. $4 \cdot 2 - 3 \cdot 7$

22. $16 \div 2 - 9 \div 3$

23. $(-2)^2 - 5(3) - 1$

24. $4 - 2 \cdot 7 - 3^2$

25. $(-1) \cdot (4 - 7)^2 \div 9 + 6 - 3 - 4(2)$

26. $(-3)^2 \cdot (5 - 7)^2 - (-9) \div 3$

27. $(1.2)^2 - 4.1(0.3)$

28. $2.4(-3) - 2.5$

29. $1.6 - (-1.6)^2$

30. $4.1(8) \div (-4.1)$

31. $(4.1 - 3.9) - 0.7^2$

32. $1.8(-2.3) - 2$

33. $-\frac{1}{2} + \frac{3}{8} \div \left(-\frac{3}{4}\right)$

34. $\left(\frac{3}{4}\right)^2 - \frac{3}{8}$

35. $\left(\frac{1}{2}\right)^2 - \left(-\frac{1}{2}\right)^2$

36. $-\frac{2}{3}\left(\frac{5}{8}\right) \div \frac{2}{7}$

37. $\frac{1}{2} - \left(\frac{3}{4} - \frac{3}{8}\right) \div \frac{1}{3}$

38. $\frac{3}{8} \div \left(-\frac{1}{2}\right)^2 + 2$

39. Which expression is equivalent to $7 - (-2^2)$?
(i) $7 + 4$ (ii) 9^2 (iii) $7 - 4$ (iv) 7×4

40. Which expression is equivalent to $3 - 5 \times 4 - 7^2$?
(i) $-2 \times (-3)^2$ (ii) $3 - 20 + 49$
(iii) $3 - 20 - 49$ (iv) $-2 \times 4 - 49$

For Exercises 41 to 56, evaluate the variable expression given $a = -2$, $b = 4$, $c = -1$, and $d = 3$.

41. $3a + 2b$

42. $6b \div (-a)$

43. $bc \div (2a)$

44. $a^2 - b^2$

45. $b^2 - c^2$

46. $2a - (c + a)^2$

47. $(b - a)^2 + 4c$

48. $\frac{b + c}{d}$

49. $\frac{d - b}{c}$

50. $\frac{2d + b}{-a}$

51. $\frac{b - d}{c - a}$

52. $\frac{bd}{a} \div c$

53. $(d - a)^2 \div 5$

54. $(b + c)^2 + (a + d)^2$

55. $(d - a)^2 - 3c$

56. $(b + d)^2 - 4a$

For Exercises 57 and 58, evaluate the expression for the given values of the variables.

57. $x - y^3z$, when $x = \frac{5}{6}$, $y = \frac{1}{2}$, and $z = \frac{8}{9}$

58. $xy^3 + z$, when $x = \frac{9}{10}$, $y = \frac{1}{3}$, and $z = \frac{7}{15}$

Applying the Concepts

59. What is the smallest integer greater than $-2^2 - (-3)^2 + 5(4) \div 10 - (-6)$?

60. Is -4 a solution of the equation $x^2 - 2x - 8 = 0$?

61. Evaluate $a \div bc$ and $a \div (bc)$ when $a = 16$, $b = 2$, and $c = -4$. Explain why the answers are not the same.

FOCUS ON PROBLEM SOLVING

Drawing Diagrams

How do you best remember something? Do you remember best what you hear? The word *aural* means "pertaining to the ear"; people with a strong aural memory remember best those things that they hear. The word *visual* means "pertaining to the sense of sight"; people with a strong visual memory remember best that which they see written down. Some people claim that their memory is in their writing hand—they remember something only if they write it down! The method by which you best remember something is probably also the method by which you can best learn something new.

In problem-solving situations, try to capitalize on your strengths. If you tend to understand the material better when you hear it spoken, read application problems aloud or have someone else read them to you. If writing helps you to organize ideas, rewrite application problems in your own words.

No matter what your main strength, visualizing a problem can be a valuable aid in problem solving. A drawing, sketch, diagram, or chart can be a useful tool in problem solving, just as calculators and computers are tools. A diagram can be helpful in gaining an understanding of the relationships inherent in a problem-solving situation. A sketch will help you to organize the given information and can lead to your being able to focus on the method by which the solution can be determined.

HOW TO • 1 A tour bus drives 5 mi south, then 4 mi west, then 3 mi north, then 4 mi east. How far is the tour bus from the starting point?

Draw a diagram of the given information.

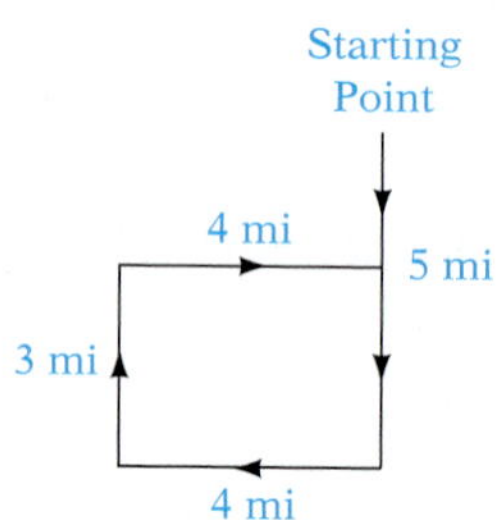

From the diagram, we can see that the solution can be determined by subtracting 3 from 5: $5 - 3 = 2$.

The bus is 2 mi from the starting point.

HOW TO • 2 If you roll two ordinary six-sided dice and multiply the two numbers that appear on top, how many different possible products are there?

Make a chart of the possible products. In the chart below, repeated products are marked with an asterisk.

$1 \cdot 1 = 1$	$2 \cdot 1 = 2$ (*)	$3 \cdot 1 = 3$ (*)	$4 \cdot 1 = 4$ (*)	$5 \cdot 1 = 5$ (*)	$6 \cdot 1 = 6$ (*)
$1 \cdot 2 = 2$	$2 \cdot 2 = 4$ (*)	$3 \cdot 2 = 6$ (*)	$4 \cdot 2 = 8$ (*)	$5 \cdot 2 = 10$ (*)	$6 \cdot 2 = 12$ (*)
$1 \cdot 3 = 3$	$2 \cdot 3 = 6$ (*)	$3 \cdot 3 = 9$	$4 \cdot 3 = 12$ (*)	$5 \cdot 3 = 15$ (*)	$6 \cdot 3 = 18$ (*)
$1 \cdot 4 = 4$	$2 \cdot 4 = 8$	$3 \cdot 4 = 12$ (*)	$4 \cdot 4 = 16$	$5 \cdot 4 = 20$ (*)	$6 \cdot 4 = 24$ (*)
$1 \cdot 5 = 5$	$2 \cdot 5 = 10$	$3 \cdot 5 = 15$	$4 \cdot 5 = 20$	$5 \cdot 5 = 25$	$6 \cdot 5 = 30$ (*)
$1 \cdot 6 = 6$	$2 \cdot 6 = 12$	$3 \cdot 6 = 18$	$4 \cdot 6 = 24$	$5 \cdot 6 = 30$	$6 \cdot 6 = 36$

By counting the products that are not repeats, we can see that there are 18 different possible products.

Look at Sections 1 and 2 of this chapter. You will notice that number lines are used to help you visualize the integers, as an aid in ordering integers, to help you understand the concepts of opposite and absolute value, and to illustrate addition of integers. As you begin your work with integers, you may find that sketching a number line proves helpful in coming to understand a problem or in working through a calculation that involves integers.

PROJECTS AND GROUP ACTIVITIES

Multiplication of Integers

The grid at the left has four regions, or quadrants, numbered counterclockwise, starting at the upper right, with the Roman numerals I, II, III, IV.

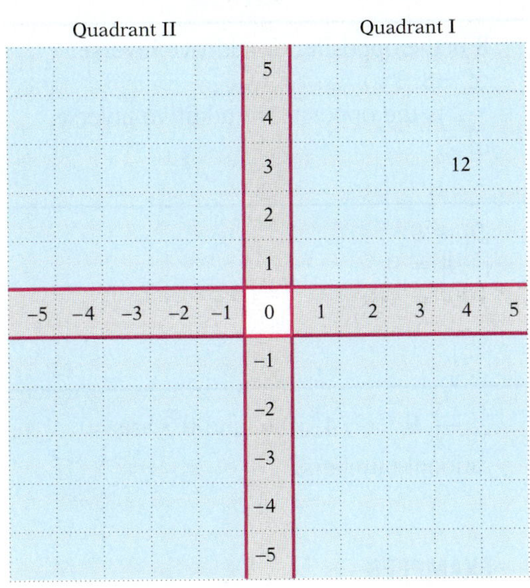

1. Complete Quadrant I by multiplying each of the horizontal numbers, 1 through 5, by each of the vertical numbers, 1 through 5. The product 4(3) has been filled in for you. Complete Quadrants II, III, and IV by again multiplying each horizontal number by each vertical number.
2. What is the sign of all the products in Quadrant I? Quadrant II? Quadrant III? Quadrant IV?
3. Describe at least three patterns that you observe in the completed table.
4. How does the table show that multiplication of integers is commutative?
5. How can you use the table to find the quotient of two integers? Provide at least two examples of division of integers.

Closure

The whole numbers are said to be *closed* with respect to addition because when two whole numbers are added, the result is a whole number. The whole numbers are not closed with respect to subtraction because, for example, 4 and 7 are whole numbers but $4 - 7 = -3$, and -3 is not a whole number. Complete the table below by entering a Y if the operation is closed for those numbers and an N if it is not closed. When we discuss whether multiplication and division are closed, zero is not included because division by zero is not defined.

Tips for Success

Three important features of this text that can be used to prepare for a test are the:
- Chapter Summary
- Chapter Review Exercises
- Chapter Test

See *AIM for Success* in the front of the book.

	Add	*Subtract*	*Multiply*	*Divide*
Whole numbers	Y	N		
Integers				
Rational numbers				

CHAPTER 3

SUMMARY

KEY WORDS	EXAMPLES
A number n is a *positive number* if $n > 0$. A number n is a *negative number* if $n < 0$. [3.1A, p. 176]	Positive numbers are numbers greater than zero. 9, 87, and 603 are positive numbers. Negative numbers are numbers less than zero. -5, -41, and -729 are negative numbers.
The *integers* are . . ., $-4, -3, -2, -1, 0, 1, 2, 3, 4, \ldots$ The integers can be defined as the whole numbers and their opposites. *Positive integers* are to the right of zero on the number line. *Negative integers* are to the left of zero on the number line. [3.1A, p. 176]	-729, -41, -5, 9, 87, and 603 are integers. 0 is an integer, but it is neither a positive nor a negative integer.
Opposite numbers are two numbers that are the same distance from zero on the number line but are on opposite sides of zero. The opposite of a number is called its *additive inverse.* [3.1B, p. 178; 3.2A, p. 188]	8 is the opposite, or additive inverse, of -8. -2 is the opposite, or additive inverse, of 2.
The *absolute value* of a number is the distance from zero to the number on the number line. The absolute value of a number is a positive number or zero. The symbol for absolute value is "$\| \|$". [3.1C, p. 179]	$\|9\| = 9$ $\|-9\| = 9$ $-\|9\| = -9$
A *rational number* is a number that can be written in the form $\frac{a}{b}$, where a and b are integers and $b \neq 0$. [3.4A, p. 214]	$\frac{3}{7}$, $-\frac{5}{8}$, 9, -2, $4\frac{1}{2}$, 0.6, and $0.\overline{3}$ are rational numbers.

ESSENTIAL RULES AND PROCEDURES	EXAMPLES
To add integers with the same sign, add the absolute values of the numbers. Then attach the sign of the addends. [3.2A, p. 187]	$6 + 4 = 10$ $-6 + (-4) = -10$
To add integers with different signs, find the absolute values of the numbers. Subtract the lesser absolute value from the greater absolute value. Then attach the sign of the addend with the greater absolute value. [3.2A, p. 187]	$-6 + 4 = -2$ $6 + (-4) = 2$
To subtract two integers, add the opposite of the second integer to the first integer. [3.2B, p. 191]	$6 - 4 = 6 + (-4) = 2$ $6 - (-4) = 6 + 4 = 10$ $-6 - 4 = -6 + (-4) = -10$ $-6 - (-4) = -6 + 4 = -2$

To multiply integers with the same sign, multiply the absolute values of the factors. The product is positive. [3.3A, p. 202]	$3 \cdot 5 = 15$ $-3(-5) = 15$
To multiply integers with different signs, multiply the absolute values of the factors. The product is negative. [3.3A, p. 202]	$-3(5) = -15$ $3(-5) = -15$
To divide two numbers with the same sign, divide the absolute values of the numbers. The quotient is positive. [3.3B, p. 205]	$15 \div 3 = 5$ $(-15) \div (-3) = 5$
To divide two numbers with different signs, divide the absolute values of the numbers. The quotient is negative. [3.3B, p. 205]	$-15 \div 3 = -5$ $15 \div (-3) = -5$
Order Relations $a > b$ if a is to the right of b on the number line. $a < b$ if a is to the left of b on the number line. [3.1A, p. 177]	$-6 > -12$ $-8 < 4$
Properties of Addition [3.2A, p. 188] **Addition Property of Zero** $a + 0 = a$ or $0 + a = a$ **Commutative Property of Addition** $a + b = b + a$ **Associative Property of Addition** $(a + b) + c = a + (b + c)$ **Inverse Property of Addition** $a + (-a) = 0$ or $-a + a = 0$	 $-6 + 0 = -6$ $-8 + 4 = 4 + (-8)$ $(-5 + 4) + 6 = -5 + (4 + 6)$ $7 + (-7) = 0$
Properties of Multiplication [3.3A, p. 203] **Multiplication Property of Zero** $a \cdot 0 = 0$ or $0 \cdot a = 0$ **Multiplication Property of One** $a \cdot 1 = a$ or $1 \cdot a = a$ **Commutative Property of Multiplication** $a \cdot b = b \cdot a$ **Associative Property of Multiplication** $(a \cdot b) \cdot c = a \cdot (b \cdot c)$	 $-9(0) = 0$ $-3(1) = -3$ $-2(6) = 6(-2)$ $(-2 \cdot 4) \cdot 5 = -2 \cdot (4 \cdot 5)$
Division Properties of Zero and One [3.3B, p. 206] If $a \neq 0$, $0 \div a = 0$. If $a \neq 0$, $a \div a = 1$. $a \div 1 = a$ $a \div 0$ is undefined.	 $0 \div (-5) = 0$ $-5 \div (-5) = 1$ $-5 \div 1 = -5$ $-5 \div 0$ is undefined.
The Order of Operations Agreement [3.5A, p. 230] **Step 1** Do all operations inside parentheses. **Step 2** Simplify any numerical expressions containing exponents. **Step 3** Do multiplication and division as they occur from left to right. **Step 4** Do addition and subtraction as they occur from left to right.	 $(-4)^2 - 3(1 - 5) = (-4)^2 - 3(-4)$ $= 16 - 3(-4)$ $= 16 - (-12)$ $= 16 + 12$ $= 28$

CHAPTER 3

CONCEPT REVIEW

Test your knowledge of the concepts presented in this chapter. Answer each question. Then check your answers against the ones provided in the Answer Section.

1. How many numbers are 6 units from 4 on the number line? What are the numbers?

2. What numbers have an absolute value of 6?

3. What is the rule for adding two integers?

4. What is the rule for subtracting two integers?

5. What operation is needed to find the change in temperature from $-5°C$ to $-14°C$?

6. Show the result on the number line: $4 - 9$.

7. If you multiply two numbers with different signs, what is the sign of the product?

8. If you divide two numbers with the same sign, what is the sign of the quotient?

9. What is the result when a number is divided by zero?

10. What is a terminating decimal?

11. What are the steps in the Order of Operations Agreement?

CHAPTER 3

REVIEW EXERCISES

1. Write the expression $8 - (-1)$ in words.

2. Evaluate $-|-36|$.

3. Find the product of -40 and -5.

4. Evaluate $-a \div b$ when $a = -27$ and $b = -3$.

5. Add: $-28 + 14$

6. Simplify: $-(-13)$

7. Graph -2 on the number line.

−6 −5 −4 −3 −2 −1 0 1 2 3 4 5 6

8. What is the sum of -65.47 and -32.91?

9. Divide: $-51 \div (-3)$

10. Find the quotient of 840 and -4.

11. Subtract: $-6 - (-7) - 15 - (-12)$

12. Evaluate $-ab$ when $a = -2$ and $b = -9$.

13. Find the sum of 18, -13, and -6.

14. Multiply: $-18(4)$

15. Simplify: $(-2)^2 - (-3)^2 \div (1 - 4)^2 \cdot 2 - 6$

16. Evaluate $-x - y$ when $x = -1$ and $y = 3$.

17. Find the difference between -15 and -28.

18. What is the quotient of $-\frac{1}{5}$ and $-\frac{7}{10}$?

19. Is -9 a solution of $-6 - t = 3$?

20. Simplify: $-9 + 16 - (-7)$

21. Divide: $\dfrac{0}{-17}$

22. Multiply: $-5(2)(-6)(-1)$

23. Add: $3 + (-9) + 4 + (-10)$

24. Evaluate $(a - b)^2 - 2a$ when $a = -2$ and $b = -3$.

25. Place the correct symbol, < or >, between the two numbers.
$-8 \quad -10$

26. Simplify: $3 \div \left(\dfrac{1}{2} - \dfrac{1}{4}\right) - 3$

27. Find the absolute value of -27.

28. Multiply: $-0.8(3.5)$

29. What is $\frac{7}{12}$ added to $-\frac{11}{16}$?

30. **Temperature** Which is colder, a temperature of $-4°C$ or $-12°C$?

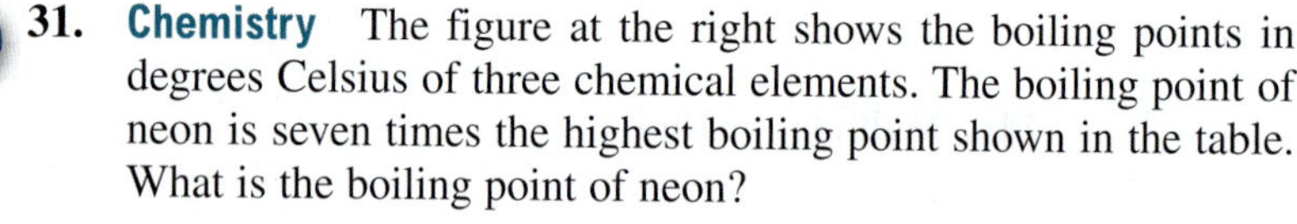

31. **Chemistry** The figure at the right shows the boiling points in degrees Celsius of three chemical elements. The boiling point of neon is seven times the highest boiling point shown in the table. What is the boiling point of neon?

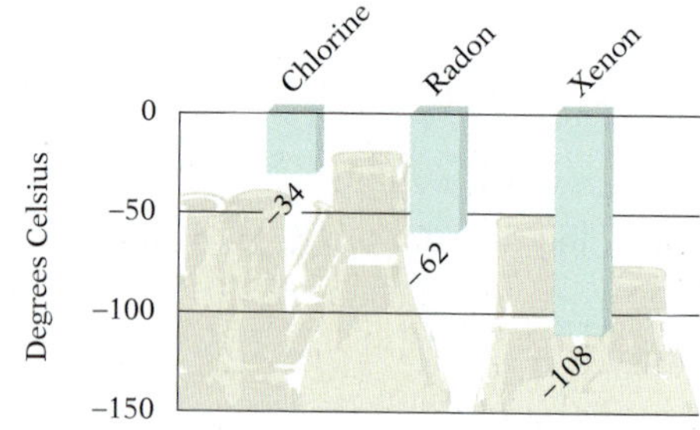

Boiling Points of Chemical Elements

32. **Temperature** Find the temperature after an increase of $5.5°C$ from $-8.5°C$.

33. **Mathematics** The distance, d, between point a and point b on the number line is given by the formula $d = |a - b|$. Find d when $a = 7$ and $b = -5$.

CHAPTER 3

TEST

1. Write the expression $-3 + (-5)$ in words.

2. Evaluate $-|-34|$.

3. What is 3 minus -15?

4. Evaluate $a + b$ when $a = -11$ and $b = -9$.

5. Evaluate $(-x)(-y)$ when $x = -4$ and $y = -6$.

6. What is $-\frac{5}{6}$ added to $\frac{4}{9}$?

7. What is -360 divided by -30?

8. Find the sum of -3, -6, and 11.

9. Place the correct symbol between the two numbers.
   16 $\quad -19$

10. Subtract: $7 - (-3) - 12$

11. Evaluate $a - b - c$ when $a = 6$, $b = -2$, and $c = 11$.

12. Simplify: $-(-49)$

13. Find the product of 50 and -5.

14. Write the given numbers in order from smallest to largest.
    $-|5|, -(-11), |-9|, -(3)$

15. Is -9 a solution of the equation $17 - x = 8$?

16. On the number line, which number is 2 units to the right of -5?

17. Divide: $\dfrac{0}{-16}$

18. Evaluate $2bc - (c + a)^3$ when $a = -2$, $b = 4$, and $c = -1$.

19. Find the opposite of 25.

20. What is 4.793 less than -6.82?

21. Subtract: $0 - 11$

22. Divide: $-96 \div (-4)$

23. Simplify: $16 \div 4 - 12 \div (-2)$

24. Evaluate $\frac{-x}{y}$ when $x = -56$ and $y = -8$.

25. Evaluate $3xy$ when $x = -2$ and $y = -10$.

26. Simplify: $7 \div \left(\frac{1}{7} - \frac{3}{14}\right) - 9$

27. Divide: $-18 \div \frac{2}{3}$

28. Evaluate xy when $x = -0.3$ and $y = 5.1$.

29. What is 14 less than 4?

30. **Temperature** Find the temperature after an increase of 11°C from -6.5°C.

31. **Environmental Science** The wind-chill factor when the temperature is -25°F and the wind is blowing at 40 mph is four times the wind-chill factor when the temperature is -5°F and the wind is blowing at 5 mph. If the wind-chill factor at -5°F with a 5-mph wind is -16°, what is the wind-chill factor at -25°F with a 40-mph wind?

32. **Temperature** The high temperature today is 8° lower than the high temperature yesterday. The high temperature today is -13°C. What was the high temperature yesterday?

33. **Mathematics** The distance, d, between point a and point b on the number line is given by the formula $d = |a - b|$. Find d when $a = 4$ and $b = -12$.

CUMULATIVE REVIEW EXERCISES

1. Find the difference between −27 and −32.

2. Estimate the product of 439 and 28.

3. Divide: $16.15 \div 0.5$

4. Simplify: $16 \div (3 + 5) \cdot 9 - 2^4$

5. Evaluate $-|-82|$.

6. Write three hundred nine thousand four hundred eighty in standard form.

7. Evaluate $5xy$ when $x = 80$ and $y = 6$.

8. What is −294 divided by −14?

9. Subtract: $-28 - (-17)$

10. Find the sum of −24, 16, and −32.

11. Find all the factors of 44.

12. Evaluate x^4y^2 when $x = \frac{1}{2}$ and $y = 4$.

13. Round 629,874 to the nearest thousand.

14. Estimate the sum of 356, 481, 294, and 117.

15. Evaluate $-a - b$ when $a = -4$ and $b = -5$.

16. Find the product of −100 and 25.

17. Find the sum of 3.97 and 4.8.

18. Add: $2\frac{1}{6} + 3\frac{1}{2}$

19. Simplify: $(1 - 5)^2 \div (-6 + 4) + 8(-3)$

20. Evaluate $-c \div d$ when $c = -32$ and $d = -8$.

21. Find the quotient of $\frac{9}{10}$ and $\frac{3}{4}$.

22. Place the correct symbol, $<$ or $>$, between the two numbers.
    −62   26

23. What is -18 multiplied by -7?

24. Divide: $(-3.312) \div (-0.8)$

25. Write $2 \cdot 2 \cdot 2 \cdot 2 \cdot 2 \cdot 7 \cdot 7$ in exponential notation.

26. Evaluate $4a + (a - b)^3$ when $a = 5$ and $b = 2$.

27. Add: $5971 + 482 + 3609$

28. What is 5 less than -21?

29. Estimate the difference between 7352 and 1986.

30. Evaluate $3^4 \cdot 5^2$.

31. **History** The land area of the United States prior to the Louisiana Purchase was 891,364 mi^2. The land area of the Louisiana Purchase, which was purchased from France in 1803, was 831,321 mi^2. What was the land area of the United States immediately after the Louisiana Purchase?

Bettmann/Corbis

Albert Einstein

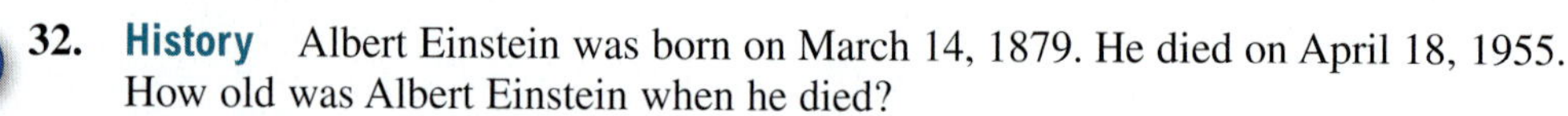

32. **History** Albert Einstein was born on March 14, 1879. He died on April 18, 1955. How old was Albert Einstein when he died?

33. **Finances** A customer makes a down payment of \$3550 on a car costing \$17,750. Find the amount that remains to be paid.

34. **Real Estate** A construction company is considering purchasing a 25-acre tract of land on which to build single-family homes. If the price is \$3690 per acre, what is the total cost of the land?

35. **Temperature** Find the temperature after an increase of 7°C from -12°C.

36. **Temperature** Record temperatures, in degrees Fahrenheit, for four states in the United States are shown at the right.

a. What is the difference between the record high and record low temperatures in Arizona?

b. For which state is the difference between the record high and record low temperatures greatest?

Record Temperatures (in degrees Fahrenheit)

State	*Lowest*	*Highest*
Alabama	−27	112
Alaska	−80	100
Arizona	−40	128
Arkansas	−29	120

Source: The World Almanac and Book of Facts 2003

37. **Sales** As a sales representative, your goal is to sell \$120,000 in merchandise during the year. You sold \$28,550 in merchandise during the first quarter of the year, \$34,850 during the second quarter, and \$31,700 during the third quarter. What must your sales for the fourth quarter be if you are to meet your goal for the year?

38. **Golf** Use the equation $S = N - P$, where S is a golfer's score relative to par in a tournament, N is the number of strokes made by the golfer, and P is par, to find a golfer's score relative to par when the golfer made 198 strokes and par is 206.

Variable Expressions

CHAPTER 4

OBJECTIVES

SECTION 4.1

A To evaluate a variable expression

SECTION 4.2

A To simplify a variable expression using the Properties of Addition

B To simplify a variable expression using the Properties of Multiplication

C To simplify a variable expression using the Distributive Property

D To simplify general variable expressions

SECTION 4.3

A To translate a verbal expression into a variable expression, given the variable

B To translate a verbal expression into a variable expression and then simplify

C To translate application problems

ARE YOU READY?

Take the Chapter 4 Prep Test to find out if you are ready to learn to:

- Evaluate a variable expression
- Simplify a variable expression
- Translate a verbal expression into a variable expression

PREP TEST

Do these exercises to prepare for Chapter 4.

1. Subtract: $-12 - (-15)$

2. Divide: $-36 \div (-9)$

3. Add: $-\frac{3}{4} + \frac{5}{6}$

4. What is the reciprocal of $-\frac{9}{4}$?

5. Divide: $-\frac{3}{4} \div \left(-\frac{5}{2}\right)$

6. Evaluate: -2^4

7. Evaluate: $\left(\frac{2}{3}\right)^3$

8. Evaluate: $3 \cdot 4^2$

9. Evaluate: $7 - 2 \cdot 3$

10. Evaluate: $5 - 7(3 - 2^2)$

SECTION 4.1 Evaluating Variable Expressions

OBJECTIVE A To evaluate a variable expression 

Tips for Success

Before you begin a new chapter, you should take some time to review previously learned skills. One way to do this is to complete the Prep Test. See page 245. This test focuses on the particular skills that will be required for the new chapter.

Point of Interest

Historical manuscripts indicate that mathematics is at least 4000 years old. Yet it was only 400 years ago that mathematicians started using variables to stand for numbers. The idea that a letter can stand for some number was a critical turning point in mathematics.

Today, *x* is used by most nations as the standard letter for a single unknown. In fact, *x*-rays were so named because the scientists who discovered them did not know what they were and thus labeled them the "unknown rays" or *x*-rays.

Often we discuss a quantity without knowing its exact value—for example, the price of gold next month, the cost of a new automobile next year, or the tuition cost for next semester. Recall that a letter of the alphabet, called a **variable,** is used to stand for a quantity that is unknown or that can change, or *vary.* An expression that contains one or more variables is called a **variable expression.**

A variable expression is shown at the right. The expression can be rewritten by writing subtraction as the addition of the opposite.

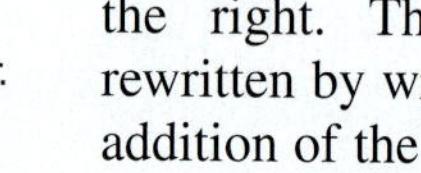

$$3x^2 - 5y + 2xy - x - 7$$
$$3x^2 + (-5y) + 2xy + (-x) + (-7)$$

Note that the expression has five addends. The **terms** of a variable expression are the addends of the expression. The expression has five terms.

Five terms

$$3x^2 \quad - \quad 5y \quad + \quad 2xy \quad - \quad x \qquad - \quad 7$$

Variable terms — Constant term

The terms $3x^2$, $-5y$, $2xy$, and $-x$ are **variable terms.**

The term -7 is a **constant term,** or simply a **constant.**

Each variable term is composed of a **numerical coefficient** and a **variable part** (the variable or variables and their exponents).

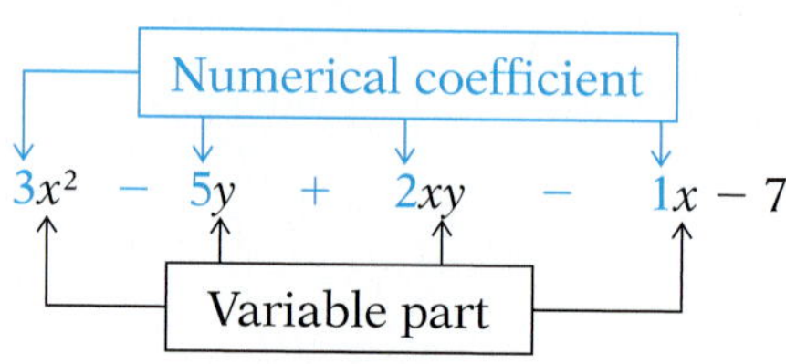

When the numerical coefficient is 1 or -1, the 1 is usually not written ($x = 1x$ and $-x = -1x$).

Variable expressions occur naturally in science. In a physics lab, a student may discover that a weight of 1 pound will stretch a spring $\frac{1}{2}$ inch. Two pounds will stretch the spring 1 inch. By experimenting, the student can discover that the distance the spring will stretch is found by multiplying the weight by $\frac{1}{2}$. By letting W represent the weight attached to the spring, the student can represent the distance the spring stretches by the variable expression $\frac{1}{2}W$.

With a weight of W pounds, the spring will stretch $\frac{1}{2} \cdot W = \frac{1}{2}W$ inches.

With a weight of 10 pounds, the spring will stretch $\frac{1}{2} \cdot 10 = 5$ inches. The number 10 is called the **value of the variable** W.

With a weight of 3 pounds, the spring will stretch $\frac{1}{2} \cdot 3 = 1\frac{1}{2}$ inches.

Integrating Technology

See the Keystroke Guide: *Evaluating Variable Expressions* for instructions on using a graphing calculator to evaluate variable expressions.

Replacing each variable by its value and then simplifying the resulting numerical expression is called **evaluating a variable expression.**

HOW TO • 1 Evaluate $ab - b^2$ when $a = 2$ and $b = -3$.

Replace each variable in the expression by its value. Then use the Order of Operations Agreement to simplify the resulting numerical expression.

$ab - b^2$

$2(-3) - (-3)^2 = -6 - 9$

$= -15$

When $a = 2$ and $b = -3$, the value of $ab - b^2$ is -15.

EXAMPLE • 1

Name the variable terms of the expression $2a^2 - 5a + 7$.

Solution

$2a^2$ and $-5a$

YOU TRY IT • 1

Name the constant term of the expression $6n^2 + 3n - 4$.

Your solution

EXAMPLE • 2

Evaluate $x^2 - 3xy$ when $x = 3$ and $y = -4$.

Solution

$x^2 - 3xy$

$3^2 - 3(3)(-4) = 9 - 3(3)(-4)$

$= 9 - 9(-4)$

$= 9 - (-36)$

$= 9 + 36 = 45$

YOU TRY IT • 2

Evaluate $2xy + y^2$ when $x = -4$ and $y = 2$.

Your solution

EXAMPLE • 3

Evaluate $\frac{a^2 - b^2}{a - b}$ when $a = 3$ and $b = -4$.

Solution

$\frac{a^2 - b^2}{a - b}$

$\frac{3^2 - (-4)^2}{3 - (-4)} = \frac{9 - 16}{3 - (-4)}$

$= \frac{-7}{7} = -1$

YOU TRY IT • 3

Evaluate $\frac{a^2 + b^2}{a + b}$ when $a = 5$ and $b = -3$.

Your solution

EXAMPLE • 4

Evaluate $x^2 - 3(x - y) - z^2$ when $x = 2$, $y = -1$, and $z = 3$.

Solution

$x^2 - 3(x - y) - z^2$

$2^2 - 3[2 - (-1)] - 3^2$

$= 2^2 - 3(3) - 3^2$

$= 4 - 3(3) - 9$

$= 4 - 9 - 9$

$= -5 - 9$

$= -14$

YOU TRY IT • 4

Evaluate $x^3 - 2(x + y) + z^2$ when $x = 2$, $y = -4$, and $z = -3$.

Your solution

Solutions on p. S11

4.1 EXERCISES

OBJECTIVE A **To evaluate a variable expression**

For Exercises 1 to 3, name the terms of the variable expression. Then underline the constant term.

1. $2x^2 + 5x - 8$

2. $-3n^2 - 4n + 7$

3. $6 - a^4$

For Exercises 4 to 6, name the variable terms of the expression. Then underline the variable part of each term.

4. $9b^2 - 4ab + a^2$

5. $7x^2y + 6xy^2 + 10$

6. $5 - 8n - 3n^2$

For Exercises 7 to 9, name the coefficients of the variable terms.

7. $x^2 - 9x + 2$

8. $12a^2 - 8ab - b^2$

9. $n^3 - 4n^2 - n + 9$

10. What is the numerical coefficient of a variable term?

11. Explain the meaning of the phrase "evaluate a variable expression."

For Exercises 12 to 32, evaluate the variable expression when $a = 2$, $b = 3$, and $c = -4$.

12. $3a + 2b$

13. $a - 2c$

14. $-a^2$

15. $2c^2$

16. $-3a + 4b$

17. $3b - 3c$

18. $b^2 - 3$

19. $-3c + 4$

20. $16 \div (2c)$

21. $6b \div (-a)$

22. $bc \div (2a)$

23. $b^2 - 4ac$

24. $a^2 - b^2$

25. $b^2 - c^2$

26. $(a + b)^2$

27. $a^2 + b^2$

28. $2a - (c + a)^2$

29. $(b - a)^2 + 4c$

30. $b^2 - \frac{ac}{8}$

31. $\frac{5ab}{6} - 3cb$

32. $(b - 2a)^2 + bc$

For Exercises 33 to 50, evaluate the variable expression when $a = -2$, $b = 4$, $c = -1$, and $d = 3$.

33. $\dfrac{b + c}{d}$

34. $\dfrac{d - b}{c}$

35. $\dfrac{2d + b}{-a}$

36. $\dfrac{b + 2d}{b}$

37. $\dfrac{b - d}{c - a}$

38. $\dfrac{2c - d}{-ad}$

39. $(b + d)^2 - 4a$

40. $(d - a)^2 - 3c$

41. $(d - a)^2 \div 5$

42. $3(b - a) - bc$

43. $\dfrac{b - 2a}{bc^2 - d}$

44. $\dfrac{b^2 - a}{ad + 3c}$

45. $\dfrac{1}{3}d^2 - \dfrac{3}{8}b^2$

46. $\dfrac{5}{8}a^4 - c^2$

47. $\dfrac{-4bc}{2a - b}$

48. $-\dfrac{3}{4}b + \dfrac{1}{2}(ac + bd)$

49. $-\dfrac{2}{3}d - \dfrac{1}{5}(bd - ac)$

50. $(b - a)^2 - (d - c)^2$

For Exercises 51 to 54, without evaluating the expression, determine whether the expression is positive or negative when $a = -25$, $b = 67$, and $c = -82$.

51. $(c - a)(-b)$

52. $(a - c) + 3b$

53. $\dfrac{b + c}{abc}$

54. $\dfrac{ac}{-b^2}$

55. The value of a is the value of $3x^2 - 4x - 5$ when $x = -2$. Find the value of $3a - 4$.

56. The value of c is the value of $a^2 + b^2$ when $a = 2$ and $b = -2$. Find the value of $c^2 - 4$.

Applying the Concepts

For Exercises 57 to 60, evaluate the expression for $x = 2$, $y = 3$, and $z = -2$.

57. $3^x - x^3$

58. z^x

59. $x^x - y^y$

60. $y^{(x^2)}$

61. For each of the following, determine the first natural number x, greater than 1, for which the second expression is larger than the first.

a. $x^3, 3^x$ **b.** $x^4, 4^x$ **c.** $x^5, 5^x$ **d.** $x^6, 6^x$

SECTION 4.2 Simplifying Variable Expressions

OBJECTIVE A To simplify a variable expression using the Properties of Addition

Like terms of a variable expression are terms with the same variable part. (Because $x^2 = x \cdot x$, x^2 and x are not like terms.)

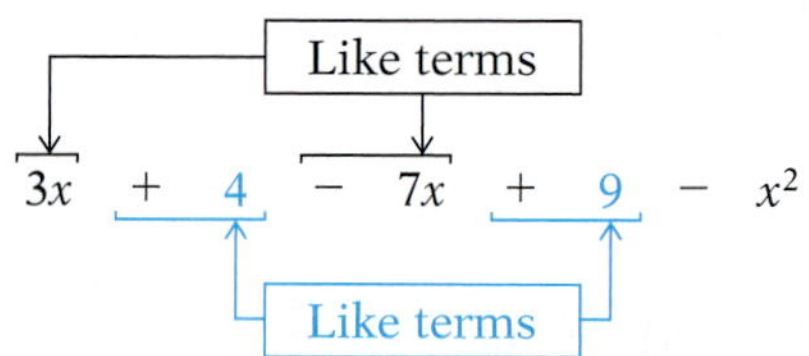

Constant terms are like terms. 4 and 9 are like terms.

To simplify a variable expression, use the Distributive Property to combine like terms by adding the numerical coefficients. The variable part remains unchanged.

> **Distributive Property**
> If a, b, and c are real numbers, then $a(b + c) = ab + ac$.

Take Note
Here is an example of the Distributive Property using just numbers.

$$2(5 + 9) = 2(5) + 2(9) = 10 + 18 = 28$$

This is the same result we would obtain using the Order of Operations Agreement.

$$2(5 + 9) = 2(14) = 28$$

The usefulness of the Distributive Property will become more apparent as we explore variable expressions.

The Distributive Property can also be written $ba + ca = (b + c)a$. This form is used to simplify a variable expression.

To simplify $2x + 3x$, use the Distributive Property to add the numerical coefficients of the like variable terms. This is called **combining like terms.**

$$2x + 3x = (2 + 3)x = 5x$$

HOW TO 1 Simplify: $5y - 11y$

$$5y - 11y = (5 - 11)y = -6y$$

• Use the Distributive Property.

Take Note
Simplifying an expression means combining like terms. The constant term 5 and the variable term $7p$ are not like terms and therefore cannot be combined.

HOW TO 2 Simplify: $5 + 7p$

The terms 5 and $7p$ are not like terms.

The expression $5 + 7p$ is in simplest form.

> **The Associative Property of Addition**
> If a, b, and c are real numbers, then $(a + b) + c = a + (b + c)$.

When three or more terms are added, the terms can be grouped (with parentheses, for example) in any order. The sum is the same. For example,

$$(5 + 7) + 15 = 5 + (7 + 15) \qquad (3x + 5x) + 9x = 3x + (5x + 9x)$$
$$12 + 15 = 5 + 22 \qquad 8x + 9x = 3x + 14x$$
$$27 = 27 \qquad 17x = 17x$$

The Commutative Property of Addition

If a and b are real numbers, then $a + b = b + a$.

When two like terms are added, the terms can be added in either order. The sum is the same. For example,

$$15 + (-28) = (-28) + 15 \qquad 2x + (-4x) = -4x + 2x$$
$$-13 = -13 \qquad -2x = -2x$$

The Addition Property of Zero

If a is a real number, then $a + 0 = 0 + a = a$.

The sum of a term and zero is the term. For example,

$$-9 + 0 = 0 + (-9) = -9 \qquad 5x + 0 = 0 + 5x = 5x$$

The Inverse Property of Addition

If a is a real number, then $a + (-a) = (-a) + a = 0$.

The sum of a term and its opposite is zero. Recall that the opposite of a number is called its **additive inverse.**

$$12 + (-12) = (-12) + 12 = 0 \qquad 7x + (-7x) = -7x + 7x = 0$$

HOW TO • 3 Simplify: $8x + 4y - 8x + y$

$8x + 4y - 8x + y$
$= (8x - 8x) + (4y + y)$ • Use the Commutative and Associative Properties of Addition to rearrange and group like terms.
$= 0 + 5y = 5y$ • Combine like terms.

HOW TO • 4 Simplify: $4x^2 + 5x - 6x^2 - 2x + 1$

$4x^2 + 5x - 6x^2 - 2x + 1$
$= (4x^2 - 6x^2) + (5x - 2x) + 1$ • Use the Commutative and Associative Properties of Addition to rearrange and group like terms.
$= -2x^2 + 3x + 1$ • Combine like terms.

EXAMPLE • 1

Simplify: $3x + 4y - 10x + 7y$

Solution

$3x + 4y - 10x + 7y = -7x + 11y$

YOU TRY IT • 1

Simplify: $3a - 2b - 5a + 6b$

Your solution

EXAMPLE • 2

Simplify: $x^2 - 7 + 4x^2 - 16$

Solution

$x^2 - 7 + 4x^2 - 16 = 5x^2 - 23$

YOU TRY IT • 2

Simplify: $-3y^2 + 7 + 8y^2 - 14$

Your solution

Solutions on p. S11

OBJECTIVE B To simplify a variable expression using the Properties of Multiplication 

In simplifying variable expressions, the following Properties of Multiplication are used.

> ✓ **Take Note**
> The Associative Property of Multiplication allows us to multiply a coefficient by a number. Without this property, the expression $2(3x)$ could not be changed.

The Associative Property of Multiplication

If a, b, and c are real numbers, then $(ab)c = a(bc)$.

When three or more factors are multiplied, the factors can be grouped in any order. The product is the same.

$$3(5 \cdot 6) = (3 \cdot 5)6 \qquad 2(3x) = (2 \cdot 3)x$$
$$3(30) = (15)6 \qquad = 6x$$
$$90 = 90$$

> ✓ **Take Note**
> The Commutative Property of Multiplication allows us to rearrange factors. This property, along with the Associative Property of Multiplication, allows us to simplify some variable expressions.

The Commutative Property of Multiplication

If a and b are real numbers, then $ab = ba$.

Two factors can be multiplied in either order. The product is the same.

$$5(-7) = -7(5) \qquad -35 = -35$$

$(5x) \cdot 3 = 3 \cdot (5x)$ • Commutative Property of Multiplication
$= (3 \cdot 5)x$ • Associative Property of Multiplication
$= 15x$

The Multiplication Property of One

If a is a real number, then $a \cdot 1 = 1 \cdot a = a$.

The product of a term and 1 is the term.

$$9 \cdot 1 = 1 \cdot 9 = 9 \qquad (8x) \cdot 1 = 1 \cdot (8x) = 8x$$

The Inverse Property of Multiplication

If a is a real number and a is not equal to zero, then

$$a \cdot \frac{1}{a} = \frac{1}{a} \cdot a = 1$$

$\frac{1}{a}$ is called the **reciprocal** of a. $\frac{1}{a}$ is also called the **multiplicative inverse** of a. The product of a number and its reciprocal is 1.

> ✓ **Take Note**
> We must state that $x \neq 0$ because division by zero is undefined.

$$7 \cdot \frac{1}{7} = \frac{1}{7} \cdot 7 = 1 \qquad x \cdot \frac{1}{x} = \frac{1}{x} \cdot x = 1, \quad x \neq 0$$

The multiplication properties are used to simplify variable expressions.

HOW TO • 5 Simplify: $2(-x)$

$2(-x) = 2(-1 \cdot x)$ • Use the Associative Property of Multiplication to group factors.
$= [2(-1)]x$
$= -2x$

HOW TO • 6 Simplify: $\frac{3}{2}\left(\frac{2x}{3}\right)$

$$\begin{aligned}\frac{3}{2}\left(\frac{2x}{3}\right) &= \frac{3}{2}\left(\frac{2}{3}x\right) && \text{• Note that } \frac{2x}{3} = \frac{2}{3}x.\\ &= \left(\frac{3}{2}\cdot\frac{2}{3}\right)x && \text{• Use the Associative Property of Multiplication to group factors.}\\ &= 1 \cdot x\\ &= x\end{aligned}$$

HOW TO • 7 Simplify: $(16x)2$

$$\begin{aligned}(16x)2 &= 2(16x) && \text{• Use the Commutative and Associative Properties of Multiplication to rearrange and group factors.}\\ &= (2 \cdot 16)x\\ &= 32x\end{aligned}$$

EXAMPLE • 3

Simplify: $-2(3x^2)$

Solution

$-2(3x^2) = -6x^2$

YOU TRY IT • 3

Simplify: $-5(4y^2)$

Your solution

EXAMPLE • 4

Simplify: $-5(-10x)$

Solution

$-5(-10x) = 50x$

YOU TRY IT • 4

Simplify: $-7(-2a)$

Your solution

EXAMPLE • 5

Simplify: $-\frac{3}{4}\left(\frac{2}{3}x\right)$

Solution

$-\frac{3}{4}\left(\frac{2}{3}x\right) = -\frac{1}{2}x$

YOU TRY IT • 5

Simplify: $-\frac{3}{5}\left(-\frac{7}{9}a\right)$

Your solution

Solutions on p. S11

OBJECTIVE C To simplify a variable expression using the Distributive Property

Recall that the Distributive Property states that if a, b, and c are real numbers, then

$$a(b + c) = ab + ac$$

The Distributive Property is used to remove parentheses from a variable expression.

HOW TO • 8 Simplify: $3(2x + 7)$

$$\begin{aligned}3(2x + 7) &= 3(2x) + 3(7) && \text{• Use the Distributive Property. Multiply each term inside the parentheses by 3.}\\ &= 6x + 21\end{aligned}$$

HOW TO 9 Simplify: $-5(4x + 6)$

$$-5(4x + 6) = -5(4x) + (-5)(6) = -20x - 30$$

• Use the Distributive Property.

HOW TO 10 Simplify: $-(2x - 4)$

$$-(2x - 4) = -1(2x - 4) = -1(2x) - (-1)(4) = -2x + 4$$

• Use the Distributive Property.

Note: When a negative sign immediately precedes the parentheses, the sign of each term inside the parentheses is changed.

HOW TO 11 Simplify: $-\frac{1}{2}(8x - 12y)$

$$-\frac{1}{2}(8x - 12y) = -\frac{1}{2}(8x) - \left(-\frac{1}{2}\right)(12y) = -4x + 6y$$

• Use the Distributive Property.

An extension of the Distributive Property is used when an expression contains more than two terms.

HOW TO 12 Simplify: $3(4x - 2y - z)$

$$3(4x - 2y - z) = 3(4x) - 3(2y) - 3(z) = 12x - 6y - 3z$$

• Use the Distributive Property.

EXAMPLE 6

Simplify: $7(4 + 2x)$

Solution

$7(4 + 2x) = 28 + 14x$

YOU TRY IT 6

Simplify: $5(3 + 7b)$

Your solution

EXAMPLE 7

Simplify: $(2x - 6)2$

Solution

$(2x - 6)2 = 4x - 12$

YOU TRY IT 7

Simplify: $(3a - 1)5$

Your solution

EXAMPLE 8

Simplify: $-3(-5a + 7b)$

Solution

$-3(-5a + 7b) = 15a - 21b$

YOU TRY IT 8

Simplify: $-8(-2a + 7b)$

Your solution

Solutions on p. S11

EXAMPLE 9

Simplify: $3(x^2 - x - 5)$

Solution

$3(x^2 - x - 5) = 3x^2 - 3x - 15$

YOU TRY IT 9

Simplify: $3(12x^2 - x + 8)$

Your solution

EXAMPLE 10

Simplify: $-2(x^2 + 5x - 4)$

Solution

$-2(x^2 + 5x - 4)$
$= -2x^2 - 10x + 8$

YOU TRY IT 10

Simplify: $3(-a^2 - 6a + 7)$

Your solution

Solutions on p. S11

OBJECTIVE D To simplify general variable expressions

When simplifying variable expressions, use the Distributive Property to remove parentheses and brackets used as grouping symbols.

HOW TO 13 Simplify: $4(x - y) - 2(-3x + 6y)$

$4(x - y) - 2(-3x + 6y)$
$= 4x - 4y + 6x - 12y$ • Use the Distributive Property.
$= 10x - 16y$ • Combine like terms.

EXAMPLE 11

Simplify: $2x - 3(2x - 7y)$

Solution

$2x - 3(2x - 7y) = 2x - 6x + 21y$
$= -4x + 21y$

YOU TRY IT 11

Simplify: $3y - 2(y - 7x)$

Your solution

EXAMPLE 12

Simplify: $7(x - 2y) - (-x - 2y)$

Solution

$7(x - 2y) - (-x - 2y)$
$= 7x - 14y + x + 2y$
$= 8x - 12y$

YOU TRY IT 12

Simplify: $-2(x - 2y) - (-x + 3y)$

Your solution

EXAMPLE 13

Simplify: $2x - 3[2x - 3(x + 7)]$

Solution

$2x - 3[2x - 3(x + 7)]$
$= 2x - 3[2x - 3x - 21]$
$= 2x - 3[-x - 21]$
$= 2x + 3x + 63$
$= 5x + 63$

YOU TRY IT 13

Simplify: $3y - 2[x - 4(2 - 3y)]$

Your solution

Solutions on p. S11

4.2 EXERCISES

OBJECTIVE A To simplify a variable expression using the Properties of Addition

1. What are *like terms*? Give an example of two like terms. Give an example of two terms that are not like terms.

2. Explain the meaning of the phrase "simplify a variable expression."

For Exercises 3 to 38, simplify.

3. $6x + 8x$
4. $12x + 13x$
5. $9a - 4a$
6. $12a - 3a$
7. $4y - 10y$
8. $8y - 6y$
9. $7 - 3b$
10. $5 + 2a$
11. $-12a + 17a$
12. $-3a + 12a$
13. $5ab - 7ab$
14. $9ab - 3ab$
15. $-12xy + 17xy$
16. $-15xy + 3xy$
17. $-3ab + 3ab$
18. $-7ab + 7ab$
19. $-\frac{1}{2}x - \frac{1}{3}x$
20. $-\frac{2}{5}y + \frac{3}{10}y$
21. $2.3x + 4.2x$
22. $6.1y - 9.2y$
23. $x - 0.55x$
24. $0.65A - A$
25. $5a - 3a + 5a$
26. $10a - 17a + 3a$
27. $-5x^2 - 12x^2 + 3x^2$
28. $-y^2 - 8y^2 + 7y^2$
29. $\frac{3}{4}x - \frac{1}{3}x - \frac{7}{8}x$
30. $-\frac{2}{5}a - \left(-\frac{3}{10}a\right) - \frac{11}{15}a$
31. $7x - 3y + 10x$
32. $8y + 8x - 8y$
33. $3a + (-7b) - 5a + b$
34. $-5b + 7a - 7b + 12a$
35. $3x + (-8y) - 10x + 4x$
36. $3y + (-12x) - 7y + 2y$
37. $x^2 - 7x + (-5x^2) + 5x$
38. $3x^2 + 5x - 10x^2 - 10x$

39. Which of the following expressions are equivalent to $-10x - 10y - 10y - 10x$?
(i) 0 (ii) $-20y$ (iii) $-20x$ (iv) $-20x - 20y$ (v) $-20y - 20x$

OBJECTIVE B **To simplify a variable expression using the Properties of Multiplication**

For Exercises 40 to 79, simplify.

40. $4(3x)$ **41.** $12(5x)$ **42.** $-3(7a)$ **43.** $-2(5a)$ **44.** $-2(-3y)$

45. $-5(-6y)$ **46.** $(4x)2$ **47.** $(6x)12$ **48.** $(3a)(-2)$ **49.** $(7a)(-4)$

50. $(-3b)(-4)$ **51.** $(-12b)(-9)$ **52.** $-5(3x^2)$ **53.** $-8(7x^2)$ **54.** $\frac{1}{3}(3x^2)$

55. $\frac{1}{6}(6x^2)$ **56.** $\frac{1}{5}(5a)$ **57.** $\frac{1}{8}(8x)$ **58.** $-\frac{1}{2}(-2x)$ **59.** $-\frac{1}{4}(-4a)$

60. $-\frac{1}{7}(-7n)$ **61.** $-\frac{1}{9}(-9b)$ **62.** $(3x)\left(\frac{1}{3}\right)$ **63.** $(12x)\left(\frac{1}{12}\right)$ **64.** $(-6y)\left(-\frac{1}{6}\right)$

65. $(-10n)\left(-\frac{1}{10}\right)$ **66.** $\frac{1}{3}(9x)$ **67.** $\frac{1}{7}(14x)$ **68.** $-0.2(10x)$ **69.** $-0.25(8x)$

70. $-\frac{2}{3}(12a^2)$ **71.** $-\frac{5}{8}(24a^2)$ **72.** $-0.5(-16y)$ **73.** $-0.75(-8y)$ **74.** $(16y)\left(\frac{1}{4}\right)$

75. $(33y)\left(\frac{1}{11}\right)$ **76.** $(-6x)\left(\frac{1}{3}\right)$ **77.** $(-10x)\left(\frac{1}{5}\right)$ **78.** $(-8a)\left(-\frac{3}{4}\right)$ **79.** $(21y)\left(-\frac{3}{7}\right)$

80. After multiplying $\frac{2}{7}x^2$ by a proper fraction, is the coefficient of x^2 greater than 1 or less than 1?

OBJECTIVE C To simplify a variable expression using the Distributive Property

For Exercises 81 to 119, simplify.

81. $2(4x - 3)$

82. $5(2x - 7)$

83. $-2(a + 7)$

84. $-5(a + 16)$

85. $-3(2y - 8)$

86. $-5(3y - 7)$

87. $-(x + 2)$

88. $-(x + 7)$

89. $(5 - 3b)7$

90. $(10 - 7b)2$

91. $\frac{1}{3}(6 - 15y)$

92. $\frac{1}{2}(-8x + 4y)$

93. $3(5x^2 + 2x)$

94. $6(3x^2 + 2x)$

95. $-2(-y + 9)$

96. $-5(-2x + 7)$

97. $(-3x - 6)5$

98. $(-2x + 7)7$

99. $2(-3x^2 - 14)$

100. $5(-6x^2 - 3)$

101. $-3(2y^2 - 7)$

102. $-8(3y^2 - 12)$

103. $3(x^2 - y^2)$

104. $5(x^2 + y^2)$

105. $-\frac{2}{3}(6x - 18y)$

106. $-\frac{1}{2}(x - 4y)$

107. $-(6a^2 - 7b^2)$

108. $3(x^2 + 2x - 6)$

109. $4(x^2 - 3x + 5)$

110. $-2(y^2 - 2y + 4)$

111. $\frac{3}{4}(2x - 6y + 8)$

112. $-\frac{2}{3}(6x - 9y + 1)$

113. $4(-3a^2 - 5a + 7)$

114. $-5(-2x^2 - 3x + 7)$

115. $-3(-4x^2 + 3x - 4)$

116. $3(2x^2 + xy - 3y^2)$

117. $5(2x^2 - 4xy - y^2)$

118. $-(3a^2 + 5a - 4)$

119. $-(8b^2 - 6b + 9)$

120. After the expression $17x - 31$ is multiplied by a negative integer, is the constant term positive or negative?

OBJECTIVE D To simplify general variable expressions

121. Which of the following expressions is equivalent to $12 - 7(y - 9)$?
(i) $5(y - 9)$ (ii) $12 - 7y - 63$ (iii) $12 - 7y + 63$ (iv) $12 - 7y - 9$

For Exercises 122 to 145, simplify.

122. $4x - 2(3x + 8)$

123. $6a - (5a + 7)$

124. $9 - 3(4y + 6)$

125. $10 - (11x - 3)$

126. $5n - (7 - 2n)$

127. $8 - (12 + 4y)$

128. $3(x + 2) - 5(x - 7)$

129. $2(x - 4) - 4(x + 2)$

130. $12(y - 2) + 3(7 - 3y)$

131. $6(2y - 7) - (3 - 2y)$

132. $3(a - b) - (a + b)$

133. $2(a + 2b) - (a - 3b)$

134. $4[x - 2(x - 3)]$

135. $2[x + 2(x + 7)]$

136. $-2[3x + 2(4 - x)]$

137. $-5[2x + 3(5 - x)]$

138. $-3[2x - (x + 7)]$

139. $-2[3x - (5x - 2)]$

140. $2x - 3[x - (4 - x)]$

141. $-7x + 3[x - (3 - 2x)]$

142. $-5x - 2[2x - 4(x + 7)] - 6$

143. $0.12(2x + 3) + x$

144. $0.05x + 0.02(4 - x)$

145. $0.03x + 0.04(1000 - x)$

Applying the Concepts

146. Determine whether the statement is true or false. If the statement is false, give an example that illustrates that it is false.
a. Division is a commutative operation.
b. Division is an associative operation.
c. Subtraction is an associative operation.
d. Subtraction is a commutative operation.

147. Give examples of two operations that occur in everyday experience that are not commutative (for example, putting on socks and then shoes).

SECTION

4.3 Translating Verbal Expressions into Variable Expressions

OBJECTIVE A To translate a verbal expression into a variable expression, given the variable

Tips for Success

Before the class meeting in which your professor begins a new section, you should read each objective statement for that section. Next, browse through the objective material. The purpose of browsing through the material is so that your brain will be prepared to accept and organize the new information when it is presented to you. See *AIM for Success* in the front of the book.

Point of Interest

The way in which expressions are symbolized has changed over time. Here are how some of the expressions shown at the right may have appeared in the early 16th century.

R p. 9 for $x + 9$. The symbol R was used for a variable to the first power. The symbol p. was used for plus.

R m. 3 for $x - 3$. The symbol R is still used for the variable. The symbol m. was used for minus.

The square of a variable was designated by Q and the cube was designated by C. The expression $x^2 + x^3$ was written Q p. C.

One of the major skills required in applied mathematics is to translate a verbal expression into a variable expression. This requires recognizing the verbal phrases that translate into mathematical operations. A partial list of the verbal phrases used to indicate the different mathematical operations follows.

Addition	added to	6 added to *y*	$y + 6$
	more than	8 more than *x*	$x + 8$
	the sum of	the sum of *x* and *z*	$x + z$
	increased by	*t* increased by 9	$t + 9$
	the total of	the total of 5 and *y*	$5 + y$
Subtraction	minus	*x* minus 2	$x - 2$
	less than	7 less than *t*	$t - 7$
	decreased by	*m* decreased by 3	$m - 3$
	the difference between	the difference between *y* and 4	$y - 4$
	subtract...from...	subtract 9 from *z*	$z - 9$
Multiplication	times	10 times *t*	$10t$
	twice	twice *w*	$2w$
	of	one-half of *x*	$\frac{1}{2}x$
	the product of	the product of *y* and *z*	yz
	multiplied by	*y* multiplied by 11	$11y$
Division	divided by	*x* divided by 12	$\frac{x}{12}$
	the quotient of	the quotient of *y* and *z*	$\frac{y}{z}$
	the ratio of	the ratio of *t* to 9	$\frac{t}{9}$
Power	the square of	the square of *x*	x^2
	the cube of	the cube of *a*	a^3

HOW TO • 1 Translate "14 less than the cube of *x*" into a variable expression.

14 less than the cube of *x* • **Identify the words that indicate the mathematical operations.**

$x^3 - 14$ • **Use the identified operations to write the variable expression.**

Translating a phrase that contains the word *sum, difference, product,* or *quotient* can be difficult. In the examples at the right, note where the operation symbol is placed.

the *sum* of x and y	$x + y$
the *difference* between x and y	$x - y$
the *product* of x and y	$x \cdot y$
the *quotient* of x and y	$\frac{x}{y}$

HOW TO • 2 Translate "the difference between the square of x and the sum of y and z" into a variable expression.

the difference between the square of x and the sum of y and z

- **Identify the words that indicate the mathematical operations.**

$x^2 - (y + z)$

- **Use the identified operations to write the variable expression.**

EXAMPLE • 1

Translate "the total of 3 times n and 5" into a variable expression.

Solution

the total of 3 times n and 5

$3n + 5$

YOU TRY IT • 1

Translate "the difference between twice n and the square of n" into a variable expression.

Your solution

EXAMPLE • 2

Translate "m decreased by the sum of n and 12" into a variable expression.

Solution

m decreased by the sum of n and 12

$m - (n + 12)$

YOU TRY IT • 2

Translate "the quotient of 7 less than b and 15" into a variable expression.

Your solution

Solutions on p. S11

OBJECTIVE B

To translate a verbal expression into a variable expression and then simplify

In most applications that involve translating phrases into variable expressions, the variable to be used is not given. To translate these phrases, a variable must be assigned to an unknown quantity before the variable expression can be written.

HOW TO • 3 Translate "a number multiplied by the total of six and the cube of the number" into a variable expression.

the unknown number: n	• **Assign a variable to one of the unknown quantities.**
the cube of the number: n^3 the total of six and the cube of the number: $6 + n^3$	• **Use the assigned variable to write an expression for any other unknown quantity.**
$n(6 + n^3)$	• **Use the assigned variable to write the variable expression.**

EXAMPLE • 3

Translate "a number added to the product of four and the square of the number" into a variable expression.

Solution

the unknown number: n
the square of the number: n^2
the product of four and the square of the number: $4n^2$
$4n^2 + n$

YOU TRY IT • 3

Translate "negative four multiplied by the total of ten and the cube of a number" into a variable expression.

Your solution

EXAMPLE • 4

Translate "four times the sum of one-half of a number and fourteen" into a variable expression. Then simplify.

Solution

the unknown number: n

one-half of the number: $\frac{1}{2}n$

the sum of one-half of the number and fourteen: $\frac{1}{2}n + 14$

$4\left(\frac{1}{2}n + 14\right)$

$= 2n + 56$

YOU TRY IT • 4

Translate "five times the difference between a number and sixty" into a variable expression. Then simplify.

Your solution

Solutions on p. S11

OBJECTIVE C To translate application problems

Many applications in mathematics require that you identify the unknown quantity, assign a variable to that quantity, and then attempt to express other unknown quantities in terms of the variable.

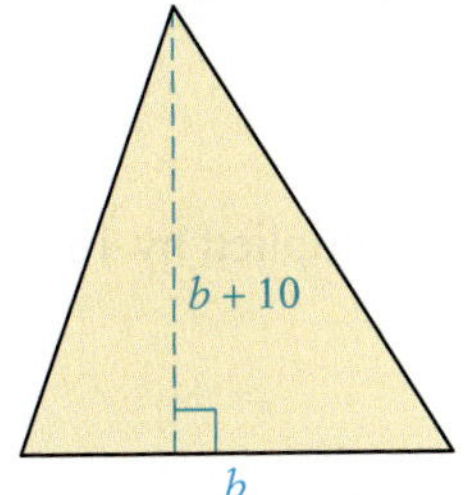

HOW TO • 4 The height of a triangle is 10 ft longer than the base of the triangle. Express the height of the triangle in terms of the base of the triangle.

the base of the triangle: b • **Assign a variable to the base of the triangle.**

the height is 10 more than the base: $b + 10$ • **Express the height of the triangle in terms of b.**

EXAMPLE • 5

The length of a swimming pool is 4 ft less than two times the width. Express the length of the pool in terms of the width.

Solution

the width of the pool: w

the length is 4 ft less than two times the width: $2w - 4$

YOU TRY IT • 5

The speed of a new jet plane is twice the speed of an older model. Express the speed of the new model in terms of the speed of the older model.

Your solution

EXAMPLE • 6

A banker divided \$5000 between two accounts, one paying 10% annual interest and the second paying 8% annual interest. Express the amount invested in the 10% account in terms of the amount invested in the 8% account.

Solution

the amount invested at 8%: x

the amount invested at 10%: $5000 - x$

YOU TRY IT • 6

A guitar string 6 ft long was cut into two pieces. Express the length of the shorter piece in terms of the length of the longer piece.

Your solution

Solutions on p. S11

4.3 EXERCISES

OBJECTIVE A To translate a verbal expression into a variable expression, given the variable

For Exercises 1 to 26, translate into a variable expression.

1. the sum of 8 and y

2. a less than 16

3. t increased by 10

4. p decreased by 7

5. z added to 14

6. q multiplied by 13

7. 20 less than the square of x

8. 6 times the difference between m and 7

9. the sum of three-fourths of n and 12

10. b decreased by the product of 2 and b

11. 8 increased by the quotient of n and 4

12. the product of -8 and y

13. the product of 3 and the total of y and 7

14. 8 divided by the difference between x and 6

15. the product of t and the sum of t and 16

16. the quotient of 6 less than n and twice n

17. 15 more than one-half of the square of x

18. 19 less than the product of n and -2

19. the total of 5 times the cube of n and the square of n

20. the ratio of 9 more than m to m

21. r decreased by the quotient of r and 3

22. four-fifths of the sum of w and 10

23. the difference between the square of x and the total of x and 17

24. s increased by the quotient of 4 and s

25. the product of 9 and the total of z and 4

26. n increased by the difference between 10 times n and 9

27. Write two different verbal phrases that translate into the variable expression $5(n^2 + 1)$.

OBJECTIVE B To translate a verbal expression into a variable expression and then simplify

For Exercises 28 to 39, translate into a variable expression.

28. twelve minus a number

29. a number divided by eighteen

30. two-thirds of a number

31. twenty more than a number

32. the quotient of twice a number and nine

33. eight less than the product of eleven and a number

34. the sum of five-eighths of a number and six

35. the quotient of seven and the total of five and a number

36. the quotient of fifteen and the sum of a number and twelve

37. the difference between forty and the quotient of a number and twenty

38. the quotient of five more than twice a number and the number

39. the sum of the square of a number and twice the number

40. Which of the following phrases translate into the variable expression $32 - \frac{a}{7}$?

(i) the difference between thirty-two and the quotient of a number and seven
(ii) thirty-two decreased by the quotient of a number and seven
(iii) thirty-two minus the ratio of a number to seven

For Exercises 41 to 56, translate into a variable expression. Then simplify.

41. ten times the difference between a number and fifty

42. nine less than the total of a number and two

43. the difference between a number and three more than the number

44. four times the sum of a number and nineteen

45. a number added to the difference between twice the number and four

46. the product of five less than a number and seven

47. a number decreased by the difference between three times the number and eight

48. the sum of eight more than a number and one-third of the number

49. a number added to the product of three and the number

50. a number increased by the total of the number and nine

51. five more than the sum of a number and six

52. a number decreased by the difference between eight and the number

53. a number minus the sum of the number and ten

54. two more than the total of a number and five

55. the sum of one-sixth of a number and four-ninths of the number

56. the difference between one-third of a number and five-eighths of the number

OBJECTIVE C To translate application problems

For Exercises 57 and 58, use the following situation: 83 more students enrolled in spring-term science classes than enrolled in fall-term science classes.

57. If s and $s + 83$ represent the quantities in this situation, what is s?

58. If n and $n - 83$ represent the quantities in this situation, what is n?

59. **Museums** In a recent year, 3.8 million more people visited the Louvre in Paris than visited the Metropolitan Museum of Art in New York City. (*Sources:* The *Art Newspaper;* museums' accounts) Express the number of visitors to the Louvre in terms of the number of visitors to the Metropolitan Museum of Art.

The Louvre

isifa Image Service s.r.o./Alamy

60. **Salaries** For an employee with a bachelor's degree in business, the average annual salary depends on experience. An employee with less than 5 years' experience is paid, on average, \$29,100 less than an employee with 10 to 20 years' experience. (*Sources:* PayScale; *The Princeton Review*) Express the salary of an employee with less than 5 years' experience in terms of the salary of an employee with 10 to 20 years' experience.

In the News

Google Websites Most Popular

During the month of February 2008, Google websites ranked number one in the world, with the highest number of unique visitors. Microsoft websites came in second, with 63 million fewer unique visitors.

Source: comScore

61. **Websites** See the news clipping at the right. Express the number of unique visitors to Microsoft websites in terms of the number of unique visitors to Google websites.

62. **Telecommunications** In 1951, phone companies began using area codes. According to information found at www.area-code.com, at the beginning of 2008 there were 183 more area codes than there were in 1951. Express the number of area codes in 2008 in terms of the number of area codes in 1951.

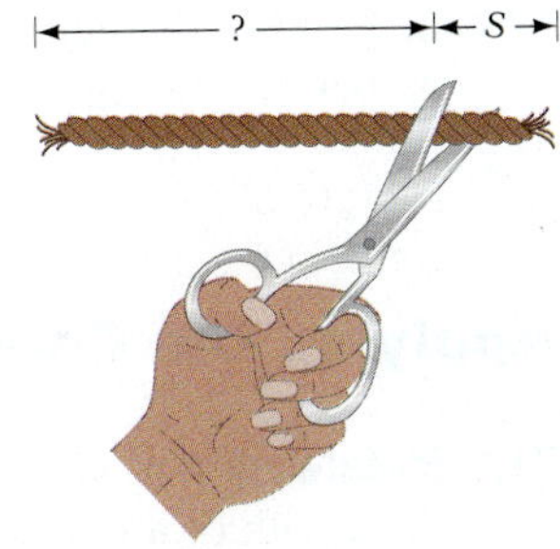

63. **Sports** A halyard 12 ft long was cut into two pieces of different lengths. Use one variable to express the lengths of the two pieces.

64. **Natural Resources** Twenty gallons of crude oil were poured into two containers of different sizes. Use one variable to express the amount of oil poured into each container.

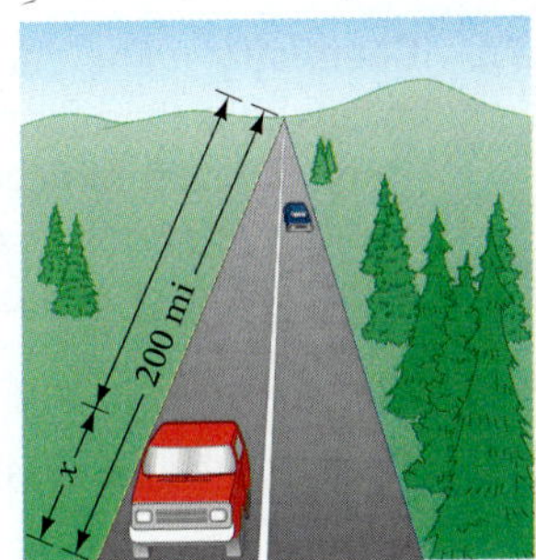

65. **Rates of Cars** Two cars start at the same place and travel at different rates in opposite directions. Two hours later the cars are 200 mi apart. Express the distance traveled by the slower car in terms of the distance traveled by the faster car.

66. **Social Networking** In June 2007, the combined number of visitors to the social networking sites Facebook and MySpace was 116,314. (*Source:* www.watblog.com) Express the number of visitors to MySpace in terms of the number of visitors to Facebook.

67. **Medicine** According to the American Podiatric Medical Association, the bones in your foot account for one-fourth of all the bones in your body. Express the number of bones in your foot in terms of the total number of bones in your body.

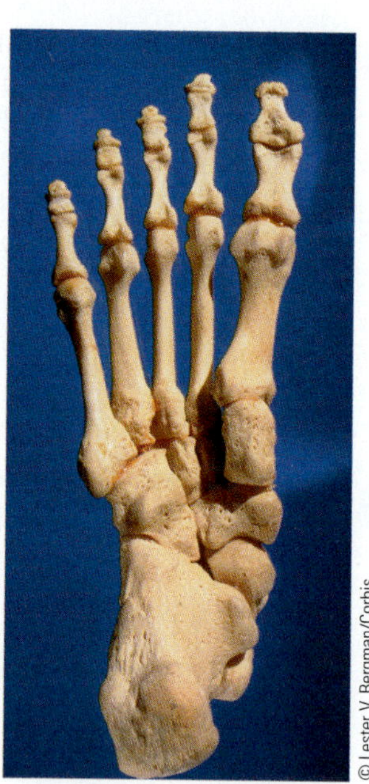
© Lester V. Bergman/Corbis

68. **Sports** The diameter of a basketball is approximately four times the diameter of a baseball. Express the diameter of a basketball in terms of the diameter of a baseball.

69. **Tax Refunds** A recent survey conducted by Turbotax.com asked, "If you receive a tax refund, what will you do?" Forty-three percent of respondents said they would pay down their debt. (*Source: USA Today,* March 27, 2008) Express the number of people who would pay down their debt in terms of the number of people surveyed.

70. **Endangered Species** Use the information in the news clipping at the right.
   a. Express the number of wild tigers in India in 2007 in terms of the number of wild tigers in India in 2002.
   b. Express the number of wild tigers in Tamil Nadu in 2007 in terms of the number of wild tigers in Tamil Nadu in 2002.

In the News

Endangered: Indian Wild Tigers

A recent survey of wild tiger populations in India shows that the number of wild tigers decreased by 2231 between 2002 and 2007. Only one state, in the southern part of the country, showed an increase in its number of tigers: the tiger population in Tamil Nadu rose by 16 tigers.

Source: news.oneindia.in

Applying the Concepts

71. **Metalwork** A wire whose length is given as x inches is bent into a square. Express the length of a side of the square in terms of x.

72. **Chemistry** The chemical formula for glucose (sugar) is $C_6H_{12}O_6$. This formula means that there are 12 hydrogen atoms for every 6 carbon atoms and 6 oxygen atoms in each molecule of glucose (see the figure at the right). If x represents the number of atoms of oxygen in a pound of sugar, express the number of hydrogen atoms in the pound of sugar in terms of x.

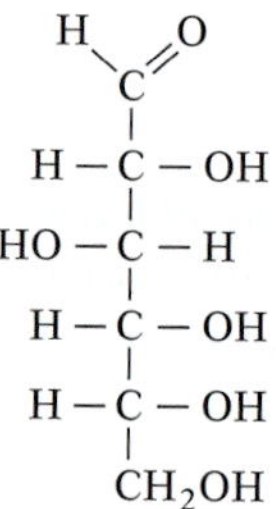

73. Translate the expressions $5x + 8$ and $5(x + 8)$ into phrases.

74. In your own words, explain how variables are used.

75. Explain the similarities and differences between the expressions "the difference between x and 5" and "5 less than x."

FOCUS ON PROBLEM SOLVING

From Concrete to Abstract

In your study of algebra, you will find that the problems are less concrete than those you studied in arithmetic. Problems that are concrete provide information pertaining to a specific instance. Algebra is more abstract. Abstract problems are theoretical; they are stated without reference to a specific instance. Let's look at an example of an abstract problem.

How many minutes are in h hours?

A strategy that can be used to solve this problem is to solve the same problem after substituting a number for the variable.

How many minutes are in 5 hours?

You know that there are 60 minutes in 1 hour. To find the number of minutes in 5 hours, multiply 5 by 60.

$$60 \cdot 5 = 300$$ There are 300 minutes in 5 hours.

Use the same procedure to find the number of minutes in h hours: multiply h by 60.

$$60 \cdot h = 60h$$ There are $60h$ minutes in h hours.

This problem might be taken a step further:

If you walk 1 mile in x minutes, how far can you walk in h hours?

Consider the same problem using numbers in place of the variables.

If you walk 1 mile in 20 minutes, how far can you walk in 3 hours?

To solve this problem, you need to calculate the number of minutes in 3 hours (multiply 3 by 60) and divide the result by the number of minutes it takes to walk 1 mile (20 minutes).

$$\frac{60 \cdot 3}{20} = \frac{180}{20} = 9$$ If you walk 1 mile in 20 minutes, you can walk 9 miles in 3 hours.

Use the same procedure to solve the related abstract problem. Calculate the number of minutes in h hours (multiply h by 60), and divide the result by the number of minutes it takes to walk 1 mile (x minutes).

$$\frac{60 \cdot h}{x} = \frac{60h}{x}$$ If you walk 1 mile in x minutes, you can walk $\frac{60h}{x}$ miles in h hours.

At the heart of the study of algebra is the use of variables. It is the variables in the problems above that make them abstract. But it is variables that enable us to generalize situations and state rules about mathematics.

Try the following problems.

1. How many hours are in d days?

2. You earn d dollars an hour. What are your wages for working h hours?

3. If p is the price of one share of stock, how many shares can you purchase with d dollars?

Jeff Greenberg/Alamy

4. A company pays a television station d dollars to air a commercial lasting s seconds. What is the cost per second?
5. After every D DVD rentals, you are entitled to one free rental. You have rented R DVDs, where $R < D$. How many more DVDs do you need to rent before you are entitled to a free rental?
6. Your car gets g miles per gallon. How many gallons of gasoline does your car consume traveling t miles?
7. If you drink j ounces of juice each day, how many days will q quarts of the juice last?
8. A TV station airs m minutes of commercials each hour. How many ads lasting s seconds each can be sold for each hour of programming?

PROJECTS AND GROUP ACTIVITIES

Prime and Composite Numbers

Recall that a prime number is a natural number greater than 1 whose only natural-number factors are itself and 1. The number 11 is a prime number because the only natural-number factors of 11 are 11 and 1.

Eratosthenes, a Greek philosopher and astronomer who lived from 270 to 190 B.C., devised a method of identifying prime numbers. It is called the **Sieve of Eratosthenes.** The procedure is illustrated below.

~~1~~	(2)	(3)	~~4~~	(5)	~~6~~	(7)	~~8~~	~~9~~	~~10~~
(11)	~~12~~	(13)	~~14~~	~~15~~	~~16~~	(17)	~~18~~	(19)	~~20~~
~~21~~	~~22~~	(23)	~~24~~	~~25~~	~~26~~	~~27~~	~~28~~	(29)	~~30~~
(31)	~~32~~	~~33~~	~~34~~	~~35~~	~~36~~	(37)	~~38~~	~~39~~	~~40~~
(41)	~~42~~	(43)	~~44~~	~~45~~	~~46~~	(47)	~~48~~	~~49~~	~~50~~
~~51~~	~~52~~	(53)	~~54~~	~~55~~	~~56~~	~~57~~	~~58~~	(59)	~~60~~
(61)	~~62~~	~~63~~	~~64~~	~~65~~	~~66~~	(67)	~~68~~	~~69~~	~~70~~
(71)	~~72~~	(73)	~~74~~	~~75~~	~~76~~	~~77~~	~~78~~	(79)	~~80~~
~~81~~	~~82~~	(83)	~~84~~	~~85~~	~~86~~	~~87~~	~~88~~	(89)	~~90~~
~~91~~	~~92~~	~~93~~	~~94~~	~~95~~	~~96~~	(97)	~~98~~	~~99~~	~~100~~

List all the natural numbers from 1 to 100. Cross out the number 1, because it is not a prime number. The number 2 is prime; circle it. Cross out all the other multiples of 2 (4, 6, 8,...) because they are not prime. The number 3 is prime; circle it. Cross out all the other multiples of 3 (6, 9, 12,...) that are not already crossed out. The number 4, the next consecutive number in the list, has already been crossed out. The number 5 is prime; circle it. Cross out all the other multiples of 5 that are not already crossed out. Continue in this manner until all the prime numbers less than 100 are circled.

A composite number is a natural number greater than 1 that has a natural-number factor other than itself and 1. The number 21 is a composite number because it has factors of 3 and 7. All the numbers crossed out in the preceding table, except the number 1, are composite numbers.

1. Use the Sieve of Eratosthenes to find the prime numbers between 100 and 200.
2. How many prime numbers are even numbers?
3. Find the "twin primes" between 100 and 200. **Twin primes** are two prime numbers whose difference is 2. For instance, 3 and 5 are twin primes; 5 and 7 are also twin primes.
4. **a.** List two prime numbers that are consecutive natural numbers.
   b. Can there be any other pairs of prime numbers that are consecutive natural numbers?
5. **a.** 4! (which is read "4 factorial") is equal to $4 \cdot 3 \cdot 2 \cdot 1$. Show that $4! + 2$, $4! + 3$, and $4! + 4$ are all composite numbers.
   b. 5! (which is read "5 factorial") is equal to $5 \cdot 4 \cdot 3 \cdot 2 \cdot 1$. Will $5! + 2$, $5! + 3$, $5! + 4$, and $5! + 5$ generate four consecutive composite numbers?
   c. Use the notation 6! to represent a list of five consecutive composite numbers.

CHAPTER 4 SUMMARY

KEY WORDS	EXAMPLES
A *variable* is a letter that is used to represent a quantity that is unknown or that can change. A *variable expression* is an expression that contains one or more variables. [4.1A, p. 246]	$4x + 2y - 6z$ is a variable expression. It contains the variables x, y, and z.
The *terms* of a variable expression are the addends of the expression. Each term is a *variable term* or a *constant term.* [4.1A, p. 246]	The expression $2a^2 - 3b^3 + 7$ has three terms: $2a^2$, $-3b^3$, and 7. $2a^2$ and $-3b^3$ are variable terms. 7 is a constant term.
A variable term is composed of a *numerical coefficient* and a *variable part.* [4.1A, p. 246]	For the expression $-7x^3y^2$, -7 is the coefficient and x^3y^2 is the variable part.
In a variable expression, replacing each variable by its value and then simplifying the resulting numerical expression is called *evaluating the variable expression.* [4.1A, p. 247]	To evaluate $2ab - b^2$ when $a = 3$ and $b = -2$, replace a by 3 and b by -2. Then simplify the numerical expression. $2(3)(-2) - (-2)^2 = -16$

Like terms of a variable expression are terms with the same variable part. Constant terms are like terms. [4.2A, p. 250]	For the expressions $3a^2 + 2b - 3$ and $2a^2 - 3a + 4$, $3a^2$ and $2a^2$ are like terms; -3 and 4 are like terms.
To simplify the sum of like variable terms, use the Distributive Property to add the numerical coefficients. This is called *combining like terms*. [4.2A, p. 250]	$5y + 3y = (5 + 3)y$ $= 8y$
The *additive inverse* of a number is the opposite of the number. [4.2A, p. 251]	-4 is the additive inverse of 4. $\frac{2}{3}$ is the additive inverse of $-\frac{2}{3}$. 0 is the additive inverse of 0.
The *multiplicative inverse* of a number is the reciprocal of the number. [4.2B, p. 252]	$\frac{3}{4}$ is the multiplicative inverse of $\frac{4}{3}$. $-\frac{1}{4}$ is the multiplicative inverse of -4.

ESSENTIAL RULES AND PROCEDURES	EXAMPLES
The Distributive Property [4.2A, p. 250] If a, b, and c are real numbers, then $a(b + c) = ab + ac$.	$5(4 + 7) = 5 \cdot 4 + 5 \cdot 7$ $= 20 + 35 = 55$
The Associative Property of Addition [4.2A, p. 250] If a, b, and c are real numbers, then $(a + b) + c = a + (b + c)$.	$-4 + (2 + 7) = -4 + 9 = 5$ $(-4 + 2) + 7 = -2 + 7 = 5$
The Commutative Property of Addition [4.2A, p. 251] If a and b are real numbers, then $a + b = b + a$.	$2 + 5 = 7$ and $5 + 2 = 7$
The Addition Property of Zero [4.2A, p. 251] If a is a real number, then $a + 0 = 0 + a = a$.	$-8 + 0 = -8$ and $0 + (-8) = -8$
The Inverse Property of Addition [4.2A, p. 251] If a is a real number, then $a + (-a) = (-a) + a = 0$.	$5 + (-5) = 0$ and $(-5) + 5 = 0$
The Associative Property of Multiplication [4.2B, p. 252] If a, b, and c are real numbers, then $(ab)c = a(bc)$.	$-3 \cdot (5 \cdot 4) = -3(20) = -60$ $(-3 \cdot 5) \cdot 4 = -15 \cdot 4 = -60$
The Commutative Property of Multiplication [4.2B, p. 252] If a and b are real numbers, then $ab = ba$.	$-3(7) = -21$ and $7(-3) = -21$
The Multiplication Property of One [4.2B, p. 252] If a is a real number, then $a \cdot 1 = 1 \cdot a = a$.	$-3(1) = -3$ and $1(-3) = -3$
The Inverse Property of Multiplication [4.2B, p. 252] If a is a real number and a is not equal to zero, then $a \cdot \frac{1}{a} = \frac{1}{a} \cdot a = 1$.	$-3 \cdot -\frac{1}{3} = 1$ and $-\frac{1}{3} \cdot -3 = 1$

CHAPTER 4

CONCEPT REVIEW

Test your knowledge of the concepts presented in this chapter. Answer each question. Then check your answers against the ones provided in the Answer Section.

1. In a term, what is the difference between the variable part and the numerical coefficient?

2. When evaluating a variable expression, what agreement must be used to simplify the resulting numerical expression?

3. What must be the same for two terms to be like terms?

4. What are like terms of a variable expression?

5. What is the difference between the Commutative Property of Multiplication and the Associative Property of Multiplication?

6. When using the Inverse Property of Addition, what is the result?

7. Which property of multiplication is needed to evaluate $6 \cdot \frac{1}{6}$?

8. What is a reciprocal?

9. Name some mathematical terms that translate into multiplication.

10. Name some mathematical terms that translate into subtraction.

CHAPTER 4

REVIEW EXERCISES

1. Simplify: $3(x^2 - 8x - 7)$

2. Simplify: $7x + 4x$

3. Simplify: $6a - 4b + 2a$

4. Simplify: $(-50n)\left(\frac{1}{10}\right)$

5. Evaluate $(5c - 4a)^2 - b$ when $a = -1$, $b = 2$, and $c = 1$.

6. Simplify: $5(2x - 7)$

7. Simplify: $2(6y^2 + 4y - 5)$

8. Simplify: $\frac{1}{4}(-24a)$

9. Simplify: $-6(7x^2)$

10. Simplify: $-9(7 + 4x)$

11. Simplify: $12y - 17y$

12. Evaluate $2bc \div (a + 7)$ when $a = 3$, $b = -5$, and $c = 4$.

13. Simplify: $7 - 2(3x + 4)$

14. Simplify: $6 + 2[2 - 5(4a - 3)]$

15. Simplify: $6(8y - 3) - 8(3y - 6)$

16. Simplify: $5c + (-2d) - 3d - (-4c)$

17. Simplify: $5(4x)$

18. Simplify: $-4(2x - 9) + 5(3x + 2)$

19. Evaluate $(b - a)^2 + c$ when $a = -2$, $b = 3$, and $c = 4$.

20. Simplify: $-9r + 2s - 6s + 12s$

21. Evaluate $(2x - y)^2 + (2x + y)^2$ when $x = -2$ and $y = -3$.

22. Evaluate $b^2 - 4ac$ when $b = -4$, $a = 1$, and $c = -3$.

23. Simplify: $4x - 3x^2 + 2x - x^2$

24. Simplify: $5[2 - 3(6x - 1)]$

25. Simplify: $0.4x + 0.6(250 - x)$

26. Simplify: $\frac{2}{3}x - \frac{3}{4}x$

27. Simplify: $(7a^2 - 2a + 3)4$

28. Simplify: $18 - (4x - 2)$

29. Evaluate $a^2 - b^2$ when $a = 3$ and $b = 4$.

30. Simplify: $-3(-12y)$

31. Translate "two-thirds of the total of x and 10" into a variable expression.

32. Translate "the product of 4 and x" into a variable expression.

33. Translate "6 less than x" into a variable expression.

34. Translate "a number plus twice the number" into a variable expression. Then simplify.

35. Translate "the difference between twice a number and one-half of the number" into a variable expression. Then simplify.

36. Translate "three times a number plus the product of five and one less than the number" into a variable expression. Then simplify.

© Guy Motil/Corbis

37. **Sports** A baseball card collection contains five times as many National League players' cards as American League players' cards. Express the number of National League players' cards in the collection in terms of the number of American League players' cards.

38. **Finance** A club treasurer has some five-dollar bills and some ten-dollar bills. The treasurer has a total of 35 bills. Express the number of five-dollar bills in terms of the number of ten-dollar bills.

39. **Nutrition** A candy bar contains eight more calories than twice the number of calories in an apple. Express the number of calories in the candy bar in terms of the number of calories in an apple.

40. **Architecture** The length of the Parthenon is approximately 1.6 times the width. Express the length of the Parthenon in terms of the width.

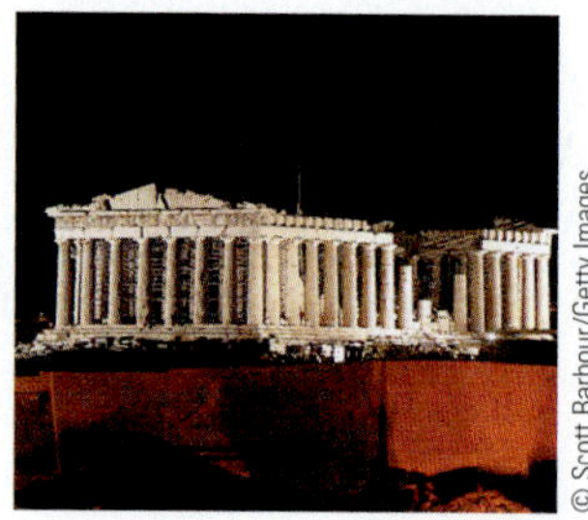
© Scott Barbour/Getty Images

41. **Anatomy** Leonardo DaVinci studied various proportions of human anatomy. One of his findings was that the standing height of a person is approximately 1.3 times the kneeling height of the same person. Represent the standing height of a person in terms of the person's kneeling height.

CHAPTER 4

TEST

1. Simplify: $3x - 5x + 7x$

2. Simplify: $-3(2x^2 - 7y^2)$

3. Simplify: $2x - 3(x - 2)$

4. Simplify: $2x + 3[4 - (3x - 7)]$

5. Simplify: $3x - 7y - 12x$

6. Evaluate $b^2 - 3ab$ when $a = 3$ and $b = -2$.

7. Simplify: $\frac{1}{5}(10x)$

8. Simplify: $5(2x + 4) - 3(x - 6)$

9. Simplify: $-5(2x^2 - 3x + 6)$

10. Simplify: $3x + (-12y) - 5x - (-7y)$

11. Evaluate $\frac{-2ab}{2b - a}$ when $a = -4$ and $b = 6$.

12. Simplify: $(12x)\left(\frac{1}{4}\right)$

13. Simplify: $-7y^2 + 6y^2 - (-2y^2)$

14. Simplify: $-2(2x - 4)$

15. Simplify: $\frac{2}{3}(-15a)$

16. Simplify: $-2[x - 2(x - y)] + 5y$

17. Simplify: $(-3)(-12y)$

18. Simplify: $5(3 - 7b)$

19. Translate "the difference between the squares of a and b" into a variable expression.

20. Translate "ten times the difference between a number and three" into a variable expression. Then simplify.

21. Translate "the sum of a number and twice the square of the number" into a variable expression.

22. Translate "three less than the quotient of six and a number" into a variable expression.

23. Translate "b decreased by the product of b and 7" into a variable expression.

24. **Sports** The speed of a pitcher's fastball is twice the speed of the catcher's return throw. Express the speed of the fastball in terms of the speed of the return throw.

25. **Metalwork** A wire is cut into two lengths. The length of the longer piece is 3 in. less than four times the length of the shorter piece. Express the length of the longer piece in terms of the length of the shorter piece.

CUMULATIVE REVIEW EXERCISES

1. Add: $-4 + 7 + (-10)$

2. Subtract: $-16 - (-25) - 4$

3. Multiply: $(-2)(3)(-4)$

4. Divide: $(-60) \div 12$

5. Find the prime factorization of 110.

6. Simplify: $\frac{7}{12} - \frac{11}{16} - \left(-\frac{1}{3}\right)$

7. Simplify: $-\frac{5}{12} \div \frac{5}{2}$

8. Simplify: $\left(-\frac{9}{16}\right) \cdot \left(\frac{8}{27}\right) \cdot \left(-\frac{3}{2}\right)$

9. Estimate the sum of 397, 516, and 408.

10. Simplify: $-2^5 \div (3 - 5)^2 - (-3)$

11. Simplify: $\left(-\frac{3}{4}\right)^2 \div \left(\frac{3}{8} - \frac{11}{12}\right)$

12. Evaluate $a^2 - 3b$ when $a = 2$ and $b = -4$.

13. Simplify: $-2x^2 - (-3x^2) + 4x^2$

14. Simplify: $5a - 10b - 12a$

15. Write eight and three hundred fifty-seven thousandths in standard form.

16. Place the correct symbol, $<$ or $>$, between the two numbers.
5.101 5.013

17. Simplify: $3(8 - 2x)$

18. Simplify: $-2(-3y + 9)$

19. Estimate the difference between 32.76 and 19.8.

20. Round 8.667 to the nearest tenth.

21. Simplify: $-4(2x^2 - 3y^2)$

22. Simplify: $-3(3y^2 - 3y - 7)$

23. Simplify: $-3x - 2(2x - 7)$

24. Simplify: $4(3x - 2) - 7(x + 5)$

25. Simplify: $2x + 3[x - 2(4 - 2x)]$

26. Simplify: $3[2x - 3(x - 2y)] + 3y$

27. Translate "the sum of one-half of *b* and *b*" into a variable expression.

28. Translate "10 divided by the difference between *y* and 2" into a variable expression.

29. Translate "the difference between eight and the quotient of a number and twelve" into a variable expression.

30. Translate "the sum of a number and two more than the number" into a variable expression. Then simplify.

31. **Geometry** The length of each side of a square is 2.25 in. Find the perimeter of the square. Use the formula $P = 4s$.

32. **Internet Connections** The speed of a DSL (Digital Subscriber Line) Internet connection is ten times faster than that of a dial-up connection. Express the speed of the DSL connection in terms of the speed of the dial-up connection.

CHAPTER

5

Solving Equations

OBJECTIVES

SECTION 5.1

A To determine whether a given number is a solution of an equation

B To solve an equation of the form $x + a = b$

C To solve an equation of the form $ax = b$

D To solve uniform motion problems

SECTION 5.2

A To solve an equation of the form $ax + b = c$

B To solve application problems using formulas

SECTION 5.3

A To solve an equation of the form $ax + b = cx + d$

B To solve an equation containing parentheses

C To solve application problems using formulas

SECTION 5.4

A To solve integer problems

B To translate a sentence into an equation and solve

SECTION 5.5

A To solve value mixture problems

B To solve uniform motion problems

ARE YOU READY?

Take the Chapter 5 Prep Test to find out if you are ready to learn to:

- Solve equations
- Solve problems using formulas
- Solve integer, mixture, and uniform motion problems

PREP TEST

Do these exercises to prepare for Chapter 5.

1. Subtract: $8 - 12$

2. Multiply: $-\frac{3}{4}\left(-\frac{4}{3}\right)$

3. Multiply: $-\frac{5}{8}(16)$

4. Simplify: $\frac{-3}{-3}$

5. Simplify: $-16 + 7y + 16$

6. Simplify: $8x - 9 - 8x$

7. Evaluate $2x + 3$ when $x = -4$.

SECTION 5.1 Introduction to Equations

OBJECTIVE A To determine whether a given number is a solution of an equation

Point of Interest

One of the most famous equations ever stated is $E = mc^2$. This equation, stated by Albert Einstein, shows that there is a relationship between mass m and energy E. As a side note, the chemical element einsteinium was named in honor of Einstein.

An **equation** expresses the equality of two mathematical expressions. The expressions can be either numerical or variable expressions.

$$\left.\begin{aligned} 9 + 3 &= 12 \\ 3x - 2 &= 10 \\ y^2 + 4 &= 2y - 1 \\ z &= 2 \end{aligned}\right\} \text{Equations}$$

The equation at the right is true if the variable is replaced by 5.

$x + 8 = 13$

$5 + 8 = 13$ A true equation

The equation is false if the variable is replaced by 7.

$7 + 8 = 13$ A false equation

A **solution of an equation** is a number that, when substituted for the variable, results in a true equation. 5 is a solution of the equation $x + 8 = 13$. 7 is not a solution of the equation $x + 8 = 13$.

HOW TO • 1 Is -2 a solution of $2x + 5 = x^2 - 3$?

$2x + 5$	$= x^2 - 3$
$2(-2) + 5$	$(-2)^2 - 3$
$-4 + 5$	$4 - 3$
1	$= 1$

- Replace x by -2.
- Evaluate the numerical expressions.
- If the results are equal, -2 is a solution of the equation. If the results are not equal, -2 is not a solution of the equation.

Yes, -2 is a solution of the equation.

Take Note

The Order of Operations Agreement applies to evaluating $2(-2) + 5$ and $(-2)^2 - 3$.

EXAMPLE • 1

Is -4 a solution of $5x - 2 = 6x + 2$?

Solution

$5x - 2$	$= 6x + 2$
$5(-4) - 2$	$6(-4) + 2$
$-20 - 2$	$-24 + 2$
-22	$= -22$

Yes, -4 is a solution.

YOU TRY IT • 1

Is $\frac{1}{4}$ a solution of $5 - 4x = 8x + 2$?

Your solution

EXAMPLE • 2

Is -4 a solution of $4 + 5x = x^2 - 2x$?

Solution

$4 + 5x$	$= x^2 - 2x$
$4 + 5(-4)$	$(-4)^2 - 2(-4)$
$4 + (-20)$	$16 - (-8)$
-16	$\neq 24$

($\neq$ means "is not equal to")

No, -4 is not a solution.

YOU TRY IT • 2

Is 5 a solution of $10x - x^2 = 3x - 10$?

Your solution

Solutions on p. S12

OBJECTIVE B To solve an equation of the form $x + a = b$

Tips for Success

To learn mathematics, you must be an active participant. Listening and watching your professor do mathematics are not enough. Take notes in class, mentally think through every question your instructor asks, and try to answer it even if you are not called on to answer it verbally. Ask questions when you have them. See *AIM for Success* in the front of the book for other ways to be an active learner.

To **solve an equation** means to find a solution of the equation. The simplest equation to solve is an equation of the form *variable* = *constant,* because the constant is the solution.

The solution of the equation $x = 5$ is 5 because $5 = 5$ is a true equation.

The solution of the equation at the right is 7 because $7 + 2 = 9$ is a true equation.

$$x + 2 = 9 \qquad 7 + 2 = 9$$

Note that if 4 is added to each side of the equation $x + 2 = 9$, the solution is still 7.

$$x + 2 = 9$$
$$x + 2 + 4 = 9 + 4$$
$$x + 6 = 13 \qquad 7 + 6 = 13$$

If -5 is added to each side of the equation $x + 2 = 9$, the solution is still 7.

$$x + 2 = 9$$
$$x + 2 + (-5) = 9 + (-5)$$
$$x - 3 = 4 \qquad 7 - 3 = 4$$

Equations that have the same solution are called **equivalent equations.** The equations $x + 2 = 9$, $x + 6 = 13$, and $x - 3 = 4$ are equivalent equations; each equation has 7 as its solution. These examples suggest that adding the same number to each side of an equation produces an equivalent equation. This is called the *Addition Property of Equations.*

Addition Property of Equations

The same number can be added to each side of an equation without changing its solution. In symbols, the equation $a = b$ has the same solution as the equation $a + c = b + c$.

In solving an equation, the goal is to rewrite the given equation in the form *variable* = *constant.* The Addition Property of Equations is used to **remove a *term*** from one side of the equation **by adding the opposite of that term to each side of the equation.**

Take Note

An equation has some properties that are similar to those of a balance scale. For instance, if a balance scale is in balance and equal weights are added to each side of the scale, then the balance scale remains in balance. If an equation is true, then adding the same number to each side of the equation produces another true equation.

HOW TO 2 Solve: $x - 4 = 2$

$x - 4 = 2$ • The goal is to rewrite the equation in the form *variable* = *constant.*

$x - 4 + 4 = 2 + 4$ • Add 4 to each side of the equation.

$x + 0 = 6$ • Simplify.

$x = 6$ • The equation is in the form *variable* = *constant.*

Check: $x - 4 = 2$

$6 - 4 \mid 2$

$2 = 2$ A true equation

The solution is 6.

Because subtraction is defined in terms of addition, the Addition Property of Equations also makes it possible to subtract the same number from each side of an equation without changing the solution of the equation.

HOW TO • 3 Solve: $y + \frac{3}{4} = \frac{1}{2}$

$$y + \frac{3}{4} = \frac{1}{2}$$

• The goal is to rewrite the equation in the form *variable* = *constant*.

$$y + \frac{3}{4} - \frac{3}{4} = \frac{1}{2} - \frac{3}{4}$$

• Subtract $\frac{3}{4}$ from each side of the equation.

$$y + 0 = \frac{2}{4} - \frac{3}{4}$$

• Simplify.

$$y = -\frac{1}{4}$$

• The equation is in the form *variable* = *constant*.

The solution is $-\frac{1}{4}$. You should check this solution.

EXAMPLE • 3

Solve: $x + \frac{2}{5} = \frac{1}{3}$

Solution

$$x + \frac{2}{5} = \frac{1}{3}$$

$$x + \frac{2}{5} - \frac{2}{5} = \frac{1}{3} - \frac{2}{5}$$

• Subtract $\frac{2}{5}$ from each side.

$$x + 0 = \frac{5}{15} - \frac{6}{15}$$

• Rewrite $\frac{1}{3}$ and $\frac{2}{5}$ with a common denominator.

$$x = -\frac{1}{15}$$

The solution is $-\frac{1}{15}$.

YOU TRY IT • 3

Solve: $\frac{5}{6} = y - \frac{3}{8}$

Your solution

Solution on p. S12

OBJECTIVE C To solve an equation of the form $ax = b$

The solution of the equation at the right is 3 because $2 \cdot 3 = 6$ is a true equation.

$$2x = 6 \qquad 2 \cdot 3 = 6$$

Note that if each side of $2x = 6$ is multiplied by 5, the solution is still 3.

$$\begin{aligned} 2x &= 6 \\ 5(2x) &= 5 \cdot 6 \\ 10x &= 30 \end{aligned} \qquad 10 \cdot 3 = 30$$

If each side of $2x = 6$ is multiplied by -4, the solution is still 3.

$$\begin{aligned} 2x &= 6 \\ (-4)(2x) &= (-4)6 \\ -8x &= -24 \end{aligned} \qquad -8 \cdot 3 = -24$$

The equations $2x = 6$, $10x = 30$, and $-8x = -24$ are equivalent equations; each equation has 3 as its solution. These examples suggest that multiplying each side of an equation by the same nonzero number produces an equivalent equation.

Multiplication Property of Equations

Each side of an equation can be multiplied by the same nonzero number without changing the solution of the equation. In symbols, if $c \neq 0$, then the equation $a = b$ has the same solutions as the equation $ac = bc$.

The Multiplication Property of Equations is used to **remove a coefficient by multiplying each side of the equation by the reciprocal of the coefficient.**

HOW TO • 4 Solve: $\frac{3}{4}z = 9$

$$\frac{3}{4}z = 9$$ • **The goal is to rewrite the equation in the form *variable = constant*.**

$$\frac{4}{3} \cdot \frac{3}{4}z = \frac{4}{3} \cdot 9$$ • **Multiply each side of the equation by $\frac{4}{3}$.**

$$1 \cdot z = 12$$ • **Simplify.**

$$z = 12$$ • **The equation is in the form *variable = constant*.**

The solution is 12. You should check this solution.

Because division is defined in terms of multiplication, each side of an equation can be divided by the same nonzero number without changing the solution of the equation.

HOW TO • 5 Solve: $6x = 14$

$$6x = 14$$ • **The goal is to rewrite the equation in the form *variable = constant*.**

$$\frac{6x}{6} = \frac{14}{6}$$ • **Divide each side of the equation by 6.**

$$x = \frac{7}{3}$$ • **Simplify. The equation is in the form *variable = constant*.**

The solution is $\frac{7}{3}$.

Take Note

Remember to check the solution.

Check:

$$6x = 14$$
$$6\left(\frac{7}{3}\right) \mid 14$$
$$14 = 14$$

When using the Multiplication Property of Equations, multiply each side of the equation by the reciprocal of the coefficient when the coefficient is a fraction. Divide each side of the equation by the coefficient when the coefficient is an integer or a decimal.

EXAMPLE • 4

Solve: $\frac{3x}{4} = -9$

Solution

$$\frac{3x}{4} = -9$$

$$\frac{4}{3} \cdot \frac{3}{4}x = \frac{4}{3}(-9)$$ • $\frac{3x}{4} = \frac{3}{4}x$

$$x = -12$$

The solution is -12.

YOU TRY IT • 4

Solve: $-\frac{2x}{5} = 6$

Your solution

EXAMPLE • 5

Solve: $5x - 9x = 12$

Solution

$$5x - 9x = 12$$

$$-4x = 12$$ • **Combine like terms.**

$$\frac{-4x}{-4} = \frac{12}{-4}$$

$$x = -3$$

The solution is -3.

YOU TRY IT • 5

Solve: $4x - 8x = 16$

Your solution

Solutions on p. S12

OBJECTIVE D To solve uniform motion problems 

Take Note
A car traveling in a *circle* at a constant speed of 45 mph is *not* in uniform motion because the direction of the car is always changing.

Any object that travels at a constant speed in a straight line is said to be in *uniform motion.* **Uniform motion** means that the speed and direction of an object do not change. For instance, a car traveling at a constant speed of 45 mph on a straight road is in uniform motion.

The solution of a uniform motion problem is based on the **uniform motion equation** $d = rt$, where d is the distance traveled, r is the rate of travel, and t is the time spent traveling. For instance, suppose a car travels at 50 mph for 3 h. Because the rate (50 mph) and time (3 h) are known, we can find the distance traveled by solving the equation $d = rt$ for d.

$$d = rt$$
$$d = 50(3) \qquad \bullet\ r = 50, t = 3$$
$$d = 150$$

The car travels a distance of 150 mi.

HOW TO • 6 A jogger runs 3 mi in 45 min. What is the rate of the jogger in miles per hour?

Strategy

- Because the answer must be in miles per *hour* and the given time is in *minutes,* convert 45 min to hours.
- To find the rate of the jogger, solve the equation $d = rt$ for r.

Solution

$$45 \text{ min} = \frac{45}{60}\text{ h} = \frac{3}{4}\text{ h}$$
$$d = rt$$
$$3 = r\left(\frac{3}{4}\right) \qquad \bullet\ d = 3, t = \frac{3}{4}$$
$$3 = \frac{3}{4}r$$
$$\left(\frac{4}{3}\right)3 = \left(\frac{4}{3}\right)\frac{3}{4}r \qquad \bullet\ \text{Multiply each side of the equation by the reciprocal of } \frac{3}{4}.$$
$$4 = r$$

The rate of the jogger is 4 mph.

If two objects are moving in opposite directions, then the rate at which the distance between them is increasing is the sum of the speeds of the two objects. For instance, in the diagram below, two cars start from the same point and travel in opposite directions. The distance between them is changing at 70 mph.

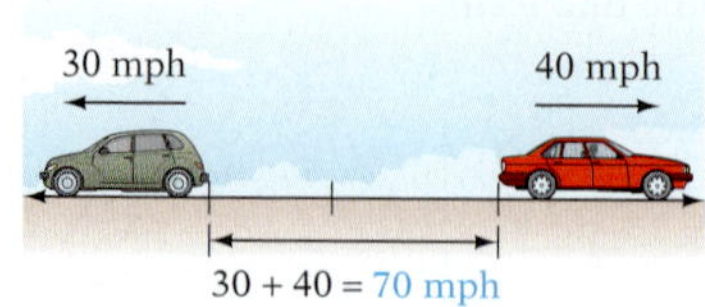

Similarly, if two objects are moving toward each other, the distance between them is decreasing at a rate that is equal to the sum of the speeds. The rate at which the two planes at the right are approaching one another is 800 mph.

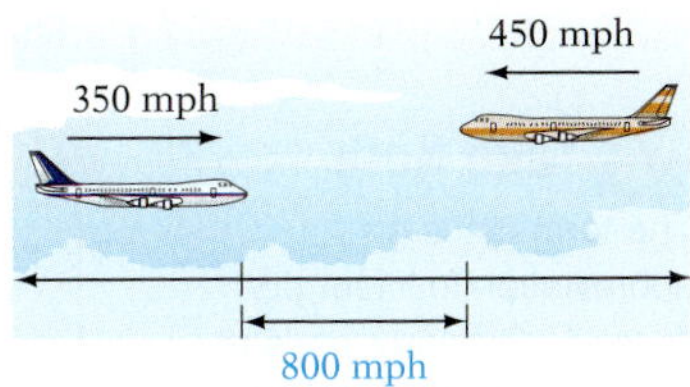

HOW TO • 7 Two cars start from the same point and move in opposite directions. The car moving west is traveling 45 mph, and the car moving east is traveling 60 mph. In how many hours will the cars be 210 mi apart?

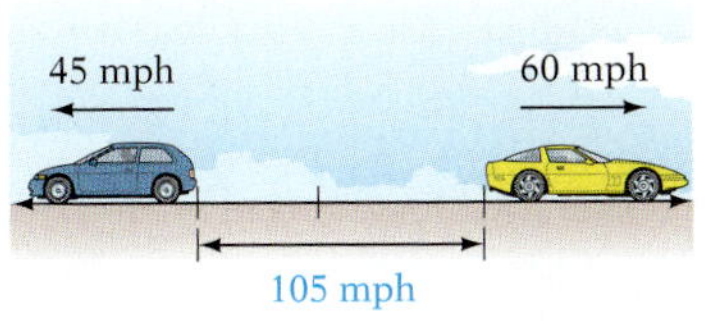

Strategy

The distance is 210 mi. Therefore, $d = 210$. The cars are moving in opposite directions, so the rate at which the distance between them is changing is the sum of the rates of each of the cars. The rate is 45 mph + 60 mph = 105 mph. Therefore, $r = 105$. To find the time, solve the equation $d = rt$ for t.

Solution

$$d = rt$$

$$210 = 105t \qquad \text{• } d = 210, r = 105$$

$$\frac{210}{105} = \frac{105t}{105} \qquad \text{• Solve for } t.$$

$$2 = t$$

In 2 h, the cars will be 210 mi apart.

If a motorboat is on a river that is flowing at a rate of 4 mph, then the boat will float down the river at a speed of 4 mph when the motor is not on. Now suppose the motor is turned on and the power adjusted so that the boat would travel 10 mph without the aid of the current. Then, if the boat is moving with the current, its effective speed is the speed of the boat using power plus the speed of the current: 10 mph + 4 mph = 14 mph. (See the figure below.)

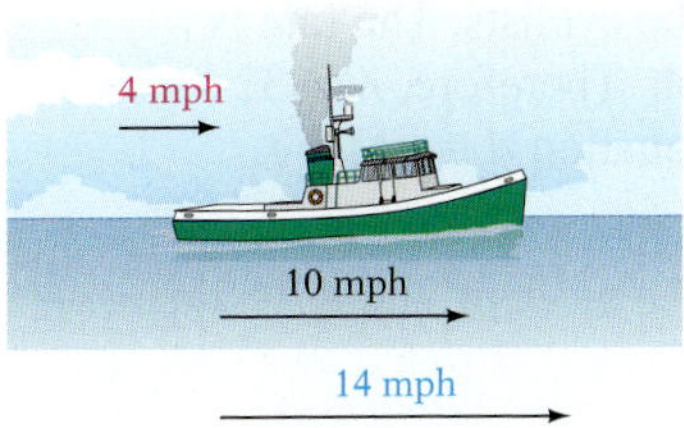

However, if the boat is moving against the current, the current slows the boat down. The effective speed of the boat is the speed of the boat using power minus the speed of the current: 10 mph − 4 mph = 6 mph. (See the figure below.)

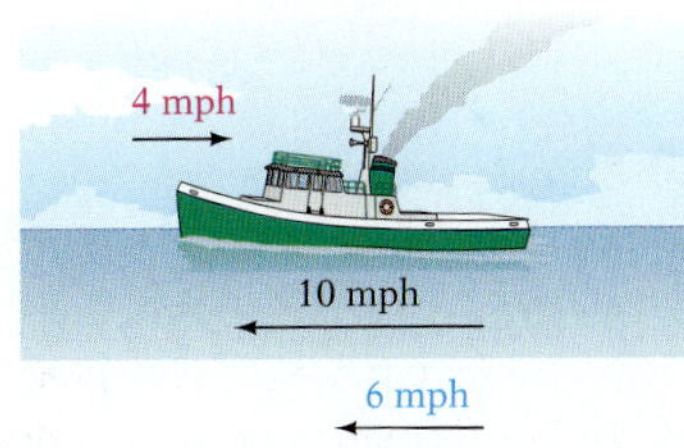

There are other situations in which the preceding concepts may be applied.

Take Note

The term ft/s is an abbreviation for "feet per second." Similarly, cm/s is "centimeters per second" and m/s is "meters per second."

Peter Titmuss/Alamy

HOW TO • 8 An airline passenger is walking between two airline terminals and decides to get on a moving sidewalk that is 150 ft long. If the passenger walks at a rate of 7 ft/s and the moving sidewalk moves at a rate of 9 ft/s, how long, in seconds, will it take for the passenger to walk from one end of the moving sidewalk to the other? Round to the nearest thousandth.

Strategy

The distance is 150 ft. Therefore, $d = 150$. The passenger is traveling at 7 ft/s and the moving sidewalk is traveling at 9 ft/s. The rate of the passenger is the sum of the two rates, or 16 ft/s. Therefore, $r = 16$. To find the time, solve the equation $d = rt$ for t.

Solution

$d = rt$

$150 = 16t$ • $d = 150, r = 16$

$\frac{150}{16} = \frac{16t}{16}$ • Solve for t.

$9.375 = t$

It will take 9.375 s for the passenger to travel the length of the moving sidewalk.

EXAMPLE • 6

Two cyclists start at the same time at opposite ends of an 80-mile course. One cyclist is traveling 18 mph, and the second cyclist is traveling 14 mph. How long after they begin cycling will they meet?

Strategy

The distance is 80 mi. Therefore, $d = 80$. The cyclists are moving toward each other, so the rate at which the distance between them is changing is the sum of the rates of each of the cyclists. The rate is 18 mph + 14 mph = 32 mph. Therefore, $r = 32$. To find the time, solve the equation $d = rt$ for t.

Solution

$d = rt$

$80 = 32t$ • $d = 80, r = 32$

$\frac{80}{32} = \frac{32t}{32}$ • Solve for t.

$2.5 = t$

The cyclists will meet in 2.5 h.

YOU TRY IT • 6

A plane that can normally travel at 250 mph in calm air is flying into a headwind of 25 mph. How far can the plane fly in 3 h?

Your strategy

Your solution

Solution on p. S12

5.1 EXERCISES

OBJECTIVE A To determine whether a given number is a solution of an equation

1. Is 4 a solution of $2x = 8$?

2. Is 3 a solution of $y + 4 = 7$?

3. Is -1 a solution of $2b - 1 = 3$?

4. Is -2 a solution of $3a - 4 = 10$?

5. Is 1 a solution of $4 - 2m = 3$?

6. Is 2 a solution of $7 - 3n = 2$?

7. Is 5 a solution of $2x + 5 = 3x$?

8. Is 4 a solution of $3y - 4 = 2y$?

9. Is -2 a solution of $3a + 2 = 2 - a$?

10. Is 3 a solution of $z^2 + 1 = 4 + 3z$?

11. Is 2 a solution of $2x^2 - 1 = 4x - 1$?

12. Is -1 a solution of $y^2 - 1 = 4y + 3$?

13. Is 4 a solution of $x(x + 1) = x^2 + 5$?

14. Is 3 a solution of $2a(a - 1) = 3a + 3$?

15. Is $-\frac{1}{4}$ a solution of $8t + 1 = -1$?

16. Is $\frac{1}{2}$ a solution of $4y + 1 = 3$?

17. Is $\frac{2}{5}$ a solution of $5m + 1 = 10m - 3$?

18. Is $\frac{3}{4}$ a solution of $8x - 1 = 12x + 3$?

19. If A is a fixed number such that $A < 0$, is a solution of the equation $5x = A$ positive or negative?

20. Can a negative number be a solution of the equation $7x - 2 = -x$?

OBJECTIVE B To solve an equation of the form $x + a = b$

21. Without solving the equation $x - \frac{11}{16} = \frac{19}{24}$, determine whether x is less than or greater than $\frac{19}{24}$. Explain your answer.

22. Without solving the equation $x + \frac{13}{15} = -\frac{21}{43}$, determine whether x is less than or greater than $-\frac{21}{43}$. Explain your answer.

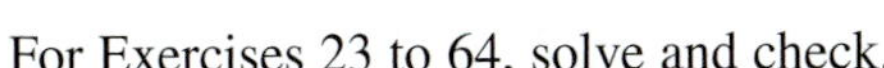

For Exercises 23 to 64, solve and check.

23. $x + 5 = 7$

24. $y + 3 = 9$

25. $b - 4 = 11$

26. $z - 6 = 10$

27. $2 + a = 8$

28. $5 + x = 12$

29. $n - 5 = -2$

30. $x - 6 = -5$

31. $b + 7 = 7$

32. $y - 5 = -5$

33. $z + 9 = 2$

34. $n + 11 = 1$

35. $10 + m = 3$

36. $8 + x = 5$

37. $9 + x = -3$

38. $10 + y = -4$

39. $2 = x + 7$

40. $-8 = n + 1$

41. $4 = m - 11$

42. $-6 = y - 5$

43. $12 = 3 + w$

44. $-9 = 5 + x$

45. $4 = -10 + b$

46. $-7 = -2 + x$

47. $m + \frac{2}{3} = -\frac{1}{3}$

48. $c + \frac{3}{4} = -\frac{1}{4}$

49. $x - \frac{1}{2} = \frac{1}{2}$

50. $x - \frac{2}{5} = \frac{3}{5}$

51. $\frac{5}{8} + y = \frac{1}{8}$

52. $\frac{4}{9} + a = -\frac{2}{9}$

53. $m + \frac{1}{2} = -\frac{1}{4}$

54. $b + \frac{1}{6} = -\frac{1}{3}$

55. $x + \frac{2}{3} = \frac{3}{4}$

56. $n + \frac{2}{5} = \frac{2}{3}$

57. $-\frac{5}{6} = x - \frac{1}{4}$

58. $-\frac{1}{4} = c - \frac{2}{3}$

59. $d + 1.3619 = 2.0148$

60. $w + 2.932 = 4.801$

61. $-0.813 + x = -1.096$

62. $-1.926 + t = -1.042$

63. $6.149 = -3.108 + z$

64. $5.237 = -2.014 + x$

OBJECTIVE C To solve an equation of the form $ax = b$

For Exercises 65 to 108, solve and check.

65. $5x = -15$

66. $4y = -28$

67. $3b = 0$

68. $2a = 0$

69. $-3x = 6$

70. $-5m = 20$

71. $-3x = -27$

72. $-\frac{1}{6}n = -30$

73. $20 = \frac{1}{4}c$

74. $18 = 2t$

75. $0 = -5x$

76. $0 = -8a$

77. $49 = -7t$

78. $\frac{x}{3} = 2$

79. $\frac{x}{4} = 3$

80. $-\frac{y}{2} = 5$

81. $-\frac{b}{3} = 6$

82. $\frac{3}{4}y = 9$

83. $\frac{2}{5}x = 6$

84. $-\frac{2}{3}d = 8$

85. $-\frac{3}{5}m = 12$

86. $\frac{2n}{3} = 0$

87. $\frac{5x}{6} = 0$

88. $\frac{-3z}{8} = 9$

89. $\frac{3x}{4} = 2$

90. $\frac{3}{4}c = \frac{3}{5}$

91. $\frac{2}{9} = \frac{2}{3}y$

92. $-\frac{6}{7} = -\frac{3}{4}b$

93. $\frac{1}{5}x = -\frac{1}{10}$

94. $-\frac{2}{3}y = -\frac{8}{9}$

95. $-1 = \frac{2n}{3}$

96. $-\frac{3}{4} = \frac{a}{8}$

97. $-\frac{2}{5}m = -\frac{6}{7}$

98. $5x + 2x = 14$

99. $3n + 2n = 20$

100. $7d - 4d = 9$

101. $10y - 3y = 21$

102. $2x - 5x = 9$

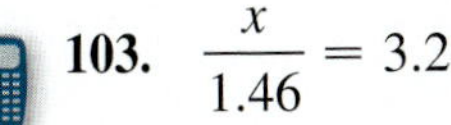

103. $\frac{x}{1.46} = 3.25$

104. $\frac{z}{2.95} = -7.88$

105. $3.47a = 7.1482$

106. $2.31m = 2.4255$

107. $-3.7x = 7.881$

108. $\dfrac{n}{2.65} = 9.08$

For Exercises 109 to 112, suppose y is a positive integer. Determine whether x is positive or negative.

109. $15x = y$

110. $-6x = y$

111. $-\dfrac{1}{4}x = y$

112. $\dfrac{2}{9}x = -y$

OBJECTIVE D To solve uniform motion problems

113. Joe and John live 2 mi apart. They leave their houses at the same time and walk toward each other until they meet. Joe walks faster than John does.

a. Is the distance walked by Joe less than, equal to, or greater than the distance walked by John?

b. Is the time spent walking by Joe less than, equal to, or greater than the time spent walking by John?

c. What is the total distance traveled by both Joe and John?

114. Morgan and Emma ride their bikes from Morgan's house to the store. Morgan begins biking 5 min before Emma begins. Emma bikes faster than Morgan and catches up with her just as they reach the store.

a. Is the distance biked by Emma less than, equal to, or greater than the distance biked by Morgan?

b. Is the time spent biking by Emma less than, equal to, or greater than the time spent biking by Morgan?

115. As part of a training program for the Boston Marathon, a runner wants to build endurance by running at a rate of 9 mph for 20 min. How far will the runner travel in that time period?

Michael Dwyer/Alamy

116. It takes a hospital dietician 40 min to drive from home to the hospital, a distance of 20 mi. What is the dietician's average rate of speed?

117. Marcella leaves home at 9:00 A.M. and drives to school, arriving at 9:45 A.M. If the distance between home and school is 27 mi, what is Marcella's average rate of speed?

118. The Ride for Health Bicycle Club has chosen a 36-mile course for this Saturday's ride. If the riders plan on averaging 12 mph while they are riding, and they have a 1-hour lunch break planned, how long will it take them to complete the trip?

© Lawrence Manning/Corbis

119. Palmer's average running speed is 3 km/h faster than his walking speed. If Palmer can run around a 30-kilometer course in 2 h, how many hours would it take for Palmer to walk the same course?

120. A shopping mall has a moving sidewalk that takes shoppers from the shopping area to the parking garage, a distance of 250 ft. If your normal walking rate is 5 ft/s and the moving sidewalk is traveling at 3 ft/s, how many seconds would it take for you to walk from one end of the moving sidewalk to the other end?

121. Two joggers start at the same time from opposite ends of an 8-mile jogging trail and begin running toward each other. One jogger is running at a rate of 5 mph, and the other jogger is running at a rate of 7 mph. How long, in minutes, after they start will the two joggers meet?

122. Two cyclists start from the same point at the same time and move in opposite directions. One cyclist is traveling at 8 mph, and the other cyclist is traveling at 9 mph. After 30 min, how far apart are the two cyclists?

© Davis Barber/PhotoEdit

123. Petra and Celine can paddle their canoe at a rate of 10 mph in calm water. How long will it take them to travel 4 mi against the 2-mph current of the river?

124. At 8:00 A.M., a train leaves a station and travels at a rate of 45 mph. At 9:00 A.M., a second train leaves the same station on the same track and travels in the direction of the first train at a speed of 60 mph. At 10:00 A.M., how far apart are the two trains?

Applying the Concepts

125. **a.** Make up an equation of the form $x + a = b$ that has 2 as a solution.
b. Make up an equation of the form $ax = b$ that has -1 as a solution.

126. Write out the steps for solving the equation $\frac{1}{2}x = -3$. Identify each Property of Real Numbers or Property of Equations as you use it.

SECTION

5.2 General Equations—Part I

OBJECTIVE A To solve an equation of the form $ax + b = c$ 

In solving an equation of the form $ax + b = c$, the goal is to rewrite the equation in the form *variable* = *constant*. This requires the application of both the Addition and the Multiplication Properties of Equations.

HOW TO 1 Solve: $\frac{3}{4}x - 2 = -11$

The goal is to write the equation in the form *variable* = *constant*.

$$\frac{3}{4}x - 2 = -11$$

$$\frac{3}{4}x - 2 + 2 = -11 + 2$$ • Add 2 to each side of the equation.

$$\frac{3}{4}x = -9$$ • Simplify.

$$\frac{4}{3} \cdot \frac{3}{4}x = \frac{4}{3}(-9)$$ • Multiply each side of the equation by $\frac{4}{3}$.

$$x = -12$$ • The equation is in the form *variable* = *constant*.

The solution is −12.

Take Note

Check:

$$\frac{3}{4}x - 2 = -11$$

$\frac{3}{4}(-12) - 2$	-11
$-9 - 2$	-11
$-11 =$	-11

A true equation

Here is an example of solving an equation that contains more than one fraction.

HOW TO 2 Solve: $\frac{2}{3}x + \frac{1}{2} = \frac{3}{4}$

$$\frac{2}{3}x + \frac{1}{2} = \frac{3}{4}$$

$$\frac{2}{3}x + \frac{1}{2} - \frac{1}{2} = \frac{3}{4} - \frac{1}{2}$$ • Subtract $\frac{1}{2}$ from each side of the equation.

$$\frac{2}{3}x = \frac{1}{4}$$ • Simplify.

$$\frac{3}{2}\left(\frac{2}{3}x\right) = \frac{3}{2}\left(\frac{1}{4}\right)$$ • Multiply each side of the equation by $\frac{3}{2}$, the reciprocal of $\frac{2}{3}$.

$$x = \frac{3}{8}$$

The solution is $\frac{3}{8}$.

It may be easier to solve an equation containing two or more fractions by multiplying each side of the equation by the least common multiple (LCM) of the denominators. For the equation above, the LCM of 3, 2, and 4 is 12. The LCM has the property that 3, 2, and 4 will divide evenly into it. Therefore, if both sides of the equation are multiplied by 12, the denominators will divide evenly into 12. The result is an equation that does not contain any fractions. Multiplying each side of an equation that contains fractions by the LCM of the denominators is called **clearing denominators.** It is an alternative method, as we show in the next example, of solving an equation that contains fractions.

Take Note

This is the same example solved on the previous page, but this time we are using the method of clearing denominators.

Observe that after we multiply both sides of the equation by the LCM of the denominators and then simplify, the equation no longer contains fractions.

HOW TO • 3 Solve: $\frac{2}{3}x + \frac{1}{2} = \frac{3}{4}$

$$\frac{2}{3}x + \frac{1}{2} = \frac{3}{4}$$

$$12\left(\frac{2}{3}x + \frac{1}{2}\right) = 12\left(\frac{3}{4}\right)$$

• Multiply each side of the equation by 12, the LCM of 3, 2, and 4.

$$12\left(\frac{2}{3}x\right) + 12\left(\frac{1}{2}\right) = 12\left(\frac{3}{4}\right)$$

• Use the Distributive Property.

$$8x + 6 = 9$$

• Simplify.

$$8x + 6 - 6 = 9 - 6$$

• Subtract 6 from each side of the equation.

$$8x = 3$$

$$\frac{8x}{8} = \frac{3}{8}$$

• Divide each side of the equation by 8.

$$x = \frac{3}{8}$$

The solution is $\frac{3}{8}$.

Note that both methods give exactly the same solution. You may use either method to solve an equation containing fractions.

EXAMPLE • 1

Solve: $3x - 7 = -5$

Solution

$$3x - 7 = -5$$

$$3x - 7 + 7 = -5 + 7$$

• Add 7 to each side.

$$3x = 2$$

$$\frac{3x}{3} = \frac{2}{3}$$

• Divide each side by 3.

$$x = \frac{2}{3}$$

The solution is $\frac{2}{3}$.

YOU TRY IT • 1

Solve: $5x + 7 = 10$

Your solution

EXAMPLE • 2

Solve: $5 = 9 - 2x$

Solution

$$5 = 9 - 2x$$

$$5 - 9 = 9 - 9 - 2x$$

• Subtract 9 from each side.

$$-4 = -2x$$

$$\frac{-4}{-2} = \frac{-2x}{-2}$$

• Divide each side by −2.

$$2 = x$$

The solution is 2.

YOU TRY IT • 2

Solve: $2 = 11 + 3x$

Your solution

Solutions on p. S12

EXAMPLE • 3

Solve: $\frac{2}{3} - \frac{x}{2} = \frac{3}{4}$

Solution

$$\frac{2}{3} - \frac{x}{2} = \frac{3}{4}$$

$$\frac{2}{3} - \frac{2}{3} - \frac{x}{2} = \frac{3}{4} - \frac{2}{3}$$ • Subtract $\frac{2}{3}$ from each side.

$$-\frac{x}{2} = \frac{1}{12}$$

$$-2\left(-\frac{x}{2}\right) = -2\left(\frac{1}{12}\right)$$ • Multiply each side by −2.

$$x = -\frac{1}{6}$$

The solution is $-\frac{1}{6}$.

YOU TRY IT • 3

Solve: $\frac{5}{8} - \frac{2x}{3} = \frac{5}{4}$

Your solution

EXAMPLE • 4

Solve $\frac{4}{5}x - \frac{1}{2} = \frac{3}{4}$ by first clearing denominators.

Solution

The LCM of 5, 2, and 4 is 20.

$$\frac{4}{5}x - \frac{1}{2} = \frac{3}{4}$$

$$20\left(\frac{4}{5}x - \frac{1}{2}\right) = 20\left(\frac{3}{4}\right)$$ • Multiply each side by 20.

$$20\left(\frac{4}{5}x\right) - 20\left(\frac{1}{2}\right) = 20\left(\frac{3}{4}\right)$$ • Use the Distributive Property.

$$16x - 10 = 15$$

$$16x - 10 + 10 = 15 + 10$$ • Add 10 to each side.

$$16x = 25$$

$$\frac{16x}{16} = \frac{25}{16}$$ • Divide each side by 16.

$$x = \frac{25}{16}$$

The solution is $\frac{25}{16}$.

YOU TRY IT • 4

Solve $\frac{2}{3}x + 3 = \frac{7}{2}$ by first clearing denominators.

Your solution

Solutions on p. S12

EXAMPLE 5

Solve: $2x + 4 - 5x = 10$

Solution

$$2x + 4 - 5x = 10$$
$$-3x + 4 = 10$$ • Combine like terms.
$$-3x + 4 - 4 = 10 - 4$$ • Subtract 4 from each side.
$$-3x = 6$$
$$\frac{-3x}{-3} = \frac{6}{-3}$$ • Divide each side by −3.
$$x = -2$$

The solution is −2.

YOU TRY IT 5

Solve: $x - 5 + 4x = 25$

Your solution

Solution on p. S13

OBJECTIVE B To solve application problems using formulas

EXAMPLE 6

To determine the total cost of production, an economist uses the equation $T = U \cdot N + F$, where T is the total cost, U is the unit cost, N is the number of units made, and F is the fixed cost. Use this equation to find the number of units made during a month in which the total cost was \$9000, the unit cost was \$25, and the fixed cost was \$3000.

Strategy

Given: $T = 9000$
$U = 25$
$F = 3000$
Unknown: N

Solution

$$T = U \cdot N + F$$
$$9000 = 25N + 3000$$
$$6000 = 25N$$
$$\frac{6000}{25} = \frac{25N}{25}$$
$$240 = N$$

There were 240 units made.

YOU TRY IT 6

The pressure at a certain depth in the ocean can be approximated by the equation $P = 15 + \frac{1}{2}D$, where P is the pressure in pounds per square inch and D is the depth in feet. Use this equation to find the depth when the pressure is 45 pounds per square inch.

Your strategy

Your solution

Solution on p. S13

5.2 EXERCISES

OBJECTIVE A To solve an equation of the form $ax + b = c$

For Exercises 1 to 80, solve and check.

1. $3x + 1 = 10$
2. $4y + 3 = 11$
3. $2a - 5 = 7$
4. $5m - 6 = 9$
5. $5 = 4x + 9$
6. $2 = 5b + 12$
7. $2x - 5 = -11$
8. $3n - 7 = -19$
9. $4 - 3w = -2$
10. $5 - 6x = -13$
11. $8 - 3t = 2$
12. $12 - 5x = 7$
13. $4a - 20 = 0$
14. $3y - 9 = 0$
15. $6 + 2b = 0$
16. $10 + 5m = 0$
17. $-2x + 5 = -7$
18. $-5d + 3 = -12$
19. $-1.2x + 3 = -0.6$
20. $-1.3 = -1.1y + 0.9$
21. $2 = 7 - 5a$
22. $3 = 11 - 4n$
23. $-35 = -6b + 1$
24. $-8x + 3 = -29$
25. $-3m - 21 = 0$
26. $-5x - 30 = 0$
27. $-4y + 15 = 15$
28. $-3x + 19 = 19$
29. $9 - 4x = 6$
30. $3t - 2 = 0$
31. $9x - 4 = 0$
32. $7 - 8z = 0$
33. $1 - 3x = 0$
34. $9d + 10 = 7$
35. $12w + 11 = 5$
36. $6y - 5 = -7$
37. $8b - 3 = -9$
38. $5 - 6m = 2$
39. $7 - 9a = 4$
40. $9 = -12c + 5$

41. $10 = -18x + 7$

42. $2y + \frac{1}{3} = \frac{7}{3}$

43. $4a + \frac{3}{4} = \frac{19}{4}$

44. $2n - \frac{3}{4} = \frac{13}{4}$

45. $3x - \frac{5}{6} = \frac{13}{6}$

46. $5y + \frac{3}{7} = \frac{3}{7}$

47. $9x + \frac{4}{5} = \frac{4}{5}$

48. $0.8 = 7d + 0.1$

49. $0.9 = 10x - 0.6$

50. $4 = 7 - 2w$

51. $7 = 9 - 5a$

52. $8t + 13 = 3$

53. $12x + 19 = 3$

54. $-6y + 5 = 13$

55. $-4x + 3 = 9$

56. $\frac{1}{2}a - 3 = 1$

57. $\frac{1}{3}m - 1 = 5$

58. $\frac{2}{5}y + 4 = 6$

59. $\frac{3}{4}n + 7 = 13$

60. $-\frac{2}{3}x + 1 = 7$

61. $-\frac{3}{8}b + 4 = 10$

62. $\frac{x}{4} - 6 = 1$

63. $\frac{y}{5} - 2 = 3$

64. $\frac{2x}{3} - 1 = 5$

65. $\frac{2}{3}x - \frac{5}{6} = -\frac{1}{3}$

66. $\frac{5}{4}x + \frac{2}{3} = \frac{1}{4}$

67. $\frac{1}{2} - \frac{2}{3}x = \frac{1}{4}$

68. $\frac{3}{4} - \frac{3}{5}x = \frac{19}{20}$

69. $\frac{3}{2} = \frac{5}{6} + \frac{3x}{8}$

70. $-\frac{1}{4} = \frac{5}{12} + \frac{5x}{6}$

71. $\frac{11}{27} = \frac{4}{9} - \frac{2x}{3}$

72. $\frac{37}{24} = \frac{7}{8} - \frac{5x}{6}$

73. $7 = \frac{2x}{5} + 4$

74. $5 - \frac{4c}{7} = 8$

75. $7 - \frac{5}{9}y = 9$

76. $6a + 3 + 2a = 11$

77. $5y + 9 + 2y = 23$

78. $7x - 4 - 2x = 6$

79. $11z - 3 - 7z = 9$

80. $2x - 6x + 1 = 9$

For Exercises 81 to 84, without solving the equation, determine whether the solution is positive or negative.

81. $15x + 73 = -347$

82. $17 = 25 - 40a$

83. $290 + 51n = 187$

84. $-72 = -86y + 49$

85. Solve $3x + 4y = 13$ when $y = -2$.

86. Solve $2x - 3y = 8$ when $y = 0$.

87. Solve $-4x + 3y = 9$ when $x = 0$.

88. Solve $5x - 2y = -3$ when $x = -3$.

89. If $2x - 3 = 7$, evaluate $3x + 4$.

90. If $3x + 5 = -4$, evaluate $2x - 5$.

91. If $4 - 5x = -1$, evaluate $x^2 - 3x + 1$.

92. If $2 - 3x = 11$, evaluate $x^2 + 2x - 3$.

OBJECTIVE B To solve application problems using formulas

Champion Trees American Forests is an organization that maintains the National Register of Big Trees, a listing of the largest trees in the United States. The formula used to award points to a tree is $P = c + h + \frac{1}{4}s$, where P is the point total for a tree with a circumference of c inches, a height of h feet, and an average crown spread of s feet. Use this formula for Exercises 93 and 94. (*Source:* www.amfor.org)

93. Find the average crown spread of the baldcypress described in the article at the right.

94. One of the smallest trees in the United States is a Florida Crossopetalum in the Key Largo Hammocks State Botanical Site. This tree stands 11 ft tall, has a circumference of just 4.8 in., and scores 16.55 points using American Forests' formula. Find the tree's average crown spread. (*Source:* www.championtrees.org)

In the News

The Senator Is a Champion

Baldcypress trees are among the most ancient of North American trees. The 3500-year-old baldcypress known as the Senator, located in Big Tree Park, Longwood, is the Florida Champion specimen of the species. With a circumference of 425 in. and a height of 118 ft, this king of the swamp forest earned a total of $557\frac{1}{4}$ points under the point system used for the National Register of Big Trees.

Source: www.championtrees.org

Nutrition The formula $C = 9f + 4p + 4c$ gives the number of calories C in a serving of food that contains f grams of fat, p grams of protein, and c grams of carbohydrate. Use this formula for Exercises 95 and 96. (*Source:* www.nutristrategy.com)

95. Find the number of grams of protein in an 8-ounce serving of vanilla yogurt that contains 174 calories, 2 g of fat, and 30 g of carbohydrate.

96. Find the number of grams of fat in a serving of granola that contains 215 calories, 42 g of carbohydrate, and 5 g of protein.

Physics The distance s, in feet, that an object will fall in t seconds is given by $s = 16t^2 + vt$, where v is the initial velocity of the object in feet per second. Use this equation for Exercises 97 and 98.

97. Find the initial velocity of an object that falls 80 ft in 2 s.

98. Find the initial velocity of an object that falls 144 ft in 3 s.

Seminole County Government

The Senator at Big Tree Park

Depreciation A company uses the equation $V = C - 6000t$ to determine the depreciated value V, after t years, of a milling machine that originally cost C dollars. Equations like this are used in accounting for straight-line depreciation. Use this equation for Exercises 99 and 100.

99. A milling machine originally cost \$50,000. In how many years will the depreciated value of the machine be \$38,000?

100. A milling machine originally cost \$78,000. In how many years will the depreciated value of the machine be \$48,000?

Anthropology Anthropologists approximate the height of a primate by the size of its humerus (the bone from the elbow to the shoulder) using the equation $H = 1.2L + 27.8$, where L is the length of the humerus and H is the height, in inches, of the primate. Use this equation for Exercises 101 and 102.

101. An anthropologist estimates the height of a primate to be 66 in. What is the approximate length of the humerus of this primate? Round to the nearest tenth of an inch.

102. An anthropologist estimates the height of a primate to be 62 in. What is the approximate length of the humerus of this primate?

Car Safety Black ice is an ice covering on roads that is especially difficult to see and therefore extremely dangerous for motorists. The distance that a car traveling 30 mph will slide after its brakes are applied is related to the outside temperature by the formula $C = \frac{1}{4}D - 45$, where C is the Celsius temperature and D is the distance in feet that the car will slide. Use this equation for Exercises 103 and 104.

103. Determine the distance a car will slide on black ice when the outside temperature is $-3°$C.

104. Determine the distance a car will slide on black ice when the outside temperature is $-11°$C.

105. If A is a positive number, is the solution of the equation $Ax + 8 = -3$ positive or negative?

106. If A is a negative number, is the solution of the equation $Ax - 2 = -5$ positive or negative?

Applying the Concepts

107. If $2x + 1 = a$ and $3x - 2 = a$, find the value of a.

108. If $1 - 4x = y$ and $2x - 5 = y$, find the value of y.

109. Solve: $x \div 15 = 25$ remainder 10

SECTION

5.3 General Equations—Part II

OBJECTIVE A To solve an equation of the form $ax + b = cx + d$

Tips for Success
Have you considered joining a study group? Getting together regularly with other students in the class to go over material and quiz each other can be very beneficial. See *AIM for Success* in the front of the book.

In solving an equation of the form $ax + b = cx + d$, the goal is to rewrite the equation in the form *variable* = *constant*. Begin by rewriting the equation so that there is only one variable term in the equation. Then rewrite the equation so that there is only one constant term.

HOW TO • 1 Solve: $2x + 3 = 5x - 9$

$$2x + 3 = 5x - 9$$

$$2x - 5x + 3 = 5x - 5x - 9$$ • Subtract $5x$ from each side of the equation.

$$-3x + 3 = -9$$ • Simplify. There is only one variable term.

$$-3x + 3 - 3 = -9 - 3$$ • Subtract 3 from each side of the equation.

$$-3x = -12$$ • Simplify. There is only one constant term.

$$\frac{-3x}{-3} = \frac{-12}{-3}$$ • Divide each side of the equation by -3.

$$x = 4$$ • The equation is in the form *variable* = *constant*.

The solution is 4. You should verify this by checking this solution.

EXAMPLE • 1

Solve: $4x - 5 = 8x - 7$

Solution

$$4x - 5 = 8x - 7$$

$$4x - 8x - 5 = 8x - 8x - 7$$ • Subtract $8x$ from each side.

$$-4x - 5 = -7$$

$$-4x - 5 + 5 = -7 + 5$$ • Add 5 to each side.

$$-4x = -2$$

$$\frac{-4x}{-4} = \frac{-2}{-4}$$ • Divide each side by -4.

$$x = \frac{1}{2}$$

The solution is $\frac{1}{2}$.

YOU TRY IT • 1

Solve: $5x + 4 = 6 + 10x$

Your solution

Solution on p. S13

EXAMPLE 2

Solve: $3x + 4 - 5x = 2 - 4x$

Solution

$3x + 4 - 5x = 2 - 4x$

$-2x + 4 = 2 - 4x$ • Combine like terms.

$-2x + 4x + 4 = 2 - 4x + 4x$ • Add $4x$ to each side.

$2x + 4 = 2$

$2x + 4 - 4 = 2 - 4$ • Subtract 4 from each side.

$2x = -2$

$\dfrac{2x}{2} = \dfrac{-2}{2}$ • Divide each side by 2.

$x = -1$

The solution is -1.

YOU TRY IT 2

Solve: $5x - 10 - 3x = 6 - 4x$

Your solution

Solution on p. S13

OBJECTIVE B To solve an equation containing parentheses 

When an equation contains parentheses, one of the steps in solving the equation requires the use of the Distributive Property. The Distributive Property is used to remove parentheses from a variable expression.

HOW TO 2 Solve: $4 + 5(2x - 3) = 3(4x - 1)$

$4 + 5(2x - 3) = 3(4x - 1)$

$4 + 10x - 15 = 12x - 3$ • Use the Distributive Property. Then simplify.

$10x - 11 = 12x - 3$

$10x - 12x - 11 = 12x - 12x - 3$ • Subtract $12x$ from each side of the equation.

$-2x - 11 = -3$ • Simplify.

$-2x - 11 + 11 = -3 + 11$ • Add 11 to each side of the equation.

$-2x = 8$ • Simplify.

$\dfrac{-2x}{-2} = \dfrac{8}{-2}$ • Divide each side of the equation by -2.

$x = -4$ • The equation is in the form *variable = constant*.

The solution is -4. You should verify this by checking this solution.

In the next example, we solve an equation with parentheses and decimals.

HOW TO • 3 Solve: $16 + 0.55x = 0.75(x + 20)$

$$16 + 0.55x = 0.75(x + 20)$$
$$16 + 0.55x = 0.75x + 15$$
- Use the Distributive Property.

$$16 + 0.55x - 0.75x = 0.75x - 0.75x + 15$$
- Subtract $0.75x$ from each side of the equation.

$$16 - 0.20x = 15$$
- Simplify.

$$16 - 16 - 0.20x = 15 - 16$$
- Subtract 16 from each side of the equation.

$$-0.20x = -1$$
- Simplify.

$$\frac{-0.20x}{-0.20} = \frac{-1}{-0.20}$$
- Divide each side of the equation by -0.20.

$$x = 5$$
- The equation is in the form *variable = constant.*

The solution is 5.

EXAMPLE • 3

Solve: $3x - 4(2 - x) = 3(x - 2) - 4$

Solution

$$3x - 4(2 - x) = 3(x - 2) - 4$$
$$3x - 8 + 4x = 3x - 6 - 4 \quad \text{• Distributive Property}$$
$$7x - 8 = 3x - 10$$
$$7x - 3x - 8 = 3x - 3x - 10 \quad \text{• Subtract } 3x.$$
$$4x - 8 = -10$$
$$4x - 8 + 8 = -10 + 8 \quad \text{• Add 8.}$$
$$4x = -2$$
$$\frac{4x}{4} = \frac{-2}{4} \quad \text{• Divide by 4.}$$
$$x = -\frac{1}{2}$$

The solution is $-\frac{1}{2}$.

YOU TRY IT • 3

Solve: $5x - 4(3 - 2x) = 2(3x - 2) + 6$

Your solution

EXAMPLE • 4

Solve: $3[2 - 4(2x - 1)] = 4x - 10$

Solution

$$3[2 - 4(2x - 1)] = 4x - 10$$
$$3[2 - 8x + 4] = 4x - 10 \quad \text{• Distributive Property}$$
$$3[6 - 8x] = 4x - 10$$
$$18 - 24x = 4x - 10 \quad \text{• Distributive Property}$$
$$18 - 24x - 4x = 4x - 4x - 10 \quad \text{• Subtract } 4x.$$
$$18 - 28x = -10$$
$$18 - 18 - 28x = -10 - 18 \quad \text{• Subtract 18.}$$
$$-28x = -28$$
$$\frac{-28x}{-28} = \frac{-28}{-28} \quad \text{• Divide by } -28.$$
$$x = 1$$

The solution is 1.

YOU TRY IT • 4

Solve: $-2[3x - 5(2x - 3)] = 3x - 8$

Your solution

Solutions on p. S13

OBJECTIVE C To solve application problems using formulas 

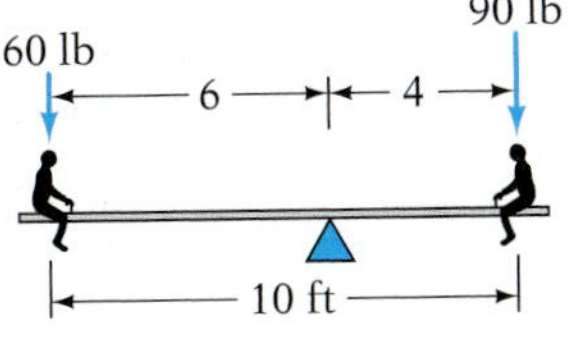

This system balances because

$$F_1x = F_2(d - x)$$
$$60(6) = 90(10 - 6)$$
$$60(6) = 90(4)$$
$$360 = 360$$

A lever system is shown at the right. It consists of a lever, or bar; a fulcrum; and two forces, F_1 and F_2. The distance d represents the length of the lever, x represents the distance from F_1 to the fulcrum, and $d - x$ represents the distance from F_2 to the fulcrum.

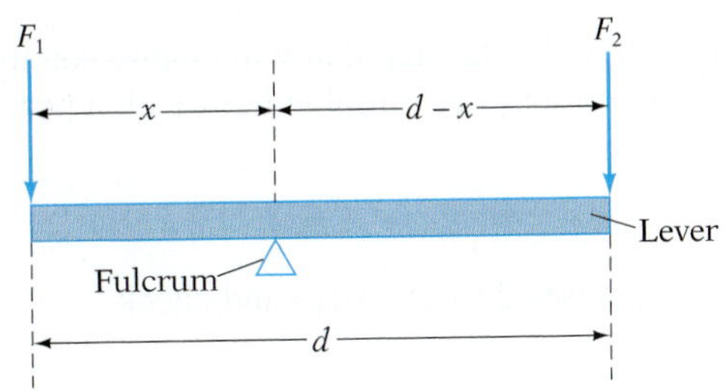

A principle of physics states that when the lever system balances, $F_1x = F_2(d - x)$.

EXAMPLE • 5

A lever is 15 ft long. A force of 50 lb is applied to one end of the lever, and a force of 100 lb is applied to the other end. Where is the fulcrum located when the system balances?

Strategy
Make a drawing.

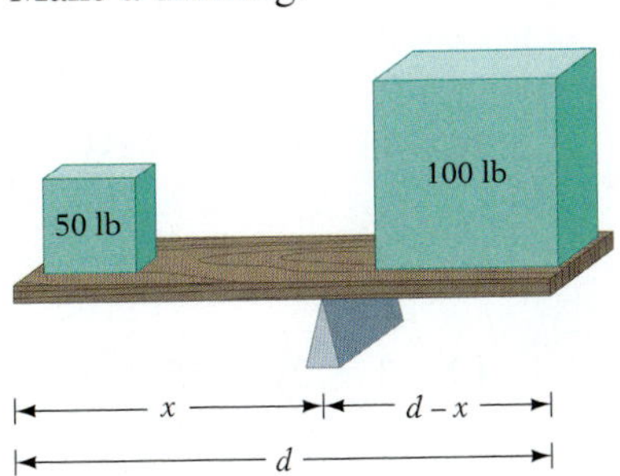

Given: $F_1 = 50$
$F_2 = 100$
$d = 15$
Unknown: x

Solution

$$F_1x = F_2(d - x)$$
$$50x = 100(15 - x)$$
$$50x = 1500 - 100x$$
$$50x + 100x = 1500 - 100x + 100x \quad \text{• Add } 100x.$$
$$150x = 1500$$
$$\frac{150x}{150} = \frac{1500}{150} \quad \text{• Divide by 150.}$$
$$x = 10$$

The fulcrum is 10 ft from the 50-pound force.

YOU TRY IT • 5

A lever is 25 ft long. A force of 45 lb is applied to one end of the lever, and a force of 80 lb is applied to the other end. Where is the location of the fulcrum when the system balances?

Your strategy

Your solution

Solution on p. S13

5.3 EXERCISES

OBJECTIVE A To solve an equation of the form $ax + b = cx + d$

1. Describe the step that will enable you to rewrite the equation $2x - 3 = 7x + 12$ so that it has one variable term with a positive coefficient.

For Exercises 2 to 28, solve and check.

2. $8x + 5 = 4x + 13$
3. $6y + 2 = y + 17$
4. $5x - 4 = 2x + 5$
5. $13b - 1 = 4b - 19$
6. $15x - 2 = 4x - 13$
7. $7a - 5 = 2a - 20$
8. $3x + 1 = 11 - 2x$
9. $n - 2 = 6 - 3n$
10. $2x - 3 = -11 - 2x$
11. $4y - 2 = -16 - 3y$
12. $0.2b + 3 = 0.5b + 12$
13. $m + 0.4 = 3m + 0.8$
14. $4y - 8 = y - 8$
15. $5a + 7 = 2a + 7$
16. $6 - 5x = 8 - 3x$
17. $10 - 4n = 16 - n$
18. $5 + 7x = 11 + 9x$
19. $3 - 2y = 15 + 4y$
20. $2x - 4 = 6x$
21. $2b - 10 = 7b$
22. $8m = 3m + 20$
23. $9y = 5y + 16$
24. $8b + 5 = 5b + 7$
25. $6y - 1 = 2y + 2$
26. $7x - 8 = x - 3$
27. $2y - 7 = -1 - 2y$
28. $2m - 1 = -6m + 5$
29. If $5x = 3x - 8$, evaluate $4x + 2$.
30. If $7x + 3 = 5x - 7$, evaluate $3x - 2$.
31. If $2 - 6a = 5 - 3a$, evaluate $4a^2 - 2a + 1$.
32. If $1 - 5c = 4 - 4c$, evaluate $3c^2 - 4c + 2$.

OBJECTIVE B To solve an equation containing parentheses

33. Without solving any of the equations, determine which of the following equations has the same solution as the equation $5 - 2(x - 1) = 8$.
(i) $3(x - 1) = 8$ (ii) $5 - 2x + 2 = 8$ (iii) $5 - 2x + 1 = 8$

For Exercises 34 to 54, solve and check.

34. $5x + 2(x + 1) = 23$

35. $6y + 2(2y + 3) = 16$

36. $9n - 3(2n - 1) = 15$

37. $12x - 2(4x - 6) = 28$

38. $7a - (3a - 4) = 12$

39. $9m - 4(2m - 3) = 11$

40. $5(3 - 2y) + 4y = 3$

41. $4(1 - 3x) + 7x = 9$

42. $5y - 3 = 7 + 4(y - 2)$

43. $0.22(x + 6) = 0.2x + 1.8$

44. $0.05(4 - x) + 0.1x = 0.32$

45. $0.3x + 0.3(x + 10) = 300$

46. $2a - 5 = 4(3a + 1) - 2$

47. $5 - (9 - 6x) = 2x - 2$

48. $7 - (5 - 8x) = 4x + 3$

49. $3[2 - 4(y - 1)] = 3(2y + 8)$

50. $5[2 - (2x - 4)] = 2(5 - 3x)$

51. $3a + 2[2 + 3(a - 1)] = 2(3a + 4)$

52. $5 + 3[1 + 2(2x - 3)] = 6(x + 5)$

53. $-2[4 - (3b + 2)] = 5 - 2(3b + 6)$

54. $-4[x - 2(2x - 3)] + 1 = 2x - 3$

55. If $4 - 3a = 7 - 2(2a + 5)$, evaluate $a^2 + 7a$.

56. If $9 - 5x = 12 - (6x + 7)$, evaluate $x^2 - 3x - 2$.

OBJECTIVE C To solve application problems using formulas

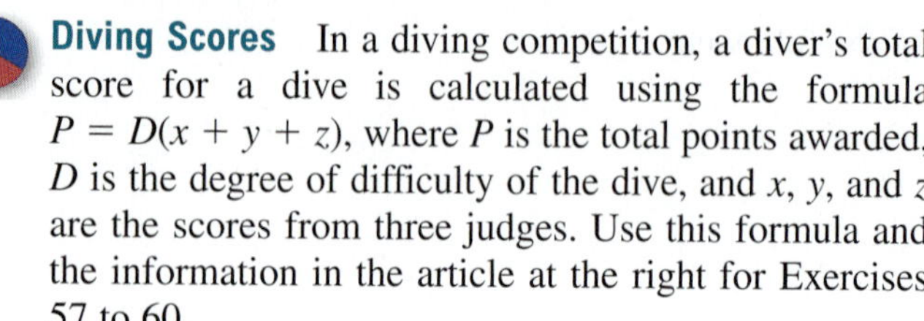

Diving Scores In a diving competition, a diver's total score for a dive is calculated using the formula $P = D(x + y + z)$, where P is the total points awarded, D is the degree of difficulty of the dive, and x, y, and z are the scores from three judges. Use this formula and the information in the article at the right for Exercises 57 to 60.

In the News

Hurricane Divers Make a Splash

University of Miami divers JJ Kinzbach and Rueben Ross took the top two spots in Men's Platform diving at the 2008 NCAA Zone B Championships. Ross also placed second in the Men's 3-meter and 1-meter events. Brittany Viola won the Women's Platform diving event and placed third in the 3-meter and 1-meter events. Statistics from some of the best dives follow.

Diver	*Event*	*Dive*	*Degree of Difficulty*	*Total Points*
Kinzbach	Platform	Inward 3½ somersault tuck	3.2	81.60
Ross	1-meter	Inward 2½ somersault tuck	3.1	77.50
Viola	Platform	Forward 3½ somersault pike	3.0	72.0
Viola	1-meter	Inward 1½ somersault pike	2.4	57.60

Source: divemeets.com

57. Two judges gave Kinzbach's platform dive scores of 8.5. Find the score given by the third judge.

58. Two judges gave Ross's 1-meter dive scores of 8 and 8.5. Find the score given by the third judge.

59. Two judges gave Viola's platform dive scores of 8. Find the score given by the third judge.

60. Two judges gave Viola's 1-meter dive scores of 8 and 8.5. Find the score given by the third judge.

61. Physics Two people sit on a seesaw that is 8 ft long. The seesaw balances when the fulcrum is 3 ft from one of the people.

a. How far is the fulcrum from the other person?

b. Which person is heavier, the person who is 3 ft from the fulcrum or the other person?

c. If the two people switch places, will the seesaw still balance?

Physics For Exercises 62 to 67, solve. Use the lever system equation $F_1x = F_2(d - x)$.

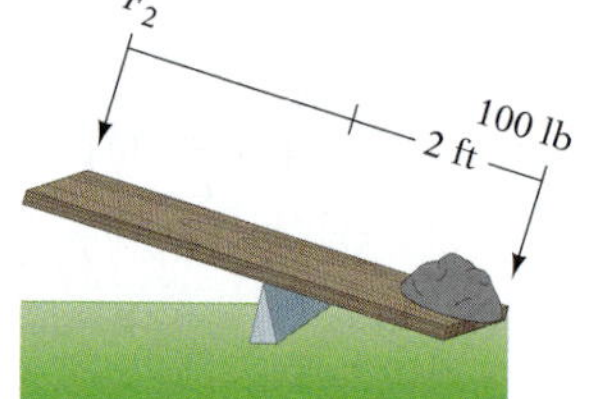

62. A lever 10 ft long is used to move a 100-pound rock. The fulcrum is placed 2 ft from the rock. What force must be applied to the other end of the lever to move the rock?

63. An adult and a child are on a seesaw 14 ft long. The adult weighs 175 lb and the child weighs 70 lb. How many feet from the child must the fulcrum be placed so that the seesaw balances?

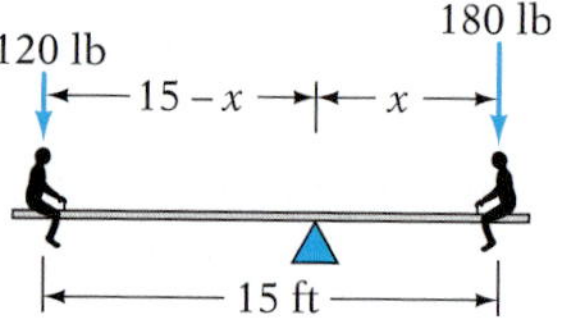

64. Two people are sitting 15 ft apart on a seesaw. One person weighs 180 lb. The second person weighs 120 lb. How far from the 180-pound person should the fulcrum be placed so that the seesaw balances?

65. Two children are sitting on a seesaw that is 12 ft long. One child weighs 60 lb. The other child weighs 90 lb. How far from the 90-pound child should the fulcrum be placed so that the seesaw balances?

66. In preparation for a stunt, two acrobats are standing on a plank 18 ft long. One acrobat weighs 128 lb and the second acrobat weighs 160 lb. How far from the 128-pound acrobat must the fulcrum be placed so that the acrobats are balanced on the plank?

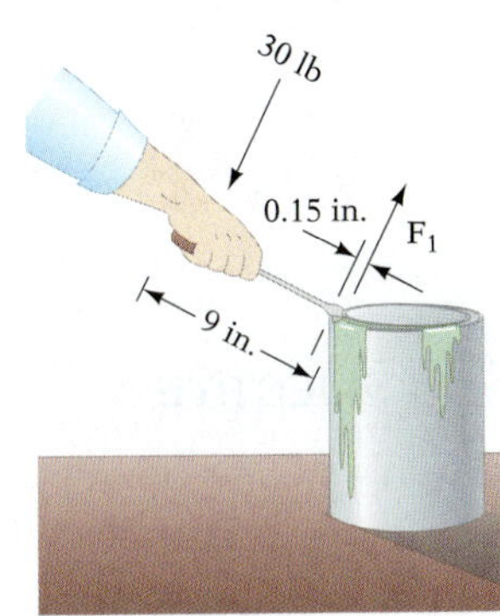

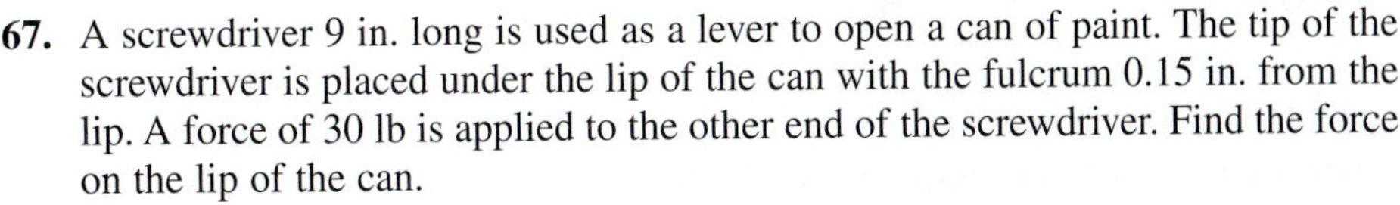

67. A screwdriver 9 in. long is used as a lever to open a can of paint. The tip of the screwdriver is placed under the lip of the can with the fulcrum 0.15 in. from the lip. A force of 30 lb is applied to the other end of the screwdriver. Find the force on the lip of the can.

Business To determine the break-even point, or the number of units that must be sold so that no profit or loss occurs, an economist uses the formula $Px = Cx + F$, where P is the selling price per unit, x is the number of units that must be sold to break even, C is the cost to make each unit, and F is the fixed cost. Use this equation for Exercises 68 to 71.

68. A business analyst has determined that the selling price per unit for a laser printer is \$1600. The cost to make one laser printer is \$950, and the fixed cost is \$211,250. Find the break-even point.

69. A business analyst has determined that the selling price per unit for a gas barbecue is \$325. The cost to make one gas barbecue is \$175, and the fixed cost is \$39,000. Find the break-even point.

Charles Mistral/Alamy

70. A manufacturer of headphones determines that the cost per unit for a pair of headphones is \$38 and that the fixed cost is \$24,400. The selling price for the headphones is \$99. Find the break-even point.

71. A manufacturing engineer determines that the cost per unit for a soprano recorder is \$12 and that the fixed cost is \$19,240. The selling price for the recorder is \$49. Find the break-even point.

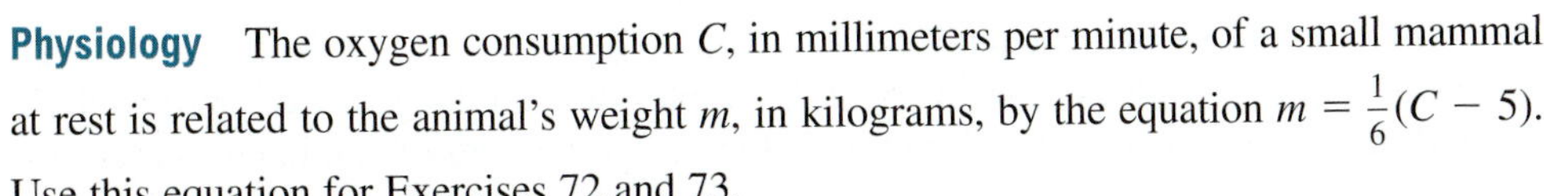

Physiology The oxygen consumption C, in millimeters per minute, of a small mammal at rest is related to the animal's weight m, in kilograms, by the equation $m = \frac{1}{6}(C - 5)$. Use this equation for Exercises 72 and 73.

72. What is the oxygen consumption of a mammal that weighs 10.4 kg?

73. What is the oxygen consumption of a mammal that weighs 8.3 kg?

Applying the Concepts

74. The equation $x = x + 1$ has no solution, whereas the solution of the equation $2x + 3 = 3$ is zero. Is there a difference between no solution and a solution of zero? Explain your answer.

SECTION

5.4 Translating Sentences into Equations

OBJECTIVE A To solve integer problems

An equation states that two mathematical expressions are equal. Therefore, to translate a sentence into an equation requires recognition of the words or phrases that mean "equals." Some of these phrases are listed below.

equals
is
is equal to
amounts to
represents

} translate to =

Once the sentence is translated into an equation, the equation can be solved by rewriting the equation in the form *variable* = *constant*.

Take Note
You can check the solution to a translation problem.

Check:

5 less than 18 is 13

$18 - 5 \mid 13$

$13 = 13$

HOW TO • 1 Translate "five less than a number is thirteen" into an equation and solve.

The unknown number: n • Assign a variable to the unknown number.

Five less than a number	is	thirteen

• Find two verbal expressions for the same value.

$n - 5 = 13$ • Write a mathematical expression for each verbal expression. Write the equals sign.

$n - 5 + 5 = 13 + 5$ • Solve the equation.

$n = 18$

The number is 18.

Recall that the integers are the numbers {..., −4, −3, −2, −1, 0, 1, 2, 3, 4, ...}. An **even integer** is an integer that is divisible by 2. Examples of even integers are −8, 0, and 22. An **odd integer** is an integer that is not divisible by 2. Examples of odd integers are −17, 1, and 39.

Consecutive integers are integers that follow one another in order. Examples of consecutive integers are shown at the right. (Assume that the variable n represents an integer.)

11, 12, 13
−8, −7, −6
$n, n + 1, n + 2$

Examples of **consecutive even integers** are shown at the right. (Assume that the variable n represents an even integer.)

24, 26, 28
−10, −8, −6
$n, n + 2, n + 4$

Take Note
Both consecutive even and consecutive odd integers are represented using $n, n + 2, n + 4, \ldots$.

Examples of **consecutive odd integers** are shown at the right. (Assume that the variable n represents an odd integer.)

19, 21, 23
−1, 1, 3
$n, n + 2, n + 4$

HOW TO • 2 The sum of three consecutive odd integers is forty-five. Find the integers.

Strategy

- First odd integer: n — • Represent three consecutive odd integers.
  Second odd integer: $n + 2$
  Third odd integer: $n + 4$
- The sum of the three odd integers is 45.

Solution

$n + (n + 2) + (n + 4) = 45$ • Write an equation.

$3n + 6 = 45$ • Solve the equation.

$3n = 39$

$n = 13$ • The first odd integer is 13.

$n + 2 = 13 + 2 = 15$ • Find the second odd integer.

$n + 4 = 13 + 4 = 17$ • Find the third odd integer.

The three consecutive odd integers are 13, 15, and 17.

EXAMPLE • 1

The sum of two numbers is sixteen. The difference between four times the smaller number and two is two more than twice the larger number. Find the two numbers.

Strategy

The smaller number: n

The larger number: $16 - n$

The difference between four times the smaller and two	is	two more than twice the larger

Solution

$$4n - 2 = 2(16 - n) + 2$$
$$4n - 2 = 32 - 2n + 2$$
$$4n - 2 = 34 - 2n$$
$$4n + 2n - 2 = 34 - 2n + 2n$$
$$6n - 2 = 34$$
$$6n - 2 + 2 = 34 + 2$$
$$6n = 36$$
$$\frac{6n}{6} = \frac{36}{6}$$
$$n = 6$$

$16 - n = 16 - 6 = 10$

The smaller number is 6.
The larger number is 10.

YOU TRY IT • 1

The sum of two numbers is twelve. The total of three times the smaller number and six amounts to seven less than the product of four and the larger number. Find the two numbers.

Your strategy

Your solution

Solution on pp. S13–S14

EXAMPLE • 2

Find three consecutive even integers such that three times the second equals four more than the sum of the first and third.

Strategy

- First even integer: n
  Second even integer: $n + 2$
  Third even integer: $n + 4$
- Three times the second equals four more than the sum of the first and third.

Solution

$$\begin{aligned} 3(n + 2) &= n + (n + 4) + 4 \\ 3n + 6 &= 2n + 8 \\ 3n - 2n + 6 &= 2n - 2n + 8 \\ n + 6 &= 8 \\ n &= 2 \end{aligned}$$

$n + 2 = 2 + 2 = 4$
$n + 4 = 2 + 4 = 6$

The three integers are 2, 4, and 6.

YOU TRY IT • 2

Find three consecutive integers whose sum is negative six.

Your strategy

Your solution

Solution on p. S14

OBJECTIVE B To translate a sentence into an equation and solve 

EXAMPLE • 3

A wallpaper hanger charges a fee of \$25 plus \$12 for each roll of wallpaper used in a room. If the total charge for hanging wallpaper is \$97, how many rolls of wallpaper were used?

Strategy

To find the number of rolls of wallpaper used, write and solve an equation using n to represent the number of rolls of wallpaper used.

Solution

\$25 plus \$12 for each roll of wallpaper	is	\$97

$$\begin{aligned} 25 + 12n &= 97 \\ 12n &= 72 \\ \frac{12n}{12} &= \frac{72}{12} \\ n &= 6 \end{aligned}$$

6 rolls of wallpaper were used.

YOU TRY IT • 3

The fee charged by a ticketing agency for a concert is \$3.50 plus \$17.50 for each ticket purchased. If your total charge for tickets is \$161, how many tickets are you purchasing?

Your strategy

Your solution

Solution on p. S14

EXAMPLE • 4

A board 20 ft long is cut into two pieces. Five times the length of the shorter piece is 2 ft more than twice the length of the longer piece. Find the length of each piece.

Strategy

Let x represent the length of the shorter piece. Then $20 - x$ represents the length of the longer piece.

Make a drawing.

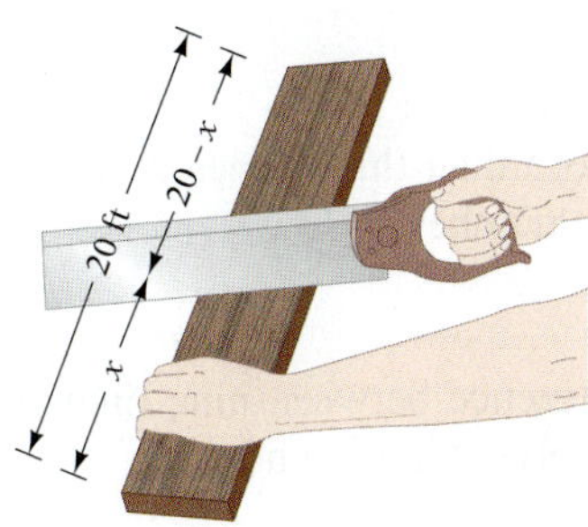

To find the lengths, write and solve an equation using x to represent the length of the shorter piece and $20 - x$ to represent the length of the longer piece.

Solution

Five times the length of the shorter piece	is	2 ft more than twice the length of the longer piece

$$\begin{aligned} 5x &= 2(20 - x) + 2 \\ 5x &= 40 - 2x + 2 \\ 5x &= 42 - 2x \\ 5x + 2x &= 42 - 2x + 2x \\ 7x &= 42 \\ \frac{7x}{7} &= \frac{42}{7} \\ x &= 6 \end{aligned}$$

$20 - x = 20 - 6 = 14$

The length of the shorter piece is 6 ft.
The length of the longer piece is 14 ft.

YOU TRY IT • 4

A wire 22 in. long is cut into two pieces. The length of the longer piece is 4 in. more than twice the length of the shorter piece. Find the length of each piece.

Your strategy

Your solution

Solution on p. S14

5.4 EXERCISES

OBJECTIVE A To solve integer problems

For Exercises 1 to 16, translate into an equation and solve.

1. The difference between a number and fifteen is seven. Find the number.

2. The sum of five and a number is three. Find the number.

3. The difference between nine and a number is seven. Find the number.

4. Three-fifths of a number is negative thirty. Find the number.

5. The difference between five and twice a number is one. Find the number.

6. Four more than three times a number is thirteen. Find the number.

7. The sum of twice a number and five is fifteen. Find the number.

8. The difference between nine times a number and six is twelve. Find the number.

9. Six less than four times a number is twenty-two. Find the number.

10. Four times the sum of twice a number and three is twelve. Find the number.

11. Three times the difference between four times a number and seven is fifteen. Find the number.

12. Twice the difference between a number and twenty-five is three times the number. Find the number.

13. The sum of two numbers is twenty. Three times the smaller is equal to two times the larger. Find the two numbers.

14. The sum of two numbers is fifteen. One less than three times the smaller is equal to the larger. Find the two numbers.

15. The sum of two numbers is fourteen. The difference between two times the smaller and the larger is one. Find the two numbers.

16. The sum of two numbers is eighteen. The total of three times the smaller and twice the larger is forty-four. Find the two numbers.

17. The sum of three consecutive odd integers is fifty-one. Find the integers.

18. Find three consecutive even integers whose sum is negative eighteen.

19. Find three consecutive odd integers such that three times the middle integer is one more than the sum of the first and third.

20. Twice the smallest of three consecutive odd integers is seven more than the largest. Find the integers.

21. Find two consecutive even integers such that three times the first equals twice the second.

22. Find two consecutive even integers such that four times the first is three times the second.

23. The sum of two numbers is seven. Twice one number is four less than the other number. Which of the following equations does *not* represent this situation?
(i) $2(7 - x) = x - 4$ (ii) $2x = (7 - x) - 4$ (iii) $2n - 4 = 7 - n$

OBJECTIVE B To translate a sentence into an equation and solve

24. **Recycling** Use the information in the article at the right to find how many tons of plastic drink bottles were stocked for sale in U.S. stores.

> **In the News**
>
> **Americans' Unquenchable Thirst**
>
> Despite efforts to increase recycling, the 2.16 million tons of plastic drink bottles that ended up in landfills this year represent four-fifths of the plastic drink bottles stocked for sale in U.S. stores.
>
> And Americans can't seem to get enough of bottled water. Last year, stores stocked 7.5 billion gallons of bottled water, an amount that is approximately the same as the volume of water that goes over Niagara Falls every three hours.
>
> *Source:* scienceline.org

25. **Robots** Kiva Systems, Inc., builds robots that companies can use to streamline order fulfillment operations in their warehouses. Salary and other benefits for one human warehouse worker can cost a company about \$64,000 a year, an amount that is 103 times the company's yearly maintenance and operation costs for one robot. Find the yearly costs for a robot. Round to the nearest hundred. (*Source: The Boston Globe*)

26. **Geometry** An isosceles triangle has two sides of equal length. The length of the third side is 1 ft less than twice the length of an equal side. Find the length of each side when the perimeter is 23 ft.

27. **Geometry** An isosceles triangle has two sides of equal length. The length of one of the equal sides is 2 m more than three times the length of the third side. If the perimeter is 46 m, find the length of each side.

28. **Union Dues** A union charges monthly dues of \$4.00 plus \$.15 for each hour worked during the month. A union member's dues for March were \$29.20. How many hours did the union member work during the month of March?

29. **Technical Support** A technical information hotline charges a customer \$15.00 plus \$2.00 per minute to answer questions about software. How many minutes did a customer who received a bill for \$37 use this service?

30. **Construction** The total cost to paint the inside of a house was \$1346. This cost included \$125 for materials and \$33 per hour for labor. How many hours of labor were required to paint the inside of the house?

31. **Telecommunications** The cellular phone service for a business executive is \$35 per month plus \$.40 per minute of phone use. For a month in which the executive's cellular phone bill was \$99.80, how many minutes did the executive use the phone?

32. **Energy** The cost of electricity in a certain city is \$.08 for each of the first 300 kWh (kilowatt-hours) and \$.13 for each kilowatt-hour over 300 kWh. Find the number of kilowatt-hours used by a family with a \$51.95 electric bill.

 **Text Messaging** For Exercises 33 and 34, use the expression $2.99 + 0.15n$, which represents the total monthly text-messaging bill for n text messages over 300 in 1 month.

33. How much does the customer pay per text message over 300 messages?

34. What is the fixed charge per month for the text-messaging service?

35. Contractors Budget Plumbing charged \$400 for a water softener and installation. The charge included \$310 for the water softener and \$30 per hour for labor. How many hours of labor were required for the job?

36. Purchasing McPherson Cement sells cement for \$75 plus \$24 for each yard of cement. How many yards of cement can be purchased for \$363?

37. Carpentry A 12-foot board is cut into two pieces. Twice the length of the shorter piece is 3 ft less than the length of the longer piece. Find the length of each piece.

38. Sports A 14-yard fishing line is cut into two pieces. Three times the length of the longer piece is four times the length of the shorter piece. Find the length of each piece.

39. Education Seven thousand dollars is divided into two scholarships. Twice the amount of the smaller scholarship is \$1000 less than the larger scholarship. What is the amount of the larger scholarship?

40. Investing An investment of \$10,000 is divided into two accounts, one for stocks and one for mutual funds. The value of the stock account is \$2000 less than twice the value of the mutual fund account. Find the amount in each account.

Applying the Concepts

41. Make up two word problems: one that requires solving the equation $6x = 123$, and one that requires solving the equation $8x + 100 = 300$, to find the answer to the problem.

42. It is always important to check the answer to an application problem to be sure that the answer makes sense. Consider the following problem. A 4-quart juice mixture is made from apple juice and cranberry juice. There are 6 more quarts of apple juice than cranberry juice. Write and solve an equation for the number of quarts of each juice in the mixture. Does the answer to this question make sense? Explain.

SECTION

5.5 Mixture and Uniform Motion Problems

OBJECTIVE A To solve value mixture problems 

A value mixture problem involves combining two ingredients that have different prices into a single blend. For example, a coffee merchant may blend two types of coffee into a single blend, or a candy manufacturer may combine two types of candy to sell as a variety pack.

The solution of a value mixture problem is based on the **value mixture equation** $AC = V$, where A is the amount of an ingredient, C is the cost per unit of the ingredient, and V is the value of the ingredient.

Take Note
The equation $AC = V$ is used to find the value of an ingredient. For example, the value of 4 lb of cashews costing \$6 per pound is

$$AC = V$$
$$4 \cdot \$6 = V$$
$$\$24 = V$$

HOW TO • 1 A coffee merchant wants to make 6 lb of a blend of coffee costing \$5 per pound. The blend is made using a \$6-per-pound grade and a \$3-per-pound grade of coffee. How many pounds of each of these grades should be used?

Strategy for Solving a Value Mixture Problem

1. For each ingredient in the mixture, write a numerical or variable expression for the amount of the ingredient used, the unit cost of the ingredient, and the value of the amount used. For the blend, write a numerical or variable expression for the amount, the unit cost of the blend, and the value of the amount. The results can be recorded in a table.

The sum of the amounts is 6 lb.

Amount of \$6 coffee: x
Amount of \$3 coffee: $6 - x$

	Amount, A	·	Unit Cost, C	=	Value, V
\$6 grade	x	·	6	=	$6x$
\$3 grade	$6 - x$	·	3	=	$3(6 - x)$
\$5 blend	6	·	5	=	$5(6)$

Take Note
Use the information given in the problem to fill in the amount and unit cost columns of the table. Fill in the value column by multiplying the two expressions you wrote in each row. Use the expressions in the last column to write the equation.

2. Determine how the values of the ingredients are related. Use the fact that the sum of the values of all the ingredients is equal to the value of the blend.

The sum of the values of the \$6 grade and the \$3 grade is equal to the value of the \$5 blend.

$$6x + 3(6 - x) = 5(6)$$
$$6x + 18 - 3x = 30$$
$$3x + 18 = 30$$
$$3x = 12$$
$$x = 4$$

$6 - x = 6 - 4 = 2$ • **Find the amount of the \$3 grade coffee.**

The merchant must use 4 lb of the \$6 coffee and 2 lb of the \$3 coffee.

EXAMPLE • 1

How many ounces of a silver alloy that costs \$4 an ounce must be mixed with 10 oz of an alloy that costs \$6 an ounce to make a mixture that costs \$4.32 an ounce?

Strategy

- Ounces of \$4 alloy: x

	Amount	*Cost*	*Value*
\$4 alloy	x	4	$4x$
\$6 alloy	10	6	6(10)
\$4.32 mixture	$10 + x$	4.32	$4.32(10 + x)$

- The sum of the values before mixing equals the value after mixing.

Solution

$$4x + 6(10) = 4.32(10 + x)$$
$$4x + 60 = 43.2 + 4.32x$$
$$-0.32x + 60 = 43.2$$
$$-0.32x = -16.8$$
$$x = 52.5$$

52.5 oz of the \$4 silver alloy must be used.

YOU TRY IT • 1

A gardener has 20 lb of a lawn fertilizer that costs \$.80 per pound. How many pounds of a fertilizer that costs \$.55 per pound should be mixed with this 20 lb of lawn fertilizer to produce a mixture that costs \$.75 per pound?

Your strategy

Your solution

Solution on p. S14

OBJECTIVE B To solve uniform motion problems 

Recall from Section 5.1 that an object traveling at a constant speed in a straight line is in *uniform motion.* The solution of a uniform motion problem is based on the equation $rt = d$, where r is the rate of travel, t is the time spent traveling, and d is the distance traveled.

HOW TO • 2 A car leaves a town traveling at 40 mph. Two hours later, a second car leaves the same town, on the same road, traveling at 60 mph. In how many hours will the second car pass the first car?

Strategy for Solving a Uniform Motion Problem

1. For each object, write a numerical or variable expression for the rate, time, and distance. The results can be recorded in a table.

The first car traveled 2 h longer than the second car.

Unknown time for the second car: t
Time for the first car: $t + 2$

Take Note

Use the information given in the problem to fill in the rate and time columns of the table. Find the expression in the distance column by multiplying the two expressions you wrote in each row.

	Rate, r	·	*Time, t*	=	*Distance, d*
First car	40	·	$t + 2$	=	$40(t + 2)$
Second car	60	·	t	=	$60t$

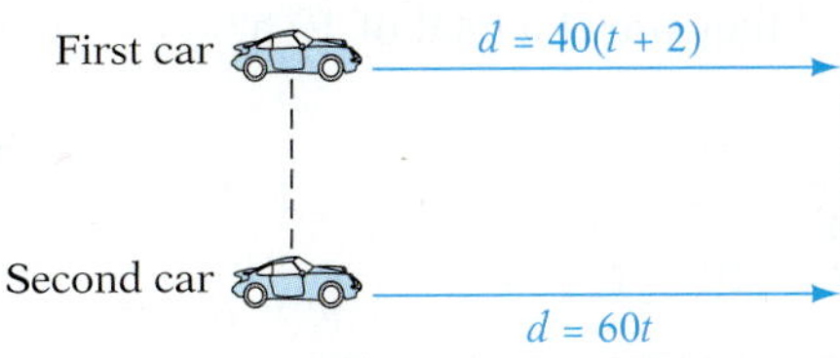

2. Determine how the distances traveled by the two objects are related. For example, the total distance traveled by both objects may be known, or it may be known that the two objects traveled the same distance.

The two cars travel the same distance.

$$40(t + 2) = 60t$$
$$40t + 80 = 60t$$
$$80 = 20t$$
$$4 = t$$

The second car will pass the first car in 4 h.

EXAMPLE • 2

Two cars, one traveling 10 mph faster than the other, start at the same time from the same point and travel in opposite directions. In 3 h, they are 300 mi apart. Find the rate of each car.

Strategy

- Rate of 1st car: r
  Rate of 2nd car: $r + 10$

	Rate	*Time*	*Distance*
1st car	r	3	$3r$
2nd car	$r + 10$	3	$3(r + 10)$

- The total distance traveled by the two cars is 300 mi.

Solution

$$3r + 3(r + 10) = 300$$
$$3r + 3r + 30 = 300$$
$$6r + 30 = 300$$
$$6r = 270$$
$$r = 45$$

$r + 10 = 45 + 10 = 55$

The first car is traveling 45 mph.
The second car is traveling 55 mph.

YOU TRY IT • 2

Two trains, one traveling at twice the speed of the other, start at the same time on parallel tracks from stations that are 288 mi apart and travel toward each other. In 3 h, the trains pass each other. Find the rate of each train.

Your strategy

Your solution

EXAMPLE • 3

How far can the members of a bicycling club ride out into the country at a speed of 12 mph and return over the same road at 8 mph if they travel a total of 10 h?

Strategy

- Time spent riding out: t
  Time spent riding back: $10 - t$

	Rate	*Time*	*Distance*
Out	12	t	$12t$
Back	8	$10 - t$	$8(10 - t)$

- The distance out equals the distance back.

Solution

$$12t = 8(10 - t)$$
$$12t = 80 - 8t$$
$$20t = 80$$
$$t = 4 \quad \text{(The time is 4 h.)}$$

The distance out $= 12t = 12(4) = 48$ mi.

The club can ride 48 mi into the country.

YOU TRY IT • 3

A pilot flew out to a parcel of land and back in 5 h. The rate out was 150 mph, and the rate returning was 100 mph. How far away was the parcel of land?

Your strategy

Your solution

Solutions on pp. S14–S15

5.5 EXERCISES

OBJECTIVE A To solve value mixture problems

1. A grocer mixes peanuts that cost \$3 per pound with almonds that cost \$7 per pound. Which of the following statements could be true about the cost per pound, C, of the mixture? There may be more than one correct answer.
(i) $C = \$10$ (ii) $C > \$7$ (iii) $C < \$7$
(iv) $C < \$3$ (v) $C > \$3$ (vi) $C = \$3$

2. An herbalist has 30 oz of herbs costing \$2 per ounce. How many ounces of herbs costing \$1 per ounce should be mixed with the 30 oz to produce a mixture costing \$1.60 per ounce?

3. The manager of a farmer's market has 500 lb of grain that costs \$1.20 per pound. How many pounds of meal costing \$.80 per pound should be mixed with the 500 lb of grain to produce a mixture that costs \$1.05 per pound?

4. Find the cost per pound of a meatloaf mixture made from 3 lb of ground beef costing \$1.99 per pound and 1 lb of ground turkey costing \$1.39 per pound.

5. Find the cost per ounce of a sunscreen made from 100 oz of a lotion that costs \$2.50 per ounce and 50 oz of a lotion that costs \$4.00 per ounce.

6. A snack food is made by mixing 5 lb of popcorn that costs \$.80 per pound with caramel that costs \$2.40 per pound. How much caramel is needed to make a mixture that costs \$1.40 per pound?

7. A wild birdseed mix is made by combining 100 lb of millet seed costing \$.60 per pound with sunflower seeds costing \$1.10 per pound. How many pounds of sunflower seeds are needed to make a mixture that costs \$.70 per pound?

8. Ten cups of a restaurant's house Italian dressing are made by blending olive oil costing \$1.50 per cup with vinegar that costs \$.25 per cup. How many cups of each are used if the cost of the blend is \$.50 per cup?

9. A high-protein diet supplement that costs \$6.75 per pound is mixed with a vitamin supplement that costs \$3.25 per pound. How many pounds of each should be used to make 5 lb of a mixture that costs \$4.65 per pound?

10. Find the cost per ounce of a mixture of 200 oz of a cologne that costs \$5.50 per ounce and 500 oz of a cologne that costs \$2.00 per ounce.

11. Find the cost per pound of a trail mix made from 40 lb of raisins that cost \$4.40 per pound and 100 lb of granola that costs \$2.30 per pound.

12. The manager of a specialty food store combined almonds that cost $4.50 per pound with walnuts that cost $2.50 per pound. How many pounds of each were used to make a 100-pound mixture that costs $3.24 per pound?

13. A goldsmith combined an alloy that cost $4.30 per ounce with an alloy that cost $1.80 per ounce. How many ounces of each were used to make a mixture of 200 oz costing $2.50 per ounce?

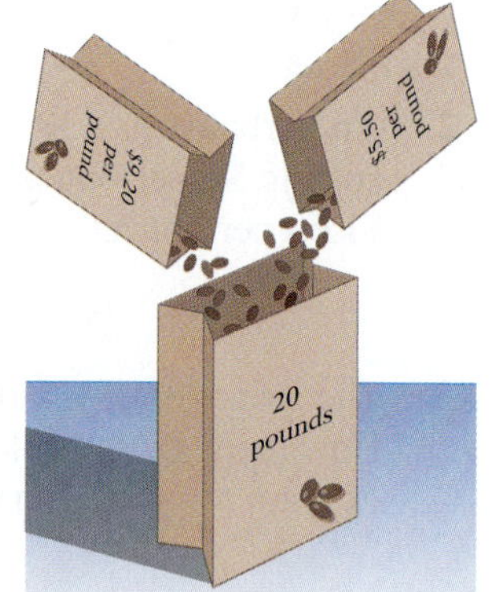

14. Find the cost per pound of sugar-coated breakfast cereal made from 40 lb of sugar that costs $1.00 per pound and 120 lb of corn flakes that cost $.60 per pound.

15. Find the cost per pound of a coffee mixture made from 8 lb of coffee that costs $9.20 per pound and 12 lb of coffee that costs $5.50 per pound.

16. Adult tickets for a play cost $6.00, and children's tickets cost $2.50. For one performance, 370 tickets were sold. Receipts for the performance totaled $1723. Find the number of adult tickets sold.

17. Tickets for a piano concert sold for $4.50 for each adult ticket. Student tickets sold for $2.00 each. The total receipts for 1720 tickets were $5980. Find the number of adult tickets sold.

18. **Tree Conservation** A town's parks department buys trees from the tree conservation program described in the news clipping at the right. The department spends $406 on 14 bundles of trees. How many bundles of seedlings and how many bundles of container-grown plants did the parks department buy?

In the News

Conservation Tree Planting Program Underway

The Kansas Forest Service is again offering its Conservation Tree Planting Program. Trees are sold in bundles of 25, in two sizes—seedlings cost $17 a bundle and larger container-grown plants cost $45 a bundle.

Source: Kansas Canopy

OBJECTIVE B To solve uniform motion problems

For Exercises 19 and 20, read the problem and state which of the following types of equations you would write to solve the problem.
(i) An equation showing two distances set equal to each other
(ii) An equation showing two distances added together and set equal to a total distance

19. Sam hiked up a mountain at a rate of 2.5 mph and returned along the same trail at a rate of 3 mph. His total hiking time was 11 h. How long was the hiking trail?

20. Sam hiked 16 mi. He hiked at one rate for the first 2 h of his hike, and then decreased his speed by 0.5 mph for the last 3 h of his hike. What was Sam's speed for the first 2 h?

21. Two small planes start from the same point and fly in opposite directions. The first plane is flying 25 mph slower than the second plane. In 2 h, the planes are 470 mi apart. Find the rate of each plane.

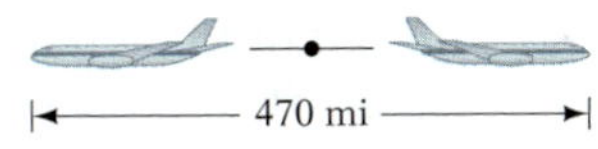

22. Two cyclists start from the same point and ride in opposite directions. One cyclist rides twice as fast as the other. In 3 h, they are 81 mi apart. Find the rate of each cyclist.

23. Two planes leave an airport at 8 A.M., one flying north at 480 km/h and the other flying south at 520 km/h. At what time will they be 3000 km apart?

24. A long-distance runner started on a course running at an average speed of 6 mph. One-half hour later, a second runner began the same course at an average speed of 7 mph. How long after the second runner started did the second runner overtake the first runner?

25. A motorboat leaves a harbor and travels at an average speed of 9 mph toward a small island. Two hours later a cabin cruiser leaves the same harbor and travels at an average speed of 18 mph toward the same island. In how many hours after the cabin cruiser leaves the harbor will it be alongside the motorboat?

26. A 555-mile, 5-hour plane trip was flown at two speeds. For the first part of the trip, the average speed was 105 mph. For the remainder of the trip, the average speed was 115 mph. How long did the plane fly at each speed?

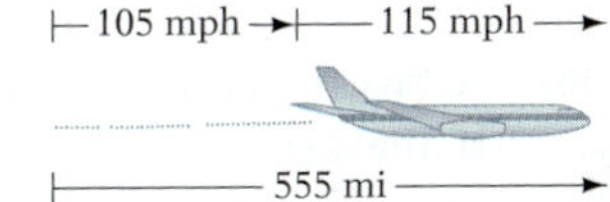

27. An executive drove from home at an average speed of 30 mph to an airport where a helicopter was waiting. The executive boarded the helicopter and flew to the corporate offices at an average speed of 60 mph. The entire distance was 150 mi. The entire trip took 3 h. Find the distance from the airport to the corporate offices.

28. After a sailboat had been on the water for 3 h, a change in the wind direction reduced the average speed of the boat by 5 mph. The entire distance sailed was 57 mi. The total time spent sailing was 6 h. How far did the sailboat travel in the first 3 h?

29. A car and a bus set out at 3 P.M. from the same point headed in the same direction. The average speed of the car is twice the average speed of the bus. In 2 h the car is 68 mi ahead of the bus. Find the rate of the car.

30. A passenger train leaves a train depot 2 h after a freight train leaves the same depot. The freight train is traveling 20 mph slower than the passenger train. Find the rate of each train if the passenger train overtakes the freight train in 3 h.

31. As part of flight training, a student pilot was required to fly to an airport and then return. The average speed on the way to the airport was 100 mph, and the average speed returning was 150 mph. Find the distance between the two airports if the total flying time was 5 h.

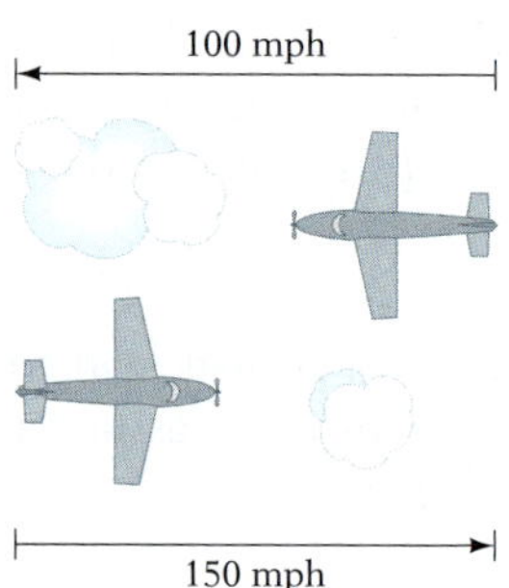

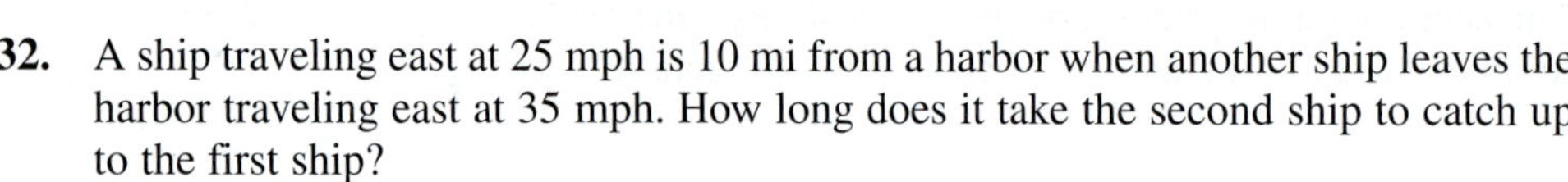

32. A ship traveling east at 25 mph is 10 mi from a harbor when another ship leaves the harbor traveling east at 35 mph. How long does it take the second ship to catch up to the first ship?

33. At 10 A.M. a plane leaves Boston, Massachusetts, for Seattle, Washington, a distance of 3000 mi. One hour later a plane leaves Seattle for Boston. Both planes are traveling at a speed of 500 mph. How many hours after the plane leaves Seattle will the planes pass each other?

34. At noon a train leaves Washington, D.C., headed for Charleston, South Carolina, a distance of 500 mi. The train travels at a speed of 60 mph. At 1 P.M. a second train leaves Charleston headed for Washington, D.C., traveling at 50 mph. How long after the train leaves Charleston will the two trains pass each other?

35. Two cyclists start at the same time from opposite ends of a course that is 51 mi long. One cyclist is riding at a rate of 16 mph, and the second cyclist is riding at a rate of 18 mph. How long after they begin will they meet?

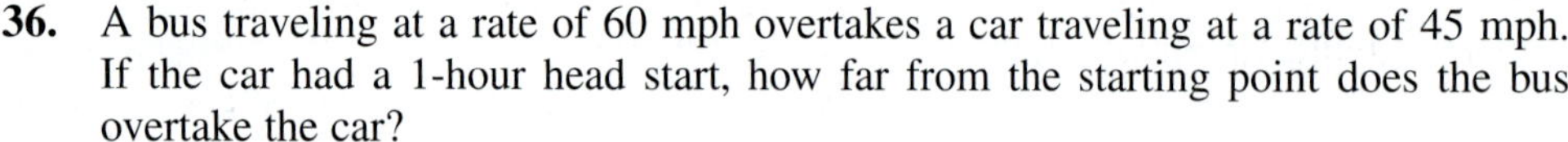

36. A bus traveling at a rate of 60 mph overtakes a car traveling at a rate of 45 mph. If the car had a 1-hour head start, how far from the starting point does the bus overtake the car?

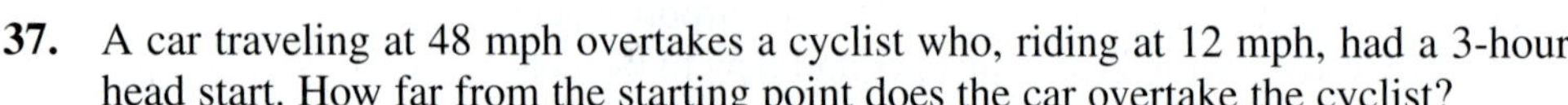

37. A car traveling at 48 mph overtakes a cyclist who, riding at 12 mph, had a 3-hour head start. How far from the starting point does the car overtake the cyclist?

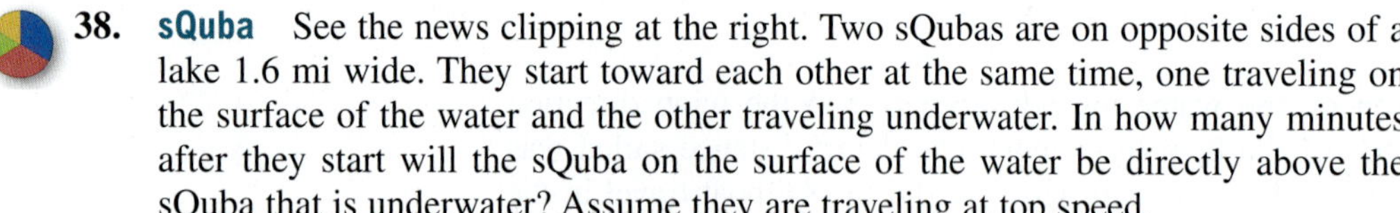

38. **sQuba** See the news clipping at the right. Two sQubas are on opposite sides of a lake 1.6 mi wide. They start toward each other at the same time, one traveling on the surface of the water and the other traveling underwater. In how many minutes after they start will the sQuba on the surface of the water be directly above the sQuba that is underwater? Assume they are traveling at top speed.

In the News

Underwater Driving—Not So Fast!

Swiss company Rinspeed, Inc., presented its new car, the sQuba, at the Geneva Auto Show. The sQuba can travel on land, on water, and underwater. With a new sQuba, you can expect top speeds of 77 mph when driving on land, 3 mph when driving on the surface of the water, and 1.8 mph when driving underwater!

Source: Seattle Times

Applying the Concepts

39. **Transportation** A bicyclist rides for 2 h at a speed of 10 mph and then returns at a speed of 20 mph. Find the cyclist's average speed for the trip.

40. **Travel** A car travels a 1-mile track at an average speed of 30 mph. At what average speed must the car travel the next mile so that the average speed for the 2 mi is 60 mph?

FOCUS ON PROBLEM SOLVING

Trial-and-Error Approach to Problem Solving

The questions below require an answer of always true, sometimes true, or never true. These problems are best solved by the trial-and-error method. The trial-and-error method of arriving at a solution to a problem involves repeated tests or experiments.

For example, consider the statement

> Both sides of an equation can be divided by the same number without changing the solution of the equation.

The solution of the equation $6x = 18$ is 3. If we divide both sides of the equation by 2, the result is $3x = 9$ and the solution is still 3. So the answer "never true" has been eliminated. We still need to determine whether there is a case for which the statement is not true. Is there a number that we could divide both sides of the equation by and the result would be an equation for which the solution is not 3?

If we divide both sides of the equation by 0, the result is $\frac{6x}{0} = \frac{18}{0}$; the solution of this equation is not 3 because the expressions on either side of the equals sign are undefined. Thus the statement is true for some numbers and not true for 0. The statement is sometimes true.

For Exercises 1 to 13, determine whether the statement is always true, sometimes true, or never true.

1. Both sides of an equation can be multiplied by the same number without changing the solution of the equation.
2. For an equation of the form $ax = b$, $a \neq 0$, multiplying both sides of the equation by the reciprocal of a will result in an equation of the form $x = constant$.
3. Adding -3 to each side of an equation yields the same result as subtracting 3 from each side of the equation.
4. An equation contains an equals sign.
5. The same variable term can be added to both sides of an equation without changing the solution of the equation.
6. An equation of the form $ax + b = c$ cannot be solved if a is a negative number.
7. The solution of the equation $\frac{x}{0} = 0$ is 0.
8. In solving an equation of the form $ax + b = cx + d$, subtracting cx from each side of the equation results in an equation with only one variable term in it.
9. If a rope 8 meters long is cut into two pieces and one of the pieces has a length of x meters, then the length of the other piece can be represented as $(x - 8)$ meters.
10. An even integer is a multiple of 2.

11. If the first of three consecutive odd integers is n, then the second and third consecutive odd integers are represented by $n + 1$ and $n + 3$.

12. If we combine an alloy that costs \$8 an ounce with an alloy that costs \$5 an ounce, the cost of the resulting mixture will be greater than \$8 an ounce.

13. If the speed of one train is 20 mph slower than that of a second train, then the speeds of the two trains can be represented as r and $20 - r$.

PROJECTS AND GROUP ACTIVITIES

Averages We often discuss temperature in terms of average high or average low temperature. Temperatures collected over a period of time are analyzed to determine, for example, the average high temperature for a given month in your city or state. The following activity is planned to help you better understand the concept of "average."

1. Choose two cities in the United States. We will refer to them as City X and City Y. Over an 8-day period, record the daily high temperature for each city.

2. Determine the average high temperature for City X for the 8-day period. (Add the eight numbers, and then divide the sum by 8.) Do not round your answer.

3. Subtract the average high temperature for City X from each of the eight daily high temperatures for City X. You should have a list of eight numbers; the list should include positive numbers, negative numbers, and possibly zero.

4. Find the sum of the list of eight differences recorded in Step 3.

5. Repeat Steps 2 through 4 for City Y.

6. Compare the two sums found in Steps 4 and 5 for City X and City Y.

7. If you were to conduct this activity again, what would you expect the outcome to be? Use the results to explain what an average high temperature is. In your own words, explain what "average" means.

CHAPTER 5

SUMMARY

KEY WORDS	EXAMPLES
An *equation* expresses the equality of two mathematical expressions. [5.1A, p. 282]	$3 + 2(4x - 5) = x + 4$ is an equation.

A *solution of an equation* is a number that, when substituted for the variable, results in a true equation. [5.1A, p. 282]	-2 is a solution of $2 - 3x = 8$ because $2 - 3(-2) = 8$ is a true equation.
To *solve an equation* means to find a solution of the equation. The goal is to rewrite the equation in the form *variable* = *constant*, because the constant is the solution. [5.1B, p. 283]	The equation $x = -3$ is in the form *variable* = *constant*. The constant, -3, is the solution of the equation.
Consecutive integers follow one another in order. [5.4A, p. 310]	5, 6, 7 are consecutive integers. $-9, -8, -7$ are consecutive integers.

ESSENTIAL RULES AND PROCEDURES	EXAMPLES
Addition Property of Equations [5.1B, p. 283] The same number can be added to each side of an equation without changing the solution of the equation.	If $a = b$, then $a + c = b + c$.
Multiplication Property of Equations [5.1C, p. 284] Each side of an equation can be multiplied by the same *nonzero* number without changing the solution of the equation.	If $a = b$ and $c \neq 0$, then $ac = bc$.
Consecutive Integers [5.4A, p. 310] $n, n + 1, n + 2, \ldots$	The sum of three consecutive integers is 33. $n + (n + 1) + (n + 2) = 33$
Consecutive Even or Consecutive Odd Integers $n, n + 2, n + 4$ [5.4A, p. 310]	The sum of three consecutive odd integers is 33. $n + (n + 2) + (n + 4) = 33$
Value Mixture Equation [5.5A, p. 317] Amount · Unit Cost = Value $AC = V$	An herbalist has 30 oz of herbs costing \$4 per ounce. How many ounces of herbs costing \$2 per ounce should be mixed with the 30 oz to produce a mixture costing \$3.20 per ounce? $30(4) + 2x = 3.20(30 + x)$
Uniform Motion Equation [5.1D, p. 286; 5.5B, p. 319] Distance = Rate · Time $d = rt$	A boat traveled from a harbor to an island at an average speed of 20 mph. The average speed on the return trip was 15 mph. The total trip took 3.5 h. How long did it take to travel to the island? $20t = 15(3.5 - t)$

CHAPTER 5

CONCEPT REVIEW

Test your knowledge of the concepts presented in this chapter. Answer each question. Then check your answers against the ones provided in the Answer Section.

1. What is the difference between an expression and an equation?

2. How do you know when a number is not a solution of an equation?

3. How is the Addition Property of Equations used to solve an equation?

4. How is the Multiplication Property of Equations used to solve an equation?

5. How do you check the solution of an equation?

6. How do you solve the equation $-14x = 28$?

7. What steps do you need to take to solve $\frac{1}{3}x - \frac{2}{9} = \frac{1}{3}$?

8. How do you solve an equation containing parentheses?

9. What formula is used to solve a uniform motion problem?

10. The solution of a value mixture problem is based on what equation?

11. What formula is used to solve a lever system problem?

12. What is the difference between consecutive integers and consecutive even integers?

CHAPTER 5

REVIEW EXERCISES

1. Solve: $x + 3 = 24$

2. Solve: $x + 5(3x - 20) = 10(x - 4)$

3. Solve: $5x - 6 = 29$

4. Is 3 a solution of $5x - 2 = 4x + 5$?

5. Solve: $\frac{3}{5}a = 12$

6. Solve: $6x + 3(2x - 1) = -27$

7. Solve: $x - 3 = -7$

8. Solve: $5x + 3 = 10x - 17$

9. Solve: $7 - [4 + 2(x - 3)] = 11(x + 2)$

10. Solve: $-6x + 16 = -2x$

11. Solve: $7 - 3x = 2 - 5x$

12. Solve: $-\frac{3}{8}x = -\frac{15}{32}$

13. Solve: $35 - 3x = 5$

14. Solve: $3x = 2(3x - 2)$

15. **Lever Systems** A lever is 12 ft long. At a distance of 2 ft from the fulcrum, a force of 120 lb is applied. How large a force must be applied to the other end so that the system will balance? Use the lever system equation $F_1x = F_2(d - x)$.

16. **Travel** A bus traveled on a level road for 2 h at an average speed that was 20 mph faster than its speed on a winding road. The time spent on the winding road was 3 h. Find the average speed on the winding road if the total trip was 200 mi.

17. **Integers** The difference between nine and twice a number is five. Find the number.

18. **Integers** The product of five and a number is fifty. Find the number.

19. **Juice Mixtures** A health food store combined cranberry juice that cost \$1.79 per quart with apple juice that cost \$1.19 per quart. How many quarts of each were used to make 10 qt of cranapple juice costing \$1.61 per quart?

20. **Integers** Four times the second of three consecutive integers equals the sum of the first and third integers. Find the integers.

21. **Integers** Translate "four less than the product of five and a number is sixteen" into an equation and solve.

Pete Seaward/Getty Images

22. **Buildings** The Empire State Building is 1472 ft tall. This is 654 ft less than twice the height of the Eiffel Tower. Find the height of the Eiffel Tower.

23. **Temperature** Find the Celsius temperature when the Fahrenheit temperature is 100°. Use the formula $F = \frac{9}{5}C + 32$, where F is the Fahrenheit temperature and C is the Celsius temperature. Round to the nearest tenth.

24. **Travel** A jet plane traveling at 600 mph overtakes a propeller-driven plane that had a 2-hour head start. The propeller-driven plane is traveling at 200 mph. How far from the starting point does the jet overtake the propeller-driven plane?

25. **Integers** The sum of two numbers is twenty-one. Three times the smaller number is two less than twice the larger number. Find the two numbers.

26. **Farming** A farmer harvested 28,336 bushels of corn. This amount represents an increase of 3036 bushels over last year's crop. How many bushels of corn did the farmer harvest last year?

CHAPTER 5

TEST

1. Solve: $3x - 2 = 5x + 8$

2. Solve: $x - 3 = -8$

3. Solve: $3x - 5 = -14$

4. Solve: $4 - 2(3 - 2x) = 2(5 - x)$

5. Is -2 a solution of $x^2 - 3x = 2x - 6$?

6. Solve: $7 - 4x = -13$

7. Solve: $5 = 3 - 4x$

8. Solve: $5x - 2(4x - 3) = 6x + 9$

9. Solve: $5x + 3 - 7x = 2x - 5$

10. Solve: $\frac{3}{4}x = -9$

11. Solve: $\frac{x}{5} - 12 = 7$

12. Solve: $8 - 3x = 2x - 8$

13. Solve: $y - 4y + 3 = 12$

14. Solve: $2x + 4(x - 3) = 5x - 1$

15. **Flour Mixtures** A baker wants to make a 15-pound blend of flour that costs \$.60 per pound. The blend is made using a rye flour that costs \$.70 per pound and a wheat flour that costs \$.40 per pound. How many pounds of each flour should be used?

16. **Manufacturing** A financial manager has determined that the cost per unit for a calculator is \$15 and that the fixed cost per month is \$2000. Find the number of calculators produced during a month in which the total cost was \$5000. Use the equation $T = U \cdot N + F$, where T is the total cost, U is the cost per unit, N is the number of units produced, and F is the fixed cost.

17. **Integers** Find three consecutive even integers whose sum is 36.

18. **Manufacturing** A clock manufacturer's fixed costs per month are \$5000. The unit cost for each clock is \$15. Find the number of clocks made during a month in which the total cost was \$65,000. Use the formula $T = U \cdot N + F$, where T is the total cost, U is the cost per unit, N is the number of units made, and F is the fixed costs.

19. **Integers** Translate "The difference between three times a number and fifteen is twenty-seven" into an equation and solve.

© Farmer/Fotolia

20. **Travel** A cross-country skier leaves a camp to explore a wilderness area. Two hours later a friend leaves the camp in a snowmobile, traveling 4 mph faster than the skier. This friend meets the skier 1 h later. Find the rate of the snowmobile.

21. **Manufacturing** A company makes 140 televisions per day. Three times the number of 15-inch TVs made equals 20 less than the number of 25-inch TVs made. Find the number of 25-inch TVs made each day.

22. **Integers** The sum of two numbers is eighteen. The difference between four times the smaller number and seven is equal to the sum of two times the larger number and five. Find the two numbers.

23. **Travel** As part of flight training, a student pilot was required to fly to an airport and then return. The average speed to the airport was 90 mph, and the average speed returning was 120 mph. Find the distance between the two airports if the total flying time was 7 h.

24. **Physics** Find the time required for a falling object to increase in velocity from 24 ft/s to 392 ft/s. Use the formula $V = V_0 + 32t$, where V is the final velocity of a falling object, V_0 is the starting velocity of the falling object, and t is the time for the object to fall.

25. **Chemistry** A chemist mixes 100 g of water at 80°C with 50 g of water at 20°C. To find the final temperature of the water after mixing, use the equation $m_1(T_1 - T) = m_2(T - T_2)$, where m_1 is the quantity of water at the hotter temperature, T_1 is the temperature of the hotter water, m_2 is the quantity of water at the cooler temperature, T_2 is the temperature of the cooler water, and T is the final temperature of the water after mixing.

CUMULATIVE REVIEW EXERCISES

1. Subtract: $-6 - (-20) - 8$

2. Multiply: $(-2)(-6)(-4)$

3. Subtract: $-\frac{5}{6} - \left(-\frac{7}{16}\right)$

4. Divide: $-2\frac{1}{3} \div 1\frac{1}{6}$

5. Simplify: $-4^2 \cdot \left(-\frac{3}{2}\right)^3$

6. Simplify: $25 - 3\frac{(5-2)^2}{2^3+1} - (-2)$

7. Evaluate $3(a - c) - 2ab$ when $a = 2$, $b = 3$, and $c = -4$.

8. Simplify: $3x - 8x + (-12x)$

9. Simplify: $2a - (-3b) - 7a - 5b$

10. Simplify: $(16x)\left(\frac{1}{8}\right)$

11. Simplify: $-4(-9y)$

12. Simplify: $-2(-x^2 - 3x + 2)$

13. Simplify: $-2(x - 3) + 2(4 - x)$

14. Simplify: $-3[2x - 4(x - 3)] + 2$

15. Is -3 a solution of $x^2 + 6x + 9 = x + 3$?

16. Is $\frac{1}{2}$ a solution of $3 - 8x = 12x - 2$?

17. Simplify: $\left(\frac{3}{8} - \frac{1}{4}\right) \div \frac{3}{4} + \frac{4}{9}$

18. Solve: $\frac{3}{5}x = -15$

19. Solve: $7x - 8 = -29$

20. Solve: $13 - 9x = -14$

21. Multiply: 9.67×0.0049

22. Find 6 less than 13.

23. Solve: $8x - 3(4x - 5) = -2x - 11$

24. Solve: $6 - 2(5x - 8) = 3x - 4$

25. Solve: $5x - 8 = 12x + 13$

26. Solve: $11 - 4x = 2x + 8$

27. **Chemistry** A chemist mixes 300 g of water at 75°C with 100 g of water at 15°C. To find the final temperature of the water after mixing, use the equation $m_1(T_1 - T) = m_2(T - T_2)$, where m_1 is the quantity of water at the hotter temperature, T_1 is the temperature of the hotter water, m_2 is the quantity of water at the cooler temperature, T_2 is the temperature of the cooler water, and T is the final temperature of the water after mixing.

28. **Integers** Translate "The difference between twelve and the product of five and a number is negative eighteen" into an equation and solve.

29. **Construction** The area of a cement foundation of a house is 2000 ft^2. This is 200 ft^2 more than three times the area of the garage. Find the area of the garage.

30. **Flour Mixtures** How many pounds of an oat flour that costs \$.80 per pound must be mixed with 40 lb of a wheat flour that costs \$.50 per pound to make a blend that costs \$.60 per pound?

31. **Integers** Translate "the sum of three times a number and four" into a mathematical expression.

32. **Integers** Three less than eight times a number is three more than five times the number. Find the number.

33. **Sprinting** A sprinter ran to the end of a track at an average rate of 8 m/s and then jogged back to the starting point at an average rate of 3 m/s. The sprinter took 55 s to run to the end of the track and jog back. Find the length of the track.

Proportion and Percent

OBJECTIVES

SECTION 6.1

A To write ratios and rates

SECTION 6.2

A To solve proportions

B To solve application problems

SECTION 6.3

A To write a percent as a fraction or a decimal

B To write a fraction or a decimal as a percent

SECTION 6.4

A To solve percent problems using the basic percent equation

B To solve percent problems using proportions

C To solve application problems

SECTION 6.5

A To solve problems involving simple interest

ARE YOU READY?

Take the Chapter 6 Prep Test to find out if you are ready to learn to:

- Write ratios and rates
- Solve proportions
- Convert fractions, decimals, and percents
- Solve percent problems using the basic percent equation
- Solve percent problems using proportions
- Solve problems involving simple interest

PREP TEST

Do these exercises to prepare for Chapter 6.

1. Simplify: $\frac{8}{10}$

2. Write as a decimal: $\frac{372}{15}$

3. Which is greater, 4×33 or 62×2?

4. Multiply: $19 \times \frac{1}{100}$

5. Multiply: 23×0.01

6. Multiply: 0.47×100

7. Multiply: $0.06 \times 47{,}500$

8. Divide: $60 \div 0.015$

9. Divide: $\frac{480}{0.06}$

10. Multiply $\frac{5}{8} \times 100$. Write the answer as a decimal.

11. Write $\frac{200}{3}$ as a mixed number.

12. Divide $28 \div 16$. Write the answer as a decimal.

SECTION

6.1 Ratios and Rates

OBJECTIVE A To write ratios and rates 

In previous work, we have used quantities with units, such as 12 ft, 3 h, 2¢, and 15 acres. In these examples, the units are feet, hours, cents, and acres.

Point of Interest

It is believed that billiards was invented in France during the reign of Louis XI (1423–1483). In the United States, the standard billiard table is 4 ft 6 in. by 9 ft. This is a ratio of 1:2. The same ratio holds for carom and snooker tables, which are 5 ft by 10 ft.

A **ratio** is the quotient or comparison of two quantities with the *same* unit. We can compare the measure of 3 ft to the measure of 8 ft by writing a quotient.

$$\frac{3 \text{ ft}}{8 \text{ ft}} = \frac{3}{8} \qquad 3 \text{ ft is } \frac{3}{8} \text{ of } 8 \text{ ft.}$$

A ratio can be written in three ways:

1. As a fraction $\frac{3}{8}$
2. As two numbers separated by a colon 3:8
3. As two numbers separated by the word *to* 3 to 8

The ratio of 15 mi to 45 mi is written as

$$\frac{15 \text{ mi}}{45 \text{ mi}} = \frac{15}{45} = \frac{1}{3} \text{ or } 1:3 \text{ or } 1 \text{ to } 3$$

A ratio is in **simplest form** when the two numbers do not have a common factor. The units are not written in a ratio.

A **rate** is the comparison of two quantities with *different* units.

A catering company prepares 9 gal of coffee for every 50 people at a reception. This rate is written

$$\frac{9 \text{ gal}}{50 \text{ people}}$$

You traveled 200 mi in 6 h. The rate is written

$$\frac{200 \text{ mi}}{6 \text{ h}} = \frac{100 \text{ mi}}{3 \text{ h}}$$

A rate is in **simplest form** when the numbers have no common factors. The units are written as part of the rate.

Many rates are written as unit rates. A **unit rate** is a rate in which the number in the denominator is 1. The word *per* generally indicates a unit rate. It means "for each" or "for every." For example,

23 miles per gallon • **The unit rate is $\frac{23 \text{ mi}}{1 \text{ gal}}$.**

65 miles per hour • **The unit rate is $\frac{65 \text{ mi}}{1 \text{ h}}$.**

\$4.78 per pound • **The unit rate is $\frac{\$4.78}{1 \text{ lb}}$.**

Unit rates make comparisons easier. For example, if you travel 37 mph and I travel 43 mph, we know that I am traveling faster than you are. It is more difficult to compare speeds if we are told that you are traveling $\frac{111 \text{ mi}}{3 \text{ h}}$ and I am traveling $\frac{172 \text{ mi}}{4 \text{ h}}$.

To find a unit rate, divide the number in the numerator of the rate by the number in the denominator of the rate. A unit rate is often written in decimal form.

HOW TO • 1 A student received $57 for working 6 h at the bookstore. Find the wage per hour (the unit rate).

$\frac{\$57}{6 \text{ h}}$ • **Write the rate as a fraction.**

$57 \div 6 = 9.5$ • **Divide the number in the numerator of the rate (57) by the number in the denominator (6).**

The unit rate is $\frac{\$9.50}{1 \text{ h}} = \$9.50/\text{h}$. This is read "$9.50 per hour."

EXAMPLE • 1

Write the comparison of 12 to 8 as a ratio in simplest form using a fraction, a colon, and the word *to*.

Solution

$\frac{12}{8} = \frac{3}{2}$

12:8 = 3:2

12 to 8 = 3 to 2

YOU TRY IT • 1

Write the comparison of 12 to 20 as a ratio in simplest form using a fraction, a colon, and the word *to*.

Your solution

EXAMPLE • 2

Write "12 hits in 26 times at bat" as a rate in simplest form.

Solution

$\frac{12 \text{ hits}}{26 \text{ at-bats}} = \frac{6 \text{ hits}}{13 \text{ at-bats}}$

YOU TRY IT • 2

Write "20 bags of grass seed for 8 acres" as a rate in simplest form.

Your solution

EXAMPLE • 3

Write "285 mi in 5 h" as a unit rate.

Solution

$\frac{285 \text{ mi}}{5 \text{ h}}$

$285 \div 5 = 57$

The unit rate is 57 mph.

YOU TRY IT • 3

Write "$8.96 for 3.5 lb" as a unit rate.

Your solution

Solutions on p. S15

6.1 EXERCISES

OBJECTIVE A To write ratios and rates

For Exercises 1 to 6, write the comparison as a ratio in simplest form using a fraction, a colon, and the word *to*.

1. 16 in. to 24 in.

2. 8 lb to 60 lb

3. 9 h to 24 h

4. \$55 to \$150

5. 9 ft to 2 ft

6. 50 min to 6 min

7. Physical Fitness The figure at the right shows the lung capacity of inactive versus athletic 45-year-olds. Write the comparison of the lung capacity of an inactive male to that of an athletic male as a ratio in simplest form using a fraction, a colon, and the word *to*.

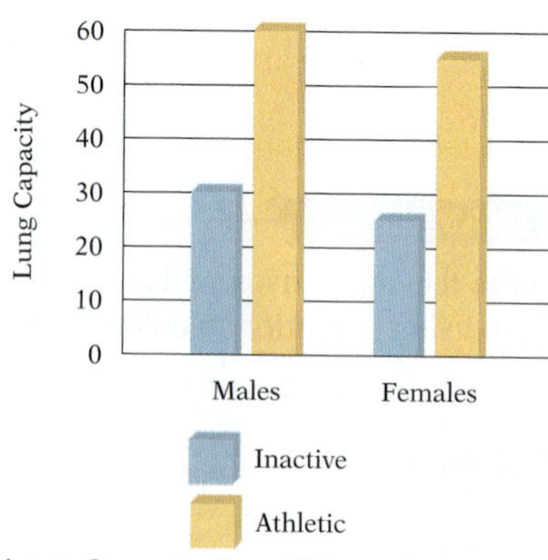

Lung Capacity (in milliliters of oxygen per kilogram of body weight per minute)

For Exercises 8 to 10, write as a ratio in simplest form using a fraction.

8. Sports A baseball player had 3 errors in 42 fielding attempts. What is the ratio of the number of times the player did not make an error to the total number of attempts?

9. Sports A basketball team won 18 games and lost 8 games during the season. What is the ratio of the number of games won to the total number of games?

10. Mechanics Find the ratio of two meshed gears if one gear has 24 teeth and the other gear has 36 teeth.

For Exercises 11 to 16, write as a rate in simplest form.

11. \$85 for 3 shirts

12. 150 mi in 6 h

13. \$76 for 8 h work

14. \$6.56 for 6 candy bars

15. 252 avocado trees on 6 acres

16. 9 children in 4 families

For Exercises 17 to 22, write as a unit rate.

17. \$460 earned for 40 h of work

18. \$38,700 earned in 12 months

19. 387.8 mi in 7 h

20. 364.8 mi on 9.5 gal of gas

21. \$19.08 for 4.5 lb

22. \$20.16 for 15 oz

23. **Sports** NCAA statistics show that for every 2800 college seniors playing college basketball, only 50 will play as rookies in the National Basketball Association. Write the ratio of the number of National Basketball Association rookies to the number of college seniors playing basketball.

24. **Energy** A transformer has 40 turns in the primary coil and 480 turns in the secondary coil. State the ratio of the number of turns in the primary coil to the number of turns in the secondary coil.

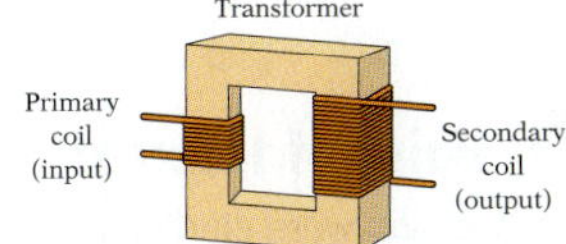

25. **Travel** An airplane flew 1155 mi in 2.5 h. Find the rate of travel.

26. **Facial Hair** Using the data in the news clipping at the right and the figure 50 million for the number of adult males in the United States, write the ratio of the number of men who participated in Movember to the number of adult males in the U.S. Write the ratio as a fraction in simplest form.

In the News

Grow a Mustache, Save a Life

Last fall, in an effort to raise money for the Prostate Cancer Foundation, approximately 2000 men participated in a month-long mustache-growing competition. The event was dubbed Movember.

Source: Time, February 18, 2008

27. **Investments** An investor purchased 100 shares of stock for \$2500. One year later the investor sold the stock for \$3200. What was the investor's profit per share?

For Exercises 28 to 30, complete the unit rate.

28. 5 miles in ___ hour

29. 15 feet in ___ second

30. 5 grams of fat in ___ serving

Applying the Concepts

31. **Compensation** You have a choice of receiving a wage of \$34,000 per year, \$2840 per month, \$650 per week, or \$18 per hour. Which pay choice would you take? Assume a 40-hour week with 52 weeks per year.

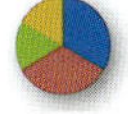

32. **Social Security** According to the Social Security Administration, the number of workers per retiree is expected to be as given in the table below.

Year	*2010*	*2020*	*2030*	*2040*
Number of workers per retiree	3.1	2.5	2.1	2.0

Why is the shrinking number of workers per retiree important to the Social Security Administration?

SECTION

6.2 Proportion

OBJECTIVE A To solve proportions 

A **proportion** is the equality of two ratios or rates.

The equality $\frac{250\text{ mi}}{5\text{ h}} = \frac{50\text{ mi}}{1\text{ h}}$ is a proportion.

Point of Interest

Proportions were studied by the earliest mathematicians. Clay tablets uncovered by archeologists show evidence of proportions in Egyptian and Babylonian cultures dating from 1800 B.C.E.

Definition of Proportion

If $\frac{a}{b}$ and $\frac{c}{d}$ are equal ratios or rates, then $\frac{a}{b} = \frac{c}{d}$ is a proportion.

Each of the four numbers in a proportion is called a **term.** Each term is numbered according to the following diagram.

first term ⟵ $\frac{a}{b} = \frac{c}{d}$ ⟶ third term
second term ⟵ ⟶ fourth term

Tips for Success

As you know, often in mathematics you learn one skill in order to perform another. This is true of this objective. You are learning to solve proportions. You will use this skill to solve application problems in the next objective and then to solve percent problems in Section 6.4.

The first and fourth terms of the proportion are called the **extremes,** and the second and third terms are called the **means.**

If we multiply the proportion by the least common multiple of the denominators, we obtain the following result:

$$\frac{a}{b} = \frac{c}{d}$$

$$bd\left(\frac{a}{b}\right) = bd\left(\frac{c}{d}\right)$$

$$ad = bc$$

***ad* is the product of the extremes.**
***bc* is the product of the means.**

In any true proportion, **the product of the means equals the product of the extremes.** This is sometimes phrased as "the cross products are equal."

In the true proportion $\frac{3}{4} = \frac{9}{12}$, the cross products are equal.

$\frac{3}{4} \times \frac{9}{12}$ ⟶ $4 \cdot 9 = 36$ ⟵ Product of the means
⟶ $3 \cdot 12 = 36$ ⟵ Product of the extremes

HOW TO • 1 Determine whether the proportion $\frac{47\text{ mi}}{2\text{ gal}} = \frac{304\text{ mi}}{13\text{ gal}}$ is a true proportion.

The product of the means:
$2 \cdot 304 = 608$

The product of the extremes:
$47 \cdot 13 = 611$

The proportion is not true because $608 \neq 611$.

When three terms of a proportion are given, the fourth term can be found. To solve a proportion for an unknown term, use the fact that the product of the means equals the product of the extremes.

Integrating Technology

To use a calculator to solve the proportion at the right, multiply the second and third terms and divide by the fourth term. Enter

5 **x** 9 **÷** 16 **=**

The display reads 2.8125.

HOW TO • 2 Solve: $\frac{n}{5} = \frac{9}{16}$

$\frac{n}{5} = \frac{9}{16}$ • Find the number (n) that will make the proportion true.

$5 \cdot 9 = n \cdot 16$ • The product of the means equals the product of the extremes.

$45 = 16n$ • Solve for n.

$\frac{45}{16} = \frac{16n}{16}$

$2.8125 = n$

EXAMPLE • 1

Determine whether $\frac{15}{3} = \frac{90}{18}$ is a true proportion.

Solution

$\frac{15}{3} \times \frac{90}{18}$ ⟶ $3 \cdot 90 = 270$; ⟶ $15 \cdot 18 = 270$

The product of the means equals the product of the extremes.

The proportion is true.

YOU TRY IT • 1

Is $\frac{50 \text{ mi}}{3 \text{ gal}} = \frac{250 \text{ mi}}{12 \text{ gal}}$ a true proportion?

Your solution

EXAMPLE • 2

Solve: $\frac{5}{9} = \frac{x}{45}$

Solution

$\frac{5}{9} = \frac{x}{45}$

$9 \cdot x = 5 \cdot 45$ • The cross products are equal.

$9x = 225$

$\frac{9x}{9} = \frac{225}{9}$

$x = 25$

YOU TRY IT • 2

Solve: $\frac{7}{12} = \frac{42}{x}$

Your solution

EXAMPLE • 3

Solve $\frac{6}{n} = \frac{45}{124}$. Round to the nearest tenth.

Solution

$\frac{6}{n} = \frac{45}{124}$

$n \cdot 45 = 6 \cdot 124$ • The cross products are equal.

$45n = 744$

$\frac{45n}{45} = \frac{744}{45}$

$n \approx 16.5$

YOU TRY IT • 3

Solve $\frac{5}{n} = \frac{3}{322}$. Round to the nearest hundredth.

Your solution

Solutions on p. S15

EXAMPLE • 4

Solve: $\frac{x+2}{3} = \frac{7}{8}$

Solution

$$\frac{x+2}{3} = \frac{7}{8}$$
$$3 \cdot 7 = (x+2)8$$ • The cross products are equal.
$$21 = 8x + 16$$
$$5 = 8x$$
$$0.625 = x$$

YOU TRY IT • 4

Solve: $\frac{4}{5} = \frac{3}{x-3}$

Your solution

Solution on p. S15

OBJECTIVE B To solve application problems 

Proportions are useful in many types of application problems. In recipes, proportions are used when a larger batch of ingredients is used than the recipe calls for. In mixing cement, the amounts of cement, sand, and rock are mixed in the same ratio. A map is drawn on a proportional basis, such as 1 in. representing 50 mi.

In setting up a proportion, keep the same units in the numerators and the same units in the denominators. For example, if *feet* is in the numerator on one side of the proportion, then *feet* must be in the numerator on the other side of the proportion.

 **Take Note**

It is also correct to write the proportion with the costs in the numerators and the number of tires in the denominators: $\frac{\$162.50}{2 \text{ tires}} = \frac{c}{5 \text{ tires}}$. The solution will be the same.

HOW TO • 3 A customer sees an ad in a newspaper advertising 2 tires for \$162.50. The customer wants to buy 5 tires and use one for the spare. How much will the 5 tires cost?

$$\frac{2 \text{ tires}}{\$162.50} = \frac{5 \text{ tires}}{c}$$ • Write a proportion. Let c = the cost of the 5 tires.
$$\$162.50 \cdot 5 = 2 \cdot c$$ • The cross products are equal.
$$812.50 = 2c$$
$$\frac{812.50}{2} = \frac{2c}{2}$$
$$406.25 = c$$

The 5 tires will cost \$406.25.

EXAMPLE • 5

During a Friday, the ratio of stocks declining in price to those advancing was 5 to 3. If 450,000 shares advanced, how many shares declined on that day?

Strategy

To find the number of shares that declined in price, write and solve a proportion using n to represent the number of shares that declined in price.

Solution

$$\frac{5 \text{ (declining)}}{3 \text{ (advancing)}} = \frac{n \text{ shares declining}}{450{,}000 \text{ shares advancing}}$$
$$3n = 5 \cdot 450{,}000$$
$$3n = 2{,}250{,}000$$
$$\frac{3n}{3} = \frac{2{,}250{,}000}{3}$$
$$n = 750{,}000$$

750,000 shares declined in price.

YOU TRY IT • 5

An automobile can travel 396 mi on 11 gal of gas. At the same rate, how many gallons of gas would be necessary to travel 832 mi? Round to the nearest tenth.

Your strategy

Your solution

EXAMPLE • 6

From previous experience, a manufacturer knows that in an average production run of 5000 calculators, 40 will be defective. What number of defective calculators can be expected from a run of 45,000 calculators?

Strategy

To find the number of defective calculators, write and solve a proportion using n to represent the number of defective calculators.

Solution

$$\frac{40 \text{ defective calculators}}{5000 \text{ calculators}} = \frac{n \text{ defective calculators}}{45{,}000 \text{ calculators}}$$
$$5000 \cdot n = 40 \cdot 45{,}000$$
$$5000n = 1{,}800{,}000$$
$$\frac{5000n}{5000} = \frac{1{,}800{,}000}{5000}$$
$$n = 360$$

The manufacturer can expect 360 defective calculators.

YOU TRY IT • 6

An automobile recall was based on tests that showed 15 defective transmissions in 1200 cars. At this rate, how many defective transmissions will be found in 120,000 cars?

Your strategy

Your solution

Solutions on p. S15

6.2 EXERCISES

OBJECTIVE A To solve proportions

For Exercises 1 to 12, determine whether the proportion is true or not true.

1. $\frac{27}{8} = \frac{9}{4}$

2. $\frac{3}{18} = \frac{4}{19}$

3. $\frac{45}{135} = \frac{3}{9}$

4. $\frac{3}{4} = \frac{54}{72}$

5. $\frac{16}{3} = \frac{48}{9}$

6. $\frac{15}{5} = \frac{3}{1}$

7. $\frac{6\text{ min}}{5\text{ cents}} = \frac{30\text{ min}}{25\text{ cents}}$

8. $\frac{7\text{ tiles}}{4\text{ ft}} = \frac{42\text{ tiles}}{20\text{ ft}}$

9. $\frac{15\text{ ft}}{3\text{ yd}} = \frac{90\text{ ft}}{18\text{ yd}}$

10. $\frac{\$65}{5\text{ days}} = \frac{\$26}{2\text{ days}}$

11. $\frac{1\text{ gal}}{4\text{ qt}} = \frac{7\text{ gal}}{28\text{ qt}}$

12. $\frac{300\text{ ft}}{4\text{ rolls}} = \frac{450\text{ ft}}{7\text{ rolls}}$

For Exercises 13 to 48, solve. Round to the nearest hundredth.

13. $\frac{2}{3} = \frac{n}{15}$

14. $\frac{7}{15} = \frac{n}{15}$

15. $\frac{n}{5} = \frac{12}{25}$

16. $\frac{n}{8} = \frac{7}{8}$

17. $\frac{3}{8} = \frac{n}{12}$

18. $\frac{5}{8} = \frac{40}{n}$

19. $\frac{3}{n} = \frac{7}{40}$

20. $\frac{7}{12} = \frac{25}{n}$

21. $\frac{16}{n} = \frac{25}{40}$

22. $\frac{15}{45} = \frac{72}{n}$

23. $\frac{120}{n} = \frac{144}{25}$

24. $\frac{65}{20} = \frac{14}{n}$

25. $\frac{0.5}{2.3} = \frac{n}{20}$

26. $\frac{1.2}{2.8} = \frac{n}{32}$

27. $\frac{0.7}{1.2} = \frac{6.4}{n}$

28. $\frac{2.5}{0.6} = \frac{165}{n}$

29. $\frac{x}{6.25} = \frac{16}{87}$

30. $\frac{x}{2.54} = \frac{132}{640}$

31. $\frac{1.2}{0.44} = \frac{y}{14.2}$

32. $\frac{12.5}{y} = \frac{102}{55}$

33. $\frac{n + 2}{5} = \frac{1}{2}$

34. $\frac{5 + n}{8} = \frac{3}{4}$

35. $\frac{4}{3} = \frac{n - 2}{6}$

36. $\frac{3}{5} = \frac{n - 7}{8}$

37. $\frac{2}{n+3} = \frac{7}{12}$

38. $\frac{5}{n+1} = \frac{7}{3}$

39. $\frac{7}{10} = \frac{3+n}{2}$

40. $\frac{3}{2} = \frac{5+n}{4}$

41. $\frac{x-4}{3} = \frac{3}{4}$

42. $\frac{x-1}{8} = \frac{5}{2}$

43. $\frac{6}{1} = \frac{x-2}{5}$

44. $\frac{7}{3} = \frac{x-4}{8}$

45. $\frac{5}{8} = \frac{2}{x-3}$

46. $\frac{5}{2} = \frac{1}{x-6}$

47. $\frac{3}{x-4} = \frac{5}{3}$

48. $\frac{8}{x-6} = \frac{5}{4}$

49. Suppose that in a true proportion you switch the numerator of the first fraction with the denominator of the second fraction. Must the result be another true proportion?

50. Write a true proportion in which the cross products are equal to 36.

OBJECTIVE B **To solve application problems**

51. Jesse walked 3 mi in 40 min. Let n be the number of miles Jesse can walk in 60 min at the same rate. To determine how many miles Jesse can walk in 60 min, a student used the proportion $\frac{40}{3} = \frac{60}{n}$. Is this a valid proportion to use in solving this problem?

For Exercises 52 to 73, solve. Round to the nearest hundredth.

52. **Nutrition** A 6-ounce package of Puffed Wheat contains 600 calories. How many calories are in a 0.5-ounce serving of the cereal?

53. **Health** Using the data at the right and a figure of 300 million for the number of Americans, determine the number of morbidly obese Americans.

54. **Fuel Efficiency** A car travels 70.5 mi on 3 gal of gas. Find the distance the car can travel on 14 gal of gas.

In the News

Number of Obese Americans Increasing

In the past 20 years, the number of obese Americans (those at least 30 pounds overweight) has doubled. The number of morbidly obese (those at least 100 pounds overweight) has quadrupled to 1 in 50.

Source: Time, July 9, 2006

55. **Landscaping** Ron Stokes uses 2 lb of fertilizer for every 100 ft^2 of lawn for landscape maintenance. At this rate, how many pounds of fertilizer did he use on a lawn that measures 3500 ft^2?

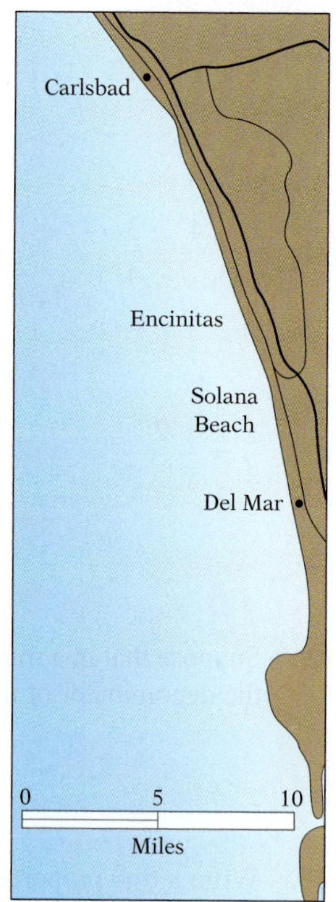

56. **Gardening** A nursery prepares a liquid plant food by adding 1 gal of water for each 2 oz of plant food. At this rate, how many gallons of water are required for 25 oz of plant food?

57. **Masonry** A brick wall 20 ft in length contains 1040 bricks. At the same rate, how many bricks would it take to build a wall 48 ft in length?

58. **Cartography** The scale on the map at the right is "1.25 inches equals 10 miles." Find the distance between Carlsbad and Del Mar, which are 2 in. apart on the map.

59. **Architecture** The scale on the plans for a new house is "1 inch equals 3 feet." Find the width and the length of a room that measures 5 in. by 8 in. on the drawing.

60. **Medicine** The dosage for a medication is $\frac{1}{3}$ oz for every 40 lb of body weight. At this rate, how many ounces of medication should a physician prescribe for a patient who weighs 150 lb? Write the answer as a decimal.

61. **Banking** A bank requires a monthly payment of \$33.45 on a \$2500 loan. At the same rate, find the monthly payment on a \$10,000 loan.

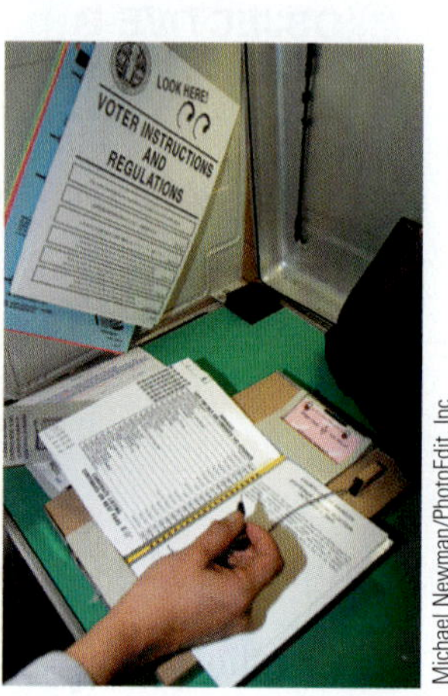

Michael Newman/PhotoEdit, Inc.

62. **Elections** A pre-election survey showed that 2 out of every 3 eligible voters would cast ballots in the county election. At this rate, how many people in a county of 240,000 eligible voters would vote in the election?

63. **Interior Design** A paint manufacturer suggests using 1 gal of paint for every 400 ft^2 of wall. At this rate, how many gallons of paint would be required for a room that has 1400 ft^2 of wall?

64. **Insurance** A 60-year-old male can obtain \$10,000 of life insurance for \$35.35 per month. At this rate, what is the monthly cost for \$50,000 of life insurance?

65. **Manufacturing** Suppose a computer chip manufacturer knows from experience that in an average production run of 2000 circuit boards, 60 will be defective. How many defective circuit boards can be expected in a run of 25,000 circuit boards?

66. **Food Waste** At the rate given in the news clipping, find the cost of food wasted yearly by **a.** the average family of three and **b.** the average family of five.

> **In the News**
>
> **How Much Food Do You Waste?**
>
> In the United States, the estimated cost of food wasted each year by the average family of four is \$590.
>
> *Source:* University of Arizona

67. **Investments** Carlos Capasso owns 50 shares of Texas Utilities that pay dividends of \$153. At this rate, what dividend would Carlos receive after buying 300 additional shares of Texas Utilities?

68. **Investments** You own 240 shares of stock in a computer company. The company declares a stock split of 5 shares for every 3 owned. How many shares of stock will you own after the stock split?

© Neil Armstrong/Corbis

69. **Physics** The ratio of weight on the moon to weight on Earth is 1:6. If a bowling ball weighs 16 lb on Earth, what would it weigh on the moon?

70. **Automobiles** When engineers designed a new car, they first built a model of the car. The ratio of the size of a part on the model to the actual size of the part is 2:5. If a door is 1.3 ft long on the model, what is the length of the door on the car?

71. **Energy** A slow-burning candle will burn 1.5 in. in 40 min. How many inches of the candle will burn in 4 h?

72. **Mixtures** A saltwater solution is made by dissolving $\frac{2}{3}$ lb of salt in 5 gal of water. At this rate, how many pounds of salt are required for 12 gal of water?

73. **Compensation** A management consulting firm recommends that the ratio of midmanagement salaries to junior management salaries be 7:5. Using this recommendation, find the yearly midmanagement salary when the junior management salary is \$90,000.

Applying the Concepts

74. **Publishing** In the first quarter of 2008, *USA Today* reported that Eckhart Tolle's *A New Earth* outsold John Grisham's *The Appeal* by 3.7 copies to 1. Explain how a proportion can be used to determine the number of copies of *A New Earth* sold given the number of copies of *The Appeal* sold.

75. **Elections** A survey of voters in a city claimed that 2 people of every 5 who voted cast a ballot in favor of city amendment A and that 3 people of every 4 who voted cast a ballot against amendment A. Is this possible? Explain your answer.

76. Write a word problem that requires solving a proportion to find the answer.

SECTION

6.3 Percent

OBJECTIVE A To write a percent as a fraction or a decimal 

Percent means "parts of 100." In the figure at the right, there are 100 parts. Because 13 of the 100 parts are shaded, 13% of the figure is shaded. The symbol % is the **percent sign.**

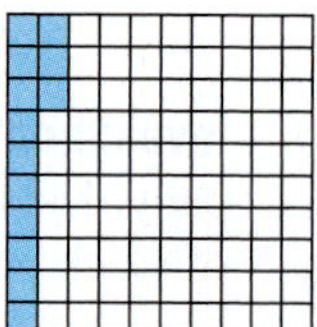

In most applied problems involving percents, it is necessary either to rewrite a percent as a fraction or a decimal or to rewrite a fraction or a decimal as a percent.

To write a percent as a fraction, remove the percent sign and multiply by $\frac{1}{100}$.

$$13\% = 13 \times \frac{1}{100} = \frac{13}{100}$$

To write a percent as a decimal, remove the percent sign and multiply by 0.01.

$$13\% = 13 \times 0.01 = 0.13$$

Move the decimal point two places to the left. Then remove the percent sign.

Take Note

Recall that division is defined as multiplication by the reciprocal. Therefore, multiplying by $\frac{1}{100}$ is equivalent to dividing by 100.

EXAMPLE 1

a. Write 120% as a fraction.
b. Write 120% as a decimal.

Solution **a.** $120\% = 120 \times \frac{1}{100} = \frac{120}{100}$

$= 1\frac{1}{5}$

b. $120\% = 120 \times 0.01 = 1.2$

Note that percents larger than 100 are greater than 1.

YOU TRY IT 1

a. Write 125% as a fraction.
b. Write 125% as a decimal.

Your solution

EXAMPLE 2

Write $16\frac{2}{3}\%$ as a fraction.

Solution $16\frac{2}{3}\% = 16\frac{2}{3} \times \frac{1}{100}$

$= \frac{50}{3} \times \frac{1}{100} = \frac{50}{300} = \frac{1}{6}$

YOU TRY IT 2

Write $33\frac{1}{3}\%$ as a fraction.

Your solution

EXAMPLE 3

Write 0.5% as a decimal.

Solution $0.5\% = 0.5 \times 0.01 = 0.005$

YOU TRY IT 3

Write 0.25% as a decimal.

Your solution

Solutions on p. S15

OBJECTIVE B To write a fraction or a decimal as a percent 

A fraction or a decimal can be written as a percent by multiplying by 100%.

HOW TO • 1 Write $\frac{3}{8}$ as a percent.

$$\frac{3}{8} = \frac{3}{8} \times 100\% = \frac{3}{8} \times \frac{100}{1}\% = \frac{300}{8}\% = 37\frac{1}{2}\% \text{ or } 37.5\%$$

HOW TO • 2 Write 0.37 as a percent.

$$0.37 = 0.37 \times 100\% = 37\%$$

Move the decimal point two places to the right. Then write the percent sign.

EXAMPLE • 4

Write 0.015, 2.15, and $0.33\frac{1}{3}$ as percents.

Solution

$$0.015 = 0.015 \times 100\% = 1.5\%$$

$$2.15 = 2.15 \times 100\% = 215\%$$

$$0.33\frac{1}{3} = 0.33\frac{1}{3} \times 100\% = 33\frac{1}{3}\%$$

YOU TRY IT • 4

Write 0.048, 3.67, and $0.62\frac{1}{2}$ as percents.

Your solution

EXAMPLE • 5

Write $\frac{2}{3}$ as a percent.
Write the remainder in fractional form.

Solution

$$\frac{2}{3} = \frac{2}{3} \times 100\% = \frac{200}{3}\% = 66\frac{2}{3}\%$$

YOU TRY IT • 5

Write $\frac{5}{6}$ as a percent.
Write the remainder in fractional form.

Your solution

EXAMPLE • 6

Write $2\frac{2}{7}$ as a percent.
Round to the nearest tenth.

Solution

$$2\frac{2}{7} = \frac{16}{7} = \frac{16}{7} \times 100\% = \frac{1600}{7}\% \approx 228.6\%$$

YOU TRY IT • 6

Write $1\frac{4}{9}$ as a percent.
Round to the nearest tenth.

Your solution

Solutions on p. S16

6.3 EXERCISES

OBJECTIVE A To write a percent as a fraction or a decimal

For Exercises 1 to 16, write as a fraction and as a decimal.

1. 25%
2. 40%
3. 130%
4. 150%
5. 100%
6. 87%
7. 73%
8. 45%
9. 383%
10. 425%
11. 70%
12. 55%
13. 88%
14. 64%
15. 32%
16. 18%

For Exercises 17 to 28, write as a fraction.

17. $66\frac{2}{3}\%$
18. $12\frac{1}{2}\%$
19. $83\frac{1}{3}\%$
20. $3\frac{1}{8}\%$
21. $11\frac{1}{9}\%$
22. $\frac{3}{8}\%$
23. $45\frac{5}{11}\%$
24. $15\frac{3}{8}\%$
25. $4\frac{2}{7}\%$
26. $5\frac{3}{4}\%$
27. $6\frac{2}{3}\%$
28. $8\frac{2}{3}\%$

For Exercises 29 to 40, write as a decimal.

29. 6.5%
30. 9.4%
31. 12.3%
32. 16.7%
33. 0.55%
34. 0.45%
35. 8.25%
36. 6.75%
37. 5.05%
38. 3.08%
39. 2%
40. 7%

41. When a certain percent is written as a fraction, the result is an improper fraction. Is the percent less than, equal to, or greater than 100%?

OBJECTIVE B To write a fraction or a decimal as a percent

For Exercises 42 to 53, write as a percent.

42. 0.16
43. 0.73
44. 0.05
45. 0.01
46. 1.07
47. 2.94
48. 0.004
49. 0.006
50. 1.012
51. 3.106
52. 0.8
53. 0.7

For Exercises 54 to 65, write as a percent. If necessary, round to the nearest tenth of a percent.

54. $\frac{27}{50}$ **55.** $\frac{37}{100}$ **56.** $\frac{1}{3}$ **57.** $\frac{2}{5}$ **58.** $\frac{5}{8}$ **59.** $\frac{1}{8}$

60. $\frac{1}{6}$ **61.** $1\frac{1}{2}$ **62.** $\frac{7}{40}$ **63.** $1\frac{2}{3}$ **64.** $1\frac{7}{9}$ **65.** $\frac{7}{8}$

For Exercises 66 to 73, write as a percent. Write the remainder in fractional form.

66. $\frac{15}{50}$ **67.** $\frac{12}{25}$ **68.** $\frac{7}{30}$ **69.** $\frac{1}{3}$

70. $2\frac{3}{8}$ **71.** $1\frac{2}{3}$ **72.** $2\frac{1}{6}$ **73.** $\frac{7}{8}$

74. Does a mixed number represent a percent greater than 100% or less than 100%?

75. A decimal number less than 0 has zeros in the tenths and hundredths places. Does the decimal represent a percent greater than 1% or less than 1%?

76. Write the part of the square that is shaded as a fraction, as a decimal, and as a percent. Write the part of the square that is not shaded as a fraction, as a decimal, and as a percent.

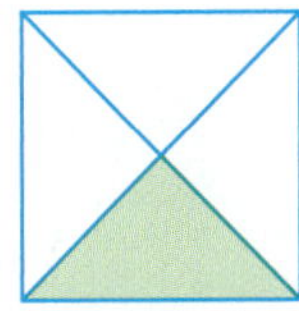

Applying the Concepts

77. The Food Industry In a survey conducted by Opinion Research Corp. for Lloyd's Barbeque Co., people were asked to name their favorite barbeque side dishes. 38% named corn on the cob, 35% named cole slaw, 11% named corn bread, and 10% named fries. What percent of those surveyed named something other than corn on the cob, cole slaw, corn bread, or fries?

David Chasey/Photodisc/Getty Images

78. Consumerism A sale on computers advertised $\frac{1}{3}$ off the regular price. What percent of the regular price does this represent?

79. Consumerism A suit was priced at 50% off the regular price. What fraction of the regular price does this represent?

80. Elections If $\frac{2}{5}$ of the population voted in an election, what percent of the population did not vote?

SECTION

6.4 The Basic Percent Equation

OBJECTIVE A To solve percent problems using the basic percent equation

A real estate broker receives a payment that is 4% of a \$285,000 sale. To find the amount the broker receives requires answering the question "4% of \$285,000 is what?"

This sentence can be written using mathematical symbols and then solved for the unknown number.

4%	of	\$285,000	is	what?
↓	↓	↓	↓	↓
Percent 4%	$\times$	base 285,000	$=$	amount n
0.04	$\times$	285,000	$=$	n
		11,400	$=$	n

of is written as $\times$ (times)
is is written as $=$ (equals)
what is written as n (the unknown number)

Note that the percent is written as a decimal.

The broker receives a payment of \$11,400.

The solution was found by solving the **basic percent equation** for amount.

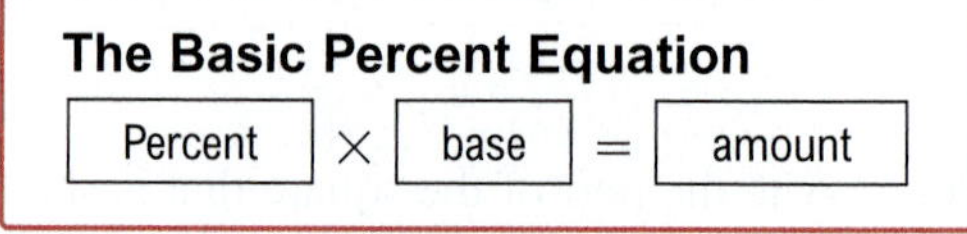

The Basic Percent Equation

Percent $\times$ base $=$ amount

Integrating Technology

The percent key **%** on a scientific calculator moves the decimal point to the left two places when pressed after a multiplication or division computation. For the example at the right, enter

800 **x** 2 **.** 5 **%** **=**

The display reads 20.

HOW TO 1 Find 2.5% of 800.

$$\text{Percent} \cdot \text{base} = \text{amount}$$
$$0.025 \cdot 800 = n$$
$$20 = n$$

• **Use the basic percent equation. Percent = 2.5% = 0.025, base = 800, amount = n**

2.5% of 800 is 20.

A recent promotional game at a grocery store listed the probability of winning a prize as "1 chance in 2." A percent can be used to describe the chance of winning. This requires answering the question "What percent of 2 is 1?"

The chance of winning can be found by solving the basic percent equation for *percent.*

What percent	of	2	is	1?
↓	↓	↓	↓	↓
Percent n	$\times$	base 2	$=$	amount 1
n	$\times$	2	$=$	1
		n	$=$	$1 \div 2$
		n	$=$	0.5
		n	$=$	50%

• **The solution must be written as a percent in order to answer the question.**

There is a 50% chance of winning a prize.

Take Note

We have written $n \cdot 20 = 32$ because that is the form of the basic percent equation. We could have written $20n = 32$.

HOW TO • 2 32 is what percent of 20?

$$\begin{aligned} \text{Percent} \cdot \text{base} &= \text{amount} \\ n \cdot 20 &= 32 \\ \frac{20n}{20} &= \frac{32}{20} \\ n &= 1.6 \\ n &= 160\% \end{aligned}$$

- Use the basic percent equation. Percent = n, base = 20, amount = 32
- Write 1.6 as a percent.

32 is 160% of 20.

In 1780, the population of Virginia was 538,000. This was 19% of the total population of the United States at that time. To find the total population at that time, you must answer the question "19% of what number is 538,000?"

19%	of	what	is	538,000?
↓	↓	↓	↓	↓
Percent 19%	×	base n	=	amount 538,000

$$\begin{aligned} 0.19 \times n &= 538{,}000 \\ n &= 538{,}000 \div 0.19 \\ n &\approx 2{,}832{,}000 \end{aligned}$$

- The population of the United States in 1780 can be found by solving the basic percent equation for the base.

The population of the United States in 1780 was approximately 2,832,000.

Take Note

The base in the basic percent equation will usually follow the word *of*. Some percent problems may use the word *find*. In this case, we can substitute *what is* for *find*. See Example 1 below.

HOW TO • 3 62% of what is 800? Round to the nearest tenth.

$$\begin{aligned} \text{Percent} \cdot \text{base} &= \text{amount} \\ 0.62 \cdot n &= 800 \\ \frac{0.62n}{0.62} &= \frac{800}{0.62} \\ n &\approx 1290.3 \end{aligned}$$

- Use the basic percent equation. Percent = 62% = 0.62, base = n, amount = 800

62% of 1290.3 is approximately 800.

Note from the HOW TO problems above that if any two parts of the basic percent equation are given, the third part can be found.

EXAMPLE • 1

Find 9.4% of 240.

Strategy

To find the amount, solve the basic percent equation.
Percent = 9.4% = 0.094, base = 240, amount = n

Solution

$$\begin{aligned} \text{Percent} \cdot \text{base} &= \text{amount} \\ 0.094 \cdot 240 &= n \\ 22.56 &= n \end{aligned}$$

22.56 is 9.4% of 240.

YOU TRY IT • 1

Find $33\frac{1}{3}\%$ of 45.

Your strategy

Your solution

Solution on p. S16

EXAMPLE • 2

What percent of 30 is 12?

Strategy

To find the percent, solve the basic percent equation.
Percent = n, base = 30, amount = 12

Solution

Percent · base = amount

$$n \cdot 30 = 12$$
$$\frac{30n}{30} = \frac{12}{30}$$
$$n = 0.4$$
$$n = 40\%$$

12 is 40% of 30.

YOU TRY IT • 2

25 is what percent of 40?

Your strategy

Your solution

EXAMPLE • 3

60 is 2.5% of what?

Strategy

To find the percent, solve the basic percent equation.
Percent = 2.5% = 0.025, base = n, amount = 60

Solution

Percent · base = amount

$$0.025 \cdot n = 60$$
$$\frac{0.025n}{0.025} = \frac{60}{0.025}$$
$$n = 2400$$

60 is 2.5% of 2400.

YOU TRY IT • 3

$16\frac{2}{3}\%$ of what is 15?

Your strategy

Your solution

Solutions on p. S16

OBJECTIVE B To solve percent problems using proportions

Percent problems can also be solved by using proportions. The proportion method is based on writing two ratios with quantities that can be found in the basic percent equation. One ratio is the percent ratio, written as $\frac{\text{percent}}{100}$. The second ratio is the amount-to-base ratio, written as $\frac{\text{amount}}{\text{base}}$. These two ratios form the proportion

$$\frac{\textbf{percent}}{\textbf{100}} = \frac{\textbf{amount}}{\textbf{base}}$$

The proportion method can be illustrated by a diagram. The rectangle at the right is divided into two parts. The whole rectangle is represented by 100 and the part by percent. On the other side, the whole rectangle is represented by the base and the part by the amount. The ratio of the percent to 100 is equal to the ratio of the *amount* to the *base*.

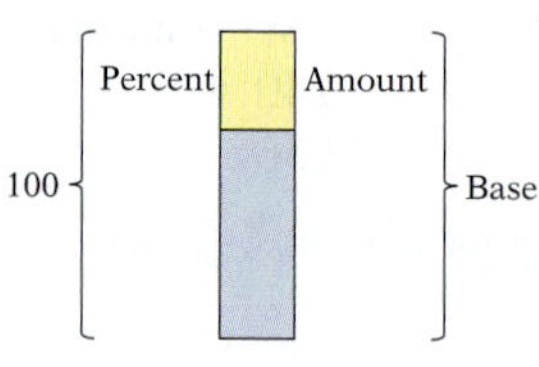

Tips for Success

Remember that a vertical red bar indicates a worked-out example. Using paper and pencil, work through the example. See *AIM for Success* in the front of the book.

HOW TO • 4 What is 32% of 85?

$$\frac{\text{percent}}{100} = \frac{\text{amount}}{\text{base}}$$

• **Sketch a diagram.**

$$\frac{32}{100} = \frac{n}{85}$$

• **Percent = 32, base = 85, amount = n**

$$100 \cdot n = 32 \cdot 85$$
$$100n = 2720$$
$$\frac{100n}{100} = \frac{2720}{100}$$
$$n = 27.2$$

32% of 85 is 27.2.

EXAMPLE • 4

24% of what is 16? Round to the nearest hundredth.

Solution

$$\frac{\text{percent}}{100} = \frac{\text{amount}}{\text{base}}$$
$$\frac{24}{100} = \frac{16}{n}$$

• **Percent = 24, amount = 16**

$$24 \cdot n = 100 \cdot 16$$
$$24n = 1600$$
$$n = \frac{1600}{24} \approx 66.67$$

16 is approximately 24% of 66.67.

YOU TRY IT • 4

8 is 25% of what?

Your solution

EXAMPLE • 5

Find 1.2% of 42.

Solution

$$\frac{\text{percent}}{100} = \frac{\text{amount}}{\text{base}}$$
$$\frac{1.2}{100} = \frac{n}{42}$$

• **Percent = 1.2, base = 42**

$$1.2 \cdot 42 = 100 \cdot n$$
$$50.4 = 100n$$
$$\frac{50.4}{100} = \frac{100n}{100}$$
$$0.504 = n$$

1.2% of 42 is 0.504.

YOU TRY IT • 5

Find 0.74% of 1200.

Your solution

EXAMPLE • 6

What percent of 52 is 13?

Solution

$$\frac{\text{percent}}{100} = \frac{\text{amount}}{\text{base}}$$
$$\frac{n}{100} = \frac{13}{52}$$

• **Amount = 13, base = 52**

$$n \cdot 52 = 100 \cdot 13$$
$$52n = 1300$$
$$\frac{52n}{52} = \frac{1300}{52}$$
$$n = 25$$

25% of 52 is 13.

YOU TRY IT • 6

What percent of 180 is 54?

Your solution

Solutions on p. S16

OBJECTIVE C To solve application problems 

HOW TO • 5

 The circle graph at the right shows the causes of death for all police officers who died while on duty during a recent year. What percent of the deaths were due to traffic accidents? Round to the nearest tenth of a percent.

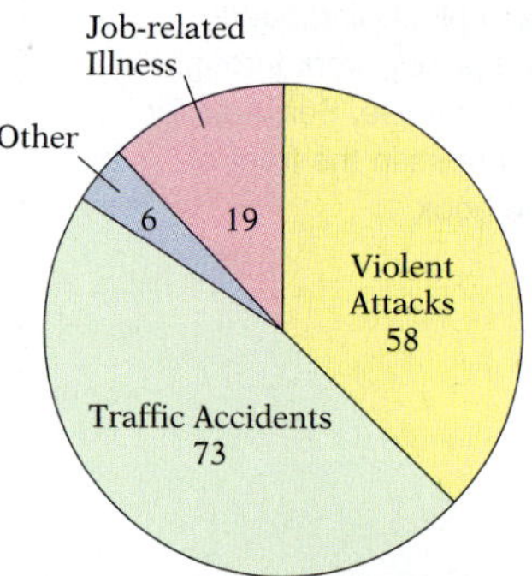

FIGURE 6.1 Causes of Death for Police Officers Killed in the Line of Duty
Source: International Union of Police Associations

Strategy To find the percent:
- Find the total number of officers who died in the line of duty.
- Use the basic percent equation. Percent = n, base = total number who died, amount = number of deaths due to traffic accidents = 73

Solution $58 + 73 + 6 + 19 = 156$

Percent · base = amount

$n \cdot 156 = 73$

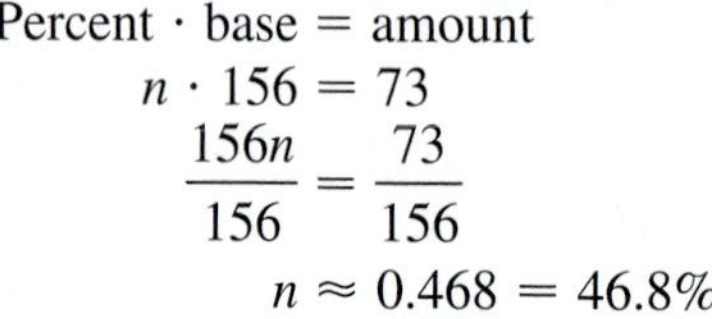

$$\frac{156n}{156} = \frac{73}{156}$$

$n \approx 0.468 = 46.8\%$

Approximately 46.8% of the deaths were due to traffic accidents.

EXAMPLE • 7

 During a recent year, 276 billion product coupons were issued by manufacturers. Shoppers redeemed 4.8 billion of these coupons. (*Source:* NCH NuWorld Consumer Behavior Study, America Coupon Council) What percent of the coupons issued were redeemed by customers? Round to the nearest tenth of a percent.

Strategy
To find the percent, use the basic percent equation. Percent = n, base = number of coupons issued = 276 billion, amount = number of coupons redeemed = 4.8 billion

Solution
Percent · base = amount

$n \cdot 276 = 4.8$

$$\frac{276n}{276} = \frac{4.8}{276}$$

$n \approx 0.017$

$n \approx 1.7\%$

Of the product coupons issued, approximately 1.7% were redeemed by customers.

YOU TRY IT • 7

An instructor receives a monthly salary of $4330, and $649.50 is deducted for income tax. Find the percent of the instructor's salary deducted for income tax.

Your strategy

Your solution

Solution on p. S16

EXAMPLE • 8

A taxpayer pays a tax rate of 35% for state and federal taxes. The taxpayer has an income of $47,500. Find the amount of state and federal taxes paid by the taxpayer.

Strategy

To find the amount, solve the basic percent equation.
Percent = 35% = 0.35, base = 47,500, amount = n

Solution

Percent · base = amount
$0.35 \cdot 47{,}500 = n$
$16{,}625 = n$

The amount of taxes paid is $16,625.

YOU TRY IT • 8

According to Board-Trac, approximately 19% of the country's 2.4 million surfers are women. Estimate the number of female surfers in this country. Write the number in standard form.

Your strategy

Your solution

EXAMPLE • 9

A department store has a blue blazer on sale for $114, which is 60% of the original price. What is the difference between the original price and the sale price?

Strategy

To find the difference between the original price and the sale price:

- Find the original price. Solve the basic percent equation.
  Percent = 60% = 0.60, amount = 114, base = n
- Subtract the sale price from the original price.

Solution

Percent · base = amount
$0.60 \cdot n = 114$
$$\frac{0.60n}{0.60} = \frac{114}{0.60}$$
$n = 190$
$190 - 114 = 76$

The difference in price is $76.

YOU TRY IT • 9

An electrician's wage this year is $30.13 per hour, which is 115% of last year's wage. What was the increase in the hourly wage over last year?

Your strategy

Your solution

Solutions on pp. S16–S17

6.4 EXERCISES

OBJECTIVE A To solve percent problems using the basic percent equation

For Exercises 1 to 22, solve. Use the basic percent equation.

1. 8% of 100 is what?

2. 16% of 50 is what?

3. 0.05% of 150 is what?

4. 0.075% of 625 is what?

5. 15 is what percent of 90?

6. 24 is what percent of 60?

7. What percent of 16 is 6?

8. What percent of 24 is 18?

9. 10 is 10% of what?

10. 37 is 37% of what?

11. 2.5% of what is 30?

12. 10.4% of what is 52?

13. Find 10.7% of 485.

14. Find 12.8% of 625.

15. 80% of 16.25 is what?

16. 26% of 19.5 is what?

17. 54 is what percent of 2000?

18. 8 is what percent of 2500?

19. 16.4 is what percent of 4.1?

20. 5.3 is what percent of 50?

21. 18 is 240% of what?

22. 24 is 320% of what?

23. True or false? If the *base* is larger than the *amount* in the basic percent equation, then the *percent* is larger than 100%.

OBJECTIVE B To solve percent problems using proportions

24. **a.** Which equation(s) below can be used to answer the question "What is 12% of 75?"
b. Which equation(s) below can be used to answer the question "75 is 12% of what?"

(i) $\frac{12}{100} = \frac{75}{n}$ (ii) $0.12 \times 75 = n$ (iii) $\frac{12}{100} = \frac{n}{75}$ (iv) $0.12 \times n = 75$

For Exercises 25 to 46, solve. Use the proportion method.

25. 26% of 250 is what?

26. Find 18% of 150.

27. 37 is what percent of 148?

28. What percent of 150 is 33?

29. 68% of what is 51?

30. 126 is 84% of what?

31. What percent of 344 is 43?

32. 750 is what percent of 50?

33. 82 is 20.5% of what?

34. 2.4% of what is 21?

35. What is 6.5% of 300?

36. Find 96% of 75.

37. 7.4 is what percent of 50?

38. What percent of 1500 is 693?

39. Find 50.5% of 124.

40. What is 87.4% of 225?

41. 120% of what is 6?

42. 14 is 175% of what?

43. What is 250% of 18?

44. 325% of 4.4 is what?

45. 87 is what percent of 29?

46. What percent of 38 is 95?

OBJECTIVE C To solve application problems

47. Read Exercise 48. Without doing any calculations, determine whether the number of people in the United States aged 18 to 24 who do not have health insurance is *less than, equal to,* or *greater than* 44 million.

48. Health Insurance Approximately 30% of the 44 million people in the United States who do not have health insurance are between the ages of 18 and 24. (*Source:* U.S. Census Bureau) About how many people in the United States aged 18 to 24 do not have health insurance?

© Galen Rowell/Corbis

49. Aviation The Federal Aviation Administration reported that 55,422 new student pilots were flying single-engine planes last year. The number of new student pilots flying single-engine planes this year is 106% of the number flying single-engine planes last year. How many new student pilots are flying single-engine planes this year?

50. Lifestyles There are 114 million households in the United States. Opposite-sex cohabitating couples comprise 4.4% of these households. (*Source:* Families and Living Arrangements) Find the number of opposite-sex cohabitating couples who maintain households in the United States. Round to the nearest million.

In the News

More Taxpayers Filing Electronically

The IRS reported that, as of May 4, it has received 128 million returns. Sixty percent of the returns were filed electronically.

Source: IRS

51. e-Filed Tax Returns See the news clipping at the right. How many of the 128 million returns were filed electronically? Round to the nearest million.

52. Prison Population The prison population in the United States is 1,596,127 people. Male prisoners comprise 91% of this population. (*Source: Time,* March 17, 2008) How many inmates are male? How many are female?

© iStockphoto.com/Paul Mckeown

53. Email The number of email messages sent each day has risen to 171 billion, of which 71% are spam. (*Source:* FeedsFarm.com) How many email messages sent per day are not spam?

54. Agriculture According to the U.S. Department of Agriculture, of the 63 billion pounds of vegetables produced in the United States in 1 year, 16 billion pounds were wasted. What percent of the vegetables produced were wasted? Round to the nearest tenth of a percent.

© iStockphoto.com/Svetlana Tebenkova

55. Wind Energy In a recent year, wind machines in the United States generated 17.8 billion kilowatt-hours of electricity, enough to serve over 1.6 million households. The nation's total electricity production that year was 4,450 billion kilowatt-hours. (*Source:* Energy Information Administration) What percent of the total energy production was generated by wind machines?

56. Diabetes Approximately 7% of the American population has diabetes. Within this group, 14.6 million are diagnosed, while 6.2 million are undiagnosed. (*Source:* The National Diabetes Education Program) What percent of Americans with diabetes have not been diagnosed with the disease? Round to the nearest tenth of a percent.

In the News

More Millionaires Audited

The Internal Revenue Service reported that 17,015 millionaires were audited this year. This figure is 33% more than last year.

Source: The Internal Revenue Service; TSN Financial Services

57. Internal Revenue Service See the news clipping at the right. Given that the number of millionaires in the United States is 9.3 million, what percent of U.S. millionaires were audited by the IRS? Round to the nearest hundredth of a percent.

58. Sociology In a survey, 1236 adults nationwide were asked, "What irks you most about the actions of other motorists?" The response "tailgaters" was given by 293 people. (*Source:* Reuters/Zogby) What percent of those surveyed were most irked by tailgaters? Round to the nearest tenth of a percent.

59. Travel Of the travelers who, during a recent year, allowed their children to miss school to go along on a trip, approximately 1.738 million allowed their children to miss school for more than a week. This represented 11% of the travelers who allowed their children to miss school. (*Source:* Travel Industry Association) About how many travelers allowed their children to miss school to go along on a trip?

60. e-Commerce Using the information in the news clipping at the right, calculate the total retail sales during the fourth quarter of last year. Round to the nearest billion.

In the News

E-commerce on the Rise

Retail e-commerce sales for the fourth quarter of last year exceeded e-commerce sales for the first three quarters of the year. E-commerce sales during October, November, and December totaled $35.3 billion, or 3.4% of total retail sales during the quarter.

Source: Service Sector Statistics

61. Marathons In 2008, 98.2% of the runners who started the Boston Marathon, or 21,963 people, crossed the finish line. (*Source:* www.bostonmarathon.org) How many runners started the Boston Marathon in 2008?

62. Education In the United States today, 23.1% of women and 27.5% of men have earned a bachelor's or graduate degree. (*Source:* Census Bureau) How many women in the United States have earned a bachelor's or graduate degree?

63. Wind-Powered Ships Using the information in the news clipping at the right, calculate the cargo ship's daily fuel bill without the kite.

Courtesy SkySails

In the News

Kite-Powered Cargo Ships

In January 2008, the first cargo ship partially powered by a giant kite set sail from Germany bound for Venezuela. The 1722-square-foot kite helped to propel the ship, which consequently used 20% less fuel, cutting approximately $1600 from the ship's daily fuel bill.

Source: The Internal Revenue Service; TSN Financial Services

64. Taxes A TurboTax online survey asked people how they planned to use their tax refunds. Seven hundred forty people, or 22% of the respondents, said they would save the money. How many people responded to the survey?

Applying the Concepts

Pets The graph at the right shows several categories of average lifetime costs of dog ownership. Use this graph for Exercises 65 to 67. Round answers to the nearest tenth of a percent.

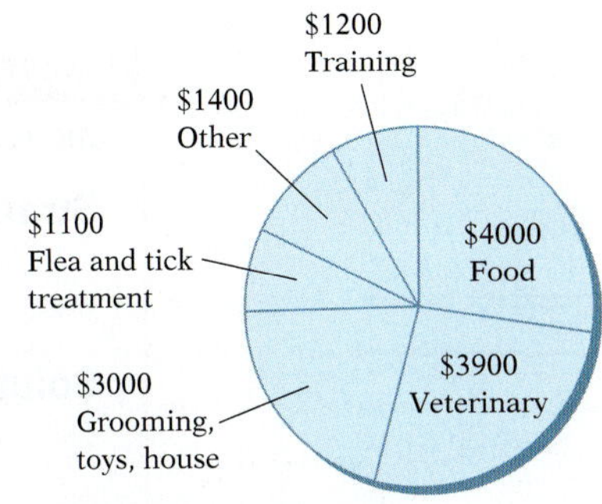

Cost of Owning a Dog

Source: Based on data from the American Kennel Club, *USA Today* research

65. What percent of the total amount is spent on food?

66. What percent of the total is spent on veterinary care?

67. What percent of the total is spent on all categories except training?

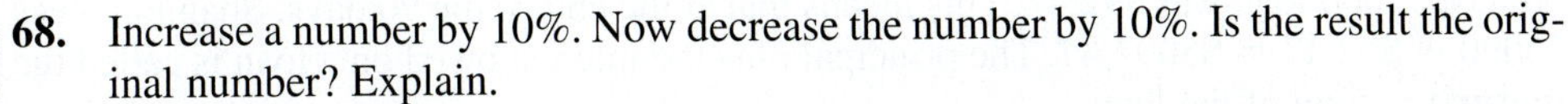

68. Increase a number by 10%. Now decrease the number by 10%. Is the result the original number? Explain.

SECTION

6.5 Simple Interest

OBJECTIVE A To solve problems involving simple interest

Don Farrall/Photodisc/Getty Images

If you deposit money in a savings account at a bank, the bank will pay you for the privilege of using that money. The amount you deposit in the savings account is called the **principal.** The amount the bank pays you for the privilege of using the money is called **interest.**

If you borrow money from the bank in order to buy a car, the amount you borrow is called the **principal.** The additional amount of money you must pay the bank, above and beyond the amount borrowed, is called **interest.**

Whether you deposit money or borrow it, the amount of interest paid is usually computed as a percent of the principal. The percent used to determine the amount of interest to be paid is the **interest rate.** Interest rates are given for specific periods of time, such as months or years.

Interest computed on the original principal is called **simple interest.** Simple interest is the cost of a loan that is for a period of about 1 year or less.

The Simple Interest Formula

$I = Prt$, where I = simple interest earned, P = principal, r = annual simple interest rate, and t = time (in years)

In the simple interest formula, t is the time in years. If a time period is given in days or months, it must be converted to years and then substituted in the formula for t. For example,

120 days is $\frac{120}{365}$ of a year. 6 months is $\frac{6}{12}$ of a year.

HOW TO 1 Shannon O'Hara borrowed \$5000 for 90 days at an annual simple interest rate of 7.5%. Find the simple interest due on the loan.

Strategy To find the simple interest owed, use the simple interest formula.

$P = 5000, r = 7.5\% = 0.075, t = \frac{90}{365}$

Solution

$$I = Prt$$

$$I = 5000(0.075)\left(\frac{90}{365}\right)$$

$$I \approx 92.47$$

The simple interest due on the loan is \$92.47.

In the example above, we calculated that the simple interest due on Shannon O'Hara's 90-day, \$5000 loan was \$92.47. This means that at the end of the 90 days, Shannon owes \$5000 + \$92.47 = \$5092.47. The principal plus the interest owed on a loan is called the **maturity value** of the loan.

Formula for the Maturity Value of a Simple Interest Loan

$M = P + I$, where M = maturity value, P = principal, and I = simple interest

The example below illustrates solving the simple interest formula for the interest rate. The solution requires the Multiplication Property of Equations.

HOW TO • 2 Ed Pabas took out a 45-day, $12,000 loan. The simple interest on the loan was $168. To the nearest tenth of a percent, what is the simple interest rate?

Strategy To find the simple interest rate, use the simple interest formula.

$P = 12{,}000,\ t = \frac{45}{365},\ I = 168$

Solution

$$I = Prt$$
$$168 = 12{,}000r\left(\frac{45}{365}\right)$$
$$168 = \frac{540{,}000}{365}r$$
$$\frac{365}{540{,}000}(168) = \frac{365}{540{,}000} \cdot \frac{540{,}000}{365}r$$
$$0.114 \approx r$$

The simple interest rate on the loan is 11.4%.

EXAMPLE • 1

You arrange for a 9-month bank loan of $9000 at an annual simple interest rate of 8.5%. Find the total amount you must repay to the bank.

Strategy

To calculate the maturity value:

- Find the simple interest due on the loan by solving the simple interest formula for I.

  $t = \frac{9}{12},\ P = 9000,\ r = 8.5\% = 0.085$
- Use the formula for the maturity value of a simple interest loan, $M = P + I$.

Solution

$I = Prt$

$I = 9000(0.085)\left(\frac{9}{12}\right)$

$I = 573.75$

$M = P + I$
$M = 9000 + 573.75$
$M = 9573.75$

The total amount owed to the bank is $9573.75.

YOU TRY IT • 1

William Carey borrowed $12,500 for 8 months at an annual simple interest rate of 9.5%. Find the maturity value of the loan.

Your strategy

Your solution

Solution on p. S17

6.5 EXERCISES

OBJECTIVE A To solve problems involving simple interest

1. Explain what each variable in the simple interest formula represents.

2. Explain the difference between interest and interest rate.

3. **a.** In the table below, the interest rate is an annual simple interest rate. Complete the table by calculating the simple interest due on the loan.

Loan Amount	*Interest Rate*	*Period*	*Interest*
\$5000	6%	1 month	
\$5000	6%	2 months	
\$5000	6%	3 months	
\$5000	6%	4 months	
\$5000	6%	5 months	

Use the pattern of your answers in the table to find the simple interest due on a \$5000 loan that has an annual simple interest rate of 6% for a period of:
b. 6 months **c.** 7 months **d.** 8 months **e.** 9 months

4. Use your solutions to Exercise 3 to answer the following questions:
   a. If you know the simple interest due on a 1-month loan, explain how you can use that figure to calculate the simple interest due on a 7-month loan for the same principal and the same interest rate.
   b. If the time period of a loan is doubled but the principal and interest rate remain the same, how many times greater is the simple interest due on the loan?

5. Kristi Yang borrowed \$15,000. The term of the loan was 90 days, and the annual simple interest rate was 7.4%. Find the simple interest due on the loan.

6. Hector Elizondo took out a 75-day loan of \$7500 at an annual interest rate of 9.5%. Find the simple interest due on the loan.

7. To finance the purchase of 15 new cars, the Lincoln Car Rental Agency borrowed \$100,000 for 9 months at an annual interest rate of 9%. What is the simple interest due on the loan?

8. A home builder obtained a preconstruction loan of \$50,000 for 8 months at an annual interest rate of 9.5%. What is the simple interest due on the loan?

9. Assume that Visa charges Francesca 1.6% per month on her unpaid balance. Find the interest owed to Visa when her unpaid balance for the month is \$1250.

10. The Mission Valley Credit Union charges its customers an interest rate of 2% per month on money that is transferred into an account that is overdrawn. Find the interest owed to the credit union for 1 month when $800 is transferred into an overdrawn account.

11. Find the simple interest Jacob Zucker owes on a 2-year loan of $8000 at an annual interest rate of 9%.

12. Find the simple interest Kara Tanamachi owes on a $1\frac{1}{2}$-year loan of $1500 at an annual interest rate of 7.5%.

13. An auto parts dealer borrowed $150,000 at a 9.5% annual simple interest rate for 1 year. Find the maturity value of the loan.

14. A corporate executive took out a $25,000 loan at an 8.2% annual simple interest rate for 1 year. Find the maturity value of the loan.

15. Capitol City Bank approves a home-improvement loan application for $14,000 at an annual simple interest rate of 10.25% for 270 days. What is the maturity value of the loan?

16. A credit union loans a member $5000 for the purchase of a used car. The loan is made for 18 months at an annual simple interest rate of 6.9%. What is the maturity value of the car loan?

17. A $12,000 investment earned $462 in interest in 6 months. Find the annual simple interest rate on the loan.

18. Michele Gabrielle borrowed $3000 for 9 months and paid $168.75 in simple interest on the loan. Find the annual simple interest rate that Michele paid on the loan.

19. An investor earned $937.50 on an investment of $50,000 in 75 days. Find the annual simple interest rate earned on the investment.

20. Don Glover borrowed $18,000 for 210 days and paid $604.80 in simple interest on the loan. What annual simple interest rate did Don pay on the loan?

21. Student A and Student B borrow the same amount of money at the same annual interest rate. Student A has a 2-year loan and Student B has a 1-year loan. In each case, state whether the first quantity is *less than*, *equal to*, or *greater than* the second quantity.
   a. Student A's principal; Student B's principal
   b. Student A's maturity value; Student B's maturity value
   c. Student A's monthly payment; Student B's monthly payment

Applying the Concepts

22. Interest has been described as a rental fee for money. Explain this description of interest.

FOCUS ON PROBLEM SOLVING

Using a Calculator as a Problem-Solving Tool

A calculator is an important tool for problem solving. Here are a few problems to solve with a calculator. You may need to research some of the questions to find information you do not know.

1. Choose any single-digit positive number. Multiply the number by 1507 and 7373. What is the answer? Choose another positive single-digit number and again multiply by 1507 and 7373. What is the answer? What pattern do you see? Why does this work?

2. The gross domestic product in 2007 was \$13,841,300,000. Is this more or less than the amount of money that would be placed on the last square of a standard checkerboard if 1 cent were placed on the first square, 2 cents were placed on the second square, 4 cents were placed on the third square, 8 cents were placed on the fourth square, and so on, until the 64th square was reached?

3. Which of the reciprocals of the first 16 natural numbers have a terminating-decimal representation and which have a repeating-decimal representation?

4. What is the largest natural number n for which $4^n > 1 \cdot 2 \cdot 3 \cdot 4 \cdot 5 \cdot \cdots \cdot n$?

5. If \$1000 bills are stacked one on top of another, is the height of \$1 billion less than or greater than the height of the Washington Monument?

6. What is the value of $1 + \cfrac{1}{1 + \cfrac{1}{1 + \cfrac{1}{1 + \cfrac{1}{1 + 1}}}}$?

7. Calculate 15^2, 35^2, 65^2, and 85^2. Study the results. Make a conjecture about a relationship between a number ending in 5 and its square. Use your conjecture to find 75^2 and 95^2. Does your conjecture work for 125^2?

8. Find the sum of the first 1000 natural numbers. (*Hint:* You could just start adding $1 + 2 + 3 + \cdots$, but even if you performed one operation every 3 s, it would take you an hour to find the sum. Instead, try pairing the numbers and then adding the pairs. Pair 1 and 1000, 2 and 999, 3 and 998, and so on. What is the sum of each pair? How many pairs are there? Use this information to answer the original question.)

9. For a borrower to qualify for a home loan, a bank requires that the monthly mortgage payment be less than 25% of the borrower's monthly take-home income. A laboratory technician has deductions for taxes, insurance, and retirement that amount to 25% of the technician's monthly gross income. What minimum monthly income must this technician earn to receive a bank loan that has a mortgage payment of \$1200 per month?

PROJECTS AND GROUP ACTIVITIES

Consumer Price Index

The consumer price index (CPI) is a percent that is written without the percent sign. For instance, a CPI of 160.1 means 160.1%. This number means that an item that cost \$100 between 1982 and 1984 (the base years) would cost \$160.10 today. Determining the cost is an application of the basic percent equation.

$$\text{Percent} \times \text{base} = \text{amount}$$

$$\text{CPI} \times \text{cost in base year} = \text{cost today}$$

$$1.601 \times 100 = 160.1 \qquad \bullet\ 160.1\% = 1.601$$

The table below gives the CPI for various products in March of 2008. You can obtain current data for the items below, as well as other items not on this list, by visiting the website of the Bureau of Labor Statistics.

Product	*CPI*
All items	213.5
Food and beverages	209.7
Housing	214.4
Clothes	120.9
Transportation	195.2
Medical care	363.0
Entertainment[1]	112.7
Education[1]	121.8

[1]Indexes on December 1997 = 100

1. Of the items listed, are there any items that in 2008 cost more than twice as much as they cost during the base year? If so, which items?

2. Of the items listed, are there any items that in 2008 cost more than one-and-one-half times as much as they cost during the base years but less than twice as much as they cost during the base years? If so, which items?

3. If the cost for textbooks for one semester was \$120 in the base years, how much did similar textbooks cost in 2008? Use the "Education" category.

4. If a new car cost \$40,000 in 2008, what would a comparable new car have cost during the base years? Use the "Transportation" category.

5. If a movie ticket cost \$10 in 2008, what would a comparable movie ticket have cost during the base years? Use the "Entertainment" category.

6. The base year for the CPI was 1967 before the change to 1982–1984. If 1967 were still used as the base year, the CPI for all items in 2008 (not just those listed above) would be 639.6.

   a. Using the base year of 1967, explain the meaning of a CPI of 639.6.

   b. Using the base year of 1967 and a CPI of 639.6, if textbooks cost \$75 for one semester in 1967, how much did similar textbooks cost in 2008?

   c. Using the base year of 1967 and a CPI of 639.6, if a family's food budget in 2008 is \$1000 per month, what would a comparable family budget have been in 1967?

CHAPTER 6

SUMMARY

KEY WORDS

KEY WORDS	EXAMPLES
A *ratio* is the comparison of two quantities with the same unit. A ratio can be written in three ways: as a fraction, as two numbers separated by a colon, or as two numbers separated by the word *to*. A ratio is in simplest form when the two quantities do not have a common factor. [6.1A, p. 336]	The comparison 16 oz to 24 oz can be written as a ratio in simplest form: $\frac{2}{3}$, 2:3, or 2 to 3
A *rate* is the comparison of two quantities with different units. A rate is in simplest form when the two quantities do not have a common factor. [6.1A, p. 336]	You earned \$63 for working 6 h. The rate is written $\frac{\$21}{2\text{ h}}$.
A *unit rate* is a rate in which the denominator is 1. [6.1A, p. 336]	You traveled 144 mi in 3 h. The unit rate is 48 mph.
A *proportion* is the equality of two ratios or rates. Each of the four members in a proportion is called a *term*. first term ⟵ a, second term ⟵ b, $\frac{a}{b} = \frac{c}{d}$, c ⟶ third term, d ⟶ fourth term. The second and third terms of the proportion are called the *means*, and the first and fourth terms are called the *extremes*. [6.2A, p. 340]	In the proportion $\frac{3}{5} = \frac{12}{20}$, 5 and 12 are the means; 3 and 20 are the extremes.
Percent means "parts of 100." [6.3A, p. 348]	23% means 23 of 100 equal parts.
Principal is the amount of money originally deposited or borrowed. *Interest* is the amount paid for the privilege of using someone else's money. The percent used to determine the amount of interest is the *interest rate*. Interest computed on the original amount is called *simple interest*. The principal plus the interest owed on a loan is called the *maturity value* of the loan. [6.5A, p. 362]	Consider a 1-year loan of \$5000 at an annual simple interest rate of 8%. The principal is \$5000. The interest rate is 8%. The interest paid on the loan is \$400. The maturity value is \$5000 + \$400 = \$5400.

ESSENTIAL RULES AND PROCEDURES

ESSENTIAL RULES AND PROCEDURES	EXAMPLES
To find a unit rate, divide the number in the numerator of the rate by the number in the denominator of the rate. [6.1A, p. 337]	You earned \$41 for working 4 h. $\frac{41}{4} = 41 \div 4 = 10.25$ The unit rate is \$10.25 per hour.
To set up a proportion, keep the same units in the numerators and the same units in the denominators. [6.2B, p. 342]	Three machines fill 5 cereal boxes per minute. How many boxes can 8 machines fill per minute? $\frac{3\text{ machines}}{5\text{ cereal boxes}} = \frac{8\text{ machines}}{x\text{ cereal boxes}}$

To solve a proportion, use the fact that the product of the means equals the product of the extremes. For the proportion $\frac{a}{b} = \frac{c}{d}$, $ad = bc$. [6.2A, p. 341]

$$\frac{6}{25} = \frac{9}{x}$$
$$25 \cdot 9 = 6 \cdot x$$
$$225 = 6x$$
$$\frac{225}{6} = \frac{6x}{6}$$
$$37.5 = x$$

To write a percent as a fraction, remove the percent sign and multiply by $\frac{1}{100}$. [6.3A, p. 348]

$$56\% = 56\left(\frac{1}{100}\right) = \frac{56}{100} = \frac{14}{25}$$

To write a percent as a decimal, remove the percent sign and multiply by 0.01. [6.3A, p. 348]

$$87\% = 87(0.01) = 0.87$$

To write a fraction as a percent, multiply by 100%. [6.3B, p. 349]

$$\frac{7}{20} = \frac{7}{20}(100\%) = \frac{700\%}{20} = 35\%$$

To write a decimal as a percent, multiply by 100%. [6.3B, p. 349]

$$0.325 = 0.325(100\%) = 32.5\%$$

The Basic Percent Equation [6.4A, p. 352]

Percent · base = amount

8% of 250 is what number?

Percent · base = amount

$$0.08 \cdot 250 = n$$
$$20 = n$$

Proportion Method of Solving a Percent Problem [6.4B, p. 354]

$$\frac{\text{percent}}{100} = \frac{\text{amount}}{\text{base}}$$

8% of 250 is what number?

$$\frac{\text{percent}}{100} = \frac{\text{amount}}{\text{base}}$$
$$\frac{8}{100} = \frac{n}{250}$$
$$100 \cdot n = 8 \cdot 250$$
$$100n = 2000$$
$$n = 20$$

Simple Interest Formula [6.5A, p. 362]

$I = Prt$, where I = simple interest earned, P = principal, r = annual simple interest rate, t = time (in years)

You borrow \$10,000 for 180 days at an annual interest rate of 8%. Find the simple interest due on the loan.

$$I = Prt$$
$$I = 10{,}000(0.08)\left(\frac{180}{365}\right)$$
$$I \approx 394.52$$

Formula for the Maturity Value of a Simple Interest Loan [6.5A, p. 363]

$M = P + I$, where M = maturity value, P = principal, I = simple interest

Suppose you paid \$400 in interest on a 1-year loan of \$5000. The maturity value of the loan is \$5000 + \$400 = \$5400.

CHAPTER 6

CONCEPT REVIEW

Test your knowledge of the concepts presented in this chapter. Answer each question. Then check your answers against the ones provided in the Answer Section.

1. If the units in a comparison are different, is it a ratio or a rate?

2. How do you find a unit rate?

3. How do you write the ratio 12:15 in simplest form?

4. How do you write the ratio 19:6 as a fraction?

5. When is a proportion true?

6. How do you solve a proportion?

7. How do the units help you to set up a proportion?

8. How do you check the solution of a proportion?

9. What is the basic percent equation?

10. What percent of 40 is 30? Did you multiply or divide?

11. Find 11.7% of 532. Did you multiply or divide?

12. 36 is 240% of what number? Did you multiply or divide?

13. How do you use the proportion method to solve a percent problem?

CHAPTER 6

REVIEW EXERCISES

1. Write the comparison 100 lb to 100 lb as a ratio in simplest form using a fraction, a colon, and the word *to*.

2. Write 18 roof supports for every 9 ft as a rate in simplest form.

3. Write \$628 earned in 40 h as a unit rate.

4. Write 8 h to 15 h as a ratio in simplest form using a fraction.

5. Solve: $\frac{n}{3} = \frac{8}{15}$

6. Write 15 lb of fertilizer for 12 trees as a rate in simplest form.

7. Write 171 mi driven in 3 h as a unit rate.

8. Solve $\frac{2}{3.5} = \frac{n}{12}$. Round to the nearest hundredth.

9. Write 32% as a fraction.

10. Write 22% as a decimal.

11. Write 25% as a fraction and as a decimal.

12. Write $3\frac{2}{5}\%$ as a fraction.

13. Write $\frac{7}{40}$ as a percent.

14. Write $1\frac{2}{7}$ as a percent. Round to the nearest tenth of a percent.

15. Write 2.8 as a percent.

16. 42% of 50 is what?

17. What percent of 3 is 15?

18. 12 is what percent of 18? Round to the nearest tenth of a percent.

19. 150% of 20 is what number?

20. Find 18% of 85.

21. 32% of what number is 180?

22. 4.5 is what percent of 80?

23. Technology In 3 years, the price of a graphing calculator went from \$125 to \$75. What is the ratio of the decrease in price to the original price?

24. Investments An investment of \$8000 earns \$520 in annual dividends. At the same rate, how much money must be invested to earn \$780 in annual dividends?

25. Lawn Care The directions on a bag of plant food recommend $\frac{1}{2}$ lb for every 50 ft^2 of lawn. How many pounds of plant food should be used on a lawn measuring 275 ft^2?

26. Profit Sharing Two attorneys share the profits of their firm in the ratio 3:2. If the attorney receiving the larger amount of this year's profits receives \$96,000, what amount does the other attorney receive?

27. Tourism The table at the right shows the countries with the highest projected numbers of tourists visiting in 2020. What percent of the tourists visiting these countries will be visiting China? Round to the nearest tenth of a percent.

Country	*Projected Number of Tourists in 2020*
China	137 million
France	93 million
Spain	71 million
USA	102 million

Source: The State of the World Atlas by Dan Smith

28. Business A company spent 7% of its \$120,000 budget for advertising. How much did the company spend for advertising?

29. Manufacturing A quality control inspector found that 1.2% of 4000 cellular telephones were defective. How many of the phones were not defective?

30. Television According to the Cabletelevision Advertising Bureau, cable households watch an average of 61.35 h of television per week. On average, what percent of the week do cable households spend watching TV? Round to the nearest tenth of a percent.

31. Finance Find the simple interest due on a 45-day loan of \$3000 at an annual simple interest rate of 8.6%.

32. Finance A realtor took out a \$10,000 loan at an 8.4% annual simple interest rate for 9 months. Find the maturity value of the loan.

33. Travel An airline knowingly overbooks flights by selling 12% more tickets than there are seats available. How many tickets would this airline sell for an airplane that has 175 seats?

CHAPTER 6

TEST

1. Write the comparison 3 yd to 24 yd as a ratio in simplest form using a fraction, a colon, and the word *to*.

2. Write 16 oz of sugar for 64 cookies as a rate in simplest form.

3. Write 120 mi driven in 200 min as a unit rate.

4. Write 200 ft to 100 ft as a ratio in simplest form using a fraction.

5. Solve: $\frac{n}{5} = \frac{3}{20}$

6. Write 8 ft walked in 4 s as a unit rate.

7. Write 2860 ft^2 mowed in 6 h as a unit rate. Round to the nearest hundredth.

8. Solve: $\frac{n}{4} = \frac{8}{9}$. Round to the nearest hundredth.

9. Write 86.4% as a decimal.

10. Write 0.4 as a percent.

11. Write $\frac{5}{4}$ as a percent.

12. Write $83\frac{1}{3}\%$ as a fraction.

13. Write 44% as a fraction.

14. Write 1.18 as a percent.

15. 18 is 20% of what number?

16. What is 68% of 73?

17. What percent of 320 is 180?

18. 28 is 14% of what number?

19. **Physical Fitness** A body builder who had been lifting weights for 2 years went from an original weight of 165 lb to 190 lb. What is the ratio of the original weight to the increased weight?

20. **Taxes** The sales tax on a $95 purchase is $7.60. Find the sales tax on a car costing $39,200.

21. **Elections** A preelection survey showed that 3 out of 4 registered voters would vote in a county election. At this rate, how many registered voters would vote in a county with 325,000 registered voters?

22. **Architecture** The scale on the architectural drawings for a new gymnasium is 1 in. equals 4 ft. How long is one of the rooms if it measures $12\frac{1}{2}$ in. on the drawing?

23. **Girl Scout Cookies** Using the information in the news clipping at the right, calculate the cash generated annually **a.** from sales of Thin Mints and **b.** from sales of Trefoil shortbread cookies.

Jeff Greenberg/age fotostock

In the News

Thin Mints Biggest Seller

Every year, sales from all the Girl Scout cookies sold by about 2.7 million girls total $700 million. The most popular cookie is Thin Mints, which earn 25% of total sales, while sales of the Trefoil shortbread cookies represent only 9% of total sales.

Source: Southwest Airlines Spirit Magazine 2007

24. **Charities** The American Red Cross spent $185,048,179 for administrative expenses. This amount was 3.16% of its total revenue. Find the American Red Cross's total revenue. Round to the nearest hundred million.

25. **Poultry** In a recent year, North Carolina produced 1,300,000,000 lb of turkey. This was 18.6% of the U.S. total in that year. Calculate the U.S. total turkey production for that year. Round to the nearest billion.

26. **Nutrition** The table at the right shows the fat, saturated fat, cholesterol, and calorie content in a 90-gram ground-beef burger and in a 90-gram soy burger. The number of fat grams in the beef burger is what percent of the number of fat grams in the soy burger?

	Beef Burger	*Soy Burger*
Fat	24 g	4 g
Saturated Fat	10 g	1.5 g
Cholesterol	75 mg	0 mg
Calories	280	140

27. **Finance** You took out a 150-day, $40,000 business loan that had an annual simple interest rate of 9.25%. Find the maturity value of the loan.

CUMULATIVE REVIEW EXERCISES

1. Simplify: $18 \div \frac{6-3}{9} - (-3)$

2. Evaluate 5^4.

3. Subtract: $7\frac{5}{12} - 3\frac{5}{9}$

4. Simplify: $\frac{4}{5} \div \frac{4}{5} + \frac{2}{3}$

5. Find the quotient of 342 and -3.

6. Evaluate $2a - 3ab$ when $a = 2$ and $b = -3$.

7. Solve: $5x - 20 = 0$

8. Solve: $3(x - 4) + 2x = 3$

9. Simplify: $-\frac{5}{8} - \left(-\frac{3}{4}\right) + \frac{5}{6}$

10. Find the product of 1.005 and 10^5.

11. Simplify: $(-5)^2 - (-8) \div (7 - 5)^2 \cdot 2 - 8$

12. Simplify: $\left(-\frac{2}{3}\right)\left(-\frac{3}{4}\right)^2$

13. Simplify: $4 - (-3) + 5 - 8$

14. Simplify: $5 - 2(1 - 3a) + 2(a - 3)$

15. Solve: $\frac{3}{4}x = -9$

16. Simplify: $-3y^2 + 3y - y^2 - 6y$

17. Divide: $3\frac{5}{8} \div 2\frac{7}{12}$

18. Write 30 cents to 1 dollar as a ratio in simplest form.

19. Write \$19,425 in 5 months as a unit rate.

20. Evaluate $a - b$ when $a = 102.5$ and $b = 77.546$.

21. Solve: $\frac{2}{3} = \frac{n}{48}$

22. Simplify: $\dfrac{\frac{1}{2} + \frac{3}{4}}{2 - \frac{5}{8}}$

23. 2.5 is what percent of 30? Round to the nearest tenth of a percent.

24. Find 42% of 60.

25. Public Transportation The figure at the right shows the average amount spent annually per household on public transportation, by region, in the United States. Find the difference between the average amount spent monthly per household in the Northeast and in the South. Round to the nearest cent.

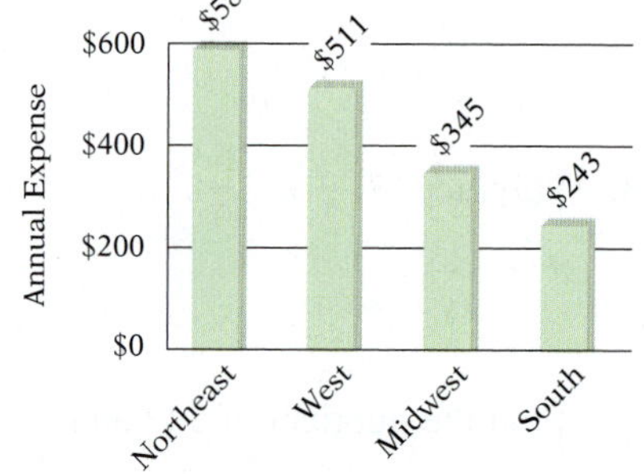

Average Annual Expense per Household on Public Transportation in the United States

Source: Bureau of Labor Statistics consumer expenditure survey

26. Five less than two-thirds of a number is three. Find the number.

27. Translate "the difference between four times a number and three times the sum of the number and two" into a variable expression. Then simplify.

28. Travel Your odometer reads 18,325 mi before you embark on a 125-mile trip. After you have driven $1\frac{1}{2}$ h, the odometer reads 18,386 mi. How many miles are left to drive?

29. Banking You had a balance of $422.89 in your checking account. You then made a deposit of $122.35 and wrote a check for $279.76. Find the new balance in your checking account.

PAYMENT/ DEBIT (−)		√ T	FEE (IF ANY) (−)	DEPOSIT/ CREDIT (+)		BALANCE	
						$ 422	89
$			$	$122	35		
279	76						

30. Jobs A data processor finished $\frac{2}{5}$ of a three-day job on the first day and $\frac{1}{3}$ on the the second day. What part of the job is to be finished on the third day?

31. Health According to the table at the right, what fraction of the population aged 75–84 is affected by Alzheimer's disease?

Age Group	*Percent Affected by Alzheimer's Disease*
65–74	4%
75–84	10%
85 +	17%

Source: Mayo Clinic Family Health Book, Encyclopedia Americana, Associated Press

32. Travel A car is driven 402.5 mi on 11.5 gal of gas. Find the number of miles traveled per gallon of gas.

33. Mechanics At a certain speed, the engine revolutions per minute (rpm) of a car in fourth gear is 2500. This is two-thirds of the rpm of the engine in third gear. Find the rpm of the engine in third gear.

CHAPTER 7

Geometry

OBJECTIVES

SECTION 7.1

A To solve problems involving lines and angles

B To solve problems involving angles formed by intersecting lines

C To solve problems involving the angles of a triangle

SECTION 7.2

A To solve problems involving the perimeter of geometric figures

B To solve problems involving the area of geometric figures

SECTION 7.3

A To solve problems using the Pythagorean Theorem

B To solve problems involving similar triangles

C To determine whether two triangles are congruent

SECTION 7.4

A To solve problems involving the volume of a solid

B To solve problems involving the surface area of a solid

ARE YOU READY?

Take the Chapter 7 Prep Test to find out if you are ready to learn to:

- Solve problems involving the angles formed by intersecting lines
- Solve problems involving the angles of a triangle
- Find the perimeter and area of geometric figures
- Solve problems using the Pythagorean Theorem
- Solve problems involving similar triangles and congruent triangles
- Find the volume and surface area of a solid

PREP TEST

Do these exercises to prepare for Chapter 7.

1. Simplify: $2(18) + 2(10)$

2. Evaluate abc when $a = 2$, $b = 3.14$, and $c = 9$.

3. Evaluate xyz^3 when $x = \frac{4}{3}$, $y = 3.14$, and $z = 3$.

4. Solve: $x + 47 = 90$

5. Solve: $32 + 97 + x = 180$

6. Solve: $\frac{5}{12} = \frac{6}{x}$

SECTION

7.1 Introduction to Geometry

OBJECTIVE A To solve problems involving lines and angles 

Tips for Success

Before you begin a new chapter, you should take some time to review previously learned skills. One way to do this is to complete the Prep Test. See page 377. This test focuses on the particular skills that will be required for the new chapter.

The word *geometry* comes from the Greek words for "earth" and "measure." The original purpose of geometry was to measure land. Today geometry is used in many fields, such as physics, medicine, and geology, and is applied in such areas as mechanical drawing and astronomy. Geometric forms are used in art and design.

Three basic concepts of geometry are point, line, and plane. A **point** is symbolized by drawing a dot. A **line** is determined by two distinct points and extends indefinitely in both directions, as the arrows on the line shown at the right indicate. This line contains points A and B and is represented by $\overleftrightarrow{AB}$. A line can also be represented by a single letter, such as ℓ.

A **ray** starts at a point and extends indefinitely in *one* direction. The point at which a ray starts is called the **endpoint** of the ray. The ray shown at the right is denoted by $\overrightarrow{AB}$. Point A is the endpoint of the ray.

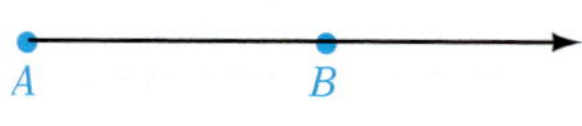

A **line segment** is part of a line and has two endpoints. The line segment shown at the right is denoted by $\overline{AB}$.

The distance between the endpoints of $\overline{AC}$ is denoted by AC. If B is a point on $\overline{AC}$, then AC (the distance from A to C) is the sum of AB (the distance from A to B) and BC (the distance from B to C).

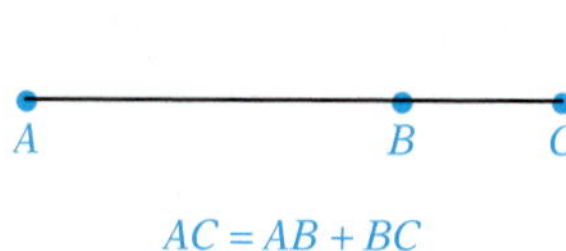

HOW TO • 1 Given $AB = 22$ cm and $AC = 31$ cm, find BC.

$AC = AB + BC$ • Write an equation for the distances between points on the line segment.

$31 = 22 + BC$ • Substitute the given distances for AB and AC into the equation.

$9 = BC$ • Solve for BC.

$BC = 9$ cm

In this section we will be discussing figures that lie in a plane. A **plane** is a flat surface and can be pictured as a table top or blackboard that extends in all directions. Figures that lie in a plane are called **plane figures.**

Point of Interest

Geometry is one of the oldest branches of mathematics. Around 350 B.C., the Greek mathematician Euclid wrote the *Elements*, which contained all of the known concepts of geometry. Euclid's contribution was to unify various concepts into a single deductive system that was based on a set of axioms.

Lines in a plane can be intersecting or parallel. **Intersecting lines** cross at a point in the plane. **Parallel lines** never intersect. The distance between them is always the same.

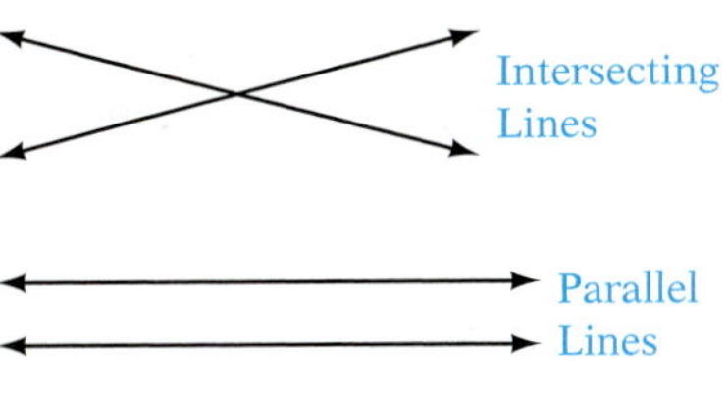

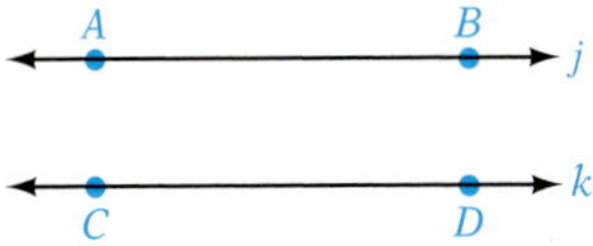

The symbol $\parallel$ means "is parallel to." In the figure at the right, $j \parallel k$ and $\overline{AB} \parallel \overline{CD}$. Note that j contains $\overline{AB}$ and k contains $\overline{CD}$. Parallel lines contain parallel line segments.

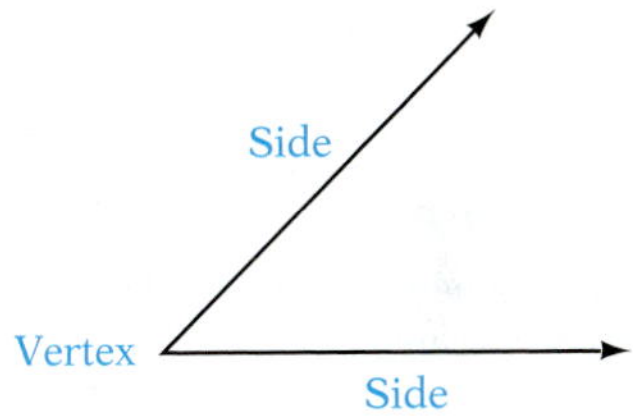

An **angle** is formed by two rays with the same endpoint. The **vertex** of the angle is the point at which the two rays meet. The rays are called the **sides** of the angle.

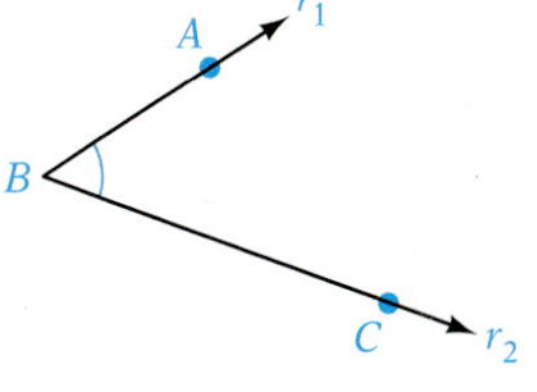

If A and C are points on rays r_1 and r_2, and B is the vertex, then the angle is called $\angle B$ or $\angle ABC$, where $\angle$ is the symbol for angle. Note that the angle is named by the vertex, or the vertex is the second point listed when the angle is named by giving three points. $\angle ABC$ could also be called $\angle CBA$.

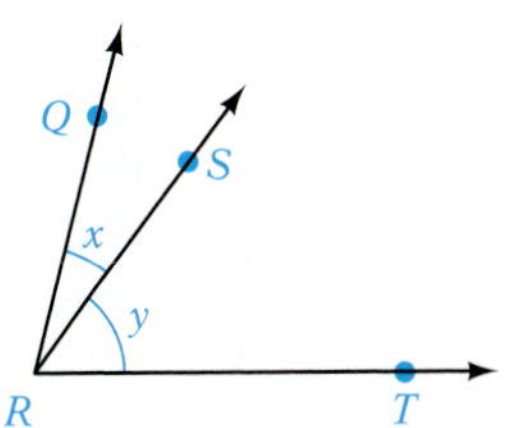

An angle can also be named by a variable written between the rays close to the vertex. In the figure at the right, $\angle x = \angle QRS$ and $\angle y = \angle SRT$. Note that in this figure, more than two rays meet at R. In this case, the vertex cannot be used to name an angle.

An angle is measured in **degrees.** The symbol for degrees is a small raised circle, °. Probably because early Babylonians believed that Earth revolves around the sun in approximately 360 days, **the angle formed by a circle has a measure of 360°** (360 degrees).

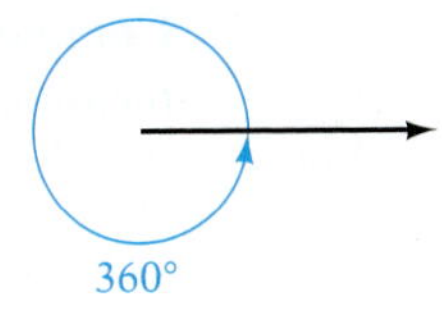

A **protractor** is used to measure an angle. Place the center of the line segment near the bottom edge of the protractor at the vertex of the angle and the line segment along a side of the angle. The angle shown in the figure below measures 58°.

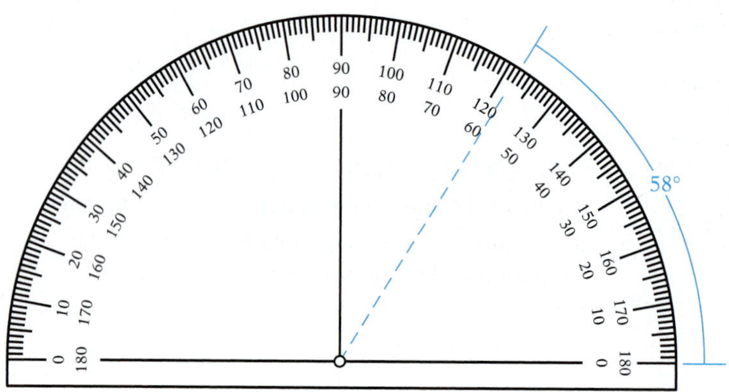

> **Take Note**
> The corner of a page of this book is a good example of a 90° angle.

A 90° angle is called a **right angle.** The symbol ∟ represents a right angle.

90°

Perpendicular lines are intersecting lines that form right angles.

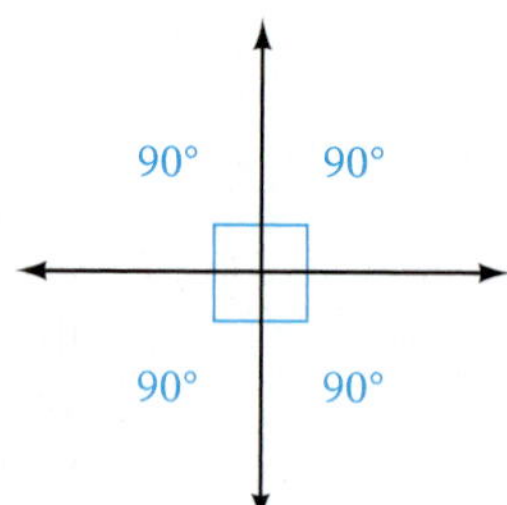

The symbol ⊥ means "is perpendicular to." In the figure at the right, $p \perp q$ and $\overline{AB} \perp \overline{CD}$. Note that line p contains $\overline{AB}$ and line q contains $\overline{CD}$. Perpendicular lines contain perpendicular line segments.

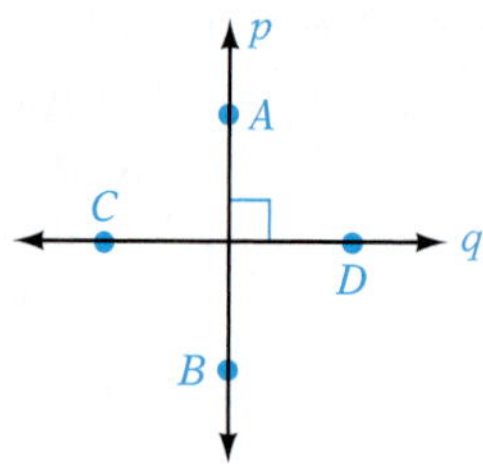

Complementary angles are two angles whose measures have the sum 90°.

$$\angle A + \angle B = 70° + 20° = 90°$$

$\angle A$ and $\angle B$ are complementary angles.

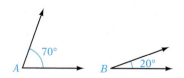

A 70° B 20°

Tips for Success
A great many new vocabulary words are introduced in this chapter. All of these terms are in **bold type.** The bold type indicates that these are concepts you must know to learn the material. Be sure to study each new term as it is presented.

A 180° angle is called a **straight angle.**

$\angle AOB$ is a straight angle.

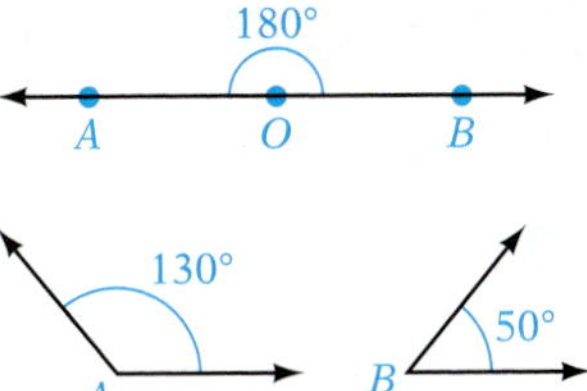

Supplementary angles are two angles whose measures have the sum 180°.

$$\angle A + \angle B = 130° + 50° = 180°$$

$\angle A$ and $\angle B$ are supplementary angles.

130°
A
50°
B

An **acute angle** is an angle whose measure is between 0° and 90°. $\angle B$ above is an acute angle. An **obtuse angle** is an angle whose measure is between 90° and 180°. $\angle A$ above is an obtuse angle.

Two angles that share a common side are **adjacent angles.** In the figure at the right, $\angle DAC$ and $\angle CAB$ are adjacent angles. $\angle DAC = 45°$ and $\angle CAB = 55°$.

$$\begin{aligned}\angle DAB &= \angle DAC + \angle CAB \\ &= 45° + 55° = 100°\end{aligned}$$

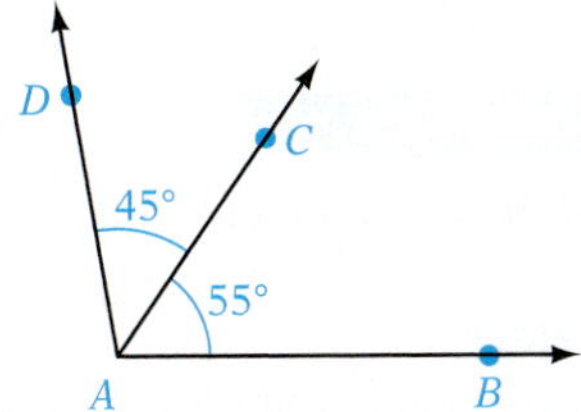

HOW TO • 2 In the figure at the right, $\angle EDG = 80°$. $\angle FDG$ is three times the measure of $\angle EDF$. Find the measure of $\angle EDF$.

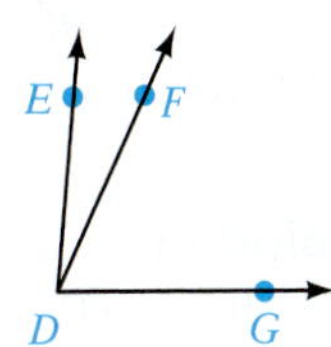

Let x = the measure of $\angle EDF$. Then $3x$ = the measure of $\angle FDG$. Write an equation and solve for x, the measure of $\angle EDF$.

$$\begin{aligned}\angle EDF + \angle FDG &= \angle EDG \\ x + 3x &= 80 \\ 4x &= 80 \\ x &= 20\end{aligned}$$

$\angle EDF = 20°$

Take Note
Answers to application problems must have units, such as degrees, feet, dollars, or hours.

EXAMPLE • 1

Given $MN = 15$ mm, $NO = 18$ mm, and $MP = 48$ mm, find OP.

Solution

$$\begin{aligned}MN + NO + OP &= MP \\ 15 + 18 + OP &= 48 \\ 33 + OP &= 48 \\ OP &= 15\end{aligned}$$

• $MN = 15$, $NO = 18$, $MP = 48$

$OP = 15$ mm

YOU TRY IT • 1

Given $QR = 24$ cm, $ST = 17$ cm, and $QT = 62$ cm, find RS.

Your solution

Solution on p. S17

EXAMPLE • 2

Given $XY = 9$ m and YZ is twice XY, find XZ.

Solution

$XZ = XY + YZ$

$XZ = XY + 2(XY)$ • YZ is twice XY.

$XZ = 9 + 2(9)$ • $XY = 9$

$XZ = 9 + 18$

$XZ = 27$

$XZ = 27$ m

YOU TRY IT • 2

Given $BC = 16$ ft and $AB = \frac{1}{4}(BC)$, find AC.

Your solution

EXAMPLE • 3

Find the complement of a 38° angle.

Strategy

Complementary angles are two angles whose sum is 90°. To find the complement, let x represent the complement of a 38° angle. Write an equation and solve for x.

Solution

$x + 38° = 90°$

$x = 52°$

The complement of a 38° angle is a 52° angle.

YOU TRY IT • 3

Find the supplement of a 129° angle.

Your strategy

Your solution

EXAMPLE • 4

Find the measure of $\angle x$.

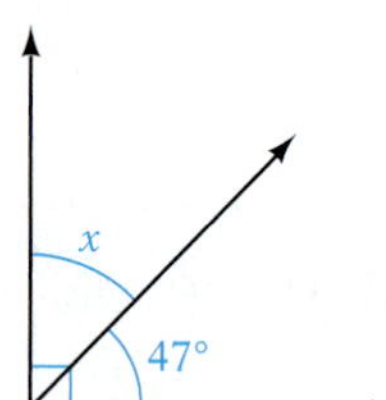

Strategy

To find the measure of $\angle x$, write an equation using the fact that the sum of the measure of $\angle x$ and 47° is 90°. Solve for $\angle x$.

Solution

$\angle x + 47° = 90°$

$\angle x = 43°$

The measure of $\angle x$ is 43°.

YOU TRY IT • 4

Find the measure of $\angle a$.

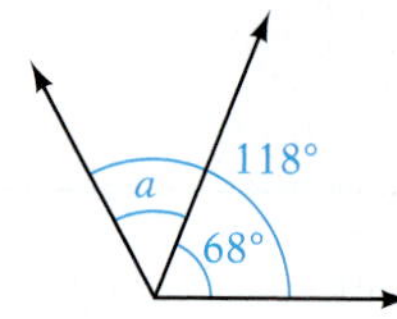

Your strategy

Your solution

Solutions on p. S17

OBJECTIVE B To solve problems involving angles formed by intersecting lines

Point of Interest

Many cities in the New World, unlike those in Europe, were designed using rectangular street grids. Washington, D.C., was planned that way except diagonal avenues were added, primarily for the purpose of enabling quick troop movement in the event that the city required defense. As an added precaution, monuments of statuary were constructed at major intersections so that attackers would not have a straight shot down a boulevard.

Four angles are formed by the intersection of two lines. If the two lines are perpendicular, each of the four angles is a right angle.

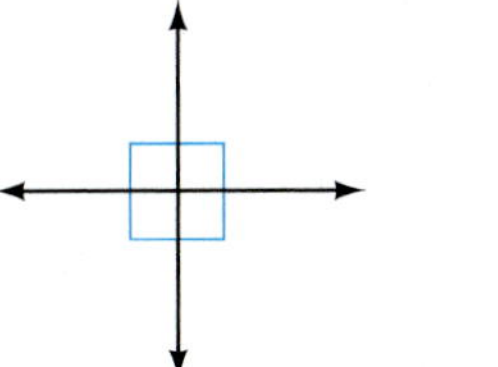

If the two lines are not perpendicular, then two of the angles formed are acute angles and two of the angles are obtuse angles. The two acute angles are always opposite each other, and the two obtuse angles are always opposite each other. In the figure at the right, $\angle w$ and $\angle y$ are acute angles. $\angle x$ and $\angle z$ are obtuse angles.

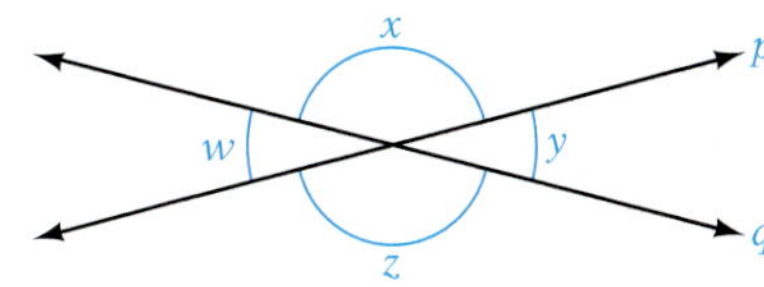

Two angles that are on opposite sides of the intersection of two lines are called **vertical angles.** Vertical angles have the same measure. $\angle w$ and $\angle y$ are vertical angles. $\angle x$ and $\angle z$ are vertical angles.

Vertical angles have the same measure.

$$\angle w = \angle y$$
$$\angle x = \angle z$$

Two angles that share a common side are called **adjacent angles.** For the figure shown above, $\angle x$ and $\angle y$ are adjacent angles, as are $\angle y$ and $\angle z$, $\angle z$ and $\angle w$, and $\angle w$ and $\angle x$. Adjacent angles of intersecting lines are supplementary angles.

Adjacent angles of intersecting lines are supplementary angles.

$$\angle x + \angle y = 180^\circ$$
$$\angle y + \angle z = 180^\circ$$
$$\angle z + \angle w = 180^\circ$$
$$\angle w + \angle x = 180^\circ$$

HOW TO • 3 Given that $\angle c = 65^\circ$, find the measures of angles a, b, and d.

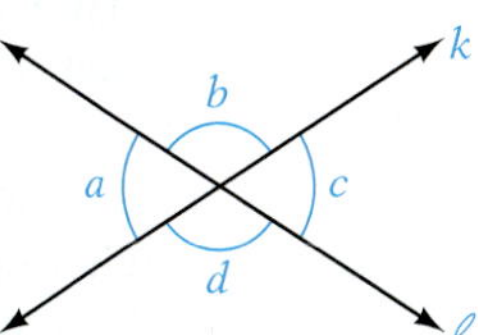

$\angle a = 65^\circ$ • $\angle a = \angle c$ because $\angle a$ and $\angle c$ are vertical angles.

$$\angle b + \angle c = 180^\circ$$
$$\angle b + 65^\circ = 180^\circ$$
$$\angle b = 115^\circ$$

• $\angle b$ is supplementary to $\angle c$ because $\angle b$ and $\angle c$ are adjacent angles of intersecting lines.

$\angle d = 115^\circ$ • $\angle d = \angle b$ because $\angle d$ and $\angle b$ are vertical angles.

A line that intersects two other lines at different points is called a **transversal.**

If the lines cut by a transversal t are parallel lines and the transversal is perpendicular to the parallel lines, all eight angles formed are right angles.

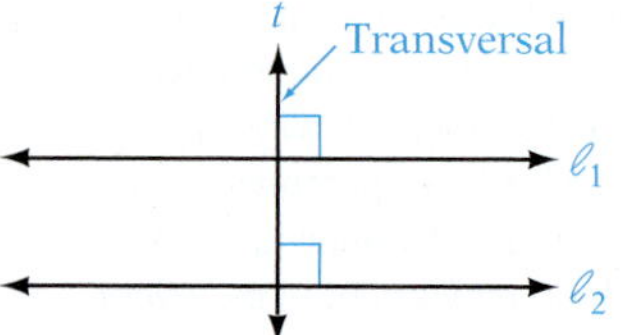

If the lines cut by a transversal t are parallel lines and the transversal is not perpendicular to the parallel lines, all four acute angles have the same measure and all four obtuse angles have the same measure. For the figure at the right,

$$\angle b = \angle d = \angle x = \angle z$$
$$\angle a = \angle c = \angle w = \angle y$$

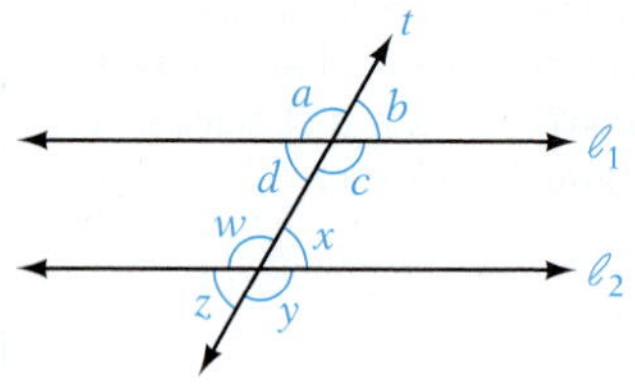

Alternate interior angles are two nonadjacent angles that are on opposite sides of the transversal and between the parallel lines. In the figure above, $\angle c$ and $\angle w$ are alternate interior angles; $\angle d$ and $\angle x$ are alternate interior angles. Alternate interior angles have the same measure.

Alternate interior angles have the same measure.

$$\angle c = \angle w$$
$$\angle d = \angle x$$

Alternate exterior angles are two nonadjacent angles that are on opposite sides of the transversal and outside the parallel lines. In the figure above, $\angle a$ and $\angle y$ are alternate exterior angles; $\angle b$ and $\angle z$ are alternate exterior angles. Alternate exterior angles have the same measure.

Alternate exterior angles have the same measure.

$$\angle a = \angle y$$
$$\angle b = \angle z$$

Corresponding angles are two angles that are on the same side of the transversal and are both acute angles or are both obtuse angles. For the figure above, the following pairs of angles are corresponding angles: $\angle a$ and $\angle w$, $\angle d$ and $\angle z$, $\angle b$ and $\angle x$, $\angle c$ and $\angle y$. Corresponding angles have the same measure.

Corresponding angles have the same measure.

$$\angle a = \angle w$$
$$\angle d = \angle z$$
$$\angle b = \angle x$$
$$\angle c = \angle y$$

HOW TO • 4 Given that $\ell_1 \parallel \ell_2$ and $\angle c = 58°$, find the measures of $\angle f$, $\angle h$, and $\angle g$.

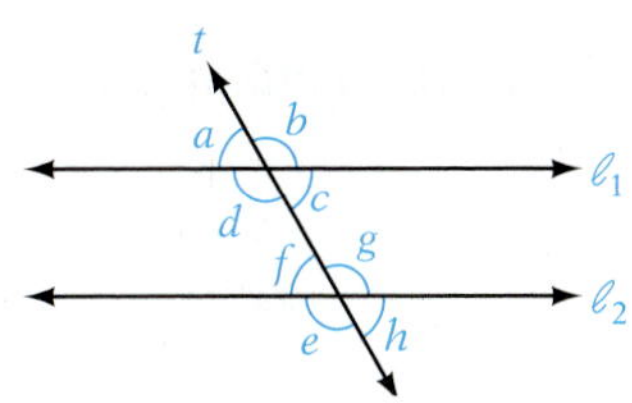

$\angle f = \angle c = 58°$ • $\angle c$ and $\angle f$ are alternate interior angles.

$\angle h = \angle c = 58°$ • $\angle c$ and $\angle h$ are corresponding angles.

$\angle g + \angle h = 180°$ • $\angle g$ is supplementary to $\angle h$.
$\angle g + 58° = 180°$
$\angle g = 122°$

EXAMPLE • 5

Find x.

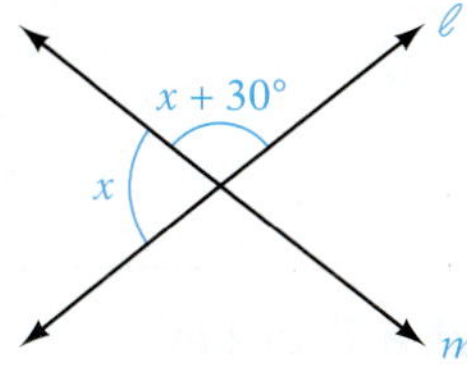

Strategy
The angles labeled are adjacent angles of intersecting lines and are therefore supplementary angles. To find x, write an equation and solve for x.

Solution

$$x + (x + 30°) = 180°$$
$$2x + 30° = 180°$$
$$2x = 150°$$
$$x = 75°$$

YOU TRY IT • 5

Find x.

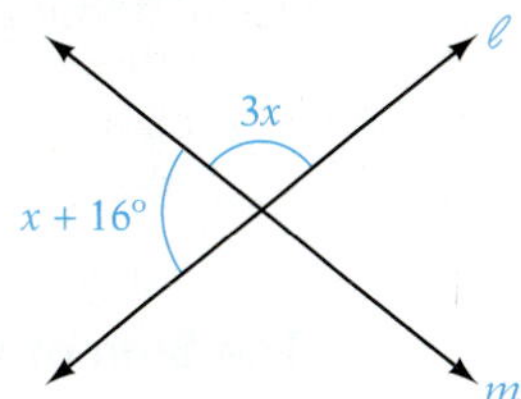

Your strategy

Your solution

EXAMPLE • 6

Given $\ell_1 \parallel \ell_2$, find x.

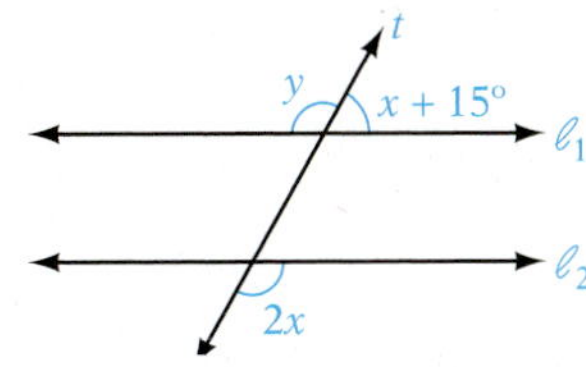

Strategy
$2x = y$ because alternate exterior angles have the same measure. $(x + 15°) + y = 180°$ because adjacent angles of intersecting lines are supplementary angles. Substitute $2x$ for y and solve for x.

Solution

$$(x + 15°) + 2x = 180°$$
$$3x + 15° = 180°$$
$$3x = 165°$$
$$x = 55°$$

YOU TRY IT • 6

Given $\ell_1 \parallel \ell_2$, find x.

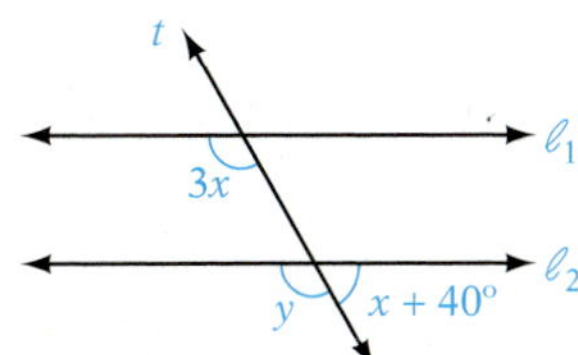

Your strategy

Your solution

Solutions on p. S17

OBJECTIVE C To solve problems involving the angles of a triangle

If the lines cut by a transversal are not parallel lines, the three lines will intersect at three points. In the figure at the right, the transversal t intersects lines p and q. The three lines intersect at points A, B, and C. These three points define three line segments, $\overline{AB}$, $\overline{BC}$, and $\overline{AC}$. The plane figure formed by these three line segments is called a **triangle.**

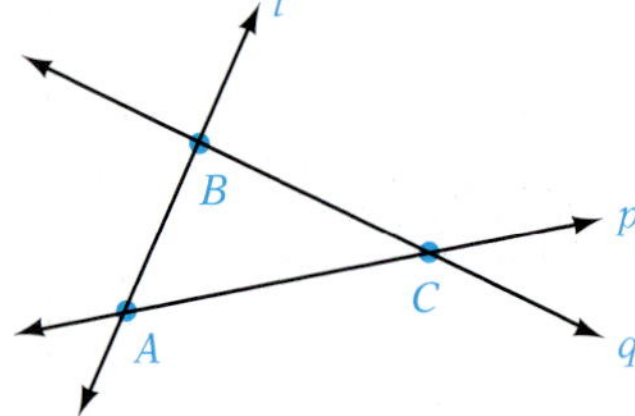

Each of the three points of intersection is the vertex of four angles. The angles within the region enclosed by the triangle are called **interior angles.** In the figure at the right, angles a, b, and c are interior angles. The sum of the measures of the interior angles of a triangle is 180°.

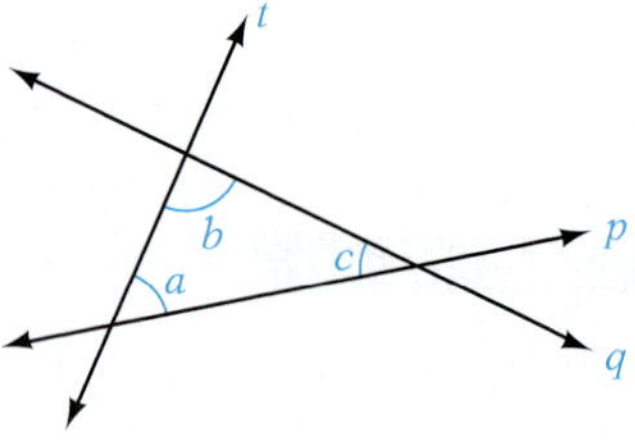

$\angle a + \angle b + \angle c = 180°$

The Sum of the Measures of the Interior Angles of a Triangle

The sum of the measures of the interior angles of a triangle is 180°.

An angle adjacent to an interior angle is an **exterior angle.** In the figure at the right, angles m and n are exterior angles for angle a. The sum of the measures of an interior and an exterior angle is 180°.

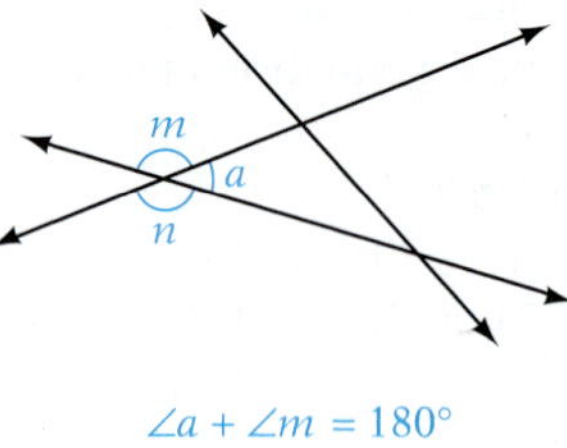

$\angle a + \angle m = 180°$
$\angle a + \angle n = 180°$

HOW TO 5 Given that $\angle c = 40°$ and $\angle d = 100°$, find the measure of $\angle e$.

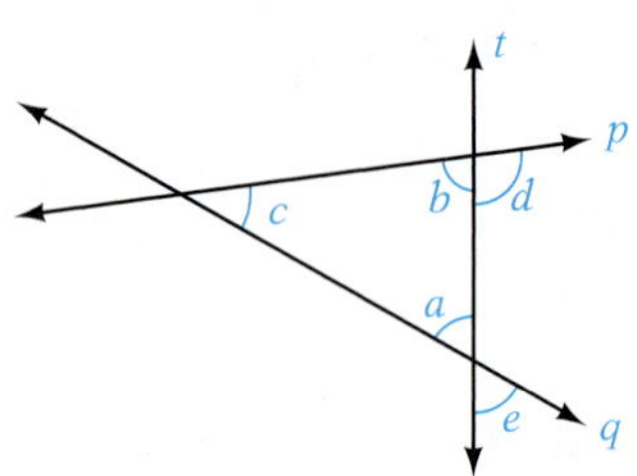

$\angle d$ and $\angle b$ are supplementary angles.

$$\begin{aligned} \angle d + \angle b &= 180° \\ 100° + \angle b &= 180° \\ \angle b &= 80° \end{aligned}$$

The sum of the interior angles is 180°.

$$\begin{aligned} \angle c + \angle b + \angle a &= 180° \\ 40° + 80° + \angle a &= 180° \\ 120° + \angle a &= 180° \\ \angle a &= 60° \end{aligned}$$

$\angle a$ and $\angle e$ are vertical angles.

$$\angle e = \angle a = 60°$$

EXAMPLE 7

Given that $\angle y = 55°$, find the measures of angles a, b, and d.

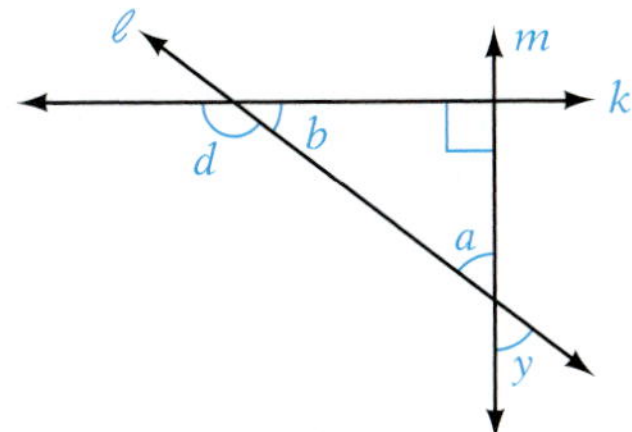

Strategy

- To find the measure of angle a, use the fact that $\angle a$ and $\angle y$ are vertical angles.
- To find the measure of angle b, use the fact that the sum of the measures of the interior angles of a triangle is 180°.
- To find the measure of angle d, use the fact that the sum of the measures of an interior and an exterior angle is 180°.

Solution

$\angle a = \angle y = 55°$

$$\begin{aligned} \angle a + \angle b + 90° &= 180° \\ 55° + \angle b + 90° &= 180° \\ \angle b + 145° &= 180° \\ \angle b &= 35° \end{aligned}$$

$$\begin{aligned} \angle d + \angle b &= 180° \\ \angle d + 35° &= 180° \\ \angle d &= 145° \end{aligned}$$

YOU TRY IT 7

Given that $\angle a = 45°$ and $\angle x = 100°$, find the measures of angles b, c, and y.

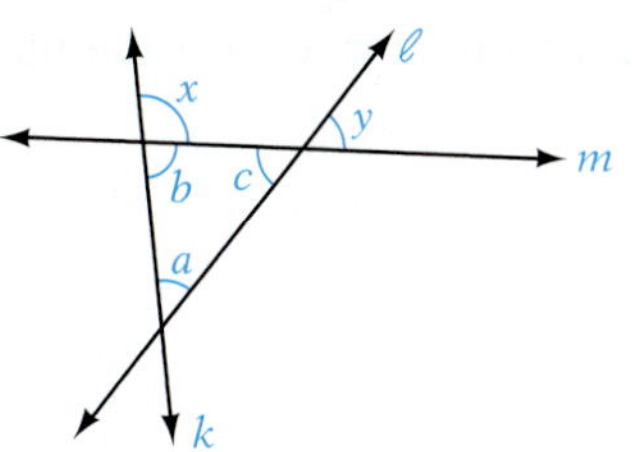

Your strategy

Your solution

EXAMPLE 8

Two angles of a triangle measure 53° and 78°. Find the measure of the third angle.

Strategy

To find the measure of the third angle, use the fact that the sum of the measures of the interior angles of a triangle is 180°. Write an equation using x to represent the measure of the third angle. Solve the equation for x.

Solution

$$\begin{aligned} x + 53° + 78° &= 180° \\ x + 131° &= 180° \\ x &= 49° \end{aligned}$$

The measure of the third angle is 49°.

YOU TRY IT 8

One angle in a triangle is a right angle, and one angle measures 34°. Find the measure of the third angle.

Your strategy

Your solution

Solutions on p. S18

7.1 EXERCISES

OBJECTIVE A To solve problems involving lines and angles

For Exercises 1 to 6, use a protractor to measure the angle. State whether the angle is acute, obtuse, or right.

1. 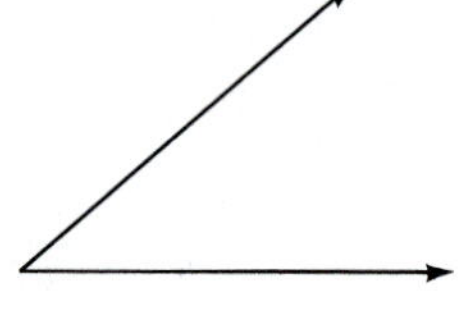

2. 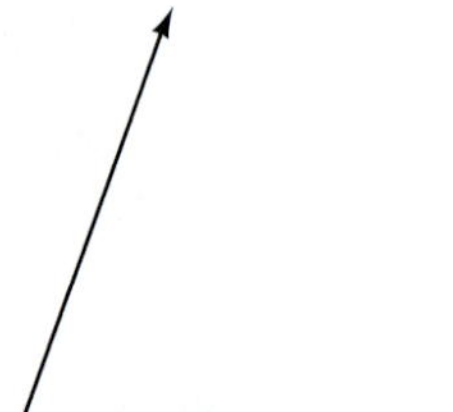

3. 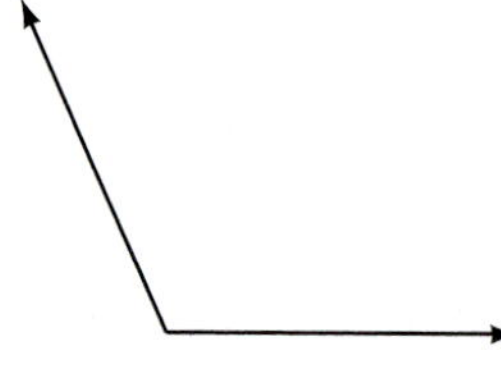

4. 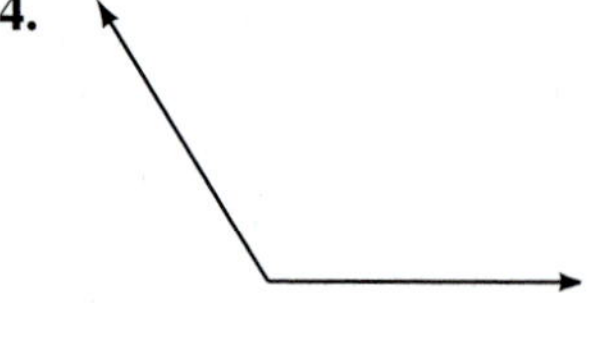

5. 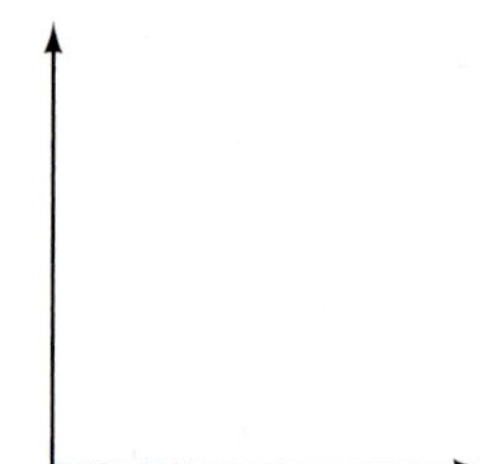

6. 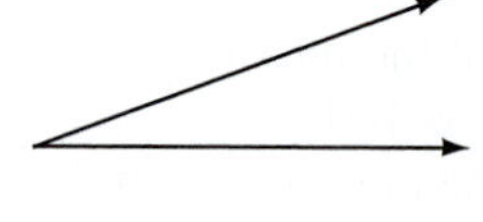

7. Find the complement of a 62° angle.

8. Find the complement of a 31° angle.

9. Find the supplement of a 162° angle.

10. Find the supplement of a 72° angle.

11. Given $AB = 12$ cm, $CD = 9$ cm, and $AD = 35$ cm, find the length of BC.

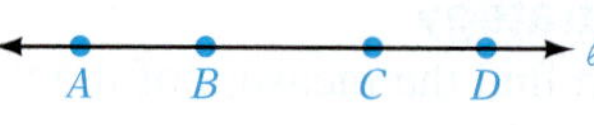

12. Given $AB = 21$ mm, $BC = 14$ mm, and $AD = 54$ mm, find the length of CD.

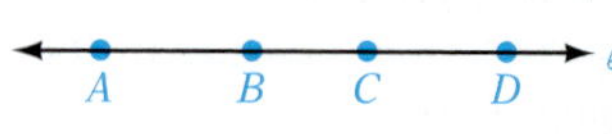

13. Given $QR = 7$ ft and RS is three times the length of QR, find the length of QS.

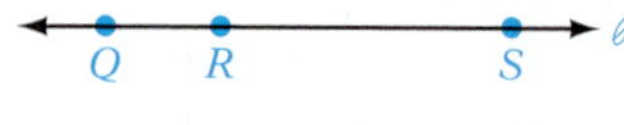

14. Given $QR = 15$ in. and RS is twice the length of QR, find the length of QS.

15. Given $EF = 20$ m and FG is one-half the length of EF, find the length of EG.

16. Given $EF = 18$ cm and FG is one-third the length of EF, find the length of EG.

17. Given $\angle LOM = 53°$ and $\angle LON = 139°$, find the measure of $\angle MON$.

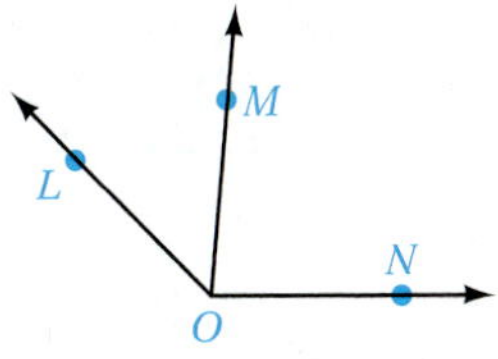

18. Given $\angle MON = 38°$ and $\angle LON = 85°$, find the measure of $\angle LOM$.

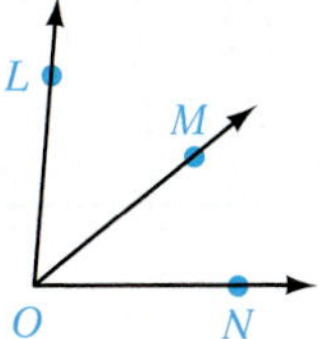

For Exercises 19 and 20, find the measure of $\angle x$.

19.

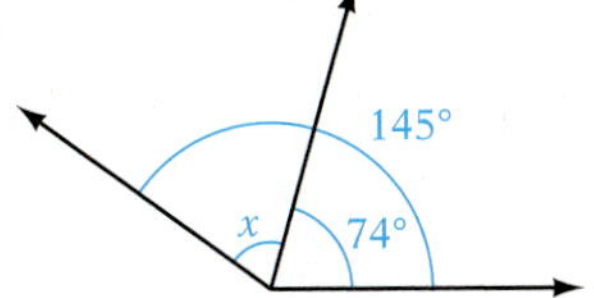

20.

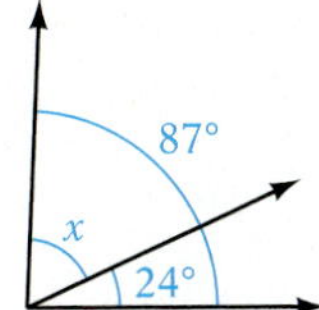

For Exercises 21 to 24, given that $\angle LON$ is a right angle, find the measure of $\angle x$.

21.

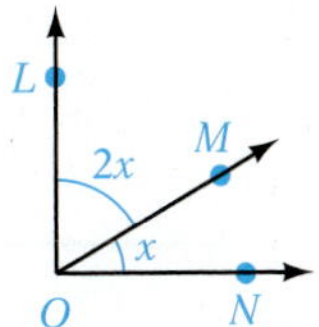

22.

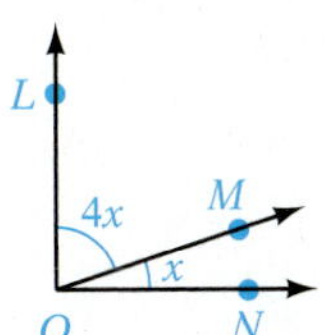

23.

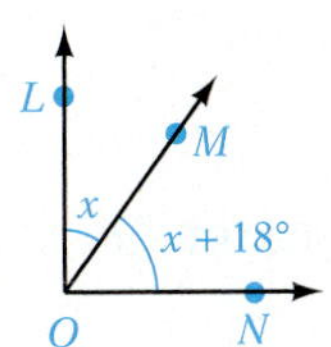

24.

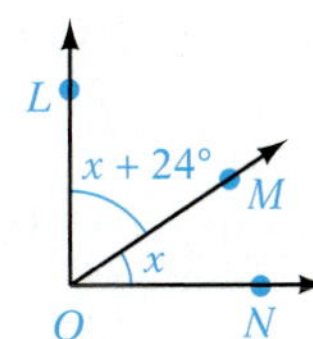

For Exercises 25 to 28, find the measure of $\angle a$.

25.

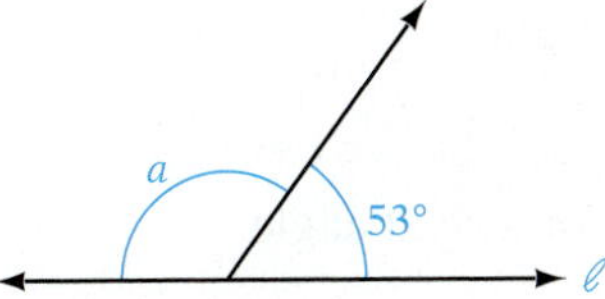

26.

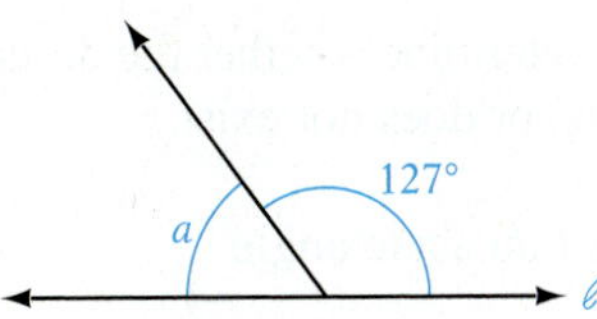

27.

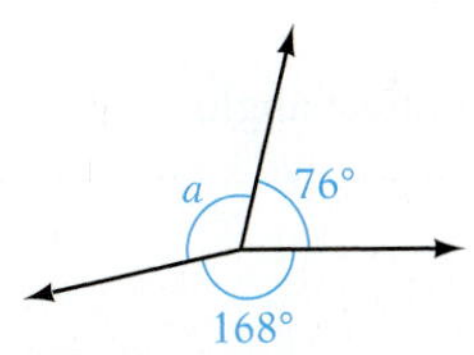

28.

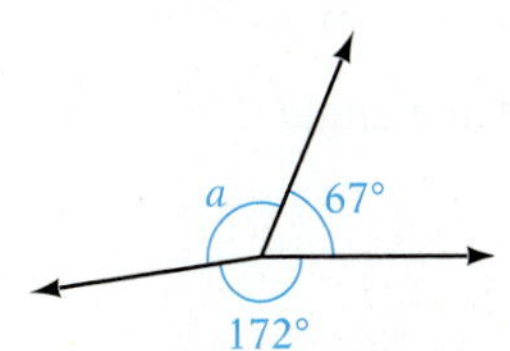

For Exercises 29 to 34, find x.

29.

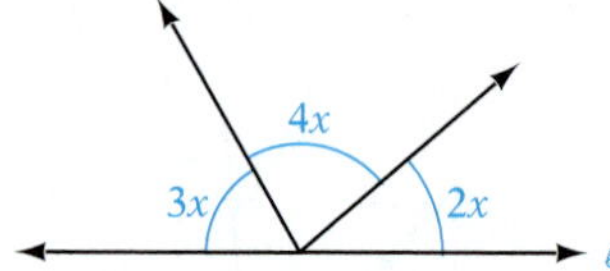

30.

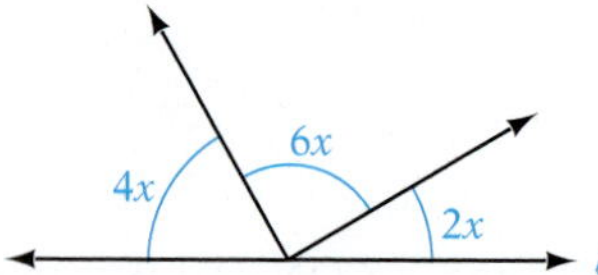

31.

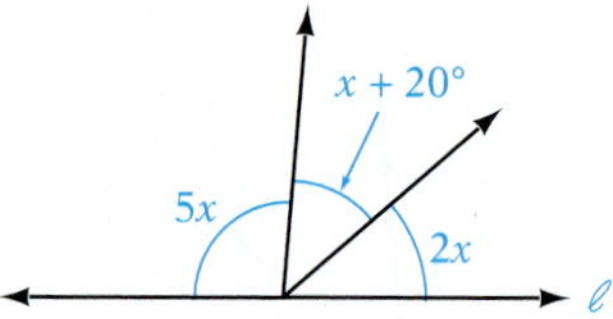

32.

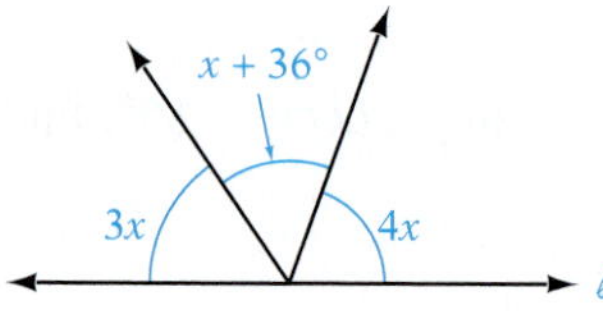

33.

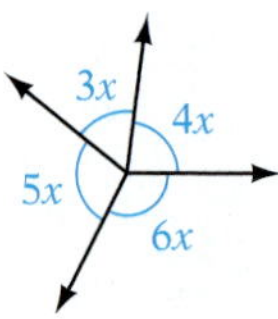

34.

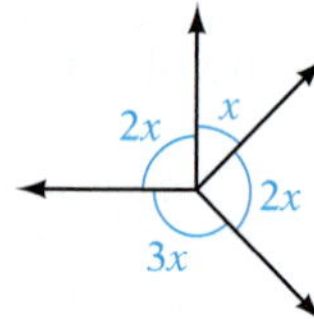

35. Given $\angle a = 51°$, find the measure of $\angle b$.

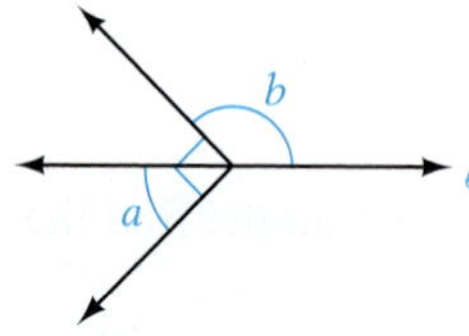

36. Given $\angle a = 38°$, find the measure of $\angle b$.

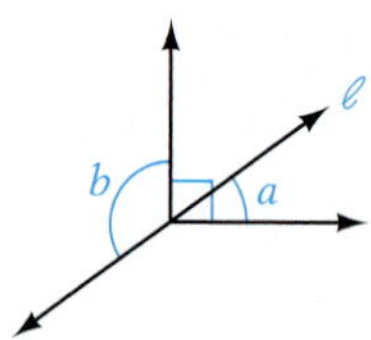

For Exercises 37 to 40, determine whether the described angle is an acute angle, is a right angle, is an obtuse angle, or does not exist.

37. The complement of an acute angle

38. The supplement of a right angle

39. The supplement of an acute angle

40. The complement of an obtuse angle

OBJECTIVE B To solve problems involving angles formed by intersecting lines

For Exercises 41 and 42, find the measure of $\angle x$.

41.

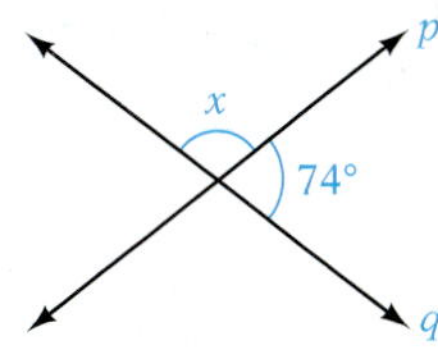

42.

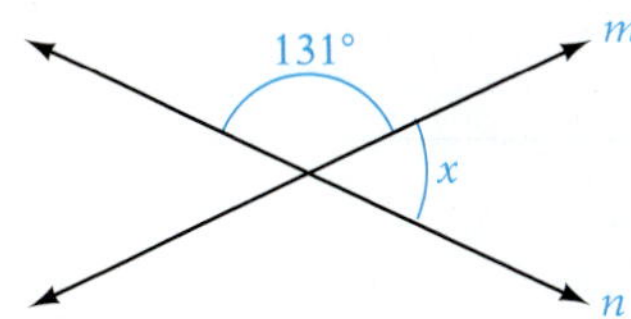

For Exercises 43 and 44, find x.

43.

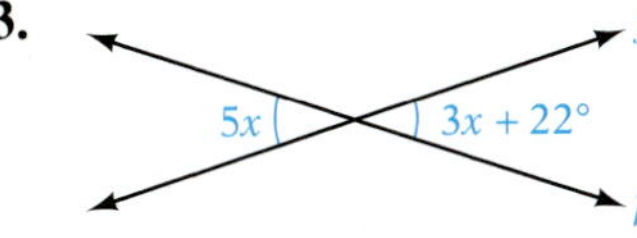

44.

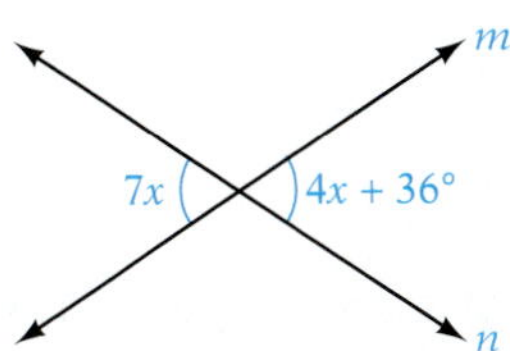

For Exercises 45 to 48, given that $\ell_1 \parallel \ell_2$, find the measures of angles a and b.

45.

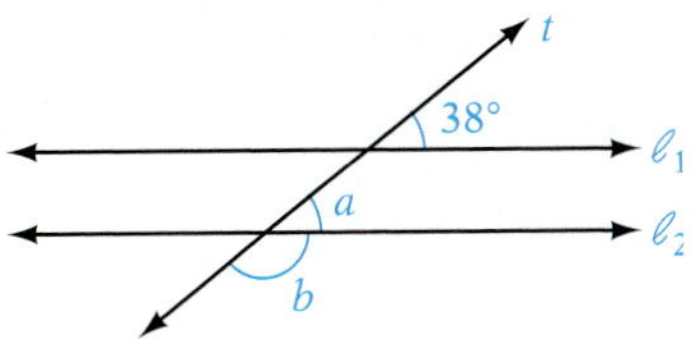

46.

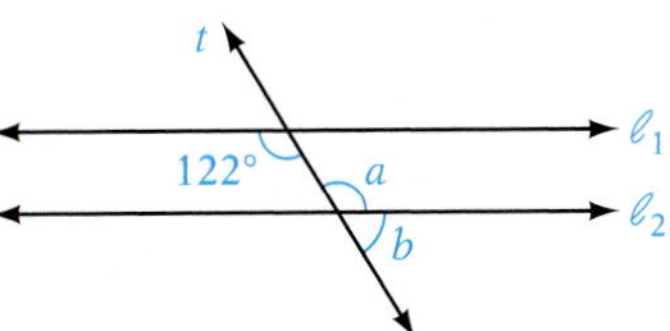

47.

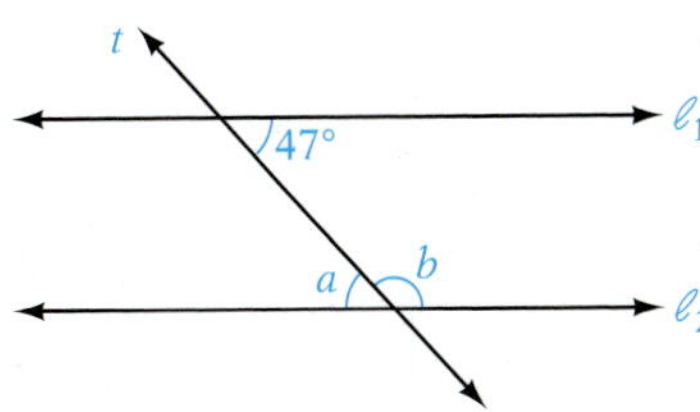

48.

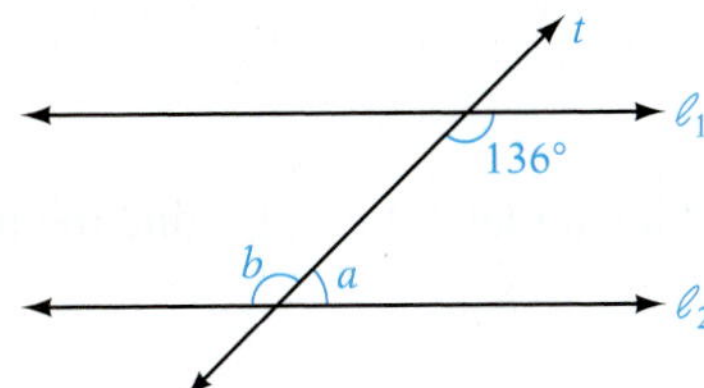

For Exercises 49 to 52, given that $\ell_1 \parallel \ell_2$, find x.

49.

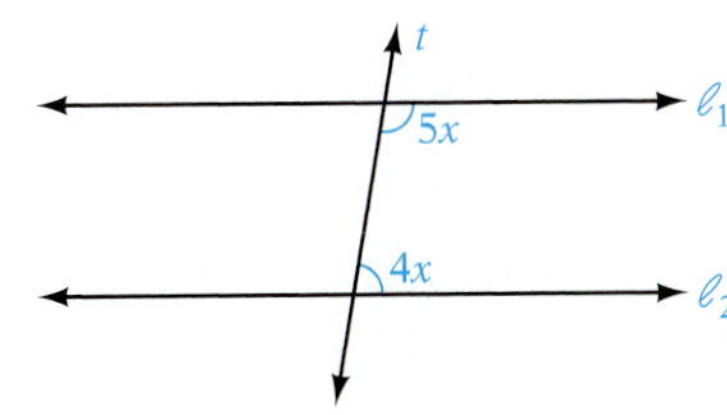

50.

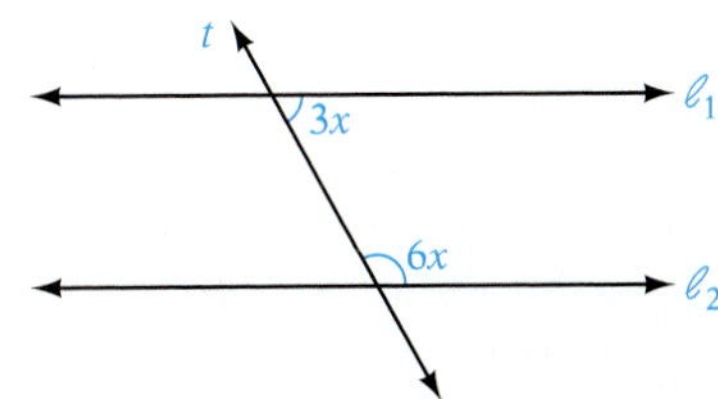

51.

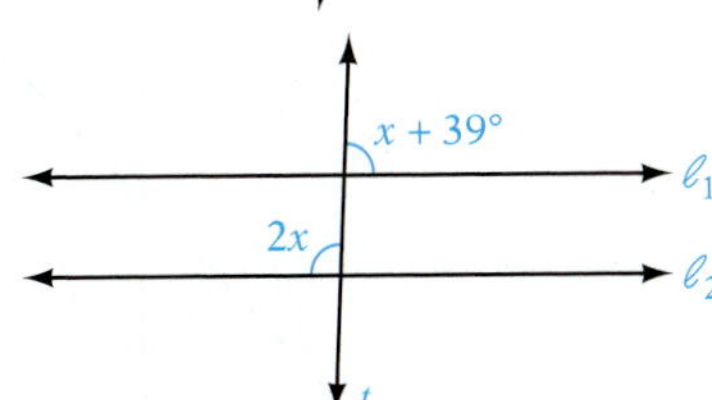

52.

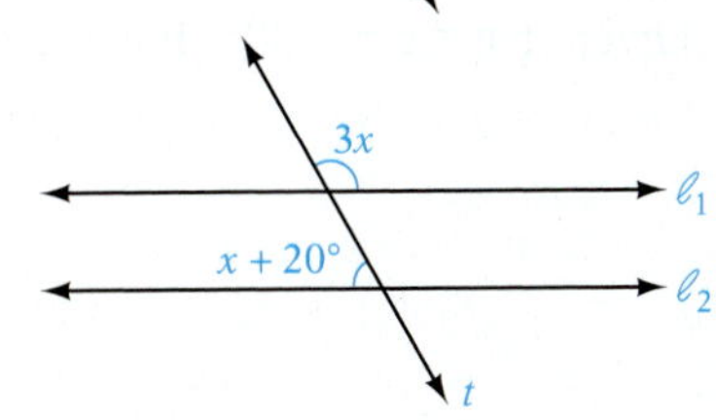

For Exercises 53 and 54, use the diagram at the right. Determine whether the given statement is *true* or *false*.

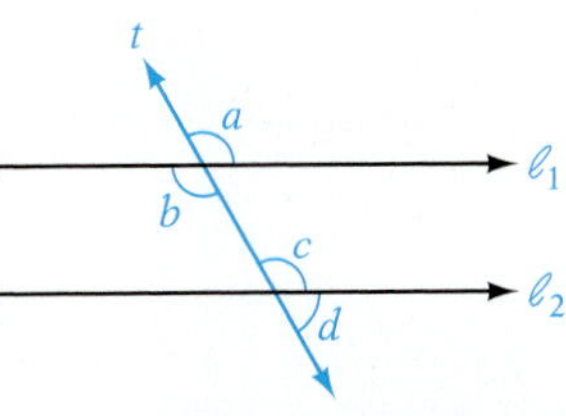

53. $\angle a$ and $\angle b$ have the same measure even if ℓ_1 and ℓ_2 are not parallel.

54. If $\angle a$ is greater than $\angle c$, then ℓ_1 and ℓ_2 are not parallel.

OBJECTIVE C To solve problems involving the angles of a triangle

55. Given that $\angle a = 95°$ and $\angle b = 70°$, find the measures of angles x and y.

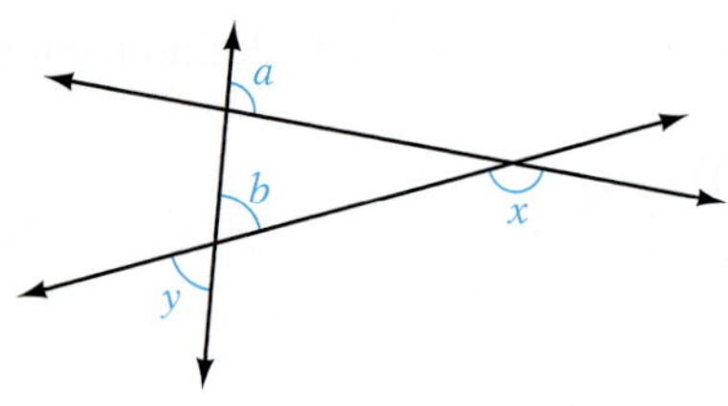

56. Given that $\angle a = 35°$ and $\angle b = 55°$, find the measures of angles x and y.

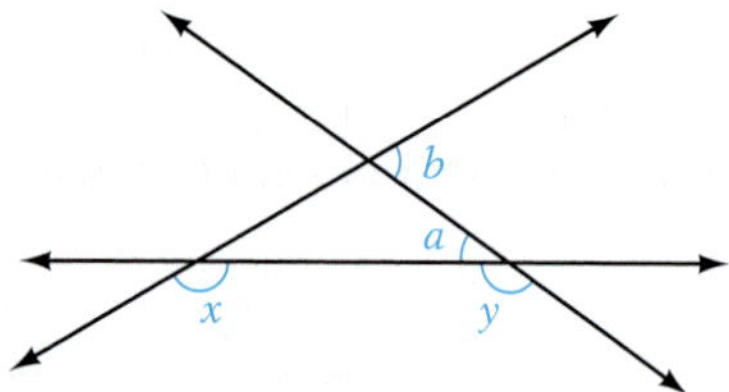

57. Given that $\angle y = 45°$, find the measures of angles a and b.

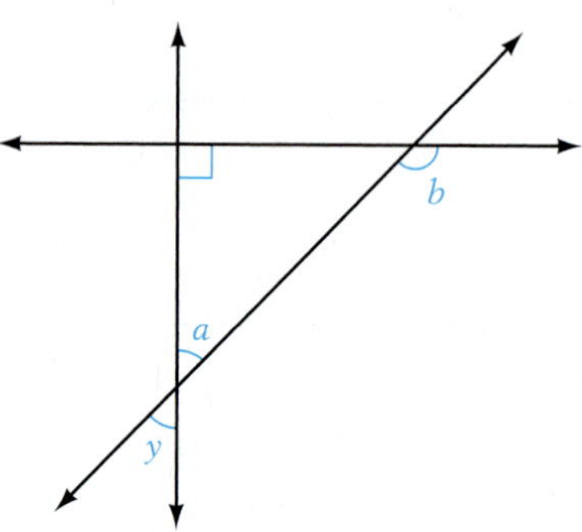

58. Given that $\angle y = 130°$, find the measures of angles a and b.

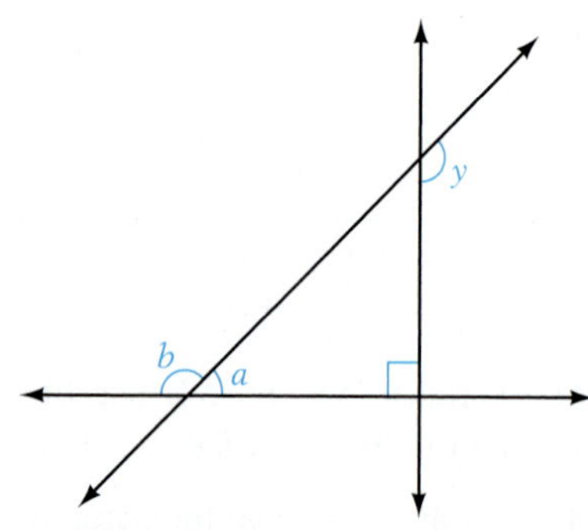

59. Given that $\overline{AO} \perp \overline{OB}$, express in terms of x the number of degrees in $\angle BOC$.

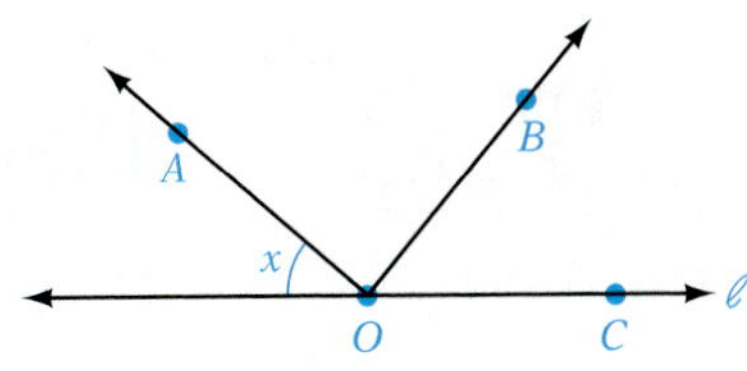

60. Given that $\overline{AO} \perp \overline{OB}$, express in terms of x the number of degrees in $\angle AOC$.

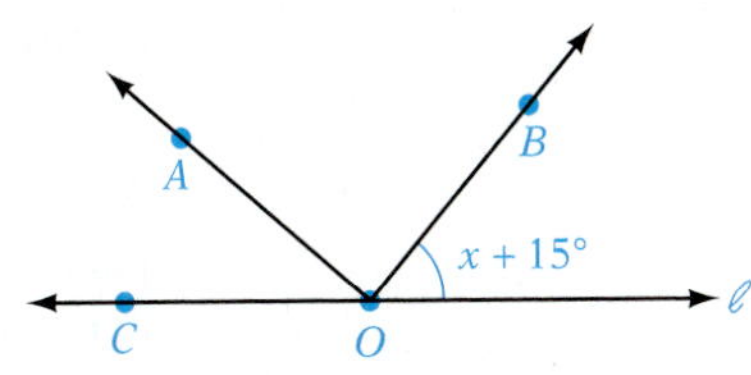

61. One angle in a triangle is a right angle, and one angle is equal to 30°. What is the measure of the third angle?

62. A triangle has a 45° angle and a right angle. Find the measure of the third angle.

63. Two angles of a triangle measure 42° and 103°. Find the measure of the third angle.

64. Two angles of a triangle measure 62° and 45°. Find the measure of the third angle.

65. A triangle has a 13° angle and a 65° angle. What is the measure of the third angle?

66. A triangle has a 105° angle and a 32° angle. What is the measure of the third angle?

For Exercises 67 and 68, determine whether the statement is *true* or *false*.

67. A triangle can have two obtuse angles.

68. If the sum of two angles of a triangle is less than 90°, then the third angle is an obtuse angle.

Applying the Concepts

69. Cut out a triangle and then tear off two of the angles, as shown at the right. Position the pieces you tore off so that angle a is adjacent to angle b and angle c is adjacent to angle b (on the other side). Describe what you observe. What does this demonstrate?

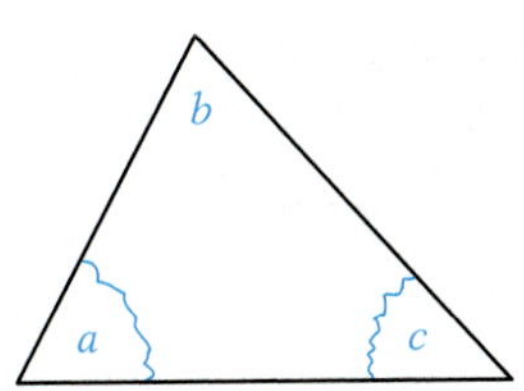

70. Determine whether the statement is always true, sometimes true, or never true.

a. Two lines that are parallel to a third line are parallel to each other.
b. A triangle contains at least two acute angles.
c. Vertical angles are complementary angles.

71. If $\overline{AB}$ and $\overline{CD}$ intersect at point O, and $\angle AOC = \angle BOC$, explain why $\overline{AB} \perp \overline{CD}$.

SECTION

7.2 Plane Geometric Figures

OBJECTIVE A To solve problems involving the perimeter of geometric figures

A **polygon** is a closed figure determined by three or more line segments that lie in a plane. The **sides of a polygon** are the line segments that form the polygon. The figures below are examples of polygons.

A

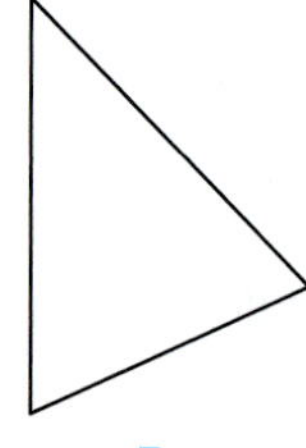
B

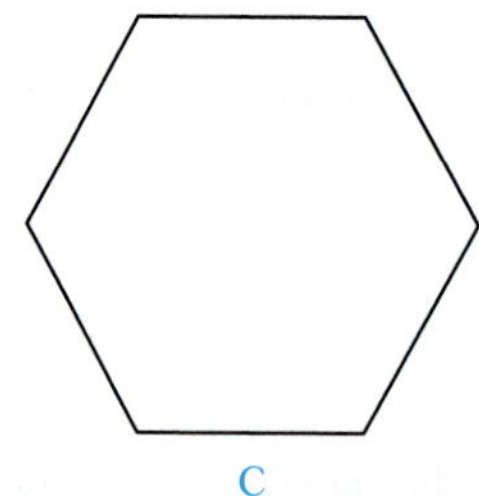
C

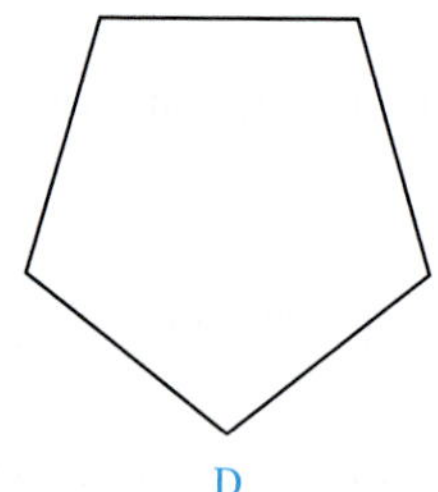
D

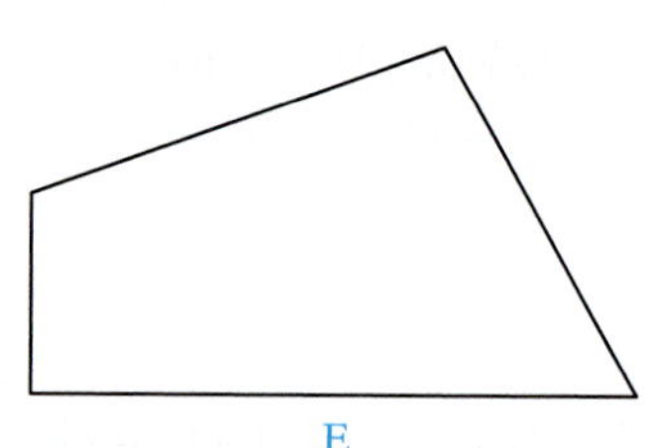
E

Point of Interest

Although a polygon is defined in terms of its *sides* (see the definition above), the word actually comes from the Latin word *polygonum*, which means "having many *angles*." This is certainly the case for many polygons.

The Pentagon in Arlington, Virginia

A **regular polygon** is one in which each side has the same length and each angle has the same measure. The polygons in Figures A, C, and D above are regular polygons.

The name of a polygon is based on the number of its sides. The table below lists the names of polygons that have from 3 to 10 sides.

Number of Sides	*Name of the Polygon*
3	Triangle
4	Quadrilateral
5	Pentagon
6	Hexagon
7	Heptagon
8	Octagon
9	Nonagon
10	Decagon

Triangles and quadrilaterals are two of the most common types of polygons. Triangles are distinguished by the number of equal sides and also by the measures of their angles.

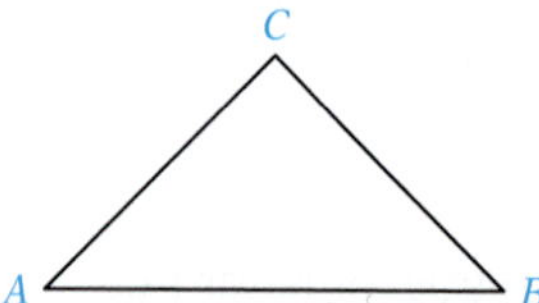

An **isosceles triangle** has two sides of equal length. The angles opposite the equal sides are of equal measure.
$AC = BC$
$\angle A = \angle B$

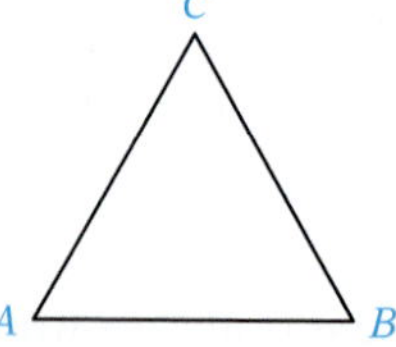

The three sides of an **equilateral triangle** are of equal length. The three angles are of equal measure.
$AB = BC = AC$
$\angle A = \angle B = \angle C$

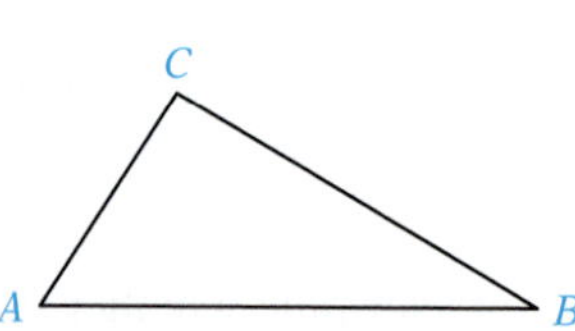

A **scalene triangle** has no two sides of equal length. No two angles are of equal measure.

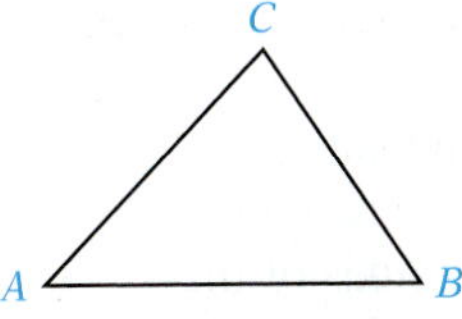

An **acute triangle** has three acute angles.

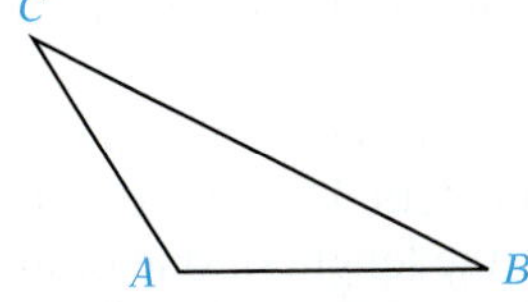

An **obtuse triangle** has one obtuse angle.

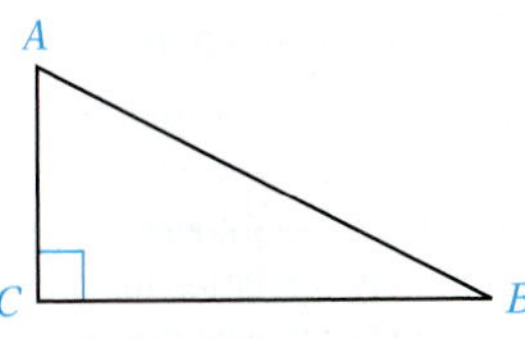

A **right triangle** has a right angle.

Quadrilaterals are also distinguished by their sides and angles, as shown below. Note that a rectangle, a square, and a rhombus are different forms of a parallelogram.

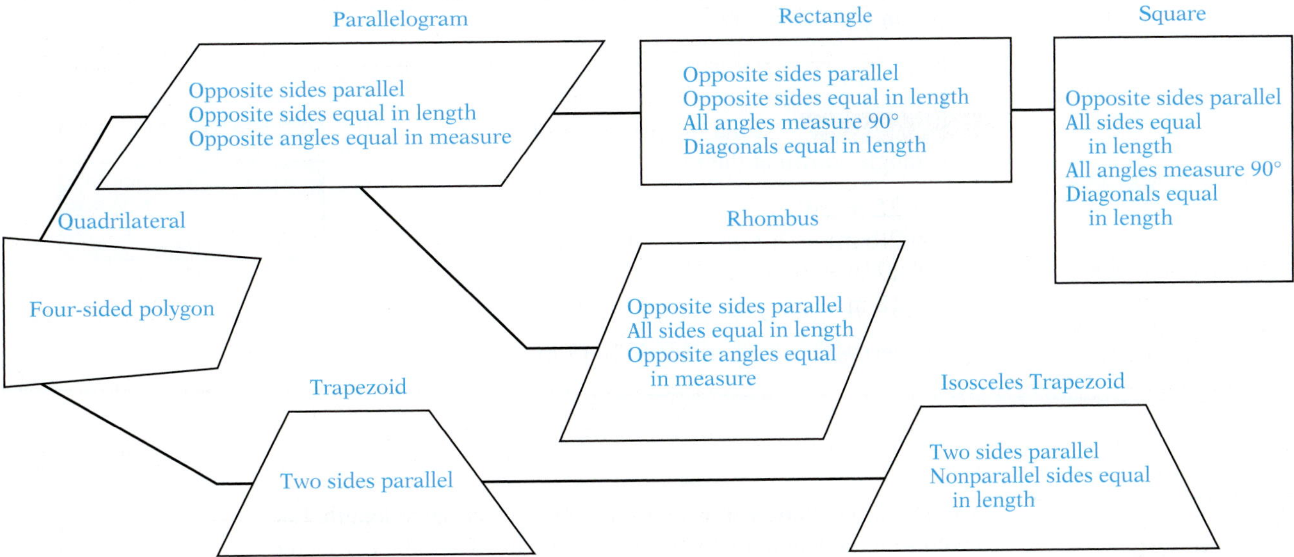

The **perimeter** of a plane geometric figure is a measure of the distance around the figure. Perimeter is used in buying fencing for a lawn or determining how much baseboard is needed for a room.

The perimeter of a triangle is the sum of the lengths of the three sides.

Perimeter of a Triangle

Let a, b, and c be the lengths of the sides of a triangle. The perimeter, P, of the triangle is given by $P = a + b + c$.

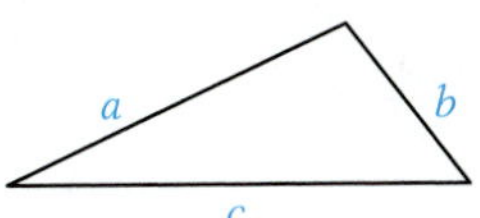

HOW TO • 1 Find the perimeter of the triangle shown at the right.

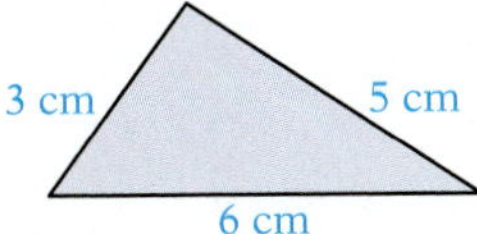

$$\begin{aligned} P &= a + b + c \\ &= 3 \text{ cm} + 5 \text{ cm} + 6 \text{ cm} \\ &= 14 \text{ cm} \end{aligned}$$

The perimeter of the triangle is 14 cm.

Point of Interest

Leonardo DaVinci painted the *Mona Lisa* on a rectangular canvas whose height was approximately 1.6 times its width. Rectangles with these proportions, called golden rectangles, were used extensively in Renaissance art.

The perimeter of a quadrilateral is the sum of the lengths of its four sides.

A rectangle has four right angles and opposite sides of equal length. Usually the length, L, of a rectangle refers to the length of one of the longer sides of the rectangle, and the width, W, refers to the length of one of the shorter sides. The perimeter can then be represented by $P = L + W + L + W$.

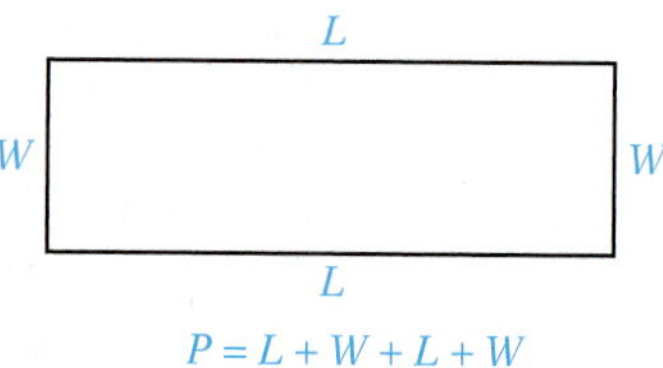

The formula for the perimeter of a rectangle is derived by combining like terms.

$$P = 2L + 2W$$

Perimeter of a Rectangle

Let L represent the length and W the width of a rectangle. The perimeter, P, of the rectangle is given by $P = 2L + 2W$.

HOW TO • 2 Find the perimeter of the rectangle shown at the right.

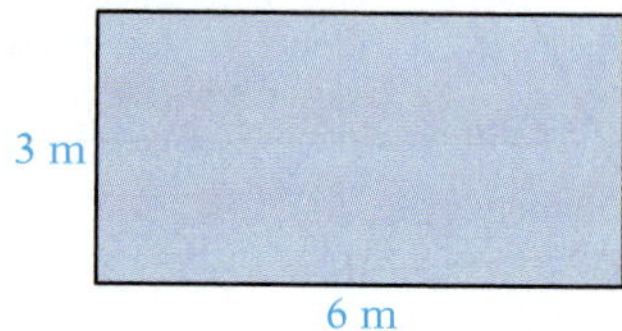

$P = 2L + 2W$

$= 2(6 \text{ m}) + 2(3 \text{ m})$ • $L = 6$ m; $W = 3$ m

$= 12 \text{ m} + 6 \text{ m}$

$= 18 \text{ m}$

The perimeter of the rectangle is 18 m.

A square is a rectangle in which each side has the same length. Let s represent the length of each side of a square. Then the perimeter of the square can be represented by $P = s + s + s + s$.

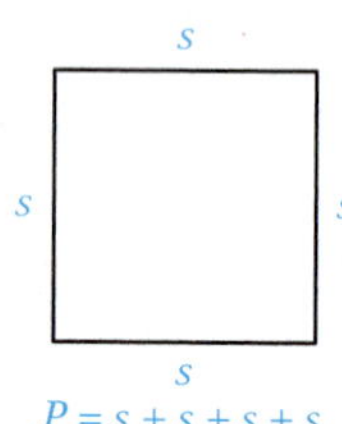

The formula for the perimeter of a square is derived by combining like terms.

$$P = 4s$$

Perimeter of a Square

Let s represent the length of a side of a square. The perimeter, P, of the square is given by $P = 4s$.

HOW TO • 3 Find the perimeter of the square shown at the right.

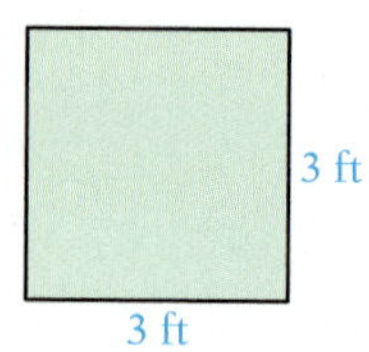

$P = 4s$

$= 4(3 \text{ ft})$ • $s = 3$ ft

$= 12 \text{ ft}$

The perimeter of the square is 12 ft.

A **circle** is a plane figure in which all points are the same distance from point O, called the **center** of the circle.

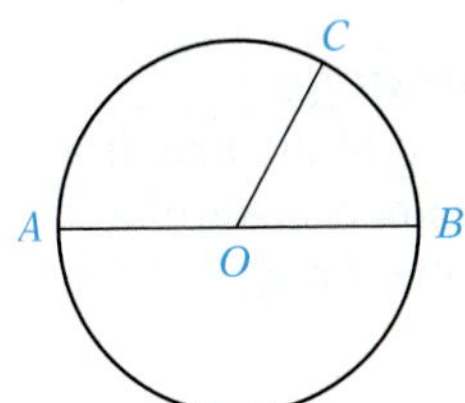

The **diameter** of a circle is a line segment across the circle through point O. AB is a diameter of the circle at the right. The variable d is used to designate a diameter of a circle.

The **radius** of a circle is a line segment from the center of the circle to a point on the circle. OC is a radius of the circle at the right. The variable r is used to designate a radius of a circle.

The length of the diameter is twice the length of the radius. $d = 2r$ or $r = \frac{1}{2}d$

Point of Interest

Archimedes (c. 287–212 B.C.E.) is the person who calculated that $\pi \approx 3\frac{1}{7}$. He actually showed that $3\frac{10}{71} < \pi < 3\frac{1}{7}$. The approximation $3\frac{10}{71}$ is a more accurate approximation of π than $3\frac{1}{7}$, but it is more difficult to use without a calculator.

The distance around a circle is called the **circumference.** The circumference, C, of a circle is equal to the product of π (pi) and the diameter. $C = \pi d$

Because $d = 2r$, the formula for the circumference can be written in terms of r. $C = 2\pi r$

Circumference of a Circle

The circumference, C, of a circle with diameter d and radius r is given by $C = \pi d$ or $C = 2\pi r$.

The formula for circumference uses the number π, which is an irrational number. The value of π can be approximated by a fraction or by a decimal. $\pi \approx \frac{22}{7}$ or $\pi \approx 3.14$

The π key on a scientific calculator gives a closer approximation of π than 3.14. Use a scientific calculator to find approximate values in calculations involving π.

Integrating Technology

The π key on your calculator can be used to find decimal approximations for expressions that contain π. To perform the calculation at the right, enter

6    .

HOW TO 4 Find the circumference of a circle with a diameter of 6 in.

$C = \pi d$ • The diameter of the circle is given. Use the circumference formula that involves the diameter. $d = 6$ in.

$C = \pi(6 \text{ in.})$

$C = 6\pi$ in. • The exact circumference of the circle is 6π in.

$C \approx 18.85$ in. • An approximate measure is found by using the π key on a calculator.

The approximate circumference is 18.85 in.

EXAMPLE • 1

A carpenter is designing a square patio with a perimeter of 44 ft. What is the length of each side?

Strategy
To find the length of each side, use the formula for the perimeter of a square. Substitute 44 for P and solve for s.

Solution

$P = 4s$
$44 = 4s$
$11 = s$

The length of each side of the patio is 11 ft.

YOU TRY IT • 1

The infield of a softball field is a square with each side of length 60 ft. Find the perimeter of the infield.

Your strategy

Your solution

EXAMPLE • 2

The dimensions of a triangular sail are 18 ft, 11 ft, and 15 ft. What is the perimeter of the sail?

Strategy
To find the perimeter, use the formula for the perimeter of a triangle. Substitute 18 ft for a, 11 ft for b, and 15 ft for c. Solve for P.

Solution

$P = a + b + c$
$P = 18 \text{ ft} + 11 \text{ ft} + 15 \text{ ft}$
$P = 44 \text{ ft}$

The perimeter of the sail is 44 ft.

YOU TRY IT • 2

What is the perimeter of a standard piece of copier paper that measures $8\frac{1}{2}$ in. by 11 in.?

Your strategy

Your solution

EXAMPLE • 3

Find the circumference of a circle with a radius of 15 cm. Round to the nearest hundredth.

Strategy
To find the circumference, use the circumference formula that involves the radius. An approximation is asked for; use the π key on a calculator. $r = 15$ cm.

Solution

$C = 2\pi r = 2\pi(15 \text{ cm}) = 30\pi \text{ cm} \approx 94.25 \text{ cm}$

The circumference is approximately 94.25 cm.

YOU TRY IT • 3

Find the circumference of a circle with a diameter of 9 in. Give the exact measure.

Your strategy

Your solution

Solutions on p. S18

OBJECTIVE B To solve problems involving the area of geometric figures 

Area is a measure of the amount of surface in a region. Area can be used to describe the size of a rug, a parking lot, a farm, or a national park. Area is measured in square units.

A square that measures 1 in. on each side has an area of 1 square inch, which is written 1 in^2.

A square that measures 1 cm on each side has an area of 1 square centimeter, which is written 1 cm^2.

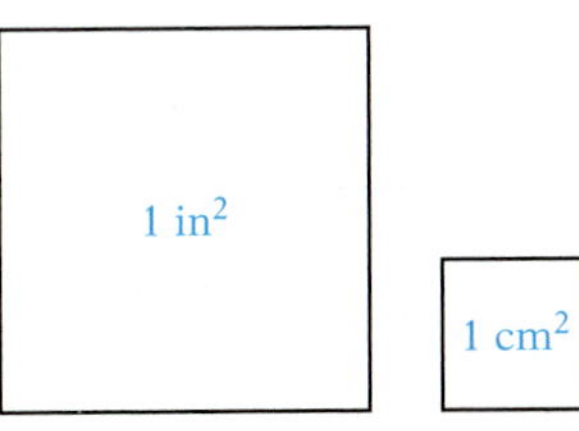

Larger areas can be measured in square feet (ft^2), square meters (m^2), square miles (mi^2), acres (43,560 ft^2), or any other square unit.

The area of a geometric figure is the number of squares that are necessary to cover the figure. In the figures below, two rectangles have been drawn and covered with squares. In the figure on the left, 12 squares, each of area 1 cm^2, were used to cover the rectangle. The area of the rectangle is 12 cm^2. In the figure on the right, 6 squares, each of area 1 in^2, were used to cover the rectangle. The area of the rectangle is 6 in^2.

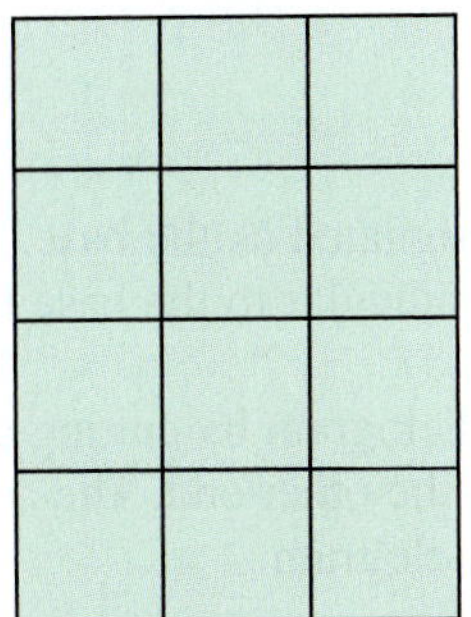

The area of the rectangle is 12 cm^2.

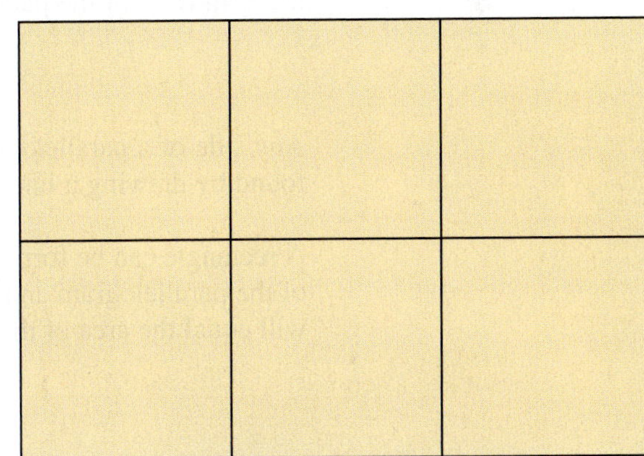

The area of the rectangle is 6 in^2.

Note from the above figures that the area of a rectangle can be found by multiplying the length of the rectangle by its width.

Area of a Rectangle

Let L represent the length and W the width of a rectangle. The area, A, of the rectangle is given by $A = LW$.

HOW TO • 5 Find the area of the rectangle shown at the right.

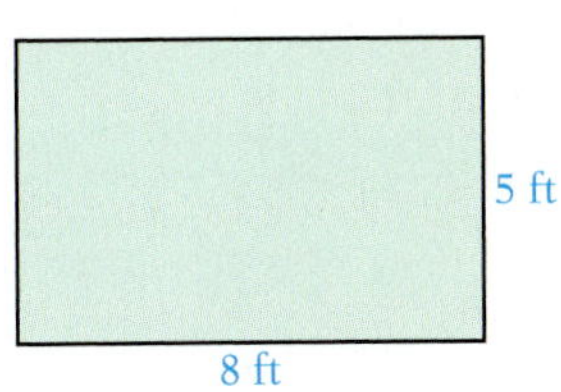

$$A = LW$$
$$= (8\text{ ft})(5\text{ ft}) \quad \bullet\ L = 8\text{ ft}, W = 5\text{ ft}$$
$$= 40\text{ ft}^2$$

The area of the rectangle is 40 ft^2.

A square is a rectangle in which all sides are the same length. Therefore, both the length and the width of a square can be represented by s, and $A = LW = s \cdot s = s^2$.

Area of a Square

Let s represent the length of a side of a square. The area, A, of the square is given by $A = s^2$.

$A = s \cdot s = s^2$

HOW TO • 6 Find the area of the square shown at the right.

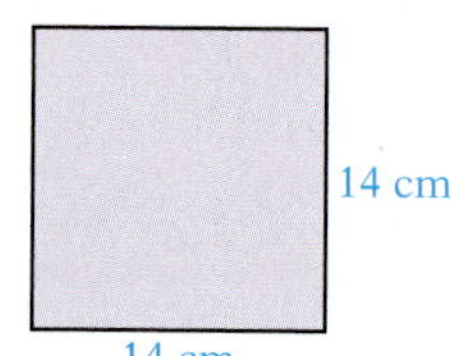

$$\begin{aligned} A &= s^2 \\ &= (14\text{ cm})^2 \quad \bullet\ s = 14\text{ cm} \\ &= 196\text{ cm}^2 \end{aligned}$$

The area of the square is 196 cm^2.

Figure $ABCD$ is a parallelogram. BC is the **base,** b, of the parallelogram. AE, perpendicular to the base, is the **height**, h, of the parallelogram.

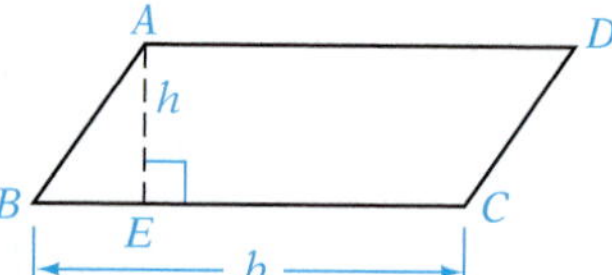

Any side of a parallelogram can be designated as the base. The corresponding height is found by drawing a line segment perpendicular to the base from the opposite side.

A rectangle can be formed from a parallelogram by cutting a right triangle from one end of the parallelogram and attaching it to the other end. The area of the resulting rectangle will equal the area of the original parallelogram.

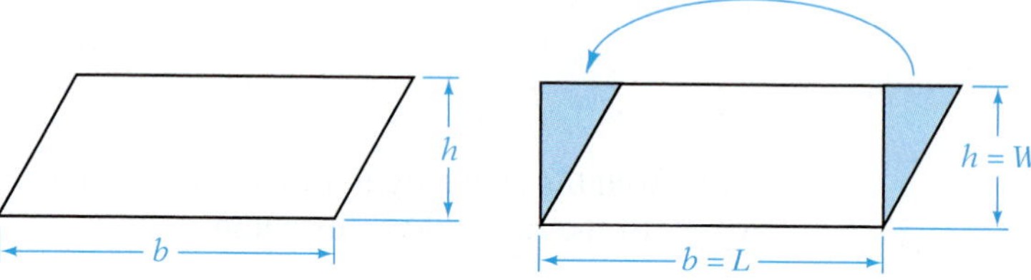

Area of a Parallelogram

Let b represent the length of the base and h the height of a parallelogram. The area, A, of the parallelogram is given by $A = bh$.

HOW TO • 7 Find the area of the parallelogram shown at the right.

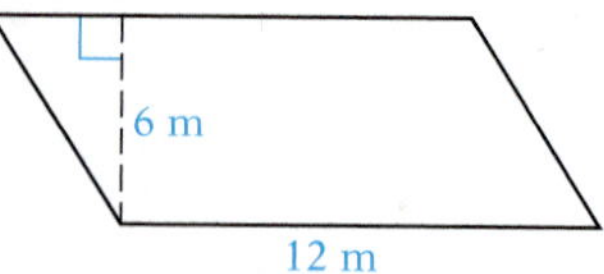

$A = bh = (12\text{ m})(6\text{ m}) = 72\text{ m}^2$

The area is 72 m^2.

Figure ABC is a triangle. AB is the **base,** b, of the triangle. CD, perpendicular to the base, is the **height**, h, of the triangle.

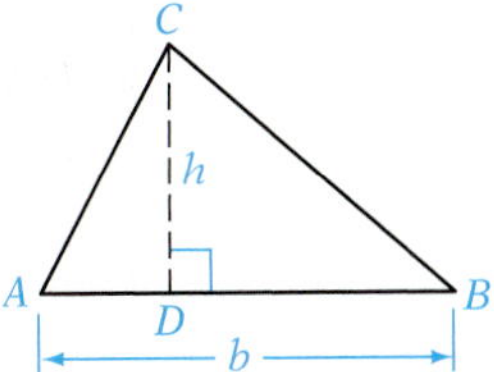

Any side of a triangle can be designated as the base. The corresponding height is found by drawing a line segment perpendicular to the base from the vertex opposite the base.

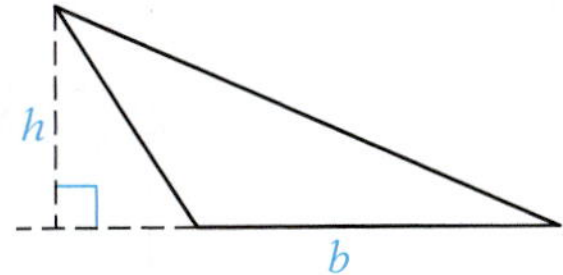

Consider the triangle with base b and height h shown at the right. By extending a line from C parallel to the base AB and equal in length to the base, a parallelogram is formed. The area of the parallelogram is bh and is twice the area of the triangle. Therefore, the area of the triangle is one-half the area of the parallelogram, or $\frac{1}{2}bh$.

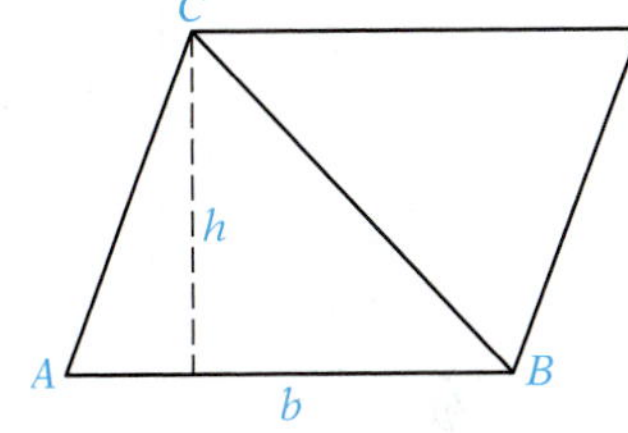

Area of a Triangle

Let b represent the length of the base and h the height of a triangle. The area, A, of the triangle is given by $A = \frac{1}{2}bh$.

Integrating Technology

To calculate the area of the triangle shown at the right, you can enter

20 **X** 5 **÷** 2 **=**

or

.5 **X** 20 **X** 5 **=**

HOW TO • 8 Find the area of the triangle shown below.

$$A = \frac{1}{2}bh$$

$$= \frac{1}{2}(20 \text{ m})(5 \text{ m}) \quad \bullet\ b = 20 \text{ m}, h = 5 \text{ m}$$

$$= 50 \text{ m}^2$$

The area of the triangle is 50 m^2.

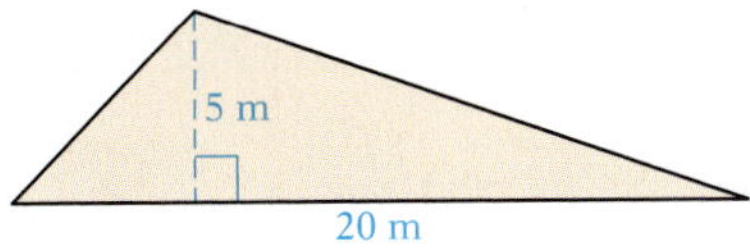

Figure $ABCD$ is a trapezoid. AB is one **base,** b_1, of the trapezoid, and CD is the other base, b_2. AE, perpendicular to the two bases, is the **height,** h.

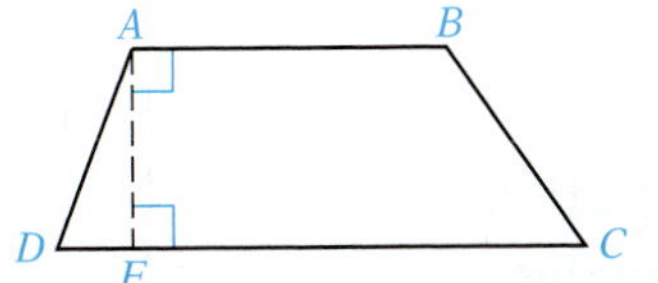

In the trapezoid at the right, the line segment BD divides the trapezoid into two triangles, ABD and BCD. In triangle ABD, b_1 is the base and h is the height. In triangle BCD, b_2 is the base and h is the height. The area of the trapezoid is the sum of the areas of the two triangles.

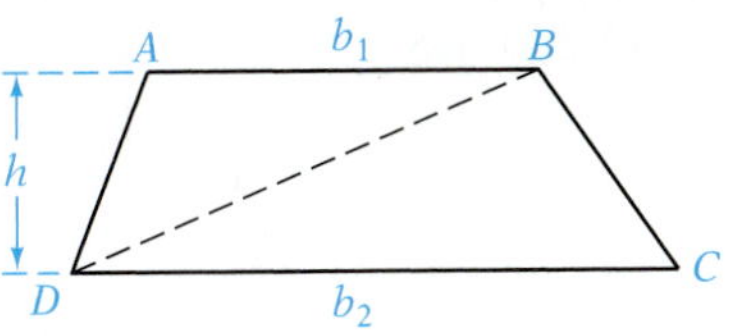

$$\text{Area of trapezoid } ABCD = \text{area of triangle } ABD + \text{area of triangle } BCD$$

$$= \frac{1}{2}b_1h + \frac{1}{2}b_2h = \frac{1}{2}h(b_1 + b_2)$$

Area of a Trapezoid

Let b_1 and b_2 represent the lengths of the bases and h the height of a trapezoid. The area, A, of the trapezoid is given by $A = \frac{1}{2}h(b_1 + b_2)$.

HOW TO • 9 Find the area of a trapezoid that has bases measuring 15 in. and 5 in. and a height of 8 in.

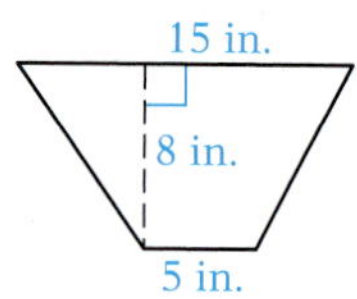

$$A = \frac{1}{2}h(b_1 + b_2)$$

$$= \frac{1}{2} \cdot (8 \text{ in.})(15 \text{ in.} + 5 \text{ in.}) = (4 \text{ in.})(20 \text{ in.}) = 80 \text{ in}^2$$

The area is 80 in^2.

The area of a circle is equal to the product of π and the square of the radius.

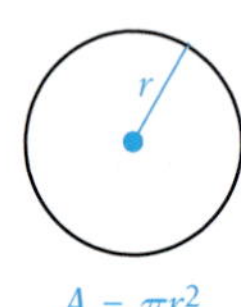

Area of a Circle

The area, A, of a circle with radius r is given by $A = \pi r^2$.

HOW TO • 10 Find the area of a circle that has a radius of 6 cm.

$A = \pi r^2$ • Use the formula for the area of a circle.
$A = \pi(6 \text{ cm})^2$ $r = 6$ cm
$A = \pi(36 \text{ cm}^2)$

$A = 36\pi \text{ cm}^2$ • The exact area of the circle is 36π cm^2.

$A \approx 113.10 \text{ cm}^2$ • An approximate measure is found by using the π key on a calculator.

The approximate area of the circle is 113.10 cm^2.

Integrating Technology

To approximate 36π on your calculator, enter 36 X π =.

For your reference, all of the formulas for the perimeter and area of the geometric figures presented in this section are listed in the Chapter Summary located at the end of this chapter.

EXAMPLE • 4

The Parks and Recreation Department of a city plans to plant grass seed in a playground that has the shape of a trapezoid, as shown below. Each bag of grass seed will seed 1500 ft². How many bags of grass seed should the department purchase?

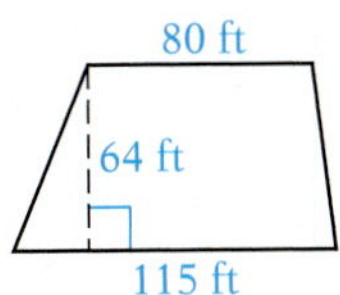

Strategy

To find the number of bags to be purchased:

- Use the formula for the area of a trapezoid to find the area of the playground.
- Divide the area of the playground by the area one bag will seed (1500).

Solution

$$A = \frac{1}{2}h(b_1 + b_2)$$

$$A = \frac{1}{2} \cdot (64 \text{ ft})(80 \text{ ft} + 115 \text{ ft})$$

$$A = 6240 \text{ ft}^2$$ • The area of the playground is 6240 ft².

$$6240 \div 1500 = 4.16$$

Because a portion of a fifth bag is needed, 5 bags of grass seed should be purchased.

YOU TRY IT • 4

An interior designer decides to wallpaper two walls of a room. Each roll of wallpaper will cover 30 ft². Each wall measures 8 ft by 12 ft. How many rolls of wallpaper should be purchased?

Your strategy

Your solution

EXAMPLE • 5

Find the area of a circle with a diameter of 5 ft. Give the exact measure.

Strategy

To find the area:

- Find the radius of the circle.
- Use the formula for the area of a circle. Leave the answer in terms of π.

Solution

$$r = \frac{1}{2}d = \frac{1}{2}(5 \text{ ft}) = 2.5 \text{ ft}$$

$$A = \pi r^2 = \pi(2.5 \text{ ft})^2 = \pi(6.25 \text{ ft}^2) = 6.25\pi \text{ ft}^2$$

The area of the circle is 6.25π ft².

YOU TRY IT • 5

Find the area of a circle with a radius of 11 cm. Round to the nearest hundredth.

Your strategy

Your solution

Solutions on p. S18

7.2 EXERCISES

OBJECTIVE A To solve problems involving the perimeter of geometric figures

For Exercises 1 to 4, name each polygon.

1. 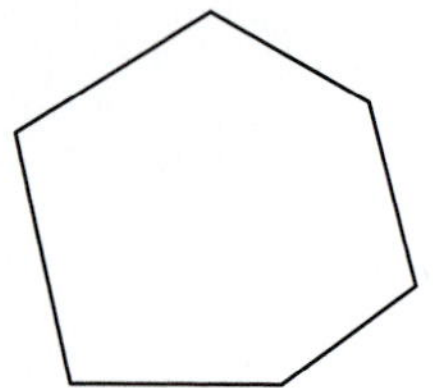

2. 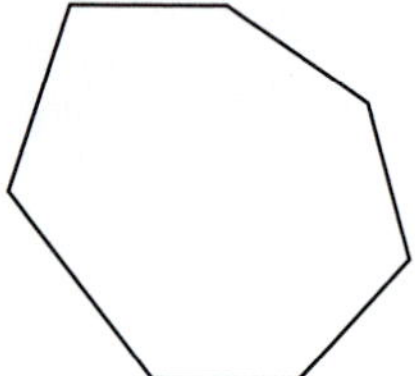

3. 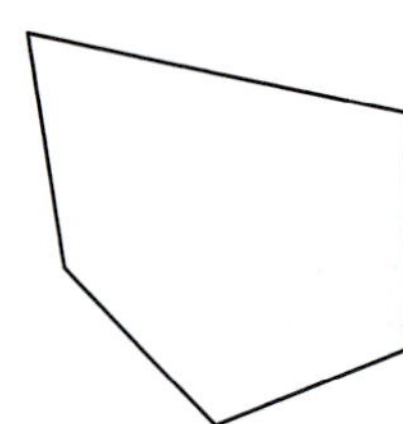

4. 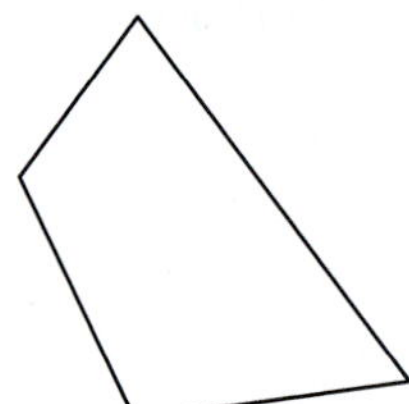

For Exercises 5 to 8, classify the triangle as isosceles, equilateral, or scalene.

5. 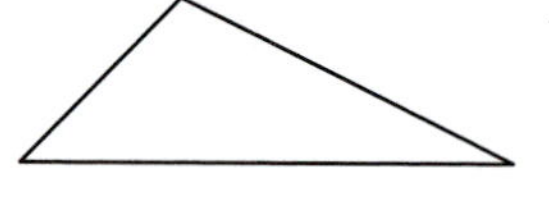

6. 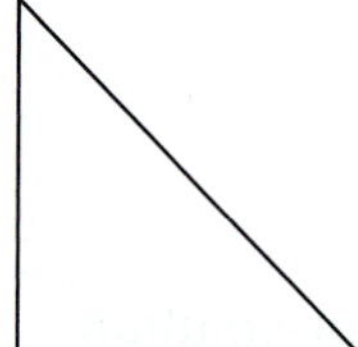

7. 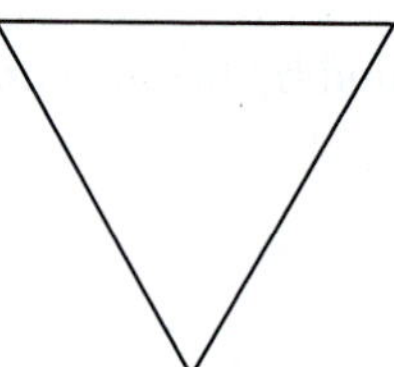

8. 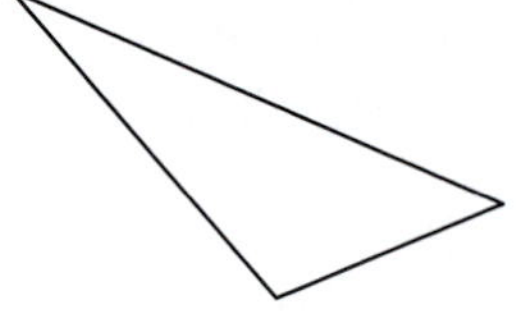

For Exercises 9 to 12, classify the triangle as acute, obtuse, or right.

9. 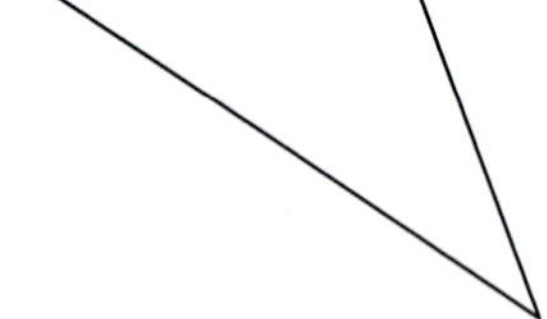

10. 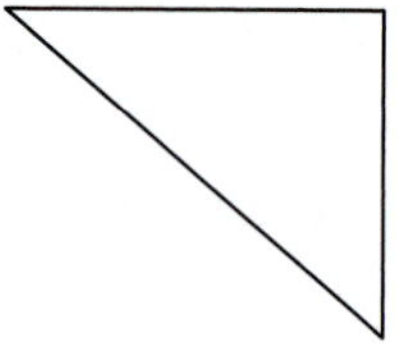

11. 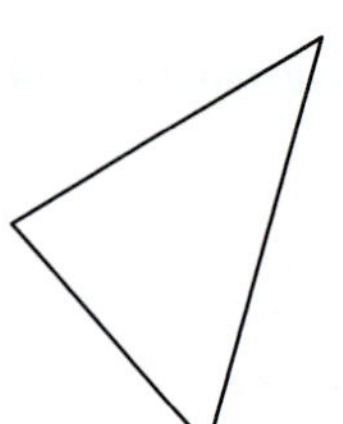

12. 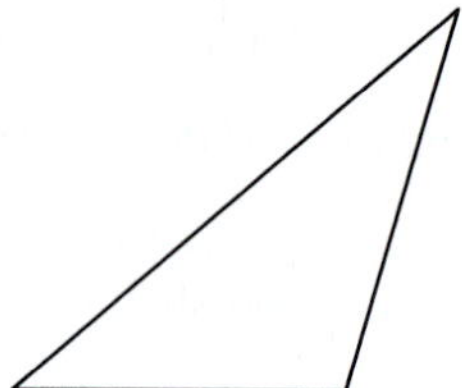

For Exercises 13 to 18, find the perimeter of the figure.

13. 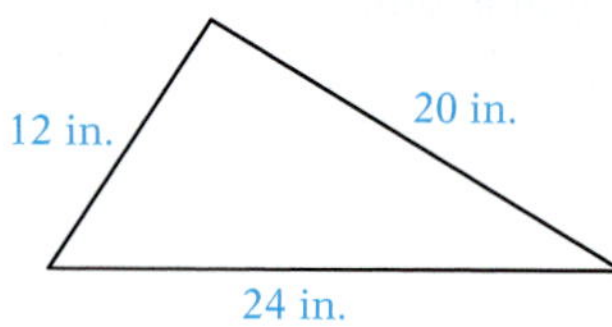

14. 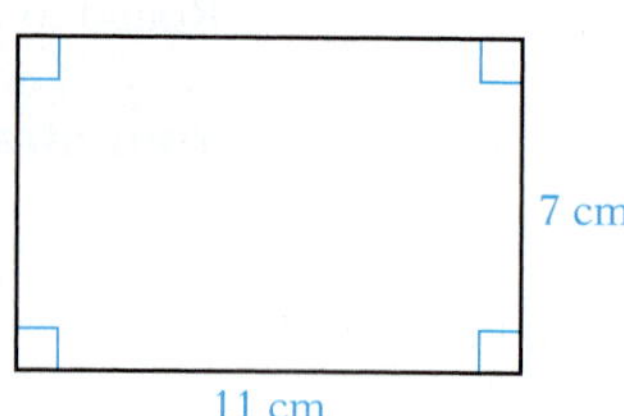

15. 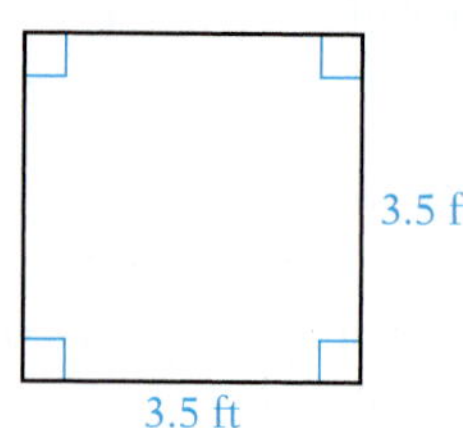

16. 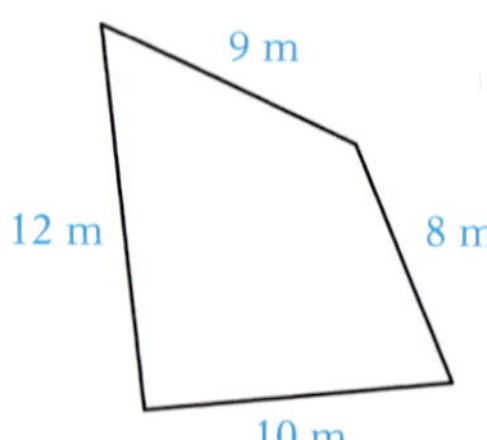

17. 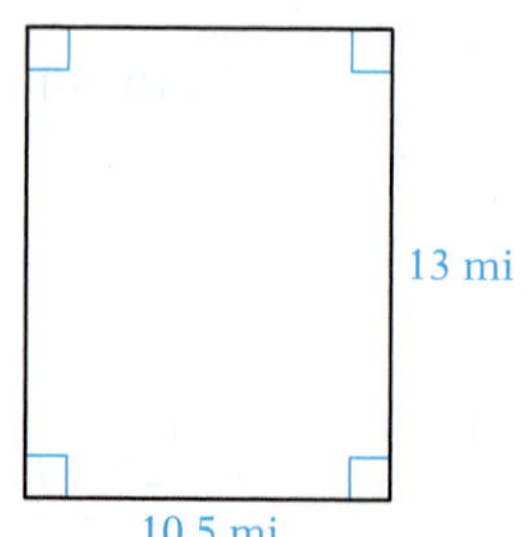

18. 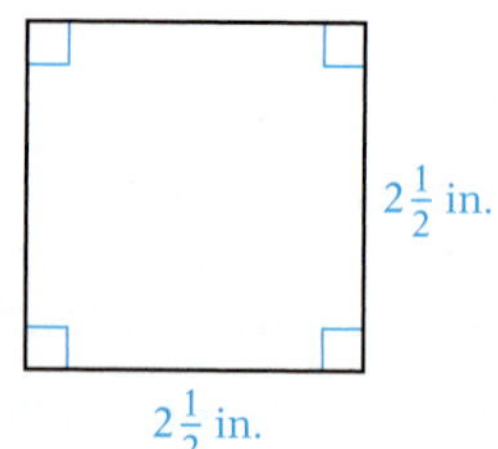

For Exercises 19 to 24, find the circumference of the figure. Give both the exact value and an approximation to the nearest hundredth.

19. 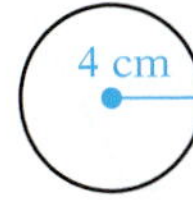

20. 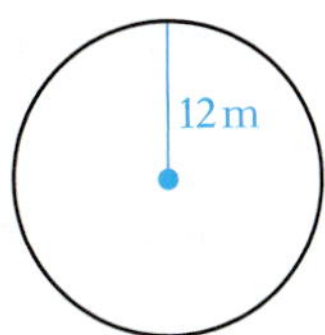

21. 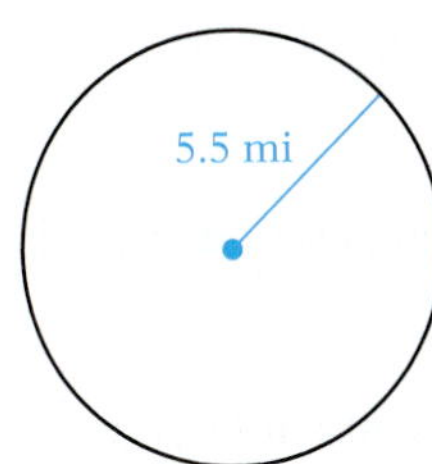

22. 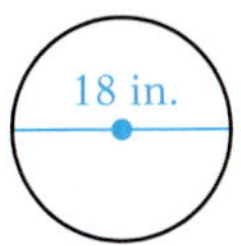

23. 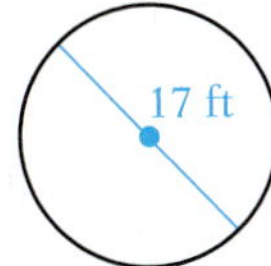

24. 

25. The lengths of the three sides of a triangle are 3.8 cm, 5.2 cm, and 8.4 cm. Find the perimeter of the triangle.

26. The lengths of the three sides of a triangle are 7.5 m, 6.1 m, and 4.9 m. Find the perimeter of the triangle.

27. The length of each of two sides of an isosceles triangle is $2\frac{1}{2}$ cm. The third side measures 3 cm. Find the perimeter of the triangle.

28. The length of each side of an equilateral triangle is $4\frac{1}{2}$ in. Find the perimeter of the triangle.

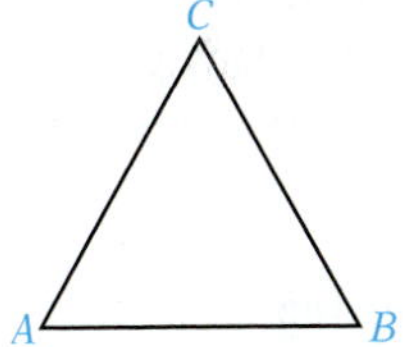

29. A rectangle has a length of 8.5 m and a width of 3.5 m. Find the perimeter of the rectangle.

30. Find the perimeter of a rectangle that has a length of $5\frac{1}{2}$ ft and a width of 4 ft.

31. The length of each side of a square is 12.2 cm. Find the perimeter of the square.

32. Find the perimeter of a square that is 0.5 m on each side.

33. Find the perimeter of a regular pentagon that measures 3.5 in. on each side.

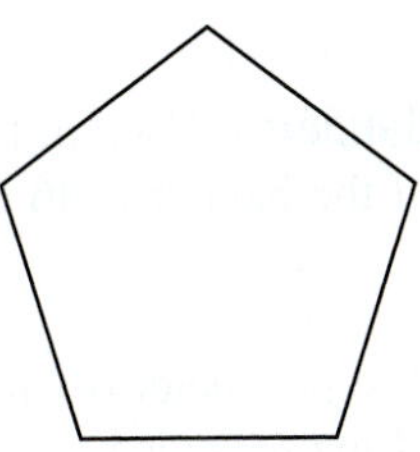

34. What is the perimeter of a regular hexagon that measures 8.5 cm on each side?

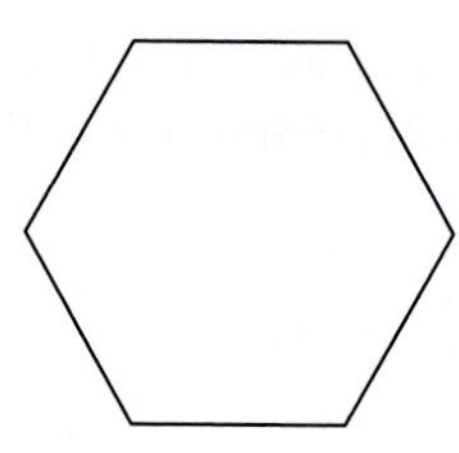

35. The radius of a circle is 4.2 cm. Find the length of a diameter of the circle.

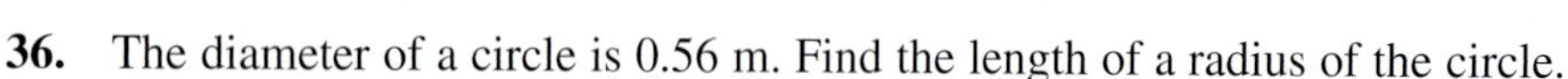

36. The diameter of a circle is 0.56 m. Find the length of a radius of the circle.

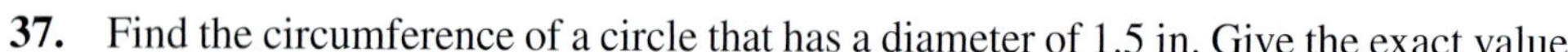

37. Find the circumference of a circle that has a diameter of 1.5 in. Give the exact value.

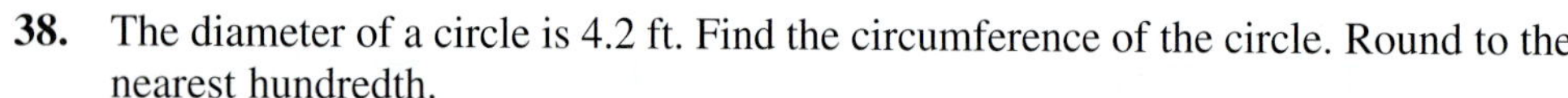

38. The diameter of a circle is 4.2 ft. Find the circumference of the circle. Round to the nearest hundredth.

39. The radius of a circle is 36 cm. Find the circumference of the circle. Round to the nearest hundredth.

40. Find the circumference of a circle that has a radius of 2.5 m. Give the exact value.

41. **Fencing** How many feet of fencing should be purchased for a rectangular garden that is 18 ft long and 12 ft wide?

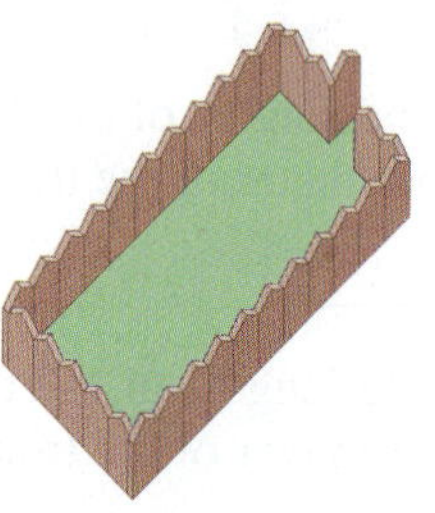

42. **Quilting** How many meters of binding are required to bind the edge of a rectangular quilt that measures 3.5 m by 8.5 m?

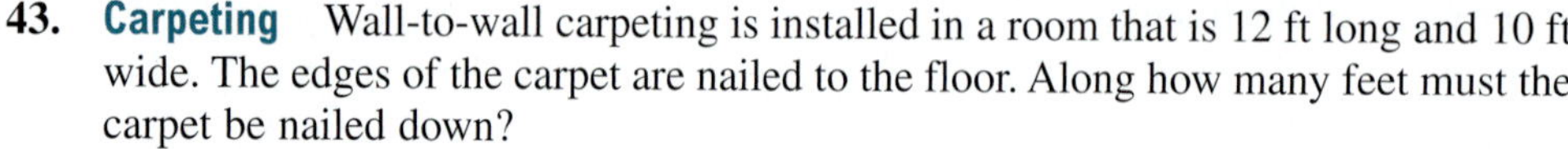

43. **Carpeting** Wall-to-wall carpeting is installed in a room that is 12 ft long and 10 ft wide. The edges of the carpet are nailed to the floor. Along how many feet must the carpet be nailed down?

44. **Fencing** The length of a rectangular park is 55 yd. The width is 47 yd. How many yards of fencing are needed to surround the park?

45. **Playgrounds** The perimeter of a rectangular playground is 440 ft. If the width is 100 ft, what is the length of the playground?

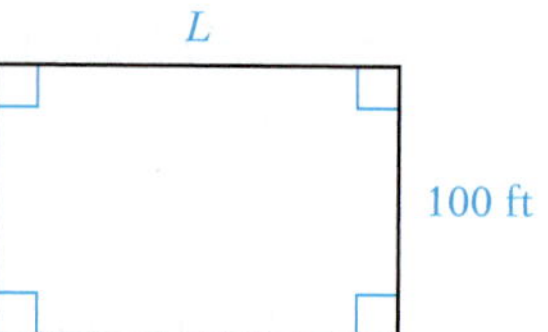

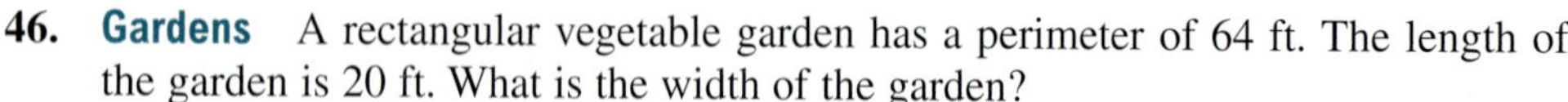

46. **Gardens** A rectangular vegetable garden has a perimeter of 64 ft. The length of the garden is 20 ft. What is the width of the garden?

47. **Banners** Each of two sides of a triangular banner measures 18 in. If the perimeter of the banner is 46 in., what is the length of the third side of the banner?

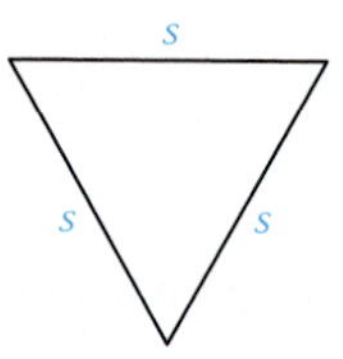

48. The perimeter of an equilateral triangle is 13.2 cm. What is the length of each side of the triangle?

49. **Framing** The perimeter of a square picture frame is 48 in. Find the length of each side of the frame.

50. **Carpeting** A square rug has a perimeter of 32 ft. Find the length of each edge of the rug.

51. The circumference of a circle is 8 cm. Find the length of a diameter of the circle. Round to the nearest hundredth.

52. The circumference of a circle is 15 in. Find the length of a radius of the circle. Round to the nearest hundredth.

53. **Carpentry** Find the length of molding needed to put around a circular table that is 4.2 ft in diameter. Round to the nearest hundredth.

54. **Carpeting** How much binding is needed to bind the edge of a circular rug that is 3 m in diameter? Round to the nearest hundredth.

55. **Cycling** A bicycle tire has a diameter of 24 in. How many feet does the bicycle travel when the wheel makes eight revolutions? Round to the nearest hundredth.

56. **Cycling** A tricycle tire has a diameter of 12 in. How many feet does the tricycle travel when the wheel makes 12 revolutions? Round to the nearest hundredth.

57. **Earth Science** The distance from the surface of Earth to its center is 6356 km. What is the circumference of Earth? Round to the nearest hundredth.

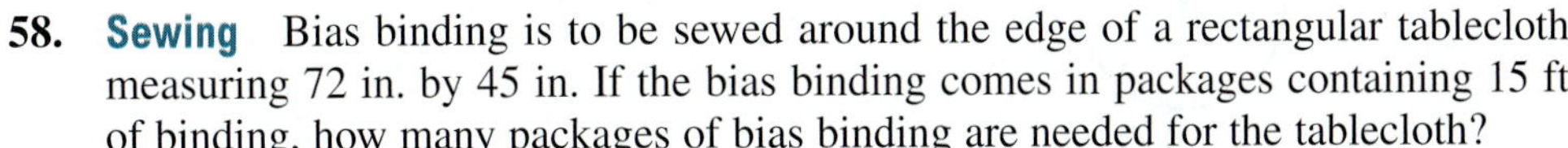

58. **Sewing** Bias binding is to be sewed around the edge of a rectangular tablecloth measuring 72 in. by 45 in. If the bias binding comes in packages containing 15 ft of binding, how many packages of bias binding are needed for the tablecloth?

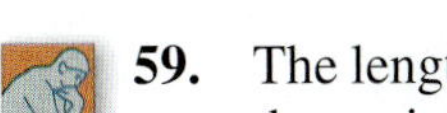

59. The length of a side of a square is equal to the diameter of a circle. Which is greater, the perimeter of the square or the circumference of the circle?

60. The length of a rectangle is equal to the diameter of a circle, and the width of the rectangle is equal to the radius of the same circle. Which is greater, the perimeter of the rectangle or the circumference of the circle?

OBJECTIVE B **To solve problems involving the area of geometric figures**

For Exercises 61 to 66, find the area of the figure.

61.

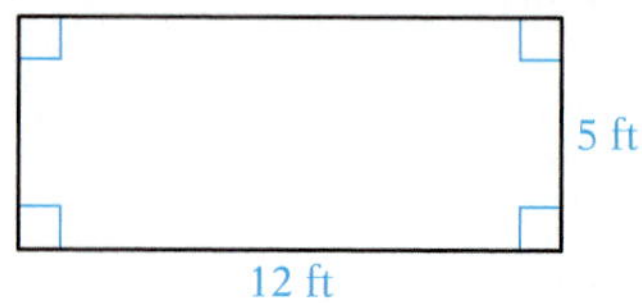

62.

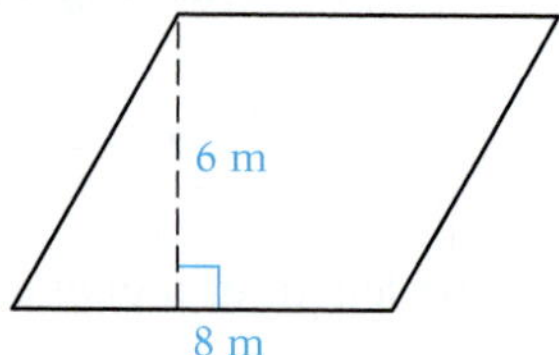

63.

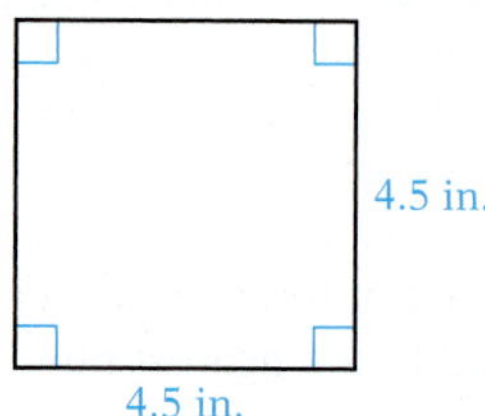

64.

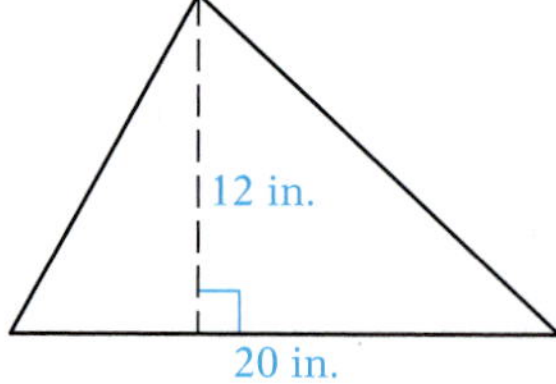

65.

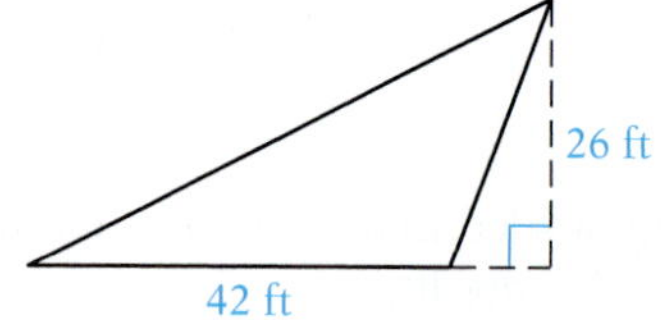

66.

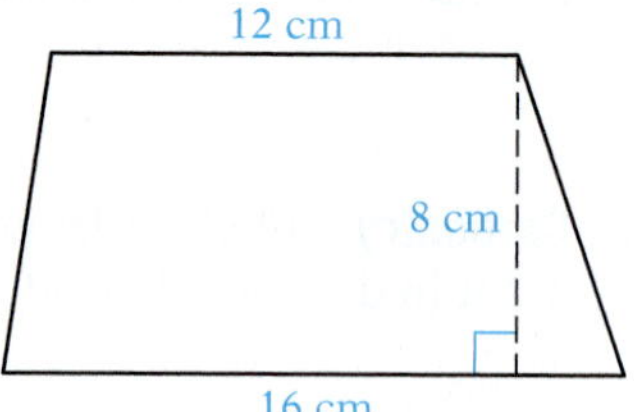

For Exercises 67 to 72, find the area of the figure. Give both the exact value and an approximation to the nearest hundredth.

67.

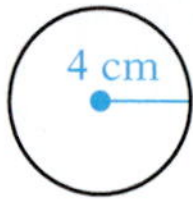

68.

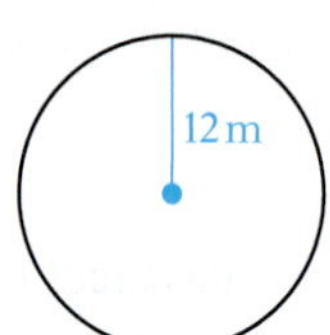

69.

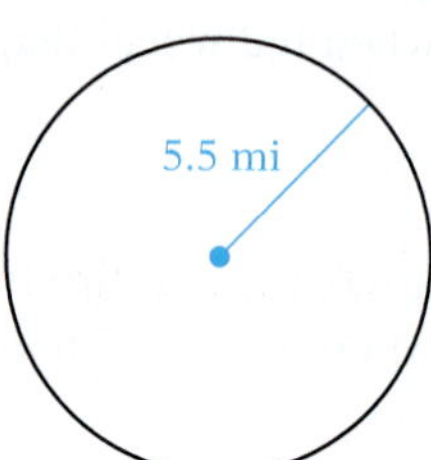

70.

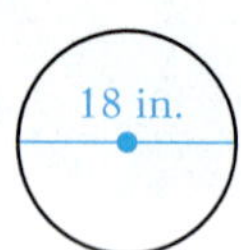

71.

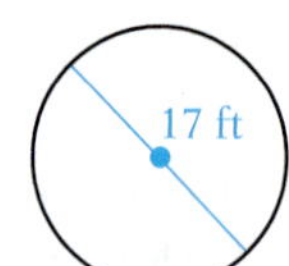

72.

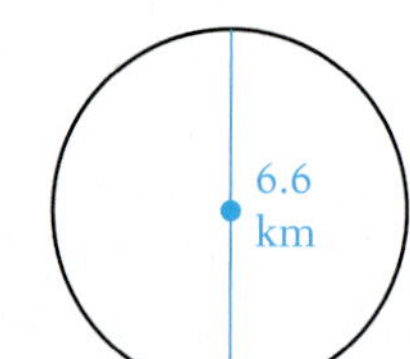

73. The length of a side of a square is 12.5 cm. Find the area of the square.

74. Each side of a square measures $3\frac{1}{2}$ in. Find the area of the square.

75. The length of a rectangle is 38 in., and the width is 15 in. Find the area of the rectangle.

76. Find the area of a rectangle that has a length of 6.5 m and a width of 3.8 m.

77. The length of the base of a parallelogram is 16 in., and the height is 12 in. Find the area of the parallelogram.

78. The height of a parallelogram is 3.4 m, and the length of the base is 5.2 m. Find the area of the parallelogram.

79. The length of the base of a triangle is 6 ft. The height is 4.5 ft. Find the area of the triangle.

80. The height of a triangle is 4.2 cm. The length of the base is 5 cm. Find the area of the triangle.

81. The length of one base of a trapezoid is 35 cm, and the length of the other base is 20 cm. If the height is 12 cm, what is the area of the trapezoid?

82. The height of a trapezoid is 5 in. The bases measure 16 in. and 18 in. Find the area of the trapezoid.

83. The radius of a circle is 5 in. Find the area of the circle. Give the exact value.

84. Find the area of a circle with a radius of 14 m. Round to the nearest hundredth.

85. Find the area of a circle with a diameter of 3.4 ft. Round to the nearest hundredth.

86. The diameter of a circle is 6.5 m. Find the area of the circle. Give the exact value.

87. **Telescopes** The lens on the Hale telescope at Mount Palomar, California, has a diameter of 200 in. Find its area. Give the exact value.

88. **Patios** What is the area of a square patio that measures 8.5 m on each side?

89. **Gardens** Find the area of a rectangular flower garden that measures 14 ft by 9 ft.

90. Irrigation An irrigation system waters a circular field that has a 50-foot radius. Find the area watered by the irrigation system. Give the exact value.

© David Frazier/Corbis

91. Athletic Fields Artificial turf is being used to cover a playing field. If the field is rectangular with a length of 100 yd and a width of 75 yd, how much artificial turf must be purchased to cover the field?

92. Interior Decorating A fabric wall hanging is to fill a space that measures 5 m by 3.5 m. Allowing for 0.1 m of the fabric to be folded back along each edge, how much fabric must be purchased for the wall hanging?

93. The area of a rectangle is 300 in^2. If the length of the rectangle is 30 in., what is the width?

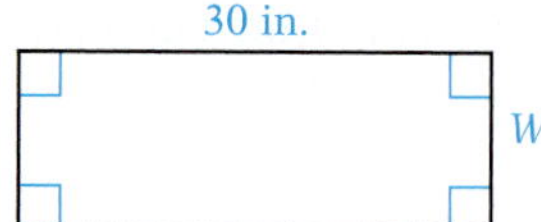

94. The width of a rectangle is 12 ft. If the area is 312 ft^2, what is the length of the rectangle?

95. The height of a triangle is 5 m. The area of the triangle is 50 m^2. Find the length of the base of the triangle.

96. The area of a parallelogram is 42 m^2. If the height of the parallelogram is 7 m, what is the length of the base?

97. Home Maintenance You plan to stain the wooden deck attached to your house. The deck measures 10 ft by 8 ft. If a quart of stain will cover 50 ft^2, how many quarts of stain should you buy?

98. Flooring You want to tile your kitchen floor. The floor measures 12 ft by 9 ft. How many tiles, each a square with side $1\frac{1}{2}$ ft, should you purchase for the job?

99. Interior Decorating You are wallpapering two walls of a child's room, one measuring 9 ft by 8 ft and the other measuring 11 ft by 8 ft. The wallpaper costs $24.50 per roll, and each roll of the wallpaper will cover 40 ft^2. What is the cost to wallpaper the two walls?

100. Parks An urban renewal project involves reseeding a park that is in the shape of a square, 60 ft on each side. Each bag of grass seed costs $9.75 and will seed 1200 ft^2. How much money should be budgeted for buying grass seed for the park?

101. A circle has a radius of 8 in. Find the increase in area when the radius is increased by 2 in. Round to the nearest hundredth.

102. A circle has a radius of 6 cm. Find the increase in area when the radius is doubled. Round to the nearest hundredth.

103. **Landscaping** A walkway 2 m wide surrounds a rectangular plot of grass. The plot is 30 m long and 20 m wide. What is the area of the walkway?

104. **Interior Decorating** You want to paint the walls of your bedroom. Two walls measure 15 ft by 9 ft, and the other two walls measure 12 ft by 9 ft. The paint you wish to purchase costs \$12.98 per gallon, and each gallon will cover 400 ft^2 of wall. Find the total amount you will spend on paint.

105. **Carpeting** You want to install wall-to-wall carpeting in your living room, which measures 15 ft by 24 ft. If the cost of the carpet you would like to purchase is \$21.95 per square yard, what is the cost of the carpeting for your living room? (*Hint:* $9\text{ ft}^2 = 1\text{ yd}^2$)

106. **Interior Design** See the news clipping at the right. What would be the cost of carpeting the entire living space if the cost of the carpet were \$36 per square yard?

STRINGER/Fotocorp

In the News

Billion-Dollar Home Built in Mumbai

The world's first billion-dollar home is a 27-story skyscraper in downtown Mumbai, India (formerly known as Bombay). It is 550 ft high with 400,000 ft^2 of living space.

Source: Forbes.com

For Exercises 107 and 108, determine whether the area of the first figure is *less than, equal to,* or *greater than* the area of the second figure.

107.

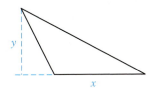

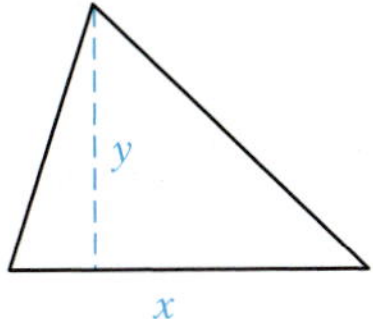

108.

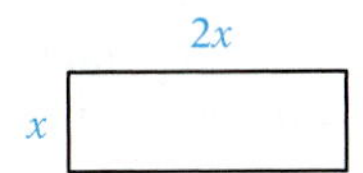

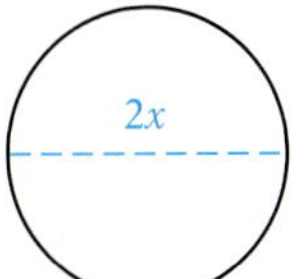

Applying the Concepts

109. If both the length and the width of a rectangle are doubled, how many times larger is the area of the resulting rectangle?

110. Suppose a circle is cut into 16 equal pieces, which are then arranged as shown at the right. The figure formed resembles a parallelogram. What variable expression could describe the base of the parallelogram? What variable could describe its height? Explain how the formula for the area of a circle is derived from this approach.

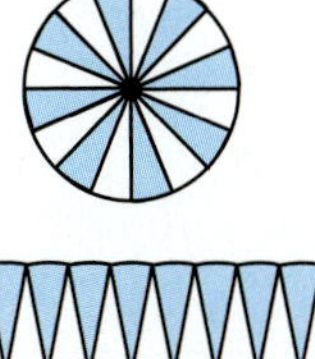

SECTION 7.3 Triangles

OBJECTIVE A To solve problems using the Pythagorean Theorem

A **right triangle** contains one right angle. The side opposite the right angle is called the **hypotenuse.** The other two sides are called **legs.**

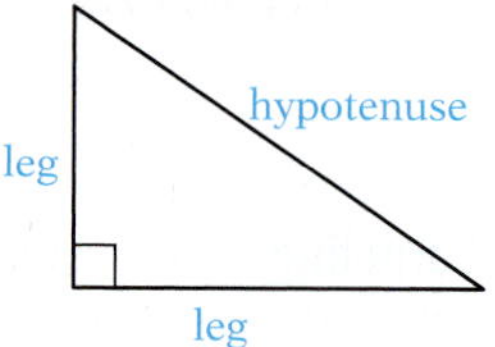

The angles in a right triangle are usually labeled with the capital letters A, B, and C, with C reserved for the right angle. The side opposite angle A is side a, the side opposite angle B is side b, and c is the hypotenuse.

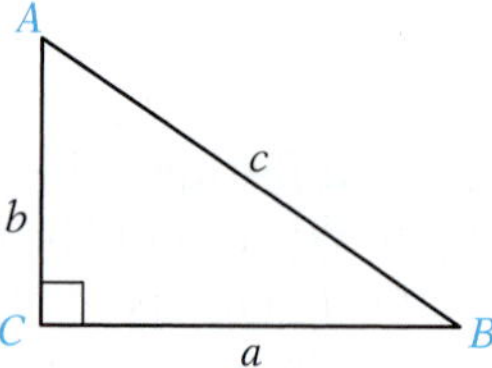

Point of Interest

The first known proof of the Pythagorean Theorem is in a Chinese textbook that dates from 150 B.C.E. The book is called *Nine Chapters on the Mathematical Art*. The diagram below is from that book and was used in the proof of the theorem.

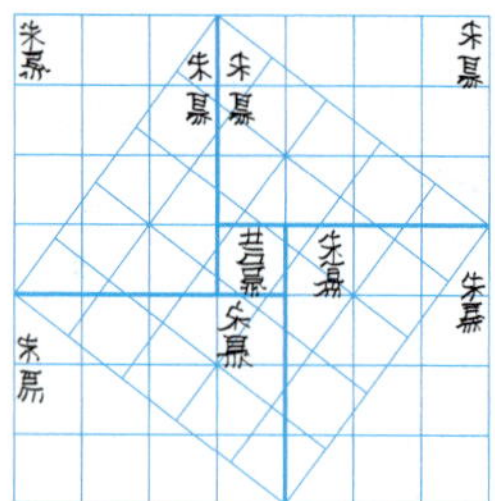

The Greek mathematician Pythagoras is generally credited with the discovery that the square of the hypotenuse of a right triangle is equal to the sum of the squares of the two legs. This is called the **Pythagorean Theorem.**

The figure at the right is a right triangle with legs measuring 3 units and 4 units and a hypotenuse measuring 5 units. Each side of the triangle is also the side of a square. The number of square units in the area of the largest square is equal to the sum of the numbers of square units in the areas of the smaller squares.

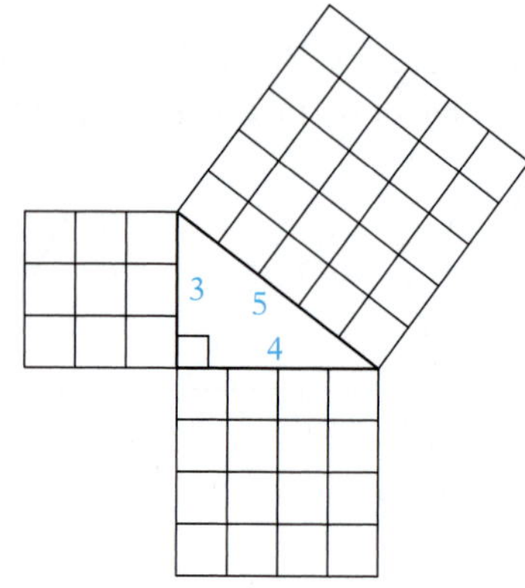

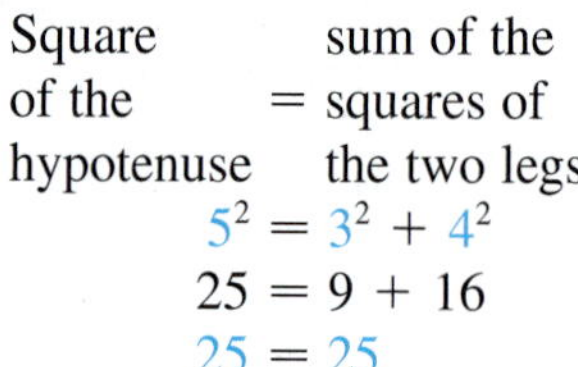

Square of the hypotenuse = sum of the squares of the two legs

$$5^2 = 3^2 + 4^2$$
$$25 = 9 + 16$$
$$25 = 25$$

> **Pythagorean Theorem**
>
> If a and b are the lengths of the legs of a right triangle and c is the length of the hypotenuse, then $c^2 = a^2 + b^2$.

If the lengths of two sides of a right triangle are known, the Pythagorean Theorem can be used to find the length of the third side.

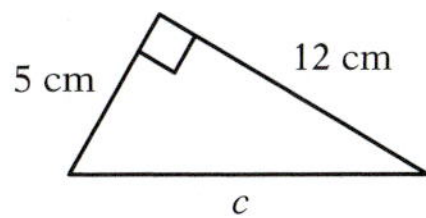

Consider a right triangle with legs that measure 5 cm and 12 cm. Use the Pythagorean Theorem, with $a = 5$ and $b = 12$, to find the length of the hypotenuse. (If you let $a = 12$ and $b = 5$, the result is the same.)

$c^2 = a^2 + b^2$
$c^2 = 5^2 + 12^2$
$c^2 = 25 + 144$
$c^2 = 169$

This equation states that the square of c is 169. Since $13^2 = 169$, $c = 13$ and the length of the hypotenuse is 13 cm. We can find c by taking the square root of 169: $\sqrt{169} = 13$. This suggests the following property.

Integrating Technology

The way in which you evaluate the square root of a number depends on the type of calculator you have. Here are two possible keystroke combinations to find $\sqrt{35}$:

35  =

or

The first method is used on many scientific calculators. The second method is used on many graphing calculators.

Principal Square Root Property

If $r^2 = s$, then $r = \sqrt{s}$, and r is called the **square root** of s.

The Principal Square Root Property and its application can be illustrated as follows: Because $5^2 = 25$, $5 = \sqrt{25}$. Therefore, if $c^2 = 25$, $c = \sqrt{25} = 5$.

Recall that numbers whose square roots are integers, such as 25, are perfect squares. If a number is not a perfect square, a calculator can be used to find an approximate square root when a decimal approximation is required.

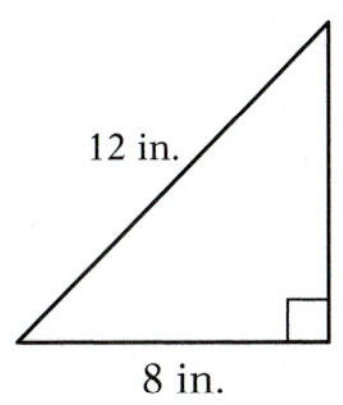

HOW TO • 1 The length of one leg of a right triangle is 8 in. The hypotenuse is 12 in. Find the length of the other leg. Round to the nearest hundredth.

$a^2 + b^2 = c^2$ • **Use the Pythagorean Theorem.**
$8^2 + b^2 = 12^2$ **$a = 8, c = 12$**
$64 + b^2 = 144$ • **Solve for b^2.**
$b^2 = 80$ **(If you let $b = 8$ and solve for a^2, the result is the same.)**
$b = \sqrt{80}$ • **Use the Principal Square Root Property.** **Since $b^2 = 80$, b is the square root of 80.**
$b \approx 8.94$ • **Use a calculator to approximate $\sqrt{80}$.**

The length of the other leg is approximately 8.94 in.

EXAMPLE • 1

The two legs of a right triangle measure 12 ft and 9 ft. Find the hypotenuse of the right triangle.

Strategy

To find the hypotenuse, use the Pythagorean Theorem. $a = 12$, $b = 9$

Solution

$c^2 = a^2 + b^2$
$c^2 = 12^2 + 9^2$
$c^2 = 144 + 81$
$c^2 = 225$
$c = \sqrt{225}$
$c = 15$

The length of the hypotenuse is 15 ft.

YOU TRY IT • 1

The hypotenuse of a right triangle measures 6 m, and one leg measures 2 m. Find the measure of the other leg. Round to the nearest hundredth.

Your strategy

Your solution

Solution on p. S19

OBJECTIVE B To solve problems involving similar triangles 

Similar objects have the same shape but not necessarily the same size. A tennis ball is similar to a basketball. A model ship is similar to an actual ship.

© Patrick Ward/Corbis

Similar objects have corresponding parts; for example, the rudder on the model ship corresponds to the rudder on the actual ship. The relationship between the sizes of each of the corresponding parts can be written as a ratio, and each ratio will be the same. If the rudder on the model ship is $\frac{1}{100}$ the size of the rudder on the actual ship, then the model wheelhouse is $\frac{1}{100}$ the size of the actual wheelhouse, the width of the model is $\frac{1}{100}$ the width of the actual ship, and so on.

The two triangles ABC and DEF shown at the right are similar. Side AB corresponds to side DE, side BC corresponds to side EF, and side AC corresponds to side DF. The ratios of corresponding sides are equal.

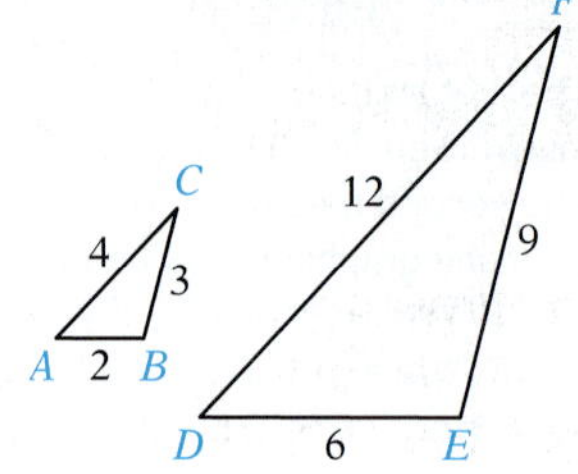

$$\frac{AB}{DE} = \frac{2}{6} = \frac{1}{3}, \frac{BC}{EF} = \frac{3}{9} = \frac{1}{3}, \text{ and } \frac{AC}{DF} = \frac{4}{12} = \frac{1}{3}$$

Since the ratios of corresponding sides are equal, three proportions can be formed.

$$\frac{AB}{DE} = \frac{BC}{EF}, \frac{AB}{DE} = \frac{AC}{DF}, \text{ and } \frac{BC}{EF} = \frac{AC}{DF}$$

The corresponding angles in similar triangles are equal. Therefore,

$$\angle A = \angle D, \angle B = \angle E, \text{ and } \angle C = \angle F$$

Triangles ABC and DEF at the right are similar triangles. AH and DK are the heights of the triangles. The ratio of the heights of similar triangles equals the ratio of corresponding sides.

Ratio of corresponding sides $= \frac{1.5}{6} = \frac{1}{4}$

Ratio of heights $= \frac{1}{4}$

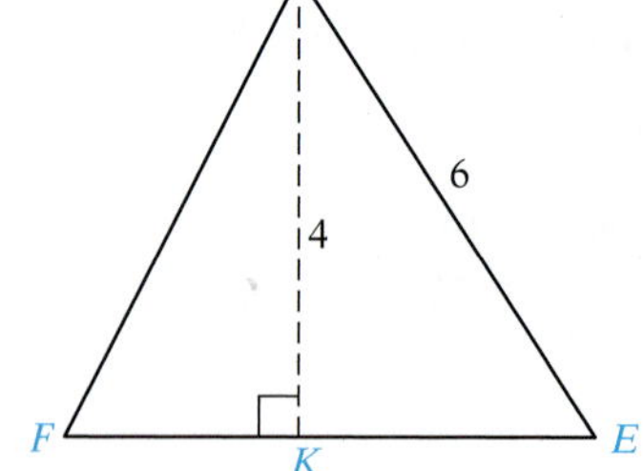

Properties of Similar Triangles

For similar triangles, the ratios of corresponding sides are equal. The ratio of corresponding heights is equal to the ratio of corresponding sides.

Point of Interest

Many mathematicians have studied similar objects. Thales of Miletus (c. 624 B.C.E.–547 B.C.E.) discovered that he could determine the heights of pyramids and other large objects by measuring a small object and the length of its shadow and then making use of similar triangles.

HOW TO • 2 The two triangles at the right are similar triangles. Find the length of side EF. Round to the nearest tenth.

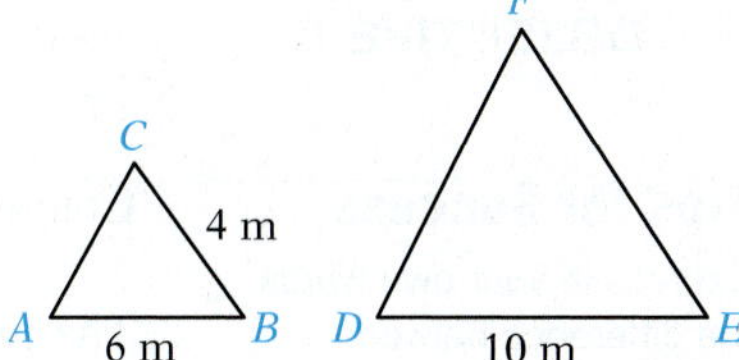

$$\frac{EF}{BC} = \frac{DE}{AB}$$ • The triangles are similar, so the ratios of corresponding sides are equal.

$$\frac{EF}{4} = \frac{10}{6}$$

$$6(EF) = 4(10)$$

$$6(EF) = 40$$

$$EF \approx 6.7$$

The length of side EF is approximately 6.7 m.

EXAMPLE • 2

Triangles ABC and DEF are similar. Find FG, the height of triangle DEF.

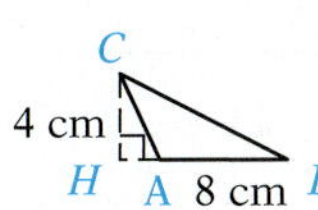

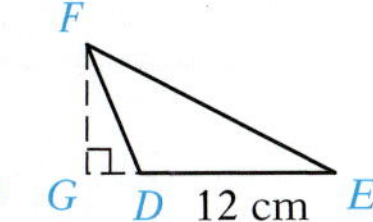

Strategy

To find FG, write a proportion using the fact that, in similar triangles, the ratio of corresponding sides equals the ratio of corresponding heights. Solve the proportion for FG.

Solution

$$\frac{AB}{DE} = \frac{CH}{FG}$$

$$\frac{8}{12} = \frac{4}{FG}$$

$$8(FG) = 12(4)$$

$$8(FG) = 48$$

$$FG = 6$$

The height FG of triangle DEF is 6 cm.

YOU TRY IT • 2

Triangles ABC and DEF are similar. Find FG, the height of triangle DEF.

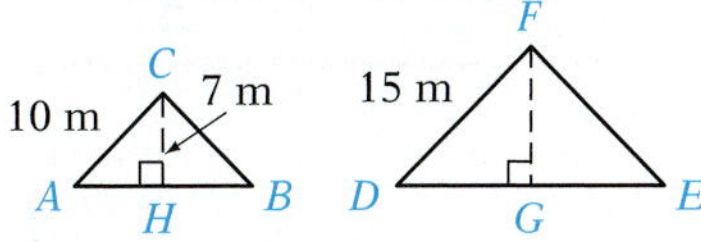

Your strategy

Your solution

Solution on p. S19

OBJECTIVE C To determine whether two triangles are congruent 

Tips for Success

Express in your own words the difference between congruent objects and similar objects. Doing so will help you to remember the difference.

Congruent objects have the same shape *and* the same size.

The two triangles at the right are congruent. They have the same size.

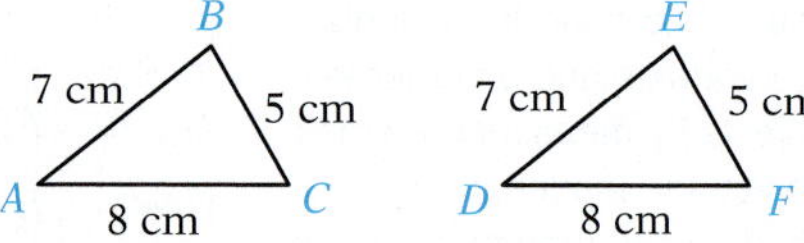

Congruent and similar triangles differ in that the corresponding sides and angles of congruent triangles must be equal, whereas for similar triangles, corresponding angles are equal, but corresponding sides are not necessarily the same length.

The three major rules used to determine whether two triangles are congruent are given below.

Side-Side-Side Rule (SSS)

Two triangles are congruent if the three sides of one triangle equal the corresponding three sides of a second triangle.

In the triangles at the right, $AC = DE$, $AB = EF$, and $BC = DF$. The corresponding sides of triangles ABC and DEF are equal. The triangles are congruent by the SSS Rule.

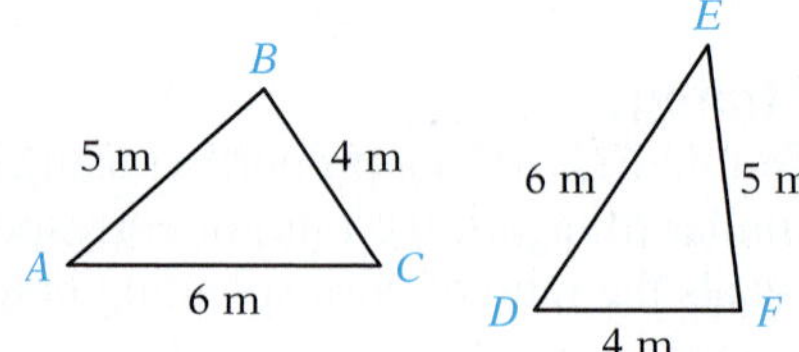

Side-Angle-Side Rule (SAS)

If two sides and the included angle of one triangle are equal to two sides and the included angle of a second triangle, the two triangles are congruent.

In the two triangles at the right, $AB = EF$, $AC = DE$, and $\angle BAC = \angle DEF$. The triangles are congruent by the SAS Rule.

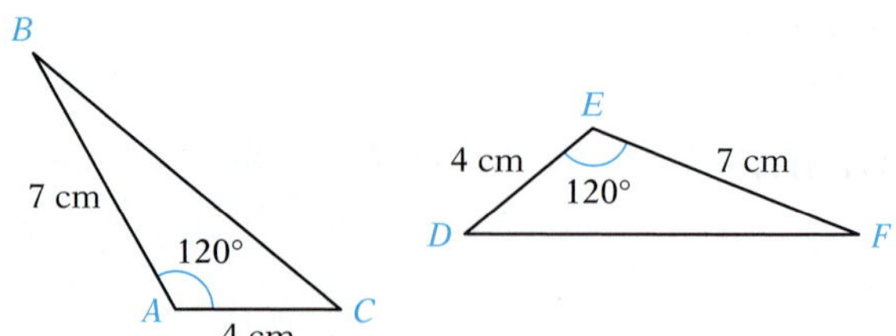

Angle-Side-Angle Rule (ASA)

If two angles and the included side of one triangle are equal to two angles and the included side of a second triangle, the two triangles are congruent.

For triangles ABC and DEF at the right, $\angle A = \angle F$, $\angle C = \angle E$, and $AC = EF$. The triangles are congruent by the ASA Rule.

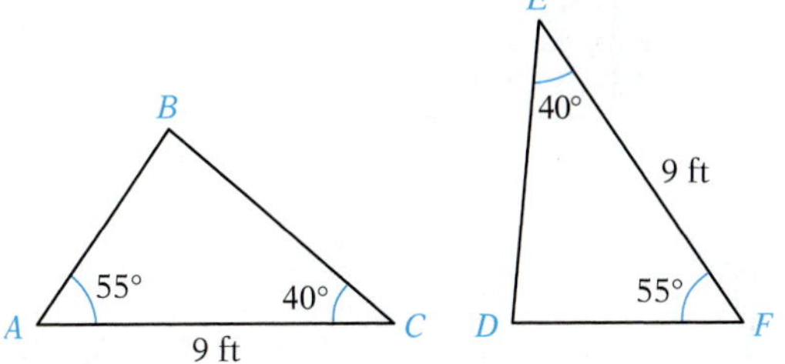

HOW TO • 3 Given triangle PQR and triangle MNO, do the conditions $\angle P = \angle O$, $\angle Q = \angle M$, and $PQ = MO$ guarantee that triangle PQR is congruent to triangle MNO?

Draw a sketch of the two triangles and determine whether one of the rules for congruence is satisfied.

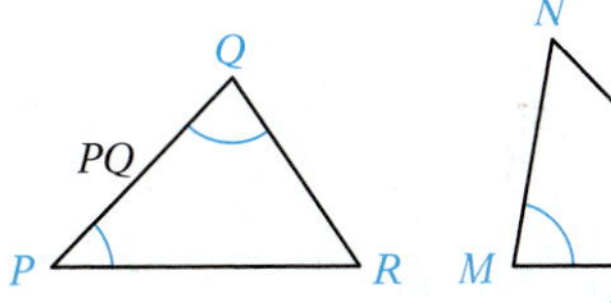

Because two angles and the included side of one triangle are equal to two angles and the included side of the second triangle, the triangles are congruent by the ASA Rule.

EXAMPLE • 3

In the figure below, is triangle ABC congruent to triangle DEF?

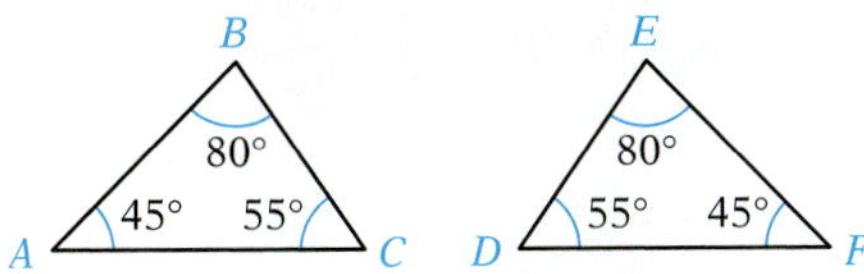

YOU TRY IT • 3

In the figure below, is triangle PQR congruent to triangle MNO?

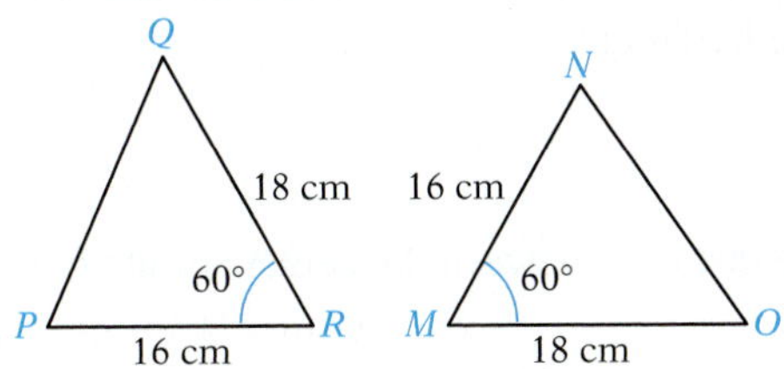

Strategy

To determine whether the triangles are congruent, determine whether one of the rules for congruence is satisfied.

Solution

The triangles do not satisfy the SSS Rule, the SAS Rule, or the ASA Rule. The triangles are not necessarily congruent.

Your strategy

Your solution

Solution on p. S19

7.3 EXERCISES

OBJECTIVE A To solve problems using the Pythagorean Theorem

For Exercises 1 to 9, find the unknown side of the triangle. Round to the nearest tenth.

1. 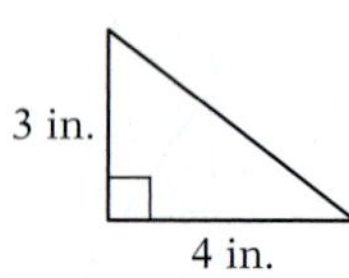

2. 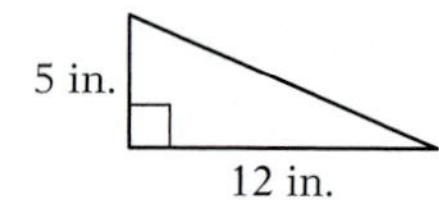

3. 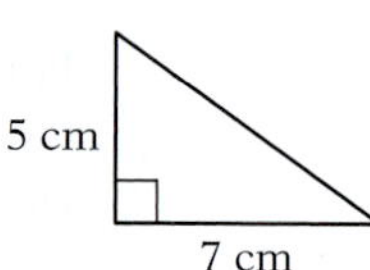

4. 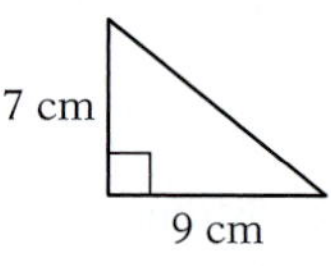

5. 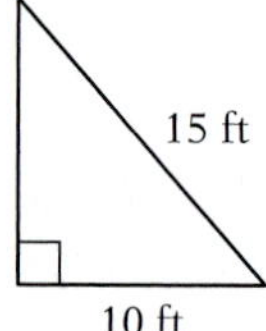

6. 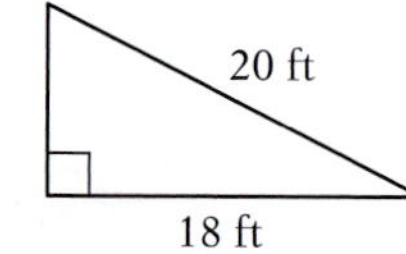

7. 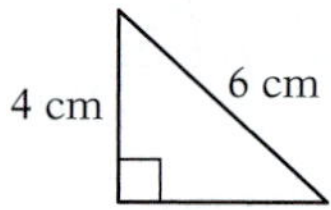

8. 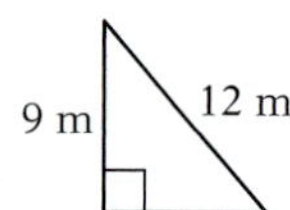

9. 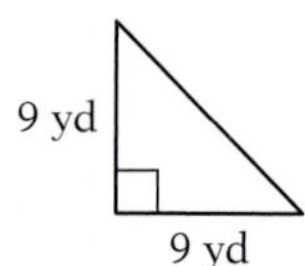

For Exercises 10 to 14, solve. Round to the nearest tenth.

10. A ladder 8 m long is leaning against a building. How high on the building will the ladder reach when the bottom of the ladder is 3 m from the building?

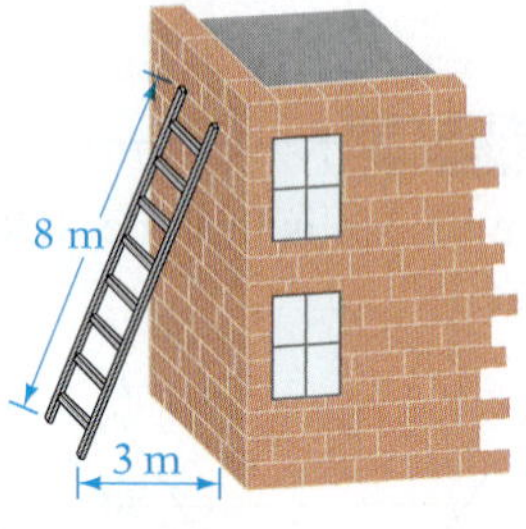

11. Find the distance between the centers of the holes in the metal plate.

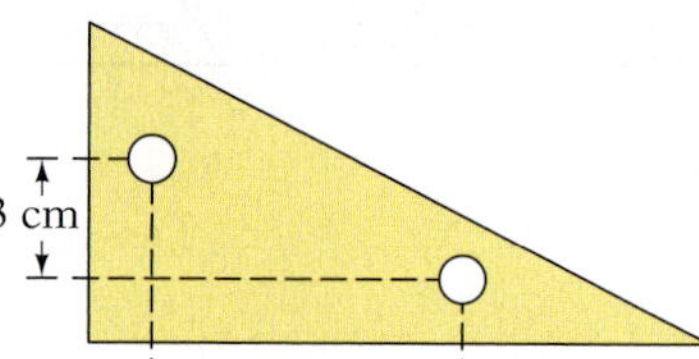

12. If you travel 18 mi east and then 12 mi north, how far are you from your starting point?

13. Find the perimeter of a right triangle with legs that measure 5 cm and 9 cm.

14. Find the perimeter of a right triangle with legs that measure 6 in. and 8 in.

For Exercises 15 and 16, use the following information. A ladder c feet long leans against the side of a building with its bottom a feet from the building. The ladder reaches a height of b feet. Refer to the following lengths: (i) 15 ft (ii) 20 ft (iii) 30 ft

15. If $c = 18$ ft, which of the given lengths is possible as a value for b?

16. If $b = 18$ ft, which of the given lengths are possible as values for c?

OBJECTIVE B To solve problems involving similar triangles

For Exercises 17 and 18, find the ratio of corresponding sides for the similar triangles.

17.

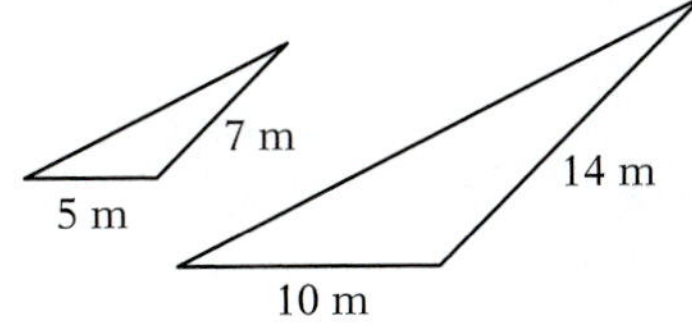

18.

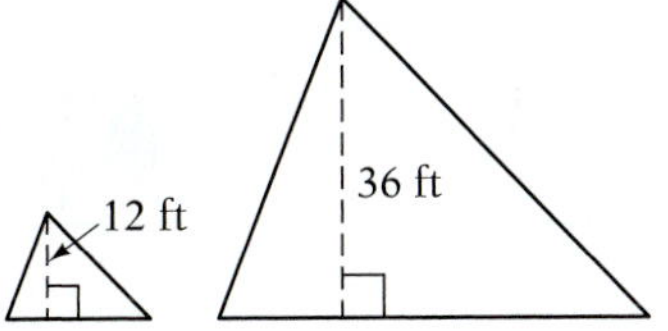

In Exercises 19 to 28, triangles ABC and DEF are similar triangles. Solve and round to the nearest tenth.

19. Find side DE.

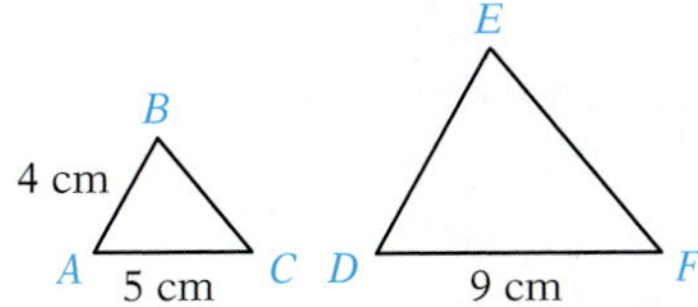

20. Find side DE.

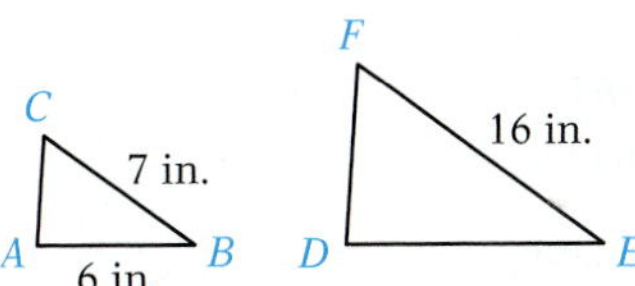

21. Find the height of triangle DEF.

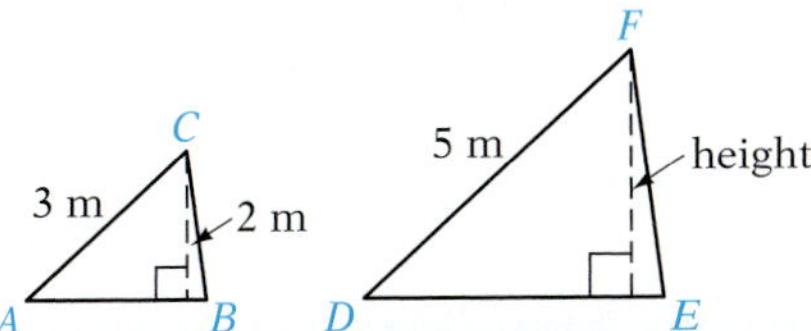

22. Find the height of triangle ABC.

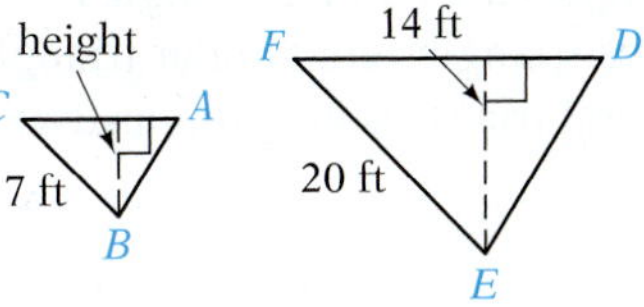

23. Find the perimeter of triangle ABC.

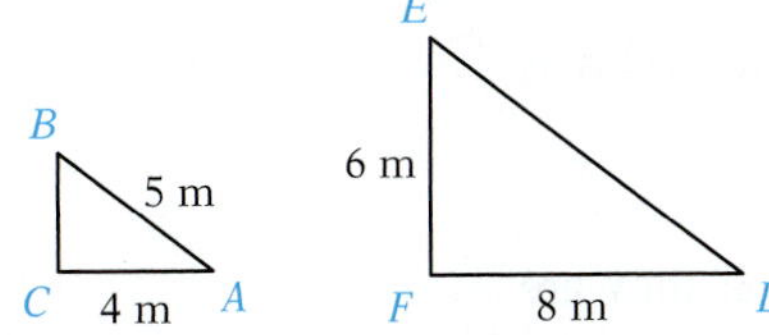

24. Find the perimeter of triangle DEF.

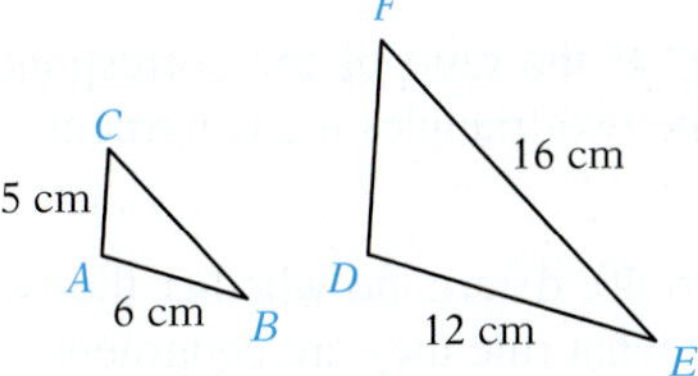

25. Find the perimeter of triangle ABC.

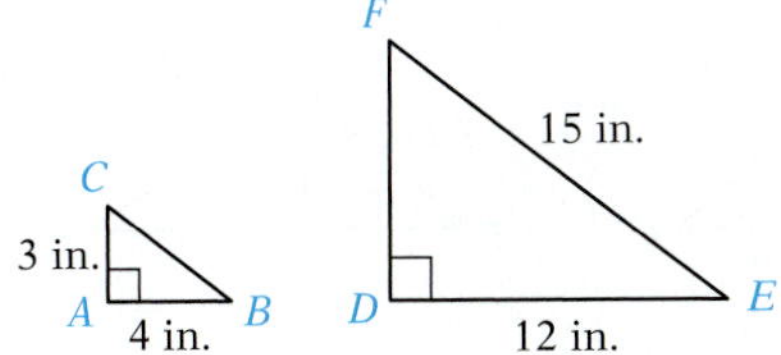

26. Find the area of triangle DEF.

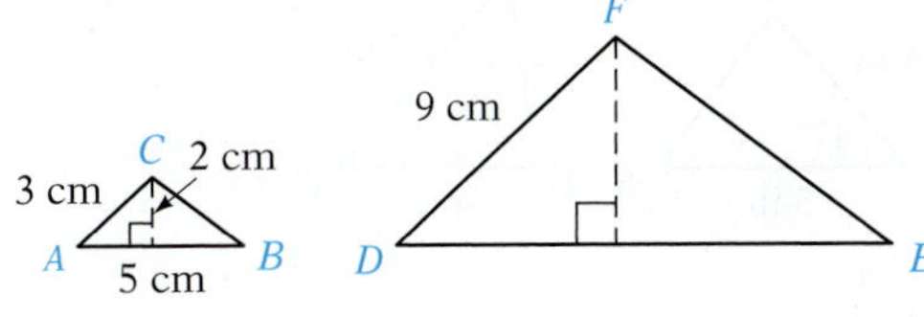

27. Find the area of triangle ABC.

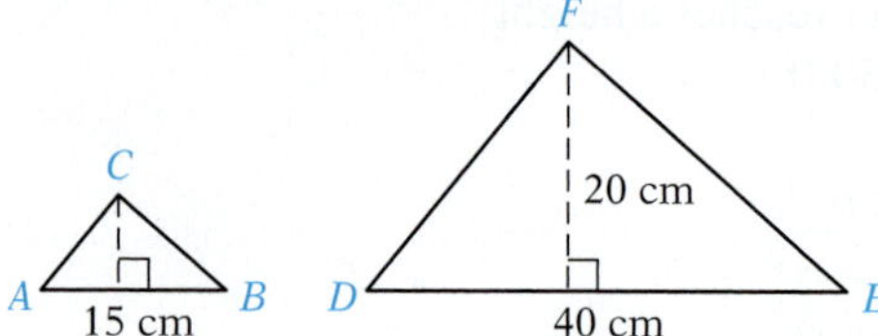

28. Find the area of triangle DEF.

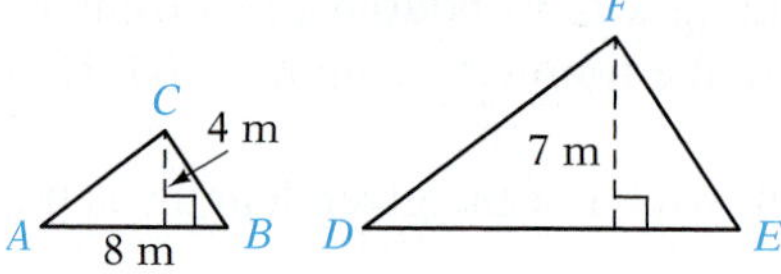

The sun's rays, objects on Earth, and the shadows cast by them form similar triangles. Use this fact for Exercises 29 to 32.

29. Find the height of the flagpole.

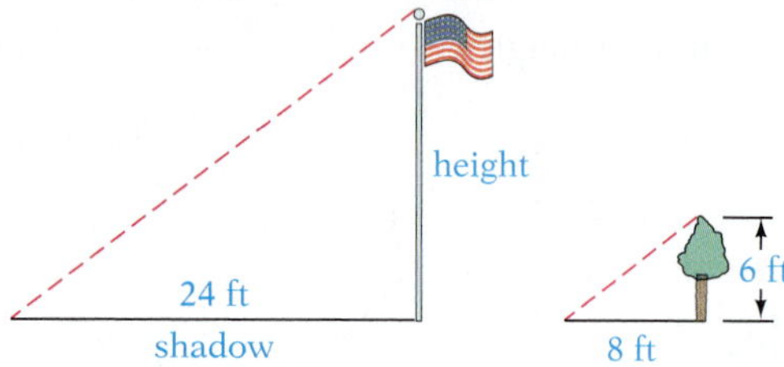

30. Find the height of the flagpole.

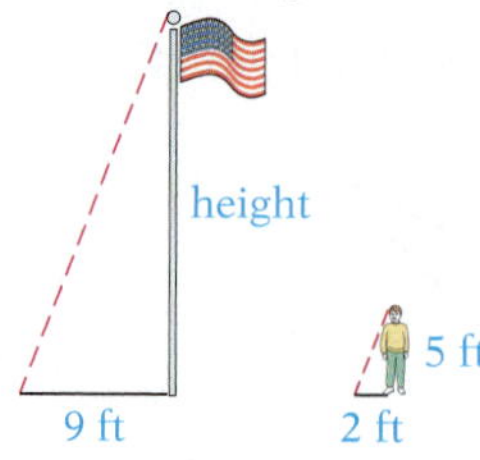

31. Find the height of the building.

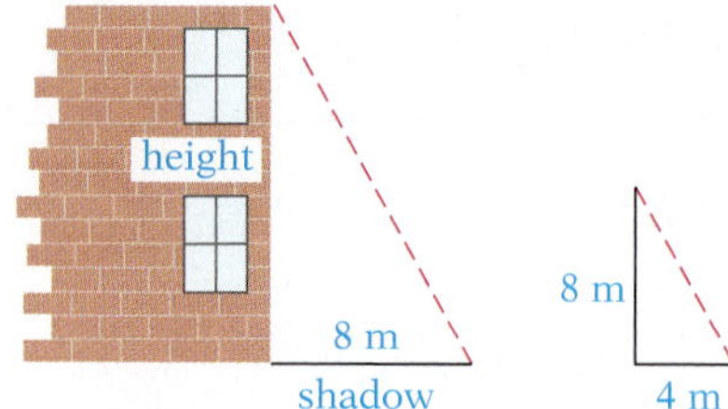

32. Find the height of the building.

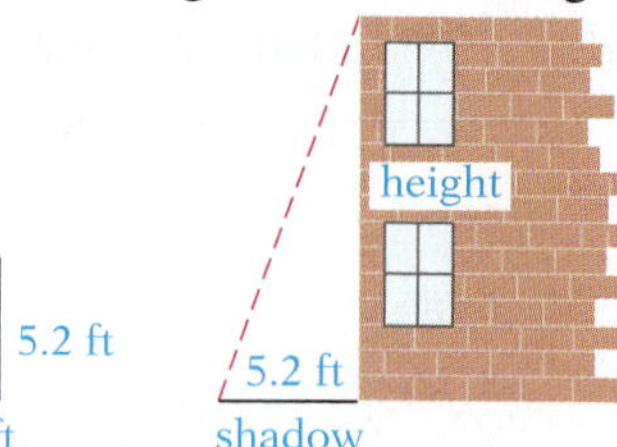

33. Determine whether the statement is always true, sometimes true, or never true.
   a. If two angles of one triangle are equal to two angles of a second triangle, then the triangles are similar triangles.
   b. Two isosceles triangles are similar triangles.
   c. Two equilateral triangles are similar triangles.

OBJECTIVE C To determine whether two triangles are congruent

34. True or false? If the ratio of the corresponding sides of two similar triangles is 1 to 1, then the two triangles are congruent.

For Exercises 35 to 40, determine whether the two triangles are congruent. If they are congruent, state by what rule they are congruent.

35.

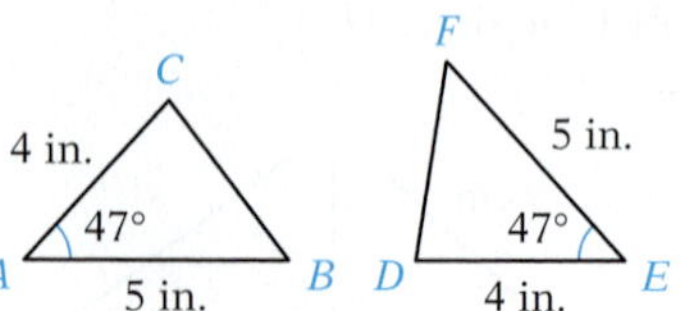

36.

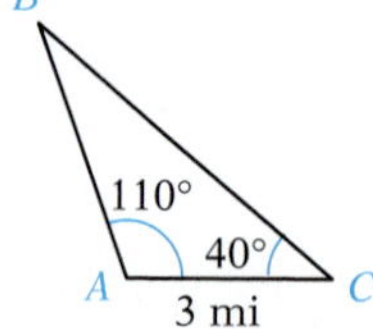

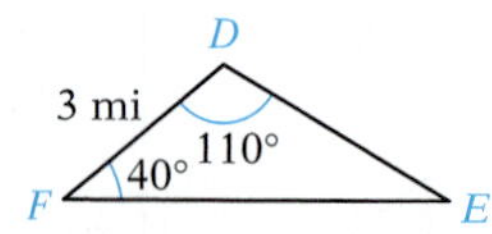

37. 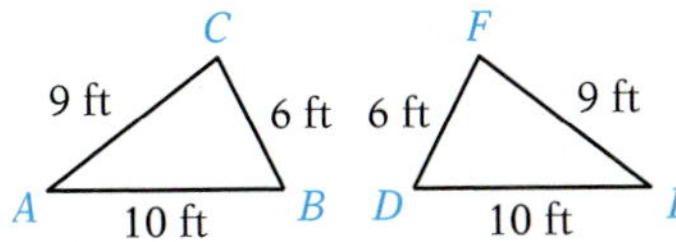

38. 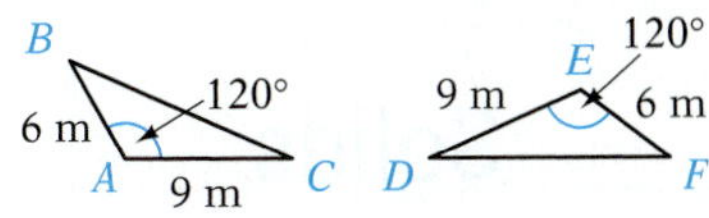

39. 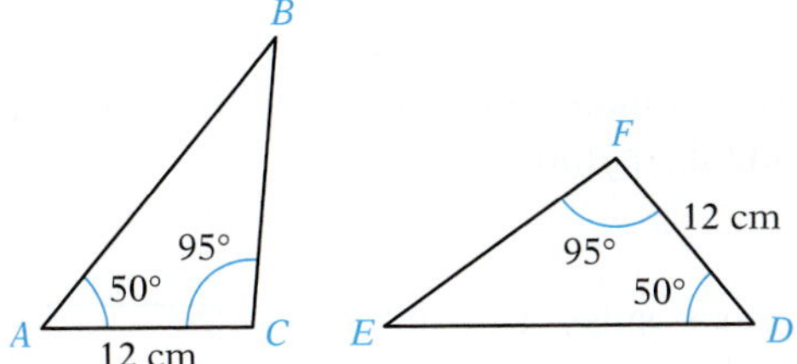

40. 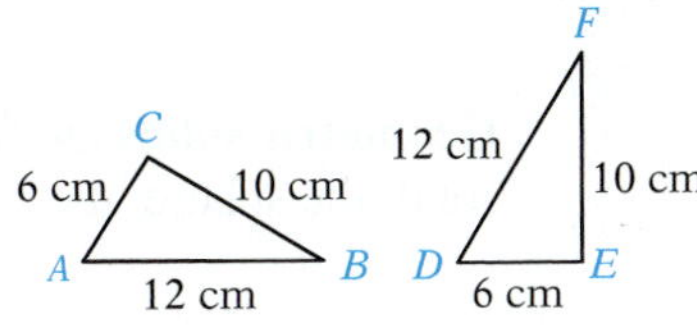

41. Given triangle *ABC* and triangle *DEF*, do the conditions $\angle C = \angle E$, $AC = EF$, and $BC = DE$ guarantee that triangle *ABC* is congruent to triangle *DEF*? If they are congruent, by what rule are they congruent?

42. Given triangle *PQR* and triangle *MNO*, do the conditions $PR = NO$, $PQ = MO$, and $QR = MN$ guarantee that triangle *PQR* is congruent to triangle *MNO*? If they are congruent, by what rule are they congruent?

43. Given triangle *LMN* and triangle *QRS*, do the conditions $\angle M = \angle S$, $\angle N = \angle Q$, and $\angle L = \angle R$ guarantee that triangle *LMN* is congruent to triangle *QRS*? If they are congruent, by what rule are they congruent?

44. Given triangle *DEF* and triangle *JKL*, do the conditions $\angle D = \angle K$, $\angle E = \angle L$, and $DE = KL$ guarantee that triangle *DEF* is congruent to triangle *JKL*? If they are congruent, by what rule are they congruent?

45. Given triangle *ABC* and triangle *PQR*, do the conditions $\angle B = \angle P$, $BC = PQ$, and $AC = QR$ guarantee that triangle *ABC* is congruent to triangle *PQR*? If they are congruent, by what rule are they congruent?

Applying the Concepts

46. **Home Maintenance** You need to clean the gutters of your home. The gutters are 24 ft above the ground. For safety, the distance a ladder reaches up a wall should be four times the distance from the bottom of the ladder to the base of the side of the house. Therefore, the ladder must be 6 ft from the base of the house. Will a 25-foot ladder be long enough to reach the gutters? Explain how you determined your answer.

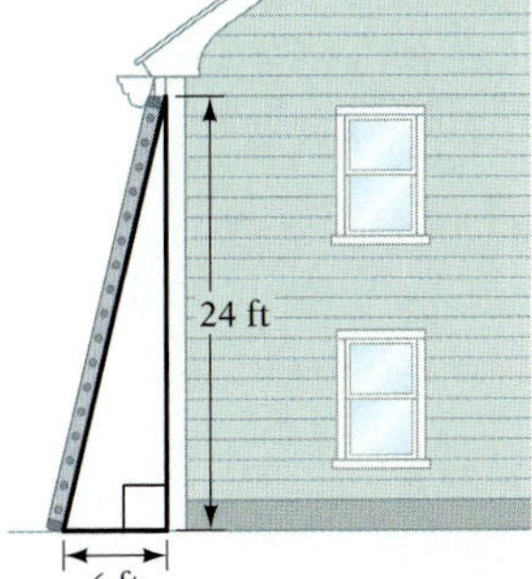

47. What is a Pythagorean triple? Provide at least three examples of Pythagorean triples.

SECTION

7.4 Solids

OBJECTIVE A To solve problems involving the volume of a solid 

Geometric solids are figures in space. Five common geometric solids are the rectangular solid, the sphere, the cylinder, the cone, and the pyramid.

A **rectangular solid** is one in which all six sides, called **faces,** are rectangles. The variable L is used to represent the length of a rectangular solid, W its width, and H its height.

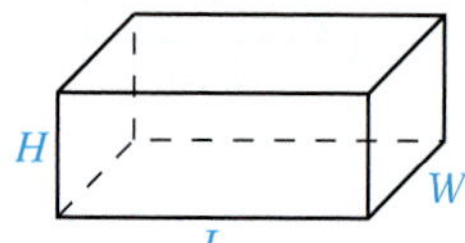

A **sphere** is a solid in which all points are the same distance from point O, called the **center** of the sphere. The **diameter,** d, of a sphere is a line across the sphere going through point O. The **radius,** r, is a line from the center to a point on the sphere. AB is a diameter and OC is a radius of the sphere shown at the right.

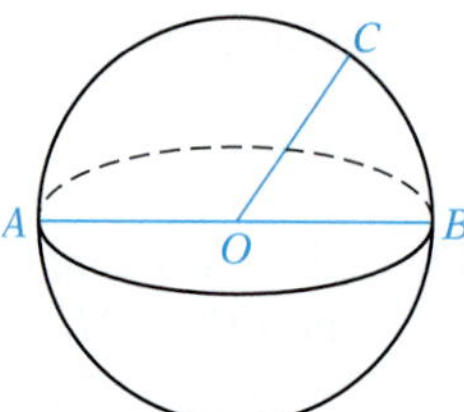

The most common cylinder, called a **right circular cylinder,** is one in which the bases are circles and are perpendicular to the height of the cylinder. The variable r is used to represent the radius of a base of a cylinder, and h represents the height. In this text, only right circular cylinders are discussed.

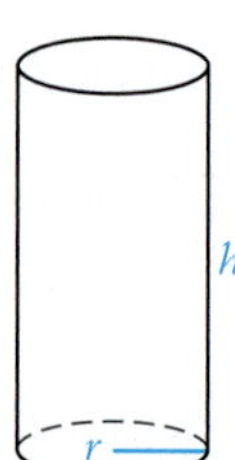

A **right circular cone** is obtained when one base of a right circular cylinder is shrunk to a point, called the **vertex,** V. The variable r is used to represent the radius of the base of the cone, and h represents the height. The variable ℓ is used to represent the **slant height,** which is the distance from a point on the circumference of the base to the vertex. In this text, only right circular cones are discussed.

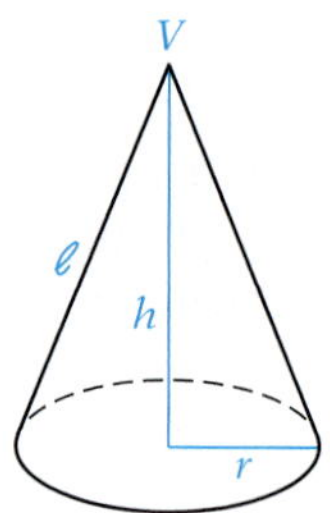

The base of a **regular pyramid** is a regular polygon, and the sides are isosceles triangles. The height, h, is the distance from the vertex, V, to the base and is perpendicular to the base. The variable ℓ is used to represent the **slant height,** which is the height of one of the isosceles triangles on the face of the pyramid. The regular square pyramid at the right has a square base. This is the only type of pyramid discussed in this text.

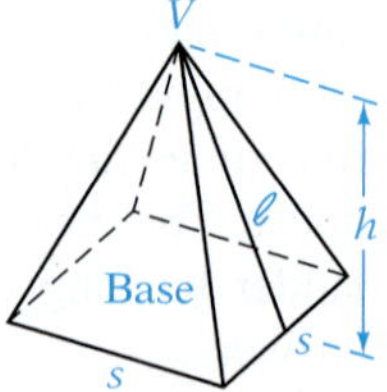

A **cube** is a special type of rectangular solid. Each of the six faces of a cube is a square. The variable s is used to represent the length of one side of a cube.

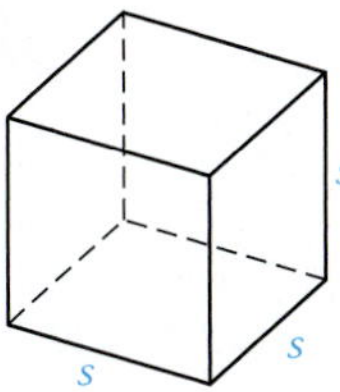

Volume is a measure of the amount of space inside a figure in space. Volume can be used to describe the amount of heating gas used for cooking, the amount of concrete delivered for the foundation of a house, or the amount of water in storage for a city's water supply.

A cube that is 1 ft on each side has a volume of 1 cubic foot, which is written 1 ft^3. A cube that measures 1 cm on each side has a volume of 1 cubic centimeter, written 1 cm^3.

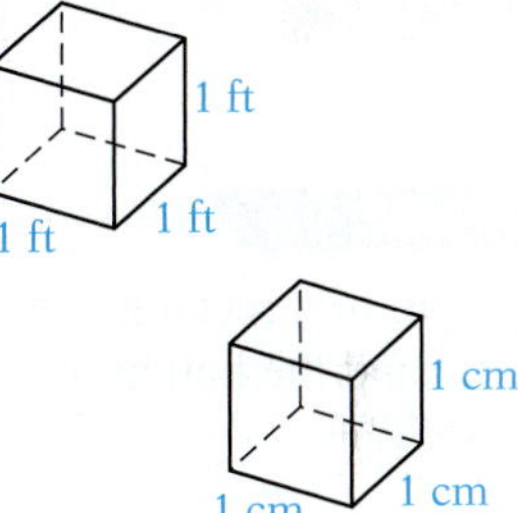

The volume of a solid is the number of cubes that are necessary to exactly fill the solid. The volume of the rectangular solid at the right is 24 cm^3 because it will hold exactly 24 cubes, each 1 cm on a side. Note that the volume can be found by multiplying the length times the width times the height.

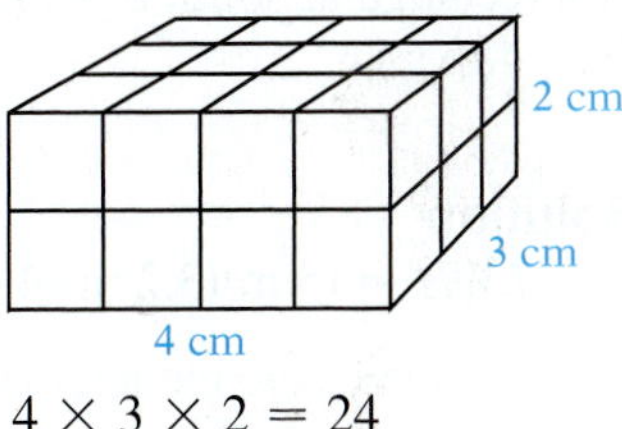

$4 \times 3 \times 2 = 24$

The formulas for the volumes of the geometric solids described above are given below.

Volumes of Geometric Solids

The volume, V, of a **rectangular solid** with length L, width W, and height H is given by $V = LWH$.

The volume, V, of a **cube** with side s is given by $V = s^3$.

The volume, V, of a **sphere** with radius r is given by $V = \frac{4}{3}\pi r^3$.

The volume, V, of a **right circular cylinder** is given by $V = \pi r^2 h$, where r is the radius of the base and h is the height.

The volume, V, of a **right circular cone** is given by $V = \frac{1}{3}\pi r^2 h$, where r is the radius of the circular base and h is the height.

The volume, V, of a **regular square pyramid** is given by $V = \frac{1}{3}s^2 h$, where s is the length of a side of the base and h is the height.

HOW TO • 1 Find the volume of a sphere with a diameter of 6 in.

$r = \frac{1}{2}d = \frac{1}{2}(6 \text{ in.}) = 3 \text{ in.}$ • First find the radius of the sphere.

$V = \frac{4}{3}\pi r^3$ • Use the formula for the volume of a sphere.

$V = \frac{4}{3}\pi(3 \text{ in.})^3$

$V = \frac{4}{3}\pi(27 \text{ in.}^3)$

$V = 36\pi \text{ in.}^3$ • The exact volume of the sphere is 36π in^3.

$V \approx 113.10 \text{ in.}^3$ • An approximate measure can be found by using the π key on a calculator.

The approximate volume is 113.10 in^3.

Integrating Technology

To approximate 36π on your calculator, enter 36 **X** **π** **=**.

EXAMPLE • 1

The length of a rectangular solid is 5 m, the width is 3.2 m, and the height is 4 m. Find the volume of the solid.

Strategy

To find the volume, use the formula for the volume of a rectangular solid. $L = 5$ m, $W = 3.2$ m, $H = 4$ m.

Solution

$V = LWH = (5 \text{ m})(3.2 \text{ m})(4 \text{ m}) = 64 \text{ m}^3$

The volume of the rectangular solid is 64 m^3.

YOU TRY IT • 1

Find the volume of a cube that measures 2.5 m on a side.

Your strategy

Your solution

EXAMPLE • 2

The radius of the base of a cone is 8 cm. The height is 12 cm. Find the volume of the cone. Round to the nearest hundredth.

Strategy

To find the volume, use the formula for the volume of a cone. An approximation is asked for; use the π key on a calculator. $r = 8$ cm, $h = 12$ cm.

Solution

$V = \frac{1}{3}\pi r^2 h$

$V = \frac{1}{3}\pi(8 \text{ cm})^2(12 \text{ cm}) = \frac{1}{3}\pi(64 \text{ cm}^2)(12 \text{ cm})$

$= 256\pi \text{ cm}^3 \approx 804.25 \text{ cm}^3$

The volume is approximately 804.25 cm^3.

YOU TRY IT • 2

The diameter of the base of a cylinder is 8 ft. The height of the cylinder is 22 ft. Find the exact volume of the cylinder.

Your strategy

Your solution

Solutions on p. S19

OBJECTIVE B To solve problems involving the surface area of a solid 

The **surface area** of a solid is the total area on the surface of the solid.

When a rectangular solid is cut open and flattened out, each face is a rectangle. The surface area, SA, of the rectangular solid is the sum of the areas of the six rectangles:

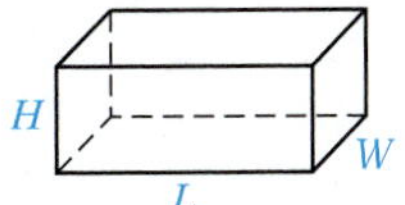

$$SA = LW + LH + WH + LW + WH + LH$$

which simplifies to

$$\mathbf{SA = 2LW + 2LH + 2WH}$$

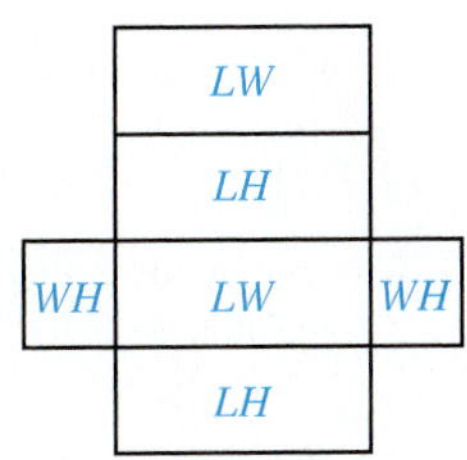

The surface area of a cube is the sum of the areas of the six faces of the cube. The area of each face is s^2. Therefore, the surface area, SA, of a cube is given by the formula $\mathbf{SA = 6s^2}$.

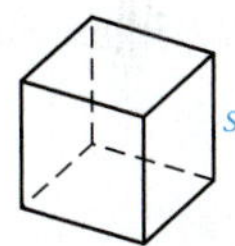

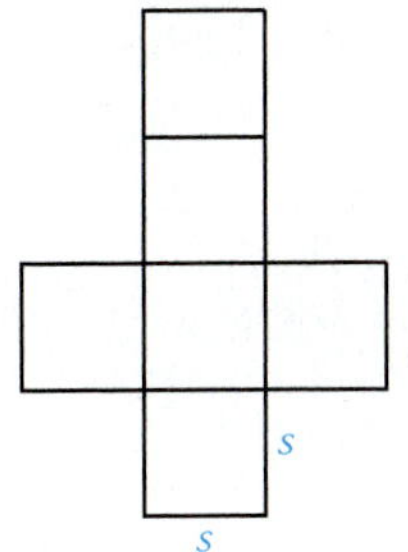

When a cylinder is cut open and flattened out, the top and bottom of the cylinder are circles. The side of the cylinder flattens out to a rectangle. The length of the rectangle is the circumference of the base, which is $2\pi r$; the width is h, the height of the cylinder. Therefore, the area of the rectangle is $2\pi rh$. The surface area, SA, of the cylinder is

$$SA = \pi r^2 + 2\pi rh + \pi r^2$$

which simplifies to

$$\mathbf{SA = 2\pi r^2 + 2\pi rh}$$

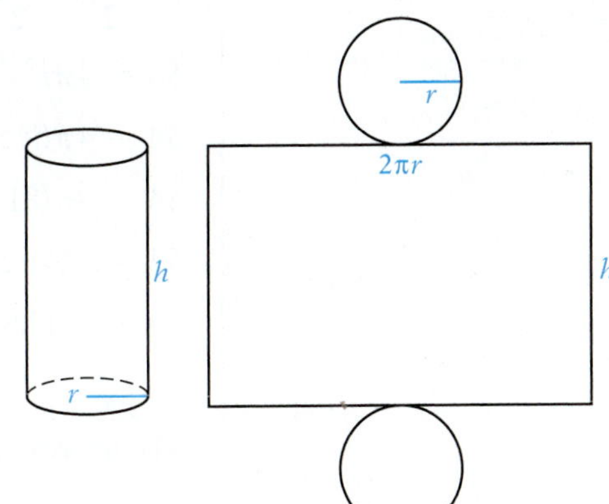

The surface area of a pyramid is the area of the base plus the area of the four isosceles triangles. A side of the square base is s; therefore, the area of the base is s^2. The slant height, ℓ, is the height of each triangle, and s is the base of each triangle. The surface area, SA, of a pyramid is

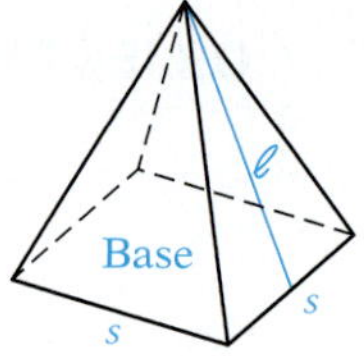

$$SA = s^2 + 4\left(\frac{1}{2}s\ell\right)$$

which simplifies to

$$\boldsymbol{SA = s^2 + 2s\ell}$$

Formulas for the surface areas of geometric solids are given below.

Surface Areas of Geometric Solids

The surface area, SA, of a **rectangular solid** with length L, width W, and height H is given by $SA = 2LW + 2LH + 2WH$.

The surface area, SA, of a **cube** with side s is given by $SA = 6s^2$.

The surface area, SA, of a **sphere** with radius r is given by $SA = 4\pi r^2$.

The surface area, SA, of a **right circular cylinder** is given by $SA = 2\pi r^2 + 2\pi rh$, where r is the radius of the base and h is the height.

The surface area, SA, of a **right circular cone** is given by $SA = \pi r^2 + \pi r\ell$, where r is the radius of the circular base and ℓ is the slant height.

The surface area, SA, of a **regular pyramid** is given by $SA = s^2 + 2s\ell$, where s is the length of a side of the base and ℓ is the slant height.

HOW TO • 2 Find the surface area of a sphere with a diameter of 18 cm.

$r = \frac{1}{2}d = \frac{1}{2}(18 \text{ cm}) = 9 \text{ cm}$ • **First find the radius of the sphere.**

$SA = 4\pi r^2$ • **Use the formula for the surface area of a sphere.**

$SA = 4\pi(9 \text{ cm})^2$

$SA = 4\pi(81 \text{ cm}^2)$

$SA = 324\pi \text{ cm}^2$ • **The exact surface area of the sphere is 324π cm².**

$SA \approx 1017.88 \text{ cm}^2$ • **An approximate measure can be found by using the π key on a calculator.**

The approximate surface area is 1017.88 cm^2.

EXAMPLE • 3

The diameter of the base of a cone is 5 m, and the slant height is 4 m. Find the surface area of the cone. Give the exact measure.

Strategy

To find the surface area of the cone:

- Find the radius of the base of the cone.
- Use the formula for the surface area of a cone. Leave the answer in terms of π.

Solution

$r = \frac{1}{2}d = \frac{1}{2}(5\text{ m}) = 2.5\text{ m}$

$SA = \pi r^2 + \pi r\ell$

$SA = \pi(2.5\text{ m})^2 + \pi(2.5\text{ m})(4\text{ m})$

$SA = 6.25\pi\text{ m}^2 + 10\pi\text{ m}^2$

$SA = 16.25\pi\text{ m}^2$

The surface area of the cone is $16.25\pi\text{ m}^2$.

YOU TRY IT • 3

The diameter of the base of a cylinder is 6 ft, and the height is 8 ft. Find the surface area of the cylinder. Round to the nearest hundredth.

Your strategy

Your solution

EXAMPLE • 4

Find the area of a label used to cover a soup can that has a radius of 4 cm and a height of 12 cm. Round to the nearest hundredth.

Strategy

To find the area of the label, use the fact that the surface area of the side of a cylinder is given by $2\pi rh$. An approximation is asked for; use the π key on a calculator. $r = 4$ cm, $h = 12$ cm.

Solution

Area of the label $= 2\pi rh$

Area of the label $= 2\pi(4\text{ cm})(12\text{ cm})$

$= 96\pi\text{ cm}^2 \approx 301.59\text{ cm}^2$

The area is approximately 301.59 cm^2.

YOU TRY IT • 4

Which has a larger surface area, a cube with a side measuring 10 cm or a sphere with a diameter measuring 8 cm?

Your strategy

Your solution

Solutions on pp. S19–S20

7.4 EXERCISES

OBJECTIVE A To solve problems involving the volume of a solid

For Exercises 1 to 6, find the volume of the figure. For calculations involving π, give both the exact value and an approximation to the nearest hundredth.

1.
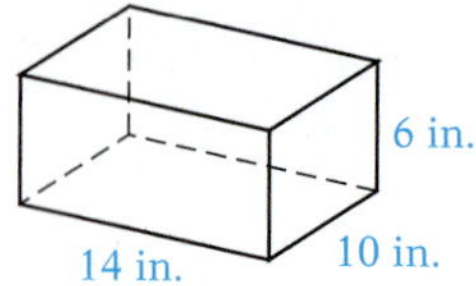

2.
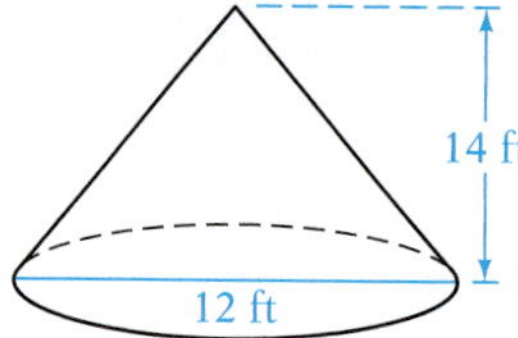

3.
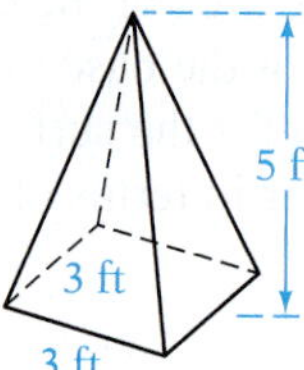

4.
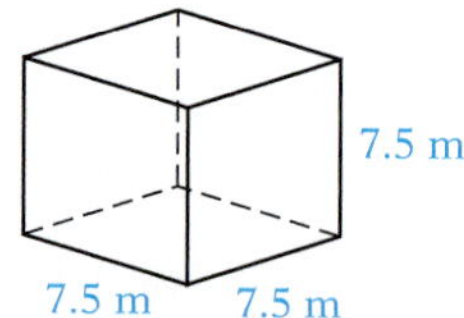

5.
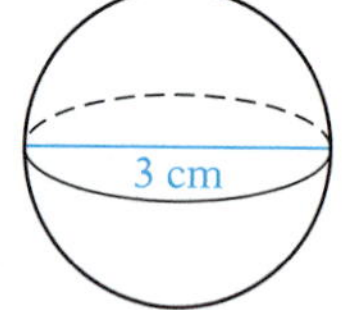

6.
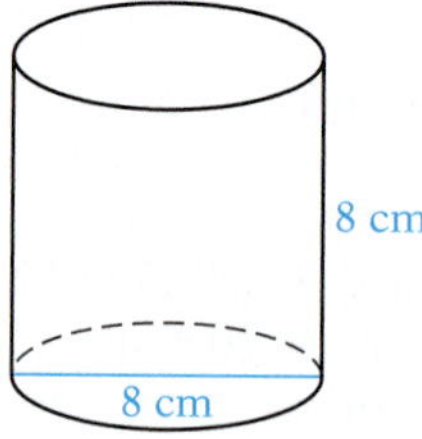

7. **Storage Units** A rectangular storage unit has a length of 6.8 m, a width of 2.5 m, and a height of 2 m. Find the volume of the storage unit.

8. **Fish Hatchery** A rectangular tank at a fish hatchery is 9 m long, 3 m wide, and 1.5 m deep. Find the volume of the water in the tank when the tank is full.

9. Find the volume of a cube whose side measures 3.5 in.

10. The length of a side of a cube is 7 cm. Find the volume of the cube.

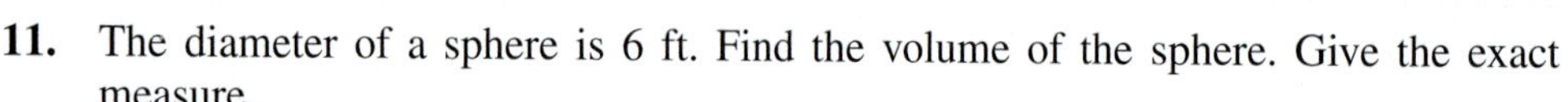

11. The diameter of a sphere is 6 ft. Find the volume of the sphere. Give the exact measure.

12. Find the volume of a sphere that has a radius of 1.2 m. Round to the nearest tenth.

13. The diameter of the base of a cylinder is 24 cm. The height of the cylinder is 18 cm. Find the volume of the cylinder. Round to the nearest hundredth.

14. The height of a cylinder is 7.2 m. The radius of the base is 4 m. Find the volume of the cylinder. Give the exact measure.

15. The radius of the base of a cone is 5 in. The height of the cone is 9 in. Find the volume of the cone. Give the exact measure.

16. The height of a cone is 15 cm. The diameter of the cone is 10 cm. Find the volume of the cone. Round to the nearest hundredth.

17. The length of a side of the base of a pyramid is 6 in., and the height is 10 in. Find the volume of the pyramid.

18. The height of a pyramid is 8 m, and the length of a side of the base is 9 m. What is the volume of the pyramid?

19. **Appliances** The volume of a freezer with a length of 7 ft and a height of 3 ft is 52.5 ft^3. Find the width of the freezer.

20. **Aquariums** The length of an aquarium is 18 in., and the width is 12 in. If the volume of the aquarium is 1836 in^3, what is the height of the aquarium?

21. The volume of a cylinder is 502.4 in^3. The diameter of the base is 10 in. Find the height of the cylinder. Round to the nearest hundredth.

22. The diameter of the base of a cylinder is 14 cm. If the volume of the cylinder is 2310 cm^3, find the height of the cylinder. Round to the nearest hundredth.

23. **Guacamole Consumption** See the news clipping at the right. What is the volume of the guacamole in cubic feet?

24. **Guacamole Consumption** See the news clipping at the right. Assuming that each person eats 1 c of guacamole, how many people could be fed from the covered football field? (1 ft^3 = 59.84 pt)

25. The length of a side of a cube is equal to the radius of a sphere. Which solid has the greater volume?

26. A sphere and a cylinder have the same radius. The height of the cylinder is equal to the radius of its base. Which solid has the greater volume?

In the News

Super Bowl Win for Guacamole

Guacamole has become the dish of choice at Super Bowl parties. If all the guacamole eaten during the Super Bowl were piled onto a football field—that's a football field which, including endzones, is 360 ft long and 160 ft wide—it would cover the field to a depth of 3 ft!

Source: www.azcentral.com

OBJECTIVE B To solve problems involving the surface area of a solid

For Exercises 27 to 29, find the surface area of the figure.

27.

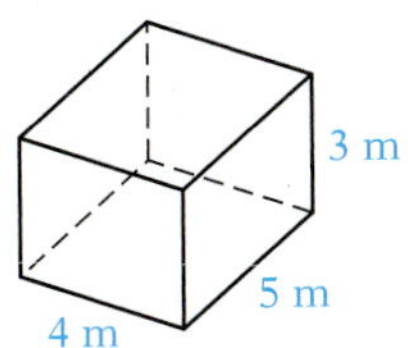

28.

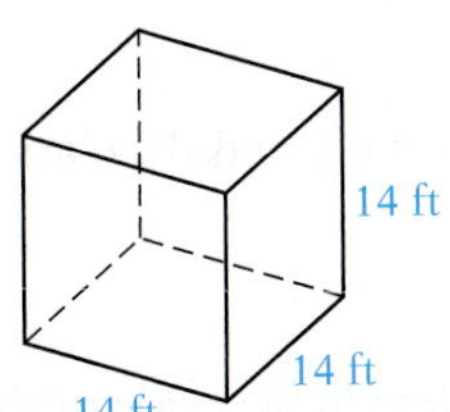

29.

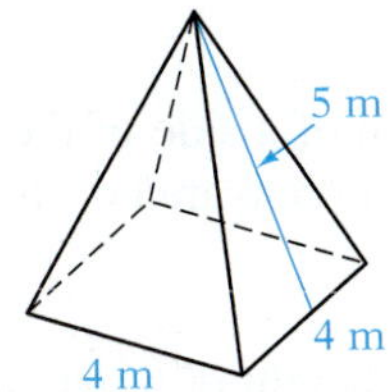

For Exercises 30 to 32, find the surface area of the figure. Give both the exact value and an approximation to the nearest hundredth.

30.

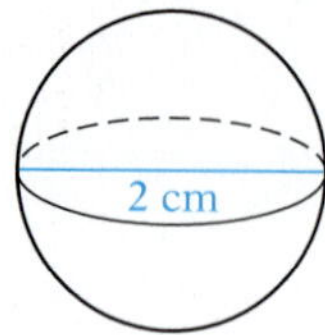

31.

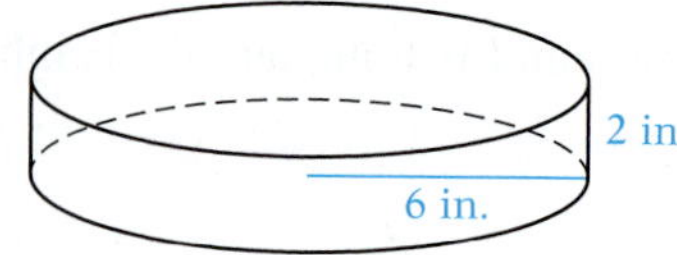

32.

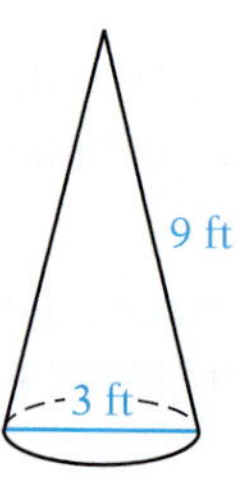

33. The height of a rectangular solid is 5 ft. The length is 8 ft, and the width is 4 ft. Find the surface area of the solid.

34. The width of a rectangular solid is 32 cm. The length is 60 cm, and the height is 14 cm. What is the surface area of the solid?

35. The side of a cube measures 3.4 m. Find the surface area of the cube.

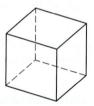

36. Find the surface area of a cube that has a side measuring 1.5 in.

37. Find the surface area of a sphere with a diameter of 15 cm. Give the exact value.

38. The radius of a sphere is 2 in. Find the surface area of the sphere. Round to the nearest hundredth.

39. The radius of the base of a cylinder is 4 in. The height of the cylinder is 12 in. Find the surface area of the cylinder. Round to the nearest hundredth.

40. The diameter of the base of a cylinder is 1.8 m. The height of the cylinder is 0.7 m. Find the surface area of the cylinder. Give the exact value.

41. The slant height of a cone is 2.5 ft. The radius of the base is 1.5 ft. Find the surface area of the cone. Give the exact value.

42. The diameter of the base of a cone is 21 in. The slant height is 16 in. What is the surface area of the cone? Round to the nearest hundredth.

43. The length of a side of the base of a pyramid is 9 in., and the slant height is 12 in. Find the surface area of the pyramid.

44. The slant height of a pyramid is 18 m, and the length of a side of the base is 16 m. What is the surface area of the pyramid?

45. The surface area of a rectangular solid is 108 cm^2. The height of the solid is 4 cm, and the length is 6 cm. Find the width of the rectangular solid.

46. The length of a rectangular solid is 12 ft. The width is 3 ft. If the surface area is 162 ft^2, find the height of the rectangular solid.

47. **Paint** A can of paint will cover 300 ft^2. How many cans of paint should be purchased in order to paint a cylinder that has a height of 30 ft and a radius of 12 ft?

48. **Ballooning** A hot air balloon is in the shape of a sphere. Approximately how much fabric was used to construct the balloon if its diameter is 32 ft? Round to the nearest whole number.

49. **Aquariums** How much glass is needed to make a fish tank that is 12 in. long, 8 in. wide, and 9 in. high? The fish tank is open at the top.

50. **Packaging** Find the area of a label used to cover a can of juice that has a diameter of 16.5 cm and a height of 17 cm. Round to the nearest hundredth.

51. The length of a side of the base of a pyramid is 5 cm, and the slant height is 8 cm. How much larger is the surface area of this pyramid than the surface area of a cone with a diameter of 5 cm and a slant height of 8 cm? Round to the nearest hundredth.

52. Half of a sphere is called a **hemisphere.** Derive formulas for the volume and surface area of a hemisphere.

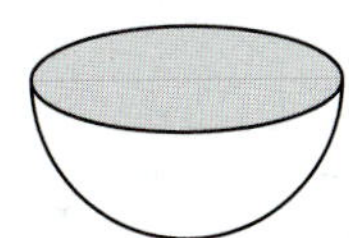

53. Determine whether the statement is always true, sometimes true, or never true.
    a. The slant height of a regular pyramid is longer than the height.
    b. The slant height of a cone is shorter than the height.
    c. The four triangular faces of a regular pyramid are equilateral triangles.

Applying the Concepts

54. **a.** What is the effect on the surface area of a rectangular solid if the width and height are doubled?
    b. What is the effect on the volume of a rectangular solid if both the length and the width are doubled?
    c. What is the effect on the volume of a cube if the length of each side of the cube is doubled?
    d. What is the effect on the surface area of a cylinder if the radius and height are doubled?

55. Explain how you could cut through a cube so that the face of the resulting solid is (a) a square, (b) an equilateral triangle, (c) a trapezoid, (d) a hexagon.

FOCUS ON PROBLEM SOLVING

Trial and Error

Some problems in mathematics are solved by using **trial and error.** The trial-and-error method of arriving at a solution to a problem involves repeated tests or experiments until a satisfactory conclusion is reached.

Many of the Applying the Concepts exercises in this text require a trial-and-error method of solution. For example, an exercise on page 431 reads as follows:

Explain how you could cut through a cube so that the face of the resulting solid is (a) a square, (b) an equilateral triangle, (c) a trapezoid, (d) a hexagon.

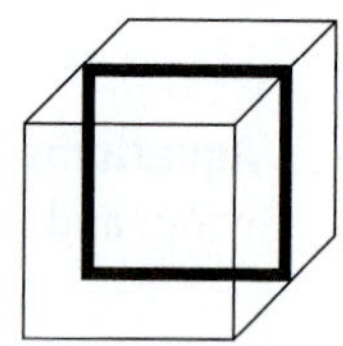

There is no formula to apply to this problem; there is no computation to perform. This problem requires picturing a cube and the results after it is cut through at different places on its surface and at different angles. For part (a), cutting perpendicular to the top and bottom of the cube and parallel to two of its sides will result in a square. The other shapes may prove more difficult.

When solving problems of this type, keep an open mind. Sometimes when using the trial-and-error method, we are hampered by narrowness of vision; we cannot expand our thinking to include other possibilities. Then, when we see someone else's solution, it appears so obvious to us! For example, for the question above, it is necessary to conceive of cutting through the cube at places other than the top surface; we need to be open to the idea of beginning the cut at one of the corner points of the cube.

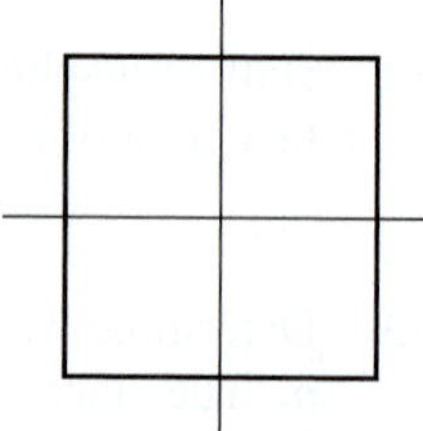

A topic of the Projects and Group Activities in this chapter is symmetry. Here again, the trial-and-error method is used to determine the lines of symmetry inherent in an object. For example, in determining lines of symmetry for a square, begin by drawing a square. The horizontal line of symmetry and the vertical line of symmetry may be immediately obvious to you.

But there are two others. Do you see that lines drawn through opposite corners of the square are also lines of symmetry?

Many of the questions in this text that require an answer of "always true," "sometimes true," or "never true" are best solved by the trial-and-error method. For example, consider the statement:

Two rectangles that have the same area have the same perimeter.

Try some numbers. Each of two rectangles, one measuring 6 units by 2 units and another measuring 4 units by 3 units, has an area of 12 square units, but the perimeter of the first is 16 units and the perimeter of the second is 14 units. So, the answer "always true" has been eliminated. We still need to determine whether there is a case for which the statement *is* true. After experimenting with a lot of numbers, you may come to realize that we are trying to determine whether it is possible for two different pairs of factors of a number to have the same sum. Is it?

Don't be afraid to perform many experiments, and remember that *errors*, or tests that "don't work," are a part of the trial-and-*error* process.

PROJECTS AND GROUP ACTIVITIES

Lines of Symmetry

Look at the letter A printed at the left. If the letter were folded along line l, the two sides of the letter would match exactly. This letter has **symmetry** with respect to line l. Line l is called the **axis of symmetry.**

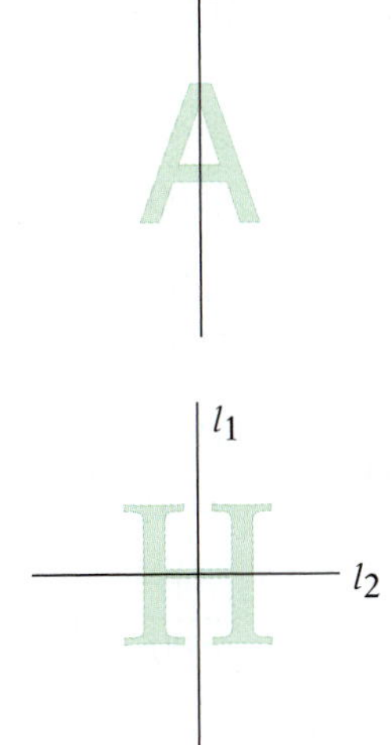

Now consider the letter H printed below at the left. Both lines l_1 and l_2 are axes of symmetry for this letter; the letter could be folded along either line and the two sides would match exactly.

1. Does the letter A have more than one axis of symmetry?
2. Find axes of symmetry for other capital letters of the alphabet.
3. Which lowercase letters have one axis of symmetry?
4. Do any of the lowercase letters have more than one axis of symmetry?
5. Find the numbers of axes of symmetry for the plane geometric figures presented in this chapter.
6. There are other types of symmetry. Look up the meaning of *point symmetry* and *rotational symmetry*. Which plane geometric figures provide examples of these types of symmetry?
7. Find examples of symmetry in nature, art, and architecture.

Preparing a Circle Graph

In Section 1 of this chapter, a protractor was used to measure angles. Preparing a circle graph requires the ability to use a protractor to draw angles.

To draw an angle of 142°, first draw a ray. Place a dot at the endpoint of the ray. This dot will be the vertex of the angle.

Place the line segment near the bottom edge of the protractor on the ray as shown in the figure at the right. Make sure the center of the line segment near the bottom edge of the protractor is located directly over the vertex point. Locate the position of the 142° mark. Place a dot next to the mark.

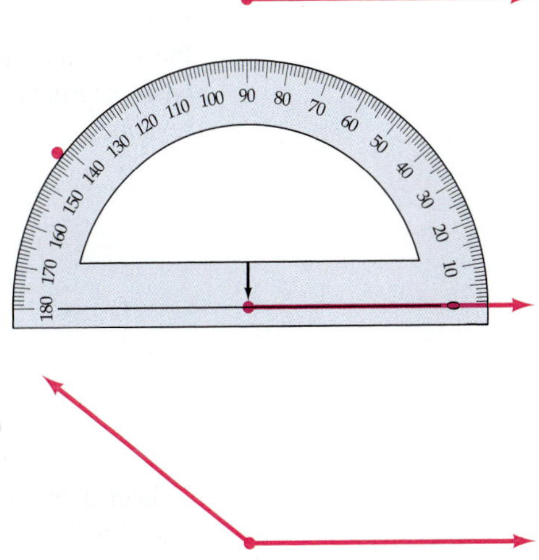

Remove the protractor and draw a ray from the vertex to the dot at the 142° mark.

An example of preparing a circle graph is given on the next page.

The revenues (in thousands of dollars) from four segments of a car dealership for the first quarter of a recent year were

New car sales:	\$2100	Used car/truck sales:	\$1500
New truck sales:	\$1200	Parts/service:	\$700

To draw a circle graph to represent the percent that each segment contributed to the total revenue from all four segments, proceed as follows.

Find the total revenue from all four segments.

$$2100 + 1200 + 1500 + 700 = 5500$$

Find what percent each segment is of the total revenue of \$5500.

New car sales: $\frac{2100}{5500} \approx 38.2\%$

New truck sales: $\frac{1200}{5500} \approx 21.8\%$

Used car/truck sales: $\frac{1500}{5500} \approx 27.3\%$

Parts/service: $\frac{700}{5500} \approx 12.7\%$

Each percent represents a sector of the circle. Because the circle contains 360°, multiply each percent by 360° to find the measure of the angle for each sector. Round to the nearest whole number.

New car sales:

$$0.382 \times 360° \approx 138°$$

New truck sales:

$$0.218 \times 360° \approx 78°$$

Used car/truck sales:

$$0.273 \times 360° \approx 98°$$

Parts/service:

$$0.127 \times 360° \approx 46°$$

Draw a circle and use a protractor to draw the sectors representing the percents that each segment contributed to the total revenue.

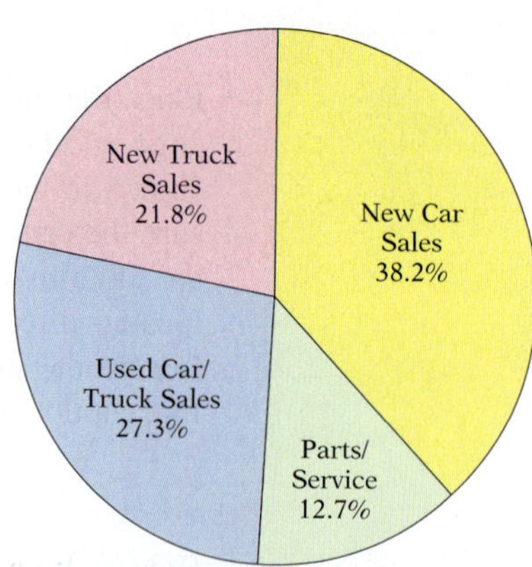

Collect data appropriate for display in a circle graph. [Some possibilities are last year's sales for the top three car manufacturers in the United States, votes cast in the last election for your state governor, the majors of the students in your math class, and the number of students enrolled in each class (senior, junior, etc.) at your college.] Then prepare the circle graph.

CHAPTER 7

SUMMARY

KEY WORDS | **EXAMPLES**

A *line* is determined by two distinct points and extends indefinitely in both directions. A *line segment* is part of a line and has two endpoints. *Parallel lines* never meet; the distance between them is always the same. *Perpendicular lines* are intersecting lines that form right angles. [7.1A, pp. 378–380]

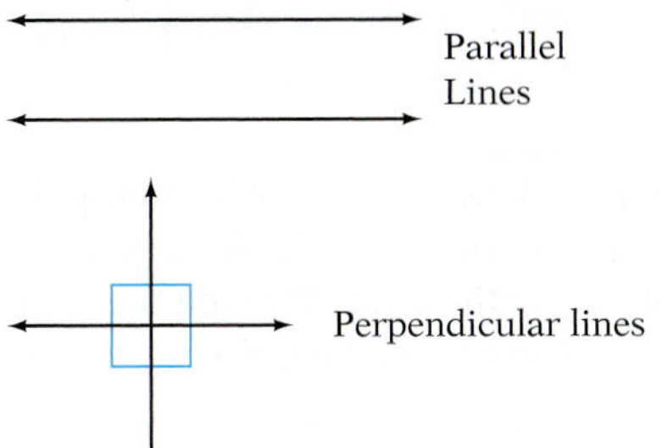

A *ray* starts at a point and extends indefinitely in one direction. The point at which a ray starts is the *endpoint* of the ray. An *angle* is formed by two rays with the same endpoint. The *vertex* of an angle is the point at which the two rays meet. An angle is measured in *degrees*. A 90° angle is a *right angle*. A 180° angle is a *straight angle*. An *acute angle* is an angle whose measure is between 0° and 90°. An *obtuse angle* is an angle whose measure is between 90° and 180°. *Complementary angles* are two angles whose measures have the sum 90°. *Supplementary angles* are two angles whose measures have the sum 180°. [7.1A, pp. 378–381]

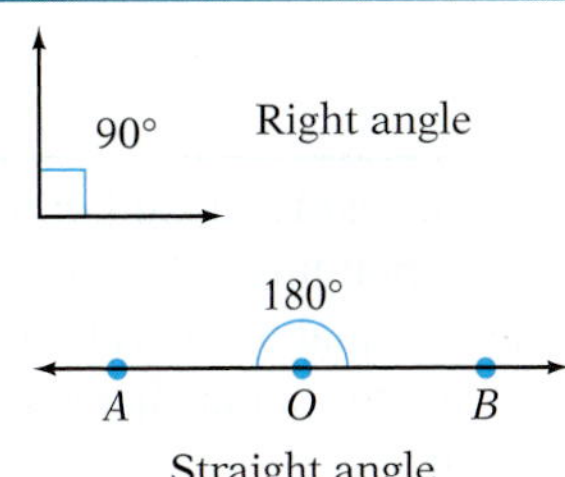

Two angles that are on opposite sides of the intersection of two lines are *vertical angles*; vertical angles have the same measure. Two angles that share a common side are *adjacent angles*; adjacent angles of intersecting lines are supplementary angles. [7.1A, p. 381; 7.1B, p. 383]

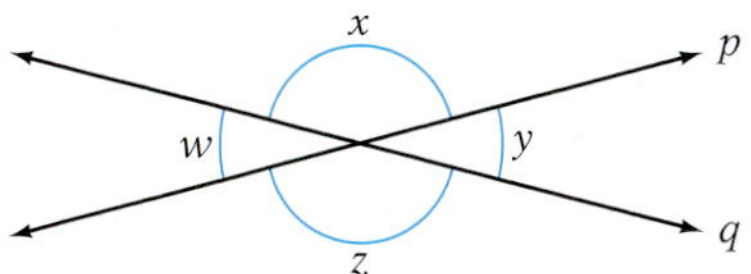

Angles w and y are vertical angles.
Angles x and y are adjacent angles.

A line that intersects two other lines at two different points is a *transversal*. If the lines cut by a transversal are parallel lines, equal angles are formed: *alternate interior angles, alternate exterior angles,* and *corresponding angles.* [7.1B, p. 384]

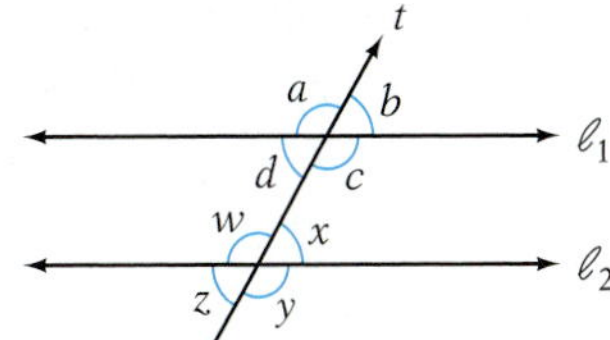

Parallel lines ℓ_1 and ℓ_2 are cut by transversal t. All four acute angles have the same measure. All four obtuse angles have the same measure.

A *polygon* is a closed figure determined by three or more line segments. The line segments that form the polygon are its *sides*. A *regular polygon* is one in which all sides have the same length and all angles have the same measure. Polygons are classified by the number of sides. A *quadrilateral* is a four-sided polygon. A parallelogram, a rectangle, a square, a rhombus, and a trapezoid are all quadrilaterals. [7.2A, pp. 394–395]

Number of Sides	*Name of the Polygon*
3	Triangle
4	Quadrilateral
5	Pentagon
6	Hexagon
7	Heptagon
8	Octagon
9	Nonagon
10	Decagon

A *triangle* is a plane figure formed by three line segments. An *isosceles triangle* has two sides of equal length. The three sides of an *equilateral triangle* are of equal length. A *scalene triangle* has no two sides of equal length. An *acute triangle* has three acute angles. An *obtuse triangle* has one obtuse angle. A *right triangle* has a right angle. [7.1C, p. 386; 7.2A, pp. 394–395]

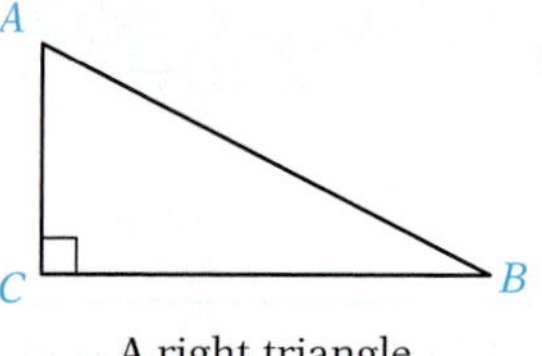

A right triangle

A *circle* is a plane figure in which all points are the same distance from the center of the circle. A *diameter* of a circle is a line segment across the circle through the center. A *radius* of a circle is a line segment from the center of the circle to a point on the circle. [7.2A, p. 397]

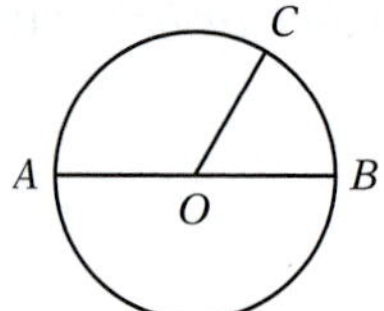

AB is a diameter of the circle.
OC is a radius.

Similar triangles have the same shape but not necessarily the same size. The ratios of corresponding sides are equal. The ratio of corresponding heights is equal to the ratio of corresponding sides. *Congruent triangles* have the same shape and the same size. [7.3B, p. 414; 7.3C, p. 416]

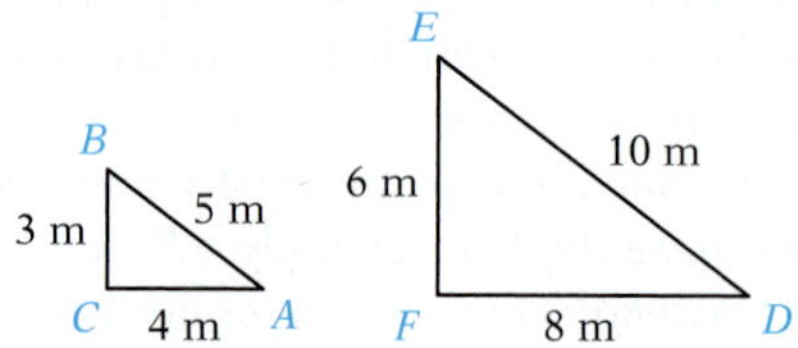

Triangles *ABC* and *DEF* are similar triangles. The ratio of corresponding sides is $\frac{1}{2}$.

ESSENTIAL RULES AND PROCEDURES

EXAMPLES

Triangles [7.1C, p. 386, 7.3C, pp. 416–417]
Sum of the measures of the interior angles = 180°
Sum of an interior angle and the corresponding exterior angle = 180°
Rules to determine congruence: SSS Rule, SAS Rule, ASA Rule

In a right triangle, the measure of one acute angle is 12°. Find the measure of the other acute angle.

$$x + 12° + 90° = 180°$$
$$x + 102° = 180°$$
$$x = 78°$$

Formulas for Perimeter (the distance around a figure) [7.2A, pp. 395–397]
Triangle: $P = a + b + c$
Rectangle: $P = 2L + 2W$
Square: $P = 4s$
Circumference of a circle: $C = \pi d$ or $C = 2\pi r$

The length of a rectangle is 8 m. The width is 5.5 m. Find the perimeter of the rectangle.

$$P = 2L + 2W$$
$$P = 2(8 \text{ m}) + 2(5.5 \text{ m})$$
$$P = 16 \text{ m} + 11 \text{ m}$$
$$P = 27 \text{ m}$$

The perimeter is 27 m.

Formulas for Area (the amount of surface in a region) [7.2B, pp. 399–402]

Triangle: $A = \frac{1}{2}bh$
Rectangle: $A = LW$
Square: $A = s^2$
Circle: $A = \pi r^2$
Parallelogram: $A = bh$
Trapezoid: $A = \frac{1}{2}h(b_1 + b_2)$

The length of the base of a parallelogram is 12 cm, and the height is 4 cm. Find the area of the parallelogram.

$A = bh$
$A = (12 \text{ cm})(4 \text{ cm})$
$A = 48 \text{ cm}^2$

The area is 48 cm^2.

Formulas for Volume (the amount of space inside a figure in space) [7.4A, p. 423]

Rectangular solid: $V = LWH$
Cube: $V = s^3$
Sphere: $V = \frac{4}{3}\pi r^3$
Right circular cylinder: $V = \pi r^2 h$
Right circular cone: $V = \frac{1}{3}\pi r^2 h$
Regular pyramid: $V = \frac{1}{3}s^2 h$

Find the volume of a cube that measures 3 in. on a side.

$V = s^3$
$V = (3 \text{ in.})^3$
$V = 27 \text{ in}^3$

The volume is 27 in^3.

Formulas for Surface Area (the total area on the surface of a solid) [7.4B, p. 426]

Rectangular solid: $SA = 2LW + 2LH + 2WH$
Cube: $SA = 6s^2$
Sphere: $SA = 4\pi r^2$
Right circular cylinder: $SA = 2\pi r^2 + 2\pi rh$
Right circular cone: $SA = \pi r^2 + \pi r\ell$
Regular pyramid: $SA = s^2 + 2s\ell$

Find the surface area of a sphere with a diameter of 10 cm. Give the exact value.

$r = \frac{1}{2}d = \frac{1}{2}(10) = 5$
$SA = 4\pi r^2$
$SA = 4\pi(5 \text{ cm})^2$
$SA = 4\pi(25 \text{ cm}^2)$
$SA = 100\pi \text{ cm}^2$

The surface area is $100\pi \text{ cm}^2$.

Pythagorean Theorem [7.3A, p. 412]

If a and b are the lengths of the legs of a right triangle and c is the length of the hypotenuse, then $c^2 = a^2 + b^2$.

Two legs of a right triangle measure 6 ft and 8 ft. Find the hypotenuse of the right triangle.

$c^2 = a^2 + b^2$
$c^2 = 6^2 + 8^2$
$c^2 = 36 + 64$
$c^2 = 100$
$c = \sqrt{100}$
$c = 10$

The length of the hypotenuse is 10 ft.

Principal Square Root Property [7.3A, p. 413]

If $r^2 = s$, then $r = \sqrt{s}$, and r is called the *square root* of s.

If $c^2 = 16$, then $c = \sqrt{16} = 4$.

CHAPTER 7

CONCEPT REVIEW

Test your knowledge of the concepts presented in this chapter. Answer each question. Then check your answers against the ones provided in the Answer Section.

1. What are perpendicular lines?

2. When are two angles complementary?

3. How are the angles in a triangle related?

4. If you know the diameter of a circle, how can you find the radius?

5. What is the formula for the perimeter of a rectangle?

6. How do you find the circumference of a circle?

7. What is the formula for the area of a triangle?

8. What three dimensions are needed to find the volume of a rectangular solid?

9. To calculate the surface area of a right circular cylinder, you are calculating the areas of what plane geometric figures?

10. How do you identify the hypotenuse of a right triangle?

11. How can you use a proportion to solve for a missing length in similar triangles?

12. What is the side-angle-side rule?

CHAPTER 7

REVIEW EXERCISES

1. Given that $\angle a = 74°$ and $\angle b = 52°$, find the measures of angles x and y.

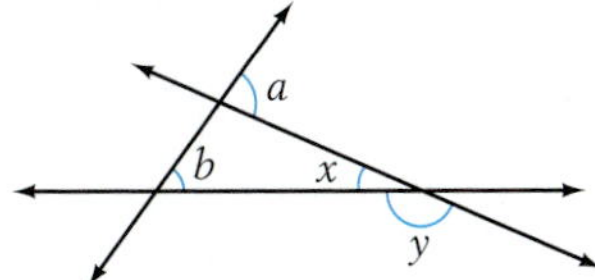

2. Triangles ABC and DEF are similar. Find the perimeter of triangle ABC.

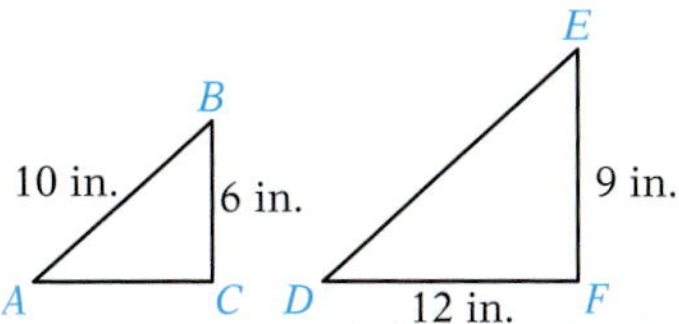

3. Given triangle ABC and triangle DEF, do the conditions $\angle B = \angle D$, $\angle A = \angle F$, and $\angle C = \angle E$ guarantee that triangle ABC is congruent to triangle DEF? If they are congruent, by what rule are they congruent?

4. Find the measure of $\angle x$.

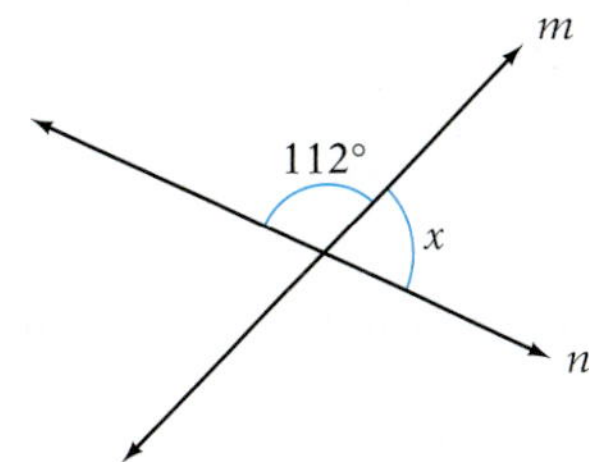

5. Determine whether the two triangles are congruent. If they are congruent, state by what rule they are congruent.

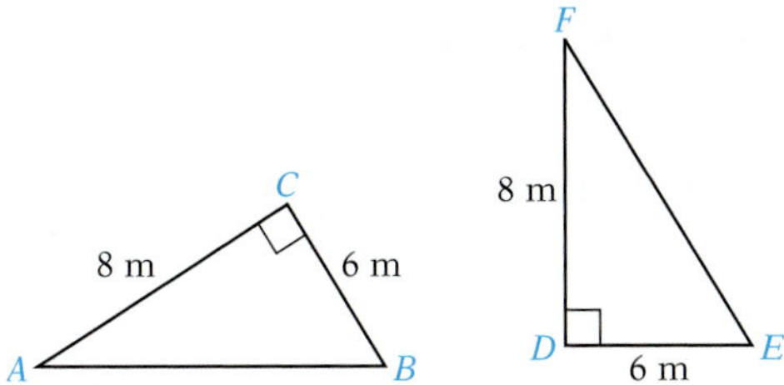

6. The two legs of a right triangle measure 4 in. and 6 in. Find the hypotenuse of the right triangle. Round to the nearest hundredth.

7. Given that $BC = 11$ cm and AB is three times the length of BC, find the length of AC.

A B C

8. Find x.

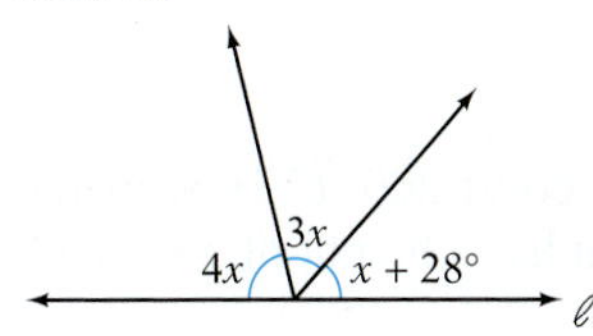

9. Find the area of the figure.

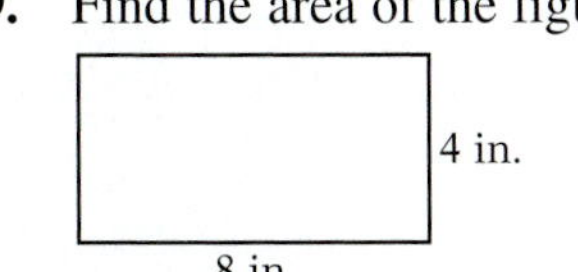

10. Find the volume of the figure.

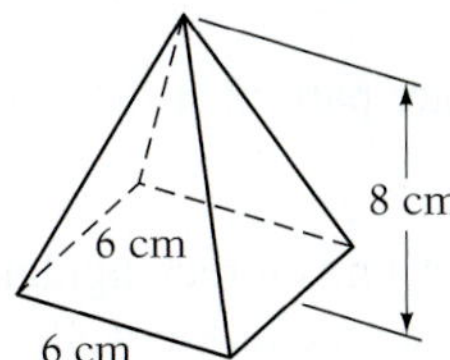

11. Find the perimeter of the figure.

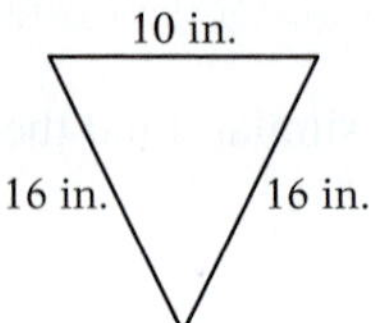

12. Given that $\ell_1 \parallel \ell_2$, find the measures of angles a and b.

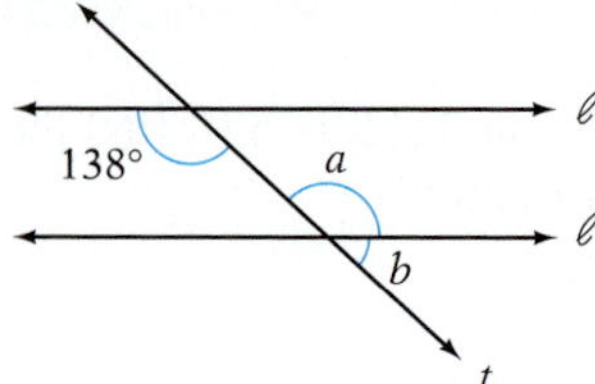

13. Find the surface area of the figure.

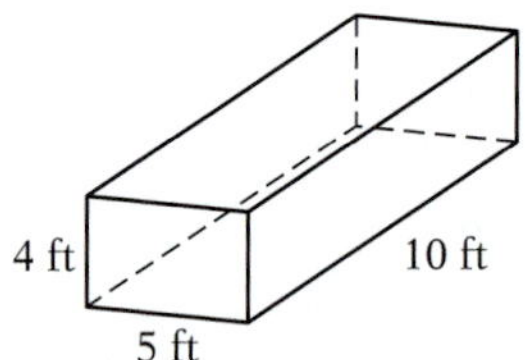

14. Find the unknown side of the triangle. Round to the nearest hundredth.

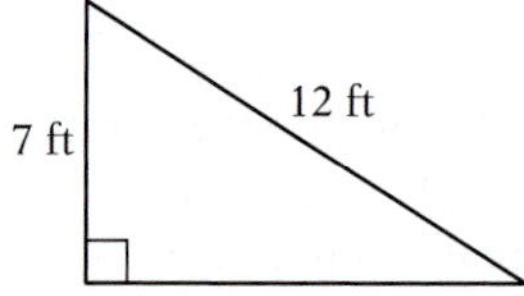

15. Find the volume of a cube whose side measures 3.5 in.

16. Find the supplement of a 32° angle.

17. Find the volume of a rectangular solid with a length of 6.5 ft, a width of 2 ft, and a height of 3 ft.

18. Two angles of a triangle measure 37° and 48°. Find the measure of the third angle.

19. The height of a triangle is 7 cm. The area of the triangle is 28 cm^2. Find the length of the base of the triangle.

20. Find the volume of a sphere that has a diameter of 12 mm. Give the exact value.

21. The perimeter of a square picture frame is 86 cm. Find the length of each side of the frame.

22. A can of paint will cover 200 ft^2. How many cans of paint should be purchased to paint a cylinder that has a height of 15 ft and a radius of 6 ft?

23. The length of a rectangular park is 56 yd. The width is 48 yd. How many yards of fencing are needed to surround the park?

24. What is the area of a square patio that measures 9.5 m on each side?

25. A walkway 2 m wide surrounds a rectangular plot of grass. The plot is 40 m long and 25 m wide. What is the area of the walkway?

CHAPTER 7

TEST

1. For the right triangle shown below, determine the length of side BC. Round to the nearest hundredth.

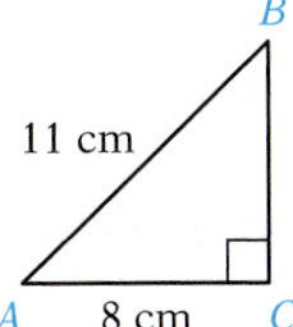

2. Determine whether the two triangles are congruent. If they are congruent, state by what rule they are congruent.

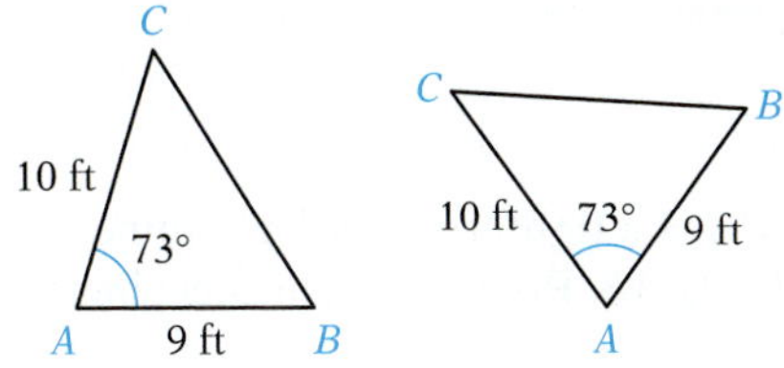

3. Determine the area of a rectangle with a length of 15 m and a width of 7.4 m.

4. Determine the area of a triangle whose base is 7 ft and whose height is 12 ft.

5. Determine the exact volume of a right circular cone whose radius is 7 cm and whose height is 16 cm.

6. Determine the exact surface area of a pyramid whose square base is 3 m on each side and whose slant height is 11 m.

7. Determine the volume of the solid shown below. Round to the nearest hundredth.

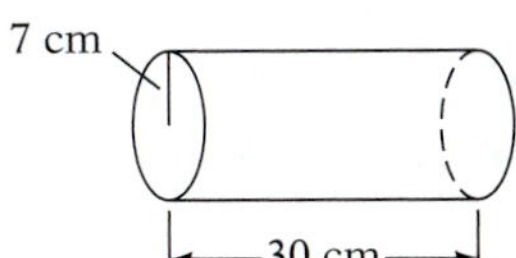

8. Determine the area of the trapezoid shown below.

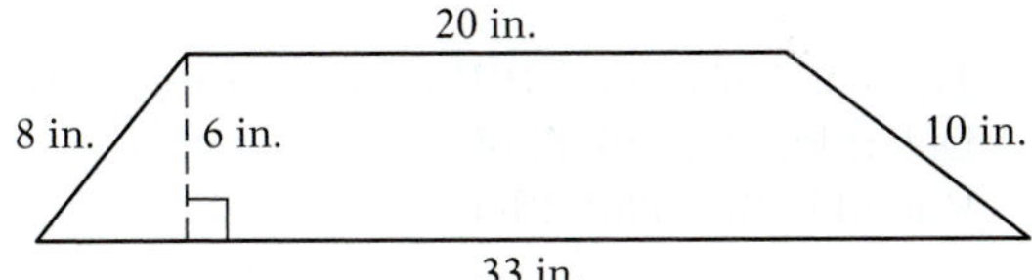

9. Determine the perimeter of the figure shown below.

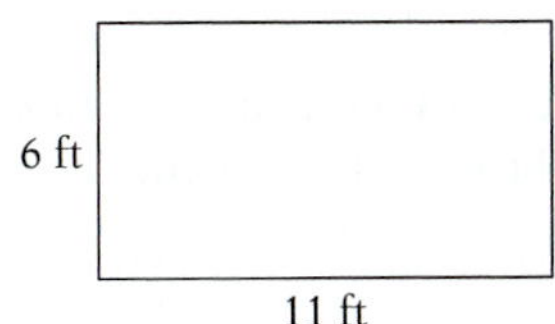

10. Two angles of a triangle measure 57° and 23°. Find the measure of the third angle.

11. Find x.

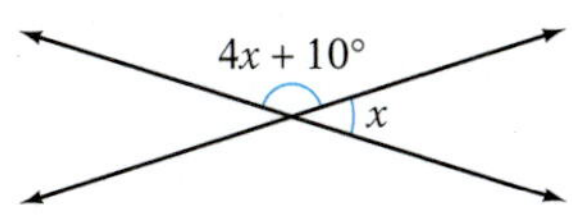

12. Name the figure shown below.

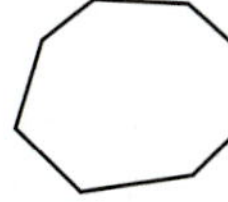

13. Determine whether the two triangles are congruent. If they are congruent, state by what rule they are congruent.

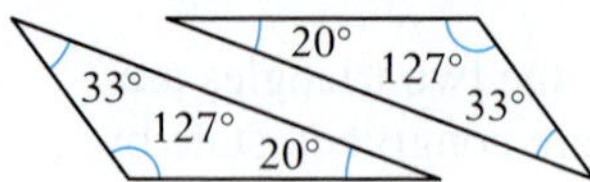

14. Determine the volume of the rectangular solid shown below.

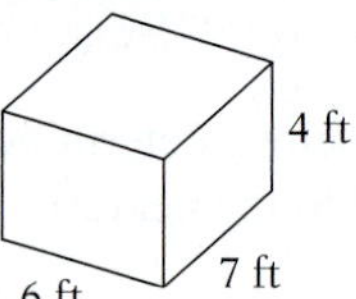

15. Figure ABC is a right triangle. Determine the length of side AB. Round to the nearest hundredth.

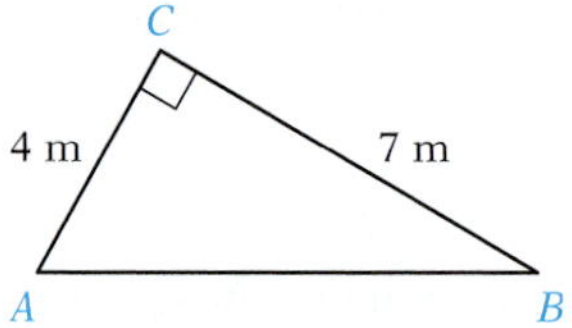

16. Given that ℓ_1 and ℓ_2 are parallel lines, determine the measure of angle a.

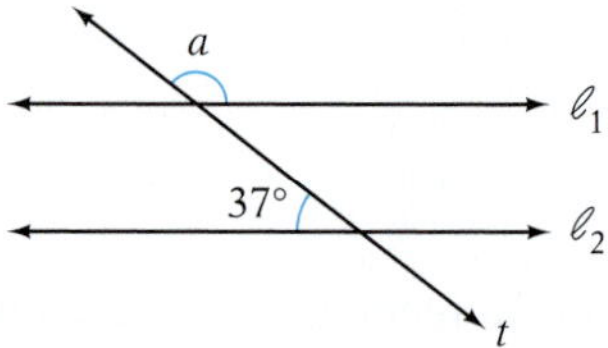

17. Determine the exact surface area of the right circular cylinder shown below.

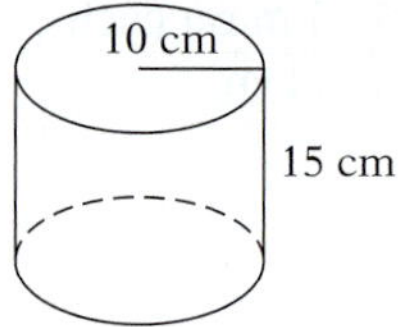

18. Determine the measure of angle a.

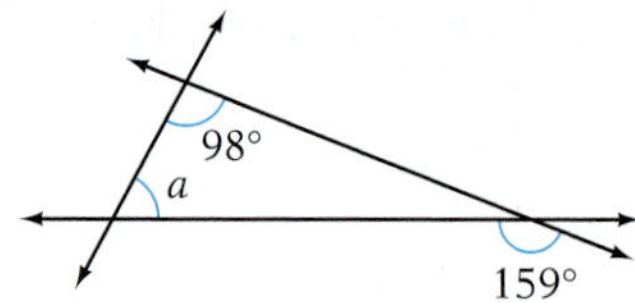

19. Triangles ABC and DEF are similar triangles. Determine the length of line segment FG. Round to the nearest hundredth.

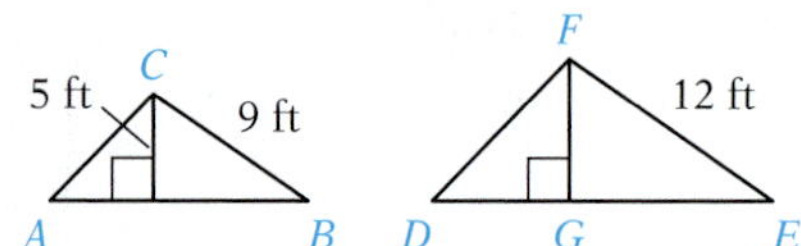

20. Triangles ABC and DEF are similar triangles. Determine the length of side BC. Round to the nearest hundredth.

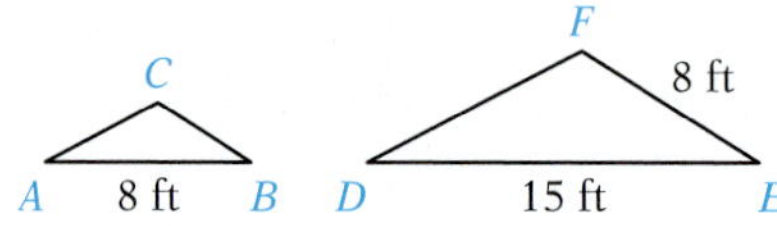

21. Determine the perimeter of a square whose side is 5 m.

22. Determine the perimeter of a rectangle whose length is 8 cm and whose width is 5 cm.

23. Find the perimeter of a right triangle with legs that measure 12 ft and 18 ft. Round to the nearest tenth.

24. Two angles of a triangle measure 41° and 37°. Find the measure of the third angle.

25. Find the supplement of a 67° angle.

CUMULATIVE REVIEW EXERCISES

1. Find 8.5% of 2400.

2. Find all the factors of 78.

3. Divide: $4\frac{2}{3} \div 5\frac{3}{5}$

4. Evaluate: $|-18|$

5. Divide and round to the nearest tenth: $82.93 \div 6.5$

6. Subtract: $-6 - (-4)$

7. Find the measure of $\angle x$.

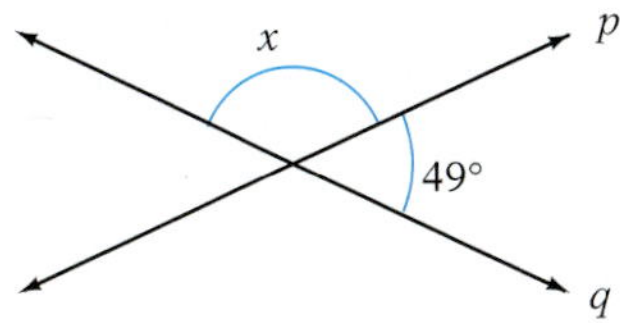

8. Find the unknown side of the triangle.

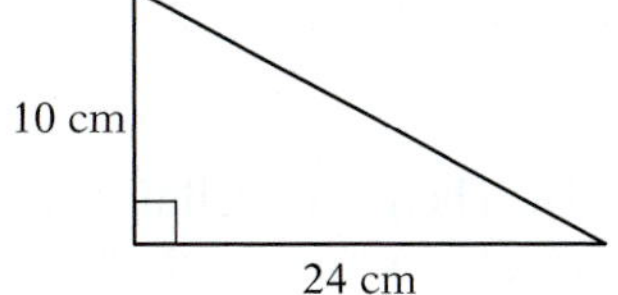

9. Add: $\frac{2}{3} + \frac{7}{10}$

10. Simplify: $-\frac{1}{4}(28b)$

11. Simplify: $3(-4) - (-1 + 5)^2$

12. Solve: $3(2x + 5) = 18$

13. Simplify: $3a + 5a - 7a + a$

14. Multiply: $-3(-25)$

15. Simplify: $5(2x + 4) - (3x + 2)$

16. Evaluate $2x + 3y^2z$ when $x = 5$, $y = -1$, and $z = -4$.

17. Evaluate $x^2y - 2z$ when $x = \frac{1}{2}$, $y = \frac{4}{5}$, and $z = -\frac{3}{10}$.

18. Find the prime factorization of 78.

19. Solve: $4x + 2 = 6x - 8$

20. Write $\frac{3}{8}$ as a percent.

21. Translate "the product of eight and twice a number" into a variable expression. Then simplify.

22. Uniform Motion Two cars, one traveling 5 mph faster than the other, start at the same time from the same point and travel in opposite directions. In 2 h they are 210 mi apart. Find the rate of each car.

23. Find the simple interest on a 270-day loan of \$20,000 at an annual interest rate of 8.875%.

24. Mixtures How many ounces of a silver alloy that costs \$3.50 per ounce must be mixed with 12 oz of an alloy that costs \$5 per ounce to make a mixture that costs \$4 per ounce?

25. Cellular Phones The charge for cellular phone service for a business executive is \$22 per month plus \$.25 per minute of phone use. In a month when the executive's phone bill was \$43.75, how many minutes did the executive use the cellular phone?

26. Taxes If the sales tax on a \$12.50 purchase is \$.75, what is the sales tax on a \$75 purchase?

27. Foreign Trade The figure at the right shows the values of the imports and exports during the first and second quarters of a recent year. Find the increase in the value of the imports from the first quarter to the second quarter.

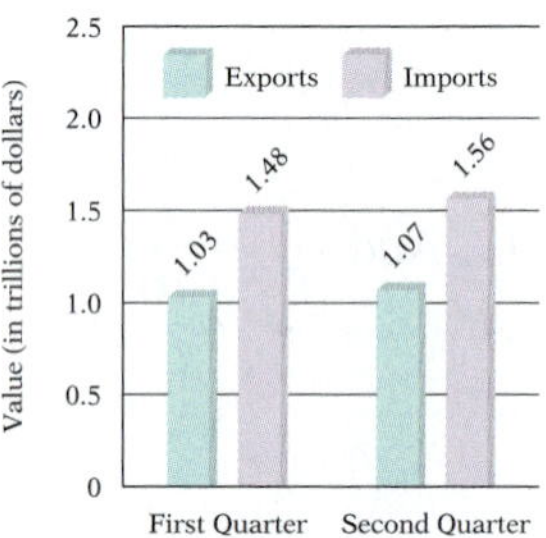

Values of Imports and Exports

Source: Bureau of Economic Analysis

28. Geometry The volume of a box is 144 ft^3. The length of the box is 12 ft, and the width is 4 ft. Find the height of the box.

29. Oceanography The pressure, P, in pounds per square inch, at a certain depth in the ocean can be approximated by the equation $P = 15 + \frac{1}{2}D$, where D is the depth in feet. Use this equation to find the depth when the pressure is 35 lb/in^2.

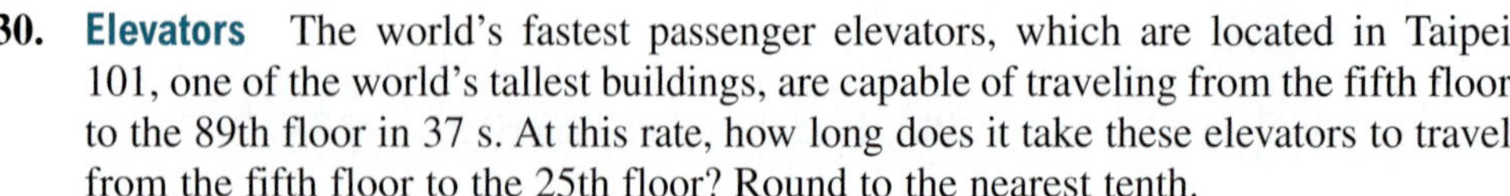

30. Elevators The world's fastest passenger elevators, which are located in Taipei 101, one of the world's tallest buildings, are capable of traveling from the fifth floor to the 89th floor in 37 s. At this rate, how long does it take these elevators to travel from the fifth floor to the 25th floor? Round to the nearest tenth.

AP Photo/Wally Santana

CHAPTER

8

Statistics and Probability

OBJECTIVES

SECTION 8.1

A To read and interpret graphs

SECTION 8.2

A To find the mean, median, and mode of a distribution

B To draw a box-and-whiskers plot

SECTION 8.3

A To calculate the probability of simple events

ARE YOU READY?

Take the Chapter 8 Prep Test to find out if you are ready to learn to:

- Read and interpret graphs
- Find the mean, median, and mode of data
- Draw a box-and-whiskers plot
- Calculate the probability of an event

PREP TEST

Do these exercises to prepare for Chapter 8.

1. Mail Bill-related mail accounted for 49 billion of the 102 billion pieces of first-class mail handled by the U.S. Postal Service during a recent year. (*Source:* U.S. Postal Service) What percent of the pieces of first-class mail handled by the U.S. Postal Service was bill-related mail? Round to the nearest tenth of a percent.

2. Education The table at the right shows the estimated costs of funding an education at a public college.

a. Between which two enrollment years is the increase in cost greatest?

b. What is the increase between these two years?

Enrollment Year	*Cost of Public College*
2005	$70,206
2006	$74,418
2007	$78,883
2008	$83,616
2009	$88,633
2010	$93,951

Source: The College Board's Annual Survey of Colleges

3. Sports During the 1924 Summer Olympics in Paris, France, the United States won 45 gold medals, 27 silver medals, and 27 bronze medals. (*Source: The Ultimate Book of Sports Lists*) Find the ratio of gold medals won by the United States to silver medals won by the United States during the 1924 Summer Olympics. Write the ratio as a fraction in simplest form.

4. The Military Approximately 198,000 women serve in the U.S. military. Six percent of these women serve in the Marine Corps. (*Source:* www.fedstats.gov) What fractional amount of women in the military are in the Marine Corps?

SECTION

8.1 Statistical Graphs

OBJECTIVE A To read and interpret graphs 

Statistics is the branch of mathematics concerned with **data,** or numerical information. Graphs are used to display numerical information in a visual format that enables the reader to quickly see relationships and trends. Three of the most common types of graphs are the bar graph, the circle graph, and the line graph. Examples of each of these types of graphs are shown below.

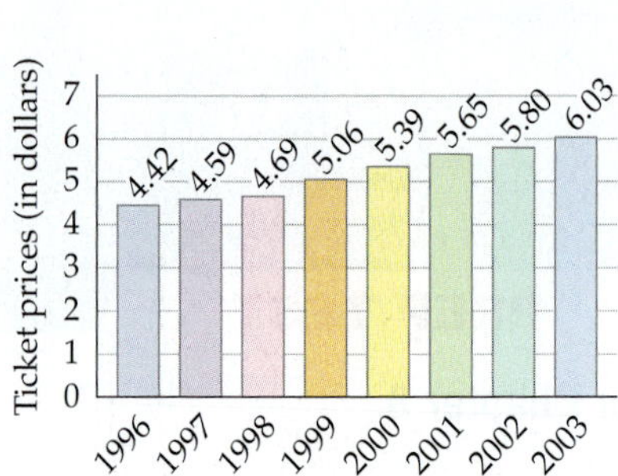

FIGURE 8.1 Average U.S. Movie Theater Ticket Prices (1996–2003)
Source: National Association of Theatre Owners (http://www.natoonline.org)

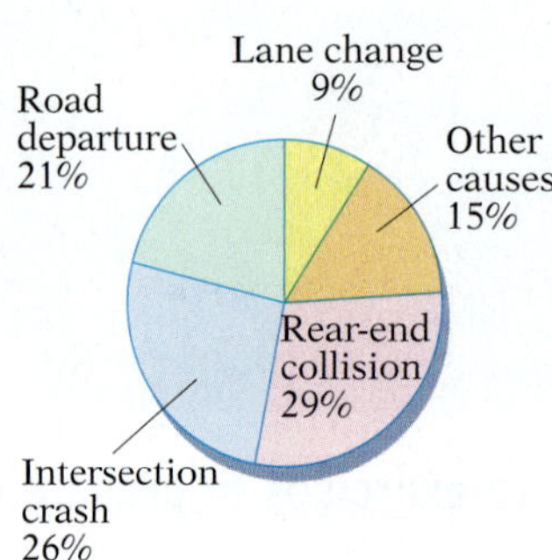

FIGURE 8.2 Types of Automobile Accidents, City of Twin Falls Year: 2005 Total Number of Accidents: 4300

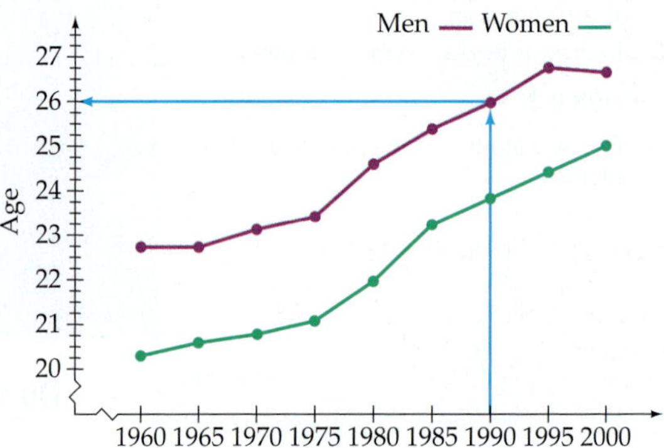

FIGURE 8.3 U.S. Average Age at First Marriage
Source: Bureau of the Census

The bar graph in Figure 8.1 displays the average U.S. movie theater ticket price for the years 1996 to 2003. Each vertical bar is used to display the average ticket price for a given year. The higher the bar, the greater the average ticket price for that year. We can see from the graph that the average movie theater ticket price has been increasing.

The circle graph in Figure 8.2 displays the percent of automobile accidents of a particular type that occurred in a given city for a given year. The largest sector of the circle corresponds to the largest percent of accidents of a given type, 29%.

Figure 8.3 shows two broken-line graphs. The upper broken-line graph displays the average age at first marriage for men for selected years from 1960 to 2000. The lower broken-line graph displays the average age at first marriage for women for selected years during the same time period. The line segments that connect the points on the graph indicate trends. Increasing trends are indicated by line segments that rise as they move to the right, and decreasing trends are indicated by line segments that fall as they move to the right. We can see from the graph that the average age at first marriage has been increasing for both men and women. The blue arrows in Figure 8.3 show that the average age at which men married for the first time in the year 1990 was about 26 years. The graph shows that, for the years shown, the average age at first marriage for men has always been greater than the average age at first marriage for women.

At the right is the circle graph shown on the previous page. Use this graph for Example 1 and You Try It 1.

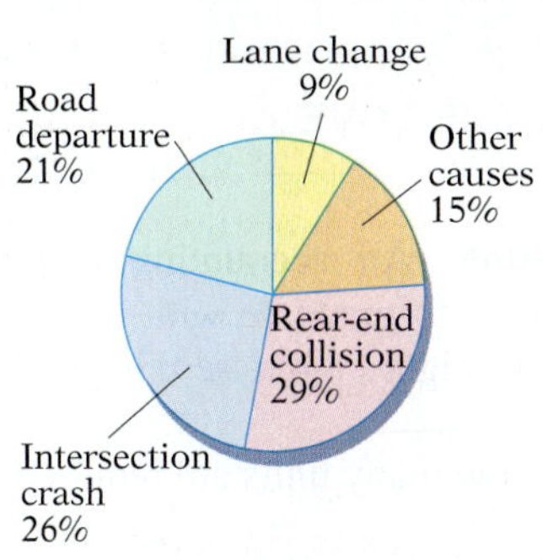

FIGURE 8.2 Types of Automobile Accidents, City of Twin Falls, Year: 2005 Total Number of Accidents: 4300

EXAMPLE • 1

a. Find the ratio, as a fraction in simplest form, of the percent of lane-change accidents to the percent of accidents resulting from "other causes."

b. Determine the number of rear-end collisions that occurred in Twin Falls in the year 2005.

Strategy

a. To find the ratio:
- From the graph, find the percent of lane-change accidents and the percent of accidents resulting from "other causes."
- Write the ratio in fractional form. Simplify.

b. To find the number of rear-end collisions:
- From the graph, find the percent of accidents that were rear-end collisions.
- Solve the basic percent equation for amount. The base is 4300.

Solution

a. Lane-change accidents: 9%

Accidents resulting from "other causes": 15%

$$\frac{9\%}{15\%} = \frac{3}{5}$$

The ratio is $\frac{3}{5}$.

b. Rear-end collisions: 29% = 0.29

Percent · base = amount
$0.29 \cdot 4300 = n$
$1247 = n$

1247 rear-end collisions occurred in Twin Falls in 2005.

YOU TRY IT • 1

a. Find the ratio, as a fraction in simplest form, of the percent of lane-change accidents to the percent of road-departure accidents.

b. Determine the number of accidents that occurred at intersections in Twin Falls in the year 2005.

Your strategy

Your solution

Solution on p. S20

8.1 EXERCISES

OBJECTIVE A To read and interpret graphs

Education An accounting major recorded the number of units required in each discipline to graduate with a degree in accounting. The results are shown in the circle graph in Figure 8.4. Use this graph for Exercises 1 to 4.

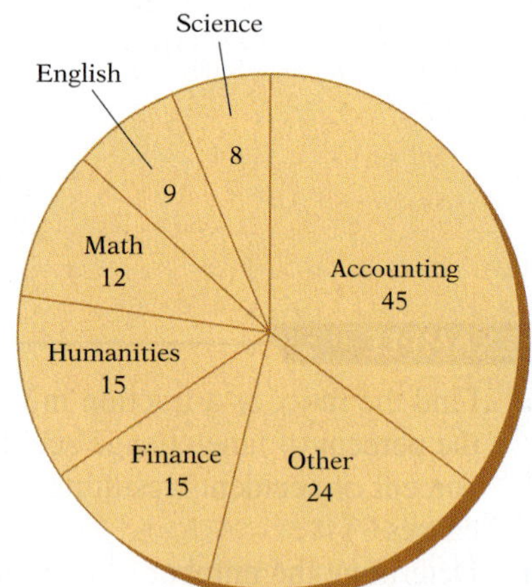

FIGURE 8.4 Number of units required to graduate with an accounting degree

1. How many units are required to graduate with a degree in accounting?

2. What is the ratio of the number of units needed in finance to the number of units needed in accounting?

3. What percent of the units required to graduate are taken in accounting? Round to the nearest tenth of a percent.

4. What percent of the units required to graduate are taken in mathematics? Round to the nearest tenth of a percent.

Video Games The circle graph in Figure 8.5 shows the breakdown of the approximately $3,100,000,000 that Americans spent on home video game equipment in one year. Use this graph for Exercises 5 to 8.

Accessories
Portable game machines
8%
9%
Game software 48%
TV game machines 35%

FIGURE 8.5 Percents of $3,100,000,000 spent annually on home video games
Source: The NPD Group, Toy Manufacturers of America

5. Find the amount of money spent on TV game machines.

6. Find the amount of money spent on portable game machines.

7. What fractional amount of the total money spent was spent on accessories?

8. Is the amount spent for TV game machines more than three times the amount spent for portable game machines?

Automobile Production The bar graph in Figure 8.6 shows the regions in which all passenger cars were produced during a recent year. Use this graph for Exercises 9 to 11.

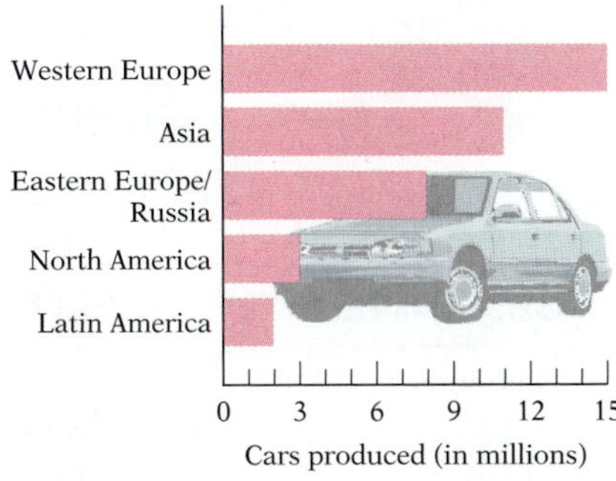

FIGURE 8.6 Number of passenger cars produced in a recent year
Source: Copyright © 2000 by the *Los Angeles Times.* Reprinted with permission.

9. How many passenger cars were produced worldwide?

10. What is the difference between the number of passenger cars produced in Western Europe and the number produced in North America?

11. What percent of the passenger cars were produced in Asia? Round to the nearest percent.

Health The double-broken-line graph in Figure 8.7 shows the number of Calories per day that should be consumed by women and men in various age groups. Use this graph for Exercises 12 to 14.

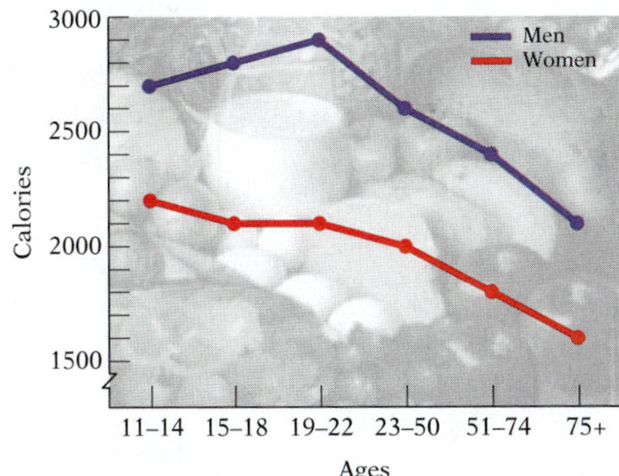

FIGURE 8.7 Recommended number of Calories per day for women and men, by age
Source: Numbers, by Andrea Sutcliffe (HarperCollins)

12. What is the difference between the number of Calories recommended for men and the number recommended for women 19 to 22 years of age?

13. People of what gender and age have the lowest recommended number of Calories?

14. Find the ratio of the number of Calories recommended for women 15 to 18 years old to the number recommended for women 51 to 74 years old.

For Exercises 15 and 16, each statement refers to a line graph (not shown) that displays the population of a particular state every 10 years between 1950 and 2000. Determine whether the statement is *true* or *false.*

15. If the population decreased between 1990 and 2000, then the segment joining the point for 1990 and the point for 2000 slants down from left to right.

16. If the points for 1960 and 1970 are connected by a horizontal line, the population in 1970 was the same as the population in 1960.

Applying the Concepts

17. The circle graph at the right shows a couple's expenditures last month. Write two observations about this couple's expenses.

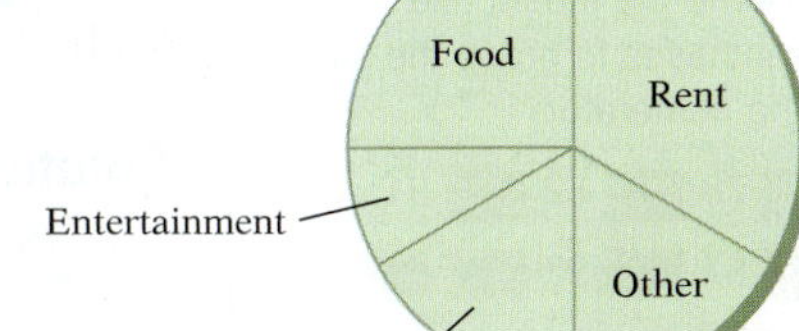

SECTION

8.2 Statistical Measures

OBJECTIVE A To find the mean, median, and mode of a distribution

The average score on the math portion of the SAT was 432. The EPA estimates that a 2010 Toyota Camry Hybrid averages 34 miles per gallon on the highway. The average rainfall for portions of Kauai is 350 inches per year. Each of these statements uses one number to describe an entire collection of numbers. Such a number is called an *average.*

In statistics there are various ways to calculate an average. Three of the most common—*mean, median,* and *mode*—are discussed here.

An automotive engineer tests the miles-per-gallon ratings of 15 cars and records the results as follows:

Miles-per-Gallon Ratings of 15 Cars														
25	22	21	27	25	35	29	31	25	26	21	39	34	32	28

The **mean** of the data is the sum of the measurements divided by the number of measurements. The symbol for the mean is $\bar{x}$.

Formula for the Mean

$$\bar{x} = \frac{\text{sum of the data values}}{\text{number of data values}}$$

To find the mean for the data above, add the numbers and then divide by 15.

$$\bar{x} = \frac{25 + 22 + 21 + 27 + 25 + 35 + 29 + 31 + 25 + 26 + 21 + 39 + 34 + 32 + 28}{15}$$

$$= \frac{420}{15} = 28$$

The mean number of miles per gallon for the 15 cars tested was 28 miles per gallon.

The mean is one of the most frequently computed averages. It is the one that is commonly used to calculate a student's performance in a class.

Integrating Technology

When using a calculator to calculate the mean, use parentheses to group the sum in the numerator.

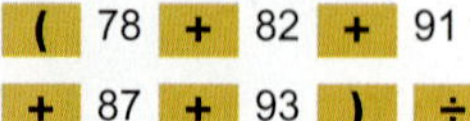

(78 + 82 + 91 + 87 + 93) ÷ 5 =

HOW TO • 1 The test scores for a student taking American history were 78, 82, 91, 87, and 93. What was the mean score for this student?

Strategy

To find the mean, divide the sum of the test scores by 5, the number of scores.

Solution

$$\bar{x} = \frac{78 + 82 + 91 + 87 + 93}{5} = \frac{431}{5} = 86.2$$

The mean score for the history student was 86.2.

The **median** of the data is the number that separates the data into two equal parts when the numbers are arranged from least to greatest (or from greatest to least). There is an equal number of values above the median and below the median.

To find the median of a set of numbers, first arrange the numbers from least to greatest. The median is the number in the middle.

The result of arranging the miles-per-gallon ratings given on the previous page from least to greatest is shown below.

21 21 22 25 25 25 26 27 28 29 31 32 34 35 39

7 values below the median — Middle number **Median** (27) — 7 values above the median

The median is 27 miles per gallon.

If data contain an *even* number of values, the median is the mean of the two middle numbers.

Tips for Success

Word problems are difficult because we must read the problem, determine the quantity we must find, think of a method to find it, actually solve the problem, and then check the answer. In short, we must devise a *strategy* and then use that strategy to find the *solution*. See *AIM for Success* at the front of the book.

HOW TO • 2 The selling prices of the last six homes sold by a real estate agent were \$275,000, \$250,000, \$350,000, \$230,000, \$345,000, and \$290,000. Find the median selling price of these homes.

Strategy

To find the median, arrange the numbers from least to greatest. Because there is an even number of values, the median is the mean of the two middle numbers.

Solution

230,000 250,000 275,000 290,000 345,000 350,000

Middle 2 numbers: 275,000 290,000

$$\text{Median} = \frac{275{,}000 + 290{,}000}{2} = 282{,}500$$

The median selling price was \$282,500.

The **mode** of a set of numbers is the value that occurs most frequently. If a set of numbers has no number occurring more than once, then the data have no mode.

Here again are the data for the gasoline mileage ratings of 15 cars.

Miles-per-Gallon Ratings of 15 Cars														
25	22	21	27	25	35	29	31	25	26	21	39	34	32	28

25 is the number that occurs most frequently.

The mode is 25 miles per gallon.

Note from the miles-per-gallon example that the mean, median, and mode may be different.

EXAMPLE • 1

Twenty students were asked the number of units in which they were enrolled. The responses were as follows:

15	12	13	15	17	18	13	20	9	16
14	10	15	12	17	16	6	14	15	12

Find the mean number of units taken by these students.

Strategy

To find the mean number of units:

- Find the sum of the 20 numbers.
- Divide the sum by 20.

Solution

$15 + 12 + 13 + 15 + 17 + 18 + 13 + 20 + 9 + 16 + 14 + 10 + 15 + 12 + 17 + 16 + 6 + 14 + 15 + 12 = 279$

$$\bar{x} = \frac{279}{20} = 13.95$$

The mean is 13.95 units.

YOU TRY IT • 1

The amounts spent by 12 customers at a McDonald's restaurant were as follows:

11.01	10.75	12.09	15.88	13.50	12.29
10.69	9.36	11.66	15.25	10.09	12.72

Find the mean amount spent by these customers. Round to the nearest cent.

Your strategy

Your solution

EXAMPLE • 2

The starting hourly wages for an apprentice electrician for six different work locations are \$12.50, \$11.25, \$10.90, \$11.56, \$13.75, and \$14.55. Find the median starting hourly wage.

Strategy

To find the median starting hourly wage:

- Arrange the numbers from least to greatest.
- Because there is an even number of values, the median is the mean of the two middle numbers.

Solution

10.90, 11.25, 11.56, 12.50, 13.75, 14.55

$$\text{Median} = \frac{11.56 + 12.50}{2} = 12.03$$

The median starting hourly wage is \$12.03.

YOU TRY IT • 2

The amounts of weight lost, in pounds, by 10 participants in a 6-month weight-reduction program were 22, 16, 31, 14, 27, 16, 29, 31, 40, and 10. Find the median weight loss for these participants.

Your strategy

Your solution

Solutions on p. S20

OBJECTIVE B To draw a box-and-whiskers plot 

Recall from the last objective that an average is one number that helps to describe all the numbers in a set of data. For example, we know from the following statement that Erie gets a lot of snow each winter.

The average annual snowfall in Erie, Pennsylvania, is 85 in.

Now look at these two statements.

San Francisco

The average annual temperature in San Francisco, California, is 57°F.

The average annual temperature in St. Louis, Missouri, is 57°F.

The average annual temperature in both cities is the same. However, we do not expect the climate in St. Louis to be like San Francisco's climate. Although both cities have the same average annual temperature, their temperature ranges differ. In fact, the difference between the average monthly high temperatures in July and January in San Francisco is 14°F, whereas the difference between the average monthly high temperatures in July and January in St. Louis is 50°F.

Note that for this example, a single number (the average annual temperature) does not provide us with a very comprehensive picture of the climate of either of these two cities.

One method used to picture an entire set of data is a box-and-whiskers plot. To prepare a box-and-whiskers plot, we begin by separating a set of data into four parts, called **quartiles.** We will illustrate this by using the average monthly high temperatures for St. Louis, in degrees Fahrenheit. These are listed below from January through December.

St. Louis

39	47	58	72	81	88	89	89	85	76	49	47

Source: The Weather Channel

First list the numbers in order from least to greatest and determine the median.

39	47	47	49	58	72	76	81	85	88	89	89

↑
Median = 74

Now find the median of the data values below the median. The median of the data values below the median is called the **first quartile,** symbolized by Q_1. Also find the median of the data values above the median. The median of the data values above the median is called the **third quartile,** symbolized by Q_3.

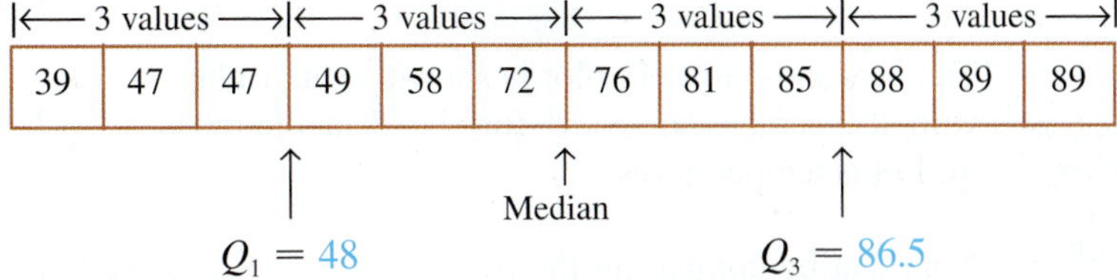

|←— 3 values —→|←— 3 values —→|←— 3 values —→|←— 3 values —→|

39	47	47	49	58	72	76	81	85	88	89	89

$Q_1 = 48$ Median $Q_3 = 86.5$

The first quartile, Q_1, is the number that one-quarter of the data lie below. This means that 25% of the data lie below the first quartile. The third quartile, Q_3, is the number that one-quarter of the data lie above. This means that 25% of the data lie above the third quartile.

The **range** of a set of numbers is the difference between the greatest number and the least number in the set. The range describes the spread of the data. For the data above,

$$\text{Range} = \text{greatest value} - \text{least value} = 89 - 39 = 50$$

Take Note
50% of the data in a distribution lie in the interquartile range.

The **interquartile range** is the difference between the third quartile, Q_3, and the first quartile, Q_1. For the data above,

$$\text{Interquartile range} = Q_3 - Q_1 = 86.5 - 48 = 38.5$$

The interquartile range is the distance that spans the "middle" 50% of the data values. Because it excludes the bottom fourth of the data values and the top fourth of the data values, it excludes any extremes in the numbers of the set.

A **box-and-whiskers plot**, or **boxplot,** is a graph that shows five numbers: the least value, the first quartile, the median, the third quartile, and the greatest value. Here are these five values for the data on St. Louis temperatures.

The least number	39
The first quartile, Q_1	48
The median	74
The third quartile, Q_3	86.5
The greatest number	89

Take Note
The number line at the right is shown as a reference only. It is not a part of the boxplot. It shows that the numbers labeled on the boxplot are plotted according to the distances on the number line above it.

Think of a number line that includes the five values listed above. With this in mind, mark off the five values. Draw a box that spans the distance from Q_1 to Q_3. Draw a vertical line the height of the box at the median.

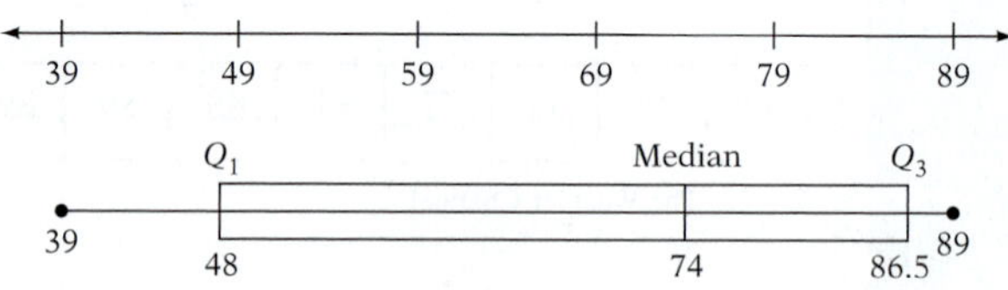

Listed below are the average monthly high temperatures for San Francisco.

57	60	61	64	68	71	71	73	74	73	60	59

Source: The Weather Channel

We can perform the same calculations on these data to determine the five values needed for the box-and-whiskers plot.

The least number	57
The first quartile, Q_1	60
The median	66
The third quartile, Q_3	72
The greatest number	74

Take Note
It is the "whiskers" on the box-and-whiskers plot that show the range of the data. The "box" on the box-and-whiskers plot shows the interquartile range of the data.

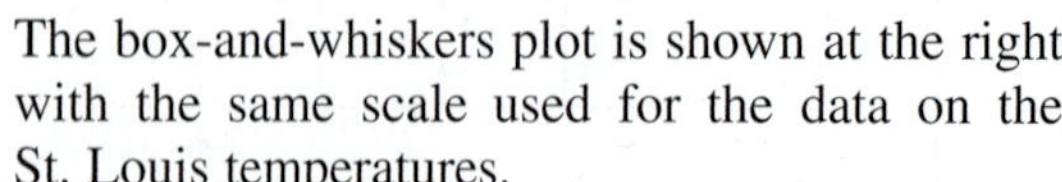

The box-and-whiskers plot is shown at the right with the same scale used for the data on the St. Louis temperatures.

Q_1 Median Q_3

57 60 66 72 74

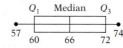

Note that by comparing the two boxplots, we can see that the range of temperatures in St. Louis is greater than the range of temperatures in San Francisco. For the St. Louis temperatures, there is a greater spread of the data below the median than above the median, whereas the data above and below the median of the San Francisco boxplot are spread nearly equally.

HOW TO • 3

The numbers of avalanche deaths in the United States during each of nine consecutive winters were 8, 24, 29, 13, 28, 30, 22, 26, and 32. (*Source:* Colorado Avalanche Information Center) Draw a box-and-whiskers plot of the data, and determine the interquartile range.

Strategy

To draw the box-and-whiskers plot, arrange the data from least to greatest. Then find the median, Q_1, and Q_3. Use the least value, Q_1, the median, Q_3, and the greatest value to draw the box-and-whiskers plot.

To find the interquartile range, find the difference between Q_3 and Q_1.

Solution

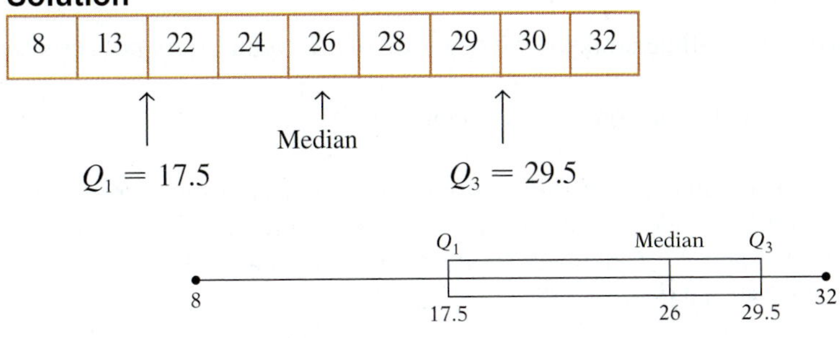

Interquartile range $= Q_3 - Q_1 = 29.5 - 17.5 = 12$

The interquartile range is 12 deaths.

Take Note

Note that the left whisker in this box-and-whiskers plot is quite long, and the length of the box from Q_1 to the median is longer than the length of the box from the median to Q_3. This illustrates a set of data in which the median is closer to the greatest data value. If the two whiskers are approximately the same length, and the distances from Q_1 to the median and from the median to Q_3 are approximately equal, then the least and greatest values are about the same distance from the median. See Example 3 below.

EXAMPLE • 3

The average monthly snowfall amounts, in inches, in Buffalo, New York, from October through April are 1, 12, 24, 25, 18, 12, and 3. (*Source:* The Weather Channel) Draw a box-and-whiskers plot of the data.

Strategy

To draw the box-and-whiskers plot:

- Arrange the data from least to greatest.
- Find the median, Q_1, and Q_3.
- Use the least value, Q_1, the median, Q_3, and the greatest value to draw the box-and-whiskers plot.

Solution

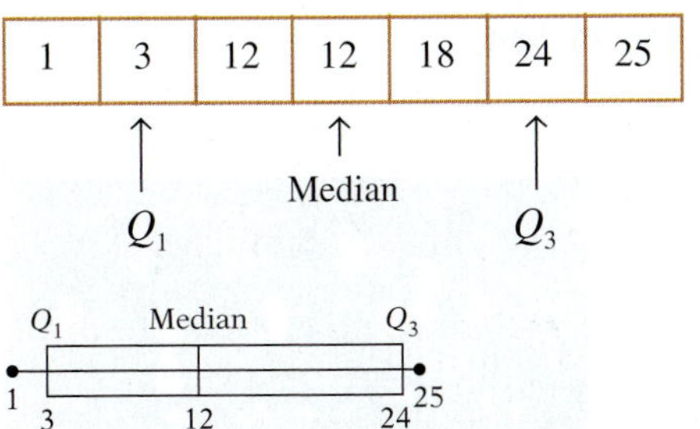

YOU TRY IT • 3

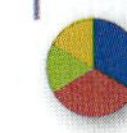

The average monthly snowfall amounts, in inches, in Denver, Colorado, from October through April are 4, 7, 7, 8, 8, 9, and 13. (*Source:* The Weather Channel)

a. Draw a box-and-whiskers plot of the data.

b. How does the spread of the data within the interquartile range compare with that in Example 3?

Your strategy

Your solution

Solution on p. S20

8.2 EXERCISES

OBJECTIVE A To find the mean, median, and mode of a distribution

1. State whether the mean, median, or mode is being used.
   a. Half of the houses in the new development are priced under $350,000.
   b. The average bill for lunch at the college union is $11.95.
   c. The college bookstore sells more green college sweatshirts than any other color.
   d. In a recent year, there were as many people age 26 and younger in the world as there were people age 26 and older.
   e. The majority of full-time students carry a load of 12 credit hours per semester.
   f. The average annual return on an investment is 6.5%.

2. **Consumerism** The number of high-definition televisions sold each month for one year was recorded by an electronics store. The results were 15, 12, 20, 20, 19, 17, 22, 24, 17, 20, 15, and 27. Calculate the mean, the median, and the mode of the number of televisions sold per month.

3. **The Airline Industry** The number of seats occupied on a jet for 16 trans–Atlantic flights was recorded. The numbers were 309, 422, 389, 412, 401, 352, 367, 319, 410, 391, 330, 408, 399, 387, 411, and 398. Calculate the mean, the median, and the mode of the number of seats occupied per flight.

4. **Sports** The times, in seconds, for a 100-meter dash at a college track meet were 10.45, 10.23, 10.57, 11.01, 10.26, 10.90, 10.74, 10.64, 10.52, and 10.78.
   a. Calculate the mean time for the 100-meter dash.
   b. Calculate the median time for the 100-meter dash.

5. **Consumerism** A consumer research group purchased identical items in eight grocery stores. The costs for the purchased items were $85.89, $92.12, $81.43, $80.67, $88.73, $82.45, $87.81, and $85.82. Calculate the mean and the median costs of the purchased items.

6. **Computers** One measure of a computer's hard-drive speed is called access time; this is measured in milliseconds (thousandths of a second). Find the mean and median access times for 11 hard drives whose access times were 5, 4.5, 4, 4.5, 5, 5.5, 6, 5.5, 3, 4.5, and 4.5. Round to the nearest tenth.

7. **Health Plans** Eight health maintenance organizations (HMOs) presented group health insurance plans to a company. The monthly rates per employee were $423, $390, $405, $396, $426, $355, $404, and $430. Calculate the mean and the median monthly rates for these eight companies.

8. **Government** The lengths of the terms, in years, of all the former Supreme Court chief justices are given in the table below. Find the mean and median length of term for the chief justices. Round to the nearest tenth.

5	0	4	34	28	8	14	21
10	8	11	4	7	15	17	19

9. Life Expectancy The life expectancies, in years, in ten selected Central and South American countries are given at the right.
a. Find the mean life expectancy in this group of countries.
b. Find the median life expectancy in this group of countries.

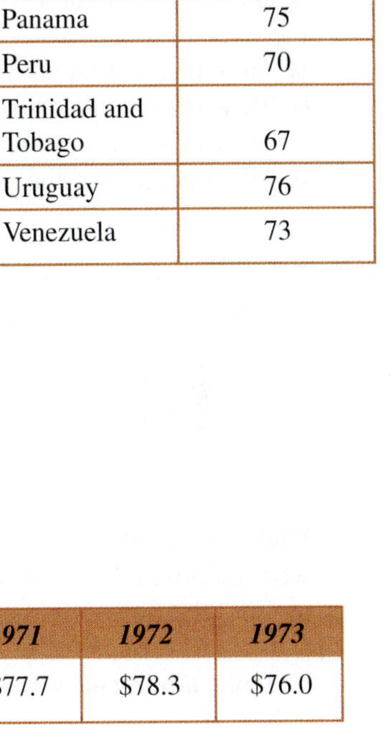

Country	*Life Expectancy*
Brazil	72
Chile	77
Costa Rica	77
Ecuador	77
Guatemala	70
Panama	75
Peru	70
Trinidad and Tobago	67
Uruguay	76
Venezuela	73

10. Education Your scores on six history tests were 78, 92, 95, 77, 94, and 88. If an "average score" of 90 receives an A for the course, which average, the mean or the median, would you prefer that the instructor use?

11. Education One student received scores of 85, 92, 86, and 89. A second student received scores of 90, 97, 91, and 94 (exactly 5 points more on each exam). Are the means of the two students the same? If not, what is the relationship between the means of the two students?

12. Defense Spending The table below shows the defense expenditures, in billions of dollars, by the federal government for 1965 through 1973, years during which the United States was actively involved in the Vietnam War.
a. Calculate the mean annual defense expenditure for these years. Round to the nearest tenth of a billion.

b. Find the median annual defense expenditure.

c. If the year 1965 were eliminated from the data, how would that affect the mean? The median?

Year	*1965*	*1966*	*1967*	*1968*	*1969*	*1970*	*1971*	*1972*	*1973*
Expenditures	$49.6	$56.8	$70.1	$80.5	$81.2	$80.3	$77.7	$78.3	$76.0

Source: Statistical Abstract of the United States

OBJECTIVE B To draw a box-and-whiskers plot

13. a. What percent of the data in a set of numbers lies above Q_3?
b. What percent of the data in a set of numbers lies above Q_1?
c. What percent of the data in a set of numbers lies below Q_3?
d. What percent of the data in a set of numbers lies below Q_1?

14. U.S. Presidents The box-and-whiskers plot below shows the distribution of the ages of presidents of the United States at the time of their inauguration.
a. What is the youngest age in the set of data?
b. What is the oldest age?
c. What is the first quartile?
d. What is the third quartile?
e. What is the median?
f. Find the range.
g. Find the interquartile range.

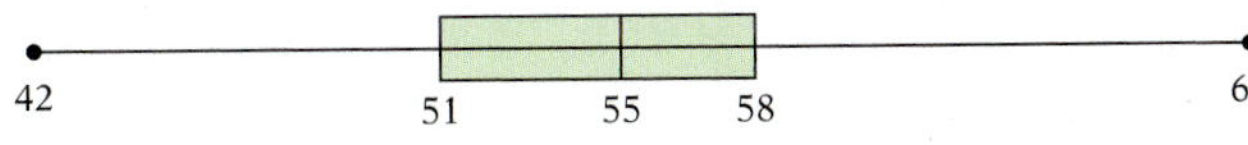

15. Compensation The box-and-whiskers plot below shows the distribution of median incomes for 50 states and the District of Columbia. What is the lowest value in the set of data? The highest value? The first quartile? The third quartile? The median? Find the range and the interquartile range.

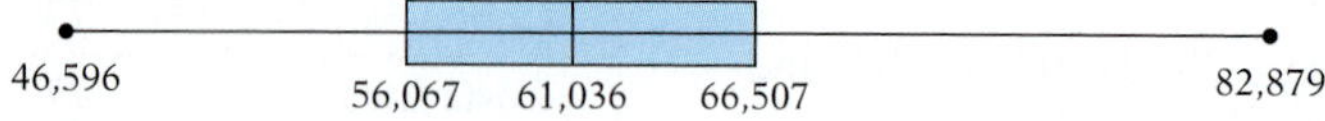

16. Education An aptitude test was taken by 200 students at the Fairfield Middle School. The box-and-whiskers plot at the right shows the distribution of their scores.

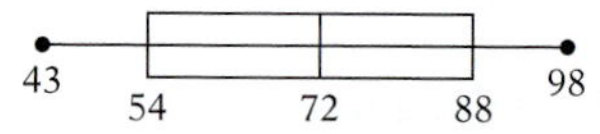

a. How many students scored over 88?
b. How many students scored below 72?
c. How many scores are represented in each quartile?
d. What percent of the students had scores of at least 54?

17. Health The cholesterol levels for 80 adults were recorded and then displayed in the box-and-whiskers plot shown at the right.

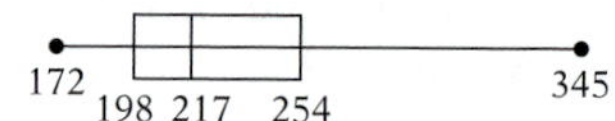

a. How many adults had a cholesterol level above 217?
b. How many adults had a cholesterol level below 254?
c. How many cholesterol levels are represented in each quartile?
d. What percent of the adults had a cholesterol level of not more than 198?

18. Fuel Efficiency The gasoline consumption of 19 cars was tested, and the results were recorded in the table below.
a. Find the range, the first quartile, the third quartile, and the interquartile range.

b. Draw a box-and-whiskers plot of the data.

c. Is the data value 21 in the interquartile range?

Miles per Gallon for 19 Cars									
33	21	30	32	20	31	25	20	16	24
22	31	30	28	26	19	21	17	26	

19. Environment Carbon dioxide is among the gases that contribute to global warming. The world's biggest emitters of carbon dioxide are listed below. The figures are emissions in millions of metric tons per year.
a. Find the range, the first quartile, the third quartile, and the interquartile range.

b. Draw a box-and-whiskers plot of the data.

c. What data value is responsible for the long whisker at the right?

Carbon Dioxide Emissions (in millions of metric tons per year)			
Canada	0.64	Japan	1.26
China	5.01	Russian Federation	1.52
Germany	0.81	South Korea	0.47
India	1.34	United Kingdom	0.59
Italy	0.45	United States	6.05

Source: U.S. Department of Energy

20. Meteorology The average monthly amounts of rainfall, in inches, from January through December for Seattle, Washington, and Houston, Texas, are listed below.

a. Is the difference between the means greater than 1 in.?

b. What is the difference between the medians?

c. Draw a box-and-whiskers plot of each set of data. Use the same scale.

d. Describe the difference between the distributions of the data for Seattle and Houston.

Seattle	6.0	4.2	3.6	2.4	1.6	1.4	0.7	1.3	2.0	3.4	5.6	6.3
Houston	3.2	3.3	2.7	4.2	4.7	4.1	3.3	3.7	4.9	3.7	3.4	3.7

Source: The Weather Channel

21. Meteorology The average monthly amounts of rainfall, in inches, from January through December for Orlando, Florida, and Portland, Oregon, are listed below.

a. Is the difference between the means greater than 1 in.?

b. What is the difference between the medians?

c. Draw a box-and-whiskers plot of each set of data. Use the same scale.

d. Describe the difference between the distributions of the data for Orlando and Portland.

Orlando	2.1	2.8	3.2	2.2	4.0	7.4	7.8	6.3	5.6	2.8	1.8	1.8
Portland	6.2	3.9	3.6	2.3	2.1	1.5	0.5	1.1	1.6	3.1	5.2	6.4

Source: The Weather Channel

22. Refer to the box-and-whiskers plot in Exercise 15. Which of the following fractions most accurately represents the fraction of states with median incomes less than $66,507?

(i) $\frac{1}{4}$ (ii) $\frac{1}{3}$ (iii) $\frac{1}{2}$ (iv) $\frac{3}{4}$

23. Write a set of data with five data values for which the mean, median, and mode are all 55.

Applying the Concepts

24. A set of data has a mean of 16, a median of 15, and a mode of 14. Which of these numbers must be a value in the data set? Explain your answer.

25. Explain each notation.

a. Q_1 **b.** Q_3 **c.** $\bar{x}$

26. The box in a box-and-whiskers plot represents 50%, or one-half, of the data in a set. Why is the box in Example 3 of this section not one-half of the entire length of the box-and-whiskers plot?

27. Create a set of data containing 25 numbers that would correspond to the box-and-whiskers plot shown at the right.

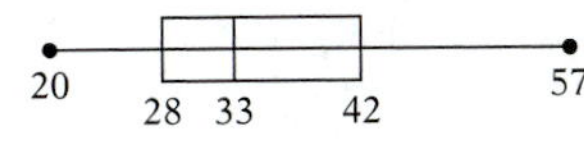

SECTION

8.3 Introduction to Probability

OBJECTIVE A To calculate the probability of simple events

Point of Interest

It was dice playing that led Antoine Gombaud, Chevalier de Mere, to ask Blaise Pascal, a French mathematician, to figure out the probability of throwing two sixes. Pascal and Pierre Fermat solved the problem, and their explorations led to the birth of probability theory.

A weather forecaster estimates that there is a 75% chance of rain. A state lottery director claims that there is a $\frac{1}{9}$ chance of winning a prize in a new game offered by the lottery. Each of these statements involves uncertainty to some extent. The degree of uncertainty is called **probability.** For the statements above, the probability of rain is 75%, and the probability of winning a prize in the new lottery game is $\frac{1}{9}$.

A probability is determined from an **experiment,** which is any activity that has an observable outcome. Examples of experiments include

Tossing a coin and observing whether it lands heads up or tails up
Interviewing voters to determine their preference for a political candidate
Drawing a card from a standard deck of 52 cards

All the possible outcomes of an experiment are called the **sample space** of the experiment. The outcomes are listed between braces. For example:

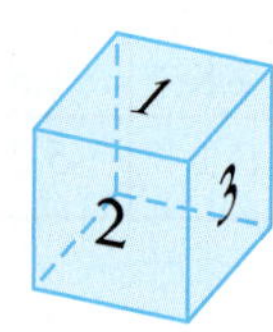

The number cube shown at the left is rolled once. Any of the numbers from 1 to 6 could show on the top of the cube. The sample space is

$$\{1, 2, 3, 4, 5, 6\}$$

A fair coin is tossed once. (A fair coin is one for which heads and tails have an equal chance of landing face up.) If H represents "heads up" and T represents "tails up," then the sample space is

$$\{H, T\}$$

An **event** is one or more outcomes of an experiment. For the experiment of rolling the six-sided cube described above, some possible events are

The number is even: $\{2, 4, 6\}$
The number is a multiple of 3: $\{3, 6\}$
The number is less than 10: $\{1, 2, 3, 4, 5, 6\}$

Note that in the last case, the event is the entire sample space.

HOW TO • 1 The spinner at the left is spun once. Assume that the spinner does not come to rest on a line.

a. What is the sample space?

The arrow could come to rest on any one of the four sectors.
The sample space is $\{1, 2, 3, 4\}$.

b. List the outcomes in the event that the spinner points to an odd number.
$\{1, 3\}$

In discussing experiments and events, it is convenient to refer to the **favorable outcomes** of an experiment. These are the outcomes of an experiment that satisfy the requirements of a particular event.

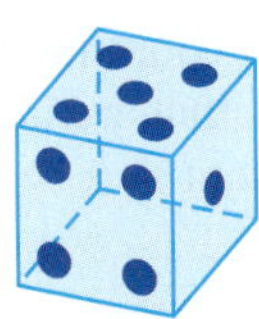

For instance, consider the experiment of rolling a fair die once. The sample space is

$$\{1, 2, 3, 4, 5, 6\}$$

and one possible event would be rolling a number that is divisible by 3. The outcomes of the experiment that are favorable to the event are 3 and 6:

$$\{3, 6\}$$

The outcomes of the experiment of tossing a fair coin are *equally likely.* Either one of the outcomes is just as likely as the other. If a fair coin is tossed once, the probability of a head is $\frac{1}{2}$, and the probability of a tail is $\frac{1}{2}$. Both events are equally likely. The theoretical probability formula, given below, applies to experiments for which the outcomes are equally likely.

Theoretical Probability Formula

The theoretical probability of an event is a fraction with the number of favorable outcomes of the experiment in the numerator and the total number of possible outcomes in the denominator.

$$\text{Probability of an event} = \frac{\text{number of favorable outcomes}}{\text{number of possible outcomes}}$$

A probability of an event is a number from 0 to 1 that tells us how likely it is that this outcome will happen.

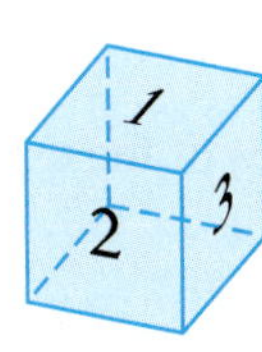

A probability of 0 means that the event is impossible.

The probability of getting a heads when rolling the die shown at the left is 0.

A probability of 1 means that the event must happen.

The probability of getting either heads or tails when tossing a coin is 1.

A probability of $\frac{1}{4}$ means that it is expected that the outcome will happen 1 in every 4 times the experiment is performed.

Take Note

The phrase **at random** means that each card has an equal chance of being drawn.

HOW TO 2 Each of the letters of the word *TENNESSEE* is written on a card, and the cards are placed in a hat. If one card is drawn at random from the hat, what is the probability that the card has the letter E on it?

Count the possible outcomes of the experiment.

There are 9 letters in *TENNESSEE.*

There are 9 possible outcomes of the experiment.

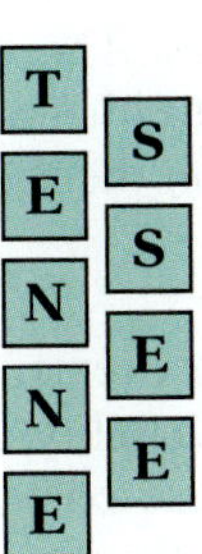

Count the number of outcomes of the experiment that are favorable to the event that a card with the letter E on it is drawn.

There are 4 cards with an E on them.

Use the probability formula.

$$\text{Probability of the event} = \frac{\text{number of favorable outcomes}}{\text{number of possible outcomes}} = \frac{4}{9}$$

The probability of drawing an E is $\frac{4}{9}$.

As just discussed, calculating the probability of an event requires counting the number of possible outcomes of an experiment and the number of outcomes that are favorable to the event. One way to do this is to list the outcomes of the experiment in a systematic way. Using a table is often helpful.

When two dice are rolled, the sample space for the experiment can be recorded systematically as in the following table.

Point of Interest

Romans called a die that was marked on four faces a *talus,* which meant "anklebone." The anklebone was considered an ideal die because it is roughly a rectangular solid and it has no marrow, so loose anklebones from sheep were more likely than other bones to be lying around after the wolves had left their prey.

Possible Outcomes from Rolling Two Dice

(1, 1)	(2, 1)	(3, 1)	(4, 1)	(5, 1)	(6, 1)
(1, 2)	(2, 2)	(3, 2)	(4, 2)	(5, 2)	(6, 2)
(1, 3)	(2, 3)	(3, 3)	(4, 3)	(5, 3)	(6, 3)
(1, 4)	(2, 4)	(3, 4)	(4, 4)	(5, 4)	(6, 4)
(1, 5)	(2, 5)	(3, 5)	(4, 5)	(5, 5)	(6, 5)
(1, 6)	(2, 6)	(3, 6)	(4, 6)	(5, 6)	(6, 6)

HOW TO • 3 Two dice are rolled once. Calculate the probability that the sum of the numbers on the two dice is 7.

Use the table above to count the number of possible outcomes of the experiment.

There are 36 possible outcomes.

Count the number of outcomes of the experiment that are favorable to the event that a sum of 7 is rolled.

There are 6 favorable outcomes: (1, 6), (2, 5), (3, 4), (4, 3), (5, 2), and (6, 1).

Use the probability formula.

$$\text{Probability of the event} = \frac{\text{number of favorable outcomes}}{\text{number of possible outcomes}} = \frac{6}{36} = \frac{1}{6}$$

The probability of a sum of 7 is $\frac{1}{6}$.

The probabilities calculated above are theoretical probabilities. The calculation of a **theoretical probability** is based on theory—for example, that either side of a coin is equally likely to land face up or that each of the six sides of a fair die is equally likely to land face up. Not all probabilities arise from such assumptions.

An **empirical probability** is based on observations of certain events. For instance, a weather forecast of a 75% chance of rain is an empirical probability. From historical records kept by the weather bureau, when a similar weather pattern existed, rain occurred 75% of the time. It is theoretically impossible to predict the weather, and only observations of past weather patterns can be used to predict future weather conditions.

Empirical Probability Formula

The empirical probability of an event is the ratio of the number of observations of the event to the total number of observations.

$$\text{Probability of an event} = \frac{\text{number of observations of the event}}{\text{total number of observations}}$$

For example, suppose the records of an insurance company show that of 2549 claims for theft filed by policy holders, 927 were claims for more than $5000. The empirical probability that the next claim for theft that this company receives will be a claim for more than $5000 is the ratio of the number of claims for over $5000 to the total number of claims.

$$\frac{927}{2549} \approx 0.36$$

The probability is approximately 0.36.

EXAMPLE • 1

There are three choices, *a, b,* or *c,* for each of the two questions on a multiple-choice quiz. If the instructor randomly chooses which questions will have an answer of *a, b,* or *c,* what is the probability that the two correct answers on this quiz will be the same letter?

Strategy

To find the probability:

- List the outcomes of the experiment in a systematic way.
- Count the number of possible outcomes of the experiment.
- Count the number of outcomes of the experiment that are favorable to the event that the two correct answers on the quiz will be the same letter.
- Use the probability formula.

Solution

Possible outcomes: (a, a) (b, a) (c, a)
(a, b) (b, b) (c, b)
(a, c) (b, c) (c, c)

There are 9 possible outcomes.

There are 3 favorable outcomes:
(a, a), (b, b), (c, c)

$$\text{Probability} = \frac{\text{number of favorable outcomes}}{\text{number of possible outcomes}}$$

$$= \frac{3}{9} = \frac{1}{3}$$

The probability that the two correct answers will be the same letter is $\frac{1}{3}$.

YOU TRY IT • 1

A professor writes three true/false questions for a quiz. If the professor randomly chooses which questions will have a true answer and which will have a false answer, what is the probability that the test will have 2 true questions and 1 false question?

Your strategy

Your solution

Solution on p. S21

8.3 EXERCISES

OBJECTIVE A To calculate the probability of simple events

1. A coin is tossed four times. List all the possible outcomes of the experiment as a sample space.

2. Three cards—one red, one green, and one blue—are to be arranged in a stack. Using R for red, G for green, and B for blue, list all the different stacks that can be formed. (Some computer monitors are called RGB monitors for the colors red, green, and blue.)

3. A tetrahedral die is one with four triangular sides. The sides show the numbers from 1 to 4. Say two tetrahedral dice are rolled. List all the possible outcomes of the experiment as a sample space.

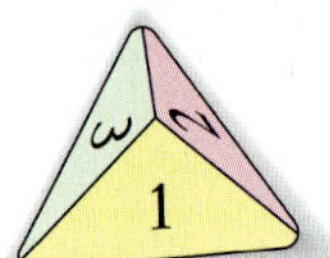

Tetrahedral die

4. A coin is tossed and then a die is rolled. List all the possible outcomes of the experiment as a sample space. [To get you started, (H, 1) is one of the possible outcomes.]

5. The spinner at the right is spun once. Assume that the spinner does not come to rest on a line.
   a. What is the sample space?
   b. List the outcomes in the event that the number is less than 4.

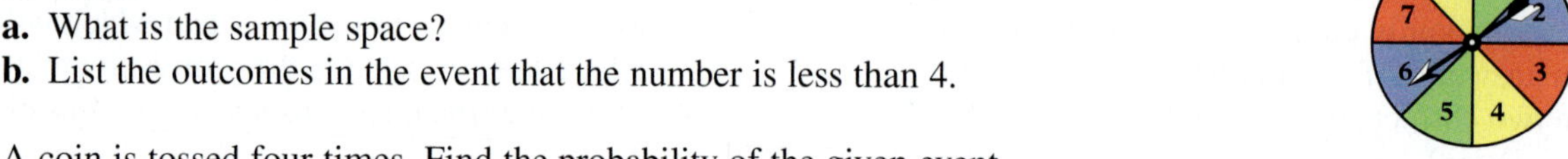

6. A coin is tossed four times. Find the probability of the given event.
   a. The outcomes are exactly in the order HHTT. (See Exercise 1.)
   b. The outcomes consist of two heads and two tails.
   c. The outcomes consist of one head and three tails.

7. Two dice are rolled. Find the probability of the given outcome.
   a. The sum of the dots on the upward faces is 5.
   b. The sum of the dots on the upward faces is 15.
   c. The sum of the dots on the upward faces is less than 15.
   d. The sum of the dots on the upward faces is 2.

8. A dodecahedral die has 12 sides numbered from 1 to 12. The die is rolled once. Find the probability of the given outcome.

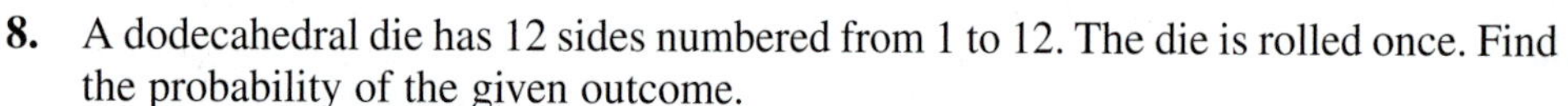

   a. The upward face shows an 11.
   b. The upward face shows a 5.

9. A dodecahedral die has 12 sides numbered from 1 to 12. The die is rolled once. Find the probability of the given outcome.

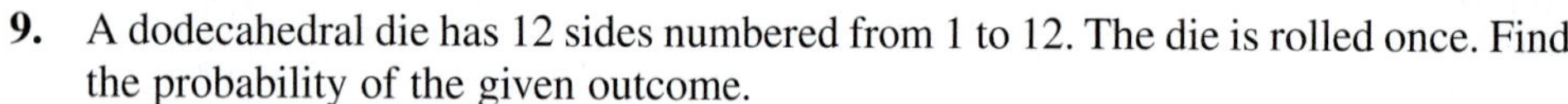

   a. The upward face shows a number that is divisible by 4.
   b. The upward face shows a number that is a multiple of 3.

Dodecahedral die

Tony Freeman/PhotoEdit, Inc.

10. Two tetrahedral dice are rolled (see Exercise 3).
   a. What is the probability that the sum on the upward faces is 4?
   b. What is the probability that the sum on the upward faces is 6?

11. Two dice are rolled. Which has the greater probability, throwing a sum of 10 or throwing a sum of 5?

12. Two dice are rolled once. Calculate the probability that the two numbers on the dice are equal.

13. Each of the letters of the word *MISSISSIPPI* is written on a card, and the cards are placed in a hat. One card is drawn at random from the hat.
   a. What is the probability that the card has the letter *I* on it?
   b. Which is greater, the probability of choosing an *S* or that of choosing a *P*?

14. Use the situation described in Exercise 12. Suppose you decide to test your result empirically by rolling a pair of dice 30 times and recording the results. Which number of "doubles" would confirm the result found in Exercise 12?
   (i) 1 (ii) 5 (iii) 6 (iv) 30

15. Use the situation described in Exercise 13. What probability does the fraction $\frac{1}{11}$ represent?

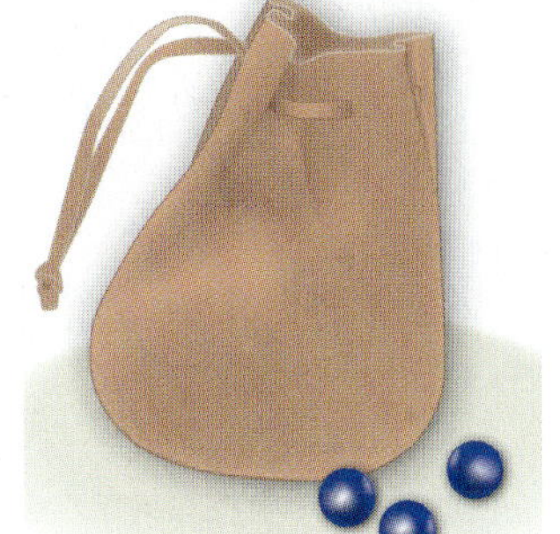

16. Three blue marbles, four green marbles, and five red marbles are placed in a bag. One marble is chosen at random.
   a. What is the probability that the marble chosen is green?
   b. Which is greater, the probability of choosing a blue marble or that of choosing a red marble?

17. Which has the greater probability, drawing a jack, queen, or king from a deck of cards or drawing a spade?

18. In a history class, a set of exams earned the following grades: 4 A's, 8 B's, 22 C's, 10 D's, and 3 F's. If a single student's exam is chosen from this class, what is the probability that it received a B?

19. A survey of 95 people showed that 37 preferred (to using a credit card) a cash discount of 2% if an item was purchased using cash or a check. Judging on the basis of this survey, what is the empirical probability that a person prefers a cash discount? Write the answer as a decimal rounded to the nearest hundredth.

20. A survey of 725 people showed that 587 had a group health insurance plan where they worked. On the basis of this survey, what is the empirical probability that an employee has a group health insurance plan? Write the answer as a decimal rounded to the nearest hundredth.

21. A television cable company surveyed some of its customers and asked them to rate the cable service as excellent, satisfactory, average, unsatisfactory, or poor. The results are recorded in the table at the right. What is the probability that a customer who was surveyed rated the service as satisfactory or excellent?

Quality of Service	*Number Who Voted*
Excellent	98
Satisfactory	87
Average	129
Unsatisfactory	42
Poor	21

Applying the Concepts

22. If the spinner at the right is spun once, is each of the numbers 1 through 5 equally likely? Why or why not?

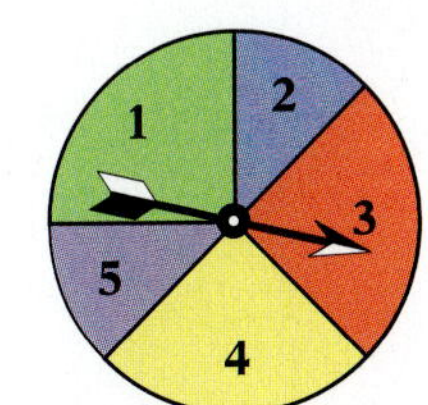

23. Why can the probability of an event not be $\frac{5}{3}$?

FOCUS ON PROBLEM SOLVING

Inductive Reasoning

Suppose that, beginning in January, you save \$25 each month. The total amount you have saved at the end of each month can be described by a list of numbers.

25	50	75	100	125	150	175	
Jan.	Feb	Mar	Apr	May	June	July	. . .

The list of numbers that indicates your total savings is an *ordered* list of numbers called a **sequence.** Each of the numbers in a sequence is called a **term** of the sequence. The list is ordered because the position of a number in the list indicates the month in which that total amount has been saved. For example, the 7th term of the sequence (indicating July) is 175. This number means that a total of \$175 has been saved by the end of the 7th month.

Now consider a person who has a different savings plan. The total amount saved by this person for the first seven months is given by the sequence

20, 35, 50, 65, 80, 95, 110, . . .

The process you use to discover the next number in the above sequence is *inductive reasoning.* **Inductive reasoning** involves making generalizations from specific examples; in other words, we reach a conclusion by making observations about particular facts or cases. In the case of the above sequence, the person saved \$15 per month after the first month.

Here is another example of inductive reasoning. Find the next two letters of the sequence A, B, E, F, I, J, ...

By trying different patterns, we can determine that a pattern for this sequence is

A, B, C, D, E, F, G, H, I, J, . . .

That is, write two letters, skip two letters, write two letters, skip two letters, and so on. The next two letters are M, N.

Use inductive reasoning to solve the following problems.

1. What is the next term of the sequence, ban, ben, bin, bon, . . . ?
2. Using a calculator, determine the decimal representation of several proper fractions that have a denominator of 99. For instance, you may use $\frac{8}{99}$, $\frac{23}{99}$, and $\frac{75}{99}$. Now use inductive reasoning to explain the pattern, and use your reasoning to find the decimal representation of $\frac{53}{99}$ without a calculator.
3. Find the next number in the sequence 1, 1, 2, 3, 5, 8, 13, 21, ...
4. The decimal representation of a number begins 0.10100100010000100000 . . . . What are the next 10 digits in this number?
5. The first seven rows of a triangle of numbers called Pascal's triangle are given below. Find the next row.

1
1 1
1 2 1
1 3 3 1
1 4 6 4 1
1 5 10 10 5 1
1 6 15 20 15 6 1

PROJECTS AND GROUP ACTIVITIES

Collecting, Organizing, Displaying, and Analyzing Data

Before standardized units of measurement became commonplace, measurements were made in terms of the human body. For example, the cubit was the distance from the end of the elbow to the tips of the fingers. The yard was the distance from the tip of the nose to the tip of the fingers on an outstretched arm.

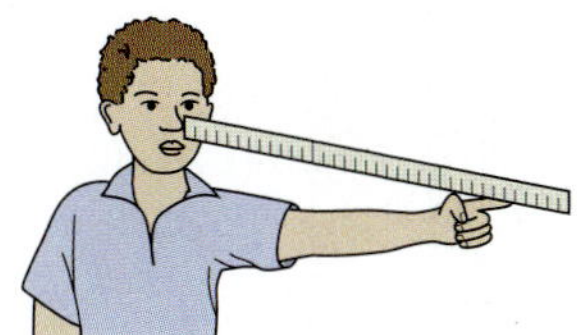

For each student in the class, find the measure from the tip of the nose to the tip of the fingers on an outstretched arm. Round each measure to the nearest centimeter. Record all the measurements on the board.

1. From the data collected, determine each of the following.

Mean __________

Median __________

Mode __________

Range __________

First quartile, Q_1 __________

Third quartile, Q_3 __________

Interquartile range __________

2. Prepare a box-and-whiskers plot of the data.

3. Write a description of the spread of the data.

4. Explain why we need standardized units of measurement.

CHAPTER 8

SUMMARY

KEY WORDS

Statistics is the branch of mathematics concerned with *data*, or numerical information. A *graph* is a pictorial representation of data. A *circle graph* represents data by the size of the sectors. [8.1A, p. 446]

EXAMPLES

The circle graph shows the results of a survey of 300 people who were asked to name their favorite sport.

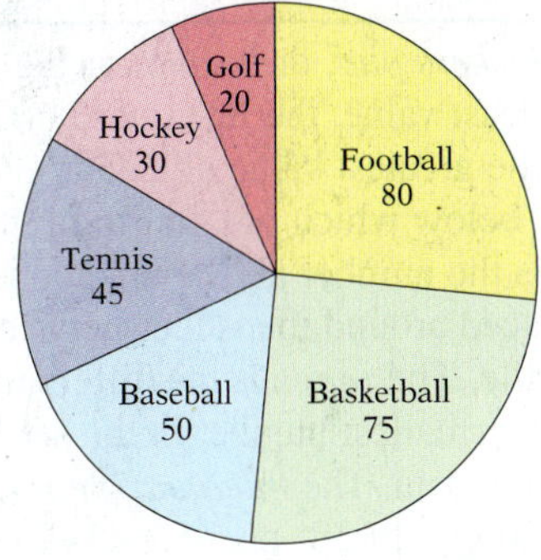

Distribution of Responses in a Survey

A *bar graph* represents data by the height of the bars. [8.1A, p. 446]

The bar graph shows the expected U.S. population aged 100 and over. *Source:* Census Bureau

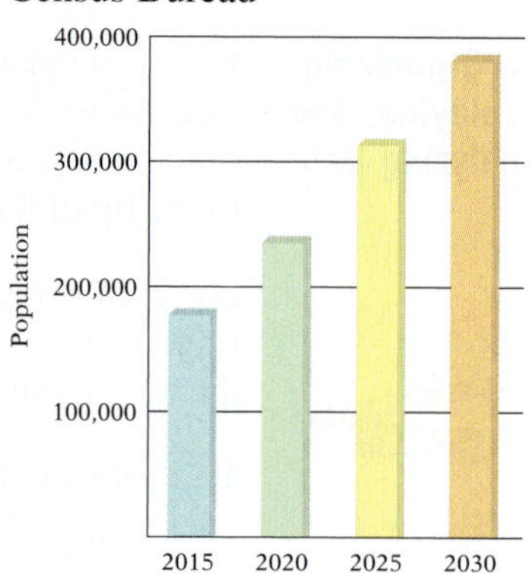

Expected U.S. Population Aged 100 and Over

A *broken-line graph* represents data by the position of the lines and shows trends or comparisons. [8.1A, p. 446]

The line graph shows a recent graduate's cumulative debt in college loans at the end of each of the four years of college.

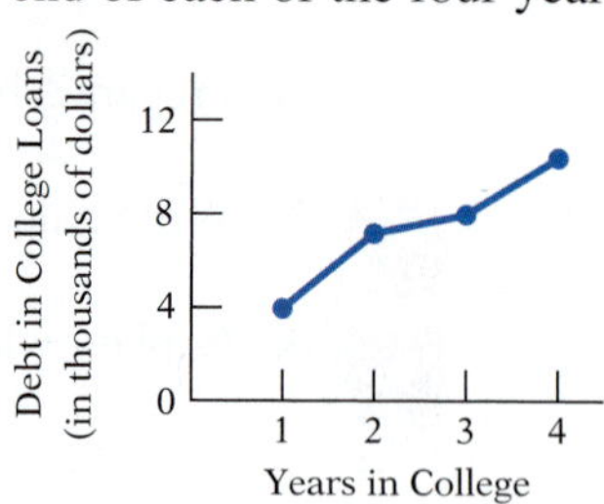

Cumulative Debt in College Loans

The *mean, median,* and *mode* are three types of averages used in statistics. The *mean* of a set of data is the sum of the data values divided by the number of values in the set. The *median* of a set of data is the number that separates the data into two equal parts when the data have been arranged from least to greatest (or greatest to least). There are an equal number of values above the median and below the median. The *mode* of a set of numbers is the value that occurs most frequently. [8.2A, pp. 450, 451]

Consider the following set of data.

24, 28, 33, 45, 45

The mean is 35.
The median is 33.
The mode is 45.

A *box-and-whiskers plot,* or *boxplot,* is a graph that shows five numbers: the least value, the first quartile, the median, the third quartile, and the greatest value of a data set. The *first quartile*, Q_1, is the number below which one-fourth of the data lie. The *third quartile*, Q_3, is the number above which one-fourth of the data lie. The box is placed around the values between the first quartile and the third quartile. The *range* is the difference between the largest number and the smallest number in the set. The range describes the spread of the data. The *interquartile range* is the difference between Q_3 and Q_1. [8.2B, pp. 453–454]

The box-and-whiskers plot for a set of test scores is shown below.

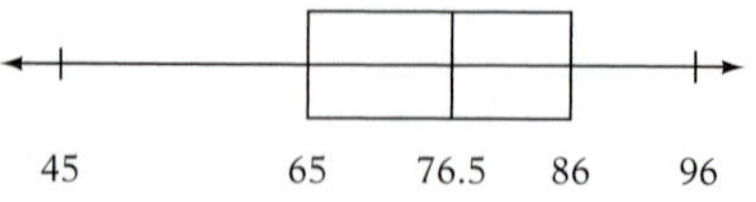

Range $= 96 - 45 = 51$
$Q_1 = 65$
$Q_3 = 86$
Interquartile range $= Q_3 - Q_1$
$= 86 - 65 = 21$

Probability is a number from 0 to 1 that tells us how likely it is that a certain outcome of an experiment will happen. An *experiment* is an activity with an observable outcome. All the possible outcomes of an experiment are called the *sample space* of the experiment. An *event* is one or more outcomes of an experiment. The *favorable outcomes* of an experiment are the outcomes that satisfy the requirements of a particular event. [8.3A, p. 460]

Tossing a single die is an example of an experiment. The sample space for this experiment is the set of possible outcomes:

$$\{1, 2, 3, 4, 5, 6\}$$

The event that the number landing face up is an odd number is represented by

$$\{1, 3, 5\}$$

ESSENTIAL RULES AND PROCEDURES / EXAMPLES

To Find the Mean of a Set of Data [8.2A, p. 450]

Divide the sum of the values by the number of values in the set.

$$\bar{x} = \frac{\text{sum of the data values}}{\text{number of data values}}$$

Consider the following set of data.

24, 28, 33, 45, 45

$$\bar{x} = \frac{24 + 28 + 33 + 45 + 45}{5} = 35$$

To Find the Median [8.2A, p. 451]

1. Arrange the numbers from least to greatest.
2. If there is an *odd* number of values in the set of data, the median is the middle number. If there is an *even* number of values in the set of data, the median is the mean of the two middle numbers.

Consider the following set of data.

24, 28, 33, 35, 45, 45

The median is $\frac{33 + 35}{2} = 34$.

To Find Q_1 [8.2B, p. 453]

Arrange the numbers from least to greatest and locate the median. Q_1 is the median of the lower half of the data.

Consider the following data.

8	10	12	14	16	19	22
	↑		↑			
	Q_1		Median			

To Find Q_3 [8.2B, p. 453]

Arrange the numbers from least to greatest and locate the median. Q_3 is the median of the upper half of the data.

Consider the following data.

8	10	12	14	16	19	22
			↑		↑	
			Median		Q_3	

Theoretical Probability Formula [8.3A, p. 461]

$$\text{Probability of an event} = \frac{\text{number of favorable outcomes}}{\text{number of possible outcomes}}$$

A die is rolled. The probability of rolling a 2 or a 4 is $\frac{2}{6} = \frac{1}{3}$.

Empirical Probability Formula [8.3A, p. 463]

$$\text{Probability of an event} = \frac{\text{number of observations of the event}}{\text{total number of observations}}$$

A thumbtack is tossed 100 times. It lands point up 15 times and lands on its side 85 times. From this experiment, the empirical probability of "point up" is $\frac{15}{100} = \frac{3}{20}$.

CHAPTER 8

CONCEPT REVIEW

Test your knowledge of the concepts presented in this chapter. Answer each question. Then check your answers against the ones provided in the Answer Section.

1. What is a sector of a circle?

2. Why is a portion of the vertical axis jagged on some bar graphs?

3. How does a broken-line graph show changes over time?

4. What is the formula for the mean?

5. To find the median, why must the data be arranged in order from least to greatest?

6. When does a set of data have no mode?

7. What five values are shown in a box-and-whiskers plot?

8. How do you find the first quartile for a set of data values?

9. What is the empirical probability formula?

10. What is the theoretical probability formula?

CHAPTER 8

REVIEW EXERCISES

Internet The circle graph in Figure 8.8 shows the approximate amount of money that government agencies spent on maintaining Internet websites for a 3-year period. Use this graph for Exercises 1 to 3.

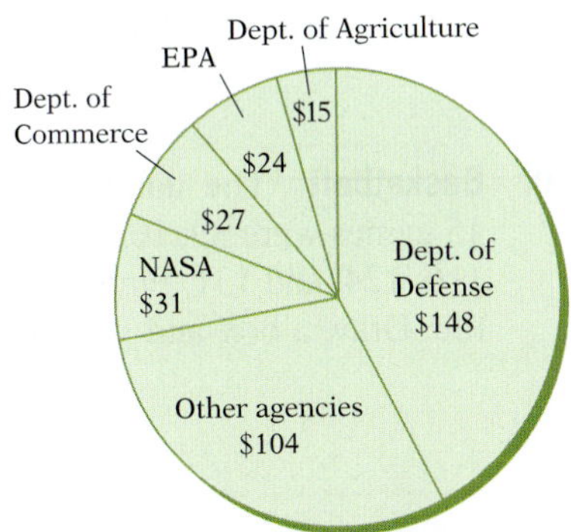

FIGURE 8.8 Amount federal agencies spent on websites (in millions of dollars)
Source: General Accounting Office

1. Find the total amount of money that these agencies spent on maintaining websites.

2. What is the ratio of the amount spent by the Department of Commerce to the amount spent by the EPA?

3. What percent of the total money spent did NASA spend? Round to the nearest tenth of a percent.

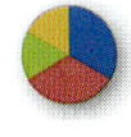

Ski Resorts The double-bar graph in Figure 8.9 shows the total days open and the days of full operation of ski resorts in different regions of the country. Use this graph for Exercises 4 to 6.

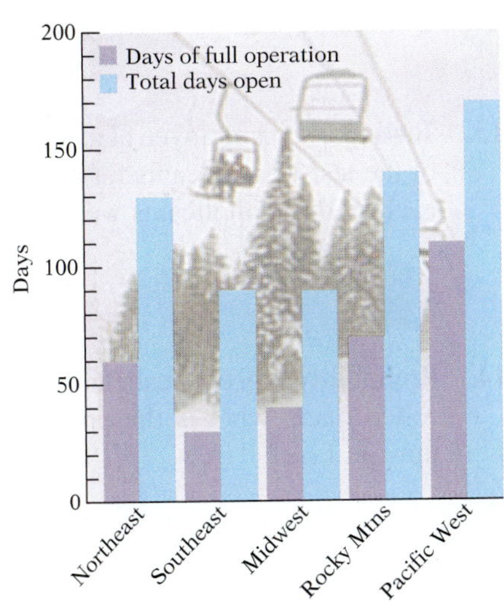

FIGURE 8.9
Source: Economic Analysis of United States Ski Areas

4. Find the difference between the total days open and the days of full operation for Midwest ski areas.

5. What percent of the total days open were the days of full operation for the Rocky Mountain ski areas?

6. **a.** Which region had the lowest number of days of full operation?
   b. How many days of full operation did this region have?

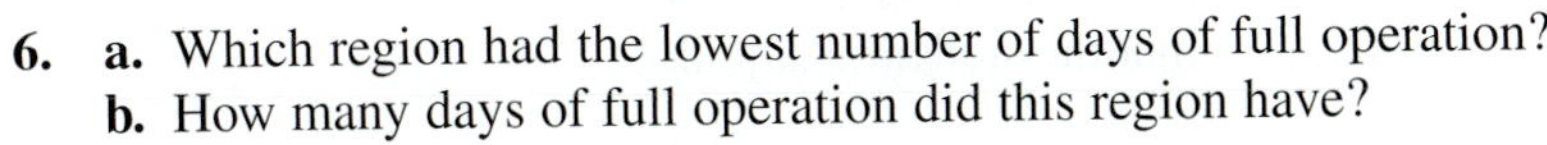

7. **Health** A health clinic administered a test for cholesterol to 11 people. The results were 180, 220, 160, 230, 280, 200, 210, 250, 190, 230, and 210. Find the mean and median of these data.

8. **Newborns** The weights, in pounds, of 10 babies born at a hospital were recorded as 6.3, 5.9, 8.1, 6.5, 7.2, 5.6, 8.9, 9.1, 6.9, and 7.2. Find the mean and median of these data.

9. **Movies** People leaving a new movie were asked to rate the movie as bad, good, very good, or excellent. The responses were bad, 28; good, 65; very good, 49; excellent, 28. What was the modal response for this survey?

10. **Basketball** The numbers of points scored by a basketball team for 15 games were 89, 102, 134, 110, 121, 124, 111, 116, 99, 120, 105, 109, 110, 124, and 131. Find the first quartile, the median, and the third quartile. Draw a box-and-whiskers plot.

11. **Probability** A coin is tossed and then a regular die is rolled. How many elements are in the sample space?

12. **Probability** A charity raffle sells 2500 raffle tickets for a big-screen television set. If you purchase 5 tickets, what is the probability that you will win the television?

13. **Testing** An employee at a department of motor vehicles analyzed the written tests of the last 10 applicants for a drivers' license. The numbers of incorrect answers for each of these applicants were 2, 0, 3, 1, 0, 4, 5, 1, 3, and 1. Find Q_1 and Q_3.

14. **Probability** A dodecahedral die has 12 sides numbered from 1 to 12. If this die is rolled once, what is the probability that a number divisible by 6 will be on the upward face?

15. **Probability** One student is randomly selected from 3 first-year students, 4 sophomores, 5 juniors, and 2 seniors. What is the probability that the student is a junior?

16. **Physical Fitness** The heart rates of 24 women tennis players were measured after each of them had run one-quarter of a mile. The results are listed in the table below.

80	82	99	91	93	87	103	94	73	96	86	80
97	94	108	81	100	109	91	84	78	96	96	100

a. Find the mean, the median, and the mode for the data. Round to the nearest tenth.
b. Find the range and the interquartile range for the data.

CHAPTER 8

TEST

The Film Industry The circle graph in Figure 8.10 categorizes the 655 films released during a recent year by their ratings. Use this graph for Exercises 1 to 3.

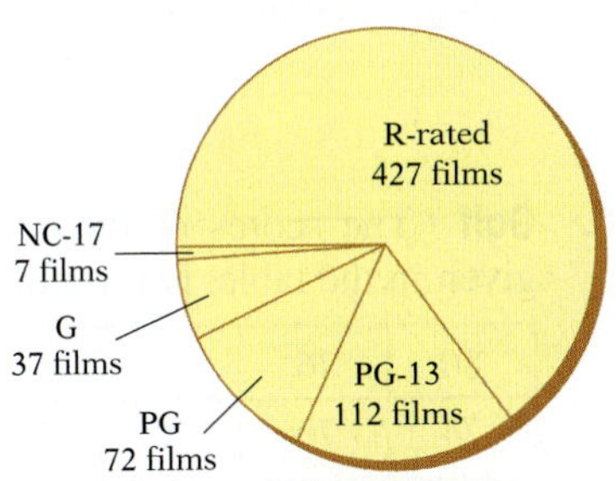

FIGURE 8.10 Ratings of films released
Source: MPA Worldwide Market Research

1. How many more R-rated films were released than PG films?

2. The number of PG-13 films released was how many times the number of NC-17 films released?

3. What percent of the films released were rated G? Round to the nearest tenth of a percent.

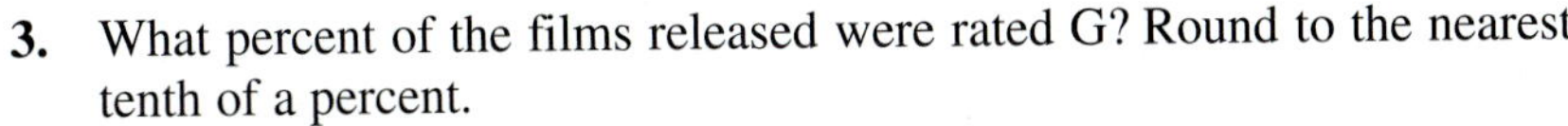

Education The broken-line graph in Figure 8.11 shows the number of students enrolled in colleges. Use this figure for Exercises 4 and 5.

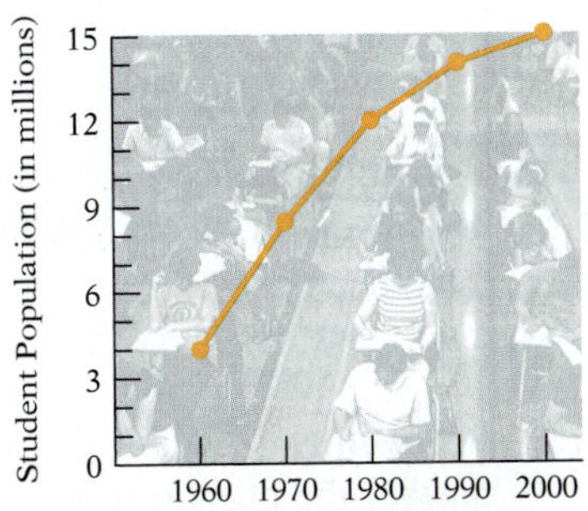

FIGURE 8.11 Student enrollment in public and private colleges
Source: National Center for Educational Statistics

4. During which decade did the student population increase the least?

5. Approximate the increase in college enrollment from 1960 to 2000.

6. **Bowling** The bowling scores for eight people were 138, 125, 162, 144, 129, 168, 184, and 173. What was the mean score for these eight people?

7. **Emergency Calls** The response times by an ambulance service to emergency calls were recorded by a public safety commission. The times (in minutes) were 17, 21, 11, 8, 22, 15, 11, 14, and 8. Determine the median response time for these calls.

8. **Education** Recent college graduates were asked to rate the quality of their education. The responses were 47, excellent; 86, very good; 32, good; 20, poor. What was the modal response?

9. **Manufacturing** The average time, in minutes, it takes a factory worker to assemble 14 different toys is given in the table. Determine the first quartile of the data.

10.5	21.0	17.3	11.2	9.3	6.5	8.6
19.8	20.3	19.6	9.8	10.5	11.9	18.5

10. **Business** The number of vacation days taken last year by each of the employees of a firm was recorded. The box-and-whiskers plot at the right represents the data. **a.** Determine the range of the data. **b.** What was the median number of vacation days taken?

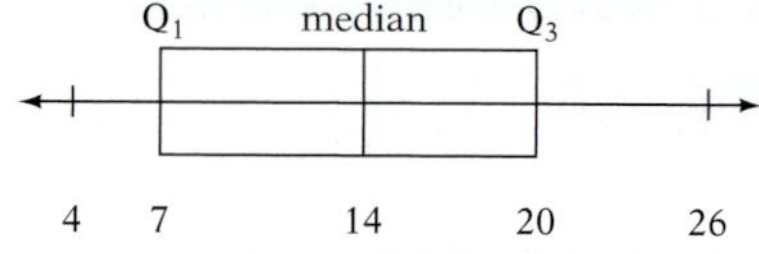

11. **Golf** The scores of the 14 leaders in a college golf tournament are given in the table. Draw a box-and-whiskers plot of the data.

80	76	70	71	74	68	72
74	70	70	73	75	69	73

12. **Probability** Three coins—a nickel, a dime, and a quarter—are stacked. List the events in the sample space.

13. **Probability** A cross-country flight has 14 passengers in first class, 32 passengers in business class, and 202 passengers in coach. If one passenger is selected at random, what is the probability that the person is in business class?

14. **Probability** Three playing cards—an ace, a king, and a queen—are randomly arranged and stacked. What is the probability that the ace is on top of the stack?

15. **Probability** A quiz contains three true/false questions. If a student attempts to answer the questions by just guessing, what is the probability that the student will answer all three questions correctly?

16. **Probability** A package of flower seeds contains 15 seeds for red flowers, 20 seeds for white flowers, and 10 seeds for pink flowers. If one seed is selected at random, what is the probability that it is not a seed for a red flower?

17. **Probability** A dodecahedral die has 12 sides. If this die is tossed once, what is the probability that the number on the upward face is less than 6?

18. **Quality Control** The length of time (in days) that various batteries operated a portable CD player continuously are given in the table below.

2.9	2.4	3.1	2.5	2.6	2.0	3.0	2.3	2.4	2.7
2.0	2.4	2.6	2.7	2.1	2.9	2.8	2.4	2.0	2.8

a. Find the mean for the data.
b. Find the median for the data.
c. Draw a box-and-whiskers plot for the data.

CUMULATIVE REVIEW EXERCISES

1. Simplify: $2^2 \cdot 3^3 \cdot 5$

2. Simplify: $3^2 \cdot (5 - 2) \div 3 + 5$

3. Find the LCM of 24 and 40.

4. Write $\frac{60}{144}$ in simplest form.

5. Find the total of $4\frac{1}{2}$, $2\frac{3}{8}$, and $5\frac{1}{5}$.

6. Subtract: $12\frac{5}{8} - 7\frac{11}{12}$

7. Multiply: $\frac{5}{8} \times 3\frac{1}{5}$

8. Find the quotient of $3\frac{1}{5}$ and $4\frac{1}{4}$.

9. Simplify: $\frac{5}{8} \div \left(\frac{3}{4} - \frac{2}{3}\right) + \frac{3}{4}$

10. Write two hundred nine and three hundred five thousandths in standard form.

11. Find the product of 4.092 and 0.69.

12. Convert $16\frac{2}{3}$ to a decimal. Round to the nearest hundredth.

13. Write “330 mi on 12.5 gal of gas” as a unit rate.

14. Solve the proportion: $\frac{n}{5} = \frac{16}{25}$

15. Write $\frac{4}{5}$ as a percent.

16. 8 is 10% of what?

17. What is 38% of 43?

18. What percent of 75 is 30?

19. Compensation Tanim Kamal, a salesperson at a department store, receives $100 per week plus 2% commission on sales. Find the income for a week in which Tanim had $27,500 in sales.

20. Insurance A life insurance policy costs $8.15 for every $1000 of insurance. At this rate, what is the cost for $50,000 of life insurance?

21. Simple Interest A contractor borrowed $125,000 for 6 months at an annual simple interest rate of 6%. Find the interest due on the loan.

22. Business A compact disc player that costs a retailer $180 is priced at 155% of the cost. Find the price of the compact disc player.

23. Finance The circle graph in Figure 8.12 shows how a family's monthly income of $3000 is budgeted. How much is budgeted for food?

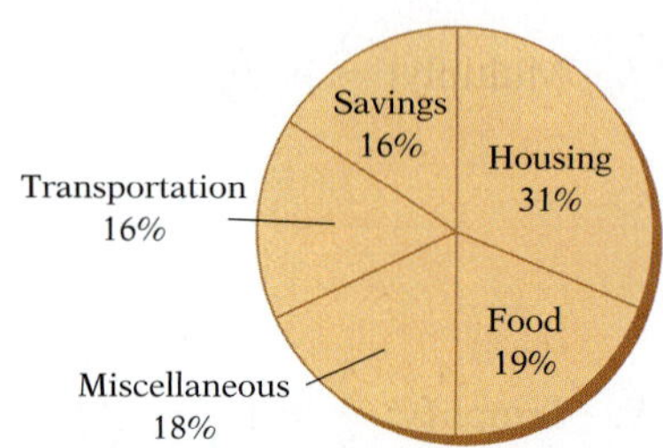

FIGURE 8.12 Budget for a monthly income of $3000

24. Education The double-broken-line graph in Figure 8.13 shows two students' scores on 5 math tests of 30 problems each. Find the difference between the scores of the two students on Test 1.

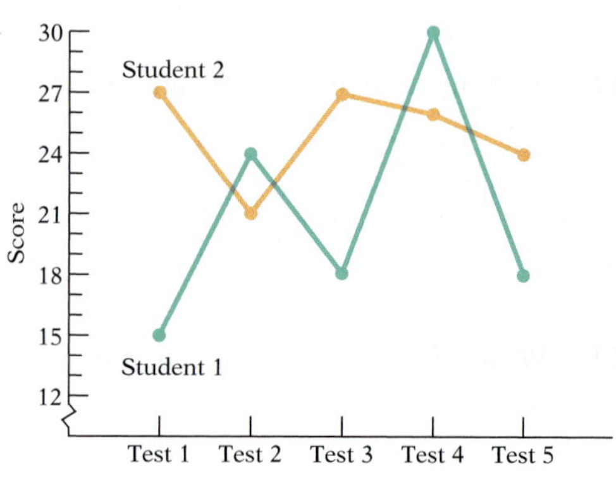

FIGURE 8.13

25. Meteorology The average daily high temperatures, in degrees Fahrenheit, for a week in Newtown were 56°, 72°, 80°, 75°, 68°, 62°, and 74°. Find the mean high temperature for the week. Round to the nearest tenth of a degree.

26. Probability Two dice are rolled. What is the probability that the sum of the dots on the upward faces is 8?

Polynomials

CHAPTER 9

OBJECTIVES

SECTION 9.1

A To add polynomials
B To subtract polynomials

SECTION 9.2

A To multiply monomials
B To simplify powers of monomials

SECTION 9.3

A To multiply a polynomial by a monomial
B To multiply two polynomials
C To multiply two binomials using the FOIL method
D To multiply binomials that have special products
E To solve application problems

SECTION 9.4

A To divide monomials
B To write a number in scientific notation

SECTION 9.5

A To divide a polynomial by a monomial
B To divide polynomials

ARE YOU READY?

Take the Chapter 9 Prep Test to find out if you are ready to learn to:

- Multiply and divide monomials
- Add, subtract, multiply, and divide polynomials
- Write a number in scientific notation

PREP TEST

Do these exercises to prepare for Chapter 9.

1. Subtract: $-2 - (-3)$

2. Multiply: $-3(6)$

3. Simplify: $-\frac{24}{-36}$

4. Evaluate $3n^4$ when $n = -2$.

5. If $\frac{a}{b}$ is a fraction in simplest form, what number is not a possible value of b?

6. Are $2x^2$ and $2x$ like terms?

7. Simplify: $3x^2 - 4x + 1 + 2x^2 - 5x - 7$

8. Simplify: $-4y + 4y$

9. Simplify: $-3(2x - 8)$

10. Simplify: $3xy - 4y - 2(5xy - 7y)$

SECTION

9.1 Addition and Subtraction of Polynomials

OBJECTIVE A To add polynomials 

Take Note

The expression $3\sqrt{x}$ is not a monomial because $\sqrt{x}$ cannot be written as a product of variables.

The expression $\frac{2x}{y}$ is not a monomial because it is a *quotient* of variables.

A **monomial** is a number, a variable, or a product of a number and variables. For instance,

7	b	$\frac{2}{3}a$	$12xy^2$
A number	A variable	A product of a number and a variable	A product of a number and variables

A **polynomial** is a variable expression in which the terms are monomials.

A polynomial of *one* term is a **monomial.** $-7x^2$ is a monomial.
A polynomial of *two* terms is a **binomial.** $4x + 2$ is a binomial.
A polynomial of *three* terms is a **trinomial.** $7x^2 + 5x - 7$ is a trinomial.

The **degree of a polynomial in one variable** is the greatest exponent on a variable. The degree of $4x^3 - 5x^2 + 7x - 8$ is 3; the degree of $2y^4 + y^2 - 1$ is 4. The degree of a nonzero constant is zero. For instance, the degree of 7 is zero.

The terms of a polynomial in one variable are usually arranged so that the exponents on the variable decrease from left to right. This is called **descending order.**

$5x^3 - 4x^2 + 6x - 1$

$7z^4 + 4z^3 + z - 6$

$2y^4 + y^3 - 2y^2 + 4y - 1$

Polynomials can be added, using either a horizontal or a vertical format, by combining like terms.

HOW TO • 1 Add $(3x^3 - 7x + 2) + (7x^2 + 2x - 7)$. Use a horizontal format.

$(3x^3 - 7x + 2) + (7x^2 + 2x - 7)$
$= 3x^3 + 7x^2 + (-7x + 2x) + (2 - 7)$ • Use the Commutative and Associative Properties of Addition to rearrange and group like terms.
$= 3x^3 + 7x^2 - 5x - 5$ • Then combine like terms.

HOW TO • 2 Add $(-4x^2 + 6x - 9) + (12 - 8x + 2x^3)$. Use a vertical format.

$$\begin{array}{r} -4x^2 + 6x - 9 \\ 2x^3 \qquad\quad - 8x + 12 \\ \hline 2x^3 - 4x^2 - 2x + 3 \end{array}$$

• Arrange the terms of each polynomial in descending order, with like terms in the same column.
• Combine the terms in each column.

EXAMPLE • 1

Use a horizontal format to add $(8x^2 - 4x - 9) + (2x^2 + 9x - 9)$.

Solution

$(8x^2 - 4x - 9) + (2x^2 + 9x - 9)$
$= (8x^2 + 2x^2) + (-4x + 9x) + (-9 - 9)$
$= 10x^2 + 5x - 18$

YOU TRY IT • 1

Use a horizontal format to add $(-4x^3 + 2x^2 - 8) + (4x^3 + 6x^2 - 7x + 5)$.

Your solution

Solution on p. S21

EXAMPLE 2

Use a vertical format to add
$(-5x^3 + 4x^2 - 7x + 9) + (2x^3 + 5x - 11)$.

Solution

$$\begin{array}{r} -5x^3 + 4x^2 - 7x + 9 \\ 2x^3 \qquad\quad + 5x - 11 \\ \hline -3x^3 + 4x^2 - 2x - 2 \end{array}$$

YOU TRY IT 2

Use a vertical format to add
$(6x^3 + 2x + 8) + (-9x^3 + 2x^2 - 12x - 8)$.

Your solution

Solution on p. S21

OBJECTIVE B To subtract polynomials

The **opposite of the polynomial** $(3x^2 - 7x + 8)$ is $-(3x^2 - 7x + 8)$.

To simplify the opposite of a polynomial, change the sign of each term to its opposite. $-(3x^2 - 7x + 8) = -3x^2 + 7x - 8$

Take Note
This is the same definition used for subtraction of integers: Subtraction is addition of the opposite.

Polynomials can be subtracted using either a horizontal or a vertical format. To subtract, add the opposite of the second polynomial to the first.

HOW TO 3 Subtract $(4y^2 - 6y + 7) - (2y^3 - 5y - 4)$. Use a horizontal format.

$(4y^2 - 6y + 7) - (2y^3 - 5y - 4)$
$= (4y^2 - 6y + 7) + (-2y^3 + 5y + 4)$ • **Add the opposite of the second polynomial to the first.**
$= -2y^3 + 4y^2 + (-6y + 5y) + (7 + 4)$
$= -2y^3 + 4y^2 - y + 11$ • **Combine like terms.**

HOW TO 4 Subtract $(9 + 4y + 3y^3) - (2y^2 + 4y - 21)$. Use a vertical format.

The opposite of $2y^2 + 4y - 21$ is $-2y^2 - 4y + 21$.

$$\begin{array}{r} 3y^3 \qquad\quad + 4y + 9 \\ -2y^2 - 4y + 21 \\ \hline 3y^3 - 2y^2 \qquad\quad + 30 \end{array}$$

• **Arrange the terms of each polynomial in descending order, with like terms in the same column.**
• **Note that $4y - 4y = 0$, but 0 is not written.**

EXAMPLE 3

Use a horizontal format to subtract
$(7c^2 - 9c - 12) - (9c^2 + 5c - 8)$.

Solution

$(7c^2 - 9c - 12) - (9c^2 + 5c - 8)$
$= (7c^2 - 9c - 12) + (-9c^2 - 5c + 8)$
$= -2c^2 - 14c - 4$

YOU TRY IT 3

Use a horizontal format to subtract
$(-4w^3 + 8w - 8) - (3w^3 - 4w^2 - 2w - 1)$.

Your solution

EXAMPLE 4

Use a vertical format to subtract
$(3k^2 - 4k + 1) - (k^3 + 3k^2 - 6k - 8)$.

Solution

$$\begin{array}{r} 3k^2 - 4k + 1 \\ -k^3 - 3k^2 + 6k + 8 \\ \hline -k^3 \qquad\quad + 2k + 9 \end{array}$$

• **Add the opposite of $(k^3 + 3k^2 - 6k - 8)$ to the first polynomial.**

YOU TRY IT 4

Use a vertical format to subtract
$(13y^3 - 6y - 7) - (4y^2 - 6y - 9)$.

Your solution

Solutions on p. S21

9.1 EXERCISES

OBJECTIVE A To add polynomials

For Exercises 1 to 8, state whether the expression is a monomial.

1. 17

2. $3x^4$

3. $\frac{17}{\sqrt{x}}$

4. xyz

5. $\frac{2}{3}y$

6. $\frac{xy}{z}$

7. $\sqrt{5}x$

8. πx

For Exercises 9 to 16, state whether the expression is a monomial, a binomial, a trinomial, or none of these.

9. $3x + 5$

10. $2y - 3\sqrt{y}$

11. $9x^2 - x - 1$

12. $x^2 + y^2$

13. $\frac{2}{x} - 3$

14. $\frac{ab}{4}$

15. $6x^2 + 7x$

16. $12a^4 - 3a + 2$

For Exercises 17 to 26, add. Use a horizontal format.

17. $(4x^2 + 2x) + (x^2 + 6x)$

18. $(-3y^2 + y) + (4y^2 + 6y)$

19. $(4x^2 - 5xy) + (3x^2 + 6xy - 4y^2)$

20. $(2x^2 - 4y^2) + (6x^2 - 2xy + 4y^2)$

21. $(2a^2 - 7a + 10) + (a^2 + 4a + 7)$

22. $(-6x^2 + 7x + 3) + (3x^2 + x + 3)$

23. $(7x + 5x^3 - 7) + (10x^2 - 8x + 3)$

24. $(4y + 3y^3 + 9) + (2y^2 + 4y - 21)$

25. $(7 - 5r + 2r^2) + (3r^3 - 6r)$

26. $(14 + 4y + 3y^3) + (-4y^2 + 21)$

For Exercises 27 to 36, add. Use a vertical format.

27. $(x^2 + 7x) + (-3x^2 - 4x)$

28. $(3y^2 - 2y) + (5y^2 + 6y)$

29. $(y^2 + 4y) + (-4y - 8)$

30. $(3x^2 + 9x) + (6x - 24)$

31. $(2x^2 + 6x + 12) + (3x^2 + x + 8)$

32. $(x^2 + x + 5) + (3x^2 - 10x + 4)$

33. $(-7x + x^3 + 4) + (2x^2 + x - 10)$

34. $(y^2 + 3y^3 + 1) + (-4y^3 - 6y - 3)$

35. $(2a^3 - 7a + 1) + (1 - 4a - 3a^2)$

36. $(5r^3 - 6r^2 + 3r) + (-3 - 2r + r^2)$

For Exercises 37 and 38, use the polynomials shown at the right. Assume that a, b, c, and d are all positive numbers. Choose the correct answer from this list:
(i) $P + Q$ (ii) $Q + R$ (iii) $P + R$ (iv) None of the above

$P = ax^3 + bx^2 - cx + d$
$Q = -ax^3 - bx^2 + cx - d$
$R = -ax^3 + bx^2 + cx + d$

37. Which sum will be a trinomial?

38. Which sum will be zero?

OBJECTIVE B To subtract polynomials

For Exercises 39 to 48, subtract. Use a horizontal format.

39. $(y^2 - 10xy) - (2y^2 + 3xy)$

40. $(x^2 - 3xy) - (-2x^2 + xy)$

41. $(3x^2 + x - 3) - (4x + x^2 - 2)$

42. $(5y^2 - 2y + 1) - (-y - 2 - 3y^2)$

43. $(-2x^3 + x - 1) - (-x^2 + x - 3)$

44. $(2x^2 + 5x - 3) - (3x^3 + 2x - 5)$

45. $(1 - 2a + 4a^3) - (a^3 - 2a + 3)$

46. $(7 - 8b + b^2) - (4b^3 - 7b - 8)$

47. $(-1 - y + 4y^3) - (3 - 3y - 2y^2)$

48. $(-3 - 2x + 3x^2) - (4 - 2x^2 + 2x^3)$

For Exercises 49 to 58, subtract. Use a vertical format.

49. $(x^2 - 6x) - (x^2 - 10x)$

50. $(y^2 + 4y) - (y^2 + 10y)$

51. $(2y^2 - 4y) - (-y^2 + 2)$

52. $(-3a^2 - 2a) - (4a^2 - 4)$

53. $(x^2 - 2x + 1) - (x^2 + 5x + 8)$

54. $(3x^2 + 2x - 2) - (5x^2 - 5x + 6)$

55. $(4x^3 + 5x + 2) - (1 + 2x - 3x^2)$

56. $(5y^2 - y + 2) - (-3 + 3y - 2y^3)$

57. $(-2y + 6y^2 + 2y^3) - (4 + y^2 + y^3)$

58. $(4 - x - 2x^2) - (-2 + 3x - x^3)$

59. What polynomial must be added to $3x^2 - 6x + 9$ so that the sum is $4x^2 + 3x - 2$?

Applying the Concepts

60. Is it possible to subtract two polynomials, each of degree 3, and have the difference be a polynomial of degree 2? If so, give an example. If not, explain why not.

61. Is it possible to add two polynomials, each of degree 3, and have the sum be a polynomial of degree 2? If so, give an example. If not, explain why not.

SECTION 9.2 Multiplication of Monomials

OBJECTIVE A To multiply monomials 

Recall that in an exponential expression such as x^6, x is the base and 6 is the exponent. The exponent indicates the number of times the base occurs as a factor.

The product of exponential expressions with the *same* base can be simplified by writing each expression in factored form and then writing the result with an exponent.

$$x^3 \cdot x^2 = \underbrace{\overbrace{(x \cdot x \cdot x)}^{3 \text{ factors}} \cdot \overbrace{(x \cdot x)}^{2 \text{ factors}}}_{5 \text{ factors}}$$
$$= x^5$$

Note that adding the exponents results in the same product.

$$x^3 \cdot x^2 = x^{3+2} = x^5$$

Rule for Multiplying Exponential Expressions

If m and n are positive integers, then $x^m \cdot x^n = x^{m+n}$.

HOW TO • 1 Simplify: $y^4 \cdot y \cdot y^3$

$y^4 \cdot y \cdot y^3 = y^{4+1+3}$ • The bases are the same. Add the exponents. Recall that $y = y^1$.

$= y^8$

HOW TO • 2 Simplify: $(-3a^4b^3)(2ab^4)$

$(-3a^4b^3)(2ab^4) = (-3 \cdot 2)(a^4 \cdot a)(b^3 \cdot b^4)$ • Use the Commutative and Associative Properties of Multiplication to rearrange and group factors.

$= -6(a^{4+1})(b^{3+4})$ • To multiply expressions with the same base, add the exponents.

$= -6a^5b^7$ • Simplify.

Take Note
The Rule for Multiplying Exponential Expressions requires that the bases be the same. The expression a^5b^7 cannot be simplified.

EXAMPLE • 1

Simplify: $(-5ab^3)(4a^5)$

Solution

$(-5ab^3)(4a^5)$
$= (-5 \cdot 4)(a \cdot a^5)b^3$ • Multiply coefficients. Add exponents with same base.
$= -20a^6b^3$

YOU TRY IT • 1

Simplify: $(8m^3n)(-3n^5)$

Your solution

EXAMPLE • 2

Simplify: $(6x^3y^2)(4x^4y^5)$

Solution

$(6x^3y^2)(4x^4y^5)$
$= (6 \cdot 4)(x^3 \cdot x^4)(y^2 \cdot y^5)$ • Multiply coefficients. Add exponents with same base.
$= 24x^7y^7$

YOU TRY IT • 2

Simplify: $(12p^4q^3)(-3p^5q^2)$

Your solution

Solutions on p. S21

OBJECTIVE B To simplify powers of monomials 

Point of Interest

One of the first symbolic representations of powers was given by Diophantus (c. 250 A.D.) in his book *Arithmetica*. He used Δ^Y for x^2 and κ^Y for x^3. The symbol Δ^Y was the first two letters of the Greek word *dunamis*, which means "power"; κ^Y was from the Greek word *kubos*, which means "cube." He also combined these symbols to denote higher powers. For instance, $\Delta\kappa^Y$ was the symbol for x^5.

The power of a monomial can be simplified by writing the power in factored form and then using the Rule for Multiplying Exponential Expressions.

$(x^4)^3 = x^4 \cdot x^4 \cdot x^4$ $\quad$ $(a^2b^3)^2 = (a^2b^3)(a^2b^3)$ • Write in factored form.

$= x^{4+4+4}$ $\quad$ $= a^{2+2}b^{3+3}$ • Use the Rule for Multiplying Exponential Expressions.

$= x^{12}$ $\quad$ $= a^4b^6$

Note that multiplying each exponent inside the parentheses by the exponent outside the parentheses results in the same product.

$(x^4)^3 = x^{4\cdot 3} = x^{12}$ $\quad$ $(a^2b^3)^2 = a^{2\cdot 2}b^{3\cdot 2} = a^4b^6$ • Multiply each exponent inside the parentheses by the exponent outside the parentheses.

Rule for Simplifying the Power of an Exponential Expression

If m and n are positive integers, then $(x^m)^n = x^{mn}$.

Rule for Simplifying the Power of a Product

If m, n, and p are positive integers, then $(x^my^n)^p = x^{mp}y^{np}$.

HOW TO • 3 Simplify: $(5x^2y^3)^3$

$(5x^2y^3)^3 = 5^{1\cdot 3}x^{2\cdot 3}y^{3\cdot 3}$ • Use the Rule for Simplifying the Power of a Product. Note that $5 = 5^1$.

$= 5^3x^6y^9$

$= 125x^6y^9$ • Evaluate 5^3.

EXAMPLE • 3

Simplify: $(-2p^3r)^4$

Solution

$(-2p^3r)^4 = (-2)^{1\cdot 4}p^{3\cdot 4}r^{1\cdot 4}$ • Use the Rule for Simplifying the Power of a Product.

$= (-2)^4p^{12}r^4 = 16p^{12}r^4$

YOU TRY IT • 3

Simplify: $(-3a^4bc^2)^3$

Your solution

EXAMPLE • 4

Simplify: $(2a^2b)(2a^3b^2)^3$

Solution

$(2a^2b)(2a^3b^2)^3$

$= (2a^2b)(2^{1\cdot 3}a^{3\cdot 3}b^{2\cdot 3})$ • Use the Rule for Simplifying the Power of a Product.

$= (2a^2b)(2^3a^9b^6)$

$= (2a^2b)(8a^9b^6) = 16a^{11}b^7$

YOU TRY IT • 4

Simplify: $(-xy^4)(-2x^3y^2)^2$

Your solution

Solutions on p. S21

9.2 EXERCISES

OBJECTIVE A To multiply monomials

For Exercises 1 and 2, state whether the expression can be simplified using the Rule for Multiplying Exponential Expressions.

1. **a.** $x^4 + x^5$ **b.** x^4x^5

2. **a.** x^4y^4 **b.** $x^4 + x^4$

For Exercises 3 to 35, simplify.

3. $(6x^2)(5x)$

4. $(-4y^3)(2y)$

5. $(7c^2)(-6c^4)$

6. $(-8z^5)(5z^8)$

7. $(-3a^3)(-3a^4)$

8. $(-5a^6)(-2a^5)$

9. $(x^2)(xy^4)$

10. $(x^2y^4)(xy^7)$

11. $(-2x^4)(5x^5y)$

12. $(-3a^3)(2a^2b^4)$

13. $(-4x^2y^4)(-3x^5y^4)$

14. $(-6a^2b^4)(-4ab^3)$

15. $(2xy)(-3x^2y^4)$

16. $(-3a^2b)(-2ab^3)$

17. $(x^2yz)(x^2y^4)$

18. $(-ab^2c)(a^2b^5)$

19. $(-a^2b^3)(-ab^2c^4)$

20. $(-x^2y^3z)(-x^3y^4)$

21. $(-5a^2b^2)(6a^3b^6)$

22. $(7xy^4)(-2xy^3)$

23. $(-6a^3)(-a^2b)$

24. $(-2a^2b^3)(-4ab^2)$

25. $(-5y^4z)(-8y^6z^5)$

26. $(3x^2y)(-4xy^2)$

27. $(x^2y)(yz)(xyz)$

28. $(xy^2z)(x^2y)(z^2y^2)$

29. $(3ab^2)(-2abc)(4ac^2)$

30. $(-2x^3y^2)(-3x^2z^2)(-5y^3z^3)$

31. $(4x^4z)(-yz^3)(-2x^3z^2)$

32. $(-a^3b^4)(-3a^4c^2)(4b^3c^4)$

33. $(-2x^2y^3)(3xy)(-5x^3y^4)$

34. $(4a^2b)(-3a^3b^4)(a^5b^2)$

35. $(3a^2b)(-6bc)(2ac^2)$

OBJECTIVE B To simplify powers of monomials

For Exercises 36 and 37, state whether the expression can be simplified using the Rule for Simplifying the Power of a Product.

36. **a.** $(xy)^3$ **b.** $(x + y)^3$

37. **a.** $(a^3 + b^4)^2$ **b.** $(a^3b^4)^2$

For Exercises 38 to 68, simplify.

38. $(z^4)^3$ **39.** $(x^3)^5$ **40.** $(y^4)^2$ **41.** $(x^7)^2$ **42.** $(-y^5)^3$

43. $(-x^2)^4$ **44.** $(-x^2)^3$ **45.** $(-y^3)^4$ **46.** $(-3y)^3$ **47.** $(-2x^2)^3$

48. $(a^3b^4)^3$ **49.** $(x^2y^3)^2$ **50.** $(2x^3y^4)^5$ **51.** $(3x^2y)^2$ **52.** $(-2ab^3)^4$

53. $(-3x^3y^2)^5$ **54.** $(3b^2)(2a^3)^4$ **55.** $(-2x)(2x^3)^2$ **56.** $(2y)(-3y^4)^3$

57. $(3x^2y)(2x^2y^2)^3$ **58.** $(a^3b)^2(ab)^3$ **59.** $(ab^2)^2(ab)^2$ **60.** $(-x^2y^3)^2(-2x^3y)^3$

61. $(-2x)^3(-2x^3y)^3$ **62.** $(-3y)(-4x^2y^3)^3$ **63.** $(-2x)(-3xy^2)^2$ **64.** $(-3y)(-2x^2y)^3$

65. $(ab^2)(-2a^2b)^3$ **66.** $(a^2b^2)(-3ab^4)^2$ **67.** $(-2a^3)(3a^2b)^3$ **68.** $(-3b^2)(2ab^2)^3$

Applying the Concepts

For Exercises 69 to 76, simplify.

69. $3x^2 + (3x)^2$ **70.** $4x^2 - (4x)^2$ **71.** $2x^6y^2 + (3x^2y)^2$ **72.** $(x^2y^2)^3 + (x^3y^3)^2$

73. $(2a^3b^2)^3 - 8a^9b^6$ **74.** $4y^2z^4 - (2yz^2)^2$ **75.** $(x^2y^4)^2 + (2xy^2)^4$ **76.** $(3a^3)^2 - 4a^6 + (2a^2)^3$

77. Evaluate $(2^3)^2$ and $2^{(3^2)}$. Are the results the same? If not, which expression has the larger value?

78. If n is a positive integer and $x^n = y^n$, does $x = y$? Explain your answer.

SECTION

9.3 Multiplication of Polynomials

OBJECTIVE A To multiply a polynomial by a monomial

To multiply a polynomial by a monomial, use the Distributive Property and the Rule for Multiplying Exponential Expressions.

HOW TO • 1 Multiply: $-3a(4a^2 - 5a + 6)$

$$-3a(4a^2 - 5a + 6) = -3a(4a^2) - (-3a)(5a) + (-3a)(6)$$
$$= -12a^3 + 15a^2 - 18a$$

• Use the Distributive Property.

EXAMPLE • 1

Multiply: $(5x + 4)(-2x)$

Solution

$(5x + 4)(-2x) = 5x(-2x) + 4(-2x) = -10x^2 - 8x$

YOU TRY IT • 1

Multiply: $(-2y + 3)(-4y)$

Your solution

EXAMPLE • 2

Multiply: $2a^2b(4a^2 - 2ab + b^2)$

Solution

$2a^2b(4a^2 - 2ab + b^2)$
$= 2a^2b(4a^2) - 2a^2b(2ab) + 2a^2b(b^2)$
$= 8a^4b - 4a^3b^2 + 2a^2b^3$

YOU TRY IT • 2

Multiply: $-a^2(3a^2 + 2a - 7)$

Your solution

Solutions on p. S21

OBJECTIVE B To multiply two polynomials

Multiplication of two polynomials requires the repeated application of the Distributive Property.

$$(y^2 - 4y - 6)(y + 2) = (y^2 - 4y - 6)y + (y^2 - 4y - 6)2$$
$$= (y^3 - 4y^2 - 6y) + (2y^2 - 8y - 12)$$
$$= y^3 - 2y^2 - 14y - 12$$

A convenient method for multiplying two polynomials is to use a vertical format similar to that used for multiplication of whole numbers.

$$\begin{array}{rll}
y^2 - 4y - 6 & & \\
y + 2 & & \\
\hline
2y^2 - 8y - 12 & = (y^2 - 4y - 6)2 & \text{• Multiply by 2.} \\
y^3 - 4y^2 - 6y \quad\quad & = (y^2 - 4y - 6)y & \text{• Multiply by } y. \\
\hline
y^3 - 2y^2 - 14y - 12 & & \text{• Add the terms in each column.}
\end{array}$$

HOW TO • 2 Multiply: $(2a^3 + a - 3)(a + 5)$

$$\begin{array}{rrrrr} & 2a^3 & & +\ a & -\ 3 \\ & & & a & +\ 5 \\ \hline & 10a^3 & & +\ 5a & -\ 15 \\ 2a^4 & & +\ a^2 & -\ 3a & \\ \hline 2a^4 & +\ 10a^3 & +\ a^2 & +\ 2a & -\ 15 \end{array}$$

- Note that spaces are provided in each product so that like terms are in the same column.
- Add the terms in each column.

EXAMPLE • 3

Multiply: $(2b^3 - b + 1)(2b + 3)$

Solution

$$\begin{array}{rrrrr} & 2b^3 & & -\ b & +\ 1 \\ & & & 2b & +\ 3 \\ \hline & 6b^3 & & -\ 3b & +\ 3 \\ 4b^4\ + & & -\ 2b^2 & +\ 2b & \\ \hline 4b^4 & +\ 6b^3 & -\ 2b^2 & -\ b & +\ 3 \end{array}$$

- $= 3(2b^3 - b + 1)$
- $= 2b(2b^3 - b + 1)$

YOU TRY IT • 3

Multiply: $(2y^3 + 2y^2 - 3)(3y - 1)$

Your solution

Solution on p. S21

OBJECTIVE C To multiply two binomials using the FOIL method

Take Note

FOIL is not really a different way of multiplying. It is based on the Distributive Property.

$(2x + 3)(x + 5)$
$= 2x(x + 5) + 3(x + 5)$
F O I L
$= 2x^2 + 10x + 3x + 15$
$= 2x^2 + 13x + 15$

It is frequently necessary to find the product of two binomials. The product can be found using a method called **FOIL,** which is based on the Distributive Property. The letters of FOIL stand for **F**irst, **O**uter, **I**nner, and **L**ast. To find the product of two binomials, add the products of the **F**irst terms, the **O**uter terms, the **I**nner terms, and the **L**ast terms.

HOW TO • 3 Multiply: $(2x + 3)(x + 5)$

Multiply the First terms.	$(2x + 3)(x + 5)$	$2x \cdot x = 2x^2$
Multiply the Outer terms.	$(2x + 3)(x + 5)$	$2x \cdot 5 = 10x$
Multiply the Inner terms.	$(2x + 3)(x + 5)$	$3 \cdot x = 3x$
Multiply the Last terms.	$(2x + 3)(x + 5)$	$3 \cdot 5 = 15$
		F O I L
Add the products.	$(2x + 3)(x + 5)$	$= 2x^2 + 10x + 3x + 15$
Combine like terms.		$= 2x^2 + 13x + 15$

HOW TO • 4 Multiply: $(4x - 3)(3x - 2)$

$$\begin{aligned}(4x - 3)(3x - 2) &= 4x(3x) + 4x(-2) + (-3)(3x) + (-3)(-2)\\ &= 12x^2 - 8x - 9x + 6\\ &= 12x^2 - 17x + 6\end{aligned}$$

HOW TO • 5 Multiply: $(3x - 2y)(x + 4y)$

$$\begin{aligned}(3x - 2y)(x + 4y) &= 3x(x) + 3x(4y) + (-2y)(x) + (-2y)(4y)\\ &= 3x^2 + 12xy - 2xy - 8y^2\\ &= 3x^2 + 10xy - 8y^2\end{aligned}$$

EXAMPLE • 4

Multiply: $(2a - 1)(3a - 2)$

Solution

$$(2a - 1)(3a - 2) = 6a^2 - 4a - 3a + 2$$
$$= 6a^2 - 7a + 2$$

YOU TRY IT • 4

Multiply: $(4y - 5)(2y - 3)$

Your solution

EXAMPLE • 5

Multiply: $(3x - 2)(4x + 3)$

Solution

$$(3x - 2)(4x + 3) = 12x^2 + 9x - 8x - 6$$
$$= 12x^2 + x - 6$$

YOU TRY IT • 5

Multiply: $(3b + 2)(3b - 5)$

Your solution

Solutions on p. S21

OBJECTIVE D To multiply binomials that have special products

Using FOIL, it is possible to find a pattern for the product of the sum and difference of two terms and for the square of a binomial.

Product of the Sum and Difference of the Same Terms

$$(a + b)(a - b) = a^2 - ab + ab - b^2$$
$$= a^2 - b^2$$

Square of the first term → a^2
Square of the second term → b^2

Square of a Binomial

$$(a + b)^2 = (a + b)(a + b) = a^2 + ab + ab + b^2$$
$$= a^2 + 2ab + b^2$$

Square of the first term → a^2
Twice the product of the two terms → $2ab$
Square of the last term → b^2

HOW TO • 6 Multiply: $(2x + 3)(2x - 3)$

$$(2x + 3)(2x - 3) = (2x)^2 - 3^2$$
$$= 4x^2 - 9$$

• This is the product of the sum and difference of the same terms.

Take Note

The word *expand* is used frequently to mean "multiply out a power."

HOW TO • 7 Expand: $(3x - 2)^2$

$$(3x - 2)^2 = (3x)^2 + 2(3x)(-2) + (-2)^2$$
$$= 9x^2 - 12x + 4$$

• This is the square of a binomial.

EXAMPLE • 6

Multiply: $(4z - 2w)(4z + 2w)$

Solution

$(4z - 2w)(4z + 2w) = 16z^2 - 4w^2$

YOU TRY IT • 6

Multiply: $(2a + 5c)(2a - 5c)$

Your solution

EXAMPLE • 7

Expand: $(2r - 3s)^2$

Solution

$(2r - 3s)^2 = 4r^2 - 12rs + 9s^2$

YOU TRY IT • 7

Expand: $(3x + 2y)^2$

Your solution

Solutions on p. S21

OBJECTIVE E To solve application problems

EXAMPLE • 8

The length of a rectangle is $(x + 7)$ m. The width is $(x - 4)$ m. Find the area of the rectangle in terms of the variable x.

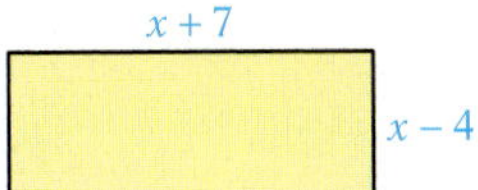

Strategy

To find the area, replace the variables L and W in the equation $A = L \cdot W$ by the given values and solve for A.

Solution

$A = L \cdot W$
$A = (x + 7)(x - 4)$
$A = x^2 - 4x + 7x - 28$
$A = x^2 + 3x - 28$

The area is $(x^2 + 3x - 28)$ m^2.

YOU TRY IT • 8

The radius of a circle is $(x - 4)$ ft. Use the equation $A = \pi r^2$, where r is the radius, to find the area of the circle in terms of x. Leave the answer in terms of π.

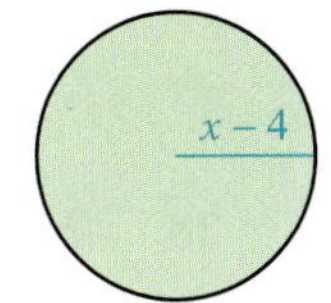

Your strategy

Your solution

Solution on p. S22

9.3 EXERCISES

OBJECTIVE A To multiply a polynomial by a monomial

For Exercises 1 to 32, multiply.

1. $x(x - 2)$

2. $y(3 - y)$

3. $-x(x + 7)$

4. $-y(7 - y)$

5. $3a^2(a - 2)$

6. $4b^2(b + 8)$

7. $-5x^2(x^2 - x)$

8. $-6y^2(y + 2y^2)$

9. $-x^3(3x^2 - 7)$

10. $-y^4(2y^2 - y^6)$

11. $2x(6x^2 - 3x)$

12. $3y(4y - y^2)$

13. $(2x - 4)3x$

14. $(3y - 2)y$

15. $(3x + 4)x$

16. $(2x + 1)2x$

17. $-xy(x^2 - y^2)$

18. $-x^2y(2xy - y^2)$

19. $x(2x^3 - 3x + 2)$

20. $y(-3y^2 - 2y + 6)$

21. $-a(-2a^2 - 3a - 2)$

22. $-b(5b^2 + 7b - 35)$

23. $x^2(3x^4 - 3x^2 - 2)$

24. $y^3(-4y^3 - 6y + 7)$

25. $2y^2(-3y^2 - 6y + 7)$

26. $4x^2(3x^2 - 2x + 6)$

27. $(a^2 + 3a - 4)(-2a)$

28. $(b^3 - 2b + 2)(-5b)$

29. $-3y^2(-2y^2 + y - 2)$

30. $-5x^2(3x^2 - 3x - 7)$

31. $xy(x^2 - 3xy + y^2)$

32. $ab(2a^2 - 4ab - 6b^2)$

33. Which of the following expressions are equivalent to $4x - x(3x - 1)$?
(i) $4x - 3x^2 - x$ (ii) $-3x^2 + 5x$ (iii) $4x - 3x^2 + x$ (iv) $9x^2 - 3x$ (v) $3x(3x - 1)$

OBJECTIVE B To multiply two polynomials

For Exercises 34 to 51, multiply.

34. $(x^2 + 3x + 2)(x + 1)$

35. $(x^2 - 2x + 7)(x - 2)$

36. $(a^2 - 3a + 4)(a - 3)$

37. $(x^2 - 3x + 5)(2x - 3)$

38. $(-2b^2 - 3b + 4)(b - 5)$

39. $(-a^2 + 3a - 2)(2a - 1)$

40. $(-2x^2 + 7x - 2)(3x - 5)$

41. $(-a^2 - 2a + 3)(2a - 1)$

42. $(x^2 + 5)(x - 3)$

43. $(y^2 - 2y)(2y + 5)$

44. $(x^3 - 3x + 2)(x - 4)$

45. $(y^3 + 4y^2 - 8)(2y - 1)$

46. $(5y^2 + 8y - 2)(3y - 8)$

47. $(3y^2 + 3y - 5)(4y - 3)$

48. $(5a^3 - 5a + 2)(a - 4)$

49. $(3b^3 - 5b^2 + 7)(6b - 1)$

50. $(y^3 + 2y^2 - 3y + 1)(y + 2)$

51. $(2a^3 - 3a^2 + 2a - 1)(2a - 3)$

52. If a polynomial of degree 3 is multiplied by a polynomial of degree 2, what is the degree of the resulting polynomial?

OBJECTIVE C To multiply two binomials using the FOIL method

For Exercises 53 to 84, multiply.

53. $(x + 1)(x + 3)$

54. $(y + 2)(y + 5)$

55. $(a - 3)(a + 4)$

56. $(b - 6)(b + 3)$

57. $(y + 3)(y - 8)$

58. $(x + 10)(x - 5)$

59. $(y - 7)(y - 3)$

60. $(a - 8)(a - 9)$

61. $(2x + 1)(x + 7)$

62. $(y + 2)(5y + 1)$

63. $(3x - 1)(x + 4)$

64. $(7x - 2)(x + 4)$

65. $(4x - 3)(x - 7)$

66. $(2x - 3)(4x - 7)$

67. $(3y - 8)(y + 2)$

68. $(5y - 9)(y + 5)$

69. $(3x + 7)(3x + 11)$

70. $(5a + 6)(6a + 5)$

71. $(7a - 16)(3a - 5)$

72. $(5a - 12)(3a - 7)$

73. $(3a - 2b)(2a - 7b)$

74. $(5a - b)(7a - b)$

75. $(a - 9b)(2a + 7b)$

76. $(2a + 5b)(7a - 2b)$

77. $(10a - 3b)(10a - 7b)$

78. $(12a - 5b)(3a - 4b)$

79. $(5x + 12y)(3x + 4y)$

80. $(11x + 2y)(3x + 7y)$

81. $(2x - 15y)(7x + 4y)$

82. $(5x + 2y)(2x - 5y)$

83. $(8x - 3y)(7x - 5y)$

84. $(2x - 9y)(8x - 3y)$

85. What polynomial has quotient $3x - 4$ when divided by $4x + 5$?

OBJECTIVE D To multiply binomials that have special products

For Exercises 86 to 93, multiply.

86. $(y - 5)(y + 5)$

87. $(y + 6)(y - 6)$

88. $(2x + 3)(2x - 3)$

89. $(4x - 7)(4x + 7)$

90. $(3x - 7)(3x + 7)$

91. $(9x - 2)(9x + 2)$

92. $(4 - 3y)(4 + 3y)$

93. $(4x - 9y)(4x + 9y)$

For Exercises 94 to 101, expand.

94. $(x + 1)^2$

95. $(y - 3)^2$

96. $(3a - 5)^2$

97. $(6x - 5)^2$

98. $(x + 3y)^2$

99. $(x - 2y)^2$

100. $(5x + 2y)^2$

101. $(2a - 9b)^2$

102. Simplify: $(a + b)^2 - (a - b)^2$

103. Expand: $(a + 3)^3$

OBJECTIVE E To solve application problems

104. Geometry The length of a rectangle is $(5x)$ ft. The width is $(2x - 7)$ ft. Find the area of the rectangle in terms of the variable x.

5x

2x − 7

105. **Geometry** The width of a rectangle is $(3x + 1)$ in. The length of the rectangle is twice the width. Find the area of the rectangle in terms of the variable x.

106. **Geometry** The length of a side of a square is $(2x + 1)$ km. Find the area of the square in terms of the variable x.

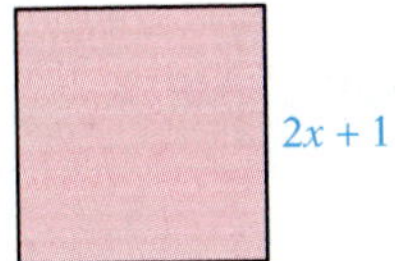

107. **Geometry** The radius of a circle is $(x + 4)$ cm. Find the area of the circle in terms of the variable x. Leave the answer in terms of π.

108. **Geometry** The base of a triangle is $(4x)$ m and the height is $(2x + 5)$ m. Find the area of the triangle in terms of the variable x.

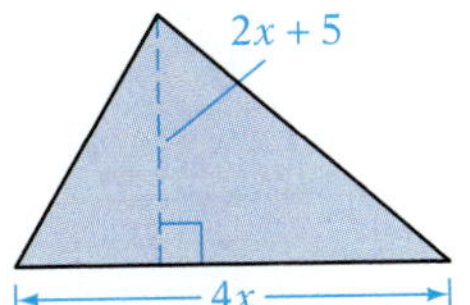

109. **Sports** A softball diamond has dimensions 45 ft by 45 ft. A base-path border x feet wide lies on both the first-base side and the third-base side of the diamond. Express the total area of the softball diamond and the base paths in terms of the variable x.

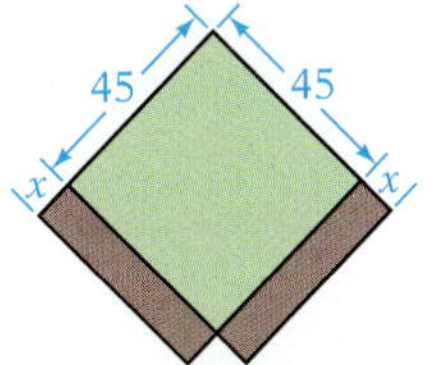

110. **Sports** An athletic field has dimensions 30 yd by 100 yd. An end zone that is w yards wide borders each end of the field. Express the total area of the field and the end zones in terms of the variable w.

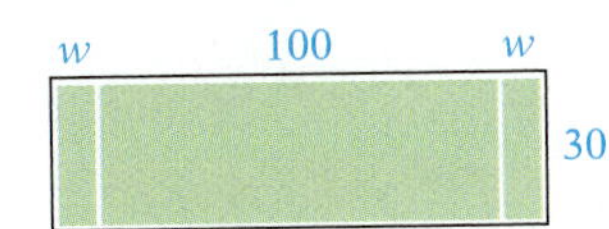

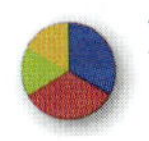

111. **The Olympics** See the news clipping at the right. The Water Cube is not actually a cube because its height is not equal to its length and width. The length of the Water Cube is 22 ft more than five times the height. (*Source:* Structurae)

a. Express the length of the Water Cube in terms of the height h.

b. Express the area of one exterior wall of the Water Cube in terms of the height h.

Christian Kober/Robert Harding World Imagery/Getty Images

The Water Cube

In the News

Olympic Water Cube Completed

The National Aquatics Center, also known as the Water Cube, was completed on the morning of December 26, 2006. Built in Beijing, China, for the 2008 Olympics, the Water Cube is designed to look like a "cube" of water molecules.

Source: Structurae

112. The expression $w(3w - 1)$ cm^2 represents the area of a rectangle of width w. Describe in words the relationship between the length and width of the rectangle.

Applying the Concepts

113. Add $x^2 + 2x - 3$ to the product of $2x - 5$ and $3x + 1$.

114. Subtract $4x^2 - x - 5$ from the product of $x^2 + x + 3$ and $x - 4$.

SECTION

9.4 Integer Exponents and Scientific Notation

OBJECTIVE A To divide monomials 

The quotient of two exponential expressions with the same base can be simplified by writing each expression in factored form, dividing by the common factors, and then writing the result with an exponent.

$$\frac{x^5}{x^2} = \frac{\overset{1}{\cancel{x}} \cdot \overset{1}{\cancel{x}} \cdot x \cdot x \cdot x}{\underset{1}{\cancel{x}} \cdot \underset{1}{\cancel{x}}} = x^3$$

Note that subtracting the exponents gives the same result.

$$\frac{x^5}{x^2} = x^{5-2} = x^3$$

To divide two monomials with the same base, subtract the exponents of the like bases.

HOW TO • 1 Simplify: $\frac{a^7}{a^3}$

$$\frac{a^7}{a^3} = a^{7-3}$$ • The bases are the same. Subtract the exponents.

$$= a^4$$

HOW TO • 2 Simplify: $\frac{r^8t^6}{r^7t}$

$$\frac{r^8t^6}{r^7t} = r^{8-7}t^{6-1}$$ • Subtract the exponents of the like bases.

$$= rt^5$$

HOW TO • 3 Simplify: $\frac{p^7}{z^4}$

Because the bases are not the same, $\frac{p^7}{z^4}$ is already in simplest form.

Consider the expression $\frac{x^4}{x^4}$, $x \neq 0$. This expression can be simplified, as shown below, by subtracting exponents or by dividing by common factors.

$$\frac{x^4}{x^4} = x^{4-4} = x^0 \qquad \frac{x^4}{x^4} = \frac{\overset{1}{\cancel{x}} \cdot \overset{1}{\cancel{x}} \cdot \overset{1}{\cancel{x}} \cdot \overset{1}{\cancel{x}}}{\underset{1}{\cancel{x}} \cdot \underset{1}{\cancel{x}} \cdot \underset{1}{\cancel{x}} \cdot \underset{1}{\cancel{x}}} = 1$$

The equations $\frac{x^4}{x^4} = x^0$ and $\frac{x^4}{x^4} = 1$ suggest the following definition of x^0.

Definition of Zero as an Exponent

If $x \neq 0$, then $x^0 = 1$. The expression 0^0 is not defined.

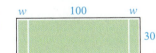

Take Note

In the example at the right, we indicate that $a \neq 0$. If we try to evaluate $(12a^3)^0$ when $a = 0$, we have

$$[12(0)^3]^0 = [12(0)]^0 = 0^0$$

However, 0^0 is not defined. Therefore, we must assume that $a \neq 0$. To avoid stating this for every example or exercise, we will assume that variables do not take on values that result in the expression 0^0.

HOW TO 4 Simplify: $(12a^3)^0, a \neq 0$

$(12a^3)^0 = 1$ • Any nonzero expression to the zero power is 1.

HOW TO 5 Simplify: $-(4x^3y^7)^0$

$-(4x^3y^7)^0 = -(1) = -1$

Consider the expression $\frac{x^4}{x^6}$, $x \neq 0$. This expression can be simplified, as shown below, by subtracting exponents or by dividing by common factors.

$$\frac{x^4}{x^6} = x^{4-6} = x^{-2} \qquad \frac{x^4}{x^6} = \frac{\overset{1}{\cancel{x}} \cdot \overset{1}{\cancel{x}} \cdot \overset{1}{\cancel{x}} \cdot \overset{1}{\cancel{x}}}{\underset{1}{\cancel{x}} \cdot \underset{1}{\cancel{x}} \cdot \underset{1}{\cancel{x}} \cdot \underset{1}{\cancel{x}} \cdot x \cdot x} = \frac{1}{x^2}$$

Point of Interest

In the 15th century, the expression $12^{2\overline{m}}$ was used to mean $12x^{-2}$. The use of $\overline{m}$ reflected an Italian influence. In Italy, *m* was used for minus and *p* was used for plus. It was understood that $2\overline{m}$ referred to an unnamed variable. Issac Newton, in the 17th century, advocated the negative exponent notation that we currently use.

The equations $\frac{x^4}{x^6} = x^{-2}$ and $\frac{x^4}{x^6} = \frac{1}{x^2}$ suggest that $x^{-2} = \frac{1}{x^2}$.

Definition of a Negative Exponent

If $x \neq 0$ and *n* is a positive integer, then

$$x^{-n} = \frac{1}{x^n} \quad \text{and} \quad \frac{1}{x^{-n}} = x^n$$

An exponential expression is in simplest form when it is written with only positive exponents.

Take Note

Note from the example at the right that 2^{-4} is a positive number. A negative exponent does not change the sign of a number.

HOW TO 6 Evaluate 2^{-4}.

$2^{-4} = \frac{1}{2^4}$ • Use the Definition of a Negative Exponent.

$= \frac{1}{16}$ • Evaluate the expression.

Take Note

For the expression $3n^{-5}$, the exponent on *n* is -5 (*negative* 5). The n^{-5} is written in the denominator as n^5. The exponent on 3 is 1 (*positive* 1). The 3 remains in the numerator. Also, we indicated that $n \neq 0$. This is done because division by zero is not defined. In this textbook, we will assume that values of the variables are chosen so that division by zero does not occur.

HOW TO 7 Simplify: $3n^{-5}, n \neq 0$

$3n^{-5} = 3 \cdot \frac{1}{n^5} = \frac{3}{n^5}$ • Use the Definition of a Negative Exponent to rewrite the expression with a positive exponent.

HOW TO 8 Simplify: $\frac{2}{5a^{-4}}$

$\frac{2}{5a^{-4}} = \frac{2}{5} \cdot \frac{1}{a^{-4}} = \frac{2}{5} \cdot a^4 = \frac{2a^4}{5}$ • Use the Definition of a Negative Exponent to rewrite the expression with a positive exponent.

The expression $\left(\frac{x^4}{y^3}\right)^2$, $y \neq 0$, can be simplified by squaring $\frac{x^4}{y^3}$ or by multiplying each exponent in the quotient by the exponent outside the parentheses.

$$\left(\frac{x^4}{y^3}\right)^2 = \left(\frac{x^4}{y^3}\right)\left(\frac{x^4}{y^3}\right) = \frac{x^4 \cdot x^4}{y^3 \cdot y^3} = \frac{x^{4+4}}{y^{3+3}} = \frac{x^8}{y^6} \qquad \left(\frac{x^4}{y^3}\right)^2 = \frac{x^{4\cdot 2}}{y^{3\cdot 2}} = \frac{x^8}{y^6}$$

Rule for Simplifying the Power of a Quotient

If m, n, and p are integers and $y \neq 0$, then $\left(\frac{x^m}{y^n}\right)^p = \frac{x^{mp}}{y^{np}}$.

Take Note

As a reminder, although it is not stated, we are assuming that $a \neq 0$ and $b \neq 0$. This assumption is made to ensure that we do not have division by zero.

HOW TO • 9 Simplify: $\left(\frac{a^3}{b^2}\right)^{-2}$

$\left(\frac{a^3}{b^2}\right)^{-2} = \frac{a^{3(-2)}}{b^{2(-2)}}$ • Use the **Rule for Simplifying the Power of a Quotient.**

$= \frac{a^{-6}}{b^{-4}} = \frac{b^4}{a^6}$ • Use the Definition of a Negative Exponent to write the expression with positive exponents.

The example above suggests the following rule.

Rule for Negative Exponents on Fractional Expressions

If $a \neq 0$, $b \neq 0$, and n is a positive integer, then

$$\left(\frac{a}{b}\right)^{-n} = \left(\frac{b}{a}\right)^n$$

Now that zero as an exponent and negative exponents have been defined, a rule for dividing exponential expressions can be stated.

Rule for Dividing Exponential Expressions

If m and n are integers and $x \neq 0$, then $\frac{x^m}{x^n} = x^{m-n}$.

HOW TO • 10 Evaluate $\frac{5^{-2}}{5}$.

$\frac{5^{-2}}{5} = 5^{-2-1} = 5^{-3}$ • Use the **Rule for Dividing Exponential Expressions.**

$= \frac{1}{5^3} = \frac{1}{125}$ • Use the Definition of a Negative Exponent to rewrite the expression with a positive exponent. Then evaluate.

HOW TO • 11 Simplify: $\frac{x^4}{x^9}$

$\frac{x^4}{x^9} = x^{4-9}$ • Use the **Rule for Dividing Exponential Expressions.**

$= x^{-5}$ • Subtract the exponents.

$= \frac{1}{x^5}$ • Use the Definition of a Negative Exponent to rewrite the expression with a positive exponent.

The rules for simplifying exponential expressions and powers of exponential expressions are true for all integers. These rules are restated here, along with the rules for dividing exponential expressions.

Rules of Exponents

If m, n, and p are integers, then

$x^m \cdot x^n = x^{m+n}$ $\quad (x^m)^n = x^{mn}$ $\quad (x^m y^n)^p = x^{mp} y^{np}$

$\frac{x^m}{x^n} = x^{m-n}, x \neq 0$ $\quad \left(\frac{x^m}{y^n}\right)^p = \frac{x^{mp}}{y^{np}}, y \neq 0$ $\quad x^{-n} = \frac{1}{x^n}, x \neq 0$

$x^0 = 1, x \neq 0$

HOW TO • 12 Simplify: $(3ab^{-4})(-2a^{-3}b^7)$

$(3ab^{-4})(-2a^{-3}b^7) = [3 \cdot (-2)](a^{1+(-3)}b^{-4+7})$ • When multiplying exponential expressions, **add the exponents on like bases.**

$= -6a^{-2}b^3$

$= -\frac{6b^3}{a^2}$

HOW TO • 13 Simplify: $\left[\frac{6m^2n^3}{8m^7n^2}\right]^{-3}$

$\left[\frac{6m^2n^3}{8m^7n^2}\right]^{-3} = \left[\frac{3m^{2-7}n^{3-2}}{4}\right]^{-3}$ • Simplify inside the brackets.

$= \left[\frac{3m^{-5}n}{4}\right]^{-3}$ • Subtract the exponents.

$= \frac{3^{-3}m^{15}n^{-3}}{4^{-3}}$ • Use the Rule for Simplifying the Power of a Quotient.

$= \frac{4^3m^{15}}{3^3n^3} = \frac{64m^{15}}{27n^3}$ • Use the Definition of a Negative Exponent to rewrite the expression with positive exponents. Then simplify.

HOW TO • 14 Simplify: $\dfrac{4a^{-2}b^5}{6a^5b^2}$

$$\frac{4a^{-2}b^5}{6a^5b^2} = \frac{2a^{-2}b^5}{3a^5b^2}$$

- Divide the coefficients by their common factor.

$$= \frac{2a^{-2-5}b^{5-2}}{3}$$

- Use the Rule for Dividing Exponential Expressions.

$$= \frac{2a^{-7}b^3}{3} = \frac{2b^3}{3a^7}$$

- Use the Definition of a Negative Exponent to rewrite the expression with positive exponents.

EXAMPLE • 1

Simplify: $(-2x)(3x^{-2})^{-3}$

Solution

$$(-2x)(3x^{-2})^{-3} = (-2x)(3^{-3}x^6)$$

- Rule for Simplifying the Power of a Product

$$= \frac{-2x^{1+6}}{3^3}$$

$$= -\frac{2x^7}{27}$$

YOU TRY IT • 1

Simplify: $(-2x^2)(x^{-3}y^{-4})^{-2}$

Your solution

EXAMPLE • 2

Simplify: $\dfrac{(2r^2t^{-1})^{-3}}{(r^{-3}t^4)^2}$

Solution

$$\frac{(2r^2t^{-1})^{-3}}{(r^{-3}t^4)^2} = \frac{2^{-3}r^{-6}t^3}{r^{-6}t^8}$$

- Rule for Simplifying the Power of a Product

$$= 2^{-3}r^{-6-(-6)}t^{3-8}$$

- Rule for Dividing Exponential Expressions

$$= 2^{-3}r^0t^{-5}$$

$$= \frac{1}{2^3t^5}$$

$$= \frac{1}{8t^5}$$

- Write the answer in simplest form.

YOU TRY IT • 2

Simplify: $\dfrac{(6a^{-2}b^3)^{-1}}{(4a^3b^{-2})^{-2}}$

Your solution

EXAMPLE • 3

Simplify: $\left[\dfrac{4a^{-2}b^3}{6a^4b^{-2}}\right]^{-3}$

Solution

$$\left[\frac{4a^{-2}b^3}{6a^4b^{-2}}\right]^{-3} = \left[\frac{2a^{-6}b^5}{3}\right]^{-3}$$

- Simplify inside brackets.

$$= \frac{2^{-3}a^{18}b^{-15}}{3^{-3}}$$

- Rule for Simplifying the Power of a Quotient

$$= \frac{27a^{18}}{8b^{15}}$$

- Write answer in simplest form.

YOU TRY IT • 3

Simplify: $\left[\dfrac{6r^3s^{-3}}{9r^3s^{-1}}\right]^{-2}$

Your solution

Solutions on p. S22

OBJECTIVE B To write a number in scientific notation

Integrating Technology

See the Keystroke Guide: *Scientific Notation* for instructions on entering a number written in scientific notation into a calculator.

Very large and very small numbers abound in the natural sciences. For example, the mass of an electron is 0.000000000000000000000000000000911 kg. Numbers such as this are difficult to read, so a more convenient system called **scientific notation** is used. In scientific notation, a number is expressed as the product of two factors, one a number between 1 and 10, and the other a power of 10.

To express a number in scientific notation, write it in the form $a \times 10^n$, where a is a number between 1 and 10, and n is an integer.

For numbers greater than or equal to 10, move the decimal point to the right of the first digit. The exponent n is positive and equal to the number of places the decimal point has been moved.

$240{,}000 = 2.4 \times 10^5$

$93{,}000{,}000 = 9.3 \times 10^7$

For numbers less than 1, move the decimal point to the right of the first nonzero digit. The exponent n is negative. The absolute value of the exponent is equal to the number of places the decimal point has been moved.

$0.0003 = 3 \times 10^{-4}$

$0.0000832 = 8.32 \times 10^{-5}$

Point of Interest

An electron microscope uses wavelengths that are approximately 4×10^{-12} meter to make images of viruses.

The human eye can detect wavelengths between 4.3×10^{-7} meter and 6.9×10^{-7} meter. Although these wavelengths are very short, they are approximately 10^5 times longer than the wavelengths used in an electron microscope.

Changing a number written in scientific notation to decimal notation also requires moving the decimal point.

When the exponent is positive, move the decimal point to the right the same number of places as the exponent.

$3.45 \times 10^6 = 3{,}450{,}000$

$2.3 \times 10^8 = 230{,}000{,}000$

When the exponent is negative, move the decimal point to the left the same number of places as the absolute value of the exponent.

$8.1 \times 10^{-3} = 0.0081$

$6.34 \times 10^{-7} = 0.000000634$

EXAMPLE 4

Write the number 824,300,000 in scientific notation.

Solution

$824{,}300{,}000 = 8.243 \times 10^8$

YOU TRY IT 4

Write the number 0.000000961 in scientific notation.

Your solution

EXAMPLE 5

Write the number 6.8×10^{-10} in decimal notation.

Solution

$6.8 \times 10^{-10} = 0.00000000068$

YOU TRY IT 5

Write the number 7.329×10^6 in decimal notation.

Your solution

Solutions on p. S22

9.4 EXERCISES

OBJECTIVE A To divide monomials

For Exercises 1 to 36, simplify.

1. $\frac{y^7}{y^3}$ **2.** $\frac{z^9}{z^2}$ **3.** $\frac{a^8}{a^5}$ **4.** $\frac{c^{12}}{c^5}$

5. $\frac{p^5}{p}$ **6.** $\frac{w^9}{w}$ **7.** $\frac{4x^8}{2x^5}$ **8.** $\frac{12z^7}{4z^3}$

9. $\frac{22k^5}{11k^4}$ **10.** $\frac{14m^{11}}{7m^{10}}$ **11.** $\frac{m^9n^7}{m^4n^5}$ **12.** $\frac{y^5z^6}{yz^3}$

13. $\frac{6r^4}{4r^2}$ **14.** $\frac{8x^9}{12x^6}$ **15.** $\frac{-16a^7}{24a^6}$ **16.** $\frac{-18b^5}{27b^4}$

17. $\frac{y^3}{y^8}$ **18.** $\frac{z^4}{z^6}$ **19.** $\frac{a^5}{a^{11}}$ **20.** $\frac{m}{m^7}$

21. $\frac{4x^2}{12x^5}$ **22.** $\frac{6y^8}{8y^9}$ **23.** $\frac{-12x}{-18x^6}$ **24.** $\frac{-24c^2}{-36c^{11}}$

25. $\frac{x^6y^5}{x^8y}$ **26.** $\frac{a^3b^2}{a^2b^3}$ **27.** $\frac{2m^6n^2}{5m^9n^{10}}$ **28.** $\frac{5r^3t^7}{6r^5t^7}$

29. $\frac{pq^3}{p^4q^4}$ **30.** $\frac{a^4b^5}{a^5b^6}$ **31.** $\frac{3x^4y^5}{6x^4y^8}$ **32.** $\frac{14a^3b^6}{21a^5b^6}$

33. $\frac{14x^4y^6z^2}{16x^3y^9z}$ **34.** $\frac{24a^2b^7c^9}{36a^7b^5c}$ **35.** $\frac{15mn^9p^3}{30m^4n^9p}$ **36.** $\frac{25x^4y^7z^2}{20x^5y^9z^{11}}$

For Exercises 37 to 44, evaluate.

37. 5^{-2} **38.** 3^{-3} **39.** $\frac{1}{8^{-2}}$ **40.** $\frac{1}{12^{-1}}$

41. $\dfrac{3^{-2}}{3}$

42. $\dfrac{5^{-3}}{5}$

43. $\dfrac{2^{-2}}{2^{-3}}$

44. $\dfrac{3^2}{3^2}$

For Exercises 45 to 92, simplify.

45. x^{-2}

46. y^{-10}

47. $\dfrac{1}{a^{-6}}$

48. $\dfrac{1}{b^{-4}}$

49. $4x^{-7}$

50. $-6y^{-1}$

51. $\dfrac{2}{3}z^{-2}$

52. $\dfrac{4}{5}a^{-4}$

53. $\dfrac{5}{b^{-8}}$

54. $\dfrac{-3}{y^{-3}}$

55. $\dfrac{1}{3x^{-2}}$

56. $\dfrac{2}{5c^{-6}}$

57. $(ab^5)^0$

58. $(32x^3y^4)^0$

59. $-(3p^2q^5)^0$

60. $-\left(\dfrac{2}{3}xy\right)^0$

61. $(-2xy^{-2})^3$

62. $(-3x^{-1}y^2)^2$

63. $(3x^{-1}y^{-2})^2$

64. $(5xy^{-3})^{-2}$

65. $(2x^{-1})(x^{-3})$

66. $(-2x^{-5})x^7$

67. $(-5a^2)(a^{-5})^2$

68. $(2a^{-3})(a^7b^{-1})^3$

69. $(-2ab^{-2})(4a^{-2}b)^{-2}$

70. $(3ab^{-2})(2a^{-1}b)^{-3}$

71. $(-5x^{-2}y)(-2x^{-2}y^2)$

72. $\dfrac{a^{-3}b^{-4}}{a^2b^2}$

73. $\dfrac{3x^{-2}y^2}{6xy^2}$

74. $\dfrac{2x^{-2}y}{8xy}$

75. $\dfrac{3x^{-2}y}{xy}$

76. $\dfrac{2x^{-1}y^4}{x^2y^3}$

77. $\dfrac{2x^{-1}y^{-4}}{4xy^2}$

78. $\dfrac{(x^{-1}y)^2}{xy^2}$

79. $\dfrac{(x^{-2}y)^2}{x^2y^3}$

80. $\dfrac{(x^{-3}y^{-2})^2}{x^6y^8}$

81. $\dfrac{(a^{-2}y^3)^{-3}}{a^2y}$

82. $\dfrac{12a^2b^3}{-27a^2b^2}$

83. $\dfrac{-16xy^4}{96x^4y^4}$

84. $\dfrac{-8x^2y^4}{44y^2z^5}$

85. $\dfrac{22a^2b^4}{-132b^3c^2}$

86. $\dfrac{-(8a^2b^4)^3}{64a^3b^8}$

87. $\dfrac{-(14ab^4)^2}{28a^4b^2}$

88. $\dfrac{(2a^{-2}b^3)^{-2}}{(4a^2b^{-4})^{-1}}$

89. $\dfrac{(3^{-1}r^4s^{-3})^{-2}}{(6r^2s^{-1}t^{-2})^2}$

90. $\left(\dfrac{6x^{-4}yz^{-1}}{14xy^{-4}z^2}\right)^{-3}$

91. $\left(\dfrac{15m^3n^{-2}p^{-1}}{25m^{-2}n^{-4}}\right)^{-3}$

92. $\left(\dfrac{18a^4b^{-2}c^4}{12ab^{-3}d^2}\right)^{-2}$

For Exercises 93 to 96, state whether the equation is true or false for all $a \neq 0$ and $b \neq 0$.

93. $\dfrac{a^{4n}}{a^n} = a^4$

94. $a^{n-m} = \dfrac{1}{a^{m-n}}$

95. $a^{-n}a^n = 1$

96. $\dfrac{a^n}{b^m} = \left(\dfrac{a}{b}\right)^{m-n}$

OBJECTIVE B To write a number in scientific notation

For Exercises 97 to 105, write in scientific notation.

97. 0.00000000324

98. 0.00000012

99. 0.000000000000000003

100. 1,800,000,000

101. 32,000,000,000,000,000

102. 76,700,000,000,000

103. 0.000000000000000000122

104. 0.00137

105. 547,000,000

For Exercises 106 to 114, write in decimal notation.

106. 2.3×10^{-12}

107. 1.67×10^{-4}

108. 2×10^{15}

109. 6.8×10^7

110. 9×10^{-21}

111. 3.05×10^{-5}

112. 9.05×10^{11}

113. 1.02×10^{-9}

114. 7.2×10^{-3}

115. If n is a negative integer, how many zeros appear after the decimal point when 1.35×10^n is written in decimal notation?

116. If n is a positive integer greater than 1, how many zeros appear before the decimal point when 1.35×10^n is written in decimal notation?

117. **Technology** See the news clipping at the right. Express in scientific notation the thickness, in meters, of the memristor.

In the News

HP Introduces the Memristor

Hewlett Packard has announced the design of the *memristor,* a new memory technology with the potential to be much smaller than the memory chips used in today's computers. HP has made a memristor with a thickness of 0.000000015 m (15 nanometers).

Source: The New York Times

AP Photo/Paul Sakuma

HP Researchers View Image of Memristor

118. **Geology** The approximate mass of the planet Earth is 5,980,000,000,000,000,000,000,000 kg. Write the mass of Earth in scientific notation.

119. **Physics** The length of an infrared light wave is approximately 0.0000037 m. Write this number in scientific notation.

120. **Electricity** The electric charge on an electron is 0.00000000000000000016 coulomb. Write this number in scientific notation.

121. **Physics** Light travels approximately 16,000,000,000 mi in 1 day. Write this number in scientific notation.

122. **Astronomy** One light-year is the distance traveled by light in 1 year. One light-year is 5,880,000,000,000 mi. Write this number in scientific notation.

123. **Astronomy** See the news clipping at the right. WASP-12b orbits a star that is 5.1156×10^{15} mi from Earth. (*Source:* news.yahoo.com) Write this number in decimal notation.

In the News

Hottest Planet Ever Discovered

A planet called WASP-12b is the hottest planet ever discovered, at about 4000°F. It orbits its star faster than any other known planet, completing a revolution once a day.

Source: news.yahoo.com

124. **Chemistry** Approximately 35 teragrams (3.5×10^{13} g) of sulfur in the atmosphere are converted to sulfate each year. Write this number in decimal notation.

Applying the Concepts

125. Evaluate 2^x when $x = -2, -1, 0, 1,$ and 2.

126. Evaluate 2^{-x} when $x = -2, -1, 0, 1,$ and 2.

SECTION

9.5 Division of Polynomials

OBJECTIVE A To divide a polynomial by a monomial 

To divide a polynomial by a monomial, divide each term in the numerator by the denominator and write the sum of the quotients.

HOW TO • 1 Divide: $\frac{6x^3 - 3x^2 + 9x}{3x}$

$$\frac{6x^3 - 3x^2 + 9x}{3x} = \frac{6x^3}{3x} - \frac{3x^2}{3x} + \frac{9x}{3x}$$

• Divide each term of the polynomial by the monomial.

$$= 2x^2 - x + 3$$

• Simplify each term.

EXAMPLE • 1

Divide: $\frac{12x^2y - 6xy + 4x^2}{2xy}$

Solution

$$\frac{12x^2y - 6xy + 4x^2}{2xy} = \frac{12x^2y}{2xy} - \frac{6xy}{2xy} + \frac{4x^2}{2xy} = 6x - 3 + \frac{2x}{y}$$

YOU TRY IT • 1

Divide: $\frac{24x^2y^2 - 18xy + 6y}{6xy}$

Your solution

Solution on p. S22

OBJECTIVE B To divide polynomials 

Tips for Success

An important element of success is practice. We cannot do anything well if we do not practice it repeatedly. Practice is crucial to success in mathematics. In this objective you are learning a new skill, how to divide polynomials. You will need to practice this skill over and over again in order to be successful at it.

The procedure for dividing two polynomials is similar to the one for dividing whole numbers. The same equation used to check division of whole numbers is used to check polynomial division.

(Quotient × divisor) + remainder = dividend

HOW TO • 2 Divide: $(x^2 - 5x + 8) \div (x - 3)$

Step 1

$$\begin{array}{r} x \phantom{{}-5x+8} \\ x - 3\overline{)x^2 - 5x + 8} \\ \underline{x^2 - 3x} \phantom{{}+8} \\ -2x + 8 \end{array}$$

• Think: $x\overline{)x^2} = \frac{x^2}{x} = x$

• Multiply: $x(x - 3) = x^2 - 3x$

• Subtract: $(x^2 - 5x) - (x^2 - 3x) = -2x$
Bring down the 8.

Step 2

$$\begin{array}{r} x - 2 \\ x - 3\overline{)x^2 - 5x + 8} \\ \underline{x^2 - 3x} \phantom{{}+8} \\ -2x + 8 \\ \underline{-2x + 6} \\ 2 \end{array}$$

• Think: $x\overline{)-2x} = \frac{-2x}{x} = -2$

• Multiply: $-2(x - 3) = -2x + 6$

• Subtract: $(-2x + 8) - (-2x + 6) = 2$

• The remainder is 2.

Check: $(x - 2)(x - 3) + 2 = x^2 - 5x + 6 + 2 = x^2 - 5x + 8$

$$(x^2 - 5x + 8) \div (x - 3) = x - 2 + \frac{2}{x - 3}$$

If a term is missing from the dividend, a zero can be inserted for that term. This helps keep like terms in the same column.

Take Note

Recall that a fraction bar means "divided by." Therefore, $6 \div 2$ can be written $\frac{6}{2}$, and $a \div b$ can be written $\frac{a}{b}$.

HOW TO • 3 Divide: $\dfrac{6x + 26 + 2x^3}{2 + x}$

$$\frac{2x^3 + 6x + 26}{x + 2}$$

• Arrange the terms of each polynomial in descending order.

$$\begin{array}{r}
2x^2 - 4x + 14 \\
x + 2\overline{)2x^3 + 0 \quad + 6x + 26} \\
\underline{2x^3 + 4x^2} \qquad\qquad\quad \\
-4x^2 + 6x \qquad \\
\underline{-4x^2 - 8x} \qquad \\
14x + 26 \\
\underline{14x + 28} \\
-2
\end{array}$$

• There is no x^2 term in $2x^3 + 6x + 26$. Insert a **zero** for the missing term.

Check:
$(2x^2 - 4x + 14)(x + 2) + (-2) = (2x^3 + 6x + 28) + (-2) = 2x^3 + 6x + 26$

$(2x^3 + 6x + 26) \div (x + 2) = 2x^2 - 4x + 14 - \dfrac{2}{x + 2}$

EXAMPLE • 2

Divide: $(8x^2 + 4x^3 + x - 4) \div (2x + 3)$

Solution

$$\begin{array}{r}
2x^2 + x - 1 \\
2x + 3\overline{)4x^3 + 8x^2 + x - 4} \\
\underline{4x^3 + 6x^2} \qquad\qquad \\
2x^2 + x \qquad \\
\underline{2x^2 + 3x} \qquad \\
-2x - 4 \\
\underline{-2x - 3} \\
-1
\end{array}$$

• Write the dividend in descending powers of x.

$(4x^3 + 8x^2 + x - 4) \div (2x + 3)$
$= 2x^2 + x - 1 - \dfrac{1}{2x + 3}$

YOU TRY IT • 2

Divide: $(2x^3 + x^2 - 8x - 3) \div (2x - 3)$

Your solution

EXAMPLE • 3

Divide: $\dfrac{x^2 - 1}{x + 1}$

Solution

$$\begin{array}{r}
x - 1 \\
x + 1\overline{)x^2 + 0 - 1} \\
\underline{x^2 + x} \qquad \\
-x - 1 \\
\underline{-x - 1} \\
0
\end{array}$$

• Insert a **zero** for the missing term.

$(x^2 - 1) \div (x + 1) = x - 1$

YOU TRY IT • 3

Divide: $\dfrac{x^3 - 2x + 1}{x - 1}$

Your solution

Solutions on p. S22

9.5 EXERCISES

OBJECTIVE A To divide a polynomial by a monomial

1. Every division problem has a related multiplication problem. What is the related multiplication problem for the division problem $\dfrac{15x^2 + 12x}{3x} = 5x + 4$?

For Exercises 2 to 22, divide.

2. $\dfrac{10a - 25}{5}$

3. $\dfrac{16b - 40}{8}$

4. $\dfrac{6y^2 + 4y}{y}$

5. $\dfrac{4b^3 - 3b}{b}$

6. $\dfrac{3x^2 - 6x}{3x}$

7. $\dfrac{10y^2 - 6y}{2y}$

8. $\dfrac{5x^2 - 10x}{-5x}$

9. $\dfrac{3y^2 - 27y}{-3y}$

10. $\dfrac{x^3 + 3x^2 - 5x}{x}$

11. $\dfrac{a^3 - 5a^2 + 7a}{a}$

12. $\dfrac{x^6 - 3x^4 - x^2}{x^2}$

13. $\dfrac{a^8 - 5a^5 - 3a^3}{a^2}$

14. $\dfrac{5x^2y^2 + 10xy}{5xy}$

15. $\dfrac{8x^2y^2 - 24xy}{8xy}$

16. $\dfrac{9y^6 - 15y^3}{-3y^3}$

17. $\dfrac{4x^4 - 6x^2}{-2x^2}$

18. $\dfrac{3x^2 - 2x + 1}{x}$

19. $\dfrac{8y^2 + 2y - 3}{y}$

20. $\dfrac{16a^2b - 20ab + 24ab^2}{4ab}$

21. $\dfrac{22a^2b - 11ab - 33ab^2}{11ab}$

22. $\dfrac{5a^2b - 15ab + 30ab^2}{5ab}$

OBJECTIVE B To divide polynomials

For Exercises 23 to 49, divide.

23. $(b^2 - 14b + 49) \div (b - 7)$

24. $(x^2 - x - 6) \div (x - 3)$

25. $(y^2 + 2y - 35) \div (y + 7)$

26. $(2x^2 + 5x + 2) \div (x + 2)$

27. $(2y^2 - 13y + 21) \div (y - 3)$

28. $(4x^2 - 16) \div (2x + 4)$

29. $\dfrac{2y^2 + 7}{y - 3}$

30. $\dfrac{x^2 + 1}{x - 1}$

31. $\dfrac{x^2 + 4}{x + 2}$

32. $\dfrac{6x^2 - 7x}{3x - 2}$

33. $\dfrac{6y^2 + 2y}{2y + 4}$

34. $\dfrac{5x^2 + 7x}{x - 1}$

35. $(6x^2 - 5) \div (x + 2)$

36. $(a^2 + 5a + 10) \div (a + 2)$

37. $(b^2 - 8b - 9) \div (b - 3)$

38. $\dfrac{2y^2 - 9y + 8}{2y + 3}$

39. $\dfrac{3x^2 + 5x - 4}{x - 4}$

40. $(8x + 3 + 4x^2) \div (2x - 1)$

41. $(10 + 21y + 10y^2) \div (2y + 3)$

42. $\dfrac{15a^2 - 8a - 8}{3a + 2}$

43. $\dfrac{12a^2 - 25a - 7}{3a - 7}$

44. $(5 - 23x + 12x^2) \div (4x - 1)$

45. $(24 + 6a^2 + 25a) \div (3a - 1)$

46. $\dfrac{5x + 3x^2 + x^3 + 3}{x + 1}$

47. $\dfrac{7x + x^3 - 6x^2 - 2}{x - 1}$

48. $(x^4 - x^2 - 6) \div (x^2 + 2)$

49. $(x^4 + 3x^2 - 10) \div (x^2 - 2)$

50. True or false? When a sixth-degree polynomial is divided by a third-degree polynomial, the quotient is a second-degree polynomial.

Applying the Concepts

51. The product of a monomial and $4b$ is $12ab^2$. Find the monomial.

52. In your own words, explain how to divide exponential expressions.

FOCUS ON PROBLEM SOLVING

Dimensional Analysis

In solving application problems, it may be useful to include the units in order to organize the problem so that the answer is in the proper units. Using units to organize and check the correctness of an application is called **dimensional analysis.** We use the operations of multiplying units and dividing units in applying dimensional analysis to application problems.

The Rule for Multiplying Exponential Expressions states that we multiply two expressions with the same base by adding the exponents.

$$x^4 \cdot x^6 = x^{4+6} = x^{10}$$

In calculations that involve quantities, the units are operated on algebraically.

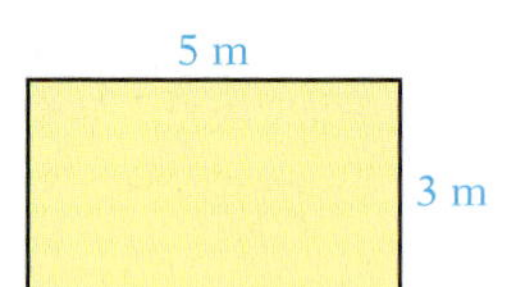

HOW TO • 1 A rectangle measures 3 m by 5 m. Find the area of the rectangle.

$A = LW = (3\text{ m})(5\text{ m}) = (3 \cdot 5)(\text{m} \cdot \text{m}) = 15\text{ m}^2$

The area of the rectangle is 15 m^2 (square meters).

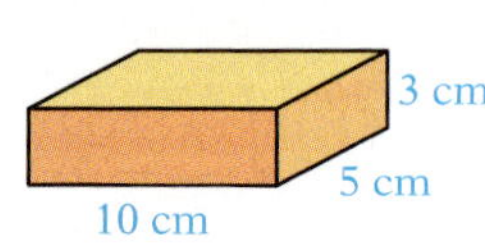

HOW TO • 2 A box measures 10 cm by 5 cm by 3 cm. Find the volume of the box.

$V = LWH = (10\text{ cm})(5\text{ cm})(3\text{ cm}) = (10 \cdot 5 \cdot 3)(\text{cm} \cdot \text{cm} \cdot \text{cm}) = 150\text{ cm}^3$

The volume of the box is 150 cm^3 (cubic centimeters).

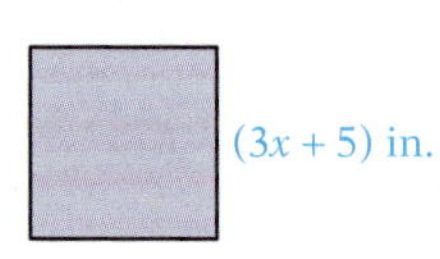

HOW TO • 3 Find the area of a square whose side measures $(3x + 5)$ in.

$A = s^2 = [(3x + 5)\text{ in.}]^2 = (3x + 5)^2\text{ in}^2 = (9x^2 + 30x + 25)\text{ in}^2$

The area of the square is $(9x^2 + 30x + 25)$ in^2 (square inches).

Dimensional analysis is used in the conversion of units.

The following example converts the unit miles to feet. The equivalent measures 1 mi = 5280 ft are used to form the following rates, which are called conversion factors: $\frac{1\text{ mi}}{5280\text{ ft}}$ and $\frac{5280\text{ ft}}{1\text{ mi}}$. Because 1 mi = 5280 ft, both of the conversion factors $\frac{1\text{ mi}}{5280\text{ ft}}$ and $\frac{5280\text{ ft}}{1\text{ mi}}$ are equal to 1.

1 mi
5280 ft

To convert 3 mi to feet, multiply 3 mi by the conversion factor $\frac{5280\text{ ft}}{1\text{ mi}}$.

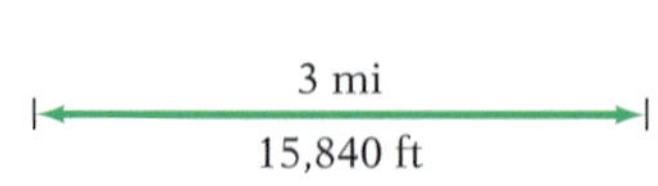

$$3\text{ mi} = 3\text{ mi} \cdot 1 = \frac{3\text{ mi}}{1} \cdot \frac{5280\text{ ft}}{1\text{ mi}} = \frac{3\text{ mi} \cdot 5280\text{ ft}}{1\text{ mi}} = 3 \cdot 5280\text{ ft} = 15{,}840\text{ ft}$$

There are two important points in the above illustration. First, **you can think of dividing the numerator and denominator by the common unit "mile" just as you would divide the numerator and denominator of a fraction by a common factor.**

Second, **the conversion factor $\frac{5280\text{ ft}}{1\text{ mi}}$ is equal to 1, and multiplying an expression by 1 does not change the value of the expression.**

In the application problem that follows, the units are kept in the problem while the problem is worked.

In 2008, a horse named Big Brown ran a 1.25-mile race in 2.02 min. Find Big Brown's average speed for that race in miles per hour. Round to the nearest tenth.

AP Photo/Rob Carr

"Big Brown"

Strategy To find the average speed, use the formula $r = \frac{d}{t}$, where r is the speed, d is the distance, and t is the time. Use the conversion factor $\frac{60 \text{ min}}{1 \text{ h}}$.

Solution

$$r = \frac{d}{t} = \frac{1.25 \text{ mi}}{2.02 \text{ min}} = \frac{1.25 \text{ mi}}{2.02 \text{ min}} \cdot \frac{60 \text{ min}}{1 \text{ h}}$$

$$= \frac{75 \text{ mi}}{2.02 \text{ h}} \approx 37.1 \text{ mph}$$

Big Brown's average speed was 37.1 mph.

Try each of the following problems. Round to the nearest tenth or nearest cent.

1. Convert 88 ft/s to miles per hour.

2. Convert 8 m/s to kilometers per hour (1 km = 1000 m).

3. A carpet is to be placed in a meeting hall that is 36 ft wide and 80 ft long. At $21.50 per square yard, how much will it cost to carpet the meeting hall?

4. A carpet is to be placed in a room that is 20 ft wide and 30 ft long. At $22.25 per square yard, how much will it cost to carpet the area?

5. Find the number of gallons of water in a fish tank that is 36 in. long and 24 in. wide and is filled to a depth of 16 in. (1 gal = 231 in^3).

6. Find the number of gallons of water in a fish tank that is 24 in. long and 18 in. wide and is filled to a depth of 12 in. (1 gal = 231 in^3).

7. A $\frac{1}{4}$-acre commercial lot is on sale for $2.15 per square foot. Find the sale price of the commercial lot (1 acre = 43,560 ft^2).

8. A 0.75-acre industrial parcel was sold for $98,010. Find the parcel's price per square foot (1 acre = 43,560 ft^2).

9. A new driveway will require 800 ft^3 of concrete. Concrete is ordered by the cubic yard. How much concrete should be ordered?

© Duomo/Corbis

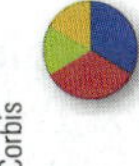

10. A piston-engined dragster traveled 440 yd in 4.936 s at Ennis, Texas, on October 9, 1988. Find the average speed of the dragster in miles per hour.

11. The Marianas Trench in the Pacific Ocean is the deepest part of the ocean. Its depth is 6.85 mi. The speed of sound under water is 4700 ft/s. Find the time it takes sound to travel from the surface of the ocean to the bottom of the Marianas Trench and back.

PROJECTS AND GROUP ACTIVITIES

Diagramming the Square of a Binomial

1. Explain why the diagram at the right represents $(a + b)^2 = a^2 + 2ab + b^2$.

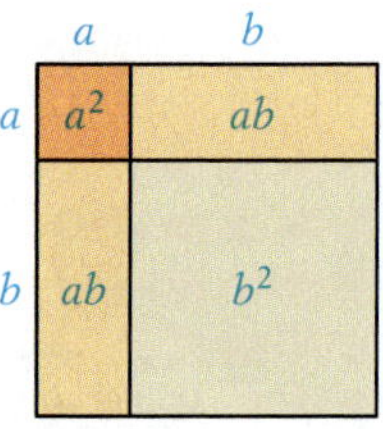

2. Draw similar diagrams representing each of the following.

$(x + 2)^2$

$(x + 4)^2$

Pascal's Triangle

Simplifying the power of a binomial is called *expanding the binomial.* The expansions of the first three powers of a binomial are shown below.

$$(a + b)^1 = a + b$$

$$(a + b)^2 = (a + b)(a + b) = a^2 + 2ab + b^2$$

$$(a + b)^3 = (a + b)^2(a + b) = (a^2 + 2ab + b^2)(a + b) = a^3 + 3a^2b + 3ab^2 + b^3$$

Point of Interest

Pascal did not invent the triangle of numbers known as Pascal's Triangle. It was known to mathematicians in China probably as early as 1050 A.D. But Pascal's *Traite du triangle arithmetique (Treatise Concerning the Arithmetical Triangle)* brought together all the different aspects of the triangle of numbers for the first time.

Find $(a + b)^4$. [*Hint*: $(a + b)^4 = (a + b)^3(a + b)$]

Find $(a + b)^5$. [*Hint*: $(a + b)^5 = (a + b)^4(a + b)$]

If we continue in this way, the results for $(a + b)^6$ are

$$(a + b)^6 = a^6 + 6a^5b + 15a^4b^2 + 20a^3b^3 + 15a^2b^4 + 6ab^5 + b^6$$

Now expand $(a + b)^8$. Before you begin, see whether you can find a pattern that will help you write the expansion of $(a + b)^8$ without having to multiply it out. Here are some hints.

1. Write out the variable terms of each binomial expansion from $(a + b)^1$ through $(a + b)^6$. Observe how the exponents on the variables change.

2. Write out the coefficients of all the terms without the variable parts. It will be helpful if you make a triangular arrangement as shown at the left. Note that each row begins and ends with a 1. Also note (in the two shaded regions, for example) that any number in a row is the sum of the two closest numbers above it. For instance, $1 + 5 = 6$ and $6 + 4 = 10$.

The triangle of numbers shown at the left is called Pascal's Triangle. To find the expansion of $(a + b)^8$, you need to find the eighth row of Pascal's Triangle. First find row seven. Then find row eight and use the patterns you have observed to write the expansion $(a + b)^8$.

Pascal's Triangle has been the subject of extensive analysis, and many patterns have been found. See whether you can find some of them.

CHAPTER 9

SUMMARY

KEY WORDS	EXAMPLES
A *monomial* is a number, a variable, or a product of a number and variables. [9.1A, p. 478]	5 is a number; y is a variable. $2a^3b^2$ is a product of a number and variables. 5, y, and $2a^3b^2$ are monomials.
A *polynomial* is a variable expression in which the terms are monomials. [9.1A, p. 478]	$5x^2y - 3xy^2 + 2$ is a polynomial. Each term of this expression is a monomial.
A polynomial of two terms is a *binomial.* [9.1A, p. 478]	$x + 2$, $y^2 - 3$, and $6a + 5b$ are binomials.
A polynomial of three terms is a *trinomial.* [9.1A, p. 478]	$x^2 - 6x + 7$ is a trinomial.
The *degree of a polynomial in one variable* is the greatest exponent on a variable. [9.1A, p. 478]	The degree of $3x - 4x^3 + 17x^2 + 25$ is 3.
A polynomial in one variable is usually written in *descending order,* where the exponents on the variable terms decrease from left to right. [9.1A, p. 478]	The polynomial $2x^4 + 3x^2 - 4x - 7$ is written in descending order.
The *opposite of a polynomial* is the polynomial with the sign of every term changed to its opposite. [9.1B, p. 479]	The opposite of the polynomial $x^2 - 3x + 4$ is $-x^2 + 3x - 4$.

ESSENTIAL RULES AND PROCEDURES	EXAMPLES
Addition of Polynomials [9.1A, p. 478] To add polynomials, add the coefficients of the like terms.	$(2x^2 + 3x - 4) + (3x^3 - 4x^2 + 2x - 5)$ $= 3x^3 + (2x^2 - 4x^2) + (3x + 2x) + (-4 - 5)$ $= 3x^3 - 2x^2 + 5x - 9$
Subtraction of Polynomials [9.1B, p. 479] To subtract polynomials, add the opposite of the second polynomial to the first.	$(3y^2 - 8y - 9) - (5y^2 - 10y + 3)$ $= (3y^2 - 8y - 9) + (-5y^2 + 10y - 3)$ $= (3y^2 - 5y^2) + (-8y + 10y) + (-9 - 3)$ $= -2y^2 + 2y - 12$
Rule for Multiplying Exponential Expressions [9.2A, p. 482] If m and n are integers, then $x^m \cdot x^n = x^{m+n}$.	$a^3 \cdot a^6 = a^{3+6} = a^9$

Rule for Simplifying the Power of an Exponential Expression [9.2B, p. 483]

If m and n are integers, then $(x^m)^n = x^{mn}$.

$(c^3)^4 = c^{3 \cdot 4} = c^{12}$

Rule for Simplifying the Power of a Product [9.2B, p. 483]

If m, n, and p are integers, then $(x^m y^n)^p = x^{mp} y^{np}$.

$(a^3b^2)^4 = a^{3 \cdot 4}b^{2 \cdot 4} = a^{12}b^8$

To multiply a polynomial by a monomial, use the Distributive Property and the Rule for Multiplying Exponential Expressions. [9.3A, p. 486]

$$\begin{aligned}&(-4y)(5y^2 + 3y - 8)\\ &\quad = (-4y)(5y^2) + (-4y)(3y) - (-4y)(8)\\ &\quad = -20y^3 - 12y^2 + 32y\end{aligned}$$

To multiply two polynomials, multiply each term of one polynomial by each term of the other polynomial. [9.3B, p. 486]

$$\begin{array}{r} x^2 - 5x + 6 \\ x + 4 \\ \hline 4x^2 - 20x + 24 \\ x^3 - 5x^2 + 6x \phantom{{}+24} \\ \hline x^3 - x^2 - 14x + 24 \end{array}$$

FOIL Method [9.3C, p. 487]

To find the product of two binomials, add the products of the **F**irst terms, the **O**uter terms, the **I**nner terms, and the **L**ast terms.

$$\begin{aligned}&(2x - 5)(3x + 4)\\ &\quad = (2x)(3x) + (2x)(4) + (-5)(3x)\\ &\qquad + (-5)(4)\\ &\quad = 6x^2 + 8x - 15x - 20\\ &\quad = 6x^2 - 7x - 20\end{aligned}$$

Product of the Sum and Difference of the Same Terms [9.3D, p. 488]

$(a + b)(a - b) = a^2 - b^2$

$$\begin{aligned}(3x + 4)(3x - 4) &= (3x)^2 - 4^2\\ &= 9x^2 - 16\end{aligned}$$

Square of a Binomial [9.3D, p. 488]

$(a + b)^2 = a^2 + 2ab + b^2$

$(a - b)^2 = a^2 - 2ab + b^2$

$$\begin{aligned}(2x + 5)^2 &= (2x)^2 + 2(2x)(5) + 5^2\\ &= 4x^2 + 20x + 25\\ (3x - 4)^2 &= (3x)^2 - 2(3x)(4) + (-4)^2\\ &= 9x^2 - 24x + 16\end{aligned}$$

Definition of Zero as an Exponent [9.4A, p. 494]

If $x \neq 0$, then $x^0 = 1$.

$17^0 = 1$; $(-6c)^0 = 1, c \neq 0$

Definition of a Negative Exponent [9.4A, p. 495]

If $x \neq 0$ and n is a positive integer, then $x^{-n} = \frac{1}{x^n}$ and $\frac{1}{x^{-n}} = x^n$.

$x^{-6} = \frac{1}{x^6}$ and $\frac{1}{x^{-6}} = x^6$

Rule for Simplifying the Power of a Quotient [9.4A, p. 496]

If m, n, and p are integers and $y \neq 0$, then $\left(\frac{x^m}{y^n}\right)^p = \frac{x^{mp}}{y^{np}}$.

$$\left(\frac{c^3}{a^5}\right)^2 = \frac{c^{3\cdot 2}}{a^{5\cdot 2}} = \frac{c^6}{a^{10}}$$

Rule for Negative Exponents on Fractional Expressions [9.4A, p. 496]

If $a \neq 0$, $b \neq 0$, and n is a positive integer, then $\left(\frac{a}{b}\right)^{-n} = \left(\frac{b}{a}\right)^n$.

$$\left(\frac{x}{y}\right)^{-3} = \left(\frac{y}{x}\right)^3$$

Rule for Dividing Exponential Expressions [9.4A, p. 496]

If m and n are integers and $x \neq 0$, then $\frac{x^m}{x^n} = x^{m-n}$.

$$\frac{a^7}{a^2} = a^{7-2} = a^5$$

To Express a Number in Scientific Notation [9.4B, p. 499]

To express a number in scientific notation, write it in the form $a \times 10^n$, where $1 \leq a < 10$ and n is an integer. If the number is greater than 10, then n is a positive integer. If the number is between 0 and 1, then n is a negative integer.

$367{,}000{,}000 = 3.67 \times 10^8$
$0.0000078 = 7.8 \times 10^{-6}$

To Change a Number in Scientific Notation to Decimal Notation [9.4B, p. 499]

To change a number in scientific notation to decimal notation, move the decimal point to the right if n is positive and to the left if n is negative. Move the decimal point the same number of places as the absolute value of the exponent on 10.

$2.418 \times 10^7 = 24{,}180{,}000$
$9.06 \times 10^{-5} = 0.0000906$

To divide a polynomial by a monomial, divide each term in the numerator by the denominator and write the sum of the quotients. [9.5A, p. 504]

$$\begin{aligned}\frac{8xy^3 - 4y^2 + 12y}{4y} &= \frac{8xy^3}{4y} - \frac{4y^2}{4y} + \frac{12y}{4y} \\ &= 2xy^2 - y + 3\end{aligned}$$

To check polynomial division, use the same equation used to check division of whole numbers:

(Quotient × divisor) + remainder = dividend

[9.5B, p. 504]

$$\begin{array}{r} x - 4 \\ x + 3\overline{)x^2 - x - 10} \\ \underline{x^2 + 3x} \\ -4x - 10 \\ \underline{-4x - 12} \\ 2 \end{array}$$

Check:

$$\begin{aligned}(x - 4)(x + 3) + 2 &= x^2 - x - 12 + 2 \\ &= x^2 - x - 10\end{aligned}$$

$$(x^2 - x - 10) \div (x + 3) = x - 4 + \frac{2}{x + 3}$$

CHAPTER 9

CONCEPT REVIEW

Test your knowledge of the concepts presented in this chapter. Answer each question. Then check your answers against the ones provided in the Answer Section.

1. Why is it important to write the terms of a polynomial in descending order before adding in a vertical format?

2. What is the opposite of $-7x^3 + 3x^2 - 4x - 2$?

3. When multiplying the terms $4p^3$ and $7p^6$, what happens to the exponents?

4. Why is the simplification of the expression $-4b(2b^2 - 3b - 5) = -8b^3 + 12b + 20$ not true?

5. How do you multiply two binomials?

6. Simplify $\frac{w^2x^4yz^6}{w^3xy^4z^0}$.

7. Simplify $\left(\frac{a^0}{b^{-2}}\right)^{-2}$.

8. How do you write a very large number in scientific notation?

9. What is wrong with this simplification? $\frac{14x^3 - 8x^2 - 6x}{2x} = 7x^2 - 8x^2 - 6x$

10. How do you check polynomial division?

CHAPTER 9

REVIEW EXERCISES

1. Multiply: $(2b - 3)(4b + 5)$

2. Add: $(12y^2 + 17y - 4) + (9y^2 - 13y + 3)$

3. Simplify: $(xy^5z^3)(x^3y^3z)$

4. Simplify: $\dfrac{8x^{12}}{12x^9}$

5. Multiply: $-2x(4x^2 + 7x - 9)$

6. Simplify: $\dfrac{3ab^4}{-6a^2b^4}$

7. Simplify: $(-2u^3v^4)^4$

8. Evaluate: $(2^3)^2$

9. Subtract: $(5x^2 - 2x - 1) - (3x^2 - 5x + 7)$

10. Simplify: $\dfrac{a^{-1}b^3}{a^3b^{-3}}$

11. Simplify: $(-2x^3)^2(-3x^4)^3$

12. Expand: $(5y - 7)^2$

13. Simplify: $(5a^7b^6)^2(4ab)$

14. Divide: $\dfrac{12b^7 + 36b^5 - 3b^3}{3b^3}$

15. Evaluate: -4^{-2}

16. Subtract: $(13y^3 - 7y - 2) - (12y^2 - 2y - 1)$

17. Divide: $\dfrac{7 - x - x^2}{x + 3}$

18. Multiply: $(2a - b)(x - 2y)$

19. Multiply: $(3y^2 + 4y - 7)(2y + 3)$

20. Divide: $(b^3 - 2b^2 - 33b - 7) \div (b - 7)$

21. Multiply: $2ab^3(4a^2 - 2ab + 3b^2)$

22. Multiply: $(2a - 5b)(2a + 5b)$

23. Multiply: $(6b^3 - 2b^2 - 5)(2b^2 - 1)$

24. Add: $(2x^3 + 7x^2 + x) + (2x^2 - 4x - 12)$

25. Divide: $\dfrac{16y^2 - 32y}{-4y}$

26. Multiply: $(a + 7)(a - 7)$

27. Write 37,560,000,000 in scientific notation.

28. Write 1.46×10^7 in decimal notation.

29. Simplify: $(2a^{12}b^3)(-9b^2c^6)(3ac)$

30. Divide: $(6y^2 - 35y + 36) \div (3y - 4)$

31. Simplify: $(-3x^{-2}y^{-3})^{-2}$

32. Multiply: $(5a - 7)(2a + 9)$

33. Write 0.000000127 in scientific notation.

34. Write 3.2×10^{-12} in decimal notation.

35. **Geometry** The length of a table-tennis table is 1 ft less than twice the width of the table. Let w represent the width of the table-tennis table. Express the area of the table in terms of the variable w.

36. **Geometry** The side of a checkerboard is $(3x - 2)$ in. Express the area of the checkerboard in terms of the variable x.

CHAPTER 9

TEST

1. Multiply: $2x(2x^2 - 3x)$

2. Divide: $\frac{12x^3 - 3x^2 + 9}{3x^2}$

3. Simplify: $\frac{12x^2}{-3x^8}$

4. Simplify: $(-2xy^2)(3x^2y^4)$

5. Divide: $(x^2 + 1) \div (x + 1)$

6. Multiply: $(x - 3)(x^2 - 4x + 5)$

7. Simplify: $(-2a^2b)^3$

8. Simplify: $\frac{(3x^{-2}y^3)^3}{3x^4y^{-1}}$

9. Multiply: $(a - 2b)(a + 5b)$

10. Divide: $\frac{16x^5 - 8x^3 + 20x}{4x}$

11. Divide: $(x^2 + 6x - 7) \div (x - 1)$

12. Multiply: $-3y^2(-2y^2 + 3y - 6)$

13. Multiply: $(-2x^3 + x^2 - 7)(2x - 3)$

14. Multiply: $(4y - 3)(4y + 3)$

15. Simplify: $(ab^2)(a^3b^5)$

16. Simplify: $\dfrac{2a^{-1}b}{2^{-2}a^{-2}b^{-3}}$

17. Divide: $\dfrac{20a - 35}{5}$

18. Subtract: $(3a^2 - 2a - 7) - (5a^3 + 2a - 10)$

19. Expand: $(2x - 5)^2$

20. Divide: $(4x^2 - 7) \div (2x - 3)$

21. Simplify: $\dfrac{-(2x^2y)^3}{4x^3y^3}$

22. Multiply: $(2x - 7y)(5x - 4y)$

23. Add: $(3x^3 - 2x^2 - 4) + (8x^2 - 8x + 7)$

24. Write 0.00000000302 in scientific notation.

25. **Geometry** The radius of a circle is $(x - 5)$ m. Use the equation $A = \pi r^2$, where r is the radius, to find the area of the circle in terms of the variable x. Leave the answer in terms of π.

$x - 5$

CUMULATIVE REVIEW EXERCISES

1. Simplify: $\frac{3}{16} - \left(-\frac{5}{8}\right) - \frac{7}{9}$

2. Evaluate $-3^2 \cdot \left(\frac{2}{3}\right)^3 \cdot \left(-\frac{5}{8}\right)$.

3. Simplify: $\left(-\frac{1}{2}\right)^3 \div \left(\frac{3}{8} - \frac{5}{6}\right) + 2$

4. Evaluate $\frac{b - (a - b)^2}{b^2}$ when $a = -2$ and $b = 3$.

5. Simplify: $-2x - (-xy) + 7x - 4xy$

6. Simplify: $(12x)\left(-\frac{3}{4}\right)$

7. Simplify: $-2[3x - 2(4 - 3x) + 2]$

8. Solve: $12 = -\frac{3}{4}x$

9. Solve: $2x - 9 = 3x + 7$

10. Solve: $2 - 3(4 - x) = 2x + 5$

11. 35.2 is what percent of 160?

12. Add: $(4b^3 - 7b^2 - 7) + (3b^2 - 8b + 3)$

13. Subtract: $(3y^3 - 5y + 8) - (-2y^2 + 5y + 8)$

14. Simplify: $(a^3b^5)^3$

15. Simplify: $(4xy^3)(-2x^2y^3)$

16. Multiply: $-2y^2(-3y^2 - 4y + 8)$

17. Multiply: $(2a - 7)(5a^2 - 2a + 3)$

18. Multiply: $(3b - 2)(5b - 7)$

19. Simplify: $\dfrac{(-2a^2b^3)^2}{8a^4b^8}$

20. Divide: $(a^2 - 4a - 21) \div (a + 3)$

21. Write 6.09×10^{-5} in decimal notation.

22. Translate "the difference between eight times a number and twice the number is eighteen" into an equation and solve.

© Brian Tolbert/Corbis

23. **Juice Mixtures** Find the cost per ounce of a fruit drink made from 200 oz of fruit juice that costs \$.25 per ounce and 300 oz of soda that costs \$.05 per ounce.

24. **Transportation** A car traveling at 50 mph overtakes a cyclist who, riding at 10 mph, has had a 2-hour head start. How far from the starting point does the car overtake the cyclist?

25. **Geometry** The width of a rectangle is 40% of the length. The perimeter of the rectangle is 42 m. Find the length and width of the rectangle.

Factoring

CHAPTER 10

OBJECTIVES

SECTION 10.1

A To factor a monomial from a polynomial

B To factor by grouping

SECTION 10.2

A To factor a trinomial of the form $x^2 + bx + c$

B To factor completely

SECTION 10.3

A To factor a trinomial of the form $ax^2 + bx + c$ by using trial factors

B To factor a trinomial of the form $ax^2 + bx + c$ by grouping

SECTION 10.4

A To factor the difference of two squares and perfect-square trinomials

B To factor completely

SECTION 10.5

A To solve equations by factoring

B To solve application problems

ARE YOU READY?

Take the Chapter 10 Prep Test to find out if you are ready to learn to:

- Factor a monomial from a polynomial
- Factor by grouping
- Factor trinomials
- Factor the difference of two squares and perfect-square trinomials
- Solve equations by factoring

PREP TEST

Do these exercises to prepare for Chapter 10.

1. Write 30 as a product of prime numbers.

2. Simplify: $-3(4y - 5)$

3. Simplify: $-(a - b)$

4. Simplify: $2(a - b) - 5(a - b)$

5. Solve: $4x = 0$

6. Solve: $2x + 1 = 0$

7. Multiply: $(x + 4)(x - 6)$

8. Multiply: $(2x - 5)(3x + 2)$

9. Simplify: $\dfrac{x^5}{x^2}$

10. Simplify: $\dfrac{6x^4y^3}{2xy^2}$

SECTION 10.1 Common Factors

OBJECTIVE A To factor a monomial from a polynomial

In Section 2.1B we discussed how to find the greatest common factor (GCF) of two or more integers. The **greatest common factor (GCF) of two or more monomials** is the product of the GCF of the coefficients and the common variable factors.

$6x^3y = 2 \cdot 3 \cdot x \cdot x \cdot x \cdot y$
$8x^2y^2 = 2 \cdot 2 \cdot 2 \cdot x \cdot x \cdot y \cdot y$
$\text{GCF} = 2 \cdot x \cdot x \cdot y = 2x^2y$

Note that **the exponent on each variable in the GCF is the same as the *smallest* exponent on that variable in either of the monomials.**

The GCF of $6x^3y$ and $8x^2y^2$ is $2x^2y$.

HOW TO • 1 Find the GCF of $12a^4b$ and $18a^2b^2c$.

The common variable factors are a^2 and b; c is not a common variable factor.

$12a^4b = 2 \cdot 2 \cdot 3 \cdot a^4 \cdot b$
$18a^2b^2c = 2 \cdot 3 \cdot 3 \cdot a^2 \cdot b^2 \cdot c$
$\text{GCF} = 2 \cdot 3 \cdot a^2 \cdot b = 6a^2b$

To **factor a polynomial** means to write the polynomial as a product of other polynomials. In the example at the right, $2x$ is the GCF of the terms $2x^2$ and $10x$.

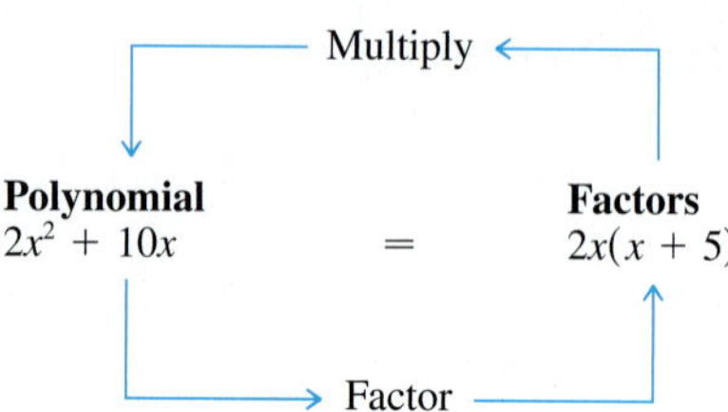

HOW TO • 2 Factor: $5x^3 - 35x^2 + 10x$

Find the GCF of the terms of the polynomial.

$5x^3 = 5 \cdot x^3$
$35x^2 = 5 \cdot 7 \cdot x^2$
$10x = 2 \cdot 5 \cdot x$

The GCF is $5x$.

Rewrite the polynomial, expressing each term as a product with the GCF as one of the factors.

$$5x^3 - 35x^2 + 10x = 5x(x^2) + 5x(-7x) + 5x(2)$$
$$= 5x(x^2 - 7x + 2)$$

• **Use the Distributive Property to write the polynomial as a product of factors.**

Take Note

At the right, the factors in parentheses are determined by dividing each term of the trinomial by the GCF, $5x$.

$\frac{5x^3}{5x} = x^2$, $\frac{-35x^2}{5x} = -7x$, and $\frac{10x}{5x} = 2$

HOW TO • 3 Factor: $21x^2y^3 - 6xy^5 + 15x^4y^2$

Find the GCF of the terms of the polynomial.

$$21x^2y^3 = 3 \cdot 7 \cdot x^2 \cdot y^3$$
$$6xy^5 = 2 \cdot 3 \cdot x \cdot y^5$$
$$15x^4y^2 = 3 \cdot 5 \cdot x^4 \cdot y^2$$

The GCF is $3xy^2$.

Rewrite the polynomial, expressing each term as a product with the GCF as one of the factors.

$$21x^2y^3 - 6xy^5 + 15x^4y^2$$
$$= 3xy^2(7xy) + 3xy^2(-2y^3) + 3xy^2(5x^3)$$
$$= 3xy^2(7xy - 2y^3 + 5x^3)$$

• **Use the Distributive Property to write the polynomial as a product of factors.**

EXAMPLE • 1

Factor: $8x^2 + 2xy$

Solution

The GCF is $2x$.

$$8x^2 + 2xy = 2x(4x) + 2x(y)$$
$$= 2x(4x + y)$$

YOU TRY IT • 1

Factor: $14a^2 - 21a^4b$

Your solution

EXAMPLE • 2

Factor: $n^3 - 5n^2 + 2n$

Solution

The GCF is n.

$$n^3 - 5n^2 + 2n = n(n^2) + n(-5n) + n(2)$$
$$= n(n^2 - 5n + 2)$$

YOU TRY IT • 2

Factor: $27b^2 + 18b + 9$

Your solution

EXAMPLE • 3

Factor: $16x^2y + 8x^4y^2 - 12x^4y^5$

Solution

The GCF is $4x^2y$.

$$16x^2y + 8x^4y^2 - 12x^4y^5$$
$$= 4x^2y(4) + 4x^2y(2x^2y) + 4x^2y(-3x^2y^4)$$
$$= 4x^2y(4 + 2x^2y - 3x^2y^4)$$

YOU TRY IT • 3

Factor: $6x^4y^2 - 9x^3y^2 + 12x^2y^4$

Your solution

Solutions on p. S22

OBJECTIVE B To factor by grouping 

A factor that has two terms is called a **binomial factor**. In the examples at the right, the binomials $a + b$ and $x - y$ are binomial factors.

$$2a(a + b)^2$$
$$3xy(x - y)$$

The Distributive Property is used to factor a common binomial factor from an expression.

The common binomial factor of the expression $6(x - 3) + y(x - 3)$ is $(x - 3)$. To factor the expression, use the Distributive Property to write the expression as a product of factors.

$$6(x - 3) + y(x - 3) = (x - 3)(6 + y)$$

Consider the following simplification of $-(a - b)$.

$$-(a - b) = -1(a - b) = -a + b = b - a$$

Thus $\qquad b - a = -(a - b)$

This equation is sometimes used to factor a common binomial from an expression.

HOW TO 4 Factor: $2x(x - y) + 5(y - x)$

$$2x(x - y) + 5(y - x) = 2x(x - y) - 5(x - y)$$
$$= (x - y)(2x - 5)$$

- $5(y - x) = 5[(-1)(x - y)] = -5(x - y)$

A polynomial can be factored by grouping if its terms can be grouped and factored in such a way that a common binomial factor is found.

A polynomial can be **factored by grouping** if its terms can be grouped and factored in such a way that a common binomial factor is found.

HOW TO 5 Factor: $ax + bx - ay - by$

$$ax + bx - ay - by = (ax + bx) - (ay + by)$$

- Group the first two terms and the last two terms. Note that $-ay - by = -(ay + by)$.

$$= x(a + b) - y(a + b)$$

- Factor each group.

$$= (a + b)(x - y)$$

- Factor the GCF, $(a + b)$, from each group.

Check: $(a + b)(x - y) = ax - ay + bx - by$
$= ax + bx - ay - by$

HOW TO 6 Factor: $6x^2 - 9x - 4xy + 6y$

$$6x^2 - 9x - 4xy + 6y = (6x^2 - 9x) - (4xy - 6y)$$

- Group the first two terms and the last two terms. Note that $-4xy + 6y = -(4xy - 6y)$.

$$= 3x(2x - 3) - 2y(2x - 3)$$

- Factor each group.

$$= (2x - 3)(3x - 2y)$$

- Factor the GCF, $(2x - 3)$, from each group.

EXAMPLE • 4

Factor: $4x(3x - 2) - 7(3x - 2)$

Solution

$4x(3x - 2) - 7(3x - 2)$ • $3x - 2$ is the common binomial factor.

$= (3x - 2)(4x - 7)$

YOU TRY IT • 4

Factor: $2y(5x - 2) - 3(2 - 5x)$

Your solution

EXAMPLE • 5

Factor: $9x^2 - 15x - 6xy + 10y$

Solution

$9x^2 - 15x - 6xy + 10y$

$= (9x^2 - 15x) - (6xy - 10y)$ • $-6xy + 10y = -(6xy - 10y)$

$= 3x(3x - 5) - 2y(3x - 5)$ • $3x - 5$ is the common factor.

$= (3x - 5)(3x - 2y)$

YOU TRY IT • 5

Factor: $a^2 - 3a + 2ab - 6b$

Your solution

EXAMPLE • 6

Factor: $3x^2y - 4x - 15xy + 20$

Solution

$3x^2y - 4x - 15xy + 20$

$= (3x^2y - 4x) - (15xy - 20)$ • $-15xy + 20 = -(15xy - 20)$

$= x(3xy - 4) - 5(3xy - 4)$ • $3xy - 4$ is the common factor.

$= (3xy - 4)(x - 5)$

YOU TRY IT • 6

Factor: $2mn^2 - n + 8mn - 4$

Your solution

EXAMPLE • 7

Factor: $4ab - 6 + 3b - 2ab^2$

Solution

$4ab - 6 + 3b - 2ab^2$

$= (4ab - 6) + (3b - 2ab^2)$

$= 2(2ab - 3) + b(3 - 2ab)$

$= 2(2ab - 3) - b(2ab - 3)$ • $3 - 2ab = -(2ab - 3)$

$= (2ab - 3)(2 - b)$ • $2ab - 3$ is the common factor.

YOU TRY IT • 7

Factor: $3xy - 9y - 12 + 4x$

Your solution

Solutions on p. S22

10.1 EXERCISES

OBJECTIVE A To factor a monomial from a polynomial

1. Explain the meaning of "a common monomial factor of a polynomial."

2. Explain the meaning of "a factor" and the meaning of "to factor."

For Exercises 3 to 41, factor.

3. $5a + 5$
4. $7b - 7$
5. $16 - 8a^2$
6. $12 + 12y^2$
7. $8x + 12$
8. $16a - 24$
9. $30a - 6$
10. $20b + 5$
11. $7x^2 - 3x$
12. $12y^2 - 5y$
13. $3a^2 + 5a^5$
14. $9x - 5x^2$
15. $14y^2 + 11y$
16. $6b^3 - 5b^2$
17. $2x^4 - 4x$
18. $3y^4 - 9y$
19. $10x^4 - 12x^2$
20. $12a^5 - 32a^2$
21. $8a^8 - 4a^5$
22. $16y^4 - 8y^7$
23. $x^2y^2 - xy$
24. $a^2b^2 + ab$
25. $3x^2y^4 - 6xy$
26. $12a^2b^5 - 9ab$
27. $x^2y - xy^3$
28. $3x^3 + 6x^2 + 9x$
29. $5y^3 - 20y^2 + 5y$
30. $2x^4 - 4x^3 + 6x^2$
31. $3y^4 - 9y^3 - 6y^2$
32. $2x^3 + 6x^2 - 14x$
33. $3y^3 - 9y^2 + 24y$
34. $2y^5 - 3y^4 + 7y^3$
35. $6a^5 - 3a^3 - 2a^2$
36. $x^3y - 3x^2y^2 + 7xy^3$
37. $2a^2b - 5a^2b^2 + 7ab^2$
38. $5y^3 + 10y^2 - 25y$
39. $4b^5 + 6b^3 - 12b$
40. $3a^2b^2 - 9ab^2 + 15b^2$
41. $8x^2y^2 - 4x^2y + x^2$

42. What is the GCF of the terms of the polynomial $x^a + x^b + x^c$ given that a, b, and c are all positive integers, and $a > b > c$?

OBJECTIVE B To factor by grouping

43. Use the three expressions at the right.
   a. Which expressions are equivalent to $x^2 - 5x + 6$?
   b. Which expression can be factored by grouping?

(i) $x^2 - 15x + 10x + 6$
(ii) $x^2 - x - 4x + 6$
(iii) $x^2 - 2x - 3x + 6$

For Exercises 44 to 70, factor.

44. $x(b + 4) + 3(b + 4)$

45. $y(a + z) + 7(a + z)$

46. $a(y - x) - b(y - x)$

47. $3r(a - b) + s(a - b)$

48. $x(x - 2) + y(2 - x)$

49. $t(m - 7) + 7(7 - m)$

50. $8c(2m - 3n) + (3n - 2m)$

51. $2y(4a + b) - (b + 4a)$

52. $2x(7 + b) - y(b + 7)$

53. $x^2 + 2x + 2xy + 4y$

54. $x^2 - 3x + 4ax - 12a$

55. $p^2 - 2p - 3rp + 6r$

56. $t^2 + 4t - st - 4s$

57. $ab + 6b - 4a - 24$

58. $xy - 5y - 2x + 10$

59. $2z^2 - z + 2yz - y$

60. $2y^2 - 10y + 7xy - 35x$

61. $8v^2 - 12vy + 14v - 21y$

62. $21x^2 + 6xy - 49x - 14y$

63. $2x^2 - 5x - 6xy + 15y$

64. $4a^2 + 5ab - 10b - 8a$

65. $3y^2 - 6y - ay + 2a$

66. $2ra + a^2 - 2r - a$

67. $3xy - y^2 - y + 3x$

68. $2ab - 3b^2 - 3b + 2a$

69. $3st + t^2 - 2t - 6s$

70. $4x^2 + 3xy - 12y - 16x$

Applying the Concepts

71. Geometry In the equation $P = 2L + 2W$, what is the effect on P when the quantity $L + W$ doubles?

72. Geometry Write an expression in factored form for the shaded portion in each of the following diagrams. Use the equation for the area of a rectangle ($A = LW$) and the equation for the area of a circle ($A = \pi r^2$).

a.

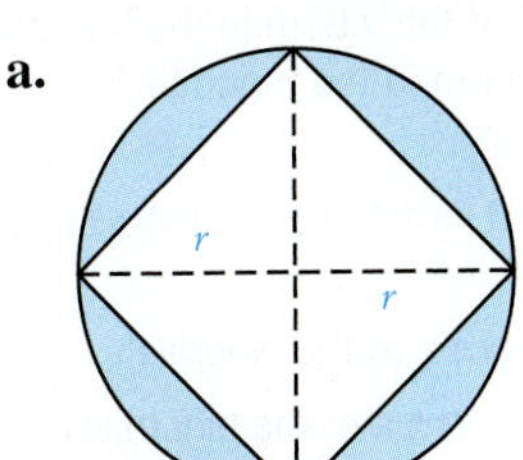

b.

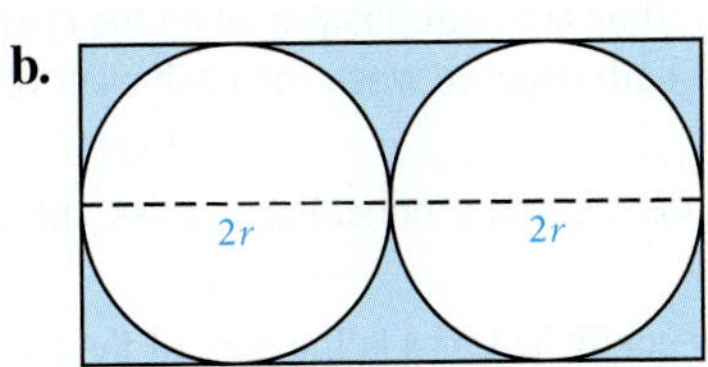

c.

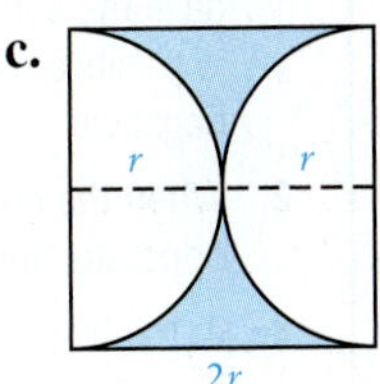

SECTION

10.2 Factoring Polynomials of the Form $x^2 + bx + c$

OBJECTIVE A To factor a trinomial of the form $x^2 + bx + c$ 

Trinomials of the form $x^2 + bx + c$, where b and c are integers, are shown at the right.

$$x^2 + 8x + 12;\ b = 8,\ c = 12$$
$$x^2 - 7x + 12;\ b = -7,\ c = 12$$
$$x^2 - 2x - 15;\ b = -2,\ c = -15$$

To factor a trinomial of this form means to express the trinomial as the product of two binomials.

Trinomials expressed as the product of binomials are shown at the right.

$$x^2 + 8x + 12 = (x + 6)(x + 2)$$
$$x^2 - 7x + 12 = (x - 3)(x - 4)$$
$$x^2 - 2x - 15 = (x + 3)(x - 5)$$

The method by which factors of a trinomial are found is based on FOIL. Consider the following binomial products, noting the relationship between the constant terms of the binomials and the terms of the trinomials.

The signs in the binomial factors are the same.

$$(x + 6)(x + 2) = x^2 + 2x + 6x + (6)(2) = x^2 + 8x + 12$$

Sum of 6 and 2 (→ 8)
Product of 6 and 2 (→ 12)

$$(x - 3)(x - 4) = x^2 - 4x - 3x + (-3)(-4) = x^2 - 7x + 12$$

Sum of −3 and −4 (→ −7)
Product of −3 and −4 (→ 12)

The signs in the binomial factors are opposites.

$$(x + 3)(x - 5) = x^2 - 5x + 3x + (3)(-5) = x^2 - 2x - 15$$

Sum of 3 and −5 (→ −2)
Product of 3 and −5 (→ −15)

$$(x - 4)(x + 6) = x^2 + 6x - 4x + (-4)(6) = x^2 + 2x - 24$$

Sum of −4 and 6 (→ 2)
Product of −4 and 6 (→ −24)

Factoring $x^2 + bx + c$: IMPORTANT RELATIONSHIPS

1. When the constant term of the trinomial is positive, the constant terms of the binomials have the same sign. They are both positive when the coefficient of the x term in the trinomial is positive. They are both negative when the coefficient of the x term in the trinomial is negative.
2. When the constant term of the trinomial is negative, the constant terms of the binomials have opposite signs.
3. In the trinomial, the coefficient of x is the sum of the constant terms of the binomials.
4. In the trinomial, the constant term is the product of the constant terms of the binomials.

HOW TO • 1 Factor: $x^2 - 7x + 10$

Because the constant term is positive and the coefficient of x is negative, the binomial constants will be negative. Find two negative factors of 10 whose sum is -7. The results can be recorded in a table.

Negative Factors of 10	*Sum*
$-1, -10$	-11
$-2, -5$	-7

• These are the correct factors.

$x^2 - 7x + 10 = (x - 2)(x - 5)$ • Write the trinomial as a product of its factors.

You can check the proposed factorization by multiplying the two binomials.

Check: $(x - 2)(x - 5) = x^2 - 5x - 2x + 10 = x^2 - 7x + 10$

Take Note
Always check your proposed factorization to ensure accuracy.

HOW TO • 2 Factor: $x^2 - 9x - 36$

The constant term is negative. The binomial constants will have opposite signs. Find two factors of -36 whose sum is -9.

Factors of −36	*Sum*
$+1, -36$	-35
$-1, +36$	35
$+2, -18$	-16
$-2, +18$	16
$+3, -12$	-9

• Once the correct factors are found, it is not necessary to try the remaining factors.

$x^2 - 9x - 36 = (x + 3)(x - 12)$ • Write the trinomial as a product of its factors.

For some trinomials it is not possible to find integer factors of the constant term whose sum is the coefficient of the middle term. A polynomial that does not factor using only integers is **nonfactorable over the integers.**

HOW TO • 3 Factor: $x^2 + 7x + 8$

The constant term is positive and the coefficient of x is positive. The binomial constants will be positive. Find two positive factors of 8 whose sum is 7.

Positive Factors of 8	*Sum*
1, 8	9
2, 4	6

• There are no positive integer factors of 8 whose sum is 7.

$x^2 + 7x + 8$ is nonfactorable over the integers.

Take Note
Just as 17 is a prime number, $x^2 + 7x + 8$ is a **prime polynomial.** Binomials of the form $x - a$ and $x + a$ are also prime polynomials.

EXAMPLE • 1

Factor: $x^2 - 8x + 15$

Solution

Find two negative factors of 15 whose sum is -8.

Factors	*Sum*
$-1, -15$	-16
$-3, -5$	-8

$x^2 - 8x + 15 = (x - 3)(x - 5)$

YOU TRY IT • 1

Factor: $x^2 + 9x + 20$

Your solution

Solution on p. S23

EXAMPLE • 2

Factor: $x^2 + 6x - 27$

Solution

Find two factors of -27 whose sum is 6.

Factors	Sum
$+1, -27$	-26
$-1, +27$	26
$+3, -9$	-6
$-3, +9$	6

$x^2 + 6x - 27 = (x - 3)(x + 9)$

YOU TRY IT • 2

Factor: $x^2 + 7x - 18$

Your solution

Solution on p. S23

OBJECTIVE B To factor completely

A polynomial is **factored completely** when it is written as a product of factors that are nonfactorable over the integers.

Take Note

The first step in *any* factoring problem is to determine whether the terms of the polynomial have a *common factor.* If they do, factor it out first.

HOW TO • 4 Factor: $4y^3 - 4y^2 - 24y$

$4y^3 - 4y^2 - 24y = 4y(y^2) - 4y(y) - 4y(6)$ • The GCF is $4y$.

$= 4y(y^2 - y - 6)$ • Use the Distributive Property to factor out the GCF.

$= 4y(y + 2)(y - 3)$ • Factor $y^2 - y - 6$. The two factors of -6 whose sum is -1 are 2 and -3.

It is always possible to check a proposed factorization by multiplying the polynomials. Here is the check for the last example.

Check: $4y(y + 2)(y - 3) = 4y(y^2 - 3y + 2y - 6)$

$= 4y(y^2 - y - 6)$

$= 4y^3 - 4y^2 - 24y$ • This is the original polynomial.

HOW TO • 5 Factor: $x^2 + 12xy + 20y^2$

There is no common factor.

Note that the variable part of the middle term is xy, and the variable part of the last term is y^2.

$x^2 + 12xy + 20y^2 = (x + 2y)(x + 10y)$ • The two factors of 20 whose sum is 12 are 2 and 10.

Take Note

The terms $2y$ and $10y$ are placed in the binomials. This is necessary so that the middle term of the trinomial contains xy and the last term contains y^2.

Note that the terms $2y$ and $10y$ are placed in the binomials. The following check shows why this is necessary.

Check: $(x + 2y)(x + 10y) = x^2 + 10xy + 2xy + 20y^2$

$= x^2 + 12xy + 20y^2$ • This is the original polynomial.

 **Take Note**

When the coefficient of the highest power in a polynomial is negative, consider factoring out a negative GCF. Example 3 is another example of this technique.

HOW TO • 6 Factor: $15 - 2x - x^2$

Because the coefficient of x^2 is -1, factor -1 from the trinomial and then write the resulting trinomial in descending order.

$15 - 2x - x^2 = -(x^2 + 2x - 15)$ • $15 - 2x - x^2 = -1(-15 + 2x + x^2) = -(x^2 + 2x - 15)$

$= -(x + 5)(x - 3)$ • Factor $x^2 + 2x - 15$. The two factors of -15 whose sum is 2 are 5 and -3.

Check: $-(x + 5)(x - 3) = -(x^2 + 2x - 15)$

$= -x^2 - 2x + 15$

$= 15 - 2x - x^2$ • This is the original polynomial.

EXAMPLE • 3

Factor: $-3x^3 + 9x^2 + 12x$

Solution

The GCF is $-3x$.

$-3x^3 + 9x^2 + 12x = -3x(x^2 - 3x - 4)$

Factor the trinomial $x^2 - 3x - 4$. Find two factors of -4 whose sum is -3.

Factors	*Sum*
+1, −4	−3

$-3x^3 + 9x^2 + 12x = -3x(x + 1)(x - 4)$

YOU TRY IT • 3

Factor: $-2x^3 + 14x^2 - 12x$

Your solution

EXAMPLE • 4

Factor: $4x^2 - 40xy + 84y^2$

Solution

The GCF is 4.

$4x^2 - 40xy + 84y^2 = 4(x^2 - 10xy + 21y^2)$

Factor the trinomial $x^2 - 10xy + 21y^2$. Find two negative factors of 21 whose sum is -10.

Factors	*Sum*
−1, −21	−22
−3, −7	−10

$4x^2 - 40xy + 84y^2 = 4(x - 3y)(x - 7y)$

YOU TRY IT • 4

Factor: $3x^2 - 9xy - 12y^2$

Your solution

Solutions on p. S23

10.2 EXERCISES

OBJECTIVE A To factor a trinomial of the form $x^2 + bx + c$

For Exercises 1 to 73, factor.

1. $x^2 + 3x + 2$

2. $x^2 + 5x + 6$

3. $x^2 - x - 2$

4. $x^2 + x - 6$

5. $a^2 + a - 12$

6. $a^2 - 2a - 35$

7. $a^2 - 3a + 2$

8. $a^2 - 5a + 4$

9. $a^2 + a - 2$

10. $a^2 - 2a - 3$

11. $b^2 - 6b + 9$

12. $b^2 + 8b + 16$

13. $b^2 + 7b - 8$

14. $y^2 - y - 6$

15. $y^2 + 6y - 55$

16. $z^2 - 4z - 45$

17. $y^2 - 5y + 6$

18. $y^2 - 8y + 15$

19. $z^2 - 14z + 45$

20. $z^2 - 14z + 49$

21. $z^2 - 12z - 160$

22. $p^2 + 2p - 35$

23. $p^2 + 12p + 27$

24. $p^2 - 6p + 8$

25. $x^2 + 20x + 100$

26. $x^2 + 18x + 81$

27. $b^2 + 9b + 20$

28. $b^2 + 13b + 40$

29. $x^2 - 11x - 42$

30. $x^2 + 9x - 70$

31. $b^2 - b - 20$

32. $b^2 + 3b - 40$

33. $y^2 - 14y - 51$

34. $y^2 - y - 72$

35. $p^2 - 4p - 21$

36. $p^2 + 16p + 39$

37. $y^2 - 8y + 32$

38. $y^2 - 9y + 81$

39. $x^2 - 20x + 75$

40. $x^2 - 12x + 11$

41. $p^2 + 24p + 63$

42. $x^2 - 15x + 56$

43. $x^2 + 21x + 38$

44. $x^2 + x - 56$

45. $x^2 + 5x - 36$

46. $a^2 - 21a - 72$

47. $a^2 - 7a - 44$

48. $a^2 - 15a + 36$

49. $a^2 - 21a + 54$

50. $z^2 - 9z - 136$

51. $z^2 + 14z - 147$

52. $c^2 - c - 90$

53. $c^2 - 3c - 180$

54. $z^2 + 15z + 44$

55. $p^2 + 24p + 135$

56. $c^2 + 19c + 34$

57. $c^2 + 11c + 18$

58. $x^2 - 4x - 96$

59. $x^2 + 10x - 75$

60. $x^2 - 22x + 112$

61. $x^2 + 21x - 100$

62. $b^2 + 8b - 105$

63. $b^2 - 22b + 72$

64. $a^2 - 9a - 36$

65. $a^2 + 42a - 135$

66. $b^2 - 23b + 102$

67. $b^2 - 25b + 126$

68. $a^2 + 27a + 72$

69. $z^2 + 24z + 144$

70. $x^2 + 25x + 156$

71. $x^2 - 29x + 100$

72. $x^2 - 10x - 96$

73. $x^2 + 9x - 112$

For Exercises 74 and 75, $x^2 + bx + c = (x + n)(x + m)$, where b and c are nonzero and n and m are positive integers.

74. Is c positive or negative?

75. Is b positive or negative?

OBJECTIVE B To factor completely

For Exercises 76 to 129, factor.

76. $2x^2 + 6x + 4$

77. $3x^2 + 15x + 18$

78. $18 + 7x - x^2$

79. $12 - 4x - x^2$

80. $ab^2 + 2ab - 15a$

81. $ab^2 + 7ab - 8a$

82. $xy^2 - 5xy + 6x$

83. $xy^2 + 8xy + 15x$

84. $z^3 - 7z^2 + 12z$

85. $-2a^3 - 6a^2 - 4a$

86. $-3y^3 + 15y^2 - 18y$

87. $4y^3 + 12y^2 - 72y$

88. $3x^2 + 3x - 36$

89. $2x^3 - 2x^2 + 4x$

90. $5z^2 - 15z - 140$

91. $6z^2 + 12z - 90$

92. $2a^3 + 8a^2 - 64a$

93. $3a^3 - 9a^2 - 54a$

94. $x^2 - 5xy + 6y^2$

95. $x^2 + 4xy - 21y^2$

96. $a^2 - 9ab + 20b^2$

97. $a^2 - 15ab + 50b^2$

98. $x^2 - 3xy - 28y^2$

99. $s^2 + 2st - 48t^2$

100. $y^2 - 15yz - 41z^2$

101. $x^2 + 85xy + 36y^2$

102. $z^4 - 12z^3 + 35z^2$

103. $z^4 + 2z^3 - 80z^2$

104. $b^4 - 22b^3 + 120b^2$

105. $b^4 - 3b^3 - 10b^2$

106. $2y^4 - 26y^3 - 96y^2$

107. $3y^4 + 54y^3 + 135y^2$

108. $-x^4 - 7x^3 + 8x^2$

109. $-x^4 + 11x^3 + 12x^2$

110. $4x^2y + 20xy - 56y$

111. $3x^2y - 6xy - 45y$

112. $c^3 + 18c^2 - 40c$

113. $-3x^3 + 36x^2 - 81x$

114. $-4x^3 - 4x^2 + 24x$

115. $x^2 - 8xy + 15y^2$

116. $y^2 - 7xy - 8x^2$

117. $a^2 - 13ab + 42b^2$

118. $y^2 + 4yz - 21z^2$

119. $y^2 + 8yz + 7z^2$

120. $y^2 - 16yz + 15z^2$

121. $3x^2y + 60xy - 63y$

122. $4x^2y - 68xy - 72y$

123. $3x^3 + 3x^2 - 36x$

124. $4x^3 + 12x^2 - 160x$

125. $2t^2 - 24ts + 70s^2$

126. $4a^2 - 40ab + 100b^2$

127. $3a^2 - 24ab - 99b^2$

128. $4x^3 + 8x^2y - 12xy^2$

129. $5x^3 + 30x^2y + 40xy^2$

130. State whether the trinomial has a factor of $x + 3$.
a. $3x^2 - 3x - 36$ **b.** $x^2y - xy - 12y$

131. State whether the trinomial has a factor of $x + y$.
a. $2x^2 - 2xy - 4y^2$ **b.** $2x^2y - 4xy - 4y$

Applying the Concepts

For Exercises 132 to 134, find all integers k such that the trinomial can be factored over the integers.

132. $x^2 + kx + 35$

133. $x^2 + kx + 18$

134. $x^2 + kx + 21$

For Exercises 135 to 140, determine the positive integer values of k for which the polynomial is factorable over the integers.

135. $y^2 + 4y + k$

136. $z^2 + 7z + k$

137. $a^2 - 6a + k$

138. $c^2 - 7c + k$

139. $x^2 - 3x + k$

140. $y^2 + 5y + k$

141. In Exercises 135 to 140, there was the stated requirement that $k > 0$. If k is allowed to be any integer, how many different values of k are possible for each polynomial?

SECTION

10.3 Factoring Polynomials of the Form $ax^2 + bx + c$

OBJECTIVE A To factor a trinomial of the form $ax^2 + bx + c$ by using trial factors 

Trinomials of the form $ax^2 + bx + c$, where a, b, and c are integers, are shown at the right.

$3x^2 - x + 4$; $a = 3$, $b = -1$, $c = 4$
$6x^2 + 2x - 3$; $a = 6$, $b = 2$, $c = -3$

These trinomials differ from those in the preceding section in that the coefficient of x^2 is not 1. There are various methods of factoring these trinomials. The method described in this objective is factoring polynomials using trial factors.

To reduce the number of trial factors that must be considered, remember the following:

1. Use the signs of the constant term and the coefficient of x in the trinomial to determine the signs of the binomial factors. **If the constant term is positive, the signs of the binomial factors will be the same as the sign of the coefficient of x in the trinomial. If the sign of the constant term is negative, the constant terms in the binomials have opposite signs.**
2. **If the terms of the trinomial do not have a common factor, then the terms of each binomial factor will not have a common factor.**

HOW TO • 1 Factor: $2x^2 - 7x + 3$

The terms have no common factor. The constant term is positive. The coefficient of x is negative. The binomial constants will be negative.

Positive Factors of 2 (coefficient of x^2)	Negative Factors of 3 (constant term)
1, 2	−1, −3

Write trial factors. Use the **O**uter and **I**nner products of FOIL to determine the middle term, $-7x$, of the trinomial.

Trial Factors	Middle Term
$(x - 1)(2x - 3)$	$-3x - 2x = -5x$
$(x - 3)(2x - 1)$	$-x - 6x = -7x$

Write the factors of the trinomial.

$$2x^2 - 7x + 3 = (x - 3)(2x - 1)$$

HOW TO • 2 Factor: $3x^2 + 14x + 15$

The terms have no common factor. The constant term is positive. The coefficient of x is positive. The binomial constants will be positive.

Positive Factors of 3 (coefficient of x^2)	Positive Factors of 15 (constant term)
1, 3	1, 15
	3, 5

Write trial factors. Use the **O**uter and **I**nner products of FOIL to determine the middle term, $14x$, of the trinomial.

Trial Factors	Middle Term
$(x + 1)(3x + 15)$	Common factor
$(x + 15)(3x + 1)$	$x + 45x = 46x$
$(x + 3)(3x + 5)$	$5x + 9x = 14x$
$(x + 5)(3x + 3)$	Common factor

Write the factors of the trinomial.

$$3x^2 + 14x + 15 = (x + 3)(3x + 5)$$

HOW TO • 3 Factor: $6x^3 + 14x^2 - 12x$

Factor the GCF, $2x$, from the terms.

$6x^3 + 14x^2 - 12x = 2x(3x^2 + 7x - 6)$

Factor the trinomial. The constant term is negative. The binomial constants will have opposite signs.

Positive Factors of 3	*Factors of −6*
1, 3	1, −6
	−1, 6
	2, −3
	−2, 3

Take Note

For this example, all the trial factors were listed. Once the correct factors have been found, however, the remaining trial factors can be omitted. For the examples and solutions in this text, all trial factors except those that have a common factor will be listed.

Write trial factors. Use the **O**uter and **I**nner products of FOIL to determine the middle term, $7x$, of the trinomial.

It is not necessary to test trial factors that have a common factor.

Trial Factors	*Middle Term*
$(x + 1)(3x - 6)$	Common factor
$(x - 6)(3x + 1)$	$x - 18x = -17x$
$(x - 1)(3x + 6)$	Common factor
$(x + 6)(3x - 1)$	$-x + 18x = 17x$
$(x + 2)(3x - 3)$	Common factor
$(x - 3)(3x + 2)$	$2x - 9x = -7x$
$(x - 2)(3x + 3)$	Common factor
$(x + 3)(3x - 2)$	$-2x + 9x = 7x$

Write the factors of the trinomial.

$6x^3 + 14x^2 - 12x = 2x(x + 3)(3x - 2)$

EXAMPLE • 1

Factor: $3x^2 + x - 2$

Solution

Positive factors of 3: 1, 3

Factors of −2: 1, −2
−1, 2

Trial Factors	*Middle Term*
$(x + 1)(3x - 2)$	$-2x + 3x = x$
$(x - 2)(3x + 1)$	$x - 6x = -5x$
$(x - 1)(3x + 2)$	$2x - 3x = -x$
$(x + 2)(3x - 1)$	$-x + 6x = 5x$

$3x^2 + x - 2 = (x + 1)(3x - 2)$

YOU TRY IT • 1

Factor: $2x^2 - x - 3$

Your solution

EXAMPLE • 2

Factor: $-12x^3 - 32x^2 + 12x$

Solution

$-12x^3 - 32x^2 + 12x = -4x(3x^2 + 8x - 3)$

Factor the trinomial.

Positive factors of 3: 1, 3

Factors of −3: 1, −3
−1, 3

Trial Factors	*Middle Term*
$(x - 3)(3x + 1)$	$x - 9x = -8x$
$(x + 3)(3x - 1)$	$-x + 9x = 8x$

$-12x^3 - 32x^2 + 12x = -4x(x + 3)(3x - 1)$

YOU TRY IT • 2

Factor: $-45y^3 + 12y^2 + 12y$

Your solution

Solutions on p. S23

OBJECTIVE B To factor a trinomial of the form $ax^2 + bx + c$ by grouping 

In the preceding objective, trinomials of the form $ax^2 + bx + c$ were factored by using trial factors. In this objective, these trinomials will be factored by grouping.

To factor $ax^2 + bx + c$, first find two factors of $a \cdot c$ whose sum is b. Then use factoring by grouping to write the factorization of the trinomial.

HOW TO • 4 Factor: $2x^2 + 13x + 15$

Find two positive factors of 30 ($a \cdot c = 2 \cdot 15 = 30$) whose sum is 13.

Positive Factors of 30	*Sum*
1, 30	31
2, 15	17
3, 10	13

• Once the required sum has been found, the remaining factors need not be checked.

$$2x^2 + 13x + 15 = 2x^2 + 3x + 10x + 15$$

• Use the factors of 30 whose sum is 13 to write $13x$ as $3x + 10x$.

$$= (2x^2 + 3x) + (10x + 15)$$

• Factor by grouping.

$$= x(2x + 3) + 5(2x + 3)$$
$$= (2x + 3)(x + 5)$$

Check: $(2x + 3)(x + 5) = 2x^2 + 10x + 3x + 15$
$= 2x^2 + 13x + 15$

HOW TO • 5 Factor: $6x^2 - 11x - 10$

Find two factors of -60 [$a \cdot c = 6(-10) = -60$] whose sum is -11.

Factors of −60	*Sum*
1, −60	−59
−1, 60	59
2, −30	−28
−2, 30	28
3, −20	−17
−3, 20	17
4, −15	−11

$$6x^2 - 11x - 10 = 6x^2 + 4x - 15x - 10$$

• Use the factors of -60 whose sum is -11 to write $-11x$ as $4x - 15x$.

$$= (6x^2 + 4x) - (15x + 10)$$

• Factor by grouping. Recall that $-15x - 10 = -(15x + 10)$.

$$= 2x(3x + 2) - 5(3x + 2)$$
$$= (3x + 2)(2x - 5)$$

Check: $(3x + 2)(2x - 5) = 6x^2 - 15x + 4x - 10$
$= 6x^2 - 11x - 10$

HOW TO • 6 Factor: $3x^2 - 2x - 4$

Find two factors of -12 $[a \cdot c = 3(-4) = -12]$ whose sum is -2.

Factors of −12	*Sum*
1, −12	−11
−1, 12	11
2, −6	−4
−2, 6	4
3, −4	−1
−3, 4	1

Because no integer factors of -12 have a sum of -2, $3x^2 - 2x - 4$ is nonfactorable over the integers.

Take Note

$3x^2 - 2x - 4$ is a prime polynomial.

EXAMPLE • 3

Factor: $2x^2 + 19x - 10$

Solution

Factors of −20 [2(−10)]	*Sum*
−1, 20	19

$$\begin{aligned} 2x^2 + 19x - 10 &= 2x^2 - x + 20x - 10 \\ &= (2x^2 - x) + (20x - 10) \\ &= x(2x - 1) + 10(2x - 1) \\ &= (2x - 1)(x + 10) \end{aligned}$$

YOU TRY IT • 3

Factor: $2a^2 + 13a - 7$

Your solution

EXAMPLE • 4

Factor: $24x^2y - 76xy + 40y$

Solution

The GCF is $4y$.

$24x^2y - 76xy + 40y = 4y(6x^2 - 19x + 10)$

Negative Factors of 60 [6(10)]	*Sum*
−1, −60	−61
−2, −30	−32
−3, −20	−23
−4, −15	−19

$$\begin{aligned} 6x^2 - 19x + 10 &= 6x^2 - 4x - 15x + 10 \\ &= (6x^2 - 4x) - (15x - 10) \\ &= 2x(3x - 2) - 5(3x - 2) \\ &= (3x - 2)(2x - 5) \end{aligned}$$

$$\begin{aligned} 24x^2y - 76xy + 40y &= 4y(6x^2 - 19x + 10) \\ &= 4y(3x - 2)(2x - 5) \end{aligned}$$

YOU TRY IT • 4

Factor: $15x^3 + 40x^2 - 80x$

Your solution

Solutions on pp. S23–S24

10.3 EXERCISES

OBJECTIVE A **To factor a trinomial of the form $ax^2 + bx + c$ by using trial factors**

For Exercises 1 to 70, factor by using trial factors.

1. $2x^2 + 3x + 1$

2. $5x^2 + 6x + 1$

3. $2y^2 + 7y + 3$

4. $3y^2 + 7y + 2$

5. $2a^2 - 3a + 1$

6. $3a^2 - 4a + 1$

7. $2b^2 - 11b + 5$

8. $3b^2 - 13b + 4$

9. $2x^2 + x - 1$

10. $4x^2 - 3x - 1$

11. $2x^2 - 5x - 3$

12. $3x^2 + 5x - 2$

13. $2t^2 - t - 10$

14. $2t^2 + 5t - 12$

15. $3p^2 - 16p + 5$

16. $6p^2 + 5p + 1$

17. $12y^2 - 7y + 1$

18. $6y^2 - 5y + 1$

19. $6z^2 - 7z + 3$

20. $9z^2 + 3z + 2$

21. $6t^2 - 11t + 4$

22. $10t^2 + 11t + 3$

23. $8x^2 + 33x + 4$

24. $7x^2 + 50x + 7$

25. $5x^2 - 62x - 7$

26. $9x^2 - 13x - 4$

27. $12y^2 + 19y + 5$

28. $5y^2 - 22y + 8$

29. $7a^2 + 47a - 14$

30. $11a^2 - 54a - 5$

31. $3b^2 - 16b + 16$

32. $6b^2 - 19b + 15$

33. $2z^2 - 27z - 14$

34. $4z^2 + 5z - 6$

35. $3p^2 + 22p - 16$

36. $7p^2 + 19p + 10$

37. $4x^2 + 6x + 2$

38. $12x^2 + 33x - 9$

39. $15y^2 - 50y + 35$

40. $30y^2 + 10y - 20$

41. $2x^3 - 11x^2 + 5x$

42. $2x^3 - 3x^2 - 5x$

43. $3a^2b - 16ab + 16b$

44. $2a^2b - ab - 21b$

45. $3z^2 + 95z + 10$

46. $8z^2 - 36z + 1$

47. $36x - 3x^2 - 3x^3$

48. $-2x^3 + 2x^2 + 4x$

49. $80y^2 - 36y + 4$

50. $24y^2 - 24y - 18$

51. $8z^3 + 14z^2 + 3z$

52. $6z^3 - 23z^2 + 20z$

53. $6x^2y - 11xy - 10y$

54. $8x^2y - 27xy + 9y$

55. $10t^2 - 5t - 50$

56. $16t^2 + 40t - 96$

57. $3p^3 - 16p^2 + 5p$

58. $6p^3 + 5p^2 + p$

59. $26z^2 + 98z - 24$

60. $30z^2 - 87z + 30$

61. $10y^3 - 44y^2 + 16y$

62. $14y^3 + 94y^2 - 28y$

63. $4yz^3 + 5yz^2 - 6yz$

64. $12a^3 + 14a^2 - 48a$

65. $42a^3 + 45a^2 - 27a$

66. $36p^2 - 9p^3 - p^4$

67. $9x^2y - 30xy^2 + 25y^3$

68. $8x^2y - 38xy^2 + 35y^3$

69. $9x^3y - 24x^2y^2 + 16xy^3$

70. $9x^3y + 12x^2y + 4xy$

For Exercises 71 and 72, let $(nx + p)$ and $(mx + q)$ be prime factors of the trinomial $ax^2 + bx + c$.

71. If n is even, must p be even or odd?

72. If p is even, must n be even or odd?

OBJECTIVE B To factor a trinomial of the form $ax^2 + bx + c$ by grouping

For Exercises 73 to 132, factor by grouping.

73. $6x^2 - 17x + 12$

74. $15x^2 - 19x + 6$

75. $5b^2 + 33b - 14$

76. $8x^2 - 30x + 25$

77. $6a^2 + 7a - 24$

78. $14a^2 + 15a - 9$

79. $4z^2 + 11z + 6$

80. $6z^2 - 25z + 14$

81. $22p^2 + 51p - 10$

82. $14p^2 - 41p + 15$

83. $8y^2 + 17y + 9$

84. $12y^2 - 145y + 12$

85. $18t^2 - 9t - 5$

86. $12t^2 + 28t - 5$

87. $6b^2 + 71b - 12$

88. $8b^2 + 65b + 8$

89. $9x^2 + 12x + 4$

90. $25x^2 - 30x + 9$

91. $6b^2 - 13b + 6$

92. $20b^2 + 37b + 15$

93. $33b^2 + 34b - 35$

94. $15b^2 - 43b + 22$

95. $18y^2 - 39y + 20$

96. $24y^2 + 41y + 12$

97. $15a^2 + 26a - 21$

98. $6a^2 + 23a + 21$

99. $8y^2 - 26y + 15$

100. $18y^2 - 27y + 4$

101. $8z^2 + 2z - 15$

102. $10z^2 + 3z - 4$

103. $15x^2 - 82x + 24$

104. $13z^2 + 49z - 8$

105. $10z^2 - 29z + 10$

106. $15z^2 - 44z + 32$

107. $36z^2 + 72z + 35$

108. $16z^2 + 8z - 35$

109. $3x^2 + xy - 2y^2$

110. $6x^2 + 10xy + 4y^2$

111. $3a^2 + 5ab - 2b^2$

112. $2a^2 - 9ab + 9b^2$

113. $4y^2 - 11yz + 6z^2$

114. $2y^2 + 7yz + 5z^2$

115. $28 + 3z - z^2$

116. $15 - 2z - z^2$

117. $8 - 7x - x^2$

118. $12 + 11x - x^2$

119. $9x^2 + 33x - 60$

120. $16x^2 - 16x - 12$

121. $24x^2 - 52x + 24$

122. $60x^2 + 95x + 20$

123. $35a^4 + 9a^3 - 2a^2$

124. $15a^4 + 26a^3 + 7a^2$

125. $15b^2 - 115b + 70$

126. $25b^2 + 35b - 30$

127. $3x^2 - 26xy + 35y^2$

128. $4x^2 + 16xy + 15y^2$

129. $216y^2 - 3y - 3$

130. $360y^2 + 4y - 4$

131. $21 - 20x - x^2$

132. $18 + 17x - x^2$

For Exercises 133 to 136, information is given about the signs of b and c in the trinomial $ax^2 + bx + c$, where $a > 0$. If you want to factor $ax^2 + bx + c$ by grouping, you look for factors of ac whose sum is b. In each case, state whether the factors of ac should be two positive numbers, two negative numbers, or one positive and one negative number.

133. $b > 0$ and $c > 0$

134. $b < 0$ and $c < 0$

135. $b < 0$ and $c > 0$

136. $b > 0$ and $c < 0$

Applying the Concepts

137. In your own words, explain how the signs of the last terms of the two binomial factors of a trinomial are determined.

For Exercises 138 to 143, factor.

138. $(x + 1)^2 - (x + 1) - 6$

139. $(x - 2)^2 + 3(x - 2) + 2$

140. $(y + 3)^2 - 5(y + 3) + 6$

141. $2(y + 2)^2 - (y + 2) - 3$

142. $3(a + 2)^2 - (a + 2) - 4$

143. $4(y - 1)^2 - 7(y - 1) - 2$

SECTION 10.4 Special Factoring

OBJECTIVE A To factor the difference of two squares and perfect-square trinomials

A polynomial of the form $a^2 - b^2$ is called a **difference of two squares.** Recall the following relationship from Objective 9.3D.

Sum and difference of the same terms		Difference of two squares
$(a + b)(a - b)$	$=$	$a^2 - b^2$

Take Note

Note that the polynomial $x^2 + y^2$ is the *sum* of two squares. The sum of two squares is nonfactorable over the integers.

Factoring the Difference of Two Squares

The difference of two squares factors as the sum and difference of the same terms.

$$a^2 - b^2 = (a + b)(a - b)$$

HOW TO • 1 Factor: $x^2 - 16$

$x^2 - 16 = (x)^2 - (4)^2$ • $x^2 - 16$ is the difference of two squares.

$= (x + 4)(x - 4)$ • Factor the difference of squares.

Check: $(x + 4)(x - 4) = x^2 - 4x + 4x - 16$

$= x^2 - 16$

HOW TO • 2 Factor: $8x^3 - 18x$

$8x^3 - 18x = 2x(4x^2 - 9)$ • The GCF is $2x$.

$= 2x[(2x)^2 - 3^2]$ • $4x^2 - 9$ is the difference of two squares.

$= 2x(2x + 3)(2x - 3)$ • Factor the difference of squares.

You should check the factorization.

HOW TO • 3 Factor: $x^2 - 10$

Because 10 cannot be written as the square of an integer, $x^2 - 10$ is nonfactorable over the integers.

A trinomial that can be written as the square of a binomial is called a **perfect-square trinomial.** Recall the pattern for finding the square of a binomial.

$$(a + b)^2 = a^2 + 2ab + b^2$$

Square of the first term (a^2)

Twice the product of the two terms ($2ab$)

Square of the last term (b^2)

Factoring a Perfect-Square Trinomial

A perfect-square trinomial factors as the square of a binomial.

$$a^2 + 2ab + b^2 = (a + b)^2$$
$$a^2 - 2ab + b^2 = (a - b)^2$$

HOW TO • 4 Factor: $4x^2 - 20x + 25$

Because the first and last terms are squares $[(2x)^2 = 4x^2; 5^2 = 25]$, try to factor this as the square of a binomial. Check the factorization.

$4x^2 - 20x + 25 = (2x - 5)^2$

Check: $(2x - 5)^2 = (2x)^2 + 2(2x)(-5) + 5^2$
$= 4x^2 - 20x + 25$ • **The factorization is correct.**

$4x^2 - 20x + 25 = (2x - 5)^2$

HOW TO • 5 Factor: $4x^2 + 37x + 9$

Because the first and last terms are squares $[(2x)^2 = 4x^2; 3^2 = 9]$, try to factor this as the square of a binomial. Check the proposed factorization.

$4x^2 + 37x + 9 = (2x + 3)^2$

Check: $(2x + 3)^2 = (2x)^2 + 2(2x)(3) + 3^2$
$= 4x^2 + 12x + 9$

Because $4x^2 + 12x + 9 \neq 4x^2 + 37x + 9$, the proposed factorization is not correct. In this case, the polynomial is not a perfect-square trinomial. It may, however, still factor. In fact, $4x^2 + 37x + 9 = (4x + 1)(x + 9)$.

EXAMPLE • 1

Factor: $16x^2 - y^2$

Solution

$16x^2 - y^2 = (4x)^2 - y^2$ • **The difference of two squares**

$= (4x + y)(4x - y)$ • **Factor.**

YOU TRY IT • 1

Factor: $25a^2 - b^2$

Your solution

EXAMPLE • 2

Factor: $z^4 - 16$

Solution

$z^4 - 16 = (z^2)^2 - 4^2$ • **The difference of two squares**

$= (z^2 + 4)(z^2 - 4)$ • **The difference of two squares**

$= (z^2 + 4)(z^2 - 2^2)$

$= (z^2 + 4)(z + 2)(z - 2)$ • **Factor.**

YOU TRY IT • 2

Factor: $n^4 - 81$

Your solution

Solutions on p. S24

EXAMPLE • 3

Factor: $9x^2 - 30x + 25$

Solution

$9x^2 = (3x)^2$, $25 = (5)^2$

$9x^2 - 30x + 25 = (3x - 5)^2$

Check: $(3x - 5)^2 = (3x)^2 + 2(3x)(-5) + 5^2$
$= 9x^2 - 30x + 25$

YOU TRY IT • 3

Factor: $16y^2 + 8y + 1$

Your solution

EXAMPLE • 4

Factor: $9x^2 + 40x + 16$

Solution

Because $9x^2 = (3x)^2$, $16 = 4^2$, and $40x \neq 2(3x)(4)$, the trinomial is not a perfect-square trinomial.

Try to factor by another method.

$9x^2 + 40x + 16 = (9x + 4)(x + 4)$

YOU TRY IT • 4

Factor: $x^2 + 15x + 36$

Your solution

Solutions on p. S24

OBJECTIVE B To factor completely

Tips for Success

You now have learned to factor many different types of polynomials. You will need to be able to recognize each of the situations described in the box at the right. To test yourself, try the exercises in the Chapter Review.

General Factoring Strategy

1. Is there a common factor? If so, factor out the common factor.
2. Is the polynomial the difference of two perfect squares? If so, factor.
3. Is the polynomial a perfect-square trinomial? If so, factor.
4. Is the polynomial a trinomial that is the product of two binomials? If so, factor.
5. Does the polynomial contain four terms? If so, try factoring by grouping.
6. Is each binomial factor nonfactorable over the integers? If not, factor the binomial.

HOW TO • 6 Factor: $z^3 + 4z^2 - 9z - 36$

$z^3 + 4z^2 - 9z - 36 = (z^3 + 4z^2) - (9z + 36)$
• Factor by grouping. Recall that $-9z - 36 = -(9z + 36)$.

$= z^2(z + 4) - 9(z + 4)$
• $z^3 + 4z^2 = z^2(z + 4)$; $9z + 36 = 9(z + 4)$

$= (z + 4)(z^2 - 9)$
• Factor out the common binomial factor $(z + 4)$.

$= (z + 4)(z + 3)(z - 3)$
• Factor the difference of squares.

EXAMPLE • 5

Factor: $3x^2 - 48$

Solution

The GCF is 3.

$3x^2 - 48 = 3(x^2 - 16)$

$= 3(x + 4)(x - 4)$ • **Factor the difference of two squares.**

YOU TRY IT • 5

Factor: $12x^3 - 75x$

Your solution

EXAMPLE • 6

Factor: $x^3 - 3x^2 - 4x + 12$

Solution

Factor by grouping.

$x^3 - 3x^2 - 4x + 12$

$= (x^3 - 3x^2) - (4x - 12)$ • **Factor by grouping.**

$= x^2(x - 3) - 4(x - 3)$ • **$x - 3$ is the common factor.**

$= (x - 3)(x^2 - 4)$ • **$x^2 - 4$ is the difference of two squares.**

$= (x - 3)(x + 2)(x - 2)$ • **Factor.**

YOU TRY IT • 6

Factor: $a^2b - 7a^2 - b + 7$

Your solution

EXAMPLE • 7

Factor: $4x^2y^2 + 12xy^2 + 9y^2$

Solution

The GCF is y^2.

$4x^2y^2 + 12xy^2 + 9y^2$

$= y^2(4x^2 + 12x + 9)$ • **Factor the GCF, y^2.**

$= y^2(2x + 3)^2$ • **Factor the perfect-square trinomial.**

YOU TRY IT • 7

Factor: $4x^3 + 28x^2 - 120x$

Your solution

Solutions on p. S24

10.4 EXERCISES

OBJECTIVE A **To factor the difference of two squares and perfect-square trinomials**

1. **a.** Provide an example of a binomial that is the difference of two squares.

   b. Provide an example of a perfect-square trinomial.

2. Explain why a binomial that is the sum of two squares is nonfactorable over the integers.

For Exercises 3 to 44, factor.

3. $x^2 - 4$ **4.** $x^2 - 9$ **5.** $a^2 - 81$ **6.** $a^2 - 49$

7. $y^2 + 2y + 1$ **8.** $y^2 + 14y + 49$ **9.** $a^2 - 2a + 1$ **10.** $x^2 - 12x + 36$

11. $4x^2 - 1$ **12.** $9x^2 - 16$ **13.** $x^6 - 9$ **14.** $y^{12} - 4$

15. $x^2 + 8x - 16$ **16.** $z^2 - 18z - 81$ **17.** $x^2 + 2xy + y^2$ **18.** $x^2 + 6xy + 9y^2$

19. $4a^2 + 4a + 1$ **20.** $25x^2 + 10x + 1$ **21.** $9x^2 - 1$ **22.** $1 - 49x^2$

23. $1 - 64x^2$ **24.** $t^2 + 36$ **25.** $x^2 + 64$ **26.** $64a^2 - 16a + 1$

27. $9a^2 + 6a + 1$ **28.** $x^4 - y^2$ **29.** $b^4 - 16a^2$ **30.** $16b^2 + 8b + 1$

31. $4a^2 - 20a + 25$ **32.** $4b^2 + 28b + 49$ **33.** $9a^2 - 42a + 49$ **34.** $9x^2 - 16y^2$

35. $25z^2 - y^2$ **36.** $x^2y^2 - 4$ **37.** $a^2b^2 - 25$ **38.** $16 - x^2y^2$

39. $25x^2 - 1$

40. $25a^2 + 30ab + 9b^2$

41. $4a^2 - 12ab + 9b^2$

42. $49x^2 + 28xy + 4y^2$

43. $4y^2 - 36yz + 81z^2$

44. $64y^2 - 48yz + 9z^2$

45. Which of the following expressions can be factored as the square of a binomial, given that a and b are positive numbers?
(i) $a^2x^2 - 2abx + b^2$ (ii) $a^2x^2 - 2abx - b^2$
(iii) $a^2x^2 + 2abx + b^2$ (iv) $a^2x^2 + 2abx - b^2$

OBJECTIVE B To factor completely

For Exercises 46 to 123, factor.

46. $8y^2 - 2$

47. $12n^2 - 48$

48. $3a^3 + 6a^2 + 3a$

49. $4rs^2 - 4rs + r$

50. $m^4 - 256$

51. $81 - t^4$

52. $9x^2 + 13x + 4$

53. $x^2 + 10x + 16$

54. $16y^4 + 48y^3 + 36y^2$

55. $36c^4 - 48c^3 + 16c^2$

56. $y^8 - 81$

57. $32s^4 - 2$

58. $25 - 20p + 4p^2$

59. $9 + 24a + 16a^2$

60. $(4x - 3)^2 - y^2$

61. $(2x + 5)^2 - 25$

62. $(x^2 - 4x + 4) - y^2$

63. $(4x^2 + 12x + 9) - 4y^2$

64. $5x^2 - 5$

65. $2x^2 - 18$

66. $x^3 + 4x^2 + 4x$

67. $y^3 - 10y^2 + 25y$

68. $x^4 + 2x^3 - 35x^2$

69. $a^4 - 11a^3 + 24a^2$

70. $5b^2 + 75b + 180$

71. $6y^2 - 48y + 72$

72. $3a^2 + 36a + 10$

73. $5a^2 - 30a + 4$

74. $2x^2y + 16xy - 66y$

75. $3a^2b + 21ab - 54b$

76. $x^3 - 6x^2 - 5x$

77. $b^3 - 8b^2 - 7b$

78. $3y^2 - 36$

79. $3y^2 - 147$

80. $20a^2 + 12a + 1$

81. $12a^2 - 36a + 27$

82. $x^2y^2 - 7xy^2 - 8y^2$

83. $a^2b^2 + 3a^2b - 88a^2$

84. $10a^2 - 5ab - 15b^2$

85. $16x^2 - 32xy + 12y^2$

86. $50 - 2x^2$

87. $72 - 2x^2$

88. $a^2b^2 - 10ab^2 + 25b^2$

89. $a^2b^2 + 6ab^2 + 9b^2$

90. $12a^3b - a^2b^2 - ab^3$

91. $2x^3y - 7x^2y^2 + 6xy^3$

92. $12a^3 - 12a^2 + 3a$

93. $18a^3 + 24a^2 + 8a$

94. $243 + 3a^2$

95. $75 + 27y^2$

96. $12a^3 - 46a^2 + 40a$

97. $24x^3 - 66x^2 + 15x$

98. $4a^3 + 20a^2 + 25a$

99. $2a^3 - 8a^2b + 8ab^2$

100. $27a^2b - 18ab + 3b$

101. $a^2b^2 - 6ab^2 + 9b^2$

102. $48 - 12x - 6x^2$

103. $21x^2 - 11x^3 - 2x^4$

104. $x^4 - x^2y^2$

105. $b^4 - a^2b^2$

106. $18a^3 + 24a^2 + 8a$

107. $32xy^2 - 48xy + 18x$

108. $2b + ab - 6a^2b$

109. $15y^2 - 2xy^2 - x^2y^2$

110. $4x^4 - 38x^3 + 48x^2$

111. $3x^2 - 27y^2$

112. $x^4 - 25x^2$

113. $y^3 - 9y$

114. $a^4 - 16$

115. $15x^4y^2 - 13x^3y^3 - 20x^2y^4$

116. $45y^2 - 42y^3 - 24y^4$

117. $a(2x - 2) + b(2x - 2)$

118. $4a(x - 3) - 2b(x - 3)$

119. $x^2(x - 2) - (x - 2)$

120. $y^2(a - b) - (a - b)$

121. $a(x^2 - 4) + b(x^2 - 4)$

122. $x(a^2 - b^2) - y(a^2 - b^2)$

123. $4(x - 5) - x^2(x - 5)$

124. The expression $x^2(x - a)(x + b)$, where a and b are positive integers, is the factored form of a polynomial P. What is the degree of the polynomial P?

Applying the Concepts

For Exercises 125 to 130, find all integers k such that the trinomial is a perfect-square trinomial.

125. $4x^2 - kx + 9$

126. $x^2 + 6x + k$

127. $64x^2 + kxy + y^2$

128. $x^2 - 2x + k$

129. $25x^2 - kx + 1$

130. $x^2 + 10x + k$

SECTION

10.5 Solving Equations

OBJECTIVE A To solve equations by factoring 

The Multiplication Property of Zero states that the product of a number and zero is zero. This property is stated below.

If a is a real number, then $a \cdot 0 = 0 \cdot a = 0$.

Now consider $a \cdot b = 0$. For this to be a true equation, then either $a = 0$ or $b = 0$.

Principle of Zero Products

If the product of two factors is zero, then at least one of the factors must be zero.

If $a \cdot b = 0$, then $a = 0$ or $b = 0$.

The Principle of Zero Products is used to solve some equations.

HOW TO 1 Solve: $(x-2)(x-3) = 0$

$(x-2)(x-3) = 0$

$x - 2 = 0 \quad x - 3 = 0$ • **Let each factor equal zero (the Principle of Zero Products).**

$x = 2 \quad x = 3$ • **Solve each equation for x.**

Check:

$(x-2)(x-3)$	0
$(2-2)(2-3)$	0
$0(-1)$	0
$0 = 0$	• **A true equation**

$(x-2)(x-3)$	0
$(3-2)(3-3)$	0
$(1)(0)$	0
$0 = 0$	• **A true equation**

The solutions are 2 and 3.

An equation that can be written in the form $ax^2 + bx + c = 0$, $a \neq 0$, is a **quadratic equation.** A quadratic equation is in **standard form** when the polynomial is in descending order and equal to zero. The quadratic equations at the right are in standard form.

$3x^2 + 2x + 1 = 0$
$a = 3, b = 2, c = 1$

$4x^2 - 3x + 2 = 0$
$a = 4, b = -3, c = 2$

HOW TO • 2 Solve: $2x^2 + x = 6$

$$2x^2 + x = 6$$

$$2x^2 + x - 6 = 0$$ • **Write the equation in standard form.**

$$(2x - 3)(x + 2) = 0$$ • **Factor.**

$$2x - 3 = 0 \qquad x + 2 = 0$$ • **Use the Principle of Zero Products.**

$$2x = 3 \qquad x = -2$$ • **Solve each equation for x.**

$$x = \frac{3}{2}$$

Check: $\frac{3}{2}$ and -2 check as solutions.

The solutions are $\frac{3}{2}$ and -2.

EXAMPLE • 1

Solve: $x(x - 3) = 0$

Solution

$$x(x - 3) = 0$$

$$x = 0 \quad x - 3 = 0$$

$$x = 3$$

• **Use the Principle of Zero Products.**

The solutions are 0 and 3.

YOU TRY IT • 1

Solve: $2x(x + 7) = 0$

Your solution

EXAMPLE • 2

Solve: $2x^2 - 50 = 0$

Solution

$$2x^2 - 50 = 0$$

$$2(x^2 - 25) = 0$$ • **Factor out the GCF, 2.**

$$2(x + 5)(x - 5) = 0$$ • **Factor the difference of two squares.**

$$x + 5 = 0 \qquad x - 5 = 0$$

$$x = -5 \qquad x = 5$$

• **Use the Principle of Zero Products.**

The solutions are -5 and 5.

YOU TRY IT • 2

Solve: $4x^2 - 9 = 0$

Your solution

EXAMPLE • 3

Solve: $(x - 3)(x - 10) = -10$

Solution

$$(x - 3)(x - 10) = -10$$

$$x^2 - 13x + 30 = -10$$ • **Multiply $(x - 3)(x - 10)$.**

$$x^2 - 13x + 40 = 0$$ • **Add 10 to each side of the equation. The equation is now in standard form.**

$$(x - 8)(x - 5) = 0$$

$$x - 8 = 0 \qquad x - 5 = 0$$

$$x = 8 \qquad x = 5$$

The solutions are 8 and 5.

YOU TRY IT • 3

Solve: $(x + 2)(x - 7) = 52$

Your solution

Solutions on p. S24

OBJECTIVE B To solve application problems 

EXAMPLE • 4

The sum of the squares of two consecutive positive even integers is equal to 100. Find the two integers.

Strategy

First positive even integer: n
Second positive even integer: $n + 2$

The sum of the square of the first positive even integer and the square of the second positive even integer is 100.

Solution

$$n^2 + (n + 2)^2 = 100$$
$$n^2 + n^2 + 4n + 4 = 100$$
$$2n^2 + 4n + 4 = 100$$
$$2n^2 + 4n - 96 = 0$$
$$2(n^2 + 2n - 48) = 0$$
$$2(n - 6)(n + 8) = 0$$

• Quadratic equation in standard form

$n - 6 = 0$ $\quad$ $n + 8 = 0$
$n = 6$ $\quad$ $n = -8$

• Principle of Zero Products

Because -8 is not a positive even integer, it is not a solution.

$n = 6$
$n + 2 = 6 + 2 = 8$

The two integers are 6 and 8.

YOU TRY IT • 4

The sum of the squares of two consecutive positive integers is 61. Find the two integers.

Your strategy

Your solution

Solution on p. S24

EXAMPLE 5

A stone is thrown into a well with an initial speed of 4 ft/s. The well is 420 ft deep. How many seconds later will the stone hit the bottom of the well? Use the equation $d = vt + 16t^2$, where d is the distance in feet that the stone travels in t seconds when its initial speed is v feet per second.

Strategy

To find the time for the stone to drop to the bottom of the well, replace the variables d and v by their given values and solve for t.

Solution

$$d = vt + 16t^2$$
$$420 = 4t + 16t^2$$
$$0 = -420 + 4t + 16t^2$$
$$0 = 16t^2 + 4t - 420$$
$$0 = 4(4t^2 + t - 105)$$
$$0 = 4(4t + 21)(t - 5)$$

• Quadratic equation in standard form

$$4t + 21 = 0 \qquad t - 5 = 0$$
$$4t = -21 \qquad t = 5$$
$$t = -\frac{21}{4}$$

• Principle of Zero Products

Because the time cannot be a negative number, $-\frac{21}{4}$ is not a solution.

The stone will hit the bottom of the well 5 s later.

YOU TRY IT 5

The length of a rectangle is 4 in. longer than twice the width. The area of the rectangle is 96 in^2. Find the length and width of the rectangle.

Your strategy

Your solution

Solution on p. S25

10.5 EXERCISES

OBJECTIVE A To solve equations by factoring

1. In your own words, explain the Principle of Zero Products.

2. Fill in the blanks. If $(x + 5)(2x - 7) = 0$, then ________ $= 0$ or ________ $= 0$.

For Exercises 3 to 60, solve.

3. $(y + 3)(y + 2) = 0$
4. $(y - 3)(y - 5) = 0$
5. $(z - 7)(z - 3) = 0$
6. $(z + 8)(z - 9) = 0$
7. $x(x - 5) = 0$
8. $x(x + 2) = 0$
9. $a(a - 9) = 0$
10. $a(a + 12) = 0$
11. $y(2y + 3) = 0$
12. $t(4t - 7) = 0$
13. $2a(3a - 2) = 0$
14. $4b(2b + 5) = 0$
15. $(b + 2)(b - 5) = 0$
16. $(b - 8)(b + 3) = 0$
17. $x^2 - 81 = 0$
18. $x^2 - 121 = 0$
19. $4x^2 - 49 = 0$
20. $16x^2 - 1 = 0$
21. $9x^2 - 1 = 0$
22. $16x^2 - 49 = 0$
23. $x^2 + 6x + 8 = 0$
24. $x^2 - 8x + 15 = 0$
25. $z^2 + 5z - 14 = 0$
26. $z^2 + z - 72 = 0$
27. $2a^2 - 9a - 5 = 0$
28. $3a^2 + 14a + 8 = 0$
29. $6z^2 + 5z + 1 = 0$
30. $6y^2 - 19y + 15 = 0$
31. $x^2 - 3x = 0$
32. $a^2 - 5a = 0$
33. $x^2 - 7x = 0$
34. $2a^2 - 8a = 0$
35. $a^2 + 5a = -4$
36. $a^2 - 5a = 24$
37. $y^2 - 5y = -6$
38. $y^2 - 7y = 8$

39. $2t^2 + 7t = 4$

40. $3t^2 + t = 10$

41. $3t^2 - 13t = -4$

42. $5t^2 - 16t = -12$

43. $x(x - 12) = -27$

44. $x(x - 11) = 12$

45. $y(y - 7) = 18$

46. $y(y + 8) = -15$

47. $p(p + 3) = -2$

48. $p(p - 1) = 20$

49. $y(y + 4) = 45$

50. $y(y - 8) = -15$

51. $x(x + 3) = 28$

52. $p(p - 14) = 15$

53. $(x + 8)(x - 3) = -30$

54. $(x + 4)(x - 1) = 14$

55. $(z - 5)(z + 4) = 52$

56. $(z - 8)(z + 4) = -35$

57. $(z - 6)(z + 1) = -10$

58. $(a + 3)(a + 4) = 72$

59. $(a - 4)(a + 7) = -18$

60. $(2x + 5)(x + 1) = -1$

For Exercises 61 and 62, the equation $ax^2 + bx + c = 0$, $a > 0$, is a quadratic equation that can be solved by factoring and then using the Principle of Zero Products.

61. If $ax^2 + bx + c = 0$ has one positive solution and one negative solution, is c greater than, less than, or equal to zero?

62. If zero is one solution of $ax^2 + bx + c = 0$, is c greater than, less than, or equal to zero?

OBJECTIVE B To solve application problems

63. **Number Sense** The square of a positive number is six more than five times the positive number. Find the number.

64. **Number Sense** The square of a negative number is fifteen more than twice the negative number. Find the number.

65. **Number Sense** The sum of two numbers is six. The sum of the squares of the two numbers is twenty. Find the two numbers.

66. **Number Sense** The sum of two numbers is eight. The sum of the squares of the two numbers is thirty-four. Find the two numbers.

For Exercises 67 and 68, use the following problem situation: The sum of the squares of two consecutive positive integers is 113. Find the two integers.

67. Which equation could be used to solve this problem?
(i) $x^2 + x^2 + 1 = 113$ (ii) $x^2 + (x + 1)^2 = 113$ (iii) $(x + x + 1)^2 = 113$

68. Suppose the solutions of the correct equation in Exercise 67 are -8 and 7. Which solution should be eliminated, and why?

69. **Number Sense** The sum of the squares of two consecutive positive integers is forty-one. Find the two integers.

70. **Number Sense** The sum of the squares of two consecutive positive even integers is one hundred. Find the two integers.

71. **Number Sense** The sum of two numbers is ten. The product of the two numbers is twenty-one. Find the two numbers.

72. **Number Sense** The sum of two numbers is thirteen. The product of the two numbers is forty. Find the two numbers.

Sum of Natural Numbers The formula $S = \frac{n^2 + n}{2}$ gives the sum S of the first n natural numbers. Use this formula for Exercises 73 and 74.

73. How many consecutive natural numbers beginning with 1 will give a sum of 78?

74. How many consecutive natural numbers beginning with 1 will give a sum of 171?

Sports The formula $N = \frac{t^2 - t}{2}$ gives the number N of football games that must be scheduled in a league with t teams if each team is to play every other team once. Use this formula for Exercises 75 and 76.

© Reuters/Corbis

75. How many teams are in a league that schedules 15 games in such a way that each team plays every other team once?

76. How many teams are in a league that schedules 45 games in such a way that each team plays every other team once?

Physics The distance s, in feet, that an object will fall (neglecting air resistance) in t seconds is given by $s = vt + 16t^2$, where v is the initial velocity of the object in feet per second. Use this formula for Exercises 77 and 78.

77. An object is released from the top of a building 192 ft high. The initial velocity is 16 ft/s, and air resistance is neglected. How many seconds later will the object hit the ground?

AP Photo/Wally Santana

78. Taipei 101 in Taipei, Taiwan, is the world's tallest inhabited building. The top of the spire is 1667 ft above ground. If an object is released from this building at a point 640 ft above the ground at an initial velocity of 48 ft/s, assuming no air resistance, how many seconds later will the object reach the ground?

Sports The height h, in feet, an object will attain (neglecting air resistance) in t seconds is given by $h = vt - 16t^2$, where v is the initial velocity of the object in feet per second. Use this formula for Exercises 79 and 80.

79. A golf ball is thrown onto a cement surface and rebounds straight up. The initial velocity of the rebound is 60 ft/s. How many seconds later will the golf ball return to the ground?

80. A foul ball leaves a bat, hits home plate, and travels straight up with an initial velocity of 64 ft/s. How many seconds later will the ball be 64 ft above the ground?

81. Geometry The length of a rectangle is 5 in. more than twice its width. Its area is 75 in^2. Find the length and width of the rectangle.

82. Geometry The width of a rectangle is 5 ft less than the length. The area of the rectangle is 176 ft^2. Find the length and width of the rectangle.

83. **Geometry** The height of a triangle is 4 m more than twice the length of the base. The area of the triangle is 35 m^2. Find the height of the triangle.

84. **Geometry** The length of each side of a square is extended 5 in. The area of the resulting square is 64 in^2. Find the length of a side of the original square.

85. **Basketball** See the news clipping at the right. If the area of the rectangular 3-second lane is 304 ft^2, find the width of the lane.

> **In the News**
>
> **New Lane for Basketball Court**
>
> The International Basketball Federation announced changes to the basketball court used in international competition. The 3-second lane, currently a trapezoid, will be a rectangle 3 ft longer than it is wide, similar to the one used in NBA games.
>
> *Source: The New York Times*

86. **Gardening** A small garden measures 8 ft by 10 ft. A uniform border around the garden increases the total area to 143 ft^2. What is the width of the border?

87. **Publishing** The page of a book measures 6 in. by 9 in. A uniform border around the page leaves 28 in^2 for type. What are the dimensions of the type area?

Anatomy The pupil is the opening in the iris that lets light into the eye. In bright light, the iris expands so that the pupil is smaller; in low light, the iris contracts so that the pupil is larger. If x is the width, in millimeters, of the iris, then the area of the iris is given by $A = (12\pi x - \pi x^2)$ mm^2. Use this formula for Exercises 88 and 89.

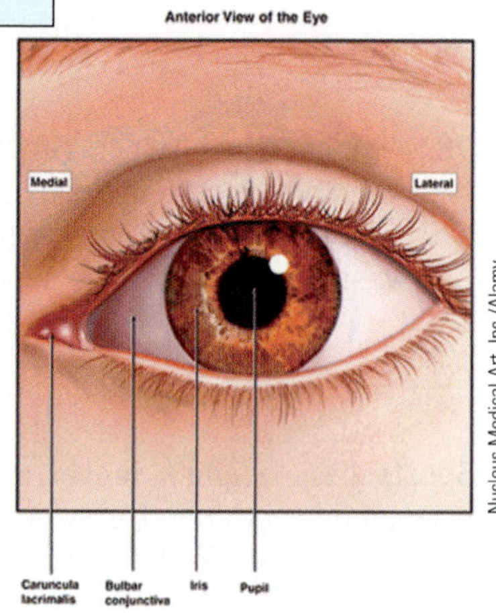

Nucleus Medical Art, Inc./Alamy

88. Find the width of the iris if the area of the iris is 20π mm^2.

89. Find the width of the iris if the area of the iris is 27π mm^2.

Applying the Concepts

90. Find $3n^2$ if $n(n + 5) = -4$.

91. Find $2n^2$ if $n(n + 3) = 4$.

For Exercises 92 to 95, solve.

92. $2y(y + 4) = -5(y + 3)$

93. $(b + 5)^2 = 16$

94. $p^3 = 9p^2$

95. $(x + 3)(2x - 1) = (3 - x)(5 - 3x)$

96. Explain the error made in solving the equation at the right. Solve the equation correctly.

$$(x + 2)(x - 3) = 6$$
$$x + 2 = 6 \quad x - 3 = 6$$
$$x = 4 \quad x = 9$$

97. Explain the error made in solving the equation at the right. Solve the equation correctly.

$$x^2 = x$$
$$\frac{x^2}{x} = \frac{x}{x}$$
$$x = 1$$

FOCUS ON PROBLEM SOLVING

Making a Table

© Bill Aron/PhotoEdit

There are six students using a gym. The wall on the gym has six lockers that are numbered 1, 2, 3, 4, 5, and 6. After a practice, the first student goes by and opens all the lockers. The second student shuts every second locker, the third student changes every third locker (opens a locker if it is shut and shuts a locker if it is open), the fourth student changes every fourth locker, the fifth student changes every fifth locker, and the sixth student changes every sixth locker. After the sixth student makes changes, which lockers are open?

One method of solving this problem would be to create a table, as shown below.

Student / *Locker*	1	2	3	4	5	6
1	O	O	O	O	O	O
2	O	C	C	C	C	C
3	O	O	C	C	C	C
4	O	C	C	O	O	O
5	O	O	O	O	C	C
6	O	C	O	O	O	C

From this table, lockers 1 and 4 are open after the sixth student passes through.

Now extend this scenario to more lockers and students. In each case, the *n*th student changes multiples of the *n*th locker. For instance, the 8th student would change the 8th, 16th, 24th, . . .

1. Suppose there are 10 lockers and 10 students. Which lockers will remain open?

2. Suppose there are 16 lockers and 16 students. Which lockers will remain open?

3. Suppose there are 25 lockers and 25 students. Which lockers will remain open?

4. Suppose there are 40 lockers and 40 students. Which lockers will remain open?

5. Suppose there are 50 lockers and 50 students. Which lockers will remain open?

6. Make a conjecture as to which lockers would be open if there were 100 lockers and 100 students.

7. Give a reason why your conjecture should be true. [*Hint:* Consider how many factors there are for the door numbers that remain open and for those that remain closed. For instance, with 40 lockers and 40 students, locker 36 (which remains open) has factors 1, 2, 3, 4, 6, 9, 12, 18, and 36—an odd number of factors. Locker 28, a closed locker, has factors 1, 2, 4, 7, 14, and 28—an even number of factors.]

PROJECTS AND GROUP ACTIVITIES

Evaluating Polynomials Using a Graphing Calculator

A graphing calculator can be used to evaluate a polynomial. To illustrate the method, consider the polynomial $2x^3 - 3x^2 + 4x - 7$. The keystrokes below are for a TI-84 Plus calculator, but the keystrokes for other calculators will closely follow these keystrokes.

Press the Y= key. You will see a screen similar to the one below. Press CLEAR to erase any expression next to Y1.

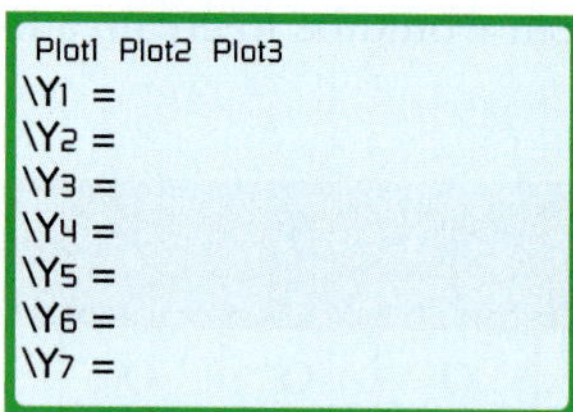

Enter the polynomial as follows. The ^ key is used to enter an exponent.

2 X,T,θ,n ^ 3 − 3 X,T,θ,n x² + 4 X,T,θ,n − 7

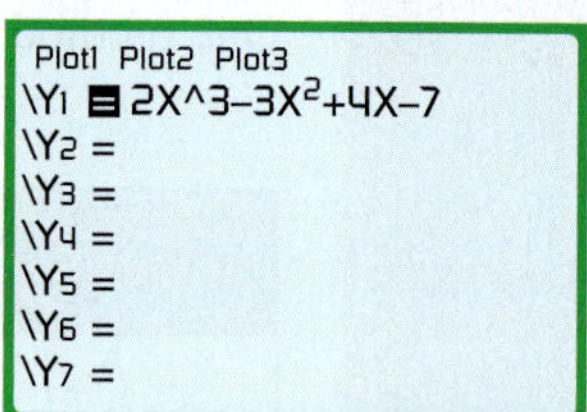

Take Note

Once the polynomial has been entered in Y1, there are several methods that can be used to evaluate it. We will show just one option.

To evaluate the polynomial when $x = 3$, first return to what is called the HOME screen by pressing 2ND QUIT.

Enter the following keystrokes. Sample screens are shown at the right.

(1) 3 STO▸ X,T,θ,n ENTER

(2) VARS ▸ ENTER

(3) ENTER

(4) ENTER

The value of the polynomial when $x = 3$ is 32.

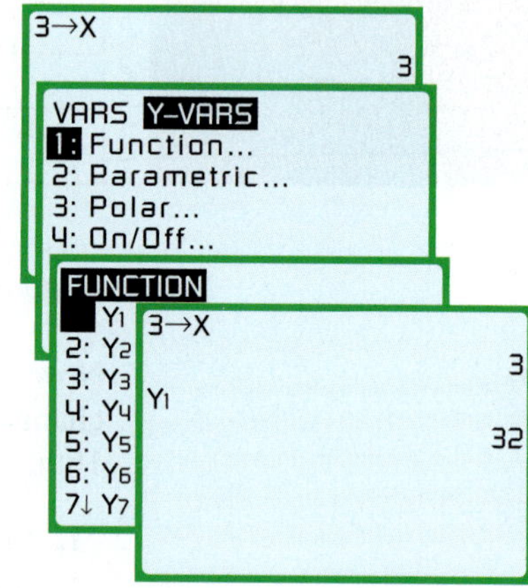

To evaluate the polynomial at a different value of x, repeat Steps 1 through 4. For instance, to evaluate the polynomial when $x = -4$, we would have

(1) (−) 4 STO▸ X,T,θ,n ENTER

(2) VARS ▸ ENTER

(3) ENTER

(4) ENTER

The value of the polynomial when $x = -4$ is -199.

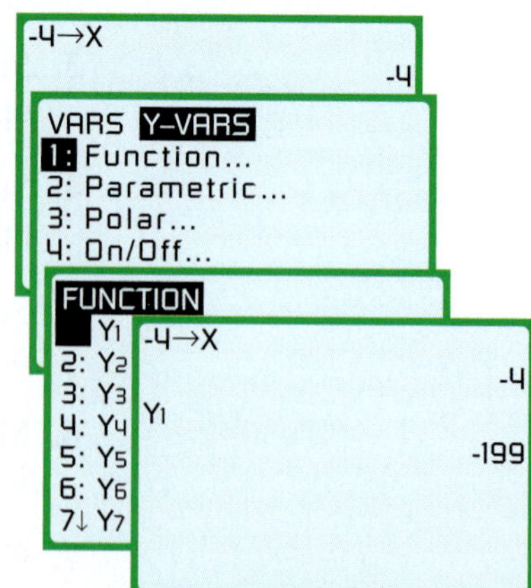

Here are some practice exercises.
Evaluate the polynomial for the given value.

1. $2x^2 - 3x + 7; x = 4$

2. $3x^2 + 7x - 12; x = -3$

3. $3x^3 - 2x^2 + 6x - 8; x = 3$

4. $2x^3 + 4x^2 - x - 2; x = 2$

5. $x^4 - 3x^3 + 6x^2 + 5x - 1; x = 2$

6. $x^5 - x^3 + 2x - 7; x = -4$

Exploring Integers *Number theory* is a branch of mathematics that focuses on integers and the relationships that exist among the integers. Some of the results from this field of study have important, practical applications for sending sensitive information such as credit card numbers over the Internet. In this project you will be asked to discover some of those relationships.

Mary Evans Picture Library/Alamy

Marin Mersenne (1588–1648) was a French mathematician, scientist, and philosopher known for his development of the Mersenne primes.

1. If n is an integer, explain why the product $n(n + 1)$ is always an even integer.
2. If n is an integer, explain why $2n$ is always an even integer.
3. If n is an integer, explain why $2n + 1$ is always an odd integer.
4. Select any odd integer greater than 1, square it, and then subtract 1. Try this for various odd integers greater than 1. Is the result always evenly divisible by 8?
5. Prove the assertion in Exercise 4. [*Suggestion:* From Exercise 3, an odd integer can be represented as $2n + 1$. Therefore, the assertion in Exercise 4 can be stated "$(2n + 1)^2 - 1$ is evenly divisible by 8." Expand this expression and explain why the result must be divisible by 8. You will need to use the result from Exercise 1.]
6. The integers 2 and 3 are consecutive prime numbers. Are there any other consecutive prime numbers? Why?
7. If n is a positive integer, for what values of n is $n^2 - 1$ a prime number?
8. A *Mersenne prime number* is a prime number that can be written in the form $2^n - 1$, where n is also a prime number. For instance, $2^5 - 1 = 32 - 1 = 31$. Because 5 and 31 are prime numbers, 31 is a Mersenne prime number. On the other hand, $2^{11} - 1 = 2048 - 1 = 2047$. In this case, although 11 is a prime number, $2047 = 23 \cdot 89$ and so is not a prime number. Find two Mersenne prime numbers other than 31.

Using the World Wide Web At the address http://www.utm.edu/research/primes/mersenne/, you can find more information on Mersenne prime numbers. By searching other websites, you can also find information about various topics in math.

The website http://mathforum.org/dr.math/ is an especially rich source of information. You can even submit math questions to this site and get an answer from Dr. Math.

CHAPTER 10 SUMMARY

KEY WORDS	EXAMPLES
The *greatest common factor (GCF) of two or more monomials* is the product of the GCF of the coefficients and the common variable factors. [10.1A, p. 522]	The GCF of $8x^2y$ and $12xyz$ is $4xy$.

To *factor a polynomial* means to write the polynomial as a product of other polynomials. [10.1A, p. 522]

To factor $x^2 + 3x + 2$ means to write it as the product $(x + 1)(x + 2)$.

A factor that has two terms is called a *binomial factor.* [10.1B, p. 524]

$(x + 1)$ is a binomial factor of $3x(x + 1)$.

A polynomial that does not factor using only integers is *nonfactorable over the integers.* [10.2A, p. 529]

The trinomial $x^2 + x + 4$ is nonfactorable over the integers. There are no integers whose product is 4 and whose sum is 1.

A polynomial is *factored completely* if it is written as a product of factors that are nonfactorable over the integers. [10.2B, p. 530]

The polynomial $3y^3 + 9y^2 - 12y$ is factored completely as $3y(y + 4)(y - 1)$.

An equation that can be written in the form $ax^2 + bx + c = 0$, $a \neq 0$, is a *quadratic equation.* A quadratic equation is in *standard form* when the polynomial is written in descending order and equal to zero. [10.5A, p. 552]

The equation $2x^2 - 3x + 7 = 0$ is a quadratic equation in standard form.

ESSENTIAL RULES AND PROCEDURES

EXAMPLES

Factoring by Grouping [10.1B, p. 524]

A polynomial can be factored by grouping if its terms can be grouped and factored in such a way that a common binomial factor is found.

$$\begin{aligned} &3a^2 - a - 15ab + 5b \\ &= (3a^2 - a) - (15ab - 5b) \\ &= a(3a - 1) - 5b(3a - 1) \\ &= (3a - 1)(a - 5b) \end{aligned}$$

Factoring $x^2 + bx + c$: IMPORTANT RELATIONSHIPS [10.2A, p. 528]

1. When the constant term of the trinomial is positive, the constant terms of the binomials have the same sign. They are both positive when the coefficient of the x term in the trinomial is positive. They are both negative when the coefficient of the x term in the trinomial is negative.

$x^2 + 6x + 8 = (x + 4)(x + 2)$

$x^2 - 6x + 5 = (x - 5)(x - 1)$

2. When the constant term of the trinomial is negative, the constant terms of the binomials have opposite signs.

$x^2 - 4x - 21 = (x + 3)(x - 7)$

3. In the trinomial, the coefficient of x is the sum of the constant terms of the binomials.

In the three examples above, note that $6 = 4 + 2$, $-6 = -5 + (-1)$, and $-4 = 3 + (-7)$.

4. In the trinomial, the constant term is the product of the constant terms of the binomials.

In the three examples above, note that $8 = 4 \cdot 2$, $5 = -5(-1)$, and $-21 = 3(-7)$.

To factor $ax^2 + bx + c$ by grouping [10.3B, p. 538]

First find two factors of $a \cdot c$ whose sum is b. Then use factoring by grouping to write the factorization of the trinomial.

$3x^2 - 11x - 20$

$a \cdot c = 3(-20) = -60$

The product of 4 and -15 is -60.

The sum of 4 and -15 is -11.

$$\begin{aligned} &3x^2 + 4x - 15x - 20 \\ &\quad = (3x^2 + 4x) - (15x + 20) \\ &\quad = x(3x + 4) - 5(3x + 4) \\ &\quad = (3x + 4)(x - 5) \end{aligned}$$

Factoring the Difference of Two Squares [10.4A, p. 544]

The difference of two squares factors as the sum and difference of the same terms.

$a^2 - b^2 = (a + b)(a - b)$

$$x^2 - 64 = (x + 8)(x - 8)$$

$$\begin{aligned} 4x^2 - 81 &= (2x)^2 - 9^2 \\ &= (2x + 9)(2x - 9) \end{aligned}$$

Factoring a Perfect-Square Trinomial [10.4A, p. 545]

A perfect-square trinomial is the square of a binomial.

$a^2 + 2ab + b^2 = (a + b)^2$

$a^2 - 2ab + b^2 = (a - b)^2$

$x^2 + 14x + 49 = (x + 7)^2$

$x^2 - 10x + 25 = (x - 5)^2$

General Factoring Strategy [10.4B, p. 546]

1. Is there a common factor? If so, factor out the common factor.

   $6x^2 - 8x = 2x(3x - 4)$

2. Is the polynomial the difference of two perfect squares? If so, factor.

   $9x^2 - 25 = (3x + 5)(3x - 5)$

3. Is the polynomial a perfect-square trinomial? If so, factor.

   $9x^2 + 6x + 1 = (3x + 1)^2$

4. Is the polynomial a trinomial that is the product of two binomials? If so, factor.

   $6x^2 + 5x - 6 = (3x - 2)(2x + 3)$

5. Does the polynomial contain four terms? If so, try factoring by grouping.

   $$\begin{aligned} &x^3 - 3x^2 + 2x - 6 \\ &\quad = (x^3 - 3x^2) + (2x - 6) \\ &\quad = x^2(x - 3) + 2(x - 3) \\ &\quad = (x - 3)(x^2 + 2) \end{aligned}$$

6. Is each binomial factor nonfactorable over the integers? If not, factor the binomial.

   $$\begin{aligned} x^4 - 16 &= (x^2 + 4)(x^2 - 4) \\ &= (x^2 + 4)(x + 2)(x - 2) \end{aligned}$$

Principle of Zero Products [10.5A, p. 552]

If the product of two factors is zero, then at least one of the factors must be zero.

If $a \cdot b = 0$, then $a = 0$ or $b = 0$.

The Principle of Zero Products is used to solve a quadratic equation by factoring.

$$\begin{aligned} x^2 + x &= 12 \\ x^2 + x - 12 &= 0 \\ (x - 3)(x + 4) &= 0 \end{aligned}$$

$$x - 3 = 0 \qquad x + 4 = 0$$

$$x = 3 \qquad x = -4$$

CHAPTER 10

CONCEPT REVIEW

Test your knowledge of the concepts presented in this chapter. Answer each question. Then check your answers against the ones provided in the Answer Section.

1. What does the GCF have to do with factoring?

2. When factoring a polynomial, do the terms of the polynomial have to be like terms?

3. After factoring a polynomial, how do you check your answer?

4. How is the GCF used in factoring by grouping?

5. When is a polynomial nonfactorable over the integers?

6. What does it mean to factor a polynomial completely?

7. When factoring a trinomial of the form $x^2 + bx + c$, why do we begin by finding the possible factors of c?

8. What are trial factors?

9. What is the middle term of a trinomial?

10. What are the binomial factors of the difference of two perfect squares?

11. What is an example of a perfect-square trinomial?

12. To solve an equation using factoring, why must the equation be set equal to zero?

CHAPTER 10

REVIEW EXERCISES

1. Factor: $b^2 - 13b + 30$

2. Factor: $4x(x - 3) - 5(3 - x)$

3. Factor $2x^2 - 5x + 6$ by using trial factors.

4. Factor: $5x^3 + 10x^2 + 35x$

5. Factor: $14y^9 - 49y^6 + 7y^3$

6. Factor: $y^2 + 5y - 36$

7. Factor $6x^2 - 29x + 28$ by using trial factors.

8. Factor: $12a^2b + 3ab^2$

9. Factor: $a^6 - 100$

10. Factor: $n^4 - 2n^3 - 3n^2$

11. Factor $12y^2 + 16y - 3$ by using trial factors.

12. Factor: $12b^3 - 58b^2 + 56b$

13. Factor: $9y^4 - 25z^2$

14. Factor: $c^2 + 8c + 12$

15. Factor $18a^2 - 3a - 10$ by grouping.

16. Solve: $4x^2 + 27x = 7$

17. Factor: $4x^3 - 20x^2 - 24x$

18. Factor: $3a^2 - 15a - 42$

19. Factor $2a^2 - 19a - 60$ by grouping.

20. Solve: $(x + 1)(x - 5) = 16$

21. Factor: $21ax - 35bx - 10by + 6ay$

22. Factor: $a^2b^2 - 1$

23. Factor: $10x^2 + 25x + 4xy + 10y$

24. Factor: $5x^2 - 5x - 30$

25. Factor: $3x^2 + 36x + 108$

26. Factor $3x^2 - 17x + 10$ by grouping.

27. **Sports** The length of the field in field hockey is 20 yd less than twice the width of the field. The area of the field in field hockey is 6000 yd^2. Find the length and width of the field.

28. **Image Projection** The size S of an image from a projector depends on the distance d of the screen from the projector and is given by $S = d^2$. Find the distance between the projector and the screen when the size of the picture is 400 ft^2.

Tony Cordoza/Alamy

29. **Photography** A rectangular photograph has dimensions 15 in. by 12 in. A picture frame around the photograph increases the total area to 270 in^2. What is the width of the frame?

30. **Gardening** The length of each side of a square garden plot is extended 4 ft. The area of the resulting square is 576 ft^2. Find the length of a side of the original garden plot.

CHAPTER 10

TEST

1. Factor: $ab + 6a - 3b - 18$

2. Factor: $2y^4 - 14y^3 - 16y^2$

3. Factor $8x^2 + 20x - 48$ by grouping.

4. Factor $6x^2 + 19x + 8$ by using trial factors.

5. Factor: $a^2 - 19a + 48$

6. Factor: $6x^3 - 8x^2 + 10x$

7. Factor: $x^2 + 2x - 15$

8. Solve: $4x^2 - 1 = 0$

9. Factor: $5x^2 - 45x - 15$

10. Factor: $p^2 + 12p + 36$

11. Solve: $x(x - 8) = -15$

12. Factor: $3x^2 + 12xy + 12y^2$

13. Factor: $b^2 - 16$

14. Factor $6x^2y^2 + 9xy^2 + 3y^2$ by grouping.

15. Factor: $p^2 + 5p + 6$

16. Factor: $a(x - 2) + b(x - 2)$

17. Factor: $x(p + 1) - (p + 1)$

18. Factor: $3a^2 - 75$

19. Factor: $2x^2 + 4x - 5$

20. Factor: $x^2 - 9x - 36$

21. Factor: $4a^2 - 12ab + 9b^2$

22. Factor: $4x^2 - 49y^2$

23. Solve: $(2a - 3)(a + 7) = 0$

24. **Number Sense** The sum of two numbers is ten. The sum of the squares of the two numbers is fifty-eight. Find the two numbers.

25. **Geometry** The length of a rectangle is 3 cm longer than twice the width. The area of the rectangle is 90 cm^2. Find the length and width of the rectangle.

$2W + 3$

W

CUMULATIVE REVIEW EXERCISES

1. Subtract: $-2 - (-3) - 5 - (-11)$

2. Simplify: $(3 - 7)^2 \div (-2) - 3 \cdot (-4)$

3. Evaluate $-2a^2 \div (2b) - c$ when $a = -4$, $b = 2$, and $c = -1$.

4. Simplify: $-\frac{3}{4}(-20x^2)$

5. Simplify: $-2[4x - 2(3 - 2x) - 8x]$

6. Solve: $-\frac{5}{7}x = -\frac{10}{21}$

7. Solve: $3x - 2 = 12 - 5x$

8. Solve: $-2 + 4[3x - 2(4 - x) - 3] = 4x + 2$

9. 120% of what number is 54?

10. Simplify: $(-3a^3b^2)^2$

11. Multiply: $(x + 2)(x^2 - 5x + 4)$

12. Divide: $(8x^2 + 4x - 3) \div (2x - 3)$

13. Simplify: $(x^{-4}y^3)^2$

14. Factor: $3a - 3b - ax + bx$

15. Factor: $15xy^2 - 20xy^4$

16. Factor: $x^2 - 5xy - 14y^2$

17. Factor: $p^2 - 9p - 10$

18. Factor: $18a^3 + 57a^2 + 30a$

19. Factor: $36a^2 - 49b^2$

20. Factor: $4x^2 + 28xy + 49y^2$

21. Factor: $9x^2 + 15x - 14$

22. Factor: $18x^2 - 48xy + 32y^2$

23. Factor: $3y(x - 3) - 2(x - 3)$

24. Solve: $3x^2 + 19x - 14 = 0$

25. Carpentry A board 10 ft long is cut into two pieces. Four times the length of the shorter piece is 2 ft less than three times the length of the longer piece. Find the length of each piece.

26. Business A portable MP3 player that regularly sells for \$165 is on sale for \$99. Find the discount rate. Use the formula $S = R - rR$, where S is the sale price, R is the regular price, and r is the discount rate. Write the answer as a percent.

27. Geometry Given that lines ℓ_1 and ℓ_2 are parallel, find the measures of angles a and b.

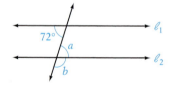

28. Travel A family drove to a resort at an average speed of 42 mph and later returned over the same road at an average speed of 56 mph. Find the distance to the resort if the total driving time was 7 h.

29. Consecutive Integers Find three consecutive even integers such that five times the middle integer is twelve more than twice the sum of the first and third integers.

30. Geometry The length of the base of a triangle is three times the height. The area of the triangle is 24 in^2. Find the length of the base of the triangle.

CHAPTER

11

Rational Expressions

OBJECTIVES

SECTION 11.1

A To simplify a rational expression
B To multiply rational expressions
C To divide rational expressions

SECTION 11.2

A To find the least common multiple (LCM) of two or more polynomials
B To express two fractions in terms of the LCM of their denominators

SECTION 11.3

A To add or subtract rational expressions with the same denominators
B To add or subtract rational expressions with different denominators

SECTION 11.4

A To simplify a complex fraction

SECTION 11.5

A To solve an equation containing fractions

SECTION 11.6

A To solve a literal equation for one of the variables

SECTION 11.7

A To solve work problems
B To use rational expressions to solve uniform motion problems

ARE YOU READY?

Take the Chapter 11 Prep Test to find out if you are ready to learn to:

- Simplify a rational expression
- Add, subtract, multiply, and divide rational expressions
- Solve an equation containing fractions
- Solve a literal equation for one of the variables
- Use rational expressions to solve work problems and uniform motion problems

PREP TEST

Do these exercises to prepare for Chapter 11.

1. Find the least common multiple (LCM) of 12 and 18.

2. Simplify: $\dfrac{9x^3y^4}{3x^2y^7}$

3. Subtract: $\dfrac{3}{4} - \dfrac{8}{9}$

4. Divide: $\left(-\dfrac{8}{11}\right) \div \dfrac{4}{5}$

5. If a is a nonzero number, are the following two quantities equal: $\frac{0}{a}$ and $\frac{a}{0}$?

6. Solve: $\dfrac{2}{3}x - \dfrac{3}{4} = \dfrac{5}{6}$

7. Factor: $x^2 - 4x - 12$

8. Factor: $2x^2 - x - 3$

9. At 9:00 A.M., Anthony begins walking on a park trail at a rate of 9 m/min. Ten minutes later his sister Jean begins walking the same trail in pursuit of her brother at a rate of 12 m/min. At what time will Jean catch up to Anthony?

SECTION

11.1 Multiplication and Division of Rational Expressions

OBJECTIVE A To simplify a rational expression

A fraction in which the numerator and denominator are polynomials is called a **rational expression.** Examples of rational expressions are shown at the right.

$$\frac{5}{z}, \quad \frac{x^2+1}{2x-1}, \quad \frac{y^2+y-1}{4y^2+1}$$

Care must be exercised with a rational expression to ensure that when the variables are replaced with numbers, the resulting denominator is not zero. Consider the rational expression at the right. The value of x cannot be 3 because the denominator would then be zero.

$$\frac{4x^2-9}{2x-6}$$

$$\frac{4(3)^2-9}{2(3)-6} = \frac{27}{0} \quad \text{Not a real number}$$

In the **simplest form of a rational expression,** the numerator and denominator have no common factors. The Multiplication Property of One is used to write a rational expression in simplest form.

HOW TO 1 Simplify: $\dfrac{x^2-4}{x^2-2x-8}$

$$\frac{x^2-4}{x^2-2x-8} = \frac{(x-2)(x+2)}{(x-4)(x+2)}$$

• Factor the numerator and denominator.

$$= \frac{x-2}{x-4}\cdot\frac{x+2}{x+2} = \frac{x-2}{x-4}\cdot 1$$

$$= \frac{x-2}{x-4}, x \neq -2, 4$$

• The restrictions, $x \neq -2$ or 4, are necessary to prevent division by zero.

This simplification is usually shown with slashes through the common factors:

$$\frac{x^2-4}{x^2-2x-8} = \frac{(x-2)\overset{1}{\cancel{(x+2)}}}{(x-4)\underset{1}{\cancel{(x+2)}}}$$

• Factor the numerator and denominator.

$$= \frac{x-2}{x-4}, x \neq -2, 4$$

• Divide by the common factors. The restrictions, $x \neq -2$ or 4, are necessary to prevent division by zero.

In summary, **to simplify a rational expression, factor the numerator and denominator. Then divide the numerator and denominator by the common factors.**

HOW TO 2 Simplify: $\dfrac{10+3x-x^2}{x^2-4x-5}$

$$\frac{10+3x-x^2}{x^2-4x-5} = \frac{-(x^2-3x-10)}{x^2-4x-5}$$

• Because the coefficient of x^2 in the numerator is -1, factor -1 from the numerator.

$$= \frac{-\overset{1}{\cancel{(x-5)}}(x+2)}{\underset{1}{\cancel{(x-5)}}(x+1)}$$

• Factor the numerator and denominator. Divide by the common factors.

$$= -\frac{x+2}{x+1}, x \neq -1, 5$$

For the remaining examples, we will omit the restrictions on the variables that prevent division by zero and assume the values of the variables are such that division by zero is not possible.

EXAMPLE • 1

Simplify: $\frac{4x^3y^4}{6x^4y}$

Solution

$\frac{4x^3y^4}{6x^4y} = \frac{2y^3}{3x}$ • Use the rules of exponents.

YOU TRY IT • 1

Simplify: $\frac{6x^5y}{12x^2y^3}$

Your solution

EXAMPLE • 2

Simplify: $\frac{x^2 + 2x - 15}{x^2 - 7x + 12}$

Solution

$$\frac{x^2 + 2x - 15}{x^2 - 7x + 12} = \frac{(x + 5)\overset{1}{\cancel{(x - 3)}}}{\underset{1}{\cancel{(x - 3)}}(x - 4)} = \frac{x + 5}{x - 4}$$

YOU TRY IT • 2

Simplify: $\frac{x^2 + 4x - 12}{x^2 - 3x + 2}$

Your solution

EXAMPLE • 3

Simplify: $\frac{9 - x^2}{x^2 + x - 12}$

Solution

$$\frac{9 - x^2}{x^2 + x - 12} = \frac{\overset{-1}{\cancel{(3 - x)}}(3 + x)}{\underset{1}{\cancel{(x - 3)}}(x + 4)} = -\frac{x + 3}{x + 4}$$

• $\frac{3 - x}{x - 3} = \frac{-1(x - 3)}{x - 3} = -1$

YOU TRY IT • 3

Simplify: $\frac{x^2 + 2x - 24}{16 - x^2}$

Your solution

Solutions on p. S25

OBJECTIVE B To multiply rational expressions

The product of two fractions is a fraction whose numerator is the product of the numerators of the two fractions and whose denominator is the product of the denominators of the two fractions.

Multiplying Rational Expressions

Multiply the numerators.
Multiply the denominators. $\frac{a}{b} \cdot \frac{c}{d} = \frac{ac}{bd}$

$$\frac{2}{3} \cdot \frac{4}{5} = \frac{8}{15} \qquad \frac{3x}{y} \cdot \frac{2}{z} = \frac{6x}{yz} \qquad \frac{x + 2}{x} \cdot \frac{3}{x - 2} = \frac{3x + 6}{x^2 - 2x}$$

HOW TO • 3 Multiply: $\frac{x^2 + 3x}{x^2 - 3x - 4} \cdot \frac{x^2 - 5x + 4}{x^2 + 2x - 3}$

$$\frac{x^2 + 3x}{x^2 - 3x - 4} \cdot \frac{x^2 - 5x + 4}{x^2 + 2x - 3}$$

$$= \frac{x(x + 3)}{(x - 4)(x + 1)} \cdot \frac{(x - 4)(x - 1)}{(x + 3)(x - 1)}$$

• Factor the numerator and denominator of each fraction.

$$= \frac{x\overset{1}{\cancel{(x + 3)}}\overset{1}{\cancel{(x - 4)}}\overset{1}{\cancel{(x - 1)}}}{\underset{1}{\cancel{(x - 4)}}(x + 1)\underset{1}{\cancel{(x + 3)}}\underset{1}{\cancel{(x - 1)}}}$$

• Multiply. Then divide by the common factors.

$$= \frac{x}{x + 1}$$

• Write the answer in simplest form.

EXAMPLE • 4

Multiply: $\frac{10x^2 - 15x}{12x - 8} \cdot \frac{3x - 2}{20x - 25}$

Solution

$$\frac{10x^2 - 15x}{12x - 8} \cdot \frac{3x - 2}{20x - 25}$$

$$= \frac{5x(2x - 3)}{4(3x - 2)} \cdot \frac{(3x - 2)}{5(4x - 5)}$$

• Factor.

$$= \frac{\overset{1}{\cancel{5}}x(2x - 3)\overset{1}{\cancel{(3x - 2)}}}{4\underset{1}{\cancel{(3x - 2)}}\underset{1}{\cancel{5}}(4x - 5)}$$

• Divide by the common factors.

$$= \frac{x(2x - 3)}{4(4x - 5)}$$

YOU TRY IT • 4

Multiply: $\frac{12x^2 + 3x}{10x - 15} \cdot \frac{8x - 12}{9x + 18}$

Your solution

EXAMPLE • 5

Multiply: $\frac{x^2 + x - 6}{x^2 + 7x + 12} \cdot \frac{x^2 + 3x - 4}{4 - x^2}$

Solution

$$\frac{x^2 + x - 6}{x^2 + 7x + 12} \cdot \frac{x^2 + 3x - 4}{4 - x^2}$$

$$= \frac{(x + 3)(x - 2)}{(x + 3)(x + 4)} \cdot \frac{(x + 4)(x - 1)}{(2 - x)(2 + x)}$$

• Factor.

$$= \frac{\overset{1}{\cancel{(x + 3)}}\overset{-1}{\cancel{(x - 2)}}\overset{1}{\cancel{(x + 4)}}(x - 1)}{\underset{1}{\cancel{(x + 3)}}\underset{1}{\cancel{(x + 4)}}\underset{1}{\cancel{(2 - x)}}(2 + x)}$$

• Divide by the common factors.

$$= -\frac{x - 1}{x + 2}$$

YOU TRY IT • 5

Multiply: $\frac{x^2 + 2x - 15}{9 - x^2} \cdot \frac{x^2 - 3x - 18}{x^2 - 7x + 6}$

Your solution

Solutions on p. S25

OBJECTIVE C To divide rational expressions

The **reciprocal of a rational expression** is the rational expression with the numerator and denominator interchanged.

Fraction $\left\{ \begin{array}{ll} \frac{a}{b} & \frac{b}{a} \\ x^2 = \frac{x^2}{1} & \frac{1}{x^2} \\ \frac{x+2}{x} & \frac{x}{x+2} \end{array} \right\}$ Reciprocal

Dividing Rational Expressions

Multiply the dividend by the reciprocal of the divisor. $\frac{a}{b} \div \frac{c}{d} = \frac{a}{b} \cdot \frac{d}{c} = \frac{ad}{bc}$

$$\frac{4}{x} \div \frac{y}{5} = \frac{4}{x} \cdot \frac{5}{y} = \frac{20}{xy} \qquad \frac{x+4}{x} \div \frac{x-2}{4} = \frac{x+4}{x} \cdot \frac{4}{x-2} = \frac{4(x+4)}{x(x-2)}$$

The basis for the division rule is shown at the right.

$$\frac{a}{b} \div \frac{c}{d} = \frac{\frac{a}{b}}{\frac{c}{d}} = \frac{\frac{a}{b} \cdot \frac{d}{c}}{\frac{c}{d} \cdot \frac{d}{c}} = \frac{\frac{a}{b} \cdot \frac{d}{c}}{1} = \frac{a}{b} \cdot \frac{d}{c}$$

EXAMPLE 6

Divide: $\frac{xy^2 - 3x^2y}{z^2} \div \frac{6x^2 - 2xy}{z^3}$

Solution

$$\frac{xy^2 - 3x^2y}{z^2} \div \frac{6x^2 - 2xy}{z^3}$$

$$= \frac{xy^2 - 3x^2y}{z^2} \cdot \frac{z^3}{6x^2 - 2xy}$$ • Multiply by the reciprocal.

$$= \frac{xy\overset{-1}{\cancel{(y - 3x)}} \cdot z^3}{z^2 \cdot 2x\underset{1}{\cancel{(3x - y)}}} = -\frac{yz}{2}$$

YOU TRY IT 6

Divide: $\frac{a^2}{4bc^2 - 2b^2c} \div \frac{a}{6bc - 3b^2}$

Your solution

EXAMPLE 7

Divide: $\frac{2x^2 + 5x + 2}{2x^2 + 3x - 2} \div \frac{3x^2 + 13x + 4}{2x^2 + 7x - 4}$

Solution

$$\frac{2x^2 + 5x + 2}{2x^2 + 3x - 2} \div \frac{3x^2 + 13x + 4}{2x^2 + 7x - 4}$$

$$= \frac{2x^2 + 5x + 2}{2x^2 + 3x - 2} \cdot \frac{2x^2 + 7x - 4}{3x^2 + 13x + 4}$$ • Multiply by the reciprocal.

$$= \frac{(2x+1)\overset{1}{\cancel{(x+2)}} \cdot \overset{1}{\cancel{(2x-1)}}\overset{1}{\cancel{(x+4)}}}{\underset{1}{\cancel{(2x-1)}}\underset{1}{\cancel{(x+2)}} \cdot (3x+1)\underset{1}{\cancel{(x+4)}}} = \frac{2x+1}{3x+1}$$

YOU TRY IT 7

Divide: $\frac{3x^2 + 26x + 16}{3x^2 - 7x - 6} \div \frac{2x^2 + 9x - 5}{x^2 + 2x - 15}$

Your solution

Solutions on p. S25

11.1 EXERCISES

OBJECTIVE A To simplify a rational expression

1. Explain the procedure for writing a rational expression in simplest form.

2. Why is the simplification at the right incorrect? $\frac{x+3}{x} = \frac{\overset{1}{\cancel{x}}+3}{\underset{1}{\cancel{x}}} = 4$

For Exercises 3 to 30, simplify.

3. $\frac{9x^3}{12x^4}$

4. $\frac{16x^2y}{24xy^3}$

5. $\frac{(x+3)^2}{(x+3)^3}$

6. $\frac{(2x-1)^5}{(2x-1)^4}$

7. $\frac{3n-4}{4-3n}$

8. $\frac{5-2x}{2x-5}$

9. $\frac{6y(y+2)}{9y^2(y+2)}$

10. $\frac{12x^2(3-x)}{18x(3-x)}$

11. $\frac{6x(x-5)}{8x^2(5-x)}$

12. $\frac{14x^3(7-3x)}{21x(3x-7)}$

13. $\frac{a^2+4a}{ab+4b}$

14. $\frac{x^2-3x}{2x-6}$

15. $\frac{4-6x}{3x^2-2x}$

16. $\frac{5xy-3y}{9-15x}$

17. $\frac{y^2-3y+2}{y^2-4y+3}$

18. $\frac{x^2+5x+6}{x^2+8x+15}$

19. $\frac{x^2+3x-10}{x^2+2x-8}$

20. $\frac{a^2+7a-8}{a^2+6a-7}$

21. $\frac{x^2+x-12}{x^2-6x+9}$

22. $\frac{x^2+8x+16}{x^2-2x-24}$

23. $\frac{x^2-3x-10}{25-x^2}$

24. $\frac{4-y^2}{y^2-3y-10}$

25. $\frac{2x^3+2x^2-4x}{x^3+2x^2-3x}$

26. $\frac{3x^3-12x}{6x^3-24x^2+24x}$

27. $\frac{6x^2-7x+2}{6x^2+5x-6}$

28. $\frac{2n^2-9n+4}{2n^2-5n-12}$

29. $\frac{x^2+3x-28}{24-2x-x^2}$

30. $\frac{x^2+7x-8}{1+x-2x^2}$

OBJECTIVE B **To multiply rational expressions**

For Exercises 31 to 54, multiply.

31. $\frac{8x^2}{9y^3} \cdot \frac{3y^2}{4x^3}$

32. $\frac{14a^2b^3}{15x^5y^2} \cdot \frac{25x^3y}{16ab}$

33. $\frac{12x^3y^4}{7a^2b^3} \cdot \frac{14a^3b^4}{9x^2y^2}$

34. $\frac{18a^4b^2}{25x^2y^3} \cdot \frac{50x^5y^6}{27a^6b^2}$

35. $\frac{3x-6}{5x-20} \cdot \frac{10x-40}{27x-54}$

36. $\frac{8x-12}{14x+7} \cdot \frac{42x+21}{32x-48}$

37. $\frac{3x^2+2x}{2xy-3y} \cdot \frac{2xy^3-3y^3}{3x^3+2x^2}$

38. $\frac{4a^2x-3a^2}{2by+5b} \cdot \frac{2b^3y+5b^3}{4ax-3a}$

39. $\frac{x^2+5x+4}{x^3y^2} \cdot \frac{x^2y^3}{x^2+2x+1}$

40. $\frac{x^2+x-2}{xy^2} \cdot \frac{x^3y}{x^2+5x+6}$

41. $\frac{x^4y^2}{x^2+3x-28} \cdot \frac{x^2-49}{xy^4}$

42. $\frac{x^5y^3}{x^2+13x+30} \cdot \frac{x^2+2x-3}{x^7y^2}$

43. $\frac{2x^2-5x}{2xy+y} \cdot \frac{2xy^2+y^2}{5x^2-2x^3}$

44. $\frac{3a^3+4a^2}{5ab-3b} \cdot \frac{3b^3-5ab^3}{3a^2+4a}$

45. $\frac{x^2-2x-24}{x^2-5x-6} \cdot \frac{x^2+5x+6}{x^2+6x+8}$

46. $\frac{x^2-8x+7}{x^2+3x-4} \cdot \frac{x^2+3x-10}{x^2-9x+14}$

47. $\frac{x^2+2x-35}{x^2+4x-21} \cdot \frac{x^2+3x-18}{x^2+9x+18}$

48. $\frac{y^2+y-20}{y^2+2y-15} \cdot \frac{y^2+4y-21}{y^2+3y-28}$

49. $\dfrac{x^2 - 3x - 4}{x^2 + 6x + 5} \cdot \dfrac{x^2 + 5x + 6}{8 + 2x - x^2}$

50. $\dfrac{25 - n^2}{n^2 - 2n - 35} \cdot \dfrac{n^2 - 8n - 20}{n^2 - 3n - 10}$

51. $\dfrac{16 + 6x - x^2}{x^2 - 10x - 24} \cdot \dfrac{x^2 - 6x - 27}{x^2 - 17x + 72}$

52. $\dfrac{x^2 - 11x + 28}{x^2 - 13x + 42} \cdot \dfrac{x^2 + 7x + 10}{20 - x - x^2}$

53. $\dfrac{2x^2 + 5x + 2}{2x^2 + 7x + 3} \cdot \dfrac{x^2 - 7x - 30}{x^2 - 6x - 40}$

54. $\dfrac{x^2 - 4x - 32}{x^2 - 8x - 48} \cdot \dfrac{3x^2 + 17x + 10}{3x^2 - 22x - 16}$

For Exercises 55 to 57, use the product $\dfrac{x^a}{y^b} \cdot \dfrac{y^c}{x^d}$, where a, b, c, and d are all positive integers.

55. If $a > d$ and $c > b$, what is the denominator of the simplified product?

56. If $a > d$ and $b > c$, which variable appears in the denominator of the simplified product?

57. If $a < d$ and $b = c$, what is the numerator of the simplified product?

OBJECTIVE C To divide rational expressions

58. What is the reciprocal of a rational expression?

59. Explain how to divide rational expressions.

For Exercises 60 to 79, divide.

60. $\dfrac{4x^2y^3}{15a^2b^3} \div \dfrac{6xy}{5a^3b^5}$

61. $\dfrac{9x^3y^4}{16a^4b^2} \div \dfrac{45x^4y^2}{14a^7b}$

62. $\dfrac{6x - 12}{8x + 32} \div \dfrac{18x - 36}{10x + 40}$

63. $\dfrac{28x + 14}{45x - 30} \div \dfrac{14x + 7}{30x - 20}$

64. $\dfrac{6x^3 + 7x^2}{12x - 3} \div \dfrac{6x^2 + 7x}{36x - 9}$

65. $\dfrac{5a^2y + 3a^2}{2x^3 + 5x^2} \div \dfrac{10ay + 6a}{6x^3 + 15x^2}$

66. $\dfrac{x^2 + 4x + 3}{x^2y} \div \dfrac{x^2 + 2x + 1}{xy^2}$

67. $\dfrac{x^3y^2}{x^2 - 3x - 10} \div \dfrac{xy^4}{x^2 - x - 20}$

68. $\dfrac{x^2 - 49}{x^4y^3} \div \dfrac{x^2 - 14x + 49}{x^4y^3}$

69. $\dfrac{x^2y^5}{x^2 - 11x + 30} \div \dfrac{xy^6}{x^2 - 7x + 10}$

70. $\dfrac{4ax - 8a}{c^2} \div \dfrac{2y - xy}{c^3}$

71. $\dfrac{3x^2y - 9xy}{a^2b} \div \dfrac{3x^2 - x^3}{ab^2}$

72. $\dfrac{x^2 - 5x + 6}{x^2 - 9x + 18} \div \dfrac{x^2 - 6x + 8}{x^2 - 9x + 20}$

73. $\dfrac{x^2 + 3x - 40}{x^2 + 2x - 35} \div \dfrac{x^2 + 2x - 48}{x^2 + 3x - 18}$

74. $\dfrac{x^2 + 2x - 15}{x^2 - 4x - 45} \div \dfrac{x^2 + x - 12}{x^2 - 5x - 36}$

75. $\dfrac{y^2 - y - 56}{y^2 + 8y + 7} \div \dfrac{y^2 - 13y + 40}{y^2 - 4y - 5}$

76. $\dfrac{8 + 2x - x^2}{x^2 + 7x + 10} \div \dfrac{x^2 - 11x + 28}{x^2 - x - 42}$

77. $\dfrac{x^2 - x - 2}{x^2 - 7x + 10} \div \dfrac{x^2 - 3x - 4}{40 - 3x - x^2}$

78. $\dfrac{2x^2 - 3x - 20}{2x^2 - 7x - 30} \div \dfrac{2x^2 - 5x - 12}{4x^2 + 12x + 9}$

79. $\dfrac{6n^2 + 13n + 6}{4n^2 - 9} \div \dfrac{6n^2 + n - 2}{4n^2 - 1}$

For Exercises 80 to 83, state whether the given division is equivalent to $\dfrac{x^2 - 3x - 4}{x^2 + 5x - 6}$.

80. $\dfrac{x - 4}{x + 6} \div \dfrac{x - 1}{x + 1}$

81. $\dfrac{x + 1}{x + 6} \div \dfrac{x - 1}{x - 4}$

82. $\dfrac{x + 1}{x - 1} \div \dfrac{x + 6}{x - 4}$

83. $\dfrac{x - 1}{x + 1} \div \dfrac{x - 4}{x + 6}$

Applying the Concepts

84. Given the expression $\dfrac{9}{x^2 + 1}$, choose some values of x and evaluate the expression for those values. Is it possible to choose a value of x for which the value of the expression is greater than 10? If so, what is that value of x? If not, explain why it is not possible.

Geometry For Exercises 85 and 86, write in simplest form the ratio of the shaded area of the figure to the total area of the figure.

85.

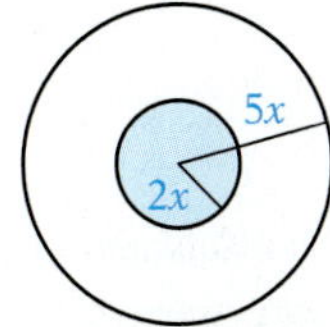

86.

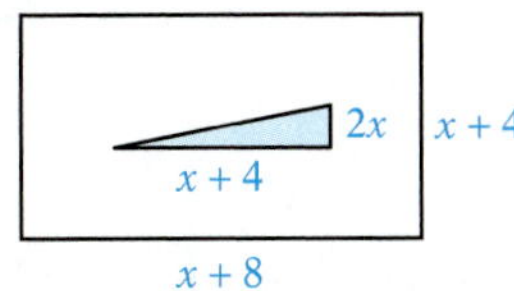

SECTION

11.2 Expressing Fractions in Terms of the Least Common Multiple (LCM)

OBJECTIVE A To find the least common multiple (LCM) of two or more polynomials

Recall that the least common multiple (LCM) of two or more numbers is the smallest number that contains the prime factorization of each number.

The LCM of 12 and 18 is 36 because 36 contains the prime factors of 12 and the prime factors of 18.

$12 = 2 \cdot 2 \cdot 3$
$18 = 2 \cdot 3 \cdot 3$

$$\text{LCM} = 36 = \overbrace{2 \cdot 2 \cdot 3}^{\text{Factors of 12}} \cdot 3 \quad (\underbrace{2 \cdot 3 \cdot 3}_{\text{Factors of 18}})$$

The **least common multiple (LCM) of two or more polynomials** is the polynomial of least degree that contains all the factors of each polynomial.

To find the LCM of two or more polynomials, first factor each polynomial completely. The LCM is the product of each factor the greatest number of times it occurs in any one factorization.

Take Note

The LCM must contain the factors of each polynomial. As shown with the braces at the right, the LCM contains the factors of $4x^2 + 4x$ and the factors of $x^2 + 2x + 1$.

HOW TO • 1 Find the LCM of $4x^2 + 4x$ and $x^2 + 2x + 1$.

The LCM of the polynomials is the product of the LCM of the numerical coefficients and each variable factor the greatest number of times it occurs in any one factorization.

$4x^2 + 4x = 4x(x + 1) = 2 \cdot 2 \cdot x(x + 1)$
$x^2 + 2x + 1 = (x + 1)(x + 1)$

$$\text{LCM} = \overbrace{2 \cdot 2 \cdot x(x + 1)}^{\text{Factors of } 4x^2 + 4x}(x + 1) = 4x(x + 1)(x + 1)$$

(Factors of $x^2 + 2x + 1$: $(x + 1)(x + 1)$)

EXAMPLE • 1

Find the LCM of $4x^2y$ and $6xy^2$.

Solution

$4x^2y = 2 \cdot 2 \cdot x \cdot x \cdot y$
$6xy^2 = 2 \cdot 3 \cdot x \cdot y \cdot y$
$\text{LCM} = 2 \cdot 2 \cdot 3 \cdot x \cdot x \cdot y \cdot y = 12x^2y^2$

YOU TRY IT • 1

Find the LCM of $8uv^2$ and $12uw$.

Your solution

EXAMPLE • 2

Find the LCM of $x^2 - x - 6$ and $9 - x^2$.

Solution

$x^2 - x - 6 = (x - 3)(x + 2)$
$9 - x^2 = -(x^2 - 9) = -(x + 3)(x - 3)$
$\text{LCM} = (x - 3)(x + 2)(x + 3)$

YOU TRY IT • 2

Find the LCM of $m^2 - 6m + 9$ and $m^2 - 2m - 3$.

Your solution

Solutions on p. S25

OBJECTIVE B To express two fractions in terms of the LCM of their denominators

When adding and subtracting fractions, it is frequently necessary to express two or more fractions in terms of a common denominator. This common denominator is the LCM of the denominators of the fractions.

HOW TO • 2 Write the fractions $\frac{x+1}{4x^2}$ and $\frac{x-3}{2x^2-4x}$ in terms of the LCM of the denominators.

Find the LCM of the denominators.

The LCM is $4x^2(x-2)$.

For each fraction, multiply the numerator and the denominator by the factors whose product with the denominator is the LCM.

$$\frac{x+1}{4x^2} = \frac{x+1}{4x^2} \cdot \frac{(x-2)}{(x-2)} = \frac{x^2-x-2}{4x^2(x-2)}$$

$$\frac{x-3}{2x^2-4x} = \frac{x-3}{2x(x-2)} \cdot \frac{2x}{2x} = \frac{2x^2-6x}{4x^2(x-2)}$$

LCM

EXAMPLE • 3

Write the fractions $\frac{x+2}{3x^2}$ and $\frac{x-1}{8xy}$ in terms of the LCM of the denominators.

Solution

The LCM is $24x^2y$.

$$\frac{x+2}{3x^2} = \frac{x+2}{3x^2} \cdot \frac{8y}{8y} = \frac{8xy+16y}{24x^2y}$$

$$\frac{x-1}{8xy} = \frac{x-1}{8xy} \cdot \frac{3x}{3x} = \frac{3x^2-3x}{24x^2y}$$

YOU TRY IT • 3

Write the fractions $\frac{x-3}{4xy^2}$ and $\frac{2x+1}{9y^2z}$ in terms of the LCM of the denominators.

Your solution

EXAMPLE • 4

Write the fractions $\frac{2x-1}{2x-x^2}$ and $\frac{x}{x^2+x-6}$ in terms of the LCM of the denominators.

Solution

$$\frac{2x-1}{2x-x^2} = \frac{2x-1}{-(x^2-2x)} = -\frac{2x-1}{x^2-2x}$$

The LCM is $x(x-2)(x+3)$.

$$\frac{2x-1}{2x-x^2} = -\frac{2x-1}{x(x-2)} \cdot \frac{x+3}{x+3} = -\frac{2x^2+5x-3}{x(x-2)(x+3)}$$

$$\frac{x}{x^2+x-6} = \frac{x}{(x-2)(x+3)} \cdot \frac{x}{x} = \frac{x^2}{x(x-2)(x+3)}$$

YOU TRY IT • 4

Write the fractions $\frac{x+4}{x^2-3x-10}$ and $\frac{2x}{25-x^2}$ in terms of the LCM of the denominators.

Your solution

Solutions on p. S25

11.2 EXERCISES

OBJECTIVE A **To find the least common multiple (LCM) of two or more polynomials**

For Exercises 1 to 30, find the LCM of the polynomials.

1. $8x^3y$
$12xy^2$

2. $6ab^2$
$18ab^3$

3. $10x^4y^2$
$15x^3y$

4. $12a^2b$
$18ab^3$

5. $8x^2$
$4x^2 + 8x$

6. $6y^2$
$4y + 12$

7. $2x^2y$
$3x^2 + 12x$

8. $4xy^2$
$6xy^2 + 12y^2$

9. $9x(x + 2)$
$12(x + 2)^2$

10. $8x^2(x - 1)^2$
$10x^3(x - 1)$

11. $3x + 3$
$2x^2 + 4x + 2$

12. $4x - 12$
$2x^2 - 12x + 18$

13. $(x - 1)(x + 2)$
$(x - 1)(x + 3)$

14. $(2x - 1)(x + 4)$
$(2x + 1)(x + 4)$

15. $(2x + 3)^2$
$(2x + 3)(x - 5)$

16. $(x - 7)(x + 2)$
$(x - 7)^2$

17. $x - 1$
$x - 2$
$(x - 1)(x - 2)$

18. $(x + 4)(x - 3)$
$x + 4$
$x - 3$

19. $x^2 - x - 6$
$x^2 + x - 12$

20. $x^2 + 3x - 10$
$x^2 + 5x - 14$

21. $x^2 + 5x + 4$
$x^2 - 3x - 28$

22. $x^2 - 10x + 21$
$x^2 - 8x + 15$

23. $x^2 - 2x - 24$
$x^2 - 36$

24. $x^2 + 7x + 10$
$x^2 - 25$

25. $2x^2 - 7x + 3$
$2x^2 + x - 1$

26. $3x^2 - 11x + 6$
$3x^2 + 4x - 4$

27. $6 + x - x^2$
$x + 2$
$x - 3$

28. $15 + 2x - x^2$
$x - 5$
$x + 3$

29. $x^2 + 3x - 18$
$3 - x$
$x + 6$

30. $x^2 - 5x + 6$
$1 - x$
$x - 6$

31. How many factors of $x - 3$ are in the LCM of each pair of expressions?
a. $x^2 + x - 12$ and $x^2 - 9$ **b.** $x^2 - x - 12$ and $x^2 + 6x + 9$ **c.** $x^2 + x - 12$ and $x^2 - 6x + 9$

OBJECTIVE B To express two fractions in terms of the LCM of their denominators

32. True or false? To write the fractions $\frac{x^2}{y(y-3)}$ and $\frac{x}{(y-3)^2}$ with a common denominator, you need only multiply the numerator and denominator of the second fraction by y.

For Exercises 33 to 52, write the fractions in terms of the LCM of the denominators.

33. $\frac{4}{x}, \frac{3}{x^2}$

34. $\frac{5}{ab^2}, \frac{6}{ab}$

35. $\frac{x}{3y^2}, \frac{z}{4y}$

36. $\frac{5y}{6x^2}, \frac{7}{9xy}$

37. $\frac{y}{x(x-3)}, \frac{6}{x^2}$

38. $\frac{a}{y^2}, \frac{6}{y(y+5)}$

39. $\frac{9}{(x-1)^2}, \frac{6}{x(x-1)}$

40. $\frac{a^2}{y(y+7)}, \frac{a}{(y+7)^2}$

41. $\frac{3}{x-3}, \frac{5}{x(3-x)}$

42. $\frac{b}{y(y-4)}, \frac{b^2}{4-y}$

43. $\frac{3}{(x-5)^2}, \frac{2}{5-x}$

44. $\frac{3}{7-y}, \frac{2}{(y-7)^2}$

45. $\frac{3}{x^2+2x}, \frac{4}{x^2}$

46. $\frac{2}{y-3}, \frac{3}{y^3-3y^2}$

47. $\frac{x-2}{x+3}, \frac{x}{x-4}$

48. $\frac{x^2}{2x-1}, \frac{x+1}{x+4}$

49. $\frac{3}{x^2+x-2}, \frac{x}{x+2}$

50. $\frac{3x}{x-5}, \frac{4}{x^2-25}$

51. $\frac{x}{x^2+x-6}, \frac{2x}{x^2-9}$

52. $\frac{x-1}{x^2+2x-15}, \frac{x}{x^2+6x+5}$

Applying the Concepts

53. When is the LCM of two polynomials equal to their product?

SECTION

11.3 Addition and Subtraction of Rational Expressions

OBJECTIVE A To add or subtract rational expressions with the same denominators

When adding rational expressions in which the denominators are the same, add the numerators. The denominator of the sum is the common denominator.

$$\frac{5x}{18} + \frac{7x}{18} = \frac{5x + 7x}{18} = \frac{12x}{18} = \frac{2x}{3}$$

$$\frac{x}{x^2 - 1} + \frac{1}{x^2 - 1} = \frac{x + 1}{x^2 - 1} = \frac{\overset{1}{\cancel{(x + 1)}}}{(x - 1)\underset{1}{\cancel{(x + 1)}}} = \frac{1}{x - 1}$$

Note that the sum is written in simplest form.

When subtracting rational expressions with like denominators, subtract the numerators. The denominator of the difference is the common denominator. Write the answer in simplest form.

$$\frac{2x}{x - 2} - \frac{4}{x - 2} = \frac{2x - 4}{x - 2} = \frac{2\overset{1}{\cancel{(x - 2)}}}{\underset{1}{\cancel{x - 2}}} = 2$$

$$\frac{3x - 1}{x^2 - 5x + 4} - \frac{2x + 3}{x^2 - 5x + 4} = \frac{(3x - 1) - (2x + 3)}{x^2 - 5x + 4} = \frac{3x - 1 - 2x - 3}{x^2 - 5x + 4}$$

$$= \frac{x - 4}{x^2 - 5x + 4} = \frac{\overset{1}{\cancel{(x - 4)}}}{\underset{1}{\cancel{(x - 4)}}(x - 1)} = \frac{1}{x - 1}$$

Adding and Subtracting Rational Expressions with the Same Denominator

Add or subtract the numerators. Place the result over the common denominator. $\frac{a}{b} + \frac{c}{b} = \frac{a + c}{b}$ $\frac{a}{b} - \frac{c}{b} = \frac{a - c}{b}$

EXAMPLE 1

Subtract: $\frac{3x^2}{x^2 - 1} - \frac{x + 4}{x^2 - 1}$

Solution

$$\frac{3x^2}{x^2 - 1} - \frac{x + 4}{x^2 - 1} = \frac{3x^2 - (x + 4)}{x^2 - 1}$$

$$= \frac{3x^2 - x - 4}{x^2 - 1}$$

$$= \frac{(3x - 4)\overset{1}{\cancel{(x + 1)}}}{(x - 1)\underset{1}{\cancel{(x + 1)}}} = \frac{3x - 4}{x - 1}$$

YOU TRY IT 1

Subtract: $\frac{2x^2}{x^2 - x - 12} - \frac{7x + 4}{x^2 - x - 12}$

Your solution

Solution on p. S25

EXAMPLE 2

Simplify:
$$\frac{2x^2+5}{x^2+2x-3}-\frac{x^2-3x}{x^2+2x-3}+\frac{x-2}{x^2+2x-3}$$

Solution

$$\frac{2x^2+5}{x^2+2x-3}-\frac{x^2-3x}{x^2+2x-3}+\frac{x-2}{x^2+2x-3}$$
$$=\frac{(2x^2+5)-(x^2-3x)+(x-2)}{x^2+2x-3}$$
$$=\frac{2x^2+5-x^2+3x+x-2}{x^2+2x-3}$$
$$=\frac{x^2+4x+3}{x^2+2x-3}$$
$$=\frac{\overset{1}{\cancel{(x+3)}}(x+1)}{\underset{1}{\cancel{(x+3)}}(x-1)}=\frac{x+1}{x-1}$$

YOU TRY IT 2

Simplify:
$$\frac{x^2-1}{x^2-8x+12}-\frac{2x+1}{x^2-8x+12}+\frac{x}{x^2-8x+12}$$

Your solution

Solution on p. S26

OBJECTIVE B To add or subtract rational expressions with different denominators

Before two fractions with unlike denominators can be added or subtracted, each fraction must be expressed in terms of a common denominator. This common denominator is the LCM of the denominators of the fractions.

HOW TO 1 Add: $\frac{x-3}{x^2-2x}+\frac{6}{x^2-4}$

The LCM is $x(x-2)(x+2)$. • **Find the LCM of the denominators.**

$$\frac{x-3}{x^2-2x}+\frac{6}{x^2-4}$$

$$=\frac{x-3}{x(x-2)}\cdot\frac{x+2}{x+2}+\frac{6}{(x-2)(x+2)}\cdot\frac{x}{x}$$ • **Write each fraction in terms of the LCM.**

$$=\frac{x^2-x-6}{x(x-2)(x+2)}+\frac{6x}{x(x-2)(x+2)}$$ • **Multiply the factors in the numerators.**

$$=\frac{(x^2-x-6)+6x}{x(x-2)(x+2)}$$ • **Add the fractions.**

$$=\frac{x^2+5x-6}{x(x-2)(x+2)}$$ • **Simplify.**

$$=\frac{(x+6)(x-1)}{x(x-2)(x+2)}$$ • **Factor.**

After combining the numerators over the common denominator, the last step is to factor the numerator to determine whether there are common factors in the numerator and denominator. For the previous example, there are no common factors, so the answer is in simplest form.

The process of adding and subtracting rational expressions is summarized below.

Adding and Subtracting Rational Expressions

1. Find the LCM of the denominators.
2. Write each fraction as an equivalent fraction using the LCM as the denominator.
3. Add or subtract the numerators and place the result over the common denominator.
4. Write the answer in simplest form.

EXAMPLE 3

Simplify: $\frac{y}{x} - \frac{4y}{3x} + \frac{3y}{4x}$

Solution

The LCM of the denominators is $12x$.

$$\frac{y}{x} - \frac{4y}{3x} + \frac{3y}{4x}$$

$$= \frac{y}{x} \cdot \frac{12}{12} - \frac{4y}{3x} \cdot \frac{4}{4} + \frac{3y}{4x} \cdot \frac{3}{3}$$ • Write each fraction using the LCM.

$$= \frac{12y}{12x} - \frac{16y}{12x} + \frac{9y}{12x}$$

$$= \frac{12y - 16y + 9y}{12x} = \frac{5y}{12x}$$ • Combine the numerators.

YOU TRY IT 3

Simplify: $\frac{z}{8y} - \frac{4z}{3y} + \frac{5z}{4y}$

Your solution

Solution on p. S26

EXAMPLE • 4

Subtract: $\frac{2x}{x-3} - \frac{5}{3-x}$

Solution

Remember that $3 - x = -(x - 3)$.

Therefore, $\frac{5}{3-x} = \frac{5}{-(x-3)} = \frac{-5}{x-3}$.

$$\frac{2x}{x-3} - \frac{5}{3-x}$$

$$= \frac{2x}{x-3} - \frac{-5}{x-3}$$ • The LCM is $x - 3$.

$$= \frac{2x-(-5)}{x-3} = \frac{2x+5}{x-3}$$ • Combine the numerators.

YOU TRY IT • 4

Add: $\frac{5x}{x-2} + \frac{3}{2-x}$

Your solution

EXAMPLE • 5

Subtract: $\frac{2x}{2x-3} - \frac{1}{x+1}$

Solution

The LCM is $(2x - 3)(x + 1)$.

$$\frac{2x}{2x-3} - \frac{1}{x+1}$$

$$= \frac{2x}{2x-3} \cdot \frac{x+1}{x+1} - \frac{1}{x+1} \cdot \frac{2x-3}{2x-3}$$

$$= \frac{2x^2+2x}{(2x-3)(x+1)} - \frac{2x-3}{(2x-3)(x+1)}$$

$$= \frac{(2x^2+2x)-(2x-3)}{(2x-3)(x+1)}$$

$$= \frac{2x^2+2x-2x+3}{(2x-3)(x+1)} = \frac{2x^2+3}{(2x-3)(x+1)}$$

YOU TRY IT • 5

Add: $\frac{4x}{3x-1} + \frac{9}{x+4}$

Your solution

EXAMPLE • 6

Add: $1 + \frac{3}{x^2}$

Solution

The LCM is x^2.

$$1 + \frac{3}{x^2} = 1 \cdot \frac{x^2}{x^2} + \frac{3}{x^2} = \frac{x^2}{x^2} + \frac{3}{x^2} = \frac{x^2+3}{x^2}$$

YOU TRY IT • 6

Subtract: $2 - \frac{1}{x-3}$

Your solution

Solutions on p. S26

EXAMPLE • 7

Add: $\frac{x+3}{x^2-2x-8}+\frac{3}{4-x}$

Solution

Recall: $\frac{3}{4-x}=\frac{-3}{x-4}$

The LCM is $(x-4)(x+2)$.

$$\frac{x+3}{x^2-2x-8}+\frac{3}{4-x}$$
$$=\frac{x+3}{(x-4)(x+2)}+\frac{(-3)}{x-4}$$
$$=\frac{x+3}{(x-4)(x+2)}+\frac{(-3)}{x-4}\cdot\frac{x+2}{x+2}$$
$$=\frac{x+3}{(x-4)(x+2)}+\frac{(-3)(x+2)}{(x-4)(x+2)}$$
$$=\frac{(x+3)+(-3)(x+2)}{(x-4)(x+2)}$$
$$=\frac{x+3-3x-6}{(x-4)(x+2)}$$
$$=\frac{-2x-3}{(x-4)(x+2)}$$

YOU TRY IT • 7

Add: $\frac{2x-1}{x^2-25}+\frac{2}{5-x}$

Your solution

EXAMPLE • 8

Simplify: $\frac{3x+2}{2x^2-x-1}-\frac{3}{2x+1}+\frac{4}{x-1}$

Solution

The LCM is $(2x+1)(x-1)$.

$$\frac{3x+2}{2x^2-x-1}-\frac{3}{2x+1}+\frac{4}{x-1}$$
$$=\frac{3x+2}{(2x+1)(x-1)}-\frac{3}{2x+1}\cdot\frac{x-1}{x-1}+\frac{4}{x-1}\cdot\frac{2x+1}{2x+1}$$
$$=\frac{3x+2}{(2x+1)(x-1)}-\frac{3x-3}{(2x+1)(x-1)}+\frac{8x+4}{(2x+1)(x-1)}$$
$$=\frac{(3x+2)-(3x-3)+(8x+4)}{(2x+1)(x-1)}$$
$$=\frac{3x+2-3x+3+8x+4}{(2x+1)(x-1)}$$
$$=\frac{8x+9}{(2x+1)(x-1)}$$

YOU TRY IT • 8

Simplify: $\frac{2x-3}{3x^2-x-2}+\frac{5}{3x+2}-\frac{1}{x-1}$

Your solution

Solutions on p. S26

11.3 EXERCISES

OBJECTIVE A **To add or subtract rational expressions with the same denominators**

For Exercises 1 to 20, simplify.

1. $\frac{3}{y^2} + \frac{8}{y^2}$

2. $\frac{6}{ab} - \frac{2}{ab}$

3. $\frac{3}{x+4} - \frac{10}{x+4}$

4. $\frac{x}{x+6} - \frac{2}{x+6}$

5. $\frac{3x}{2x+3} + \frac{5x}{2x+3}$

6. $\frac{6y}{4y+1} - \frac{11y}{4y+1}$

7. $\frac{2x+1}{x-3} + \frac{3x+6}{x-3}$

8. $\frac{4x+3}{2x-7} + \frac{3x-8}{2x-7}$

9. $\frac{5x-1}{x+9} - \frac{3x+4}{x+9}$

10. $\frac{6x-5}{x-10} - \frac{3x-4}{x-10}$

11. $\frac{x-7}{2x+7} - \frac{4x-3}{2x+7}$

12. $\frac{2n}{3n+4} - \frac{5n-3}{3n+4}$

13. $\frac{x}{x^2+2x-15} - \frac{3}{x^2+2x-15}$

14. $\frac{3x}{x^2+3x-10} - \frac{6}{x^2+3x-10}$

15. $\frac{2x+3}{x^2-x-30} - \frac{x-2}{x^2-x-30}$

16. $\frac{3x-1}{x^2+5x-6} - \frac{2x-7}{x^2+5x-6}$

17. $\frac{4y+7}{2y^2+7y-4} - \frac{y-5}{2y^2+7y-4}$

18. $\frac{x+1}{2x^2-5x-12} + \frac{x+2}{2x^2-5x-12}$

19. $\frac{2x^2+3x}{x^2-9x+20} + \frac{2x^2-3}{x^2-9x+20} - \frac{4x^2+2x+1}{x^2-9x+20}$

20. $\frac{2x^2+3x}{x^2-2x-63} - \frac{x^2-3x+21}{x^2-2x-63} - \frac{x-7}{x^2-2x-63}$

21. Which expressions are equivalent to $\frac{3}{y-5} - \frac{y-2}{y-5}$?

(i) $\frac{5-y}{y-5}$ (ii) $\frac{1-y}{y-5}$ (iii) $\frac{5-y}{2y-10}$ (iv) -1 (v) $\frac{1-y}{-10}$

OBJECTIVE B To add or subtract rational expressions with different denominators

22. True or false? $\frac{3}{x-8} + \frac{3}{8-x} = 0$

For Exercises 23 to 80, simplify.

23. $\frac{4}{x} + \frac{5}{y}$

24. $\frac{7}{a} + \frac{5}{b}$

25. $\frac{12}{x} - \frac{5}{2x}$

26. $\frac{5}{3a} - \frac{3}{4a}$

27. $\frac{1}{2x} - \frac{5}{4x} + \frac{7}{6x}$

28. $\frac{7}{4y} + \frac{11}{6y} - \frac{8}{3y}$

29. $\frac{5}{3x} - \frac{2}{x^2} + \frac{3}{2x}$

30. $\frac{6}{y^2} + \frac{3}{4y} - \frac{2}{5y}$

31. $\frac{2}{x} - \frac{3}{2y} + \frac{3}{5x} - \frac{1}{4y}$

32. $\frac{5}{2a} + \frac{7}{3b} - \frac{2}{b} - \frac{3}{4a}$

33. $\frac{2x+1}{3x} + \frac{x-1}{5x}$

34. $\frac{4x-3}{6x} + \frac{2x+3}{4x}$

35. $\frac{x-3}{6x} + \frac{x+4}{8x}$

36. $\frac{2x-3}{2x} + \frac{x+3}{3x}$

37. $\frac{2x+9}{9x} - \frac{x-5}{5x}$

38. $\frac{3y-2}{12y} - \frac{y-3}{18y}$

39. $\frac{x+4}{2x} - \frac{x-1}{x^2}$

40. $\frac{x-2}{3x^2} - \frac{x+4}{x}$

41. $\frac{x-10}{4x^2} + \frac{x+1}{2x}$

42. $\frac{x+5}{3x^2} + \frac{2x+1}{2x}$

43. $\frac{4}{x+4} - x$

44. $2x + \frac{1}{x}$

45. $5 - \frac{x - 2}{x + 1}$

46. $3 + \frac{x - 1}{x + 1}$

47. $\frac{x + 3}{6x} - \frac{x - 3}{8x^2}$

48. $\frac{x + 2}{xy} - \frac{3x - 2}{x^2y}$

49. $\frac{3x - 1}{xy^2} - \frac{2x + 3}{xy}$

50. $\frac{4x - 3}{3x^2y} + \frac{2x + 1}{4xy^2}$

51. $\frac{5x + 7}{6xy^2} - \frac{4x - 3}{8x^2y}$

52. $\frac{x - 2}{8x^2} - \frac{x + 7}{12xy}$

53. $\frac{3x - 1}{6y^2} - \frac{x + 5}{9xy}$

54. $\frac{4}{x - 2} + \frac{5}{x + 3}$

55. $\frac{2}{x - 3} + \frac{5}{x - 4}$

56. $\frac{6}{x - 7} - \frac{4}{x + 3}$

57. $\frac{3}{y + 6} - \frac{4}{y - 3}$

58. $\frac{2x}{x + 1} + \frac{1}{x - 3}$

59. $\frac{3x}{x - 4} + \frac{2}{x + 6}$

60. $\frac{4x}{2x - 1} - \frac{5}{x - 6}$

61. $\frac{6x}{x + 5} - \frac{3}{2x + 3}$

62. $\frac{2a}{a - 7} + \frac{5}{7 - a}$

63. $\frac{4x}{6 - x} + \frac{5}{x - 6}$

64. $\frac{x}{x^2 - 9} + \frac{3}{x - 3}$

65. $\frac{y}{y^2 - 16} + \frac{1}{y - 4}$

66. $\frac{2x}{x^2 - x - 6} - \frac{3}{x + 2}$

67. $\frac{(x - 1)^2}{(x + 1)^2} - 1$

68. $1 - \frac{(y-2)^2}{(y+2)^2}$

69. $\frac{x}{1-x^2} - 1 + \frac{x}{1+x}$

70. $\frac{y}{x-y} + 2 - \frac{x}{y-x}$

71. $\frac{3x-1}{x^2-10x+25} - \frac{3}{x-5}$

72. $\frac{2a+3}{a^2-7a+12} - \frac{2}{a-3}$

73. $\frac{x+4}{x^2-x-42} + \frac{3}{7-x}$

74. $\frac{x+3}{x^2-3x-10} + \frac{2}{5-x}$

75. $\frac{1}{x+1} + \frac{x}{x-6} - \frac{5x-2}{x^2-5x-6}$

76. $\frac{x}{x-4} + \frac{5}{x+5} - \frac{11x-8}{x^2+x-20}$

77. $\frac{3x+1}{x-1} - \frac{x-1}{x-3} + \frac{x+1}{x^2-4x+3}$

78. $\frac{4x+1}{x-8} - \frac{3x+2}{x+4} - \frac{49x+4}{x^2-4x-32}$

79. $\frac{2x+9}{3-x} + \frac{x+5}{x+7} - \frac{2x^2+3x-3}{x^2+4x-21}$

80. $\frac{3x+5}{x+5} - \frac{x+1}{2-x} - \frac{4x^2-3x-1}{x^2+3x-10}$

Applying the Concepts

81. **Transportation** Suppose that you drive about 12,000 mi per year and that the cost of gasoline averages $3.70 per gallon.

 a. Let x represent the number of miles per gallon your car gets. Write a variable expression for the amount you spend on gasoline in 1 year.
 b. Write and simplify a variable expression for the amount of money you will save each year if you can increase your gas mileage by 5 mi/gal.
 c. If you currently get 25 mi/gal and you increase your gas mileage by 5 mi/gal, how much will you save in 1 year?

82. Explain the process of adding rational expressions with different denominators.

SECTION

11.4 Complex Fractions

OBJECTIVE A To simplify a complex fraction

Point of Interest

There are many instances of complex fractions in application problems. The fraction $\dfrac{1}{\dfrac{1}{r_1}+\dfrac{1}{r_2}}$ is used to determine the total resistance in certain electric circuits.

Take Note

You may use either method to simplify a complex fraction. The result will be the same.

A **complex fraction** is a fraction in which the numerator or denominator contains one or more fractions. Examples of complex fractions are shown at the right.

$$\frac{3}{2-\dfrac{1}{2}},\quad \frac{4+\dfrac{1}{x}}{3+\dfrac{2}{x}},\quad \frac{\dfrac{1}{x-1}+x+3}{x-3+\dfrac{1}{x+4}}$$

To simplify a complex fraction, use one of the following methods.

Simplifying Complex Fractions

Method 1: Multiply by 1 in the form $\dfrac{\text{LCM}}{\text{LCM}}$.

1. Determine the LCM of the denominators of the fractions in the numerator and denominator of the complex fraction.
2. Multiply the numerator and denominator of the complex fraction by the LCM.
3. Simplify.

Method 2: Multiply the numerator by the reciprocal of the denominator.

1. Simplify the numerator to a single fraction and simplify the denominator to a single fraction.
2. Using the definition for dividing fractions, multiply the numerator by the reciprocal of the denominator.
3. Simplify.

Here is an example using Method 1.

HOW TO • 1 Simplify: $\dfrac{9-\dfrac{4}{x^2}}{3+\dfrac{2}{x}}$

The LCM of x and x^2 is x^2.

• Find the LCM of the denominators of the fractions in the numerator and the denominator.

$$\frac{9-\dfrac{4}{x^2}}{3+\dfrac{2}{x}}=\frac{9-\dfrac{4}{x^2}}{3+\dfrac{2}{x}}\cdot\frac{x^2}{x^2}$$

• Multiply the numerator and denominator by the LCM.

$$=\frac{9\cdot x^2-\dfrac{4}{x^2}\cdot x^2}{3\cdot x^2+\dfrac{2}{x}\cdot x^2}=\frac{9x^2-4}{3x^2+2x}$$

• Use the Distributive Property.

$$=\frac{(3x-2)\overset{1}{\cancel{(3x+2)}}}{x\underset{1}{\cancel{(3x+2)}}}=\frac{3x-2}{x}$$

• Simplify.

Here is the same example using Method 2.

HOW TO • 2 Simplify: $\dfrac{9-\dfrac{4}{x^2}}{3+\dfrac{2}{x}}$

$$\frac{9-\dfrac{4}{x^2}}{3+\dfrac{2}{x}}=\frac{\dfrac{9x^2}{x^2}-\dfrac{4}{x^2}}{\dfrac{3x}{x}+\dfrac{2}{x}}=\frac{\dfrac{9x^2-4}{x^2}}{\dfrac{3x+2}{x}}$$

• Simplify the numerator to a single fraction and simplify the denominator to a single fraction.

$$=\frac{9x^2-4}{x^2}\cdot\frac{x}{3x+2}$$

• Multiply the numerator by the reciprocal of the denominator.

$$=\frac{x(3x-2)\overset{1}{\cancel{(3x+2)}}}{x^2\underset{1}{\cancel{(3x+2)}}}$$

• Simplify.

$$=\frac{3x-2}{x}$$

For the examples below, we will use the first method.

EXAMPLE • 1

Simplify: $\dfrac{\dfrac{1}{x}+\dfrac{1}{2}}{\dfrac{1}{x^2}-\dfrac{1}{4}}$

Solution

The LCM of x, 2, x^2, and 4 is $4x^2$.

$$\frac{\dfrac{1}{x}+\dfrac{1}{2}}{\dfrac{1}{x^2}-\dfrac{1}{4}}=\frac{\dfrac{1}{x}+\dfrac{1}{2}}{\dfrac{1}{x^2}-\dfrac{1}{4}}\cdot\frac{4x^2}{4x^2}$$

• Multiply by the LCM.

$$=\frac{\dfrac{1}{x}\cdot 4x^2+\dfrac{1}{2}\cdot 4x^2}{\dfrac{1}{x^2}\cdot 4x^2-\dfrac{1}{4}\cdot 4x^2}$$

• Distributive Property

$$=\frac{4x+2x^2}{4-x^2}$$

• Simplify.

$$=\frac{2x\overset{1}{\cancel{(2+x)}}}{(2-x)\underset{1}{\cancel{(2+x)}}}$$

$$=\frac{2x}{2-x}$$

YOU TRY IT • 1

Simplify: $\dfrac{\dfrac{1}{3}-\dfrac{1}{x}}{\dfrac{1}{9}-\dfrac{1}{x^2}}$

Your solution

Solution on p. S26

EXAMPLE • 2

Simplify: $\dfrac{1 - \dfrac{2}{x} - \dfrac{15}{x^2}}{1 - \dfrac{11}{x} + \dfrac{30}{x^2}}$

Solution

The LCM of x and x^2 is x^2.

$$\frac{1 - \frac{2}{x} - \frac{15}{x^2}}{1 - \frac{11}{x} + \frac{30}{x^2}} = \frac{1 - \frac{2}{x} - \frac{15}{x^2}}{1 - \frac{11}{x} + \frac{30}{x^2}} \cdot \frac{x^2}{x^2}$$

• Multiply by the LCM.

$$= \frac{1 \cdot x^2 - \frac{2}{x} \cdot x^2 - \frac{15}{x^2} \cdot x^2}{1 \cdot x^2 - \frac{11}{x} \cdot x^2 + \frac{30}{x^2} \cdot x^2}$$

• Distributive Property

$$= \frac{x^2 - 2x - 15}{x^2 - 11x + 30}$$

$$= \frac{\overset{1}{\cancel{(x - 5)}}(x + 3)}{\underset{1}{\cancel{(x - 5)}}(x - 6)} = \frac{x + 3}{x - 6}$$

• Simplify.

YOU TRY IT • 2

Simplify: $\dfrac{1 + \dfrac{4}{x} + \dfrac{3}{x^2}}{1 + \dfrac{10}{x} + \dfrac{21}{x^2}}$

Your solution

EXAMPLE • 3

Simplify: $\dfrac{x - 8 + \dfrac{20}{x + 4}}{x - 10 + \dfrac{24}{x + 4}}$

Solution

The LCM is $x + 4$.

$$\frac{x - 8 + \frac{20}{x + 4}}{x - 10 + \frac{24}{x + 4}}$$

$$= \frac{x - 8 + \frac{20}{x + 4}}{x - 10 + \frac{24}{x + 4}} \cdot \frac{x + 4}{x + 4}$$

• Multiply by the LCM.

$$= \frac{(x - 8)(x + 4) + \frac{20}{x + 4} \cdot (x + 4)}{(x - 10)(x + 4) + \frac{24}{x + 4} \cdot (x + 4)}$$

• Distributive Property

$$= \frac{x^2 - 4x - 32 + 20}{x^2 - 6x - 40 + 24} = \frac{x^2 - 4x - 12}{x^2 - 6x - 16}$$

• Simplify.

$$= \frac{(x - 6)\overset{1}{\cancel{(x + 2)}}}{(x - 8)\underset{1}{\cancel{(x + 2)}}} = \frac{x - 6}{x - 8}$$

YOU TRY IT • 3

Simplify: $\dfrac{x + 3 - \dfrac{20}{x - 5}}{x + 8 + \dfrac{30}{x - 5}}$

Your solution

Solutions on p. S27

11.4 EXERCISES

OBJECTIVE A To simplify a complex fraction

For Exercises 1 to 30, simplify.

1. $\dfrac{1 + \dfrac{3}{x}}{1 - \dfrac{9}{x^2}}$

2. $\dfrac{1 + \dfrac{4}{x}}{1 - \dfrac{16}{x^2}}$

3. $\dfrac{2 - \dfrac{8}{x + 4}}{3 - \dfrac{12}{x + 4}}$

4. $\dfrac{5 - \dfrac{25}{x + 5}}{1 - \dfrac{3}{x + 5}}$

5. $\dfrac{1 + \dfrac{5}{y - 2}}{1 - \dfrac{2}{y - 2}}$

6. $\dfrac{2 - \dfrac{11}{2x - 1}}{3 - \dfrac{17}{2x - 1}}$

7. $\dfrac{4 - \dfrac{2}{x + 7}}{5 + \dfrac{1}{x + 7}}$

8. $\dfrac{5 + \dfrac{3}{x - 8}}{2 - \dfrac{1}{x - 8}}$

9. $\dfrac{1 - \dfrac{1}{x} - \dfrac{6}{x^2}}{1 - \dfrac{9}{x^2}}$

10. $\dfrac{1 + \dfrac{4}{x} + \dfrac{4}{x^2}}{1 - \dfrac{2}{x} - \dfrac{8}{x^2}}$

11. $\dfrac{1 - \dfrac{5}{x} - \dfrac{6}{x^2}}{1 + \dfrac{6}{x} + \dfrac{5}{x^2}}$

12. $\dfrac{1 - \dfrac{7}{a} + \dfrac{12}{a^2}}{1 + \dfrac{1}{a} - \dfrac{20}{a^2}}$

13. $\dfrac{1 - \dfrac{6}{x} + \dfrac{8}{x^2}}{\dfrac{4}{x^2} + \dfrac{3}{x} - 1}$

14. $\dfrac{1 + \dfrac{3}{x} - \dfrac{18}{x^2}}{\dfrac{21}{x^2} - \dfrac{4}{x} - 1}$

15. $\dfrac{x - \dfrac{4}{x + 3}}{1 + \dfrac{1}{x + 3}}$

16. $\dfrac{y + \dfrac{1}{y - 2}}{1 + \dfrac{1}{y - 2}}$

17. $\dfrac{1 - \dfrac{x}{2x + 1}}{x - \dfrac{1}{2x + 1}}$

18. $\dfrac{1 - \dfrac{2x - 2}{3x - 1}}{x - \dfrac{4}{3x - 1}}$

19. $\dfrac{x - 5 + \dfrac{14}{x + 4}}{x + 3 - \dfrac{2}{x + 4}}$

20. $\dfrac{a + 4 + \dfrac{5}{a - 2}}{a + 6 + \dfrac{15}{a - 2}}$

21. $\dfrac{x + 3 - \dfrac{10}{x - 6}}{x + 2 - \dfrac{20}{x - 6}}$

22. $\dfrac{x - 7 + \dfrac{5}{x - 1}}{x - 3 + \dfrac{1}{x - 1}}$

23. $\dfrac{y - 6 + \dfrac{22}{2y + 3}}{y - 5 + \dfrac{11}{2y + 3}}$

24. $\dfrac{x + 2 - \dfrac{12}{2x - 1}}{x + 1 - \dfrac{9}{2x - 1}}$

25. $\dfrac{x - \dfrac{2}{2x - 3}}{2x - 1 - \dfrac{8}{2x - 3}}$

26. $\dfrac{x + 3 - \dfrac{18}{2x + 1}}{x - \dfrac{6}{2x + 1}}$

27. $\dfrac{\dfrac{1}{x} - \dfrac{2}{x - 1}}{\dfrac{3}{x} + \dfrac{1}{x - 1}}$

28. $\dfrac{\dfrac{3}{n + 1} + \dfrac{1}{n}}{\dfrac{2}{n + 1} + \dfrac{3}{n}}$

29. $\dfrac{\dfrac{3}{2x - 1} - \dfrac{1}{x}}{\dfrac{4}{x} + \dfrac{2}{2x - 1}}$

30. $\dfrac{\dfrac{4}{3x + 1} + \dfrac{3}{x}}{\dfrac{6}{x} - \dfrac{2}{3x + 1}}$

31. True or false? If the denominator of a complex fraction is the reciprocal of the numerator, then the complex fraction is equal to the square of its numerator.

Applying the Concepts

For Exercises 32 to 37, simplify.

32. $1 + \dfrac{1}{1 + \dfrac{1}{2}}$

33. $1 + \dfrac{1}{1 + \dfrac{1}{1 + \dfrac{1}{2}}}$

34. $\dfrac{a^{-1} - b^{-1}}{a^{-2} - b^{-2}}$

35. $\dfrac{1 + x^{-1}}{1 - x^{-1}}$

36. $\dfrac{x + x^{-1}}{x - x^{-1}}$

37. $\dfrac{x^{-1}}{y^{-1}} + \dfrac{x}{y}$

SECTION 11.5 Solving Equations Containing Fractions

OBJECTIVE A To solve an equation containing fractions 

Recall that to solve an equation containing fractions, clear denominators by multiplying each side of the equation by the LCM of the denominators. Then solve for the variable.

HOW TO • 1 Solve: $\frac{3x-1}{4} + \frac{2}{3} = \frac{7}{6}$

$$\frac{3x-1}{4} + \frac{2}{3} = \frac{7}{6}$$

$$12\left(\frac{3x-1}{4} + \frac{2}{3}\right) = 12 \cdot \frac{7}{6}$$

• The LCM is 12. To clear denominators, multiply each side of the equation by the LCM.

$$12\left(\frac{3x-1}{4}\right) + 12 \cdot \frac{2}{3} = 12 \cdot \frac{7}{6}$$

• Simplify using the Distributive Property and the Properties of Fractions.

$$\frac{\overset{3}{\cancel{12}}}{1}\left(\frac{3x-1}{\underset{1}{\cancel{4}}}\right) + \frac{\overset{4}{\cancel{12}}}{1} \cdot \frac{2}{\underset{1}{\cancel{3}}} = \frac{\overset{2}{\cancel{12}}}{1} \cdot \frac{7}{\underset{1}{\cancel{6}}}$$

$$9x - 3 + 8 = 14$$

• Solve for x.

$$9x + 5 = 14$$
$$9x = 9$$
$$x = 1$$

1 checks as a solution. The solution is 1.

Occasionally, a value of the variable that appears to be a solution of an equation will make one of the denominators zero. In such a case, that value is not a solution of the equation.

Take Note

The example at the right illustrates the importance of checking a solution of a rational equation when each side is multiplied by a variable expression. As shown in this example, a proposed solution may not check when it is substituted into the original equation.

HOW TO • 2 Solve: $\frac{2x}{x-2} = 1 + \frac{4}{x-2}$

$$\frac{2x}{x-2} = 1 + \frac{4}{x-2}$$

$$(x-2)\frac{2x}{x-2} = (x-2)\left(1 + \frac{4}{x-2}\right)$$

• The LCM is $x - 2$. Multiply each side of the equation by the LCM.

$$(x-2)\frac{2x}{x-2} = (x-2) \cdot 1 + (x-2)\frac{4}{x-2}$$

• Simplify using the Distributive Property and the Properties of Fractions.

$$\frac{\overset{1}{\cancel{(x-2)}}}{1} \cdot \frac{2x}{\underset{1}{\cancel{x-2}}} = (x-2) \cdot 1 + \frac{\overset{1}{\cancel{(x-2)}}}{1} \cdot \frac{4}{\underset{1}{\cancel{x-2}}}$$

$$2x = x - 2 + 4$$

• Solve for x.

$$2x = x + 2$$
$$x = 2$$

When x is replaced by 2, the denominators of $\frac{2x}{x-2}$ and $\frac{4}{x-2}$ are zero. Therefore, the equation has no solution.

EXAMPLE • 1

Solve: $\dfrac{x}{x+4} = \dfrac{2}{x}$

Solution

The LCM is $x(x+4)$.

$$\frac{x}{x+4} = \frac{2}{x}$$

$$x(x+4)\left(\frac{x}{x+4}\right) = x(x+4)\left(\frac{2}{x}\right)$$ • **Multiply by the LCM.**

$$\frac{\overset{1}{\cancel{x(x+4)}}}{1} \cdot \frac{x}{\underset{1}{\cancel{x+4}}} = \frac{\overset{1}{\cancel{x}}(x+4)}{1} \cdot \frac{2}{\underset{1}{\cancel{x}}}$$ • **Divide by the common factors.**

$$x^2 = (x+4)2$$ • **Simplify.**

$$x^2 = 2x + 8$$

Solve the quadratic equation by factoring.

$$x^2 - 2x - 8 = 0$$ • **Write in standard form.**

$$(x-4)(x+2) = 0$$ • **Factor.**

$$x - 4 = 0 \qquad x + 2 = 0$$ • **Principle of Zero Products**

$$x = 4 \qquad x = -2$$

Both 4 and -2 check as solutions.

The solutions are 4 and -2.

YOU TRY IT • 1

Solve: $\dfrac{x}{x+6} = \dfrac{3}{x}$

Your solution

EXAMPLE • 2

Solve: $\dfrac{3x}{x-4} = 5 + \dfrac{12}{x-4}$

Solution

The LCM is $x - 4$.

$$\frac{3x}{x-4} = 5 + \frac{12}{x-4}$$

$$(x-4)\left(\frac{3x}{x-4}\right) = (x-4)\left(5 + \frac{12}{x-4}\right)$$ • **Clear denominators.**

$$\frac{\overset{1}{\cancel{(x-4)}}}{1} \cdot \frac{3x}{\underset{1}{\cancel{x-4}}} = (x-4)5 + \frac{\overset{1}{\cancel{(x-4)}}}{1} \cdot \frac{12}{\underset{1}{\cancel{x-4}}}$$

$$3x = (x-4)5 + 12$$ • **Solve for x.**

$$3x = 5x - 20 + 12$$

$$3x = 5x - 8$$

$$-2x = -8$$

$$x = 4$$

4 does not check as a solution.

The equation has no solution.

YOU TRY IT • 2

Solve: $\dfrac{5x}{x+2} = 3 - \dfrac{10}{x+2}$

Your solution

Solutions on p. S27

11.5 EXERCISES

OBJECTIVE A To solve an equation containing fractions

When a proposed solution of a rational equation does not check in the original equation, it is because the proposed solution results in an expression that involves division by zero. For Exercises 1 to 3, state the values of x that would result in division by zero when substituted into the original equation.

1. $\frac{6x}{x+1} - \frac{x}{x-2} = 4$

2. $\frac{1}{x+5} = \frac{x}{x-3} + \frac{2}{x^2+2x-15}$

3. $\frac{3}{x-9} = \frac{1}{x^2-9x} + 2$

For Exercises 4 to 36, solve.

4. $\frac{2x}{3} - \frac{5}{2} = -\frac{1}{2}$

5. $\frac{x}{3} - \frac{1}{4} = \frac{1}{12}$

6. $\frac{x}{3} - \frac{1}{4} = \frac{x}{4} - \frac{1}{6}$

7. $\frac{2y}{9} - \frac{1}{6} = \frac{y}{9} + \frac{1}{6}$

8. $\frac{2x-5}{8} + \frac{1}{4} = \frac{x}{8} + \frac{3}{4}$

9. $\frac{3x+4}{12} - \frac{1}{3} = \frac{5x+2}{12} - \frac{1}{2}$

10. $\frac{6}{2a+1} = 2$

11. $\frac{12}{3x-2} = 3$

12. $\frac{9}{2x-5} = -2$

13. $\frac{6}{4-3x} = 3$

14. $2 + \frac{5}{x} = 7$

15. $3 + \frac{8}{n} = 5$

16. $1 - \frac{9}{x} = 4$

17. $3 - \frac{12}{x} = 7$

18. $\frac{2}{y} + 5 = 9$

19. $\frac{6}{x} + 3 = 11$

20. $\frac{3}{x-2} = \frac{4}{x}$

21. $\frac{5}{x+3} = \frac{3}{x-1}$

22. $\frac{2}{3x-1} = \frac{3}{4x+1}$

23. $\frac{5}{3x-4} = \frac{-3}{1-2x}$

24. $\frac{-3}{2x+5} = \frac{2}{x-1}$

25. $\frac{4}{5y - 1} = \frac{2}{2y - 1}$

26. $\frac{4x}{x - 4} + 5 = \frac{5x}{x - 4}$

27. $\frac{2x}{x + 2} - 5 = \frac{7x}{x + 2}$

28. $2 + \frac{3}{a - 3} = \frac{a}{a - 3}$

29. $\frac{x}{x + 4} = 3 - \frac{4}{x + 4}$

30. $\frac{x}{x - 1} = \frac{8}{x + 2}$

31. $\frac{x}{x + 12} = \frac{1}{x + 5}$

32. $\frac{2x}{x + 4} = \frac{3}{x - 1}$

33. $\frac{5}{3n - 8} = \frac{n}{n + 2}$

34. $\frac{x}{x + 4} = \frac{11}{x^2 - 16} + 2$

35. $x - \frac{6}{x - 3} = \frac{2x}{x - 3}$

36. $\frac{8}{r} + \frac{3}{r - 1} = 3$

Applying the Concepts

37. Explain the procedure for solving an equation containing fractions. Include in your discussion an explanation of how the LCM of the denominators is used to eliminate fractions in the equation.

For Exercises 38 to 43, solve.

38. $\frac{3}{5}y - \frac{1}{3}(1 - y) = \frac{2y - 5}{15}$

39. $\frac{3}{4}a = \frac{1}{2}(3 - a) + \frac{a - 2}{4}$

40. $\frac{b + 2}{5} = \frac{1}{4}b - \frac{3}{10}(b - 1)$

41. $\frac{x}{2x^2 - x - 1} = \frac{3}{x^2 - 1} + \frac{3}{2x + 1}$

42. $\frac{x + 1}{x^2 + x - 2} = \frac{x + 2}{x^2 - 1} + \frac{3}{x + 2}$

43. $\frac{y + 2}{y^2 - y - 2} + \frac{y + 1}{y^2 - 4} = \frac{1}{y + 1}$

SECTION

11.6 Literal Equations

OBJECTIVE A To solve a literal equation for one of the variables

A **literal equation** is an equation that contains more than one variable. Examples of literal equations are shown at the right.

$$2x + 3y = 6$$
$$4w - 2x + z = 0$$

Formulas are used to express a relationship among physical quantities. A **formula** is a literal equation that states a rule about measurements. Examples of formulas are shown at the right.

$$\frac{1}{R_1} + \frac{1}{R_2} = \frac{1}{R} \quad \text{(Physics)}$$
$$s = a + (n - 1)d \quad \text{(Mathematics)}$$
$$A = P + Prt \quad \text{(Business)}$$

The Addition and Multiplication Properties can be used to solve a literal equation for one of the variables. The goal is to rewrite the equation so that the variable being solved for is alone on one side of the equation and all the other numbers and variables are on the other side.

HOW TO • 1 Solve $A = P(1 + i)$ for i.

The goal is to rewrite the equation so that i is on one side of the equation and all other variables are on the other side.

$$A = P(1 + i)$$
$$A = P + Pi$$ • Use the Distributive Property to remove parentheses.
$$A - P = P - P + Pi$$ • Subtract P from each side of the equation.
$$A - P = Pi$$
$$\frac{A - P}{P} = \frac{Pi}{P}$$ • Divide each side of the equation by P.
$$\frac{A - P}{P} = i$$

EXAMPLE • 1

Solve $3x - 4y = 12$ for y.

Solution

$$3x - 4y = 12$$
$$3x - 3x - 4y = -3x + 12$$ • Subtract $3x$.
$$-4y = -3x + 12$$
$$\frac{-4y}{-4} = \frac{-3x + 12}{-4}$$ • Divide by -4.
$$y = \frac{3}{4}x - 3$$

YOU TRY IT • 1

Solve $5x - 2y = 10$ for y.

Your solution

Solution on p. S27

EXAMPLE • 2

Solve $I = \dfrac{E}{R + r}$ for R.

Solution

$$I = \frac{E}{R + r}$$

$$(R + r)I = (R + r)\frac{E}{R + r}$$ • Multiply by $(R + r)$.

$$RI + rI = E$$

$$RI + rI - rI = E - rI$$ • Subtract rI.

$$RI = E - rI$$

$$\frac{RI}{I} = \frac{E - rI}{I}$$ • Divide by I.

$$R = \frac{E - rI}{I}$$

YOU TRY IT • 2

Solve $s = \dfrac{A + L}{2}$ for L.

Your solution

EXAMPLE • 3

Solve $L = a(1 + ct)$ for c.

Solution

$$L = a(1 + ct)$$

$$L = a + act$$ • Distributive Property

$$L - a = a - a + act$$ • Subtract a.

$$L - a = act$$

$$\frac{L - a}{at} = \frac{act}{at}$$ • Divide by at.

$$\frac{L - a}{at} = c$$

YOU TRY IT • 3

Solve $S = a + (n - 1)d$ for n.

Your solution

EXAMPLE • 4

Solve $S = C - rC$ for C.

Solution

$$S = C - rC$$

$$S = (1 - r)C$$ • Factor.

$$\frac{S}{1 - r} = \frac{(1 - r)C}{1 - r}$$ • Divide by $(1 - r)$.

$$\frac{S}{1 - r} = C$$

YOU TRY IT • 4

Solve $S = rS + C$ for S.

Your solution

Solutions on p. S27

11.6 EXERCISES

OBJECTIVE A To solve a literal equation for one of the variables

For Exercises 1 to 15, solve for y.

1. $3x + y = 10$ **2.** $2x + y = 5$ **3.** $4x - y = 3$ **4.** $5x - y = 7$

5. $3x + 2y = 6$ **6.** $2x + 3y = 9$ **7.** $2x - 5y = 10$ **8.** $5x - 2y = 4$

9. $2x + 7y = 14$ **10.** $6x - 5y = 10$ **11.** $x + 3y = 6$ **12.** $x + 2y = 8$

13. $y - 2 = 3(x + 2)$ **14.** $y + 4 = -2(x - 3)$ **15.** $y - 1 = -\frac{2}{3}(x + 6)$

For Exercises 16 to 23, solve for x.

16. $x + 3y = 6$ **17.** $x + 6y = 10$ **18.** $3x - y = 3$ **19.** $2x - y = 6$

20. $2x + 5y = 10$ **21.** $4x + 3y = 12$ **22.** $x - 2y + 1 = 0$ **23.** $x - 4y - 3 = 0$

24. Two students are working with the equation $A = P(1 + i)$. State whether the two students' answers are equivalent.

a. When asked to solve the equation for i, one student answered $i = \frac{A}{P} - 1$ and the other student answered $i = \frac{A - P}{P}$.

b. When asked to solve the equation for i, one student answered $i = -\frac{P - A}{P}$ and the other student answered $i = \frac{A - P}{P}$.

For Exercises 25 to 40, solve the formula for the given variable.

25. $d = rt$; t (Physics) **26.** $E = IR$; R (Physics)

27. $PV = nRT$; T (Chemistry) **28.** $A = bh$; h (Geometry)

29. $P = 2l + 2w; l$ (Geometry)

30. $F = \frac{9}{5}C + 32; C$ (Temperature conversion)

31. $A = \frac{1}{2}h(b_1 + b_2); b_1$ (Geometry)

32. $s = a(x - vt); t$ (Physics)

33. $V = \frac{1}{3}Ah; h$ (Geometry)

34. $P = R - C; C$ (Business)

35. $R = \frac{C - S}{t}; S$ (Business)

36. $P = \frac{R - C}{n}; R$ (Business)

37. $A = P + Prt; P$ (Business)

38. $T = fm - gm; m$ (Engineering)

39. $A = Sw + w; w$ (Physics)

40. $a = S - Sr; S$ (Mathematics)

Applying the Concepts

Business Break-even analysis is a method used to determine the sales volume required for a company to break even, or experience neither a profit nor a loss on the sale of a product. The break-even point represents the number of units that must be made and sold for income from sales to equal the cost of the product. The break-even point can be calculated using the formula $B = \frac{F}{S - V}$, where F is the fixed costs, S is the selling price per unit, and V is the variable costs per unit. Use this information for Exercise 41.

41. **a.** Solve the formula $B = \frac{F}{S - V}$ for S.

b. Use your answer to part (a) to find the selling price per unit required for a company to break even. The fixed costs are $20,000, the variable costs per unit are $80, and the company plans to make and sell 200 underwater cameras.

c. Use your answer to part (a) to find the selling price per unit required for a company to break even. The fixed costs are $15,000, the variable costs per unit are $50, and the company plans to make and sell 600 pen scanners.

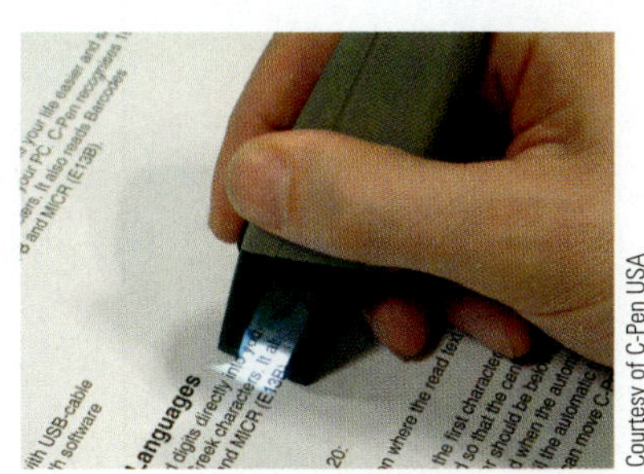

Courtesy of C-Pen USA

SECTION

11.7 Application Problems

OBJECTIVE A To solve work problems 

If a painter can paint a room in 4 h, then in 1 h the painter can paint $\frac{1}{4}$ of the room. The painter's rate of work is $\frac{1}{4}$ of the room each hour. The **rate of work** is the part of a task that is completed in 1 unit of time.

A pipe can fill a tank in 30 min. This pipe can fill $\frac{1}{30}$ of the tank in 1 min. The rate of work is $\frac{1}{30}$ of the tank each minute. If a second pipe can fill the tank in x min, the rate of work for the second pipe is $\frac{1}{x}$ of the tank each minute.

In solving a work problem, the goal is to determine the time it takes to complete a task. The basic equation that is used to solve work problems is

Rate of work × time worked = part of task completed

For example, if a faucet can fill a sink in 6 min, then in 5 min the faucet will fill $\frac{1}{6} \times 5 = \frac{5}{6}$ of the sink. In 5 min the faucet completes $\frac{5}{6}$ of the task.

Tips for Success

Note in the examples in this section that solving a word problem includes stating a strategy and using the strategy to find a solution. If you have difficulty with a word problem, write down the known information. Be very specific. Write out a phrase or sentence that states what you are trying to find. See *AIM for Success* at the front of the book.

Take Note

Use the information given in the problem to fill in the "Rate" and "Time" columns of the table. Fill in the "Part Completed" column by multiplying the two expressions you wrote in each row.

HOW TO • 1 A painter can paint a wall in 20 min. The painter's apprentice can paint the same wall in 30 min. How long will it take them to paint the wall when they work together?

Strategy for Solving a Work Problem

1. For each person or machine, write a numerical or variable expression for the rate of work, the time worked, and the part of the task completed. The results can be recorded in a table.

Unknown time to paint the wall working together: t

	Rate of Work	·	*Time Worked*	=	*Part of Task Completed*
Painter	$\frac{1}{20}$	·	t	=	$\frac{t}{20}$
Apprentice	$\frac{1}{30}$	·	t	=	$\frac{t}{30}$

2. Determine how the parts of the task completed are related. Use the fact that the sum of the parts of the task completed must equal 1, the complete task.

$$\frac{t}{20} + \frac{t}{30} = 1$$

• The sum of the part of the task completed by the painter and the part of the task completed by the apprentice is 1.

$$60\left(\frac{t}{20} + \frac{t}{30}\right) = 60 \cdot 1$$

• Multiply by the LCM of 20 and 30.

$$3t + 2t = 60$$

• Distributive Property

$$5t = 60$$
$$t = 12$$

Working together, they will paint the wall in 12 min.

EXAMPLE • 1

A small water pipe takes three times longer to fill a tank than does a large water pipe. With both pipes open it takes 4 h to fill the tank. Find the time it would take the small pipe, working alone, to fill the tank.

Strategy

- Time for large pipe to fill the tank: t
  Time for small pipe to fill the tank: $3t$

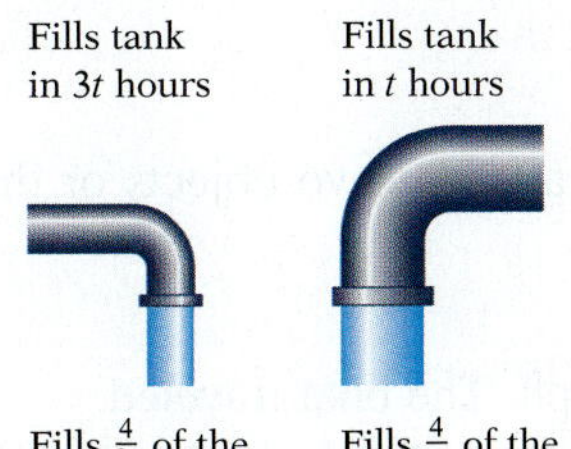

	Rate	*Time*	*Part*
Small pipe	$\frac{1}{3t}$	4	$\frac{4}{3t}$
Large pipe	$\frac{1}{t}$	4	$\frac{4}{t}$

- The sum of the parts of the task completed by each pipe must equal 1.

Solution

$$\frac{4}{3t} + \frac{4}{t} = 1$$

$$3t\left(\frac{4}{3t} + \frac{4}{t}\right) = 3t \cdot 1$$ • **Multiply by the LCM of 3*t* and *t*.**

$$4 + 12 = 3t$$ • **Distributive Property**

$$16 = 3t$$

$$\frac{16}{3} = t$$

$$3t = 3\left(\frac{16}{3}\right) = 16$$

The small pipe, working alone, takes 16 h to fill the tank.

YOU TRY IT • 1

Two computer printers that work at the same rate are working together to print the payroll checks for a large corporation. After they work together for 2 h, one of the printers quits. The second printer requires 3 h more to complete the payroll checks. Find the time it would take one printer, working alone, to print the payroll.

Your strategy

Your solution

Solution on p. S28

OBJECTIVE B To use rational expressions to solve uniform motion problems 

A car that travels constantly in a straight line at 30 mph is in uniform motion. **Uniform motion** means that the speed or direction of an object does not change.

The basic equation used to solve uniform motion problems is

$$\textbf{Distance} = \textbf{rate} \times \textbf{time}$$

An alternative form of this equation can be written by solving the equation for time.

$$\frac{\textbf{Distance}}{\textbf{Rate}} = \textbf{time}$$

This form of the equation is useful when the total time of travel for two objects or the time of travel between two points is known.

HOW TO • 2 The speed of a boat in still water is 20 mph. The boat traveled 75 mi down a river in the same amount of time it took to travel 45 mi up the river. Find the rate of the river's current.

Strategy for Solving a Uniform Motion Problem

1. For each object, write a numerical or variable expression for the distance, rate, and time. The results can be recorded in a table.

The unknown rate of the river's current: r

	Distance	÷	*Rate*	=	*Time*
Down river	75	÷	$20 + r$	=	$\frac{75}{20 + r}$
Up river	45	÷	$20 - r$	=	$\frac{45}{20 - r}$

Take Note

Use the information given in the problem to fill in the "Distance" and "Rate" columns of the table. Fill in the "Time" column by dividing the two expressions you wrote in each row.

2. Determine how the times traveled by each object are related. For example, it may be known that the times are equal, or the total time may be known.

$$\frac{75}{20 + r} = \frac{45}{20 - r}$$

• The time down the river is equal to the time up the river.

$$(20 + r)(20 - r)\frac{75}{20 + r} = (20 + r)(20 - r)\frac{45}{20 - r}$$

• Multiply by the LCM.

$$(20 - r)75 = (20 + r)45$$

$$1500 - 75r = 900 + 45r$$

• Distributive Property

$$-120r = -600$$

$$r = 5$$

The rate of the river's current is 5 mph.

EXAMPLE • 2

A cyclist rode the first 20 mi of a trip at a constant rate. For the next 16 mi, the cyclist reduced the speed by 2 mph. The total time for the 36 mi was 4 h. Find the rate of the cyclist for each leg of the trip.

Strategy

- Rate for the first 20 mi: r
  Rate for the next 16 mi: $r - 2$

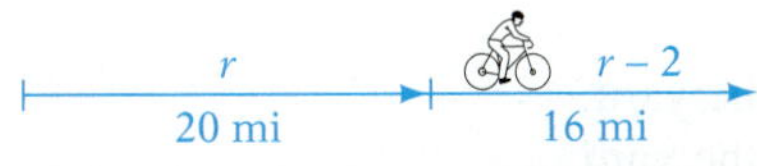

	Distance	*Rate*	*Time*
First 20 mi	20	r	$\frac{20}{r}$
Next 16 mi	16	$r - 2$	$\frac{16}{r-2}$

- The total time for the trip was 4 h.

Solution

$$\frac{20}{r} + \frac{16}{r-2} = 4$$ • The total time was 4 h.

$$r(r-2)\left[\frac{20}{r} + \frac{16}{r-2}\right] = r(r-2) \cdot 4$$ • Multiply by the LCM.

$$(r-2)20 + 16r = 4r^2 - 8r$$ • Distributive Property

$$20r - 40 + 16r = 4r^2 - 8r$$

$$36r - 40 = 4r^2 - 8r$$

Solve the quadratic equation by factoring.

$$0 = 4r^2 - 44r + 40$$ • Standard form

$$0 = 4(r^2 - 11r + 10)$$

$$0 = 4(r - 10)(r - 1)$$ • Factor.

$$r - 10 = 0 \qquad r - 1 = 0$$ • Principle of Zero Products

$$r = 10 \qquad r = 1$$

The solution $r = 1$ mph is not possible, because the rate on the last 16 mi would then be -1 mph.

10 mph was the rate for the first 20 mi.
8 mph was the rate for the next 16 mi.

YOU TRY IT • 2

The total time it took for a sailboat to sail back and forth across a lake 6 km wide was 2 h. The rate sailing back was three times the rate sailing across. Find the rate sailing out across the lake.

Your strategy

Your solution

Solution on p. S28

11.7 EXERCISES

OBJECTIVE A To solve work problems

1. Explain the meaning of the phrase "rate of work."

2. If $\frac{2}{5}$ of a room can be painted in 1 h, what is the rate of work? At the same rate, how long will it take to paint the entire room?

3. It takes Sam h hours to rake the yard, and it takes Emma k hours to rake the yard, where $h > k$. Let t be the amount of time it takes Sam and Emma to rake the yard working together. Is t less than k, between k and h, or greater than k?

4. One grocery clerk can stock a shelf in 20 min, whereas a second clerk requires 30 min to stock the same shelf. How long would it take to stock the shelf if the two clerks worked together?

5. One person with a skiploader requires 12 h to remove a large quantity of earth. A second, larger skiploader can remove the same amount of earth in 4 h. How long would it take to remove the earth with both skiploaders working together?

© Tom Carter/PhotoEdit

6. An experienced painter can paint a fence twice as fast as an inexperienced painter. Working together, the painters require 4 h to paint the fence. How long would it take the experienced painter, working alone, to paint the fence?

7. A new machine can make 10,000 aluminum cans three times faster than an older machine. With both machines working, 10,000 cans can be made in 9 h. How long would it take the new machine, working alone, to make the 10,000 cans?

8. A small air conditioner can cool a room 5° in 75 min. A larger air conditioner can cool the room 5° in 50 min. How long would it take to cool the room 5° with both air conditioners working?

9. One printing press can print the first edition of a book in 55 min, whereas a second printing press requires 66 min to print the same number of copies. How long would it take to print the first edition with both presses operating?

© Sheldan Collins/Corbis

10. Two oil pipelines can fill a small tank in 30 min. One of the pipelines would require 45 min to fill the tank. How long would it take the second pipeline, working alone, to fill the tank?

11. A mason can construct a retaining wall in 10 h. With the mason's apprentice assisting, the task takes 6 h. How long would it take the apprentice, working alone, to construct the wall?

© David Frazier/PhotoEdit

12. A mechanic requires 2 h to repair a transmission, whereas an apprentice requires 6 h to make the same repairs. The mechanic worked alone for 1 h and then stopped. How long will it take the apprentice, working alone, to complete the repairs?

13. One technician can wire a security alarm in 4 h, whereas it takes 6 h for a second technician to do the same job. After working alone for 2 h, the first technician quit. How long will it take the second technician to complete the wiring?

14. A wallpaper hanger requires 2 h to hang the wallpaper on one wall of a room. A second wallpaper hanger requires 4 h to hang the same amount of paper. The first wallpaper hanger worked alone for 1 h and then quit. How long will it take the second wallpaper hanger, working alone, to complete the wall?

15. Two welders who work at the same rate are welding the girders of a building. After they work together for 10 h, one of the welders quits. The second welder requires 20 more hours to complete the welds. Find the time it would have taken one of the welders, working alone, to complete the welds.

© Galen Rowell/Corbis

16. A large and a small heating unit are being used to heat the water of a pool. The large unit, working alone, requires 8 h to heat the pool. After both units have been operating for 2 h, the large unit is turned off. The small unit requires 9 h more to heat the pool. How long would it take the small unit, working alone, to heat the pool?

17. Two machines that fill cereal boxes work at the same rate. After they work together for 7 h, one machine breaks down. The second machine requires 14 h more to finish filling the boxes. How long would it have taken one of the machines, working alone, to fill the boxes?

18. A large and a small drain are opened to drain a pool. The large drain can empty the pool in 6 h. After both drains have been open for 1 h, the large drain becomes clogged and is closed. The smaller drain remains open and requires 9 h more to empty the pool. How long would it have taken the small drain, working alone, to empty the pool?

19. Zachary and Eli picked a row of peas together in m minutes. It would have taken Zachary n minutes to pick the row of peas by himself. What fraction of the row of peas did Zachary pick? What fraction of the row of peas did Eli pick?

OBJECTIVE B To use rational expressions to solve uniform motion problems

20. Running at a constant speed, a jogger ran 24 mi in 3 h. How far did the jogger run in 2 h?

21. For uniform motion, distance = rate · time. How is time related to distance and rate? How is rate related to distance and time?

22. Commuting from work to home, a lab technician traveled 10 mi at a constant rate through congested traffic. On reaching the expressway, the technician increased the speed by 20 mph. An additional 20 mi was traveled at the increased speed. The total time for the trip was 1 h. Find the rate of travel through the congested traffic.

10 mi r | 20 mi $r + 20$

23. The president of a company traveled 1800 mi by jet and 300 mi on a prop plane. The rate of the jet was four times the rate of the prop plane. The entire trip took a total of 5 h. Find the rate of the jet plane.

24. As part of a conditioning program, a jogger ran 8 mi in the same amount of time a cyclist rode 20 mi. The rate of the cyclist was 12 mph faster than the rate of the jogger. Find the rate of the jogger and that of the cyclist.

8 mi r

20 mi $r + 12$

25. An express train travels 600 mi in the same amount of time it takes a freight train to travel 360 mi. The rate of the express train is 20 mph faster than that of the freight train. Find the rate of each train.

© Billy E. Barnes/PhotoEdit

26. To assess the damage done by a fire, a forest ranger traveled 1080 mi by jet and then an additional 180 mi by helicopter. The rate of the jet was four times the rate of the helicopter. The entire trip took a total of 5 h. Find the rate of the jet.

27. As part of an exercise plan, Camille Ellison walked for 40 min and then ran for 20 min. If Camille runs 3 mph faster than she walks and covered 5 mi during the 1-hour exercise period, what is her walking speed?

28. A car and a bus leave a town at 1 P.M. and head for a town 300 mi away. The rate of the car is twice the rate of the bus. The car arrives 5 h ahead of the bus. Find the rate of the car.

29. A car is traveling at a rate that is 36 mph faster than the rate of a cyclist. The car travels 384 mi in the same amount of time it takes the cyclist to travel 96 mi. Find the rate of the car.

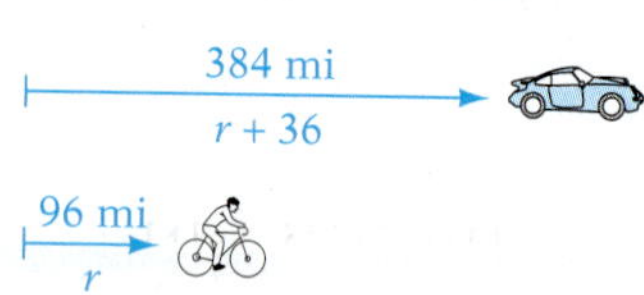

30. On a recent trip, a trucker traveled 330 mi at a constant rate. Because of road construction, the trucker then had to reduce the speed by 25 mph. An additional 30 mi was traveled at the reduced rate. The total time for the entire trip was 7 h. Find the rate of the trucker for the first 330 mi.

© Lee Cohen/Corbis

31. A backpacker hiking into a wilderness area walked 9 mi at a constant rate and then reduced this rate by 1 mph. Another 4 mi was hiked at the reduced rate. The time required to hike the 4 mi was 1 h less than the time required to walk the 9 mi. Find the rate at which the hiker walked the first 9 mi.

32. A plane can fly 180 mph in calm air. Flying with the wind, the plane can fly 600 mi in the same amount of time it takes to fly 480 mi against the wind. Find the rate of the wind.

33. A commercial jet can fly 550 mph in calm air. Traveling with the jet stream, the plane flew 2400 mi in the same amount of time it takes to fly 2000 mi against the jet stream. Find the rate of the jet stream.

2400 mi
$550 + r$

2000 mi
$550 - r$

34. A cruise ship can sail at 28 mph in calm water. Sailing with the gulf current, the ship can sail 170 mi in the same amount of time that it can sail 110 mi against the gulf current. Find the rate of the gulf current.

© Joel Bit, 2009. Used under license from Shutterstock.com

35. Rowing with the current of a river, a rowing team can row 25 mi in the same amount of time it takes to row 15 mi against the current. The rate of the rowing team in calm water is 20 mph. Find the rate of the current.

For Exercises 36 and 37, use the following problem situation: A plane can fly 380 mph in calm air. In the time it takes the plane to fly 1440 mi against a headwind, it could fly 1600 mi with the wind. Use the equation $\frac{1440}{380 - r} = \frac{1600}{380 + r}$ to find the rate r of the wind.

36. Explain the meaning of $380 - r$ and $380 + r$ in terms of the problem situation.

37. Explain the meaning of $\frac{1440}{380 - r}$ and $\frac{1600}{380 + r}$ in terms of the problem situation.

Applying the Concepts

38. **Work** One pipe can fill a tank in 2 h, a second pipe can fill the tank in 4 h, and a third pipe can fill the tank in 5 h. How long will it take to fill the tank with all three pipes working?

39. **Transportation** Because of bad weather, a bus driver reduced the usual speed along a 150-mile bus route by 10 mph. The bus arrived only 30 min later than its usual arrival time. How fast does the bus usually travel?

FOCUS ON PROBLEM SOLVING

Negations

The sentence "George Washington was the first president of the United States" is a true sentence. The **negation** of that sentence is "George Washington was **not** the first president of the United States." That sentence is false. In general, the negation of a true sentence is a false sentence.

© Francis G. Mayer/Corbis

The negation of a false sentence is a true sentence. For instance, the sentence "The moon is made of green cheese" is a false sentence. The negation of that sentence, "The moon is **not** made of green cheese," is true.

The words *all, no* (or *none*), and *some* are called **quantifiers.** Writing the negation of a sentence that contains these words requires special attention. Consider the sentence "All pets are dogs." This sentence is not true because there are pets that are not dogs; cats, for example, are pets. Because the sentence is false, its negation must be true. You might be tempted to write "All pets are not dogs," but that sentence is not true because some pets are dogs. The correct negation of "All pets are dogs" is "Some pets are not dogs." Note the use of the word *some* in the negation.

Now consider the sentence "Some computers are portable." Because that sentence is true, its negation must be false. Writing "Some computers are not portable" as the negation is not correct, because that sentence is true. The negation of "Some computers are portable" is "No computers are portable."

The sentence "No flowers have red blooms" is false, because there is at least one flower (some roses, for example) that has red blooms. Because the sentence is false, its negation must be true. The negation is "Some flowers have red blooms."

Sentence	*Negation*
All *A* are *B*.	Some *A* are not *B*.
No *A* are *B*.	Some *A* are *B*.
Some *A* are *B*.	No *A* are *B*.
Some *A* are not *B*.	All *A* are *B*.

Write the negation of the sentence.

1. All cats like milk.

2. All computers need people.

3. Some trees are tall.

4. No politicians are honest.

5. No houses have kitchens.

6. All police officers are tall.

7. All lakes are not polluted.

8. Some drivers are unsafe.

9. Some speeches are interesting.

10. All laws are good.

11. All businesses are not profitable.

12. All motorcycles are not large.

13. Some vegetables are good for you to eat.

14. Some banks are not open on Sunday.

PROJECTS AND GROUP ACTIVITIES

Body Mass Index

Body mass index, or **BMI,** expresses the relationship between a person's height and weight. It is a measurement for gauging a person's weight-related level of risk for high blood pressure, heart disease, and diabetes. A BMI value of 25 or less indicates a very low to low risk; a BMI value of 25 to 30 indicates low to moderate risk; a BMI of 30 or more indicates a moderate to very high risk.

The formula for body mass index is $B = \frac{705W}{H^2}$, where B is the BMI, W is the weight in pounds, and H is height in inches.

1. Amy is 140 lb and 5'8" tall. Calculate Amy's BMI. Round to the nearest tenth. Would you rank Amy as a low, moderate, or high risk for weight-related disease?
2. Roger is 5'11". How much should he weigh in order to have a BMI of 25? Round to the nearest whole number.
3. Brenda is 5'3". What should she weigh in order to have a BMI of 24? Round to the nearest whole number.
4. Carlos weighs 185 lb and is 5'9". How many pounds must Carlos lose in order to reach a BMI of 23? Round to the nearest whole number.
5. Pat is 6'3" and weighs 245 lb. Calculate the number of pounds Pat must lose in order to reach a BMI of 22. Round to the nearest whole number.
6. Zack weighs 205 lb and is 6'0". He would like to lower his BMI to 20. **a.** By how many points must Zack lower his BMI? Round to the nearest tenth. **b.** How many pounds must Zack lose in order to reach a BMI of 20? Round to the nearest whole number.
7. Felicia weighs 160 lb and is 5'7". She would like to lower her BMI to 20. **a.** By how many points must Felicia lower her BMI? Round to the nearest tenth. **b.** How many pounds must Felicia lose in order to reach a BMI of 20? Round to the nearest whole number.

CHAPTER 11

SUMMARY

KEY WORDS	EXAMPLES
A *rational expression* is a fraction in which the numerator and denominator are polynomials. A rational expression is in *simplest form* when the numerator and denominator have no common factors. [11.1A, p. 574]	$\frac{2x+1}{x^2+4}$ is a rational expression in simplest form.
The *reciprocal of a rational expression* is the rational expression with the numerator and denominator interchanged. [11.1C, p. 577]	The reciprocal of $\frac{3x-y}{x+4}$ is $\frac{x+4}{3x-y}$.
The *least common multiple (LCM) of two or more polynomials* is the polynomial of least degree that contains all the factors of each polynomial. [11.2A, p. 582]	The LCM of $3x^2 - 6x$ and $x^2 - 4$ is $3x(x-2)(x+2)$, because it contains the factors of $3x^2 - 6x = 3x(x-2)$ and the factors of $x^2 - 4 = (x-2)(x+2)$.

A *complex fraction* is a fraction whose numerator or denominator contains one or more fractions. [11.4A, p. 595]

$\dfrac{x - \dfrac{2}{x+1}}{1 - \dfrac{4}{x}}$ is a complex fraction.

A *literal equation* is an equation that contains more than one variable. A *formula* is a literal equation that states a rule about measurements. [11.6A, p. 604]

$3x - 4y = 12$ is a literal equation.
$A = LW$ is a literal equation that is also the formula for the area of a rectangle.

ESSENTIAL RULES AND PROCEDURES

EXAMPLES

Simplifying Rational Expressions [11.1A, p. 574]
Factor the numerator and denominator. Divide the numerator and denominator by the common factors.

$$\frac{x^2 - 3x - 10}{x^2 - 25} = \frac{(x+2)(x-5)}{(x+5)(x-5)} = \frac{x+2}{x+5}$$

Multiplying Rational Expressions [11.1B, p. 575]
Multiply the numerators. Multiply the denominators. Write the answer in simplest form.

$$\frac{a}{b} \cdot \frac{c}{d} = \frac{ac}{bd}$$

$$\begin{aligned}\frac{x^2 - 3x}{x^2 + x} \cdot \frac{x^2 + 5x + 4}{x^2 - 4x + 3} &= \frac{x(x-3)}{x(x+1)} \cdot \frac{(x+1)(x+4)}{(x-3)(x-1)} \\ &= \frac{x(x-3)(x+1)(x+4)}{x(x+1)(x-3)(x-1)} \\ &= \frac{x+4}{x-1}\end{aligned}$$

Dividing Rational Expressions [11.1C, p. 577]
Multiply the dividend by the reciprocal of the divisor. Write the answer in simplest form.

$$\frac{a}{b} \div \frac{c}{d} = \frac{a}{b} \cdot \frac{d}{c} = \frac{ad}{bc}$$

$$\begin{aligned}\frac{4x + 16}{3x - 6} \div \frac{x^2 + 6x + 8}{x^2 - 4} &= \frac{4x + 16}{3x - 6} \cdot \frac{x^2 - 4}{x^2 + 6x + 8} \\ &= \frac{4(x+4)}{3(x-2)} \cdot \frac{(x-2)(x+2)}{(x+4)(x+2)} \\ &= \frac{4}{3}\end{aligned}$$

Adding and Subtracting Rational Expressions [11.3B, p. 588]

1. Find the LCM of the denominators.
2. Write each fraction as an equivalent fraction using the LCM as the denominator.
3. Add or subtract the numerators and place the result over the common denominator.
4. Write the answer in simplest form.

$$\frac{a}{b} + \frac{c}{b} = \frac{a+c}{b} \qquad \frac{a}{b} - \frac{c}{b} = \frac{a-c}{b}$$

$$\begin{aligned}\frac{x}{x+1} - \frac{x+3}{x-2} &= \frac{x}{x+1} \cdot \frac{x-2}{x-2} - \frac{x+3}{x-2} \cdot \frac{x+1}{x+1} \\ &= \frac{x(x-2)}{(x+1)(x-2)} - \frac{(x+3)(x+1)}{(x+1)(x-2)} \\ &= \frac{x(x-2) - (x+3)(x+1)}{(x+1)(x-2)} \\ &= \frac{(x^2 - 2x) - (x^2 + 4x + 3)}{(x+1)(x-2)} \\ &= \frac{-6x - 3}{(x+1)(x-2)}\end{aligned}$$

Simplifying Complex Fractions [11.4A, p. 595]

Method 1: Multiply by 1 in the form $\frac{\text{LCM}}{\text{LCM}}$.

1. Determine the LCM of the denominators of the fractions in the numerator and denominator of the complex fraction.
2. Multiply the numerator and denominator of the complex fraction by the LCM.
3. Simplify.

Method 1:

$$\frac{\frac{1}{x}+\frac{1}{y}}{\frac{1}{x}-\frac{1}{y}} = \frac{\frac{1}{x}+\frac{1}{y}}{\frac{1}{x}-\frac{1}{y}} \cdot \frac{xy}{xy}$$
$$= \frac{\frac{1}{x}\cdot xy + \frac{1}{y}\cdot xy}{\frac{1}{x}\cdot xy - \frac{1}{y}\cdot xy}$$
$$= \frac{y+x}{y-x}$$

Method 2: Multiply the numerator by the reciprocal of the denominator.

1. Simplify the numerator to a single fraction and simplify the denominator to a single fraction.
2. Using the definition for dividing fractions, multiply the numerator by the reciprocal of the denominator.
3. Simplify.

Method 2:

$$\frac{\frac{1}{x}+\frac{1}{y}}{\frac{1}{x}-\frac{1}{y}} = \frac{\frac{y+x}{xy}}{\frac{y-x}{xy}}$$
$$= \frac{y+x}{xy} \cdot \frac{xy}{y-x}$$
$$= \frac{y+x}{y-x}$$

Solving Equations Containing Fractions [11.5A, p. 600]

Clear denominators by multiplying each side of the equation by the LCM of the denominators. Then solve for the variable.

$$\frac{1}{2a} = \frac{2}{a} - \frac{3}{8}$$
$$8a\left(\frac{1}{2a}\right) = 8a\left(\frac{2}{a}\right) - 8a\left(\frac{3}{8}\right)$$
$$4 = 16 - 3a$$
$$-12 = -3a$$
$$4 = a$$

Solving Literal Equations [11.6A, p. 604]

Rewrite the equation so that the letter being solved for is alone on one side of the equation and all numbers and other variables are on the other side.

$$2x + ax = 5$$
• **Solve for x.**
$$x(2+a) = 5$$
$$\frac{x(2+a)}{2+a} = \frac{5}{2+a}$$
$$x = \frac{5}{2+a}$$

Work Problems [11.7A, p. 608]

Rate of work × time worked = part of task completed

Pat can do a certain job in 3 h. Chris can do the same job in 5 h. How long would it take them, working together, to get the job done?

$$\frac{t}{3} + \frac{t}{5} = 1$$

Uniform Motion Problems with Rational Expressions [11.7B, p. 610]

$$\frac{\text{Distance}}{\text{Rate}} = \text{time}$$

Train A's speed is 15 mph faster than train B's speed. Train A travels 150 mi in the same amount of time it takes train B to travel 120 mi. Find the rate of train B.

$$\frac{120}{r} = \frac{150}{r+15}$$

CHAPTER 11

CONCEPT REVIEW

Test your knowledge of the concepts presented in this chapter. Answer each question. Then check your answers against the ones provided in the Answer Section.

1. When is a rational expression in simplest form?

2. How is the reciprocal useful when dividing rational expressions?

3. How do you find the LCM of two polynomials?

4. When subtracting two rational expressions, what must be the same about both expressions before subtraction can take place?

5. What are the steps for adding rational expressions?

6. What are the steps used to simplify a complex fraction? Use either method.

7. When solving an equation that contains fractions, why do we first clear the denominators?

8. What is the goal when you solve a literal equation for a particular variable?

9. What is the rate of work if a job is completed in x hours?

10. Suppose the speed of a boat in calm water is r mph, and the speed of the current in a river is c mph. Express the rate of the boat when traveling down this river and the rate of the boat when traveling up this river.

CHAPTER 11

REVIEW EXERCISES

1. Divide: $\dfrac{6a^2b^7}{25x^3y} \div \dfrac{12a^3b^4}{5x^2y^2}$

2. Add: $\dfrac{x+7}{15x} + \dfrac{x-2}{20x}$

3. Multiply: $\dfrac{3x^3 + 9x^2}{6xy^2 - 18y^2} \cdot \dfrac{4xy^3 - 12y^3}{5x^2 + 15x}$

4. Divide: $\dfrac{2x(x-y)}{x^2y(x+y)} \div \dfrac{3(x-y)}{x^2y^2}$

5. Simplify: $\dfrac{x - \dfrac{16}{5x-2}}{3x - 4 - \dfrac{88}{5x-2}}$

6. Simplify: $\dfrac{x^2 + x - 30}{15 + 2x - x^2}$

7. Simplify: $\dfrac{16x^5y^3}{24xy^{10}}$

8. Solve: $\dfrac{20}{x+2} = \dfrac{5}{16}$

9. Divide: $\dfrac{10 - 23y + 12y^2}{6y^2 - y - 5} \div \dfrac{4y^2 - 13y + 10}{18y^2 + 3y - 10}$

10. Solve $3ax - x = 5$ for x.

11. Solve: $\dfrac{2}{x} + \dfrac{3}{4} = 1$

12. Add: $\dfrac{x}{y} + \dfrac{3}{x}$

13. Solve $5x + 4y = 20$ for y.

14. Multiply: $\dfrac{8ab^2}{15x^3y} \cdot \dfrac{5xy^4}{16a^2b}$

15. Simplify: $\dfrac{1 - \dfrac{1}{x}}{1 - \dfrac{8x-7}{x^2}}$

16. Write each fraction in terms of the LCM of the denominators.
$\dfrac{x}{12x^2 + 16x - 3}, \dfrac{4x^2}{6x^2 + 7x - 3}$

17. Solve $T = 2(ab + bc + ca)$ for a.

18. Solve: $\dfrac{5}{7} + \dfrac{x}{2} = 2 - \dfrac{x}{7}$

19. Simplify: $\dfrac{2 + \dfrac{1}{x}}{3 - \dfrac{2}{x}}$

20. Subtract: $\dfrac{2x}{x-5} - \dfrac{x+1}{x-2}$

21. Solve $i = \frac{100m}{c}$ for c.

22. Solve: $\frac{x+8}{x+4} = 1 + \frac{5}{x+4}$

23. Divide: $\frac{20x^2 - 45x}{6x^3 + 4x^2} \div \frac{40x^3 - 90x^2}{12x^2 + 8x}$

24. Add: $\frac{2y}{5y-7} + \frac{3}{7-5y}$

25. Subtract: $\frac{5x+3}{2x^2+5x-3} - \frac{3x+4}{2x^2+5x-3}$

26. Find the LCM of $10x^2 - 11x + 3$ and $20x^2 - 17x + 3$.

27. Solve $4x + 9y = 18$ for y.

28. Multiply: $\frac{2x^2 - 5x - 3}{3x^2 - 7x - 6} \cdot \frac{3x^2 + 8x + 4}{x^2 + 4x + 4}$

29. Solve: $\frac{20}{2x+3} = \frac{17x}{2x+3} - 5$

30. Add: $\frac{x-1}{x+2} + \frac{3x-2}{5-x} + \frac{5x^2+15x-11}{x^2-3x-10}$

31. Solve: $\frac{6}{x-7} = \frac{8}{x-6}$

32. Solve: $\frac{3}{20} = \frac{x}{80}$

33. **Work** One hose can fill a pool in 15 h. A second hose can fill the pool in 10 h. How long would it take to fill the pool using both hoses?

34. **Travel** A car travels 315 mi in the same amount of time in which a bus travels 245 mi. The rate of the car is 10 mph faster than that of the bus. Find the rate of the car.

35. **Travel** The rate of a jet is 400 mph in calm air. Traveling with the wind, the jet can fly 2100 mi in the same amount of time it takes to fly 1900 mi against the wind. Find the rate of the wind.

CHAPTER 11

TEST

1. Subtract: $\dfrac{x}{x+3} - \dfrac{2x-5}{x^2+x-6}$

2. Solve: $\dfrac{3}{x+4} = \dfrac{5}{x+6}$

3. Multiply: $\dfrac{x^2+2x-3}{x^2+6x+9} \cdot \dfrac{2x^2-11x+5}{2x^2+3x-5}$

4. Simplify: $\dfrac{16x^5y}{24x^2y^4}$

5. Solve $d = s + rt$ for t.

6. Solve: $\dfrac{6}{x} - 2 = 1$

7. Simplify: $\dfrac{x^2+4x-5}{1-x^2}$

8. Find the LCM of $6x - 3$ and $2x^2 + x - 1$.

9. Subtract: $\dfrac{2}{2x-1} - \dfrac{3}{3x+1}$

10. Divide: $\dfrac{x^2+3x+2}{x^2+5x+4} \div \dfrac{x^2-x-6}{x^2+2x-15}$

11. Simplify: $\dfrac{1+\dfrac{1}{x}-\dfrac{12}{x^2}}{1+\dfrac{2}{x}-\dfrac{8}{x^2}}$

12. Write each fraction in terms of the LCM of the denominators.
$\dfrac{3}{x^2-2x}, \dfrac{x}{x^2-4}$

13. Subtract: $\dfrac{2x}{x^2 + 3x - 10} - \dfrac{4}{x^2 + 3x - 10}$

14. Solve $3x - 8y = 16$ for y.

15. Solve: $\dfrac{2x}{x + 1} - 3 = \dfrac{-2}{x + 1}$

16. Multiply: $\dfrac{x^3y^4}{x^2 - 4x + 4} \cdot \dfrac{x^2 - x - 2}{x^6y^4}$

17. Divide: $\dfrac{8a^2b^5}{3xy^4} \div \dfrac{4a^3b}{9x^2y}$

18. Add: $\dfrac{4}{5x^2y} + \dfrac{1}{5x^2y}$

19. **Work** A ski resort can manufacture enough machine-made snow to open its beginners' run in 4 h, whereas naturally falling snow would take 12 h to provide enough snow. If the resort makes snow at the same time it is snowing naturally, how long will it take until the run can be opened?

20. **Work** A pool can be filled with one pipe in 6 h, whereas a second pipe requires 12 h to fill the pool. How long would it take to fill the pool with both pipes turned on?

21. **Travel** A small plane can fly at 110 mph in calm air. Flying with the wind, the plane can fly 260 mi in the same amount of time it takes to fly 180 mi against the wind. Find the rate of the wind.

22. **Travel** A jet ski can comfortably travel across calm water at 35 mph. If a rider traveled 4 mi down a river in the same amount of time it took to travel 3 mi back up the river, find the rate of the river's current.

CUMULATIVE REVIEW EXERCISES

1. Evaluate: $\left(\frac{2}{3}\right)^2 \div \left(\frac{3}{2} - \frac{2}{3}\right) + \frac{1}{2}$

2. Evaluate $-a^2 + (a - b)^2$ when $a = -2$ and $b = 3$.

3. Simplify: $-2x - (-3y) + 7x - 5y$

4. Simplify: $2[3x - 7(x - 3) - 8]$

5. Solve: $4 - \frac{2}{3}x = 7$

6. Solve: $3[x - 2(x - 3)] = 2(3 - 2x)$

7. Find $16\frac{2}{3}\%$ of 60.

8. Simplify: $(a^2b^5)(ab^2)$

9. Multiply: $(a - 3b)(a + 4b)$

10. Divide: $\frac{15b^4 - 5b^2 + 10b}{5b}$

11. Divide: $(x^3 - 8) \div (x - 2)$

12. Factor: $12x^2 - x - 1$

13. Factor: $y^2 - 7y + 6$

14. Factor: $2a^3 + 7a^2 - 15a$

15. Factor: $4b^2 - 100$

16. Solve: $(x + 3)(2x - 5) = 0$

17. Simplify: $\frac{12x^4y^2}{18xy^7}$

18. Simplify: $\frac{x^2 - 7x + 10}{25 - x^2}$

19. Divide: $\dfrac{x^2 - x - 56}{x^2 + 8x + 7} \div \dfrac{x^2 - 13x + 40}{x^2 - 4x - 5}$

20. Subtract: $\dfrac{2}{2x - 1} - \dfrac{1}{x + 1}$

21. Simplify: $\dfrac{1 - \dfrac{2}{x} - \dfrac{15}{x^2}}{1 - \dfrac{25}{x^2}}$

22. Solve: $\dfrac{3x}{x - 3} - 2 = \dfrac{10}{x - 3}$

23. Solve: $\dfrac{2}{x - 2} = \dfrac{12}{x + 3}$

24. Solve $f = v + at$ for t.

25. **Number Sense** Translate "the difference between five times a number and thirteen is the opposite of eight" into an equation and solve.

26. **Home-Schooling** According to the National Center for Education Statistics, 1.1 million students are home-schooled. This number is 2.2% of the school-age population in the United States. What is the school-age population in the United States?

27. **Geometry** The length of the base of a triangle is 2 in. less than twice the height. The area of the triangle is 30 in^2. Find the base and height of the triangle.

28. **Insurance** A life insurance policy costs \$16 for every \$1000 of coverage. At this rate, how much money would a policy of \$5000 cost?

29. **Work** One water pipe can fill a tank in 9 min, whereas a second pipe requires 18 min to fill the tank. How long would it take both pipes, working together, to fill the tank?

30. **Travel** The rower of a boat can row at a rate of 5 mph in calm water. Rowing with the current, the boat travels 14 mi in the same amount of time it takes to travel 6 mi against the current. Find the rate of the current.

Linear Equations in Two Variables

OBJECTIVES

SECTION 12.1

- A To graph points in a rectangular coordinate system
- B To determine ordered-pair solutions of an equation in two variables
- C To determine whether a set of ordered pairs is a function
- D To evaluate a function

SECTION 12.2

- A To graph an equation of the form $y = mx + b$
- B To graph an equation of the form $Ax + By = C$
- C To solve application problems

SECTION 12.3

- A To find the x- and y-intercepts of a straight line
- B To find the slope of a straight line
- C To graph a line using the slope and the y-intercept

SECTION 12.4

- A To find the equation of a line given a point and the slope
- B To find the equation of a line given two points
- C To solve application problems

ARE YOU READY?

Take the Chapter 12 Prep Test to find out if you are ready to learn to:

- Evaluate a function
- Graph equations of the form $y = mx + b$ and of the form $Ax + By = C$
- Find the x- and y-intercepts of a straight line
- Find the slope of a straight line
- Find the equation of a line given a point and the slope or given two points

PREP TEST

Do these exercises to prepare for Chapter 12.

1. Simplify: $-\frac{5 - (-7)}{4 - 8}$

2. Evaluate $\frac{a - b}{c - d}$ when $a = 3$, $b = -2$, $c = -3$, and $d = 2$.

3. Simplify: $-3(x - 4)$

4. Solve: $3x + 6 = 0$

5. Solve $4x + 5y = 20$ when $y = 0$.

6. Solve $3x - 7y = 11$ when $x = -1$.

7. Divide: $\frac{12x - 15}{-3}$

8. Solve: $\frac{2x + 1}{3} = \frac{3x}{4}$

9. Solve $3x - 5y = 15$ for y.

10. Solve $y + 3 = -\frac{1}{2}(x + 4)$ for y.

SECTION

12.1 The Rectangular Coordinate System

OBJECTIVE A To graph points in a rectangular coordinate system

Before the 15th century, geometry and algebra were considered separate branches of mathematics. That all changed when René Descartes, a French mathematician who lived from 1596 to 1650, founded **analytic geometry.** In this geometry, a *coordinate system* is used to study relationships between variables.

A **rectangular coordinate system** is formed by two number lines, one horizontal and one vertical, that intersect at the zero point of each line. The point of intersection is called the **origin.** The two lines are called **coordinate axes,** or simply **axes.** The axes determine a **plane,** which can be thought of as a large, flat sheet of paper. The two axes divide the plane into four regions called **quadrants,** which are numbered counterclockwise from I to IV.

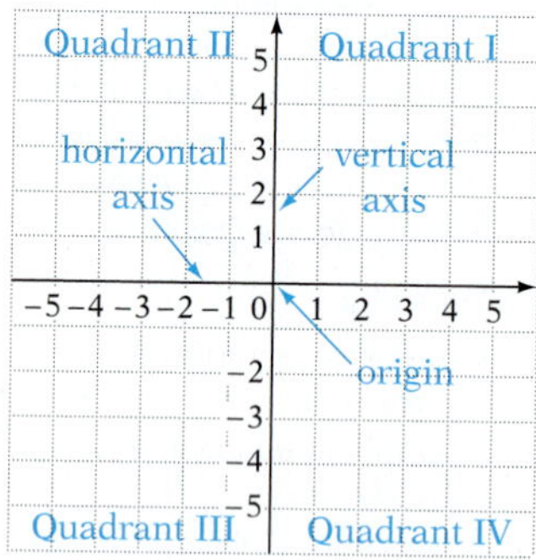

Each point in the plane can be identified by a pair of numbers called an **ordered pair.** The first number of the pair measures a horizontal distance and is called the **abscissa.** The second number of the pair measures a vertical distance and is called the **ordinate.** The **coordinates of a point** are the numbers in the ordered pair associated with the point. The abscissa is also called the **first coordinate** of the ordered pair, and the ordinate is also called the **second coordinate** of the ordered pair.

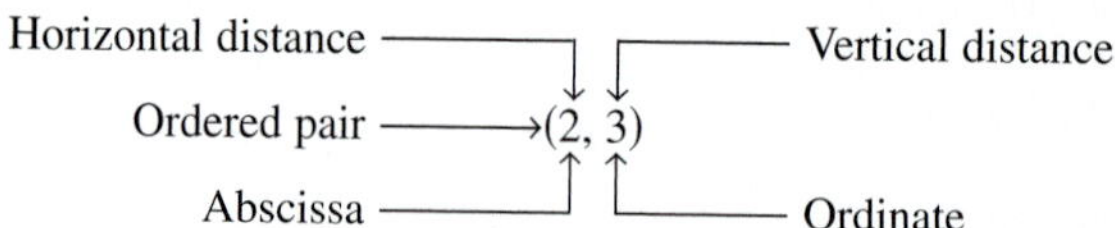

When drawing a rectangular coordinate system, we often label the horizontal axis x and the vertical axis y. In this case, the coordinate system is called an ***xy*-coordinate system.** The coordinates of the points are given by ordered pairs (x, y), where the abscissa is called the ***x*-coordinate** and the ordinate is called the ***y*-coordinate.**

To **graph or plot a point in the plane,** place a dot at the location given by the ordered pair. The **graph of an ordered pair** (x, y) is the dot drawn at the coordinates of the point in the plane. The points whose coordinates are $(3, 4)$ and $(-2.5, -3)$ are graphed in the figures below.

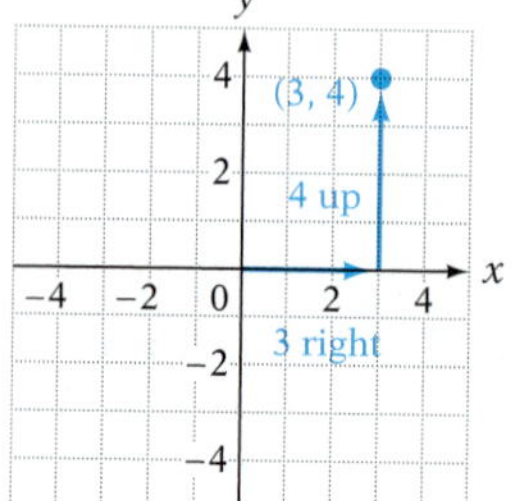

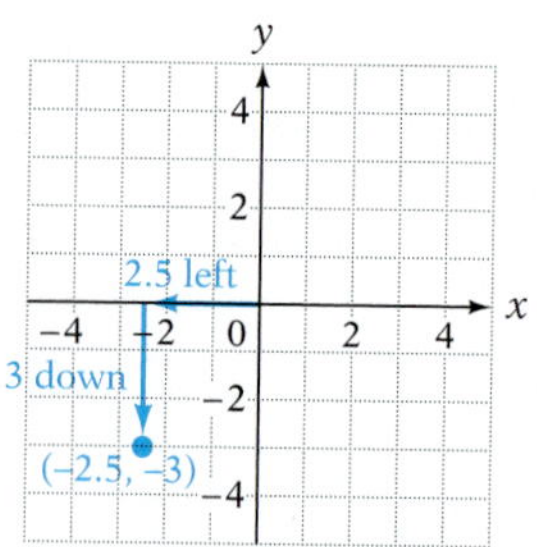

Take Note

This concept is very important. An **ordered pair** is a *pair* of coordinates, and the *order* in which the coordinates appear is crucial.

The points whose coordinates are $(3, -1)$ and $(-1, 3)$ are graphed at the right. Note that the graphed points are in different locations. *The order of the coordinates of an ordered pair is important.*

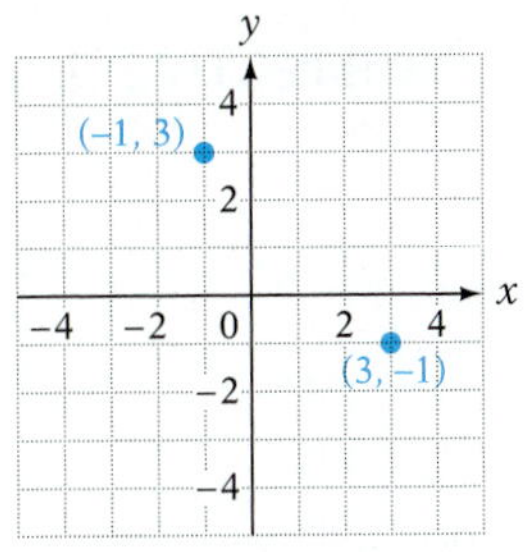

Each point in the plane is associated with an ordered pair, and each ordered pair is associated with a point in the plane. Although only the labels for integers are given on a coordinate grid, the graph of any ordered pair can be approximated. For example, the points whose coordinates are $(-2.3, 4.1)$ and $(\pi, 1)$ are shown on the graph at the right.

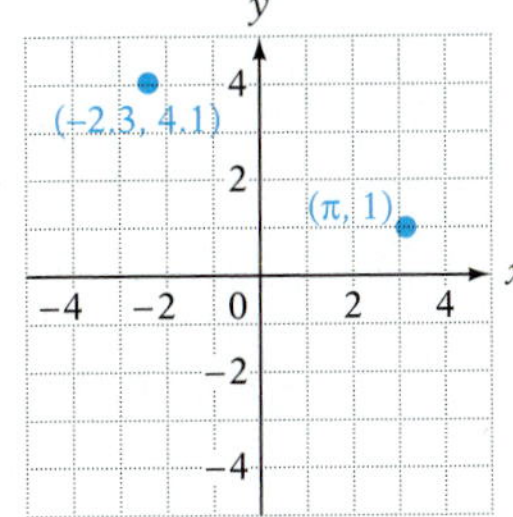

EXAMPLE • 1

Graph the ordered pairs $(-2, -3)$, $(3, -2)$, $(0, -2)$, and $(3, 0)$.

Solution

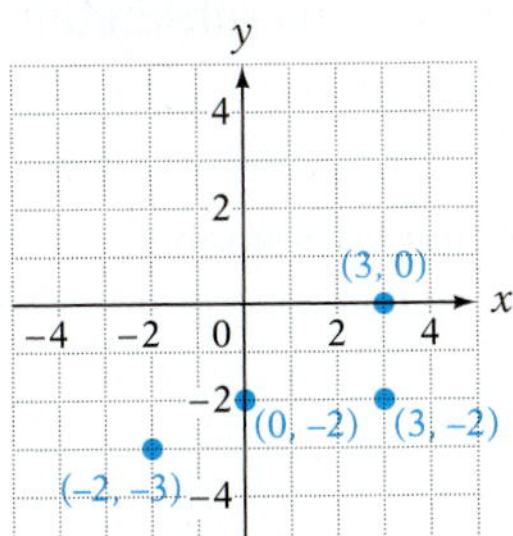

YOU TRY IT • 1

Graph the ordered pairs $(-4, 1)$, $(3, -3)$, $(0, 4)$, and $(-3, 0)$.

Your solution

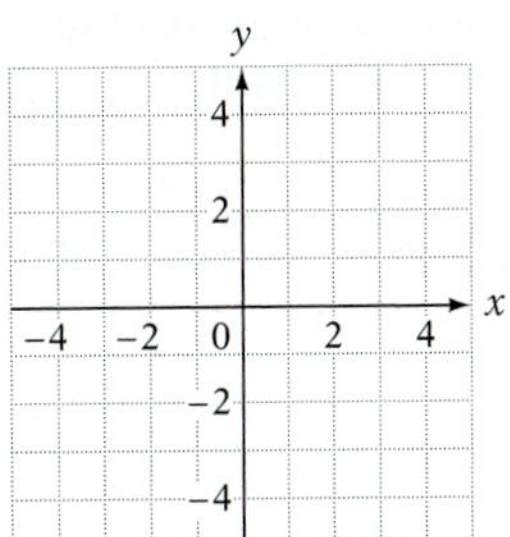

EXAMPLE • 2

Give the coordinates of the points labeled *A* and *B*. Give the abscissa of point *C* and the ordinate of point *D*.

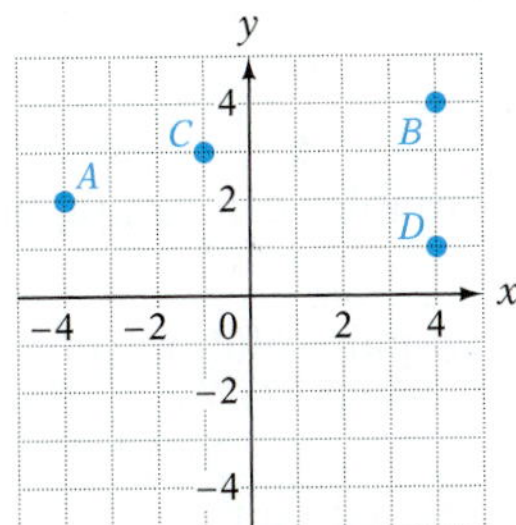

Solution

The coordinates of *A* are $(-4, 2)$.
The coordinates of *B* are $(4, 4)$.
The abscissa of *C* is -1.
The ordinate of *D* is 1.

YOU TRY IT • 2

Give the coordinates of the points labeled *A* and *B*. Give the abscissa of point *D* and the ordinate of point *C*.

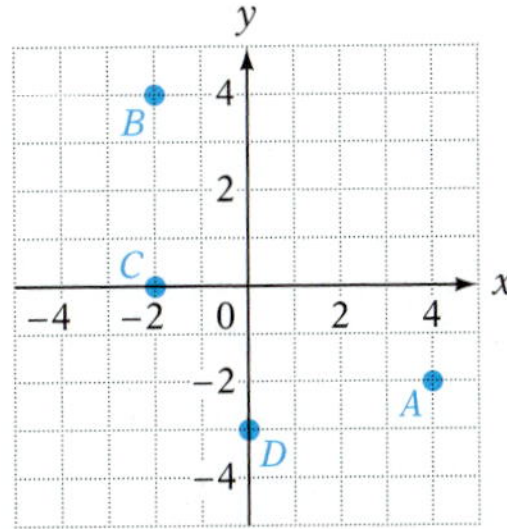

Your solution

Solutions on p. S28

OBJECTIVE B To determine ordered-pair solutions of an equation in two variables 

An *xy*-coordinate system is used to study the relationship between two variables. Frequently this relationship is given by an equation. Examples of equations in two variables include

$$y = 2x - 3 \qquad 3x + 2y = 6 \qquad x^2 - y = 0$$

A **solution of an equation in two variables** is an ordered pair (x, y) whose coordinates make the equation a true statement.

HOW TO • 1 Is $(-3, 7)$ a solution of $y = -2x + 1$?

$$\begin{array}{c|l} y & = -2x + 1 \\ \hline 7 & -2(-3) + 1 \\ & 6 + 1 \end{array}$$
$7 = 7$

• **Replace x by -3; replace y by 7.**

• **The results are equal.**

$(-3, 7)$ is a solution of the equation $y = -2x + 1$.

Besides $(-3, 7)$, there are many other ordered-pair solutions of $y = -2x + 1$. For example, $(0, 1)$, $\left(-\frac{3}{2}, 4\right)$, and $(4, -7)$ are also solutions. In general, an equation in two variables has an infinite number of solutions. By choosing any value of x and substituting that value into the equation, we can calculate a corresponding value of y.

HOW TO • 2 Find the ordered-pair solution of $y = \frac{2}{3}x - 3$ that corresponds to $x = 6$.

$$y = \frac{2}{3}x - 3$$

$$= \frac{2}{3}(6) - 3 \qquad \text{• Replace } x \text{ by } 6.$$

$$= 4 - 3 = 1 \qquad \text{• Simplify.}$$

The ordered-pair solution is $(6, 1)$.

The solutions of an equation in two variables can be graphed in an *xy*-coordinate system.

HOW TO • 3 Graph the ordered-pair solutions of $y = -2x + 1$ when $x = -2, -1, 0, 1,$ and 2.

Use the values of x to determine ordered-pair solutions of the equation. It is convenient to record these in a table.

x	$y = -2x + 1$	y	(x, y)
-2	$-2(-2) + 1$	5	$(-2, 5)$
-1	$-2(-1) + 1$	3	$(-1, 3)$
0	$-2(0) + 1$	1	$(0, 1)$
1	$-2(1) + 1$	-1	$(1, -1)$
2	$-2(2) + 1$	-3	$(2, -3)$

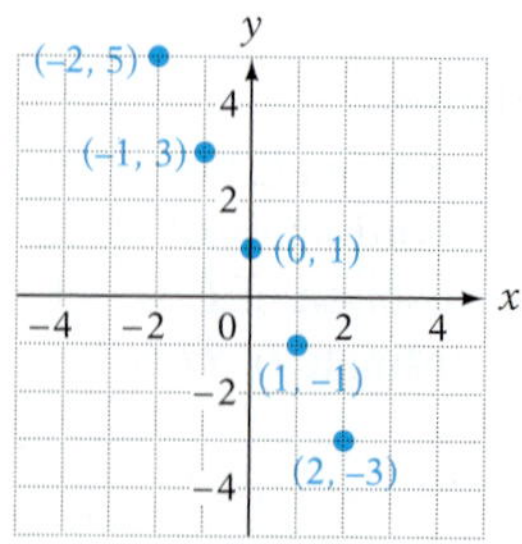

EXAMPLE • 3

Is $(3, -2)$ a solution of $3x - 4y = 15$?

Solution

$$\begin{array}{r|l} 3x - 4y = 15 & \\ \hline 3(3) - 4(-2) & 15 \\ 9 + 8 & \\ 17 \neq 15 & \end{array}$$

• **Replace x by 3 and y by -2.**

No. $(3, -2)$ is not a solution of $3x - 4y = 15$.

YOU TRY IT • 3

Is $(-2, 4)$ a solution of $x - 3y = -14$?

Your solution

EXAMPLE • 4

Graph the ordered-pair solutions of $2x - 3y = 6$ when $x = -3, 0, 3,$ and 6.

Solution

$$2x - 3y = 6$$
$$-3y = -2x + 6$$
$$y = \frac{2}{3}x - 2$$

• **Solve $2x - 3y = 6$ for y.**

Replace x in $y = \frac{2}{3}x - 2$ by $-3, 0, 3,$ and 6. For each value of x, determine the value of y.

x	$y = \frac{2}{3}x - 2$	y	(x, y)
-3	$\frac{2}{3}(-3) - 2$	-4	$(-3, -4)$
0	$\frac{2}{3}(0) - 2$	-2	$(0, -2)$
3	$\frac{2}{3}(3) - 2$	0	$(3, 0)$
6	$\frac{2}{3}(6) - 2$	2	$(6, 2)$

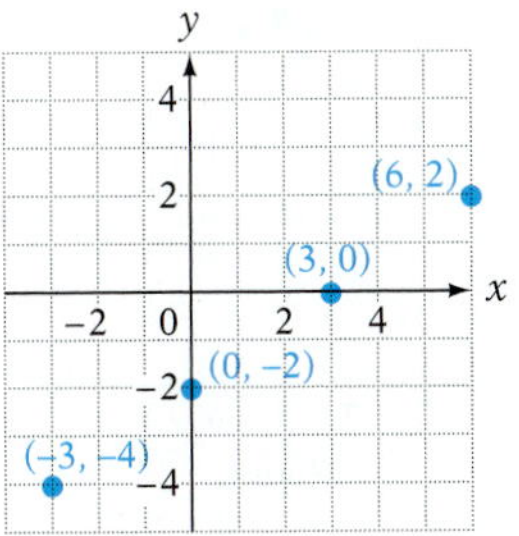

YOU TRY IT • 4

Graph the ordered-pair solutions of $x + 2y = 4$ when $x = -4, -2, 0,$ and 2.

Your solution

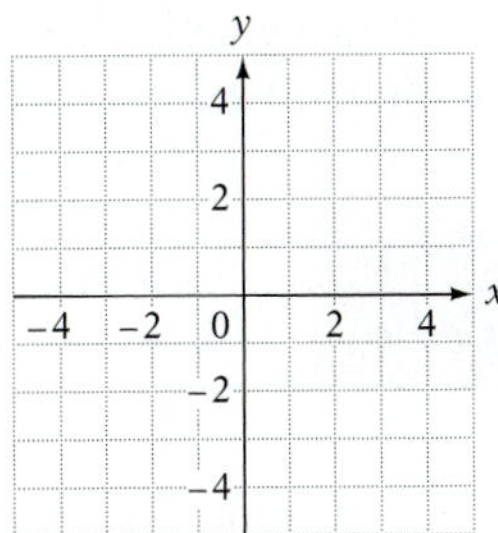

Solutions on p. S28

OBJECTIVE C To determine whether a set of ordered pairs is a function 

© Craig Tuttle/Corbis

Discovering a relationship between two variables is an important task in the application of mathematics. Here are some examples.

- Botanists study the relationship between the number of bushels of wheat yielded per acre and the amount of watering per acre.
- Environmental scientists study the relationship between the incidents of skin cancer and the amount of ozone in the atmosphere.
- Business analysts study the relationship between the price of a product and the number of products that are sold at that price.

Each of these relationships can be described by a set of ordered pairs.

Definition of a Relation

A **relation** is any set of ordered pairs.

The following table shows the number of hours that each of nine students spent studying for a midterm exam and the grade that each of these nine students received.

Hours	3	3.5	2.75	2	4	4.5	3	2.5	5
Grade	78	75	70	65	85	85	80	75	90

This information can be written as the relation

$$\{(3, 78), (3.5, 75), (2.75, 70), (2, 65), (4, 85), (4.5, 85), (3, 80), (2.5, 75), (5, 90)\}$$

where the first coordinate of the ordered pair is the hours spent studying and the second coordinate is the score on the midterm.

The **domain** of a relation is the set of first coordinates of the ordered pairs; the **range** is the set of second coordinates. For the relation above,

$$\text{Domain} = \{2, 2.5, 2.75, 3, 3.5, 4, 4.5, 5\} \qquad \text{Range} = \{65, 70, 75, 78, 80, 85, 90\}$$

The **graph of a relation** is the graph of the ordered pairs that belong to the relation. The graph of the relation given above is shown at the right. The horizontal axis represents the hours spent studying (the domain); the vertical axis represents the test score (the range). The axes could be labeled H for hours studied and S for test score.

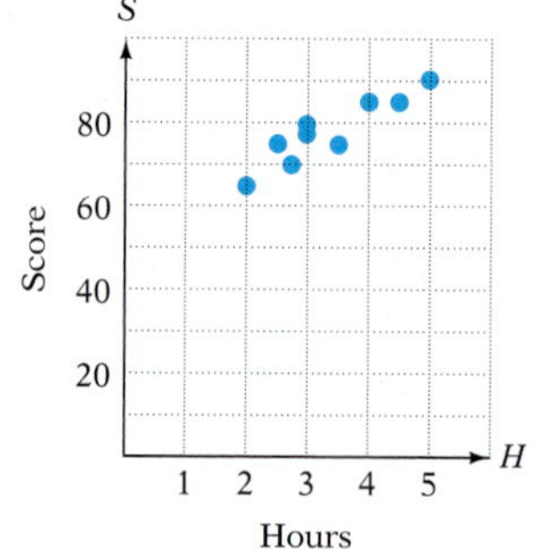

A *function* is a special type of relation in which no two ordered pairs have the same first coordinate.

Definition of a Function

A **function** is a relation in which no two ordered pairs have the same first coordinate.

The table at the right is the grading scale for a 100-point test. This table defines a relationship between the *score* on the test and a *letter grade.* Some of the ordered pairs of this function are (78, C), (97, A), (84, B), and (82, B).

Score	*Grade*
90–100	A
80–89	B
70–79	C
60–69	D
0–59	F

The grading-scale table defines a function because no two ordered pairs can have the same first coordinate and different second coordinates. For instance, it is not possible to have the ordered pairs (72, C), and (72, B)—same first coordinate (test score) but different second coordinates (test grade). The domain of this function is {0, 1, 2,…, 99, 100}. The range is {A, B, C, D, F}.

The example of hours spent studying and test score given earlier is *not* a function, because (3, 78) and (3, 80) are ordered pairs of the relation that have the *same* first coordinate but *different* second coordinates.

Consider, again, the grading-scale example. Note that (84, B) and (82, B) are ordered pairs of the function. Ordered pairs of a function may have the same *second* coordinates but not the same first coordinates.

Although relations and functions can be given by tables, they are frequently given by an equation in two variables.

The equation $y = 2x$ expresses the relationship between a number, x, and twice the number, y. For instance, if $x = 3$, then $y = 6$, which is twice 3. To indicate exactly which ordered pairs are determined by the equation, the domain (values of x) is specified. If $x \in \{-2, -1, 0, 1, 2\}$, then the ordered pairs determined by the equation are $\{(-2, -4), (-1, -2), (0, 0), (1, 2), (2, 4)\}$. This relation is a function because no two ordered pairs have the same first coordinate.

The graph of the function $y = 2x$ with domain $\{-2, -1, 0, 1, 2\}$ is shown at the right. The horizontal axis (domain) is labeled x; the vertical axis (range) is labeled y.

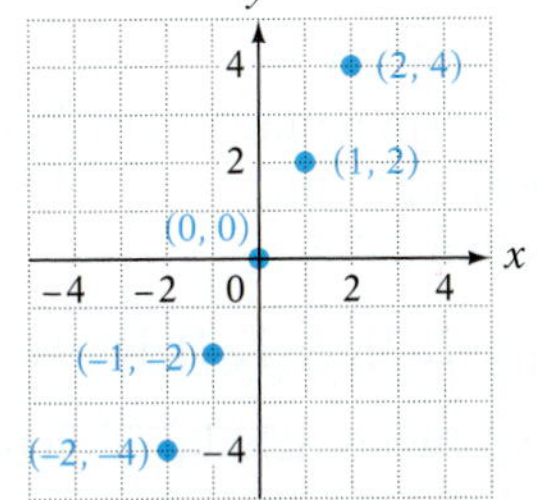

The domain $\{-2, -1, 0, 1, 2\}$ was chosen arbitrarily. Other domains could have been selected. The type of application usually influences the choice of the domain.

For the equation $y = 2x$, we say that "y is a function of x" because the set of ordered pairs is a function.

Not all equations, however, define a function. For instance, the equation $|y| = x + 2$ does not define y as a function of x. The ordered pairs (2, 4) and (2, −4) both satisfy the equation. Thus there are two ordered pairs with the same first coordinate but different second coordinates.

EXAMPLE • 5

The table below shows the amount of money invested in college savings plans and the amount invested in prepaid college tuition plans over a six-year period. (*Sources:* Investment Company Institute and College Savings Plan Network)

Year	*Assets in College Savings Plans (in billions of dollars)*	*Assets in Prepaid Tuition Plans (in billions of dollars)*
1	9	7
2	19	8
3	35	11
4	52	13
5	69	14
6	90	16

Write a relation in which the first coordinate is the amount of money in college savings plans and the second coordinate is the amount of money in prepaid tuition plans (both in billions of dollars). Is the relation a function?

Solution

The relation is

{(9, 7), (19, 8), (35, 11), (52, 13), (69, 14), (90, 16)}

There are no two ordered pairs with the same first coordinate. The relation is a function.

YOU TRY IT • 5

Six students decided to go on a diet and fitness program over the summer. Their weights (in pounds) at the beginning and end of the program are given in the table below.

Beginning	*End*
145	140
140	125
150	130
165	150
140	130
165	160

Write a relation wherein the first coordinate is the weight at the beginning of the summer and the second coordinate is the weight at the end of the summer. Is the relation a function?

Your solution

EXAMPLE • 6

Does $y = x^2 + 3$, where $x \in \{-2, -1, 1, 3\}$, define y as a function of x?

Solution

Determine the ordered pairs defined by the equation. Replace x in $y = x^2 + 3$ by the given values and solve for y.

$\{(-2, 7), (-1, 4), (1, 4), (3, 12)\}$

No two ordered pairs have the same first coordinate. Therefore, the relation is a function and the equation $y = x^2 + 3$ defines y as a function of x.

Note that $(-1, 4)$ and $(1, 4)$ are ordered pairs that belong to this function. Ordered pairs of a function may have the same *second* coordinate but not the same *first* coordinate.

YOU TRY IT • 6

Does $y = \frac{1}{2}x + 1$, where $x \in \{-4, 0, 2\}$, define y as a function of x?

Your solution

Solutions on p. S28

OBJECTIVE D To evaluate a function 

When an equation defines y as a function of x, **function notation** is frequently used to emphasize that the relation is a function. In this case, it is common to replace y in the function's equation with the symbol $f(x)$, where

$f(x)$ is read "f of x" or "the value of f at x."

For instance, the equation $y = x^2 + 3$ from Example 6 defined y as a function of x. The equation can also be written

$$f(x) = x^2 + 3$$

where y has been replaced by $f(x)$.

The symbol $f(x)$ is called the **value of a function at x** because it is the result of evaluating a variable expression. For instance, $f(4)$ means to replace x by 4 and then simplify the resulting numerical expression.

$$f(x) = x^2 + 3$$
$$f(4) = 4^2 + 3$$
$$= 16 + 3 = 19$$

Replace x by 4.

This process is called **evaluating a function.**

HOW TO • 4 Given $f(x) = x^2 + x - 3$, find $f(-2)$.

$$f(x) = x^2 + x - 3$$
$$f(-2) = (-2)^2 + (-2) - 3$$
$$= 4 - 2 - 3 = -1$$

• **Replace x by -2.**

$$f(-2) = -1$$

In this example, $f(-2)$ is the second coordinate of an ordered pair of the function; the first coordinate is -2. Therefore, an ordered pair of this function is $(-2, f(-2))$, or, because $f(-2) = -1$, $(-2, -1)$.

For the function given by $y = f(x) = x^2 + x - 3$, y is called the **dependent variable** because its value depends on the value of x. The **independent variable** is x.

Functions can be written using other letters or even combinations of letters. For instance, some calculators use $ABS(x)$ for the absolute-value function. Thus the equation $y = |x|$ would be written $ABS(x) = |x|$, where $ABS(x)$ replaces y.

EXAMPLE • 7

Given $G(t) = \frac{3t}{t + 4}$, find $G(1)$.

Solution

$$G(t) = \frac{3t}{t + 4}$$

$$G(1) = \frac{3(1)}{1 + 4}$$

• **Replace t by 1. Then simplify.**

$$G(1) = \frac{3}{5}$$

YOU TRY IT • 7

Given $H(x) = \frac{x}{x - 4}$, find $H(8)$.

Your solution

Solution on p. S28

12.1 EXERCISES

OBJECTIVE A To graph points in a rectangular coordinate system

1. Graph $(-2, 1)$, $(3, -5)$, $(-2, 4)$, and $(0, 3)$.

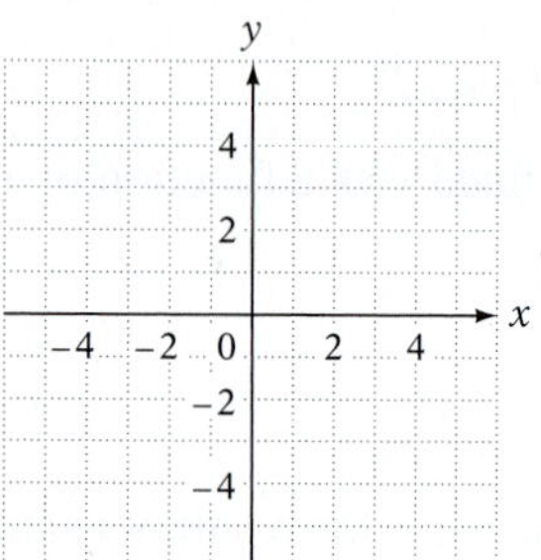

2. Graph $(5, -1)$, $(-3, -3)$, $(-1, 0)$, and $(1, -1)$.

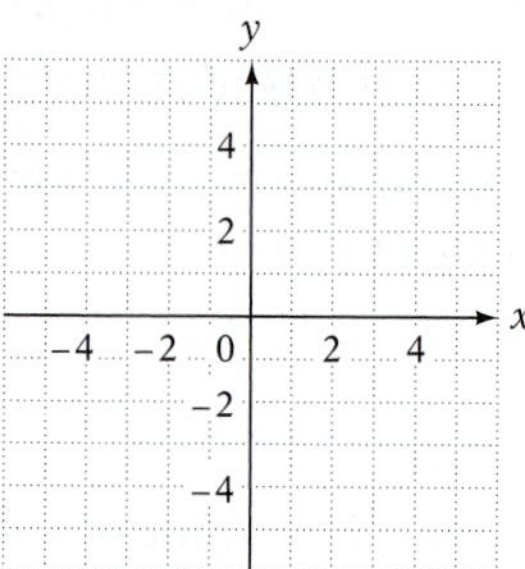

3. Graph $(0, 0)$, $(0, -5)$, $(-3, 0)$, and $(0, 2)$.

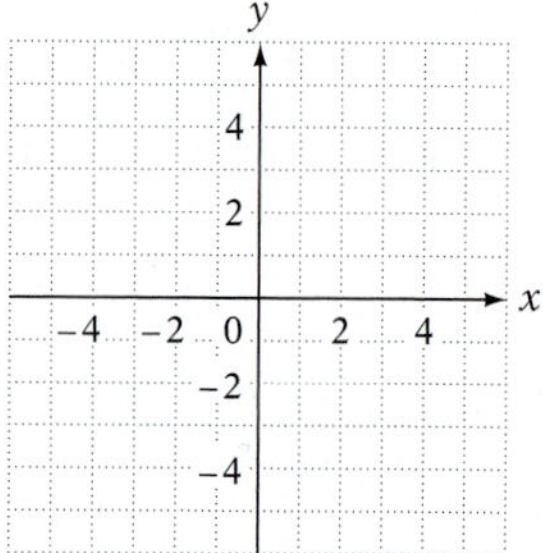

4. Graph $(-4, 5)$, $(-3, 1)$, $(3, -4)$, and $(5, 0)$.

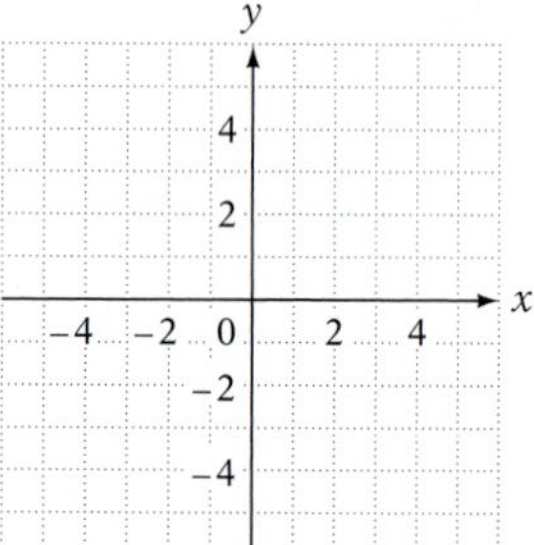

5. Graph $(-1, 4)$, $(-2, -3)$, $(0, 2)$, and $(4, 0)$.

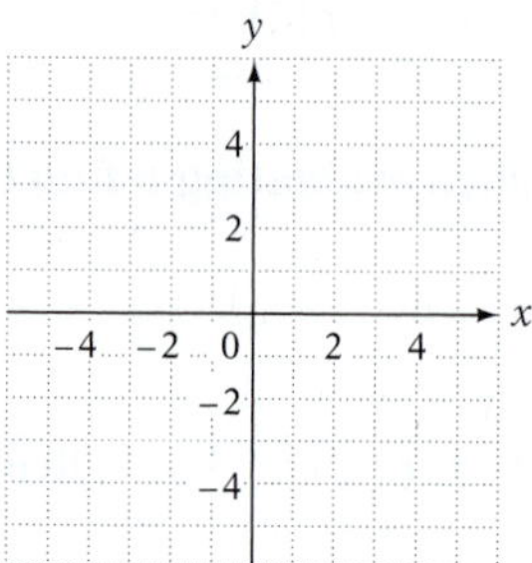

6. Graph $(5, 2)$, $(-4, -1)$, $(0, 0)$, and $(0, 3)$.

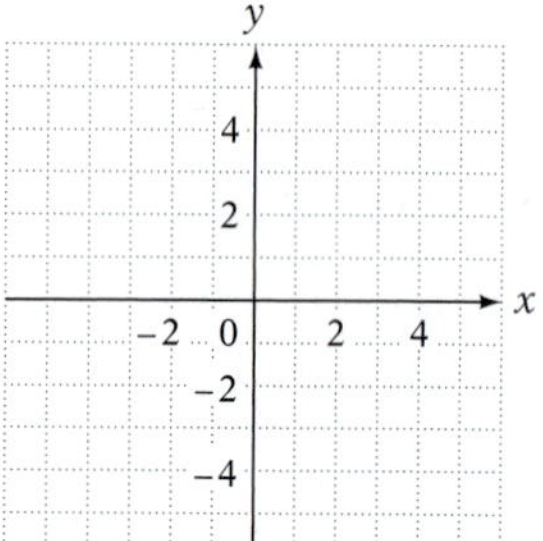

7. Find the coordinates of each of the points.

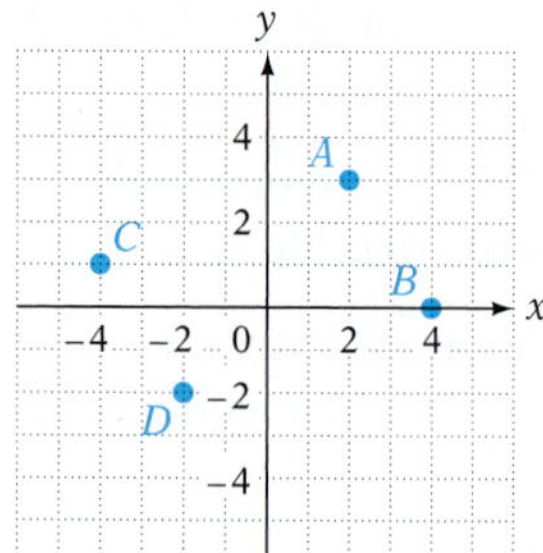

8. Find the coordinates of each of the points.

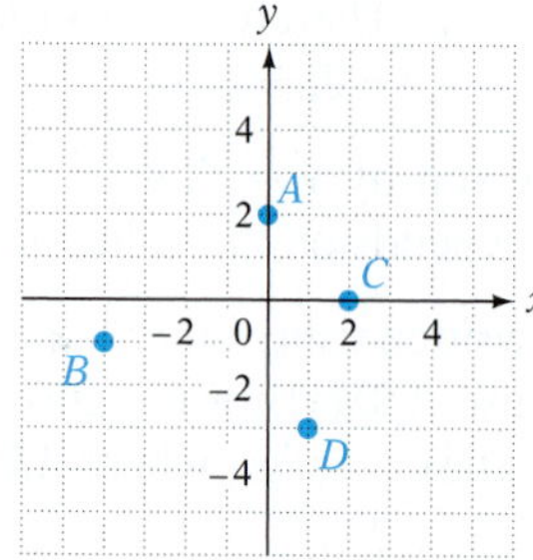

9. Find the coordinates of each of the points.

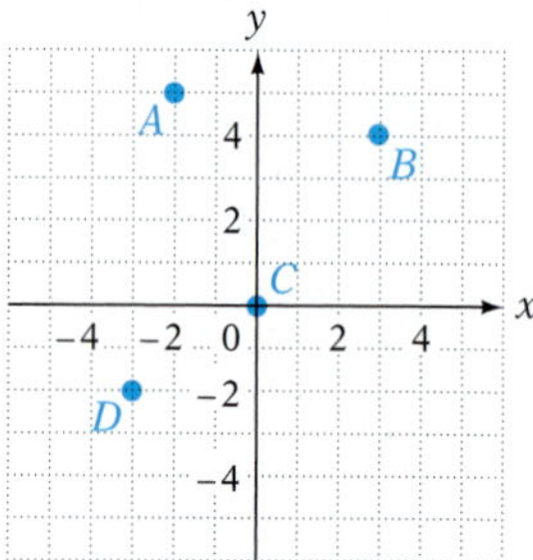

10. Find the coordinates of each of the points.

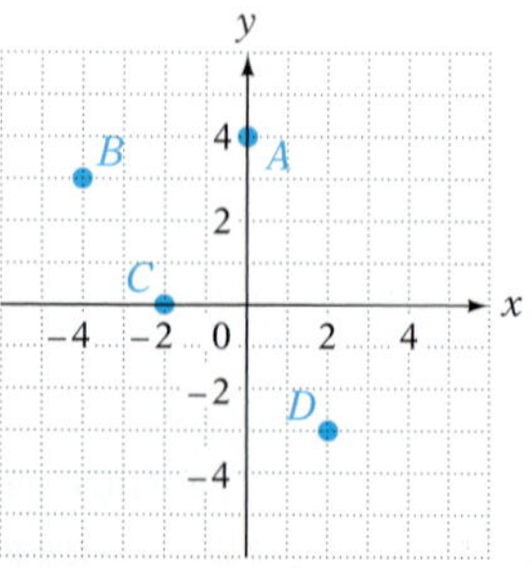

11. **a.** Name the abscissas of points *A* and *C*.
b. Name the ordinates of points *B* and *D*.

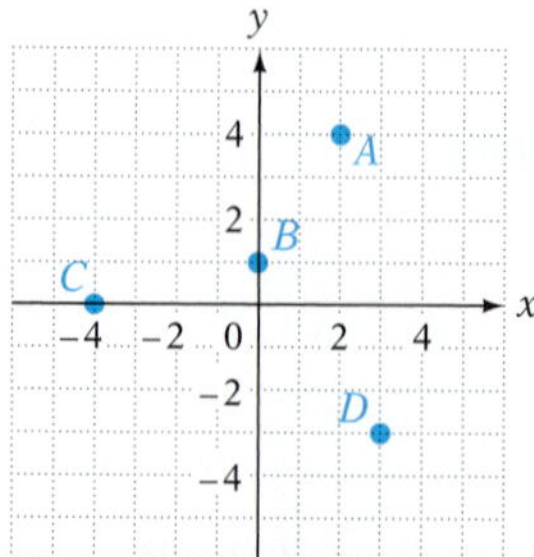

12. **a.** Name the abscissas of points *A* and *C*.
b. Name the ordinates of points *B* and *D*.

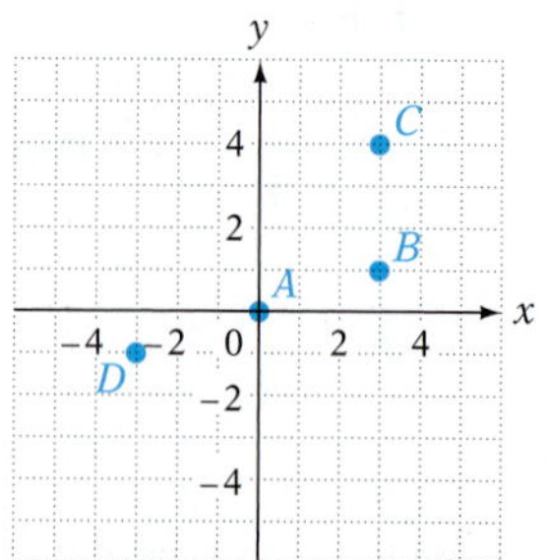

13. **a.** On an xy-coordinate system, what is the name of the axis for which all the x-coordinates are zero?
b. On an xy-coordinate system, what is the name of the axis for which all the y-coordinates are zero?

14. Let a and b be positive numbers such that $a < b$. In which quadrant is each point located?
a. (a, b) **b.** $(-a, b)$ **c.** $(-a, -b)$ **d.** $(b - a, -b)$

OBJECTIVE B To determine ordered-pair solutions of an equation in two variables

15. Is (3, 4) a solution of $y = -x + 7$?

16. Is $(2, -3)$ a solution of $y = x + 5$?

17. Is $(-1, 2)$ a solution of $y = \frac{1}{2}x - 1$?

18. Is $(1, -3)$ a solution of $y = -2x - 1$?

19. Is (4, 1) a solution of $2x - 5y = 4$?

20. Is $(-5, 3)$ a solution of $3x - 2y = 9$?

21. Suppose (x, y) is a solution of the equation $y = -3x + 6$, where $x > 2$. Is y positive or negative?

22. Suppose (x, y) is a solution of the equation $y = 4x - 8$, where $y > 0$. Is x less than or greater than 2?

For Exercises 23 to 28, graph the ordered-pair solutions of the equation for the given values of x.

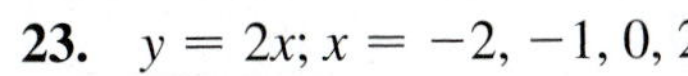

23. $y = 2x$; $x = -2, -1, 0, 2$

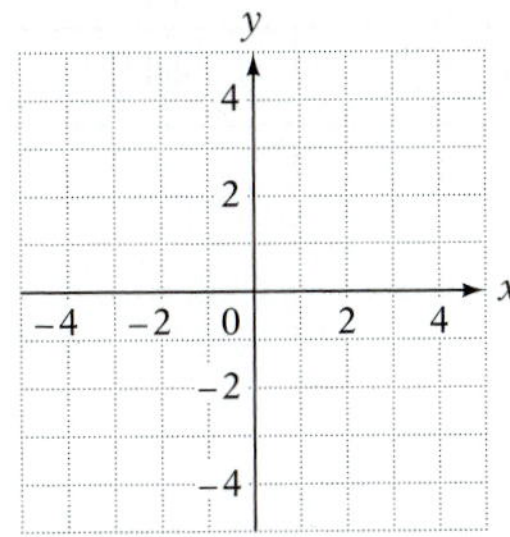

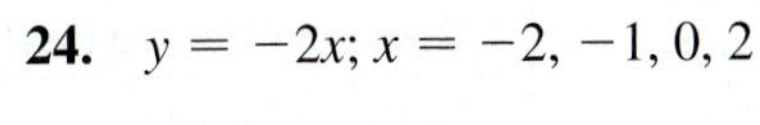

24. $y = -2x$; $x = -2, -1, 0, 2$

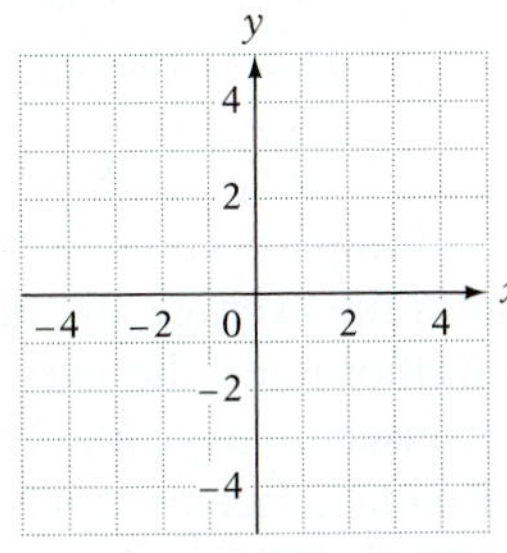

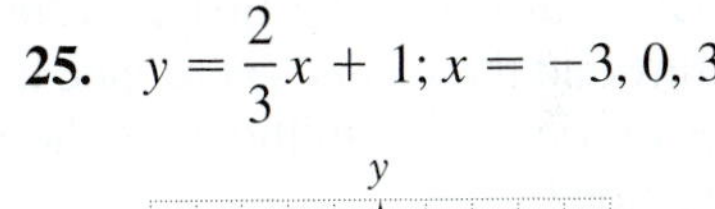

25. $y = \frac{2}{3}x + 1$; $x = -3, 0, 3$

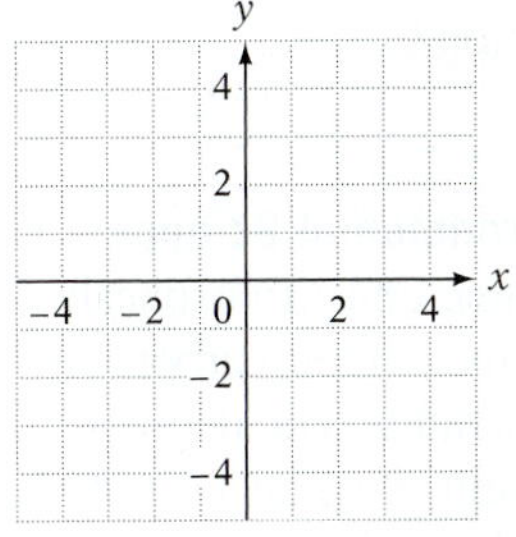

26. $y = -\frac{1}{3}x - 2$; $x = -3, 0, 3$

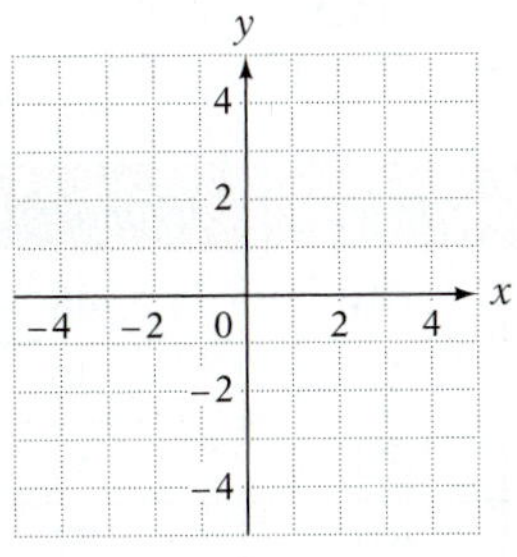

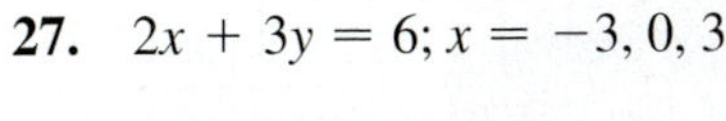

27. $2x + 3y = 6$; $x = -3, 0, 3$

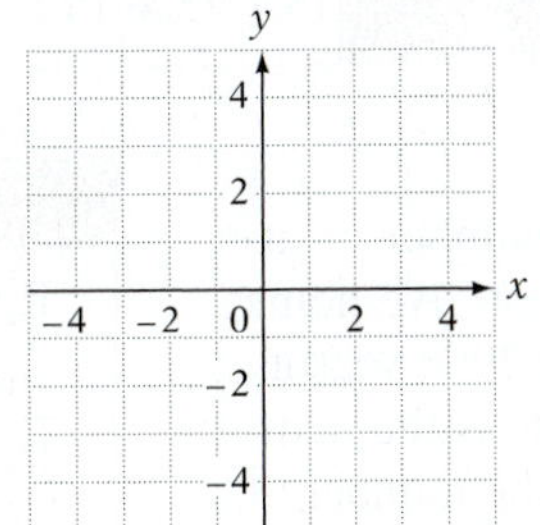

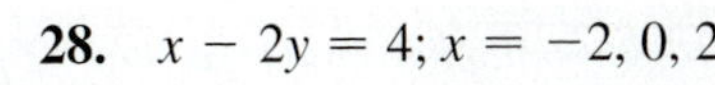

28. $x - 2y = 4$; $x = -2, 0, 2$

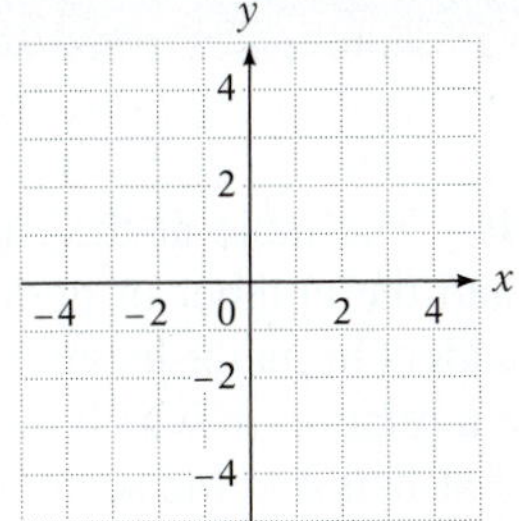

OBJECTIVE C To determine whether a set of ordered pairs is a function

For Exercises 29 and 30, D is the set of all dates of the year {January 1, January 2, January 3, ...}, and P is the set of all the people in the world.

29. A relation has domain D and range P. Ordered pairs in the relation are of the form (date, person born on that date). Is the relation a function?

30. A relation has domain P and range D. Ordered pairs in the relation are of the form (person, birth date of that person). Is this relation a function?

31. Biology The table below shows the length, in centimeters, of the humerus (the long bone of the forelimb, from shoulder to elbow) and the total wingspan, in centimeters, of several pterosaurs, which are extinct flying reptiles of the order Pterosauria. Write a relation in which the first coordinate is the length of the humerus and the second is the wingspan. Is the relation a function?

Humerus (in centimeters)	24	32	22	15	4.4	17	15	4.4
Wingspan (in centimeters)	600	750	430	300	68	370	310	55

Pterosaur

32. Nielsen Ratings The ratings (each rating point is equivalent to 1,145,000 households) and the numbers of viewers for selected television shows for a week in October 2008 are shown in the table at the right. Write a relation in which the first coordinate is the rating and the second coordinate is the number of viewers in millions. Is the relation a function?

Television Show	*Rating*	*Number of Viewers (in millions)*
Dancing with the Stars	12	18
CSI	11.7	19
Desperate Housewives	10.1	16
Criminal Minds	9.5	15
Grey's Anatomy	9.5	14

Source: www.nielsenmedia.com

33. Environmental Science The table below, based in part on data from the National Oceanic and Atmospheric Administration, shows the average annual concentration of atmospheric carbon dioxide (in parts per million) and the average sea surface temperature (in degrees Celsius) for eight consecutive years. Write a relation wherein the first coordinate is the carbon dioxide concentration and the second coordinate is the average sea surface temperature. Is the relation a function?

Carbon dioxide concentration (in parts per million)	352	353	354	355	356	358	360	361
Surface sea temperature (in degrees Celsius)	15.4	15.4	15.1	15.1	15.2	15.4	15.3	15.5

34. Sports The table at the right shows the number of at-bats and the number of home runs for the top five home run leaders in major league baseball for the 2008 season. Write a relation in which the first coordinate is the number of at-bats and the second coordinate is the number of home runs per at-bat rounded to the nearest thousandth. Is the relation a function?

Player	*At-bats*	*Home runs*
Ryan Howard	610	48
Adam Dunn	517	40
Carlos Delgado	598	38
Miguel Cabrera	616	37
Manny Ramirez	548	37

35. Marathons See the news clipping at the right. The table below shows the ages and finishing times of the top eight finishers in the Manhattan Island Marathon Swim. Write a relation in which the first coordinate is the age of a swimmer and the second coordinate is the swimmer's finishing time. Is the relation a function?

Age (in years)	35	45	38	24	47	51	35	48
Time (in hours)	7.50	7.58	7.63	7.78	7.80	7.86	7.89	7.92

In the News

Swimmers Go the Distance

Twenty-three swimmers competed in this year's Manhattan Island Marathon Swim. The race started at Battery Park City–South Cove at 9:05 A.M., with the first swimmer finishing the 28.5-mile swim around Manhattan Island 7 hours, 30 minutes, and 15 seconds later.

Source: www.nycswim.org

36. Does $y = -2x - 3$, where $x \in \{-2, -1, 0, 3\}$, define y as a function of x?

37. Does $y = 2x + 3$, where $x \in \{-2, -1, 1, 4\}$, define y as a function of x?

38. Does $|y| = x - 1$, where $x \in \{1, 2, 3, 4\}$, define y as a function of x?

39. Does $y = x^2$, where $x \in \{-2, -1, 0, 1, 2\}$, define y as a function of x?

OBJECTIVE D To evaluate a function

40. Given $f(x) = 3x - 4$, find $f(4)$.

41. Given $f(x) = 5x + 1$, find $f(2)$.

42. Given $f(x) = x^2$, find $f(3)$.

43. Given $f(x) = x^2 - 1$, find $f(1)$.

44. Given $G(x) = x^2 + x$, find $G(-2)$.

45. Given $H(x) = x^2 - x$, find $H(-2)$.

46. Given $s(t) = \frac{3}{t - 1}$, find $s(-2)$.

47. Given $P(x) = \frac{4}{2x + 1}$, find $P(-2)$.

48. Given $h(x) = 3x^2 - 2x + 1$, find $h(3)$.

49. Given $Q(r) = 4r^2 - r - 3$, find $Q(2)$.

50. Given $f(x) = \frac{x}{x + 5}$, find $f(-3)$.

51. Given $v(t) = \frac{2t}{2t + 1}$, find $v(3)$.

For Exercises 52 to 55, use the function $f(x) = x^2 - 4$. For the given condition on a, determine whether $f(a)$ *must be positive, must be negative,* or *could be either positive or negative.*

52. $a > 2$

53. $a < 0$

54. $a > -2$

55. $a < -2$

Applying the Concepts

56. Write a few sentences that describe the similarities and differences between relations and functions.

SECTION

12.2 Linear Equations in Two Variables

OBJECTIVE A To graph an equation of the form $y = mx + b$ 

The **graph of an equation in two variables** is a graph of the ordered-pair solutions of the equation.

Consider $y = 2x + 1$. Choosing $x = -2, -1, 0, 1,$ and 2 and determining the corresponding values of y produces some of the ordered pairs of the equation. These are recorded in the table at the right. See the graph of the ordered pairs in Figure 1.

x	$y = 2x + 1$	y	(x, y)
-2	$2(-2) + 1$	-3	$(-2, -3)$
-1	$2(-1) + 1$	-1	$(-1, -1)$
0	$2(0) + 1$	1	$(0, 1)$
1	$2(1) + 1$	3	$(1, 3)$
2	$2(2) + 1$	5	$(2, 5)$

Choosing values of x that are not integers produces more ordered pairs to graph, such as $\left(-\frac{5}{2}, -4\right)$ and $\left(\frac{3}{2}, 4\right)$, as shown in Figure 2. Choosing still other values of x would result in more and more ordered pairs being graphed. The result would be so many dots that the graph would appear as the straight line shown in Figure 3, which is the graph of $y = 2x + 1$.

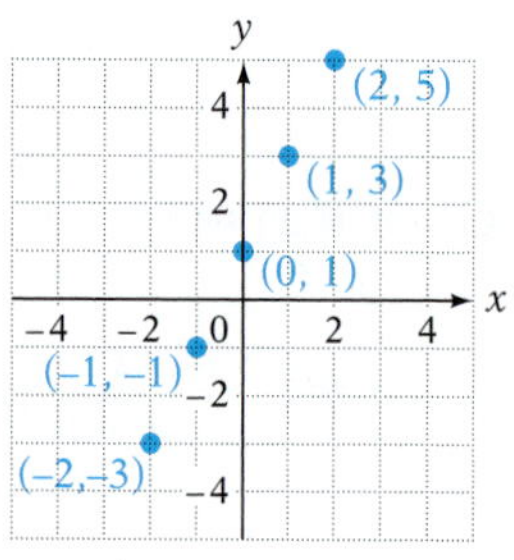

FIGURE 1

FIGURE 2

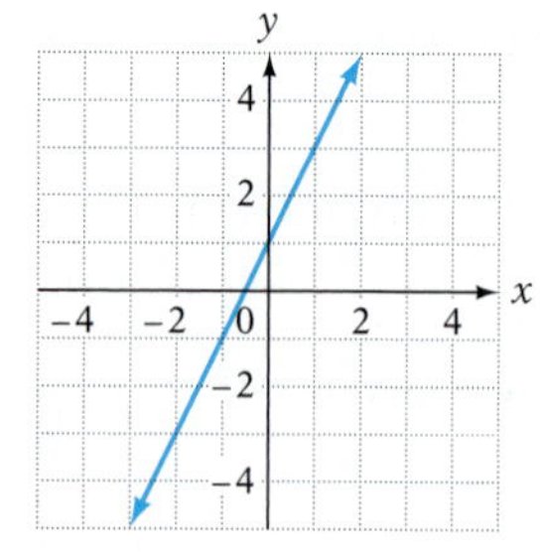

FIGURE 3

Equations in two variables have characteristic graphs. The equation $y = 2x + 1$ is an example of a *linear equation*, or *linear function*, because its graph is a straight line. It is also called a *first-degree equation* in two variables because the exponent on each variable is 1.

Linear Equation in Two Variables

Any equation of the form $y = mx + b$, where m is the coefficient of x and b is a constant, is a **linear equation in two variables**, or a **first-degree equation in two variables**, or a **linear function.** The graph of a linear equation in two variables is a straight line.

Examples of linear equations are shown at the right. These equations represent linear functions because there is only one possible y for each x. Note that for $y = 3 - 2x$, m is the coefficient of x and b is the constant.

$y = 2x + 1 \quad (m = 2, b = 1)$

$y = x - 4 \quad (m = 1, b = -4)$

$y = -\frac{3}{4}x \quad \left(m = -\frac{3}{4}, b = 0\right)$

$y = 3 - 2x \quad (m = -2, b = 3)$

The equation $y = x^2 + 4x + 3$ is not a linear equation in two variables because there is a term with a variable squared. The equation $y = \frac{3}{x - 4}$ is not a linear equation because a variable occurs in the denominator of a fraction.

Integrating Technology

The Projects and Group Activities feature at the end of this chapter contains information on using a calculator to graph an equation.

To graph a linear equation, choose some values of x and then find the corresponding values of y. Because a straight line is determined by two points, it is sufficient to find only two ordered-pair solutions. However, it is recommended that at least three ordered-pair solutions be found to ensure accuracy.

HOW TO • 1 Graph $y = -\frac{3}{2}x + 2$.

This is a linear equation with $m = -\frac{3}{2}$ and $b = 2$. Find at least three solutions. Because m is a fraction, choose values of x that will simplify the calculations. We have chosen -2, 0, and 4 for x. (Any values of x could have been selected.)

x	$y = -\frac{3}{2}x + 2$	y	(x, y)
-2	$-\frac{3}{2}(-2) + 2$	5	$(-2, 5)$
0	$-\frac{3}{2}(0) + 2$	2	$(0, 2)$
4	$-\frac{3}{2}(4) + 2$	-4	$(4, -4)$

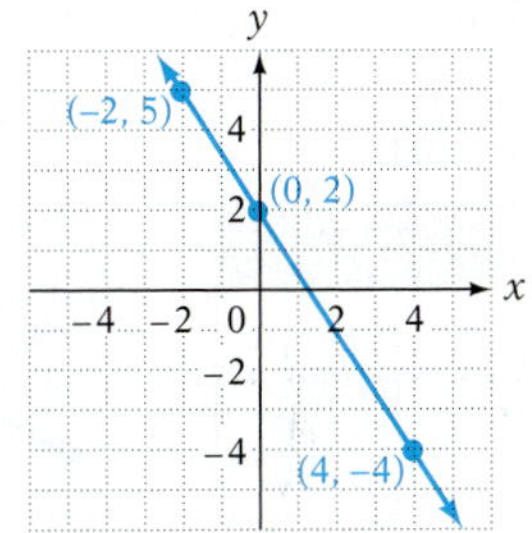

The graph of $y = -\frac{3}{2}x + 2$ is shown at the right.

Remember that a graph is a drawing of the ordered-pair solutions of an equation. Therefore, every point on the graph is a solution of the equation, and every solution of the equation is a point on the graph.

The graph at the right is the graph of $y = x + 2$. Note that $(-4, -2)$ and $(1, 3)$ are points on the graph and that these points are solutions of $y = x + 2$. The point whose coordinates are $(4, 1)$ is not a point on the graph and is not a solution of the equation.

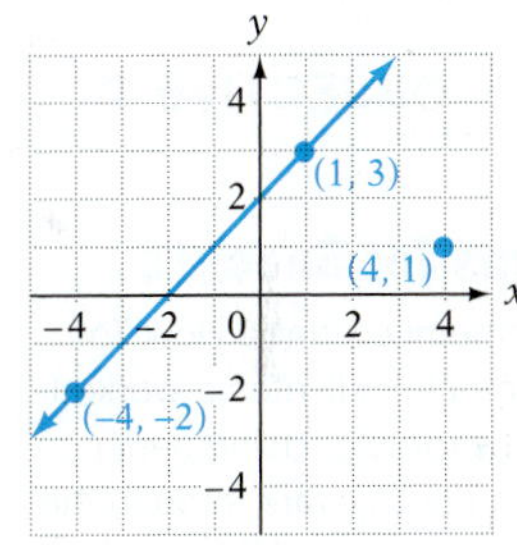

EXAMPLE • 1

Graph $y = 3x - 2$.

Solution

x	y
0	-2
-1	-5
2	4

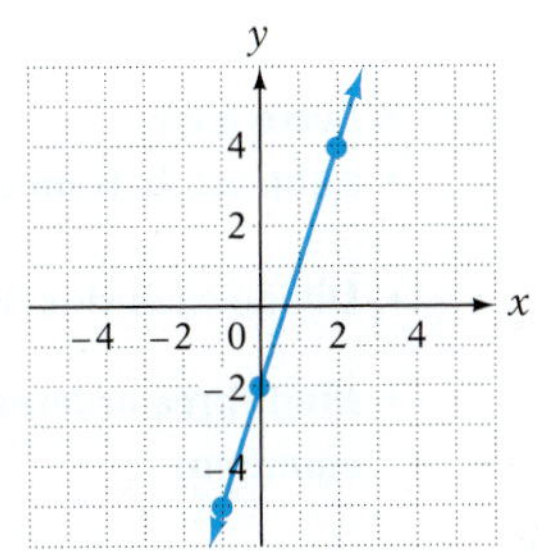

YOU TRY IT • 1

Graph $y = 3x + 1$.

Your solution

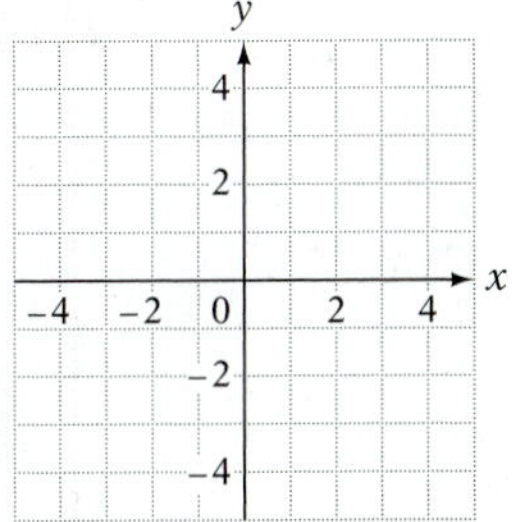

Solution on p. S29

EXAMPLE • 2

Graph $y = 2x$.

Solution

x	y
0	0
2	4
−2	−4

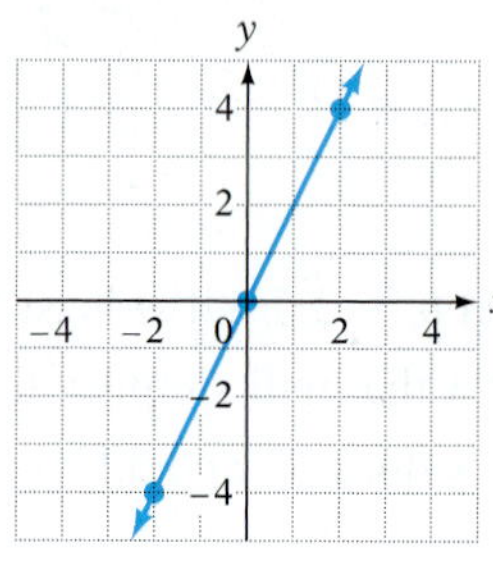

YOU TRY IT • 2

Graph $y = -2x$.

Your solution

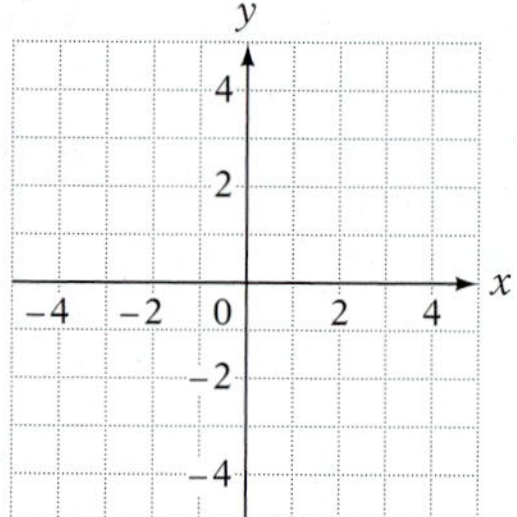

EXAMPLE • 3

Graph $y = \frac{1}{2}x - 1$.

Solution

x	y
0	−1
2	0
−2	−2

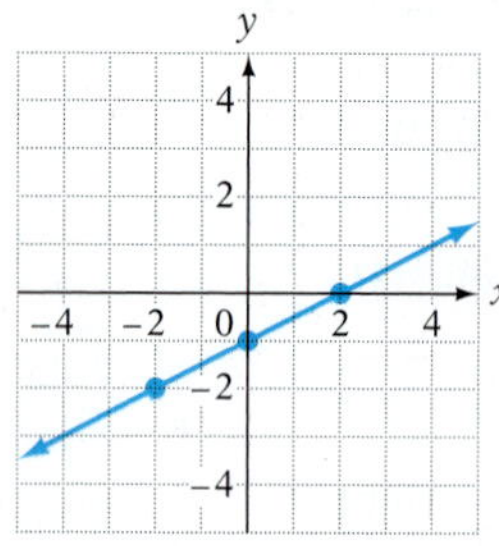

YOU TRY IT • 3

Graph $y = \frac{1}{3}x - 3$.

Your solution

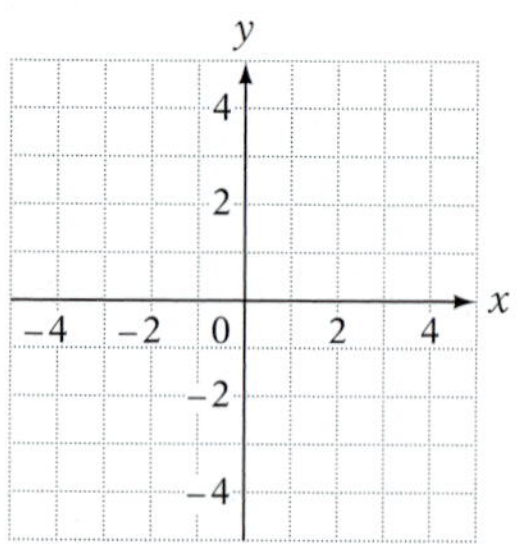

Solutions on p. S29

OBJECTIVE B To graph an equation of the form $Ax + By = C$

Tips for Success

Remember that a How To example indicates a worked-out example. Using paper and pencil, work through the example. See *AIM for Success* at the front of the book.

The equation $Ax + By = C$, where A and B are coefficients and C is a constant, is called the **standard form of a linear equation in two variables.** Examples are shown at the right.

$2x + 3y = 6$ $(A = 2, B = 3, C = 6)$

$x - 2y = -4$ $(A = 1, B = -2, C = -4)$

$2x + y = 0$ $(A = 2, B = 1, C = 0)$

$4x - 5y = 2$ $(A = 4, B = -5, C = 2)$

To graph an equation of the form $Ax + By = C$, first solve the equation for y. Then follow the same procedure used for graphing $y = mx + b$.

HOW TO • 2 Graph $3x + 4y = 12$.

$3x + 4y = 12$

$4y = -3x + 12$

$y = -\frac{3}{4}x + 3$

- **Solve for y.**
- **Subtract $3x$ from each side of the equation.**
- **Divide each side of the equation by 4.**
- **Find three ordered-pair solutions of the equation.**

x	y
0	3
4	0
−4	6

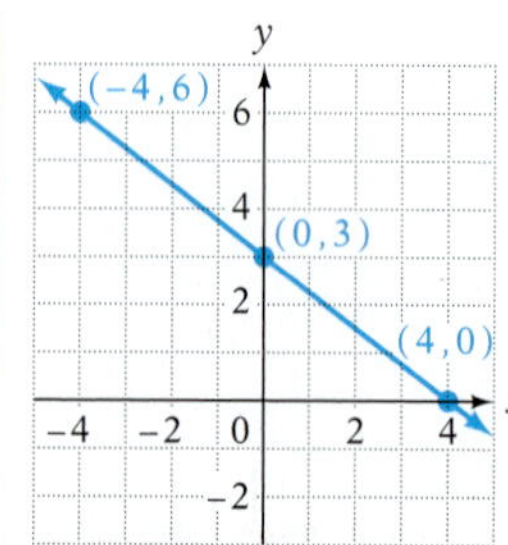

- **Graph the ordered pairs and then draw a line through the points.**

The graph of a linear equation with one of the variables missing is either a horizontal or a vertical line.

The equation $y = 2$ could be written $0 \cdot x + y = 2$. Because $0 \cdot x = 0$ for any value of x, the value of y is always 2 no matter what value of x is chosen. For instance, replace x by -4, by -1, by 0, and by 3. In each case, $y = 2$.

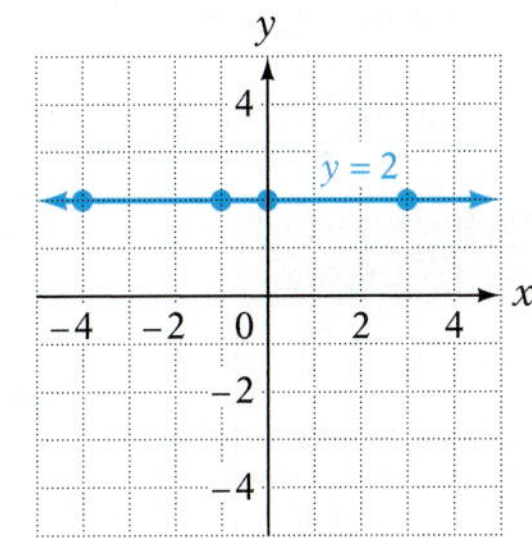

$$0x + y = 2$$

$0(-4) + y = 2$	$(-4, 2)$ is a solution.
$0(-1) + y = 2$	$(-1, 2)$ is a solution.
$0(0) + y = 2$	$(0, 2)$ is a solution.
$0(3) + y = 2$	$(3, 2)$ is a solution.

The solutions are plotted in the graph at the right, and a line is drawn through the plotted points. Note that the line is horizontal.

Graph of a Horizontal Line

The graph of $y = b$ is a horizontal line passing through $(0, b)$.

The equation $x = -2$ could be written $x + 0 \cdot y = -2$. Because $0 \cdot y = 0$ for any value of y, the value of x is always -2 no matter what value of y is chosen. For instance, replace y by -2, by 0, by 2, and by 3. In each case, $x = -2$.

$$x + 0y = -2$$

$x + 0(-2) = -2$	$(-2, -2)$ is a solution.
$x + 0(0) = -2$	$(-2, 0)$ is a solution.
$x + 0(2) = -2$	$(-2, 2)$ is a solution.
$x + 0(3) = -2$	$(-2, 3)$ is a solution.

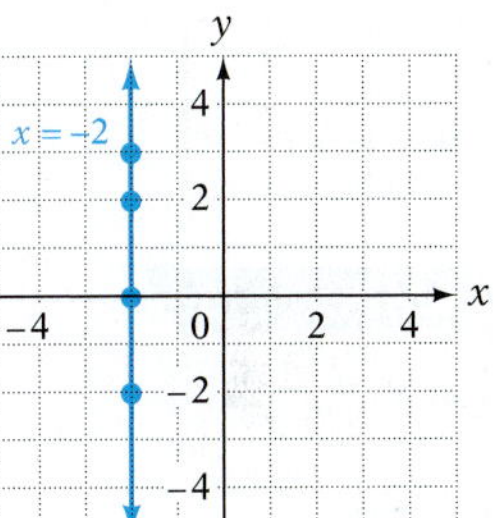

The solutions are plotted in the graph at the right, and a line is drawn through the plotted points. Note that the line is vertical.

Graph of a Vertical Line

The graph of $x = a$ is a vertical line passing through $(a, 0)$.

HOW TO • 3 Graph $x = -3$ and $y = 1$ on the same coordinate grid.

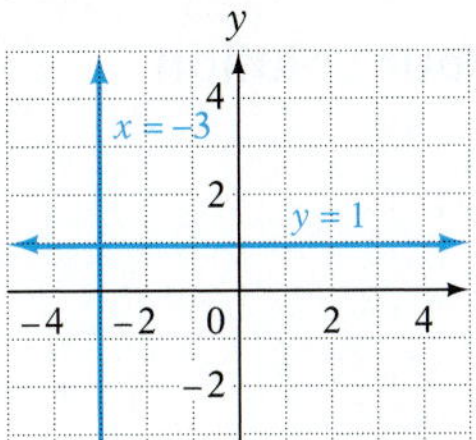

- **The graph of $x = -3$ is a vertical line passing through $(-3, 0)$.**
- **The graph of $y = 1$ is a horizontal line passing through $(0, 1)$.**

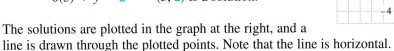

EXAMPLE • 4

Graph $2x - 5y = 10$.

Solution Solve $2x - 5y = 10$ for y.

$$2x - 5y = 10$$
$$-5y = -2x + 10$$
$$y = \frac{2}{5}x - 2$$

x	y
0	−2
5	0
−5	−4

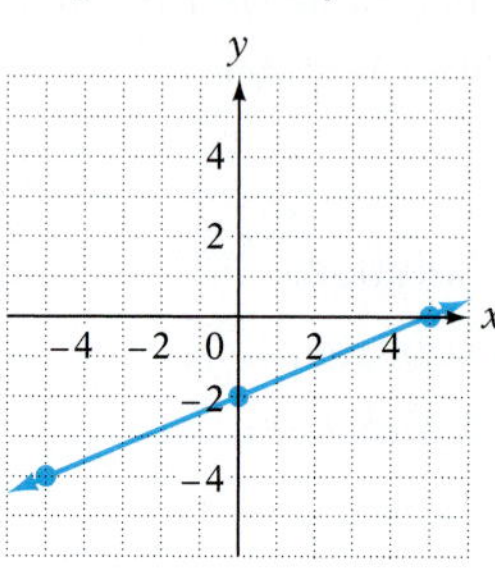

YOU TRY IT • 4

Graph $5x - 2y = 10$.

Your solution

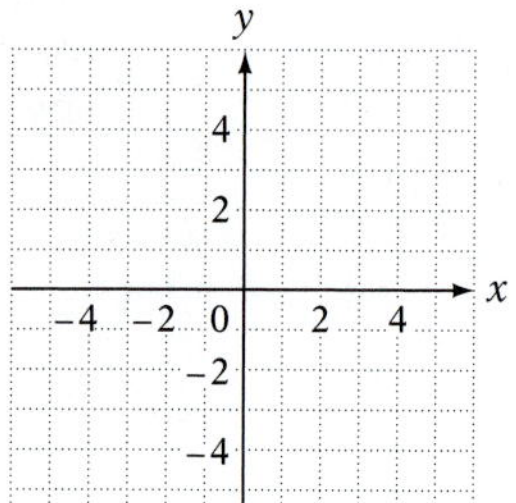

EXAMPLE • 5

Graph $x + 2y = 6$.

Solution Solve $x + 2y = 6$ for y.

$$x + 2y = 6$$
$$2y = -x + 6$$
$$y = -\frac{1}{2}x + 3$$

x	y
0	3
−2	4
4	1

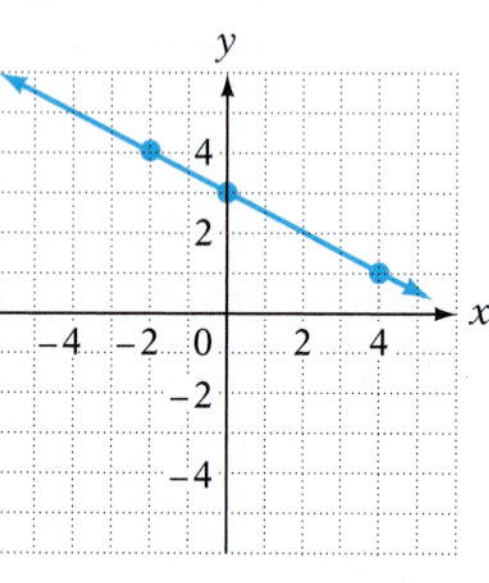

YOU TRY IT • 5

Graph $x - 3y = 9$.

Your solution

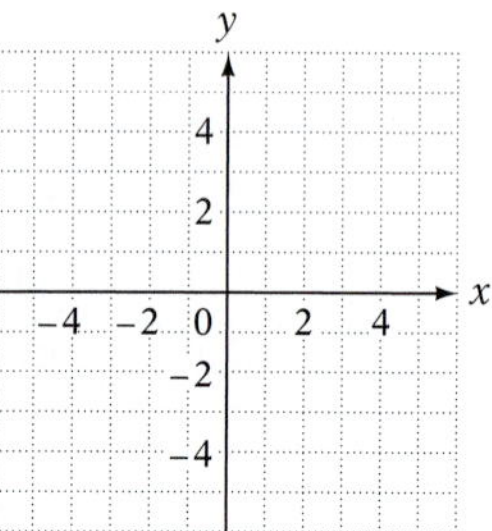

EXAMPLE • 6

Graph $y = -2$.

Solution

The graph of an equation of the form $y = b$ is a horizontal line passing through the point $(0, b)$.

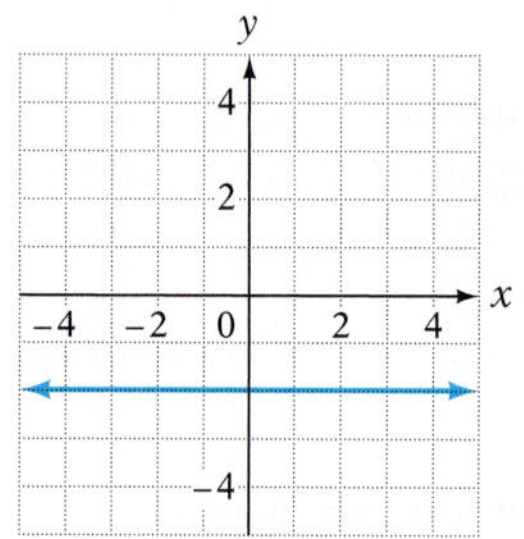

YOU TRY IT • 6

Graph $y = 3$.

Your solution

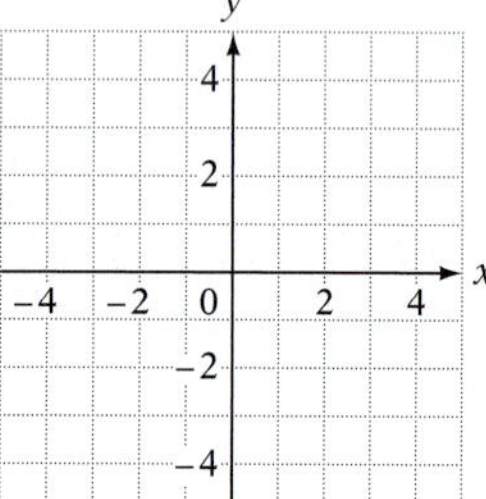

EXAMPLE • 7

Graph $x = 3$.

Solution

The graph of an equation of the form $x = a$ is a vertical line passing through the point $(a, 0)$.

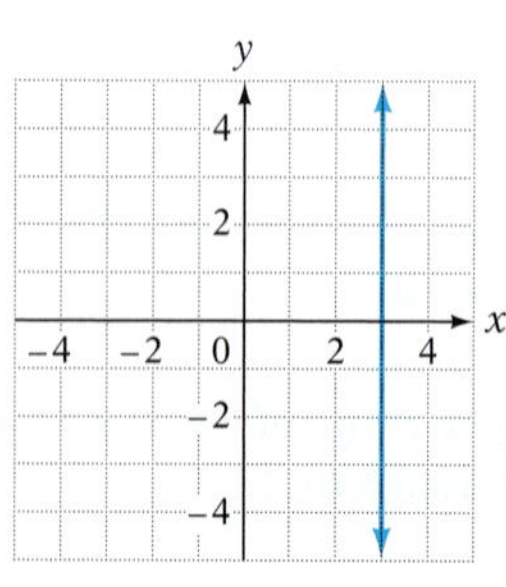

YOU TRY IT • 7

Graph $x = -4$.

Your solution

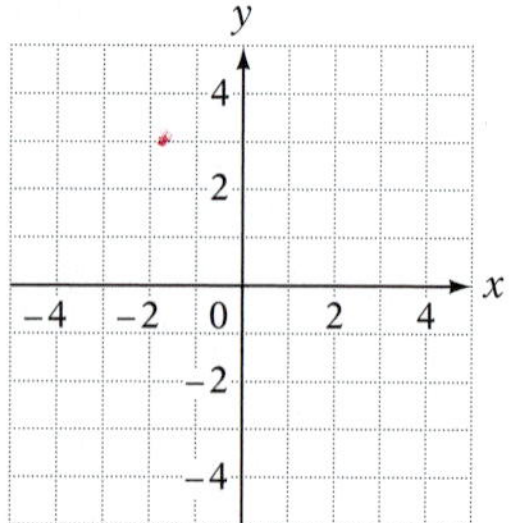

Solutions on p. S29

OBJECTIVE C To solve application problems 

There are a variety of applications of linear functions.

HOW TO • 4 The temperature of a cup of water that has been placed in a microwave oven to be heated can be approximated by the equation $T = 0.7s + 65$, where T is the temperature (in degrees Fahrenheit) of the water s seconds after the microwave oven is turned on.

a. Graph this equation for values of s from 0 to 200. (*Note:* In many applications, the domain of the variable is given so that the equation makes sense. For instance, it would not be sensible to have values of s that are less than 0. This would correspond to negative time. The choice of 200 is somewhat arbitrary and was chosen so that the water would not boil over.)

b. The point whose coordinates are (120, 149) is on the graph of this equation. Write a sentence that describes the meaning of this ordered pair.

Solution

a.

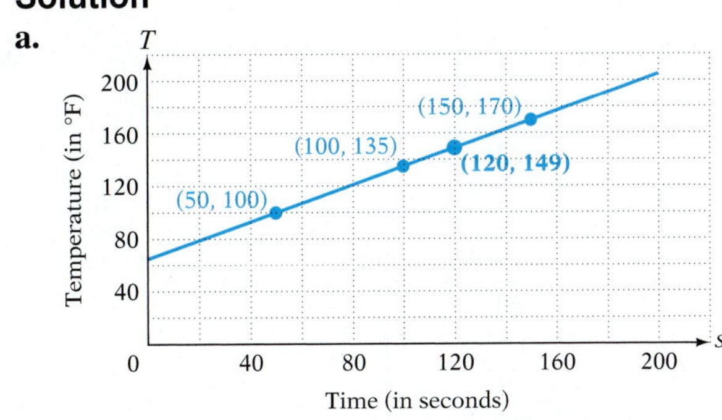

• **By choosing $s = 50, 100,$ and 150, you can find the corresponding ordered pairs (50, 100), (100, 135), and (150, 170). Plot these points and draw a line through the points.**

b. The point whose coordinates are (120, 149) means that 120 s (2 min) after the oven is turned on, the water temperature is 149°F.

EXAMPLE • 8

The number of kilobytes K of an MP3 file that remain to be downloaded t seconds after starting the download is given by $K = 935 - 5.5t$. Graph this equation for values of t from 0 to 170. The point whose coordinates are (50, 660) is on this graph. Write a sentence that describes the meaning of this ordered pair.

Solution

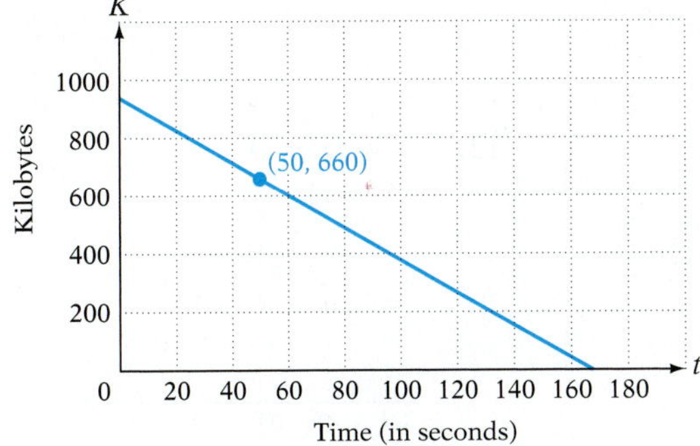

The ordered pair (50, 660) means that after 50 s, there are 660 K remaining to be downloaded.

YOU TRY IT • 8

A car is traveling at a uniform speed of 40 mph. The distance d the car travels in t hours is given by $d = 40t$. Graph this equation for values of t from 0 to 5. The point whose coordinates are (3, 120) is on the graph. Write a sentence that describes the meaning of this ordered pair.

Your solution

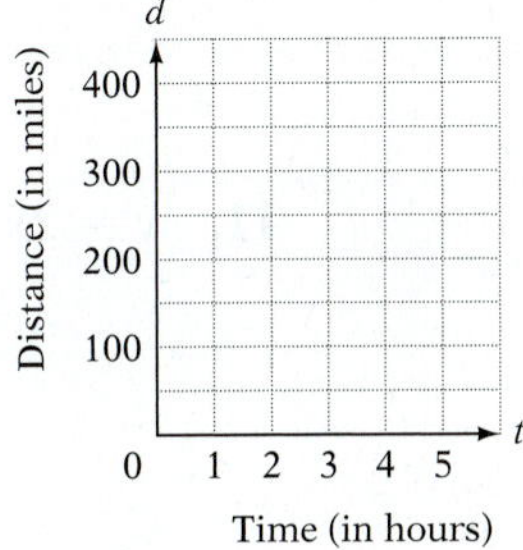

Solution on p. S29

12.2 EXERCISES

OBJECTIVE A To graph an equation of the form $y = mx + b$

For Exercises 1 to 18, graph.

1. $y = 2x - 3$

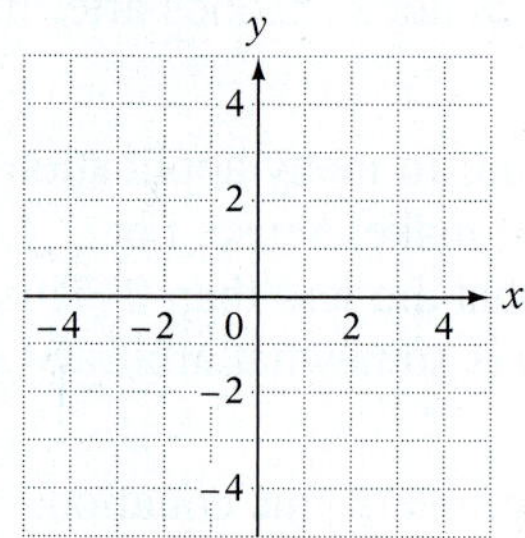

2. $y = -2x + 2$

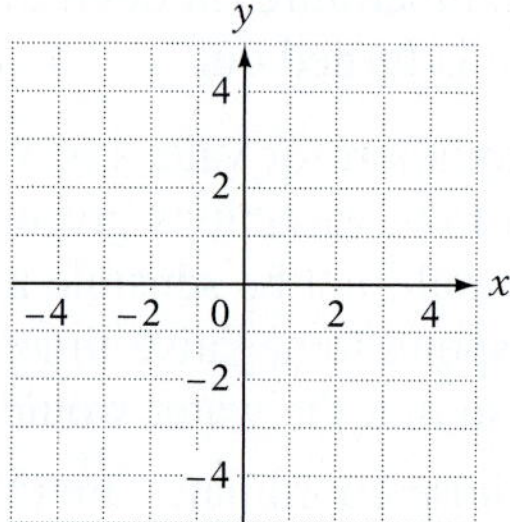

3. $y = \frac{1}{3}x$

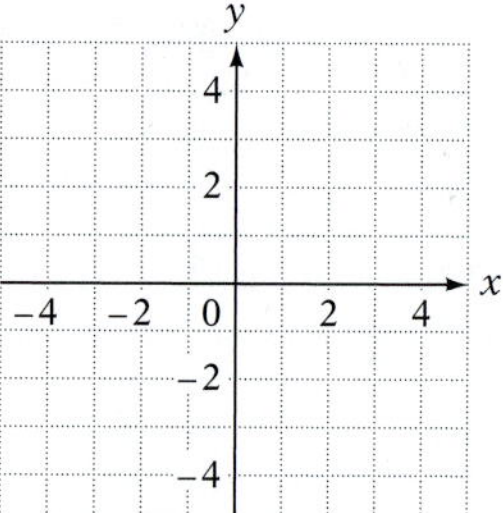

4. $y = -3x$

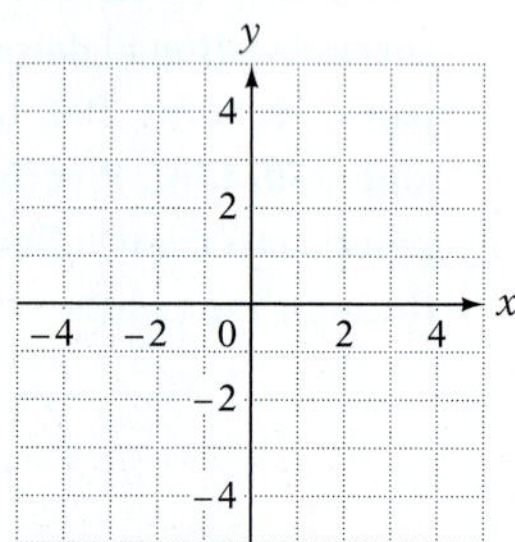

5. $y = \frac{2}{3}x - 1$

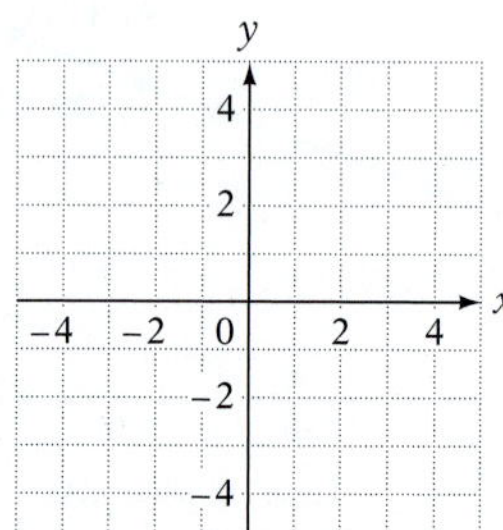

6. $y = \frac{3}{4}x + 2$

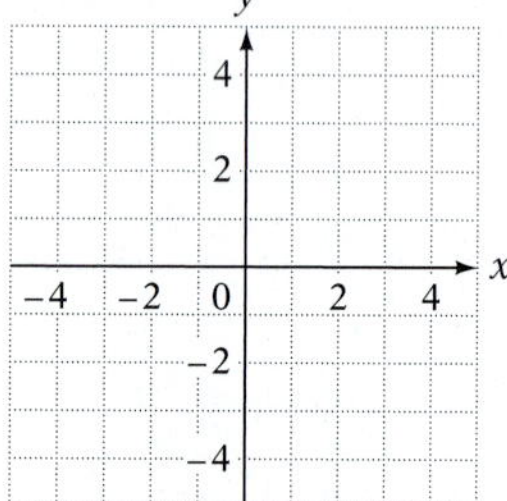

7. $y = -\frac{1}{4}x + 2$

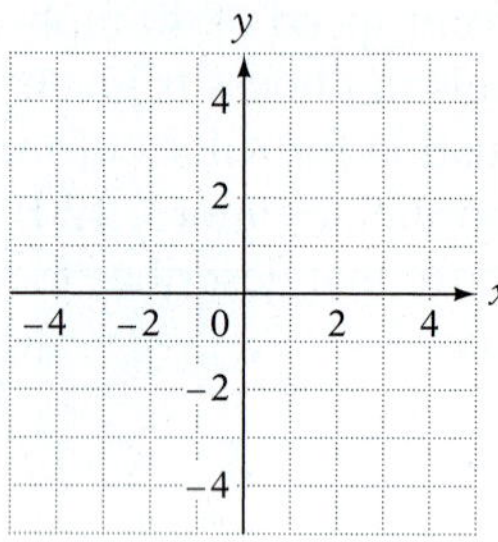

8. $y = -\frac{1}{3}x + 1$

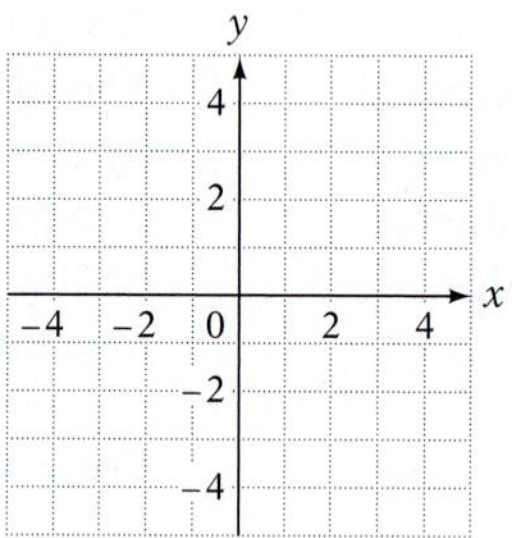

9. $y = -\frac{2}{5}x + 1$

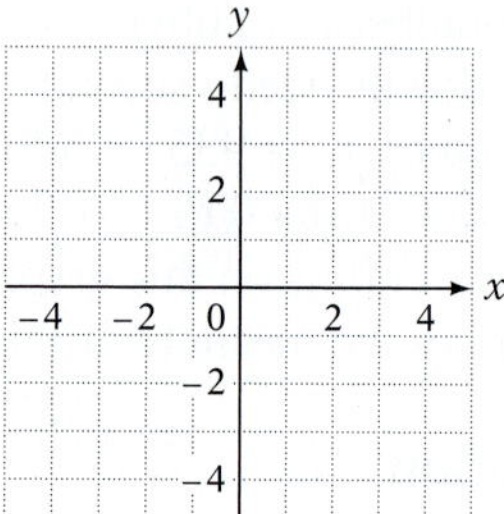

10. $y = -\frac{1}{2}x + 3$

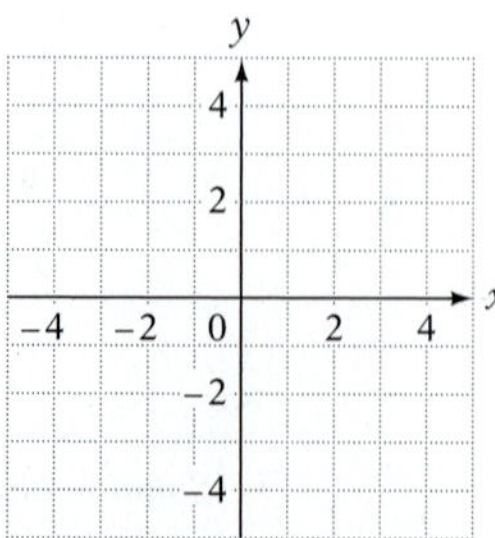

11. $y = 2x - 4$

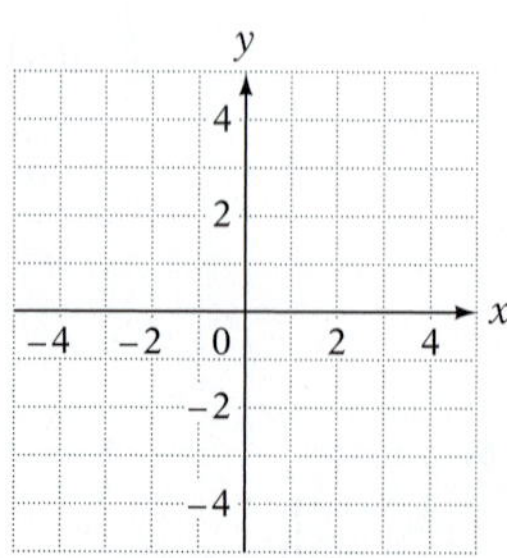

12. $y = 3x - 4$

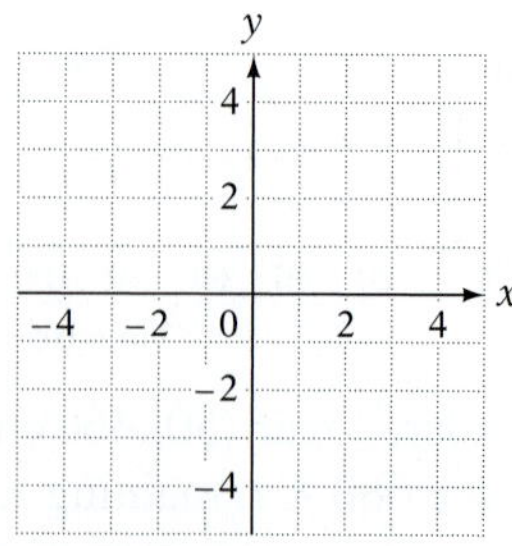

13. $y = x - 3$

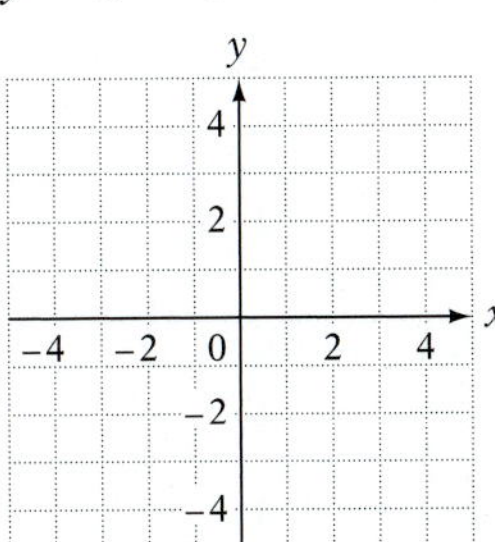

14. $y = x + 2$

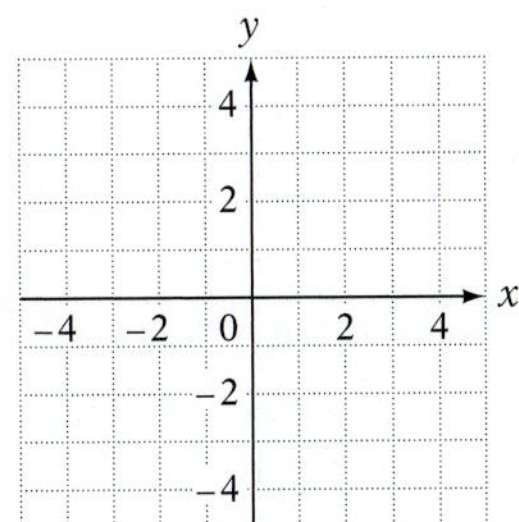

15. $y = -x + 2$

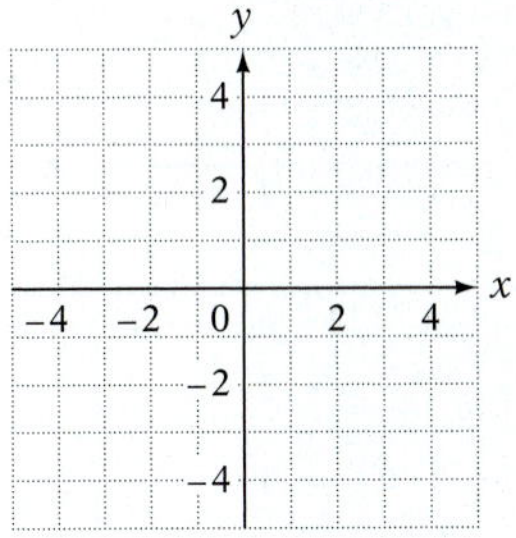

16. $y = -x - 1$

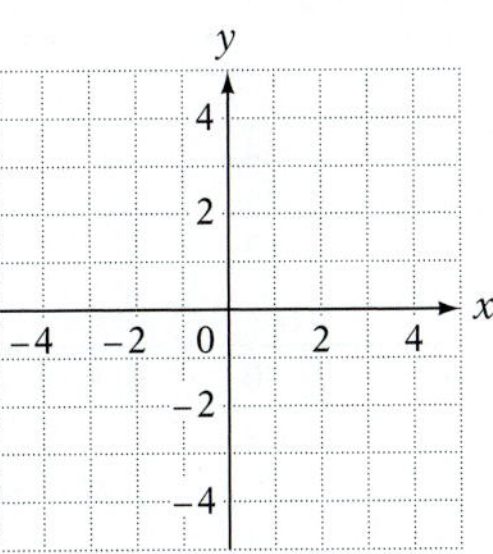

17. $y = -\frac{2}{3}x + 1$

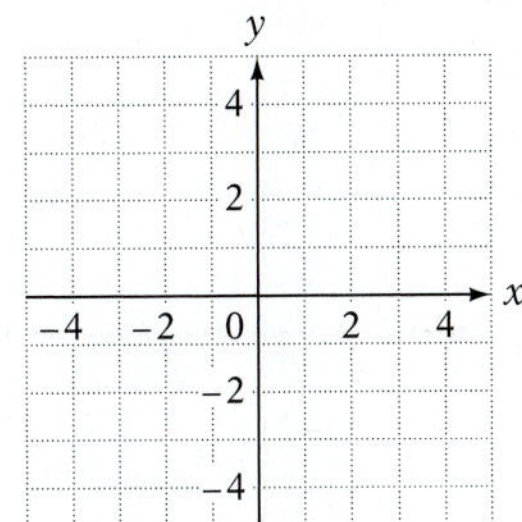

18. $y = 5x - 4$

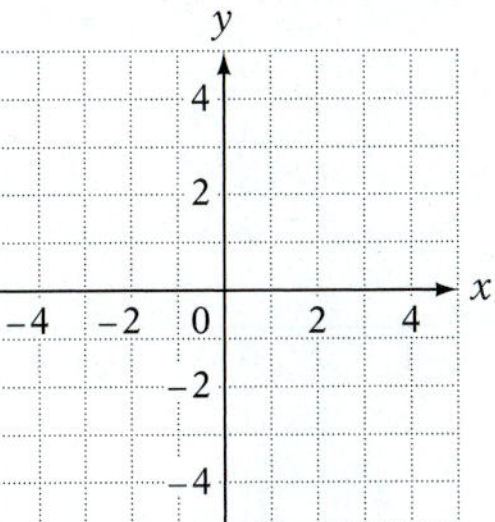

19. If the graph of $y = mx + b$ passes through the origin, (0, 0), what is the value of b?

OBJECTIVE B To graph an equation of the form $Ax + By = C$

For Exercises 20 to 37, graph.

20. $3x + y = 3$

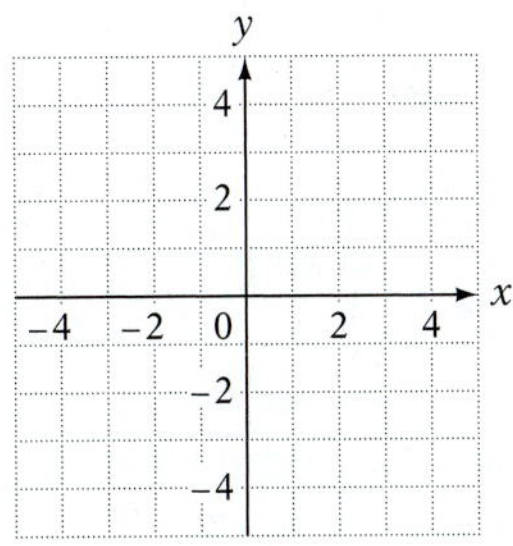

21. $2x + y = 4$

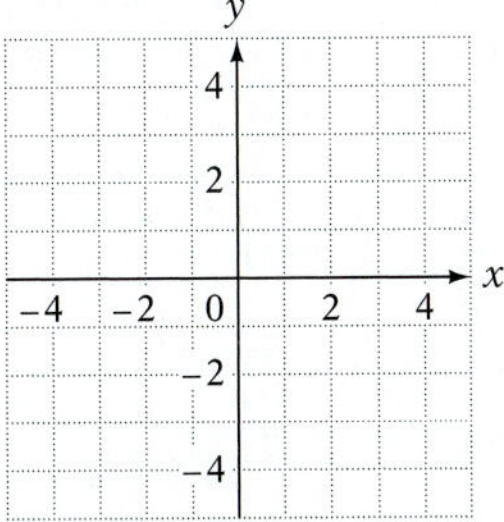

22. $2x + 3y = 6$

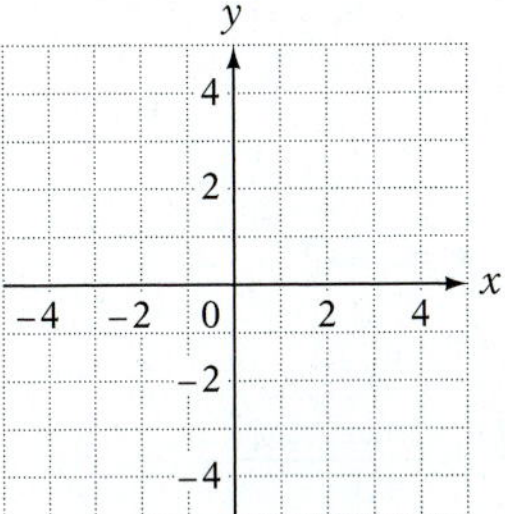

23. $3x + 2y = 4$

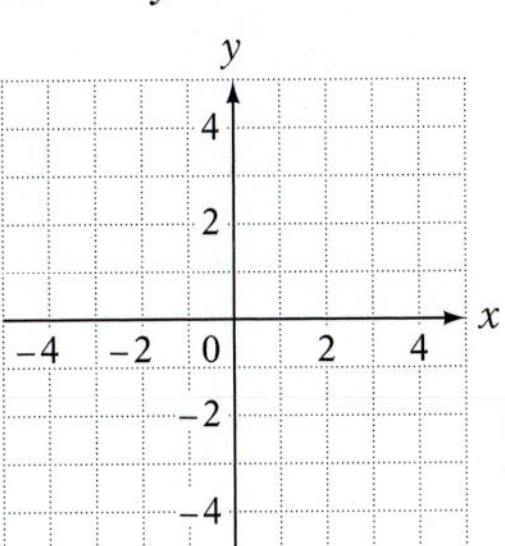

24. $x - 2y = 4$

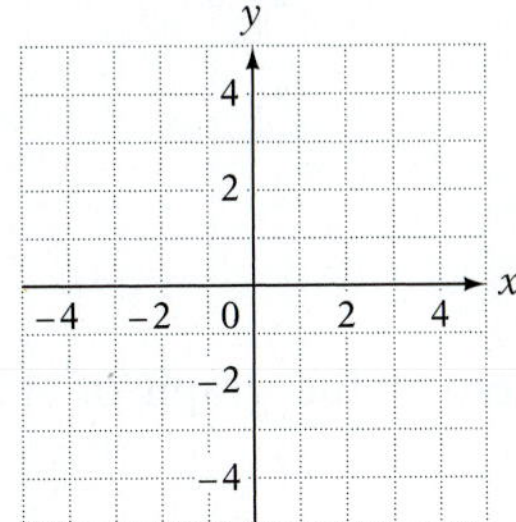

25. $x - 3y = 6$

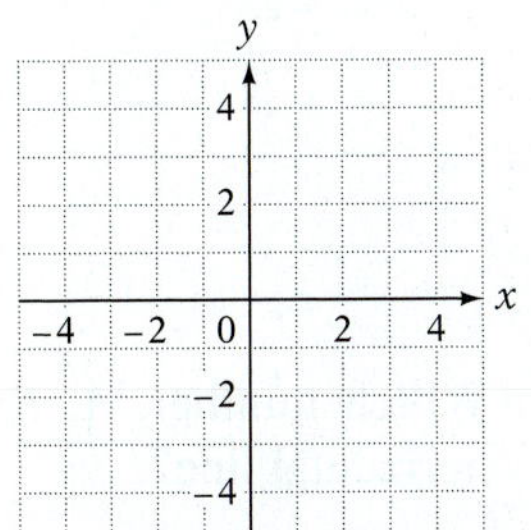

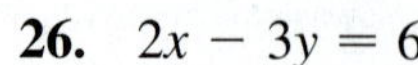

26. $2x - 3y = 6$

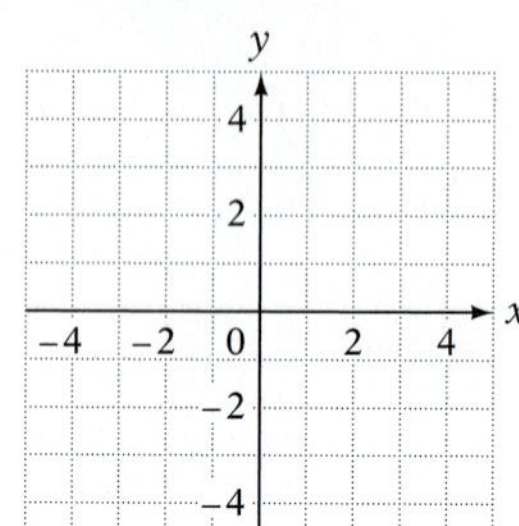

27. $3x - 2y = 8$

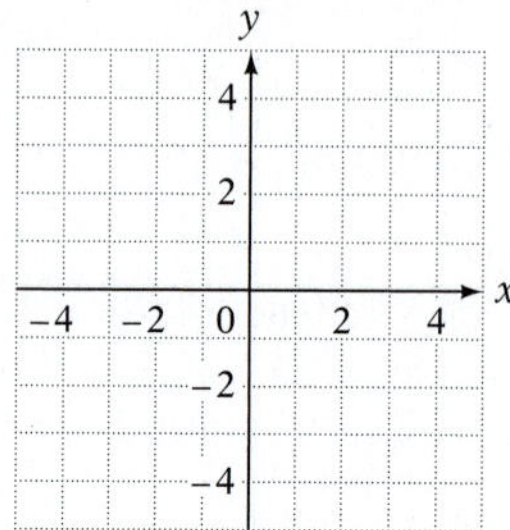

28. $2x + 5y = 10$

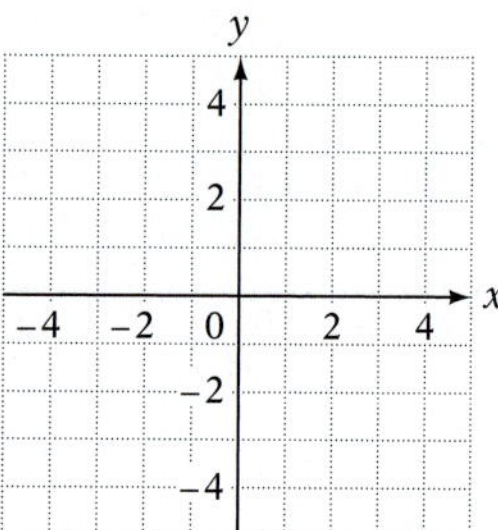

29. $3x + 4y = 12$

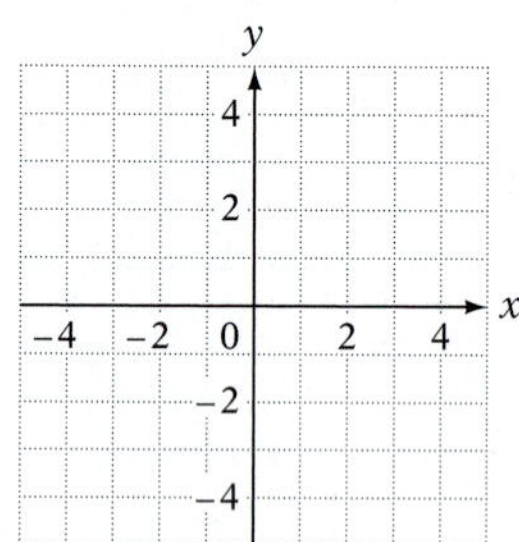

30. $x = 3$

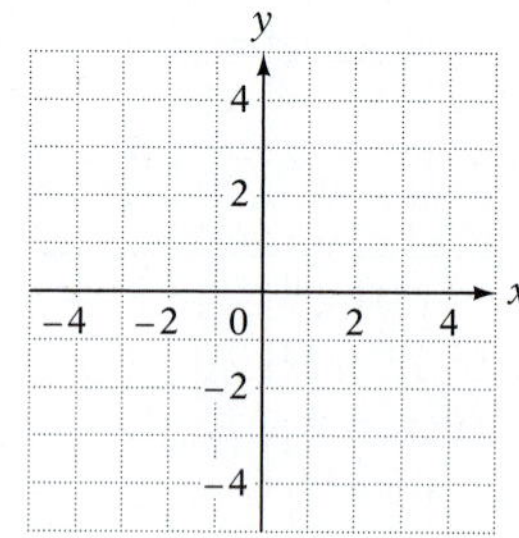

31. $y = -4$

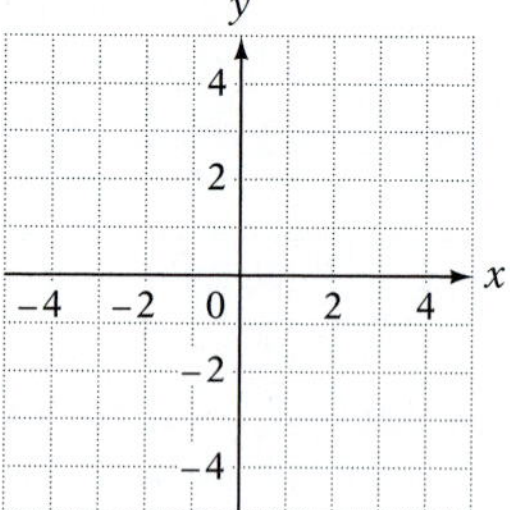

32. $x + 4y = 4$

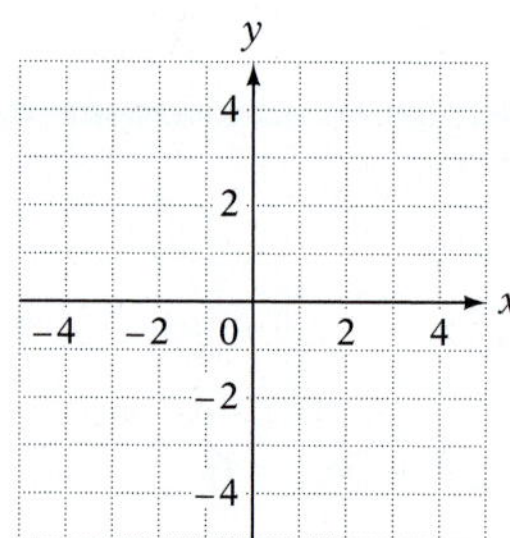

33. $4x - 3y = 12$

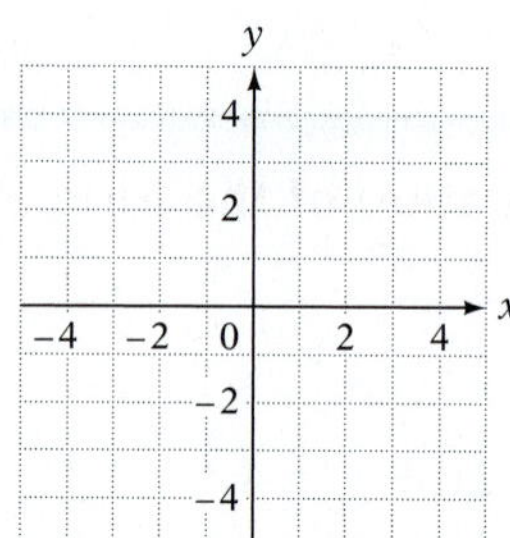

34. $y = 4$

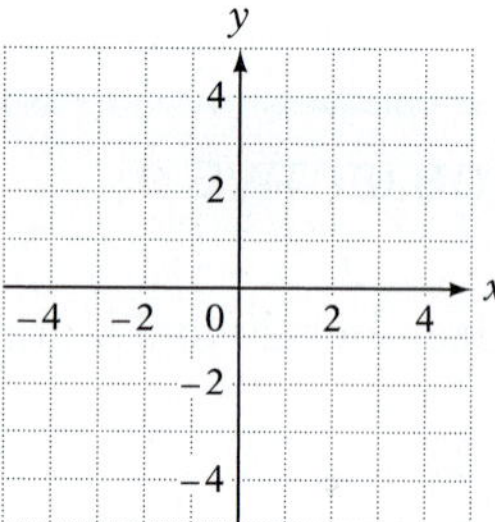

35. $x = -2$

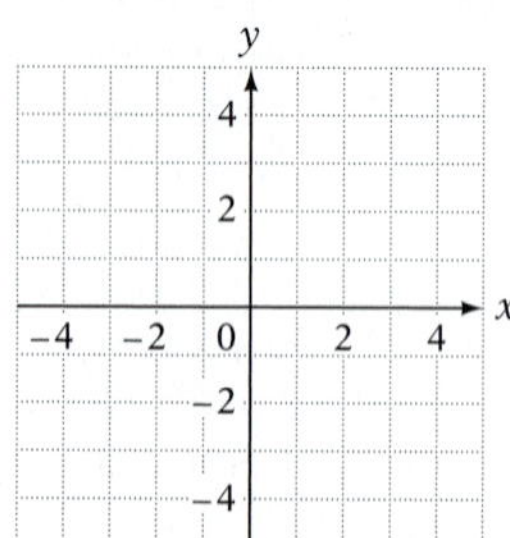

36. $\frac{x}{5} + \frac{y}{4} = 1$

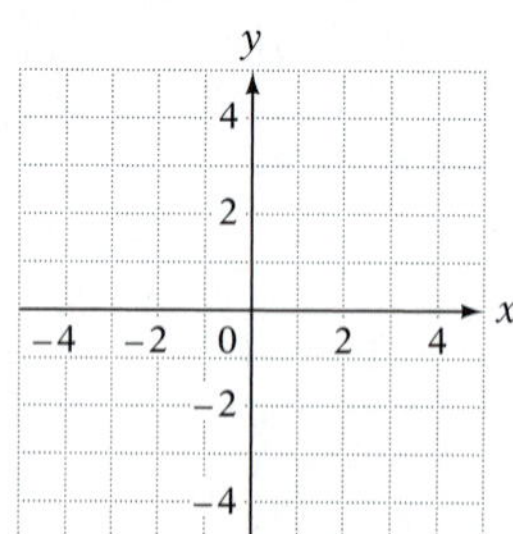

37. $\frac{x}{4} - \frac{y}{3} = 1$

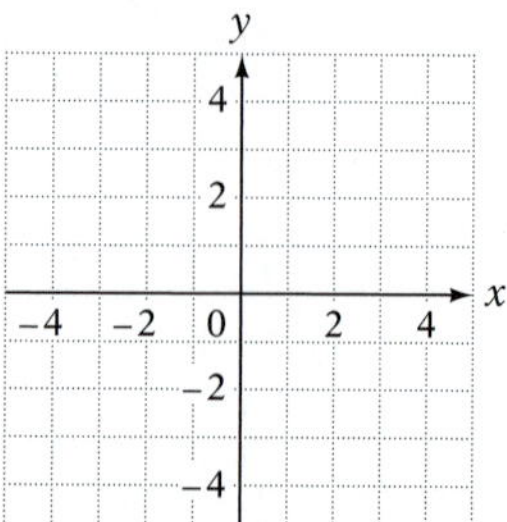

38. Which number, A, B, or C, must be zero if the graph of $Ax + By = C$ is a horizontal line?

OBJECTIVE C To solve application problems

39. Use the oven temperature graph on page 645 to determine whether the statement is true or false.

Sixty seconds after the oven is turned on, the temperature is still below 100°F.

40. Business A custom-illustrated sign or banner can be commissioned for a cost of \$25 for the material and \$10.50 per square foot for the artwork. The equation that represents this cost is given by $y = 10.50x + 25$, where y is the cost and x is the number of square feet in the sign. Graph this equation for values of x from 0 to 20. The point (15, 182.5) is on the graph. Write a sentence that describes the meaning of this ordered pair.

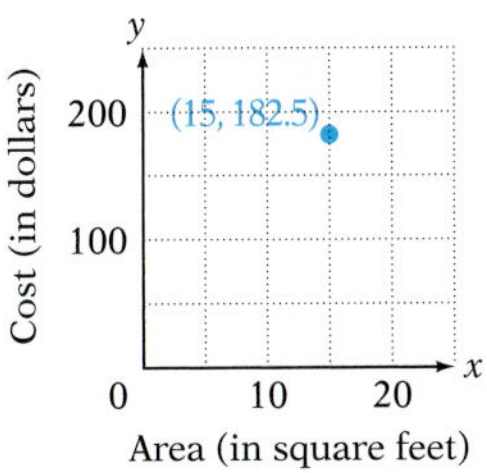

41. Emergency Response A rescue helicopter is rushing at a constant speed of 150 mph to reach several people stranded in the ocean 11 mi away after their boat sank. The rescuers can determine how far they are from the victims by using the equation $D = 11 - 2.5t$, where D is the distance in miles and t is the time elapsed in minutes. Graph this equation for values of t from 0 to 4. The point (3, 3.5) is on the graph. Write a sentence that describes the meaning of this ordered pair.

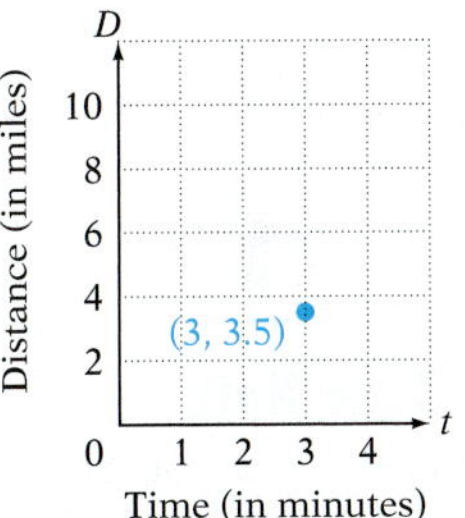

42. Veterinary Science According to some veterinarians, the age x of a dog can be translated to "human years" by using the equation $H = 4x + 16$, where H is the human equivalent age for the dog. Graph this equation for values of x from 2 to 21. The point whose coordinates are (6, 40) is on this graph. Write a sentence that explains the meaning of this ordered pair.

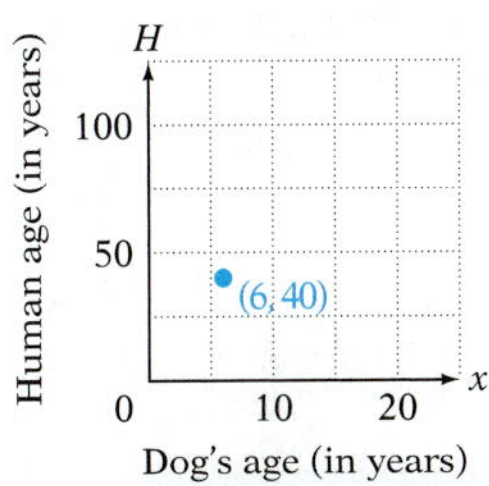

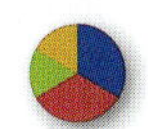

43. Taxi Fares See the news clipping at the right. You can use the equation $F = 2.80M + 2.20$ to calculate the fare F, in dollars, for a ride of M miles. Graph this equation for values of M from 1 to 5. The point (3, 10.6) is on the graph. Write a sentence that describes the meaning of this ordered pair.

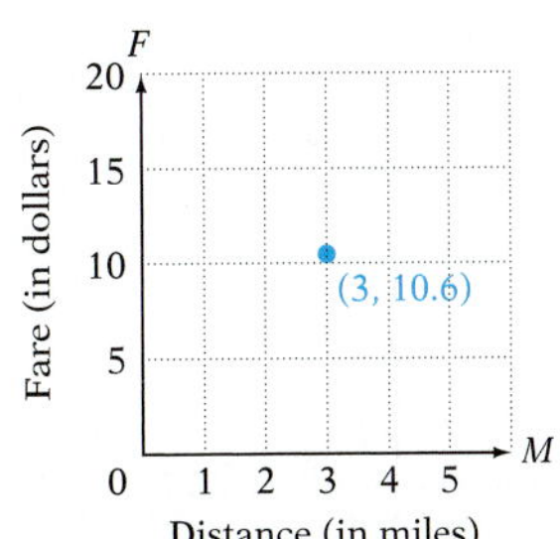

In the News

Rate Hike for Boston Cab Rides

Taxi drivers soon will be raising their rates, perhaps in an effort to help pay for their required switch to hybrid vehicles by 2015. In the near future, a passenger will have to pay \$5.00 for the first mile of a taxi ride and \$2.80 for each additional mile.

Source: The Boston Globe

Applying the Concepts

44. Graph $y = 2x - 2$, $y = 2x$, and $y = 2x + 3$. What observation can you make about the graphs?

45. Graph $y = x + 3$, $y = 2x + 3$, and $y = -\frac{1}{2}x + 3$. What observation can you make about the graphs?

SECTION

12.3 Intercepts and Slopes of Straight Lines

OBJECTIVE A To find the x- and y-intercepts of a straight line

The graph of the equation $2x + 3y = 6$ is shown at the right. The graph crosses the x-axis at the point (3, 0) and crosses the y-axis at the point (0, 2). The point at which a graph crosses the x-axis is called the **x-intercept.** At the x-intercept, the y-coordinate is 0. The point at which a graph crosses the y-axis is called the **y-intercept.** At the y-intercept, the x-coordinate is 0.

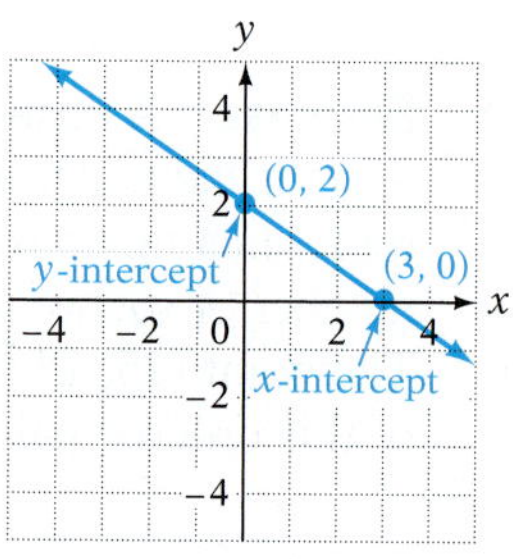

HOW TO • 1 Find the x- and y-intercepts of the graph of the equation $2x - 3y = 12$.

To find the x-intercept, let $y = 0$. (Any point on the x-axis has y-coordinate 0.)

$$\begin{aligned} 2x - 3y &= 12 \\ 2x - 3(0) &= 12 \\ 2x &= 12 \\ x &= 6 \end{aligned}$$

The x-intercept is (6, 0).

To find the y-intercept, let $x = 0$. (Any point on the y-axis has x-coordinate 0.)

$$\begin{aligned} 2x - 3y &= 12 \\ 2(0) - 3y &= 12 \\ -3y &= 12 \\ y &= -4 \end{aligned}$$

The y-intercept is (0, −4).

Take Note

To find the x-intercept, let $y = 0$ and solve for x. To find the y-intercept, let $x = 0$ and solve for y.

Some linear equations can be graphed by finding the x- and y-intercepts and then drawing a line through these two points.

EXAMPLE • 1

Find the x- and y-intercepts for $x - 2y = 4$. Graph the line.

Solution

To find the x-intercept, let $y = 0$ and solve for x.

$$\begin{aligned} x - 2y &= 4 \\ x - 2(0) &= 4 \\ x &= 4 \quad (4, 0) \end{aligned}$$

To find the y-intercept, let $x = 0$ and solve for y.

$$\begin{aligned} x - 2y &= 4 \\ 0 - 2y &= 4 \\ -2y &= 4 \\ y &= -2 \quad (0, -2) \end{aligned}$$

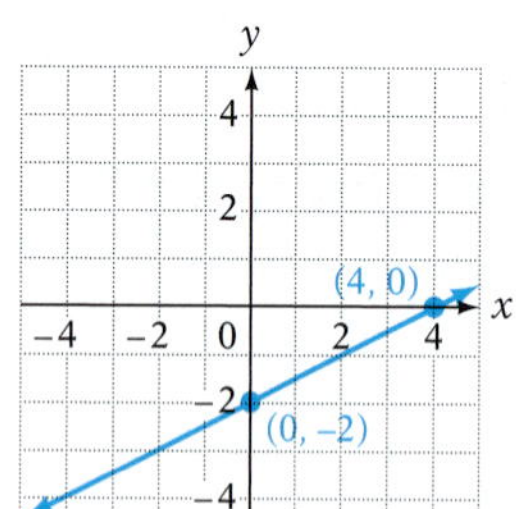

YOU TRY IT • 1

Find the x- and y-intercepts for $2x - y = 4$. Graph the line.

Your solution

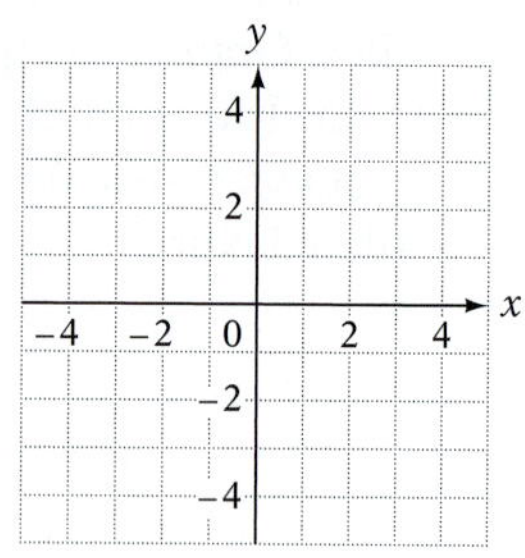

Solution on p. S29

OBJECTIVE B To find the slope of a straight line

The graphs of $y = \frac{2}{3}x + 1$ and $y = 2x + 1$ are shown in Figure 1. Each graph crosses the y-axis at the point (0, 1), but the graphs have different slants. The **slope** of a line is a measure of the slant of the line. The symbol for slope is m.

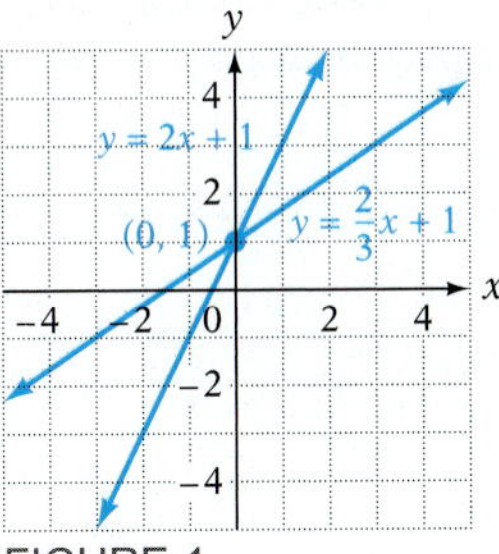

FIGURE 1

Take Note

The change in the y values can be thought of as the *rise* of the line, and the change in the x values can be thought of as the *run*. Then

$$\text{Slope} = m = \frac{\text{rise}}{\text{run}}$$

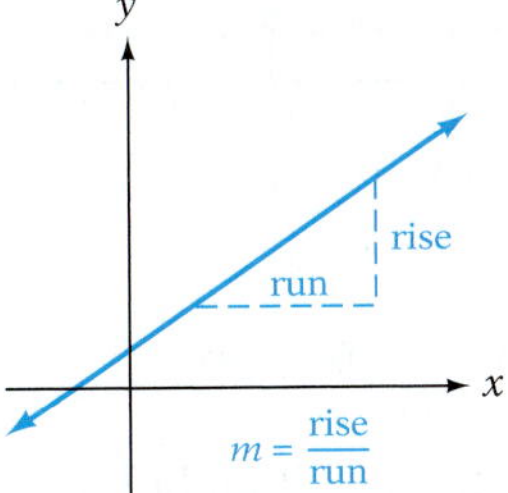
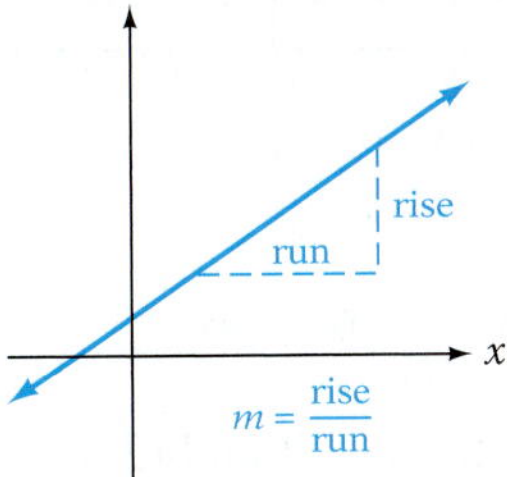

The slope of a line containing two points is the ratio of the change in the y values of the two points to the change in the x values. The line containing the points $(-2, -3)$ and $(6, 1)$ is graphed in Figure 2. The change in the y values is the difference between the two ordinates.

$$\text{Change in } y = 1 - (-3) = 4$$

FIGURE 2

The change in the x values is the difference between the two abscissas (Figure 3).

$$\text{Change in } x = 6 - (-2) = 8$$

$$\text{Slope} = m = \frac{\text{change in } y}{\text{change in } x} = \frac{4}{8} = \frac{1}{2}$$

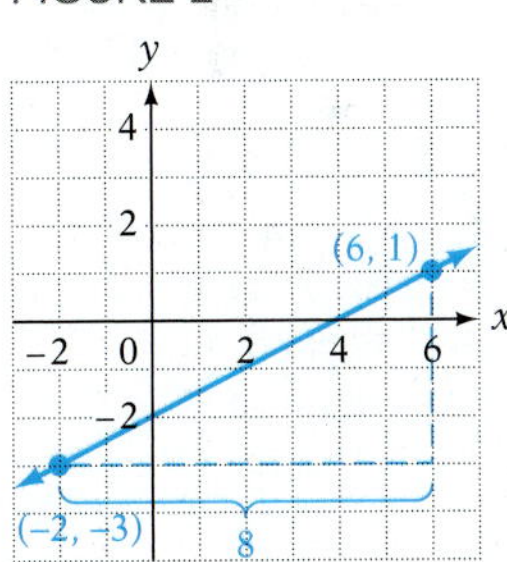

FIGURE 3

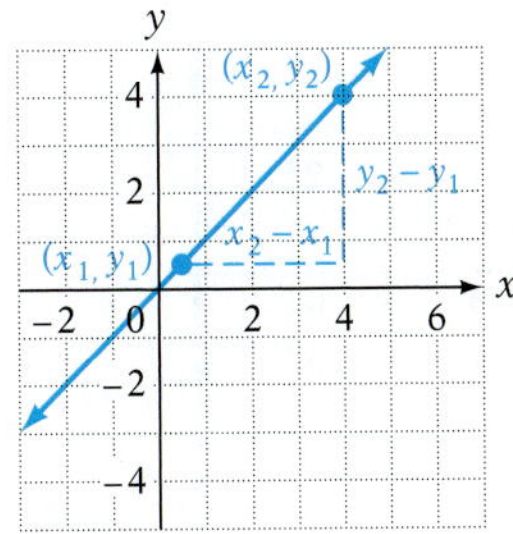

FIGURE 4

Slope Formula

If $P_1(x_1, y_1)$ and $P_2(x_2, y_2)$ are two points on a line and $x_1 \neq x_2$, then $m = \frac{y_2 - y_1}{x_2 - x_1}$ (Figure 4). If $x_1 = x_2$, the slope is undefined.

HOW TO 2 Find the slope of the line containing the points $(-1, 1)$ and $(2, 3)$.

Let P_1 be $(-1, 1)$ and P_2 be $(2, 3)$. Then $x_1 = -1$, $y_1 = 1$, $x_2 = 2$, and $y_2 = 3$.

$$m = \frac{y_2 - y_1}{x_2 - x_1} = \frac{3 - 1}{2 - (-1)} = \frac{2}{3}$$

The slope is $\frac{2}{3}$.

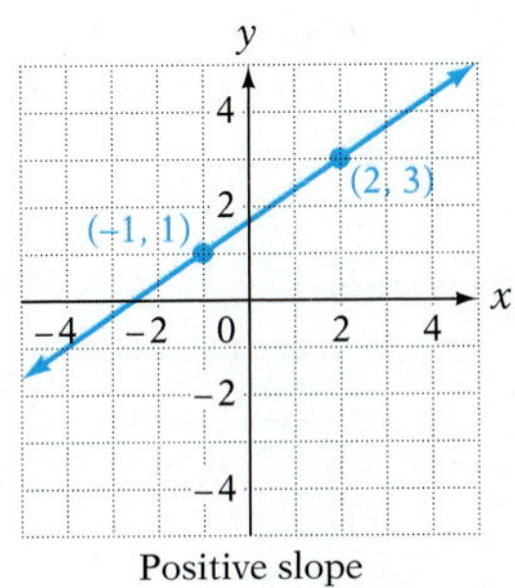

Positive slope

Take Note

Positive slope means that the value of y increases as the value of x increases.

A line that slants upward to the right always has a **positive slope.**

Note that you obtain the same results if the points are named oppositely. Let P_1 be (2, 3) and P_2 be $(-1, 1)$. Then $x_1 = 2$, $y_1 = 3$, $x_2 = -1$, and $y_2 = 1$.

$$m = \frac{y_2 - y_1}{x_2 - x_1} = \frac{1 - 3}{-1 - 2} = \frac{-2}{-3} = \frac{2}{3}$$

The slope is $\frac{2}{3}$. Therefore, it does not matter which point is named P_1 and which is named P_2; the slope remains the same.

Take Note

Negative slope means that the value of y decreases as x increases. Compare this to positive slope.

HOW TO • 3 Find the slope of the line containing the points $(-3, 4)$ and $(2, -2)$.

Let P_1 be $(-3, 4)$ and P_2 be $(2, -2)$.

$$m = \frac{y_2 - y_1}{x_2 - x_1} = \frac{-2 - 4}{2 - (-3)} = \frac{-6}{5} = -\frac{6}{5}$$

The slope is $-\frac{6}{5}$.

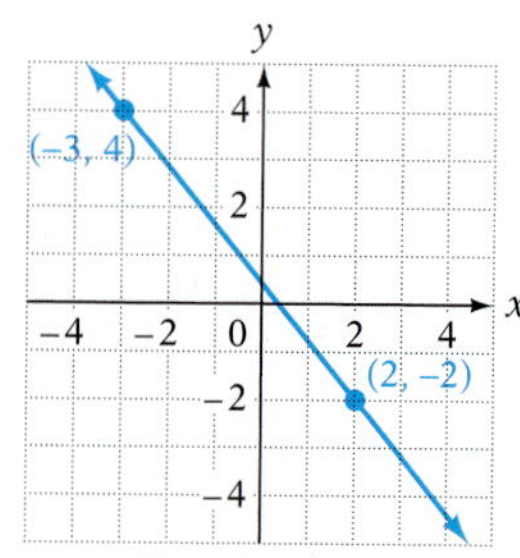

Negative slope

A line that slants downward to the right always has a **negative slope.**

HOW TO • 4 Find the slope of the line containing the points $(-1, 3)$ and $(4, 3)$.

Let P_1 be $(-1, 3)$ and P_2 be $(4, 3)$.

$$m = \frac{y_2 - y_1}{x_2 - x_1} = \frac{3 - 3}{4 - (-1)} = \frac{0}{5} = 0$$

The slope is 0.

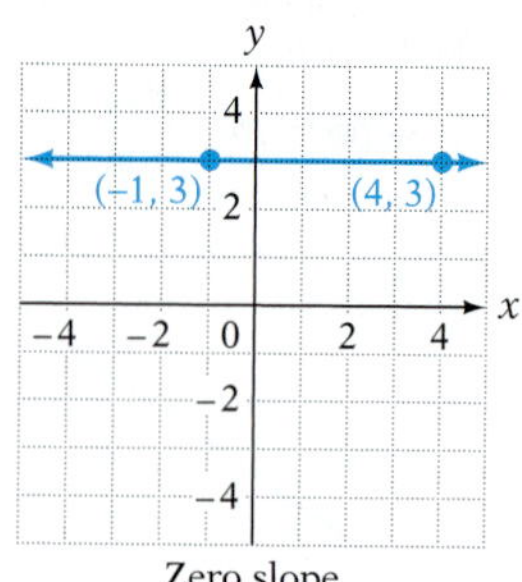

Zero slope

A horizontal line has **zero slope.**

HOW TO • 5 Find the slope of the line containing the points $(2, -2)$ and $(2, 4)$.

Let P_1 be $(2, -2)$ and P_2 be $(2, 4)$.

$$m = \frac{y_2 - y_1}{x_2 - x_1} = \frac{4 - (-2)}{2 - 2} = \frac{6}{0}$$

Division by zero is not defined.

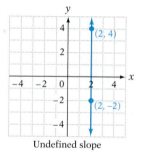

Undefined slope

A vertical line has **undefined slope.**

Two lines in the plane that never intersect are called parallel lines. The lines l_1 and l_2 in the figure at the right are parallel. Calculating the slope of each line, we have

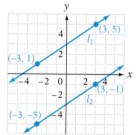

$$\text{Slope of } l_1\text{: } m_1 = \frac{y_2 - y_1}{x_2 - x_1} = \frac{5 - 1}{3 - (-3)} = \frac{4}{6} = \frac{2}{3}$$

$$\text{Slope of } l_2\text{: } m_2 = \frac{y_2 - y_1}{x_2 - x_1} = \frac{-1 - (-5)}{3 - (-3)} = \frac{4}{6} = \frac{2}{3}$$

Note that these parallel lines have the same slope. This is always true for parallel lines.

Take Note

We must separate the description of parallel lines at the right into two parts because vertical lines in the plane are parallel, but their slopes are undefined.

Parallel Lines

Two nonvertical lines in the plane are parallel if and only if they have the same slope. Vertical lines in the plane are parallel.

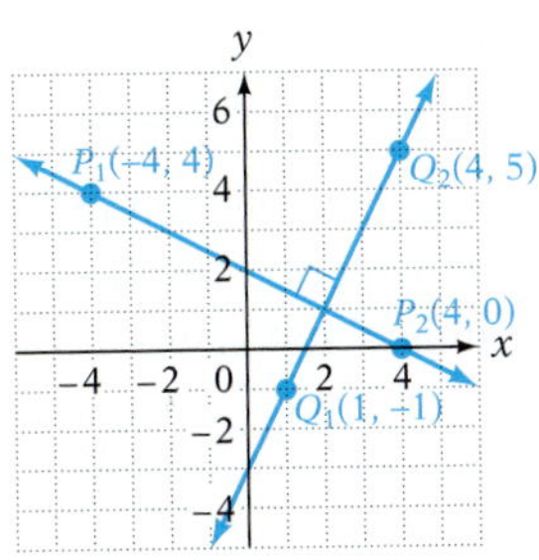

Two lines that intersect at a 90° angle (right angle) are perpendicular lines. The lines at the left are perpendicular.

> **Perpendicular Lines**
>
> Two nonvertical lines in the plane are perpendicular if and only if the product of their slopes is −1. A vertical and a horizontal line are perpendicular.

The slope of the line between P_1 and P_2 is $m_1 = \frac{0-4}{4-(-4)} = -\frac{4}{8} = -\frac{1}{2}$. The slope of the line between Q_1 and Q_2 is $m_2 = \frac{5-(-1)}{4-1} = \frac{6}{3} = 2$. The product of the slopes is $\left(-\frac{1}{2}\right)2 = -1$. Because the product of the slopes is **−1**, the graphs are perpendicular.

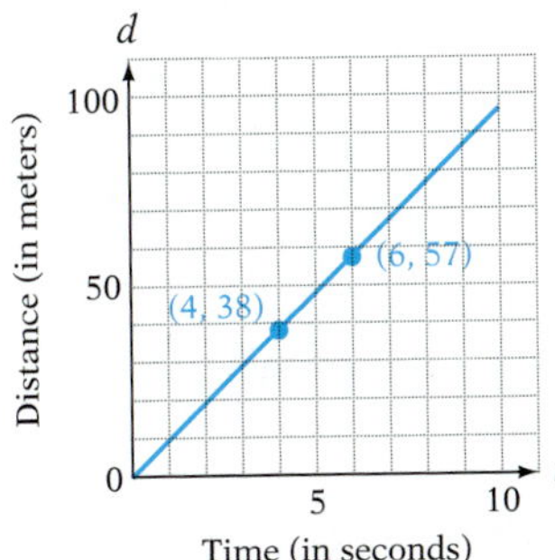

There are many applications of the concept of slope. Here is an example.

When Florence Griffith-Joyner set the world record for the 100-meter dash, her average rate of speed was approximately 9.5 m/s. The graph at the left shows the distance she ran during her record-setting run. From the graph, note that after 4 s she had traveled 38 m and that after 6 s she had traveled 57 m. The slope of the line between these two points is

$$m = \frac{57-38}{6-4} = \frac{19}{2} = 9.5$$

Note that the slope of the line is the same as the rate at which she was running, 9.5 m/s. The average speed of an object is related to slope.

EXAMPLE 2

Find the slope of the line containing the points $(-2, -3)$ and $(3, 4)$.

Solution

Let $P_1 = (-2, -3)$ and $P_2 = (3, 4)$.

$$m = \frac{y_2 - y_1}{x_2 - x_1} = \frac{4-(-3)}{3-(-2)}$$

• $y_2 = 4, y_1 = -3$
• $x_2 = 3, x_1 = -2$

$$= \frac{7}{5}$$

The slope is $\frac{7}{5}$.

YOU TRY IT 2

Find the slope of the line containing the points $(1, 4)$ and $(-3, 8)$.

Your solution

EXAMPLE 3

Find the slope of the line containing the points $(-1, 4)$ and $(-1, 0)$.

Solution

Let $P_1 = (-1, 4)$ and $P_2 = (-1, 0)$.

$$m = \frac{y_2 - y_1}{x_2 - x_1} = \frac{0-4}{-1-(-1)}$$

• $y_2 = 0, y_1 = 4$
• $x_2 = -1, x_1 = -1$

$$= \frac{-4}{0}$$

The slope is undefined.

YOU TRY IT 3

Find the slope of the line containing the points $(-1, 2)$ and $(4, 2)$.

Your solution

Solutions on p. S29

EXAMPLE 4

The graph below shows the height of a plane above an airport during its 30-minute descent from cruising altitude to landing. Find the slope of the line. Write a sentence that explains the meaning of the slope.

Solution

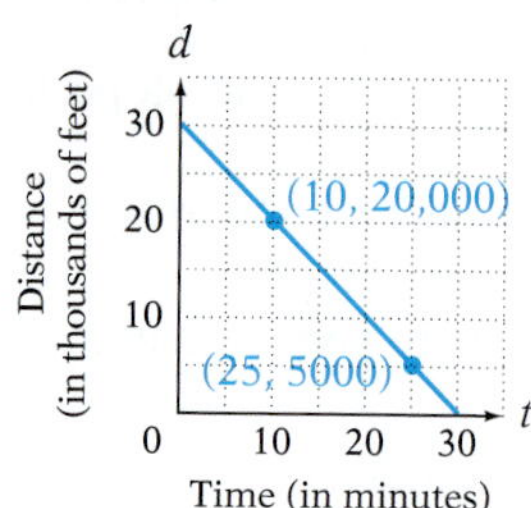

$$m = \frac{5000 - 20{,}000}{25 - 10} = \frac{-15{,}000}{15}$$
$$= -1000$$

A slope of -1000 means that the height of the plane is *decreasing* at the rate of 1000 ft/min.

YOU TRY IT 4

The graph below shows the approximate decline in the value of a used car over a 5-year period. Find the slope of the line. Write a sentence that states the meaning of the slope.

Your solution

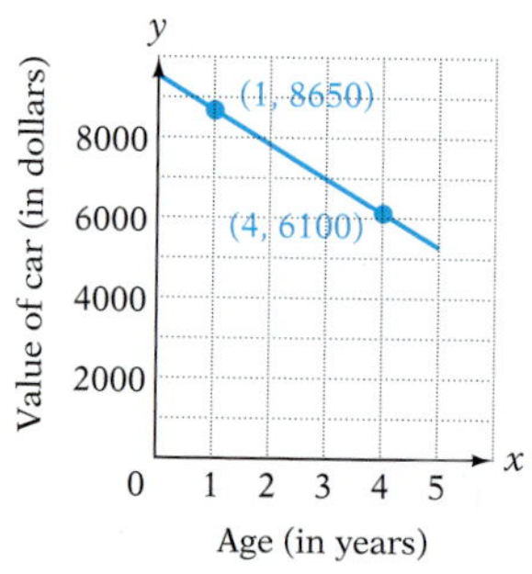

Solution on p. S29

OBJECTIVE C To graph a line using the slope and the y-intercept

HOW TO 6 Find the y-intercept of $y = 3x + 4$.

$y = 3x + 4 = 3(0) + 4 = 4$ • **Let $x = 0$.**

The y-intercept is (0, 4).

For any equation of the form $y = mx + b$, the y-intercept is $(0, b)$.

The graph of the equation $y = \frac{2}{3}x + 1$ is shown at the right. The points $(-3, -1)$ and $(3, 3)$ are on the graph. The slope of the line between the two points is

$$m = \frac{3 - (-1)}{3 - (-3)} = \frac{4}{6} = \frac{2}{3}$$

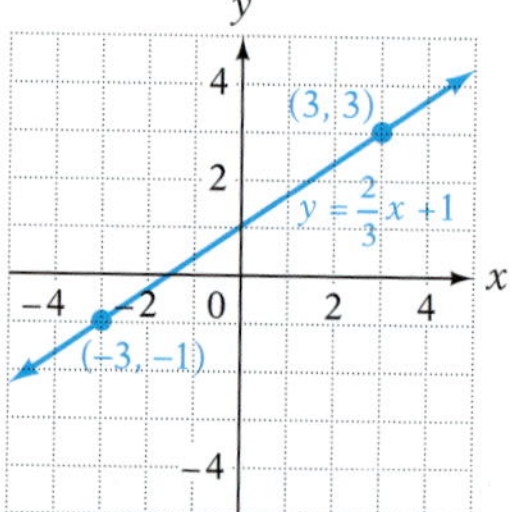

Observe that the slope of the line is the coefficient of x in the equation $y = \frac{2}{3}x + 1$. Also recall that the y-intercept is (0, 1), where 1 is the constant term of the equation.

Take Note

Here are some equations in slope-intercept form.

$y = 2x - 3$: Slope is 2; y-intercept is $(0, -3)$.

$y = -x + 2$: Slope is -1 (recall that $-x = -1x$); y-intercept is (0, 2).

$y = \frac{x}{2}$: Because $\frac{x}{2} = \frac{1}{2}x$, slope is $\frac{1}{2}$; y-intercept is (0, 0).

Slope-Intercept Form of a Linear Equation

An equation of the form $y = mx + b$ is called the **slope-intercept form** of a straight line. The slope of the line is m, the coefficient of x. The y-intercept is $(0, b)$, where b is the constant term of the equation.

When an equation of a line is in slope-intercept form, the graph can be drawn using the slope and the y-intercept. First locate the y-intercept. Use the slope to find a second point on the line. Then draw a line through the two points.

HOW TO • 7 Graph $y = 2x - 3$.

y-intercept $= (0, b) = (0, -3)$

$m = 2 = \frac{2}{1} = \frac{\text{change in } y}{\text{change in } x}$

Beginning at the y-intercept, move right 1 unit (change in x) and then up 2 units (change in y).

$(1, -1)$ is a second point on the graph.

Draw a line through the two points $(0, -3)$ and $(1, -1)$.

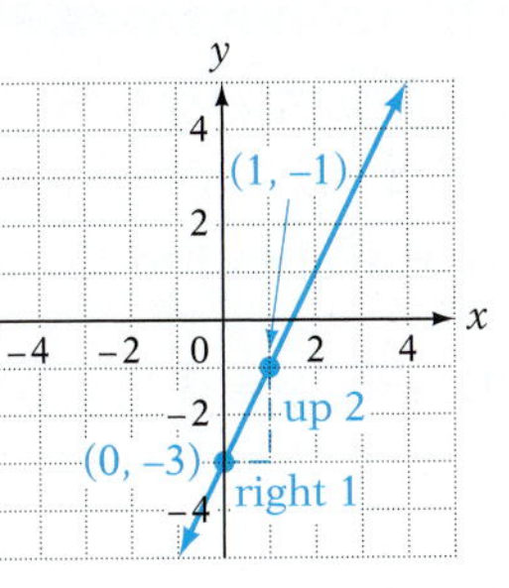

EXAMPLE • 5

Graph $y = -\frac{2}{3}x + 1$ by using the slope and y-intercept.

Solution

y-intercept $= (0, b) = (0, 1)$

$m = -\frac{2}{3} = \frac{-2}{3} = \frac{\text{change in } y}{\text{change in } x}$

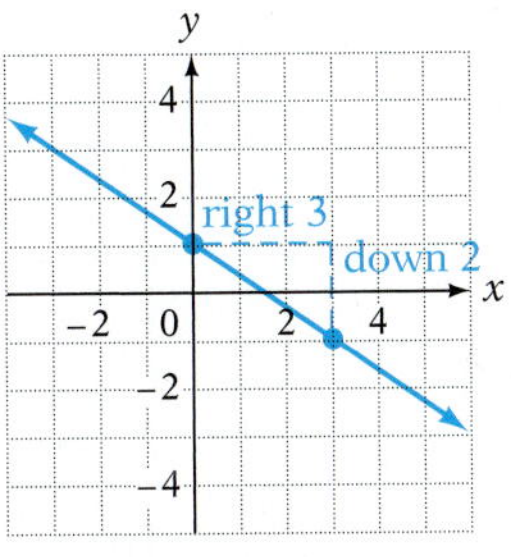

YOU TRY IT • 5

Graph $y = -\frac{1}{4}x - 1$ by using the slope and y-intercept.

Your solution

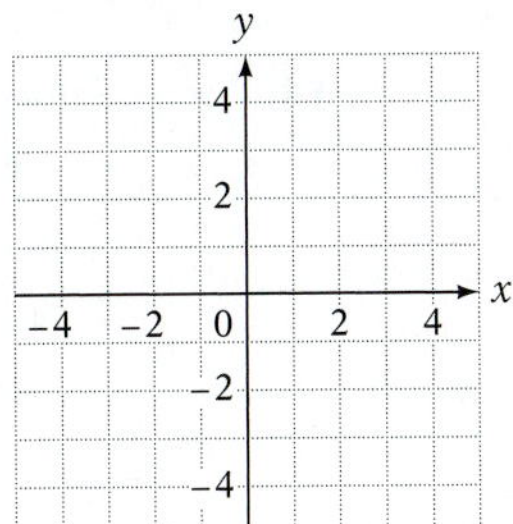

EXAMPLE • 6

Graph $2x - 3y = 6$ by using the slope and y-intercept.

Solution

The equation is in the form $Ax + By = C$. Rewrite it in slope-intercept form by solving it for y.

$$2x - 3y = 6$$
$$-3y = -2x + 6$$
$$y = \frac{2}{3}x - 2$$

y-intercept $= (0, -2)$; $m = \frac{2}{3}$

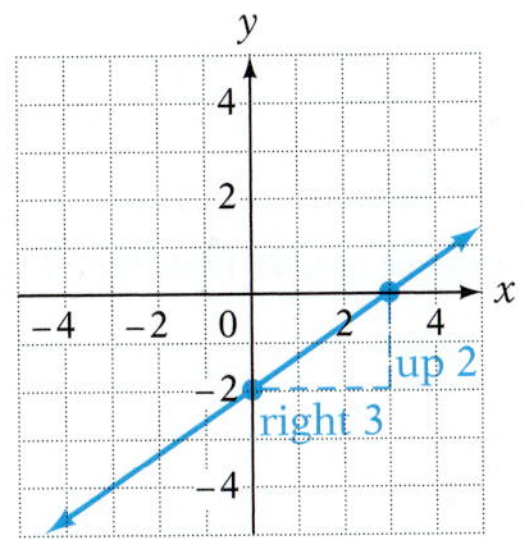

YOU TRY IT • 6

Graph $x - 2y = 4$ by using the slope and y-intercept.

Your solution

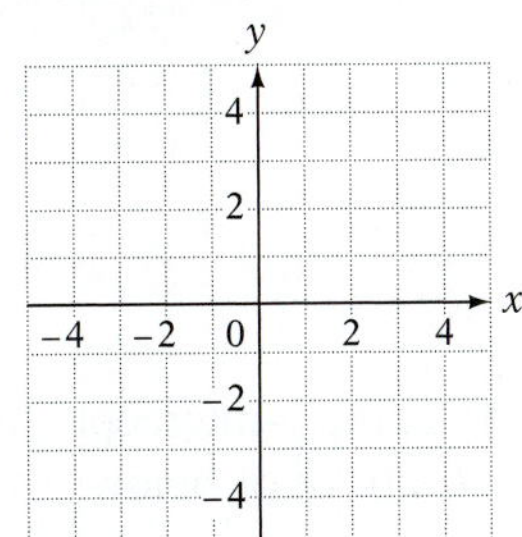

Solutions on pp. S29–S30

12.3 EXERCISES

OBJECTIVE A To find the x- and y-intercepts of a straight line

For Exercises 1 to 12, find the x- and y-intercepts.

1. $x - y = 3$

2. $3x + 4y = 12$

3. $3x - y = 6$

4. $2x - y = -10$

5. $x - 5y = 10$

6. $3x + 2y = 12$

7. $3x - y = -12$

8. $5x - y = -10$

9. $2x - 3y = 0$

10. $3x + 4y = 0$

11. $x + 2y = 6$

12. $2x - 3y = 12$

For Exercises 13 to 18, find the x- and y-intercepts, and then graph.

13. $5x + 2y = 10$

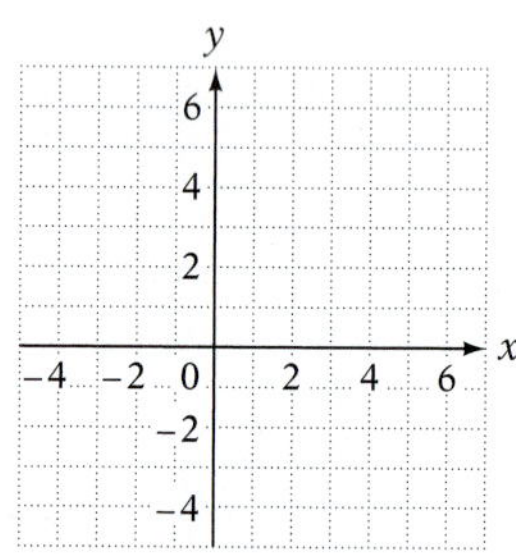

14. $x - 3y = 6$

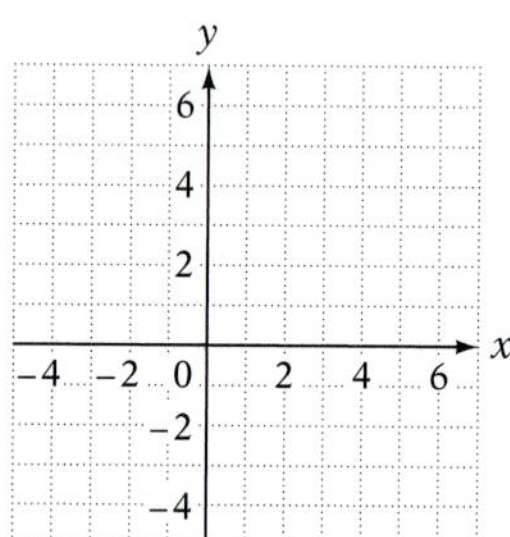

15. $3x - 4y = 12$

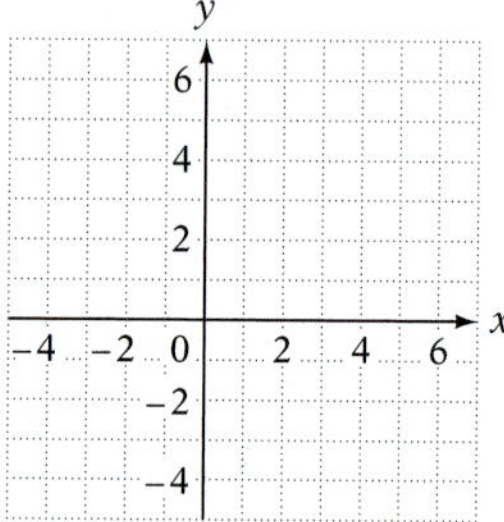

16. $2x - 5y = 10$

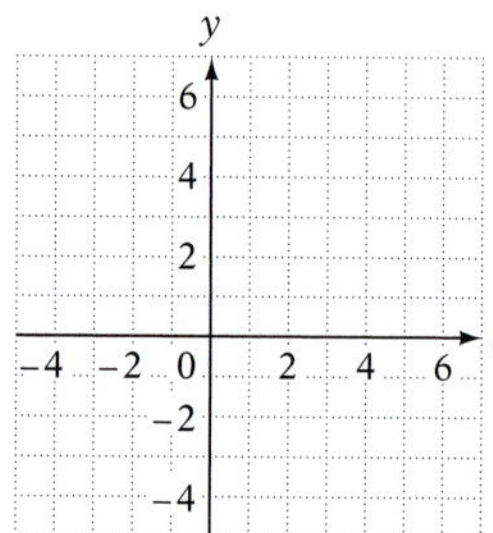

17. $5y - 3x = 15$

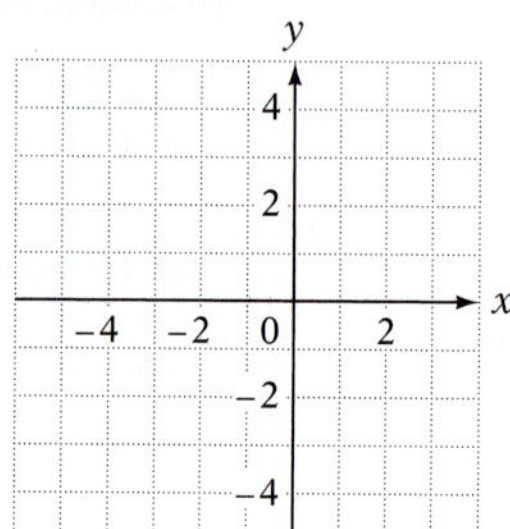

18. $9y - 4x = 18$

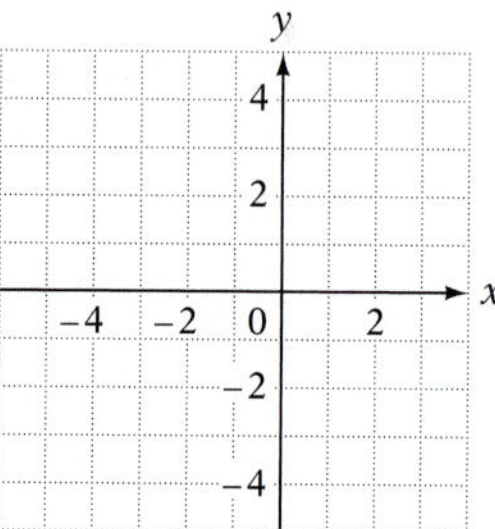

19. If $A > 0$, $B > 0$, and $C > 0$, is the y-intercept of the graph of $Ax + By = C$ above or below the x-axis?

20. If $A > 0$, $B > 0$, and $C > 0$, is the x-intercept of the graph of $Ax + By = C$ to the left or to the right of the y-axis?

OBJECTIVE B To find the slope of a straight line

21. What is the difference between a line that has zero slope and one that has undefined slope?

For Exercises 22 to 33, find the slope of the line containing the given points.

22. $P_1(4, 2), P_2(3, 4)$ **23.** $P_1(2, 1), P_2(3, 4)$ **24.** $P_1(-1, 3), P_2(2, 4)$ **25.** $P_1(-2, 1), P_2(2, 2)$

26. $P_1(2, 4), P_2(4, -1)$ **27.** $P_1(1, 3), P_2(5, -3)$ **28.** $P_1(3, -4), P_2(3, 5)$ **29.** $P_1(-1, 2), P_2(-1, 3)$

30. $P_1(4, -2), P_2(3, -2)$ **31.** $P_1(5, 1), P_2(-2, 1)$ **32.** $P_1(0, -1), P_2(3, -2)$ **33.** $P_1(3, 0), P_2(2, -1)$

For Exercises 34 and 35, l is a line passing through two distinct points (a, b) and (c, d).

34. Describe any relationships that must exist among a, b, c, and d in order for the slope of l to be undefined.

35. Describe any relationships that must exist among a, b, c, and d in order for the slope of l to be zero.

For Exercises 36 to 43, determine whether the line through P_1 and P_2 is parallel, perpendicular, or neither parallel nor perpendicular to the line through Q_1 and Q_2.

36. $P_1(-3, 4), P_2(2, -5); Q_1(3, 6), Q_2(-2, -3)$

37. $P_1(4, -5), P_2(6, -9); Q_1(5, -4), Q_2(1, 4)$

38. $P_1(0, 1), P_2(2, 4); Q_1(-4, -7), Q_2(2, 5)$

39. $P_1(5, 1), P_2(3, -2); Q_1(0, -2), Q_2(3, -4)$

40. $P_1(-2, 4), P_2(2, 4); Q_1(-3, 6), Q_2(4, 6)$

41. $P_1(1, -1), P_2(3, -2); Q_1(-4, 1), Q_2(2, -5)$

42. $P_1(7, -1), P_2(-4, 6); Q_1(3, 0), Q_2(-5, 3)$

43. $P_1(5, -2), P_2(-1, 3); Q_1(3, 4), Q_2(-2, -2)$

44. Deep-Sea Diving The pressure, in pounds per square inch, on a diver is shown in the graph at the right. Find the slope of the line. Write a sentence that explains the meaning of the slope.

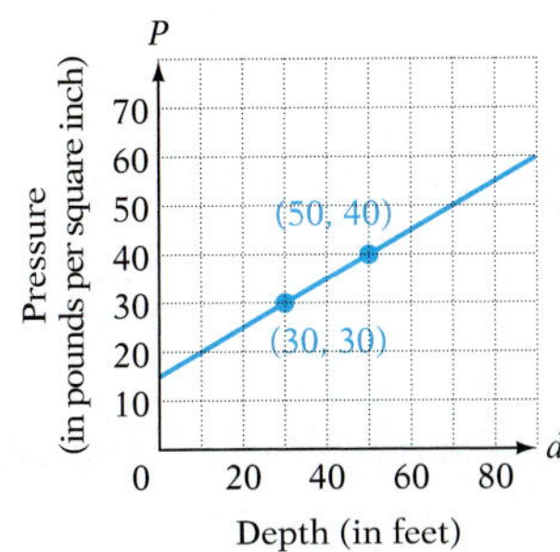

45. Panama Canal Ships in the Panama Canal are lowered through a series of locks. A ship is lowered as the water in a lock is discharged. The graph at the right shows the number of gallons of water N remaining in a lock t minutes after the valves are opened to discharge the water. Find the slope of the line. Write a sentence that explains the meaning of the slope.

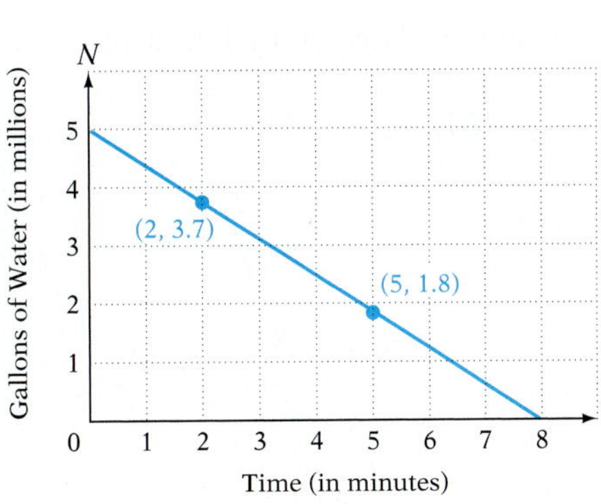

Traffic Safety See the news clipping below. Use the information in the clipping for Exercises 46 and 47.

In the News

Buckling Up Saves Lives

Annual surveys conducted by the National Highway Safety Administration show that Americans' steady increase in seat belt use has been accompanied by a steady decrease in deaths due to motor vehicle accidents.

Source: National Highway Traffic Safety Association

Seat Belt Use

S

100

75

50

25

0

'01

'03

'05

t

Seat Belt Use (in percent)

Year

(2001, 73)

(2005, 82)

Passenger Deaths

D

200

150

100

50

0

'01

'03

'05

t

Deaths per 10 Billion Miles Traveled

Year

(2001, 127)

(2005, 115)

46. Find the slope of the line in the Seat Belt Use graph. Write a sentence that states the meaning of the slope in the context of the article.

47. Find the slope of the line in the Passenger Deaths graph. Write a sentence that states the meaning of the slope in the context of the article.

OBJECTIVE C To graph a line using the slope and the y-intercept

For Exercises 48 to 55, find the slope and y-intercept of the graph of the equation.

48. $y = -\frac{3}{8}x + 5$ **49.** $y = -x + 7$ **50.** $2x - 3y = 6$ **51.** $4x + 3y = 12$

52. $2x + 5y = 10$ **53.** $2x + y = 0$ **54.** $x - 4y = 0$ **55.** $2x + 3y = 8$

For Exercises 56 to 70, graph by using the slope and y-intercept.

56. $y = 3x + 1$

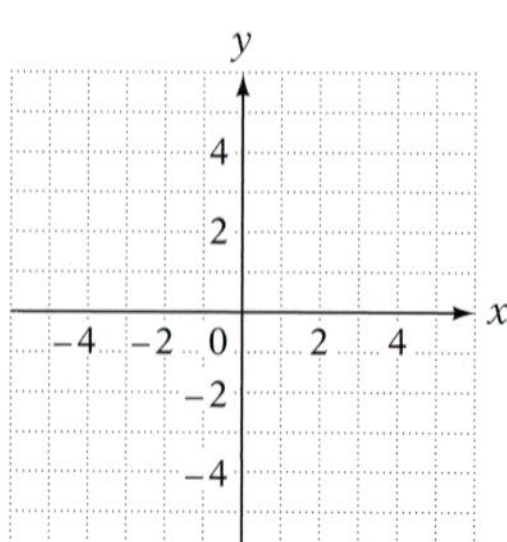

57. $y = -2x - 1$

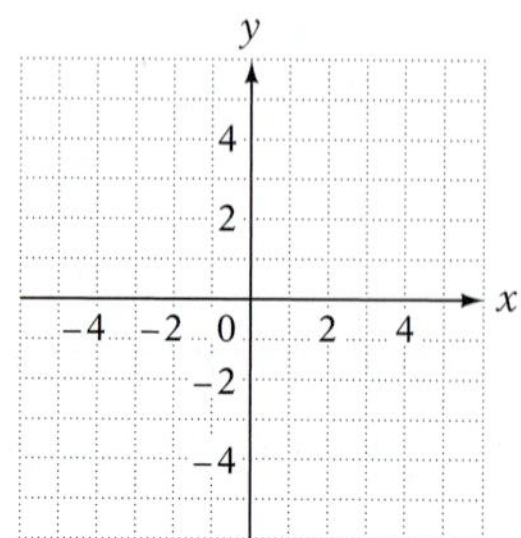

58. $y = \frac{2}{5}x - 2$

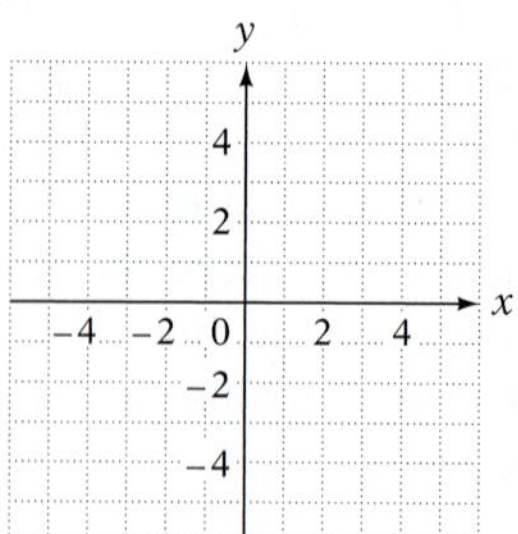

59. $y = \frac{3}{4}x + 1$

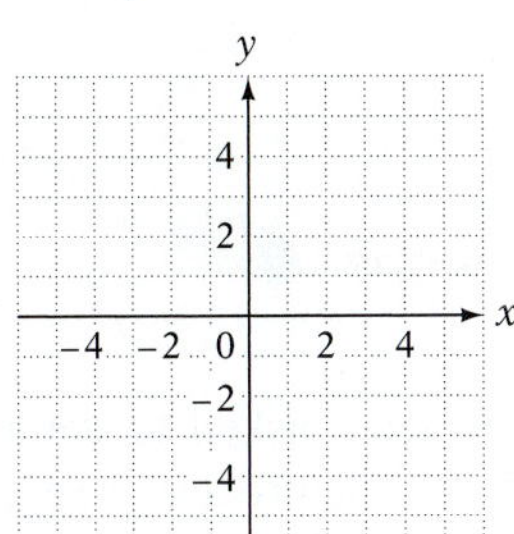

60. $2x + y = 3$

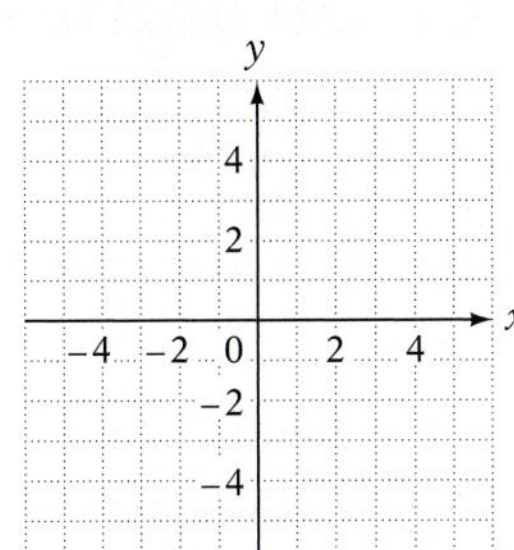

61. $3x - y = 1$

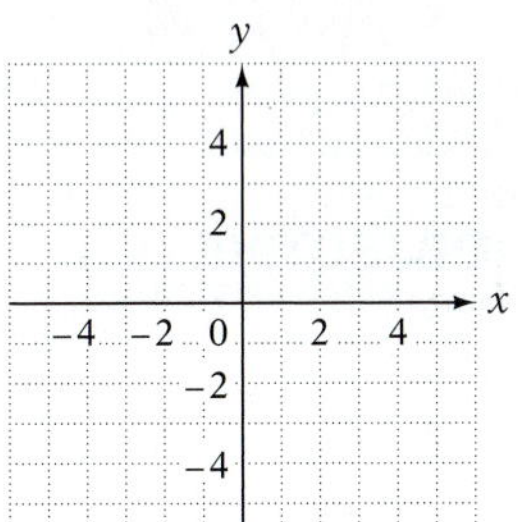

62. $x - 2y = 4$

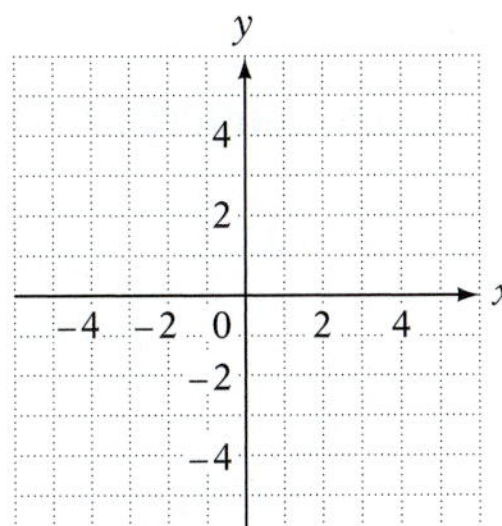

63. $x + 3y = 6$

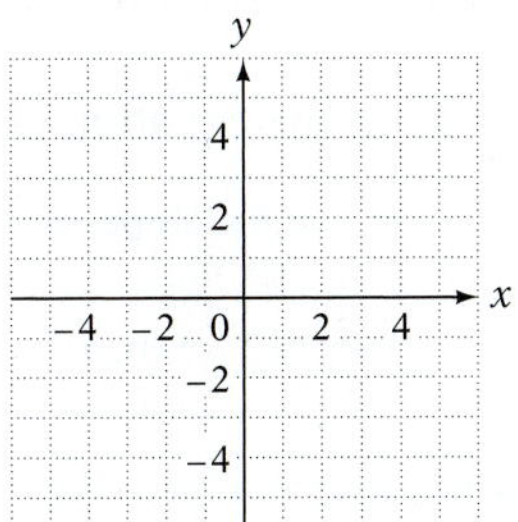

64. $y = \frac{2}{3}x$

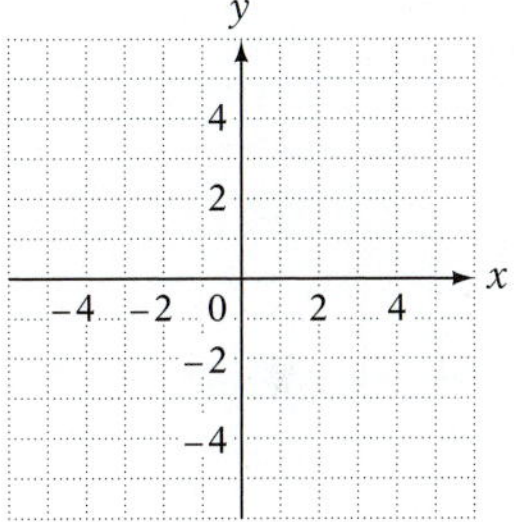

65. $y = \frac{1}{2}x$

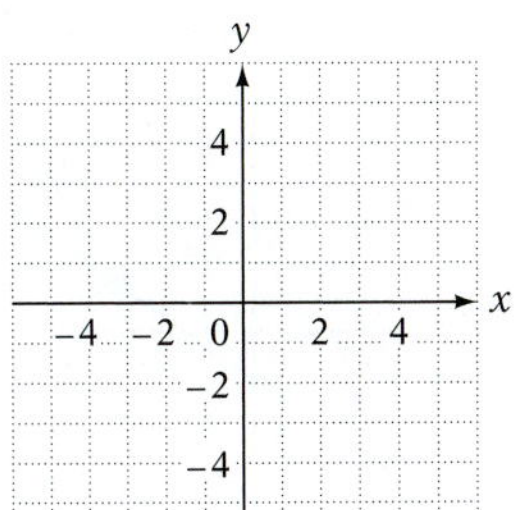

66. $y = -x + 1$

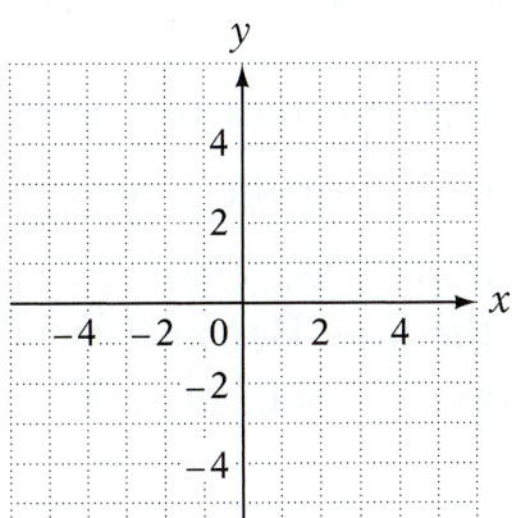

67. $y = -x - 3$

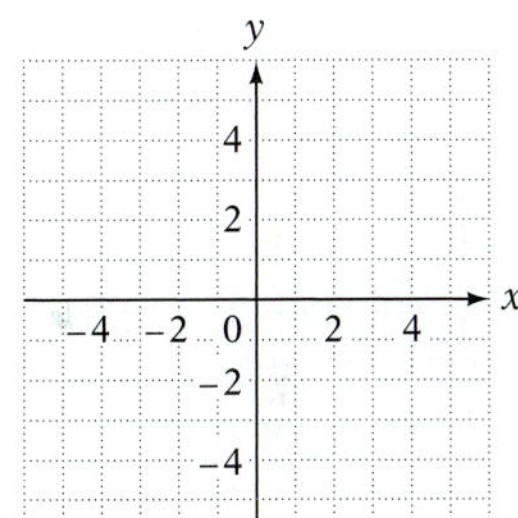

68. $3x - 4y = 12$

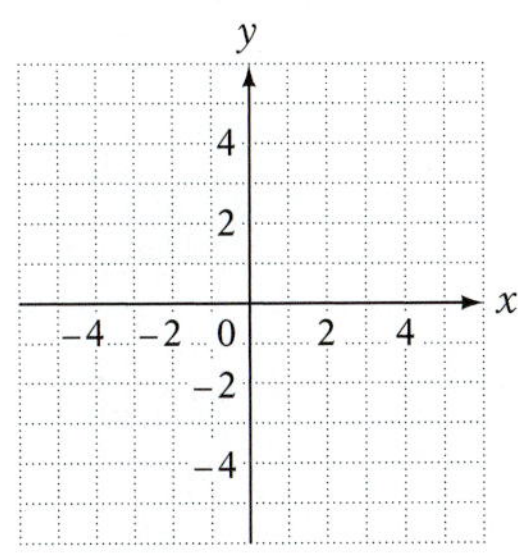

69. $5x - 2y = 10$

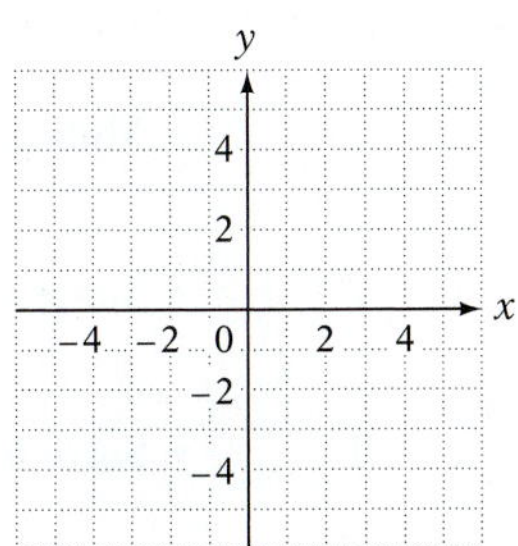

70. $y = -4x + 2$

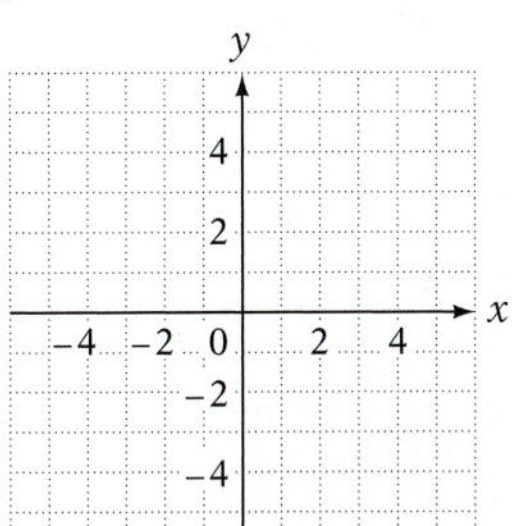

For Exercises 71 and 72, use the given conditions on A, B, and C to determine whether the graph of $Ax + By = C$ slants upward to the right or downward to the right.

71. $A > 0$, $B > 0$, and $C > 0$

72. $A < 0$, $B > 0$, and $C > 0$

Applying the Concepts

73. Do all straight lines have a y-intercept? If not, give an example of a line that does not.

74. If two lines have the same slope and the same y-intercept, must the graphs of the lines be the same? If not, give an example.

SECTION

12.4 Equations of Straight Lines

OBJECTIVE A To find the equation of a line given a point and the slope

In earlier sections, the equation of a line was given and you were asked to determine some properties of the line, such as its intercepts and slope. Here, the process is reversed. Given properties of a line, you will determine its equation.

If the slope and y-intercept of a line are known, the equation of the line can be determined by using the slope-intercept form of a straight line.

HOW TO 1 Find the equation of the line with slope $-\frac{1}{2}$ and y-intercept $(0, 3)$.

$y = mx + b$ • **Use the slope-intercept form.**

$y = -\frac{1}{2}x + 3$ • $m = -\frac{1}{2}$; $(0, b) = (0, 3)$, so $b = 3$.

The equation of the line is $y = -\frac{1}{2}x + 3$.

When the slope and the coordinates of a point other than the y-intercept are known, the equation of the line can be found by using the formula for slope.

Suppose a line passes through the point $(3, 1)$ and has a slope of $\frac{2}{3}$. The equation of the line with these properties is determined by letting (x, y) be the coordinates of an unknown point on the line. Because the slope of the line is known, use the slope formula to write an equation. Then solve for y.

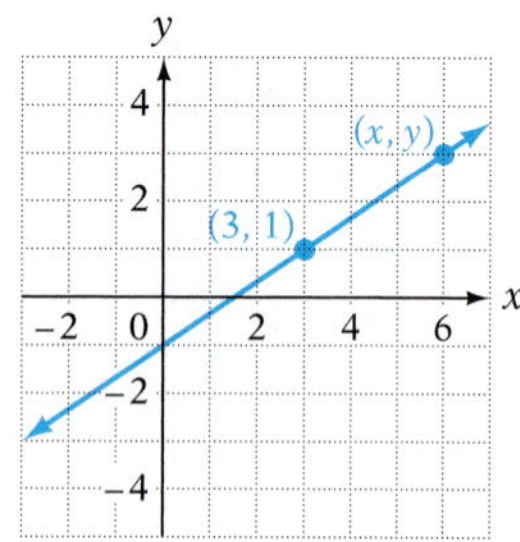

$\frac{y - 1}{x - 3} = \frac{2}{3}$ • $\frac{y_2 - y_1}{x_2 - x_1} = m$; $m = \frac{2}{3}$; $(x_2, y_2) = (x, y)$; $(x_1, y_1) = (3, 1)$

$\frac{y - 1}{x - 3}(x - 3) = \frac{2}{3}(x - 3)$ • **Multiply each side by $(x - 3)$.**

$y - 1 = \frac{2}{3}x - 2$ • **Simplify.**

$y = \frac{2}{3}x - 1$ • **Solve for y.**

The equation of the line is $y = \frac{2}{3}x - 1$.

The same procedure that was used above is used to derive the *point-slope formula.* We use this formula to determine the equation of a line when we are given the coordinates of a point on the line and the slope of the line.

Let (x_1, y_1) be the given coordinates of a point on a line, m the given slope of the line, and (x, y) the coordinates of an unknown point on the line. Then

$\frac{y - y_1}{x - x_1} = m$ • **Formula for slope.**

$\frac{y - y_1}{x - x_1}(x - x_1) = m(x - x_1)$ • **Multiply each side by $x - x_1$.**

$y - y_1 = m(x - x_1)$ • **Simplify.**

Point-Slope Formula

If (x_1, y_1) is a point on a line with slope m, then $y - y_1 = m(x - x_1)$.

HOW TO • 2 Find the equation of the line that passes through the point (2, 3) and has slope −2.

$y - y_1 = m(x - x_1)$ • Use the point-slope formula.

$y - 3 = -2(x - 2)$ • $m = -2$; $(x_1, y_1) = (2, 3)$

$y - 3 = -2x + 4$ • Solve for y.

$y = -2x + 7$

The equation of the line is $y = -2x + 7$.

EXAMPLE • 1

Find the equation of the line that contains the point (0, −1) and has slope $-\frac{2}{3}$.

Solution

Because the slope and y-intercept are known, use the slope-intercept formula, $y = mx + b$.

$y = -\frac{2}{3}x - 1$ • $m = -\frac{2}{3}$; $b = -1$

YOU TRY IT • 1

Find the equation of the line that contains the point (0, 2) and has slope $\frac{5}{3}$.

Your solution

EXAMPLE • 2

Use the point-slope formula to find the equation of the line that passes through the point (−2, −1) and has slope $\frac{3}{2}$.

Solution

$y - y_1 = m(x - x_1)$

$y - (-1) = \frac{3}{2}[x - (-2)]$ • $m = \frac{3}{2}$; $(x_1, y_1) = (-2, -1)$

$y + 1 = \frac{3}{2}(x + 2)$

$y + 1 = \frac{3}{2}x + 3$

$y = \frac{3}{2}x + 2$

YOU TRY IT • 2

Use the point-slope formula to find the equation of the line that passes through the point (4, −2) and has slope $\frac{3}{4}$.

Your solution

Solutions on p. S30

OBJECTIVE B To find the equation of a line given two points

The point-slope formula is used to find the equation of a line when a point on the line and the slope of the line are known. But this formula can also be used **to find the equation of a line given two points on the line.** In this case,

1. **Use the slope formula to determine the slope of the line between the points.**
2. **Use the point-slope formula, the slope you just calculated, and one of the given points to find the equation of the line.**

HOW TO • 3 Find the equation of the line that passes through the points $(-3, -1)$ and $(3, 3)$.

Use the slope formula to determine the slope of the line between the points.

$$m = \frac{y_2 - y_1}{x_2 - x_1} = \frac{3 - (-1)}{3 - (-3)} = \frac{4}{6} = \frac{2}{3} \qquad \bullet\ (x_1, y_1) = (-3, -1); (x_2, y_2) = (3, 3)$$

Use the point-slope formula, the slope you just calculated, and one of the given points to find the equation of the line.

$$y - y_1 = m(x - x_1) \qquad \bullet \text{ Point-slope formula}$$

$$y - (-1) = \frac{2}{3}[x - (-3)] \qquad \bullet\ m = \frac{2}{3}; (x_1, y_1) = (-3, -1)$$

$$y + 1 = \frac{2}{3}(x + 3)$$

$$y + 1 = \frac{2}{3}x + 2$$

$$y = \frac{2}{3}x + 1$$

Take Note
You can verify that the equation $y = \frac{2}{3}x + 1$ passes through the points $(-3, -1)$ and $(3, 3)$ by substituting the coordinates of these points into the equation.

Check:

$y = \frac{2}{3}x + 1$

-1	$\frac{2}{3}(-3) + 1$
-1	$-2 + 1$

$-1 = -1$ • $(x, y) = (-3, -1)$

$y = \frac{2}{3}x + 1$

3	$\frac{2}{3}(3) + 1$
3	$2 + 1$

$3 = 3$ • $(x, y) = (3, 3)$

The equation of the line that passes through the two points is $y = \frac{2}{3}x + 1$.

EXAMPLE • 3

Find the equation of the line that passes through the points $(-4, 0)$ and $(2, -3)$.

Solution

Find the slope of the line between the two points.

$$m = \frac{y_2 - y_1}{x_2 - x_1} = \frac{-3 - 0}{2 - (-4)} = \frac{-3}{6} = -\frac{1}{2}$$

Use the point-slope formula.

$$y - y_1 = m(x - x_1) \qquad \bullet \text{ Point-slope formula}$$

$$y - 0 = -\frac{1}{2}[x - (-4)] \qquad \bullet\ m = -\frac{1}{2}; (x_1, y_1) = (-4, 0)$$

$$y = -\frac{1}{2}(x + 4)$$

$$y = -\frac{1}{2}x - 2$$

The equation of the line is $y = -\frac{1}{2}x - 2$.

YOU TRY IT • 3

Find the equation of the line that passes through the points $(-6, -2)$ and $(3, 1)$.

Your solution

Solution on p. S30

OBJECTIVE C To solve application problems

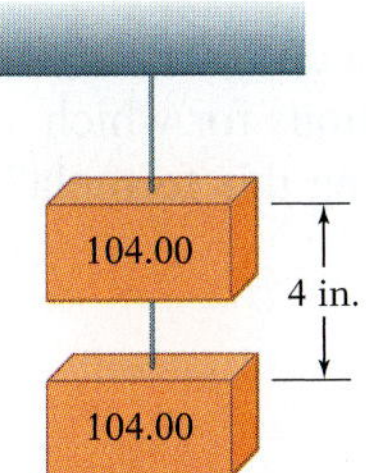

A **linear model** is a first-degree equation that is used to describe a relationship between quantities. In many cases, a linear model is used to approximate collected data. The data are graphed as points in a coordinate system, and then a line is drawn that approximates the data. The graph of the points is called a **scatter diagram;** the line is called the **line of best fit.**

Consider an experiment to determine the weight required to stretch a spring a certain distance. Data from such an experiment are shown in the table below.

Distance (in inches)	2.5	4	2	3.5	1	4.5
Weight (in pounds)	63	104	47	85	27	115

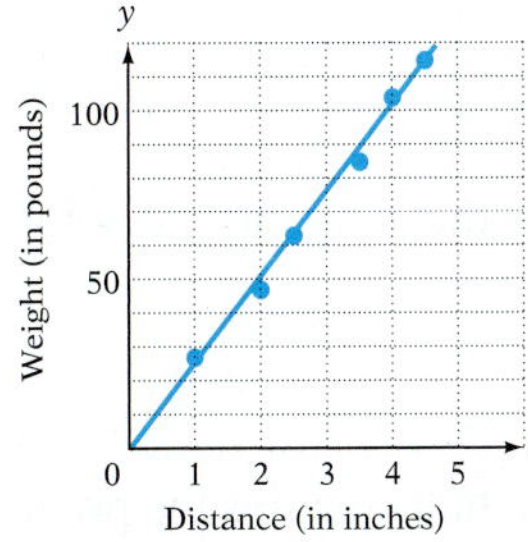

The accompanying graph shows the scatter diagram, which is the plotted points, and the line of best fit, which is the line that approximately goes through the plotted points. The equation of the line of best fit is $y = 25.6x - 1.3$, where x is the number of inches the spring is stretched and y is the weight in pounds.

The table below shows the values that the model would predict to the nearest tenth. Good linear models should predict values that are close to the actual values. A more thorough analysis of lines of best fit is undertaken in statistics courses.

Distance, x	2.5	4	2	3.5	1	4.5
Weight predicted using $y = 25.6x - 1.3$	62.7	101.1	49.9	88.3	24.3	113.9

EXAMPLE 4

The data in the table below show the growth in defense spending by the U.S. government. (*Source:* Office of Management and Budget) The line of best fit is $y = 49x + 220.3$, where x is the year (with 2005 corresponding to $x = 5$) and y is the defense spending in billions of dollars.

Year	5	6	7	8
Defense Spending (in billions of dollars)	475	490	530	625

Graph the data and the line of best fit in the coordinate system below. Write a sentence that describes the meaning of the slope of the line.

Solution

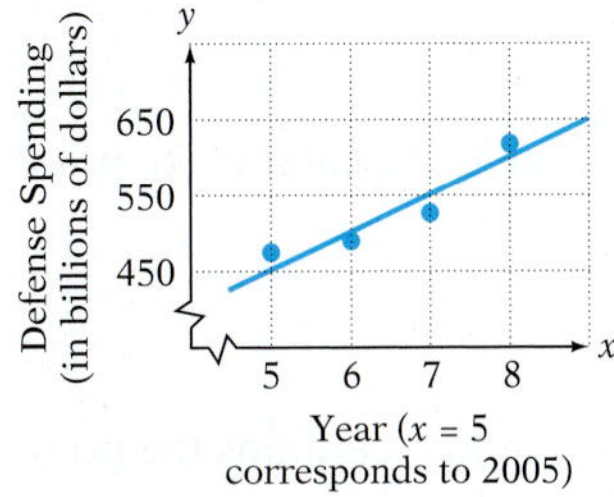

The slope of the line means that the amount spent on defense increased by $49 billion per year.

YOU TRY IT 4

The data in the table below show a reading test grade and the final exam grade in a history class. The line of best fit is $y = 8.3x - 7.8$, where x is the reading test score and y is the history test score.

Reading	8.5	9.4	10.0	11.4	12.0
History	64	68	76	87	92

Graph the data and the line of best fit in the coordinate system below. Write a sentence that describes the meaning of the slope of the line of best fit.

Your solution

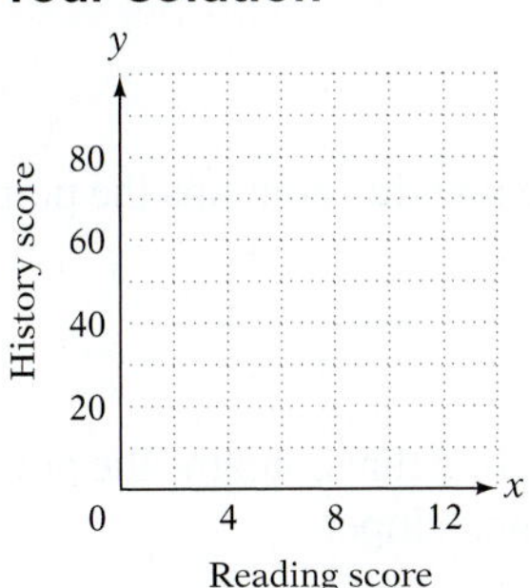

Solution on p. S30

12.4 EXERCISES

OBJECTIVE A To find the equation of a line given a point and the slope

1. What is the point-slope formula and how is it used?

2. Can the point-slope formula be used to find the equation of any line? If not, equations for which types of lines cannot be found using this formula?

For Exercises 3 to 6, sketch the line described in the indicated exercise. Use your graph to determine whether the *b*-value of the equation of the line is positive or negative.

3. Exercise 8

4. Exercise 10

5. Exercise 12

6. Exercise 14

7. Find the equation of the line that contains the point $(0, 2)$ and has slope 2.

8. Find the equation of the line that contains the point $(0, -1)$ and has slope -2.

9. Find the equation of the line that contains the point $(-1, 2)$ and has slope -3.

10. Find the equation of the line that contains the point $(2, -3)$ and has slope 3.

11. Find the equation of the line that contains the point $(3, 1)$ and has slope $\frac{1}{3}$.

12. Find the equation of the line that contains the point $(-2, 3)$ and has slope $\frac{1}{2}$.

13. Find the equation of the line that contains the point $(4, -2)$ and has slope $\frac{3}{4}$.

14. Find the equation of the line that contains the point $(2, 3)$ and has slope $-\frac{1}{2}$.

15. Find the equation of the line that contains the point $(5, -3)$ and has slope $-\frac{3}{5}$.

16. Find the equation of the line that contains the point $(5, -1)$ and has slope $\frac{1}{5}$.

17. Find the equation of the line that contains the point $(2, 3)$ and has slope $\frac{1}{4}$.

18. Find the equation of the line that contains the point $(-1, 2)$ and has slope $-\frac{1}{2}$.

19. Find the equation of the line that contains the point $(2, -2)$ and has slope 0.

20. Find the equation of the line that contains the point $(-4, 5)$ and has slope 0.

21. Find the equation of the line that contains the point $(-3, 1)$ and has undefined slope.

22. Find the equation of the line that contains the point $(6, -8)$ and has undefined slope.

23. Use the point-slope formula to write the equation of the line with slope m and y-intercept $(0, b)$. Does your answer simplify to the slope-intercept form of a straight line with slope m and y-intercept $(0, b)$?

24. Use the point-slope formula to write the equation of the line that goes through the point $(0, b)$ and has slope 0. Does your answer simplify to the equation of a horizontal line through $(0, b)$?

OBJECTIVE B To find the equation of a line given two points

For Exercises 25 to 28, sketch the line described in the indicated exercise. Use your graph to determine whether the m-value of the equation of the line is positive or negative.

25. Exercise 31

26. Exercise 32

27. Exercise 35

28. Exercise 36

29. Find the equation of the line that passes through the points $(1, -1)$ and $(-2, -7)$.

30. Find the equation of the line that passes through the points $(2, 3)$ and $(3, 2)$.

31. Find the equation of the line that passes through the points $(-2, 1)$ and $(1, -5)$.

32. Find the equation of the line that passes through the points $(-1, -3)$ and $(2, -12)$.

33. Find the equation of the line that passes through the points $(0, 0)$ and $(-3, -2)$.

34. Find the equation of the line that passes through the points $(0, 0)$ and $(-5, 1)$.

35. Find the equation of the line that passes through the points $(2, 3)$ and $(-4, 0)$.

36. Find the equation of the line that passes through the points $(3, -1)$ and $(0, -3)$.

37. Find the equation of the line that passes through the points $(-4, 1)$ and $(4, -5)$.

38. Find the equation of the line that passes through the points $(-5, 0)$ and $(10, -3)$.

39. Find the equation of the line that passes through the points $(-2, 1)$ and $(2, 4)$.

40. Find the equation of the line that passes through the points $(3, -2)$ and $(-3, -3)$.

41. Find the equation of the line that passes through the points $(4, -3)$ and $(-1, -3)$.

42. Find the equation of the line that passes through the points $(-1, 4)$ and $(2, 4)$.

43. Find the equation of the line that passes through the points $(-2, 6)$ and $(-2, -7)$.

44. Find the equation of the line that passes through the points $(5, -1)$ and $(5, 3)$.

45. If (x_1, y_1) and (x_2, y_2) are the coordinates of two points on the graph of $y = 2x - 3$, what is the value of $\frac{y_2 - y_1}{x_2 - x_1}$?

OBJECTIVE C To solve application problems

46. Refer to Example 4 on page 663. Use the points for Year 5 and Year 6. Is the slope of the line between these two points greater than or less than the slope of the line of best fit?

47. Sports The data in the table below show the number of carbohydrates used for various amounts of time during a strenuous tennis workout. The line of best fit is $y = 1.55x + 1.45$, where x is the time of the workout in minutes and y is the number of carbohydrates used in grams.

Time of workout, x (in minutes)	5	10	20	30	60
Carbohydrates used, y (in grams)	10	15	33	49	94

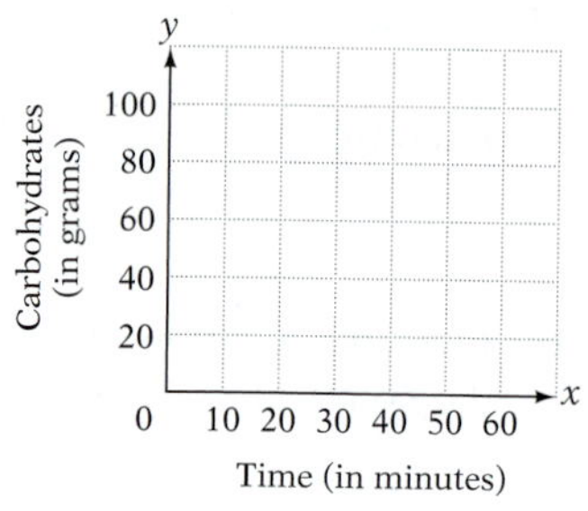

Graph the data and the line of best fit in the coordinate system at the right. Write a sentence that describes the meaning of the slope of the line of best fit in the context of this problem.

48. Sports The data in the table below show the amount of water a professional tennis player loses for various times during a tennis match. The line of best fit is $y = 34.6x + 207$, where x is the time of the workout in minutes and y is the milliliters of water lost during the match.

Time of workout, x (in minutes)	10	20	30	40	50	60
Water lost, y (in milliliters)	600	900	1200	1500	2000	2300

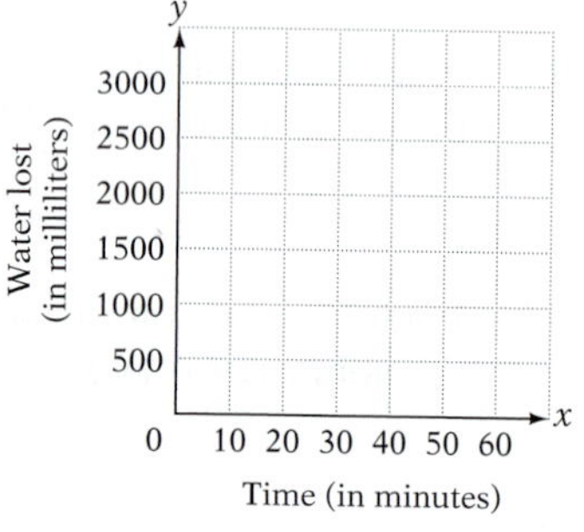

Graph the data and the line of best fit in the coordinate system at the right. Write a sentence that describes the meaning of the slope of the line of best fit in the context of this problem.

49. Evaporation The data in the table below show the amount of water that evaporates from swimming pools of various surface areas. The line of best fit is $y = 0.17x - 1$, where x is the surface area of the swimming pool in square feet and y is the number of gallons of water that evaporate in one day.

Surface area, x (in square feet)	100	200	300	400	600	1000
Water evaporated, y (in gallons)	25	30	45	60	100	170

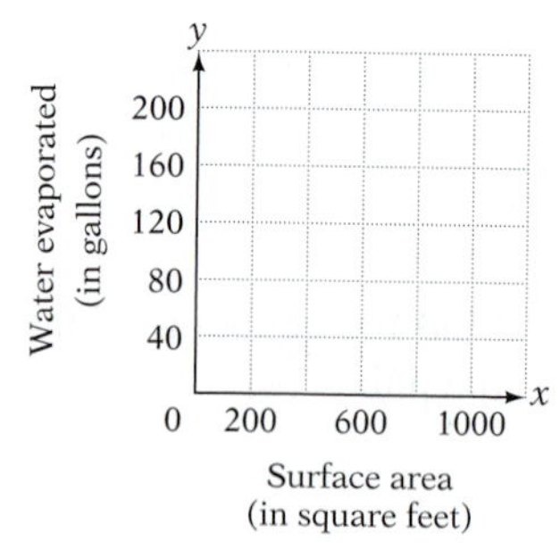

Graph the data and the line of best fit in the coordinate system at the right. Write a sentence that describes the meaning of the slope of the line of best fit in the context of this problem.

50. Alternative Energy Read the following news clipping.

In the News

GWEC Issues Annual Global Wind Report

In its recently released Global Wind Report, the Global Wind Energy Council predicts continued worldwide growth of new installations of wind turbines. The Council's predictions for the energy-producing capacity, in gigawatts, of new installations for the years 2007 to 2012 are shown in the table.

Year	1	2	3	4	5	6
Capacity (in gW)	19.9	23.1	26.0	28.9	32.3	36.1

Source: Global Wind Energy Council, Global Wind 2007 Report

The line of best fit for the data in the article is $y = 3.19x + 16.57$, where x is the year (with $x = 0$ corresponding to 2006) and y is the energy producing capacity, in gigawatts (gW), of the new installations. Graph the data and the line of best fit in the coordinate system at the right. Write a sentence that describes the meaning of the slope of the line of best fit in the context of this problem.

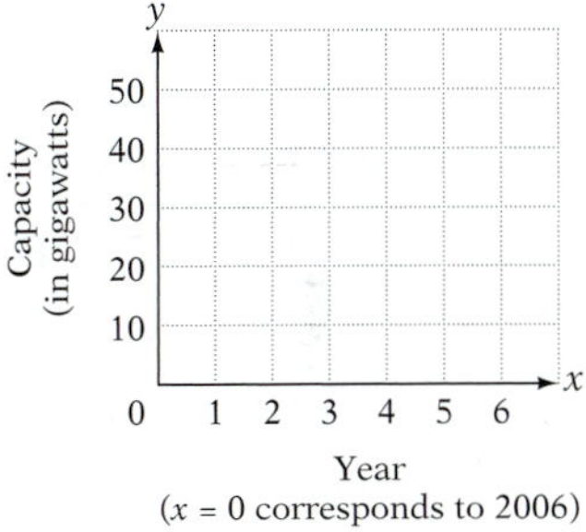

Applying the Concepts

51. For the equation $y = 3x + 2$, when the value of x changes from 1 to 2, does the value of y increase or decrease? What is the change in y? Suppose that the value of x changes from 13 to 14. What is the change in y?

52. For the equation $y = -2x + 1$, when the value of x changes from 1 to 2, does the value of y increase or decrease? What is the change in y? Suppose that the value of x changes from 13 to 14. What is the change in y?

In Exercises 53 to 56, the first two given points are on a line. Determine whether the third point is on the line.

53. $(-3, 2), (4, 1); (-1, 0)$

54. $(2, -2), (3, 4); (-1, 5)$

55. $(-3, -5), (1, 3); (4, 9)$

56. $(-3, 7), (0, -2); (1, -5)$

57. If $(-2, 4)$ are the coordinates of a point on the line whose equation is $y = mx + 1$, what is the slope of the line?

58. If $(3, 1)$ are the coordinates of a point on the line whose equation is $y = mx - 3$, what is the slope of the line?

59. If $(0, -3)$, $(6, -7)$, and $(3, n)$ are coordinates of points on the same line, determine n.

60. If $(-4, 11)$, $(2, -4)$, and $(6, n)$ are coordinates of points on the same line, determine n.

FOCUS ON PROBLEM SOLVING

Counterexamples

Some of the exercises in this text ask you to determine whether a statement is true or false. For instance, the statement "Every real number has a reciprocal" is false because 0 is a real number and 0 does not have a reciprocal.

Finding an example, such as "0 has no reciprocal," to show that a statement is not always true is called finding a counterexample. A **counterexample** is an example that shows that a statement is not always true.

Here are some counterexamples to the statement "The square of a number is always larger than the number."

$$\left(\frac{1}{2}\right)^2 = \frac{1}{4} \quad \text{but} \quad \frac{1}{4} < \frac{1}{2} \qquad 1^2 = 1 \quad \text{but} \quad 1 = 1$$

For Exercises 1 to 7, answer true if the statement is always true. If there is an instance when the statement is false, give a counterexample.

1. The product of two integers is always a positive number.
2. The sum of two prime numbers is never a prime number.
3. For all real numbers, $|x + y| = |x| + |y|$.
4. If x and y are nonzero real numbers and $x > y$, then $x^2 > y^2$.
5. The quotient of any two nonzero real numbers is less than either one of the numbers.
6. The reciprocal of a positive number is always smaller than the number.
7. If $x < 0$, then $|x| = -x$.

PROJECTS AND GROUP ACTIVITIES

Graphing Linear Equations with a Graphing Utility

A computer or graphing calculator screen is divided into *pixels*. There are approximately 6000 to 790,000 pixels available on the screen (depending on the computer or calculator). The greater the number of pixels, the smoother a graph will appear. A portion of a screen is shown at the left. Each little rectangle represents one pixel.

The graphing utilities that are used by computers or calculators to graph an equation do basically what we have shown in the text: They choose values of x and, for each, calculate the corresponding value of y. The pixel corresponding to the ordered pair is then turned on. The graph is jagged because pixels are much larger than the dots we draw on paper.

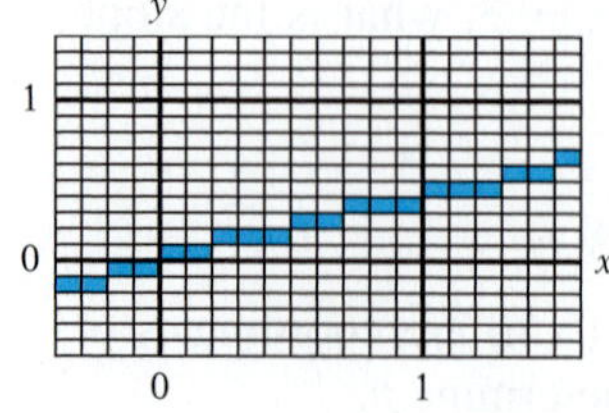

The graph of $y = 0.45x$ is shown at the left as the calculator drew it (jagged). The x- and y-axes have been chosen so that each pixel represents $\frac{1}{10}$ of a unit. Consider the region of the graph where $x = 1$, 1.1, and 1.2.

The corresponding values of y are 0.45, 0.495, and 0.54. Because the y-axis is in tenths, the numbers 0.45, 0.495, and 0.54 are rounded to the nearest tenth before plotting. Rounding 0.45, 0.495, and 0.54 to the nearest tenth results in 0.5 for each number. Thus the ordered pairs (1, 0.45), (1.1, 0.495), and (1.2, 0.54) are graphed as (1, 0.5), (1.1, 0.5), and (1.2, 0.5). These points appear as three illuminated horizontal pixels. However, if you use the TRACE feature of the calculator (see the Appendix), the actual y-coordinate for each value of x is displayed.

Take Note

Xmin and Xmax are the smallest and largest values of x that will be shown on the screen. Ymin and Ymax are the smallest and largest values of y that will be shown on the screen.

Here are the keystrokes to graph $y = \frac{2}{3}x + 1$ on a TI-84 calculator. First the equation is entered. Then the domain (Xmin to Xmax) and the range (Ymin to Ymax) are entered. This is called the **viewing window.**

10 ENTER 1 ENTER GRAPH

Integrating Technology

See the Keystroke Guide:  and WINDOW for assistance.

By changing the keystrokes 2 X,T,θ,n ÷ 3 + 1, you can graph different equations.

For Exercises 1 to 4, graph on a graphing calculator.

1. $y = 2x + 1$ **2.** $y = -\frac{1}{2}x - 2$ **3.** $3x + 2y = 6$ **4.** $4x + 3y = 75$

CHAPTER 12

SUMMARY

KEY WORDS

EXAMPLES

A *rectangular coordinate system* is formed by two number lines, one horizontal and one vertical, that intersect at the zero point of each line. The number lines that make up a rectangular coordinate system are called the *coordinate axes,* or simply *axes.* The *origin* is the point of intersection of the two coordinate axes. Generally, the horizontal axis is labeled the x-axis and the vertical axis is labeled the y-axis. The coordinate system divides the plane into four regions called *quadrants.* The *coordinates of a point* in the plane are given by an *ordered pair* (x, y). The first number in the ordered pair is called the *abscissa* or x*-coordinate.* The second number in the ordered pair is the *ordinate* or y*-coordinate.* The *graph of an ordered pair* (x, y) is the dot drawn at the coordinates of the point in the plane. [12.1A, p. 628]

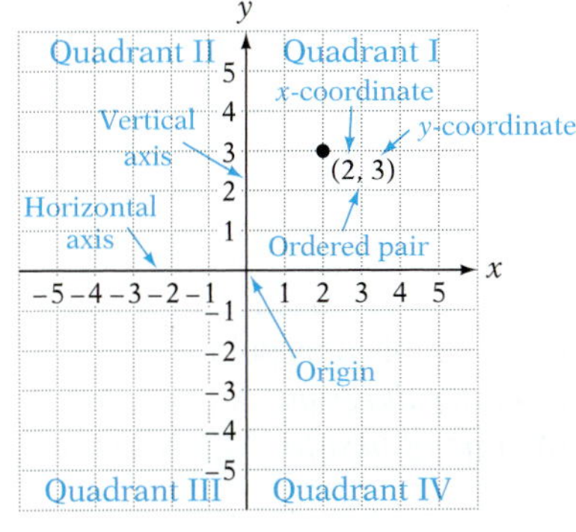

A *solution of an equation in two variables* is an ordered pair (x, y) that makes the equation a true statement. [12.1B, p. 630]

The ordered pair $(-1, 1)$ is a solution of the equation $y = 2x + 3$ because when -1 is substituted for x and 1 is substituted for y, the result is a true equation.

A *relation* is any set of ordered pairs. The *domain* of a relation is the set of first coordinates of the ordered pairs. The *range* is the set of second coordinates of the ordered pairs. [12.1C, p. 632]

For the relation $\{(-1, 2), (2, 4), (3, 5), (3, 7)\}$, the domain is $\{-1, 2, 3\}$; the range is $\{2, 4, 5, 7\}$.

A *function* is a relation in which no two ordered pairs have the same first coordinate. [12.1C, p. 632]

The relation {(−2, −3), (0, 4), (1, 5)} is a function. No two ordered pairs have the same first coordinate.

The *graph of an equation in two variables* is a graph of the ordered-pair solutions of the equation. An equation of the form $y = mx + b$ is a *linear equation in two variables.* [12.2A, p. 640]

$y = 2x + 3$ is a linear equation in two variables. Its graph is shown at the right.

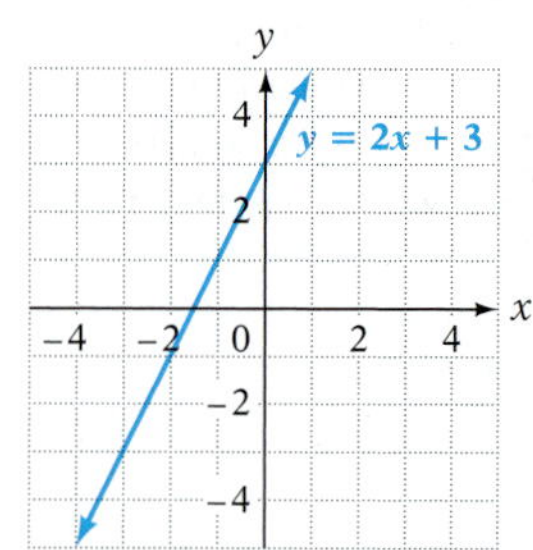

An equation written in the form $Ax + By = C$ is the *standard form of a linear equation in two variables.* [12.2B, p. 642]

$2x + 7y = 10$ is an example of a linear equation in two variables written in standard form.

The point at which a graph crosses the x-axis is called the *x-intercept.* At the x-intercept, the y-coordinate is 0. The point at which a graph crosses the y-axis is called the *y-intercept.* At the y-intercept, the x-coordinate is 0. [12.3A, p. 650]

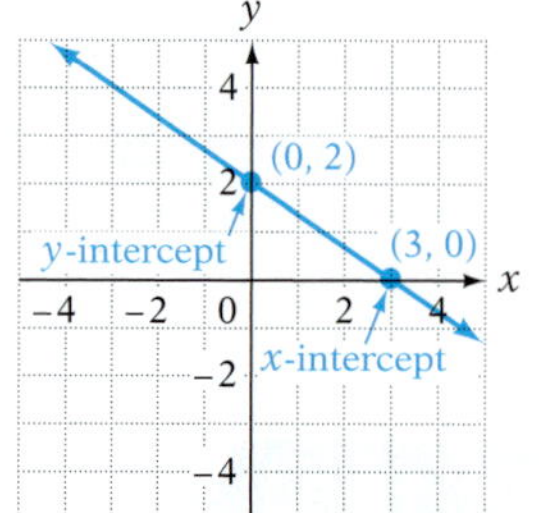

The *slope* of a line is a measure of the slant of the line. The symbol for slope is m. A line with *positive slope* slants upward to the right. A line with *negative slope* slants downward to the right. A horizontal line has *zero slope.* A vertical line has an *undefined slope.* [12.3B, pp. 651–652]

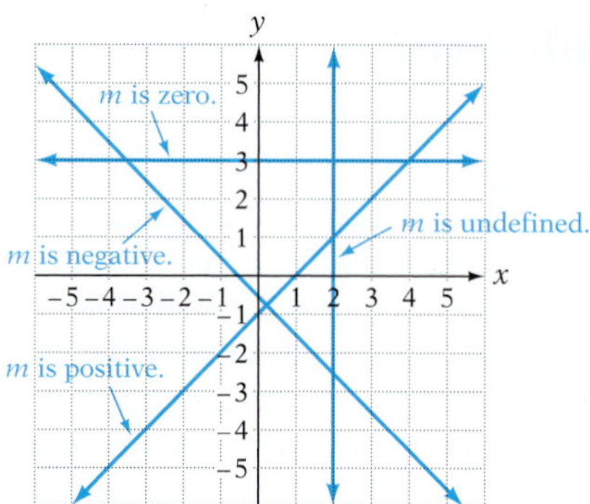

When data are graphed as points in a coordinate system, the graph is called a *scatter diagram.* A line drawn to approximate the data is called the *line of best fit.* [12.4C, p. 663]

The graph shown at the right is the scatter diagram and line of best fit for the spring data on page 663.

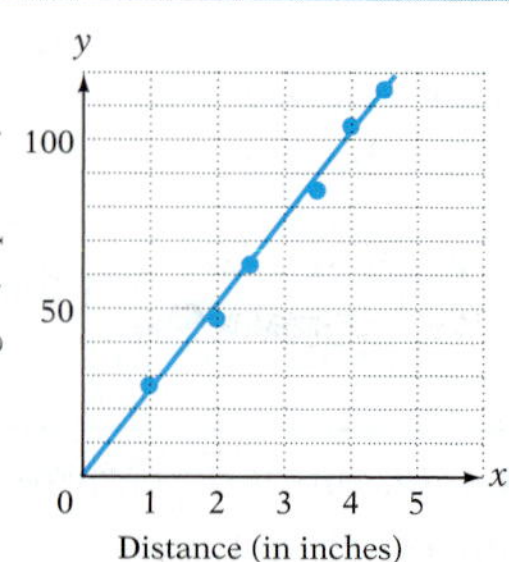

ESSENTIAL RULES AND PROCEDURES

EXAMPLES

Function Notation [12.1D, p. 635]
The equation of a function is written in function notation when y is replaced by the symbol $f(x)$, where $f(x)$ is read "f of x" or "the value of f at x." To evaluate a function at a given value of x, replace x by the given value and then simplify the resulting numerical expression to find the value of $f(x)$.

$y = x^2 + 2x - 1$ is written in function notation as $f(x) = x^2 + 2x - 1$. To evaluate $f(x) = x^2 + 2x - 1$ at $x = -3$, find $f(-3)$.

$$f(-3) = (-3)^2 + 2(-3) - 1$$
$$= 9 - 6 - 1 = 2$$

Horizontal and Vertical Lines [12.2B, p. 643]
The graph of $y = b$ is a horizontal line passing through $(0, b)$. The graph of $x = a$ is a vertical line passing through $(a, 0)$.

The graph of $y = -2$ is a horizontal line passing through $(0, -2)$. The graph of $x = 3$ is a vertical line passing through $(3, 0)$.

To find the *x*-intercept, let $y = 0$ and solve for x.
To find the *y*-intercept, let $x = 0$ and solve for y.
[12.3A, p. 650]

To find the x-intercept of $4x - 5y = 20$, let $y = 0$ and solve for x. To find the y-intercept, let $x = 0$ and solve for y.

$$\begin{aligned} 4x - 5y &= 20 \\ 4x - 5(0) &= 20 \\ 4x &= 20 \\ x &= 5 \end{aligned} \qquad \begin{aligned} 4x - 5y &= 20 \\ 4(0) - 5y &= 20 \\ -5y &= 20 \\ y &= -4 \end{aligned}$$

The x-intercept is $(5, 0)$. The y-intercept is $(0, -4)$.

Slope Formula [12.3B, p. 651]
If $P_1(x_1, y_1)$ and $P_2(x_2, y_2)$ are two points on a line and $x_1 \neq x_2$, then

$$m = \frac{y_2 - y_1}{x_2 - x_1}$$

To find the slope of the line between the points $(1, -2)$ and $(-3, -1)$, let $P_1 = (1, -2)$ and $P_2 = (-3, -1)$. Then $m = \frac{y_2 - y_1}{x_2 - x_1} = \frac{-1 - (-2)}{-3 - 1} = \frac{1}{-4} = -\frac{1}{4}$.

Parallel Lines [12.3B, p. 652]
Two nonvertical lines in the plane are parallel if and only if they have the same slope. Vertical lines in the plane are parallel.

The slope of the line through $P_1(3, -6)$ and $P_2(5, -10)$ is $m_1 = \frac{-10 - (-6)}{5 - 3} = -2$.

The slope of the line through $Q_1(4, -5)$ and $Q_2(0, 3)$ is $m_2 = \frac{3 - (-5)}{0 - 4} = -2$.

Because $m_1 = m_2$, the lines are parallel.

Perpendicular Lines [12.3B, p. 653]
Two nonvertical lines in the plane are perpendicular if and only if the product of their slopes is -1. A vertical and a horizontal line are perpendicular.

The slope of the line through $P_1(5, -3)$ and $P_2(2, -1)$ is $m_1 = \frac{-1 - (-3)}{2 - 5} = -\frac{2}{3}$.

The slope of the line through $Q_1(1, -4)$ and $Q_2(3, -1)$ is $m_2 = \frac{-1 - (-4)}{3 - 1} = \frac{3}{2}$.

Because $m_1 m_2 = \left(-\frac{2}{3}\right)\left(\frac{3}{2}\right) = -1$, the lines are perpendicular.

Slope-Intercept Form of a Linear Equation [12.3C, p. 654]
An equation of the form $y = mx + b$ is called the slope-intercept form of a straight line. The slope of the line is m, the coefficient of x. The y-intercept is $(0, b)$, where b is the constant term of the equation.

For the line with equation $y = -3x + 2$, the slope is -3 and the y-intercept is $(0, 2)$.

Point-Slope Formula [12.4A, p. 661]
If (x_1, y_1) is a point on a line with slope m, then

$$y - y_1 = m(x - x_1)$$

The equation of the line that passes through the point $(5, -3)$ and has slope -2 is:

$$\begin{aligned} y - y_1 &= m(x - x_1) \\ y - (-3) &= -2(x - 5) \\ y + 3 &= -2x + 10 \\ y &= -2x + 7 \end{aligned}$$

CHAPTER 12

CONCEPT REVIEW

Test your knowledge of the concepts presented in this chapter. Answer each question. Then check your answers against the ones provided in the Answer Section.

1. How is the ordinate different from the abscissa?

2. How many ordered-pair solutions are there for a linear equation in two variables?

3. When is a relation a function?

4. What is the difference between an independent variable and a dependent variable?

5. In the general equation $y = mx + b$, what do m and b represent?

6. How many ordered-pair solutions of a linear function should be found to ensure the accuracy of a graph?

7. How is the equation of a vertical line different from the equation of a horizontal line?

8. How are the ordered pairs different for an x-intercept and a y-intercept?

9. What does it mean for a line to have an undefined slope?

10. Given two ordered pairs on a line, how do you find the slope of the line?

11. What is the difference between parallel and perpendicular lines?

12. What is the point-slope formula?

CHAPTER 12

REVIEW EXERCISES

1. **a.** Graph the ordered pairs $(-2, 4)$ and $(3, -2)$.
   b. Name the abscissa of point A.
   c. Name the ordinate of point B.

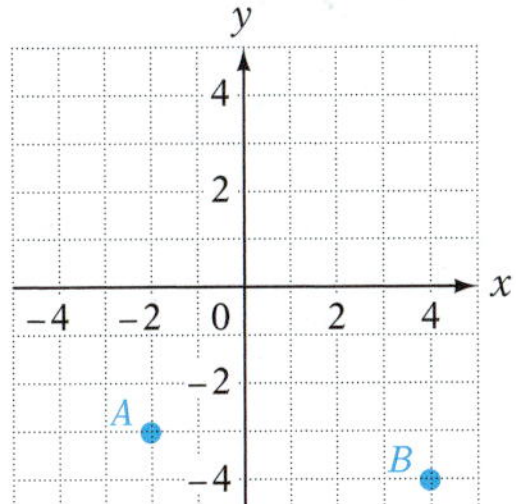

2. Graph the ordered-pair solutions of $y = -\frac{1}{2}x - 2$ when $x \in \{-4, -2, 0, 2\}$.

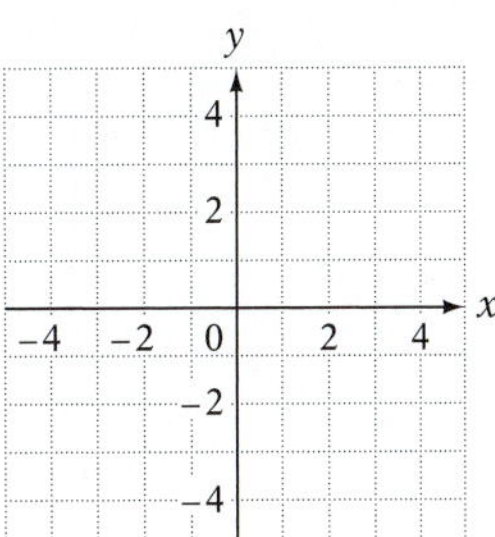

3. Determine the equation of the line that passes through the points $(-1, 3)$ and $(2, -5)$.

4. Determine the equation of the line that passes through the point $(6, 1)$ and has slope $-\frac{5}{2}$.

5. Graph $y = \frac{1}{4}x + 3$.

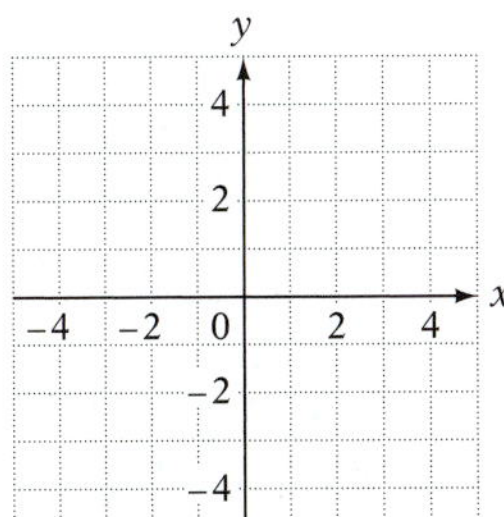

6. Graph $5x + 3y = 15$.

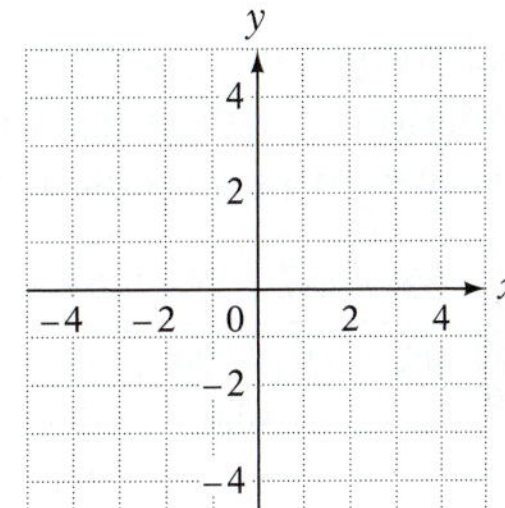

7. Is the line that passes through $(7, -5)$ and $(6, -1)$ parallel, perpendicular, or neither parallel nor perpendicular to the line that passes through $(4, 5)$ and $(2, -3)$?

8. Given $f(x) = x^2 - 2$, find $f(-1)$.

9. Does $y = -x + 3$, where $x \in \{-2, 0, 3, 5\}$, define y as a function of x?

10. Find the slope of the line containing the points $(9, 8)$ and $(-2, 1)$.

11. Find the x- and y-intercepts of $3x - 2y = 24$.

12. Find the slope of the line containing the points $(-2, -3)$ and $(4, -3)$.

13. Graph the line that has slope $\frac{1}{2}$ and y-intercept $(0, -1)$.

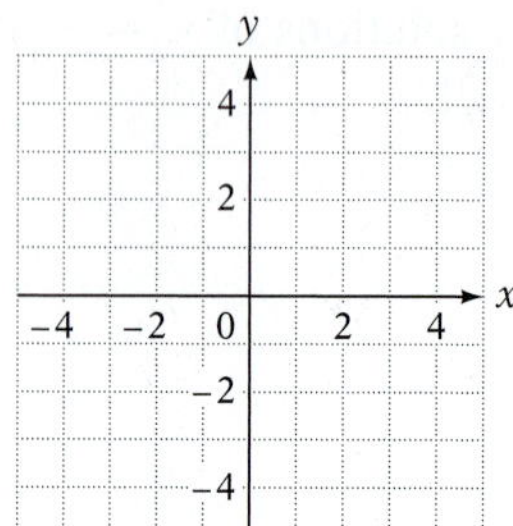

14. Graph $x = -3$.

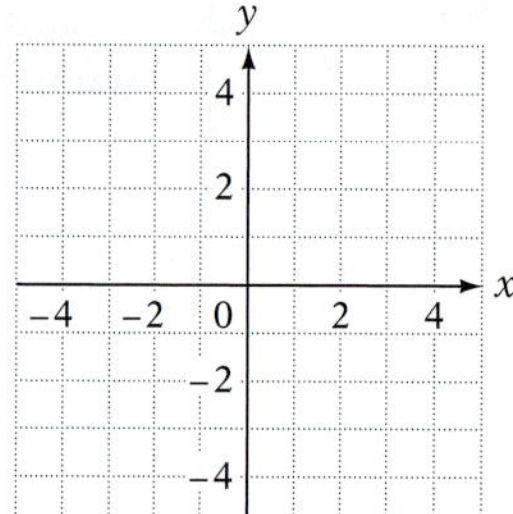

15. Graph the line that has slope $-\frac{2}{3}$ and y-intercept $(0, 2)$.

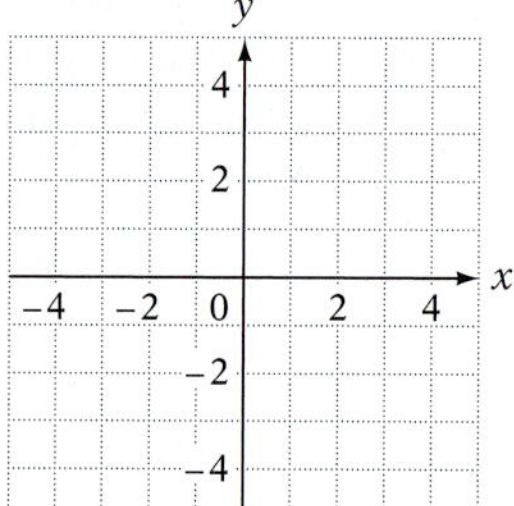

16. Graph $y = -2x - 1$.

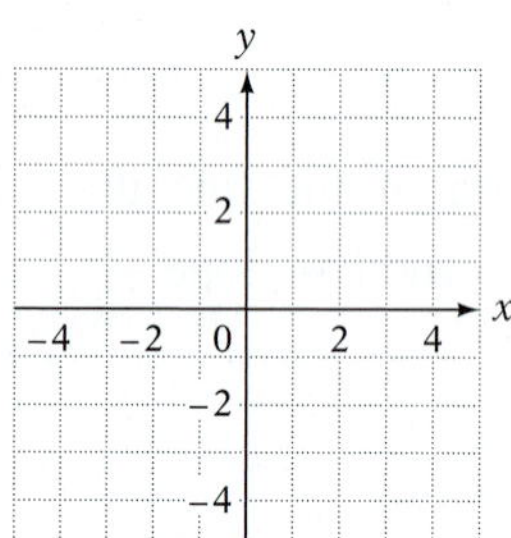

17. Graph the line that has slope 2 and y-intercept $(0, -4)$.

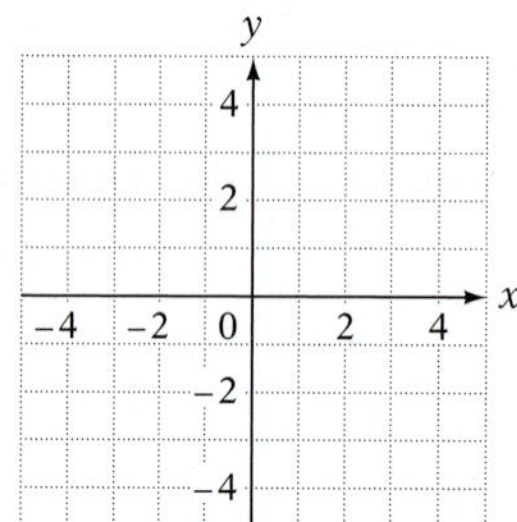

18. Graph $3x - 2y = -6$.

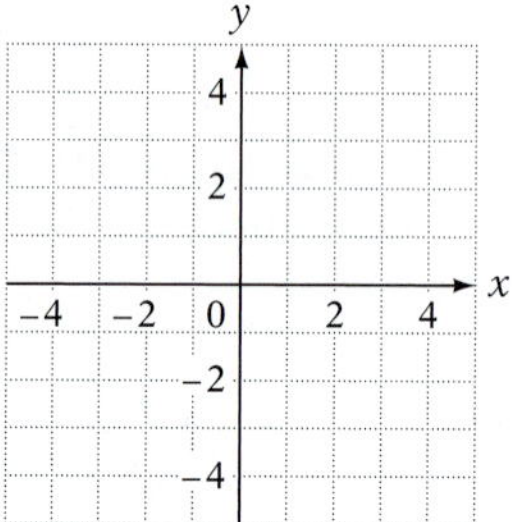

19. **Health** The height and weight of 8 seventh-grade students are shown in the following table. Write a relation in which the first coordinate is height in inches and the second coordinate is weight in pounds. Is the relation a function?

Height (in inches)	55	57	53	57	60	61	58	54
Weight (in pounds)	95	101	94	98	100	105	97	95

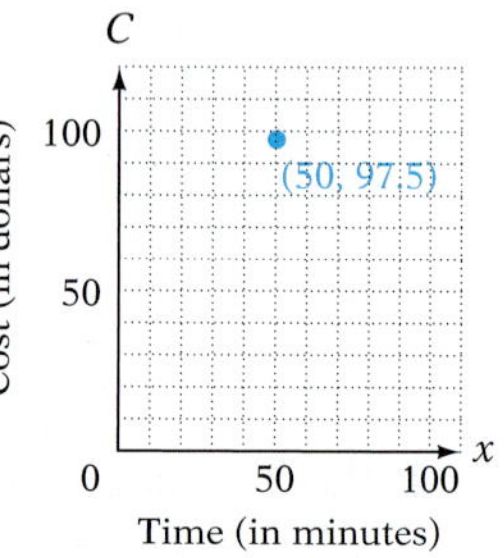

20. **Business** An online research service charges a monthly access fee of \$75 plus \$.45 per minute to use the service. An equation that represents the monthly cost to use this service is $C = 0.45x + 75$, where C is the monthly cost and x is the number of minutes of access used. Graph this equation for values of x from 0 to 100. The point $(50, 97.5)$ is on the graph. Write a sentence that describes the meaning of this ordered pair.

21. **Telecommunications** The data in the table below show the annual costs of telephone bills for a family for 6 years. The line of best fit is $y = 34x + 657$, where x is the year and y is the annual cost, in dollars, of telephone bills.

Year, x	1	2	3	4	5	6
Cost of telephone bills, y (in dollars)	690	708	772	809	830	849

Graph the data and the line of best fit in the coordinate system at the right. Write a sentence that describes the meaning of the slope of the line of best fit.

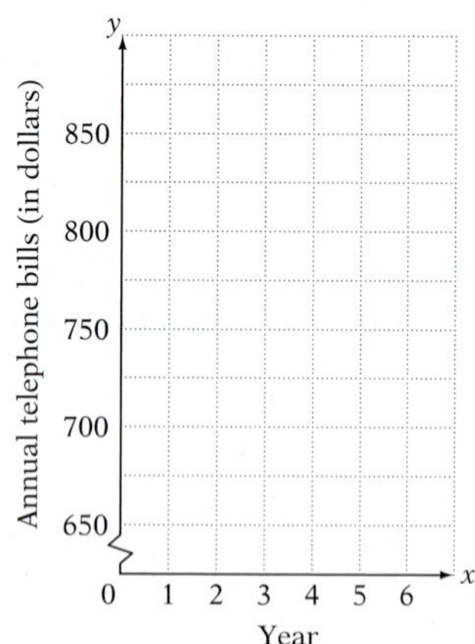

CHAPTER 12

TEST

1. Find the ordered-pair solution of $2x - 3y = 15$ corresponding to $x = 3$.

2. Graph the ordered-pair solutions of $y = -\frac{3}{2}x + 1$ when $x = -2, 0,$ and 4.

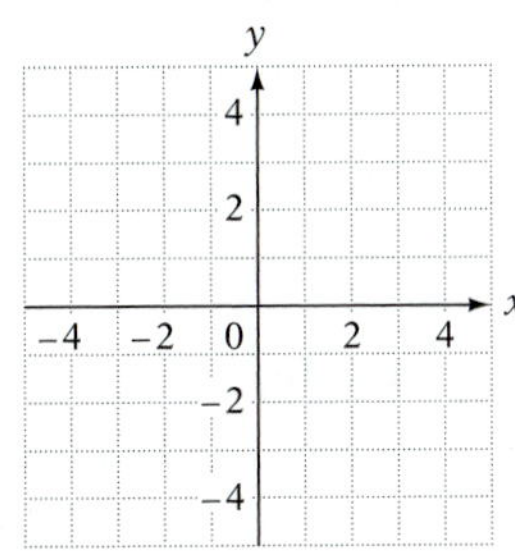

3. Does $y = \frac{1}{2}x - 3$ define y as a function of x for $x \in \{-2, 0, 4\}$?

4. Given $f(t) = t^2 + t$, find $f(2)$.

5. Given $f(x) = x^2 - 2x$, find $f(-1)$.

6. **Emergency Response** The distance a house is from a fire station and the amount of damage that the house sustained in a fire are given in the following table. Write a relation wherein the first coordinate of the ordered pair is the distance, in miles, from the fire station and the second coordinate is the amount of damage in thousands of dollars. Is the relation a function?

Distance (in miles)	3.5	4.0	5.2	5.0	4.0	6.3	5.4
Damage (in thousands of dollars)	25	30	45	38	42	12	34

7. Graph $y = 3x + 1$.

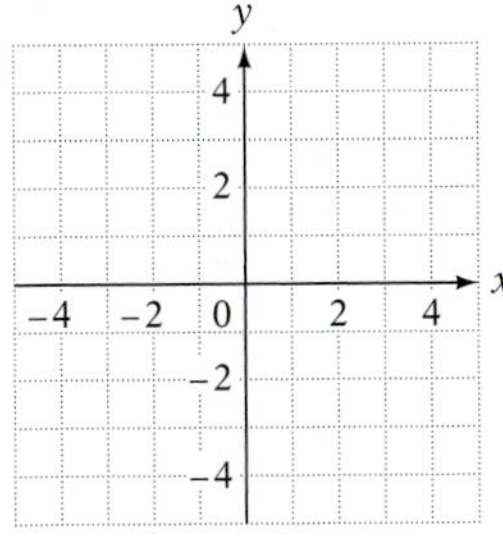

8. Graph $y = -\frac{3}{4}x + 3$.

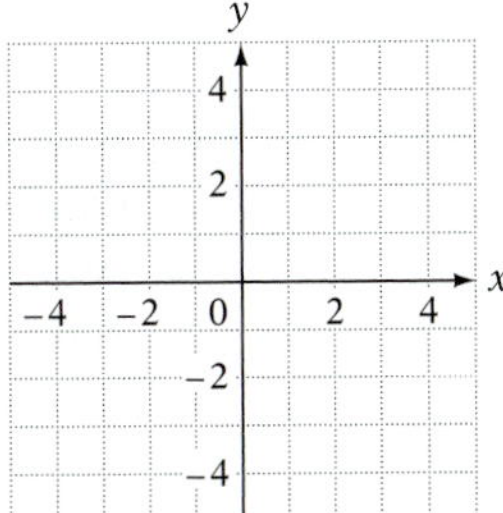

9. Graph $3x - 2y = 6$.

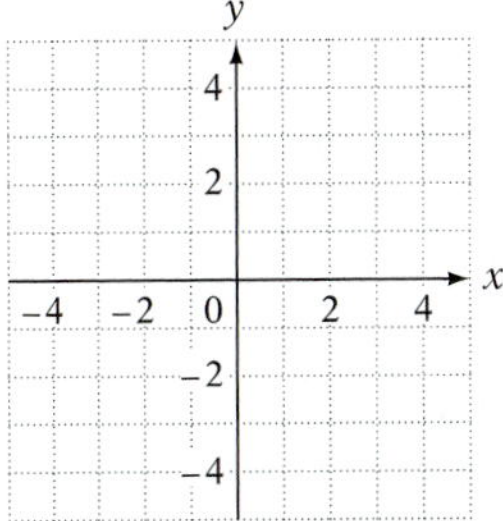

10. Graph $x + 3 = 0$.

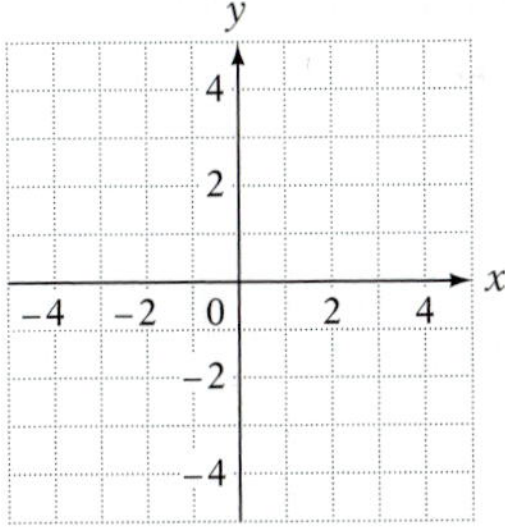

11. Graph the line that has slope $-\frac{2}{3}$ and y-intercept $(0, 4)$.

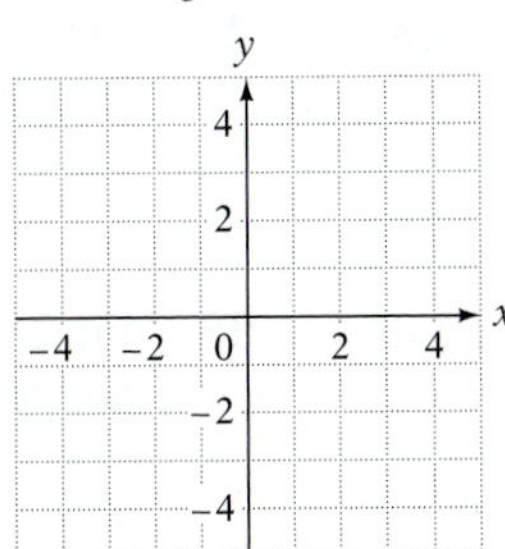

12. Graph the line that has slope 2 and y-intercept -2.

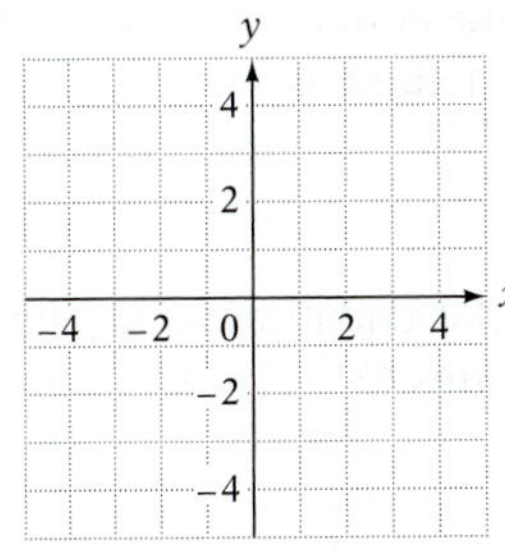

13. **Sports** The equation for the speed of a ball that is thrown straight up with an initial speed of 128 ft/s is $v = 128 - 32t$, where v is the speed of the ball after t seconds. Graph this equation for values of t from 0 to 4. The point whose coordinates are (1, 96) is on the graph. Write a sentence that describes the meaning of this ordered pair.

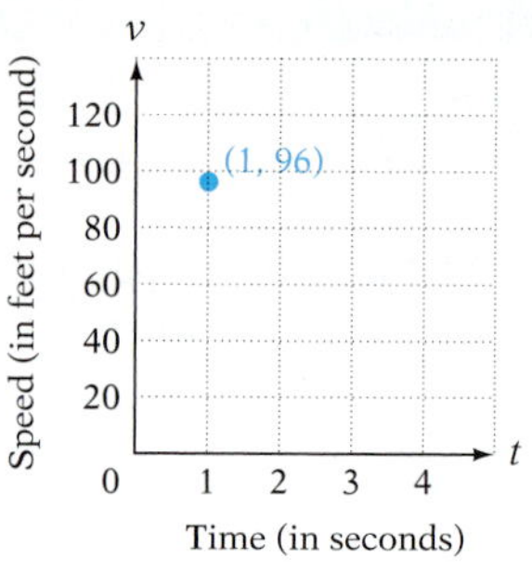

14. **Health** The graph at the right shows the relationship between distance walked and calories burned. Find the slope of the line. Write a sentence that explains the meaning of the slope.

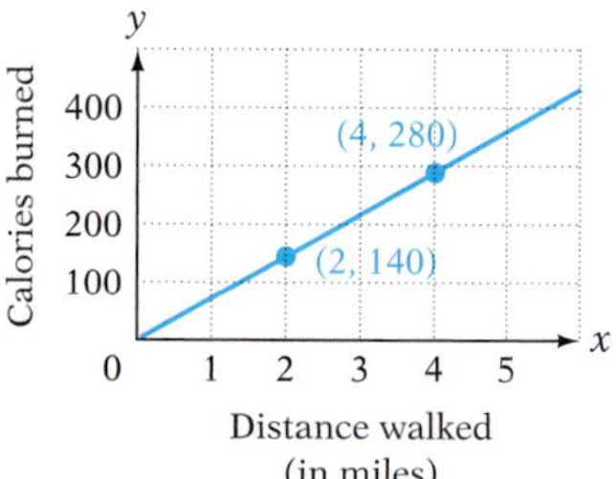

15. **Tuition** The data in the table below show the annual tuition costs at a 4-year college over a 6-year period. The line of best fit is $y = 809x + 11{,}390$, where x is the year and y is the annual tuition cost in dollars.

Year, x	1	2	3	4	5	6
Tuition Costs, y (in dollars)	12,400	12,800	13,700	14,700	15,400	16,300

Graph the data and the line of best fit in the coordinate system at the right. Write a sentence that describes the meaning of the slope of the line of best fit.

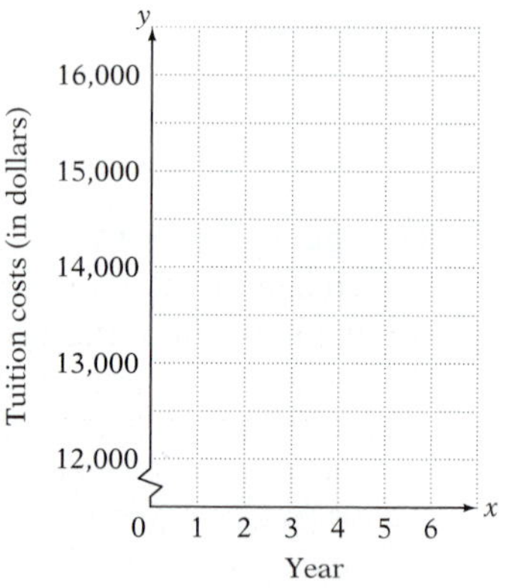

16. Find the x- and y-intercepts for $6x - 4y = 12$.

17. Find the x- and y-intercepts for $y = \frac{1}{2}x + 1$.

18. Find the slope of the line containing the points $(2, -3)$ and $(4, 1)$.

19. Is the line that passes through $(2, 5)$ and $(-1, 1)$ parallel, perpendicular, or neither parallel nor perpendicular to the line that passes through $(-2, 3)$ and $(4, 11)$?

20. Find the slope of the line containing the points $(-5, 2)$ and $(-5, 7)$.

21. Find the slope of the line whose equation is $2x + 3y = 6$.

22. Find the equation of the line that contains the point $(0, -1)$ and has slope 3.

23. Find the equation of the line that contains the point $(-3, 1)$ and has slope $\frac{2}{3}$.

24. Find the equation of the line that passes through the points $(5, -4)$ and $(-3, 1)$.

25. Find the equation of the line that passes through the points $(-2, 0)$ and $(5, -2)$.

CUMULATIVE REVIEW EXERCISES

1. Simplify: $12 - 18 \div 3 \cdot (-2)^2$

2. Evaluate $\frac{a-b}{a^2-c}$ when $a = -2$, $b = 3$, and $c = -4$.

3. Given $f(x) = \frac{2}{x-1}$, find $f(-2)$.

4. Solve: $2x - \frac{2}{3} = \frac{7}{3}$

5. Solve: $3x - 2[x - 3(2 - 3x)] = x - 7$

6. Write $6\frac{2}{3}\%$ as a fraction.

7. Simplify: $(-2x^2y)^3(2xy^2)^2$

8. Simplify: $\frac{-15x^7}{5x^5}$

9. Divide: $(x^2 - 4x - 21) \div (x - 7)$

10. Factor: $5x^2 + 15x + 10$

11. Factor: $x(a + 2) + y(a + 2)$

12. Solve: $x(x - 2) = 8$

13. Multiply: $\frac{x^5y^3}{x^2 - x - 6} \cdot \frac{x^2 - 9}{x^2y^4}$

14. Subtract: $\frac{3x}{x^2 + 5x - 24} - \frac{9}{x^2 + 5x - 24}$

15. Solve: $3 - \frac{1}{x} = \frac{5}{x}$

16. Solve $4x - 5y = 15$ for y.

17. Find the ordered-pair solution of $y = 2x - 1$ corresponding to $x = -2$.

18. Find the slope of the line that contains the points (2, 3) and (−2, 3).

19. Find the equation of the line that contains the point (2, −1) and has slope $\frac{1}{2}$.

20. Find the equation of the line that contains the point (0, 2) and has slope −3.

21. Find the equation of the line that contains the point (−1, 0) and has slope 2.

22. Find the equation of the line that contains the point (6, 1) and has slope $\frac{2}{3}$.

23. **Probability** Four blue marbles, three red marbles, and two green marbles are placed in a bag. One marble is chosen at random. What is the probability that the marble chosen is not red?

24. **Geometry** The measure of the first angle of a triangle is 3° more than the measure of the second angle. The measure of the third angle is 5° more than twice the measure of the second angle. Find the measure of each angle.

25. **Taxes** The real estate tax for a home that costs \$500,000 is \$6250. At this rate, what is the value of a home for which the real estate tax is \$13,750?

26. **Business** An electrician requires 6 h to wire a garage. An apprentice can do the same job in 10 h. How long would it take to wire the garage if both the electrician and the apprentice worked together?

27. Graph $y = \frac{1}{2}x - 1$.

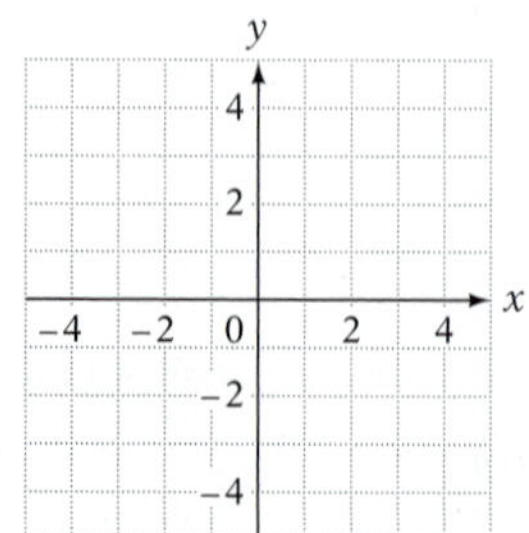

28. Graph the line that has slope $-\frac{2}{3}$ and y-intercept 2.

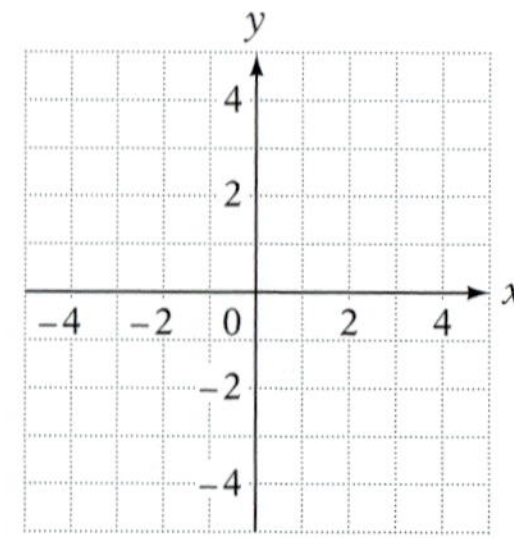

CHAPTER 13

Systems of Linear Equations

OBJECTIVES

SECTION 13.1

A To solve a system of linear equations by graphing

SECTION 13.2

A To solve a system of linear equations by the substitution method

B To solve investment problems

SECTION 13.3

A To solve a system of linear equations by the addition method

SECTION 13.4

A To solve rate-of-wind or rate-of-current problems

B To solve application problems using two variables

ARE YOU READY?

Take the Chapter 13 Prep Test to find out if you are ready to learn to:

- Solve a system of linear equations by graphing, by the substitution method, or by the addition method
- Solve investment problems and rate-of-wind or rate-of-current problems

PREP TEST

Do these exercises to prepare for Chapter 13.

1. Solve $3x - 4y = 24$ for y.

2. Solve: $50 + 0.07x = 0.05(x + 1400)$

3. Simplify: $-3(2x - 7y) + 3(2x + 4y)$

4. Simplify: $4x + 2(3x - 5)$

5. Is $(-4, 2)$ a solution of $3x - 5y = -22$?

6. Find the x- and y-intercepts for $3x - 4y = 12$.

7. Are the graphs of $3x + y = 6$ and $y = -3x - 4$ parallel?

8. Graph: $y = \frac{5}{4}x - 2$

y
4
2
x
–4 –2 0 2 4
–2
–4

9. Hiking One hiker starts along a trail walking at 3 mph. One-half hour later, another hiker starts on the same walking trail at a speed of 4 mph. How long after the second hiker starts will the two hikers be side-by-side?

SECTION

13.1 Solving Systems of Linear Equations by Graphing

OBJECTIVE A To solve a system of linear equations by graphing

Two or more equations considered together are called a **system of equations.** Three examples of *linear* systems of equations in *two* variables are shown below, along with the graphs of the equations of each system.

System I

$x - 2y = -8$
$2x + 5y = 11$

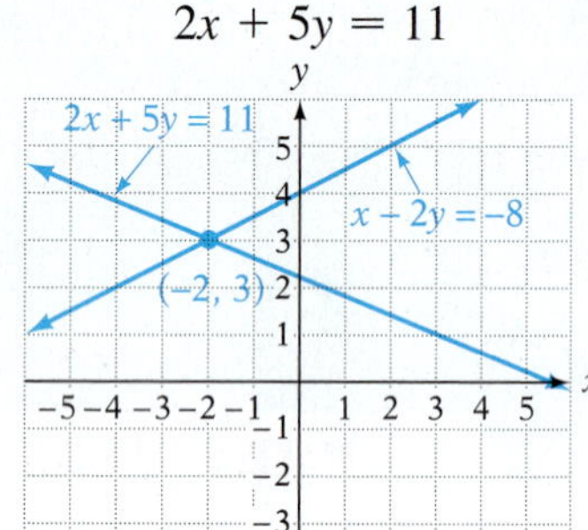

System II

$4x + 2y = 6$
$y = -2x + 3$

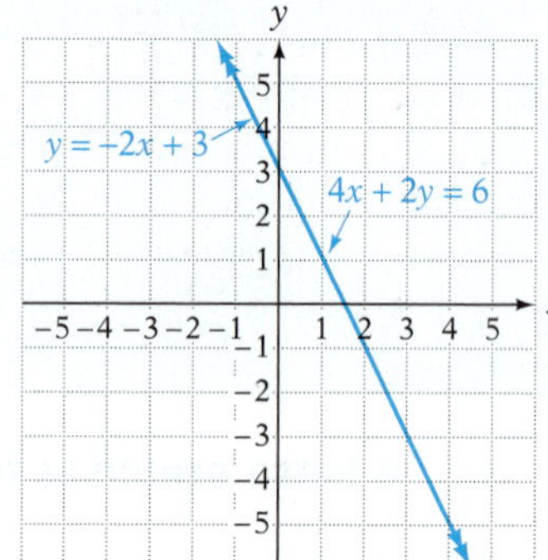

System III

$4x + 6y = 12$
$6x + 9y = -9$

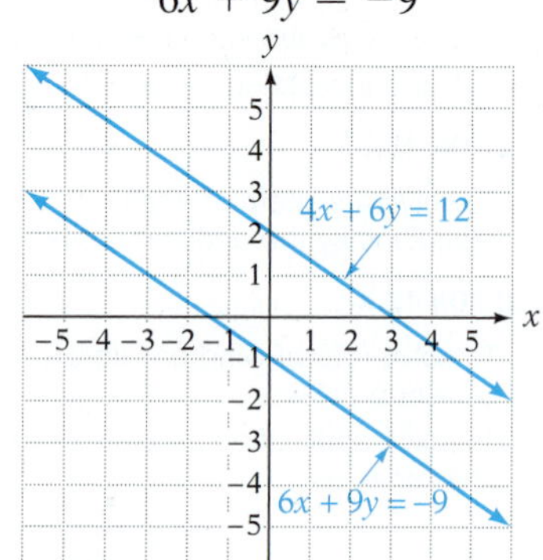

Take Note

The systems of equations above are *linear systems of equations* because each of the equations in the system has a graph that is a line. Also, each equation has two variables. In future math courses, you will study equations that contain more than two variables.

For system I, the two lines intersect at a single point, $(-2, 3)$. Because this point lies on both lines, it is a solution of each equation of the system of equations. We can check this by replacing x by -2 and y by 3. The check is shown below.

$$\begin{array}{r|l} \multicolumn{2}{c}{x - 2y = -8} \\ \hline -2 - 2(3) & -8 \\ -2 - 6 & -8 \\ -8 = -8 & \checkmark \end{array} \qquad \begin{array}{r|l} \multicolumn{2}{c}{2x + 5y = 11} \\ \hline 2(-2) + 5(3) & 11 \\ -4 + 15 & 11 \\ 11 = 11 & \checkmark \end{array}$$

• Replace x by -2 and replace y by 3.

A **solution of a system of equations in two variables** is an ordered pair that is a solution of each equation of the system. The ordered pair $(-2, 3)$ is a solution of system I.

HOW TO 1 Is $(-1, 4)$ a solution of the system of equations? $7x + 3y = 5$, $3x - 2y = 12$

$$\begin{array}{r|l} \multicolumn{2}{c}{7x + 3y = 5} \\ \hline 7(-1) + 3(4) & 5 \\ -7 + 12 & 5 \\ 5 = 5 & \checkmark \end{array} \qquad \begin{array}{r|l} \multicolumn{2}{c}{3x - 2y = 12} \\ \hline 3(-1) - 2(4) & 12 \\ -3 - 8 & 12 \\ -11 \neq 12 & \end{array}$$

• Replace x by -1 and replace y by 4.

• Does not check

Because $(-1, 4)$ is not a solution of both equations, $(-1, 4)$ is not a solution of the system of equations.

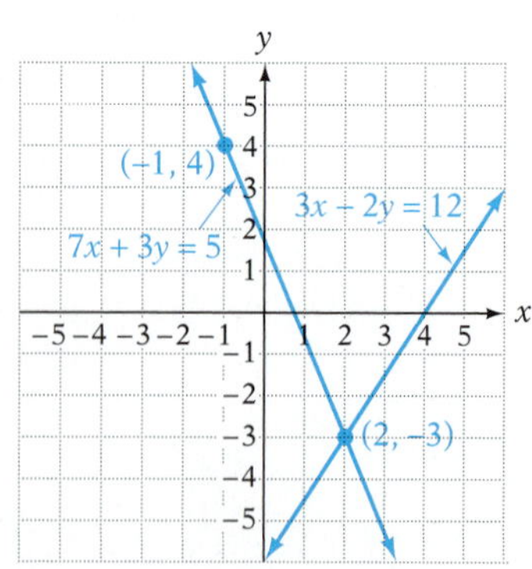

Using the system of equations above and the graph at the right, note that the graph of the ordered pair $(-1, 4)$ lies on the graph of $7x + 3y = 5$ but not on *both* lines. The ordered pair $(-1, 4)$ is *not* a solution of the system of equations. The graph of the ordered pair $(2, -3)$ does lie on both lines and therefore the ordered pair $(2, -3)$ is a solution of the system of equations.

Take Note

The fact that there is an infinite number of ordered pairs that are solutions of the system at the right does not mean *every* ordered pair is a solution. For instance, (0, 3), (−2, 7), and (2, −1) are solutions. However, (3, 1), (−1, 4), and (1, 6) are not solutions. You should verify these statements.

System II from the preceding page and the graph of the equations of that system are shown again at the right. Note that the graph of $y = -2x + 3$ lies directly on top of the graph of $4x + 2y = 6$. Thus the two lines intersect at an infinite number of points. The graphs intersect at an infinite number of points, so there are an infinite number of solutions of this system of equations. Because each equation represents the same set of points, the solutions of the system of equations can be stated by using the ordered pairs of either one of the equations. Therefore, we can say, "The solutions are the ordered pairs that satisfy $4x + 2y = 6$," or we can say "The solutions are the ordered pairs that satisfy $y = -2x + 3$."

$$4x + 2y = 6$$
$$y = -2x + 3$$

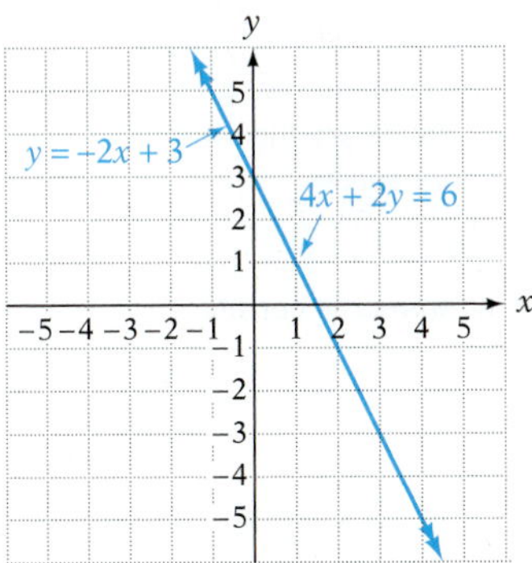

System III from the preceding page and the graph of the equations of that system are shown again at the right. Note that in this case, the graphs of the lines are parallel and do not intersect. Because the graphs do not intersect, there is no point that is on both lines. Therefore, the system of equations has no solution.

$$4x + 6y = 12$$
$$6x + 9y = -9$$

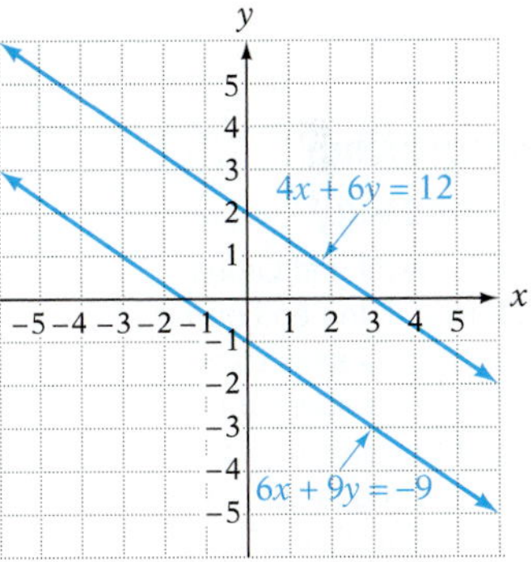

The preceding examples illustrate three types of systems of linear equations. An **independent system** has exactly one solution—the graphs intersect at one point. A **dependent system** has an infinite number of solutions—the graphs are the same line. An **inconsistent system** has no solution—the graphs are parallel lines.

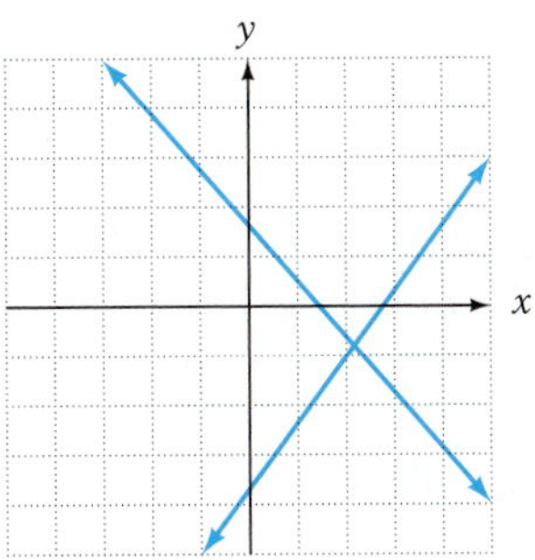

Independent: one solution

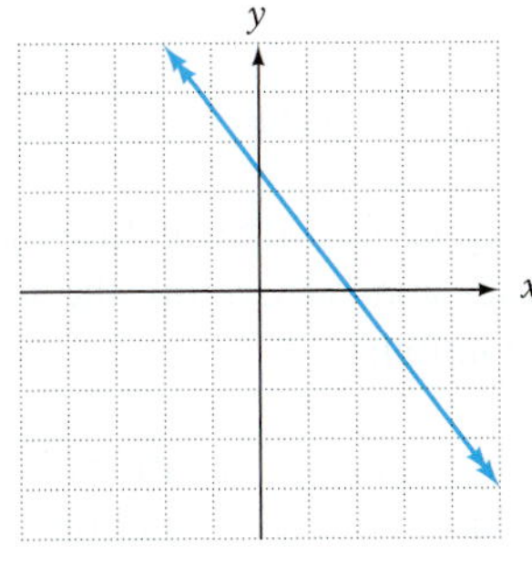

Dependent: infinitely many solutions

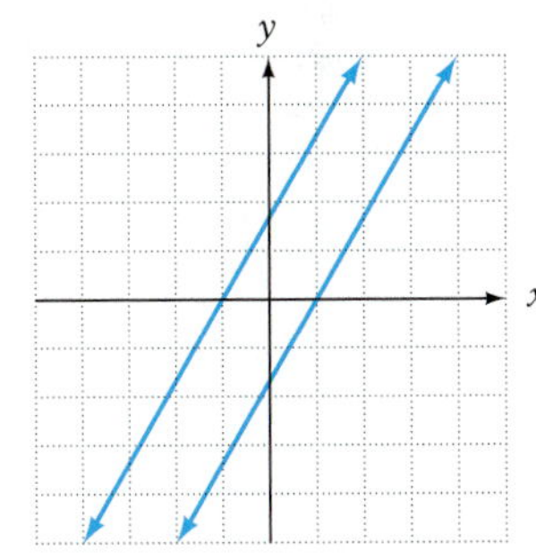

Inconsistent: no solution

HOW TO 2 The graphs of the equations in the system of equations below are shown at the right. What is the solution of the system of equations?

$$2x + 3y = 6$$
$$2x + y = -2$$

The graphs intersect at $(-3, 4)$. This is an *independent* system of equations. The solution of the system of equations is $(-3, 4)$.

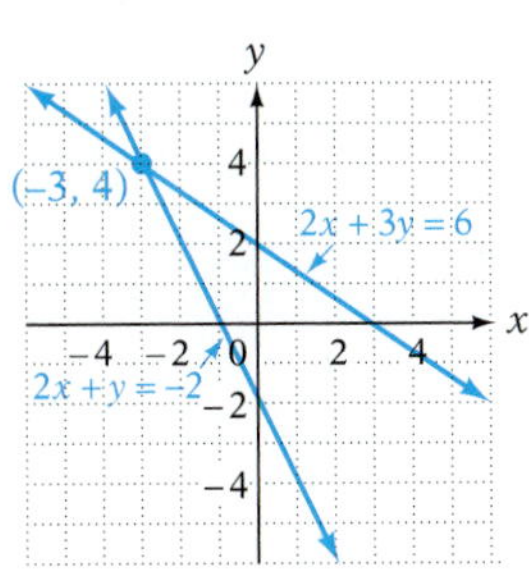

Take Note

Because both equations represent the same ordered pairs, we can also say that the solutions of the system of equations are the ordered pairs that satisfy $x = \frac{1}{2}y + 1$.

Either answer is correct.

HOW TO • 3 The graphs of the equations in the system of equations at the right are shown below. What is the solution of the system of equations?

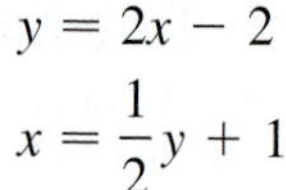

$$y = 2x - 2$$
$$x = \frac{1}{2}y + 1$$

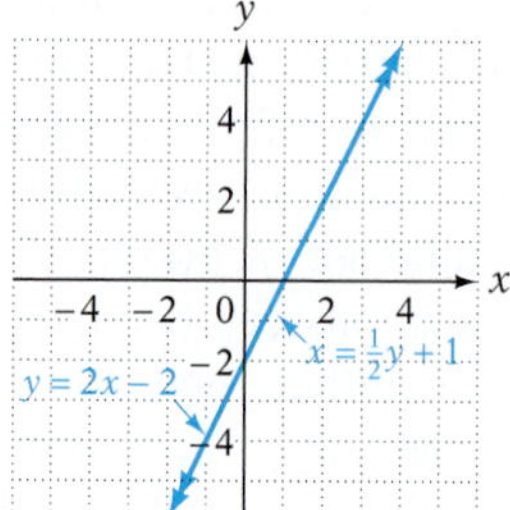

The two graphs lie directly on top of one another. Thus the two lines intersect at an infinite number of points, and the system of equations has an infinite number of solutions. This is a *dependent* system of equations. The solutions of the system of equations are the ordered pairs that satisfy $y = 2x - 2$.

Integrating Technology

The Projects and Group Activities at the end of this chapter discusses using a calculator to approximate the solution of an independent system of equations. Also see the Keystroke Guide: *Intersect*.

Solving a system of equations means finding the ordered-pair solutions of the system. One way to do this is to draw the graphs of the equations in the system of equations and determine where the graphs intersect.

To solve a system of linear equations in two variables by graphing, graph each equation on the same coordinate system, and then determine the points of intersection.

HOW TO • 4 Solve by graphing: $2x - y = -1$
$x + 2y = 7$

Graph each line.

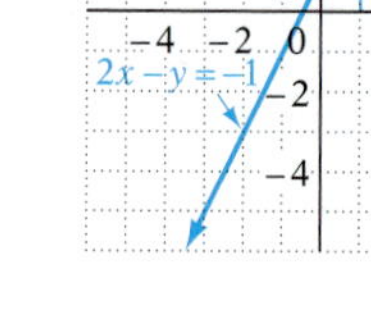

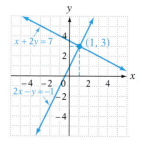

The point of intersection of the two graphs lies on both lines and is therefore the solution of the system of equations.

The system of equations is independent. (1, 3) is a solution of each equation.

The solution is (1, 3).

HOW TO • 5 Solve by graphing: $y = 2x + 2$
$4x - 2y = 4$

Graph each line.

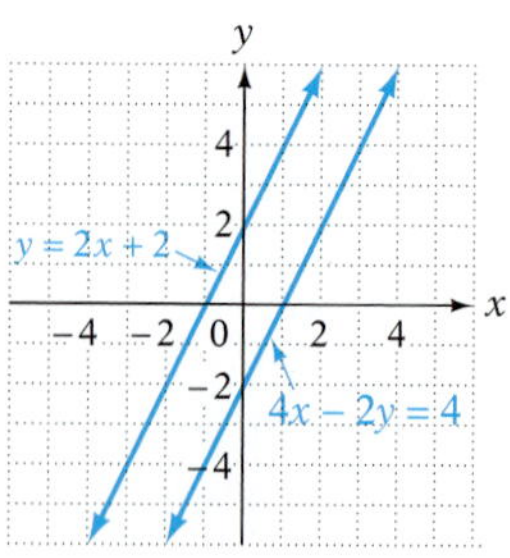

The graphs do not intersect. The system of equations is inconsistent.

The system of equations has no solution.

EXAMPLE • 1

Is $(1, -3)$ a solution of the following system?

$$3x + 2y = -3$$
$$x - 3y = 6$$

Solution

Replace x by 1 and y by -3.

$$\begin{array}{r|l} \multicolumn{2}{c}{3x + 2y = -3} \\ \hline 3 \cdot 1 + 2(-3) & -3 \\ 3 + (-6) & -3 \\ \multicolumn{2}{r}{-3 = -3} \end{array} \qquad \begin{array}{r|l} \multicolumn{2}{c}{x - 3y = 6} \\ \hline 1 - 3(-3) & 6 \\ 1 - (-9) & 6 \\ \multicolumn{2}{r}{10 \neq 6} \end{array}$$

No, $(1, -3)$ is not a solution of the system of equations.

YOU TRY IT • 1

Is $(-1, -2)$ a solution of the following system?

$$2x - 5y = 8$$
$$-x + 3y = -5$$

Your solution

EXAMPLE • 2

Solve by graphing:

$$x - 2y = 2$$
$$x + y = 5$$

Solution

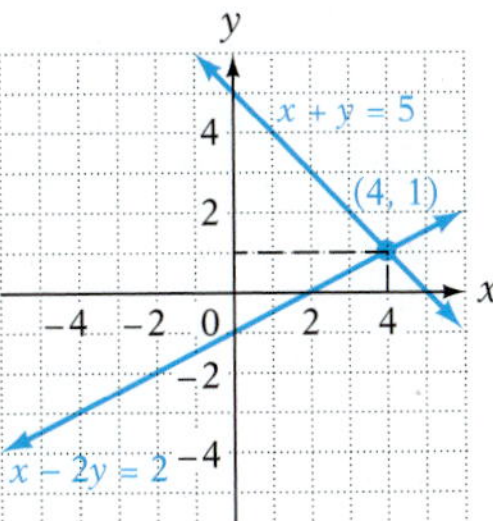

The solution is $(4, 1)$.

YOU TRY IT • 2

Solve by graphing:

$$x + 3y = 3$$
$$-x + y = 5$$

Your solution

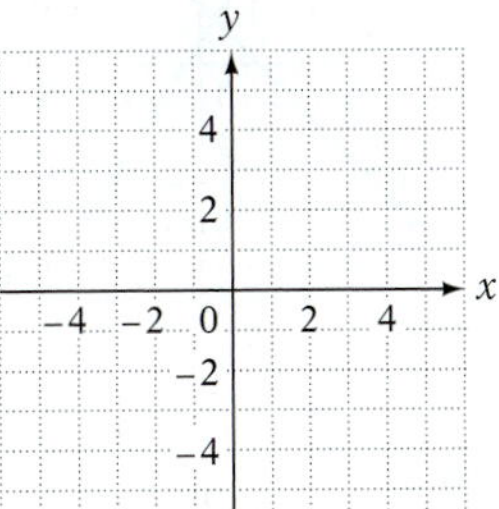

EXAMPLE • 3

Solve by graphing:

$$4x - 2y = 6$$
$$y = 2x - 3$$

Solution

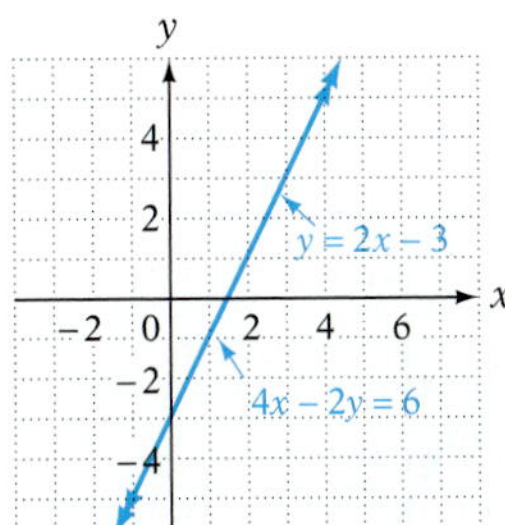

The solutions are the ordered pairs that satisfy the equation $y = 2x - 3$.

YOU TRY IT • 3

Solve by graphing:

$$y = 3x - 1$$
$$6x - 2y = -6$$

Your solution

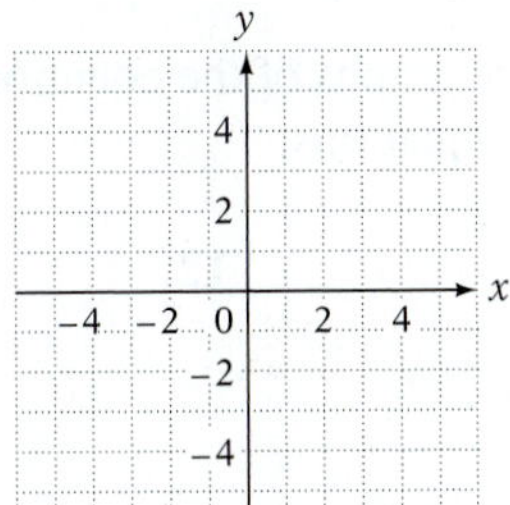

Solutions on p. S30

13.1 EXERCISES

OBJECTIVE A To solve a system of linear equations by graphing

1. Is (2, 3) a solution of $\begin{aligned} 3x + 4y &= 18 \\ 2x - y &= 1 \end{aligned}$?

2. Is (2, −1) a solution of $\begin{aligned} x - 2y &= 4 \\ 2x + y &= 3 \end{aligned}$?

3. Is (4, 3) a solution of $\begin{aligned} 5x - 2y &= 14 \\ x + y &= 8 \end{aligned}$?

4. Is (2, 5) a solution of $\begin{aligned} 3x + 2y &= 16 \\ 2x - 3y &= 4 \end{aligned}$?

5. Is (2, −3) a solution of $\begin{aligned} y &= 2x - 7 \\ 3x - y &= 9 \end{aligned}$?

6. Is (−1, −2) a solution of $\begin{aligned} 3x - 4y &= 5 \\ y &= x - 1 \end{aligned}$?

7. Is (0, 0) a solution of $\begin{aligned} 3x + 4y &= 0 \\ y &= x \end{aligned}$?

8. Is (3, −4) a solution of $\begin{aligned} 5x - 2y &= 23 \\ 2x - 5y &= 25 \end{aligned}$?

For Exercises 9 and 10, label each system of equations (systems I, II, and III) as (a) independent, (b) dependent, or (c) inconsistent.

9. I

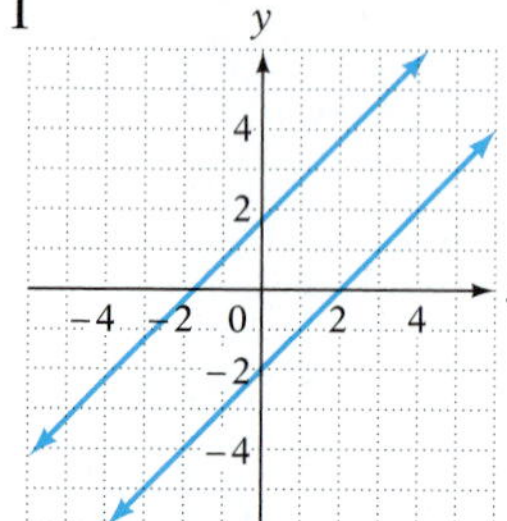

II

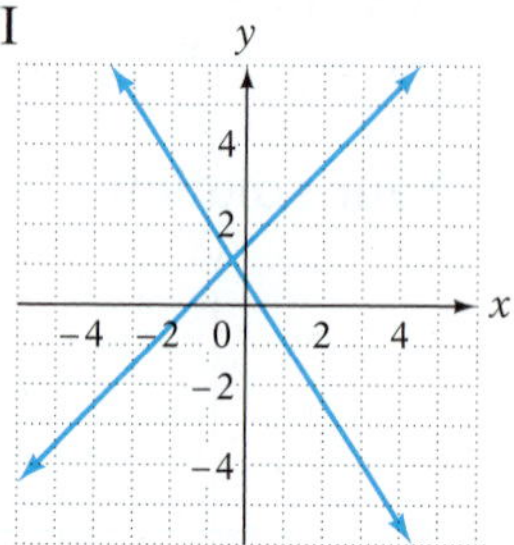

III

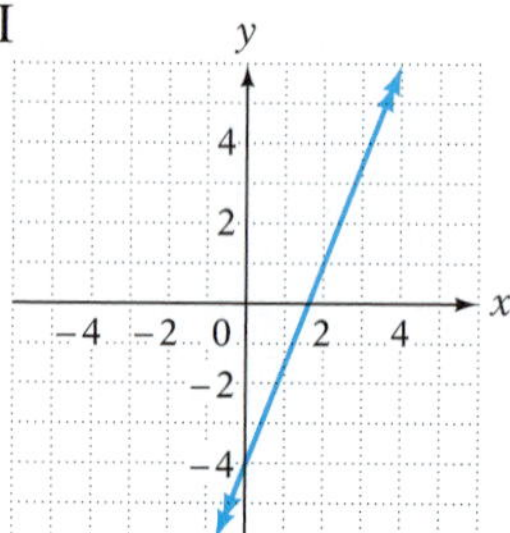

10. I

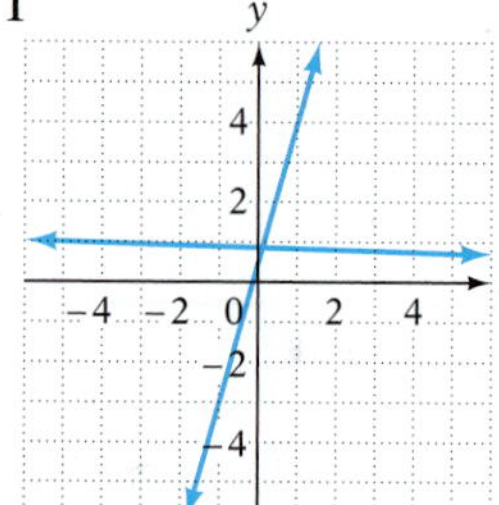

II

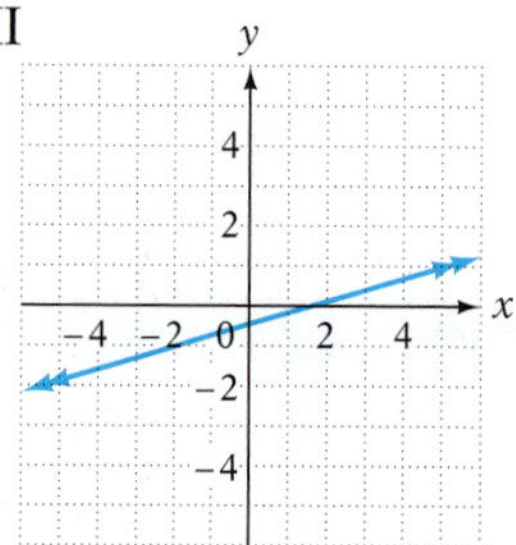

III

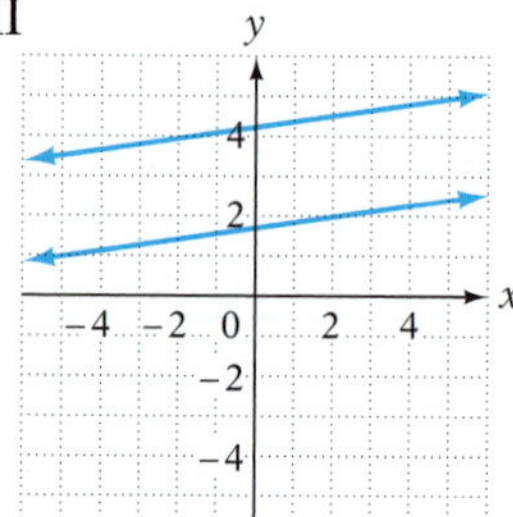

For Exercises 11 to 19, use the graphs of the equations of the system of equations to find the solution of the system of equations.

11.

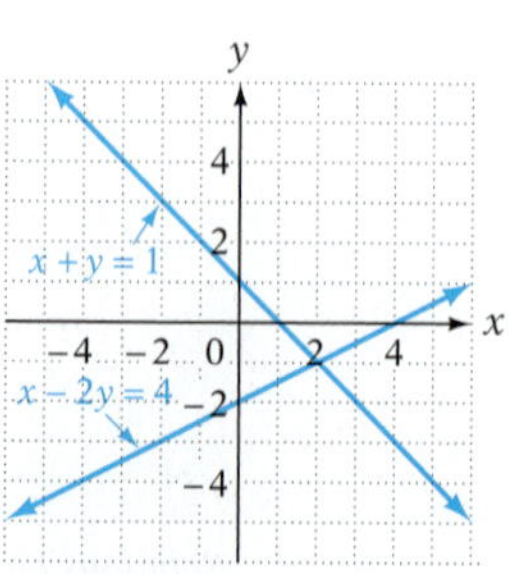

12.

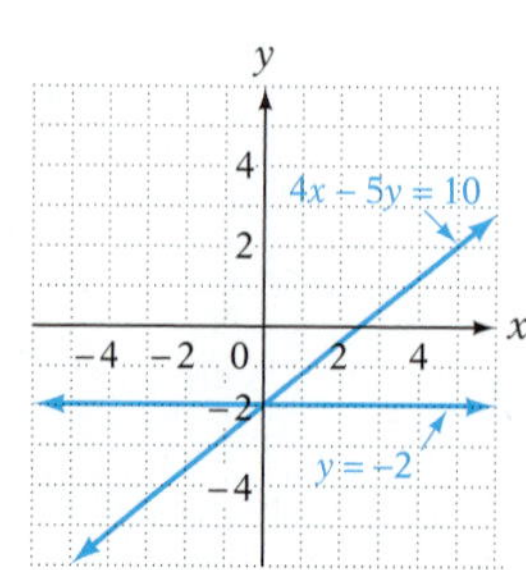

13.

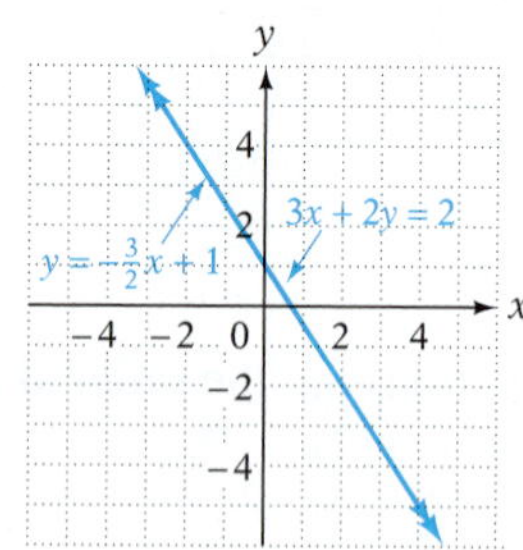

14.

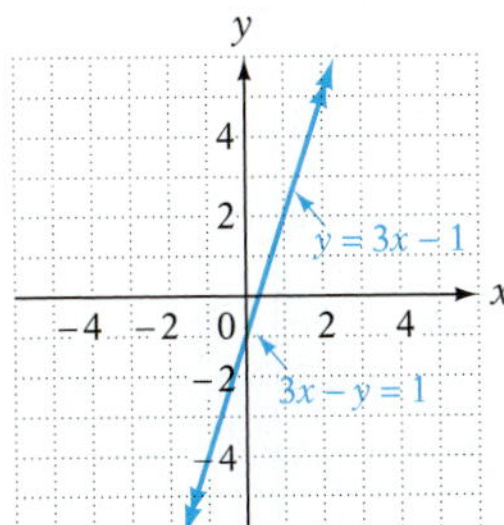

15.

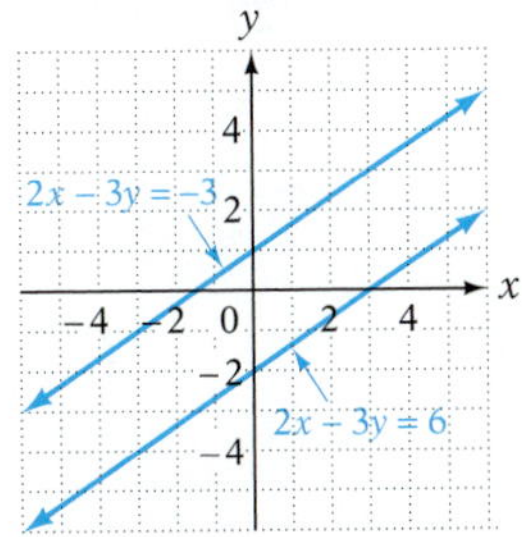

16.

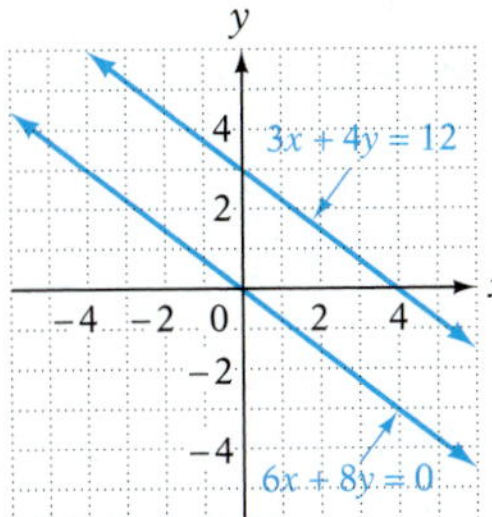

17.

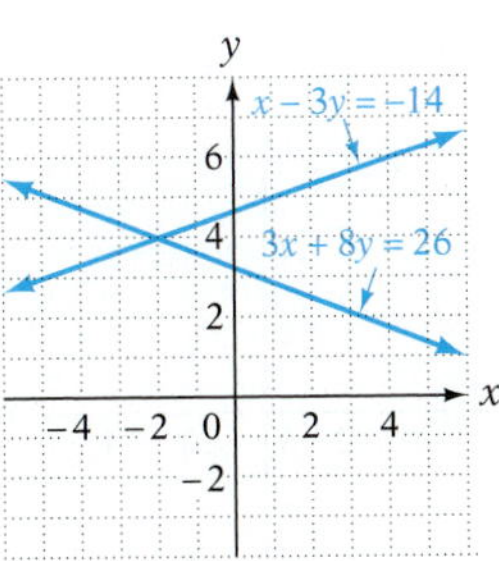

18.

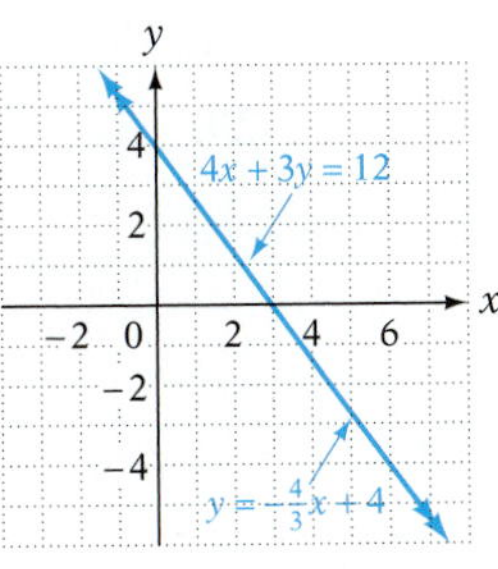

19.

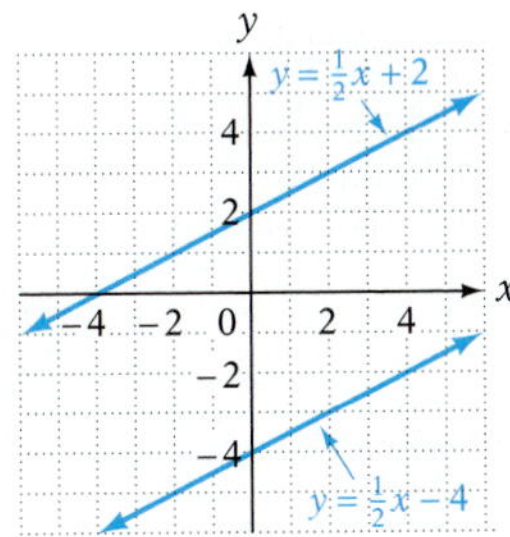

For Exercises 20 to 39, solve by graphing.

20. $x - y = 3$
$x + y = 5$

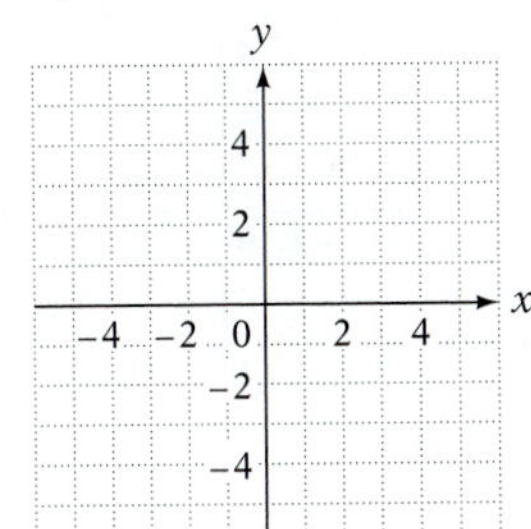

21. $2x - y = 4$
$x + y = 5$

22. $x + 2y = 6$
$x - y = 3$

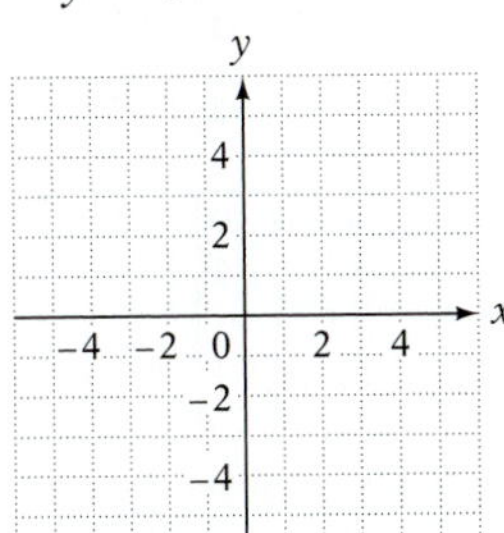

23. $3x - y = 3$
$2x + y = 2$

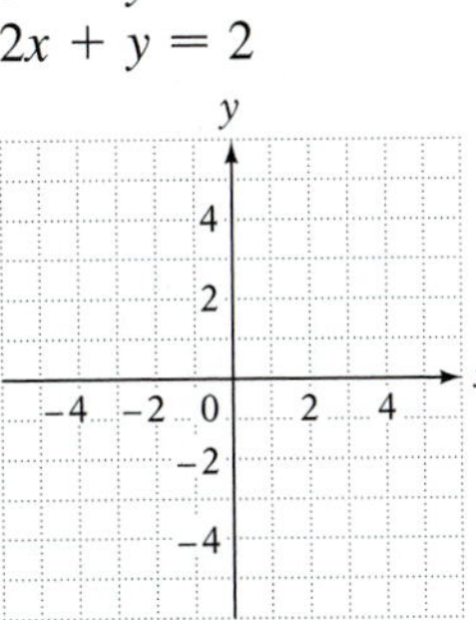

24. $3x - 2y = 6$
$y = 3$

25. $x = 2$
$3x + 2y = 4$

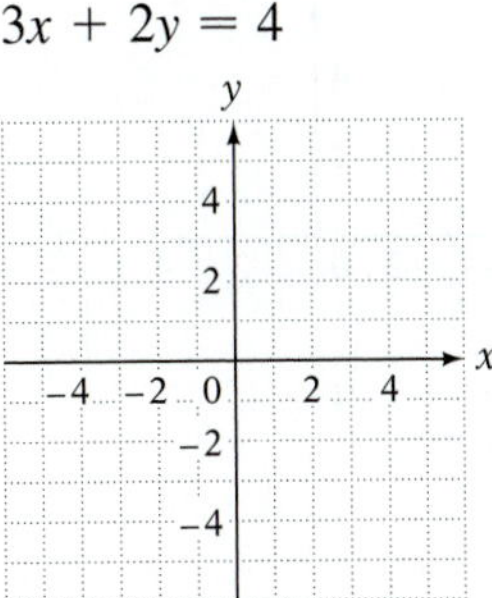

26. 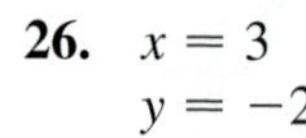
$x = 3$
$y = -2$

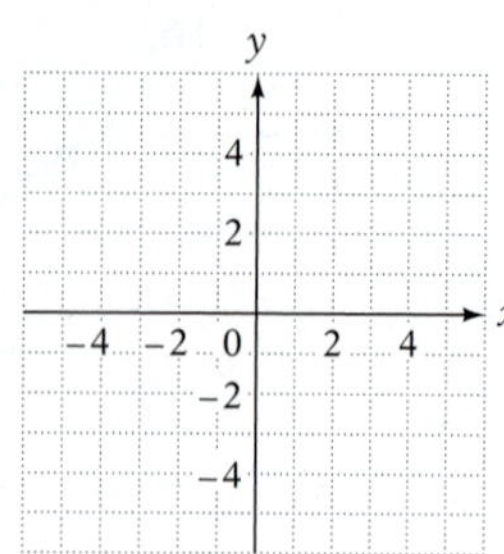

27. $x + 1 = 0$
$y - 3 = 0$

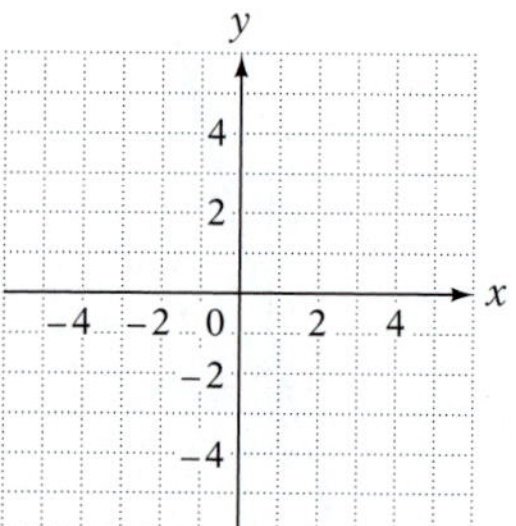

28. $y = 2x - 6$
$x + y = 0$

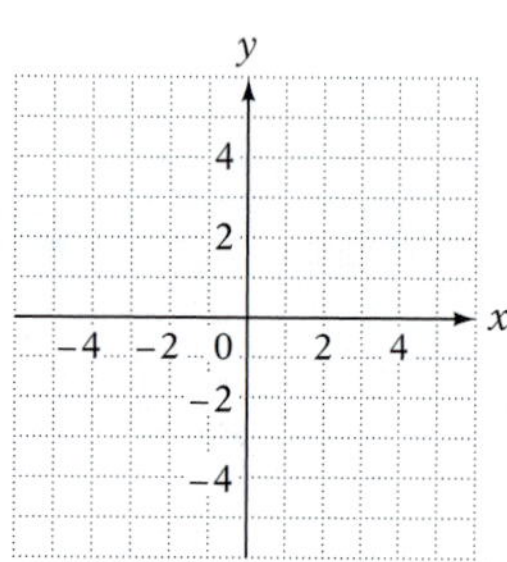

29. $5x - 2y = 11$
$y = 2x - 5$

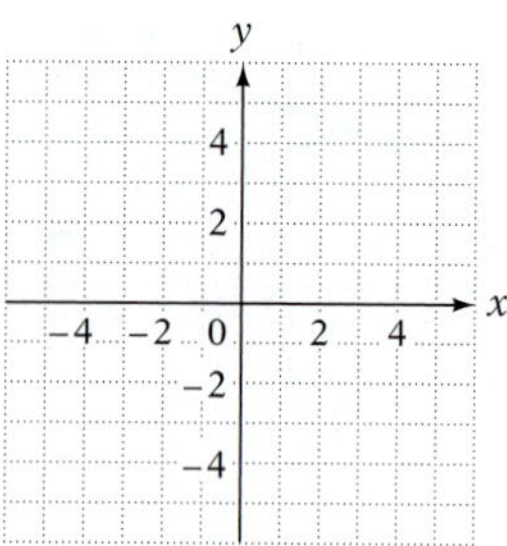

30. $2x + y = -2$
$6x + 3y = 6$

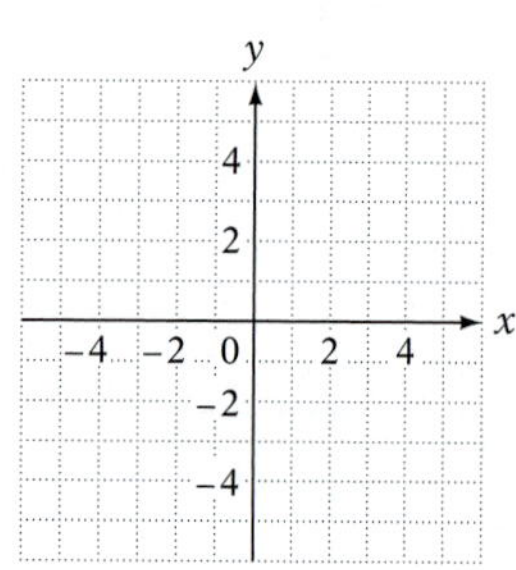

31. $x + y = 5$
$3x + 3y = 6$

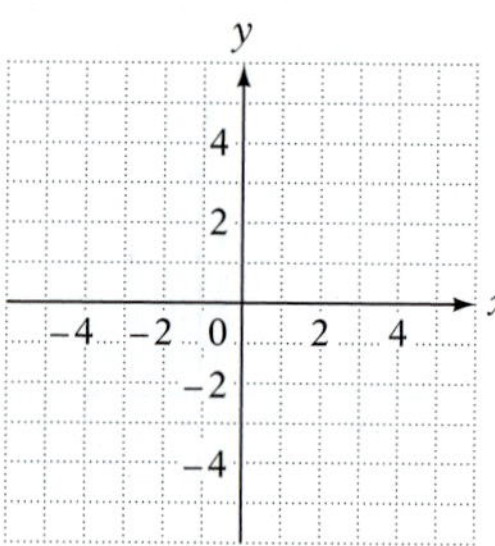

32. $y = 2x - 2$
$4x - 2y = 4$

33. 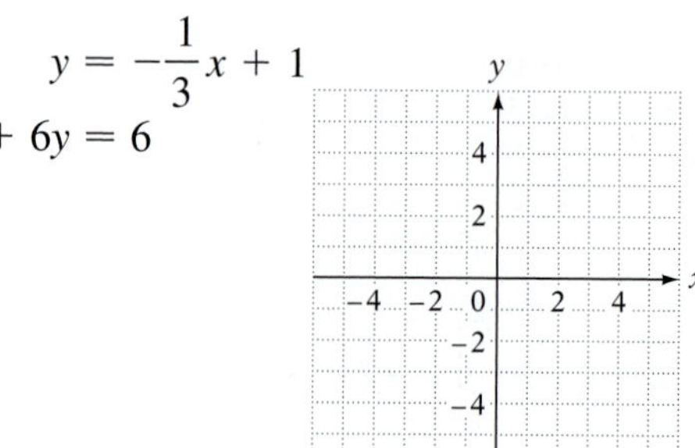
$y = -\frac{1}{3}x + 1$
$2x + 6y = 6$

34. $x - y = 5$
$2x - y = 6$

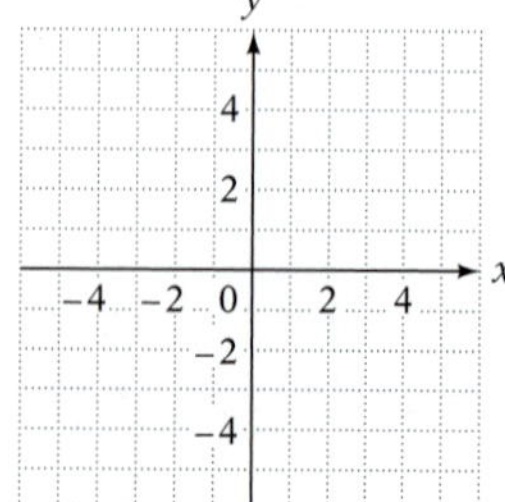

35. $5x - 2y = 10$
$3x + 2y = 6$

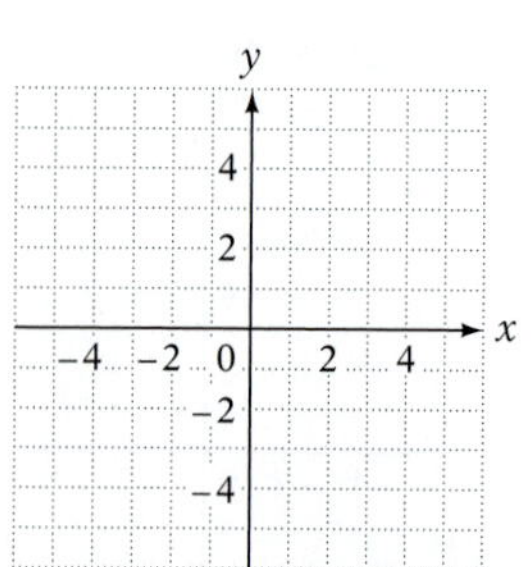

36. $3x + 4y = 0$
$2x - 5y = 0$

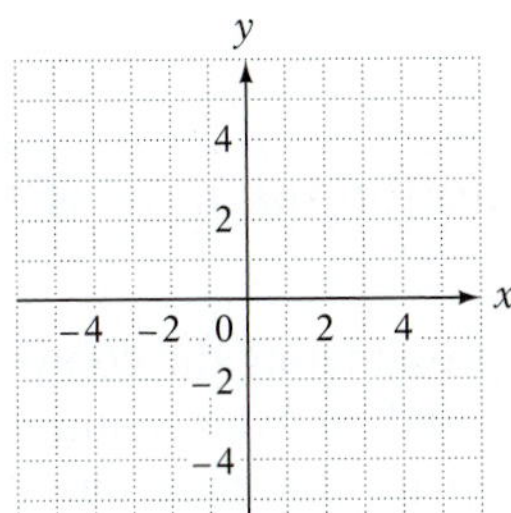

37. $2x - 3y = 0$
$y = -\frac{1}{3}x$

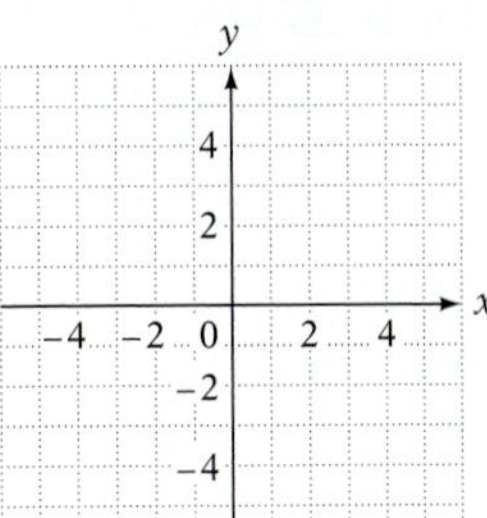

38. $x - 3y = 3$
$2x - 6y = 12$

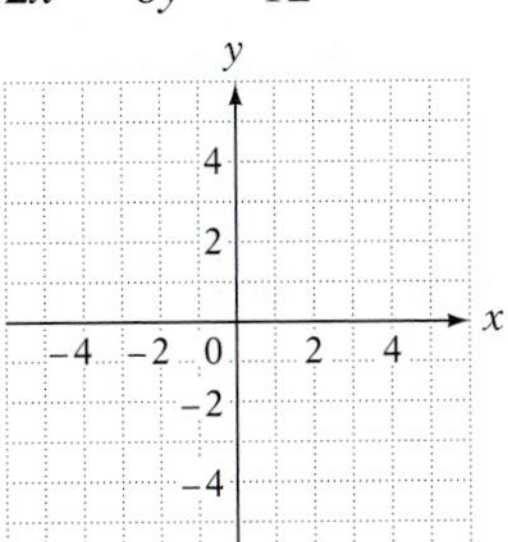

39. $4x + 6y = 12$
$6x + 9y = 18$

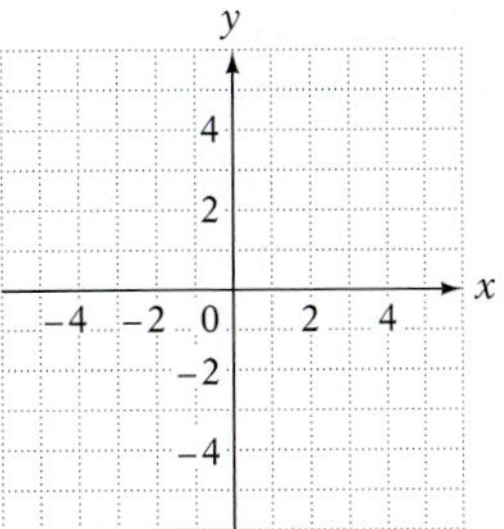

In Exercises 40 and 41, *A, B, C,* and *D* are nonzero real numbers. State whether the system of equations is independent, inconsistent, or dependent.

40. $y = Ax + B$
$y = Ax + C, B \neq C$

41. $x = C$
$y = D$

Applying the Concepts

42. Determine whether the statement is always true, sometimes true, or never true.
a. A solution of a system of two equations in two variables is a point in the plane.
b. Two parallel lines have the same slope.
c. Two different lines with the same *y*-intercept are parallel.
d. Two different lines with the same slope are parallel.

43. Write a system of equations that has $(-2, 4)$ as its only solution.

44. Write a system of equations for which there is no solution.

45. Write a system of equations that is a dependent system of equations.

SECTION

13.2 Solving Systems of Linear Equations by the Substitution Method

OBJECTIVE A To solve a system of linear equations by the substitution method

A graphical solution of a system of equations is found by approximating the coordinates of a point of intersection. Algebraic methods can be used to find an exact solution of a system of equations. The **substitution method** can be used to eliminate one of the variables in one of the equations so that we have one equation in one unknown.

HOW TO • 1 Solve by the substitution method: (1) $2x + 5y = -11$
(2) $y = 3x - 9$

Equation (2) states that $y = 3x - 9$. Substitute $3x - 9$ for y in Equation (1). Then solve for x.

$$\begin{aligned} 2x + 5y &= -11 \\ 2x + 5(3x - 9) &= -11 \\ 2x + 15x - 45 &= -11 \\ 17x - 45 &= -11 \\ 17x &= 34 \\ x &= 2 \end{aligned}$$

- This is Equation (1).
- From Equation (2), substitute $3x - 9$ for y.
- Solve for x.

Now substitute the value of x into Equation (2) and solve for y.

$$\begin{aligned} y &= 3x - 9 \\ y &= 3(2) - 9 \\ y &= 6 - 9 = -3 \end{aligned}$$

- This is Equation (2).
- Substitute 2 for x.

The solution is the ordered pair $(2, -3)$.

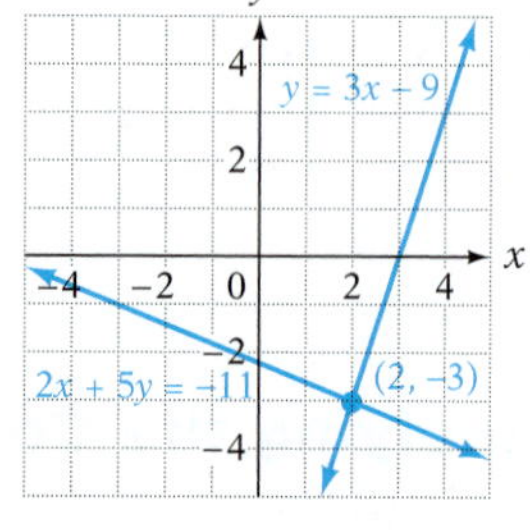

The graph of the equations in this system of equations is shown at the right. Note that the lines intersect at the point whose coordinates are $(2, -3)$, which is the algebraic solution we determined by the substitution method.

To solve a system of equations by the substitution method, we may need to solve one of the equations in the system of equations for one of its variables. For instance, the first step in solving the system of equations

$$\begin{aligned} (1)\quad x + 2y &= -3 \\ (2)\quad 2x - 3y &= 5 \end{aligned}$$

is to solve an equation of the system for one of its variables. Either equation can be used.

Solving Equation (1) for x:

$$\begin{aligned} x + 2y &= -3 \\ x &= -2y - 3 \end{aligned}$$

Solving Equation (2) for x:

$$\begin{aligned} 2x - 3y &= 5 \\ 2x &= 3y + 5 \\ x &= \frac{3y + 5}{2} = \frac{3}{2}y + \frac{5}{2} \end{aligned}$$

Because solving Equation (1) for x does not result in fractions, it is the easier of the two equations to use.

Here is the solution of the system of equations given on the preceding page.

HOW TO • 2 Solve by the substitution method: (1) $x + 2y = -3$
(2) $2x - 3y = 5$

To use the substitution method, we must solve one equation of the system for one of its variables. We used Equation (1) because solving it for x does not result in fractions.

$$x + 2y = -3$$
(3) $x = -2y - 3$ • **Solve for x. This is Equation (3).**

Now substitute $-2y - 3$ for x in Equation (2) and solve for y.

$2x - 3y = 5$ • **This is Equation (2).**
$2(-2y - 3) - 3y = 5$ • **From Equation (3), substitute $-2y - 3$ for x.**
$-4y - 6 - 3y = 5$ • **Solve for y.**
$-7y - 6 = 5$
$-7y = 11$
$y = -\frac{11}{7}$

Substitute the value of y into Equation (3) and solve for x.

$x = -2y - 3$ • **This is Equation (3).**
$= -2\left(-\frac{11}{7}\right) - 3$ • **Substitute $-\frac{11}{7}$ for y.**
$= \frac{22}{7} - 3 = \frac{22}{7} - \frac{21}{7} = \frac{1}{7}$

The solution is $\left(\frac{1}{7}, -\frac{11}{7}\right)$.

The graph of the system of equations given above is shown at the right. It would be difficult to determine the exact solution of this system of equations from the graphs of the equations.

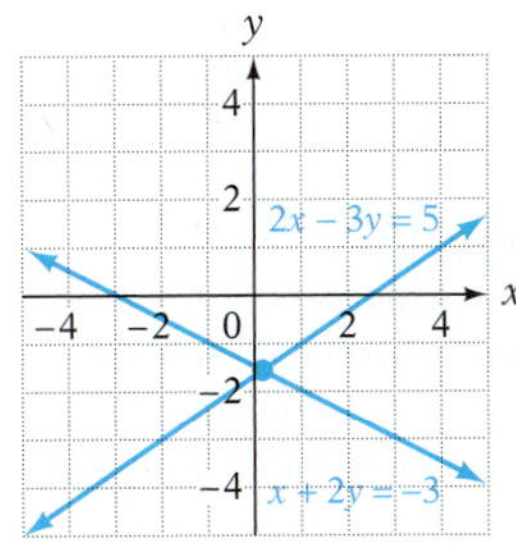

HOW TO • 3 Solve by the substitution method: (1) $y = 3x - 1$
(2) $y = -2x - 6$

$y = -2x - 6$
$3x - 1 = -2x - 6$ • **Substitute $3x - 1$ for y in Equation (2).**
$5x = -5$ • **Solve for x.**
$x = -1$

Substitute this value of x into Equation (1) or Equation (2) and solve for y. Equation (1) is used here.

$y = 3x - 1$
$y = 3(-1) - 1 = -4$ • **Substitute -1 for x.**

The solution is $(-1, -4)$.

The substitution method can be used to analyze inconsistent and dependent systems of equations. **If, when solving a system of equations algebraically, the variable is eliminated and the result is a false equation,** such as $0 = 4$, **the system of equations is inconsistent. If the variable is eliminated and the result is a true equation,** such as $12 = 12$, **the system of equations is dependent.**

HOW TO • 4 Solve by the substitution method: (1) $2x + 3y = 3$ (2) $y = -\frac{2}{3}x + 3$

$2x + 3y = 3$ • This is Equation (1).

$2x + 3\left(-\frac{2}{3}x + 3\right) = 3$ • From Equation (2), replace y with $-\frac{2}{3}x + 3$.

$2x - 2x + 9 = 3$ • Solve for x.

$9 = 3$ • This is a false equation.

Because $9 = 3$ is a false equation, the system of equations has no solution. The system is inconsistent.

Solving Equation (1) above for y, we have $y = -\frac{2}{3}x + 1$.

Comparing this with Equation (2) reveals that the slopes are equal and the y-intercepts are different. The graphs of the equations that make up this system of equations are parallel and thus never intersect. Because the graphs do not intersect, there are no solutions of the system of equations. The system of equations is inconsistent.

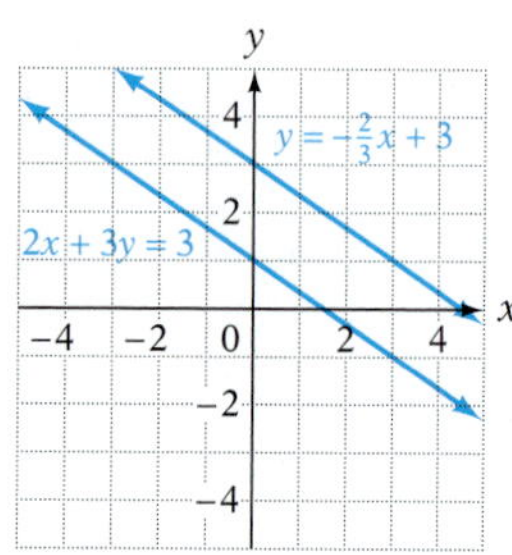

HOW TO • 5 Solve by the substitution method: (1) $x = 2y + 3$ (2) $4x - 8y = 12$

$4x - 8y = 12$ • This is Equation (2).

$4(2y + 3) - 8y = 12$ • From Equation (1), replace x by $2y + 3$.

$8y + 12 - 8y = 12$ • Solve for y.

$12 = 12$ • This is a true equation.

The true equation $12 = 12$ indicates that any ordered pair (x, y) that satisfies one equation of the system satisfies the other equation. Therefore, the system of equations has an infinite number of solutions. The system is dependent. The solutions are the ordered pairs (x, y) that are solutions of $x = 2y + 3$.

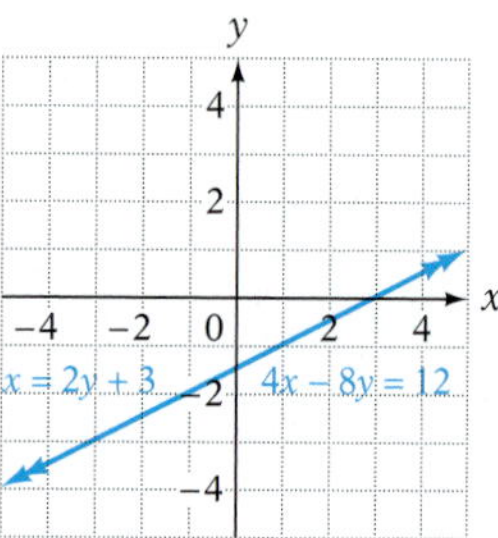

Take Note

As we mentioned in the previous section, when a system of equations is dependent, either equation can be used to write the ordered-pair solutions. Thus we could have said, "The solutions are the ordered pairs (x, y) that are solutions of $4x - 8y = 12$." Also note that, as we show at the right, if we solve each equation for y, the equations have the same slope-intercept form. This means we could also say, "The solutions are the ordered pairs (x, y) that are solutions of $y = \frac{1}{2}x - \frac{3}{2}$." When a system of equations is dependent, there are many ways in which the solutions can be stated.

If we write Equation (1) and Equation (2) in slope-intercept form, we have

$$\begin{aligned} x &= 2y + 3 \\ -2y &= -x + 3 \\ y &= \frac{1}{2}x - \frac{3}{2} \end{aligned} \qquad \begin{aligned} 4x - 8y &= 12 \\ -8y &= -4x + 12 \\ y &= \frac{1}{2}x - \frac{3}{2} \end{aligned}$$

The slope-intercept forms of the equations are the same, and therefore the graphs are the same. If we graph these two equations, we essentially graph one over the other, so the graphs intersect at an infinite number of points.

EXAMPLE • 1

Solve by substitution:

(1) $3x + 4y = -2$

(2) $-x + 2y = 4$

Solution

$-x + 2y = 4$ • Solve Equation (2) for x.

$-x = -2y + 4$

$x = 2y - 4$

Substitute in Equation (1).

(1) $3x + 4y = -2$

$3(2y - 4) + 4y = -2$ • $x = 2y - 4$

$6y - 12 + 4y = -2$ • Solve for y.

$10y - 12 = -2$

$10y = 10$

$y = 1$

Substitute in $x = 2y - 4$.

$x = 2y - 4$

$x = 2(1) - 4$ • $y = 1$

$x = 2 - 4$

$x = -2$

The solution is $(-2, 1)$.

YOU TRY IT • 1

Solve by substitution:

(1) $7x - y = 4$

(2) $3x + 2y = 9$

Your solution

EXAMPLE • 2

Solve by substitution:

$4x + 2y = 5$

$y = -2x + 1$

Solution

$4x + 2y = 5$

$4x + 2(-2x + 1) = 5$ • $y = -2x + 1$

$4x - 4x + 2 = 5$ • Solve for x.

$2 = 5$ • A false equation

The system of equations is inconsistent and therefore does not have a solution.

YOU TRY IT • 2

Solve by substitution:

$3x - y = 4$

$y = 3x + 2$

Your solution

EXAMPLE • 3

Solve by substitution:

$y = 3x - 2$

$6x - 2y = 4$

Solution

$6x - 2y = 4$

$6x - 2(3x - 2) = 4$ • $y = 3x - 2$

$6x - 6x + 4 = 4$ • Solve for x.

$4 = 4$ • A true equation

The system of equations is dependent. The solutions are the ordered pairs that satisfy the equation $y = 3x - 2$.

YOU TRY IT • 3

Solve by substitution:

$y = -2x + 1$

$6x + 3y = 3$

Your solution

Solutions on pp. S30–S31

OBJECTIVE B To solve investment problems 

The annual simple interest that an investment earns is given by the equation $Pr = I$, where P is the principal, or the amount invested, r is the simple interest rate, and I is the simple interest.

For instance, if you invest \$750 at a simple interest rate of 6%, then the interest earned after 1 year is calculated as follows:

$$Pr = I$$
$$750(0.06) = I \quad \text{• Replace } P \text{ by 750 and } r \text{ by 0.06 (6\%).}$$
$$45 = I \quad \text{• Simplify.}$$

The amount of interest earned is \$45.

Tips for Success

Word problems are difficult because we must read the problem, determine the quantity we must find, think of a method to find it, actually solve the problem, and then check the answer. In short, we must devise a *strategy* and then use that strategy to find the *solution*. See *AIM for Success* at the front of the book.

HOW TO 6 A medical lab technician decides to open an Individual Retirement Account (IRA) by placing \$2000 in two simple interest accounts. On one account, a corporate bond fund, the annual simple interest rate is 7.5%. On the second account, a real estate investment trust, the annual simple interest rate is 9%. If the technician wants to have annual earnings of \$168 from these two investments, how much must be invested in each account?

Strategy for Solving Simple-Interest Investment Problems

1. For each amount invested, use the equation $Pr = I$. Write a numerical or variable expression for the principal, the interest rate, and the interest earned.

Amount invested at 7.5%: x
Amount invested at 9%: y

	Principal, P	·	*Interest rate, r*	=	*Interest earned, I*
Amount at 7.5%	x	·	0.075	=	$0.075x$
Amount at 9%	y	·	0.09	=	$0.09y$

2. Write a system of equations. One equation will express the relationship between the amounts invested. The second equation will express the relationship between the amounts of interest earned by the investments.

The total amount invested is \$2000: $x + y = 2000$
The total annual interest earned is \$168: $0.075x + 0.09y = 168$

Solve the system of equations.

(1) $x + y = 2000$
(2) $0.075x + 0.09y = 168$

Solve Equation (1) for y and substitute into Equation (2).

(3) $y = -x + 2000$

$$0.075x + 0.09(-x + 2000) = 168 \quad \text{• Substitute } -x + 2000 \text{ for } y.$$
$$0.075x - 0.09x + 180 = 168$$
$$-0.015x = -12$$
$$x = 800$$

Substitute the value of x into Equation (3) and solve for y.

$$y = -x + 2000$$
$$y = -800 + 2000 = 1200 \quad \text{• Substitute 800 for } x.$$

The amount invested at 7.5% is \$800. The amount invested at 9% is \$1200.

EXAMPLE • 4

A hair stylist invested some money at an annual simple interest rate of 5.2%. A second investment, \$1000 more than the first, was invested at an annual simple interest rate of 7.2%. The total annual interest earned was \$320. How much was invested in each account?

Strategy

- Amount invested at 5.2%: x
  Amount invested at 7.2%: y

	Principal	*Rate*	*Interest*
Amount at 5.2%	x	0.052	$0.052x$
Amount at 7.2%	y	0.072	$0.072y$

- The second investment is \$1000 more than the first investment:

  $y = x + 1000$

  The sum of the interest earned at 5.2% and the interest earned at 7.2% equals \$320.

  $0.052x + 0.072y = 320$

Solution

(1) $y = x + 1000$
(2) $0.052x + 0.072y = 320$

Replace y in Equation (2) by $x + 1000$ from Equation (1). Then solve for x.

$$
\begin{aligned}
0.052x + 0.072y &= 320 \\
0.052x + 0.072(x + 1000) &= 320 && \bullet\ y = x + 1000 \\
0.052x + 0.072x + 72 &= 320 && \bullet\ \text{Solve for } x. \\
0.124x + 72 &= 320 \\
0.124x &= 248 \\
x &= 2000
\end{aligned}
$$

$$
\begin{aligned}
y &= x + 1000 \\
&= 2000 + 1000 && \bullet\ x = 2000 \\
&= 3000
\end{aligned}
$$

\$2000 was invested at an annual simple interest rate of 5.2%; \$3000 was invested at 7.2%.

YOU TRY IT • 4

The manager of a city's investment income wishes to place \$330,000 in two simple interest accounts. The first account earns 6.5% annual interest, and the second account earns 4.5%. How much should be invested in each account so that both accounts earn the same annual interest?

Your strategy

Your solution

Solution on p. S31

13.2 EXERCISES

OBJECTIVE A To solve a system of linear equations by the substitution method

1. When you solve a system of equations by the substitution method, how do you determine whether the system of equations is inconsistent?

2. When you solve a system of equations by the substitution method, how do you determine whether the system of equations is dependent?

For Exercises 3 to 32, solve by substitution.

3. $2x + 3y = 7$
$x = 2$

4. $y = 3$
$3x - 2y = 6$

5. $y = x - 3$
$x + y = 5$

6. $y = x + 2$
$x + y = 6$

7. $x = y - 2$
$x + 3y = 2$

8. $x = y + 1$
$x + 2y = 7$

9. $y = 4 - 3x$
$3x + y = 5$

10. $y = 2 - 3x$
$6x + 2y = 7$

11. $x = 3y + 3$
$2x - 6y = 12$

12. $x = 2 - y$
$3x + 3y = 6$

13. $3x + 5y = -6$
$x = 5y + 3$

14. $y = 2x + 3$
$4x - 3y = 1$

15. $3x + y = 4$
$4x - 3y = 1$

16. $x - 4y = 9$
$2x - 3y = 11$

17. $3x - y = 6$
$x + 3y = 2$

18. $4x - y = -5$
$2x + 5y = 13$

19. $3x - y = 5$
$2x + 5y = -8$

20. $3x + 4y = 18$
$2x - y = 1$

21. $4x + 3y = 0$
$2x - y = 0$

22. $5x + 2y = 0$
$x - 3y = 0$

23. $2x - y = 2$
$6x - 3y = 6$

24. $3x + y = 4$
$9x + 3y = 12$

25. $x = 3y + 2$
$y = 2x + 6$

26. $x = 4 - 2y$
$y = 2x - 13$

27. $y = 2x + 11$
$y = 5x - 19$

28. $y = 2x - 8$
$y = 3x - 13$

29. $y = -4x + 2$
$y = -3x - 1$

30. $x = 3y + 7$
$x = 2y - 1$

31. $x = 4y - 2$
$x = 6y + 8$

32. $x = 3 - 2y$
$x = 5y - 10$

For Exercises 33 and 34, assume that A, B, and C are nonzero real numbers. State whether the system of equations is independent, inconsistent, or dependent.

33. $x + y = A$
$x = A - y$

34. $x + y = B$
$y = -x + C, C \neq B$

OBJECTIVE B To solve investment problems

For Exercises 35 and 36, use the system of equations at the right, which represents the following situation. Owen Marshall places \$10,000 in two simple interest accounts. One account earns 8% annual simple interest, and the second account earns 6.5% annual simple interest.

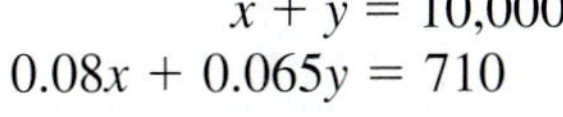

$$x + y = 10{,}000$$
$$0.08x + 0.065y = 710$$

35. What do the variables x and y represent? Explain the meaning of each equation in terms of the problem situation.

36. Write a question that could be answered by solving the system of equations.

37. An investment of \$3500 is divided between two simple interest accounts. On one account, the annual simple interest rate is 5%, and on the second account, the annual simple interest rate is 7.5%. How much should be invested in each account so that the total interest earned from the two accounts is \$215?

38. A mortgage broker purchased two trust deeds for a total of \$250,000. One trust deed earns 7% simple annual interest, and the second one earns 8% simple annual interest. If the total annual interest earned from the two trust deeds is \$18,500, what was the purchase price of each trust deed?

39. When Sara Whitehorse changed jobs, she rolled over the \$6000 in her retirement account into two simple interest accounts. On one account, the annual simple interest rate is 9%; on the second account, the annual simple interest rate is 6%. How much must be invested in each account if the accounts earn the same amount of annual interest?

40. An animal trainer decided to take the \$15,000 won on a game show and deposit it in two simple interest accounts. Part of the winnings were placed in an account paying 7% annual simple interest, and the remainder was used to purchase a government bond that earns 6.5% annual simple interest. The amount of interest earned for 1 year was \$1020. How much was invested in each account?

41. A police officer has chosen a high-yield stock fund that earns 8% annual simple interest for part of a \$6000 investment. The remaining portion is used to purchase a preferred stock that earns 11% annual simple interest. How much should be invested in each account so that the amount earned on the 8% account is twice the amount earned on the 11% account?

42. To plan for the purchase of a new car, a deposit was made into an account that earns 7% annual simple interest. Another deposit, \$1500 less than the first deposit, was placed in a certificate of deposit earning 9% annual simple interest. The total interest earned on both accounts for 1 year was \$505. How much money was deposited in the certificate of deposit?

43. The Pacific Investment Group invested some money in a certificate of deposit (CD) that earns 6.5% annual simple interest. Twice the amount invested at 6.5% was invested in a second CD that earns 8.5% annual simple interest. If the total annual interest earned from the two investments was \$4935, how much was invested at 6.5%?

44. A corporation gave a university \$300,000 to support product safety research. The university deposited some of the money in a 10% simple interest account and the remainder in an 8.5% simple interest account. How much should be deposited in each account so that the annual interest earned is \$28,500?

45. Ten co-workers formed an investment club, and each deposited \$2000 in the club's account. They decided to take the total amount and invest some of it in preferred stock that pays 8% annual simple interest and the remainder in a municipal bond that pays 7% annual simple interest. The amount of interest earned each year from the investments was \$1520. How much was invested in each?

46. A financial consultant advises a client to invest part of \$30,000 in municipal bonds that earn 6.5% annual simple interest and the remainder of the money in 8.5% corporate bonds. How much should be invested in each so that the total interest earned each year is \$2190?

47. Alisa Rhodes placed some money in a real estate investment trust that earns 7.5% annual simple interest. A second investment, which was one-half the amount placed in the real estate investment trust, was used to purchase a trust deed that earns 9% annual simple interest. If the total annual interest earned from the two investments was \$900, how much was invested in the trust deed?

Applying the Concepts

For Exercises 48 to 50, find the value of k for which the system of equations has no solution.

48. $2x - 3y = 7$
$kx - 3y = 4$

49. $8x - 4y = 1$
$2x - ky = 3$

50. $x = 4y + 4$
$kx - 8y = 4$

51. The following was offered as a solution of the system of equations.

(1) $y = \frac{1}{2}x + 2$

(2) $2x + 5y = 10$

$2x + 5y = 10$ • Equation (2)

$2x + 5\left(\frac{1}{2}x + 2\right) = 10$ • Substitute $\frac{1}{2}x + 2$ for y.

$2x + \frac{5}{2}x + 10 = 10$ • Solve for x.

$\frac{9}{2}x = 0$

$x = 0$

At this point the student stated that because $x = 0$, the system of equations has no solution. If this assertion is correct, is the system of equations independent, dependent, or inconsistent? If the assertion is not correct, what is the correct solution?

52. **Investments** A plant manager invested \$3000 more in stocks than in bonds. The stocks paid 8% annual simple interest, and the bonds paid 9.5% annual simple interest. Both investments yielded the same income. Find the total annual interest received on both investments.

53. **Compound Interest** The exercises in this objective were based on annual *simple* interest, r, which means that the amount of interest earned after 1 year is given by $I = Pr$. For **compound interest,** the interest earned for a certain period of time (usually daily or monthly) is added to the principal before the interest for the next period is calculated. The compound interest earned in 1 year is given by the formula $I = P\left[\left(1 + \frac{r}{n}\right)^n - 1\right]$, where n is the number of times per year the interest is compounded. For instance, if interest is compounded daily, then $n = 365$; if interest is compounded monthly, then $n = 12$. Suppose an investment of \$5000 is made into three different accounts. The first account earns 8% annual simple interest, the second earns 8% compounded monthly ($n = 12$), and the third earns 8% compounded daily ($n = 365$). Find the amount of interest earned from each account.

SECTION

13.3 Solving Systems of Linear Equations by the Addition Method

OBJECTIVE A To solve a system of linear equations by the addition method 

Another method of solving a system of equations is called the **addition method.** This method is based on the Addition Property of Equations.

Note, for the system of equations at the right, the effect of adding Equation (2) to Equation (1). Because $2y$ and $-2y$ are opposites, adding the equations results in an equation with only one variable.

$$\begin{array}{lr} (1) & 5x + 2y = 11 \\ (2) & 3x - 2y = 13 \\ \hline & 8x + 0y = 24 \\ & 8x = 24 \end{array}$$

Solving $8x = 24$ for x gives the first coordinate of the ordered-pair solution of the system of equations.

$$\frac{8x}{8} = \frac{24}{8}$$
$$x = 3$$

The second coordinate is found by substituting the value of x into Equation (1) or Equation (2) and then solving for y. Equation (1) is used here.

$$\begin{array}{lr} (1) & 5x + 2y = 11 \\ & 5(3) + 2y = 11 \\ & 15 + 2y = 11 \\ & 2y = -4 \\ & y = -2 \end{array}$$

The solution is $(3, -2)$.

Sometimes adding the two equations does not eliminate one of the variables. In this case, use the Multiplication Property of Equations to rewrite one or both of the equations so that the coefficients of one variable are opposites. Then add the equations and solve for the variables.

HOW TO • 1 Solve by the addition method: (1) $4x + y = 5$
(2) $2x - 5y = 19$

Multiply Equation (2) by -2. The coefficients of x will then be opposites.

$-2(2x - 5y) = -2 \cdot 19$ • **Multiply Equation (2) by -2.**
(3) $-4x + 10y = -38$ • **Simplify. This is Equation (3).**

Add Equation (1) to Equation (3). Then solve for y.

$$\begin{array}{lr} (1) & 4x + y = 5 \\ (3) & -4x + 10y = -38 \\ \hline & 11y = -33 \\ & y = -3 \end{array}$$

• **Note that the coefficients of x are opposites.**
• **Add the two equations.**
• **Solve for y.**

Substitute the value of y into Equation (1) or Equation (2) and solve for x. Equation (1) is used here.

$$\begin{array}{lr} (1) & 4x + y = 5 \\ & 4x + (-3) = 5 \\ & 4x - 3 = 5 \\ & 4x = 8 \\ & x = 2 \end{array}$$

• **Substitute -3 for y.**
• **Solve for x.**

The solution is $(2, -3)$.

Sometimes each equation of a system of equations must be multiplied by a constant so that the coefficients of one variable are opposites.

HOW TO • 2 Solve by the addition method: (1) $3x + 7y = 2$
(2) $5x - 3y = -26$

To eliminate x, multiply Equation (1) by 5 and Equation (2) by -3. Note at the right how the constants are chosen.

$$5(3x + 7y) = 5 \cdot 2$$
$$-3(5x - 3y) = -3(-26)$$

• **The negative is used so that the coefficients will be opposites.**

$$\begin{aligned} 15x + 35y &= 10 && \text{• 5 times Equation (1)} \\ -15x + 9y &= 78 && \text{• } -3 \text{ times Equation (2)} \\ \hline 44y &= 88 && \text{• Add the equations.} \\ y &= 2 && \text{• Solve for } y. \end{aligned}$$

Substitute the value of y into Equation (1) or Equation (2) and solve for x. Equation (1) is used here.

$$\begin{aligned} (1) \quad 3x + 7y &= 2 \\ 3x + 7(2) &= 2 && \text{• Substitute 2 for } y. \\ 3x + 14 &= 2 && \text{• Solve for } x. \\ 3x &= -12 \\ x &= -4 \end{aligned}$$

The solution is $(-4, 2)$.

For the above system of equations, the value of x was determined by substitution. This value can also be determined by eliminating y from the system.

$$\begin{aligned} 9x + 21y &= 6 && \text{• 3 times Equation (1)} \\ 35x - 21y &= -182 && \text{• 7 times Equation (2)} \\ \hline 44x \quad &= -176 && \text{• Add the equations.} \\ x &= -4 && \text{• Solve for } x. \end{aligned}$$

Note that this is the same value of x as was determined by using substitution.

Take Note

When you use the addition method to solve a system of equations and the result is an equation that is always true (like the one at the right), the system of equations is dependent. Compare this result with the following example.

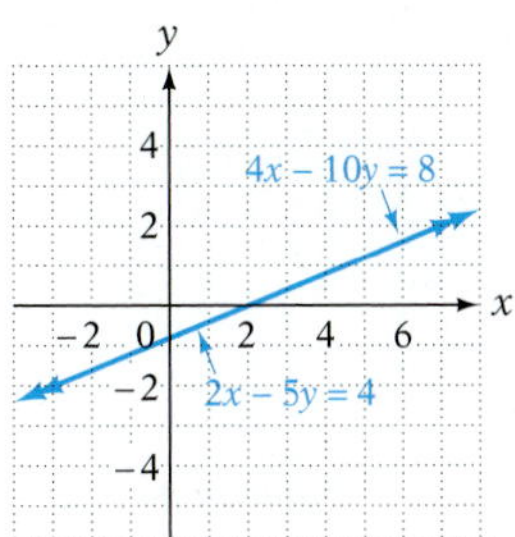

HOW TO • 3 Solve by the addition method: (1) $2x - 5y = 4$
(2) $4x - 10y = 8$

Eliminate x. Multiply Equation (1) by -2.

$$\begin{aligned} -2(2x - 5y) &= -2(4) && \text{• } -2 \text{ times Equation (1)} \\ (3) \quad -4x + 10y &= -8 && \text{• This is Equation (3).} \end{aligned}$$

Add Equation (3) to Equation (2) and solve for y.

$$\begin{aligned} (2) \quad 4x - 10y &= 8 \\ (3) \quad -4x + 10y &= -8 \\ \hline 0x + 0y &= 0 \\ 0 &= 0 \end{aligned}$$

The equation $0 = 0$ means that the system of equations is dependent. Therefore, the solutions of the system of equations are the ordered pairs that satisfy $2x - 5y = 4$.

The graphs of the two equations in the system of equations above are shown at the left. One line is on top of the other; therefore, the lines intersect infinitely often.

HOW TO • 4 Solve by the addition method: (1) $2x + y = 2$
(2) $4x + 2y = -5$

Eliminate y. Multiply Equation (1) by -2.

$$\begin{aligned} (1)\quad -2(2x + y) &= -2 \cdot 2 \\ (3)\quad -4x - 2y &= -4 \end{aligned}$$

• **-2 times Equation (1)**
• **This is Equation (3).**

Add Equation (2) to Equation (3) and solve for x.

$$\begin{aligned} (3)\quad -4x - 2y &= -4 \\ (2)\quad 4x + 2y &= -5 \\ \hline 0x + 0y &= -9 \\ 0 &= -9 \end{aligned}$$

• **Add Equation (2) to Equation (3).**
• **This is a false equation.**

The system of equations is inconsistent and therefore does not have a solution.

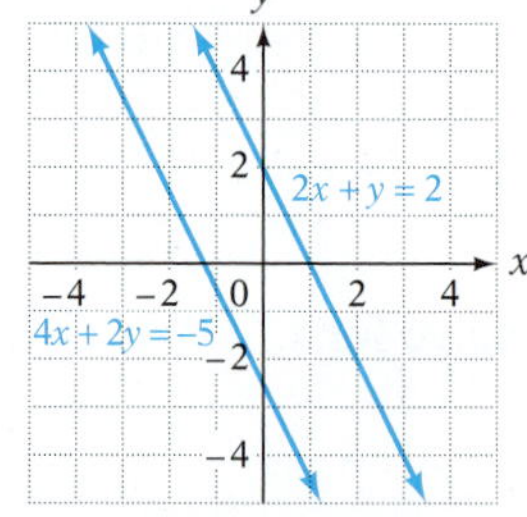

The graphs of the two equations in the system of equations above are shown at the left. Note that the graphs are parallel and therefore do not intersect. Thus the system of equations has no solution.

EXAMPLE • 1

Solve by the addition method:
(1) $2x + 4y = 7$
(2) $5x - 3y = -2$

Solution

Eliminate x.

$$\begin{aligned} 5(2x + 4y) &= 5 \cdot 7 \\ -2(5x - 3y) &= -2(-2) \end{aligned}$$

• **5 times Equation (1)**
• **-2 times Equation (2)**

$$\begin{aligned} 10x + 20y &= 35 \\ -10x + 6y &= 4 \\ \hline 26y &= 39 \\ y &= \frac{39}{26} = \frac{3}{2} \end{aligned}$$

• **Add the equations.**
• **Solve for y.**

Substitute $\frac{3}{2}$ for y in Equation (1).

$$\begin{aligned} (1)\quad 2x + 4y &= 7 \\ 2x + 4\left(\frac{3}{2}\right) &= 7 \\ 2x + 6 &= 7 \\ 2x &= 1 \\ x &= \frac{1}{2} \end{aligned}$$

• **Replace y by $\frac{3}{2}$.**
• **Solve for x.**

The solution is $\left(\frac{1}{2}, \frac{3}{2}\right)$.

YOU TRY IT • 1

Solve by the addition method:
(1) $2x - 3y = 1$
(2) $-3x + 4y = 6$

Your solution

Solution on p. S31

EXAMPLE • 2

Solve by the addition method:

(1) $6x + 9y = 15$

(2) $4x + 6y = 10$

Solution

Eliminate x.

$4(6x + 9y) = 4 \cdot 15$ • 4 times Equation (1)

$-6(4x + 6y) = -6 \cdot 10$ • −6 times Equation (2)

$24x + 36y = 60$

$-24x - 36y = -60$

$0x + 0y = 0$ • Add the equations.

$0 = 0$

The system of equations is dependent. The solutions are the ordered pairs that satisfy the equation $6x + 9y = 15$.

YOU TRY IT • 2

Solve by the addition method:

$2x - 3y = 4$

$-4x + 6y = -8$

Your solution

EXAMPLE • 3

Solve by the addition method:

(1) $2x = y + 8$

(2) $3x + 2y = 5$

Solution

Write Equation (1) in the form $Ax + By = C$.

$2x = y + 8$

(3) $2x - y = 8$ • This is Equation (3).

Eliminate y.

$2(2x - y) = 2 \cdot 8$ • 2 times Equation (3)

$3x + 2y = 5$ • This is Equation (2).

$4x - 2y = 16$

$3x + 2y = 5$

$7x = 21$ • Add the equations.

$x = 3$

Replace x in Equation (1).

(1) $2x = y + 8$

$2 \cdot 3 = y + 8$ • Replace x by 3.

$6 = y + 8$

$-2 = y$

The solution is $(3, -2)$.

YOU TRY IT • 3

Solve by the addition method:

$4x + 5y = 11$

$3y = x + 10$

Your solution

Solutions on pp. S31–S32

13.3 EXERCISES

OBJECTIVE A To solve a system of linear equations by the addition method

For Exercises 1 to 36, solve by the addition method.

1. $x + y = 4$
   $x - y = 6$

2. $2x + y = 3$
   $x - y = 3$

3. $x + y = 4$
   $2x + y = 5$

4. $x - 3y = 2$
   $x + 2y = -3$

5. $2x - y = 1$
   $x + 3y = 4$

6. $x - 2y = 4$
   $3x + 4y = 2$

7. $4x - 5y = 22$
   $x + 2y = -1$

8. $3x - y = 11$
   $2x + 5y = 13$

9. $2x - y = 1$
   $4x - 2y = 2$

10. $x + 3y = 2$
    $3x + 9y = 6$

11. $4x + 3y = 15$
    $2x - 5y = 1$

12. $3x - 7y = 13$
    $6x + 5y = 7$

13. $2x - 3y = 1$
    $4x - 6y = 2$

14. $2x + 4y = 6$
    $3x + 6y = 9$

15. $3x - 6y = -1$
    $6x - 4y = 2$

16. $5x + 2y = 3$
    $3x - 10y = -1$

17. $5x + 7y = 10$
    $3x - 14y = 6$

18. $7x + 10y = 13$
    $4x + 5y = 6$

19. $3x - 2y = 0$
    $6x + 5y = 0$

20. $5x + 2y = 0$
    $3x + 5y = 0$

21. $2x - 3y = 16$
    $3x + 4y = 7$

22. $3x + 4y = 10$
    $4x + 3y = 11$

23. $5x + 3y = 7$
    $2x + 5y = 1$

24. $-2x + 7y = 9$
    $3x + 2y = -1$

25. $3x + 4y = 4$
$5x + 12y = 5$

26. $2x + 5y = 2$
$3x + 3y = 1$

27. $8x - 3y = 11$
$6x - 5y = 11$

28. $4x - 8y = 36$
$3x - 6y = 15$

29. $5x + 15y = 20$
$2x + 6y = 12$

30. $y = 2x - 3$
$3x + 4y = -1$

31. $3x = 2y + 7$
$5x - 2y = 13$

32. $2y = 4 - 9x$
$9x - y = 25$

33. $2x + 9y = 16$
$5x = 1 - 3y$

34. $3x - 4 = y + 18$
$4x + 5y = -21$

35. $2x + 3y = 7 - 2x$
$7x + 2y = 9$

36. $5x - 3y = 3y + 4$
$4x + 3y = 11$

In Exercises 37 to 39, assume that A, B, and C are nonzero real numbers, where $A \neq B \neq C$. State whether the system of equations is independent, inconsistent, or dependent.

37. $Ax + By = C$
$2Ax + 2By = 2C$

38. $x - Ay = B$
$3x - 3Ay = 3C$

39. $Ax + By = C$
$Bx + Ay = 2C$

Applying the Concepts

40. The point of intersection of the graphs of the equations $Ax + 2y = 2$ and $2x + By = 10$ is $(2, -2)$. Find A and B.

41. The point of intersection of the graphs of the equations $Ax - 4y = 9$ and $4x + By = -1$ is $(-1, -3)$. Find A and B.

42. For what value of k is the system of equations dependent?

a. $2x + 3y = 7$
$4x + 6y = k$

b. $y = \frac{2}{3}x - 3$
$y = kx - 3$

c. $x = ky - 1$
$y = 2x + 2$

43. For what value of k is the system of equations inconsistent?

a. $x + y = 7$
$kx + y = 3$

b. $x + 2y = 4$
$kx + 3y = 2$

c. $2x + ky = 1$
$x + 2y = 2$

SECTION

13.4 Application Problems in Two Variables

OBJECTIVE A To solve rate-of-wind or rate-of-current problems 

We normally need two variables to solve motion problems that involve an object moving with or against a wind or current.

HOW TO • 1 Flying with the wind, a small plane can fly 600 mi in 3 h. Against the wind, the plane can fly the same distance in 4 h. Find the rate of the plane in calm air and the rate of the wind.

Strategy for Solving Rate-of-Wind or Rate-of-Current Problems

1. Choose one variable to represent the rate of the object in calm conditions and a second variable to represent the rate of the wind or current. Using these variables, express the rate of the object traveling with and against the wind or current. Use the equation $rt = d$ to write expressions for the distance traveled by the object. The results can be recorded in a table.

Rate of plane in calm air: p
Rate of wind: w

	Rate	·	*Time*	=	*Distance*
With the wind	$p + w$	·	3	=	$3(p + w)$
Against the wind	$p - w$	·	4	=	$4(p - w)$

2. Determine how the expressions for distance are related.

The distance traveled with the wind is 600 mi. $3(p + w) = 600$
The distance traveled against the wind is 600 mi. $4(p - w) = 600$

Solve the system of equations.

$$3(p + w) = 600 \quad \longrightarrow \quad \frac{1}{3} \cdot 3(p + w) = \frac{1}{3} \cdot 600 \quad \longrightarrow \quad p + w = 200$$

$$4(p - w) = 600 \quad \quad \frac{1}{4} \cdot 4(p - w) = \frac{1}{4} \cdot 600 \quad \quad p - w = 150$$

$$2p = 350$$
$$p = 175$$

$$p + w = 200$$
$$175 + w = 200 \quad \text{• } p = 175$$
$$w = 25$$

The rate of the plane in calm air is 175 mph.
The rate of the wind is 25 mph.

EXAMPLE • 1

A 450-mile trip from one city to another takes 3 h when a plane is flying with the wind. The return trip, against the wind, takes 5 h. Find the rate of the plane in still air and the rate of the wind.

Strategy

- Rate of the plane in still air: p
  Rate of the wind: w

	Rate	Time	Distance
With wind	$p + w$	3	$3(p + w)$
Against wind	$p - w$	5	$5(p - w)$

- The distance traveled with the wind is 450 mi.
  The distance traveled against the wind is 450 mi.

Solution

$$3(p + w) = 450 \qquad \frac{1}{3} \cdot 3(p + w) = \frac{1}{3} \cdot 450$$

$$5(p - w) = 450 \qquad \frac{1}{5} \cdot 5(p - w) = \frac{1}{5} \cdot 450$$

$$p + w = 150$$
$$p - w = 90$$

$$2p = 240$$
$$p = 120$$

$$p + w = 150$$
$$120 + w = 150 \qquad \bullet\ p = 120$$
$$w = 30$$

The rate of the plane in still air is 120 mph.
The rate of the wind is 30 mph.

YOU TRY IT • 1

A canoeist paddling with the current can travel 15 mi in 3 h. Against the current, it takes the canoeist 5 h to travel the same distance. Find the rate of the current and the rate of the canoeist in calm water.

Your strategy

Your solution

Solution on p. S32

OBJECTIVE B To solve application problems using two variables 

The application problems in this section are varieties of those problems solved earlier in the text. Each of the strategies for the problems in this section will result in a system of equations.

© Spencer Grant/PhotoEdit

Point of Interest

The Babylonians had a method for solving systems of equations. Here is an adaptation of a problem from an ancient (around 1500 B.C.E.) Babylonian text. "There are two silver blocks. The sum of $\frac{1}{7}$ of the first block and $\frac{1}{11}$ of the second block is one sheqel (a weight). The first block diminished by $\frac{1}{7}$ of its weight equals the second diminished by $\frac{1}{11}$ of its weight. What are the weights of the two blocks?"

HOW TO • 2 A jeweler purchased 5 oz of a gold alloy and 20 oz of a silver alloy for a total cost of \$540. The next day, at the same prices per ounce, the jeweler purchased 4 oz of the gold alloy and 25 oz of the silver alloy for a total cost of \$450. Find the cost per ounce of the gold and silver alloys.

Strategy for Solving an Application Problem in Two Variables

1. Choose one variable to represent one of the unknown quantities and a second variable to represent the other unknown quantity. Write numerical or variable expressions for all of the remaining quantities. These results can be recorded in two tables, one for each of the conditions.

Cost per ounce of gold: g
Cost per ounce of silver: s

First day:

	Amount	·	*Unit Cost*	=	*Value*
Gold	5	·	g	=	$5g$
Silver	20	·	s	=	$20s$

Second day:

	Amount	·	*Unit Cost*	=	*Value*
Gold	4	·	g	=	$4g$
Silver	25	·	s	=	$25s$

2. Determine a system of equations. Each table will give one equation of the system.

The total value of the purchase on the first day was \$540. $5g + 20s = 540$

The total value of the purchase on the second day was \$450. $4g + 25s = 450$

Solve the system of equations.

$$5g + 20s = 540 \qquad 4(5g + 20s) = 4 \cdot 540 \qquad 20g + 80s = 2160$$
$$4g + 25s = 450 \qquad -5(4g + 25s) = -5 \cdot 450 \qquad -20g - 125s = -2250$$
$$-45s = -90$$
$$s = 2$$

$$5g + 20s = 540$$
$$5g + 20(2) = 540 \qquad \bullet\ s = 2$$
$$5g + 40 = 540$$
$$5g = 500$$
$$g = 100$$

The cost per ounce of the gold alloy was \$100.
The cost per ounce of the silver alloy was \$2.

EXAMPLE • 2

A store owner purchased 20 halogen light bulbs and 30 fluorescent bulbs for a total cost of \$630. A second purchase, at the same prices, included 30 halogen bulbs and 10 fluorescent bulbs for a total cost of \$560. Find the cost of a halogen bulb and of a fluorescent bulb.

Strategy

Cost of a halogen bulb: h
Cost of a fluorescent bulb: f

First purchase:

	Amount	*Unit Cost*	*Value*
Halogen	20	h	$20h$
Fluorescent	30	f	$30f$

Second purchase:

	Amount	*Unit Cost*	*Value*
Halogen	30	h	$30h$
Fluorescent	10	f	$10f$

The total cost of the first purchase was \$630.
The total cost of the second purchase was \$560.

Solution

$$20h + 30f = 630 \qquad 20h + 30f = 630$$
$$30h + 10f = 560 \qquad -3(30h + 10f) = -3(560)$$

$$\begin{aligned} 20h + 30f &= 630 \\ -90h - 30f &= -1680 \\ \hline -70h &= -1050 \\ h &= 15 \end{aligned}$$

$$\begin{aligned} 20h + 30f &= 630 \\ 20(15) + 30f &= 630 \quad \bullet\ h = 15 \\ 300 + 30f &= 630 \\ 30f &= 330 \\ f &= 11 \end{aligned}$$

The cost of a halogen light bulb is \$15.
The cost of a fluorescent light bulb is \$11.

YOU TRY IT • 2

A citrus grower purchased 25 orange trees and 20 grapefruit trees for \$2900. The next week, at the same prices, the grower bought 20 orange trees and 30 grapefruit trees for \$3300. Find the cost of an orange tree and the cost of a grapefruit tree.

Your strategy

Your solution

Solution on p. S32

13.4 EXERCISES

OBJECTIVE A To solve rate-of-wind or rate-of-current problems

1. Traveling with the wind, a plane flies m miles in h hours. Traveling against the wind, the plane flies n miles in h hours. Is m less than, equal to, or greater than n?

2. Traveling against the current, it takes a boat h hours to go m miles. Traveling with the current, the boat takes k hours to go m miles. Is k less than, equal to, or greater than h?

3. A rowing team rowing with the current traveled 40 km in 2 h. Rowing against the current, the team could travel only 16 km in 2 h. Find the rowing rate in calm water and the rate of the current.

4. A plane flying with the jet stream flew from Los Angeles to Chicago, a distance of 2250 mi, in 5 h. Flying against the jet stream, the plane could fly only 1750 mi in the same amount of time. Find the rate of the plane in calm air and the rate of the wind.

5. A whale swimming against an ocean current traveled 60 mi in 2 h. Swimming in the opposite direction, with the current, the whale was able to travel the same distance in 1.5 h. Find the speed of the whale in calm water and the rate of the ocean current.

WILDLIFE/Peter Arnold Inc.

6. The bird capable of the fastest flying speed is the swift. A swift flying with the wind to a favorite feeding spot traveled 26 mi in 0.2 h. On returning, now against the wind, the swift was able to travel only 16 mi in the same amount of time. What is the rate of the swift in calm air, and what was the rate of the wind?

7. A private Learjet 31A was flying with a tailwind and traveled 1120 mi in 2 h. Flying against the wind on the return trip, the jet was able to travel only 980 mi in 2 h. Find the speed of the jet in calm air and the rate of the wind.

8. A plane flying with a tailwind flew 300 mi in 2 h. Against the wind, it took 3 h to travel the same distance. Find the rate of the plane in calm air and the rate of the wind.

9. A Boeing Apache Longbow military helicopter traveling directly into a strong headwind was able to travel 450 mi in 2.5 h. The return trip, now with a tailwind, took 1 h 40 min. Find the speed of the helicopter in calm air and the rate of the wind.

10. Rowing with the current, a canoeist paddled 14 mi in 2 h. Against the current, the canoeist could paddle only 10 mi in the same amount of time. Find the rate of the canoeist in calm water and the rate of the current.

11. A motorboat traveling with the current went 35 mi in 3.5 h. Traveling against the current, the boat went 12 mi in 3 h. Find the rate of the boat in calm water and the rate of the current.

12. With the wind, a quarterback passes a football 140 ft in 2 s. Against the wind, the same pass would have traveled 80 ft in 2 s. Find the rate of the pass and the rate of the wind.

OBJECTIVE B To solve application problems using two variables

For Exercises 13 and 14, use the system of equations at the right, which represents the following situation. You spent \$320 on theater tickets for 4 adults and 2 children. For the same performance, your neighbor spent \$240 on tickets for 2 adults and 3 children.

$$4x + 2y = 320$$
$$2x + 3y = 240$$

13. What do the variables x and y represent? Explain the meaning of each equation in terms of the problem situation.

14. Write a question that could be answered by solving the system of equations.

Sandra Baker/Alamy

15. **Flour Mixtures** A baker purchased 12 lb of wheat flour and 15 lb of rye flour for a total cost of \$18.30. A second purchase, at the same prices, included 15 lb of wheat flour and 10 lb of rye flour. The cost of the second purchase was \$16.75. Find the cost per pound of the wheat flour and of the rye flour.

16. **Consumerism** For using a computerized financial news network for 50 min during prime time and 70 min during non-prime time, a customer was charged \$10.75. A second customer was charged \$13.35 for using the network for 60 min of prime time and 90 min of non-prime time. Find the cost per minute for using the financial news network during prime time.

17. **Consumerism** The employees of a hardware store ordered lunch from a local delicatessen. The lunch consisted of 4 turkey sandwiches and 7 orders of french fries, for a total cost of \$38.30. The next day, the employees ordered 5 turkey sandwiches and 5 orders of french fries totaling \$40.75. What does the delicatessen charge for a turkey sandwich? What is the charge for an order of french fries?

18. **Fuel Mixtures** An octane number of 87 on gasoline means that it will fight engine "knock" as effectively as a reference fuel that is 87% isooctane, a type of gas. Suppose you want to fill an empty 18-gallon tank with some 87-octane gasoline and some 93-octane fuel to produce a mixture that is 89-octane. How much of each type of gasoline must you use?

19. Food Mixtures A pastry chef created a 50-ounce sugar solution that was 34% sugar from a 20% sugar solution and a 40% sugar solution. How much of the 20% sugar solution and how much of the 40% sugar solution were used?

© Jeff Greenberg/PhotoEdit

Ideal Body Weight There are various formulas for calculating ideal body weight. In each of the formulas in Exercises 20 and 21, W is ideal body weight in kilograms, and x is height in inches above 60 in.

20. J. D. Robinson gave the following formula for men: $W = 52 + 1.9x$. D. R. Miller published a slightly different formula for men: $W = 56.2 + 1.41x$. At what height do both formulas give the same ideal body weight? Round to the nearest whole number.

21. J. D. Robinson gave the following formula for women: $W = 49 + 1.7x$. D. R. Miller published a slightly different formula for women: $W = 53.1 + 1.36x$. At what height do both formulas give the same ideal body weight? Round to the nearest whole number.

22. Fuel Economy Read the article at the right. Suppose you use 10 gal of gas to drive a 2007 Ford Taurus 208 mi. Using the new miles-per-gallon estimates given in the article, find the number of city miles and the number of highway miles you drove.

In the News

New Miles-per-Gallon Estimates

Beginning with model year 2008, the Environmental Protection Agency is using a new method to estimate miles-per-gallon ratings for motor vehicles. In general, estimates will be lower than before. For example, under the new method, ratings for a 2007 Ford Taurus would be lowered to 18 mpg in the city and 25 mpg on the highway.

Source: www.fueleconomy.gov

23. Stamps Stolen in 1967, the famous "Ice House" envelope (named for the address shown on the envelope) was recovered in 2006. The envelope displays a Lincoln stamp, a Thomas Jefferson stamp, and a Henry Clay stamp.

a. The original postage value of three Lincoln stamps and five Jefferson stamps was \$3.20. The original postage value of two Lincoln stamps and three Jefferson stamps was \$2.10. Find the original value of the Lincoln stamp and the Jefferson stamp.

b. The total postage on the Ice House envelope was \$1.12. What was the original postage value of the Henry Clay stamp?

Applying the Concepts

24. Geometry Two angles are supplementary. The measure of the larger angle is 15° more than twice the measure of the smaller angle. Find the measures of the two angles. (Supplementary angles are two angles whose sum is 180°.)

25. Investments An investor has \$5000 to invest in two accounts. The first account earns 8% annual simple interest, and the second account earns 10% annual simple interest. How much money should be invested in each account so that the total annual simple interest earned is \$600?

FOCUS ON PROBLEM SOLVING

Relevant Information

One of the challenges of problem solving is to separate the information that is relevant to the problem from other information. Following is an example:

A lawyer drove 8 mi to the train station. After a 35-minute ride of 18 mi, the lawyer walked 10 min to the office. Find the total time it took the lawyer to get to work.

From this situation, answer the following questions before reading on.

a. What is asked for?

b. Is there enough information to answer the question?

c. Is information given that is not needed?

Here are the answers.

a. We want the total time for the lawyer to get to work.

b. No. We do not know the time it takes the lawyer to get to the train station.

c. Yes. Neither the distance to the train station nor the distance of the train ride is necessary to answer the question.

For each of the following problems, answer these questions:

a. What is asked for?

b. Is there enough information to answer the question?

c. Is information given that is not needed?

1. A customer bought six boxes of strawberries and paid with a \$20 bill. What was the change?

2. A board is cut into two pieces. One piece is 3 feet longer than the other piece. What is the length of the original board?

3. A family rented a car for their vacation and drove 680 miles. The cost of the rental car was \$21 per day with 150 free miles per day and \$.15 for each mile driven above the number of free miles allowed. How many miles did the family drive per day?

4. An investor bought 8 acres of land for \$80,000. One and one-half acres were set aside for a park, and the remainder of the land was developed into one-half-acre lots. How many lots were available for sale?

5. You wrote checks of \$43.67, \$122.88, and \$432.22 after making a deposit of \$768.55. How much do you have left in your checking account?

PROJECTS AND GROUP ACTIVITIES

Solving a System of Equations with a Graphing Calculator

A graphing calculator can be used to approximate the solution of a system of equations in two variables. Graph each equation of the system of equations, and then approximate the coordinates of the point of intersection. The process by which you approximate the solution depends on what model of calculator you have. In all cases, however, you must first solve each equation in the system of equations for y.

Solve: $2x - 5y = 9$
$4x + 3y = 2$

$$\begin{aligned} 2x - 5y &= 9 \\ -5y &= -2x + 9 \\ y &= \frac{2}{5}x - \frac{9}{5} \end{aligned} \qquad \begin{aligned} 4x + 3y &= 2 \\ 3y &= -4x + 2 \\ y &= -\frac{4}{3}x + \frac{2}{3} \end{aligned}$$

• Solve each equation for y.

Take Note

The graphing calculator screens shown here are taken from a TI-84. Similar screens would display if we used a different model of graphing calculator.

For the TI-84, press Y=. Enter one equation as Y1 and the other as Y2. The result should be similar to the screen at the left below. Press GRAPH. The graphs of the two equations should appear on the screen, as shown at the right below. If the point of intersection is not on the screen, adjust the viewing window by pressing the WINDOW key.

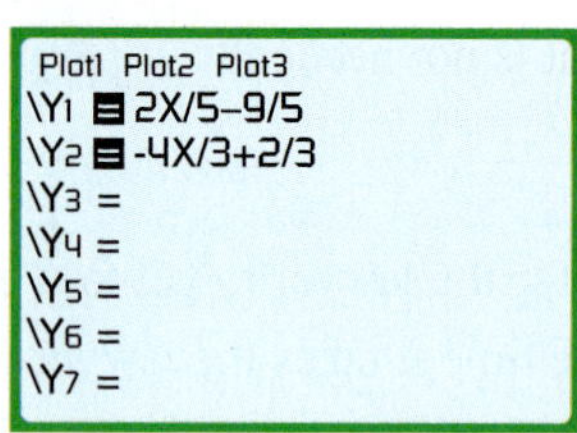

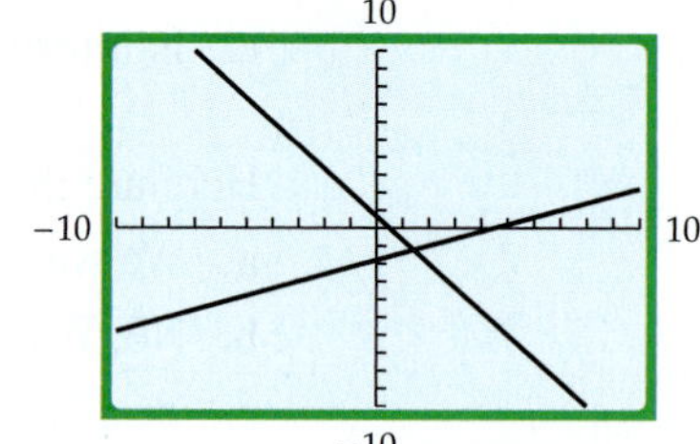

Integrating Technology

See the Keystroke Guide: *Intersect* for instructions on using a graphing calculator to solve systems of equations.

Press  CALC 5  . After a few seconds, the point of intersection will show on the bottom of the screen as X = 1.4230769, Y = −1.230769.

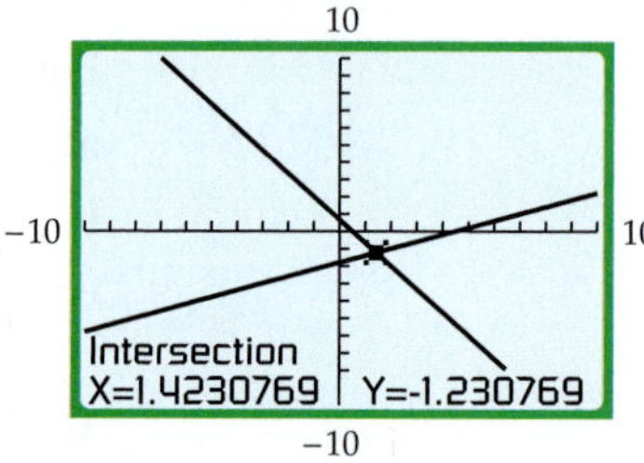

For Exercises 1 to 4, solve by using a graphing calculator.

1. $4x - 5y = 8$
   $5x + 7y = 7$
2. $3x + 2y = 11$
   $7x - 6y = 13$
3. $x = 3y + 2$
   $y = 4x - 2$
4. $x = 2y - 5$
   $x = 3y + 2$

CHAPTER 13

SUMMARY

KEY WORDS	EXAMPLES
Two or more equations considered together are called a *system of equations.* [13.1A, p. 680]	An example of a system of equations is $2x - 3y = 9$, $3x + 4y = 5$
A *solution of a system of equations in two variables* is an ordered pair that is a solution of each equation of the system. [13.1A, p. 680]	The solution of the system of equations shown above is the ordered pair $(3, -1)$ because it is a solution of each equation of the system of equations.
An *independent system* of linear equations has exactly one solution. The graphs of the equations in an independent system of linear equations intersect at one point. [13.1A, p. 681]	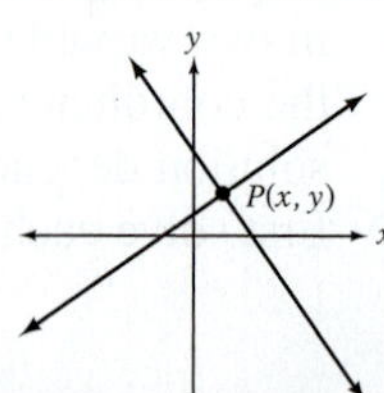

A *dependent system* of linear equations has an infinite number of solutions. The graphs of the equations in a dependent system of linear equations are the same line. [13.1A, p. 681]

If, when solving a system of equations algebraically, the variable is eliminated and the result is a true equation, such as $5 = 5$, the system of equations is dependent. [13.2A, p. 690]

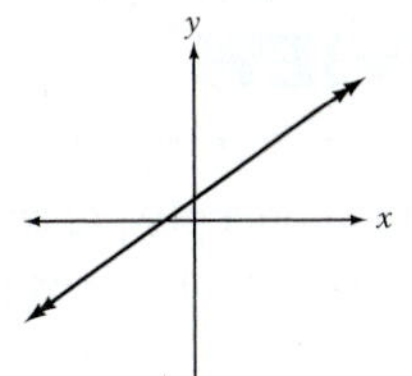

An *inconsistent system* of linear equations has no solution. The graphs of the equations of an inconsistent system of linear equations are parallel lines. [13.1A, p. 681]

If, when solving a system of equations algebraically, the variable is eliminated and the result is a false equation, such as $0 = 4$, the system of equations is inconsistent. [13.2A, p. 690]

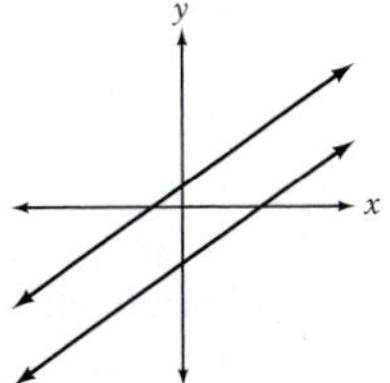

ESSENTIAL RULES AND PROCEDURES

To solve a system of linear equations in two variables by graphing, graph each equation on the same coordinate system, and then determine the point of intersection. [13.1A, p. 682]

EXAMPLES

Solve by graphing: $x + 2y = 4$
$2x + y = -1$

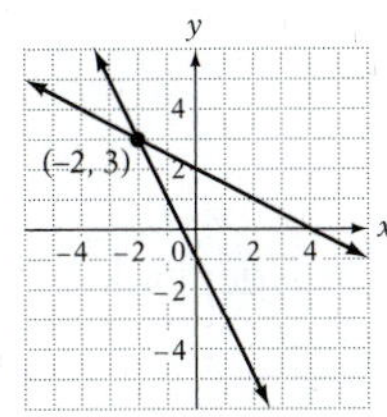

The solution is $(-2, 3)$.

To solve a system of linear equations by the substitution method, one variable must be written in terms of the other variable. [13.2A, p. 688]

Solve by substitution: $2x + y = 5$ (1)
$3x - 2y = 11$ (2)

$$2x + y = 5$$
$$y = -2x + 5 \quad \text{• Solve Equation (1) for } y.$$
$$3x - 2y = 11$$
$$3x - 2(-2x + 5) = 11 \quad \text{• Substitute for } y \text{ in Equation (2).}$$
$$3x + 4x - 10 = 11$$
$$7x - 10 = 11$$
$$7x = 21$$
$$x = 3$$

$$y = -2x + 5$$
$$y = -2(3) + 5$$
$$y = -1$$

The solution is $(3, -1)$.

To solve a system of linear equations by the addition method, use the Multiplication Property of Equations to rewrite one or both of the equations so that the coefficients of one variable are opposites. Then add the equations and solve for the variables. [13.3A, p. 698]

Solve by the addition method:
$2x + 5y = 8$ (1)
$3x - 4y = -11$ (2)

$$6x + 15y = 24 \quad \text{• 3 times Equation (1)}$$
$$-6x + 8y = 22 \quad \text{• −2 times Equation (2)}$$
$$23y = 46 \quad \text{• Add the equations.}$$
$$y = 2 \quad \text{• Solve for } y.$$

$$2x + 5y = 8$$
$$2x + 5(2) = 8 \quad \text{• Replace } y \text{ by 2 in Equation (1).}$$
$$2x + 10 = 8 \quad \text{• Solve for } x.$$
$$2x = -2$$
$$x = -1$$

The solution is $(-1, 2)$.

CHAPTER 13

CONCEPT REVIEW

Test your knowledge of the concepts presented in this chapter. Answer each question. Then check your answers against the ones provided in the Answer Section.

1. After graphing a system of linear equations, how is the solution determined?

2. What is the difference between an independent system and a dependent system of equations?

3. What does the graph of an inconsistent system of equations look like?

4. What does the graph of a dependent system of equations look like?

5. What steps are used to solve a system of linear equations by the substitution method?

6. What formula is used to solve a simple interest problem?

7. What steps are used to solve a system of linear equations by the addition method?

8. When using the addition method, after adding the two equations in a system of equations, what type of resulting equation tells you that the system of equations is dependent?

9. In a rate-of-wind problem, what do the expressions $p + w$ and $p - w$ represent?

10. In application problems in two variables, why are two equations written?

CHAPTER 13

REVIEW EXERCISES

1. Is $(-1, -3)$ a solution of this system of equations?
$$\begin{aligned} 5x + 4y &= -17 \\ 2x - y &= 1 \end{aligned}$$

2. Is $(-2, 0)$ a solution of this system of equations?
$$\begin{aligned} -x + 9y &= 2 \\ 6x - 4y &= 12 \end{aligned}$$

3. Solve by graphing:
$$\begin{aligned} 3x - y &= 6 \\ y &= -3 \end{aligned}$$

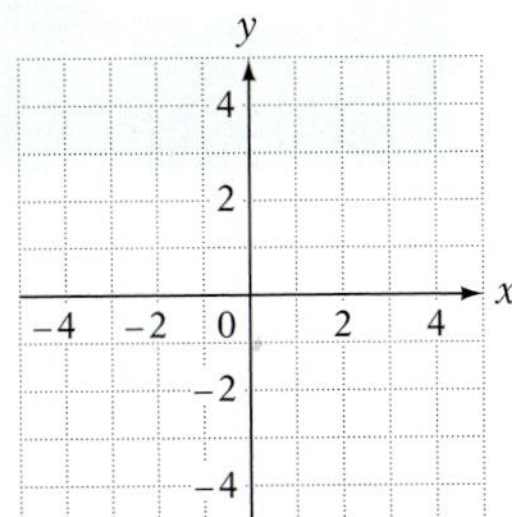

4. Solve by graphing:
$$\begin{aligned} 4x - 2y &= 8 \\ y &= 2x - 4 \end{aligned}$$

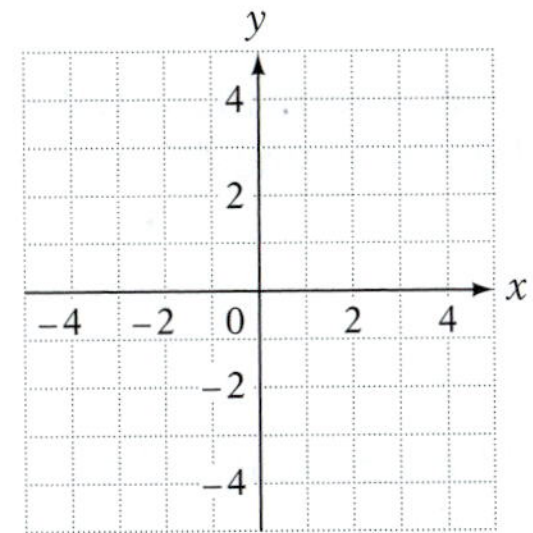

5. Solve by graphing:
$$\begin{aligned} x + 2y &= 3 \\ y &= -\frac{1}{2}x + 1 \end{aligned}$$

6. Solve by substitution:
$$\begin{aligned} 4x + 7y &= 3 \\ x &= y - 2 \end{aligned}$$

7. Solve by substitution:
$$\begin{aligned} 6x - y &= 0 \\ 7x - y &= 1 \end{aligned}$$

8. Solve by the addition method:
$$\begin{aligned} 3x + 8y &= -1 \\ x - 2y &= -5 \end{aligned}$$

9. Solve by the addition method:
$$\begin{aligned} 6x + 4y &= -3 \\ 12x - 10y &= -15 \end{aligned}$$

10. Solve by substitution:
$$\begin{aligned} 12x - 9y &= 18 \\ y &= \frac{4}{3}x - 3 \end{aligned}$$

11. Solve by substitution:
$$\begin{aligned} 8x - y &= 2 \\ y &= 5x + 1 \end{aligned}$$

12. Solve by the addition method:
$$\begin{aligned} 4x - y &= 9 \\ 2x + 3y &= -13 \end{aligned}$$

13. Solve by the addition method:
$$\begin{aligned} 5x + 7y &= 21 \\ 20x + 28y &= 63 \end{aligned}$$

14. Solve by substitution:
$$\begin{aligned} 4x + 3y &= 12 \\ y &= -\frac{4}{3}x + 4 \end{aligned}$$

15. Solve by substitution:
$7x + 3y = -16$
$x - 2y = 5$

16. Solve by the addition method:
$3x + y = -2$
$-9x - 3y = 6$

17. Solve by the addition method:
$6x - 18y = 7$
$9x + 24y = 2$

18. **Sculling** A sculling team rowing with the current went 24 mi in 2 h. Rowing against the current, the sculling team went 18 mi in 3 h. Find the rate of the sculling team in calm water and the rate of the current.

© Joel W. Rogers/Corbis

19. **Investments** An investor bought 1500 shares of stock, some at \$6 per share and the rest at \$25 per share. If \$12,800 worth of stock was purchased, how many shares of each kind did the investor buy?

20. **Travel** A flight crew flew 420 km in 3 h with a tailwind. Flying against the wind, the flight crew flew 440 km in 4 h. Find the rate of the flight crew in calm air and the rate of the wind.

21. **Travel** A small plane flying with the wind flew 360 mi in 3 h. Against a headwind, the plane took 4 h to fly the same distance. Find the rate of the plane in calm air and the rate of the wind.

22. **Consumerism** A computer online service charges one hourly rate for regular use and a higher hourly rate for designated "premium" services. A customer was charged \$14.00 for 9 h of basic use and 2 h of premium use. Another customer was charged \$13.50 for 6 h of regular use and 3 h of premium use. What is the service charge per hour for regular and premium services?

23. **Investments** Terra Cotta Art Center receives an annual income of \$915 from two simple interest investments. One investment, in a corporate bond fund, earns 8.5% annual simple interest. The second investment, in a real estate investment trust, earns 7% annual simple interest. If the total amount invested in the two accounts is \$12,000, how much is invested in each account?

24. **Grain Mixtures** A silo contains a mixture of lentils and corn. If 50 bushels of lentils were added, there would be twice as many bushels of lentils as of corn. If 150 bushels of corn were added instead, there would be the same amount of corn as of lentils. How many bushels of each were originally in the silo?

25. **Investments** Mosher Children's Hospital received a \$300,000 donation that it invested in two simple interest accounts, one earning 5.4% and the other earning 6.6%. If each account earned the same amount of annual interest, how much was invested in each account?

CHAPTER 13

TEST

1. Is $(-2, 3)$ a solution of this system?
$2x + 5y = 11$
$x + 3y = 7$

2. Is $(1, -3)$ a solution of this system?
$3x - 2y = 9$
$4x + y = 1$

3. Solve by graphing: $3x + 2y = 6$
$5x + 2y = 2$

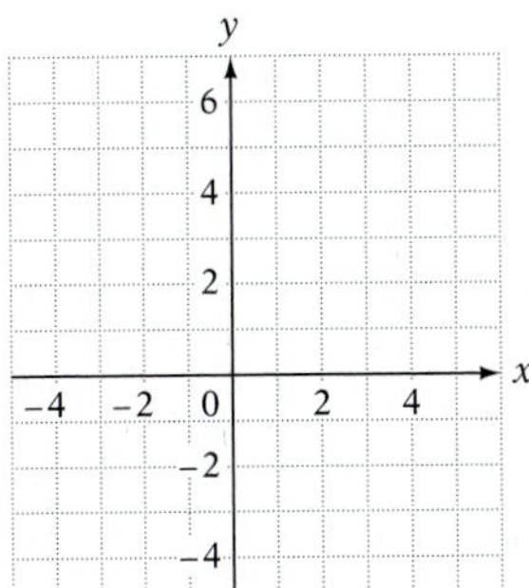

4. Solve by substitution:
$4x - y = 11$
$y = 2x - 5$

5. Solve by substitution:
$x = 2y + 3$
$3x - 2y = 5$

6. Solve by substitution:
$3x + 5y = 1$
$2x - y = 5$

7. Solve by substitution:
$3x - 5y = 13$
$x + 3y = 1$

8. Solve by substitution:
$2x - 4y = 1$
$y = \frac{1}{2}x + 3$

9. Solve by the addition method:
$4x + 3y = 11$
$5x - 3y = 7$

10. Solve by the addition method:
$2x - 5y = 6$
$4x + 3y = -1$

11. Solve by the addition method:
$x + 2y = 8$
$3x + 6y = 24$

12. Solve by the addition method:
$7x + 3y = 11$
$2x - 5y = 9$

13. Solve by the addition method:
$5x + 6y = -7$
$3x + 4y = -5$

14. **Travel** With the wind, a plane flies 240 mi in 2 h. Against the wind, the plane requires 3 h to fly the same distance. Find the rate of the plane in calm air and the rate of the wind.

15. **Entertainment** For the first performance of a play in a community theater, 50 reserved-seat tickets and 80 general-admission tickets were sold. The total receipts were \$980. For the second performance, 60 reserved-seat tickets and 90 general-admission tickets were sold. The total receipts were \$1140. Find the price of a reserved-seat ticket and the price of a general-admission ticket.

16. **Investments** Bernardo Community Library received a \$28,000 donation that it invested in two accounts, one earning 7.6% simple interest and the other earning 6.4% simple interest. If both accounts earned the same amount of annual interest, how much was invested in each account?

CUMULATIVE REVIEW EXERCISES

1. Evaluate $\frac{a^2 - b^2}{2a}$ when $a = 4$ and $b = -2$.

2. Solve: $-\frac{3}{4}x = \frac{9}{8}$

3. Given $f(x) = x^2 + 2x - 1$, find $f(2)$.

4. Multiply: $(2a^2 - 3a + 1)(2 - 3a)$

5. Simplify: $\frac{(-2x^2y)^4}{-8x^3y^2}$

6. Divide: $(4b^2 - 8b + 4) \div (2b - 3)$

7. Simplify: $\frac{8x^{-2}y^5}{-2xy^4}$

8. Factor: $4x^2y^4 - 64y^2$

9. Solve: $(x - 5)(x + 2) = -6$

10. Divide: $\frac{x^2 - 6x + 8}{2x^3 + 6x^2} \div \frac{2x - 8}{4x^3 + 12x^2}$

11. Add: $\frac{x - 1}{x + 2} + \frac{2x + 1}{x^2 + x - 2}$

12. Simplify: $\frac{x + 4 - \frac{7}{x - 2}}{x + 8 + \frac{21}{x - 2}}$

13. Solve: $\frac{x}{2x - 3} + 2 = \frac{-7}{2x - 3}$

14. Solve $A = P + Prt$ for r.

15. Find the x- and y-intercepts for $2x - 3y = 12$.

16. Find the slope of the line that passes through the points $(2, -3)$ and $(-3, 4)$.

17. Find the equation of the line that passes through the point $(-2, 3)$ and has slope $-\frac{3}{2}$.

18. Is $(2, 0)$ a solution of this system?

$$5x - 3y = 10$$
$$4x + 7y = 8$$

19. Solve by substitution:
$$3x - 5y = -23$$
$$x + 2y = -4$$

20. Solve by the addition method:
$$5x - 3y = 29$$
$$4x + 7y = -5$$

21. **Investments** A total of \$8750 is invested in two accounts. On one account, the annual simple interest rate is 9.6%; on the second account, the annual simple interest rate is 7.2%. How much should be invested in each account so that both accounts earn the same interest?

22. **Travel** A passenger train leaves a train depot $\frac{1}{2}$ h after a freight train leaves the same depot. The freight train is traveling 8 mph slower than the passenger train. Find the rate of each train if the passenger train overtakes the freight train in 3 h.

23. **Geometry** The length of each side of a square is extended 4 in. The area of the resulting square is 144 in^2. Find the length of a side of the original square.

24. **Travel** A plane can travel 160 mph in calm air. Flying with the wind, the plane can fly 570 mi in the same amount of time as it takes to fly 390 mi against the wind. Find the rate of the wind.

25. Graph $2x - 3y = 6$.

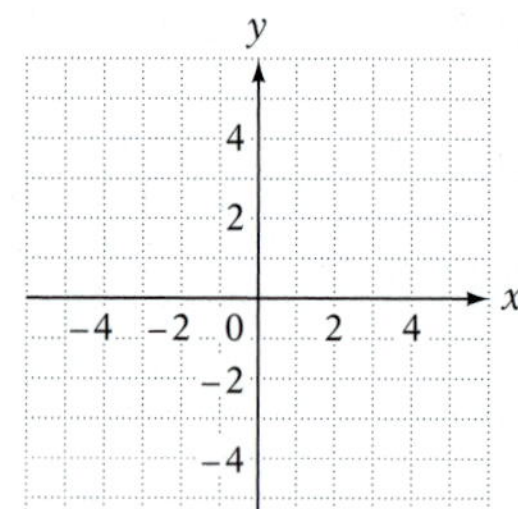

26. Solve by graphing: $3x + 2y = 6$
$3x - 2y = 6$

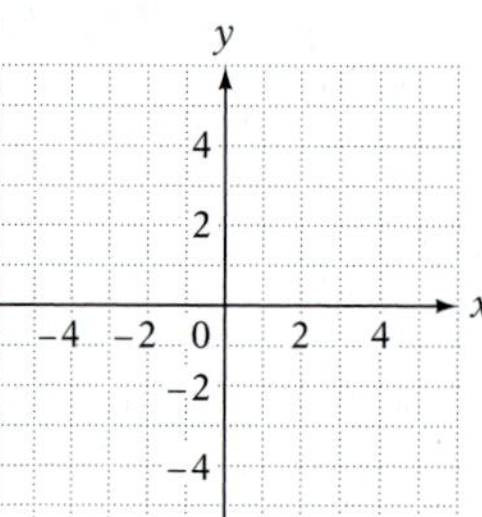

27. **Travel** With the current, a motorboat can travel 48 mi in 3 h. Against the current, the boat requires 4 h to travel the same distance. Find the rate of the boat in calm water.

28. **Registered Voters** In a recent year, the U.S. voting-age population was 205 million people, but only 156 million Americans were registered to vote. (*Source:* The Election Center) What percent of the voting-age population was registered to vote? Round to the nearest percent.

Inequalities

CHAPTER 14

OBJECTIVES

SECTION 14.1

A To write a set using the roster method

B To write and graph sets of real numbers

SECTION 14.2

A To solve an inequality using the Addition Property of Inequalities

B To solve an inequality using the Multiplication Property of Inequalities

C To solve application problems

SECTION 14.3

A To solve general inequalities

B To solve application problems

SECTION 14.4

A To graph an inequality in two variables

ARE YOU READY?

Take the Chapter 14 Prep Test to find out if you are ready to learn to:

- Write a set using the roster method, set-builder notation, and interval notation
- Graph an inequality on the number line
- Solve an inequality
- Graph an inequality in two variables

PREP TEST

Do these exercises to prepare for Chapter 14.

1. Place the correct symbol, < or >, between the two numbers.
$-45 \quad -27$

2. Simplify: $3x - 5(2x - 3)$

3. State the Addition Property of Equations.

4. State the Multiplication Property of Equations.

5. **Nutrition** A certain grade of hamburger contains 15% fat. How many pounds of fat are in 3 lb of this hamburger?

6. Solve: $4x - 5 = -7$

7. Solve: $4 = 2 - \frac{3}{4}x$

8. Solve: $7 - 2(2x - 3) = 3x - 1$

9. Graph: $y = \frac{2}{3}x - 3$

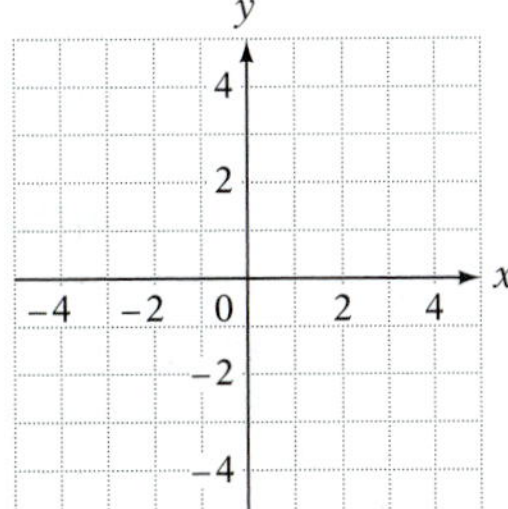

10. Graph: $3x + 4y = 12$

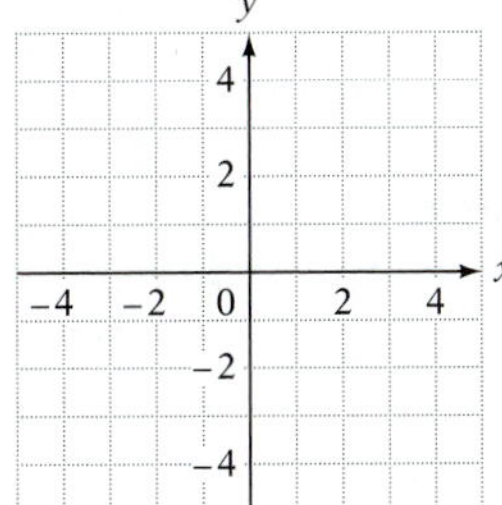

SECTION

14.1 Sets

OBJECTIVE A To write a set using the roster method 

Recall that a *set* is a collection of objects, which are called the *elements* of the set. The roster method of writing a set encloses a list of the elements in braces.

The set of the positive integers less than 5 is written $\{1, 2, 3, 4\}$.

HOW TO • 1 Use the roster method to write the set of integers between 0 and 10.

$A = \{1, 2, 3, 4, 5, 6, 7, 8, 9\}$ • **A set can be designated by a capital letter. Note that 0 and 10 are not elements of the set.**

HOW TO • 2 Use the roster method to write the set of natural numbers.

$A = \{1, 2, 3, 4, \ldots\}$ • **The three dots mean that the pattern of numbers continues without end.**

The **empty set,** or **null set,** is the set that contains no elements. The symbol $\varnothing$ or $\{\ \}$ is used to represent the empty set.

The set of people who have run a 2-minute mile is an empty set.

Union of Two Sets

The **union** of two sets, written $A \cup B$, is the set of all elements that belong to either set A *or* set B.

HOW TO • 3 Find $A \cup B$, given $A = \{1, 2, 3, 4\}$ and $B = \{3, 4, 5, 6\}$.

$A \cup B = \{1, 2, 3, 4, 5, 6\}$ • **The union of A and B contains all the elements of A and all the elements of B. Elements in both sets are listed only once.**

Intersection of Two Sets

The **intersection** of two sets, written $A \cap B$, is the set that contains the elements that are common to both A and B.

HOW TO • 4 Find $A \cap B$, given $A = \{1, 2, 3, 4\}$ and $B = \{3, 4, 5, 6\}$.

$A \cap B = \{3, 4\}$ • **The intersection of A and B contains the elements common to A and B.**

EXAMPLE • 1

Use the roster method to write the set of the odd positive integers less than 12.

Solution

$A = \{1, 3, 5, 7, 9, 11\}$

YOU TRY IT • 1

Use the roster method to write the set of the odd negative integers greater than -10.

Your solution

Solution on p. S32

EXAMPLE • 2

Use the roster method to write the set of the even positive integers.

Solution

$A = \{2, 4, 6, \ldots\}$

YOU TRY IT • 2

Use the roster method to write the set of the odd positive integers.

Your solution

EXAMPLE • 3

Find $D \cup E$, given $D = \{6, 8, 10, 12\}$ and $E = \{-8, -6, 10, 12\}$.

Solution

$D \cup E = \{-8, -6, 6, 8, 10, 12\}$

YOU TRY IT • 3

Find $A \cup B$, given $A = \{-2, -1, 0, 1, 2\}$ and $B = \{0, 1, 2, 3, 4\}$.

Your solution

EXAMPLE • 4

Find $A \cap B$, given $A = \{5, 6, 9, 11\}$ and $B = \{5, 9, 13, 15\}$.

Solution

$A \cap B = \{5, 9\}$

YOU TRY IT • 4

Find $C \cap D$, given $C = \{10, 12, 14, 16\}$ and $D = \{10, 16, 20, 26\}$.

Your solution

EXAMPLE • 5

Find $A \cap B$, given $A = \{1, 2, 3, 4\}$ and $B = \{8, 9, 10, 11\}$.

Solution

$A \cap B = \emptyset$

YOU TRY IT • 5

Find $A \cap B$, given $A = \{-5, -4, -3, -2\}$ and $B = \{2, 3, 4, 5\}$.

Your solution

Solutions on p. S32

OBJECTIVE B To write and graph sets of real numbers

Point of Interest

The symbol $\in$ was first used in the book *Arithmeticae Principia,* published in 1889. It is the first letter of the Greek word *εστι*, which means "is." The symbols for union and intersection were also introduced around the same time.

Another method of representing sets is called **set-builder notation.** This method of writing sets uses a rule to describe the elements of the set. Using set-builder notation, we represent the set of all positive integers less than 10 as

$\{x \mid x < 10, x \in \text{positive integers}\}$, which is read "the set of all positive integers x that are less than 10."

HOW TO • 5 Use set-builder notation to write the set of integers less than or equal to 12.

$\{x \mid x \leq 12, x \in \text{integers}\}$

- **This is read "the set of all integers x that are less than or equal to 12."**

HOW TO • 6 Use set-builder notation to write the set of real numbers greater than 4.

$\{x \mid x > 4, x \in \text{real numbers}\}$

- **This is read "the set of all real numbers x that are greater than 4."**

For the remainder of this section, all variables will represent real numbers. Given this convention, $\{x|x > 4, x \in \text{real numbers}\}$ is written $\{x|x > 4\}$.

Some sets of real numbers written in set-builder notation can be written in **interval notation.** For instance, the interval notation $[-3, 2)$ represents the set of real numbers between -3 and 2. The bracket means that -3 is included in the set, and the parenthesis means that 2 is *not* included in the set. Using set-builder notation, the interval $[-3, 2)$ is written

$\{x|-3 \le x < 2\}$ • **This is read "the set of all real numbers x between -3 and 2, including -3 but excluding 2."**

To indicate an interval that extends forever in the positive direction, we use the **infinity symbol, ∞;** to indicate an interval that extends forever in the negative direction, we use the **negative infinity symbol, −∞.**

Take Note

When writing a set in interval notation, note that we always use a parenthesis to the right of ∞ and to the left of −∞. Infinity is not a real number, so it cannot be represented as belonging to the set of real numbers by using a bracket.

HOW TO • 7 Write $\{x|x > 1\}$ in interval notation.

$\{x|x > 1\}$ is the set of real numbers greater than 1. This set extends forever in the positive direction. In interval notation, this set is written $(1, \infty)$.

HOW TO • 8 Write $\{x|x \le -2\}$ in interval notation.

$\{x|x \le -2\}$ is the set of real numbers less than or equal to -2. This set extends forever in the negative direction. In interval notation, this set is written $(-\infty, -2]$.

HOW TO • 9 Write $[1, 3]$ in set-builder notation.

This is the set of real numbers between 1 and 3, including 1 and 3. In set-builder notation, this set is written $\{x|1 \le x \le 3\}$.

We can graph sets of real numbers given in set-builder notation or in interval notation.

HOW TO • 10 Graph: $(-\infty, -1)$

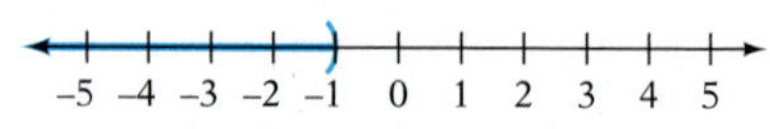

This is the set of real numbers less than -1, excluding -1. The parenthesis at -1 indicates that -1 is excluded from the set.

HOW TO • 11 Graph: $\{x|x \ge 1\}$

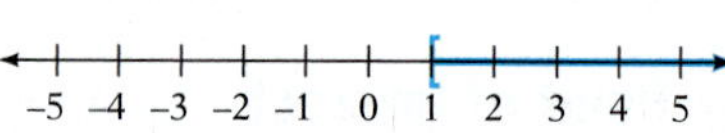

This is the set of real numbers greater than or equal to 1. The bracket at 1 indicates that 1 is included in the set.

EXAMPLE • 6

Write $\{x|x \ge 2\}$ in interval notation.

Solution

$\{x|x \ge 2\}$ is the set of real numbers greater than or equal to 2. This set extends forever in the positive direction. In interval notation, this set is written $[2, \infty)$.

YOU TRY IT • 6

Write $\{x|x \le 3\}$ in interval notation.

Your solution

Solution on p. S32

EXAMPLE • 7

Write $\{x|0 \le x \le 1\}$ in interval notation.

Solution

$\{x|0 \le x \le 1\}$ is the set of real numbers between 0 and 1, including 0 and 1. In interval notation, this set is written $[0, 1]$.

YOU TRY IT • 7

Write $\{x|-5 \le x \le -3\}$ in interval notation.

Your solution

EXAMPLE • 8

Write $(-\infty, 0]$ in set-builder notation.

Solution

The interval $(-\infty, 0]$ is the set of real numbers less than or equal to 0. In set-builder notation, this set is written $\{x|x \le 0\}$.

YOU TRY IT • 8

Write $(-3, \infty)$ in set-builder notation.

Your solution

EXAMPLE • 9

Write $(-3, 3)$ in set-builder notation.

Solution

The interval $(-3, 3)$ is the set of real numbers between -3 and 3, excluding -3 and 3. In set-builder notation, this set is written $\{x|-3 < x < 3\}$.

YOU TRY IT • 9

Write $[0, 4)$ in set-builder notation.

Your solution

EXAMPLE • 10

Graph: $\{x|-2 < x < 1\}$

Solution

The graph is the set of real numbers between -2 and 1, excluding -2 and 1. Use parentheses at -2 and 1.

−5 −4 −3 −2 −1 0 1 2 3 4 5

YOU TRY IT • 10

Graph: $\{x|-4 \le x \le 4\}$

Your solution

−5 −4 −3 −2 −1 0 1 2 3 4 5

EXAMPLE • 11

Graph: $\{x|x < 4\}$

Solution

The graph is the set of real numbers less than 4. Use a parenthesis at 4.

−5 −4 −3 −2 −1 0 1 2 3 4 5

YOU TRY IT • 11

Graph: $\{x|x > -3\}$

Your solution

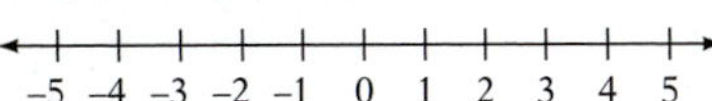

EXAMPLE • 12

Graph: $(-\infty, 5)$

Solution

The graph is the set of real numbers less than 5. Use a parenthesis at 5.

YOU TRY IT • 12

Graph: $(-2, \infty)$

Your solution

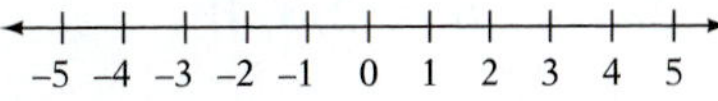

EXAMPLE • 13

Graph: $[-4, 3)$

Solution

The graph is the set of real numbers between -4 and 3, including -4 and excluding 3.

−5 −4 −3 −2 −1 0 1 2 3 4 5

YOU TRY IT • 13

Graph: $[2, 5]$

Your solution

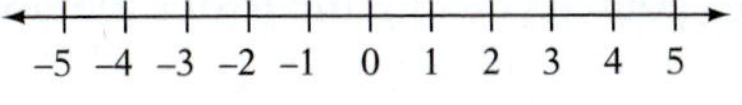

Solutions on pp. S32–S33

14.1 EXERCISES

OBJECTIVE A To write a set using the roster method

1. Explain how to find the union of two sets.

2. Explain how to find the intersection of two sets.

For Exercises 3 to 8, use the roster method to write the set.

3. The integers between 15 and 22

4. The integers between -10 and -4

5. The odd integers between 8 and 18

6. The even integers between -11 and -1

7. The letters of the alphabet between a and d

8. The letters of the alphabet between p and v

For Exercises 9 to 16, find $A \cup B$.

9. $A = \{3, 4, 5\}$ $B = \{4, 5, 6\}$

10. $A = \{-3, -2, -1\}$ $B = \{-2, -1, 0\}$

11. $A = \{-10, -9, -8\}$ $B = \{8, 9, 10\}$

12. $A = \{a, b, c\}$ $B = \{x, y, z\}$

13. $A = \{a, b, d, e\}$ $B = \{c, d, e, f\}$

14. $A = \{m, n, p, q\}$ $B = \{m, n, o\}$

15. $A = \{1, 3, 7, 9\}$ $B = \{7, 9, 11, 13\}$

16. $A = \{-3, -2, -1\}$ $B = \{-1, 1, 2\}$

For Exercises 17 to 22, find $A \cap B$.

17. $A = \{3, 4, 5\}$ $B = \{4, 5, 6\}$

18. $A = \{-4, -3, -2\}$ $B = \{-6, -5, -4\}$

19. $A = \{-4, -3, -2\}$ $B = \{2, 3, 4\}$

20. $A = \{1, 2, 3, 4\}$ $B = \{1, 2, 3, 4\}$

21. $A = \{a, b, c, d, e\}$ $B = \{c, d, e, f, g\}$

22. $A = \{m, n, o, p\}$ $B = \{k, l, m, n\}$

23. Make up sets A and B such that $A \cup B$ has five elements and $A \cap B$ has two elements. Write your sets using the roster method.

24. True or false? If $A \cup B = A$, then $A \cap B = B$.

OBJECTIVE B To write and graph sets of real numbers

For Exercises 25 to 30, use set-builder notation to write the set.

25. The negative integers greater than -5

26. The positive integers less than 5

27. The integers greater than 30

28. The integers less than -70

29. The real numbers greater than 8

30. The real numbers less than 57

For Exercises 31 to 39, write the set in interval notation.

31. $\{x|1 < x < 2\}$

32. $\{x|-2 < x \le 4\}$

33. $\{x|x > 3\}$

34. $\{x|x \le 0\}$

35. $\{x|-4 \le x < 5\}$

36. $\{x|-3 \le x \le 0\}$

37. $\{x|x \le 2\}$

38. $\{x|x \ge -3\}$

39. $\{x|-3 \le x \le 1\}$

For Exercises 40 to 48, write the interval in set-builder notation.

40. $[-4, 5]$

41. $(-5, -3)$

42. $(4, \infty)$

43. $(-\infty, -2]$

44. $(4, 9]$

45. $[-3, -2]$

46. $[0, \infty)$

47. $(-\infty, 6]$

48. $(-\infty, \infty)$

For Exercises 49 to 64, graph the set.

49. $[-5, 4]$

-5 -4 -3 -2 -1 0 1 2 3 4 5

50. $(-3, 5]$

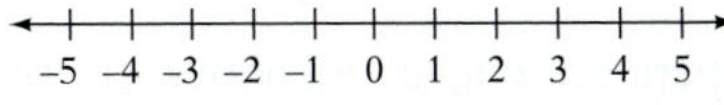

51. $\{x|x < 4\}$

–5 –4 –3 –2 –1 0 1 2 3 4 5

52. $\{x|x \geq -3\}$

–5 –4 –3 –2 –1 0 1 2 3 4 5

53. $\{x|x \leq -4\}$

–5 –4 –3 –2 –1 0 1 2 3 4 5

54. $\{x|x > 0\}$

–5 –4 –3 –2 –1 0 1 2 3 4 5

55. $(-\infty, 3]$

–5 –4 –3 –2 –1 0 1 2 3 4 5

56. $(4, \infty)$

–5 –4 –3 –2 –1 0 1 2 3 4 5

57. $[-1, 3)$

–5 –4 –3 –2 –1 0 1 2 3 4 5

58. $(-3, 0]$

–5 –4 –3 –2 –1 0 1 2 3 4 5

59. $\{x|-3 < x < 3\}$

–5 –4 –3 –2 –1 0 1 2 3 4 5

60. $\{x|0 \leq x < 4\}$

–5 –4 –3 –2 –1 0 1 2 3 4 5

61. $\{x|2 \leq x \leq 4\}$

–5 –4 –3 –2 –1 0 1 2 3 4 5

62. $\{x|-4 < x < 1\}$

–5 –4 –3 –2 –1 0 1 2 3 4 5

63. $\{x|-\infty < x < \infty\}$

–5 –4 –3 –2 –1 0 1 2 3 4 5

64. $(-\infty, \infty)$

–5 –4 –3 –2 –1 0 1 2 3 4 5

65. How many elements are in the set given in interval notation as $(4, 4)$?

66. How many elements are in the set given by $\{x|4 \leq x \leq 4\}$?

Applying the Concepts

For Exercises 67 and 68, write an inequality that describes the situation.

67. To avoid shipping charges, one must spend a minimum m of \$250.

68. The temperature t never got above freezing (32°F).

SECTION

14.2 The Addition and Multiplication Properties of Inequalities

OBJECTIVE A To solve an inequality using the Addition Property of Inequalities

The inequality at the right is true if the variable is replaced by 7, 9.3, or $\frac{15}{2}$.

$$x + 5 > 8$$
$$\left.\begin{array}{r} 7 + 5 > 8 \\ 9.3 + 5 > 8 \\ \frac{15}{2} + 5 > 8 \end{array}\right\} \text{True inequalities}$$

The inequality $x + 5 > 8$ is false if the variable is replaced by 2, 1.5, or $-\frac{1}{2}$.

$$\left.\begin{array}{r} 2 + 5 > 8 \\ 1.5 + 5 > 8 \\ -\frac{1}{2} + 5 > 8 \end{array}\right\} \text{False inequalities}$$

The **solution set of an inequality** is the set of numbers each element of which, when substituted for the variable, results in a true inequality. The values of x that will make the inequality $x + 5 > 8$ true are the numbers greater than 3. The solution set of $x + 5 > 8$ is $\{x \mid x > 3\}$. This set could also be written in interval notation as $(3, \infty)$.

At the right is the graph of the solution set of $x + 5 > 8$.

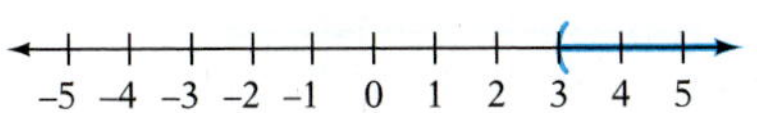

In solving an inequality, the goal is to rewrite the given inequality in the form *variable* > *constant* or *variable* < *constant*. The Addition Property of Inequalities is used to rewrite an inequality in this form.

Addition Property of Inequalities

The same term can be added to each side of an inequality without changing the solution set of the inequality.

If $a > b$, then $a + c > b + c$.

If $a < b$, then $a + c < b + c$.

The Addition Property of Inequalities also holds true for an inequality containing the symbol $\geq$ or $\leq$. **The Addition Property of Inequalities is used when,** in order to rewrite an inequality in the form *variable* > *constant* or *variable* < *constant*, **we must remove a term from one side of the inequality. Add the opposite of that term to each side of the inequality.**

HOW TO 1 Solve and write the answer in set-builder notation: $x - 4 < -3$

$$x - 4 < -3$$
$$x - 4 + 4 < -3 + 4$$ • **Add 4 to each side of the inequality.**
$$x < 1$$ • **Simplify.**
$$\{x \mid x < 1\}$$ • **Write in set-builder notation.**

At the right is the graph of the solution set of $x - 4 < -3$.

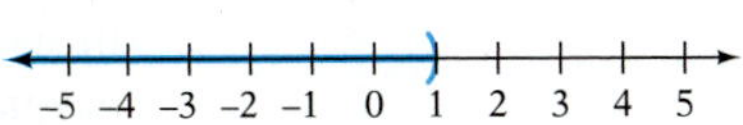

Because subtraction is defined in terms of addition, the Addition Property of Inequalities allows the same term to be subtracted from each side of an inequality.

HOW TO • 2 Solve and write the answer in set-builder notation: $5x - 6 \le 4x - 4$

$5x - 6 \le 4x - 4$

$5x - 4x - 6 \le 4x - 4x - 4$ • **Subtract 4x from each side of the inequality.**

$x - 6 \le -4$ • **Simplify.**

$x - 6 + 6 \le -4 + 6$ • **Add 6 to each side of the inequality.**

$x \le 2$ • **Simplify.**

$\{x \mid x \le 2\}$ • **Write in set-builder notation.**

EXAMPLE • 1

Solve $3 < x + 5$ and write the answer in interval notation. Graph the solution set.

Solution

$3 < x + 5$

$3 - 5 < x + 5 - 5$ • **Subtract 5.**

$-2 < x$

$(-2, \infty)$

YOU TRY IT • 1

Solve $x + 2 < -2$ and write the answer in interval notation. Graph the solution set.

Your solution

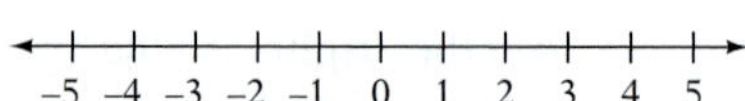

EXAMPLE • 2

Solve and write the answer in set-builder notation: $7x - 14 \le 6x - 16$

Solution

$7x - 14 \le 6x - 16$

$7x - 6x - 14 \le 6x - 6x - 16$ • **Subtract 6x.**

$x - 14 \le -16$

$x - 14 + 14 \le -16 + 14$ • **Add 14.**

$x \le -2$

$\{x \mid x \le -2\}$

YOU TRY IT • 2

Solve and write the answer in set-builder notation: $5x + 3 > 4x + 5$

Your solution

Solutions on p. S33

OBJECTIVE B

To solve an inequality using the Multiplication Property of Inequalities

Consider the two inequalities below and the effect of multiplying each inequality by 2, a *positive* number.

$$-3 < 7 \qquad 6 > 4$$
$$2(-3) < 2(7) \qquad 2(6) > 2(4)$$
$$-6 < 14 \qquad 12 > 8$$

In each case, the inequality symbol remains the same. Multiplying each side of an inequality by a **positive** number does not change the inequality.

Now consider the same inequalities and the effect of multiplying by -2, a *negative* number.

$$-3 < 7 \qquad\qquad 6 > 4$$
$$-2(-3) > -2(7) \qquad\qquad -2(6) < -2(4)$$
$$6 > -14 \qquad\qquad -12 < -8$$

In order for the inequality to be true, the inequality symbol must be reversed. If each side of an inequality is multiplied by a **negative** number, the inequality symbol must be reversed in order for the inequality to remain a true inequality.

Take Note

Any time an inequality is multiplied or divided by a negative number, the inequality symbol must be reversed. Compare the next two examples.

$2x < -4$	Divide each side by *positive* 2.
$\frac{2x}{2} < \frac{-4}{2}$	
$x < -2$	Inequality *is not* reversed.
$-2x < 4$	Divide each side by *negative* 2.
$\frac{-2x}{-2} > \frac{4}{-2}$	
$x > -2$	Inequality *is* reversed.

Multiplication Property of Inequalities—Part 1

Each side of an inequality can be multiplied by the same **positive** number without changing the solution set of the inequality. In symbols, this is stated as follows.

If $a < b$ and $c > 0$, then $ac < bc$. If $a > b$ and $c > 0$, then $ac > bc$.

Multiplication Property of Inequalities—Part 2

Multiplying each side of an inequality by the same **negative** number and reversing the inequality symbol does not change the solution set of the inequality. In symbols, this is stated as follows.

If $a < b$ and $c < 0$, then $ac > bc$. If $a > b$ and $c < 0$, then $ac < bc$.

In solving an inequality, the goal is to rewrite the given inequality in the form *variable* > *constant* or *variable* < *constant*. **The Multiplication Property of Inequalities is used when,** in order to rewrite an inequality in this form, **we must remove a coefficient from one side of the inequality.**

The Multiplication Property of Inequalities also holds true for an inequality containing the symbol ≥ or ≤.

HOW TO • 3 Solve $-\frac{3}{2}x \le 6$ and write the answer in set-builder notation. Graph the solution set.

$$-\frac{3}{2}x \le 6$$

$$-\frac{2}{3}\left(-\frac{3}{2}x\right) \ge -\frac{2}{3}(6)$$

• **Multiply each side of the inequality by $-\frac{2}{3}$. Because $-\frac{2}{3}$ is a negative number, the inequality symbol must be reversed.**

$$x \ge -4$$

$$\{x|x \ge -4\}$$

• **Write in set-builder notation.**

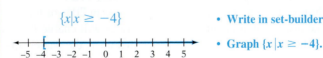

-5 -4 -3 -2 -1 0 1 2 3 4 5

• **Graph $\{x|x \ge -4\}$.**

Because division is defined in terms of multiplication, the Multiplication Property of Inequalities allows each side of an inequality to be divided by a nonzero constant.

Take Note

As shown in the example at the right, the goal in solving an inequality can be *constant* < *variable* or *constant* > *variable*. We could have written the third line of this example as $x > -\frac{2}{3}$.

HOW TO • 4 Solve and write the answer in set-builder notation: $-4 < 6x$

$$-4 < 6x$$

$$\frac{-4}{6} < \frac{6x}{6}$$

• **Divide each side of the inequality by 6.**

$$-\frac{2}{3} < x$$

• **Simplify: $\frac{-4}{6} = -\frac{2}{3}$**

$$\left\{x\middle|x > -\frac{2}{3}\right\}$$

• **Write in set-builder notation.**

EXAMPLE • 3

Solve $-7x > 14$ and write the answer in interval notation. Graph the solution set.

Solution

$$-7x > 14$$

$$\frac{-7x}{-7} < \frac{14}{-7} \quad \text{• Divide by } -7.$$

$$x < -2$$

$(-\infty, -2)$

-5 -4 -3 -2 -1 0 1 2 3 4 5

YOU TRY IT • 3

Solve $-3x > -9$ and write the answer in interval notation. Graph the solution set.

Your solution

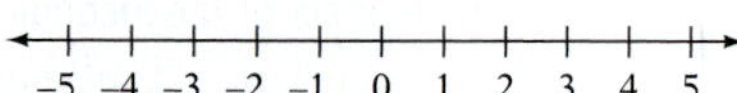

EXAMPLE • 4

Solve and write the answer in set-builder notation: $-\frac{5}{8}x \le \frac{5}{12}$

Solution

$$-\frac{5}{8}x \le \frac{5}{12}$$

$$-\frac{8}{5}\left(-\frac{5}{8}x\right) \ge -\frac{8}{5}\left(\frac{5}{12}\right) \quad \text{• Multiply by } -\frac{8}{5}.$$

$$x \ge -\frac{2}{3}$$

$$\left\{x \mid x \ge -\frac{2}{3}\right\}$$

YOU TRY IT • 4

Solve and write the answer in set-builder notation: $-\frac{3}{4}x \ge 18$

Your solution

Solutions on p. S33

OBJECTIVE C To solve application problems

EXAMPLE • 5

A student must have at least 450 points out of 500 points on five tests to receive an A in a course. One student's results on the first four tests were 94, 87, 77, and 95. What scores on the last test will enable this student to receive an A in the course?

Strategy

To find the scores, write and solve an inequality using N to represent the possible scores on the last test.

Solution

Total number of points on the five tests	is greater than or equal to	450

$$94 + 87 + 77 + 95 + N \ge 450$$

$$353 + N \ge 450 \quad \text{• Simplify.}$$

$$353 - 353 + N \ge 450 - 353 \quad \text{• Subtract 353.}$$

$$N \ge 97$$

The student's score on the last test must be greater than or equal to 97.

YOU TRY IT • 5

A consumer electronics dealer will make a profit on the sale of an LCD HDTV if the cost of the TV is less than 70% of the selling price. What selling prices will enable the dealer to make a profit on a TV that costs the dealer $942?

Your strategy

Your solution

Solution on p. S33

14.2 EXERCISES

OBJECTIVE A To solve an inequality using the Addition Property of Inequalities

For Exercises 1 to 8, solve the inequality and write the answer in set-builder notation. Graph the solution set.

1. $x + 1 < 3$

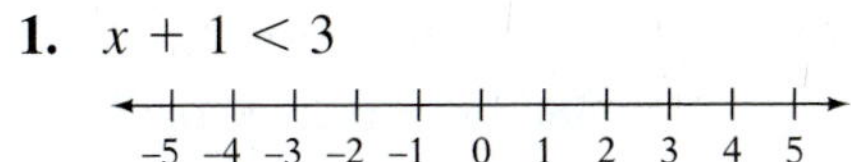

2. $y + 2 < 2$

–5 –4 –3 –2 –1 0 1 2 3 4 5

3. $x - 5 > -2$

–5 –4 –3 –2 –1 0 1 2 3 4 5

4. $x - 3 > -2$

–5 –4 –3 –2 –1 0 1 2 3 4 5

5. $7 \le n + 4$

–5 –4 –3 –2 –1 0 1 2 3 4 5

6. $3 \le 5 + x$

–5 –4 –3 –2 –1 0 1 2 3 4 5

7. $x - 6 \le -10$

–5 –4 –3 –2 –1 0 1 2 3 4 5

8. $y - 8 \le -11$

–5 –4 –3 –2 –1 0 1 2 3 4 5

For Exercises 9 to 20, solve and write the answer in interval notation.

9. $y - 3 \ge -12$

10. $x + 8 \ge -14$

11. $3x - 5 < 2x + 7$

12. $5x + 4 < 4x - 10$

13. $8x - 7 \ge 7x - 2$

14. $3n - 9 \ge 2n - 8$

15. $2x + 4 < x - 7$

16. $9x + 7 < 8x - 7$

17. $4x - 8 \le 2 + 3x$

18. $5b - 9 < 3 + 4b$

19. $6x + 4 \ge 5x - 2$

20. $7x - 3 \ge 6x - 2$

For Exercises 21 to 38, solve and write the answer in set-builder notation.

21. $2x - 12 > x - 10$

22. $3x + 9 > 2x + 7$

23. $d + \frac{1}{2} < \frac{1}{3}$

24. $x - \frac{3}{8} < \frac{5}{6}$

25. $x + \frac{5}{8} \ge -\frac{2}{3}$

26. $y + \frac{5}{12} \ge -\frac{3}{4}$

27. $x - \frac{3}{8} < \frac{1}{4}$

28. $y + \frac{5}{9} \le \frac{5}{6}$

29. $2x - \frac{1}{2} < x + \frac{3}{4}$

30. $6x - \frac{1}{3} \le 5x - \frac{1}{2}$

31. $3x + \frac{5}{8} > 2x + \frac{5}{6}$

32. $4b - \frac{7}{12} \ge 3b - \frac{9}{16}$

33. $3.8x < 2.8x - 3.8$

34. $1.2x < 0.2x - 7.3$

35. $x + 5.8 \le 4.6$

36. $n - 3.82 \le 3.95$

37. $x - 3.5 < 2.1$

38. $x - 0.23 \le 0.47$

For Exercises 39 to 42, assume that n and a are both positive numbers. State whether the solution set of an inequality in the given form contains only negative numbers, only positive numbers, or both negative and positive numbers.

39. $x + n < a$, where $n > a$

40. $x + n > a$, where $n < a$

41. $x + n < a$, where $n < a$

42. $x + n > a$, where $n > a$

OBJECTIVE B To solve an inequality using the Multiplication Property of Inequalities

For Exercises 43 to 52, solve and write the answer in set-builder notation. Graph the solution set.

43. $3x < 12$

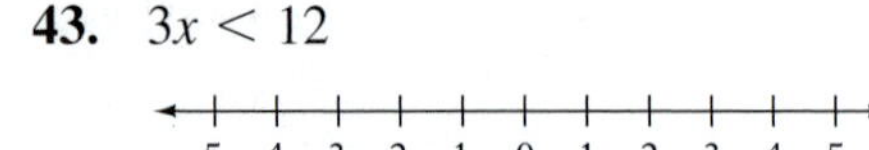

44. $8x \le -24$

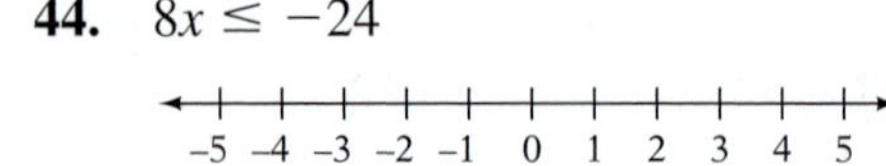

45. $15 \le 5y$

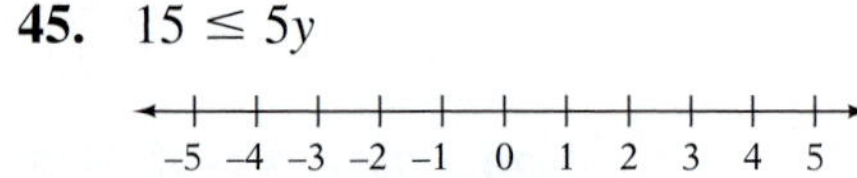

46. $-48 < 24x$

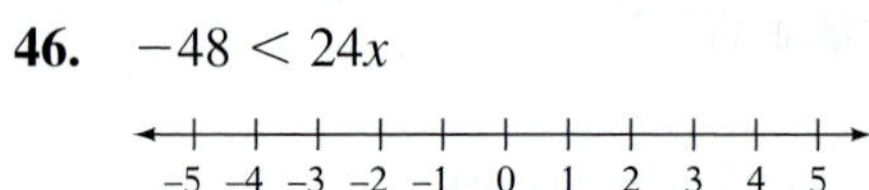

47. $16x \le 16$

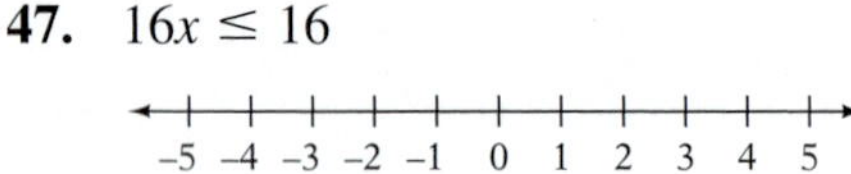

48. $3x > 0$

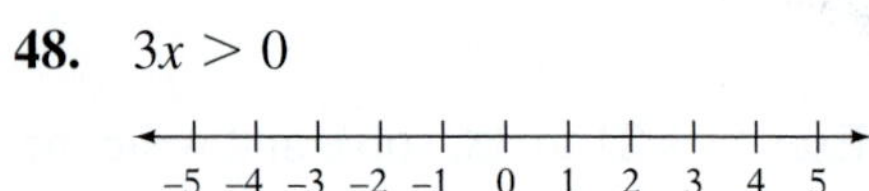

49. $-8x > 8$

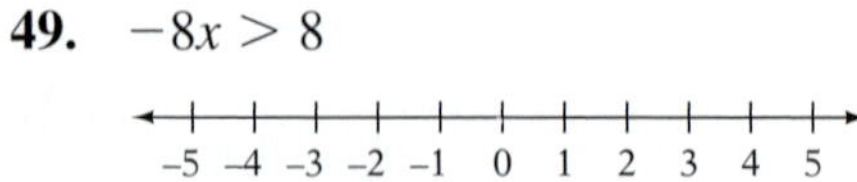

50. $-2n \le -8$

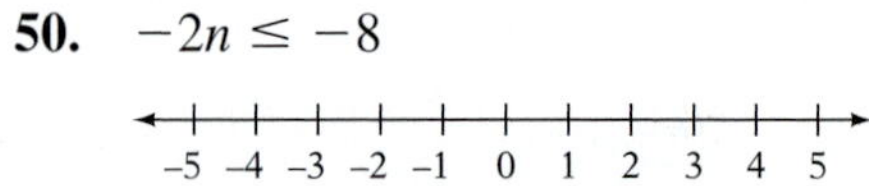

51. $-6b > 24$

52. $-4x < 8$

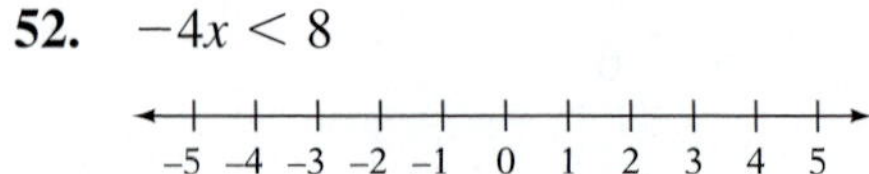

For Exercises 53 to 68, solve and write the answer in interval notation.

53. $-5y \ge 0$

54. $-3z < 0$

55. $7x > 2$

56. $6x \le -1$

57. $2x \le -5$
58. $-x \ge 3$
59. $-y < 4$
60. $-b > -7$

61. $2 > -y$
62. $-5 \le -x$
63. $\frac{5}{6}n < 15$
64. $\frac{3}{4}x < 12$

65. $\frac{2}{3}y \ge 4$
66. $10 \le \frac{5}{8}x$
67. $4 \ge \frac{2}{3}x$
68. $-\frac{3}{7}x \le 6$

For Exercises 69 to 84, solve and write the answer in set-builder notation.

69. $-\frac{2}{11}b \ge -6$
70. $-\frac{4}{7}x \ge -12$
71. $\frac{2}{3}n < \frac{1}{2}$
72. $-\frac{3}{5}x < 0$

73. $-\frac{2}{3}x \ge 0$
74. $-\frac{3}{8}x \ge \frac{9}{14}$
75. $-\frac{3}{5}x < -\frac{6}{7}$
76. $-\frac{4}{5}x < -\frac{8}{15}$

77. $-\frac{3}{4}y \ge -\frac{5}{8}$
78. $-\frac{8}{9}x \ge -\frac{16}{27}$
79. $1.5x \le 6.30$
80. $2.3x \le 5.29$

81. $-3.5d > 7.35$
82. $-0.24x > 0.768$
83. $4.25m > -34$
84. $-3.9x \ge -19.5$

For Exercises 85 to 87, without actually solving the inequality or using a calculator, determine which of the following statements is true.
(i) n must be positive. (ii) n must be negative. (iii) n can be positive, negative, or zero.

85. $-0.8157n > 7.304$
86. $3.978n \le -0.615$
87. $-917n \ge -10{,}512$

OBJECTIVE C To solve application problems

88. Consider the following statement: Today's high temperature will be at least 10 degrees lower than yesterday's high temperature. If the inequality $T \le t - 10$ correctly represents this statement, what does the variable t represent?

89. **Mortgages** See the news clipping at the right. Suppose a couple's mortgage application is approved. Their monthly mortgage payment is \$2050. What is the couple's monthly household income? Round to the nearest dollar.

In the News

New Federal Standard for Mortgages

A new federal regulation states that the purchaser of a house is not to be approved for a monthly mortgage payment that is more than 38% of the purchaser's monthly household income.

Source: US News & World Report

90. Sports To be eligible for a basketball tournament, a basketball team must win at least 60% of its remaining games. If the team has 17 games remaining, how many games must the team win to qualify for the tournament?

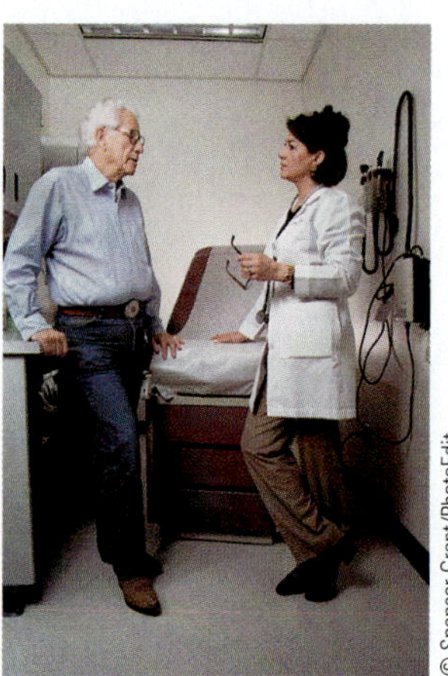
© Spencer Grant/PhotoEdit

91. Recycling A service organization will receive a bonus of \$200 for collecting more than 1850 lb of aluminum cans during its four collection drives. On the first three drives, the organization collected 505 lb, 493 lb, and 412 lb. How many pounds of cans must the organization collect on the fourth drive to receive the bonus?

92. Health A health official recommends a maximum cholesterol level of 200 units. How many units must a patient with a cholesterol level of 275 units reduce his cholesterol level to satisfy the recommended maximum level?

93. Grading To pass a course with a B grade, a student must have an average of 80 points on five tests. The student's grades on the first four tests were 75, 83, 86, and 78. What scores can the student receive on the fifth test to earn a B grade?

Alternative Energy For Exercises 94 to 96, use the information in the article at the right.

94. a. A couple living in a town that has not changed the set-back requirement wants to install an 80-foot wind turbine on their property. How far back from the property line must the turbine be set?

b. Suppose the town lowers the 150% requirement to 125%. How far back from the property line must the turbine be set?

95. You live in a town that has not changed the set-back requirement. You want to install a wind turbine 68 ft from your property line. To the nearest foot, what is the height of the tallest wind turbine you can install?

96. You live in a town that has changed the set-back requirement to 115%. A good spot for a wind turbine on your property is 75 ft from the property line. To the nearest foot, what is the height of the tallest wind turbine you can install?

In the News

New Law Eases Restrictions on Small Wind Systems

A research project created by students in the University of New Hampshire's Environmental Politics class has led to a change in New Hampshire state law. Under the new law, a small wind turbine installed on a residential property must be set back from the property line by a distance greater than 150% of the turbine height. Individual towns may lower the 150% requirement, but they may not increase it.

Source: www.gencourt.state.nh.us

Applying the Concepts

For Exercises 97 to 102, given that $a > b$ and that a and b are real numbers, determine for which real numbers c the statement is true. Use set-builder notation to write the answer.

97. $ac > bc$

98. $ac < bc$

99. $a + c > b + c$

100. $a + c < b + c$

101. $\frac{a}{c} > \frac{b}{c}$

102. $\frac{a}{c} < \frac{b}{c}$

SECTION 14.3 General Inequalities

OBJECTIVE A To solve general inequalities 

Solving an inequality frequently requires application of both the Addition and Multiplication Properties of Inequalities.

HOW TO • 1 Solve and write the answer in interval notation: $4y - 3 \geq 6y + 5$

$4y - 3 \geq 6y + 5$

$4y - 6y - 3 \geq 6y - 6y + 5$ • Subtract $6y$ from each side of the inequality.

$-2y - 3 \geq 5$ • Simplify.

$-2y - 3 + 3 \geq 5 + 3$ • Add 3 to each side of the inequality.

$-2y \geq 8$ • Simplify.

$\dfrac{-2y}{-2} \leq \dfrac{8}{-2}$ • Divide each side of the inequality by -2. Because -2 is a negative number, the inequality symbol must be reversed.

$y \leq -4$

$(-\infty, -4]$ • Write in interval notation.

Take Note

When an inequality contains parentheses, one of the steps in solving the inequality requires the use of the Distributive Property.

HOW TO • 2 Solve and write the answer in set-builder notation:
$-2(x - 7) > 8 - 4(2x - 3)$

$-2(x - 7) > 8 - 4(2x - 3)$

$-2x + 14 > 8 - 8x + 12$ • Use the Distributive Property to remove parentheses.

$-2x + 14 > -8x + 20$ • Simplify.

$-2x + 8x + 14 > -8x + 8x + 20$ • Add $8x$ to each side of the inequality.

$6x + 14 > 20$ • Simplify.

$6x + 14 - 14 > 20 - 14$ • Subtract 14 from each side of the inequality.

$6x > 6$ • Simplify.

$\dfrac{6x}{6} > \dfrac{6}{6}$ • Divide each side of the inequality by 6.

$x > 1$

$\{x \mid x > 1\}$ • Write in set-builder notation.

EXAMPLE • 1

Solve and write the answer in interval notation:
$7x - 3 \leq 3x + 17$

Solution

$7x - 3 \leq 3x + 17$

$7x - 3x - 3 \leq 3x - 3x + 17$ • Subtract $3x$.

$4x - 3 \leq 17$

$4x - 3 + 3 \leq 17 + 3$ • Add 3.

$4x \leq 20$

$\dfrac{4x}{4} \leq \dfrac{20}{4}$ • Divide by 4.

$x \leq 5$

$(-\infty, 5]$

YOU TRY IT • 1

Solve and write the answer in interval notation:
$5 - 4x > 9 - 8x$

Your solution

Solution on p. S33

EXAMPLE • 2

Solve and write the answer in set-builder notation:
$3(3 - 2x) \geq -5x - 2(3 - x)$

Solution

$$3(3 - 2x) \geq -5x - 2(3 - x)$$
$$9 - 6x \geq -5x - 6 + 2x \quad \text{• Distributive Property}$$
$$9 - 6x \geq -3x - 6$$
$$9 - 6x + 3x \geq -3x + 3x - 6 \quad \text{• Add } 3x.$$
$$9 - 3x \geq -6$$
$$9 - 9 - 3x \geq -6 - 9 \quad \text{• Subtract 9.}$$
$$-3x \geq -15$$
$$\frac{-3x}{-3} \leq \frac{-15}{-3} \quad \text{• Divide by } -3.$$
$$x \leq 5$$
$$\{x|x \leq 5\}$$

YOU TRY IT • 2

Solve and write the answer in set-builder notation:
$8 - 4(3x + 5) \leq 6(x - 8)$

Your solution

Solution on p. S33

OBJECTIVE B To solve application problems

EXAMPLE • 3

A rectangle is 10 ft wide and $(2x + 4)$ ft long. Express as an integer the maximum length of the rectangle when the area is less than 200 ft². (The area of a rectangle is equal to its length times its width.)

Strategy

To find the maximum length:

- Replace the variables in the area formula by the given values and solve for x.
- Replace the variable in the expression $2x + 4$ with the value found for x.

Solution

Length times width	is less than	200 ft²

$$10(2x + 4) < 200$$
$$20x + 40 < 200 \quad \text{• Distributive Property}$$
$$20x + 40 - 40 < 200 - 40 \quad \text{• Subtract 40.}$$
$$20x < 160$$
$$\frac{20x}{20} < \frac{160}{20} \quad \text{• Divide by 20.}$$
$$x < 8$$

The length is $(2x + 4)$ ft. Because $x < 8$, $2x + 4 < 2(8) + 4 = 20$. Therefore, the length is less than 20 ft. The maximum length is 19 ft.

YOU TRY IT • 3

Company A rents cars for \$8 a day and \$.10 for every mile driven. Company B rents cars for \$10 a day and \$.08 per mile driven. You want to rent a car for 1 week. What is the maximum number of miles you can drive a Company A car if it is to cost you less than a Company B car?

Your strategy

Your solution

Solution on pp. S33–S34

14.3 EXERCISES

OBJECTIVE A To solve general inequalities

For Exercises 1 to 9, solve and write the answer in interval notation.

1. $4x - 8 < 2x$

2. $7x - 4 < 3x$

3. $2x - 8 > 4x$

4. $3y + 2 > 7y$

5. $8 - 3x \leq 5x$

6. $10 - 3x \leq 7x$

7. $3x + 2 > 5x - 8$

8. $2n - 9 \geq 5n + 4$

9. $5x - 2 < 3x - 2$

For Exercises 10 to 20, solve and write the answer in set-builder notation.

10. $8x - 9 > 3x - 9$

11. $0.1(180 + x) > x$

12. $x > 0.2(50 + x)$

13. $2(2y - 5) \leq 3(5 - 2y)$

14. $2(5x - 8) \leq 7(x - 3)$

15. $5(2 - x) > 3(2x - 5)$

16. $4(3d - 1) > 3(2 - 5d)$

17. $4 - 3(3 - n) \leq 3(2 - 5n)$

18. $15 - 5(3 - 2x) \leq 4(x - 3)$

19. $2x - 3(x - 4) \geq 4 - 2(x - 7)$

20. $4 + 2(3 - 2y) \leq 4(3y - 5) - 6y$

21. Which of the following inequalities are equivalent to the inequality $-7x - 2 > -4x + 1$?
(i) $-3 > -11x$ (ii) $3x > 3$ (iii) $-3 > 3x$ (iv) $3x < -3$

OBJECTIVE B To solve application problems

22. An automatic garage door opener costs \$325 plus an installation labor charge of \$30 per hour, with a minimum of 1 h and a maximum of 3 h of labor. Which of the following are *not* possible amounts for the total cost of the door and installation? There may be more than one correct answer.
(i) \$355 (ii) \$450 (iii) \$325 (iv) \$415 (v) \$350

23. **Wages** A sales agent for a jewelry company is offered a flat monthly salary of \$3200 or a salary of \$1000 plus an 11% commission on the selling price of each item sold by the agent. If the agent chooses the \$3200, what dollar amount does the agent expect to sell in 1 month?

24. **Sports** A baseball player is offered an annual salary of \$200,000 or a base salary of \$100,000 plus a bonus of \$1000 for each hit over 100 hits. How many hits must the baseball player make to earn more than \$200,000?

25. **Comparing Services** A site licensing fee for a computer program is \$1500. Paying this fee allows the company to use the program at any computer terminal within the company. Alternatively, the company can choose to pay \$200 for each individual computer it has. How many individual computers must a company have for the site license to be more economical for the company?

26. **Transportation** A shuttle service taking skiers to a ski area charges \$8 per person each way. Four skiers are debating whether to take the shuttle bus or rent a car for \$45 plus \$.25 per mile. Assuming that the skiers will share the cost of the car and that they want the least expensive method of transportation, find how far away the ski area is if they choose the shuttle service.

27. **Health** For a product to be labeled orange juice, a state agency requires that at least 80% of the drink be real orange juice. How many ounces of artificial flavors can be added to 32 oz of real orange juice and have it still be legal to label the drink orange juice?

28. **Health** Grade A hamburger cannot contain more than 20% fat. How much fat can a butcher mix with 300 lb of lean meat to meet the 20% requirement?

29. **Oceanography** Read the news clipping at the right. The inequality $d + rm \leq 5000$ describes the fact that the current depth d of the *Sentry,* plus the product of the rate of descent r and the number of minutes spent descending m, is not more than 5000 ft. Suppose the *Sentry* is hovering at a depth of 1230 m and is ready to descend further. Use the given inequality and a rate of 58 m/min to determine for how long, in minutes, the *Sentry* can descend before stopping again.

In the News

Sentry Completes First Mission

Woods Hole Oceanographic Institute announced the debut of the autonomous underwater vehicle (AUV) *Sentry,* an unmanned research robot that can dive to a depth of 5000 m. Unlike many earlier AUVs, the *Sentry* can start, stop, and change direction, all with no input from scientists working on shore.

Source: www.whoi.edu

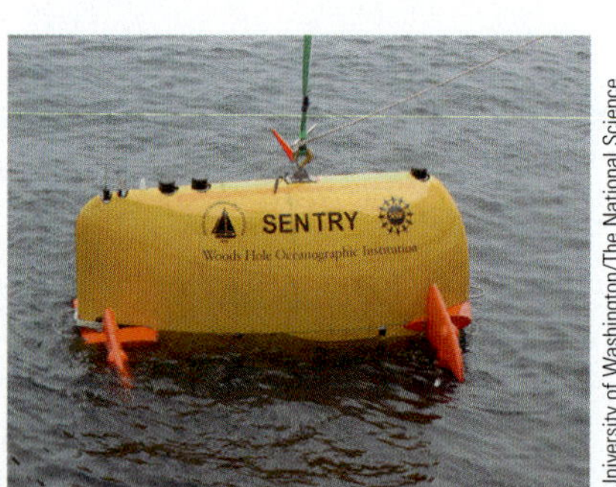

The autonomous underwater vehicle Sentry

Applying the Concepts

30. Determine whether the statement is always true, sometimes true, or never true, given that a, b, and c are real numbers.
    a. If $a > b$, then $-a > -b$.
    b. If $a < b$, then $ac < bc$.
    c. If $a > b$, then $a + c > b + c$.
    d. If $a \neq 0$, $b \neq 0$, and $a > b$, then $\frac{1}{a} > \frac{1}{b}$.

SECTION

14.4 Graphing Linear Inequalities

OBJECTIVE A To graph an inequality in two variables

Point of Interest

Linear inequalities play an important role in applied mathematics. They are used in a branch of mathematics called *linear programming,* which was developed during World War II to solve problems in supplying the Air Force with the machine parts necessary to keep planes flying. Today, linear programming applications extend to many other disciplines.

The graph of the linear equation $y = x - 2$ separates a plane into three sets:

the set of points on the line,
the set of points above the line, and
the set of points below the line.

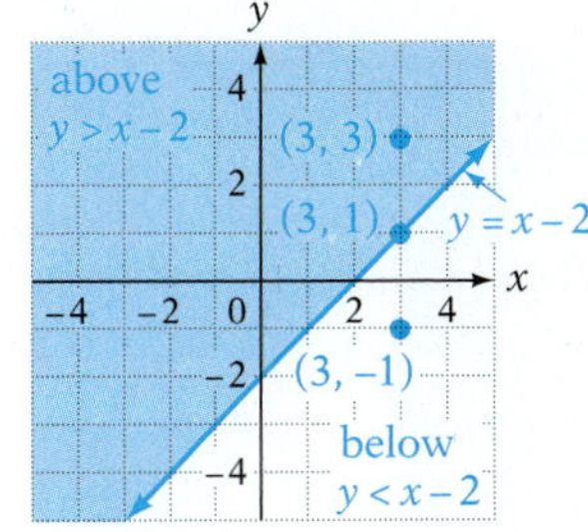

The point (3, 1) is a solution of $y = x - 2$.

$$\begin{array}{c|c} \multicolumn{2}{c}{y = x - 2} \\ \hline 1 & 3 - 2 \\ \multicolumn{2}{c}{1 = 1} \end{array}$$

The point (3, 3) is a solution of $y > x - 2$.

$$\begin{array}{c|c} \multicolumn{2}{c}{y > x - 2} \\ \hline 3 & 3 - 2 \\ \multicolumn{2}{c}{3 > 1} \end{array}$$

Any point above the line is a solution of $y > x - 2$.

Tips for Success

Be sure to do all you need to do in order to be successful at graphing linear inequalities: Read through the introductory material, work through the How To examples, study the paired examples, do the You Try Its, and check your solutions against those in the back of the book. See *AIM for Success* at the front of the book.

The point (3, −1) is a solution of $y < x - 2$.

$$\begin{array}{c|c} \multicolumn{2}{c}{y < x - 2} \\ \hline -1 & 3 - 2 \\ \multicolumn{2}{c}{-1 < 1} \end{array}$$

Any point below the line is a solution of $y < x - 2$.

The solution set of $y = x - 2$ is all points on the line. The solution set of $y > x - 2$ is all points above the line. The solution set of $y < x - 2$ is all points below the line. The solution set of an inequality in two variables is a **half-plane.**

The following example illustrates the procedure for graphing a linear inequality.

HOW TO • 1 Graph the solution set of $2x + 3y \le 6$.

Solve the inequality for y.

$$2x + 3y \le 6$$

$$2x - 2x + 3y \le -2x + 6 \quad \text{• Subtract } 2x \text{ from each side.}$$

$$3y \le -2x + 6 \quad \text{• Simplify.}$$

$$\frac{3y}{3} \le \frac{-2x + 6}{3} \quad \text{• Divide each side by 3.}$$

$$y \le -\frac{2}{3}x + 2 \quad \text{• Simplify.}$$

Change the inequality to an equality and graph $y = -\frac{2}{3}x + 2$. If the inequality is $\ge$ or $\le$, the line is part of the solution set and is shown by a solid line. If the inequality is $>$ or $<$, the line is not part of the solution set and is shown by a dashed line.

If the inequality is $>$ or $\ge$, shade the upper half-plane. If the inequality is $<$ or $\le$, shade the lower half-plane.

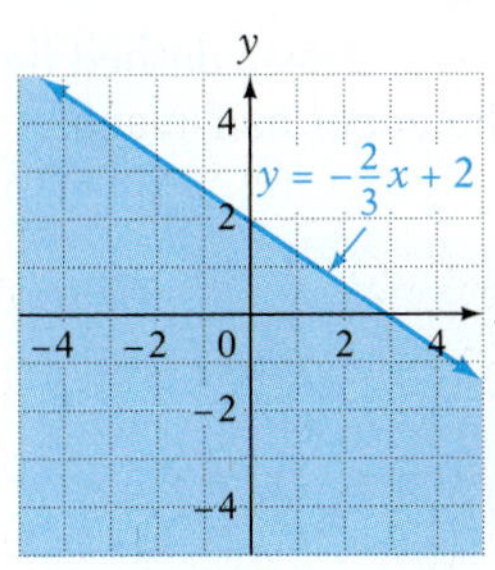

EXAMPLE 1

Graph the solution set of $3x + y > -2$.

Solution

$$3x + y > -2$$
$$3x - 3x + y > -3x - 2 \quad \text{• Subtract } 3x.$$
$$y > -3x - 2$$

Graph $y = -3x - 2$ as a dashed line.
Shade the upper half-plane.

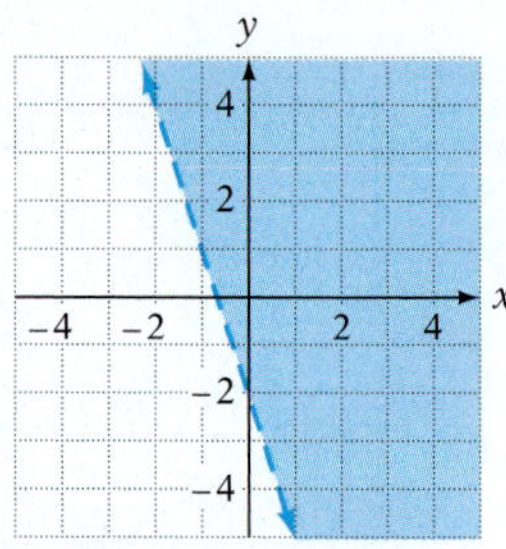

YOU TRY IT 1

Graph the solution set of $x - 3y < 2$.

Your solution

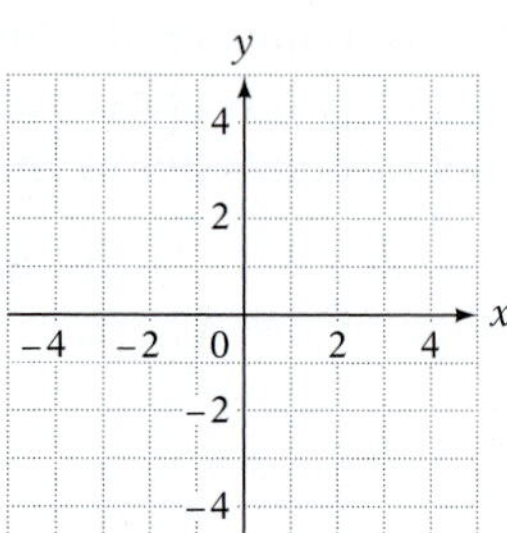

EXAMPLE 2

Graph the solution set of $2x - y \geq 2$.

Solution

$$2x - y \geq 2$$
$$2x - 2x - y \geq -2x + 2 \quad \text{• Subtract } 2x.$$
$$-y \geq -2x + 2$$
$$-1(-y) \leq -1(-2x + 2) \quad \text{• Multiply by } -1.$$
$$y \leq 2x - 2$$

Graph $y = 2x - 2$ as a solid line.
Shade the lower half-plane.

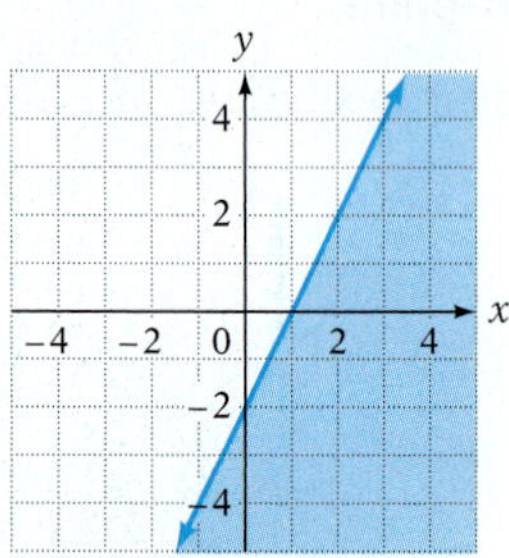

YOU TRY IT 2

Graph the solution set of $2x - 4y \leq 8$.

Your solution

EXAMPLE 3

Graph the solution set of $y > -1$.

Solution

Graph $y = -1$ as a dashed line.
Shade the upper half-plane.

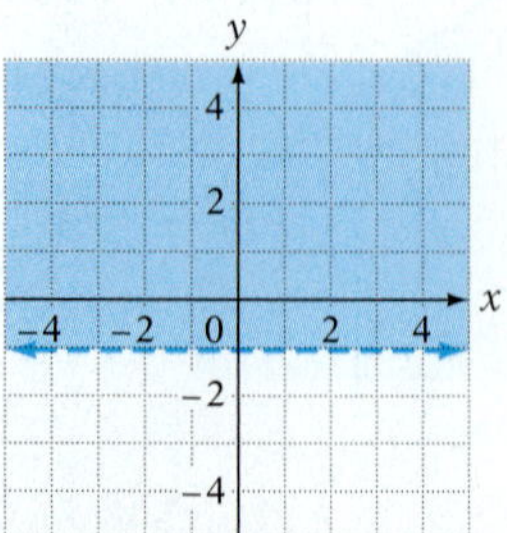

YOU TRY IT 3

Graph the solution set of $x < 3$.

Your solution

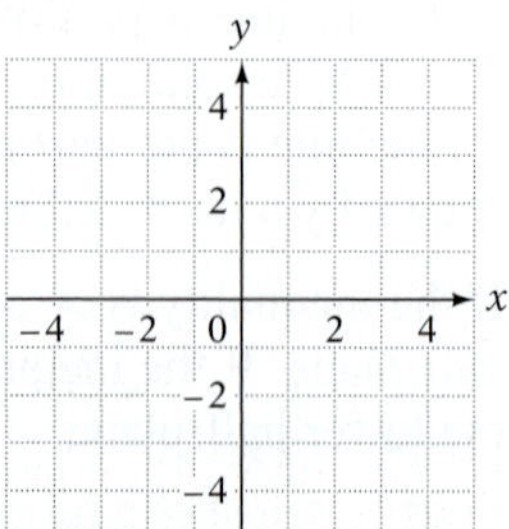

Solutions on p. S34

14.4 EXERCISES

OBJECTIVE A To graph an inequality in two variables

For Exercises 1 to 12, graph the solution set of the inequality.

1. $y > -x + 4$

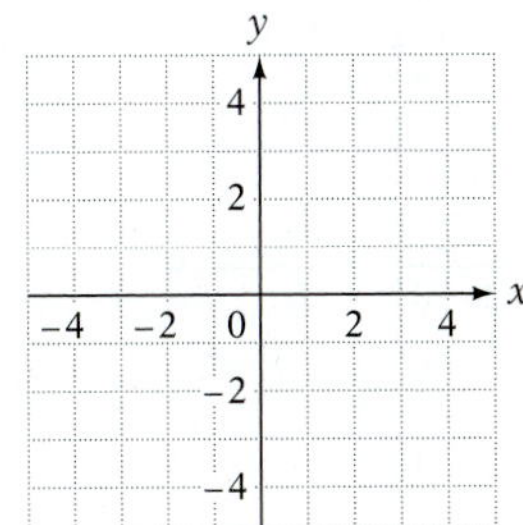

2. $y < x + 3$

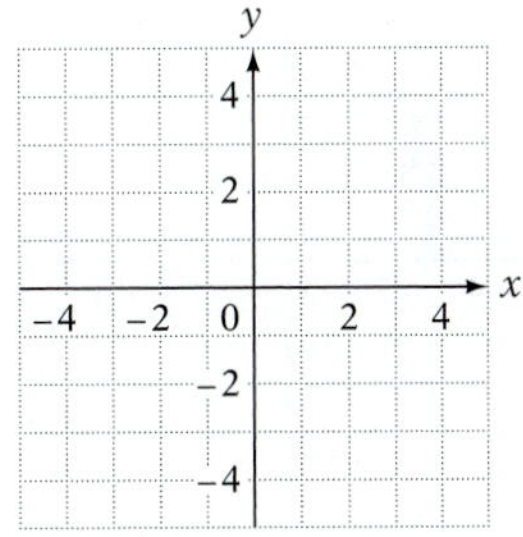

3. $y > 2x + 3$

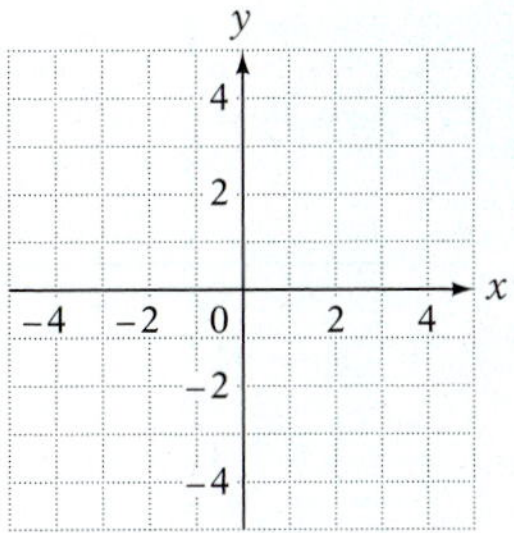

4. $y > 3x - 9$

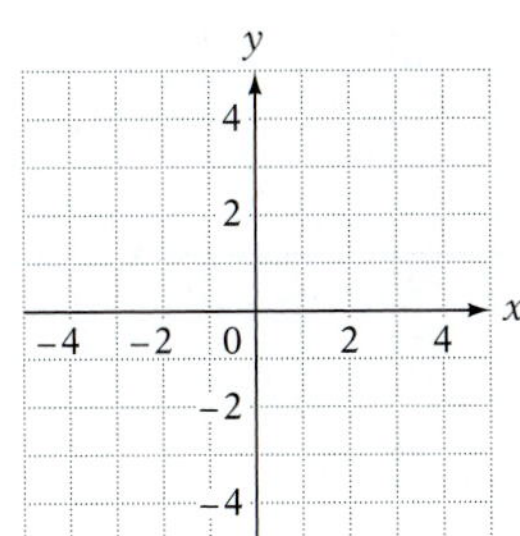

5. $2x + y \geq 4$

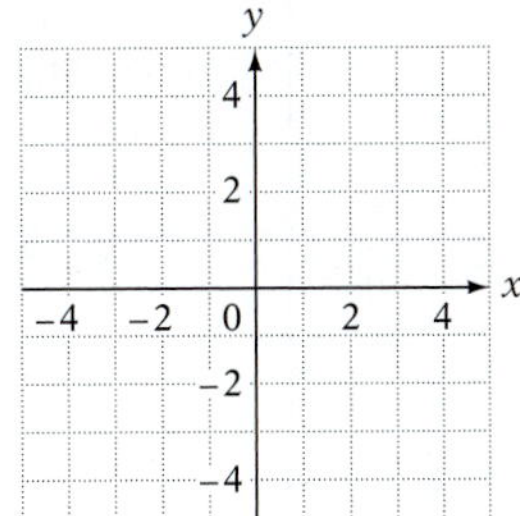

6. $3x + y \geq 6$

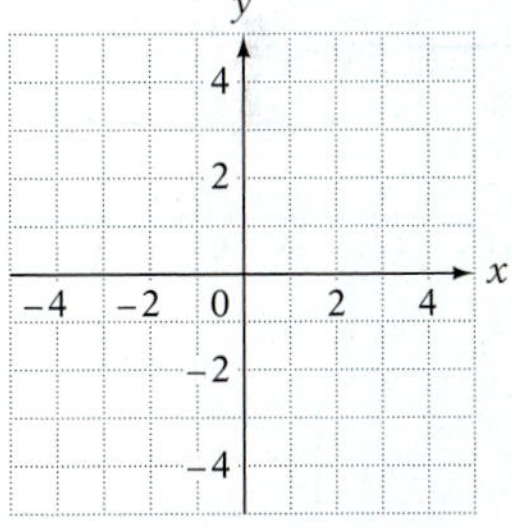

7. $y \leq -2$

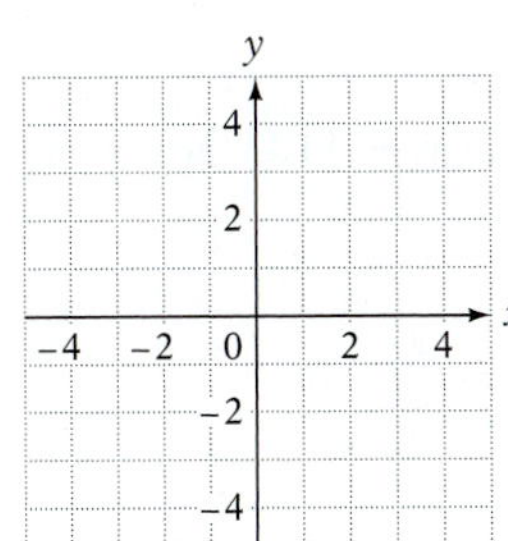

8. $y > 3$

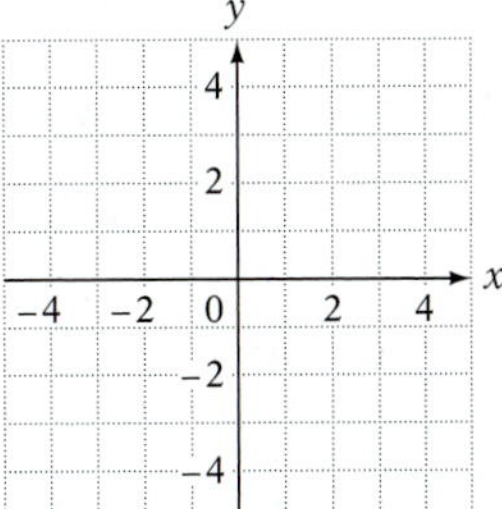

9. $3x - 2y < 8$

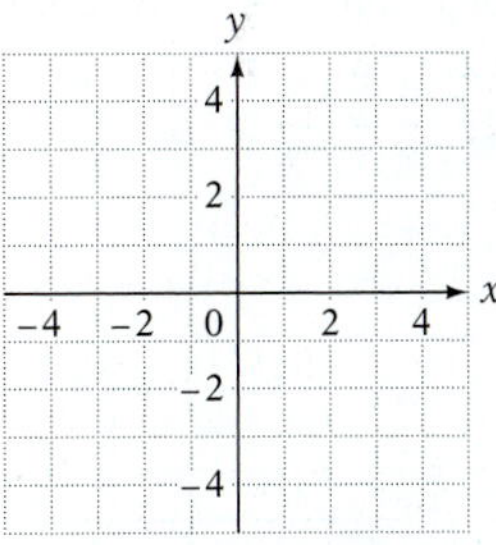

10. $5x + 4y > 4$

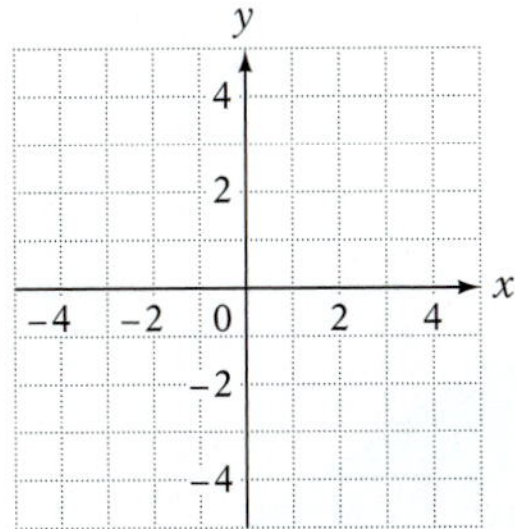

11. $-3x - 4y \geq 4$

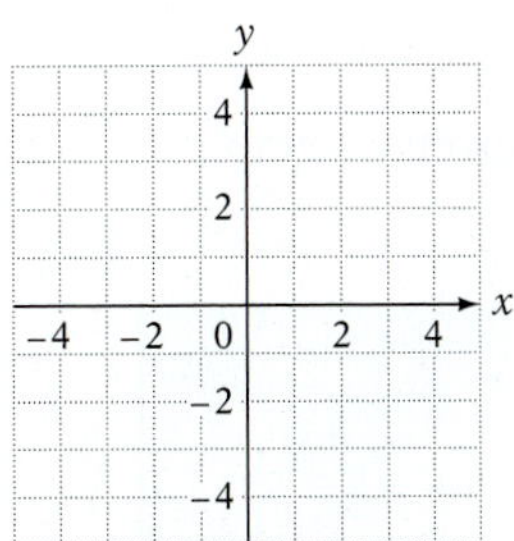

12. $-5x - 2y \geq 8$

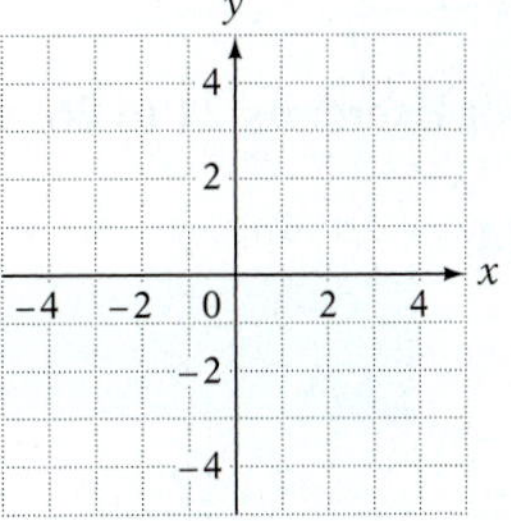

13. Suppose (0, 0) is a point on the graph of the linear inequality $Ax + By > C$, where C is not zero. Is C positive or negative?

14. Suppose $Ax + By < C$, where C is a negative number. Is (0, 0) a point on the graph of $Ax + By < C$?

For Exercises 15 to 20, graph the solution set of the inequality.

15. $6x + 5y \leq -10$

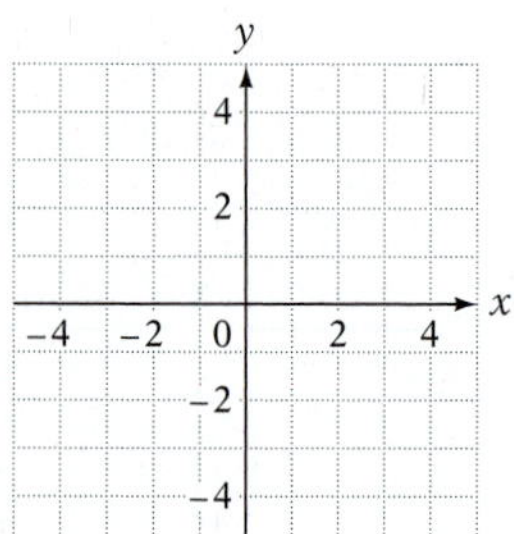

16. $2x + 2y \leq -4$

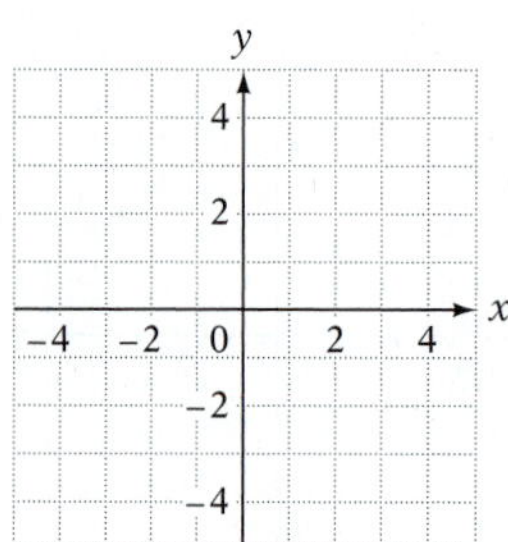

17. $-4x + 3y < -12$

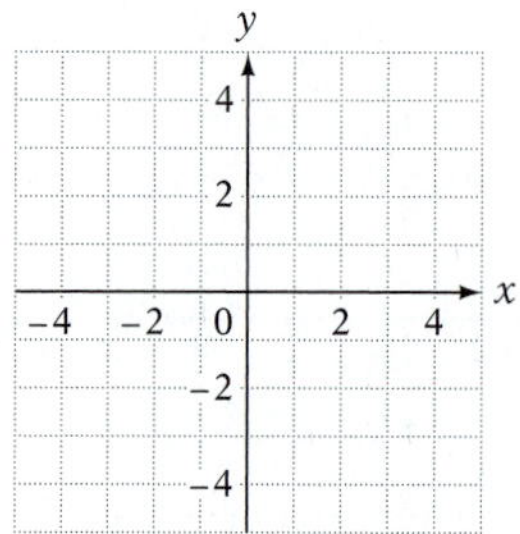

18. $-4x + 5y < 15$

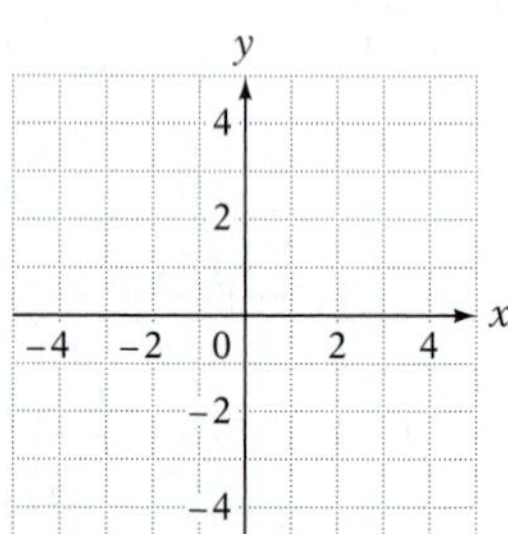

19. $-2x + 3y \leq 6$

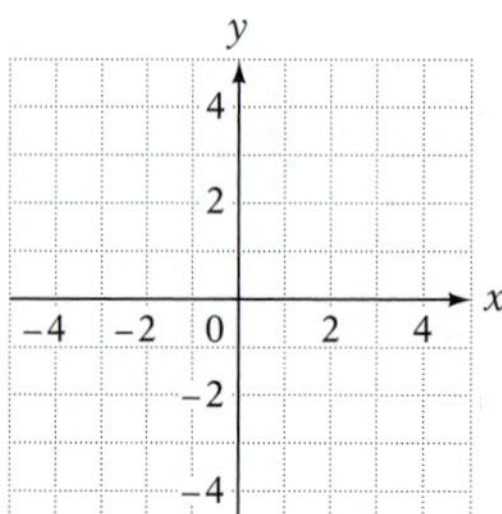

20. $3x - 4y > 12$

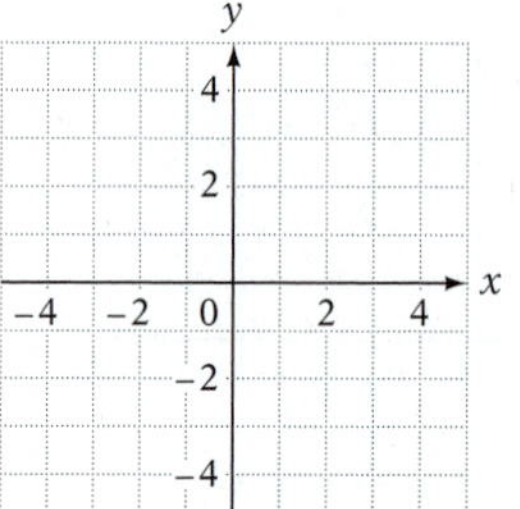

Applying the Concepts

For Exercises 21 to 23, graph the solution set of the inequality.

21. $\frac{x}{4} + \frac{y}{2} > 1$

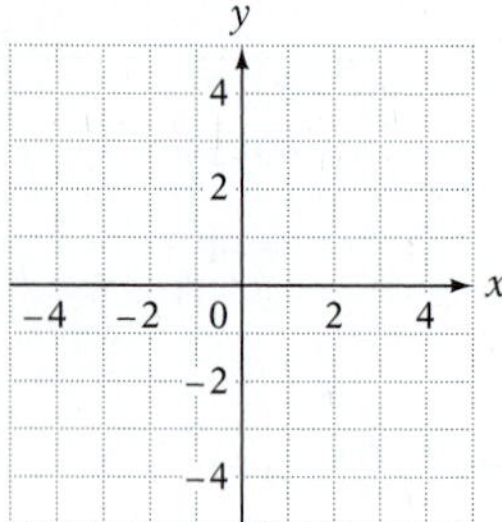

22. $2x - 3(y + 1) > y - (4 - x)$

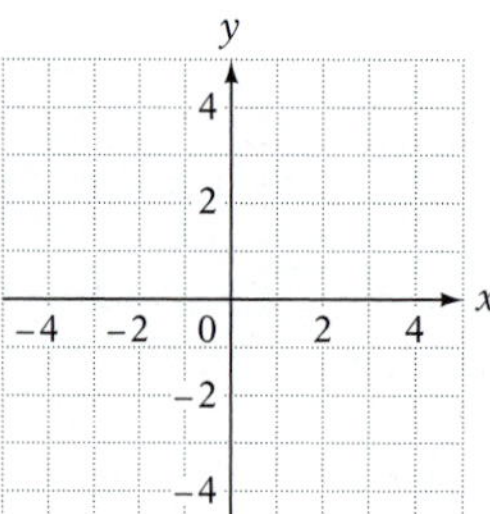

23. $4y - 2(x + 1) \geq 3(y - 1) + 3$

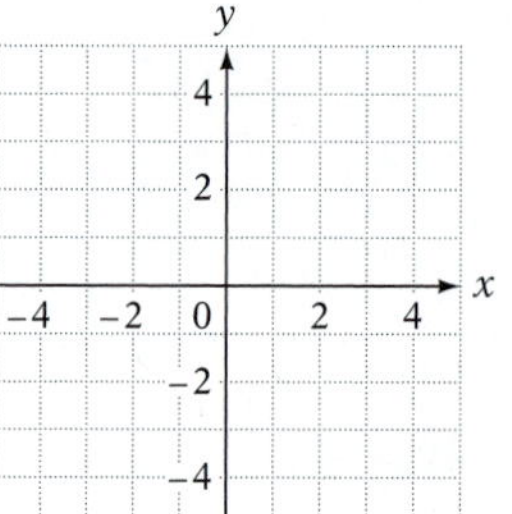

For Exercises 24 to 26, write the inequality given its graph.

24.

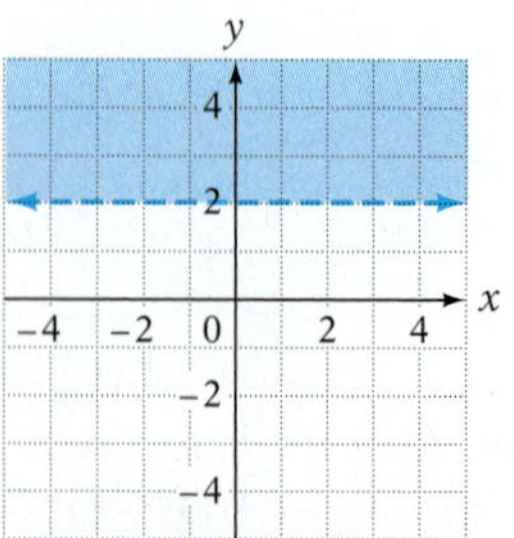

25.

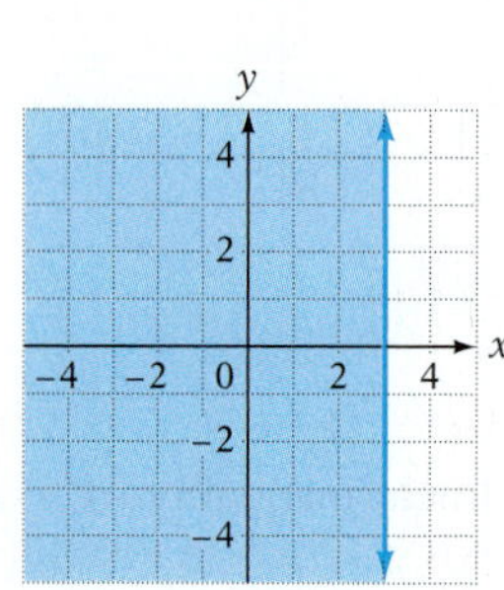

26.

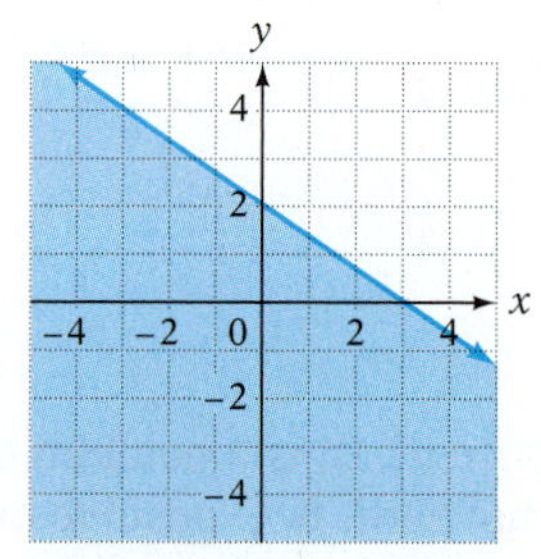

FOCUS ON PROBLEM SOLVING

Graphing Data

Graphs are very useful in displaying data. By studying a graph, we can reach various conclusions about the data.

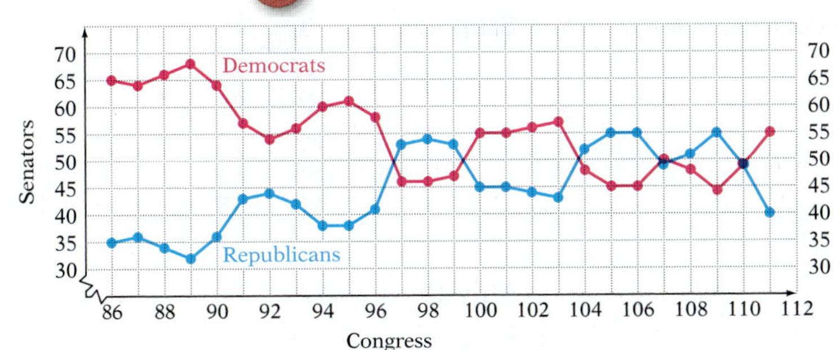

The double-line graph at the left shows the number of Democrats and the number of Republicans in the U.S. Senate for the 86th Congress through the 111th Congress.

1. How many Democratic and how many Republican senators were in the 90th Congress?
2. In which Congress was the difference between the numbers of Democrats and Republicans the greatest?
3. In which Congress did the majority first change from Democratic to Republican?
4. Between which two Congresses did the number of Republican senators increase but the number of Democratic senators remain the same?
5. In what percent of the Congresses did the number of Democrats exceed the number of Republicans? Round to the nearest tenth.
6. In which Congresses were there a greater number of Republican senators than Democratic senators?

PROJECTS AND GROUP ACTIVITIES

Mean and Standard Deviation

The **mean** of a set of data is the sum of the measurements divided by the number of measurements. The symbol for the mean is $\bar{x}$.

$$\textbf{Mean} = \bar{x} = \frac{\textbf{sum of all data values}}{\textbf{number of data values}}$$

The mean is one of the most frequently computed averages. It is the one that is commonly used to calculate a student's performance in a class.

Consider two students, each of whom has taken five exams.

Scores for student A

84	86	83	85	87

Scores for student B

90	75	94	68	98

$$\bar{x} = \frac{84 + 86 + 83 + 85 + 87}{5} = \frac{425}{5} = 85$$

The mean for student A is 85.

$$\bar{x} = \frac{90 + 75 + 94 + 68 + 98}{5} = \frac{425}{5} = 85$$

The mean for student B is 85.

For each of these students, the mean (average) for the five exams is 85. However, student A has a more consistent record of scores than student B. One way to measure the consistency, or "clustering" near the mean, of data is to use the **standard deviation.**

To calculate the standard deviation:

Step 1. Sum the squares of the differences between each data value and the mean.

Step 2. Divide the result in Step 1 by the number of items in the set of data.

Step 3. Take the square root of the result in Step 2.

The calculation for student A is shown at the right.

Step 1:

x	$x - \bar{x}$	$(x - \bar{x})^2$
84	84 − 85	$(-1)^2 = 1$
86	86 − 85	$1^2 = 1$
83	83 − 85	$(-2)^2 = 4$
85	85 − 85	$0^2 = 0$
87	87 − 85	$2^2 = 4$
		Total = 10

The symbol for standard deviation is the lowercase Greek letter *sigma*, σ.

Step 2: $\frac{10}{5} = 2$

Step 3: $\sigma = \sqrt{2} \approx 1.414$

The standard deviation of student A's scores is approximately 1.414.

Following a similar procedure for student B shows that the standard deviation of student B's scores is approximately 11.524. Because the standard deviation of student B's scores is greater than that of student A's (11.524 > 1.414), student B's scores are not as consistent as those of student A.

1. The weights in ounces of six newborn infants were recorded by a hospital. The weights were 96, 105, 84, 90, 102, and 99. Find the standard deviation of the weights. Round to the nearest hundredth.

2. Seven coins were each tossed 100 times. The numbers of heads recorded for each coin were 56, 63, 49, 50, 48, 53, and 52. Find the standard deviation of the numbers of heads. Round to the nearest hundredth.

3. The high temperatures, in degrees Fahrenheit, for 11 consecutive days at a desert resort were 95°, 98°, 98°, 104°, 97°, 100°, 96°, 97°, 108°, 93°, and 104°. For the same days, the high temperatures in Antarctica were 27°, 28°, 28°, 30°, 28°, 27°, 30°, 25°, 24°, 26°, and 21°. Which location has the greater standard deviation of high temperatures?

4. The scores for five college basketball games were 56, 68, 60, 72, and 64. The scores for five professional basketball games were 106, 118, 110, 122, and 114. Which set of scores has the greater standard deviation?

5. One student received test scores of 85, 92, 86, and 89. A second student received scores of 90, 97, 91, and 94 (exactly 5 points more on each test). Are the mean scores of the two students the same? If not, what is the relationship between the mean scores of the two students? Are the standard deviations of the scores of the two students the same? If not, what is the relationship between the standard deviations of the scores of the two students?

6. A company is negotiating with its employees the terms of a raise in salary. One proposal would add $500 a year to each employee's salary. The second proposal would give each employee a 4% raise. Explain how each of these proposals would affect the current mean and standard deviation of salaries for the company.

CHAPTER 14
SUMMARY

KEY WORDS	EXAMPLES
The *empty set* or *null set,* written ∅, is the set that contains no elements. [14.1A, p. 722]	The set of cars that can travel faster than 1000 mph is an empty set.
The *union* of two sets, written $A \cup B$, is the set that contains the elements of A and the elements of B. [14.1A, p. 722]	Let $A = \{2, 4, 6, 8\}$ and $B = \{0, 1, 2, 3, 4\}$. Then $A \cup B = \{0, 1, 2, 3, 4, 6, 8\}$.
The *intersection* of two sets, written $A \cap B$, is the set that contains the elements that are common to both A and B. [14.1A, p. 722]	Let $A = \{2, 4, 6, 8\}$ and $B = \{0, 1, 2, 3, 4\}$. Then $A \cap B = \{2, 4\}$.
Set-builder notation and *interval notation* are used to describe the elements of a set. [14.1B, pp. 723–724]	The set of real numbers greater than 2 is written in set-builder notation as $\{x \mid x > 2, x \in \text{real numbers}\}$ and in interval notation as $(2, \infty)$.
The *solution set of an inequality* is a set of numbers each element of which, when substituted for the variable, results in a true inequality. The solution set of an inequality can be graphed on a number line. [14.2A, p. 729]	The solution set of $3x - 1 < 5$ is $\{x \mid x < 2\}$. The graph of the solution set is −5 −4 −3 −2 −1 0 1 2 3 4 5
The solution set of a linear inequality in two variables is a *half-plane.* [14.4A, p. 741]	The solution set of $3x + 4y \geq 12$ is the half-plane shown at the right. 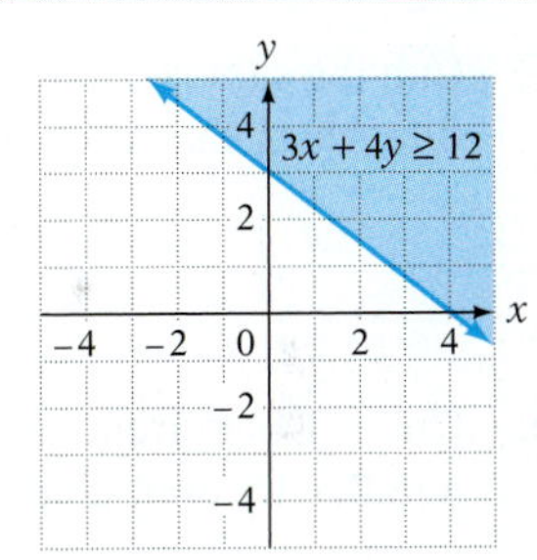

ESSENTIAL RULES AND PROCEDURES	EXAMPLES

Addition Property of Inequalities [14.2A, p. 729]
The same term can be added to each side of an inequality without changing the solution set of the inequality.

If $a > b$, then $a + c > b + c$.
If $a < b$, then $a + c < b + c$.

$$x - 3 < -7$$
$$x - 3 + 3 < -7 + 3$$
$$x < -4$$

Multiplication Property of Inequalities [14.2B, p. 731]
Each side of an inequality can be multiplied by the same positive number without changing the solution set of the inequality.

If $a > b$ and $c > 0$, then $ac > bc$.
If $a < b$ and $c > 0$, then $ac < bc$.

$$4x > -8$$
$$\frac{4x}{4} > \frac{-8}{4}$$
$$x > -2$$

If each side of an inequality is multiplied by the same negative number and the inequality symbol is reversed, then the solution set of the inequality is not changed.

If $a > b$ and $c < 0$, then $ac < bc$.
If $a < b$ and $c < 0$, then $ac > bc$.

$$-2x < 6$$
$$\frac{-2x}{-2} > \frac{6}{-2}$$
$$x > -3$$

CHAPTER 14

CONCEPT REVIEW

Test your knowledge of the concepts presented in this chapter. Answer each question. Then check your answers against the ones provided in the Answer Section.

1. How is the empty set or null set represented?

2. What do $A \cup B$ and $A \cap B$ mean?

3. What is the difference between the roster method and set-builder notation?

4. How is the solution set of an inequality represented on a number line?

5. Can the same term be added to each side of an inequality without changing the solution set of the inequality?

6. Under what circumstances can each side of an inequality be multiplied by the same number without changing the direction of the inequality symbol?

7. How is the Multiplication Property of Inequalities different from the Multiplication Property of Equations?

8. How is solving a general first-degree inequality different from solving a general first-degree equation?

9. How is graphing a linear inequality in two variables different from graphing a linear equation in two variables?

10. When graphing an inequality in two variables, when is a dashed line used?

CHAPTER 14

REVIEW EXERCISES

1. Solve and write the solution in set-builder notation: $2x - 3 > x + 15$

2. Find $A \cap B$, given $A = \{0, 2, 4, 6, 8\}$ and $B = \{-2, -4\}$.

3. Use set-builder notation to write the set of odd integers greater than -8.

4. Find $A \cup B$, given $A = \{6, 8, 10\}$ and $B = \{2, 4, 6\}$.

5. Use the roster method to write the set of odd positive integers less than 8.

6. Solve and write the solution set in interval notation: $12 - 4(x - 1) \le 5(x - 4)$

7. Graph: $\{x \mid x > 3\}$

8. Solve and write the solution set in set-builder notation: $3x + 4 \ge -8$

9. Graph: $3x + 2y \le 12$

10. Graph: $5x + 2y < 6$

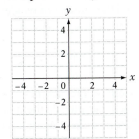

11. Write the set $\{x \mid x > -4\}$ in interval notation.

12. Solve $x - 3 > -1$ and write the solution set in interval notation. Graph the solution set.

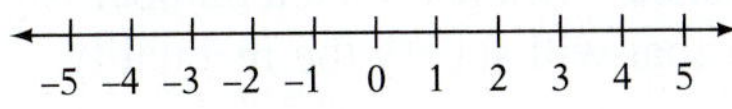

13. Find $A \cap B$, given $A = \{1, 5, 9, 13\}$ and $B = \{1, 3, 5, 7, 9\}$.

14. Graph the interval $[1, 4]$.

15. Graph: $\{x | -1 < x \leq 2\}$

−5 −4 −3 −2 −1 0 1 2 3 4 5

16. Solve and write the solution set in set-builder notation: $-15x \leq 45$

17. Solve and write the solution set in interval notation: $6x - 9 < 4x + 3(x + 3)$

18. Solve and write the solution set in set-builder notation: $5 - 4(x + 9) > 11(12x - 9)$

19. Solve and write the solution set in set-builder notation: $-\frac{3}{4}x > \frac{2}{3}$

20. Solve and write the solution set in interval notation: $7x - 2(x + 3) \geq x + 10$

21. Graph: $2x - 3y < 9$

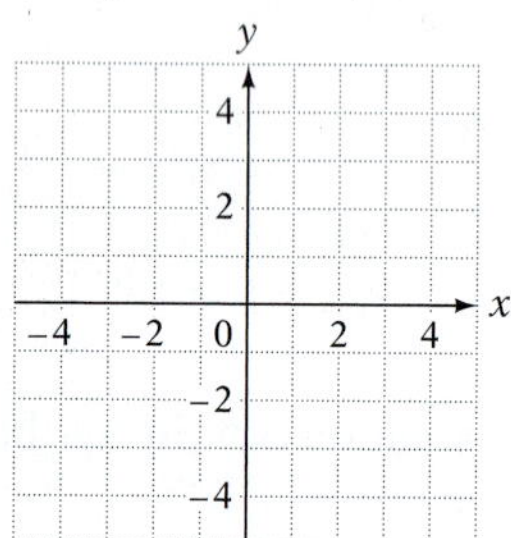

22. **Floral Delivery** Florist A charges a \$6 delivery fee plus \$57 per bouquet delivered. Florist B charges a \$15 delivery fee plus \$48 per bouquet delivered. A church wants to supply each resident of a small nursing home with a bouquet for Grandparents Day. Find the number of residents of the nursing home if using florist B is more economical than using florist A.

23. **Landscaping** The width of a rectangular garden is 12 ft. The length of the garden is $(3x + 5)$ ft. Express as an integer the minimum length of the garden when the area is greater than 276 ft^2. (The area of a rectangle is equal to its length times its width.)

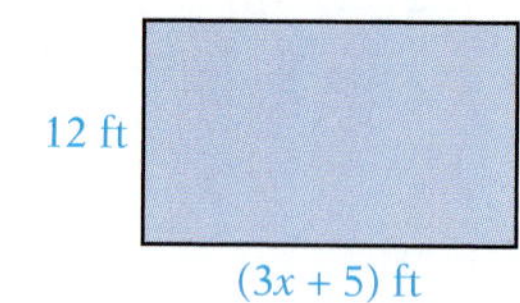

24. **Number Sense** Six less than a number is greater than twenty-five. Find the smallest integer that will satisfy the inequality.

25. **Grading** A student's grades on five sociology tests were 68, 82, 90, 73, and 95. What is the lowest score the student can receive on the next test and still be able to attain a minimum of 480 points?

CHAPTER 14
TEST

1. Graph the interval (0, 5).

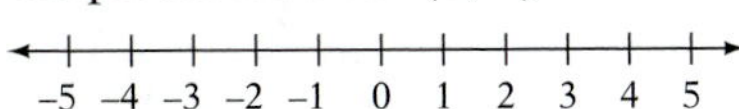

2. Use set-builder notation to write the set of positive integers less than 50.

3. Use the roster method to write the set of the even positive integers between 3 and 9.

4. Solve and write the solution set in interval notation: $3(2x - 5) \geq 8x - 9$

5. Solve and write the solution set in set-builder notation: $x + \frac{1}{2} > \frac{5}{8}$

6. Graph: $\{x \mid x > -2\}$

-5 -4 -3 -2 -1 0 1 2 3 4 5

7. Solve and write the solution set in interval notation: $5 - 3x > 8$

8. Use set-builder notation to write the set of real numbers greater than -23.

9. Graph the solution set of $3x + y > 4$.

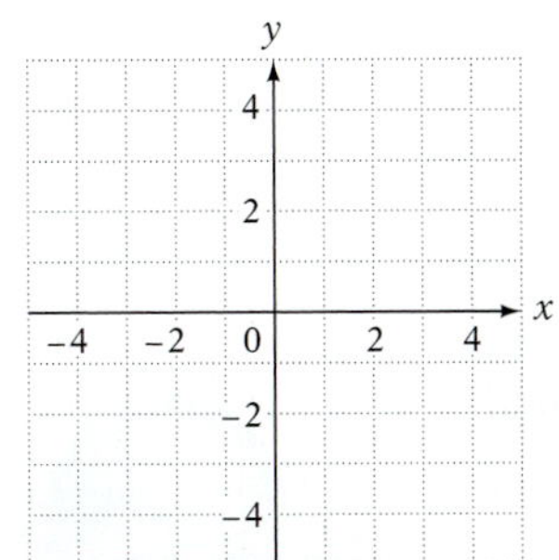

10. Graph the solution set of $4x - 5y \geq 15$.

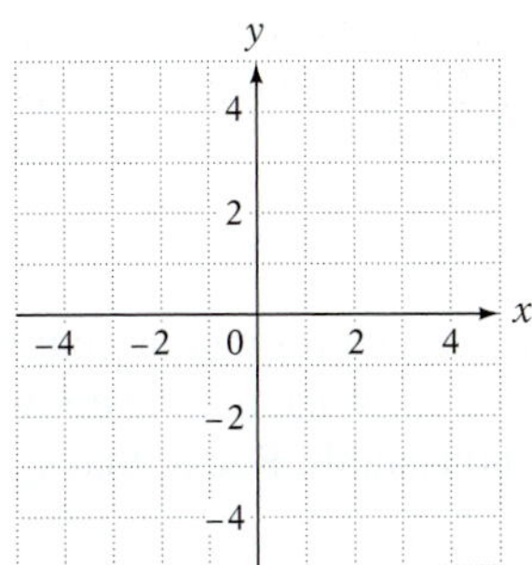

11. Find $A \cap B$, given $A = \{6, 8, 10, 12\}$ and $B = \{12, 14, 16\}$.

12. Solve $4 + x < 1$ and write the solution set in set-builder notation. Graph the solution set.

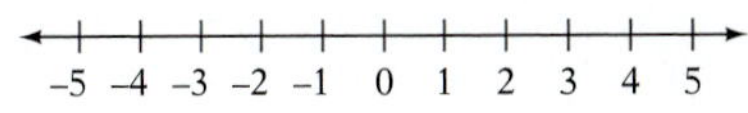

13. Solve and write the solution set in set-builder notation: $-\frac{3}{8}x \leq 5$

14. Solve and write the solution set in interval notation: $6x - 3(2 - 3x) < 4(2x - 7)$

15. Solve $\frac{2}{3}x \geq 2$ and write the solution set in interval notation. Graph the solution set.

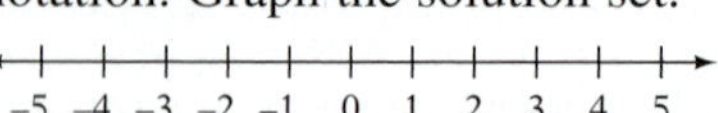

16. Solve and write the solution set in set-builder notation: $2x - 7 \leq 6x + 9$

17. **Safety** To ride a certain roller coaster at an amusement park, a person must be at least 48 in. tall. How many inches must a child who is 43 in. tall grow to be eligible to ride the roller coaster?

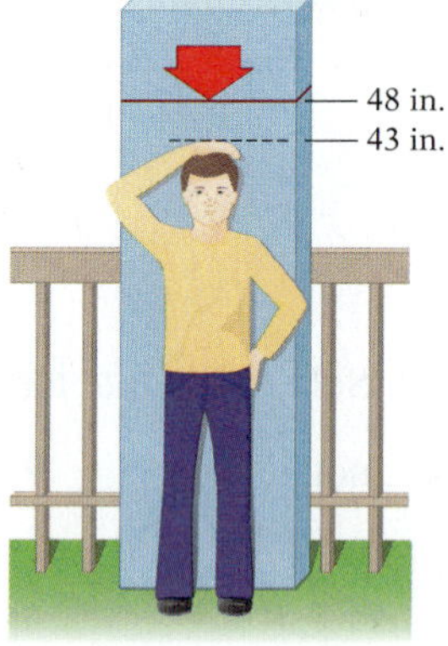

18. **Geometry** A rectangle is 15 ft long and $(2x - 4)$ ft wide. Express as an integer the maximum width of the rectangle if the area is less than 180 ft^2. (The area of a rectangle is equal to its length times its width.)

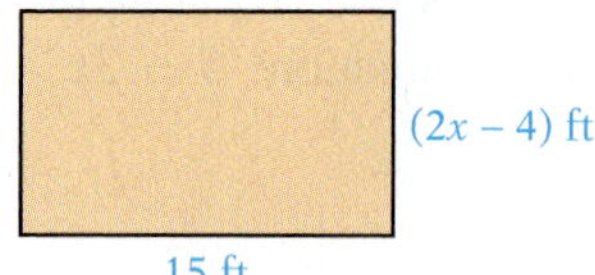

19. **Machining** A ball bearing for a rotary engine must have a circumference between 0.1220 in. and 0.1240 in. What are the allowable diameters for the bearing? Round to the nearest ten-thousandth. Recall that $C = \pi d$.

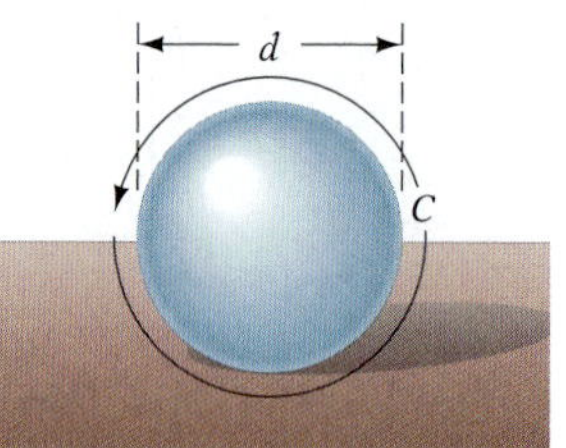

20. **Wages** A stockbroker receives a monthly salary that is the greater of \$2500 or \$1000 plus 2% of the total value of all stock transactions the broker processes during the month. What dollar amounts of transactions did the broker process in a month for which the broker's salary was \$2500?

CUMULATIVE REVIEW EXERCISES

1. Simplify: $2[5a - 3(2 - 5a) - 8]$

2. Solve: $\frac{5}{8} - 4x = \frac{1}{8}$

3. Solve: $2x - 3[x - 2(x - 3)] = 2$

4. Simplify: $(-3a)(-2a^3b^2)^2$

5. Simplify: $\frac{27a^3b^2}{(-3ab^2)^3}$

6. Divide: $(16x^2 - 12x - 2) \div (4x - 1)$

7. Given $f(x) = x^2 - 4x - 5$, find $f(-1)$.

8. Factor: $27a^2x^2 - 3a^2$

9. Divide: $\frac{x^2 - 2x}{x^2 - 2x - 8} \div \frac{x^3 - 5x^2 + 6x}{x^2 - 7x + 12}$

10. Subtract: $\frac{4a}{2a - 3} - \frac{2a}{a + 3}$

11. Solve: $\frac{5y}{6} - \frac{5}{9} = \frac{y}{3} - \frac{5}{6}$

12. Solve $R = \frac{C - S}{t}$ for C.

13. Find the slope of the line that passes through the points $(2, -3)$ and $(-1, 4)$.

14. Find the equation of the line that passes through the point $(1, -3)$ and has slope $-\frac{3}{2}$.

15. Solve by substitution.
$$x = 3y + 1$$
$$2x + 5y = 13$$

16. Solve by the addition method.
$$9x - 2y = 17$$
$$5x + 3y = -7$$

17. Find $A \cup B$, given $A = \{0, 1, 2\}$ and $B = \{-10, -2\}$.

18. Use set-builder notation to write the set of real numbers less than 48.

19. Write $\{x|x < 4\}$ in interval notation.

20. Graph the solution set of $\frac{3}{8}x > -\frac{3}{4}$.

−5 −4 −3 −2 −1 0 1 2 3 4 5

21. Solve: $-\frac{4}{5}x > 12$

22. Solve: $15 - 3(5x - 7) < 2(7 - 2x)$

23. **Number Sense** Three-fifths of a number is less than negative fifteen. What integers satisfy this inequality? Write the answer in set-builder notation.

24. **Rental Agencies** Company A rents cars for \$6 a day and \$.25 for every mile driven. Company B rents cars for \$15 a day and \$.10 per mile. You want to rent a car for 6 days. What is the maximum number of miles you can drive a Company A car if it is to cost you less than a Company B car?

25. **Conservation** In a lake, 100 fish are caught, tagged, and then released. Later, 150 fish are caught. Three of these 150 fish are found to have tags. Estimate the number of fish in the lake.

26. **Geometry** The measure of the first angle of a triangle is 30° more than the measure of the second angle. The measure of the third angle is 10° more than twice the measure of the second angle. Find the measure of each angle.

27. Graph: $y = 2x - 1$

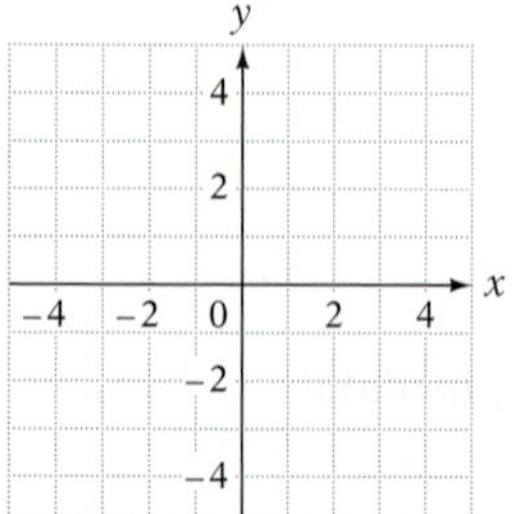

28. Graph the solution set of $6x - 3y \geq 6$.

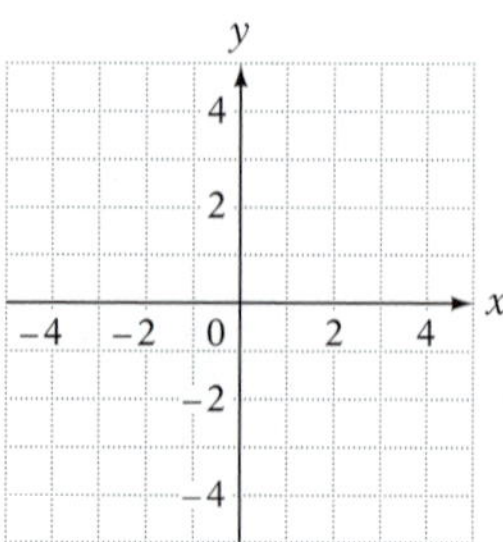

CHAPTER

15

Radical Expressions

OBJECTIVES

SECTION 15.1

A To simplify numerical radical expressions

B To simplify variable radical expressions

SECTION 15.2

A To add and subtract radical expressions

SECTION 15.3

A To multiply radical expressions

B To divide radical expressions

SECTION 15.4

A To solve an equation containing a radical expression

B To solve application problems

ARE YOU READY?

Take the Chapter 15 Prep Test to find out if you are ready to learn to:

- Simplify numerical and variable radical expressions
- Add, subtract, multiply, and divide radical expressions
- Solve an equation containing a radical expression

PREP TEST

Do these exercises to prepare for Chapter 15.

1. Evaluate: $-|-14|$

2. Simplify: $3x^2y - 4xy^2 - 5x^2y$

3. Solve: $1.5h = 21$

4. Solve: $3x - 2 = 5 - 2x$

5. Simplify: $x^3 \cdot x^3$

6. Expand: $(x + y)^2$

7. Expand: $(2x - 3)^2$

8. Multiply: $(2 - 3v)(2 + 3v)$

9. Multiply: $(a - 5)(a + 5)$

10. Simplify: $\dfrac{2x^4y^3}{18x^2y}$

SECTION

15.1 Introduction to Radical Expressions

OBJECTIVE A To simplify numerical radical expressions

Point of Interest

The radical symbol was first used in 1525 but was written as √. Some historians suggest that the radical symbol also developed into the symbols for "less than" and "greater than." Because typesetters of that time did not want to make additional symbols, the radical was rotated to the position > and used as a "greater than" symbol and rotated to < and used as the "less than" symbol. Other evidence, however, suggests that the "less than" and "greater than" symbols were developed independently of the radical symbol.

A **square root** of a positive number *a* is a number whose square is *a*.

A square root of 16 is 4 because $4^2 = 16$.

A square root of 16 is -4 because $(-4)^2 = 16$.

Every positive number has two square roots, one a positive and one a negative number. The symbol $\sqrt{\ }$, called a **radical sign,** is used to indicate the positive or **principal square root** of a number. For example, $\sqrt{16} = 4$ (the principal square root of 16 is 4) and $\sqrt{25} = 5$ (the principal square root of 25 is 5). The number under the radical sign is called the **radicand.**

When the negative square root of a number is to be found, a negative sign is placed in front of the radical. For example, $-\sqrt{16} = -4$ and $-\sqrt{25} = -5$.

The square of an integer is a **perfect square.** For instance, 49, 81, and 144 are perfect squares. The principal square root of a perfect-square integer is a positive integer.

$7^2 = 49 \qquad \sqrt{49} = 7$

$9^2 = 81 \qquad \sqrt{81} = 9$

$12^2 = 144 \qquad \sqrt{144} = 12$

If a number is not a perfect square, its square root can only be approximated. For example, 2 and 7 are not perfect squares. The square roots of these numbers are *irrational numbers.* Their decimal approximations never terminate or repeat.

$$\sqrt{2} \approx 1.4142135\ldots \qquad \sqrt{7} \approx 2.6457513\ldots$$

Take Note

Recall that a factor of a number divides the number evenly. For instance, 6 is a factor of 18. The perfect square 9 is also a factor of 18. It is a *perfect-square factor* of 18, whereas 6 is not a perfect-square factor of 18.

A radical expression is in *simplest form* when the radicand contains no factor greater than 1 that is a perfect square. For instance, $\sqrt{50}$ is not in simplest form because 25 is a perfect-square factor of 50. The radical expression $\sqrt{15}$ is in simplest form because there are no perfect-square factors of 15 that are greater than 1.

The Product Property of Square Roots and a knowledge of perfect squares are used to simplify radicands that are not perfect squares.

The Product Property of Square Roots

If *a* and *b* are positive real numbers, then $\sqrt{ab} = \sqrt{a} \cdot \sqrt{b}$.

The chart below shows the square roots of some perfect squares.

Square Roots of Perfect Squares			
$\sqrt{1} = 1$	$\sqrt{16} = 4$	$\sqrt{49} = 7$	$\sqrt{100} = 10$
$\sqrt{4} = 2$	$\sqrt{25} = 5$	$\sqrt{64} = 8$	$\sqrt{121} = 11$
$\sqrt{9} = 3$	$\sqrt{36} = 6$	$\sqrt{81} = 9$	$\sqrt{144} = 12$

HOW TO • 1 Simplify: $\sqrt{72}$

$\sqrt{72} = \sqrt{36 \cdot 2}$ • Write the radicand as the product of a perfect square and a factor that does not contain a perfect square.

$= \sqrt{36}\sqrt{2}$ • Use the Product Property of Square Roots to write the expression as a product.

$= 6\sqrt{2}$ • Simplify.

Note that 72 must be written as the product of a perfect square and *a factor that does not contain a perfect square.* Therefore, it would not be correct to simplify $\sqrt{72}$ as $\sqrt{9 \cdot 8}$. Although 9 is a perfect-square factor of 72, 8 also contains a perfect-square factor ($8 = 4 \cdot 2$). Therefore, $\sqrt{8}$ is not in simplest form. Remember to find the largest perfect-square factor of the radicand.

$$\sqrt{72} = \sqrt{9 \cdot 8} = \sqrt{9} \cdot \sqrt{8} = 3\sqrt{8}$$

Not in simplest form

HOW TO • 2 Simplify: $\sqrt{147}$

$\sqrt{147} = \sqrt{49 \cdot 3}$ • **Write the radicand as the product of a perfect square and a factor that does not contain a perfect square.**

$= \sqrt{49}\sqrt{3}$ • **Use the Product Property of Square Roots to write the expression as a product.**

$= 7\sqrt{3}$ • **Simplify.**

HOW TO • 3 Simplify: $\sqrt{360}$

$\sqrt{360} = \sqrt{36 \cdot 10}$ • **Write the radicand as the product of a perfect square and a factor that does not contain a perfect square.**

$= \sqrt{36}\sqrt{10}$ • **Use the Product Property of Square Roots to write the expression as a product.**

$= 6\sqrt{10}$ • **Simplify.**

From the last example, note that $\sqrt{360} = 6\sqrt{10}$. The two expressions are different representations of the same number. Using a calculator, we find that $\sqrt{360} \approx 18.973666$ and $6\sqrt{10} \approx 6(3.1622777) = 18.9736662$.

HOW TO • 4 Simplify: $\sqrt{-16}$

Because the square of any real number is positive, there is no real number whose square is -16. $\sqrt{-16}$ is not a real number.

EXAMPLE • 1

Simplify: $3\sqrt{90}$

Solution

$3\sqrt{90} = 3\sqrt{9 \cdot 10}$ • **9 is a perfect-square factor.**

$= 3\sqrt{9}\sqrt{10}$ • **Product Property of Square Roots**

$= 3 \cdot 3\sqrt{10}$

$= 9\sqrt{10}$

YOU TRY IT • 1

Simplify: $\sqrt{216}$

Your solution

EXAMPLE • 2

Simplify: $\sqrt{252}$

Solution

$\sqrt{252} = \sqrt{36 \cdot 7}$ • **36 is a perfect-square factor.**

$= \sqrt{36}\sqrt{7}$ • **Product Property of Square Roots**

$= 6\sqrt{7}$

YOU TRY IT • 2

Simplify: $-5\sqrt{32}$

Your solution

Solutions on p. S34

OBJECTIVE B To simplify variable radical expressions 

Variable expressions that contain radicals do not always represent real numbers. For example, if $a = -4$, then

$$\sqrt{a^3} = \sqrt{(-4)^3} = \sqrt{-64}$$

and $\sqrt{-64}$ is not a real number.

Now consider the expression $\sqrt{x^2}$. Evaluate this expression for $x = -2$ and $x = 2$.

$$\sqrt{x^2} \qquad\qquad \sqrt{x^2}$$
$$\sqrt{(-2)^2} = \sqrt{4} = 2 = |-2| \qquad \sqrt{2^2} = \sqrt{4} = 2 = |2|$$

This suggests the following:

For any real number a, $\sqrt{a^2} = |a|$. If $a \geq 0$, then $\sqrt{a^2} = a$.

In order to avoid variable expressions that do not represent real numbers, and so that absolute value signs are not needed for certain expressions, the variables in this chapter will represent *positive* numbers unless otherwise stated.

A variable or a product of variables written in exponential form is a perfect square when each exponent is an even number.

To find the square root of a perfect square, remove the radical sign and multiply each exponent by $\frac{1}{2}$.

HOW TO • 5 Simplify: $\sqrt{a^6}$

$\sqrt{a^6} = a^3$ • Remove the radical sign and multiply the exponent by $\frac{1}{2}$.

A variable radical expression is in simplest form when the radicand contains no factor greater than 1 that is a perfect square.

HOW TO • 6 Simplify: $\sqrt{x^7}$

$\sqrt{x^7} = \sqrt{x^6 \cdot x}$ • Write x^7 as the product of a perfect square and x.

$= \sqrt{x^6}\sqrt{x}$ • Use the Product Property of Square Roots.

$= x^3\sqrt{x}$ • Simplify the perfect square.

HOW TO • 7 Simplify: $3x\sqrt{8x^3y^{13}}$

$3x\sqrt{8x^3y^{13}} = 3x\sqrt{4x^2y^{12}(2xy)}$ • Write the radicand as the product of perfect squares and factors that do not contain a perfect square.

$= 3x\sqrt{4x^2y^{12}}\sqrt{2xy}$ • Use the Product Property of Square Roots.

$= 3x \cdot 2xy^6\sqrt{2xy}$ • Simplify.

$= 6x^2y^7\sqrt{2xy}$

HOW TO • 8 Simplify: $\sqrt{25(x+2)^2}$

$$\sqrt{25(x+2)^2} = 5(x+2)$$
$$= 5x + 10$$

EXAMPLE • 3

Simplify: $\sqrt{b^{15}}$

Solution

$\sqrt{b^{15}} = \sqrt{b^{14} \cdot b}$ • b^{14} is a perfect square.

$= \sqrt{b^{14}} \cdot \sqrt{b} = b^7\sqrt{b}$

YOU TRY IT • 3

Simplify: $\sqrt{y^{19}}$

Your solution

EXAMPLE • 4

Simplify: $\sqrt{24x^5}$

Solution

$\sqrt{24x^5} = \sqrt{4x^4(6x)}$ • 4 and x^4 are perfect squares.

$= \sqrt{4x^4}\sqrt{6x}$

$= 2x^2\sqrt{6x}$

YOU TRY IT • 4

Simplify: $\sqrt{45b^7}$

Your solution

EXAMPLE • 5

Simplify: $2a\sqrt{18a^3b^{10}}$

Solution

$2a\sqrt{18a^3b^{10}}$

$= 2a\sqrt{9a^2b^{10}(2a)}$ • 9, a^2, and b^{10} are perfect squares.

$= 2a\sqrt{9a^2b^{10}}\sqrt{2a}$

$= 2a \cdot 3ab^5\sqrt{2a}$

$= 6a^2b^5\sqrt{2a}$

YOU TRY IT • 5

Simplify: $3a\sqrt{28a^9b^{18}}$

Your solution

EXAMPLE • 6

Simplify: $\sqrt{16(x+5)^2}$

Solution

$\sqrt{16(x+5)^2} = 4(x+5) = 4x + 20$

YOU TRY IT • 6

Simplify: $\sqrt{25(a+3)^2}$

Your solution

EXAMPLE • 7

Simplify: $\sqrt{x^2 + 10x + 25}$

Solution

$\sqrt{x^2 + 10x + 25} = \sqrt{(x+5)^2} = x + 5$

YOU TRY IT • 7

Simplify: $\sqrt{x^2 + 14x + 49}$

Your solution

Solutions on p. S34

15.1 EXERCISES

OBJECTIVE A To simplify numerical radical expressions

1. Describe in your own words how to simplify a radical expression.

2. Explain why $2\sqrt{2}$ is in simplest form and $\sqrt{8}$ is not in simplest form.

For Exercises 3 to 26, simplify.

3. $\sqrt{16}$
4. $\sqrt{64}$
5. $\sqrt{49}$
6. $\sqrt{144}$
7. $\sqrt{32}$
8. $\sqrt{50}$
9. $\sqrt{8}$
10. $\sqrt{12}$
11. $-6\sqrt{18}$
12. $-3\sqrt{48}$
13. $5\sqrt{40}$
14. $2\sqrt{28}$
15. $\sqrt{15}$
16. $\sqrt{21}$
17. $\sqrt{29}$
18. $\sqrt{13}$
19. $-9\sqrt{72}$
20. $-11\sqrt{80}$
21. $\sqrt{45}$
22. $\sqrt{225}$
23. $\sqrt{0}$
24. $\sqrt{210}$
25. $6\sqrt{128}$
26. $9\sqrt{288}$

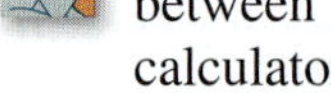
For Exercises 27 to 30, find consecutive integers m and n such that the given number is between m and n, or state that the given number is not a real number. Do not use a calculator.

27. $-\sqrt{115}$
28. $-\sqrt{-90}$
29. $\sqrt{\sqrt{64}}$
30. $\sqrt{200}$

For Exercises 31 to 36, find the decimal approximation rounded to the nearest thousandth.

31. $\sqrt{240}$
32. $\sqrt{300}$
33. $\sqrt{288}$
34. $\sqrt{600}$
35. $\sqrt{350}$
36. $\sqrt{500}$

OBJECTIVE B To simplify variable radical expressions

For Exercises 37 to 76, simplify.

37. $\sqrt{x^{14}}$
38. $\sqrt{x^{12}}$
39. $\sqrt{y^{15}}$
40. $\sqrt{y^{11}}$
41. $\sqrt{a^{20}}$
42. $\sqrt{a^{16}}$
43. $\sqrt{x^4y^4}$
44. $\sqrt{x^{12}y^8}$
45. $\sqrt{4x^4}$
46. $\sqrt{25y^8}$
47. $\sqrt{24x^2}$
48. $\sqrt{x^3y^{15}}$
49. $\sqrt{60x^5}$
50. $\sqrt{72y^7}$
51. $\sqrt{49a^4b^8}$
52. $\sqrt{144x^2y^8}$

53. $\sqrt{18x^5y^7}$
54. $\sqrt{32a^5b^{15}}$
55. $\sqrt{40x^{11}y^7}$
56. $\sqrt{72x^9y^3}$

57. $\sqrt{80a^9b^{10}}$
58. $\sqrt{96a^5b^7}$
59. $2\sqrt{16a^2b^3}$
60. $5\sqrt{25a^4b^7}$

61. $x\sqrt{x^4y^2}$
62. $y\sqrt{x^3y^6}$
63. $4\sqrt{20a^4b^7}$
64. $5\sqrt{12a^3b^4}$

65. $3x\sqrt{12x^2y^7}$
66. $4y\sqrt{18x^5y^4}$
67. $2x^2\sqrt{8x^2y^3}$
68. $3y^2\sqrt{27x^4y^3}$

69. $\sqrt{25(a + 4)^2}$
70. $\sqrt{81(x + y)^4}$
71. $\sqrt{4(x + 2)^4}$
72. $\sqrt{9(x + 2)^8}$

73. $\sqrt{x^2 + 4x + 4}$
74. $\sqrt{b^2 + 8b + 16}$
75. $\sqrt{y^2 + 2y + 1}$
76. $\sqrt{a^2 + 6a + 9}$

For Exercises 77 to 80, assume that a is a positive integer that is not a perfect square. State whether the expression represents a rational number or an irrational number.

77. $\sqrt{100a^6}$
78. $\sqrt{9a^9}$
79. $\sqrt{\sqrt{25a^{16}}}$
80. $\sqrt{\sqrt{81a^8}}$

Applying the Concepts

Shaun Egan/Getty Images

81. **Automotive Safety** Traffic accident investigators can estimate the speed S, in miles per hour, of a car from the length of its skid mark by using the formula $S = \sqrt{30fl}$, where f is the coefficient of friction (which depends on the type of road surface) and l is the length, in feet, of the skid mark. Say the coefficient of friction is 1.2 and the length of a skid mark is 60 ft.
    a. Determine the speed of the car as a radical expression in simplest form.

    b. Write the answer to part (a) as a decimal rounded to the nearest integer.

82. **Travel** The distance a passenger in an airplane can see to the horizon can be approximated by $d = 1.2\sqrt{h}$, where d is the distance to the horizon in miles and h is the height of the plane in feet. To the nearest tenth of a mile, what is the distance to the horizon of a passenger who is flying at an altitude of 5000 ft?

© Jeff Greenberg/PhotoEdit

83. If a and b are positive real numbers, does $\sqrt{a + b} = \sqrt{a} + \sqrt{b}$? If not, give an example in which the expressions are not equal.

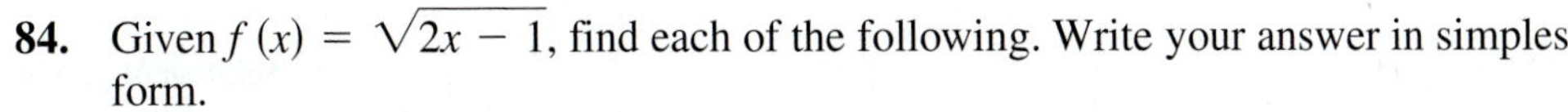

84. Given $f(x) = \sqrt{2x - 1}$, find each of the following. Write your answer in simplest form.
    a. $f(1)$ **b.** $f(5)$ **c.** $f(14)$

SECTION

15.2 Addition and Subtraction of Radical Expressions

OBJECTIVE A To add and subtract radical expressions 

The Distributive Property is used to simplify the sum or difference of radical expressions with like radicands.

$5\sqrt{2} + 3\sqrt{2} = (5 + 3)\sqrt{2} = 8\sqrt{2}$

$6\sqrt{2x} - 4\sqrt{2x} = (6 - 4)\sqrt{2x} = 2\sqrt{2x}$

Radical expressions that are in simplest form and have unlike radicands cannot be simplified by the Distributive Property.

$2\sqrt{3} + 4\sqrt{2}$ cannot be simplified by the Distributive Property.

To simplify the sum or difference of radical expressions, first simplify each radical expression.

HOW TO • 1 Simplify: $4\sqrt{8} - 10\sqrt{2}$

$$\begin{aligned}4\sqrt{8} - 10\sqrt{2} &= 4\sqrt{4 \cdot 2} - 10\sqrt{2}\\ &= 4\sqrt{4}\sqrt{2} - 10\sqrt{2}\\ &= 4 \cdot 2\sqrt{2} - 10\sqrt{2}\\ &= 8\sqrt{2} - 10\sqrt{2}\\ &= (8 - 10)\sqrt{2}\\ &= -2\sqrt{2}\end{aligned}$$

- Use the Product Property of Square Roots.
- Simplify the expression by using the Distributive Property.

HOW TO • 2 Simplify: $8\sqrt{18x} - 2\sqrt{32x}$

$$\begin{aligned}8\sqrt{18x} - 2\sqrt{32x} &= 8\sqrt{9 \cdot 2x} - 2\sqrt{16 \cdot 2x}\\ &= 8\sqrt{9}\sqrt{2x} - 2\sqrt{16}\sqrt{2x}\\ &= 8 \cdot 3\sqrt{2x} - 2 \cdot 4\sqrt{2x}\\ &= 24\sqrt{2x} - 8\sqrt{2x}\\ &= (24 - 8)\sqrt{2x}\\ &= 16\sqrt{2x}\end{aligned}$$

- Use the Product Property of Square Roots.
- Simplify the expression by using the Distributive Property.

EXAMPLE • 1

Simplify: $5\sqrt{2} - 3\sqrt{2} + 12\sqrt{2}$

Solution

$$\begin{aligned}&5\sqrt{2} - 3\sqrt{2} + 12\sqrt{2}\\ &= (5 - 3 + 12)\sqrt{2}\\ &= 14\sqrt{2}\end{aligned}$$

- Distributive Property

YOU TRY IT • 1

Simplify: $9\sqrt{3} + 3\sqrt{3} - 18\sqrt{3}$

Your solution

Solution on p. S34

EXAMPLE • 2

Simplify: $3\sqrt{12} - 5\sqrt{27}$

Solution

$3\sqrt{12} - 5\sqrt{27}$

$= 3\sqrt{4 \cdot 3} - 5\sqrt{9 \cdot 3}$ • Simplify $\sqrt{12}$ and $\sqrt{27}$.

$= 3\sqrt{4}\sqrt{3} - 5\sqrt{9}\sqrt{3}$

$= 3 \cdot 2\sqrt{3} - 5 \cdot 3\sqrt{3}$

$= 6\sqrt{3} - 15\sqrt{3}$

$= (6 - 15)\sqrt{3}$ • Distributive Property

$= -9\sqrt{3}$

YOU TRY IT • 2

Simplify: $2\sqrt{50} - 5\sqrt{32}$

Your solution

EXAMPLE • 3

Simplify: $3\sqrt{12x^3} - 2x\sqrt{3x}$

Solution

$3\sqrt{12x^3} - 2x\sqrt{3x}$

$= 3\sqrt{4x^2 \cdot 3x} - 2x\sqrt{3x}$ • Simplify $\sqrt{12x^3}$.

$= 3\sqrt{4x^2}\sqrt{3x} - 2x\sqrt{3x}$

$= 3 \cdot 2x\sqrt{3x} - 2x\sqrt{3x}$

$= 6x\sqrt{3x} - 2x\sqrt{3x}$

$= (6x - 2x)\sqrt{3x}$ • Distributive Property

$= 4x\sqrt{3x}$

YOU TRY IT • 3

Simplify: $y\sqrt{28y} + 7\sqrt{63y^3}$

Your solution

EXAMPLE • 4

Simplify: $2x\sqrt{8y} - 3\sqrt{2x^2y} + 2\sqrt{32x^2y}$

Solution

$2x\sqrt{8y} - 3\sqrt{2x^2y} + 2\sqrt{32x^2y}$

$= 2x\sqrt{4 \cdot 2y} - 3\sqrt{x^2 \cdot 2y} + 2\sqrt{16x^2 \cdot 2y}$

$= 2x\sqrt{4}\sqrt{2y} - 3\sqrt{x^2}\sqrt{2y} + 2\sqrt{16x^2}\sqrt{2y}$

$= 2x \cdot 2\sqrt{2y} - 3 \cdot x\sqrt{2y} + 2 \cdot 4x\sqrt{2y}$

$= 4x\sqrt{2y} - 3x\sqrt{2y} + 8x\sqrt{2y}$

$= 9x\sqrt{2y}$

YOU TRY IT • 4

Simplify: $2\sqrt{27a^5} - 4a\sqrt{12a^3} + a^2\sqrt{75a}$

Your solution

Solutions on pp. S34–S35

15.2 EXERCISES

OBJECTIVE A To add and subtract radical expressions

1. Which of the numbers 2, 9, 20, 25, 50, 81, and 100 are *not* perfect squares?

2. Write down a number that has a perfect-square factor that is greater than 1.

For Exercises 3 to 58, simplify.

3. $2\sqrt{2} + \sqrt{2}$
4. $3\sqrt{5} + 8\sqrt{5}$
5. $-3\sqrt{7} + 2\sqrt{7}$
6. $4\sqrt{5} - 10\sqrt{5}$
7. $-3\sqrt{11} - 8\sqrt{11}$
8. $-3\sqrt{3} - 5\sqrt{3}$
9. $2\sqrt{x} + 8\sqrt{x}$
10. $3\sqrt{y} + 2\sqrt{y}$
11. $8\sqrt{y} - 10\sqrt{y}$
12. $-5\sqrt{2a} + 2\sqrt{2a}$
13. $-2\sqrt{3b} - 9\sqrt{3b}$
14. $-7\sqrt{5a} - 5\sqrt{5a}$
15. $3x\sqrt{2} - x\sqrt{2}$
16. $2y\sqrt{3} - 9y\sqrt{3}$
17. $2a\sqrt{3a} - 5a\sqrt{3a}$
18. $-5b\sqrt{3x} - 2b\sqrt{3x}$
19. $3\sqrt{xy} - 8\sqrt{xy}$
20. $-4\sqrt{xy} + 6\sqrt{xy}$
21. $\sqrt{45} + \sqrt{125}$
22. $\sqrt{32} - \sqrt{98}$
23. $2\sqrt{2} + 3\sqrt{8}$
24. $4\sqrt{128} - 3\sqrt{32}$
25. $5\sqrt{18} - 2\sqrt{75}$
26. $5\sqrt{75} - 2\sqrt{18}$
27. $5\sqrt{4x} - 3\sqrt{9x}$
28. $-3\sqrt{25y} + 8\sqrt{49y}$
29. $3\sqrt{3x^2} - 5\sqrt{27x^2}$
30. $-2\sqrt{8y^2} + 5\sqrt{32y^2}$
31. $2x\sqrt{xy^2} - 3y\sqrt{x^2y}$
32. $4a\sqrt{b^2a} - 3b\sqrt{a^2b}$
33. $3x\sqrt{12x} - 5\sqrt{27x^3}$
34. $2a\sqrt{50a} + 7\sqrt{32a^3}$
35. $4y\sqrt{8y^3} - 7\sqrt{18y^5}$
36. $2a\sqrt{8ab^2} - 2b\sqrt{2a^3}$
37. $b^2\sqrt{a^5b} + 3a^2\sqrt{ab^5}$
38. $y^2\sqrt{x^5y} + x\sqrt{x^3y^5}$
39. $4\sqrt{2} - 5\sqrt{2} + 8\sqrt{2}$
40. $3\sqrt{3} + 8\sqrt{3} - 16\sqrt{3}$
41. $5\sqrt{x} - 8\sqrt{x} + 9\sqrt{x}$

42. $\sqrt{x} - 7\sqrt{x} + 6\sqrt{x}$

43. $8\sqrt{2} - 3\sqrt{y} - 8\sqrt{2}$

44. $8\sqrt{3} - 5\sqrt{2} - 5\sqrt{3}$

45. $8\sqrt{8} - 4\sqrt{32} - 9\sqrt{50}$

46. $2\sqrt{12} - 4\sqrt{27} + \sqrt{75}$

47. $-2\sqrt{3} + 5\sqrt{27} - 4\sqrt{45}$

48. $-2\sqrt{8} - 3\sqrt{27} + 3\sqrt{50}$

49. $4\sqrt{75} + 3\sqrt{48} - \sqrt{99}$

50. $2\sqrt{75} - 5\sqrt{20} + 2\sqrt{45}$

51. $\sqrt{25x} - \sqrt{9x} + \sqrt{16x}$

52. $\sqrt{4x} - \sqrt{100x} - \sqrt{49x}$

53. $3\sqrt{3x} + \sqrt{27x} - 8\sqrt{75x}$

54. $5\sqrt{5x} + 2\sqrt{45x} - 3\sqrt{80x}$

55. $2a\sqrt{75b} - a\sqrt{20b} + 4a\sqrt{45b}$

56. $2b\sqrt{75a} - 5b\sqrt{27a} + 2b\sqrt{20a}$

57. $x\sqrt{3y^2} - 2y\sqrt{12x^2} + xy\sqrt{3}$

58. $a\sqrt{27b^2} + 3b\sqrt{147a^2} - ab\sqrt{3}$

59. Determine whether the statement is true or false.
a. $7x\sqrt{x} + x\sqrt{x} = 7x^2\sqrt{x}$ **b.** $\sqrt{9 + y^2} = 3 + y$

60. Which expression is equivalent to $\sqrt{2ab} + \sqrt{2ab}$?
(i) $2\sqrt{ab}$ (ii) $\sqrt{4ab}$ (iii) $2ab$ (iv) $\sqrt{8ab}$

Applying the Concepts

61. Given $G(x) = \sqrt{x + 5} + \sqrt{5x + 3}$, write $G(3)$ in simplest form.

62. For each equation, write "ok" if the equation is correct. If the equation is incorrect, correct the right-hand side.
a. $3\sqrt{ab} + 5\sqrt{ab} = 8\sqrt{2ab}$
b. $7\sqrt{x^3} - 3x\sqrt{x} - x\sqrt{16x} = 0$
c. $5 - 2\sqrt{y} = 3\sqrt{y}$

63. Write a sentence or two that you could email to a friend to explain the concept of a perfect-square factor.

SECTION

15.3 Multiplication and Division of Radical Expressions

OBJECTIVE A To multiply radical expressions

The Product Property of Square Roots is used to multiply radical expressions.

$$\sqrt{2x}\sqrt{3y} = \sqrt{2x \cdot 3y} = \sqrt{6xy}$$

HOW TO • 1 Simplify: $\sqrt{2x^2}\sqrt{32x^5}$

$\sqrt{2x^2}\sqrt{32x^5} = \sqrt{2x^2 \cdot 32x^5}$ • Use the Product Property of Square Roots.

$= \sqrt{64x^7}$ • Multiply the radicands.

$= \sqrt{64x^6 \cdot x}$ • Simplify.

$= \sqrt{64x^6}\sqrt{x} = 8x^3\sqrt{x}$

HOW TO • 2 Simplify: $\sqrt{2x}(x + \sqrt{2x})$

$\sqrt{2x}(x + \sqrt{2x}) = \sqrt{2x}(x) + \sqrt{2x}\sqrt{2x}$ • Use the Distributive Property to remove parentheses.

$= x\sqrt{2x} + \sqrt{4x^2}$

$= x\sqrt{2x} + 2x$ • Simplify.

Use FOIL to multiply radical expressions with two terms.

HOW TO • 3 Simplify: $(\sqrt{2} - 3x)(\sqrt{2} + x)$

$(\sqrt{2} - 3x)(\sqrt{2} + x) = \sqrt{2 \cdot 2} + x\sqrt{2} - 3x\sqrt{2} - 3x^2$ • Use the FOIL method to remove parentheses.

$= \sqrt{4} + (x - 3x)\sqrt{2} - 3x^2$

$= 2 - 2x\sqrt{2} - 3x^2$

The expressions $a + b$ and $a - b$, which differ only in the sign of one term, are called **conjugates.** Recall that $(a + b)(a - b) = a^2 - b^2$.

Take Note

For $x > 0$, $(\sqrt{x})^2 = x$ because $(\sqrt{x})^2 = \sqrt{x} \cdot \sqrt{x} = \sqrt{x^2} = x$.

HOW TO • 4 Simplify: $(2 + \sqrt{7})(2 - \sqrt{7})$

$(2 + \sqrt{7})(2 - \sqrt{7}) = 2^2 - (\sqrt{7})^2$ • $(2 + \sqrt{7})(2 - \sqrt{7})$ is the product of conjugates.

$= 4 - 7 = -3$

HOW TO • 5 Simplify: $(3 + \sqrt{y})(3 - \sqrt{y})$

$(3 + \sqrt{y})(3 - \sqrt{y}) = 3^2 - (\sqrt{y})^2$ • $(3 + \sqrt{y})(3 - \sqrt{y})$ is the product of conjugates.

$= 9 - y$

EXAMPLE • 1

Simplify: $\sqrt{3x^4}\sqrt{2x^2y}\sqrt{6xy^2}$

Solution

$\sqrt{3x^4}\sqrt{2x^2y}\sqrt{6xy^2}$

$= \sqrt{36x^7y^3}$ • Product Property of Square Roots

$= \sqrt{36x^6y^2 \cdot xy}$ • Simplify.

$= \sqrt{36x^6y^2}\sqrt{xy}$

$= 6x^3y\sqrt{xy}$

YOU TRY IT • 1

Simplify: $\sqrt{5a}\sqrt{15a^3b^4}\sqrt{20b^5}$

Your solution

Solution on p. S35

EXAMPLE • 2

Simplify: $\sqrt{3ab}(\sqrt{3a} + \sqrt{9b})$

Solution

$\sqrt{3ab}(\sqrt{3a} + \sqrt{9b})$

$= \sqrt{9a^2b} + \sqrt{27ab^2}$ • Distributive Property

$= \sqrt{9a^2 \cdot b} + \sqrt{9b^2 \cdot 3a}$ • Simplify.

$= \sqrt{9a^2}\sqrt{b} + \sqrt{9b^2}\sqrt{3a}$

$= 3a\sqrt{b} + 3b\sqrt{3a}$

YOU TRY IT • 2

Simplify: $\sqrt{5x}(\sqrt{5x} - \sqrt{25y})$

Your solution

EXAMPLE • 3

Simplify: $(\sqrt{x} - 2\sqrt{y})(4\sqrt{x} + \sqrt{y})$

Solution

$(\sqrt{x} - 2\sqrt{y})(4\sqrt{x} + \sqrt{y})$

$= 4(\sqrt{x})^2 + \sqrt{xy} - 8\sqrt{xy} - 2(\sqrt{y})^2$ • FOIL

$= 4x - 7\sqrt{xy} - 2y$

YOU TRY IT • 3

Simplify: $(3\sqrt{x} - \sqrt{y})(5\sqrt{x} - 2\sqrt{y})$

Your solution

EXAMPLE • 4

Simplify: $(\sqrt{a} - \sqrt{b})(\sqrt{a} + \sqrt{b})$

Solution

$(\sqrt{a} - \sqrt{b})(\sqrt{a} + \sqrt{b})$ • Product of conjugates

$= (\sqrt{a})^2 - (\sqrt{b})^2$

$= a - b$

YOU TRY IT • 4

Simplify: $(2\sqrt{x} + 7)(2\sqrt{x} - 7)$

Your solution

Solutions on p. S35

OBJECTIVE B To divide radical expressions 

The Quotient Property of Square Roots

If a and b are positive real numbers, then $\sqrt{\frac{a}{b}} = \frac{\sqrt{a}}{\sqrt{b}}$ and $\frac{\sqrt{a}}{\sqrt{b}} = \sqrt{\frac{a}{b}}$.

This property states that the square root of a quotient is equal to the quotient of the square roots.

HOW TO • 6 Simplify: $\sqrt{\frac{4x^2}{z^6}}$

$\sqrt{\frac{4x^2}{z^6}} = \frac{\sqrt{4x^2}}{\sqrt{z^6}}$ • Rewrite the radical expression as the quotient of the square roots.

$= \frac{2x}{z^3}$ • Simplify.

Point of Interest

A radical expression that occurs in Einstein's Theory of Relativity is

$$\frac{1}{\sqrt{1-\frac{v^2}{c^2}}}$$

where v is the velocity of an object and c is the speed of light.

HOW TO • 7 Simplify: $\sqrt{\dfrac{24x^3y^7}{3x^7y^2}}$

$$\sqrt{\frac{24x^3y^7}{3x^7y^2}} = \sqrt{\frac{8y^5}{x^4}}$$ • Simplify the radicand.

$$= \frac{\sqrt{8y^5}}{\sqrt{x^4}}$$ • Rewrite the radical expression as the quotient of the square roots.

$$= \frac{\sqrt{4y^4 \cdot 2y}}{\sqrt{x^4}}$$ • Simplify.

$$= \frac{\sqrt{4y^4}\sqrt{2y}}{\sqrt{x^4}}$$

$$= \frac{2y^2\sqrt{2y}}{x^2}$$

The Quotient Property of Square Roots is used to divide radical expressions.

HOW TO • 8 Simplify: $\dfrac{\sqrt{4x^2y}}{\sqrt{xy}}$

$$\frac{\sqrt{4x^2y}}{\sqrt{xy}} = \sqrt{\frac{4x^2y}{xy}}$$ • Use the Quotient Property of Square Roots.

$$= \sqrt{4x}$$ • Simplify the radicand.

$$= \sqrt{4}\sqrt{x}$$ • Simplify the radical expression.

$$= 2\sqrt{x}$$

The previous examples all result in radical expressions written in simplest form.

Simplest Form of a Radical Expression

For a radical expression to be in simplest form, three conditions must be met:

1. The radicand contains no factor greater than 1 that is a perfect square.
2. There is no fraction under the radical sign.
3. There is no radical in the denominator of a fraction.

The procedure used to remove a radical from a denominator is called **rationalizing the denominator.**

HOW TO • 9 Simplify: $\dfrac{2}{\sqrt{3}}$

$$\frac{2}{\sqrt{3}} = \frac{2}{\sqrt{3}} \cdot \frac{\sqrt{3}}{\sqrt{3}}$$ • To rationalize the denominator, multiply the expression by $\frac{\sqrt{3}}{\sqrt{3}}$, which equals 1.

$$= \frac{2\sqrt{3}}{(\sqrt{3})^2}$$

$$= \frac{2\sqrt{3}}{3}$$ • Simplify.

When the denominator contains a radical expression with two terms, rationalize the denominator by multiplying the numerator and denominator by the conjugate of the denominator.

HOW TO • 10 Simplify: $\dfrac{\sqrt{2y}}{\sqrt{y}+3}$

$$\frac{\sqrt{2y}}{\sqrt{y}+3} = \frac{\sqrt{2y}}{\sqrt{y}+3}\cdot\frac{\sqrt{y}-3}{\sqrt{y}-3}$$

• Multiply the numerator and denominator by $\sqrt{y}-3$, the conjugate of $\sqrt{y}+3$.

$$= \frac{\sqrt{2y^2}-3\sqrt{2y}}{(\sqrt{y})^2-3^2} = \frac{y\sqrt{2}-3\sqrt{2y}}{y-9}$$

EXAMPLE • 5

Simplify: $\dfrac{\sqrt{4x^2y^5}}{\sqrt{3x^4y}}$

Solution

$$\frac{\sqrt{4x^2y^5}}{\sqrt{3x^4y}} = \sqrt{\frac{4x^2y^5}{3x^4y}} = \sqrt{\frac{4y^4}{3x^2}} = \frac{\sqrt{4y^4}}{\sqrt{3x^2}}$$

$$= \frac{2y^2}{x\sqrt{3}} = \frac{2y^2}{x\sqrt{3}}\cdot\frac{\sqrt{3}}{\sqrt{3}}$$

• Rationalize the denominator.

$$= \frac{2y^2\sqrt{3}}{3x}$$

YOU TRY IT • 5

Simplify: $\dfrac{\sqrt{15x^6y^7}}{\sqrt{3x^7y^9}}$

Your solution

EXAMPLE • 6

Simplify: $\dfrac{\sqrt{2}}{\sqrt{2}+\sqrt{6}}$

Solution

$$\frac{\sqrt{2}}{\sqrt{2}+\sqrt{6}}$$

$$= \frac{\sqrt{2}}{\sqrt{2}+\sqrt{6}}\cdot\frac{\sqrt{2}-\sqrt{6}}{\sqrt{2}-\sqrt{6}}$$

• Multiply the numerator and denominator by the conjugate of the denominator.

$$= \frac{(\sqrt{2})^2-\sqrt{12}}{2-6} = \frac{2-2\sqrt{3}}{-4}$$

$$= \frac{2(1-\sqrt{3})}{-4} = \frac{1-\sqrt{3}}{-2} = -\frac{1-\sqrt{3}}{2}$$

YOU TRY IT • 6

Simplify: $\dfrac{\sqrt{3}}{\sqrt{3}-\sqrt{6}}$

Your solution

EXAMPLE • 7

Simplify: $\dfrac{3-\sqrt{y}}{2+3\sqrt{y}}$

Solution

$$\frac{3-\sqrt{y}}{2+3\sqrt{y}} = \frac{3-\sqrt{y}}{2+3\sqrt{y}}\cdot\frac{2-3\sqrt{y}}{2-3\sqrt{y}}$$

• Rationalize the denominator.

$$= \frac{6-9\sqrt{y}-2\sqrt{y}+3(\sqrt{y})^2}{4-9y}$$

$$= \frac{6-11\sqrt{y}+3y}{4-9y}$$

YOU TRY IT • 7

Simplify: $\dfrac{5+\sqrt{y}}{1-2\sqrt{y}}$

Your solution

Solutions on p. S35

15.3 EXERCISES

OBJECTIVE A To multiply radical expressions

For Exercises 1 to 36, simplify.

1. $\sqrt{5} \cdot \sqrt{5}$

2. $\sqrt{11} \cdot \sqrt{11}$

3. $\sqrt{3} \cdot \sqrt{12}$

4. $\sqrt{2} \cdot \sqrt{8}$

5. $\sqrt{x} \cdot \sqrt{x}$

6. $\sqrt{y} \cdot \sqrt{y}$

7. $\sqrt{xy^3} \cdot \sqrt{x^5y}$

8. $\sqrt{a^3b^5} \cdot \sqrt{ab^5}$

9. $\sqrt{3a^2b^5} \cdot \sqrt{6ab^7}$

10. $\sqrt{5x^3y} \cdot \sqrt{10x^2y}$

11. $\sqrt{6a^3b^2} \cdot \sqrt{24a^5b}$

12. $\sqrt{8ab^5} \cdot \sqrt{12a^7b}$

13. $\sqrt{2}(\sqrt{2} - \sqrt{3})$

14. $3(\sqrt{12} - \sqrt{3})$

15. $\sqrt{x}(\sqrt{x} - \sqrt{y})$

16. $\sqrt{b}(\sqrt{a} - \sqrt{b})$

17. $\sqrt{5}(\sqrt{10} - \sqrt{x})$

18. $\sqrt{6}(\sqrt{y} - \sqrt{18})$

19. $\sqrt{3a}(\sqrt{3a} - \sqrt{3b})$

20. $\sqrt{5x}(\sqrt{10x} - \sqrt{x})$

21. $\sqrt{2ac} \cdot \sqrt{5ab} \cdot \sqrt{10cb}$

22. $\sqrt{3xy} \cdot \sqrt{6x^3y} \cdot \sqrt{2y^2}$

23. $(\sqrt{x} - 3)^2$

24. $(2\sqrt{a} - y)^2$

25. $(\sqrt{5} + 3)(2\sqrt{5} - 4)$

26. $(2 - 3\sqrt{7})(5 + 2\sqrt{7})$

27. $(4 + \sqrt{8})(3 + \sqrt{2})$

28. $(6 - \sqrt{27})(2 + \sqrt{3})$

29. $(2\sqrt{x} + 4)(3\sqrt{x} - 1)$

30. $(5 + \sqrt{y})(6 - 3\sqrt{y})$

31. $(3\sqrt{x} - 2y)(5\sqrt{x} - 4y)$

32. $(5\sqrt{x} + 2\sqrt{y})(3\sqrt{x} - \sqrt{y})$

33. $(3 + \sqrt{5})(3 - \sqrt{5})$

34. $(1 + \sqrt{6})(1 - \sqrt{6})$

35. $(3\sqrt{x} - 4)(3\sqrt{x} + 4)$

36. $(\sqrt{x} - y)(\sqrt{x} + y)$

37. For $a > 0$, is $(\sqrt{a} - 1)(\sqrt{a} + 1)$ less than, equal to, or greater than a?

38. For $a > 0$, is $\sqrt{a}(\sqrt{2a} - \sqrt{a})$ less than, equal to, or greater than a?

OBJECTIVE B To divide radical expressions

39. Why is $\frac{\sqrt{3}}{3}$ in simplest form but $\frac{1}{\sqrt{3}}$ not in simplest form?

40. Why can we multiply $\frac{2}{\sqrt{5}}$ by $\frac{\sqrt{5}}{\sqrt{5}}$ without changing the value of $\frac{2}{\sqrt{5}}$?

For Exercises 41 to 70, simplify.

41. $\frac{\sqrt{32}}{\sqrt{2}}$ **42.** $\frac{\sqrt{45}}{\sqrt{5}}$ **43.** $\frac{\sqrt{98}}{\sqrt{2}}$ **44.** $\frac{\sqrt{48}}{\sqrt{3}}$ **45.** $\frac{\sqrt{27a}}{\sqrt{3a}}$

46. $\frac{\sqrt{72x^5}}{\sqrt{2x}}$ **47.** $\frac{\sqrt{15x^3y}}{\sqrt{3xy}}$ **48.** $\frac{\sqrt{40x^5y^2}}{\sqrt{5xy}}$ **49.** $\frac{\sqrt{2a^5b^4}}{\sqrt{98ab^4}}$ **50.** $\frac{\sqrt{48x^5y^2}}{\sqrt{3x^3y}}$

51. $\frac{\sqrt{9xy^2}}{\sqrt{27x}}$ **52.** $\frac{\sqrt{4x^2y}}{\sqrt{3xy^3}}$ **53.** $\frac{\sqrt{16x^3y^2}}{\sqrt{8x^3y}}$ **54.** $\frac{\sqrt{2}}{\sqrt{8}+4}$

55. $\frac{1}{\sqrt{2}-3}$ **56.** $\frac{5}{\sqrt{7}-3}$ **57.** $\frac{3}{5+\sqrt{5}}$ **58.** $\frac{\sqrt{3}}{5-\sqrt{27}}$

59. $\frac{7}{\sqrt{2}-7}$ **60.** $\frac{3-\sqrt{6}}{5-2\sqrt{6}}$ **61.** $\frac{6-2\sqrt{3}}{4+3\sqrt{3}}$ **62.** $\frac{-6}{4+\sqrt{2}}$

63. $\frac{-\sqrt{15}}{3-\sqrt{12}}$ **64.** $\frac{-12}{\sqrt{6}-3}$ **65.** $\frac{\sqrt{2}+2\sqrt{6}}{2\sqrt{2}-3\sqrt{6}}$ **66.** $\frac{2\sqrt{3}-\sqrt{6}}{5\sqrt{3}+2\sqrt{6}}$

67. $\frac{3+\sqrt{x}}{2-\sqrt{x}}$ **68.** $\frac{\sqrt{a}-4}{2\sqrt{a}+2}$ **69.** $\frac{\sqrt{xy}}{\sqrt{x}-\sqrt{y}}$ **70.** $\frac{\sqrt{x}}{\sqrt{x}-\sqrt{y}}$

71. For $a > 0$, is $\frac{a}{\sqrt{a}}$ less than, equal to, or greater than $\sqrt{a}$?

72. For $a > 0$ and $b > 0$, is $\frac{a-b}{\sqrt{a}-\sqrt{b}}$ less than, equal to, or greater than $\sqrt{a}$?

Applying the Concepts

For Exercises 73 to 76, answer true or false. If the equation is false, correct it.

73. $(\sqrt{y})^4 = y^2$ **74.** $(2\sqrt{x})^3 = 8x\sqrt{x}$ **75.** $(\sqrt{x}+1)^2 = x+1$ **76.** $\frac{1}{2-\sqrt{3}} = 2+\sqrt{3}$

SECTION

15.4 Solving Equations Containing Radical Expressions

OBJECTIVE A To solve an equation containing a radical expression 

An equation that contains a variable expression in a radicand is a **radical equation.**

$\sqrt{x} = 4$
$\sqrt{x+2} = \sqrt{x-7}$ } Radical equations

The following property of equality, which states that if two numbers are equal, the squares of the numbers are equal, is used to solve radical equations.

Property of Squaring Both Sides of an Equation

If a and b are real numbers and $a = b$, then $a^2 = b^2$.

To solve a radical equation containing one radical, use the following procedure.

Solving a Radical Equation

1. Write the equation with the radical alone on one side.
2. Square both sides of the equation.
3. Solve for the variable.
4. Check the solution(s) in the original equation.

Tips for Success

When we suggest that you check a solution, you should substitute the solution into the original equation. Below is the check for the equation at the right.

Check:

$\sqrt{x-2} - 7 = 0$

$\sqrt{51-2} - 7$	0
$\sqrt{49} - 7$	0
$7 - 7$	0
$0 = 0$	

A true equation

HOW TO • 1 Solve: $\sqrt{x-2} - 7 = 0$

$\sqrt{x-2} - 7 = 0$ • Isolate the radical by adding 7 to both sides of the equation.
$\sqrt{x-2} = 7$
$(\sqrt{x-2})^2 = 7^2$ • Square both sides of the equation.
$x - 2 = 49$ • Solve the resulting equation.
$x = 51$

The check is shown at the left.

The solution is 51.

When both sides of an equation are squared, the resulting equation may have a solution that is not a solution of the original equation. Checking a proposed solution of a radical equation, as we did at the left, is a necessary step.

HOW TO • 2 Solve: $\sqrt{2x-5} + 3 = 0$

$\sqrt{2x-5} + 3 = 0$ • Isolate the radical by subtracting 3 from both sides of the equation.
$\sqrt{2x-5} = -3$
$(\sqrt{2x-5})^2 = (-3)^2$ • Square both sides of the equation.
$2x - 5 = 9$ • Solve for x.
$2x = 14$
$x = 7$

Take Note

Any time each side of an equation is squared, you must check the proposed solution of the equation.

Here is the check for the equation on the preceding page.

Check:

$$\begin{array}{r|l} \sqrt{2x-5}+3 & 0 \\ \hline \sqrt{2\cdot 7-5}+3 & 0 \\ \sqrt{14-5}+3 & 0 \\ \sqrt{9}+3 & 0 \\ 3+3 & 0 \\ 6 \neq 0 & \end{array}$$

7 does not check as a solution. The equation has no solution.

EXAMPLE • 1

Solve: $\sqrt{3x}+2=5$

Solution

$$\begin{aligned} \sqrt{3x}+2&=5 \\ \sqrt{3x}&=3 &&\text{• Isolate } \sqrt{3x}. \\ (\sqrt{3x})^2&=3^2 &&\text{• Square both sides.} \\ 3x&=9 \\ x&=3 &&\text{• Solve for } x. \end{aligned}$$

Check:

$$\begin{array}{r|l} \sqrt{3x}+2 & 5 \\ \hline \sqrt{3\cdot 3}+2 & 5 \\ \sqrt{9}+2 & 5 \\ 3+2 & 5 \\ 5 = 5 & \end{array}$$

The solution is 3.

YOU TRY IT • 1

Solve: $\sqrt{4x}+3=7$

Your solution

EXAMPLE • 2

Solve: $1=\sqrt{x}-\sqrt{x-5}$

Solution

When an equation contains two radicals, isolate the radicals one at a time.

$$\begin{aligned} 1&=\sqrt{x}-\sqrt{x-5} \\ 1+\sqrt{x-5}&=\sqrt{x} &&\text{• Isolate } \sqrt{x}. \\ (1+\sqrt{x-5})^2&=(\sqrt{x})^2 &&\text{• Square both sides.} \\ 1+2\sqrt{x-5}+(x-5)&=x &&\text{• Expand the left side.} \\ 2\sqrt{x-5}&=4 &&\text{• Simplify.} \\ \sqrt{x-5}&=2 &&\text{• Isolate } \sqrt{x-5}. \\ (\sqrt{x-5})^2&=2^2 &&\text{• Square both sides.} \\ x-5&=4 \\ x&=9 &&\text{• Solve for } x. \end{aligned}$$

Check:

$$\begin{array}{l|l} 1 & \sqrt{x}-\sqrt{x-5} \\ \hline 1 & \sqrt{9}-\sqrt{9-5} \\ 1 & \sqrt{9}-\sqrt{4} \\ 1 & 3-2 \\ 1=1 & \end{array}$$

The solution is 9.

YOU TRY IT • 2

Solve: $\sqrt{x}+\sqrt{x+9}=9$

Your solution

Solutions on p. S35

OBJECTIVE B To solve application problems 

Pythagoras
(c. 580 B.C.E.–520 B.C.E.)

A **right triangle** is a triangle that contains a 90° angle. The side opposite the 90° angle is called the **hypotenuse.** The other two sides are called **legs.**

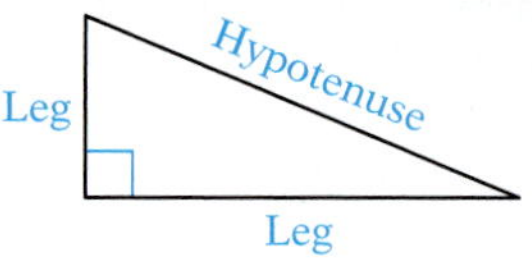

Pythagoras, a Greek mathematician who lived around 550 B.C.E., is given credit for the Pythagorean Theorem. It states that the square of the hypotenuse of a right triangle is equal to the sum of the squares of the two legs. Actually, this theorem was known to the Babylonians around 1200 B.C.E.

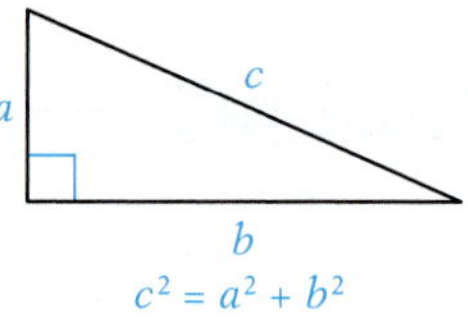

Point of Interest

The first known proof of this theorem occurs in a Chinese text, *Arithmetic Classic,* which was first written around 600 B.C.E. (but there are no existing copies) and revised over a period of 500 years. The earliest known copy of this text dates from approximately 100 B.C.E.

Pythagorean Theorem

If a and b are the lengths of the legs of a right triangle and c is the length of the hypotenuse, then $c^2 = a^2 + b^2$.

Using this theorem, we can find the hypotenuse of a right triangle when we know the two legs. Use the formula

$$\begin{aligned}\text{Hypotenuse} &= \sqrt{(\text{leg})^2 + (\text{leg})^2}\\ c &= \sqrt{a^2 + b^2}\\ &= \sqrt{(5)^2 + (12)^2}\\ &= \sqrt{25 + 144}\\ &= \sqrt{169}\\ &= 13\end{aligned}$$

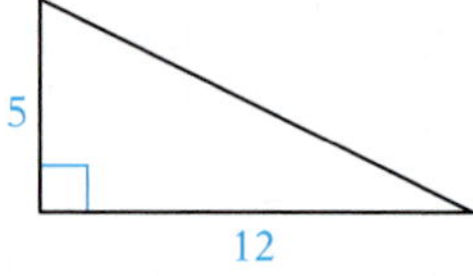

The leg of a right triangle can be found when one leg and the hypotenuse are known. Use the formula

$$\begin{aligned}\text{Leg} &= \sqrt{(\text{hypotenuse})^2 - (\text{leg})^2}\\ a &= \sqrt{c^2 - b^2}\\ &= \sqrt{(25)^2 - (20)^2}\\ &= \sqrt{625 - 400}\\ &= \sqrt{225}\\ &= 15\end{aligned}$$

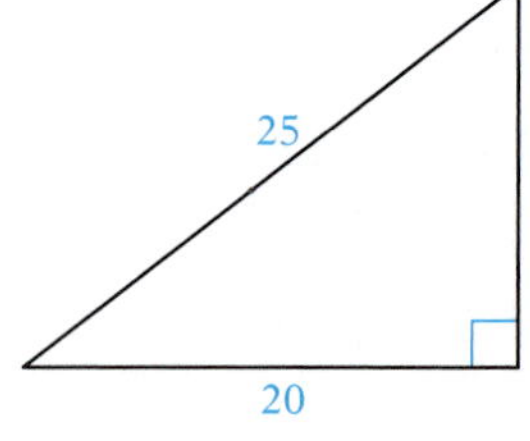

Example 3 and You Try It 3 on the following page illustrate the use of the Pythagorean Theorem. Example 4 and You Try It 4 illustrate other applications of radical equations.

EXAMPLE • 3

A guy wire is attached to a point 20 m above the ground on a telephone pole. The wire is anchored to the ground at a point 8 m from the base of the pole. Find the length of the guy wire. Round to the nearest tenth.

Strategy

To find the length of the guy wire, use the Pythagorean Theorem. One leg is 20 m. The other leg is 8 m. The guy wire is the hypotenuse. Solve the Pythagorean Theorem for the hypotenuse.

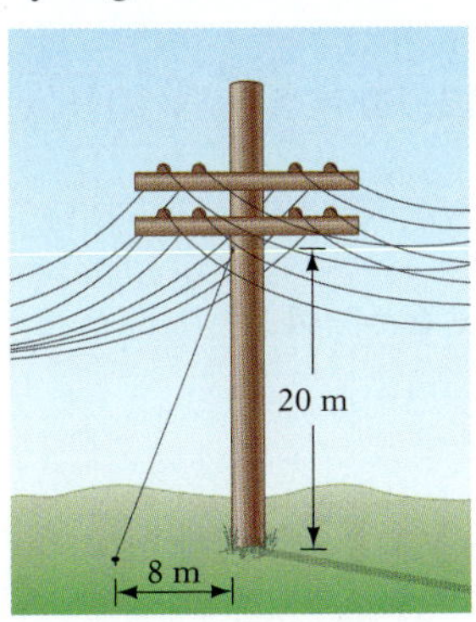

Solution

$$c = \sqrt{a^2 + b^2}$$
$$= \sqrt{(20)^2 + (8)^2} \quad \bullet\ a = 20, b = 8$$
$$= \sqrt{400 + 64} = \sqrt{464} \approx 21.5$$

The guy wire has a length of approximately 21.5 m.

YOU TRY IT • 3

A ladder 8 ft long is resting against a building. How high on the building will the ladder reach when the bottom of the ladder is 3 ft from the building? Round to the nearest hundredth.

Your strategy

Your solution

EXAMPLE • 4

How far above the water would a submarine periscope have to be to locate a ship 4 mi away? The equation for the distance in miles that the lookout can see is $d = \sqrt{1.5h}$, where h is the height in feet above the surface of the water. Round to the nearest hundredth.

Strategy

To find the height above the water, replace d in the equation with the given value and solve for h.

Solution

$$d = \sqrt{1.5h}$$
$$4 = \sqrt{1.5h} \quad \bullet\ d = 4$$
$$4^2 = (\sqrt{1.5h})^2$$
$$16 = 1.5h$$
$$10.67 \approx h$$

The periscope must be approximately 10.67 ft above the water.

YOU TRY IT • 4

Find the length of a pendulum that makes one swing in 2.5 s. The equation for the time for one swing is $T = 2\pi\sqrt{\frac{L}{32}}$, where T is the time in seconds and L is the length in feet. Use 3.14 for π. Round to the nearest hundredth.

Your strategy

Your solution

Solutions on p. S36

15.4 EXERCISES

OBJECTIVE A To solve an equation containing a radical expression

For Exercises 1 to 21, solve and check.

1. $\sqrt{x} = 5$

2. $\sqrt{y} = 7$

3. $\sqrt{a} = 12$

4. $\sqrt{a} = 9$

5. $\sqrt{4x} + 5 = 2$

6. $\sqrt{3x} + 9 = 4$

7. $\sqrt{3x - 2} = 4$

8. $\sqrt{5x + 6} = 1$

9. $\sqrt{2x + 1} = 7$

10. $\sqrt{5x + 4} = 3$

11. $0 = 2 - \sqrt{3 - x}$

12. $0 = 5 - \sqrt{10 + x}$

13. $\sqrt{5x + 2} = 0$

14. $\sqrt{3x - 7} = 0$

15. $\sqrt{3x} - 6 = -4$

16. $\sqrt{x^2 + 5} = x + 1$

17. $\sqrt{x^2 - 5} = 5 - x$

18. $\sqrt{x} + \sqrt{x + 7} = 1$

19. $\sqrt{x} + \sqrt{x - 12} = 2$

20. $\sqrt{2x + 1} - \sqrt{2x - 4} = 1$

21. $\sqrt{3x + 1} - \sqrt{3x - 2} = 1$

22. Without solving the equations, identify which equation has no solution.
(i) $-\sqrt{2x - 5} = -3$ (ii) $\sqrt{2x} - 5 = -3$ (iii) $\sqrt{2x - 5} = -3$

OBJECTIVE B To solve application problems

23. Is the given equation equivalent to the equation given in Exercise 24? Assume C and H are positive numbers.

a. $\dfrac{C^2}{H} = 32$ **b.** $C = 4\sqrt{2H}$ **c.** $\dfrac{C^2}{32} = H$ **d.** $\left(\dfrac{C}{4}\right)^2 = 2H$

24. **Physics** A formula used in the study of shallow-water wave motion is $C = \sqrt{32H}$, where C is the wave velocity in feet per second and H is the depth in feet. Use this formula to find the depth of the water when the wave velocity is 20 ft/s.

25. **Physics** See the news clipping at the right. The time it takes an object to fall a certain distance is given by the equation $t = \sqrt{\dfrac{d}{16}}$, where t is the time in seconds and d is the distance in feet. Use this equation to find the height from which the hay was dropped.

> **In the News**
>
> **Hay Drop for Stranded Cattle**
>
> The Wyoming and Colorado National Guards have come to the aid of thousands of cattle stranded by the blizzard that has paralyzed southeastern Colorado. Flying low over the cattle, the guardsmen drop bales of hay that 6 s later smash into the ground, break apart, and provide food for the animals, which would otherwise starve.
>
> *Sources: The Denver Post;* www.af.mil

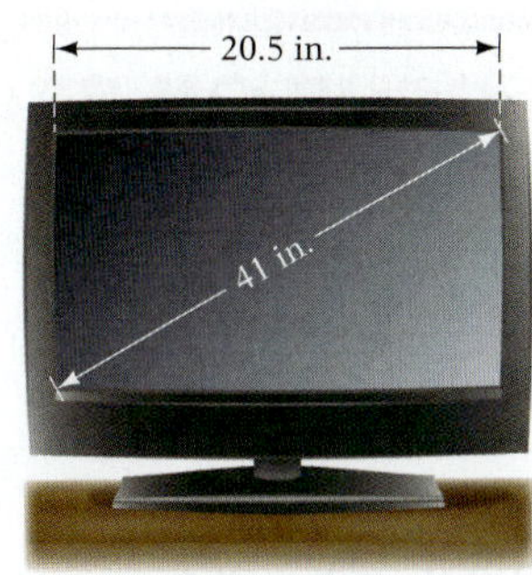

26. Home Entertainment The measure of a television screen is given by the length of a diagonal across the screen. A 41-inch television has a width of 20.5 in. Find the height of the screen to the nearest tenth of an inch.

27. Recreation The speed of a child riding a merry-go-round at a carnival is given by the equation $v = \sqrt{12r}$, where v is the speed in feet per second and r is the distance in feet from the center of the merry-go-round to the rider. If a child is moving at 15 ft/s, how far is the child from the center of the merry-go-round?

28. Time Find the length of a pendulum that makes one swing in 1.5 s. The equation for the time of one swing of a pendulum is $T = 2\pi\sqrt{\frac{L}{32}}$, where T is the time in seconds and L is the length in feet. Use 3.14 for π. Round to the nearest hundredth.

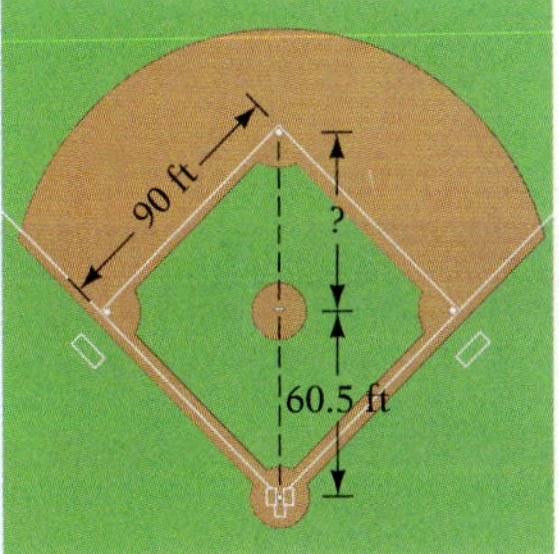

29. Sports The infield of a baseball diamond is a square. The distance between successive bases is 90 ft. The pitcher's mound is on the diagonal between home plate and second base at a distance of 60.5 ft from home plate. (See the figure at the right.) Is the pitcher's mound more or less than halfway between home plate and second base?

30. Periscopes How far above the water would a submarine periscope have to be to locate a ship 5 mi away? The equation for the distance in miles that the lookout can see is $d = \sqrt{1.5h}$, where h is the height in feet above the surface of the water. Round to the nearest hundredth.

31. Credit Cards See the news clipping at the right. The equation $N = 2.3\sqrt{S}$, where S is a student's year in college, can be used to find the average number of credit cards N that a student has. Use this equation to find the average number of credit cards for **a.** a first-year student, **b.** a sophomore, **c.** a junior, and **d.** a senior. Round to the nearest tenth.

> **In the News**
>
> **Student Credit Card Debt Grows**
>
> With each advancing year in college, students get more credit cards and accumulate more debt. The average credit card balance for a first-year student is $1585, for a sophomore is $1581, for a junior is $2000, and for a senior or fifth-year student is $2864.
>
> *Source:* Nellie Mae

Applying the Concepts

32. Geometry In the coordinate plane, a triangle is formed by drawing lines between the points (0, 0) and (5, 0), (5, 0) and (5, 12), and (5, 12) and (0, 0). Find the perimeter of the triangle.

33. Geometry The hypotenuse of a right triangle is $5\sqrt{2}$ cm, and one leg is $4\sqrt{2}$ cm.
a. Find the perimeter of the triangle.
b. Find the area of the triangle.

34. Geometry Can the Pythagorean Theorem be used to find the length of side c of the triangle at the right? If so, determine c. If not, explain why the theorem cannot be used.

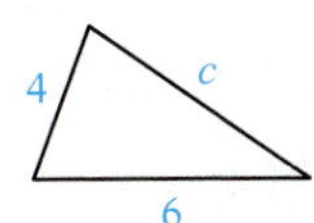

FOCUS ON PROBLEM SOLVING

Deductive Reasoning

Deductive reasoning uses a rule or statement of fact to reach a conclusion. For instance, if two angles of one triangle are equal to two angles of another triangle, then the two triangles are similar. Thus any time we establish this fact about two triangles, we know that the triangles are similar. Below are two examples of deductive reasoning.

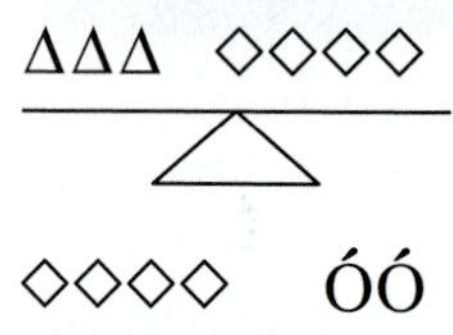

Given that ΔΔΔ = ◇◇◇◇ and ◇◇◇◇ = ÓÓ, then ΔΔΔΔΔΔ is equivalent to how many Ós?

Because 3 Δs = 4 ◇s and 4 ◇s = 2 Ós, 3 Δs = 2 Ós.

6 Δs is twice 3 Δs. We need to find twice 2 Ós, which is 4 Ós.

Therefore, ΔΔΔΔΔΔ = ÓÓÓÓ.

Lomax, Parish, Thorpe, and Wong are neighbors. Each drives a different type of vehicle: a compact car, a sedan, a sports car, or a hybrid. From the following statements, determine which type of vehicle each of the neighbors drives.

1. Although the vehicle owned by Lomax has more mileage on it than does either the sedan or the sports car, it does not have the highest mileage of all four cars. (Use X1 in the chart below to eliminate the possibilities that this statement rules out.)
2. Wong and the owner of the sports car live on one side of the street, and Thorpe and the owner of the compact car live on the other side of the street. (Use X2 to eliminate the possibilities that this statement rules out.)
3. Thorpe owns the vehicle with the most mileage on it. (Use X3 to eliminate the possibilities that this statement rules out.)

Take Note
To use the chart to solve this problem, write an X in a box to indicate that a possibility has been eliminated. Write a √ to show that a match has been found. When a row or column has three X's, a √ is written in the remaining open box in that row or column of the chart.

	Compact	*Sedan*	*Sports Car*	*Hybrid*
Lomax	√	X1	X1	X2
Parish	X2	X2	√	X2
Thorpe	X2	X3	X2	√
Wong	X2		X2	

Lomax drives the compact car, Parish drives the sports car, Thorpe drives the hybrid, and Wong drives the sedan.

1. Given that ‡‡ = ••••• and ••••• = ΛΛ, then ‡‡‡‡‡ = how many Λs?
2. Given that □□□□□□ = ÓÓÓÓ and ÓÓÓÓ = ÎÎ, then □□□ = how many Îs?
3. Given that ¤ ¤ ¤ ¤ = ΩΩΩ and ΩΩΩ = ΔΔ, then ΔΔΔΔ = how many ¤ s?
4. Given that ¥¥¥¥¥ = §§ and §§ = ÂÂÂ, then ÂÂÂÂÂÂ = how many ¥s?

5. Anna, Kay, Megan, and Nicole decide to travel together during spring break, but they need to find a destination where each of them will be able to participate in her favorite sport (golf, horseback riding, sailing, or tennis). From the following statements, determine the favorite sport of each student.
   a. Anna and the student whose favorite sport is sailing both like to swim, whereas Nicole and the student whose favorite sport is tennis would prefer to scuba dive.
   b. Megan and the student whose favorite sport is sailing are roommates. Nicole and the student whose favorite sport is golf live by themselves in singles.

6. Chang, Nick, Pablo, and Saul each take a different form of transportation (bus, car, subway, or taxi) from the office to the airport. From the following statements, determine which form of transportation each takes.
   a. Chang spent more on transportation than the fellow who took the bus but less than the fellow who took the taxi.
   b. Pablo, who did not travel by bus and who spent the least on transportation, arrived at the airport after Nick but before the fellow who took the subway.
   c. Saul spent less on transportation than either Chang or Nick.

PROJECTS AND GROUP ACTIVITIES

Distance to the Horizon

The formula $d = \sqrt{1.5h}$ can be used to calculate the approximate distance d (in miles) that a person can see who uses a periscope h feet above the water. The formula is derived by using the Pythagorean Theorem.

Consider the diagram (not to scale) at the right, which shows Earth as a sphere and the periscope as extending h feet above its surface. From geometry, because AB is tangent to the circle and OA is a radius, triangle AOB is a right triangle. Therefore,

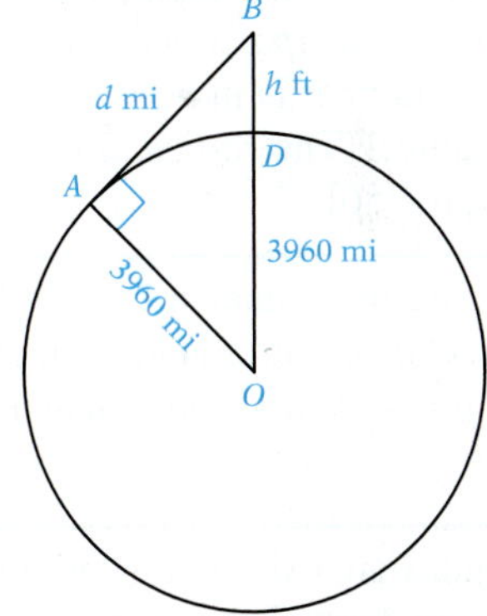

$$(OA)^2 + (AB)^2 = (OB)^2$$

Substituting into this formula, we have

$$3960^2 + d^2 = \left(3960 + \frac{h}{5280}\right)^2$$

• **Because h is in feet, $\frac{h}{5280}$ is in miles.**

$$3960^2 + d^2 = 3960^2 + \frac{2 \cdot 3960}{5280}h + \left(\frac{h}{5280}\right)^2$$

$$d^2 = \frac{3}{2}h + \left(\frac{h}{5280}\right)^2$$

$$d = \sqrt{\frac{3}{2}h + \left(\frac{h}{5280}\right)^2}$$

At this point, an assumption is made that $\sqrt{\frac{3}{2}h + \left(\frac{h}{5280}\right)^2} \approx \sqrt{1.5h}$, where we have written $\frac{3}{2}$ as 1.5. Thus $d \approx \sqrt{1.5h}$ is used to approximate the distance that can be seen using a periscope h feet above the water.

1. Write a paragraph that justifies the assumption that

$$\sqrt{\frac{3}{2}h + \left(\frac{h}{5280}\right)^2} \approx \sqrt{1.5h}$$

(*Suggestion:* Evaluate each expression for various values of h. Because h is the height of a periscope above water, it is unlikely that $h > 25$ ft.)

2. The distance d is the distance from the top of the periscope to A. The distance along the surface of the water is given by arc AD. This distance can be approximated by the equation

$$L \approx \sqrt{1.5h} + 0.306186\left(\sqrt{\frac{h}{5280}}\right)^3$$

Using this formula, calculate L when $h = 10$.

CHAPTER 15

SUMMARY

KEY WORDS	EXAMPLES
A *square root* of a positive number a is a number whose square is a. Every positive number has two square roots, one a positive and one a negative number. The square root of a negative number is not a real number. [15.1A, pp. 756–757]	A square root of 49 is 7 because $7^2 = 49$. A square root of 49 is -7 because $(-7)^2 = 49$. $\sqrt{-9}$ is not a real number.
The symbol $\sqrt{\ }$ is called a *radical sign* and is used to indicate the positive or *principal square root* of a number. The negative square root of a number is indicated by placing a negative sign in front of the radical. The *radicand* is the expression under the radical sign. [15.1A, p. 756]	$\sqrt{49} = 7$ $-\sqrt{49} = -7$ In the expression $\sqrt{49xy}$, $49xy$ is the radicand.
The square of an integer is a *perfect square.* If a number is not a perfect square, its square root can only be approximated. Such square roots are *irrational numbers.* Their decimal representations never terminate or repeat. [15.1A, p. 756]	1, 4, 9, 16, 25, 36, 49, 64, . . . are examples of perfect squares. 7 is not a perfect square. $\sqrt{7}$ is an irrational number.
Conjugates are expressions with two terms that differ only in the sign of one term. The expressions $a + b$ and $a - b$ are conjugates. [15.3A, p. 766]	$-5 + \sqrt{11}$ and $-5 - \sqrt{11}$ are conjugates. $\sqrt{x} - 3$ and $\sqrt{x} + 3$ are conjugates.
A *radical equation* is an equation that contains a variable expression in a radicand. [15.4A, p. 772]	$\sqrt{2x} + 5 = 9$ is a radical equation. $2x + \sqrt{5} = 9$ is not a radical equation.
A *right triangle* is a triangle that contains a 90° angle. The side opposite the 90° angle is the *hypotenuse.* The other two sides are called *legs.* [15.4B, p. 774]	Hypotenuse Leg Leg

ESSENTIAL RULES AND PROCEDURES	EXAMPLES
The Product Property of Square Roots [15.1A, p. 756] If a and b are positive real numbers, then $\sqrt{ab} = \sqrt{a} \cdot \sqrt{b}$. Use the Product Property of Square Roots and a knowledge of perfect squares to simplify radicands that are not perfect squares.	$\sqrt{28} = \sqrt{4 \cdot 7} = \sqrt{4} \cdot \sqrt{7} = 2\sqrt{7}$ $\sqrt{9x^7} = \sqrt{9x^6 \cdot x} = \sqrt{9x^6}\sqrt{x} = 3x^3\sqrt{x}$

Adding or Subtracting Radical Expressions [15.2A, p. 762]
The Distributive Property is used to simplify the sum or difference of radical expressions with like radicands.

$8\sqrt{2x} - 3\sqrt{2x} = (8 - 3)\sqrt{2x} = 5\sqrt{2x}$

Multiplying Radical Expressions [15.3A, p. 766]
The Product Property of Square Roots is used to multiply radical expressions.

Use FOIL to multiply radical expressions with two terms.

$\sqrt{2y}(\sqrt{3} - \sqrt{x}) = \sqrt{6y} - \sqrt{2xy}$

$(3 - \sqrt{x})(5 + \sqrt{x})$
$= 15 + 3\sqrt{x} - 5\sqrt{x} - (\sqrt{x})^2$
$= 15 - 2\sqrt{x} - x$

The Quotient Property of Square Roots [15.3B, p. 767]

If a and b are positive real numbers, then $\sqrt{\frac{a}{b}} = \frac{\sqrt{a}}{\sqrt{b}}$ and $\frac{\sqrt{a}}{\sqrt{b}} = \sqrt{\frac{a}{b}}$.

The Quotient Property of Square Roots is used to divide radical expressions.

$\frac{\sqrt{27}}{\sqrt{3}} = \sqrt{\frac{27}{3}} = \sqrt{9} = 3$

$\frac{\sqrt{3x^5y}}{\sqrt{75xy^3}} = \sqrt{\frac{3x^5y}{75xy^3}} = \sqrt{\frac{x^4}{25y^2}} = \frac{x^2}{5y}$

Simplest Form of a Radical Expression [15.3B, p. 768]
For a radical expression to be in simplest form, three conditions must be met:

1. The radicand contains no factor greater than 1 that is a perfect square.
2. There is no fraction under the radical sign.
3. There is no radical in the denominator of a fraction.

$\sqrt{12}$, $\sqrt{\frac{3}{4}}$, and $\frac{1}{\sqrt{3}}$ are not in simplest form.

$5\sqrt{3}$ and $\frac{\sqrt{3}}{3}$ are in simplest form.

Rationalizing the Denominator [15.3B, p. 768]
The procedure used to remove a radical from a denominator is called **rationalizing the denominator.**

$\frac{5}{\sqrt{7}} = \frac{5}{\sqrt{7}} \cdot \frac{\sqrt{7}}{\sqrt{7}} = \frac{5\sqrt{7}}{7}$

Property of Squaring Both Sides of an Equation [15.4A, p. 772]
If a and b are real numbers and $a = b$, then $a^2 = b^2$.

$\sqrt{x} = 5$
$(\sqrt{x})^2 = 5^2$
$x = 25$

Solving a Radical Equation Containing One Radical [15.4A, p. 772]

1. Write the equation with the radical alone on one side.
2. Square both sides of the equation.
3. Solve for the variable.
4. Check the solution(s) in the original equation.

$\sqrt{2x} - 1 = 5$
$\sqrt{2x} = 6$ • **Isolate the radical.**
$(\sqrt{2x})^2 = 6^2$ • **Square both sides.**
$2x = 36$
$x = 18$ • **Solve for *x*.**

The solution checks.

Pythagorean Theorem [15.4B, p. 774]
If a and b are the lengths of the legs of a right triangle and c is the length of the hypotenuse, then $c^2 = a^2 + b^2$.

Two legs of a right triangle measure 4 cm and 7 cm. Find the length of the hypotenuse.

$c = \sqrt{a^2 + b^2}$
$c = \sqrt{4^2 + 7^2}$ • $a = 4, b = 7$
$c = \sqrt{16 + 49}$
$c = \sqrt{65}$

The length of the hypotenuse is $\sqrt{65}$ cm.

CHAPTER 15

CONCEPT REVIEW

Test your knowledge of the concepts presented in this chapter. Answer each question. Then check your answers against the ones provided in the Answer Section.

1. What is the principal square root of a number?

2. When are square roots irrational numbers?

3. What does the Product Property of Square Roots state?

4. How can you tell when a radical expression is simplified?

5. When can you add two radical expressions?

6. Given the expression $3 + \sqrt{8}$, what is its conjugate?

7. How do you rationalize the denominator of $\frac{5}{\sqrt{2x}}$?

8. When are conjugates used to rationalize a denominator?

9. What does the Property of Squaring Both Sides of an Equation state?

10. Why is it important to check your solution to a radical equation?

11. What is a right triangle?

12. What does the Pythagorean Theorem state?

CHAPTER 15

REVIEW EXERCISES

1. Simplify: $\sqrt{3}(\sqrt{12} - \sqrt{3})$

2. Simplify: $3\sqrt{18a^5b}$

3. Simplify: $2\sqrt{36}$

4. Simplify: $\sqrt{6a}(\sqrt{3a} + \sqrt{2a})$

5. Simplify: $\dfrac{12}{\sqrt{6}}$

6. Simplify: $2\sqrt{8} - 3\sqrt{32}$

7. Simplify: $(3 - \sqrt{7})(3 + \sqrt{7})$

8. Solve: $\sqrt{x+3} - \sqrt{x} = 1$

9. Simplify: $\dfrac{2x}{\sqrt{3} - \sqrt{5}}$

10. Simplify: $-3\sqrt{120}$

11. Solve: $\sqrt{5x} = 10$

12. Simplify: $5\sqrt{48}$

13. Simplify: $\dfrac{\sqrt{98x^7y^9}}{\sqrt{2x^3y}}$

14. Solve: $3 - \sqrt{7x} = 5$

15. Simplify: $6a\sqrt{80b} - \sqrt{180a^2b} + 5a\sqrt{b}$

16. Simplify: $4\sqrt{250}$

17. Simplify: $2x\sqrt{60x^3y^3} + 3x^2y\sqrt{15xy}$

18. Simplify: $(4\sqrt{y} - \sqrt{5})(2\sqrt{y} + 3\sqrt{5})$

19. Simplify: $3\sqrt{12x} + 5\sqrt{48x}$

20. Solve: $\sqrt{2x - 3} + 4 = 0$

21. Simplify: $\dfrac{8}{\sqrt{x} - 3}$

22. Simplify: $4y\sqrt{243x^{17}y^9}$

23. Simplify: $y\sqrt{24y^6}$

24. Solve: $2x + 4 = \sqrt{x^2 + 3}$

25. Simplify:
$2x^2\sqrt{18x^2y^5} + 6y\sqrt{2x^6y^3} - 9xy^2\sqrt{8x^4y}$

26. Simplify: $\dfrac{16}{\sqrt{a}}$

27. **Surveying** To find the distance across a pond, a surveyor constructs a right triangle as shown at the right. Find the distance d across the pond. Round to the nearest foot.

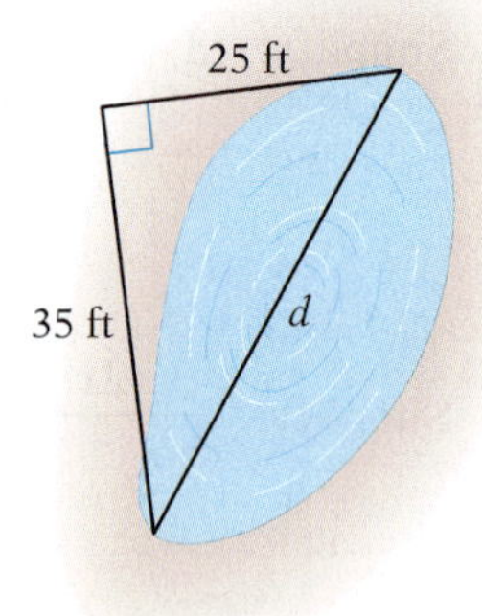

28. **Space Exploration** The weight of an object is related to the distance the object is above the surface of Earth. An equation for this relationship is $d = 4000\sqrt{\dfrac{W_0}{W_d}} - 4000$, where W_0 is an object's weight on the surface of Earth and W_d is the object's weight at a distance of d miles above Earth's surface. If a space explorer weighs 36 lb at a distance of 4000 mi above the surface of Earth, how much does the explorer weigh on the surface of Earth?

29. **Tsunamis** A tsunami is a great sea wave produced by underwater earthquakes or volcanic eruption. The velocity of a tsunami as it approaches land depends on the depth of the water and can be approximated by the equation $v = 3\sqrt{d}$, where d is the depth of the water in feet and v is the velocity of the tsunami in feet per second. Find the depth of the water if the velocity is 30 ft/s.

30. **Bicycle Safety** A bicycle will overturn if it rounds a corner too sharply or too fast. An equation for the maximum velocity at which a cyclist can turn a corner without tipping over is $v = 4\sqrt{r}$, where v is the velocity of the bicycle in miles per hour and r is the radius of the corner in feet. What is the radius of the sharpest corner that a cyclist can safely turn while riding at 20 mph?

CHAPTER 15

TEST

1. Simplify: $\sqrt{121x^8y^2}$

2. Simplify: $\sqrt{3x^2y}\sqrt{6xy^2}\sqrt{2x}$

3. Simplify: $5\sqrt{8} - 3\sqrt{50}$

4. Simplify: $\sqrt{45}$

5. Simplify: $\dfrac{\sqrt{162}}{\sqrt{2}}$

6. Solve: $\sqrt{9x} + 3 = 18$

7. Simplify: $\sqrt{32a^5b^{11}}$

8. Simplify: $\dfrac{\sqrt{98a^6b^4}}{\sqrt{2a^3b^2}}$

9. Simplify: $\dfrac{2}{\sqrt{3} - 1}$

10. Simplify: $\sqrt{8x^3y}\sqrt{10xy^4}$

11. Solve: $\sqrt{x-5} + \sqrt{x} = 5$

12. Simplify: $3\sqrt{8y} - 2\sqrt{72x} + 5\sqrt{18y}$

13. Simplify: $\sqrt{72x^7y^2}$

14. Simplify: $(\sqrt{y} - 3)(\sqrt{y} + 5)$

15. Simplify: $2x\sqrt{3xy^3} - 2y\sqrt{12x^3y} - 3xy\sqrt{xy}$

16. Simplify: $\dfrac{2 - \sqrt{5}}{6 + \sqrt{5}}$

17. Simplify: $\sqrt{a}(\sqrt{a} - \sqrt{b})$

18. Simplify: $\sqrt{75}$

19. **Time** Find the length of a pendulum that makes one swing in 3 s. The equation for the time of one swing of a pendulum is $T = 2\pi\sqrt{\frac{L}{32}}$, where T is the time in seconds and L is the length in feet. Use 3.14 for π. Round to the nearest hundredth.

20. **Camping** A support rope for a tent is attached to the top of a pole and then secured to the ground as shown in the figure at the right. If the rope is 8 ft long and the pole is 4 ft high, how far x from the base of the pole should the rope be secured? Round to the nearest foot.

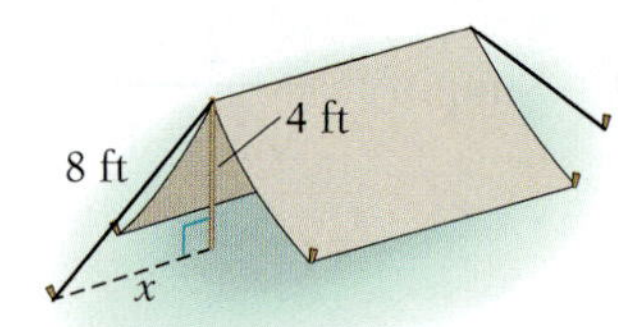

CUMULATIVE REVIEW EXERCISES

1. Simplify:
$\left(\frac{2}{3}\right)^2 \cdot \left(\frac{3}{4} - \frac{3}{2}\right) + \left(\frac{1}{2}\right)^2$

2. Simplify:
$-3[x - 2(3 - 2x) - 5x] + 2x$

3. Solve:
$2x - 4[3x - 2(1 - 3x)] = 2(3 - 4x)$

4. Simplify: $(-3x^2y)(-2x^3y^4)$

5. Simplify: $\dfrac{12b^4 - 6b^2 + 2}{-6b^2}$

6. Given $f(x) = \dfrac{2x}{x - 3}$, find $f(-3)$.

7. Factor: $2a^3 - 16a^2 + 30a$

8. Multiply: $\dfrac{3x^3 - 6x^2}{4x^2 + 4x} \cdot \dfrac{3x - 9}{9x^3 - 45x^2 + 54x}$

9. Subtract: $\dfrac{x + 2}{x - 4} - \dfrac{6}{(x - 4)(x - 3)}$

10. Solve: $\dfrac{x}{2x - 5} - 2 = \dfrac{3x}{2x - 5}$

11. Find the equation of the line that contains the point $(-2, -3)$ and has slope $\frac{1}{2}$.

12. Solve by substitution:
$4x - 3y = 1$
$2x + y = 3$

13. Solve by the addition method:
$5x + 4y = 7$
$3x - 2y = 13$

14. Solve: $3(x - 7) \geq 5x - 12$

15. Simplify: $\sqrt{108}$

16. Simplify: $3\sqrt{32} - 2\sqrt{128}$

17. Simplify: $2a\sqrt{2ab^3} + b\sqrt{8a^3b} - 5ab\sqrt{ab}$

18. Simplify: $\sqrt{2a^9b}\sqrt{98ab^3}\sqrt{2a}$

19. Simplify: $\sqrt{3}(\sqrt{6} - \sqrt{x^2})$

20. Simplify: $\dfrac{\sqrt{320}}{\sqrt{5}}$

21. Simplify: $\dfrac{3}{2 - \sqrt{5}}$

22. Solve: $\sqrt{3x - 2} - 4 = 0$

23. **Business** The selling price of a book is \$59.40. The markup rate used by the bookstore is 20%. Find the cost of the book. Use the formula $S = C + rC$, where S is the selling price, C is the cost, and r is the markup rate.

24. **Travel** Two cyclists start from the same point and ride in opposite directions. One cyclist rides 4 mph faster than the other. In 2 h, they are 52 mi apart. Find the rate of the faster cyclist.

25. **Number Sense** The sum of two numbers is twenty-one. The product of the two numbers is one hundred four. Find the two numbers.

26. **Work** A small water pipe takes twice as long to fill a tank as does a larger water pipe. With both pipes open, it takes 16 h to fill the tank. Find the time it would take the small pipe, working alone, to fill the tank.

27. Solve by graphing: $3x - 2y = 8$
$4x + 5y = 3$

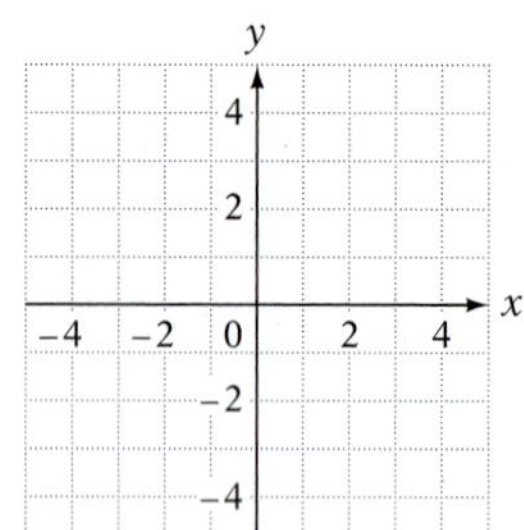

28. Graph the solution set of $3x + y \leq 2$.

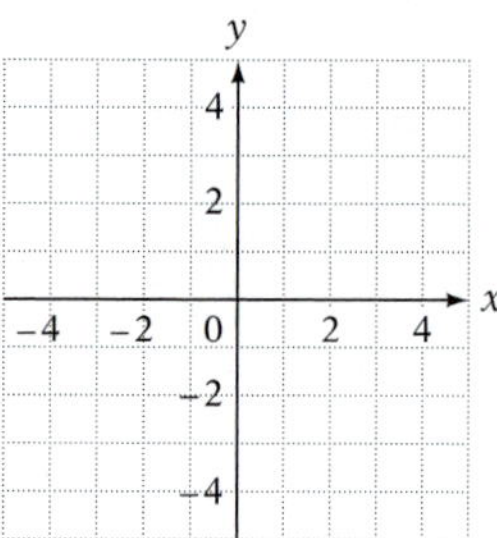

29. **Number Sense** The square root of the sum of two consecutive integers is equal to 9. Find the smaller integer.

30. **Physics** A stone is dropped from a building and hits the ground 5 s later. How high is the building? The equation for the distance an object falls in T seconds is $T = \sqrt{\dfrac{d}{16}}$, where d is the distance in feet.

CHAPTER 16

Quadratic Equations

OBJECTIVES

SECTION 16.1

A To solve a quadratic equation by factoring

B To solve a quadratic equation by taking square roots

SECTION 16.2

A To solve a quadratic equation by completing the square

SECTION 16.3

A To solve a quadratic equation by using the quadratic formula

SECTION 16.4

A To graph a quadratic equation of the form $y = ax^2 + bx + c$

SECTION 16.5

A To solve application problems

ARE YOU READY?

Take the Chapter 16 Prep Test to find out if you are ready to learn to:

- Solve a quadratic equation by factoring, by taking square roots, by completing the square, and by using the quadratic formula
- Graph a quadratic equation of the form $y = ax^2 + bx + c$

PREP TEST

Do these exercises to prepare for Chapter 16.

1. Evaluate $b^2 - 4ac$ when $a = 2$, $b = -3$, and $c = -4$.

2. Solve: $5x + 4 = 3$

3. Factor: $x^2 + x - 12$

4. Factor: $4x^2 - 12x + 9$

5. Is $x^2 - 10x + 25$ a perfect square trinomial?

6. Solve: $\dfrac{5}{x-2} = \dfrac{15}{x}$

7. Graph: $y = -2x + 3$

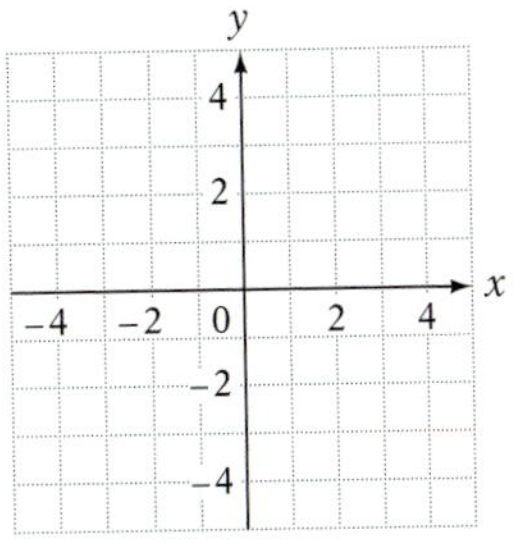

8. Simplify: $\sqrt{28}$

9. If a is *any* real number, simplify $\sqrt{a^2}$.

10. Exercising Walking at a constant speed of 4.5 mph, Lucy and Sam walked from the beginning to the end of a hiking trail. When they reached the end, they immediately started back along the same path at a constant speed of 3 mph. If the round-trip took 2 h, what is the length of the hiking trail?

Lori Adamski Peek/Stone/Getty Images

SECTION

16.1 Solving Quadratic Equations by Factoring or by Taking Square Roots

OBJECTIVE A To solve a quadratic equation by factoring

An equation of the form $ax^2 + bx + c = 0$, where a, b, and c are real numbers and $a \neq 0$, is a **quadratic equation.**

$4x^2 - 3x + 1 = 0, a = 4, b = -3, c = 1$
$3x^2 - 4 = 0, \quad a = 3, b = 0, c = -4$
$\frac{x^2}{2} - 2x + 4 = 0, \quad a = \frac{1}{2}, b = -2, c = 4$

A quadratic equation is also called a **second-degree equation.**

A quadratic equation is in **standard form** when the polynomial is in descending order and equal to zero. $3x^2 + 5x - 2 = 0$ is a quadratic equation in standard form.

Recall that the Principle of Zero Products states that if the product of two factors is zero, then at least one of the factors must be zero.

If $a \cdot b = 0$,
then $a = 0$ or $b = 0$.

The Principle of Zero Products can be used to solve quadratic equations by factoring. **Write the equation in standard form, factor the polynomial, apply the Principle of Zero Products, and solve for the variable.**

HOW TO 1 Solve by factoring: $2x^2 - x = 1$

$2x^2 - x = 1$

$2x^2 - x - 1 = 0$ • Write the equation in standard form.

$(2x + 1)(x - 1) = 0$ • Factor.

$2x + 1 = 0 \qquad x - 1 = 0$ • Use the Principle of Zero Products to set each factor equal to zero.

$2x = -1 \qquad x = 1$ • Rewrite each equation in the form *variable = constant.*

$x = -\frac{1}{2}$

Check:

$2x^2 - x = 1$	
$2\left(-\frac{1}{2}\right)^2 - \left(-\frac{1}{2}\right)$	1
$2 \cdot \frac{1}{4} + \frac{1}{2}$	1
$\frac{1}{2} + \frac{1}{2}$	1
$1 = 1$	

$2x^2 - x = 1$	
$2(1)^2 - 1$	1
$2 \cdot 1 - 1$	1
$2 - 1$	1
$1 = 1$	

The solutions are $-\frac{1}{2}$ and 1.

Take Note
You should always check your solutions by substituting the proposed solutions back into the *original* equation.

HOW TO • 2 Solve by factoring: $3x^2 - 4x + 8 = (4x + 1)(x - 2)$

$3x^2 - 4x + 8 = (4x + 1)(x - 2)$

$3x^2 - 4x + 8 = 4x^2 - 7x - 2$ • Multiply the factors on the right side of the equation.

$0 = x^2 - 3x - 10$ • Write the equation in standard form.

$0 = (x - 5)(x + 2)$ • Factor.

$x - 5 = 0 \qquad x + 2 = 0$ • Use the Principle of Zero Products to set each factor equal to zero.

$x = 5 \qquad x = -2$ • Rewrite each equation in the form *variable = constant.*

Check:

$3x^2 - 4x + 8$	$(4x + 1)(x - 2)$	$3x^2 - 4x + 8$	$(4x + 1)(x - 2)$
$3(5)^2 - 4(5) + 8$	$(4[5] + 1)(5 - 2)$	$3(-2)^2 - 4(-2) + 8$	$(4[-2] + 1)(-2 - 2)$
$3(25) - 4(5) + 8$	$(20 + 1)(3)$	$3(4) - 4(-2) + 8$	$(-8 + 1)(-4)$
$75 - 20 + 8$	$(21)(3)$	$12 + 8 + 8$	$(-7)(-4)$
$63 = 63$		$28 = 28$	

The solutions are 5 and −2.

HOW TO • 3 Solve by factoring: $x^2 - 10x + 25 = 0$

$x^2 - 10x + 25 = 0$

$(x - 5)(x - 5) = 0$ • Factor.

$x - 5 = 0 \qquad x - 5 = 0$ • Use the Principle of Zero Products.

$x = 5 \qquad x = 5$ • Solve each equation for x.

The solution is 5.

In this last example, 5 is called a **double root** of the quadratic equation.

EXAMPLE • 1

Solve by factoring: $\frac{z^2}{2} - \frac{z}{4} - \frac{1}{4} = 0$

Solution

$\frac{z^2}{2} - \frac{z}{4} - \frac{1}{4} = 0$

$4\left(\frac{z^2}{2} - \frac{z}{4} - \frac{1}{4}\right) = 4(0)$ • Multiply each side by 4.

$2z^2 - z - 1 = 0$

$(2z + 1)(z - 1) = 0$ • Factor.

$2z + 1 = 0 \qquad z - 1 = 0$ • Principle of Zero Products

$2z = -1 \qquad z = 1$

$z = -\frac{1}{2}$

The solutions are $-\frac{1}{2}$ and 1.

YOU TRY IT • 1

Solve by factoring: $\frac{3y^2}{2} + y - \frac{1}{2} = 0$

Your solution

Solution on p. S36

OBJECTIVE B To solve a quadratic equation by taking square roots

Consider a quadratic equation of the form $x^2 = a$. This equation can be solved by factoring.

$$
\begin{aligned}
x^2 &= 25 \\
x^2 - 25 &= 0 \\
(x-5)(x+5) &= 0
\end{aligned}
$$

$$
\begin{array}{ll}
x - 5 = 0 & x + 5 = 0 \\
x = 5 & x = -5
\end{array}
$$

The solutions are 5 and −5. The solutions are plus or minus the same number, which is frequently written by using ±; for example, "the solutions are ±5." An alternative method of solving this equation is suggested by the fact that ±5 can be written as $\pm\sqrt{25}$.

Take Note

Recall that the solution of the equation $|x| = 5$ is ± 5. This principle is used when solving an equation by taking square roots. Remember that $\sqrt{x^2} = |x|$. Therefore,

$$
\begin{aligned}
x^2 &= 25 \\
\sqrt{x^2} &= \sqrt{25} \\
|x| &= 5 && \bullet\ \sqrt{x^2} = |x| \\
x &= \pm 5 && \bullet\ \text{If } |x| = 5, \text{ then } x = \pm 5.
\end{aligned}
$$

Principle of Taking the Square Root of Each Side of an Equation

If $x^2 = a$, then $x = \pm\sqrt{a}$.

HOW TO 4 Solve by taking square roots: $x^2 = 25$

$$
\begin{aligned}
x^2 &= 25 \\
\sqrt{x^2} &= \sqrt{25} && \bullet\ \text{Take the square root of each side of the equation. Then simplify.} \\
x &= \pm\sqrt{25} = \pm 5
\end{aligned}
$$

The solutions are 5 and −5.

Take Note

Here is a check for the example at the right.

Check:

$$
\begin{array}{r|l}
3x^2 = & 36 \\
\hline
3(2\sqrt{3})^2 & 36 \\
3(12) & 36 \\
36 = & 36
\end{array}
$$

$$
\begin{array}{r|l}
3x^2 = & 36 \\
\hline
3(-2\sqrt{3})^2 & 36 \\
3(12) & 36 \\
36 = & 36
\end{array}
$$

HOW TO 5 Solve by taking square roots: $3x^2 = 36$

$$
\begin{aligned}
3x^2 &= 36 \\
x^2 &= 12 && \bullet\ \text{Solve for } x^2. \\
\sqrt{x^2} &= \sqrt{12} && \bullet\ \text{Take the square root of each side.} \\
x &= \pm\sqrt{12} && \bullet\ \text{Simplify.} \\
x &= \pm 2\sqrt{3}
\end{aligned}
$$

The solutions are $2\sqrt{3}$ and $-2\sqrt{3}$.

HOW TO 6 Solve by taking square roots: $49y^2 - 25 = 0$

$$
\begin{aligned}
49y^2 - 25 &= 0 \\
49y^2 &= 25 \\
y^2 &= \frac{25}{49} && \bullet\ \text{Solve for } y^2. \\
\sqrt{y^2} &= \sqrt{\frac{25}{49}} && \bullet\ \text{Take the square root of each side.} \\
y &= \pm\frac{5}{7} && \bullet\ \text{Simplify.}
\end{aligned}
$$

The solutions are $\frac{5}{7}$ and $-\frac{5}{7}$.

An equation that contains the square of a binomial can be solved by taking square roots.

Take Note

Here is a check for one of the solutions in the example at the right. You should check all solutions.

Check:

$2(x-1)^2 - 36 = 0$	
$2(1+3\sqrt{2}-1)^2 - 36$	0
$2(3\sqrt{2})^2 - 36$	0
$2(18) - 36$	0
$36 - 36$	0
$0 = 0$	

HOW TO • 7 Solve by taking square roots: $2(x-1)^2 - 36 = 0$

$$2(x-1)^2 - 36 = 0$$
$$2(x-1)^2 = 36$$
$$(x-1)^2 = 18 \quad \text{• Solve for } (x-1)^2.$$
$$\sqrt{(x-1)^2} = \sqrt{18} \quad \text{• Take the square root of each side of the equation.}$$
$$x - 1 = \pm\sqrt{18}$$
$$x - 1 = \pm 3\sqrt{2} \quad \text{• Simplify.}$$
$$x = 1 \pm 3\sqrt{2} \quad \text{• Solve for } x.$$

The solutions are $1 + 3\sqrt{2}$ and $1 - 3\sqrt{2}$.

EXAMPLE • 2

Solve by taking square roots:
$x^2 + 16 = 0$

Solution

$$x^2 + 16 = 0$$
$$x^2 = -16 \quad \text{• Solve for } x^2.$$
$$\sqrt{x^2} = \sqrt{-16} \quad \text{• Take square roots.}$$

$\sqrt{-16}$ is not a real number.

The equation has no real number solution.

YOU TRY IT • 2

Solve by taking square roots:
$x^2 + 81 = 0$

Your solution

EXAMPLE • 3

Solve by taking square roots:
$5(y-4)^2 = 25$

Solution

$$5(y-4)^2 = 25$$
$$(y-4)^2 = 5 \quad \text{• Solve for } (y-4)^2.$$
$$\sqrt{(y-4)^2} = \sqrt{5} \quad \text{• Take square roots.}$$
$$y - 4 = \pm\sqrt{5} \quad \text{• Simplify.}$$
$$y = 4 \pm\sqrt{5} \quad \text{• Solve for } y.$$

The solutions are $4 + \sqrt{5}$ and $4 - \sqrt{5}$.

YOU TRY IT • 3

Solve by taking square roots:
$7(z+2)^2 = 21$

Your solution

Solutions on p. S36

16.1 EXERCISES

OBJECTIVE A To solve a quadratic equation by factoring

For Exercises 1 to 4, solve for x.

1. $(x + 3)(x - 5) = 0$

2. $x(x - 7) = 0$

3. $(2x + 5)(3x - 1) = 0$

4. $(x - 4)(2x - 7) = 0$

For Exercises 5 to 34, solve by factoring.

5. $x^2 + 2x - 15 = 0$

6. $t^2 + 3t - 10 = 0$

7. $z^2 - 4z + 3 = 0$

8. $s^2 - 5s + 4 = 0$

9. $p^2 + 3p + 2 = 0$

10. $v^2 + 6v + 5 = 0$

11. $x^2 - 6x + 9 = 0$

12. $y^2 - 8y + 16 = 0$

13. $12y^2 + 8y = 0$

14. $6x^2 - 9x = 0$

15. $r^2 - 10 = 3r$

16. $t^2 - 12 = 4t$

17. $3v^2 - 5v + 2 = 0$

18. $2p^2 - 3p - 2 = 0$

19. $3s^2 + 8s = 3$

20. $3x^2 + 5x = 12$

21. $\frac{3}{4}z^2 - z = -\frac{1}{3}$

22. $\frac{r^2}{2} = 1 - \frac{r}{12}$

23. $4t^2 = 4t + 3$

24. $5y^2 + 11y = 12$

25. $4v^2 - 4v + 1 = 0$

26. $9s^2 - 6s + 1 = 0$

27. $x^2 - 9 = 0$

28. $t^2 - 16 = 0$

29. $4y^2 - 1 = 0$

30. $9z^2 - 4 = 0$

31. $x + 15 = x(x - 1)$

32. $p + 18 = p(p - 2)$

33. $r^2 - r - 2 = (2r - 1)(r - 3)$

34. $s^2 + 5s - 4 = (2s + 1)(s - 4)$

35. Let a be a positive integer. Which equation has a positive double root?
(i) $x^2 - a^2 = 0$ (ii) $x^2 + 2ax + a^2 = 0$ (iii) $x^2 - 2ax + a^2 = 0$

OBJECTIVE B To solve a quadratic equation by taking square roots

For Exercises 36 to 62, solve by taking square roots.

36. $x^2 = 36$

37. $y^2 = 49$

38. $v^2 - 1 = 0$

39. $z^2 - 64 = 0$

40. $4x^2 - 49 = 0$

41. $9w^2 - 64 = 0$

42. $9y^2 = 4$

43. $4z^2 = 25$

44. $16v^2 - 9 = 0$

45. $25x^2 - 64 = 0$

46. $y^2 + 81 = 0$

47. $z^2 + 49 = 0$

48. $w^2 - 24 = 0$

49. $v^2 - 48 = 0$

50. $(x - 1)^2 = 36$

51. $(y + 2)^2 = 49$

52. $2(x + 5)^2 = 8$

53. $4(z - 3)^2 = 100$

54. $9(x - 1)^2 - 16 = 0$

55. $4(y + 3)^2 - 81 = 0$

56. $49(v + 1)^2 - 25 = 0$

57. $81(y - 2)^2 - 64 = 0$

58. $(x - 4)^2 - 20 = 0$

59. $(y + 5)^2 - 50 = 0$

60. $(x + 1)^2 + 36 = 0$

61. $2\left(z - \frac{1}{2}\right)^2 = 12$

62. $3\left(v + \frac{3}{4}\right)^2 = 36$

For Exercises 63 to 66, assume that a and b are both positive numbers. In each case, state how many real number solutions the equation has.

63. $(x + a)^2 = 0$

64. $ax^2 - b = 0$

65. $(x + a)^2 = b$

66. $ax^2 + b = 0$

Applying the Concepts

67. **Investments** The value A of an initial investment of P dollars after 2 years is given by $A = P(1 + r)^2$, where r is the annual percentage rate earned by the investment. If an initial investment of \$1500 grew to a value of \$1782.15 in 2 years, what was the annual percentage rate?

68. **Automotive Safety** On a certain type of street surface, the equation $d = 0.0074v^2$ can be used to approximate the distance d, in feet, a car traveling v miles per hour will slide when its brakes are applied. After applying the brakes, the owner of a car involved in an accident skidded 40 ft. Did the traffic officer investigating the accident issue the car owner a ticket for speeding if the speed limit is 65 mph?

SECTION

16.2 Solving Quadratic Equations by Completing the Square

OBJECTIVE A To solve a quadratic equation by completing the square 

Recall that a perfect-square trinomial is the square of a binomial.

Perfect-Square Trinomial		Square of a Binomial
$x^2 + 6x + 9$	=	$(x + 3)^2$
$x^2 - 10x + 25$	=	$(x - 5)^2$
$x^2 + 8x + 16$	=	$(x + 4)^2$

For each perfect-square trinomial, the square of $\frac{1}{2}$ of the coefficient of x equals the constant term.

$$x^2 + 6x + 9, \quad \left(\frac{1}{2} \cdot 6\right)^2 = 9$$

$$x^2 - 10x + 25, \quad \left[\frac{1}{2}(-10)\right]^2 = 25$$

$$x^2 + 8x + 16, \quad \left(\frac{1}{2} \cdot 8\right)^2 = 16$$

Adding to a binomial the constant term that makes it a perfect-square trinomial is called **completing the square.**

HOW TO • 1 Complete the square of $x^2 - 8x$. Write the resulting perfect-square trinomial as the square of a binomial.

$\left[\frac{1}{2}(-8)\right]^2 = 16$ • **Find the constant term.**

$x^2 - 8x + 16$ • **Complete the square on $x^2 - 8x$ by adding the constant term.**

$x^2 - 8x + 16 = (x - 4)^2$ • **Write the resulting perfect-square trinomial as the square of a binomial.**

HOW TO • 2 Complete the square of $y^2 + 5y$. Write the resulting perfect-square trinomial as the square of a binomial.

$\left(\frac{1}{2} \cdot 5\right)^2 = \left(\frac{5}{2}\right)^2 = \frac{25}{4}$ • **Find the constant term.**

$y^2 + 5y + \frac{25}{4}$ • **Complete the square on $y^2 + 5y$ by adding the constant term.**

$y^2 + 5y + \frac{25}{4} = \left(y + \frac{5}{2}\right)^2$ • **Write the resulting perfect-square trinomial as the square of a binomial.**

Point of Interest

Early mathematicians solved quadratic equations by literally *completing the square.* For these mathematicians, all equations had geometric interpretations. They found that a quadratic equation could be solved by making certain figures into squares. See the Projects and Group Activities at the end of this chapter for an idea of how this was done.

A quadratic equation that cannot be solved by factoring can be solved by completing the square. **When the quadratic equation is in the form $x^2 + bx = c$, add to each side of the equation the term that completes the square on $x^2 + bx$. Factor the perfect-square trinomial, and write it as the square of a binomial. Take the square root of each side of the equation, and then solve for x.**

Tips for Success

This is a new skill and one that is difficult for many students. Be sure to do all you need to do in order to be successful at solving quadratic equations by completing the square: Read through the introductory material, work through the How To examples, study the paired examples, and do the You Try Its and check your solutions against the ones given in the back of the book. See *AIM for Success* at the front of the book.

HOW TO 3 Solve by completing the square: $x^2 + 8x - 2 = 0$

$$x^2 + 8x - 2 = 0$$

$$x^2 + 8x = 2$$
• Add 2 to each side of the equation.

$$x^2 + 8x + \left(\frac{1}{2} \cdot 8\right)^2 = 2 + \left(\frac{1}{2} \cdot 8\right)^2$$
• Complete the square on $x^2 + 8x$. Add $\left(\frac{1}{2} \cdot 8\right)^2$ to each side of the equation.

$$x^2 + 8x + 16 = 2 + 16$$
• Simplify.

$$(x + 4)^2 = 18$$
• Factor the perfect-square trinomial.

$$\sqrt{(x + 4)^2} = \sqrt{18}$$
• Take the square root of each side of the equation.

$$x + 4 = \pm\sqrt{18}$$
• Solve for x.

$$x + 4 = \pm 3\sqrt{2}$$

$$x = -4 \pm 3\sqrt{2}$$

Check:

$x^2 + 8x - 2 = 0$	
$(-4 + 3\sqrt{2})^2 + 8(-4 + 3\sqrt{2}) - 2$	0
$16 - 24\sqrt{2} + 18 - 32 + 24\sqrt{2} - 2$	0
$0 = 0$	

$x^2 + 8x - 2 = 0$	
$(-4 - 3\sqrt{2})^2 + 8(-4 - 3\sqrt{2}) - 2$	0
$16 + 24\sqrt{2} + 18 - 32 - 24\sqrt{2} - 2$	0
$0 = 0$	

The solutions are $-4 + 3\sqrt{2}$ and $-4 - 3\sqrt{2}$.

If the coefficient of the second-degree term is not 1, a necessary step in completing the square is to multiply each side of the equation by the reciprocal of that coefficient.

HOW TO 4 Solve by completing the square: $2x^2 - 3x + 1 = 0$

$$2x^2 - 3x + 1 = 0$$

$$2x^2 - 3x = -1$$
• Subtract 1 from each side of the equation.

$$\frac{1}{2}(2x^2 - 3x) = \frac{1}{2} \cdot (-1)$$
• In order to complete the square, the coefficient of x^2 must be 1. Multiply each side of the equation by $\frac{1}{2}$.

$$x^2 - \frac{3}{2}x = -\frac{1}{2}$$

$$x^2 - \frac{3}{2}x + \left[\frac{1}{2}\left(-\frac{3}{2}\right)\right]^2 = -\frac{1}{2} + \left[\frac{1}{2}\left(-\frac{3}{2}\right)\right]^2$$
• Complete the square. Add $\left[\frac{1}{2}\left(-\frac{3}{2}\right)\right]^2$ to each side of the equation.

$$x^2 - \frac{3}{2}x + \frac{9}{16} = -\frac{1}{2} + \frac{9}{16}$$
• Simplify.

$$\left(x - \frac{3}{4}\right)^2 = \frac{1}{16}$$
• Factor the perfect-square trinomial.

$$\sqrt{\left(x - \frac{3}{4}\right)^2} = \sqrt{\frac{1}{16}}$$
• Take the square root of each side of the equation.

$$x - \frac{3}{4} = \pm\frac{1}{4}$$
• Solve for x.

$$x = \frac{3}{4} \pm \frac{1}{4}$$

$$x = \frac{3}{4} + \frac{1}{4} = 1 \qquad x = \frac{3}{4} - \frac{1}{4} = \frac{1}{2}$$

The solutions are $\frac{1}{2}$ and 1.

EXAMPLE • 1

Solve by completing the square:
$2x^2 - 4x - 1 = 0$

Solution

$$2x^2 - 4x - 1 = 0$$

$$2x^2 - 4x = 1 \qquad \text{• Add 1.}$$

$$\frac{1}{2}(2x^2 - 4x) = \frac{1}{2} \cdot 1 \qquad \text{• Multiply by } \frac{1}{2}.$$

$$x^2 - 2x = \frac{1}{2} \qquad \text{• The coefficient of } x^2 \text{ is 1.}$$

Complete the square.

$$x^2 - 2x + 1 = \frac{1}{2} + 1 \qquad \text{• } \left[\frac{1}{2} \cdot (-2)\right]^2 = [-1]^2 = 1$$

$$(x - 1)^2 = \frac{3}{2} \qquad \text{• Factor.}$$

$$\sqrt{(x - 1)^2} = \sqrt{\frac{3}{2}} \qquad \text{• Take square roots.}$$

$$x - 1 = \pm\frac{\sqrt{6}}{2} \qquad \text{• Simplify.}$$

$$x = 1 \pm \frac{\sqrt{6}}{2}$$

$$x = 1 + \frac{\sqrt{6}}{2} = \frac{2 + \sqrt{6}}{2} \qquad x = 1 - \frac{\sqrt{6}}{2} = \frac{2 - \sqrt{6}}{2}$$

Check:

$2x^2 - 4x - 1$	$= 0$
$2\left(\frac{2 + \sqrt{6}}{2}\right)^2 - 4\left(\frac{2 + \sqrt{6}}{2}\right) - 1$	0
$2\left(\frac{4 + 4\sqrt{6} + 6}{4}\right) - 2(2 + \sqrt{6}) - 1$	0
$2 + 2\sqrt{6} + 3 - 4 - 2\sqrt{6} - 1$	0
$0 = 0$	

$2x^2 - 4x - 1$	$= 0$
$2\left(\frac{2 - \sqrt{6}}{2}\right)^2 - 4\left(\frac{2 - \sqrt{6}}{2}\right) - 1$	0
$2\left(\frac{4 - 4\sqrt{6} + 6}{4}\right) - 2(2 - \sqrt{6}) - 1$	0
$2 - 2\sqrt{6} + 3 - 4 + 2\sqrt{6} - 1$	0
$0 = 0$	

The solutions are $\frac{2 + \sqrt{6}}{2}$ and $\frac{2 - \sqrt{6}}{2}$.

YOU TRY IT • 1

Solve by completing the square:
$3x^2 - 6x - 2 = 0$

Your solution

Solution on p. S36

EXAMPLE 2

Solve by completing the square:
$x^2 + 4x + 5 = 0$

Solution

$x^2 + 4x + 5 = 0$

$x^2 + 4x = -5$ • **Subtract 5.**

Complete the square.

$x^2 + 4x + 4 = -5 + 4$ • $\left(\frac{1}{2} \cdot 4\right)^2 = 2^2 = 4$

$(x + 2)^2 = -1$ • **Factor.**

$\sqrt{(x + 2)^2} = \sqrt{-1}$ • **Take square roots.**

$\sqrt{-1}$ is not a real number.

The quadratic equation has no real number solution.

YOU TRY IT 2

Solve by completing the square:
$x^2 + 6x + 12 = 0$

Your solution

EXAMPLE 3

Solve $x^2 = -6x - 4$ by completing the square. Approximate the solutions to the nearest thousandth.

Solution

$x^2 = -6x - 4$

$x^2 + 6x = -4$ • **Add 6x.**

Complete the square.

$x^2 + 6x + 9 = -4 + 9$ • $\left(\frac{1}{2} \cdot 6\right)^2 = 3^2 = 9$

$(x + 3)^2 = 5$ • **Factor.**

$\sqrt{(x + 3)^2} = \sqrt{5}$ • **Take square roots.**

$x + 3 = \pm\sqrt{5}$

$x + 3 = \sqrt{5}$ $\qquad$ $x + 3 = -\sqrt{5}$

$x = -3 + \sqrt{5}$ $\qquad$ $x = -3 - \sqrt{5}$

$\approx -3 + 2.236$ $\qquad$ $\approx -3 - 2.236$

≈ -0.764 $\qquad$ ≈ -5.236

The solutions are approximately -0.764 and -5.236.

YOU TRY IT 3

Solve $x^2 + 8x + 8 = 0$ by completing the square. Approximate the solutions to the nearest thousandth.

Your solution

Solutions on p. S37

16.2 EXERCISES

OBJECTIVE A To solve a quadratic equation by completing the square

For Exercises 1 to 4, complete the square on each binomial. Write the resulting trinomial as the square of a binomial.

1. $x^2 - 8x$

2. $x^2 + 6x$

3. $x^2 + 5x$

4. $x^2 - 3x$

For Exercises 5 to 45, solve by completing the square.

5. $x^2 + 2x - 3 = 0$

6. $y^2 + 4y - 5 = 0$

7. $z^2 - 6z - 16 = 0$

8. $w^2 + 8w - 9 = 0$

9. $x^2 = 4x - 4$

10. $z^2 = 8z - 16$

11. $v^2 - 6v + 13 = 0$

12. $x^2 + 4x + 13 = 0$

13. $y^2 + 5y + 4 = 0$

14. $v^2 - 5v - 6 = 0$

15. $w^2 + 7w = 8$

16. $y^2 + 5y = -4$

17. $v^2 + 4v + 1 = 0$

18. $y^2 - 2y - 5 = 0$

19. $x^2 + 6x = 5$

20. $w^2 - 8w = 3$

21. $\frac{z^2}{2} = z + \frac{1}{2}$

22. $\frac{y^2}{10} = y - 2$

23. $p^2 + 3p = 1$

24. $r^2 + 5r = 2$

25. $t^2 - 3t = -2$

26. $z^2 - 5z = -3$

27. $v^2 + v - 3 = 0$

28. $x^2 - x = 1$

29. $y^2 = 7 - 10y$

30. $v^2 = 14 + 16v$

31. $r^2 - 3r = 5$

32. $s^2 + 3s = -1$

33. $t^2 - t = 4$

34. $y^2 + y - 4 = 0$

35. $x^2 - 3x + 5 = 0$

36. $z^2 + 5z + 7 = 0$

37. $2t^2 - 3t + 1 = 0$

38. $2x^2 - 7x + 3 = 0$ **39.** $2r^2 + 5r = 3$ **40.** $2y^2 - 3y = 4$ **41.** $2s^2 = 7s - 1$

42. $4v^2 + 4v - 1 = 0$ **43.** $6s^2 + s = 3$ **44.** $6z^2 = z + 2$ **45.** $6p^2 = 5p + 4$

For Exercises 46 and 47, without using a calculator, determine if the given solutions are both negative, both positive, or one negative and one positive.

46. A quadratic equation has solutions $-3 \pm \sqrt{5}$. **47.** A quadratic equation has solutions $2 \pm \sqrt{7}$.

For Exercises 48 to 51, solve by completing the square. Approximate the solutions to the nearest thousandth.

48. $y^2 + 3y = 5$ **49.** $w^2 + 5w = 2$ **50.** $2z^2 - 3z = 7$ **51.** $2x^2 + 3x = 11$

Applying the Concepts

52. Explain why the equation $(x - 2)^2 = -4$ does not have a real number solution.

For Exercises 53 to 58, solve.

53. $\sqrt{x + 2} = x - 4$ **54.** $\sqrt{3x + 4} - x = 2$ **55.** $\frac{x + 1}{2} + \frac{3}{x - 1} = 4$

56. $\frac{x - 2}{3} + \frac{2}{x + 2} = 4$ **57.** $4\sqrt{x + 1} - x = 4$ **58.** $3\sqrt{x - 1} + 3 = x$

59. **Sports** A basketball player shoots at a basket 25 ft away. The height of the ball above the ground at time t is given by $h = -16t^2 + 32t + 6.5$. How many seconds after the ball is released does it hit the basket? *Hint:* When the ball hits the basket, $h = 10$ ft.

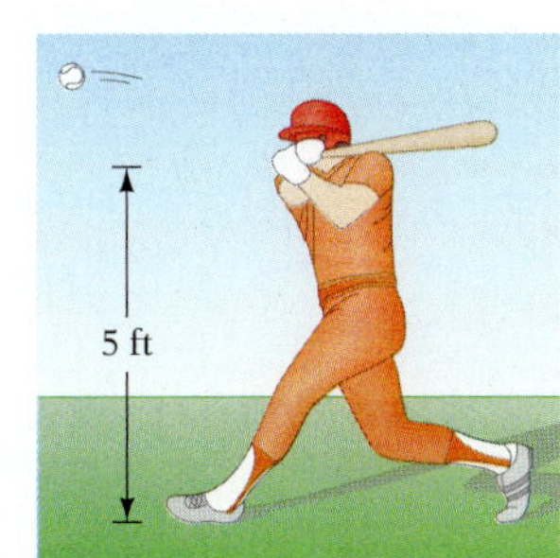

60. **Sports** A ball player hits a ball. The height of the ball above the ground can be approximated by the equation $h = -16t^2 + 76t + 5$. When will the ball hit the ground? *Hint:* The ball strikes the ground when $h = 0$ ft.

SECTION

16.3 Solving Quadratic Equations by Using the Quadratic Formula

OBJECTIVE A To solve a quadratic equation by using the quadratic formula 

Any quadratic equation can be solved by completing the square. Applying this method to the standard form of a quadratic equation produces a formula that can be used to solve any quadratic equation.

Solve $ax^2 + bx + c = 0$ by completing the square.

$$ax^2 + bx + c = 0$$

Add the opposite of the constant term to each side of the equation.

$$ax^2 + bx + c + (-c) = 0 + (-c)$$

$$ax^2 + bx = -c$$

Multiply each side of the equation by the reciprocal of a, the coefficient of x^2.

$$\frac{1}{a}(ax^2 + bx) = \frac{1}{a}(-c)$$

$$x^2 + \frac{b}{a}x = -\frac{c}{a}$$

Complete the square by adding $\left(\frac{1}{2} \cdot \frac{b}{a}\right)^2$ to each side of the equation.

$$x^2 + \frac{b}{a}x + \left(\frac{1}{2} \cdot \frac{b}{a}\right)^2 = \left(\frac{1}{2} \cdot \frac{b}{a}\right)^2 - \frac{c}{a}$$

$$x^2 + \frac{b}{a}x + \frac{b^2}{4a^2} = \frac{b^2}{4a^2} - \frac{c}{a}$$

Simplify the right side of the equation.

$$x^2 + \frac{b}{a}x + \frac{b^2}{4a^2} = \frac{b^2}{4a^2} - \left(\frac{c}{a} \cdot \frac{4a}{4a}\right)$$

$$x^2 + \frac{b}{a}x + \frac{b^2}{4a^2} = \frac{b^2}{4a^2} - \frac{4ac}{4a^2}$$

$$x^2 + \frac{b}{a}x + \frac{b^2}{4a^2} = \frac{b^2 - 4ac}{4a^2}$$

Factor the perfect-square trinomial on the left side of the equation.

$$\left(x + \frac{b}{2a}\right)^2 = \frac{b^2 - 4ac}{4a^2}$$

Take the square root of each side of the equation.

$$\sqrt{\left(x + \frac{b}{2a}\right)^2} = \sqrt{\frac{b^2 - 4ac}{4a^2}}$$

$$x + \frac{b}{2a} = \pm\frac{\sqrt{b^2 - 4ac}}{2a}$$

Solve for x.

$$x + \frac{b}{2a} = \frac{\sqrt{b^2 - 4ac}}{2a} \qquad x + \frac{b}{2a} = -\frac{\sqrt{b^2 - 4ac}}{2a}$$

$$x = -\frac{b}{2a} + \frac{\sqrt{b^2 - 4ac}}{2a} \qquad x = -\frac{b}{2a} - \frac{\sqrt{b^2 - 4ac}}{2a}$$

$$= \frac{-b + \sqrt{b^2 - 4ac}}{2a} \qquad = \frac{-b - \sqrt{b^2 - 4ac}}{2a}$$

The Quadratic Formula

The solutions of $ax^2 + bx + c = 0$, $a \neq 0$, are

$$x = \frac{-b \pm \sqrt{b^2 - 4ac}}{2a}$$

HOW TO • 1 Solve by using the quadratic formula: $2x^2 = 4x - 1$

$$2x^2 = 4x - 1$$
$$2x^2 - 4x + 1 = 0$$ • Write the equation in standard form. Subtract $4x$ from, and add 1 to, each side of the equation.
$$x = \frac{-b \pm \sqrt{b^2 - 4ac}}{2a}$$ • The quadratic formula
$$= \frac{-(-4) \pm \sqrt{(-4)^2 - (4 \cdot 2 \cdot 1)}}{2 \cdot 2}$$ • $a = 2, b = -4, c = 1$. Replace a, b, and c by their values.
$$= \frac{4 \pm \sqrt{16 - 8}}{4} = \frac{4 \pm \sqrt{8}}{4}$$ • Simplify.
$$= \frac{4 \pm 2\sqrt{2}}{4} = \frac{2 \pm \sqrt{2}}{2}$$

The solutions are $\frac{2 + \sqrt{2}}{2}$ and $\frac{2 - \sqrt{2}}{2}$.

Take Note
$$\frac{4 \pm 2\sqrt{2}}{4} = \frac{2(2 \pm \sqrt{2})}{2 \cdot 2} = \frac{2 \pm \sqrt{2}}{2}$$

EXAMPLE • 1

Solve by using the quadratic formula:
$2x^2 - 3x + 1 = 0$

Solution

$$2x^2 - 3x + 1 = 0$$ • Standard form
$$x = \frac{-(-3) \pm \sqrt{(-3)^2 - 4(2)(1)}}{2 \cdot 2}$$ • $a = 2, b = -3, c = 1$
$$= \frac{3 \pm \sqrt{9 - 8}}{4} = \frac{3 \pm \sqrt{1}}{4} = \frac{3 \pm 1}{4}$$
$$x = \frac{3 + 1}{4} = \frac{4}{4} = 1 \qquad x = \frac{3 - 1}{4} = \frac{2}{4} = \frac{1}{2}$$

The solutions are 1 and $\frac{1}{2}$.

YOU TRY IT • 1

Solve by using the quadratic formula:
$3x^2 + 4x - 4 = 0$

Your solution

EXAMPLE • 2

Solve by using the quadratic formula: $\frac{x^2}{2} = 2x - \frac{5}{4}$

Solution

$$\frac{x^2}{2} = 2x - \frac{5}{4}$$
$$4\left(\frac{x^2}{2}\right) = 4\left(2x - \frac{5}{4}\right)$$ • Multiply by 4.
$$2x^2 = 8x - 5$$
$$2x^2 - 8x + 5 = 0$$ • Standard form
$$x = \frac{-(-8) \pm \sqrt{(-8)^2 - 4(2)(5)}}{2 \cdot 2}$$ • $a = 2, b = -8, c = 5$
$$= \frac{8 \pm \sqrt{64 - 40}}{4} = \frac{8 \pm \sqrt{24}}{4}$$
$$= \frac{8 \pm 2\sqrt{6}}{4} = \frac{4 \pm \sqrt{6}}{2}$$

The solutions are $\frac{4 + \sqrt{6}}{2}$ and $\frac{4 - \sqrt{6}}{2}$.

YOU TRY IT • 2

Solve by using the quadratic formula: $\frac{x^2}{4} + \frac{x}{2} = \frac{1}{4}$

Your solution

Solutions on p. S37

16.3 EXERCISES

OBJECTIVE A **To solve a quadratic equation by using the quadratic formula**

For Exercises 1 to 30, solve by using the quadratic formula.

1. $x^2 - 4x - 5 = 0$

2. $y^2 + 3y + 2 = 0$

3. $y^2 = 2y + 3$

4. $w^2 = 3w + 18$

5. $2y^2 - y - 1 = 0$

6. $2t^2 - 5t + 3 = 0$

7. $w^2 + 3w + 5 = 0$

8. $x^2 - 2x + 6 = 0$

9. $4y^2 + 4y = 15$

10. $6y^2 + 5y - 4 = 0$

11. $2x^2 + x + 1 = 0$

12. $3r^2 - r + 2 = 0$

13. $\frac{1}{2}t^2 - t = \frac{5}{2}$

14. $y^2 - 4y = 6$

15. $\frac{1}{3}t^2 + 2t - \frac{1}{3} = 0$

16. $z^2 + 4z + 1 = 0$

17. $w^2 = 4w + 9$

18. $y^2 = 8y + 3$

19. $9y^2 + 6y - 1 = 0$

20. $9s^2 - 6s - 2 = 0$

21. $4p^2 + 4p + 1 = 0$

22. $9z^2 + 12z + 4 = 0$

23. $\frac{x^2}{2} = x - \frac{5}{4}$

24. $r^2 = \frac{5}{3}r - 2$

25. $4p^2 + 16p = -11$

26. $4y^2 - 12y = -1$

27. $4x^2 = 4x + 11$

28. $4s^2 + 12s = 3$

29. $9v^2 = -30v - 23$

30. $9t^2 = 30t + 17$

31. True or false? If you use the quadratic formula to solve $ax^2 + bx + c = 0$ and get rational solutions, then you could have solved the equation by factoring.

32. True or false? If the value of $b^2 - 4ac$ in the quadratic formula is 0, then $ax^2 + bx + c = 0$ has only one solution, a double root.

For Exercises 33 to 41, solve by using the quadratic formula. Approximate the solutions to the nearest thousandth.

33. $x^2 - 2x - 21 = 0$

34. $y^2 + 4y - 11 = 0$

35. $s^2 - 6s - 13 = 0$

36. $w^2 + 8w - 15 = 0$

37. $2p^2 - 7p - 10 = 0$

38. $3t^2 - 8t - 1 = 0$

39. $4z^2 + 8z - 1 = 0$

40. $4x^2 + 7x + 1 = 0$

41. $5v^2 - v - 5 = 0$

Applying the Concepts

42. Factoring, completing the square, and using the quadratic formula are three methods of solving quadratic equations. Describe each method, and cite the advantages and disadvantages of each.

For Exercises 43 to 48, solve.

43. $\sqrt{x + 3} = x - 3$

44. $\sqrt{x + 4} = x + 4$

45. $\sqrt{x + 1} = x - 1$

46. $\sqrt{x^2 + 2x + 1} = x - 1$

47. $\frac{x}{4} + \frac{3}{x} = \frac{5}{2}$

48. $\frac{x + 1}{5} - \frac{4}{x - 1} = 2$

49. True or False?

a. The equations $x = \sqrt{12 - x}$ and $x^2 = 12 - x$ have the same solutions.

b. If $\sqrt{a} + \sqrt{b} = c$, then $a + b = c^2$.

c. $\sqrt{9} = \pm 3$

d. $\sqrt{x^2} = |x|$

50. **Distance** An L-shaped sidewalk from the parking lot to a memorial is shown in the figure at the right. The distance directly across the grass to the memorial is 650 ft. The distance to the corner is 600 ft. Find the distance from the corner to the memorial.

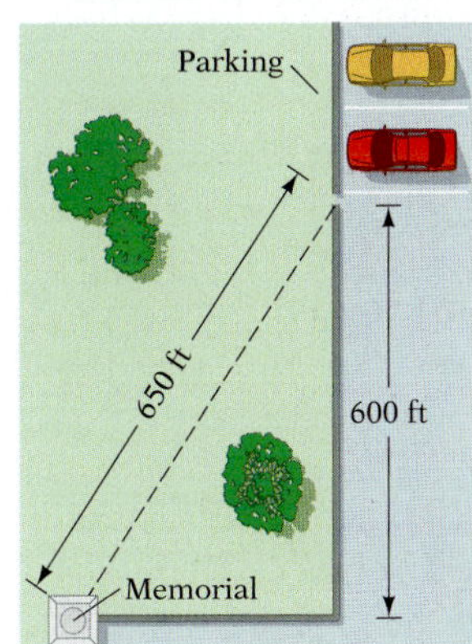

51. **Travel** A commuter plane leaves an airport traveling due south at 400 mph. Another plane leaving at the same time travels due east at 300 mph. Find the distance between the two planes after 2 h.

SECTION

16.4 Graphing Quadratic Equations in Two Variables

OBJECTIVE A To graph a quadratic equation of the form $y = ax^2 + bx + c$

Take Note

For the equation $y = 3x^2 - x + 1$, $a = 3$, $b = -1$, and $c = 1$.

Point of Interest

Mirrors in some telescopes are ground into the shape of a parabola. The mirror at the Palomar Mountain Observatory is 2 ft thick at the ends and weighs 14.75 tons. The mirror has been ground to a true paraboloid (the three-dimensional version of a parabola) to within 0.0000015 in. A possible equation of the mirror is $y = 2640x^2$.

Take Note

One of the equations at the right was written as $y = 2x^2 + 3x - 2$, and the other was written using function notation as $f(x) = -x^2 + 3x + 2$. Remember that y and $f(x)$ are different symbols for the same quantity.

An equation of the form $y = ax^2 + bx + c$, $a \neq 0$, is a **quadratic equation in two variables.** Examples of quadratic equations in two variables are shown at the right.

$y = 3x^2 - x + 1$
$y = -x^2 - 3$
$y = 2x^2 - 5x$

For these equations, y is a function of x, and we can write $f(x) = ax^2 + bx + c$. This equation represents a **quadratic function.**

HOW TO • 1 Evaluate $f(x) = 2x^2 - 3x + 4$ when $x = -2$.

$f(x) = 2x^2 - 3x + 4$
$f(-2) = 2(-2)^2 - 3(-2) + 4$ • **Replace x by -2.**
$= 2(4) + 6 + 4 = 18$ • **Simplify.**

The value of the function when $x = -2$ is 18.

The graph of $y = ax^2 + bx + c$ or $f(x) = ax^2 + bx + c$ is a **parabola. The graph is U-shaped, and opens up when a is positive and down when a is negative.** The graphs of two parabolas are shown below.

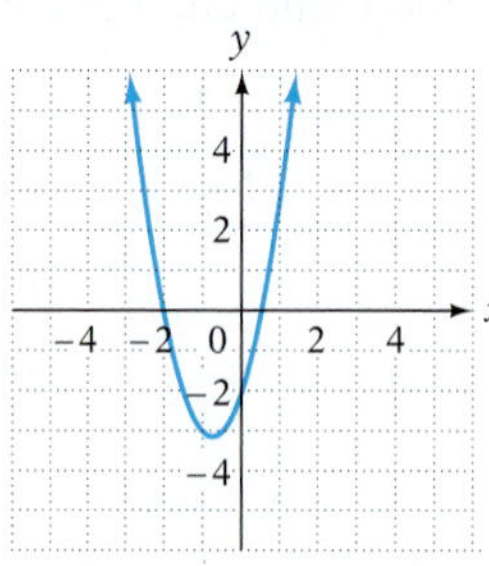

$y = 2x^2 + 3x - 2$
$a = 2$, a positive number
Parabola opens up.

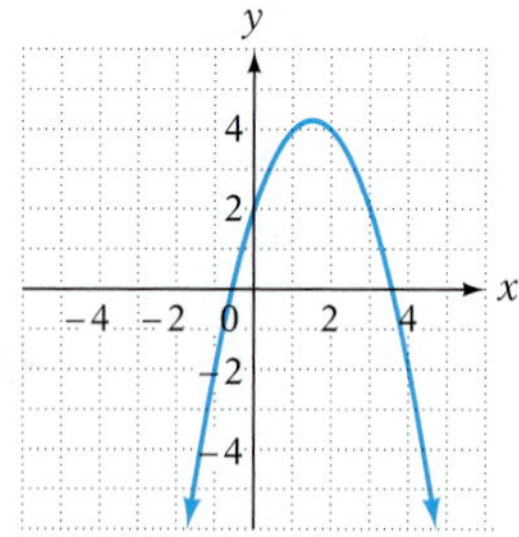

$f(x) = -x^2 + 3x + 2$
$a = -1$, a negative number
Parabola opens down.

HOW TO • 2 Graph $y = x^2 - 2x - 3$.

x	y
−2	5
−1	0
0	−3
1	−4
2	−3
3	0
4	5

• **Find several solutions of the equation. Because the graph is not a straight line, several solutions must be found in order to determine the U-shape. Record the ordered pairs in a table.**

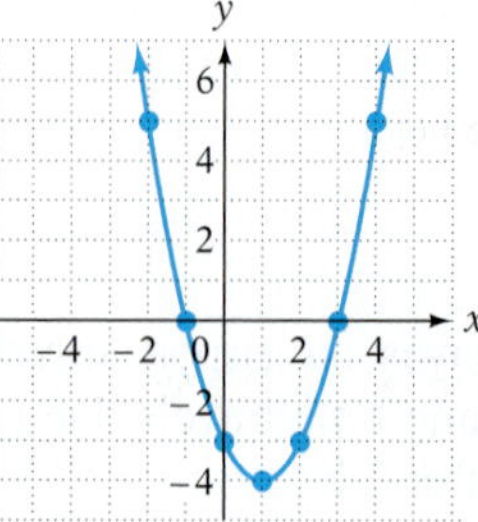

• **Graph the ordered-pair solutions on a rectangular coordinate system. Draw a parabola through the points.**

Note that the graph of $y = x^2 - 2x - 3$, shown again below, crosses the x-axis at $(-1, 0)$ and $(3, 0)$. This is also confirmed from the table for the graph. From the table, note that $y = 0$ when $x = -1$ and when $x = 3$. The x-intercepts of the graph are $(-1, 0)$ and $(3, 0)$.

x	y
-2	5
-1	0
0	-3
1	-4
2	-3
3	0
4	5

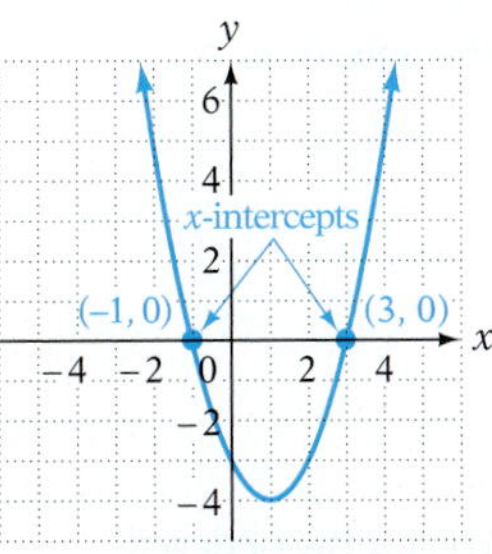

The y-intercept is the point at which the graph crosses the y-axis. At this point, $x = 0$. From the graph, we can see that the y-intercept is $(0, -3)$.

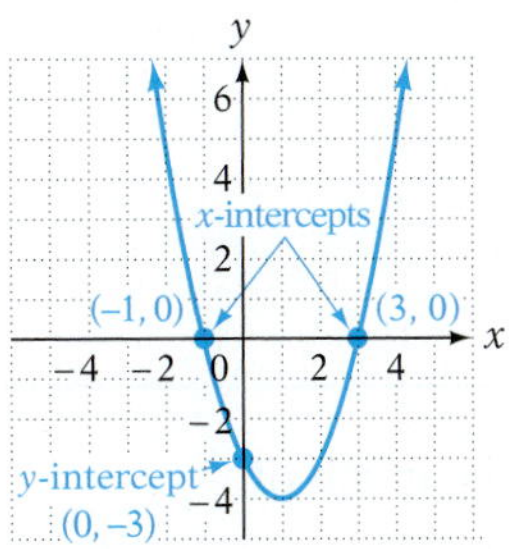

We can find the x-intercepts algebraically by letting $y = 0$ and solving for x.

$y = x^2 - 2x - 3$
$0 = x^2 - 2x - 3$ • **Replace y by 0 and solve for x.**
$0 = (x + 1)(x - 3)$ • **This equation can be solved by factoring. However, it will be necessary to use the quadratic formula to solve some quadratic equations.**
$x + 1 = 0 \qquad x - 3 = 0$
$x = -1 \qquad x = 3$

The x-intercepts are $(-1, 0)$ and $(3, 0)$.

We can find the y-intercept algebraically by letting $x = 0$ and solving for y.

$y = x^2 - 2x - 3$
$y = 0^2 - 2(0) - 3$ • **Replace x by 0 and simplify.**
$= -3$

The y-intercept is $(0, -3)$.

Graph of a Quadratic Equation in Two Variables

To graph a quadratic equation in two variables, find several solutions of the equation. Graph the ordered-pair solutions on a rectangular coordinate system. Draw a parabola through the points.

To find the x-intercepts of the graph of a quadratic equation in two variables, let $y = 0$ and solve for x.

To find the y-intercept, let $x = 0$ and solve for y.

EXAMPLE • 1

Graph $y = x^2 - 2x$.

Solution

x	y
-1	3
0	0
1	-1
2	0
3	3

• **Find several solutions of the equation.**

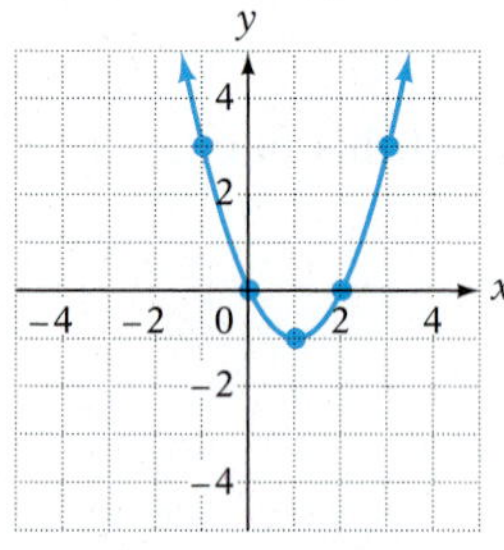

• **Graph the ordered-pair solutions. Draw a parabola through the points.**

YOU TRY IT • 1

Graph $y = x^2 + 2$.

Your solution

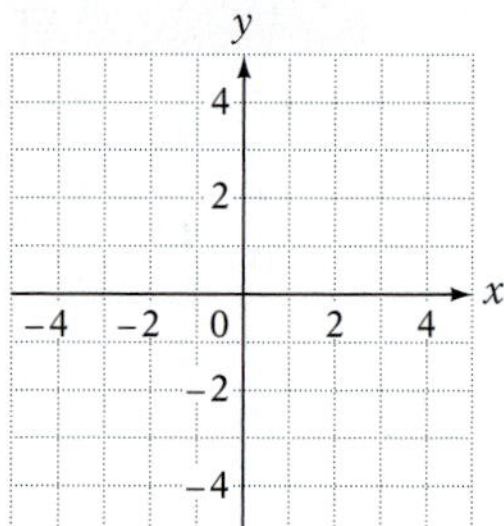

EXAMPLE • 2

Find the x- and y-intercepts of the graph of $y = x^2 - 2x - 5$.

Solution

To find the x-intercepts, let $y = 0$ and solve for x. This gives the equation $0 = x^2 - 2x - 5$, which is not factorable over the integers. Use the quadratic formula.

$$
\begin{aligned}
x &= \frac{-b \pm \sqrt{b^2 - 4ac}}{2a} \\
&= \frac{-(-2) \pm \sqrt{(-2)^2 - 4(1)(-5)}}{2(1)} \\
&= \frac{2 \pm \sqrt{24}}{2} \\
&= \frac{2 \pm 2\sqrt{6}}{2} \\
&= 1 \pm \sqrt{6}
\end{aligned}
$$

• **$a = 1, b = -2, c = -5$**

The x-intercepts are $(1 - \sqrt{6}, 0)$ and $(1 + \sqrt{6}, 0)$.

To find the y-intercept, let $x = 0$ and solve for y.

$$
\begin{aligned}
y &= x^2 - 2x - 5 \\
&= 0^2 - 2(0) - 5 \\
&= -5
\end{aligned}
$$

• **Replace x by 0.**

The y-intercept is $(0, -5)$.

YOU TRY IT • 2

Find the x- and y-intercepts of the graph of $f(x) = x^2 - 6x + 9$.

Your solution

Solutions on p. S37

16.4 EXERCISES

OBJECTIVE A To graph a quadratic equation of the form $y = ax^2 + bx + c$

For Exercises 1 to 4, determine whether the graph of the equation opens up or down.

1. $y = -\frac{1}{3}x^2$

2. $y = x^2 - 2x - 3$

3. $y = 2x^2 - 4$

4. $f(x) = 3 - 2x - x^2$

For Exercises 5 to 10, evaluate the function for the given value of x.

5. $f(x) = x^2 - 2x + 1; x = 3$

6. $f(x) = 2x^2 + x - 1; x = -2$

7. $f(x) = 4 - x^2; x = -3$

8. $f(x) = x^2 + 6x + 9; x = -3$

9. $f(x) = -x^2 + 5x - 6; x = -4$

10. $f(x) = -2x^2 + 2x - 1; x = -3$

For Exercises 11 to 25, graph.

11. $y = x^2$

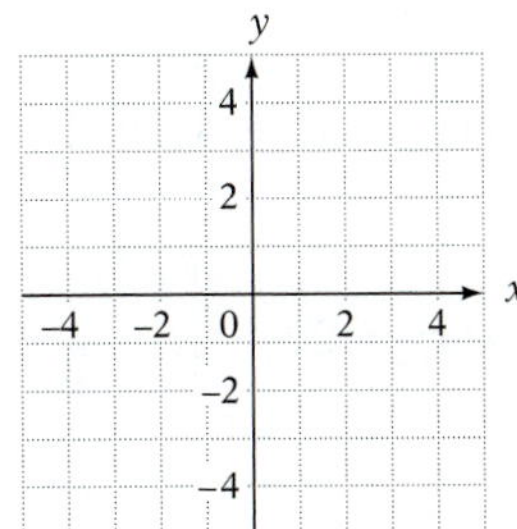

12. $y = -x^2$

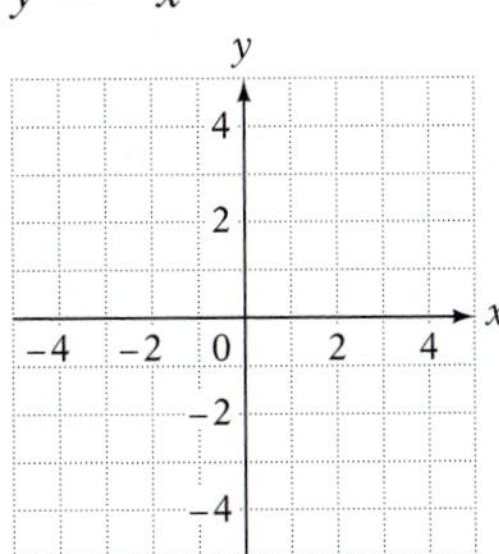

13. $y = -x^2 + 1$

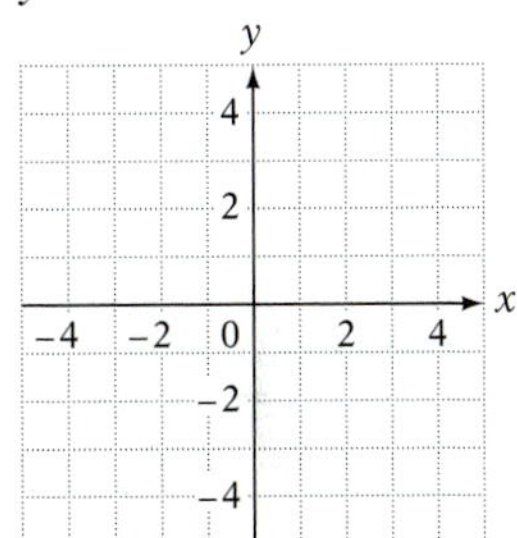

14. $y = x^2 - 1$

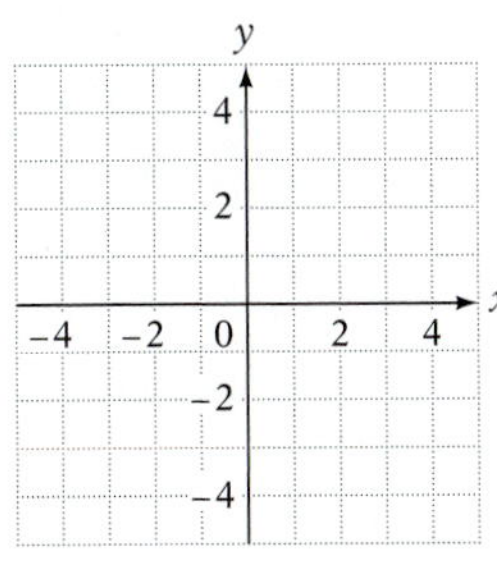

15. $f(x) = 2x^2$

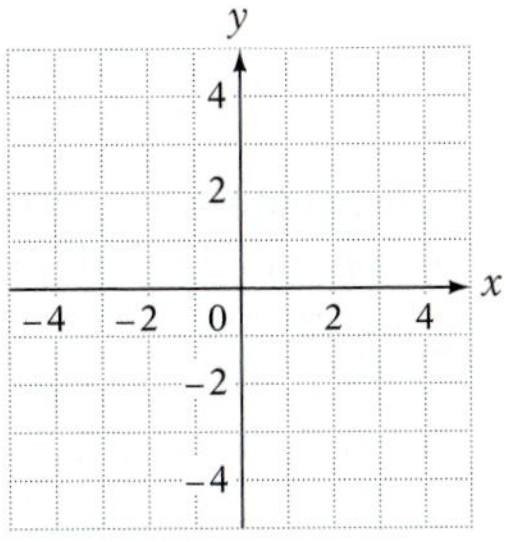

16. $f(x) = \frac{1}{2}x^2$

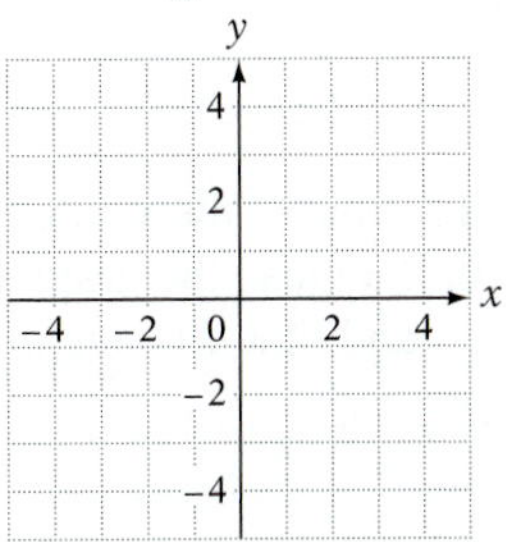

17. $f(x) = -\frac{1}{2}x^2 + 1$

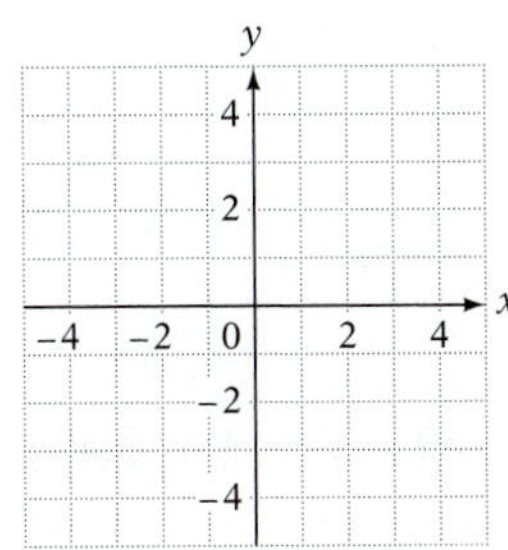

18. $f(x) = 2x^2 - 1$

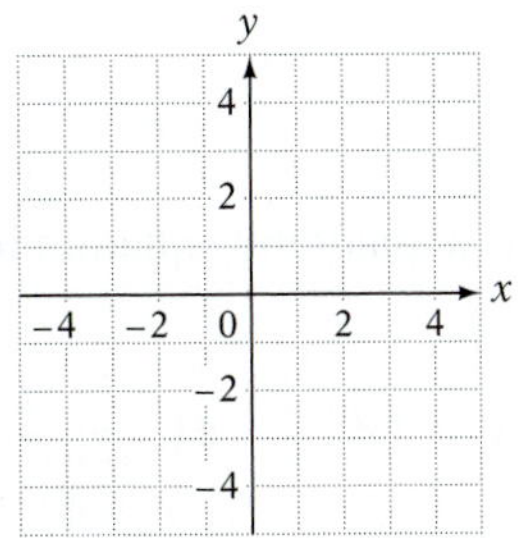

19. $y = x^2 - 4x$

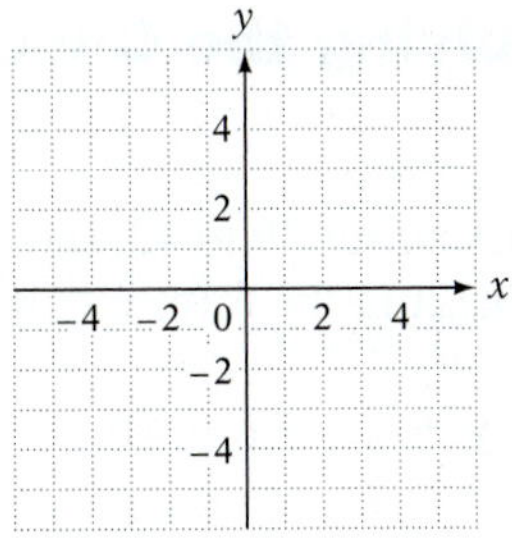

20. $y = x^2 + 4x$

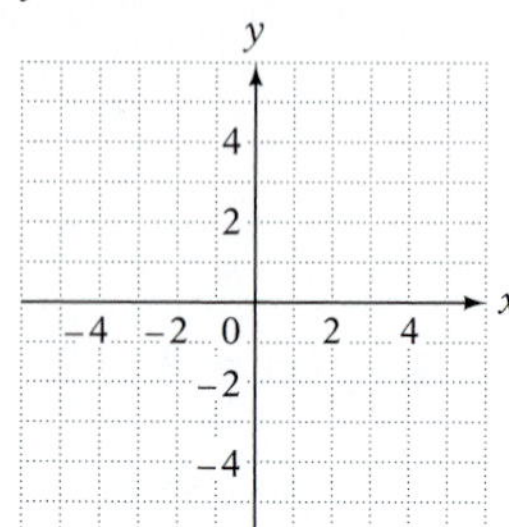

21. $y = x^2 - 2x + 3$

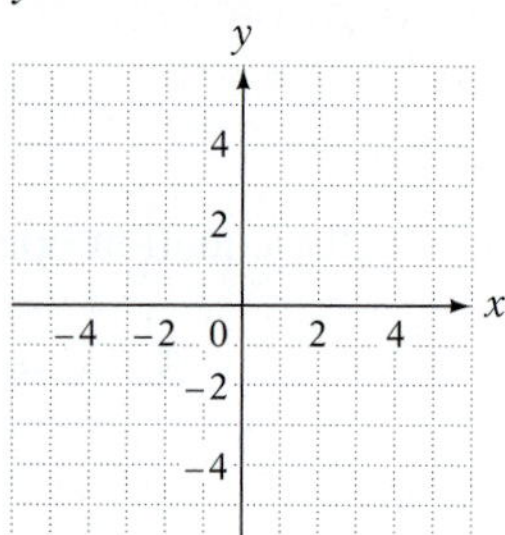

22. $y = x^2 - 4x + 2$

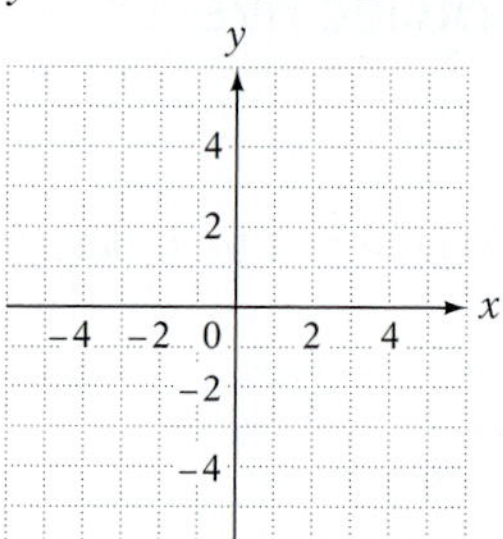

23. $y = -x^2 + 2x + 3$

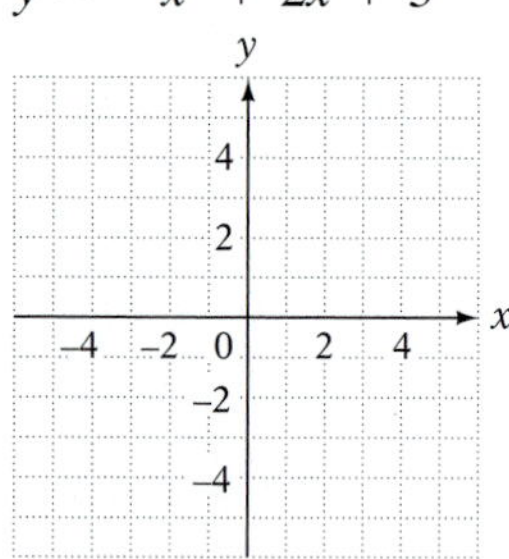

24. $y = -x^2 - 2x + 3$

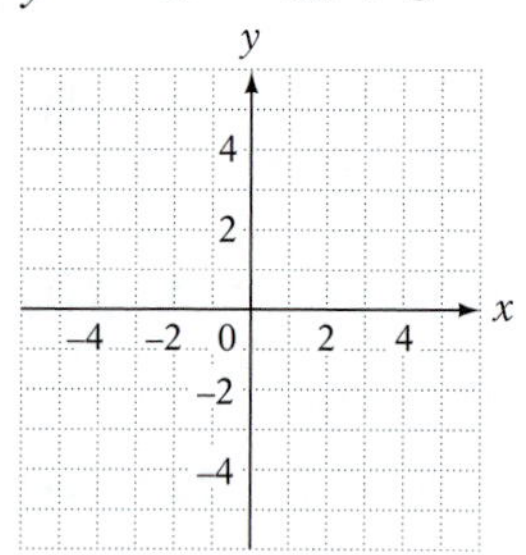

25. $y = -x^2 + 4x - 4$

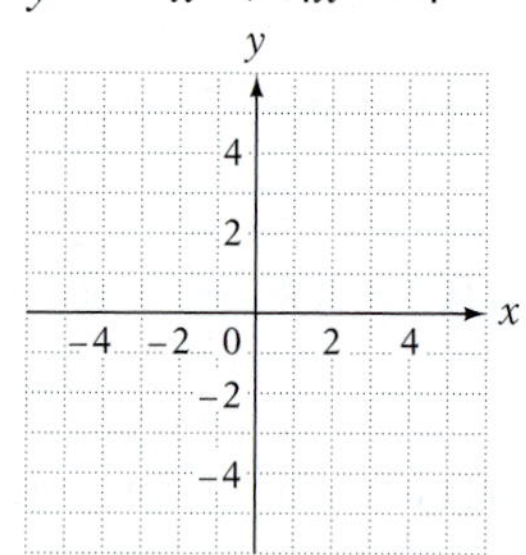

For Exercises 26 to 37, determine the x- and y-intercepts.

26. $y = x^2 - 5x + 6$

27. $y = x^2 + 5x - 6$

28. $f(x) = 9 - x^2$

29. $f(x) = x^2 + 12x + 36$

30. $y = x^2 + 2x - 6$

31. $f(x) = x^2 + 4x - 2$

32. $y = x^2 + 2x + 3$

33. $y = x^2 - x + 1$

34. $f(x) = 2x^2 - x - 3$

35. $f(x) = 2x^2 - 13x + 15$

36. $y = 4 - x - x^2$

37. $y = 2 - 3x - 3x^2$

38. **a.** What is the y-intercept of the parabola with equation $y = ax^2 + bx + c$?
b. Suppose the graph of $y = ax^2 + bx + c$ is a parabola with only one x-intercept, and that a is negative. Is c positive or negative?

Applying the Concepts

For Exercises 39 to 42, show that the equation is a quadratic equation in two variables by writing it in the form $y = ax^2 + bx + c$.

39. $y + 1 = (x - 4)^2$

40. $y - 2 = 3(x + 1)^2$

41. $y - 4 = 2(x - 3)^2$

42. $y + 3 = 3(x - 1)^2$

SECTION

16.5 Application Problems

OBJECTIVE A To solve application problems 

The application problems in this section are varieties of those problems solved earlier in the text. Each of the strategies for the problems in this section will result in a quadratic equation.

© Rudi Von Briel/PhotoEdit

HOW TO 1 In 5 h, two campers rowed 12 mi down a stream and then rowed back to their campsite. The rate of the stream's current was 1 mph. Find the rate at which the campers rowed.

Strategy for Solving an Application Problem

1. Determine the type of problem. For example, is it a distance-rate problem, a geometry problem, or a work problem?

The problem is a distance-rate problem.

2. Choose a variable to represent the unknown quantity. Write numerical or variable expressions for all the remaining quantities. These results can be recorded in a table.

The unknown rate of the campers: r

	Distance	÷	*Rate*	=	*Time*
Downstream	12	÷	$r+1$	=	$\frac{12}{r+1}$
Upstream	12	÷	$r-1$	=	$\frac{12}{r-1}$

3. Determine how the quantities are related.

The total time of the trip was 5 h.

$$\frac{12}{r+1}+\frac{12}{r-1}=5$$

$$(r+1)(r-1)\left(\frac{12}{r+1}+\frac{12}{r-1}\right)=(r+1)(r-1)5$$

$$(r-1)12+(r+1)12=(r^2-1)5$$

$$12r-12+12r+12=5r^2-5$$

$$24r=5r^2-5$$

$$0=5r^2-24r-5$$

$$0=(5r+1)(r-5)$$

$$5r+1=0 \qquad r-5=0$$

$$5r=-1 \qquad r=5$$

$$r=-\frac{1}{5}$$

The rowing rate was 5 mph.

Take Note

The time going downstream plus the time going upstream is equal to the time of the entire trip.

Take Note

The solution $r=-\frac{1}{5}$ is not possible because the rate cannot be a negative number.

EXAMPLE • 1

Working together, a painter and the painter's apprentice can paint a room in 2 h. Working alone, the apprentice requires 3 more hours to paint the room than the painter requires working alone. How long does it take the painter, working alone, to paint the room?

Strategy

- This is a work problem.
- Time for the painter to paint the room: t
  Time for the apprentice to paint the room: $t + 3$

	Rate	*Time*	*Part*
Painter	$\frac{1}{t}$	2	$\frac{2}{t}$
Apprentice	$\frac{1}{t+3}$	2	$\frac{2}{t+3}$

- The sum of the parts of the task completed must equal 1.

Solution

$$\frac{2}{t} + \frac{2}{t+3} = 1$$

$$t(t+3)\left(\frac{2}{t} + \frac{2}{t+3}\right) = t(t+3) \cdot 1$$

$$(t+3)2 + t(2) = t(t+3)$$

$$2t + 6 + 2t = t^2 + 3t$$

$$4t + 6 = t^2 + 3t$$

$$0 = t^2 - t - 6$$

$$0 = (t-3)(t+2)$$

$t - 3 = 0 \qquad t + 2 = 0$

$t = 3 \qquad t = -2$

The solution $t = -2$ is not possible.

The time is 3 h.

YOU TRY IT • 1

The length of a rectangle is 2 m more than the width. The area is 15 m^2. Find the width.

Your strategy

Your solution

Solution on p. S38

16.5 EXERCISES

OBJECTIVE A To solve application problems

1. **Geometry** The height of a triangle is 2 m more than twice the length of the base. The area of the triangle is 20 m^2. Find the height of the triangle and the length of the base.

In the News

Long Board for the Longhorns

The University of Texas Longhorns have replaced their stadium's old 2800-square-foot scoreboard with a new, state-of-the-art, 7370-square-foot scoreboard designed and built by local business Daktronics, Inc.

Sources: Business Wire, www.engadget.com

2. **Geometry** The length of a rectangle is 4 ft more than twice the width. The area of the rectangle is 160 ft^2. Find the length and width of the rectangle.

3. **Sports** Read the article at the right. The Longhorns' old scoreboard was a rectangle with a length 30 ft greater than its width. Find the length and width of the old scoreboard.

4. **Sports** The area of the batter's box on a major-league baseball field is 24 ft^2. The length of the batter's box is 2 ft more than the width. Find the length and width of the batter's box.

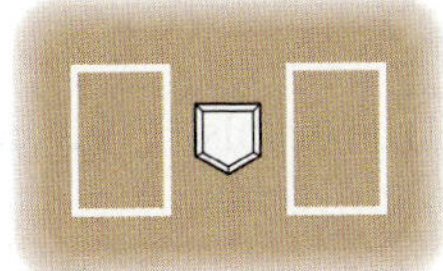

5. **Sports** The length of the batter's box on a softball field is 1 ft less than twice the width. The area of the batter's box is 15 ft^2. Find the length and width of the batter's box.

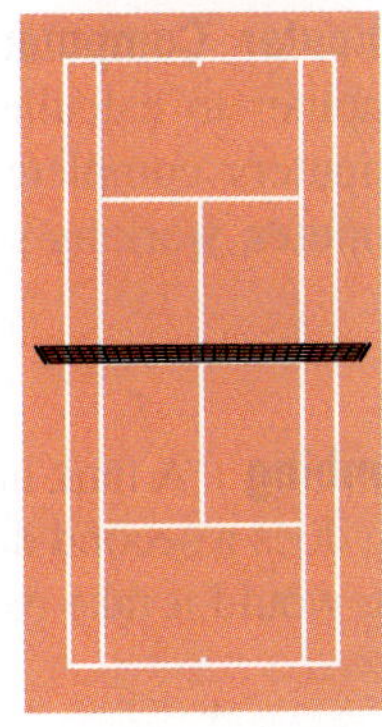

6. **Sports** The length of a swimming pool is twice the width. The area of the pool is 5000 ft^2. Find the length and width of the pool.

7. **Sports** The length of a singles tennis court is 24 ft more than twice the width. The area of the tennis court is 2106 ft^2. Find the length and width of the court.

8. **Sports** The hang time of a football that is kicked on the opening kickoff is given by $s = -16t^2 + 88t + 4$, where s is the height of the football t seconds after leaving the kicker's foot. What is the hang time of a kickoff that hits the ground without being caught? Round to the nearest tenth.

9. **Manufacturing** A square piece of cardboard is to be formed into a box to transport pizzas. The box is formed by cutting 2-inch square corners from the cardboard and folding them up as shown in the figure at the right. If the volume of the box is 512 in^3, what are the dimensions of the cardboard?

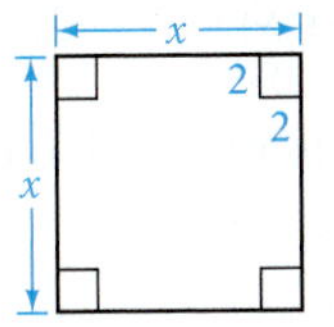

10. **Landscaping** The perimeter of a rectangular garden is 54 ft. The area of the garden is 180 ft^2. Find the length and width of the garden.

11. **Food Preparation** The radius of a large pizza is 1 in. less than twice the radius of a small pizza. The difference between the areas of the two pizzas is 33π in^2. Find the radius of the large pizza.

12. **Botany** Botanists have determined that some species of weeds grow in a circular pattern. For one such weed, the area A, in square meters, can be approximated by $A(t) = 0.005\pi t^2$, where t is the time in days after the growth of the weed first can be observed. How many days after the growth is first observed will this weed cover an area of 10 m²? Round to the nearest whole number.

Hans Pfletschinger/ Peter Arnold Inc.

13. **Geometry** The hypotenuse of a right triangle is $\sqrt{13}$ cm. One leg is 1 cm shorter than twice the length of the other leg. Find the lengths of the legs of the right triangle.

For Exercises 14 and 15, answer *without* writing and solving an equation. Use the following situation: A small pipe takes 12 min longer to fill a tank than does a larger pipe. Working together, the pipes can fill the tank in 4 min.

14. True or false? The amount of time needed for the larger pipe to fill the tank is less than 4 min.

15. True or false? The amount of time needed for the small pipe to fill the tank is greater than 16 min.

16. **Computer Computations** One computer takes 21 min longer than a second computer to calculate the value of a complex equation. Working together, these computers complete the calculation in 10 min. How long would it take each computer, working separately, to calculate the value?

17. **Plumbing** A tank has two drains. One drain takes 16 min longer to empty the tank than does a second drain. With both drains open, the tank is emptied in 6 min. How long would it take each drain, working alone, to empty the tank?

18. **Transportation** Using one engine of a ferryboat, it takes 6 h longer to cross a channel than it does using a second engine alone. With both engines operating, the ferryboat can make the crossing in 4 h. How long would it take each engine, working alone, to power the ferryboat across the channel?

© Bonnie Kamin/PhotoEdit

19. **Masonry** An apprentice mason takes 8 h longer to build a small fireplace than an experienced mason. Working together, they can build the fireplace in 3 h. How long would it take each mason, working alone, to complete the fireplace?

20. **Travel** It took a small plane 2 h longer to fly 375 mi against the wind than to fly the same distance with the wind. The rate of the wind was 25 mph. Find the rate of the plane in calm air.

21. **Travel** It took a motorboat 1 h longer to travel 36 mi against the current than to go 36 mi with the current. The rate of the current was 3 mph. Find the rate of the boat in calm water.

22. **Physics** The kinetic energy of a moving body is given by $E = \frac{1}{2}mv^2$, where E is the kinetic energy, m is the mass, and v is the velocity in meters per second. What is the velocity of a moving body whose mass is 5 kg and whose kinetic energy is 250 newton-meters?

> **In the News**
> **Boomers Turn 65**
> By the time the last baby boomer turns 65, the population of people aged 65 and older will have more than doubled, from 35 million to 71 million.
> *Source:* Census Bureau

23. **Demography** See the news clipping at the right. Approximate the year in which there will be 50 million people aged 65 and older in the United States. Use the equation $y = 0.03x^2 + 0.36x + 34.6$, where y is the population, in millions, in year x, where $x = 0$ corresponds to the year 2000.

> **In the News**
> **Alzheimer's Diagnoses Rising**
> As the population of senior citizens grows, so will the number of people diagnosed with Alzheimer's, the disease that afflicted former president Ronald Reagan for the last 10 years of his life.
> *Source:* The Alzheimer's Association

24. **Alzheimer's** See the news clipping at the right. Find the year in which 15 million Americans are expected to have Alzheimer's. Use the equation $y = 0.002x^2 + 0.05x + 2$, where y is the population, in millions, with Alzheimer's in year x, where $x = 0$ corresponds to the year 1980.

25. **Automotive Safety** The distance s, in feet, a car needs to come to a stop on a certain surface depends on the velocity v, in feet per second, of the car when the brakes are applied. The equation is given by $s = 0.0344v^2 - 0.758v$. What is the maximum velocity a car can have when the brakes are applied and stop within 150 ft?

> **In the News**
> **Spike in Internet Traffic**
> Alarm is growing over the increase on the Internet of video clips, movies, social networks, and multiplayer games, all of which require more bandwidth. One reaserch firm projects that user demand for the Internet will outpace capacity by 2011.
> *Source: New York Times*

26. **The Internet** See the news clipping at the right. Find the year in which consumer Internet traffic will reach 10 million terabytes. Use the equation $y = 0.27x^2 - 2.6x + 7.6$, where y is consumer Internet traffic, in millions of terabytes, in year x, where $x = 5$ corresponds to the year 2005.

Applying the Concepts

27. **Food Preparation** If a pizza with a diameter of 8 in. costs $10, what should be the cost of a pizza with a diameter of 16 in. if both pizzas cost the same amount per square inch?

28. **Geometry** A wire 8 ft long is cut into two pieces. A circle is formed from one piece, and a square is formed from the other. The total area of both figures is given by $A = \frac{1}{16}(8 - x)^2 + \frac{x^2}{4\pi}$. What is the length of each piece of wire if the total area is 4.5 ft^2?

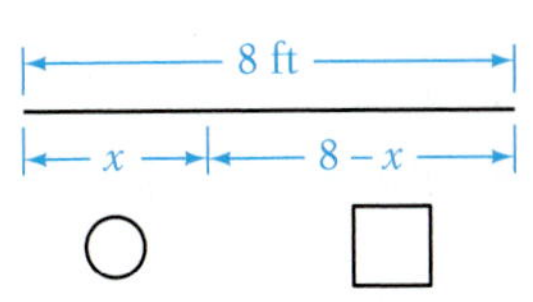

FOCUS ON PROBLEM SOLVING

Algebraic Manipulation and Graphing Techniques

Problem solving is often easier when we have both algebraic manipulation and graphing techniques at our disposal. Solving quadratic equations and graphing quadratic equations in two variables are used here to solve problems involving profit.

A company's **revenue** R is the total amount of money the company earned by selling its products. The **cost** C is the total amount of money the company spent to manufacture and sell its products. A company's **profit** P is the difference between the revenue and the cost: $P = R - C$. A company's revenue and cost may be represented by equations.

A company manufactures and sells woodstoves. The total weekly cost, in dollars, to produce n woodstoves is $C = 30n + 2000$. Write a variable expression for the company's weekly profit if the revenue, in dollars, obtained from selling all n woodstoves is $R = 150n - 0.4n^2$.

$$P = R - C$$
$$P = 150n - 0.4n^2 - (30n + 2000)$$
$$P = -0.4n^2 + 120n - 2000$$

• Replace R by $150n - 0.4n^2$ and C by $30n + 2000$. Then simplify.

How many woodstoves must the company manufacture and sell in order to make a profit of \$6000 a week?

$$P = -0.4n^2 + 120n - 2000$$
$$6000 = -0.4n^2 + 120n - 2000$$ • Substitute 6000 for P.
$$0 = -0.4n^2 + 120n - 8000$$ • Write the equation in standard form.
$$0 = n^2 - 300n + 20{,}000$$ • Divide each side of the equation by -0.4.
$$0 = (n - 100)(n - 200)$$ • Factor.

$n - 100 = 0 \qquad n - 200 = 0$ • Solve for n.
$n = 100 \qquad n = 200$

The company will make a weekly profit of \$6000 if either 100 or 200 woodstoves are manufactured and sold.

The graph of $P = -0.4n^2 + 120n - 2000$ is shown at the right. Note that when $P = 6000$, the values of n are 100 and 200.

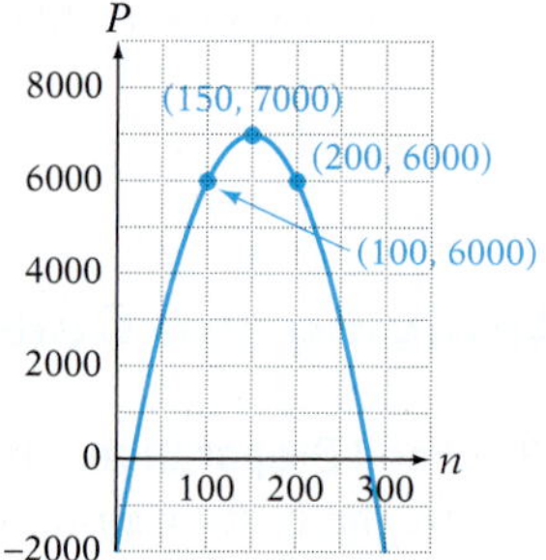

Also note that the coordinates of the highest point on the graph are (150, 7000). This means that the company makes a *maximum* profit of \$7000 per week when 150 woodstoves are manufactured and sold.

1. The total cost, in dollars, for a company to produce and sell n guitars per week is $C = 240n + 1200$. The company's revenue, in dollars, from selling all n guitars is $R = 400n - 2n^2$.
   a. How many guitars must the company produce and sell each week in order to make a weekly profit of \$1200?
   b. Graph the profit equation. What is the maximum weekly profit that the company can make?

Review Topics

CHAPTER R

OBJECTIVES

SECTION R.1

A To add and subtract fractions
B To multiply and divide fractions

SECTION R.2

A To add and subtract integers
B To multiply and divide integers

SECTION R.3

A To add and subtract rational numbers
B To multiply and divide rational numbers
C To evaluate exponential expressions
D To use the Order of Operations Agreement to simplify expressions

SECTION R.4

A To solve a first-degree equation in one variable

SECTION

R.1 Fractions

OBJECTIVE A To add and subtract fractions 

2.3A Add fractions

When we add two fractions with the same denominator, we add the numerators. The denominator remains the same. Then we write the answer in simplest form.

$$\frac{2}{9} + \frac{4}{9} = \frac{2+4}{9} = \frac{6}{9} = \frac{2}{3}$$

Addition of Fractions

To add fractions with the same denominator, add the numerators and place the sum over the common denominator.

$$\frac{a}{b} + \frac{c}{b} = \frac{a+c}{b}, \quad \text{where} \quad b \neq 0$$

2.1A Find the least common multiple (LCM)

Before two fractions can be added, the fractions must have the same denominator. To add fractions with different denominators, first rewrite the fractions as equivalent fractions with a common denominator. The common denominator is the least common multiple (LCM) of the denominators of the fractions. The LCM of denominators is sometimes called the least common denominator (LCD).

HOW TO • 1 Add: $\frac{2}{3} + \frac{1}{4}$

$$\frac{2}{3} + \frac{1}{4} = \frac{8}{12} + \frac{3}{12}$$

- **The common denominator is the LCM of 3 and 4, which is 12. Write the fractions as equivalent fractions with the common denominator.**

$$= \frac{8+3}{12} = \frac{11}{12}$$

- **Add the fractions. $\frac{11}{12}$ is in simplest form.**

2.3B Subtract fractions

When we subtract two fractions with the same denominator, we subtract the numerators. The denominator remains the same. Then we write the answer in simplest form.

$$\frac{7}{9} - \frac{4}{9} = \frac{7-4}{9} = \frac{3}{9} = \frac{1}{3}$$

Subtraction of Fractions

To subtract fractions with the same denominator, subtract the numerators and place the difference over the common denominator.

$$\frac{a}{b} - \frac{c}{b} = \frac{a-c}{b}, \quad \text{where} \quad b \neq 0$$

To subtract fractions with different denominators, first rewrite the fractions as equivalent fractions with a common denominator. As with addition of fractions, the common denominator is the least common multiple (LCM) of the denominators of the fractions.

HOW TO • 2 Subtract: $\frac{2}{3} - \frac{5}{8}$

$\frac{2}{3} - \frac{5}{8} = \frac{16}{24} - \frac{15}{24}$

- The common denominator is the LCM of 3 and 8, which is 24. Write the fractions as equivalent fractions with the common denominator.

$= \frac{16 - 15}{24} = \frac{1}{24}$

- Subtract the fractions. $\frac{1}{24}$ is in simplest form.

EXAMPLE • 1

Add: $\frac{3}{8} + \frac{7}{12}$

Solution

$\frac{3}{8} + \frac{7}{12} = \frac{9}{24} + \frac{14}{24}$

- The LCM of 8 and 12 is 24.

$= \frac{23}{24}$

- Add the numerators. Place the sum over the common denominator.

YOU TRY IT • 1

Add: $\frac{5}{12} + \frac{9}{16}$

Your solution

EXAMPLE • 2

Add: $\frac{2}{3} + \frac{3}{5} + \frac{5}{6}$

Solution

$\frac{2}{3} + \frac{3}{5} + \frac{5}{6}$

- The LCM of 3, 5, and 6 is 30.

$= \frac{20}{30} + \frac{18}{30} + \frac{25}{30}$

- Write equivalent fractions using the LCM.

$= \frac{63}{30}$

- Add the numerators. Place the sum over the common denominator.

$= \frac{21}{10} = 2\frac{1}{10}$

- Simplify.

YOU TRY IT • 2

Add: $\frac{3}{4} + \frac{4}{5} + \frac{5}{8}$

Your solution

EXAMPLE • 3

Subtract: $\frac{11}{16} - \frac{5}{12}$

Solution

$\frac{11}{16} - \frac{5}{12} = \frac{33}{48} - \frac{20}{48}$

- The LCM of 16 and 12 is 48.

$= \frac{13}{48}$

- Subtract the numerators. Place the difference over the common denominator.

YOU TRY IT • 3

Subtract: $\frac{5}{6} - \frac{1}{4}$

Your solution

Solutions on p. S38

OBJECTIVE B To multiply and divide fractions 

To multiply two fractions, multiply the numerators and multiply the denominators.

2.4A Multiply fractions

Multiplication of Fractions

The product of two fractions is the product of the numerators over the product of the denominators.

$$\frac{a}{b} \cdot \frac{c}{d} = \frac{ac}{bd}, \quad \text{where} \quad b \neq 0 \quad \text{and} \quad d \neq 0$$

Note that fractions do not need to have the same denominator in order to be multiplied.

HOW TO • 3 Multiply: $\frac{2}{3} \cdot \frac{4}{5}$

$\frac{2}{3} \cdot \frac{4}{5} = \frac{2 \cdot 4}{3 \cdot 5}$ • Multiply the numerators. Multiply the denominators.

$= \frac{8}{15}$

After multiplying two fractions, write the product in simplest form, as illustrated in the example below.

HOW TO • 4 Multiply: $\frac{3}{4} \cdot \frac{14}{15}$

$\frac{3}{4} \cdot \frac{14}{15} = \frac{3 \cdot 14}{4 \cdot 15}$ • Multiply the numerators. Multiply the denominators.

$= \frac{3 \cdot 2 \cdot 7}{2 \cdot 2 \cdot 3 \cdot 5}$ • Express the fraction in simplest form by first writing the prime factorization of each number.

$= \frac{\overset{1}{\cancel{3}} \cdot \overset{1}{\cancel{2}} \cdot 7}{\underset{1}{\cancel{2}} \cdot 2 \cdot \underset{1}{\cancel{3}} \cdot 5}$ • Divide by the common factors.

$= \frac{7}{10}$ • Write the product in simplest form.

Division is defined as multiplication by the reciprocal. Fractions are divided by applying this definition.

2.4B Divide fractions

Division of Fractions

To divide two fractions, multiply the first fraction by the reciprocal of the second fraction.

$$\frac{a}{b} \div \frac{c}{d} = \frac{a}{b} \cdot \frac{d}{c}, \quad \text{where} \quad b \neq 0, \quad c \neq 0, \quad \text{and} \quad d \neq 0$$

HOW TO • 5 Divide: $\frac{2}{3} \div \frac{3}{4}$

$$\frac{2}{3} \div \frac{3}{4} = \frac{2}{3} \cdot \frac{4}{3}$$

• Multiply the first fraction by the reciprocal of the second fraction.

$$= \frac{2 \cdot 4}{3 \cdot 3}$$

• Multiply the numerators. Multiply the denominators.

$$= \frac{2 \cdot 2 \cdot 2}{3 \cdot 3} = \frac{8}{9}$$

• There are no common factors in the numerator and denominator.

EXAMPLE • 4

Multiply: $\frac{4}{15} \cdot \frac{5}{28}$

Solution

$$\frac{4}{15} \cdot \frac{5}{28} = \frac{4 \cdot 5}{15 \cdot 28}$$

• Multiply the numerators. Multiply the denominators.

$$= \frac{2 \cdot 2 \cdot 5}{3 \cdot 5 \cdot 2 \cdot 2 \cdot 7}$$

• Write the prime factorization of each number.

$$= \frac{\overset{1}{\cancel{2}} \cdot \overset{1}{\cancel{2}} \cdot \overset{1}{\cancel{5}}}{3 \cdot \underset{1}{\cancel{5}} \cdot \underset{1}{\cancel{2}} \cdot \underset{1}{\cancel{2}} \cdot 7}$$

• Divide by the common factors.

$$= \frac{1}{21}$$

• Write the product in simplest form.

YOU TRY IT • 4

Multiply: $\frac{4}{21} \cdot \frac{7}{44}$

Your solution

EXAMPLE • 5

Multiply: $\frac{3}{16} \cdot 4$

Solution

$$\frac{3}{16} \cdot 4 = \frac{3}{16} \cdot \frac{4}{1}$$

• Write 4 as $\frac{4}{1}$.

$$= \frac{3 \cdot 4}{16 \cdot 1}$$

• Multiply the fractions.

$$= \frac{3 \cdot \overset{1}{\cancel{2}} \cdot \overset{1}{\cancel{2}}}{\underset{1}{\cancel{2}} \cdot \underset{1}{\cancel{2}} \cdot 2 \cdot 2 \cdot 1}$$

• Divide by the common factors.

$$= \frac{3}{4}$$

• Write the product in simplest form.

YOU TRY IT • 5

Multiply: $\frac{2}{15} \cdot 5$

Your solution

EXAMPLE • 6

Divide: $\frac{3}{4} \div \frac{9}{10}$

Solution

$$\frac{3}{4} \div \frac{9}{10} = \frac{3}{4} \cdot \frac{10}{9}$$

• Multiply the first fraction by the reciprocal of the second fraction.

$$= \frac{3 \cdot 10}{4 \cdot 9}$$

$$= \frac{\overset{1}{\cancel{3}} \cdot \overset{1}{\cancel{2}} \cdot 5}{\underset{1}{\cancel{2}} \cdot 2 \cdot \underset{1}{\cancel{3}} \cdot 3}$$

$$= \frac{5}{6}$$

YOU TRY IT • 6

Divide: $\frac{1}{6} \div \frac{4}{9}$

Your solution

Solutions on p. S38

R.1 EXERCISES

OBJECTIVE A To add and subtract fractions

For Exercises 1 to 36, add or subtract.

1. $\frac{2}{7} + \frac{1}{7}$

2. $\frac{3}{11} + \frac{5}{11}$

3. $\frac{1}{2} + \frac{1}{2}$

4. $\frac{1}{3} + \frac{2}{3}$

5. $\frac{3}{8} + \frac{7}{8} + \frac{1}{8}$

6. $\frac{5}{12} + \frac{7}{12} + \frac{1}{12}$

7. $\frac{1}{2} + \frac{2}{3}$

8. $\frac{2}{3} + \frac{1}{4}$

9. $\frac{3}{14} + \frac{5}{7}$

10. $\frac{7}{10} + \frac{3}{5}$

11. $\frac{8}{15} + \frac{7}{20}$

12. $\frac{1}{6} + \frac{7}{9}$

13. $\frac{3}{20} + \frac{7}{30}$

14. $\frac{5}{12} + \frac{7}{30}$

15. $\frac{1}{3} + \frac{5}{6} + \frac{7}{9}$

16. $\frac{2}{3} + \frac{5}{6} + \frac{7}{12}$

17. $\frac{2}{3} + \frac{1}{5} + \frac{7}{12}$

18. $\frac{3}{4} + \frac{4}{5} + \frac{7}{12}$

19. $\frac{11}{12} - \frac{7}{12}$

20. $\frac{13}{15} - \frac{4}{15}$

21. $\frac{3}{4} - \frac{1}{8}$

22. $\frac{2}{3} - \frac{1}{6}$

23. $\frac{5}{9} - \frac{4}{15}$

24. $\frac{11}{12} - \frac{2}{3}$

25. $\frac{9}{20} - \frac{7}{20}$

26. $\frac{48}{55} - \frac{13}{55}$

27. $\frac{11}{24} - \frac{5}{24}$

28. $\frac{23}{30} - \frac{13}{30}$

29. $\frac{5}{7} - \frac{3}{14}$

30. $\frac{5}{9} - \frac{7}{15}$

31. $\frac{8}{15} - \frac{7}{20}$

32. $\frac{7}{9} - \frac{1}{6}$

33. $\frac{5}{6} - \frac{3}{4}$

34. $\frac{4}{5} - \frac{2}{3}$

35. $\frac{7}{8} - \frac{2}{3}$

36. $\frac{5}{12} - \frac{1}{3}$

OBJECTIVE B To multiply and divide fractions

For Exercises 37 to 72, multiply or divide.

37. $\frac{2}{3} \cdot \frac{7}{8}$

38. $\frac{1}{2} \cdot \frac{2}{3}$

39. $\frac{15}{16} \cdot \frac{7}{15}$

40. $\frac{3}{8} \cdot \frac{6}{7}$

41. $\frac{2}{5} \cdot \frac{5}{6}$

42. $\frac{11}{12} \cdot \frac{3}{5}$

43. $\frac{3}{5} \cdot \frac{10}{11}$

44. $\frac{6}{7} \cdot \frac{14}{15}$

45. $\frac{8}{9} \cdot \frac{27}{4}$

46. $\frac{3}{5} \cdot \frac{3}{10}$

47. $\frac{3}{8} \cdot \frac{5}{12}$

48. $\frac{3}{2} \cdot \frac{4}{9}$

49. $\frac{7}{8} \cdot \frac{3}{14}$

50. $\frac{5}{12} \cdot \frac{6}{7}$

51. $\frac{5}{6} \cdot \frac{4}{15}$

52. $\frac{5}{7} \cdot \frac{14}{15}$

53. $\frac{2}{3} \cdot \frac{5}{4} \cdot \frac{1}{9}$

54. $\frac{3}{4} \cdot \frac{5}{6} \cdot \frac{8}{9}$

55. $\frac{2}{3} \cdot 6$

56. $14 \cdot \frac{5}{7}$

57. $\frac{3}{7} \div \frac{3}{2}$

58. $\frac{3}{7} \div \frac{3}{7}$

59. $0 \div \frac{1}{2}$

60. $\frac{5}{24} \div \frac{15}{36}$

61. $\frac{2}{15} \div \frac{3}{5}$

62. $\frac{1}{9} \div \frac{2}{3}$

63. $\frac{2}{5} \div \frac{4}{7}$

64. $\frac{3}{8} \div \frac{5}{12}$

65. $\frac{1}{2} \div \frac{1}{4}$

66. $\frac{1}{3} \div \frac{1}{9}$

67. $\frac{4}{15} \div \frac{2}{5}$

68. $\frac{7}{15} \div \frac{14}{5}$

69. $4 \div \frac{2}{3}$

70. $\frac{2}{3} \div 4$

71. $\frac{3}{2} \div 3$

72. $3 \div \frac{3}{2}$

SECTION

R.2 Integers

OBJECTIVE A To add and subtract integers 

The rule for adding two integers depends on whether the signs of the integers are the same or different.

3.2A Add integers

Rule for Adding Two Integers

To add two integers with the same sign, add the absolute values of the numbers. Then attach the sign of the addends.

To add two integers with different signs, find the absolute values of the numbers. Subtract the smaller absolute value from the larger absolute value. Then attach the sign of the addend with the larger absolute value.

HOW TO 1 Add: $(-5) + (-11)$

$(-5) + (-11) = -16$

• The signs of the addends are the same. Add the absolute values of the numbers. $|-5| = 5, |-11| = 11, 5 + 11 = 16$ Attach the sign of the addends. (Both addends are negative. The sum is negative.)

HOW TO 2 Add: $-16 + (-32)$

$-16 + (-32) = -48$

• The signs of the addends are the same. Add the absolute values of the numbers. Attach the sign of the addends.

HOW TO 3 Add: $7 + (-20)$

$7 + (-20) = -13$

• The signs of the addends are different. Find the absolute values of the numbers. $|7| = 7, |-20| = 20$ Subtract the smaller absolute value from the larger absolute value. $20 - 7 = 13$ Attach the sign of the number with the larger absolute value. ($|-20| > |7|$. Attach the negative sign.)

HOW TO 4 Add: $82 + (-136)$

$82 + (-136) = -54$

• The signs are different. Find the difference between the absolute values of the numbers. $136 - 82 = 54$ Attach the sign of the number with the larger absolute value.

Opposites are used to rewrite subtraction problems as related addition problems. Notice below that the subtraction of two whole numbers is the same as addition of the opposite number.

Subtraction		*Addition of the Opposite*	
$9 - 5$	$=$	$9 + (-5)$	$= 4$
$7 - 4$	$=$	$7 + (-4)$	$= 3$
$8 - 3$	$=$	$8 + (-3)$	$= 5$

Subtraction of integers can be written as the addition of the opposite number. To subtract two integers, rewrite the subtraction expression as the first number plus the opposite of the second number.

3.2B Subtract integers

Rule for Subtracting Two Integers

To subtract two integers, add the opposite of the second integer to the first integer.

HOW TO • 5 Subtract: $(-14) - 62$

$(-14) - 62$

$= (-14) + (-62)$ • **Rewrite the subtraction operation as the first number plus the opposite of the second number. The opposite of 62 is −62.**

$= -76$ • **Add.**

HOW TO • 6 Subtract: $7 - (-5)$

$7 - (-5)$

$= 7 + 5$ • **Rewrite the subtraction operation as the first number plus the opposite of the second number. The opposite of −5 is 5.**

$= 12$ • **Add.**

HOW TO • 7 Subtract: $9 - 18$

$9 - 18$

$= 9 + (-18)$ • **Rewrite the subtraction operation as the first number plus the opposite of the second number. The opposite of 18 is −18.**

$= -9$ • **Add.**

When subtraction occurs several times in an expression, rewrite each subtraction as addition of the opposite, and then add.

HOW TO • 8 Subtract: $-23 - 7 - (-5)$

$-23 - 7 - (-5)$ • **Rewrite each subtraction as addition of the opposite.**

$= -23 + (-7) + 5$

$= -30 + 5$ • **Add.**

$= -25$

EXAMPLE • 1

Add.

a. $-6 + 17$ **b.** $-15 + (-8)$ **c.** $19 + (-26)$

Solution

a. $-6 + 17 = 11$ • The signs are different. Subtract the absolute values. The sum has the same sign as the number with the larger absolute value.

b. $-15 + (-8) = -23$ • The signs are the same. Add the absolute values. The sum has the same sign as the addends.

c. $19 + (-26) = -7$ • The signs are different. Subtract the absolute values. The sum has the same sign as the number with the larger absolute value.

YOU TRY IT • 1

Add.

a. $-25 + 13$ **b.** $-41 + (-60)$ **c.** $37 + (-9)$

Your solution

EXAMPLE • 2

Subtract.

a. $-9 - 18$ **b.** $-25 - (-13)$
c. $17 - (-40)$ **d.** $10 - (-3) - 7 + 6$

Solution

a. $-9 - 18 = -9 + (-18)$
$= -27$
• -9 minus 18 = -9 plus the opposite of 18. The opposite of 18 is -18.

b. $-25 - (-13) = -25 + 13$
$= -12$
• -25 minus -13 = -25 plus the opposite of -13. The opposite of -13 is 13.

c. $17 - (-40) = 17 + 40$
$= 57$
• 17 minus -40 = 17 plus the opposite of -40. The opposite of -40 is 40.

d. $10 - (-3) - 7 + 6$
$= 10 + 3 + (-7) + 6$ • Rewrite each subtraction as addition of the opposite.
$= 13 + (-7) + 6$ • Add the numbers.
$= 6 + 6$
$= 12$

YOU TRY IT • 2

Subtract.

a. $-27 - 18$ **b.** $-34 - (-90)$
c. $8 - 42$ **d.** $-12 + 9 - (-5) - 4$

Your solution

Solutions on pp. S38–S39

OBJECTIVE B To multiply and divide integers 

The rule for multiplying two integers depends on whether the signs of the integers are the same or different.

3.3A Multiply integers

Rule for Multiplying Two Integers

To multiply two integers with the same sign, multiply the absolute values of the numbers. The product is positive.

To multiply two integers with different signs, multiply the absolute values of the numbers. The product is negative.

HOW TO • 9 Multiply: $-3(-12)$

$-3(-12) = 36$

• The signs of the factors are the same (they are both negative). Multiply the absolute values of the factors. $|-3| = 3, |-12| = 12, 3(12) = 36$ The product is positive.

HOW TO • 10 Multiply: $-6(30)$

$-6(30) = -180$

• The signs of the factors are different. Multiply the absolute values of the factors. The product is negative.

The rule for dividing two integers depends on whether the signs of the integers are the same or different.

3.3B Divide integers

Rule for Dividing Two Integers

To divide two integers with the same sign, divide the absolute values of the numbers. The quotient is positive.

To divide two integers with different signs, divide the absolute values of the numbers. The quotient is negative.

HOW TO • 11 Divide: $(-24) \div (-8)$

$(-24) \div (-8) = 3$

• The signs of the numbers are the same. Divide the absolute values of the numbers. $|-24| = 24, |-8| = 8, 24 \div 8 = 3$ The quotient is positive.

HOW TO • 12 Divide: $(-44) \div 11$

$(-44) \div 11 = -4$

• The signs of the numbers are different. Divide the absolute values of the numbers. The quotient is negative.

EXAMPLE • 3

Multiply.

a. $(-5)(-12)$ **b.** $4(-9)$ **c.** $(-2)(3)(8)(-10)$

Solution

a. $(-5)(-12) = 60$

• **The signs are the same. The product is positive.**

b. $4(-9) = -36$

• **The signs are different. The product is negative.**

c. $(-2)(3)(8)(-10)$
$= -6(8)(-10)$
$= -48(-10)$
$= 480$

• **Multiply the first two numbers. Then multiply the product by the third number. Continue until all the numbers have been multiplied.**

YOU TRY IT • 3

Multiply.

a. $-25(4)$ **b.** $-4(-61)$ **c.** $(-4)(5)(-3)(-1)$

Your solution

EXAMPLE • 4

Divide.

a. $-18 \div (-18)$ **b.** $16 \div (-4)$ **c.** $\frac{-20}{-5}$

Solution

a. $-18 \div (-18) = 1$

• **The signs are the same. The quotient is positive.**

b. $16 \div (-4) = -4$

• **The signs are different. The quotient is negative.**

c. $\frac{-20}{-5} = 4$

• **The fraction bar can be read "divided by."**
$\frac{-20}{-5} = (-20) \div (-5)$
The signs are the same. The quotient is positive.

YOU TRY IT • 4

Divide.

a. $(-30) \div 6$ **b.** $(-50) \div (-25)$ **c.** $\frac{32}{-8}$

Your solution

Solutions on p. S39

R.2 EXERCISES

OBJECTIVE A To add and subtract integers

For Exercises 1 to 46, add or subtract.

1. $4 + (-9)$

2. $6 + (-7)$

3. $(-5) + (-12)$

4. $(-8) + (-11)$

5. $-5 + 8$

6. $-8 + 5$

7. $-14 + (-6)$

8. $-17 + (-3)$

9. $-6 + 6$

10. $-19 + 19$

11. $64 + (-43)$

12. $-78 + 51$

13. $8 - 15$

14. $7 - 10$

15. $-8 - 3$

16. $-10 - 5$

17. $7 - (-1)$

18. $4 - (-5)$

19. $-9 - (-9)$

20. $-13 - (-13)$

21. $-10 - 15$

22. $-8 - 7$

23. $(-11) - (-2)$

24. $(-8) - (-5)$

25. $6 - (-16)$

26. $4 - (-26)$

27. $(-12) - (-6)$

28. $-3 - (-17)$

29. $8 - (-8)$

30. $(-32) - 46$

31. $45 - 77$

32. $-82 - (-16)$

33. $0 + (-15)$

34. $-18 + 0$

35. $(-21) - (-7)$

36. $-13 - (-4)$

37. $5 - (-6)$

38. $12 - (-2)$

39. $6 - (-10)$

40. $13 - (-5)$

41. $7 + 3 - (-3)$

42. $(-9) - 8 + (-6)$

43. $-2 + (-5) - (-12)$

44. $-3 - 5 + 8 - 1$

45. $-4 + 6 - 9 - 2$

46. $7 - (-3) - 6 + 5$

OBJECTIVE B To multiply and divide integers

For Exercises 47 to 94, multiply or divide.

47. $-3 \cdot 7$

48. $-6 \cdot 8$

49. $-5(-7)$

50. $-9(-2)$

51. $4(-9)$

52. $3(-11)$

53. $-10(5)$

54. $-8(4)$

55. $(-7)(-7)$

56. $(-4)(-7)$

57. $(-9)(0)$

58. $-16(1)$

59. $15(4)$

60. $42(3)$

61. $-21(6)$

62. $-14(2)$

63. $(-3)(-27)$

64. $(-6)(-32)$

65. $8(-24)$

66. $7(-30)$

67. $-5 \cdot (17)$

68. $-6 \cdot (22)$

69. $-7(-14)$

70. $-4(-62)$

71. $2 \cdot (-8) \cdot 5$

72. $5 \cdot 6 \cdot (-1)$

73. $-2(-6)(-3)(4)$

74. $-1(4)(-9)(-2)$

75. $12 \div (-4)$

76. $18 \div (-6)$

77. $(-81) \div (-9)$

78. $(-48) \div (-6)$

79. $0 \div (-4)$

80. $-36 \div 1$

81. $77 \div (-7)$

82. $-50 \div (-10)$

83. $\dfrac{36}{-4}$

84. $\dfrac{40}{-8}$

85. $\dfrac{-66}{-3}$

86. $\dfrac{-100}{-20}$

87. $-84 \div (-6)$

88. $-112 \div (-7)$

89. $-48 \div 0$

90. $(-210) \div (-210)$

91. $-126 \div 6$

92. $-160 \div (-5)$

93. $(-240) \div 6$

94. $(-96) \div (-8)$

SECTION

R.3 Rational Numbers

OBJECTIVE A To add and subtract rational numbers

In this section, operations with rational numbers are discussed. A **rational number** is the quotient of two integers.

Rational Numbers

A rational number is a number that can be written in the form $\frac{a}{b}$, where a and b are integers and $b \neq 0$.

3.4A Add or subtract rational numbers

We begin by reviewing addition of rational numbers in fractional form. If an addend is a fraction containing a negative sign, rewrite the fraction with the negative sign in the numerator. Then add the numerators and place the sum over the common denominator.

HOW TO • 1 Add: $-\frac{5}{6} + \frac{3}{8}$

$$-\frac{5}{6} + \frac{3}{8} = -\frac{20}{24} + \frac{9}{24}$$

• The LCM of the denominators 6 and 8 is 24. Write each fraction with a denominator of 24.

$$= \frac{-20}{24} + \frac{9}{24}$$

• Write the negative sign in the numerator.

$$= \frac{-20 + 9}{24}$$

• Add the fractions.

$$= \frac{-11}{24}$$

• Simplify the numerator.

$$= -\frac{11}{24}$$

• Write the negative sign in front of the fraction.

HOW TO • 2 Add: $-\frac{3}{5} + \left(-\frac{1}{3}\right)$

$$-\frac{3}{5} + \left(-\frac{1}{3}\right) = -\frac{9}{15} + \left(-\frac{5}{15}\right)$$

• The LCM of the denominators 5 and 3 is 15. Write each fraction with a denominator of 15.

$$= \frac{-9}{15} + \frac{-5}{15}$$

• Write the negative signs in the numerators.

$$= \frac{-9 + (-5)}{15}$$

• Add the fractions.

$$= \frac{-14}{15}$$

• Simplify the numerator.

$$= -\frac{14}{15}$$

• Write the negative sign in front of the fraction.

To subtract fractions with negative signs, rewrite the fractions with the negative signs in the numerators.

HOW TO • 3 Subtract: $-\frac{2}{3} - \frac{5}{8}$

$-\frac{2}{3} - \frac{5}{8} = -\frac{16}{24} - \frac{15}{24}$
• The LCM of the denominators 3 and 8 is 24. Write each fraction with a denominator of 24.

$= \frac{-16}{24} + \frac{-15}{24}$
• Rewrite subtraction as addition of the opposite. Write the negative signs in the numerators.

$= \frac{-16 + (-15)}{24}$
• Add the fractions.

$= \frac{-31}{24}$
• Simplify the numerator.

$= -\frac{31}{24} = -1\frac{7}{24}$
• Write the negative sign in front of the fraction.

HOW TO • 4 Subtract: $-\frac{1}{6} - \left(-\frac{2}{9}\right)$

$-\frac{1}{6} - \left(-\frac{2}{9}\right) = -\frac{1}{6} + \frac{2}{9}$
• Rewrite subtraction as addition of the opposite.

$= -\frac{3}{18} + \frac{4}{18}$
• Write the fractions as equivalent fractions with a common denominator.

$= \frac{-3}{18} + \frac{4}{18}$
• Write the negative sign in the numerator.

$= \frac{-3 + 4}{18}$
• Add the fractions.

$= \frac{1}{18}$
• Simplify the numerator.

The sign rules for adding and subtracting decimals are the same rules used to add and subtract integers.

HOW TO • 5 Simplify: $-29.871 + 34.06$

$34.06 - 29.871 = 4.189$
• The signs of the addends are different. Subtract the smaller absolute value from the larger absolute value.

$|34.06| > |-29.871|$
$-29.871 + 34.06 = 4.189$
• Attach the sign of the number with the larger absolute value. The sum is positive.

Recall that the opposite of n is $-n$ and the opposite of $-n$ is n. To find the opposite of a number, change the sign of the number.

HOW TO • 6 Simplify: $-3.92 - 21.7$

$$-3.92 - 21.7$$
$$= -3.92 + (-21.7)$$
$$= -25.62$$

- Rewrite subtraction as addition of the opposite. The opposite of 21.7 is −21.7.
- The signs of the addends are the same. Add the absolute values of the numbers. Attach the sign of the addends.

EXAMPLE • 1

Add: $-\frac{17}{20} + \frac{4}{5}$

Solution

$$-\frac{17}{20} + \frac{4}{5} = -\frac{17}{20} + \frac{16}{20}$$

- Write the fractions with a common denominator.

$$= \frac{-17}{20} + \frac{16}{20}$$

- Write the negative sign in the numerator.

$$= \frac{-17 + 16}{20}$$

- Add the fractions.

$$= \frac{-1}{20}$$

- Simplify the numerator.

$$= -\frac{1}{20}$$

- Write the negative sign in front of the fraction.

YOU TRY IT • 1

Add: $-\frac{1}{4} + \left(-\frac{3}{8}\right)$

Your solution

EXAMPLE • 2

Subtract: $-\frac{7}{20} - \frac{1}{5}$

Solution

$$-\frac{7}{20} - \frac{1}{5} = -\frac{7}{20} - \frac{4}{20}$$

- Write the fractions with a common denominator.

$$= \frac{-7}{20} + \frac{-4}{20}$$

- Rewrite subtraction as addition of the opposite. Write the negative signs in the numerators.

$$= \frac{-7 + (-4)}{20}$$

- Add the fractions.

$$= \frac{-11}{20}$$

- Simplify the numerator.

$$= -\frac{11}{20}$$

- Write the negative sign in front of the fraction.

YOU TRY IT • 2

Subtract: $-\frac{3}{8} - \left(-\frac{1}{6}\right)$

Your solution

Solutions on p. S39

EXAMPLE • 3

Simplify: $-3.97 - (-10.8)$

Solution

$-3.97 - (-10.8)$

$= -3.97 + 10.8$ • Rewrite subtraction as addition of the opposite.

$= 6.83$ • Subtract the absolute values of the numbers. The sum has the same sign as the number with the larger absolute value.

YOU TRY IT • 3

Simplify: $4.69 - 12.5$

Your solution

Solution on p. S39

OBJECTIVE B To multiply and divide rational numbers

3.4B Multiply or divide rational numbers

The product of two rational numbers written in fractional form is the product of the numerators over the product of the denominators. The sign rules are the same rules used to multiply integers.

The product of two numbers with the same sign is positive.

The product of two numbers with different signs is negative.

HOW TO • 7 Multiply: $-\frac{3}{8} \cdot \frac{4}{15}$

$-\frac{3}{8} \cdot \frac{4}{15} = -\left(\frac{3}{8} \cdot \frac{4}{15}\right)$ • The signs are different. The product is negative.

$= -\frac{3 \cdot 4}{8 \cdot 15}$ • Multiply the numerators. Multiply the denominators.

$= -\frac{3 \cdot 2 \cdot 2}{2 \cdot 2 \cdot 2 \cdot 3 \cdot 5}$ • Write the product is simplest form.

$= -\frac{1}{10}$

The sign rules for dividing rational numbers are the same rules used to divide integers.

The quotient of two numbers with the same sign is positive.

The quotient of two numbers with different signs is negative.

HOW TO • 8 Divide: $-\frac{3}{8} \div \left(-\frac{4}{5}\right)$

$$-\frac{3}{8} \div \left(-\frac{4}{5}\right) = \frac{3}{8} \div \frac{4}{5}$$

- The signs are the same. The quotient is positive.

$$= \frac{3}{8} \cdot \frac{5}{4}$$

- Rewrite division as multiplication by the reciprocal.

$$= \frac{3 \cdot 5}{8 \cdot 4}$$

- Multiply the fractions.

$$= \frac{3 \cdot 5}{2 \cdot 2 \cdot 2 \cdot 2 \cdot 2}$$

$$= \frac{15}{32}$$

The sign rules for multiplying and dividing decimals are the same rules used to multiply and divide integers.

HOW TO • 9 Multiply: $(-3.25)(-10.1)$

$$(-3.25)(-10.1) = 32.825$$

- The signs are the same. The product is positive. Multiply the absolute values of the numbers.

HOW TO • 10 Divide: $-29.4 \div 3.5$

$$-29.4 \div 3.5 = -8.4$$

- The signs are different. The quotient is negative. Divide the absolute values of the numbers.

EXAMPLE • 4

Multiply: $\left(-\frac{5}{8}\right)\left(-\frac{3}{5}\right)$

Solution

$$\left(-\frac{5}{8}\right)\left(-\frac{3}{5}\right) = \left(\frac{5}{8}\right)\left(\frac{3}{5}\right)$$

- The signs are the same. The product is positive.

$$= \frac{5 \cdot 3}{8 \cdot 5}$$

- Multiply the numerators. Multiply the denominators.

$$= \frac{5 \cdot 3}{2 \cdot 2 \cdot 2 \cdot 5}$$

- Write the product in simplest form.

$$= \frac{3}{8}$$

YOU TRY IT • 4

Multiply: $\frac{10}{11}\left(-\frac{2}{5}\right)$

Your solution

Solutions on p. S39

EXAMPLE • 5

Divide: $-\frac{9}{16} \div \frac{3}{4}$

Solution

$$-\frac{9}{16} \div \frac{3}{4} = -\left(\frac{9}{16} \div \frac{3}{4}\right)$$

$$= -\left(\frac{9}{16} \cdot \frac{4}{3}\right)$$

$$= -\frac{9 \cdot 4}{16 \cdot 3}$$

$$= -\frac{3 \cdot 3 \cdot 2 \cdot 2}{2 \cdot 2 \cdot 2 \cdot 2 \cdot 3}$$

$$= -\frac{3}{4}$$

- The signs are different. The quotient is negative.
- Rewrite division as multiplication by the reciprocal.
- Multiply the fractions.

YOU TRY IT • 5

Divide: $-\frac{3}{8} \div \left(-\frac{1}{2}\right)$

Your solution

EXAMPLE • 6

Multiply: $(-8.9)(0.25)$

Solution

$(-8.9)(0.25) = -2.225$

- The signs are different. The product is negative. Multiply the absolute values of the numbers.

YOU TRY IT • 6

Multiply: $(-3.6)(-1.45)$

Your solution

EXAMPLE • 7

Divide: $(-16.2) \div (-3.6)$

Solution

$(-16.2) \div (-3.6) = 4.5$

- The signs are the same. The quotient is positive. Divide the absolute values of the numbers.

YOU TRY IT • 7

Divide: $5.04 \div (-8.4)$

Your solution

Solutions on pp. S39–S40

OBJECTIVE C To evaluate exponential expressions

Recall that an exponent indicates repeated multiplication of the same factor. For example,

$$3^5 = 3 \cdot 3 \cdot 3 \cdot 3 \cdot 3$$

The **exponent,** 5, indicates how many times the **base,** 3, occurs as a factor in the multiplication.

The base of an exponential expression can be any rational number, for example, 0.5^4. To evaluate this expression, write the factor as many times as indicated by the exponent and then multiply.

$$0.5^4 = 0.5(0.5)(0.5)(0.5) = 0.25(0.5)(0.5) = 0.125(0.5) = 0.0625$$

EXAMPLE 8

Simplify: $\left(-\frac{3}{4}\right)^3 \cdot 8^2$

Solution

$\left(-\frac{3}{4}\right)^3 \cdot 8^2$

$= \left(-\frac{3}{4}\right)\left(-\frac{3}{4}\right)\left(-\frac{3}{4}\right) \cdot 8 \cdot 8$ • Write each factor the number of times indicated by the exponent.

$= -\left(\frac{3}{4} \cdot \frac{3}{4} \cdot \frac{3}{4} \cdot \frac{8}{1} \cdot \frac{8}{1}\right)$ • The product is negative.

$= -\frac{3 \cdot 3 \cdot 3 \cdot 8 \cdot 8}{4 \cdot 4 \cdot 4 \cdot 1 \cdot 1}$ • Multiply the fractions.

$= -27$ • Simplify.

YOU TRY IT 8

Simplify: $\left(\frac{2}{9}\right)^2 \cdot (-3)^4$

Your solution

Solution on p. S40

OBJECTIVE D To use the Order of Operations Agreement to simplify expressions

Whenever an expression contains more than one operation, the operations must be performed in a specified order, as listed on the next page in the Order of Operations Agreement.

3.5A Use the Order of Operations Agreement

The Order of Operations Agreement

Step 1 Do all operations inside grouping symbols. Grouping symbols include parentheses ( ), brackets [ ], and absolute value symbols | |.

Step 2 Simplify any numerical expressions containing exponents.

Step 3 Do multiplication and division as they occur from left to right.

Step 4 Do addition and subtraction as they occur from left to right.

The Order of Operations Agreement is used to simplify the expression in the following example.

HOW TO • 11 Simplify: $0.2(2.5 - 5.6) + (1.4)^2$

$0.2(2.5 - 5.6) + (1.4)^2$

$= 0.2(-3.1) + (1.4)^2$ • Perform operations inside parentheses.

$= 0.2(-3.1) + 1.96$ • Simplify the exponential expression.

$= -0.62 + 1.96$ • Do the multiplication.

$= 1.34$ • Do the addition.

EXAMPLE • 9

Simplify: $3 \div \left(\frac{1}{4} - \frac{1}{2}\right)^2 - 5$

Solution

$3 \div \left(\frac{1}{4} - \frac{1}{2}\right)^2 - 5$ • Use the Order of Operations Agreement.

$= 3 \div \left(-\frac{1}{4}\right)^2 - 5$ • Perform the operation inside the parentheses.

$= 3 \div \frac{1}{16} - 5$ • Simplify the exponential expression.

$= 3(16) - 5$ • Rewrite division as multiplication by the reciprocal.

$= 48 - 5$ • Do the multiplication.

$= 43$ • Do the subtraction.

YOU TRY IT • 9

Simplify: $7 \div \left(\frac{1}{7} - \frac{3}{14}\right) - 9$

Your solution

Solution on p. S40

R.3 EXERCISES

OBJECTIVE A To add and subtract rational numbers

For Exercises 1 to 27, simplify.

1. $-\frac{3}{4} + \frac{2}{3}$

2. $-\frac{5}{12} + \frac{3}{8}$

3. $\frac{2}{5} + \left(-\frac{11}{15}\right)$

4. $\frac{1}{4} + \left(-\frac{1}{7}\right)$

5. $-\frac{1}{2} - \frac{3}{8}$

6. $-\frac{5}{6} - \frac{1}{9}$

7. $-\frac{3}{10} - \frac{4}{5}$

8. $-\frac{5}{12} - \left(-\frac{2}{3}\right)$

9. $-\frac{5}{8} - \left(-\frac{7}{12}\right)$

10. $-\frac{3}{4} - \left(-\frac{5}{16}\right)$

11. $-\frac{2}{3} + \left(-\frac{1}{12}\right)$

12. $-\frac{2}{5} + \left(-\frac{4}{15}\right)$

13. $\frac{3}{8} + \left(-\frac{1}{2}\right) + \frac{7}{12}$

14. $-\frac{7}{12} + \frac{2}{3} + \left(-\frac{4}{5}\right)$

15. $\frac{2}{3} + \left(-\frac{5}{6}\right) + \frac{1}{4}$

16. $-\frac{5}{8} + \frac{3}{4} + \frac{1}{2}$

17. $-42.1 - 8.6$

18. $-6.57 - 8.933$

19. $5.73 - 9.042$

20. $-31.894 + 7.5$

21. $1.09 - (-8.3)$

22. $-8 - (-10.37)$

23. $-19 - (-2.65)$

24. $3.18 - 5.72 - 6.4$

25. $-12.3 - 4.07 + 6.82$

26. $-8.9 + 7.36 - 14.2$

27. $-5.6 - (-3.82) - 17.409$

28. Without simplifying, which is greater, $\frac{5}{8} - \left(-\frac{5}{6}\right)$ or $-\frac{5}{6} - \frac{5}{9}$?

29. Without simplifying, which is greater, $-\frac{1}{8} - \frac{3}{4}$ or $\frac{11}{12} - \left(-\frac{1}{4}\right)$?

OBJECTIVE B To multiply and divide rational numbers

For Exercises 30 to 68, simplify.

30. $-\frac{6}{7} \cdot \frac{11}{12}$

31. $\frac{3}{8} \cdot \left(-\frac{2}{3}\right)$

32. $\frac{5}{6} \cdot \left(-\frac{2}{5}\right)$

33. $\left(-\frac{4}{15}\right)\left(-\frac{3}{8}\right)$

34. $\left(-\frac{3}{4}\right)\left(-\frac{2}{9}\right)$

35. $-\frac{3}{4} \cdot \frac{1}{2}$

36. $-\frac{8}{15} \cdot \frac{5}{12}$

37. $-\frac{7}{12} \cdot \frac{5}{8} \cdot \frac{16}{25}$

38. $\frac{5}{12} \cdot \left(-\frac{1}{3}\right) \cdot \left(-\frac{8}{15}\right)$

39. $\left(-\frac{3}{5}\right) \cdot \frac{1}{2} \cdot \left(-\frac{5}{8}\right)$

40. $\frac{5}{6} \cdot \left(-\frac{2}{3}\right) \cdot \frac{3}{25}$

41. $12 \cdot \left(-\frac{5}{8}\right)$

42. $24\left(-\frac{3}{8}\right)$

43. $-9 \cdot \frac{7}{15}$

44. $\frac{1}{3} \cdot (-9)$

45. $-\frac{5}{2} \cdot 4$

46. $\frac{4}{7} \div \left(-\frac{4}{7}\right)$

47. $\left(-\frac{3}{8}\right) \div \frac{7}{8}$

48. $-\frac{5}{16} \div \left(-\frac{3}{8}\right)$

49. $\left(-\frac{3}{4}\right) \div \left(-\frac{5}{6}\right)$

50. $\frac{3}{4} \div (-6)$

51. $-\frac{2}{3} \div 8$

52. $\frac{5}{12} \div \left(-\frac{15}{32}\right)$

53. $\frac{3}{8} \div \left(-\frac{5}{12}\right)$

54. $-5.2(0.8)$

55. $(-2.1)(-0.7)$

56. $(-6.3)(-2.4)$

57. $(1.9)(-3.7)$

58. $-1.3(4.2)$

59. $-8.1(-7.5)$

60. $1.31(-0.006)$

61. $-10(0.59)$

62. $(-100)(4.73)$

63. $27.08 \div (-0.4)$

64. $-8.919 \div 0.9$

65. $(-3.312) \div (-0.8)$

66. $84.66 \div (-1.7)$

67. $-2.501 \div 0.41$

68. $1.003 \div (-0.59)$

For Exercises 69 to 71, divide. Round to the nearest tenth.

69. $-6.824 \div 0.053$

70. $0.0416 \div (-0.53)$

71. $(-31.792) \div (-0.86)$

72. Without simplifying, which is greater, $\left(-\frac{8}{9}\right)\left(-\frac{3}{4}\right)$ or $-\frac{5}{16} \div \frac{3}{8}$?

73. Without simplifying, which is greater, $-\frac{5}{6} \div (-5)$ or $-\frac{3}{4}\left(\frac{2}{9}\right)$?

OBJECTIVE C To evaluate exponential expressions

For Exercises 74 to 81, simplify.

74. $\left(-\frac{1}{6}\right)^3$

75. $\left(-\frac{2}{7}\right)^3$

76. $(2.25)^2$

77. $(3.5)^2$

78. $\left(\frac{4}{5}\right)^4 \cdot \left(-\frac{5}{8}\right)^3$

79. $\left(-\frac{9}{11}\right)^2 \cdot \left(\frac{1}{3}\right)^4$

80. $-4 \cdot \left(\frac{4}{7}\right)^2 \cdot \left(-\frac{3}{4}\right)^3$

81. $-3 \cdot \left(\frac{2}{5}\right)^2 \cdot \left(-\frac{1}{6}\right)^2$

OBJECTIVE D To use the Order of Operations Agreement to simplify expressions

For Exercises 82 to 93, simplify.

82. $(0.2)^2 \cdot (-0.5) + 1.72$

83. $0.3(1.7 - 4.8) + (1.2)^2$

84. $(1.8)^2 - 2.52 \div (1.8)$

85. $(1.65 - 1.05)^2 \div 0.4 + 0.8$

86. $\frac{7}{12} + \frac{5}{6}\left(\frac{1}{6} - \frac{2}{3}\right)$

87. $-\frac{3}{4}\left(\frac{11}{12} - \frac{7}{8}\right) + \frac{5}{16}$

88. $\frac{11}{16} - \left(-\frac{3}{4}\right)^2 + \frac{7}{8}$

89. $\left(-\frac{2}{3}\right)^2 - \frac{7}{18} + \frac{5}{6}$

90. $\left(-\frac{1}{3}\right)^2 \cdot \left(-\frac{9}{4}\right) + \frac{3}{4}$

91. $\left(-\frac{2}{3}\right)^2 + \left(-\frac{1}{6}\right) \div \frac{3}{8}$

92. $\left(\frac{1}{3} - \frac{5}{6}\right) + \frac{7}{8} \div \left(-\frac{1}{2}\right)^3$

93. $\left(-\frac{1}{4}\right)^2 \div \left(\frac{1}{2} - \frac{3}{4}\right) + \frac{3}{8}$

94. Arrange the expressions in order from greatest value to least value.

$16 - 3(3 - 8) \div 5$
$4(-3) \div [2(6 - 7)^2]$
$18 \div (-2) + (-3)^2 - (-15)$

95. Arrange the expressions in order from greatest value to least value.

$20 \div (6 - 2^4) + (-5)$
$18 \div |2^3 - 9| + (-3)$
$16 + 15 \div (-5) - (-4)$

SECTION

R.4 Equations

OBJECTIVE A To solve a first-degree equation in one variable 

An **equation** expresses the equality of two mathematical expressions. Each of the equations below is a **first-degree equation in one variable.** *First degree* means that the variable has an exponent of 1.

$$x + 11 = 14$$
$$3a + 5 = 8a$$
$$2(6y - 1) = 3$$

A **solution** of an equation is a number that, when substituted for the variable, results in a true equation.

3 is a solution of the equation $x + 4 = 7$ because $3 + 4 = 7$.

9 is not a solution of the equation $x + 4 = 7$ because $9 + 4 \neq 7$.

5.1B, 5.1C, 5.2A, 5.3A, 5.3B
Solve first-degree equations in one variable

To **solve an equation** means to find a solution of the equation. In solving an equation, the goal is to rewrite the given equation with the variable alone on one side of the equation and a constant term on the other side of the equation; the constant term is the solution of the equation. The following properties of equations are used to rewrite equations in this form.

Properties of Equations

Addition Property of Equations

The same number can be added to each side of an equation without changing the solution of the equation. In symbols, the equation $a = b$ has the same solution as the equation $a + c = b + c$.

Multiplication Property of Equations

Each side of an equation can be multiplied by the same nonzero number without changing the solution of the equation. In symbols, if $c \neq 0$, then the equation $a = b$ has the same solution as the equation $ac = bc$.

Take Note
Subtraction is defined as addition of the opposite.

$$a - b = a + (-b)$$

The Addition Property of Equations is used to remove a term from one side of an equation by adding the opposite of that term to each side of the equation. Because subtraction is defined in terms of addition, the Addition Property of Equations also makes it possible to subtract the same number from each side of an equation without changing the solution of the equation.

For example, to solve the equation $t + 9 = -4$, subtract the constant term (9) from each side of the equation.

$$\begin{aligned} t + 9 &= -4 \\ t + 9 - 9 &= -4 - 9 \\ t &= -13 \end{aligned}$$

Now the variable is alone on one side of the equation and a constant term (-13) is on the other side. The solution is the constant. The solution is -13.

 **Take Note**

Division is defined as multiplication by the reciprocal.

$$a \div b = a \cdot \frac{1}{b}$$

The Multiplication Property of Equations is used to remove a coefficient by multiplying each side of the equation by the reciprocal of the coefficient. Because division is defined in terms of multiplication, each side of an equation can be divided by the same nonzero number without changing the solution of the equation.

 **Take Note**

When using the Multiplication Property of Equations, multiply each side of the equation by the reciprocal of the coefficient when the coefficient is a fraction. Divide each side of the equation by the coefficient when the coefficient is an integer or a decimal.

For example, to solve the equation $-5q = 120$, divide each side of the equation by the coefficient -5.

$$\begin{aligned} -5q &= 120 \\ \frac{-5q}{-5} &= \frac{120}{-5} \\ q &= -24 \end{aligned}$$

Now the variable is alone on one side of the equation and a constant (-24) is on the other side. The solution is the constant. The solution is -24.

In solving more complicated first-degree equations in one variable, use the following sequence of steps.

Steps for Solving a First-Degree Equation in One Variable

Step 1 Use the Distributive Property to remove parentheses.

Step 2 Combine any like terms on the right side of the equation and any like terms on the left side of the equation.

Step 3 Use the Addition Property to rewrite the equation with only one variable term.

Step 4 Use the Addition Property to rewrite the equation with only one constant term.

Step 5 Use the Multiplication Property to rewrite the equation with the variable alone on one side of the equation and a constant on the other side of the equation.

If one of the above steps is not needed to solve a given equation, proceed to the next step.

EXAMPLE • 1

Solve: $3x + 5 - 4x = 6$

Solution

$$
\begin{aligned}
3x + 5 - 4x &= 6 \\
5 - x &= 6 && \text{• Step 2} \\
5 - 5 - x &= 6 - 5 && \text{• Step 4} \\
-x &= 1 \\
\frac{-x}{-1} &= \frac{1}{-1} && \text{• Step 5} \\
x &= -1
\end{aligned}
$$

The solution is -1.

YOU TRY IT • 1

Solve: $5x + 3 - 7x = 9$

Your solution

EXAMPLE • 2

Solve: $5x + 9 = 23 - 2x$

Solution

$$
\begin{aligned}
5x + 9 &= 23 - 2x \\
5x + 2x + 9 &= 23 - 2x + 2x && \text{• Step 3} \\
7x + 9 &= 23 \\
7x + 9 - 9 &= 23 - 9 && \text{• Step 4} \\
7x &= 14 \\
\frac{7x}{7} &= \frac{14}{7} && \text{• Step 5} \\
x &= 2
\end{aligned}
$$

The solution is 2.

YOU TRY IT • 2

Solve: $4x + 3 = 7x + 9$

Your solution

EXAMPLE • 3

Solve: $8x - 3(4x - 5) = -2x + 6$

Solution

$$
\begin{aligned}
8x - 3(4x - 5) &= -2x + 6 \\
8x - 12x + 15 &= -2x + 6 && \text{• Step 1} \\
-4x + 15 &= -2x + 6 && \text{• Step 2} \\
-4x + 2x + 15 &= -2x + 2x + 6 && \text{• Step 3} \\
-2x + 15 &= 6 \\
-2x + 15 - 15 &= 6 - 15 && \text{• Step 4} \\
-2x &= -9 \\
\frac{-2x}{-2} &= \frac{-9}{-2} && \text{• Step 5} \\
x &= \frac{9}{2}
\end{aligned}
$$

The solution is $\frac{9}{2}$.

YOU TRY IT • 3

Solve: $4 - (5x - 8) = 4x + 3$

Your solution

Solutions on p. S40

R.4 EXERCISES

OBJECTIVE A To solve a first-degree equation in one variable

For Exercises 1 to 36, solve.

1. $x + 7 = -5$

2. $9 + b = 21$

3. $-9 = z - 8$

4. $b - 11 = 11$

5. $-48 = 6z$

6. $-9a = -108$

7. $-\frac{3}{4}x = 15$

8. $\frac{5}{2}x = -10$

9. $-\frac{x}{4} = -2$

10. $\frac{2x}{5} = -8$

11. $3x + 8 = 17$

12. $2 + 5a = 12$

13. $5 = 3x - 10$

14. $4 = 3 - 5x$

15. $\frac{2}{3}x + 5 = 3$

16. $-\frac{1}{2}x + 4 = 1$

17. $2b + 6 - 3b = 4$

18. $3x + 4 - 5x = 8$

19. $4 - 2b = 2 - 4b$

20. $4y - 10 = 6 + 2y$

21. $5x - 3 = 9x - 7$

22. $5x + 7 = 8x + 5$

23. $2 - 6y = 5 - 7y$

24. $4b + 15 = 3 - 2b$

25. $2(x + 1) + 5x = 23$

26. $9n - 15 = 3(2n - 1)$

27. $7a - (3a - 4) = 12$

28. $5(3 - 2y) = 3 - 4y$

29. $9 - 7x = 4(1 - 3x)$

30. $2(3b + 5) - 1 = 10b + 1$

31. $2z - 2 = 5 - (9 - 6z)$

32. $4a + 3 = 7 - (5 - 8a)$

33. $5(6 - 2x) = 2(5 - 3x)$

34. $4(3y + 1) = 2(y - 8)$

35. $2(3b - 5) = 4(6b - 2)$

36. $3(x - 4) = 1 - (2x - 7)$

Appendix

The Metric System of Measurement

International trade, or trade among nations, is a vital and growing segment of business in the world today. The opening of McDonald's restaurants around the globe is testimony to the expansion of international business.

Point of Interest

To learn more about these tests, go to **www.ed.gov.** Use the search feature and enter TIMSS.

The United States, as a nation, is dependent on world trade. And world trade is dependent on internationally standardized units of measurement: the metric system. The Third International Mathematics and Science Study (TIMSS) compared the performance of half a million students from 41 countries at five different grade levels on tests of their mathematics and science knowledge. One area of mathematics in which the U.S. average was below the international average was measurement, due in large part to the fact that the units cited in the questions were metric units. Because the United States has not yet converted to the metric system, its citizens are less familiar with it.

In this Appendix, we present the metric system of measurement and explain how to convert between different units.

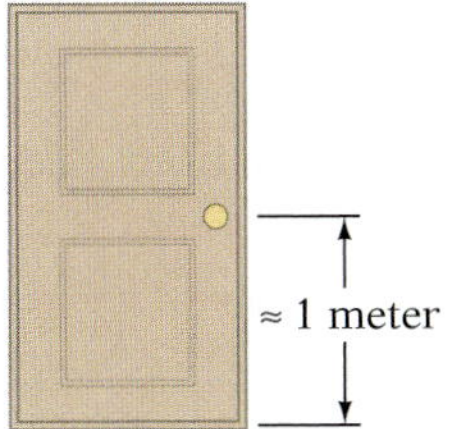

The basic unit of *length,* or distance, in the metric system is the **meter** (m). One meter is approximately the distance from a doorknob to the floor. All units of length in the metric system are derived from the meter. Prefixes to the basic unit denote the length of each unit. For example, the prefix *centi-* means "one-hundredth"; therefore, 1 centimeter is 1 one-hundredth of a meter (0.01 m).

Point of Interest

Originally, the meter (spelled *metre* in some countries) was defined as $\frac{1}{10,000,000}$ of the distance from the equator to the North Pole. Modern scientists have redefined the meter as 1,650,753.73 wavelengths of the orange-red light given off by the element krypton.

kilo-	= 1 000	1 kilometer (km)	= 1 000 meters (m)
hecto-	= 100	1 hectometer (hm)	= 100 m
deca-	= 10	1 decameter (dam)	= 10 m
		1 meter (m)	= 1 m
deci-	= 0.1	1 decimeter (dm)	= 0.1 m
centi-	= 0.01	1 centimeter (cm)	= 0.01 m
milli-	= 0.001	1 millimeter (mm)	= 0.001 m

Note in this list that 1000 is written as 1 000, with a space between the 1 and the zeros. **When writing numbers using metric units, each group of three numbers is separated by a space instead of a comma.** A space is also used after each group of three numbers to the right of a decimal point. For example, 31,245.2976 is written 31 245.297 6 in metric notation.

Mass and weight are closely related. *Weight* is a measure of how strongly gravity is pulling on an object. Therefore, an object's weight is less in space than on Earth's surface. However, the amount of material in the object, its *mass,* remains the same. On the surface of Earth, the terms *mass* and *weight* can be used interchangeably.

The basic unit of mass in the metric system is the **gram** (g). If a box that is 1 centimeter long on each side is filled with water, the mass of that water is 1 gram.

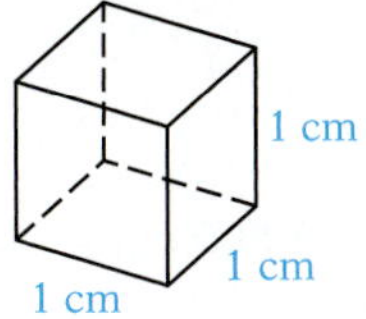

1 gram = the mass of water in a box that is 1 centimeter long on each side

The units of mass in the metric system have the same prefixes as the units of length.

1 kilogram (kg) = 1 000 grams (g)
1 hectogram (hg) = 100 g
1 decagram (dag) = 10 g
1 gram (g) = 1 g
1 decigram (dg) = 0.1 g
1 centigram (cg) = 0.01 g
1 milligram (mg) = 0.001 g

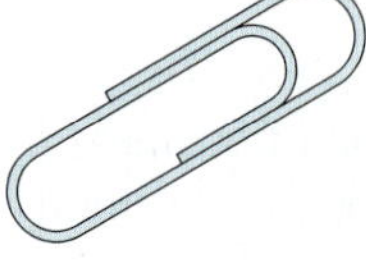
Weight ≈ 1 gram

The gram is a very small unit of mass. A paperclip weighs about 1 gram. In applications, the kilogram (1 000 grams) is a more useful unit of mass. This textbook weighs about 1 kilogram.

Liquid substances are measured in units of *capacity*.

The basic unit of capacity in the metric system is the **liter** (L). One liter is defined as the capacity of a box that is 10 centimeters long on each side.

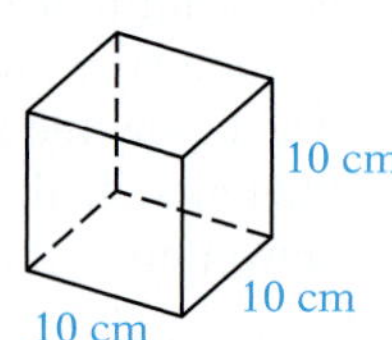

1 liter = the capacity of a box that is 10 centimeters long on each side

The units of capacity in the metric system have the same prefixes as the units of length.

1 kiloliter (kl) = 1 000 liters (L)
1 hectoliter (hl) = 100 L
1 decaliter (dal) = 10 L
1 liter (L) = 1 L
1 deciliter (dl) = 0.1 L
1 centiliter (cl) = 0.01 L
1 milliliter (ml) = 0.001 L

Point of Interest

The definition of 1 inch has been changed as a consequence of the wide acceptance of the metric system. One inch is now exactly 25.4 mm.

Converting between units in the metric system involves moving the decimal point to the right or to the left. Listing the units in order from largest to smallest will indicate how many places to move the decimal point in which direction.

To convert 3 800 cm to meters, write the units of length in order from largest to smallest.

km hm dam m dm cm mm

2 positions

- **Converting from centimeters to meters requires moving two places to the left.**

3 800 cm = 38.00 m

2 places

- **Move the decimal point the same number of places in the same direction.**

HOW TO • 1 Convert 27 kg to grams.

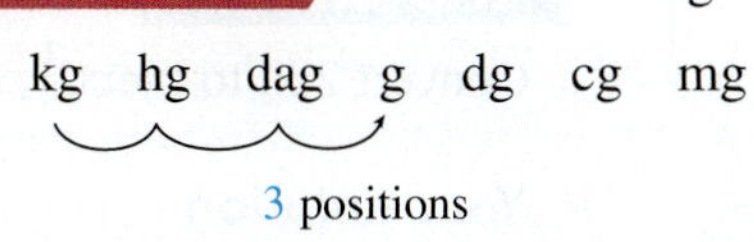

- Write the units of mass in order from largest to smallest.
- Converting kilograms to grams requires moving three positions to the right.

27 kg = 27 000 g

3 places

- Move the decimal point the same number of places in the same direction.

EXAMPLE • 1

Convert 4.08 m to centimeters.

Solution

Write the units of length from largest to smallest.

km hm dam (m) dm (cm) mm

Converting meters to centimeters requires moving two positions to the right.

4.08 m = 408 cm

YOU TRY IT • 1

Convert 1 295 m to kilometers.

Your solution

EXAMPLE • 2

Convert 5.93 g to milligrams.

Solution

Write the units of mass from largest to smallest.

kg hg dag (g) dg cg (mg)

Converting grams to milligrams requires moving three positions to the right.

5.93 g = 5 930 mg

YOU TRY IT • 2

Convert 7 543 g to kilograms.

Your solution

EXAMPLE • 3

Convert 82 ml to liters.

Solution

Write the units of capacity from largest to smallest.

kl hl dal (L) dl cl (ml)

Converting milliliters to liters requires moving three positions to the left.

82 ml = 0.082 L

YOU TRY IT • 3

Convert 6.3 L to milliliters.

Your solution

Solutions on p. S40

EXAMPLE 4

Convert 9 kl to liters.

Solution

Write the units of capacity from largest to smallest.

(kl) hl dal (L) dl cl ml

Converting kiloliters to liters requires moving three positions to the right.

9 kl = 9 000 L

YOU TRY IT 4

Convert 2 kl to liters.

Your solution

Solution on p. S40

Other prefixes in the metric system are becoming more common as a result of technological advances in the computer industry. For example:

tera- = 1 000 000 000 000
giga- = 1 000 000 000
mega- = 1 000 000

micro- = 0.000 001
nano- = 0.000 000 001
pico- = 0.000 000 000 001

A **bit** is the smallest unit of code that computers can read; it is a binary digit, either a 0 or a 1. Usually bits are grouped into **bytes** of 8 bits. Each byte stands for a letter, a number, or any other symbol we might use in communicating information. For example, the letter W can be represented by 01010111.

The amount of memory in a computer hard drive is measured in terabytes, gigabytes, and megabytes. The speed of a computer used to be measured in microseconds and then nanoseconds, but now speeds are measured in picoseconds.

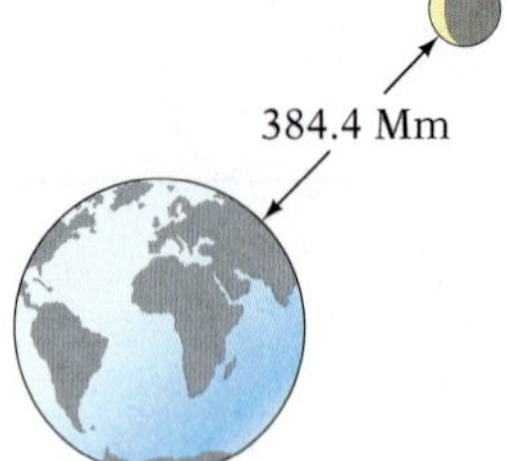

Here are a few more examples of how these prefixes are used.

The mass of Earth gains 40 Gg (gigagrams) each year from captured meteorites and cosmic dust.

The average distance from Earth to the moon is 384.4 Mm (megameters), and the average distance from Earth to the sun is 149.5 Gm (gigameters).

The wavelength of yellow light is 590 nm (nanometers).

The diameter of a hydrogen atom is about 70 pm (picometers).

There are additional prefixes in the metric system, indicating both larger and smaller numbers. We may hear them more and more often as computer chips hold more and more information, as computers get faster and faster, and as we learn more and more about objects in our universe that are great distances away.

The U.S. Metric Association (USMA) advocates U.S. conversion to the metric system, which is also referred to as the International System of Units, abbreviated SI. The process of changing measurement units to the metric system is called **metric transition** or **metrication.**

Exercises

1. In the metric system, what is the basic unit of length? of liquid measure? of weight?

2. **a.** Explain how to convert meters to centimeters.
   b. Explain how to convert milliliters to liters.

For Exercises 3 to 26, name the unit in the metric system that would be used to measure each.

3. The distance from New York to London
4. The weight of a truck
5. A person's waist
6. The amount of coffee in a mug
7. The weight of a thumbtack
8. The amount of water in a swimming pool
9. The distance a baseball player hits a baseball
10. A person's hat size
11. The amount of fat in a slice of cheddar cheese
12. A person's weight
13. The maple syrup served with pancakes
14. The amount of water in a water cooler
15. The amount of vitamin C in a vitamin tablet
16. A serving of cereal
17. The width of a hair
18. A person's height
19. The amount of medication in an aspirin
20. The weight of a lawnmower
21. The weight of a slice of bread
22. The contents of a bottle of salad dressing
23. The amount of water a family uses monthly
24. The newspapers collected at a recycling center
25. The amount of liquid in a bowl of soup
26. The distance to the bank

For Exercises 27 to 56, convert the given measure.

27. 42 cm = _____ mm
28. 91 cm = _____ mm
29. 360 g = _____ kg
30. 1 856 g = _____ kg
31. 5 194 ml = _____ L
32. 7 285 ml = _____ L
33. 2 m = _____ mm
34. 8 m = _____ mm
35. 217 mg = _____ g
36. 34 mg = _____ g
37. 4.52 L = _____ ml
38. 0.029 7 L = _____ ml
39. 8 406 m = _____ km
40. 7 530 m = _____ km
41. 2.4 kg = _____ g
42. 9.2 kg = _____ g
43. 6.18 kl = _____ L
44. 0.036 kl = _____ L
45. 9.612 km = _____ m
46. 2.35 km = _____ m
47. 0.24 g = _____ mg
48. 0.083 g = _____ mg
49. 298 cm = _____ m
50. 71.6 cm = _____ m
51. 2 431 L = _____ kl
52. 6 302 L = _____ kl
53. 0.66 m = _____ cm
54. 4.58 m = _____ cm
55. 243 mm = _____ cm
56. 92 mm = _____ cm

57. a. Complete the table.

Metric System Prefix	*Symbol*	*Magnitude*	*Means Multiply the Basic Unit By:*
tera-	T	10^{12}	1 000 000 000 000
giga-	G	?	1 000 000 000
mega-	M	10^{6}	?
kilo-	?	?	1 000
hecto-	h	?	100
deca-	da	10^{1}	?
deci-	d	$\frac{1}{10}$	?
centi-	?	$\frac{1}{10^2}$	?
milli-	?	?	0.001
micro-	μ (mu)	$\frac{1}{10^6}$	?
nano-	n	$\frac{1}{10^9}$	?
pico-	p	?	0.000 000 000 001

b. How can the magnitude column in the table above be used to determine how many places to move the decimal point when converting to the basic unit in the metric system?

58. The Olympics

a. One of the events in the summer Olympics is the 50 000-meter walk. How many kilometers do the entrants in this event walk?

b. One of the events in the winter Olympic Games is the 10 000-meter speed skating event. How many kilometers do the entrants in this event skate?

59. Gemstones A carat is a unit of weight equal to 200 mg. Find the weight in grams of a 10-carat precious stone.

60. Fabric How many pieces of material, each 75 cm long, can be cut from a bolt of fabric that is 6 m long?

61. Swimming Pools An athletic club uses 800 ml of chlorine each day for its swimming pool. How many liters of chlorine are used in a month of 30 days?

62. **Carpentry** Each of the four shelves in a bookcase measures 175 cm. Find the cost of the shelves when the price of lumber is $15.75 per meter.

63. **Containers of Milk** The printed label from a container of milk is shown at the right. To the nearest whole number, how many 230-milliliter servings are in the container?

Dairy Hill
Skim Milk
Vitamin A & D Added
Pasteurized • Homogenized
INGREDIENTS: PASTEURIZED SKIM MILK, NONFAT MILK SOLIDS, VITAMIN A PALMITATE AND VITAMIN D3 ADDED.
0 15400 20209 1
1 GAL. (3.78 L)

64. **Cereal** A 1.19-kilogram container of Quaker Oats contains 30 servings. Find the number of grams in one serving of the oatmeal. Round to the nearest gram.

65. **Nutritional Supplements** A patient is advised to supplement her diet with 2 g of calcium per day. The calcium tablets she purchases contain 500 mg of calcium per tablet. How many tablets per day should the patient take?

66. **Education** A laboratory assistant is in charge of ordering acid for three chemistry classes of 30 students each. Each student requires 80 ml of acid. How many liters of acid should be ordered? The assistant must order by the whole liter.

67. **Unit Cost** A case of 12 one-liter bottles of apple juice costs $19.80. A case of 24 cans, each can containing 340 ml of apple juice, costs $14.50. Which case of apple juice costs less per milliliter?

68. **Construction** A column assembly is being constructed in a building. The components are shown in the diagram at the right. What length column must be cut?

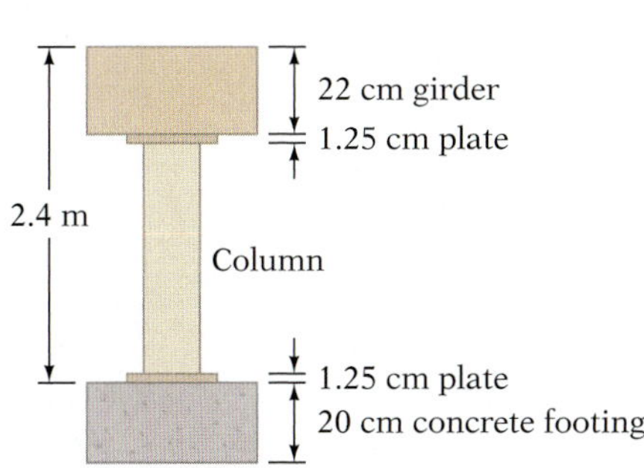

69. **Speed of Light** The distance between Earth and the sun is 150 000 000 km. Light travels 300 000 000 m in 1 s. How long does it take for light to reach Earth from the sun?

70. Why is it necessary to have internationally standardized units of measurement?

Applying the Concepts

71. **Business** A service station operator bought 85 kl of gasoline for $38,500. The gasoline was sold for $.658 per liter. Find the profit on the 85 kl of gasoline.

72. **Business** For $149.50, a cosmetician buys 5 L of moisturizer and repackages it in 125-milliliter jars. Each jar costs the cosmetician $.55. Each jar of moisturizer is sold for $8.95. Find the profit on the 5 L of moisturizer.

73. **Business** A health food store buys nuts in 10-kilogram containers and repackages the nuts for resale. The store packages the nuts in 200-gram bags, costing $.06 each, and sells them for $2.89 per bag. Find the profit on a 10-kilogram container of nuts costing $75.

74. Form two debating teams. One team should argue in favor of changing to the metric system in the United States, and the other should argue against it.

Solutions to "You Try It"

SOLUTIONS TO CHAPTER 1 "YOU TRY IT"

SECTION 1.1

You Try It 1

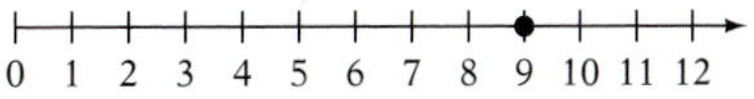

You Try It 2

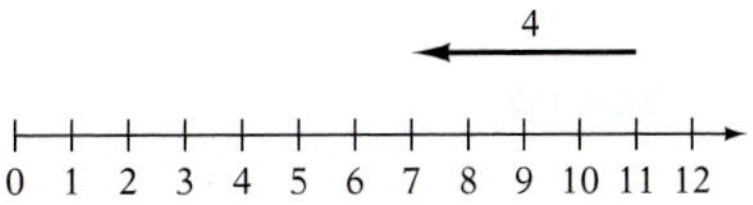

7 is 4 units to the left of 11.

You Try It 3 **a.** $47 > 19$ **b.** $26 > 0$

You Try It 4 0, 3, 17, 52, 68, 94

You Try It 5 Forty-six million thirty-two thousand seven hundred fifteen

You Try It 6 920,008

You Try It 7 70,000 + 6000 + 200 + 40 + 5

You Try It 8

Given place value
529,374
$9 > 5$

529,374 rounded to the nearest ten-thousand is 530,000.

You Try It 9

Given place value
7985
$8 > 5$

7985 rounded to the nearest hundred is 8000.

You Try It 10

Strategy To find the sport named by the greatest number of people, find the largest number given in the circle graph.

Solution The largest number given in the graph is 80.
The sport named by the greatest number of people was football.

You Try It 11

Strategy To find the shorter distance, compare the numbers 347 and 387.

Solution $347 < 387$

The shorter distance is between Los Angeles and San Jose.

You Try It 12

Strategy To determine which state has fewer sanctioned league bowlers, compare the numbers 239,951 and 239,010.

Solution $239{,}010 < 239{,}951$

Ohio has fewer sanctioned league bowlers.

You Try It 13

Strategy To find the land area to the nearest thousand square miles, round 3,851,809 to the nearest thousand.

Solution 3,851,809 rounded to the nearest thousand is 3,852,000.

To the nearest thousand, the land area of Canada is 3,852,000 mi^2.

SECTION 1.2

You Try It 1

$$\begin{array}{rcr} 6285 & \longrightarrow & 6000 \\ 3972 & \longrightarrow & 4000 \\ 5140 & \longrightarrow & +\ 5000 \\ \hline & & 15{,}000 \end{array}$$

You Try It 2 The Addition Property of Zero

You Try It 3

$$\begin{array}{r} 111{,}100{,}000 \\ 61{,}600{,}000 \\ 24{,}100{,}000 \\ +\ \ 1{,}600{,}000 \\ \hline 198{,}400{,}000 \end{array}$$

A total of 198,400,000 cases of eggs were produced during the year.

You Try It 4

$x + y + z$
$1692 + 4783 + 5046$

$$\begin{array}{r} \scriptstyle 1\,2\,1 \\ 1692 \\ 4783 \\ +\ 5046 \\ \hline 11{,}521 \end{array}$$

You Try It 5

$$\begin{array}{l|l} 13 = b + 6 & \\ \hline 13 & 7 + 6 \\ 13 = 13 & \end{array}$$

Yes, 7 is a solution of the equation.

You Try It 6

$$\begin{array}{r} \scriptstyle 8\ \ 9\ 9\ 12 \\ 4\not{9},\not{0}\not{0}\not{2} \\ -3\,1,8\,6\,5 \\ \hline 1\,7,1\,3\,7 \end{array} \qquad \textit{Check:} \begin{array}{r} 31{,}865 \\ +17{,}137 \\ \hline 49{,}002 \end{array}$$

You Try It 7

$$\begin{array}{l} 8544 \longrightarrow \\ 3621 \longrightarrow \end{array} \begin{array}{r} 9000 \\ -4000 \\ \hline 5000 \end{array} \quad \begin{array}{r} 8544 \\ -3621 \\ \hline 4923 \end{array}$$

You Try It 8

2020: 612 quadrillion Btu
1990: 346 quadrillion Btu

$$\begin{array}{r} 612 \\ -346 \\ \hline 266 \end{array}$$

The difference is 266 quadrillion Btu.

You Try It 9

$x - y$
$7061 - 3229$

$$\begin{array}{r} \scriptstyle 6\ 10\ 5\ 11 \\ \not{7}\not{0}\not{6}\not{1} \\ -\ 3\,2\,2\,9 \\ \hline 3\,8\,3\,2 \end{array}$$

You Try It 10

$$\begin{array}{l|l} 46 = 58 - p & \\ \hline 46 & 58 - 11 \\ 46 \neq 47 & \end{array}$$

No, 11 is not a solution of the equation.

You Try It 11

Strategy To find the total number of fatal accidents during 1991 through 1999:

- Find the number of fatal accidents each year.
- Add the nine numbers.

Solution

1991: 3 1994: 2 1997: 4
1992: 2 1995: 3 1998: 5
1993: 4 1996: 3 1999: 6

$3 + 2 + 4 + 2 + 3 + 3 + 4 + 5 + 6$
$= 32$

During 1991 through 1999, there were 32 fatal accidents on amusement rides.

You Try It 12

Strategy To find the price, replace C by 148 and M by 74 in the given formula and solve for P.

Solution

$P = C + M$
$P = 148 + 74$
$P = 222$

The price of the leather jacket is \$222.

You Try It 13

Strategy Draw a diagram.

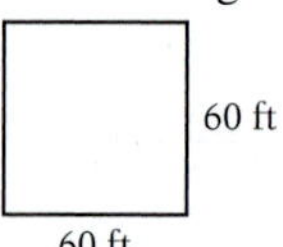

To find the length of fencing needed, use the formula for the perimeter of a rectangle, $P = L + W + L + W$. $L = 60$ and $W = 60$.

Solution

$P = L + W + L + W$
$P = 60 + 60 + 60 + 60$
$P = 240$

240 ft of fencing is needed.

SECTION 1.3

You Try It 1 The average monthly savings in France is \$175.

$$\begin{array}{r} 175 \\ \times\ 12 \\ \hline 350 \\ 175 \\ \hline 2100 \end{array}$$

The average annual savings of individuals in France is \$2100.

You Try It 2

$8704 \longrightarrow 9000$
$93 \longrightarrow 90$

$9000 \cdot 90 = 810{,}000$

You Try It 3

$5xy$
$5(20)(60) = 100(60)$
$= 6000$

You Try It 4 $90(7000) = 630{,}000$

You Try It 5 $0 \cdot 10 = 0$

You Try It 6

$$\begin{array}{r|l} 7a = 77 & \\ \hline 7 \cdot 11 & 77 \\ 77 = 77 & \end{array}$$

Yes, 11 is a solution of the equation.

You Try It 7 $2 \cdot 2 \cdot 2 \cdot 3 \cdot 3 \cdot 3 \cdot 3 = 2^3 \cdot 3^4$

You Try It 8 $6^4 = 6 \cdot 6 \cdot 6 \cdot 6 = 36 \cdot 6 \cdot 6$
$= 216 \cdot 6 = 1296$

You Try It 9 $10^8 = 100{,}000{,}000$

You Try It 10 $2^4 \cdot 3^2 = (2 \cdot 2 \cdot 2 \cdot 2) \cdot (3 \cdot 3)$
$= 16 \cdot 9 = 144$

You Try It 11 x^4y^2
$1^4 \cdot 3^2 = (1 \cdot 1 \cdot 1 \cdot 1) \cdot (3 \cdot 3)$
$= 1 \cdot 9$
$= 9$

You Try It 12

$$\begin{array}{r} 320\text{ r}14 \\ 24\overline{)7694} \\ -72 \\ \hline 49 \\ -48 \\ \hline 14 \\ -0 \\ \hline 14 \end{array}$$

Check: $(320 \cdot 24) + 14 = 7680 + 14$
$= 7694$

You Try It 13 The annual expense for food is \$7200.

$7200 \div 12 = 600$

The monthly expense for food is \$600.

You Try It 14 $216{,}936 \longrightarrow 200{,}000$
$207 \longrightarrow 200$

$200{,}000 \div 200 = 1000$

You Try It 15 $\dfrac{x}{y}$

$\dfrac{672}{8} = 84$

You Try It 16 $\dfrac{60}{y} = 2$

$$\begin{array}{c|c} \frac{60}{12} & 2 \end{array}$$

$5 \neq 2$

No, 12 is not a solution of the equation.

You Try It 17
$30 \div 1 = 30$
$30 \div 2 = 15$
$30 \div 3 = 10$
$30 \div 4$ Does not divide evenly.
$30 \div 5 = 6$
$30 \div 6 = 5$ The factors are repeating.

The factors of 30 are 1, 2, 3, 5, 6, 10, 15, and 30.

You Try It 18

$$\begin{array}{r} 11 \\ 2\overline{)22} \\ 2\overline{)44} \\ 2\overline{)88} \end{array}$$

$88 = 2 \cdot 2 \cdot 2 \cdot 11 = 2^3 \cdot 11$

You Try It 19

$$\begin{array}{r} 59 \\ 5\overline{)295} \end{array}$$

$295 = 5 \cdot 59$

You Try It 20

Strategy To find how many times more expensive a stamp was, divide the cost in 1997 (32) by the cost in 1960 (4).

Solution $32 \div 4 = 8$

A stamp was 8 times more expensive in 1997.

You Try It 21

Strategy Draw a diagram.

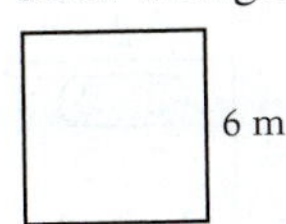

To find the amount of carpet that should be purchased, use the formula for the area of a square, $A = s^2$, with $s = 6$.

Solution $A = s^2$
$A = 6^2$
$A = 36$

36 m^2 of carpet should be purchased.

You Try It 22

Strategy To find the speed, replace d by 486 and t by 9 in the given formula and solve for r.

Solution $r = \dfrac{d}{t}$

$r = \dfrac{486}{9}$

$r = 54$

You would need to travel at a speed of 54 mph.

SECTION 1.4

You Try It 1 $4 \cdot (8 - 3) \div 5 - 2 = 4 \cdot 5 \div 5 - 2$
$= 20 \div 5 - 2$
$= 4 - 2$
$= 2$

You Try It 2 $16 + 3(6 - 1)^2 \div 5 = 16 + 3(5)^2 \div 5$
$= 16 + 3(25) \div 5$
$= 16 + 75 \div 5$
$= 16 + 15$
$= 31$

You Try It 3 $(a - b)^2 + 5c$
$(7 - 2)^2 + 5(4) = 5^2 + 5(4)$
$= 25 + 5(4)$
$= 25 + 20$
$= 45$

SOLUTIONS TO CHAPTER 2 "YOU TRY IT"

SECTION 2.1

You Try It 1

	2	3	5
12 =	(2 · 2)	3	
27 =		(3 · 3 · 3)	
50 =	2		(5 · 5)

The LCM $= 2 \cdot 2 \cdot 3 \cdot 3 \cdot 3 \cdot 5 \cdot 5 = 2700$

You Try It 2

	2	3	5
36 =	(2 · 2)	3 · 3	
60 =	2 · 2	(3)	5
72 =	2 · 2 · 2	3 · 3	

The GCF $= 2 \cdot 2 \cdot 3 = 12$.

You Try It 3

	2	3	5	11
11 =				11
24 =	2 · 2 · 2	3		
30 =	2	3	5	

Because no numbers are circled, the GCF = 1.

SECTION 2.2

You Try It 1 $\frac{19}{6}; 3\frac{1}{6}$

You Try It 2 $\begin{array}{r} 8 \\ 3\overline{)26} \\ -24 \\ \hline 2 \end{array}$ $\quad \frac{26}{3} = 8\frac{2}{3}$

You Try It 3 $\begin{array}{r} 9 \\ 4\overline{)36} \\ -36 \\ \hline 0 \end{array}$ $\quad \frac{36}{4} = 9$

You Try It 4 $9\frac{4}{7} = \frac{(7 \cdot 9) + 4}{7} = \frac{63 + 4}{7} = \frac{67}{7}$

You Try It 5 $3 = \frac{3}{1}$

You Try It 6 $48 \div 8 = 6$

$\frac{5}{8} = \frac{5 \cdot 6}{8 \cdot 6} = \frac{30}{48}$

$\frac{30}{48}$ is equivalent to $\frac{5}{8}$.

You Try It 7 $8 = \frac{8}{1}$ $\quad 12 \div 1 = 12$

$8 = \frac{8}{1} = \frac{8 \cdot 12}{1 \cdot 12} = \frac{96}{12}$

$\frac{96}{12}$ is equivalent to 8.

You Try It 8 $\frac{21}{84} = \frac{3 \cdot 7}{2 \cdot 2 \cdot 3 \cdot 7} = \frac{1}{4}$

You Try It 9 $\frac{32}{12} = \frac{2 \cdot 2 \cdot 2 \cdot 2 \cdot 2}{2 \cdot 2 \cdot 3} = \frac{8}{3}$

You Try It 10 $\frac{11t}{11} = \frac{11 \cdot t}{11} = t$

You Try It 11 The LCM of 9 and 21 is 63.

$\frac{4}{9} = \frac{28}{63} \qquad \frac{8}{21} = \frac{24}{63}$

$\frac{28}{63} > \frac{24}{63}$

$\frac{4}{9} > \frac{8}{21}$

You Try It 12 The LCM of 24 and 9 is 72.

$\frac{17}{24} = \frac{51}{72} \qquad \frac{7}{9} = \frac{56}{72}$

$\frac{51}{72} < \frac{56}{72}$

$\frac{17}{24} < \frac{7}{9}$

SECTION 2.3

You Try It 1 $\frac{7}{12} + \frac{3}{8} = \frac{14}{24} + \frac{9}{24} = \frac{23}{24}$

You Try It 2 $\frac{3}{5} + \frac{2}{3} + \frac{5}{6} = \frac{18}{30} + \frac{20}{30} + \frac{25}{30} = \frac{63}{30}$

$= 2\frac{3}{30} = 2\frac{1}{10}$

You Try It 3 $16 + 8\frac{5}{9} = 24\frac{5}{9}$

You Try It 4

$$\begin{array}{c|c} \multicolumn{2}{c}{\frac{2}{3} + z = \frac{23}{24}} \\ \hline \frac{2}{3} + \frac{3}{8} & \frac{23}{24} \\ \frac{16}{24} + \frac{9}{24} & \frac{23}{24} \\ \multicolumn{2}{c}{\frac{25}{24} \neq \frac{23}{25}} \end{array}$$

No, $\frac{3}{8}$ is not a solution of $\frac{2}{3} + z = \frac{23}{24}$.

You Try It 5 $x + y + z$

$3\frac{5}{6} + 2\frac{1}{9} + 5\frac{5}{12} = 3\frac{30}{36} + 2\frac{4}{36} + 5\frac{15}{36}$

$= 10\frac{49}{36}$

$= 11\frac{13}{36}$

You Try It 6 $\frac{5}{6} - \frac{7}{9} = \frac{15}{18} - \frac{14}{18} = \frac{1}{18}$

You Try It 7 $9\frac{7}{8} - 5\frac{2}{3} = 9\frac{21}{24} - 5\frac{16}{24} = 4\frac{5}{24}$

You Try It 8 $6 - 4\frac{2}{11} = 5\frac{11}{11} - 4\frac{2}{11} = 1\frac{9}{11}$

You Try It 9

Strategy To find the fraction of the respondents who did not name glazed, filled, or frosted:

- Add the three fractions to find the fraction who named glazed, filled, or frosted.
- Subtract the fraction who named glazed, filled, or frosted from 1, the entire group surveyed.

Solution

$$\frac{2}{5} + \frac{8}{25} + \frac{3}{20} = \frac{40}{100} + \frac{32}{100} + \frac{15}{100} = \frac{87}{100}$$

$$1 - \frac{87}{100} = \frac{100}{100} - \frac{87}{100} = \frac{13}{100}$$

$\frac{13}{100}$ of the respondents did not name glazed, filled, or frosted as their favorite type of doughnut.

SECTION 2.4

You Try It 1 $\frac{y}{10} \cdot \frac{z}{7} = \frac{y \cdot z}{10 \cdot 7} = \frac{yz}{70}$

You Try It 2

$$\frac{5}{12} \cdot \frac{9}{35} \cdot \frac{7}{8} = \frac{5 \cdot 9 \cdot 7}{12 \cdot 35 \cdot 8} = \frac{5 \cdot 3 \cdot 3 \cdot 7}{2 \cdot 2 \cdot 3 \cdot 5 \cdot 7 \cdot 2 \cdot 2 \cdot 2} = \frac{3}{32}$$

You Try It 3

$$\frac{8}{9} \cdot 6 = \frac{8}{9} \cdot \frac{6}{1} = \frac{8 \cdot 6}{9 \cdot 1} = \frac{2 \cdot 2 \cdot 2 \cdot 2 \cdot 3}{3 \cdot 3 \cdot 1} = \frac{16}{3} = 5\frac{1}{3}$$

You Try It 4

$$3\frac{6}{7} \cdot 2\frac{4}{9} = \frac{27}{7} \cdot \frac{22}{9} = \frac{27 \cdot 22}{7 \cdot 9} = \frac{3 \cdot 3 \cdot 3 \cdot 2 \cdot 11}{7 \cdot 3 \cdot 3} = \frac{66}{7} = 9\frac{3}{7}$$

You Try It 5 x^4y^3

$$\left(2\frac{1}{3}\right)^4 \cdot \left(\frac{3}{7}\right)^3 = \left(\frac{7}{3}\right)^4 \cdot \left(\frac{3}{7}\right)^3 = \frac{7}{3} \cdot \frac{7}{3} \cdot \frac{7}{3} \cdot \frac{7}{3} \cdot \frac{3}{7} \cdot \frac{3}{7} \cdot \frac{3}{7} = \frac{7 \cdot 7 \cdot 7 \cdot 7 \cdot 3 \cdot 3 \cdot 3}{3 \cdot 3 \cdot 3 \cdot 3 \cdot 7 \cdot 7 \cdot 7} = \frac{7}{3} = 2\frac{1}{3}$$

You Try It 6

$$\frac{5}{6} \div \frac{10}{27} = \frac{5}{6} \cdot \frac{27}{10} = \frac{5 \cdot 27}{6 \cdot 10} = \frac{5 \cdot 3 \cdot 3 \cdot 3}{2 \cdot 3 \cdot 2 \cdot 5} = \frac{9}{4} = 2\frac{1}{4}$$

You Try It 7

$$\frac{x}{8} \div \frac{y}{6} = \frac{x}{8} \cdot \frac{6}{y} = \frac{x \cdot 6}{8 \cdot y} = \frac{x \cdot 2 \cdot 3}{2 \cdot 2 \cdot 2 \cdot y} = \frac{3x}{4y}$$

You Try It 8

$$4\frac{3}{8} \div 3\frac{1}{2} = \frac{35}{8} \div \frac{7}{2} = \frac{35}{8} \cdot \frac{2}{7} = \frac{35 \cdot 2}{8 \cdot 7} = \frac{5 \cdot 7 \cdot 2}{2 \cdot 2 \cdot 2 \cdot 7} = \frac{5}{4} = 1\frac{1}{4}$$

You Try It 9 $x \div y$

$$2\frac{1}{4} \div 9 = \frac{9}{4} \div \frac{9}{1} = \frac{9}{4} \cdot \frac{1}{9} = \frac{9 \cdot 1}{4 \cdot 9} = \frac{3 \cdot 3 \cdot 1}{2 \cdot 2 \cdot 3 \cdot 3} = \frac{1}{4}$$

You Try It 10

$$\frac{2y + 3}{y} = 2$$

$\frac{2\left(\frac{1}{2}\right) + 3}{\frac{1}{2}}$	2
$\frac{1 + 3}{\frac{1}{2}}$	2
$\frac{4}{\frac{1}{2}}$	2
$4(2)$	2
$8 \neq 2$	

No, $\frac{1}{2}$ is not a solution of the equation.

You Try It 11

$$\frac{x}{y-z}$$

$$\frac{2\frac{4}{9}}{3-1\frac{1}{3}} = \frac{\frac{22}{9}}{\frac{5}{3}} = \frac{22}{9} \div \frac{5}{3} = \frac{22}{9} \cdot \frac{3}{5}$$
$$= \frac{22}{15} = 1\frac{7}{15}$$

You Try It 12

Strategy To find the area, use the formula for the area of a triangle, $A = \frac{1}{2}bh$.

$b = 18$ and $h = 9$.

Solution

$$A = \frac{1}{2}bh$$
$$A = \frac{1}{2}(18)(9)$$
$$A = 81$$

81 in^2 of felt are needed.

You Try It 13

Strategy To find the total cost:

- Multiply the amount of material per sash $\left(1\frac{3}{8}\right)$ by the number of sashes (22) to find the total number of yards of material needed.
- Multiply the total number of yards of material needed by the cost per yard (12).

Solution

$$1\frac{3}{8} \cdot 22 = \frac{11}{8} \cdot \frac{22}{1} = \frac{11 \cdot 22}{8 \cdot 1} = \frac{11 \cdot 2 \cdot 11}{2 \cdot 2 \cdot 2 \cdot 1}$$
$$= \frac{121}{4} = 30\frac{1}{4}$$
$$30\frac{1}{4} \cdot 12 = \frac{121}{4} \cdot \frac{12}{1} = \frac{121 \cdot 12}{4 \cdot 1}$$
$$= \frac{11 \cdot 11 \cdot 2 \cdot 2 \cdot 3}{2 \cdot 2 \cdot 1} = 363$$

The total cost of the material is $363.

SECTION 2.5

You Try It 1 The digit 4 is in the thousandths place.

You Try It 2 $\frac{501}{1000} = 0.501$

(five hundred one thousandths)

You Try It 3 $0.67 = \frac{67}{100}$ (sixty-seven hundredths)

You Try It 4 Fifty-five and six thousand eighty-three ten-thousandths

You Try It 5 806.00491

You Try It 6

$$0.065 = 0.0650$$
$$0.0650 < 0.0802$$
$$0.065 < 0.0802$$

You Try It 7

3.03, 0.33, 0.30, 3.30, 0.03
0.03, 0.30, 0.33, 3.03, 3.30
0.03, 0.3, 0.33, 3.03, 3.3

You Try It 8

3.675849 (Given place value: 4; 4 < 5)

3.675849 rounded to the nearest ten-thousandth is 3.6758.

You Try It 9

48.907 (Given place value: 9; 0 < 5)

48.907 rounded to the nearest tenth is 48.9.

You Try It 10

31.8652 (Given place value: 1; 8 > 5)

31.8652 rounded to the nearest whole number is 32.

You Try It 11

Strategy To determine who had more home runs for every 100 times at bat, compare the numbers 7.03 and 7.09.

Solution $7.09 > 7.03$

Ralph Kiner had more home runs for every 100 times at bat.

You Try It 12

Strategy To determine the average annual precipitation to the nearest inch, round the number 2.65 to the nearest whole number.

Solution 2.65 rounded to the nearest whole number is 3.

To the nearest inch, the average annual precipitation in Yuma is 3 in.

SECTION 2.6

You Try It 1

```
 11 1
  8.64
 52.7
+ 0.39105
 61.73105
```

You Try It 2

```
  4 9 10
 25.00          Check:   4.91
-  4.91                +20.09
 20.09                  25.00
```

You Try It 3

$$\begin{array}{lcr} 6.514 & \longrightarrow & 7 \\ 8.903 & \longrightarrow & 9 \\ 2.275 & \longrightarrow & +\ 2 \\ \hline & & 18 \end{array}$$

You Try It 4

$x + y + z$

$7.84 + 3.05 + 2.19$

$= 10.89 + 2.19$

$= 13.08$

You Try It 5

$$\begin{array}{r} 0.000081 \\ \times \quad 0.025 \\ \hline 405 \\ 162\ \ \\ \hline 0.000002025 \end{array}$$

You Try It 6

$$\begin{array}{lcr} 6.407 & \longrightarrow & 6 \\ 0.959 & \longrightarrow & \times 1 \\ \hline & & 6 \end{array}$$

You Try It 7

$1.756 \cdot 10^4 = 17{,}560$

You Try It 8

$25xy$

$25(0.8)(0.6) = 20(0.6) = 12$

You Try It 9

$$\begin{array}{r} 48.2 \\ 6.53\overline{)314.74.6} \\ -\ 261\ 2 \\ \hline 53\ 54 \\ -\ 52\ 24 \\ \hline 1\ 30\ 6 \\ -\ 1\ 30\ 6 \\ \hline 0 \end{array}$$

You Try It 10

$$\begin{array}{lcr} 62.7 & \longrightarrow & 60 \\ 3.45 & \longrightarrow & 3 \end{array}$$

$60 \div 3 = 20$

You Try It 11

$$\begin{array}{r} 6.0391 \approx 6.039 \\ 86\overline{)519.3700} \\ -516 \\ \hline 3\ 3 \\ -\ \ 0 \\ \hline 3\ 37 \\ -\ 2\ 58 \\ \hline 790 \\ -\ 774 \\ \hline 160 \\ -\ \ 86 \\ \hline 74 \end{array}$$

You Try It 12

$63.7 \div 100 = 0.637$

You Try It 13

$\dfrac{x}{y}$

$\dfrac{40.6}{0.7} = 58$

You Try It 14

$$\begin{array}{r} 0.8 \\ 5\overline{)4.0} \end{array} \qquad \frac{4}{5} = 0.8$$

You Try It 15

$$\begin{array}{r} 0.8333 \\ 6\overline{)5.0000} \end{array} \qquad 1\frac{5}{6} = 1.8\overline{3}$$

You Try It 16

$6.2 = 6\dfrac{2}{10} = 6\dfrac{1}{5}$

You Try It 17

$\dfrac{7}{12} \approx 0.5833$

$0.5880 > 0.5833$

$0.588 > \dfrac{7}{12}$

You Try It 18

Strategy

To find the change you receive:

- Multiply the number of stamps (12) by the cost of each stamp (44¢) to find the total cost of the stamps.
- Convert the total cost of the stamps to dollars and cents.
- Subtract the total cost of the stamps from $10.

Solution

$12(44) = 528$ The stamps cost 528¢.

$528¢ = \$5.28$ The stamps cost $5.28.

$10.00 - 5.28 = \$4.72$

You receive $4.72 in change.

You Try It 19

Strategy

To find the profit:

- Divide the number of pounds per 100-pound container (100) by the number of pounds packaged in each bag (2) to find the number of bags sold.
- Multiply the number of bags sold by the selling price per bag (12.50) to find the income from selling the nuts.
- Multiply the number of bags sold by the cost for each bag (.06) to find the total cost of the bags.
- Subtract the cost of the bags and the cost of the nuts (475) from the income.

Solution

$100 \div 2 = 50$ Each container makes 50 bags of nuts.

$50(12.50) = 625$ The income from the 50 bags is $625.

$50(.06) = 3$ The total cost of the bags is $3.

$625 - 3 - 475 = 147$

The profit is $147.

You Try It 20

Strategy

To find the insurance premium due, replace B by 276.25 and F by 1.8 in the given formula and solve for P.

Solution

$P = BF$

$P = 276.25(1.8)$

$P = 497.25$

The insurance premium due is $497.25.

SECTION 2.7

You Try It 1

$(1.2 - 0.8)^2 + (1.5)(6)$
$= (0.4)^2 + (1.5)(6)$
$= 0.16 + (1.5)(6)$
$= 0.16 + 9$
$= 9.16$

You Try It 2

$\left(\frac{1}{2}\right)^3 \cdot \frac{7-3}{9-4} + \frac{4}{5}$

$= \left(\frac{1}{2}\right)^3 \cdot \frac{4}{5} + \frac{4}{5}$

$= \frac{1}{8} \cdot \frac{4}{5} + \frac{4}{5}$

$= \frac{1}{10} + \frac{4}{5} = \frac{1}{10} + \frac{8}{10} = \frac{9}{10}$

SOLUTIONS TO CHAPTER 3 "YOU TRY IT"

SECTION 3.1

You Try It 1

4 units

–4 –3 –2 –1 0 1 2 3 4

-3 is 4 units to the left of 1.

You Try It 2

–5 –4 –3 –2 –1 0 1 2 3

A is -5, and C is -3.

You Try It 3

a. 2 is to the right of -5 on the number line.
$2 > -5$

b. -4 is to the left of 3 on the number line.
$-4 < 3$

You Try It 4 $-7, -1, 0, 4, 8$

You Try It 5 **a.** -24 **b.** 13 **c.** b

You Try It 6 **a.** Negative three minus twelve **b.** Eight plus negative five

You Try It 7 **a.** $-(-59) = 59$ **b.** $-(y) = -y$

You Try It 8 **a.** $|-8| = 8$ **b.** $|12| = 12$

You Try It 9 **a.** $|0| = 0$ **b.** $-|35| = -35$

You Try It 10 $|-y| = |-2| = 2$

You Try It 11 $|6| = 6, |-2| = 2, -(-1) = 1, -|-8| = -8$
$-8, -4, 1, 2, 6$
$-|-8|, -4, -(-1), |-2|, |6|$

You Try It 12

Strategy To determine which is closer to blastoff, find the absolute value of each number. The number with the smaller absolute value is closer to zero and, therefore, closer to blastoff.

Solution $|-9| = 9, |-7| = 7$
$7 < 9$
-7 s and counting is closer to blastoff than -9 s and counting.

SECTION 3.2

You Try It 1 $-38 + (-62) = -100$ • The signs of the addends are the same.

You Try It 2 $47 + (-53) = -6$ • The signs of the addends are different.

You Try It 3 $-36 + 17 + (-21) = -19 + (-21)$
$= -40$

You Try It 4 $-154 + (-37) = -191$

You Try It 5 $-x + y$
$-(-3) + (-10) = 3 + (-10) = -7$

You Try It 6

$2 = 11 + a$
$2 \mid 11 + (-9)$ • Replace a by -9.
$2 = 2$

Yes, -9 is a solution of the equation.

You Try It 7 $-35 - (-34)$
$= -35 + 34$
$= -1$
• Rewrite "–" as "+". The opposite of -34 is 34.

You Try It 8 $83 - (-29)$
$= 83 + 29$
$= 112$
• Rewrite "–" as "+". The opposite of -29 is 29.

You Try It 9 The boiling point of xenon is -108.
The melting point of xenon is -112.
$-108 - (-112) = -108 + 112$
$= 4$

The difference is 4°C.

You Try It 10 $-8 - 14$
$= -8 + (-14)$
$= -22$
• Rewrite "–" as "+". The opposite of 14 is -14.

You Try It 11 $25 - 68$
$= 25 + (-68)$
$= -43$
• Rewrite "–" as "+". The opposite of 68 is -68.

You Try It 12 $-4 - (-3) + 12 - (-7) - 20$
$= -4 + 3 + 12 + 7 + (-20)$
$= -1 + 12 + 7 + (-20)$
$= 11 + 7 + (-20)$
$= 18 + (-20)$
$= -2$

You Try It 13 $x - y$
$-9 - 7 = -9 + (-7)$
$= -16$

You Try It 14

$a - 5 = -8$
$-3 - 5 \mid -8$ • Replace a by -3.
$-3 + (-5) \mid -8$
$-8 = -8$

Yes, -3 is a solution of the equation.

You Try It 15

Strategy To find the difference, subtract the lowest melting point shown (−259) from the highest melting point shown (181).

Solution

$$181 - (-259) = 181 + 259$$
$$= 440$$

The difference is 440°C.

You Try It 16

Strategy To find the temperature, add the increase (10) to the previous temperature (−3).

Solution $-3 + 10 = 7$

The temperature is 7°C.

You Try It 17

Strategy To find the difference, subtract the lower temperature (−70) from the higher temperature (57).

Solution

$$57 - (-70) = 57 + 70$$
$$= 127$$

The difference between the average temperatures is 127°F.

You Try It 18

Strategy To find d, replace a by −6 and b by 5 in the given formula and solve for d.

Solution

$$d = |a - b|$$
$$d = |-6 - 5|$$
$$d = |-11|$$
$$d = 11$$

The distance between the two points is 11 units.

SECTION 3.3

You Try It 1 $-38(51) = -1938$ • **The signs are different. The product is negative.**

You Try It 2

$$-7(-8)(9)(-2) = 56(9)(-2)$$
$$= 504(-2)$$
$$= -1008$$

You Try It 3

$-9y$

$-9(20) = -180$

You Try It 4

$12 = -4a$

$12 \mid -4(-3)$

$12 = 12$

Yes, −3 is a solution of the equation.

You Try It 5 $0 \div (-17) = 0$ • **If $a \neq 0$, $\frac{0}{a} = 0$.**

You Try It 6 $\frac{84}{-6} = -14$ • **The signs are different. The quotient is negative.**

You Try It 7 Any number divided by 1 is the number.

$x \div 1 = x$

You Try It 8

$$\frac{a}{-b}$$

$$\frac{-14}{-(-7)} = \frac{-14}{7} = -2$$

You Try It 9

$$\frac{-6}{y} = -2$$

$\frac{-6}{-3} \mid -2$ • **Replace y by −3.**

$2 \neq -2$

No, −3 is not a solution of the equation.

You Try It 10

Strategy To find the average daily high temperature:

- Add the seven temperature readings.
- Divide by 7.

Solution

$-7 + (-8) + 0 + (-1) + (-6) + (-11) + (-2) = -35$

$-35 \div 7 = -5$

The average daily high temperature was −5°.

SECTION 3.4

You Try It 1

$$-\frac{5}{12} + \frac{5}{8} + \left(-\frac{1}{6}\right)$$

$= \frac{-5}{12} + \frac{5}{8} + \frac{-1}{6}$ • **Rewrite with negative signs in the numerators.**

$= \frac{-10}{24} + \frac{15}{24} + \frac{-4}{24}$ • **The LCM of the denominators is 24.**

$= \frac{-10 + 15 + (-4)}{24}$ • **Add the numerators.**

$= \frac{1}{24}$

You Try It 2

$$-\frac{5}{6}-\frac{7}{9}=\frac{-5}{6}-\frac{7}{9}$$
$$=\frac{-15}{18}-\frac{14}{18}$$
$$=\frac{-15-14}{18}$$
$$=\frac{-29}{18}$$
$$=-\frac{29}{18}=-1\frac{11}{18}$$

You Try It 3

$$4.002-9.378=4.002+(-9.378)$$
$$=-5.376$$

You Try It 4

$$x+y+z$$
$$-7.84+(-3.05)+2.19$$
$$=-10.89+2.19$$
$$=-8.7$$

You Try It 5

$$\frac{2}{3}-y=\frac{11}{12}$$

$\frac{2}{3}-\left(-\frac{1}{4}\right)$	$\frac{11}{12}$
$\frac{2}{3}+\frac{1}{4}$	$\frac{11}{12}$
$\frac{8}{12}+\frac{3}{12}$	$\frac{11}{12}$

• Replace y by $-\frac{1}{4}$.

$$\frac{11}{12}=\frac{11}{12}$$

Yes, $-\frac{1}{4}$ is a solution of the equation.

You Try It 6

$$-\frac{1}{3}\left(-\frac{5}{12}\right)\left(\frac{8}{15}\right)=\frac{1}{3}\cdot\frac{5}{12}\cdot\frac{8}{15}$$
$$=\frac{1\cdot5\cdot8}{3\cdot12\cdot15}$$
$$=\frac{1\cdot5\cdot2\cdot2\cdot2}{3\cdot2\cdot2\cdot3\cdot3\cdot5}$$
$$=\frac{2}{27}$$

• The product of two negative fractions is positive.

You Try It 7

$$\left(3\frac{6}{7}\right)\left(-\frac{4}{9}\right)=-\left(3\frac{6}{7}\cdot\frac{4}{9}\right)$$
$$=-\left(\frac{27}{7}\cdot\frac{4}{9}\right)$$
$$=-\left(\frac{27\cdot4}{7\cdot9}\right)$$
$$=-\left(\frac{3\cdot3\cdot3\cdot2\cdot2}{7\cdot3\cdot3}\right)$$
$$=-\frac{12}{7}=-1\frac{5}{7}$$

• The signs are different. The product is negative.

You Try It 8

$$4\div\left(-\frac{6}{7}\right)=-\left(\frac{4}{1}\div\frac{6}{7}\right)$$
$$=-\left(\frac{4}{1}\cdot\frac{7}{6}\right)$$
$$=-\frac{4\cdot7}{1\cdot6}$$
$$=-\frac{2\cdot2\cdot7}{1\cdot2\cdot3}=-\frac{14}{3}=-4\frac{2}{3}$$

• The signs are different. The quotient is negative.

You Try It 9 $(-0.7)(-5.8)=4.06$

• The signs are the same. The product is positive.

You Try It 10 $-25.7\div0.31\approx-82.9$

• The signs are different. The quotient is negative.

You Try It 11

$$xy$$
$$\left(-5\frac{1}{8}\right)\left(-\frac{2}{3}\right)=\frac{41}{8}\cdot\frac{2}{3}$$
$$=\frac{41\cdot2}{8\cdot3}$$
$$=\frac{41\cdot2}{2\cdot2\cdot2\cdot3}$$
$$=\frac{41}{12}=3\frac{5}{12}$$

• The signs are the same. The product is positive.

You Try It 12

$$\frac{x}{y}$$
$$\frac{-40.6}{-0.7}=58$$

You Try It 13

$$25xy$$
$$25(-0.8)(0.6)=-20(0.6)=-12$$

You Try It 14

$$-2=\frac{d}{-0.6}$$

-2	$\frac{-1.2}{-0.6}$

• Replace d by -1.2.

$$-2\neq2$$

No, -1.2 is not a solution of the equation.

You Try It 15

Strategy To find how many degrees the temperature fell, subtract the lower temperature (-13.33) from the higher temperature (12.78).

Solution

$$12.78-(-13.33)=12.78+13.33$$
$$=26.11$$

The temperature fell 26.11°C in the 15-minute period.

SECTION 3.5

You Try It 1

$$\begin{aligned} 8 \div 4 \cdot 4 - (-2)^2 &= 8 \div 4 \cdot 4 - 4 && \text{• Exponents} \\ &= 2 \cdot 4 - 4 && \text{• Division} \\ &= 8 - 4 && \text{• Multiplication} \\ &= 4 && \text{• Subtraction} \end{aligned}$$

You Try It 2

$$\begin{aligned} &\left(-\frac{1}{2}\right)^3 \cdot \frac{7-3}{4-9} + \frac{4}{5} \\ &= \left(-\frac{1}{2}\right)^3 \cdot \frac{4}{-5} + \frac{4}{5} && \text{• Simplify above and below the fraction bar.} \\ &= -\frac{1}{8} \cdot \frac{4}{-5} + \frac{4}{5} && \text{• Exponents} \\ &= \frac{1}{10} + \frac{4}{5} && \text{• Multiplication} \\ &= \frac{1}{10} + \frac{8}{10} = \frac{9}{10} && \text{• Addition} \end{aligned}$$

You Try It 3

$$\begin{aligned} &3a - 4b \\ &3(-2) - 4(5) = -6 - 4(5) \\ &\qquad = -6 - 20 \\ &\qquad = -6 + (-20) \\ &\qquad = -26 \end{aligned}$$

SOLUTIONS TO CHAPTER 4 "YOU TRY IT"

SECTION 4.1

You Try It 1 -4 is the constant term.

You Try It 2

$$\begin{aligned} &2xy + y^2 \\ &2(-4)(2) + (2)^2 \\ &\quad = 2(-4)(2) + 4 \\ &\quad = (-8)(2) + 4 \\ &\quad = (-16) + 4 \\ &\quad = -12 \end{aligned}$$

You Try It 3

$$\begin{aligned} &\frac{a^2 + b^2}{a + b} \\ &\frac{5^2 + (-3)^2}{5 + (-3)} = \frac{25 + 9}{5 + (-3)} \\ &\qquad = \frac{34}{2} \\ &\qquad = 17 \end{aligned}$$

You Try It 4

$$\begin{aligned} &x^3 - 2(x + y) + z^2 \\ &(2)^3 - 2[2 + (-4)] + (-3)^2 \\ &\quad = 8 - 2(-2) + 9 \\ &\quad = 8 + 4 + 9 \\ &\quad = 12 + 9 \\ &\quad = 21 \end{aligned}$$

SECTION 4.2

You Try It 1 $3a - 2b - 5a + 6b = -2a + 4b$

You Try It 2 $-3y^2 + 7 + 8y^2 - 14 = 5y^2 - 7$

You Try It 3 $-5(4y^2) = -20y^2$

You Try It 4 $-7(-2a) = 14a$

You Try It 5 $-\frac{3}{5}\left(-\frac{7}{9}a\right) = \frac{7}{15}a$

You Try It 6 $5(3 + 7b) = 15 + 35b$

You Try It 7 $(3a - 1)5 = 15a - 5$

You Try It 8 $-8(-2a + 7b) = 16a - 56b$

You Try It 9 $3(12x^2 - x + 8) = 36x^2 - 3x + 24$

You Try It 10 $3(-a^2 - 6a + 7) = -3a^2 - 18a + 21$

You Try It 11

$$\begin{aligned} 3y - 2(y - 7x) &= 3y - 2y + 14x \\ &= y + 14x \end{aligned}$$

You Try It 12

$$\begin{aligned} -2(x - 2y) - (-x + 3y) &= -2x + 4y + x - 3y \\ &= -x + y \end{aligned}$$

You Try It 13

$$\begin{aligned} 3y - 2[x - 4(2 - 3y)] &= 3y - 2[x - 8 + 12y] \\ &= 3y - 2x + 16 - 24y \\ &= -2x - 21y + 16 \end{aligned}$$

SECTION 4.3

You Try It 1 the difference between twice n and the square of n

$2n - n^2$

You Try It 2 the quotient of 7 less than b and 15

$\frac{b - 7}{15}$

You Try It 3 the unknown number: n

the cube of the number: n^3

the total of ten and the cube of the number: $10 + n^3$

$-4(10 + n^3)$

You Try It 4 the unknown number: x

the difference between the number and sixty: $x - 60$

$$\begin{aligned} &5(x - 60) \\ &\quad = 5x - 300 \end{aligned}$$

You Try It 5 the speed of the older model: s

the speed of the new jet plane is twice the speed of the older model: $2s$

You Try It 6 the length of the longer piece: y

the length of the shorter piece: $6 - y$

SOLUTIONS TO CHAPTER 5 "YOU TRY IT"

SECTION 5.1

You Try It 1

$$5 - 4x = 8x + 2$$

$5 - 4\left(\frac{1}{4}\right)$	$8\left(\frac{1}{4}\right) + 2$
$5 - 1$	$2 + 2$

$$4 = 4$$

Yes, $\frac{1}{4}$ is a solution.

You Try It 2

$$10x - x^2 = 3x - 10$$

$10(5) - (5)^2$	$3(5) - 10$
$50 - 25$	$15 - 10$

$$25 \neq 5$$

No, 5 is not a solution.

You Try It 3

$$\frac{5}{6} = y - \frac{3}{8}$$
$$\frac{5}{6} + \frac{3}{8} = y - \frac{3}{8} + \frac{3}{8}$$
$$\frac{29}{24} = y$$

The solution is $\frac{29}{24}$.

You Try It 4

$$-\frac{2x}{5} = 6$$
$$\left(-\frac{5}{2}\right)\left(-\frac{2}{5}x\right) = \left(-\frac{5}{2}\right)(6)$$
$$x = -15$$

The solution is -15.

You Try It 5

$$4x - 8x = 16$$
$$-4x = 16$$
$$\frac{-4x}{-4} = \frac{16}{-4}$$
$$x = -4$$

The solution is -4.

You Try It 6

Strategy To find the distance, solve the equation $d = rt$ for d. The time is 3 h. Therefore, $t = 3$. The plane is moving against the wind, which means the headwind is slowing the actual speed of the plane. 250 mph − 25 mph = 225 mph. Thus $r = 225$.

Solution

$$d = rt$$
$$d = 225(3) \quad \text{• } r = 225, t = 3$$
$$= 675$$

The plane travels 675 mi in 3 h.

SECTION 5.2

You Try It 1

$$5x + 7 = 10$$
$$5x + 7 - 7 = 10 - 7 \quad \text{• Subtract 7.}$$
$$5x = 3$$
$$\frac{5x}{5} = \frac{3}{5} \quad \text{• Divide by 5.}$$
$$x = \frac{3}{5}$$

The solution is $\frac{3}{5}$.

You Try It 2

$$2 = 11 + 3x$$
$$2 - 11 = 11 - 11 + 3x \quad \text{• Subtract 11.}$$
$$-9 = 3x$$
$$\frac{-9}{3} = \frac{3x}{3} \quad \text{• Divide by 3.}$$
$$-3 = x$$

The solution is -3.

You Try It 3

$$\frac{5}{8} - \frac{2x}{3} = \frac{5}{4}$$
$$\frac{5}{8} - \frac{5}{8} - \frac{2}{3}x = \frac{5}{4} - \frac{5}{8} \quad \text{• Recall that } \frac{2x}{3} = \frac{2}{3}x.$$
$$-\frac{2}{3}x = \frac{5}{8}$$
$$-\frac{3}{2}\left(-\frac{2}{3}x\right) = -\frac{3}{2}\left(\frac{5}{8}\right) \quad \text{• Multiply by } -\frac{3}{2}.$$
$$x = -\frac{15}{16}$$

The solution is $-\frac{15}{16}$.

You Try It 4

$$\frac{2}{3}x + 3 = \frac{7}{2}$$
$$6\left(\frac{2}{3}x + 3\right) = 6\left(\frac{7}{2}\right)$$
$$6\left(\frac{2}{3}x\right) + 6(3) = 6\left(\frac{7}{2}\right) \quad \text{• Distributive Property}$$
$$4x + 18 = 21$$
$$4x + 18 - 18 = 21 - 18 \quad \text{• Subtract 18.}$$
$$4x = 3$$
$$\frac{4x}{4} = \frac{3}{4} \quad \text{• Divide by 4.}$$
$$x = \frac{3}{4}$$

The solution is $\frac{3}{4}$.

You Try It 5

$$x - 5 + 4x = 25$$
$$5x - 5 = 25$$
$$5x - 5 + 5 = 25 + 5$$
$$5x = 30$$
$$\frac{5x}{5} = \frac{30}{5}$$
$$x = 6$$

The solution is 6.

You Try It 6

Strategy Given: $P = 45$
Unknown: D

Solution

$$P = 15 + \frac{1}{2}D$$
$$45 = 15 + \frac{1}{2}D$$
$$45 - 15 = 15 - 15 + \frac{1}{2}D$$
$$30 = \frac{1}{2}D$$
$$2(30) = 2 \cdot \frac{1}{2}D$$
$$60 = D$$

The depth is 60 ft.

SECTION 5.3

You Try It 1

$$5x + 4 = 6 + 10x$$
$$5x - 10x + 4 = 6 + 10x - 10x \quad \text{• Subtract } 10x.$$
$$-5x + 4 = 6$$
$$-5x + 4 - 4 = 6 - 4 \quad \text{• Subtract 4.}$$
$$-5x = 2$$
$$\frac{-5x}{-5} = \frac{2}{-5} \quad \text{• Divide by } -5.$$
$$x = -\frac{2}{5}$$

The solution is $-\frac{2}{5}$.

You Try It 2

$$5x - 10 - 3x = 6 - 4x$$
$$2x - 10 = 6 - 4x \quad \text{• Combine like terms.}$$
$$2x + 4x - 10 = 6 - 4x + 4x \quad \text{• Add } 4x.$$
$$6x - 10 = 6$$
$$6x - 10 + 10 = 6 + 10 \quad \text{• Add 10.}$$
$$6x = 16$$
$$\frac{6x}{6} = \frac{16}{6} \quad \text{• Divide by 6.}$$
$$x = \frac{8}{3}$$

The solution is $\frac{8}{3}$.

You Try It 3

$$5x - 4(3 - 2x) = 2(3x - 2) + 6$$
$$5x - 12 + 8x = 6x - 4 + 6 \quad \text{• Distributive Property}$$
$$13x - 12 = 6x + 2$$
$$13x - 6x - 12 = 6x - 6x + 2 \quad \text{• Subtract } 6x.$$
$$7x - 12 = 2$$
$$7x - 12 + 12 = 2 + 12 \quad \text{• Add 12.}$$
$$7x = 14$$
$$\frac{7x}{7} = \frac{14}{7} \quad \text{• Divide by 7.}$$
$$x = 2$$

The solution is 2.

You Try It 4

$$-2[3x - 5(2x - 3)] = 3x - 8$$
$$-2[3x - 10x + 15] = 3x - 8 \quad \text{• Distributive Property}$$
$$-2[-7x + 15] = 3x - 8$$
$$14x - 30 = 3x - 8$$
$$14x - 3x - 30 = 3x - 3x - 8 \quad \text{• Subtract } 3x.$$
$$11x - 30 = -8$$
$$11x - 30 + 30 = -8 + 30 \quad \text{• Add 30.}$$
$$11x = 22$$
$$\frac{11x}{11} = \frac{22}{11} \quad \text{• Divide by 11.}$$
$$x = 2$$

The solution is 2.

You Try It 5

Strategy Given: $F_1 = 45$
$F_2 = 80$
$d = 25$
Unknown: x

Solution

$$F_1x = F_2(d - x)$$
$$45x = 80(25 - x)$$
$$45x = 2000 - 80x$$
$$45x + 80x = 2000 - 80x + 80x$$
$$125x = 2000$$
$$\frac{125x}{125} = \frac{2000}{125}$$
$$x = 16$$

The fulcrum is 16 ft from the 45-pound force.

SECTION 5.4

You Try It 1

The smaller number: n
The larger number: $12 - n$

The total of three times the smaller number and six	amounts to	seven less than the product of four and the larger number

(continued)

(continued)

$$\begin{aligned}3n + 6 &= 4(12 - n) - 7\\ 3n + 6 &= 48 - 4n - 7\\ 3n + 6 &= 41 - 4n\\ 3n + 4n + 6 &= 41 - 4n + 4n\\ 7n + 6 &= 41\\ 7n + 6 - 6 &= 41 - 6\\ 7n &= 35\\ \frac{7n}{7} &= \frac{35}{7}\\ n &= 5\end{aligned}$$

$12 - n = 12 - 5 = 7$

The smaller number is 5.

The larger number is 7.

You Try It 2

Strategy

- First integer: n
  Second integer: $n + 1$
  Third integer: $n + 2$
- The sum of the three integers is -6.

Solution

$$\begin{aligned}n + (n + 1) + (n + 2) &= -6\\ 3n + 3 &= -6\\ 3n &= -9\\ n &= -3\end{aligned}$$

$n + 1 = -3 + 1 = -2$
$n + 2 = -3 + 2 = -1$

The three consecutive integers are -3, -2, and -1.

You Try It 3

Strategy To find the number of tickets that you are purchasing, write and solve an equation using x to represent the number of tickets purchased.

Solution

$3.50 plus $17.50 for each ticket	is	$161

$$\begin{aligned}3.50 + 17.50x &= 161\\ 3.50 - 3.50 + 17.50x &= 161 - 3.50\\ 17.50x &= 157.50\\ \frac{17.50x}{17.50} &= \frac{157.50}{17.50}\\ x &= 9\end{aligned}$$

You are purchasing 9 tickets.

You Try It 4

Strategy

To find the length, write and solve an equation using x to represent the length of the shorter piece and $22 - x$ to represent the length of the longer piece.

Solution

The length of the longer piece	is	4 in. more than twice the length of the shorter piece

$$\begin{aligned}22 - x &= 2x + 4\\ 22 - x - 2x &= 2x - 2x + 4\\ 22 - 3x &= 4\\ 22 - 22 - 3x &= 4 - 22\\ -3x &= -18\\ \frac{-3x}{-3} &= \frac{-18}{-3}\\ x &= 6\end{aligned}$$

$22 - x = 22 - 6 = 16$

The length of the shorter piece is 6 in.

The length of the longer piece is 16 in.

SECTION 5.5

You Try It 1

Strategy

- Pounds of $.55 fertilizer: x

	Amount	*Cost*	*Value*
$.80 fertilizer	20	0.80	0.80(20)
$.55 fertilizer	x	0.55	$0.55x$
$.75 fertilizer	$20 + x$	0.75	$0.75(20 + x)$

- The sum of the values before mixing equals the value after mixing.

Solution

$$\begin{aligned}0.80(20) + 0.55x &= 0.75(20 + x)\\ 16 + 0.55x &= 15 + 0.75x\\ 16 - 0.20x &= 15\\ -0.20x &= -1\\ x &= 5\end{aligned}$$

5 lb of the $.55 fertilizer must be added.

You Try It 2

Strategy

- Rate of the first train: r
  Rate of the second train: $2r$

	Rate	*Time*	*Distance*
1st train	r	3	$3r$
2nd train	$2r$	3	$3(2r)$

- The sum of the distances traveled by the two trains equals 288 mi.

Solution

$$3r + 3(2r) = 288$$
$$3r + 6r = 288$$
$$9r = 288$$
$$r = 32$$

$2r = 2(32) = 64$

The first train is traveling at 32 mph.

The second train is traveling at 64 mph.

You Try It 3

Strategy

- Time spent flying out: t
  Time spent flying back: $5 - t$

	Rate	*Time*	*Distance*
Out	150	t	$150t$
Back	100	$5 - t$	$100(5 - t)$

- The distance out equals the distance back.

Solution

$$150t = 100(5 - t)$$
$$150t = 500 - 100t$$
$$250t = 500$$
$$t = 2 \quad \text{(The time out was 2 h.)}$$

The distance out $= 150t = 150(2)$
$= 300$ mi

The parcel of land was 300 mi away.

SOLUTIONS TO CHAPTER 6 "YOU TRY IT"

SECTION 6.1

You Try It 1

$$\frac{12}{20} = \frac{3}{5}$$

12:20 = 3:5

12 to 20 = 3 to 5

You Try It 2

$$\frac{20 \text{ bags}}{8 \text{ acres}} = \frac{5 \text{ bags}}{2 \text{ acres}}$$

You Try It 3

$$\frac{\$8.96}{3.5 \text{ lb}}$$

$8.96 \div 3.5 = 2.56$

The unit rate is \$2.56/lb.

SECTION 6.2

You Try It 1

$$\frac{50}{3} \times \frac{250}{12} \quad 3 \cdot 250 = 750 \quad 50 \cdot 12 = 600$$

$750 \neq 600$

The proportion is not true.

You Try It 2

$$\frac{7}{12} = \frac{42}{x}$$
$$12 \cdot 42 = 7 \cdot x$$ • The cross products are equal.
$$504 = 7x$$
$$72 = x$$

You Try It 3

$$\frac{5}{n} = \frac{3}{322}$$
$$n \cdot 3 = 5 \cdot 322$$ • The cross products are equal.
$$3n = 1610$$
$$\frac{3n}{3} = \frac{1610}{3}$$
$$n \approx 536.67$$

You Try It 4

$$\frac{4}{5} = \frac{3}{x - 3}$$
$$5 \cdot 3 = 4(x - 3)$$ • The cross products are equal.
$$15 = 4x - 12$$
$$27 = 4x$$
$$6.75 = x$$

You Try It 5

Strategy To find the number of gallons, write and solve a proportion using n to represent the number of gallons needed to travel 832 mi.

Solution

$$\frac{396 \text{ mi}}{11 \text{ gal}} = \frac{832 \text{ mi}}{n \text{ gal}}$$
$$11 \cdot 832 = 396 \cdot n$$
$$9152 = 396n$$
$$23.1 \approx n$$

To travel 832 mi, approximately 23.1 gal of gas are needed.

You Try It 6

Strategy To find the number of defective transmissions, write and solve a proportion using n to represent the number of defective transmissions in 120,000 cars.

Solution

$$\frac{15 \text{ defective transmissions}}{1200 \text{ cars}} = \frac{n \text{ defective transmissions}}{120{,}000 \text{ cars}}$$
$$1200 \cdot n = 15 \cdot 120{,}000$$
$$1200n = 1{,}800{,}000$$
$$n = 1500$$

1500 defective transmissions would be found in 120,000 cars.

SECTION 6.3

You Try It 1

a. $125\% = 125 \times \frac{1}{100} = \frac{125}{100} = 1\frac{1}{4}$

b. $125\% = 125 \times 0.01 = 1.25$

You Try It 2

$$33\frac{1}{3}\% = 33\frac{1}{3} \times \frac{1}{100}$$
$$= \frac{100}{3} \times \frac{1}{100}$$
$$= \frac{100}{300} = \frac{1}{3}$$

You Try It 3

$0.25\% = 0.25 \times 0.01 = 0.0025$

You Try It 4

$0.048 = 0.048 \times 100\% = 4.8\%$
$3.67 = 3.67 \times 100\% = 367\%$
$0.62\frac{1}{2} = 0.62\frac{1}{2} \times 100\%$
$= 62\frac{1}{2}\%$

You Try It 5 $\frac{5}{6} = \frac{5}{6} \times 100\% = \frac{500}{6}\% = 83\frac{1}{3}\%$

You Try It 6 $1\frac{4}{9} = \frac{13}{9} = \frac{13}{9} \times 100\%$
$= \frac{1300}{9}\% \approx 144.4\%$

SECTION 6.4

You Try It 1

Strategy To find the amount, solve the basic percent equation.

Percent $= 33\frac{1}{3}\% = \frac{1}{3}$, base $= 45$, amount $= n$

Solution Percent · base = amount

$\frac{1}{3}(45) = n$
$15 = n$

15 is $33\frac{1}{3}\%$ of 45.

You Try It 2

Strategy To find the percent, solve the basic percent equation. Percent $= n$, base $= 40$, amount $= 25$

Solution Percent · base = amount

$n \cdot 40 = 25$
$\frac{40n}{40} = \frac{25}{40}$
$n = 0.625 = 62.5\%$

25 is 62.5% of 40.

You Try It 3

Strategy To find the base, solve the basic percent equation.

Percent $= 16\frac{2}{3}\% = \frac{1}{6}$, base $= n$, amount $= 15$

Solution Percent · base = amount

$\frac{1}{6} \cdot n = 15$
$6 \cdot \frac{1}{6}n = 15 \cdot 6$
$n = 90$

$16\frac{2}{3}\%$ of 90 is 15.

You Try It 4 Percent $= 25$, base $= n$, amount $= 8$

$\frac{25}{100} = \frac{8}{n}$
$25 \cdot n = 100 \cdot 8$
$25n = 800$
$\frac{25n}{25} = \frac{800}{25}$
$n = 32$

8 is 25% of 32.

You Try It 5 Percent $= 0.74$, base $= 1200$, amount $= n$

$\frac{0.74}{100} = \frac{n}{1200}$
$100 \cdot n = 0.74 \cdot 1200$
$100n = 888$
$\frac{100n}{100} = \frac{888}{100}$
$n = 8.88$

0.74% of 1200 is 8.88.

You Try It 6 Percent $= n$, base $= 180$, amount $= 54$

$\frac{n}{100} = \frac{54}{180}$
$n \cdot 180 = 100 \cdot 54$
$180n = 5400$
$\frac{180n}{180} = \frac{5400}{180}$
$n = 30$

30% of 180 is 54.

You Try It 7

Strategy To find the percent, use the basic percent equation. Percent $= n$, base $= 4330$, amount $= 649.50$

Solution Percent · base = amount

$n \cdot 4330 = 649.50$
$\frac{4330n}{4330} = \frac{649.50}{4330}$
$n = 0.15$

15% of the instructor's salary is deducted for income tax.

You Try It 8

Strategy To find the number, solve the basic percent equation. Percent $= 19\% = 0.19$, base $= 2.4$ million, amount $= n$

Solution Percent · base = amount

$0.19 \cdot 2.4 = n$
$0.456 = n$

0.456 million = 456,000

There are approximately 456,000 female surfers in this country.

You Try It 9

Strategy To find the increase in the hourly wage:

- Find last year's wage. Solve the basic percent equation.
  Percent = 115% = 1.15, base = n, amount = 30.13
- Subtract last year's wage from this year's wage.

Solution

$$\text{Percent} \cdot \text{base} = \text{amount}$$
$$1.15 \cdot n = 30.13$$
$$\frac{1.15n}{1.15} = \frac{30.13}{1.15}$$
$$n = 26.20$$

$$30.13 - 26.20 = 3.93$$

The increase in the hourly wage was $3.93.

SECTION 6.5

You Try It 1

Strategy To calculate the maturity value:

- Find the simple interest due on the loan by solving the simple interest formula for I.

$t = \frac{8}{12}$, $P = 12{,}500$, $r = 9.5\% = 0.095$

- Use the formula for the maturity value of a simple interest loan, $M = P + I$.

Solution

$$I = Prt$$
$$I = 12{,}500(0.095)\left(\frac{8}{12}\right)$$
$$I \approx 791.67$$

$$M = P + I$$
$$M = 12{,}500 + 791.67$$
$$M = 13{,}291.67$$

The maturity value of the loan is $13,291.67.

SOLUTIONS TO CHAPTER 7 "YOU TRY IT"

SECTION 7.1

You Try It 1

$$QR + RS + ST = QT$$
$$24 + RS + 17 = 62$$
$$41 + RS = 62$$
$$RS = 21$$

$RS = 21$ cm

You Try It 2

$$AC = AB + BC$$
$$AC = \frac{1}{4}(BC) + BC$$
$$AC = \frac{1}{4}(16) + 16$$
$$AC = 4 + 16$$
$$AC = 20$$
$AC = 20$ ft

You Try It 3

Strategy Supplementary angles are two angles whose sum is 180°. To find the supplement, let x represent the supplement of a 129° angle. Write an equation and solve for x.

Solution

$$x + 129^\circ = 180^\circ$$
$$x = 51^\circ$$

The supplement of a 129° angle is a 51° angle.

You Try It 4

Strategy To find the measure of $\angle a$, write an equation using the fact that the sum of the measure of $\angle a$ and 68° is 118°. Solve for $\angle a$.

Solution

$$\angle a + 68^\circ = 118^\circ$$
$$\angle a = 50^\circ$$

The measure of $\angle a$ is 50°.

You Try It 5

Strategy The angles labeled are adjacent angles of intersecting lines and are therefore supplementary angles. To find x, write an equation and solve for x.

Solution

$$(x + 16^\circ) + 3x = 180^\circ$$
$$4x + 16^\circ = 180^\circ$$
$$4x = 164^\circ$$
$$x = 41^\circ$$

You Try It 6

Strategy $3x = y$ because corresponding angles have the same measure. $y + (x + 40^\circ) = 180^\circ$ because adjacent angles of intersecting lines are supplementary angles. Substitute $3x$ for y and solve for x.

Solution

$$3x + (x + 40^\circ) = 180^\circ$$
$$4x + 40^\circ = 180^\circ$$
$$4x = 140^\circ$$
$$x = 35^\circ$$

You Try It 7

Strategy

- To find the measure of angle b, use the fact that $\angle b$ and $\angle x$ are supplementary angles.
- To find the measure of angle c, use the fact that the sum of the interior angles of a triangle is 180°.
- To find the measure of angle y, use the fact that $\angle c$ and $\angle y$ are vertical angles.

Solution

$$\angle b + \angle x = 180^\circ$$
$$\angle b + 100^\circ = 180^\circ$$
$$\angle b = 80^\circ$$

$$\angle a + \angle b + \angle c = 180^\circ$$
$$45^\circ + 80^\circ + \angle c = 180^\circ$$
$$125^\circ + \angle c = 180^\circ$$
$$\angle c = 55^\circ$$

$$\angle y = \angle c = 55^\circ$$

You Try It 8

Strategy

To find the measure of the third angle, use the fact that the measure of a right angle is 90° and the fact that the sum of the measures of the interior angles of a triangle is 180°. Write an equation using x to represent the measure of the third angle. Solve the equation for x.

Solution

$$x + 90^\circ + 34^\circ = 180^\circ$$
$$x + 124^\circ = 180^\circ$$
$$x = 56^\circ$$

The measure of the third angle is 56°.

SECTION 7.2

You Try It 1

Strategy

To find the perimeter, use the formula for the perimeter of a square. Substitute 60 ft for s and solve for P.

Solution

$$P = 4s$$
$$P = 4(60 \text{ ft})$$
$$P = 240 \text{ ft}$$

The perimeter of the infield is 240 ft.

You Try It 2

Strategy

To find the perimeter, use the formula for the perimeter of a rectangle. Substitute 11 in. for L and $8\frac{1}{2}$ in. for W and solve for P.

Solution

$$P = 2L + 2W$$
$$P = 2(11 \text{ in.}) + 2\left(8\frac{1}{2} \text{ in.}\right)$$
$$P = 2(11 \text{ in.}) + 2\left(\frac{17}{2} \text{ in.}\right)$$
$$P = 22 \text{ in.} + 17 \text{ in.}$$
$$P = 39 \text{ in.}$$

The perimeter of a standard piece of copier paper is 39 in.

You Try It 3

Strategy

To find the circumference, use the circumference formula that involves the diameter. Leave the answer in terms of π.

Solution

$$C = \pi d$$
$$C = \pi(9 \text{ in.})$$
$$C = 9\pi \text{ in.}$$

The circumference is 9π in.

You Try It 4

Strategy

To find the number of rolls of wallpaper to be purchased:

- Use the formula for the area of a rectangle to find the area of one wall.
- Multiply the area of one wall by the number of walls to be covered (2).
- Divide the area of wall to be covered by the area one roll of wallpaper will cover (30).

Solution

$A = LW$

$A = (12 \text{ ft})(8 \text{ ft}) = 96 \text{ ft}^2$ • **The area of one wall is 96 ft^2.**

$2(96 \text{ ft}^2) = 192 \text{ ft}^2$ • **The area of the two walls is 192 ft^2.**

$192 \div 30 = 6.4$

Because a portion of a seventh roll is needed, 7 rolls of wallpaper should be purchased.

You Try It 5

Strategy

To find the area, use the formula for the area of a circle. An approximation is asked for; use the π key on a calculator. $r = 11$ cm.

Solution

$$A = \pi r^2$$
$$A = \pi(11 \text{ cm})^2$$
$$A = 121\pi \text{ cm}^2$$
$$A \approx 380.13 \text{ cm}^2$$

The area is approximately 380.13 cm^2.

SECTION 7.3

You Try It 1

Strategy To find the measure of the other leg, use the Pythagorean Theorem.
$a = 2, c = 6$

Solution
$a^2 + b^2 = c^2$
$2^2 + b^2 = 6^2$
$4 + b^2 = 36$
$b^2 = 32$
$b = \sqrt{32}$
$b \approx 5.66$

The measure of the other leg is approximately 5.66 m.

You Try It 2

Strategy To find FG, write a proportion using the fact that, in similar triangles, the ratio of corresponding sides equals the ratio of corresponding heights. Solve the proportion for FG.

Solution
$\frac{AC}{DF} = \frac{CH}{FG}$
$\frac{10}{15} = \frac{7}{FG}$
$10(FG) = 15(7)$
$10(FG) = 105$
$FG = 10.5$

The height FG of triangle DEF is 10.5 m.

You Try It 3

Strategy To determine whether the triangles are congruent, determine whether one of the rules for congruence is satisfied.

Solution $PR = MN$, $QR = MO$, and $\angle QRP = \angle OMN$. Two sides and the included angle of one triangle are equal to two sides and the included angle of the other triangle.

The triangles are congruent by the SAS Rule.

SECTION 7.4

You Try It 1

Strategy To find the volume, use the formula for the volume of a cube. $s = 2.5$ m.

Solution
$V = s^3$
$V = (2.5\text{ m})^3 = 15.625\text{ m}^3$

The volume of the cube is 15.625 m^3.

You Try It 2

Strategy To find the volume:
- Find the radius of the base of the cylinder. $d = 8$ ft.
- Use the formula for the volume of a cylinder. Leave the answer in terms of π.

Solution
$r = \frac{1}{2}d = \frac{1}{2}(8\text{ ft}) = 4\text{ ft}$
$V = \pi r^2 h = \pi(4\text{ ft})^2(22\text{ ft})$
$= \pi(16\text{ ft}^2)(22\text{ ft}) = 352\pi\text{ ft}^3$

The volume of the cylinder is 352π ft^3.

You Try It 3

Strategy To find the surface area of the cylinder.
- Find the radius of the base of the cylinder. $d = 6$ ft.
- Use the formula for the surface area of a cylinder. An approximation is asked for; use the π key on a calculator.

Solution
$r = \frac{1}{2}d = \frac{1}{2}(6\text{ ft}) = 3\text{ ft}$
$SA = 2\pi r^2 + 2\pi rh$
$SA = 2\pi(3\text{ ft})^2 + 2\pi(3\text{ ft})(8\text{ ft})$
$SA = 2\pi(9\text{ ft}^2) + 2\pi(3\text{ ft})(8\text{ ft})$
$SA = 18\pi\text{ ft}^2 + 48\pi\text{ ft}^2$
$SA = 66\pi\text{ ft}^2$
$SA \approx 207.35\text{ ft}^2$

The surface area of the cylinder is approximately 207.35 ft^2.

You Try It 4

Strategy To find which solid has the larger surface area:
- Use the formula for the surface area of a cube to find the surface area of the cube. $s = 10$ cm.
- Find the radius of the sphere. $d = 8$ cm.
- Use the formula for the surface area of a sphere to find the surface area of the sphere. Because this number is to be compared to another number, use the π key on a calculator to approximate the surface area.
- Compare the two numbers.

(continued)

(continued)

Solution

$SA = 6s^2$
$SA = 6(10 \text{ cm})^2 = 6(100 \text{ cm}^2)$
$= 600 \text{ cm}^2$

The surface area of the cube is 600 cm^2.

$r = \frac{1}{2}d = \frac{1}{2}(8 \text{ cm}) = 4 \text{ cm}$
$SA = 4\pi r^2$
$SA = 4\pi(4 \text{ cm})^2 = 4\pi(16 \text{ cm}^2)$
$= 64\pi \text{ cm}^2 \approx 201.06 \text{ cm}^2$

The surface area of the sphere is approximately 201.06 cm^2.

$600 > 201.06$

The cube has a larger surface area than the sphere.

SOLUTIONS TO CHAPTER 8 "YOU TRY IT"

SECTION 8.1

You Try It 1

Strategy

a. To find the ratio:
- From the graph, find the percent of lane-change accidents and the percent of road-departure accidents.
- Write the ratio in fractional form. Simplify.

b. To find the number of accidents that occurred at intersections:
- From the graph, find the percent of accidents that occurred at intersections.
- Solve the basic percent equation for amount. The base is 4300.

Solution

a. Lane-change accidents: 9%
Road-departure accidents: 21%

$\frac{9\%}{21\%} = \frac{3}{7}$

The ratio is $\frac{3}{7}$.

b. Accidents that occurred at intersections: 26% = 0.26

Percent · base = amount
0.26 · 4300 = n
1118 = n

1118 accidents occurred at intersections in Twin Falls in 2005.

SECTION 8.2

You Try It 1

Strategy

To find the mean amount spent by the 12 customers:
- Find the sum of the numbers.
- Divide the sum by the number of customers (12).

Solution

11.01 + 10.75 + 12.09 + 15.88 + 13.50 + 12.29 + 10.69 + 9.36 + 11.66 + 15.25 + 10.09 + 12.72 = 145.29

$\bar{x} = \frac{145.29}{12} \approx 12.11$

The mean amount spent by the 12 customers was $12.11.

You Try It 2

Strategy

To find the median weight loss:
- Arrange the weight losses from least to greatest.
- Because there is an even number of values, the median is the mean of the middle two numbers.

Solution

10, 14, 16, 16, 22, 27, 29, 31, 31, 40

$\text{Median} = \frac{22 + 27}{2} = \frac{49}{2} = 24.5$

The median weight loss was 24.5 lb.

You Try It 3

Strategy

To draw the box-and-whiskers plot:
- Find the median, Q_1, and Q_3.
- Use the least value, Q_1, the median, Q_3, and the greatest value to draw the box-and-whiskers plot.

Solution

4 7 7 8 8 9 13
↑ Q_1 (at 7), ↑ Median (at 8), ↑ Q_3 (at 9)

a.

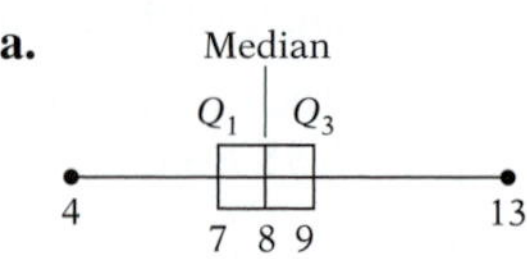

b. Answers about the spread of the data will vary. For example, in You Try It 3, the values in the interquartile range are all very close to the median. They are not so close to the median in Example 3. The whiskers are long with respect to the box in You Try It 3, whereas they are short with respect to the box in Example 3. This shows that the data values outside the interquartile range are closer together in Example 3 than in You Try It 3.

SECTION 8.3

You Try It 1

Strategy To find the probability:

- List the outcomes of the experiment in a systematic way. We will use a table.
- Use the table to count the number of possible outcomes of the experiment.
- Count the number of outcomes of the experiment that are favorable to the event of two true questions and one false question.
- Use the probability formula.

Solution

Question 1	Question 2	Question 3
T	T	T
T	T	F
T	F	T
T	F	F
F	T	T
F	T	F
F	F	T
F	F	F

There are 8 possible outcomes:

S = {TTT, TTF, TFT, TFF, FTT, FTF, FFT, FFF}

There are 3 outcomes favorable to the event:

{TTF, TFT, FTT}

Probability of an event

$$= \frac{\text{number of favorable outcomes}}{\text{number of possible outcomes}} = \frac{3}{8}$$

The probability of two true questions and one false question is $\frac{3}{8}$.

SOLUTIONS TO CHAPTER 9 "YOU TRY IT"

SECTION 9.1

You Try It 1

$$\begin{aligned}&(-4x^3 + 2x^2 - 8) + (4x^3 + 6x^2 - 7x + 5)\\ &\quad = (-4x^3 + 4x^3) + (2x^2 + 6x^2) + (-7x) + (-8 + 5)\\ &\quad = 8x^2 - 7x - 3\end{aligned}$$

You Try It 2

$$\begin{array}{r} 6x^3 \qquad\quad + \ 2x + 8 \\ -9x^3 + 2x^2 - 12x - 8 \\ \hline -3x^3 + 2x^2 - 10x \qquad \end{array}$$

You Try It 3

$$\begin{aligned}&(-4w^3 + 8w - 8) - (3w^3 - 4w^2 - 2w - 1)\\ &\quad = (-4w^3 + 8w - 8)\\ &\qquad + (-3w^3 + 4w^2 + 2w + 1)\\ &\quad = -7w^3 + 4w^2 + 10w - 7\end{aligned}$$

You Try It 4

$$\begin{array}{r} 13y^3 \qquad\qquad - 6y - 7 \\ - 4y^2 + 6y + 9 \\ \hline 13y^3 - 4y^2 \qquad\quad + 2 \end{array}$$

SECTION 9.2

You Try It 1

$$\begin{aligned}&(8m^3n)(-3n^5)\\ &\quad = [8(-3)](m^3)(n \cdot n^5)\\ &\quad = -24m^3n^6\end{aligned}$$

• Multiply coefficients. Add exponents with same base.

You Try It 2

$$\begin{aligned}&(12p^4q^3)(-3p^5q^2)\\ &\quad = [12(-3)](p^4 \cdot p^5)(q^3 \cdot q^2)\\ &\quad = -36p^9q^5\end{aligned}$$

• Multiply coefficients. Add exponents with same base.

You Try It 3

$$\begin{aligned}(-3a^4bc^2)^3 &= (-3)^{1\cdot 3}a^{4\cdot 3}b^{1\cdot 3}c^{2\cdot 3}\\ &= (-3)^3a^{12}b^3c^6\\ &= -27a^{12}b^3c^6\end{aligned}$$

• Rule for Simplifying the Power of a Product

You Try It 4

$$\begin{aligned}(-xy^4)(-2x^3y^2)^2 &= (-xy^4)[(-2)^{1\cdot 2}x^{3\cdot 2}y^{2\cdot 2}]\\ &= (-xy^4)[(-2)^2x^6y^4]\\ &= (-xy^4)(4x^6y^4)\\ &= -4x^7y^8\end{aligned}$$

• Rule for Simplifying the Power of a Product

SECTION 9.3

You Try It 1

$$(-2y + 3)(-4y) = -2y(-4y) + 3(-4y) = 8y^2 - 12y$$

You Try It 2

$$\begin{aligned}-a^2(3a^2 + 2a - 7) &= -a^2(3a^2) + (-a^2)(2a) - (-a^2)(7)\\ &= -3a^4 - 2a^3 + 7a^2\end{aligned}$$

You Try It 3

$$\begin{array}{rl} 2y^3 + 2y^2 \qquad - 3 & \\ 3y - 1 & \\ \hline -2y^3 - 2y^2 \qquad + 3 & = -1(2y^3 + 2y^2 - 3) \\ 6y^4 + 6y^3 \qquad - 9y \qquad & = 3y(2y^3 + 2y^2 - 3) \\ \hline 6y^4 + 4y^3 - 2y^2 - 9y + 3 & \end{array}$$

You Try It 4

$$\begin{aligned}(4y - 5)(2y - 3) &= 8y^2 - 12y - 10y + 15\\ &= 8y^2 - 22y + 15\end{aligned}$$

You Try It 5

$$\begin{aligned}(3b + 2)(3b - 5) &= 9b^2 - 15b + 6b - 10\\ &= 9b^2 - 9b - 10\end{aligned}$$

You Try It 6

$$(2a + 5c)(2a - 5c) = 4a^2 - 25c^2$$

You Try It 7

$$(3x + 2y)^2 = 9x^2 + 12xy + 4y^2$$

You Try It 8

Strategy To find the area, replace the variable r in the equation $A = \pi r^2$ by $(x - 4)$ and solve for A.

Solution

$$A = \pi r^2$$
$$A = \pi(x - 4)^2$$
$$A = \pi(x^2 - 8x + 16)$$
$$A = \pi x^2 - 8\pi x + 16\pi$$

The area of the circle is $(\pi x^2 - 8\pi x + 16\pi)$ ft^2.

SECTION 9.4

You Try It 1

$$(-2x^2)(x^{-3}y^{-4})^{-2}$$
$$= (-2x^2)(x^6y^8) \quad \text{• Rule for Simplifying the Power of a Product}$$
$$= -2x^8y^8$$

You Try It 2

$$\frac{(6a^{-2}b^3)^{-1}}{(4a^3b^{-2})^{-2}}$$
$$= \frac{6^{-1}a^2b^{-3}}{4^{-2}a^{-6}b^4} \quad \text{• Rule for Simplifying the Power of a Product}$$
$$= 4^2(6^{-1}a^8b^{-7}) \quad \text{• Rule for Dividing Exponential Expressions}$$
$$= \frac{16a^8}{6b^7} = \frac{8a^8}{3b^7}$$

You Try It 3

$$\left[\frac{6r^3s^{-3}}{9r^3s^{-1}}\right]^{-2} = \left[\frac{2r^0s^{-2}}{3}\right]^{-2}$$
$$= \frac{2^{-2}s^4}{3^{-2}} = \frac{9s^4}{4}$$

You Try It 4 $0.000000961 = 9.61 \times 10^{-7}$

You Try It 5 $7.329 \times 10^6 = 7{,}329{,}000$

SECTION 9.5

You Try It 1

$$\frac{24x^2y^2 - 18xy + 6y}{6xy} = \frac{24x^2y^2}{6xy} - \frac{18xy}{6xy} + \frac{6y}{6xy}$$
$$= 4xy - 3 + \frac{1}{x}$$

You Try It 2

$$\begin{array}{r} x^2 + 2x - 1 \\ 2x - 3\overline{)2x^3 + x^2 - 8x - 3} \\ \underline{2x^3 - 3x^2} \qquad\qquad \\ 4x^2 - 8x \qquad \\ \underline{4x^2 - 6x} \qquad \\ -2x - 3 \\ \underline{-2x + 3} \\ -6 \end{array}$$

$$(2x^3 + x^2 - 8x - 3) \div (2x - 3)$$
$$= x^2 + 2x - 1 - \frac{6}{2x - 3}$$

You Try It 3

$$\begin{array}{r} x^2 + x - 1 \\ x - 1\overline{)x^3 + 0 - 2x + 1} \\ \underline{x^3 - x^2} \qquad\qquad \\ x^2 - 2x \qquad \\ \underline{x^2 - x} \qquad \\ -x + 1 \\ \underline{-x + 1} \\ 0 \end{array}$$

$$(x^3 - 2x + 1) \div (x - 1) = x^2 + x - 1$$

SOLUTIONS TO CHAPTER 10 "YOU TRY IT"

SECTION 10.1

You Try It 1 The GCF is $7a^2$.

$$14a^2 - 21a^4b = 7a^2(2) + 7a^2(-3a^2b)$$
$$= 7a^2(2 - 3a^2b)$$

You Try It 2 The GCF is 9.

$$27b^2 + 18b + 9$$
$$= 9(3b^2) + 9(2b) + 9(1)$$
$$= 9(3b^2 + 2b + 1)$$

You Try It 3

The GCF is $3x^2y^2$.

$$6x^4y^2 - 9x^3y^2 + 12x^2y^4$$
$$= 3x^2y^2(2x^2) + 3x^2y^2(-3x) + 3x^2y^2(4y^2)$$
$$= 3x^2y^2(2x^2 - 3x + 4y^2)$$

You Try It 4

$$2y(5x - 2) - 3(2 - 5x)$$
$$= 2y(5x - 2) + 3(5x - 2) \quad \text{• } 5x - 2 \text{ is the common factor.}$$
$$= (5x - 2)(2y + 3)$$

You Try It 5

$$a^2 - 3a + 2ab - 6b$$
$$= (a^2 - 3a) + (2ab - 6b)$$
$$= a(a - 3) + 2b(a - 3) \quad \text{• } a - 3 \text{ is the common factor.}$$
$$= (a - 3)(a + 2b)$$

You Try It 6

$$2mn^2 - n + 8mn - 4$$
$$= (2mn^2 - n) + (8mn - 4)$$
$$= n(2mn - 1) + 4(2mn - 1) \quad \text{• } 2mn - 1 \text{ is the common factor.}$$
$$= (2mn - 1)(n + 4)$$

You Try It 7

$$3xy - 9y - 12 + 4x$$
$$= (3xy - 9y) - (12 - 4x) \quad \text{• } -12 + 4x = -(12 - 4x)$$
$$= 3y(x - 3) - 4(3 - x) \quad \text{• } -(3 - x) = (x - 3)$$
$$= 3y(x - 3) + 4(x - 3) \quad \text{• } x - 3 \text{ is the common factor.}$$
$$= (x - 3)(3y + 4)$$

SECTION 10.2

You Try It 1

Find the positive factors of 20 whose sum is 9.

Factors	Sum
1, 20	21
2, 10	12
4, 5	9

$x^2 + 9x + 20 = (x + 4)(x + 5)$

You Try It 2

Find the factors of -18 whose sum is 7.

Factors	Sum
+1, −18	−17
−1, +18	17
+2, −9	−7
−2, +9	7
+3, −6	−3
−3, +6	3

$x^2 + 7x - 18 = (x + 9)(x - 2)$

You Try It 3

The GCF is $-2x$.

$-2x^3 + 14x^2 - 12x = -2x(x^2 - 7x + 6)$

Factor the trinomial $x^2 - 7x + 6$. Find two negative factors of 6 whose sum is -7.

Factors	Sum
−1, −6	−7
−2, −3	−5

$-2x^3 + 14x^2 - 12x = -2x(x - 6)(x - 1)$

You Try It 4

The GCF is 3.

$3x^2 - 9xy - 12y^2 = 3(x^2 - 3xy - 4y^2)$

Factor the trinomial.

Find the factors of -4 whose sum is -3.

Factors	Sum
+1, −4	−3
−1, +4	3
+2, −2	0

$3x^2 - 9xy - 12y^2 = 3(x + y)(x - 4y)$

SECTION 10.3

You Try It 1

Factor the trinomial $2x^2 - x - 3$.

Positive factors of 2: 1, 2

Factors of -3: +1, −3; −1, +3

Trial Factors	Middle Term
$(x + 1)(2x - 3)$	$-3x + 2x = -x$
$(x - 3)(2x + 1)$	$x - 6x = -5x$
$(x - 1)(2x + 3)$	$3x - 2x = x$
$(x + 3)(2x - 1)$	$-x + 6x = 5x$

$2x^2 - x - 3 = (x + 1)(2x - 3)$

You Try It 2

The GCF is $-3y$.

$-45y^3 + 12y^2 + 12y = -3y(15y^2 - 4y - 4)$

Factor the trinomial $15y^2 - 4y - 4$.

Positive factors of 15: 1, 15; 3, 5

Factors of -4: 1, −4; −1, 4; 2, −2

Trial Factors	Middle Term
$(y + 1)(15y - 4)$	$-4y + 15y = 11y$
$(y - 4)(15y + 1)$	$y - 60y = -59y$
$(y - 1)(15y + 4)$	$4y - 15y = -11y$
$(y + 4)(15y - 1)$	$-y + 60y = 59y$
$(y + 2)(15y - 2)$	$-2y + 30y = 28y$
$(y - 2)(15y + 2)$	$2y - 30y = -28y$
$(3y + 1)(5y - 4)$	$-12y + 5y = -7y$
$(3y - 4)(5y + 1)$	$3y - 20y = -17y$
$(3y - 1)(5y + 4)$	$12y - 5y = 7y$
$(3y + 4)(5y - 1)$	$-3y + 20y = 17y$
$(3y + 2)(5y - 2)$	$-6y + 10y = 4y$
$(3y - 2)(5y + 2)$	$6y - 10y = -4y$

$-45y^3 + 12y^2 + 12y = -3y(3y - 2)(5y + 2)$

You Try It 3

Factors of −14 [2(−7)]	Sum
+1, −14	−13
−1, +14	13
+2, −7	−5
−2, +7	5

$$\begin{aligned} 2a^2 + 13a - 7 &= 2a^2 - a + 14a - 7 \\ &= (2a^2 - a) + (14a - 7) \\ &= a(2a - 1) + 7(2a - 1) \\ &= (2a - 1)(a + 7) \end{aligned}$$

$2a^2 + 13a - 7 = (2a - 1)(a + 7)$

You Try It 4

The GCF is $5x$.

$15x^3 + 40x^2 - 80x = 5x(3x^2 + 8x - 16)$

Factors of -48 [$3(-16)$]	Sum
$+1, -48$	-47
$-1, +48$	47
$+2, -24$	-22
$-2, +24$	22
$+3, -16$	-13
$-3, +16$	13
$+4, -12$	-8
$-4, +12$	8

$$\begin{aligned}3x^2 + 8x - 16 &= 3x^2 - 4x + 12x - 16\\ &= (3x^2 - 4x) + (12x - 16)\\ &= x(3x - 4) + 4(3x - 4)\\ &= (3x - 4)(x + 4)\end{aligned}$$

$$\begin{aligned}15x^3 + 40x^2 - 80x &= 5x(3x^2 + 8x - 16)\\ &= 5x(3x - 4)(x + 4)\end{aligned}$$

SECTION 10.4

You Try It 1

$$\begin{aligned}25a^2 - b^2 &= (5a)^2 - b^2\\ &= (5a + b)(5a - b)\end{aligned}$$

• Difference of two squares

You Try It 2

$n^4 - 81 = (n^2)^2 - 9^2$ • Difference of two squares

$= (n^2 + 9)(n^2 - 9)$ • Difference of two squares

$= (n^2 + 9)(n + 3)(n - 3)$

You Try It 3

Because $16y^2 = (4y)^2$, $1 = 1^2$, and $8y = 2(4y)(1)$, the trinomial is a perfect-square trinomial.

$16y^2 + 8y + 1 = (4y + 1)^2$

You Try It 4

Because $x^2 = (x)^2$, $36 = 6^2$, and $15x \neq 2(x)(6)$, the trinomial is not a perfect-square trinomial. Try to factor the trinomial by another method.

$x^2 + 15x + 36 = (x + 3)(x + 12)$

You Try It 5

The GCF is $3x$.

$$\begin{aligned}12x^3 - 75x &= 3x(4x^2 - 25)\\ &= 3x(2x + 5)(2x - 5)\end{aligned}$$

You Try It 6

Factor by grouping.

$a^2b - 7a^2 - b + 7$

$= (a^2b - 7a^2) - (b - 7)$

$= a^2(b - 7) - (b - 7)$ • $b - 7$ is the common factor.

$= (b - 7)(a^2 - 1)$ • $a^2 - 1$ is the difference of two squares.

$= (b - 7)(a + 1)(a - 1)$

You Try It 7

The GCF is $4x$.

$4x^3 + 28x^2 - 120x$

$= 4x(x^2 + 7x - 30)$ • Factor out the GCF, $4x$.

$= 4x(x + 10)(x - 3)$ • Factor the trinomial.

SECTION 10.5

You Try It 1

$2x(x + 7) = 0$

$2x = 0 \quad x + 7 = 0$ • Principle of Zero Products

$x = 0 \quad x = -7$

The solutions are 0 and -7.

You Try It 2

$4x^2 - 9 = 0$ • Difference of two squares

$(2x - 3)(2x + 3) = 0$

$2x - 3 = 0 \quad 2x + 3 = 0$ • Principle of Zero Products

$2x = 3 \quad 2x = -3$

$x = \frac{3}{2} \quad x = -\frac{3}{2}$

The solutions are $\frac{3}{2}$ and $-\frac{3}{2}$.

You Try It 3

$(x + 2)(x - 7) = 52$

$x^2 - 5x - 14 = 52$

$x^2 - 5x - 66 = 0$

$(x + 6)(x - 11) = 0$

$x + 6 = 0 \quad x - 11 = 0$ • Principle of Zero Products

$x = -6 \quad x = 11$

The solutions are -6 and 11.

You Try It 4

Strategy First consecutive positive integer: n
Second consecutive positive integer: $n + 1$
The sum of the squares of the two consecutive positive integers is 61.

Solution

$n^2 + (n + 1)^2 = 61$

$n^2 + n^2 + 2n + 1 = 61$

$2n^2 + 2n + 1 = 61$

$2n^2 + 2n - 60 = 0$

$2(n^2 + n - 30) = 0$

$2(n - 5)(n + 6) = 0$

$n - 5 = 0 \quad n + 6 = 0$ • Principle of Zero Products

$n = 5 \quad n = -6$

Because -6 is not a positive integer, it is not a solution.

$n = 5$

$n + 1 = 5 + 1 = 6$

The two integers are 5 and 6.

You Try It 5

Strategy Width = x

Length = $2x + 4$

The area of the rectangle is 96 in^2.

Use the equation $A = L \cdot W$.

Solution

$$\begin{aligned} A &= L \cdot W \\ 96 &= (2x+4)x \\ 96 &= 2x^2 + 4x \\ 0 &= 2x^2 + 4x - 96 \\ 0 &= 2(x^2 + 2x - 48) \\ 0 &= 2(x+8)(x-6) \end{aligned}$$

$x + 8 = 0 \quad x - 6 = 0$ • Principle of Zero Products

$x = -8 \quad x = 6$

Because the width cannot be a negative number, -8 is not a solution.

$x = 6$

$2x + 4 = 2(6) + 4 = 12 + 4 = 16$

The length is 16 in. The width is 6 in.

SOLUTIONS TO CHAPTER 11 "YOU TRY IT"

SECTION 11.1

You Try It 1 $\dfrac{6x^5y}{12x^2y^3} = \dfrac{\overset{1}{\cancel{2}} \cdot \overset{1}{\cancel{3}} \cdot x^5y}{\underset{1}{\cancel{2}} \cdot 2 \cdot \underset{1}{\cancel{3}} \cdot x^2y^3} = \dfrac{x^3}{2y^2}$

You Try It 2

$$\frac{x^2+4x-12}{x^2-3x+2} = \frac{\overset{1}{\cancel{(x-2)}}(x+6)}{(x-1)\underset{1}{\cancel{(x-2)}}} = \frac{x+6}{x-1}$$

You Try It 3

$$\frac{x^2+2x-24}{16-x^2} = \frac{\overset{-1}{\cancel{(x-4)}}(x+6)}{\underset{1}{\cancel{(4-x)}}(4+x)} = -\frac{x+6}{x+4}$$

• $\dfrac{x-4}{4-x} = \dfrac{x-4}{-1(x-4)} = -1$

You Try It 4

$$\frac{12x^2+3x}{10x-15} \cdot \frac{8x-12}{9x+18} = \frac{3x(4x+1)}{5(2x-3)} \cdot \frac{4(2x-3)}{9(x+2)}$$

$$= \frac{\overset{1}{\cancel{3}}x(4x+1) \cdot 2 \cdot 2\overset{1}{\cancel{(2x-3)}}}{5\underset{1}{\cancel{(2x-3)}} \cdot \underset{1}{\cancel{3}} \cdot 3(x+2)} = \frac{4x(4x+1)}{15(x+2)}$$

You Try It 5

$$\frac{x^2+2x-15}{9-x^2} \cdot \frac{x^2-3x-18}{x^2-7x+6}$$

$$= \frac{(x-3)(x+5)}{(3-x)(3+x)} \cdot \frac{(x+3)(x-6)}{(x-1)(x-6)}$$ • Factor.

$$= \frac{\overset{-1}{\cancel{(x-3)}}(x+5) \cdot \overset{1}{\cancel{(x+3)}}\overset{1}{\cancel{(x-6)}}}{\underset{1}{\cancel{(3-x)}}\underset{1}{\cancel{(3+x)}} \cdot (x-1)\underset{1}{\cancel{(x-6)}}} = -\frac{x+5}{x-1}$$

You Try It 6

$$\frac{a^2}{4bc^2-2b^2c} \div \frac{a}{6bc-3b^2}$$

$$= \frac{a^2}{4bc^2-2b^2c} \cdot \frac{6bc-3b^2}{a}$$ • Multiply by the reciprocal.

$$= \frac{a^2 \cdot 3\overset{1}{\cancel{b}}\overset{1}{\cancel{(2c-b)}}}{2\underset{1}{\cancel{b}}c\underset{1}{\cancel{(2c-b)}} \cdot a} = \frac{3a}{2c}$$

You Try It 7

$$\frac{3x^2+26x+16}{3x^2-7x-6} \div \frac{2x^2+9x-5}{x^2+2x-15}$$

$$= \frac{3x^2+26x+16}{3x^2-7x-6} \cdot \frac{x^2+2x-15}{2x^2+9x-5}$$ • Multiply by the reciprocal.

$$= \frac{\overset{1}{\cancel{(3x+2)}}(x+8) \cdot \overset{1}{\cancel{(x+5)}}\overset{1}{\cancel{(x-3)}}}{\underset{1}{\cancel{(3x+2)}}\underset{1}{\cancel{(x-3)}} \cdot (2x-1)\underset{1}{\cancel{(x+5)}}} = \frac{x+8}{2x-1}$$

SECTION 11.2

You Try It 1

$8uv^2 = 2 \cdot 2 \cdot 2 \cdot u \cdot v \cdot v$

$12uw = 2 \cdot 2 \cdot 3 \cdot u \cdot w$

LCM $= 2 \cdot 2 \cdot 2 \cdot 3 \cdot u \cdot v \cdot v \cdot w = 24uv^2w$

You Try It 2

$m^2 - 6m + 9 = (m-3)(m-3)$

$m^2 - 2m - 3 = (m+1)(m-3)$

LCM $= (m-3)(m-3)(m+1)$

You Try It 3 The LCM is $36xy^2z$.

$$\frac{x-3}{4xy^2} = \frac{x-3}{4xy^2} \cdot \frac{9z}{9z} = \frac{9xz-27z}{36xy^2z}$$

$$\frac{2x+1}{9y^2z} = \frac{2x+1}{9y^2z} \cdot \frac{4x}{4x} = \frac{8x^2+4x}{36xy^2z}$$

You Try It 4

The LCM is $(x+2)(x-5)(x+5)$.

$$\frac{x+4}{x^2-3x-10} = \frac{x+4}{(x+2)(x-5)} \cdot \frac{x+5}{x+5} = \frac{x^2+9x+20}{(x+2)(x-5)(x+5)}$$

$$\frac{2x}{25-x^2} = \frac{2x}{-(x^2-25)} = -\frac{2x}{(x-5)(x+5)} \cdot \frac{x+2}{x+2} = -\frac{2x^2+4x}{(x+2)(x-5)(x+5)}$$

SECTION 11.3

You Try It 1

$$\frac{2x^2}{x^2-x-12} - \frac{7x+4}{x^2-x-12}$$

$$= \frac{2x^2-(7x+4)}{x^2-x-12} = \frac{2x^2-7x-4}{x^2-x-12}$$

$$= \frac{(2x+1)\overset{1}{\cancel{(x-4)}}}{(x+3)\underset{1}{\cancel{(x-4)}}} = \frac{2x+1}{x+3}$$

You Try It 2

$$\frac{x^2-1}{x^2-8x+12}-\frac{2x+1}{x^2-8x+12}+\frac{x}{x^2-8x+12}$$
$$=\frac{(x^2-1)-(2x+1)+x}{x^2-8x+12}=\frac{x^2-1-2x-1+x}{x^2-8x+12}$$
$$=\frac{x^2-x-2}{x^2-8x+12}=\frac{(x+1)\overset{1}{\cancel{(x-2)}}}{\underset{1}{\cancel{(x-2)}}(x-6)}=\frac{x+1}{x-6}$$

You Try It 3

The LCM of the denominators is $24y$.

$$\frac{z}{8y}-\frac{4z}{3y}+\frac{5z}{4y}$$
$$=\frac{z}{8y}\cdot\frac{3}{3}-\frac{4z}{3y}\cdot\frac{8}{8}+\frac{5z}{4y}\cdot\frac{6}{6}$$ • Write each fraction using the LCM.
$$=\frac{3z}{24y}-\frac{32z}{24y}+\frac{30z}{24y}$$
$$=\frac{3z-32z+30z}{24y}=\frac{z}{24y}$$ • Combine the numerators.

You Try It 4

$2-x=-(x-2)$; therefore, $\frac{3}{2-x}=\frac{-3}{x-2}$.

$$\frac{5x}{x-2}+\frac{3}{2-x}=\frac{5x}{x-2}+\frac{-3}{x-2}$$ • The LCM is $x-2$.
$$=\frac{5x+(-3)}{x-2}=\frac{5x-3}{x-2}$$ • Combine the numerators.

You Try It 5

The LCM is $(3x-1)(x+4)$.

$$\frac{4x}{3x-1}+\frac{9}{x+4}=\frac{4x}{3x-1}\cdot\frac{x+4}{x+4}+\frac{9}{x+4}\cdot\frac{3x-1}{3x-1}$$
$$=\frac{4x^2+16x}{(3x-1)(x+4)}+\frac{27x-9}{(3x-1)(x+4)}$$
$$=\frac{(4x^2+16x)+(27x-9)}{(3x-1)(x+4)}$$
$$=\frac{4x^2+16x+27x-9}{(3x-1)(x+4)}$$
$$=\frac{4x^2+43x-9}{(3x-1)(x+4)}$$

You Try It 6 The LCM is $x-3$.

$$2-\frac{1}{x-3}=2\cdot\frac{x-3}{x-3}-\frac{1}{x-3}$$
$$=\frac{2x-6}{x-3}-\frac{1}{x-3}$$
$$=\frac{2x-6-1}{x-3}$$
$$=\frac{2x-7}{x-3}$$

You Try It 7

$$\frac{2}{5-x}=\frac{-2}{x-5}$$

The LCM is $(x+5)(x-5)$.

$$\frac{2x-1}{x^2-25}+\frac{2}{5-x}=\frac{2x-1}{(x+5)(x-5)}+\frac{-2}{x-5}$$
$$=\frac{2x-1}{(x+5)(x-5)}+\frac{-2}{x-5}\cdot\frac{x+5}{x+5}$$
$$=\frac{2x-1}{(x+5)(x-5)}+\frac{-2(x+5)}{(x+5)(x-5)}$$
$$=\frac{2x-1+(-2)(x+5)}{(x+5)(x-5)}$$
$$=\frac{2x-1-2x-10}{(x+5)(x-5)}$$
$$=\frac{-11}{(x+5)(x-5)}$$
$$=-\frac{11}{(x+5)(x-5)}$$

You Try It 8

The LCM is $(3x+2)(x-1)$.

$$\frac{2x-3}{3x^2-x-2}+\frac{5}{3x+2}-\frac{1}{x-1}$$
$$=\frac{2x-3}{(3x+2)(x-1)}+\frac{5}{3x+2}\cdot\frac{x-1}{x-1}-\frac{1}{x-1}\cdot\frac{3x+2}{3x+2}$$
$$=\frac{2x-3}{(3x+2)(x-1)}+\frac{5x-5}{(3x+2)(x-1)}-\frac{3x+2}{(3x+2)(x-1)}$$
$$=\frac{(2x-3)+(5x-5)-(3x+2)}{(3x+2)(x-1)}$$
$$=\frac{2x-3+5x-5-3x-2}{(3x+2)(x-1)}$$
$$=\frac{4x-10}{(3x+2)(x-1)}=\frac{2(2x-5)}{(3x+2)(x-1)}$$

SECTION 11.4

You Try It 1

The LCM of 3, x, 9, and x^2 is $9x^2$.

$$\frac{\frac{1}{3}-\frac{1}{x}}{\frac{1}{9}-\frac{1}{x^2}}=\frac{\frac{1}{3}-\frac{1}{x}}{\frac{1}{9}-\frac{1}{x^2}}\cdot\frac{9x^2}{9x^2}=\frac{\frac{1}{3}\cdot 9x^2-\frac{1}{x}\cdot 9x^2}{\frac{1}{9}\cdot 9x^2-\frac{1}{x^2}\cdot 9x^2}$$ • Multiply by the LCM.
$$=\frac{3x^2-9x}{x^2-9}=\frac{3x\overset{1}{\cancel{(x-3)}}}{\underset{1}{\cancel{(x-3)}}(x+3)}=\frac{3x}{x+3}$$

You Try It 2

The LCM of x and x^2 is x^2.

$$\frac{1+\frac{4}{x}+\frac{3}{x^2}}{1+\frac{10}{x}+\frac{21}{x^2}}=\frac{1+\frac{4}{x}+\frac{3}{x^2}}{1+\frac{10}{x}+\frac{21}{x^2}}\cdot\frac{x^2}{x^2}$$ • Multiply by the LCM.

$$=\frac{1\cdot x^2+\frac{4}{x}\cdot x^2+\frac{3}{x^2}\cdot x^2}{1\cdot x^2+\frac{10}{x}\cdot x^2+\frac{21}{x^2}\cdot x^2}$$ • Distributive Property

$$=\frac{x^2+4x+3}{x^2+10x+21}=\frac{(x+1)\overset{1}{\cancel{(x+3)}}}{\underset{1}{\cancel{(x+3)}}(x+7)}$$

$$=\frac{x+1}{x+7}$$

You Try It 3

The LCM is $x-5$.

$$\frac{x+3-\frac{20}{x-5}}{x+8+\frac{30}{x-5}}=\frac{x+3-\frac{20}{x-5}}{x+8+\frac{30}{x-5}}\cdot\frac{x-5}{x-5}$$

$$=\frac{(x+3)(x-5)-\frac{20}{x-5}\cdot(x-5)}{(x+8)(x-5)+\frac{30}{x-5}\cdot(x-5)}$$

$$=\frac{x^2-2x-15-20}{x^2+3x-40+30}=\frac{x^2-2x-35}{x^2+3x-10}$$

$$=\frac{\overset{1}{\cancel{(x+5)}}(x-7)}{(x-2)\underset{1}{\cancel{(x+5)}}}=\frac{x-7}{x-2}$$

SECTION 11.5

You Try It 1

$$\frac{x}{x+6}=\frac{3}{x}$$ • The LCM is $x(x+6)$.

$$\frac{x\overset{1}{\cancel{(x+6)}}}{1}\cdot\frac{x}{\underset{1}{\cancel{x+6}}}=\frac{x(x+6)}{1}\cdot\frac{3}{x}$$ • Multiply by the LCM.

$$x^2=(x+6)3$$ • Simplify.

$$x^2=3x+18$$

$$x^2-3x-18=0$$

$$(x+3)(x-6)=0$$ • Factor.

$$x+3=0 \qquad x-6=0$$ • Principle of Zero Products

$$x=-3 \qquad x=6$$

Both -3 and 6 check as solutions.

The solutions are -3 and 6.

You Try It 2

$$\frac{5x}{x+2}=3-\frac{10}{x+2}$$ • The LCM is $x+2$.

$$\frac{(x+2)}{1}\cdot\frac{5x}{x+2}=\frac{(x+2)}{1}\left(3-\frac{10}{x+2}\right)$$ • Clear denominators.

$$\frac{\overset{1}{\cancel{x+2}}}{1}\cdot\frac{5x}{\underset{1}{\cancel{x+2}}}=\frac{x+2}{1}\cdot3-\frac{\overset{1}{\cancel{x+2}}}{1}\cdot\frac{10}{\underset{1}{\cancel{x+2}}}$$

$$5x=(x+2)3-10$$ • Solve for x.

$$5x=3x+6-10$$

$$5x=3x-4$$

$$2x=-4$$

$$x=-2$$

-2 does not check as a solution.

The equation has no solution.

SECTION 11.6

You Try It 1

$$5x-2y=10$$

$$5x-5x-2y=-5x+10$$ • Subtract $5x$.

$$-2y=-5x+10$$

$$\frac{-2y}{-2}=\frac{-5x+10}{-2}$$ • Divide by -2.

$$y=\frac{5}{2}x-5$$

You Try It 2

$$s=\frac{A+L}{2}$$

$$2\cdot s=2\left(\frac{A+L}{2}\right)$$ • Multiply by 2.

$$2s=A+L$$ • Subtract A.

$$2s-A=A-A+L$$

$$2s-A=L$$

You Try It 3

$$S=a+(n-1)d$$

$$S=a+nd-d$$

$$S-a=a-a+nd-d$$ • Subtract a.

$$S-a=nd-d$$

$$S-a+d=nd-d+d$$ • Add d.

$$S-a+d=nd$$

$$\frac{S-a+d}{d}=\frac{nd}{d}$$ • Divide by d.

$$\frac{S-a+d}{d}=n$$

You Try It 4

$$S=rS+C$$

$$S-rS=rS-rS+C$$ • Subtract rS.

$$S-rS=C$$

$$(1-r)S=C$$ • Factor.

$$\frac{(1-r)S}{1-r}=\frac{C}{1-r}$$ • Divide by $1-r$.

$$S=\frac{C}{1-r}$$

SECTION 11.7

You Try It 1

Strategy • Time for one printer to complete the job: t

	Rate	Time	Part
1st printer	$\frac{1}{t}$	2	$\frac{2}{t}$
2nd printer	$\frac{1}{t}$	5	$\frac{5}{t}$

• The sum of the parts of the task completed must equal 1.

Solution

$$\frac{2}{t}+\frac{5}{t}=1$$

$$t\left(\frac{2}{t}+\frac{5}{t}\right)=t\cdot 1$$

$$2+5=t$$

$$7=t$$

Working alone, one printer takes 7 h to print the payroll.

You Try It 2

Strategy • Rate sailing across the lake: r
Rate sailing back: $3r$

	Distance	Rate	Time
Across	6	r	$\frac{6}{r}$
Back	6	$3r$	$\frac{6}{3r}$

• The total time for the trip was 2 h.

Solution

$$\frac{6}{r}+\frac{6}{3r}=2$$

$$3r\left(\frac{6}{r}+\frac{6}{3r}\right)=3r(2)$$ • Multiply by the LCM, $3r$.

$$3r\cdot\frac{6}{r}+3r\cdot\frac{6}{3r}=6r$$

$$18+6=6r$$ • Solve for r.

$$24=6r$$

$$4=r$$

The rate sailing across the lake was 4 km/h.

SOLUTIONS TO CHAPTER 12 "YOU TRY IT"

SECTION 12.1

You Try It 1

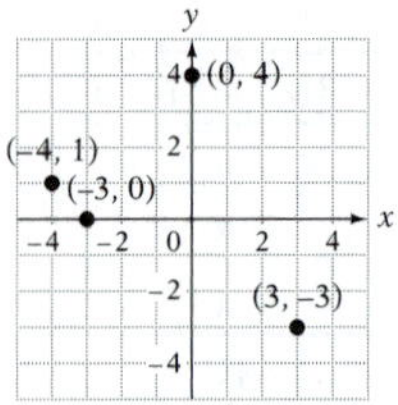

You Try It 2

$A(4, -2), B(-2, 4)$
The abscissa of D is 0.
The ordinate of C is 0.

You Try It 3

$$\begin{array}{c|c} x-3y & -14 \\ \hline -2-3(4) & -14 \\ -2-12 & -14 \\ -14 & -14 \end{array}$$

$-14 = -14$

Yes, $(-2, 4)$ is a solution of $x-3y=-14$.

You Try It 4

$$x+2y=4$$

$$2y=-x+4$$

$$y=-\frac{1}{2}x+2$$

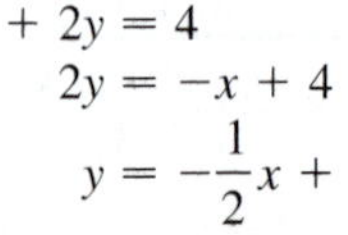

x	y
−4	4
−2	3
0	2
2	1

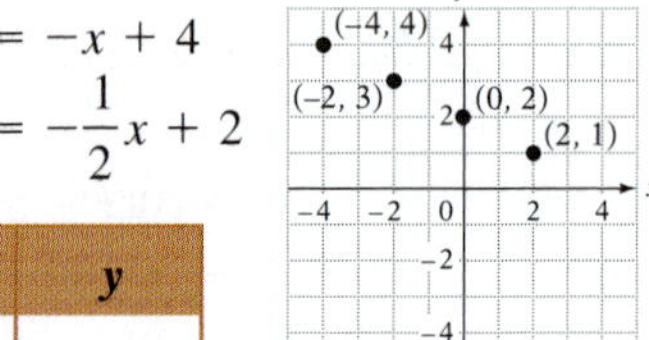

You Try It 5

{(145, 140), (140, 125), (150, 130), (165, 150), (140, 130), (165, 160)}

No, the relation is not a function. The two ordered pairs (140, 125) and (140, 130) have the same first coordinate but different second coordinates.

You Try It 6

Determine the ordered pairs defined by the equation. Replace x in $y=\frac{1}{2}x+1$ by the given values and solve for y: $\{(-4, -1), (0, 1), (2, 2)\}$.
Yes, y is a function of x.

You Try It 7

$$H(x)=\frac{x}{x-4}$$

$$H(8)=\frac{8}{8-4}$$ • Replace x by 8.

$$H(8)=\frac{8}{4}=2$$

SECTION 12.2

You Try It 1

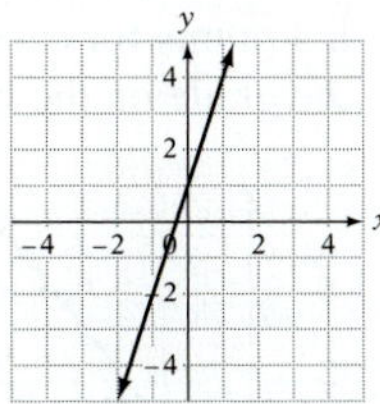

You Try It 2

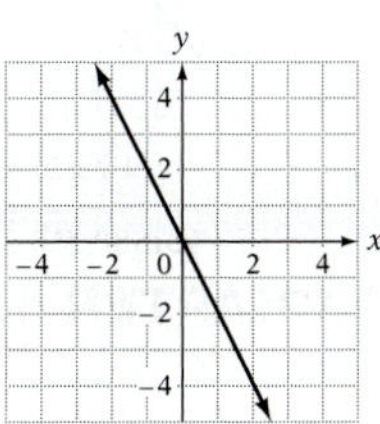

You Try It 3

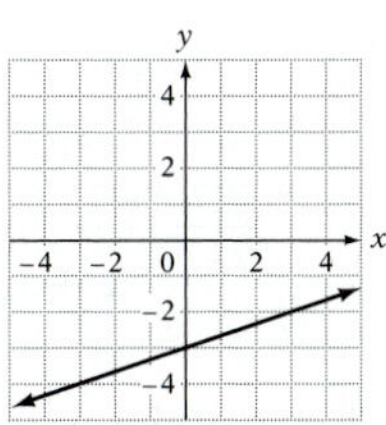

You Try It 4

$$5x - 2y = 10 \qquad \text{• Solve for } y.$$
$$-2y = -5x + 10$$
$$y = \frac{5}{2}x - 5$$

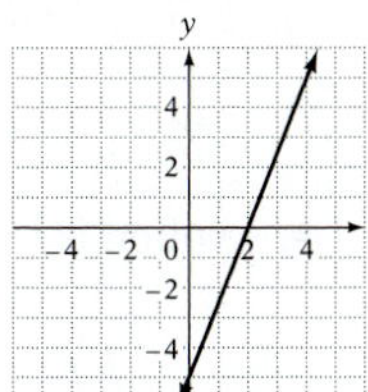

You Try It 5

$$x - 3y = 9 \qquad \text{• Solve for } y.$$
$$-3y = -x + 9$$
$$y = \frac{1}{3}x - 3$$

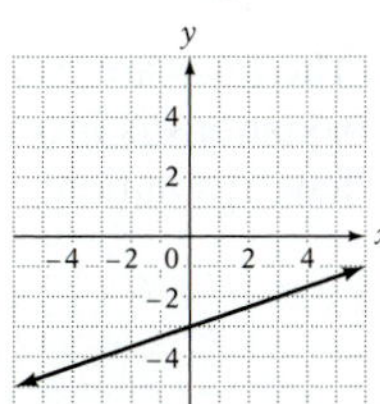

You Try It 6

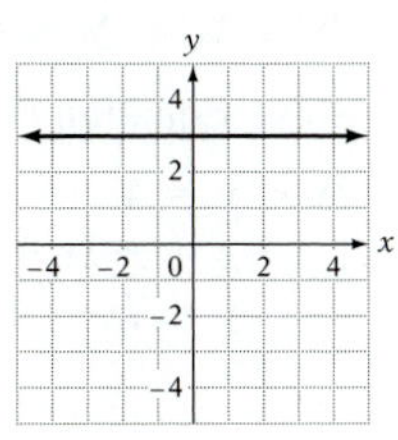

You Try It 7

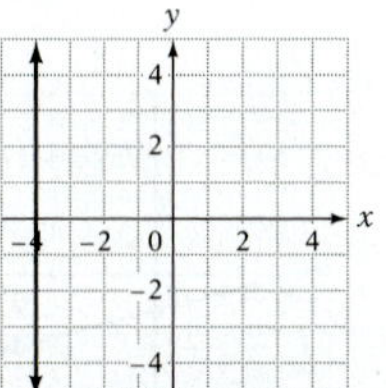

You Try It 8 The ordered pair (3, 120) means that in 3 h the car will travel 120 mi.

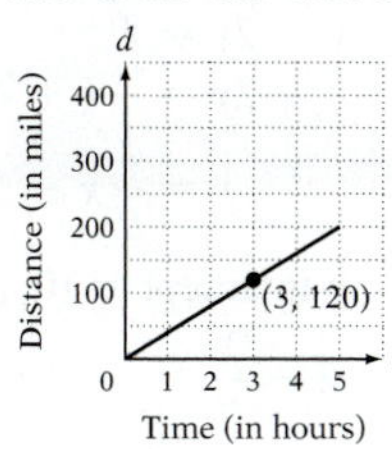

SECTION 12.3

You Try It 1

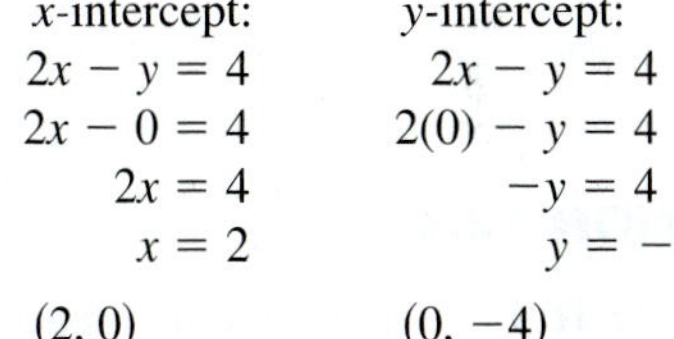

x-intercept:	y-intercept:
$2x - y = 4$	$2x - y = 4$
$2x - 0 = 4$	$2(0) - y = 4$
$2x = 4$	$-y = 4$
$x = 2$	$y = -4$

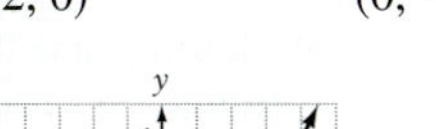

(2, 0) (0, −4)

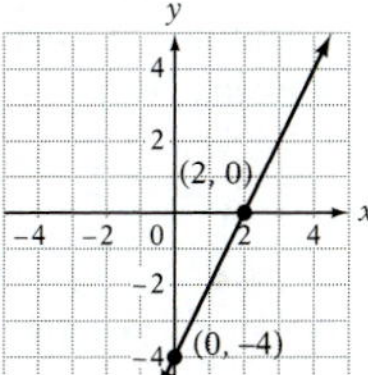

You Try It 2 Let $P_1 = (1, 4)$ and $P_2 = (-3, 8)$.

$$m = \frac{y_2 - y_1}{x_2 - x_1} = \frac{8 - 4}{-3 - 1} = \frac{4}{-4} = -1$$

The slope is −1.

You Try It 3 Let $P_1 = (-1, 2)$ and $P_2 = (4, 2)$.

$$m = \frac{y_2 - y_1}{x_2 - x_1} = \frac{2 - 2}{4 - (-1)} = \frac{0}{5} = 0$$

The slope is 0.

You Try It 4

$$m = \frac{8650 - 6100}{1 - 4} = \frac{2550}{-3}$$
$$m = -850$$

A slope of −850 means that the value of the car is decreasing at a rate of \$850 per year.

You Try It 5 y-intercept $= (0, b) = (0, -1)$

$$m = -\frac{1}{4}$$

(continued)

(continued)

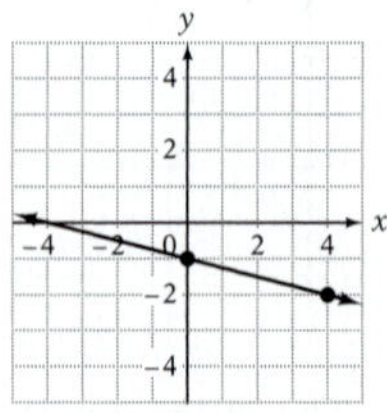

You Try It 6 Solve the equation for y.

$$x - 2y = 4$$
$$-2y = -x + 4$$
$$y = \frac{1}{2}x - 2$$

y-intercept $= (0, b) = (0, -2)$

$m = \frac{1}{2}$

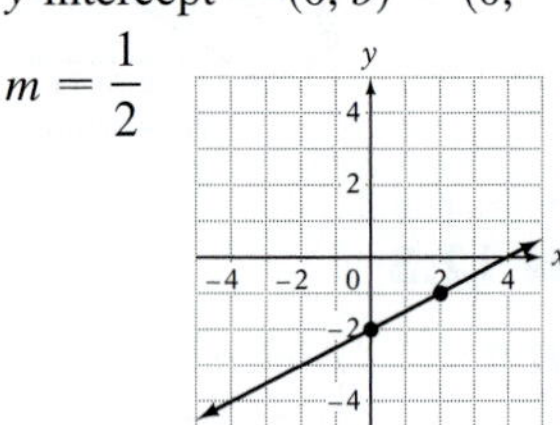

SECTION 12.4

You Try It 1 Because the slope and y-intercept are known, use the slope-intercept formula, $y = mx + b$.

$$y = mx + b$$
$$y = \frac{5}{3}x + 2 \quad \bullet\ m = \frac{5}{3};\ b = 2$$

You Try It 2 $m = \frac{3}{4}$ $(x_1, y_1) = (4, -2)$

$$y - y_1 = m(x - x_1)$$
$$y - (-2) = \frac{3}{4}(x - 4)$$
$$y + 2 = \frac{3}{4}x - 3$$
$$y = \frac{3}{4}x - 5$$

The equation of the line is $y = \frac{3}{4}x - 5$.

You Try It 3 Find the slope of the line between the two points.

$P_1 = (-6, -2), P_2 = (3, 1)$

$$m = \frac{y_2 - y_1}{x_2 - x_1} = \frac{1 - (-2)}{3 - (-6)} = \frac{3}{9} = \frac{1}{3}$$

Use the point-slope formula.

$$y - y_1 = m(x - x_1)$$
$$y - (-2) = \frac{1}{3}[x - (-6)] \quad \bullet\ y_1 = -2;\ x_1 = -6$$
$$y + 2 = \frac{1}{3}x + 2$$
$$y = \frac{1}{3}x$$

You Try It 4 The slope of the line means that the grade on the history test increases 8.3 points for each 1-point increase in the grade on the reading test.

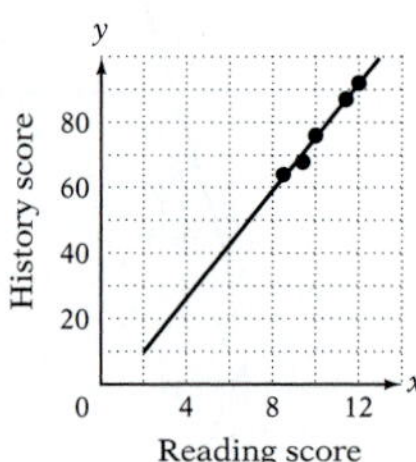

SOLUTIONS TO CHAPTER 13 "YOU TRY IT"

SECTION 13.1

You Try It 1

$2x - 5y = 8$	
$2(-1) - 5(-2)$	8
$-2 + 10$	8
$8 = 8$	

$-x + 3y = -5$	
$-(-1) + 3(-2)$	-5
$1 + (-6)$	-5
$-5 = -5$	

Yes, $(-1, -2)$ is a solution of the system of equations.

You Try It 2

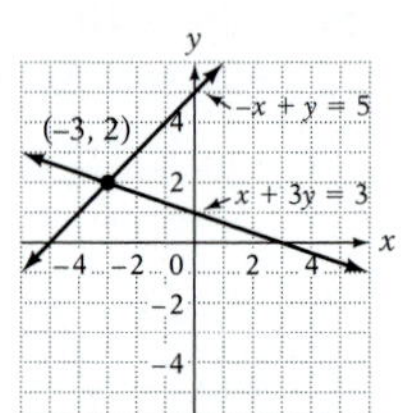

The solution is $(-3, 2)$.

You Try It 3

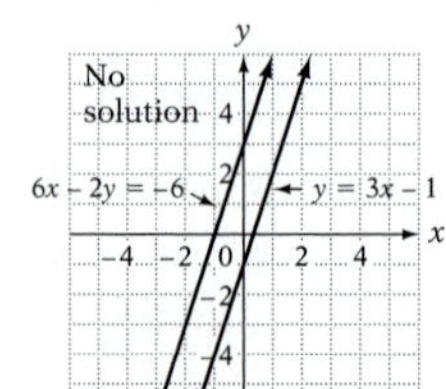

The lines are parallel. The system of equations is inconsistent and does not have a solution.

SECTION 13.2

You Try It 1

(1) $7x - y = 4$

(2) $3x + 2y = 9$

Solve Equation (1) for y.

$$7x - y = 4$$
$$-y = -7x + 4$$
$$y = 7x - 4$$

You Try It 3

$y\sqrt{28y} + 7\sqrt{63y^3}$

$= y\sqrt{4 \cdot 7y} + 7\sqrt{9y^2 \cdot 7y}$ • Simplify the radicands.

$= y\sqrt{4}\sqrt{7y} + 7\sqrt{9y^2}\sqrt{7y}$

$= y \cdot 2\sqrt{7y} + 7 \cdot 3y\sqrt{7y}$

$= 2y\sqrt{7y} + 21y\sqrt{7y}$

$= (2y + 21y)\sqrt{7y}$ • Distributive Property

$= 23y\sqrt{7y}$

You Try It 4

$2\sqrt{27a^5} - 4a\sqrt{12a^3} + a^2\sqrt{75a}$

$= 2\sqrt{9a^4 \cdot 3a} - 4a\sqrt{4a^2 \cdot 3a} + a^2\sqrt{25 \cdot 3a}$

$= 2\sqrt{9a^4}\sqrt{3a} - 4a\sqrt{4a^2}\sqrt{3a} + a^2\sqrt{25}\sqrt{3a}$

$= 2 \cdot 3a^2\sqrt{3a} - 4a \cdot 2a\sqrt{3a} + a^2 \cdot 5\sqrt{3a}$

$= 6a^2\sqrt{3a} - 8a^2\sqrt{3a} + 5a^2\sqrt{3a} = 3a^2\sqrt{3a}$

SECTION 15.3

You Try It 1

$\sqrt{5a}\sqrt{15a^3b^4}\sqrt{20b^5}$

$= \sqrt{1500a^4b^9} = \sqrt{100a^4b^8 \cdot 15b}$

$= \sqrt{100a^4b^8} \cdot \sqrt{15b}$

$= 10a^2b^4\sqrt{15b}$

You Try It 2

$\sqrt{5x}(\sqrt{5x} - \sqrt{25y})$

$= \sqrt{25x^2} - \sqrt{125xy}$ • Distributive Property

$= \sqrt{25x^2} - \sqrt{25 \cdot 5xy} = \sqrt{25x^2} - \sqrt{25}\sqrt{5xy}$

$= 5x - 5\sqrt{5xy}$

You Try It 3

$(3\sqrt{x} - \sqrt{y})(5\sqrt{x} - 2\sqrt{y})$

$= 15(\sqrt{x})^2 - 6\sqrt{xy} - 5\sqrt{xy} + 2(\sqrt{y})^2$ • FOIL

$= 15(\sqrt{x})^2 - 11\sqrt{xy} + 2(\sqrt{y})^2$

$= 15x - 11\sqrt{xy} + 2y$

You Try It 4

$(2\sqrt{x} + 7)(2\sqrt{x} - 7)$

$= 4(\sqrt{x})^2 - 7^2$ • Product of conjugates

$= 4x - 49$

You Try It 5

$\dfrac{\sqrt{15x^6y^7}}{\sqrt{3x^7y^9}} = \sqrt{\dfrac{15x^6y^7}{3x^7y^9}} = \sqrt{\dfrac{5}{xy^2}} = \dfrac{\sqrt{5}}{\sqrt{xy^2}}$

$= \dfrac{\sqrt{5}}{y\sqrt{x}} = \dfrac{\sqrt{5}}{y\sqrt{x}} \cdot \dfrac{\sqrt{x}}{\sqrt{x}}$ • Rationalize the denominator.

$= \dfrac{\sqrt{5x}}{xy}$

You Try It 6

$\dfrac{\sqrt{3}}{\sqrt{3} - \sqrt{6}} = \dfrac{\sqrt{3}}{\sqrt{3} - \sqrt{6}} \cdot \dfrac{\sqrt{3} + \sqrt{6}}{\sqrt{3} + \sqrt{6}}$ • Rationalize the denominator.

$= \dfrac{3 + \sqrt{18}}{3 - 6} = \dfrac{3 + 3\sqrt{2}}{-3}$

$= \dfrac{3(1 + \sqrt{2})}{-3} = -1(1 + \sqrt{2})$

$= -1 - \sqrt{2}$

You Try It 7

$\dfrac{5 + \sqrt{y}}{1 - 2\sqrt{y}} = \dfrac{5 + \sqrt{y}}{1 - 2\sqrt{y}} \cdot \dfrac{1 + 2\sqrt{y}}{1 + 2\sqrt{y}}$ • Rationalize the denominator.

$= \dfrac{5 + 10\sqrt{y} + \sqrt{y} + 2(\sqrt{y})^2}{1 - 4y}$

$= \dfrac{5 + 11\sqrt{y} + 2y}{1 - 4y}$

SECTION 15.4

You Try It 1

$\sqrt{4x} + 3 = 7$

$\sqrt{4x} = 4$ • Isolate $\sqrt{4x}$.

$(\sqrt{4x})^2 = 4^2$ • Square both sides.

$4x = 16$

$x = 4$ • Solve for x.

Check:

$\sqrt{4x} + 3 = 7$	
$\sqrt{4 \cdot 4} + 3$	7
$\sqrt{16} + 3$	7
$4 + 3$	7
$7 = 7$	

The solution is 4.

You Try It 2

$\sqrt{x} + \sqrt{x + 9} = 9$

$\sqrt{x} = 9 - \sqrt{x + 9}$ • Isolate $\sqrt{x}$.

$(\sqrt{x})^2 = (9 - \sqrt{x + 9})^2$ • Square both sides.

$x = 81 - 18\sqrt{x + 9} + (x + 9)$

$-90 = -18\sqrt{x + 9}$

$5 = \sqrt{x + 9}$ • Isolate $\sqrt{x + 9}$.

$5^2 = (\sqrt{x + 9})^2$ • Square both sides.

$25 = x + 9$

$16 = x$ • Solve for x.

Check:

$\sqrt{x} + \sqrt{x + 9} = 9$	
$\sqrt{16} + \sqrt{16 + 9}$	9
$\sqrt{16} + \sqrt{25}$	9
$4 + 5$	9
$9 = 9$	

The solution is 16.

You Try It 3

Strategy To find the distance, use the Pythagorean Theorem. The hypotenuse is the length of the ladder. One leg is the distance from the bottom of the ladder to the base of the building. The distance along the building from the ground to the top of the ladder is the unknown leg.

Solution

$$\begin{aligned} a &= \sqrt{c^2 - b^2} \\ &= \sqrt{(8)^2 - (3)^2} && \bullet\ c = 8, b = 3 \\ &= \sqrt{64 - 9} \\ &= \sqrt{55} \\ &\approx 7.42 \end{aligned}$$

The distance is approximately 7.42 ft.

You Try It 4

Strategy To find the length of the pendulum, replace T in the equation with the given value and solve for L.

Solution

$$\begin{aligned} T &= 2\pi\sqrt{\frac{L}{32}} \\ 2.5 &= 2(3.14)\sqrt{\frac{L}{32}} && \bullet\ T = 2.5 \\ 2.5 &= 6.28\sqrt{\frac{L}{32}} \\ \frac{2.5}{6.28} &= \sqrt{\frac{L}{32}} \\ \left(\frac{2.5}{6.28}\right)^2 &= \left(\sqrt{\frac{L}{32}}\right)^2 \\ \frac{6.25}{39.4384} &= \frac{L}{32} \\ (32)\left(\frac{6.25}{39.4384}\right) &= (32)\left(\frac{L}{32}\right) \\ \frac{200}{39.4384} &= L \\ 5.07 &\approx L \end{aligned}$$

The length of the pendulum is approximately 5.07 ft.

SOLUTIONS TO CHAPTER 16 "YOU TRY IT"

SECTION 16.1

You Try It 1

$$\begin{aligned} \frac{3y^2}{2} + y - \frac{1}{2} &= 0 \\ 2\left(\frac{3y^2}{2} + y - \frac{1}{2}\right) &= 2(0) && \bullet\ \text{Multiply each side by 2.} \\ 3y^2 + 2y - 1 &= 0 \\ (3y - 1)(y + 1) &= 0 && \bullet\ \text{Factor.} \end{aligned}$$

$$\begin{aligned} 3y - 1 &= 0 \qquad & y + 1 &= 0 && \bullet\ \text{Principle of Zero Products} \\ 3y &= 1 & y &= -1 \\ y &= \frac{1}{3} \end{aligned}$$

The solutions are $\frac{1}{3}$ and -1.

You Try It 2

$$\begin{aligned} x^2 + 81 &= 0 \\ x^2 &= -81 && \bullet\ \text{Solve for } x^2. \\ \sqrt{x^2} &= \sqrt{-81} && \bullet\ \text{Take square roots.} \end{aligned}$$

$\sqrt{-81}$ is not a real number.

The equation has no real number solution.

You Try It 3

$$\begin{aligned} 7(z + 2)^2 &= 21 \\ (z + 2)^2 &= 3 && \bullet\ \text{Solve for } (z + 2)^2. \\ \sqrt{(z + 2)^2} &= \sqrt{3} && \bullet\ \text{Take square roots.} \\ z + 2 &= \pm\sqrt{3} \\ z &= -2 \pm \sqrt{3} && \bullet\ \text{Solve for } z. \end{aligned}$$

The solutions are $-2 + \sqrt{3}$ and $-2 - \sqrt{3}$.

SECTION 16.2

You Try It 1

$$\begin{aligned} 3x^2 - 6x - 2 &= 0 \\ 3x^2 - 6x &= 2 && \bullet\ \text{Add 2.} \\ \frac{1}{3}(3x^2 - 6x) &= \frac{1}{3} \cdot 2 && \bullet\ \text{Multiply by } \frac{1}{3}. \\ x^2 - 2x &= \frac{2}{3} \end{aligned}$$

Complete the square.

$$\begin{aligned} x^2 - 2x + 1 &= \frac{2}{3} + 1 && \bullet\ \left[\frac{1}{2}(-2)\right]^2 = [-1]^2 = 1 \\ (x - 1)^2 &= \frac{5}{3} && \bullet\ \text{Factor.} \\ \sqrt{(x - 1)^2} &= \sqrt{\frac{5}{3}} && \bullet\ \text{Take square roots.} \\ x - 1 &= \pm\sqrt{\frac{5}{3}} && \bullet\ \text{Simplify.} \\ x &= 1 \pm \sqrt{\frac{5}{3}} \\ x &= 1 \pm \frac{\sqrt{15}}{3} \\ x &= \frac{3 \pm \sqrt{15}}{3} \end{aligned}$$

The solutions are $\frac{3 + \sqrt{15}}{3}$ and $\frac{3 - \sqrt{15}}{3}$.

You Try It 2

$$x^2 + 6x + 12 = 0$$
$$x^2 + 6x = -12 \quad \text{• Subtract 12.}$$
$$x^2 + 6x + 9 = -12 + 9 \quad \text{• } \left(\frac{1}{2} \cdot 6\right)^2 = 3^2 = 9$$
$$(x + 3)^2 = -3 \quad \text{• Factor.}$$
$$\sqrt{(x + 3)^2} = \sqrt{-3} \quad \text{• Take square roots.}$$

$\sqrt{-3}$ is not a real number.

The quadratic equation has no real number solution.

You Try It 3

$$x^2 + 8x + 8 = 0$$
$$x^2 + 8x = -8 \quad \text{• Subtract 8.}$$
$$x^2 + 8x + 16 = -8 + 16 \quad \text{• } \left(\frac{1}{2} \cdot 8\right)^2 = 4^2 = 16$$
$$(x + 4)^2 = 8 \quad \text{• Factor.}$$
$$\sqrt{(x + 4)^2} = \sqrt{8} \quad \text{• Take square roots.}$$
$$x + 4 = \pm\sqrt{8}$$
$$x + 4 = \pm 2\sqrt{2}$$
$$x = -4 \pm 2\sqrt{2}$$

$$x = -4 + 2\sqrt{2} \approx -4 + 2(1.414) \approx -4 + 2.828 \approx -1.172$$
$$x = -4 - 2\sqrt{2} \approx -4 - 2(1.414) \approx -4 - 2.828 \approx -6.828$$

The solutions are approximately -1.172 and -6.828.

SECTION 16.3

You Try It 1

$$3x^2 + 4x - 4 = 0$$
$$a = 3, b = 4, c = -4$$
$$x = \frac{-(4) \pm \sqrt{(4)^2 - 4(3)(-4)}}{2 \cdot 3}$$
$$= \frac{-4 \pm \sqrt{16 + 48}}{6}$$
$$= \frac{-4 \pm \sqrt{64}}{6} = \frac{-4 \pm 8}{6}$$
$$x = \frac{-4 + 8}{6} = \frac{4}{6} = \frac{2}{3} \qquad x = \frac{-4 - 8}{6} = \frac{-12}{6} = -2$$

The solutions are $\frac{2}{3}$ and -2.

You Try It 2

$$\frac{x^2}{4} + \frac{x}{2} = \frac{1}{4}$$
$$4\left(\frac{x^2}{4} + \frac{x}{2}\right) = 4\left(\frac{1}{4}\right) \quad \text{• Multiply by 4.}$$
$$x^2 + 2x = 1$$
$$x^2 + 2x - 1 = 0 \quad \text{• Standard form}$$
$$a = 1, b = 2, c = -1$$
$$x = \frac{-(2) \pm \sqrt{(2)^2 - 4(1)(-1)}}{2 \cdot 1}$$
$$= \frac{-2 \pm \sqrt{4 + 4}}{2} = \frac{-2 \pm \sqrt{8}}{2}$$
$$= \frac{-2 \pm 2\sqrt{2}}{2} = -1 \pm \sqrt{2}$$

The solutions are $-1 + \sqrt{2}$ and $-1 - \sqrt{2}$.

SECTION 16.4

You Try It 1

$$y = x^2 + 2$$

x	y
-2	6
-1	3
0	2
1	3
2	6

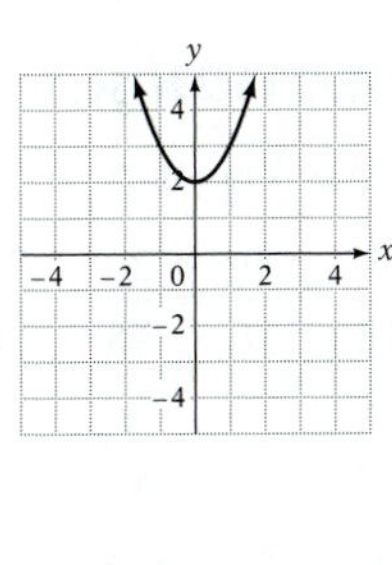

You Try It 2

To find the x-intercept, let $f(x) = 0$ and solve for x.

$$f(x) = x^2 - 6x + 9$$
$$0 = x^2 - 6x + 9$$
$$0 = (x - 3)(x - 3) \quad \text{• Factor.}$$
$$x - 3 = 0 \qquad x - 3 = 0 \quad \text{• Principle of Zero Products}$$
$$x = 3 \qquad x = 3$$

The x-intercept is $(3, 0)$.

There is only one x-intercept. The equation has a double root.

To find the y-intercept, evaluate the function at $x = 0$.

$$f(x) = x^2 - 6x + 9$$
$$f(0) = 0^2 - 6(0) + 9 = 9$$

The y-intercept is $(0, 9)$.

SECTION 16.5

You Try It 1

Strategy

- This is a geometry problem.
- Width of the rectangle: W
  Length of the rectangle: $W + 2$
- Use the equation $A = L \cdot W$.

Solution

$A = L \cdot W$
$15 = (W + 2)W$ • $A = 15, L = W + 2$
$15 = W^2 + 2W$
$0 = W^2 + 2W - 15$
$0 = (W + 5)(W - 3)$ • Factor.

$W + 5 = 0 \qquad W - 3 = 0$ • Principle of Zero Products
$W = -5 \qquad W = 3$

The solution -5 is not possible.
The width is 3 m.

SOLUTIONS TO CHAPTER R "YOU TRY IT"

SECTION R.1

You Try It 1

$\frac{5}{12} + \frac{9}{16} = \frac{20}{48} + \frac{27}{48}$ • The LCM of 12 and 16 is 48.

$= \frac{47}{48}$ • Add the numerators. Place the sum over the common denominator.

You Try It 2

$\frac{3}{4} + \frac{4}{5} + \frac{5}{8}$ • The LCM of 4, 5, and 8 is 40. Write equivalent fractions using the LCM.

$= \frac{30}{40} + \frac{32}{40} + \frac{25}{40}$

$= \frac{87}{40}$ • Add the numerators. Place the sum over the common denominator.

$= 2\frac{7}{40}$

You Try It 3

$\frac{5}{6} - \frac{1}{4} = \frac{10}{12} - \frac{3}{12}$ • The LCM of 6 and 4 is 12.

$= \frac{7}{12}$ • Subtract the numerators. Place the difference over the common denominator.

You Try It 4

$\frac{4}{21} \cdot \frac{7}{44} = \frac{4 \cdot 7}{21 \cdot 44}$ • Multiply the numerators. Multiply the denominators.

$= \frac{2 \cdot 2 \cdot 7}{3 \cdot 7 \cdot 2 \cdot 2 \cdot 11}$ • Write the prime factorization of each number.

$= \frac{\overset{1}{\cancel{2}} \cdot \overset{1}{\cancel{2}} \cdot \overset{1}{\cancel{7}}}{3 \cdot \underset{1}{\cancel{7}} \cdot \underset{1}{\cancel{2}} \cdot \underset{1}{\cancel{2}} \cdot 11}$ • Divide by the common factors.

$= \frac{1}{33}$ • Write the fraction in simplest form.

You Try It 5

$\frac{2}{15} \cdot 5 = \frac{2}{15} \cdot \frac{5}{1}$ • Write 5 as $\frac{5}{1}$.

$= \frac{2 \cdot 5}{15 \cdot 1}$ • Multiply the fractions.

$= \frac{2 \cdot \overset{1}{\cancel{5}}}{3 \cdot \underset{1}{\cancel{5}} \cdot 1}$ • Divide by the common factors.

$= \frac{2}{3}$ • Write the fraction in simplest form.

You Try It 6

$\frac{1}{6} \div \frac{4}{9} = \frac{1}{6} \cdot \frac{9}{4}$ • Multiply the first fraction by the reciprocal of the second fraction.

$= \frac{1 \cdot 9}{6 \cdot 4}$

$= \frac{1 \cdot \overset{1}{\cancel{3}} \cdot 3}{2 \cdot \underset{1}{\cancel{3}} \cdot 2 \cdot 2}$

$= \frac{3}{8}$

SECTION R.2

You Try It 1

a. $-25 + 13 = -12$ • The signs are different. Subtract the absolute values. The sum has the same sign as the number with the larger absolute value.

b. $-41 + (-60) = -101$ • The signs are the same. Add the absolute values. The sum has the same sign as the addends.

c. $37 + (-9) = 28$ • The signs are different. Subtract the absolute values. The sum has the same sign as the number with the larger absolute value.

You Try It 2

a. $-27 - 18 = -27 + (-18)$
$= -45$

• -27 minus 18 = -27 plus the opposite of 18. The opposite of 18 is -18.

b. $-34 - (-90) = -34 + 90$
$= 56$

• -34 minus -90 = -34 plus the opposite of -90. The opposite of -90 is 90.

c. $8 - 42 = 8 + (-42)$
$= -34$

• 8 minus 42 = 8 plus the opposite of 42. The opposite of 42 is -42.

d. $-12 + 9 - (-5) - 4$
$= -12 + 9 + 5 + (-4)$

• Rewrite each subtraction as addition of the opposite.

$= -3 + 5 + (-4)$
$= 2 + (-4)$
$= -2$

• Add the numbers.

You Try It 3

a. $-25(4) = -100$

• The signs are different. The product is negative.

b. $-4(-61) = 244$

• The signs are the same. The product is positive.

c. $(-4)(5)(-3)(-1)$
$= (-20)(-3)(-1)$
$= 60(-1)$
$= -60$

• Multiply the first two numbers. Then multiply the product by the third number. Continue until all the numbers have been multiplied.

You Try It 4

a. $(-30) \div 6 = -5$

• The signs are different. The quotient is negative.

b. $(-50) \div (-25) = 2$

• The signs are the same. The quotient is positive.

c. $\dfrac{32}{-8} = -4$

• The fraction bar can be read "divided by." $\dfrac{32}{-8} = (32) \div (-8)$ The signs are different. The quotient is negative.

SECTION R.3

You Try It 1

$$-\frac{1}{4} + \left(-\frac{3}{8}\right) = -\frac{2}{8} + \left(-\frac{3}{8}\right)$$

• Write the fractions with a common denominator.

$$= \frac{-2}{8} + \frac{-3}{8}$$

• Write the negative signs in the numerators.

$$= \frac{-2 + (-3)}{8}$$

• Add the fractions.

$$= \frac{-5}{8}$$

• Simplify the numerator.

$$= -\frac{5}{8}$$

• Write the negative sign in front of the fraction.

You Try It 2

$$-\frac{3}{8} - \left(-\frac{1}{6}\right) = -\frac{3}{8} + \frac{1}{6}$$

• Rewrite subtraction as addition of the opposite.

$$= -\frac{9}{24} + \frac{4}{24}$$

• Write the fractions with a common denominator.

$$= \frac{-9}{24} + \frac{4}{24}$$

• Write the negative sign in the numerator.

$$= \frac{-9 + 4}{24}$$

• Add the fractions.

$$= \frac{-5}{24}$$

• Simplify the numerator.

$$= -\frac{5}{24}$$

• Write the negative sign in front of the fraction.

You Try It 3

$4.69 - 12.5 = 4.69 + (-12.5)$

• Rewrite subtraction as addition of the opposite.

$= -7.81$

• Subtract the absolute values of the numbers. The sum has the same sign as the number with the larger absolute value.

You Try It 4

$$\frac{10}{11}\left(-\frac{2}{5}\right) = -\left(\frac{10}{11} \cdot \frac{2}{5}\right)$$

• The signs are different. The product is negative.

$$= -\frac{10 \cdot 2}{11 \cdot 5}$$

• Multiply the numerators. Multiply the denominators.

$$= -\frac{2 \cdot 5 \cdot 2}{11 \cdot 5}$$

• Write the product in simplest form.

$$= -\frac{4}{11}$$

You Try It 5

$$-\frac{3}{8} \div \left(-\frac{1}{2}\right) = \frac{3}{8} \div \frac{1}{2}$$

• The signs are the same. The quotient is positive.

$$= \frac{3}{8} \cdot \frac{2}{1}$$

• Rewrite division as multiplication by the reciprocal.

$$= \frac{3 \cdot 2}{8 \cdot 1}$$

• Multiply the fractions.

$$= \frac{3 \cdot 2}{2 \cdot 2 \cdot 2 \cdot 1}$$

$$= \frac{3}{4}$$

You Try It 6

$(-3.6)(-1.45) = 5.22$

• The signs are the same. The product is positive. Multiply the absolute values of the numbers.

You Try It 7

$5.04 \div (-8.4) = -0.6$ • The signs are different. The quotient is negative. Divide the absolute values of the numbers.

You Try It 8

$\left(\frac{2}{9}\right)^2 \cdot (-3)^4$

$= \left(\frac{2}{9}\right)\left(\frac{2}{9}\right) \cdot (-3)(-3)(-3)(-3)$ • Write each factor the number of times indicated by the exponent.

$= \left(\frac{2}{9}\right)\left(\frac{2}{9}\right) \cdot (3)(3)(3)(3)$ • The product is positive.

$= \frac{2 \cdot 2 \cdot 3 \cdot 3 \cdot 3 \cdot 3}{3 \cdot 3 \cdot 3 \cdot 3}$ • Multiply.

$= 4$

You Try It 9

$7 \div \left(\frac{1}{7} - \frac{3}{14}\right) - 9$ • Use the Order of Operations Agreement.

$= 7 \div \left(-\frac{1}{14}\right) - 9$ • Perform the operation inside the parentheses.

$= 7(-14) - 9$ • Rewrite division as multiplication by the reciprocal.

$= -98 - 9$ • Do the multiplication.

$= -98 + (-9)$ • Do the subtraction.

$= -107$

SECTION R.4

You Try It 1

$5x + 3 - 7x = 9$

$3 - 2x = 9$ • Step 2

$3 - 3 - 2x = 9 - 3$ • Step 4

$-2x = 6$

$\frac{-2x}{-2} = \frac{6}{-2}$ • Step 5

$x = -3$

The solution is -3.

You Try It 2

$4x + 3 = 7x + 9$

$4x - 7x + 3 = 7x - 7x + 9$ • Step 3

$-3x + 3 = 9$

$-3x + 3 - 3 = 9 - 3$ • Step 4

$-3x = 6$

$\frac{-3x}{-3} = \frac{6}{-3}$ • Step 5

$x = -2$

The solution is -2.

You Try It 3

$4 - (5x - 8) = 4x + 3$

$4 - 5x + 8 = 4x + 3$ • Step 1

$-5x + 12 = 4x + 3$ • Step 2

$-5x - 4x + 12 = 4x - 4x + 3$ • Step 3

$-9x + 12 = 3$

$-9x + 12 - 12 = 3 - 12$ • Step 4

$-9x = -9$

$\frac{-9x}{-9} = \frac{-9}{-9}$ • Step 5

$x = 1$

The solution is 1.

SOLUTIONS TO APPENDIX "YOU TRY IT"

You Try It 1 1 295 m = 1.295 km

You Try It 2 7 543 g = 7.543 kg

You Try It 3 6.3 L = 6 300 ml

You Try It 4 2 kl = 2 000 L

Answers to Selected Exercises

ANSWERS TO CHAPTER 1 SELECTED EXERCISES

PREP TEST

1. 8 **2.** 1 2 3 4 5 6 7 8 9 10 **3.** a and D; b and E; c and A; d and B; e and F; f and C **4.** 0 **5.** Fifty

SECTION 1.1

3. 0 1 2 3 4 5 6 7 8 9 10 11 12 **5.** 0 1 2 3 4 5 6 7 8 9 10 11 12

7. 0 1 2 3 4 5 6 7 8 9 10 11 12 **9.** 5 **11.** 5 **13.** 0 **15.** $27 < 39$ **17.** $0 < 52$ **19.** $273 > 194$

21. $2761 < 3857$ **23.** $4610 > 4061$ **25.** $8005 < 8050$ **27.** Yes **29.** 18, 27, 35, 60, 71 **31.** 28, 45, 54, 63, 109 **33.** 155, 271, 358, 496, 505 **35.** 400, 404, 440, 444, 4000 **37.** Seven hundred four **39.** Three hundred seventy-four **41.** Two thousand eight hundred sixty-one **43.** Forty-eight thousand two hundred ninety-seven **45.** Five hundred sixty-three thousand seventy-eight **47.** Six million three hundred seventy-nine thousand four hundred eighty-two **49.** 75 **51.** 2851 **53.** 130,212 **55.** 8073 **57.** 603,132 **59.** 3,004,008 **61.** Millions **63.** 7000 + 200 + 40 + 5 **65.** 500,000 + 30,000 + 2000 + 700 + 90 + 1 **67.** 5000 + 60 + 4 **69.** 20,000 + 300 + 90 + 7 **71.** 400,000 + 2000 + 700 + 8 **73.** 8,000,000 + 300 + 10 + 6 **75.** 3050 **77.** 1600 **79.** 17,600 **81.** 5000 **83.** 85,000 **85.** 390,000 **87.** 750,000 **89.** 37,000,000 **91.** False **93.** Billy Hamilton **95.** *Fiddler on the Roof* **97.** Two tablespoons of peanut butter **99.** St. Louis to San Diego **101.** Neptune **103.** 571,000 mi^2 **105. a.** July **b.** July **107. a.** 1985 **b.** Decrease **109.** 300,000 km/s **111.** 999; 10,000 **113. a.** True **b.** False

SECTION 1.2

3. 1,383,659 **5.** 6043 **7.** 112,152 **9.** 12,548 **11.** 199,556 **13.** 327,473 **15.** 168,574 **17.** 7947 **19.** 99,637 **21.** 1872 students **23.** 15,000; 15,040 **25.** 1,400,000; 1,388,917 **27.** 2000; 1998 **29.** 307,000; 329,801 **31.** 1272 **33.** 12,150 **35.** 89,900 **37.** 1572 **39.** 14,591 **41.** 56,010 **43.** The Commutative Property of Addition **45.** The Associative Property of Addition **47.** The Addition Property of Zero **49.** 28 **51.** 4 **53.** 15 **55.** The Commutative Property of Addition **57.** No **59.** Yes **61.** No **65.** 353 **67.** 467 **69.** 103 **71.** 658 **73.** 2786 **75.** 2127 **77.** 4738 **79.** 61,757 **81.** 1618 **83.** 7378 **85.** 17,548 **87.** 15 ft **89.** 2000; 2136 **91.** 40,000; 38,283 **93.** 35,000; 31,195 **95.** 100,000; 125,665 **97.** 13 **99.** 643 **101.** 355 **103.** 5211 **105.** 766 **107.** 18,231 **109.** Yes **111.** No **113.** Yes **115.** 210 **117.** 901 **119.** The difference represents the increase in the number of people aged 100 and over from 2014 to 2018. **121.** 550 more calories **123.** 60 ft **125.** 144 cm **127.** 68 ft **129.** \$1924 **131.** \$544 **133.** 280,000 mi^2 **135.** Car sales decreased the most between February and March. The amount of decrease was 19 cars. **137.** 2011 **139.** \$9284 **141.** \$188,800 **143.** 410 mph **145.** No. The data tell us how many motorists were driving between 66 mph and 70 mph. **147.** More people are driving at or below the posted speed limit. **149.** There are 90 two-digit numbers. There are 900 three-digit numbers.

SECTION 1.3

3. 1143 **5.** 46,963 **7.** 470,152 **9.** 48,493 **11.** 324,438 **13.** 3,206,160 **15.** 1500 **17.** 2000 **19.** 0 **21.** qrs **23.** 1,200,000; 1,244,653 **25.** 1,200,000; 1,138,134 **27.** 42,000; 46,935 **29.** 6,300,000; 6,491,166 **31.** 14,880 **33.** 3255 **35.** 1800 **37.** 3082 **39.** Answers will vary. One example is 5 and 20. **41.** The Associative Property of Multiplication **43.** The Multiplication Property of Zero **45.** 5 **47.** 1 **49.** No **51.** Yes **53.** No **55.** $3^6 \cdot 5^3$ **57.** $7^2 \cdot 11^3 \cdot 19^4$ **59.** d^3 **61.** a^2b^4 **63.** 64 **65.** 1,000,000,000 **67.** 288 **69.** 1600 **71.** 0 **73.** 4050 **75.** 144 **77.** 512 **79.** a^4 **81.** 24 **83.** 320 **85.** 225 **89.** 307 **91.** 309 r4 **93.** 2550 **95.** 21 r9 **97.** 147 r38 **99.** 200 r8 **101.** 404 r34 **103.** 16 r97 **105.** 907 **107.** 881 r1 **109.** $\frac{c}{d}$ **111.** 800; 776 **113.** 5000; 5129 **115.** 500; 493 r37 **117.** 1500; 1516 **119.** 48 **121.** Undefined **123.** 9800 **125.** False **127.** Yes **129.** No **131.** 1, 2, 4, 5, 10, 20 **133.** 1, 3, 9 **135.** 1, 2, 4, 8, 16 **137.** 1, 17 **139.** 1, 2, 3, 4, 6, 8, 12, 24 **141.** 1, 2, 3, 4, 6, 9, 12, 18, 36 **143.** 1, 3, 5, 9, 15, 45 **145.** 1, 2, 4, 8, 16, 32 **147.** 1, 2, 4, 8, 16, 32, 64 **149.** 1, 3, 5, 15, 25, 75 **151.** 2^4 **153.** $2^2 \cdot 3$ **155.** $3 \cdot 5$ **157.** $2^3 \cdot 5$ **159.** Prime **161.** $5 \cdot 13$ **163.** $2^2 \cdot 7$ **165.** $2 \cdot 3 \cdot 7$ **167.** $3 \cdot 17$ **169.** $2 \cdot 23$ **171.** 460 calories **173.** 4325 gal **175. a.** 78 m **b.** 360 m^2 **177.** 96 ft **179.** 576 ft^2 **181.** 59,136 cm^2 **183.** $16,000 **185.** $6840 **187.** 9 h **189.** $21 **191.** (iv) **193.** 87,312

SECTION 1.4

3. 4 **5.** 29 **7.** 13 **9.** 19 **11.** 11 **13.** 6 **15.** 61 **17.** 54 **19.** 19 **21.** 24 **23.** 186 **25.** 39 **27.** 18 **29.** 14 **31.** 14 **33.** 2 **35.** 57 **37.** 8 **39.** 68 **41.** 16 **43.** $12 + (9 - 5) \cdot 3 > 11 + (8 + 4) \div 6$ **45.** $8 - (2 \cdot 3) + 1$ **47.** $(8 - 2) \cdot (3 + 1)$ **49.** 97

CHAPTER 1 CONCEPT REVIEW*

1. The symbol < means "is less than." A number that appears to the left of a given number on the number line is less than (<) the given number. For example, $4 < 9$. The symbol > means "is greater than." A number that appears to the right of a given number on the number line is greater than (>) the given number. For example, $5 > 2$. [1.1A]
2. To round a four-digit whole number to the nearest hundred, look at the digit in the tens place. If the digit in the tens place is less than 5, that digit and the digit in the ones place are replaced by zeros. If the digit in the tens place is greater than or equal to 5, increase the digit in the hundreds place by 1 and replace the digits in the tens place and the ones place by zeros. [1.1C]
3. The Commutative Property of Addition states that two numbers can be added in either order; the sum is the same. For example, $3 + 5 = 5 + 3$. The Associative Property of Addition states that changing the grouping of three or more addends does not change their sum. For example, $3 + (4 + 5) = (3 + 4) + 5$. Note that in the Commutative Property of Addition, the order in which the numbers appear changes, while in the Associative Property of Addition, the order in which the numbers appear does not change. [1.2A]
4. To estimate the sum of two numbers, round each number to the highest place value of the number. Then add the numbers. For example, to estimate the sum of 562,397 and 41,086, round the numbers to 600,000 and 40,000. Then add $600{,}000 + 40{,}000 = 640{,}000$. [1.2A]
5. It is necessary to borrow when performing subtraction if, in any place value, the lower digit is larger than the upper digit. [1.2B]
6. The Multiplication Property of Zero states that the product of a number and zero is zero. For example, $8 \times 0 = 0$. The Multiplication Property of One states that the product of a number and one is the number. For example, $8 \times 1 = 8$. [1.3A]
7. To multiply a whole number by 100, write two zeros to the right of the number. For example, $64 \times 100 = 6400$. [1.3A]
8. To estimate the product of two numbers, round each number so that it contains only one nonzero digit. Then multiply. For example, to estimate the product of 87 and 43, round the two numbers to 90 and 40; then multiply $90 \times 40 = 3600$. [1.3A]
9. $0 \div 9 = 0$. Zero divided by any whole number except zero is zero. $9 \div 0$ is undefined. Division by zero is not allowed. [1.3C]
10. To check the answer to a division problem that has a remainder, multiply the quotient by the divisor. Add the remainder to the product. The result should be the dividend. For example, $16 \div 5 = 3$ r1. Check: $(3 \times 5) + 1 = 16$, the dividend. [1.3C]
11. The steps in the Order of Operations Agreement are:
    1. Do all operations inside parentheses.
    2. Simplify any numerical expressions containing exponents.
    3. Do multiplication and division as they occur from left to right.
    4. Do addition and subtraction as they occur from left to right. [1.4A]

**Note:* The numbers in brackets following the answers in the Concept Review are a reference to the objective that corresponds to that problem. For example, the reference [1.2A] stands for Section 1.2, Objective A. This notation will be used for all Prep Tests, Concept Reviews, Chapter Reviews, Chapter Tests, and Cumulative Reviews throughout the text.

12. A number is a factor of another number if it divides that number evenly (there is no remainder). For example, 7 is a factor of 21 because 21 ÷ 7 = 3, with a remainder of 0. [1.3D]

13. Three is a factor of a number if the sum of the digits of the number is divisible by 3. For the number 285, 2 + 8 + 5 = 15, which is divisible by 3. Thus 285 is divisible by 3. [1.3D]

CHAPTER 1 REVIEW EXERCISES

1. 0 1 2 3 4 5 6 7 8 9 10 11 12 [1.1A] **2.** 10,000 [1.3B] **3.** 2583 [1.2B] **4.** $3^2 \cdot 5^4$ [1.3B]
5. 1389 [1.2A] **6.** 38,700 [1.1C] **7.** 247 > 163 [1.1A] **8.** 32,509 [1.1B] **9.** 700 [1.3A]
10. 2607 [1.3C] **11.** 4048 [1.2B] **12.** 1500 [1.2A] **13.** 1, 2, 5, 10, 25, 50 [1.3D] **14.** Yes [1.2B]
15. 18 [1.4A] **16.** The Commutative Property of Addition [1.2A] **17.** Four million nine hundred twenty-seven thousand thirty-six [1.1B] **18.** 675 [1.3B] **19. a.** 16 times more **b.** 61 times more [1.3E] **20.** 67 r70 [1.3C]
21. 2636 [1.3A] **22.** 137 [1.2B] **23.** $2 \cdot 3^2 \cdot 5$ [1.3D] **24.** 80 [1.3C] **25.** 1 [1.3A] **26.** 10 [1.4A]
27. 932 [1.2A] **28.** 432 [1.3A] **29.** 56 [1.4A] **30.** Kareem Abdul-Jabbar [1.1D] **31.** $182,000 [1.3E]
32. a. 74 m **b.** 300 m^2 [1.2C, 1.3E] **33.** 42 mi [1.3E] **34.** $449 [1.2C] **35.** 6 h [1.3E]

CHAPTER 1 TEST

1. 329,700 [1.3A, How To 4] **2.** 16,000 [1.3B, Example 10] **3.** 4029 [1.2B, Example 6] **4.** x^4y^3 [1.3B, Example 7]
5. Yes [1.2A, You Try It 5] **6.** 3000 [1.1C, Example 9] **7.** 7177 < 7717 [1.1A, Example 3]
8. 8490 [1.1B, Example 6] **9.** Three hundred eighty-two thousand nine hundred four [1.1B, Example 5]
10. 2000 [1.2A, Example 1] **11.** 11,008 [1.3A, How To 1] **12.** 2,400,000 [1.3A, Example 2]
13. 1, 2, 4, 23, 46, 92 [1.3D, How To 17] **14.** $2^4 \cdot 3 \cdot 5$ [1.3D, How To 18] **15.** 30,866 [1.2B, How To 14]
16. The Commutative Property of Addition [1.2A, Example 2] **17.** 897 [1.3C, Example 15] **18.** 26 [1.4A, Example 1]
19. $13,900 [1.2C, Example 11] **20.** 7 [1.4A, How To 1] **21.** 3000 + 900 + 70 + 2 [1.1B, Example 7]
22. 56 [1.4A, How To 3] **23.** 7 [1.2A, Example 2] **24.** 720 [1.3E, Example 21] **25.** $556 [1.2C, Example 11]
26. a. 96 cm **b.** 576 cm^2 [1.3E, How To 21] **27.** $4456 [1.2C, Example 11] **28. a.** $51,184 **b.** $121,468 **c.** $70,284 [1.2C, Example 11; 1.3E, Example 20] **29.** $960 [1.3E, Example 22] **30.** $11 [1.3E, Example 22]

ANSWERS TO CHAPTER 2 SELECTED EXERCISES

PREP TEST

1. 20 [1.3A] **2.** 120 [1.3A] **3.** 9 [1.3A] **4.** 10 [1.2A] **5.** 7 [1.2B] **6.** 2 r3 [1.3C]
7. 1, 2, 3, 4, 6, 12 [1.3C] **8.** 59 [1.4A] **9.** 7 [1.2A] **10.** 44 < 48 [1.1A] **11.** 36,900 [1.1C]

SECTION 2.1

1. 40 **3.** 24 **5.** 30 **7.** 12 **9.** 24 **11.** 60 **13.** 56 **15.** 9 **17.** 32 **19.** 36 **21.** 660
23. 9384 **25.** 24 **27.** 30 **29.** 24 **31.** 576 **33.** 1680 **35.** True **37.** 1 **39.** 3 **41.** 5 **43.** 25
45. 1 **47.** 4 **49.** 4 **51.** 6 **53.** 4 **55.** 1 **57.** 7 **59.** 5 **61.** 8 **63.** 1 **65.** 25 **67.** 7
69. 8 **71.** True **73.** They will have another day off together in 12 days.

SECTION 2.2

1. $\frac{4}{5}$ **3.** $\frac{1}{4}$ **5.** $\frac{4}{3}$; $1\frac{1}{3}$ **7.** $\frac{13}{5}$; $2\frac{3}{5}$ **9.** $3\frac{1}{4}$ **11.** 4 **13.** $2\frac{7}{10}$ **15.** 7 **17.** $1\frac{8}{9}$ **19.** $2\frac{2}{5}$ **21.** 18
23. $2\frac{2}{15}$ **25.** 1 **27.** $9\frac{1}{3}$ **29.** $\frac{9}{4}$ **31.** $\frac{11}{2}$ **33.** $\frac{14}{5}$ **35.** $\frac{47}{6}$ **37.** $\frac{7}{1}$ **39.** $\frac{33}{4}$ **41.** $\frac{31}{3}$
43. $\frac{55}{12}$ **45.** $\frac{8}{1}$ **47.** $\frac{64}{5}$ **49.** True **51.** $\frac{6}{12}$ **53.** $\frac{9}{24}$ **55.** $\frac{6}{51}$ **57.** $\frac{24}{32}$ **59.** $\frac{108}{18}$ **61.** $\frac{30}{90}$ **63.** $\frac{14}{21}$
65. $\frac{42}{49}$ **67.** $\frac{8}{18}$ **69.** $\frac{28}{4}$ **71.** $\frac{1}{4}$ **73.** $\frac{3}{4}$ **75.** $\frac{1}{6}$ **77.** $\frac{8}{33}$ **79.** 0 **81.** $\frac{7}{6}$ **83.** 1 **85.** $\frac{3}{5}$
87. $\frac{4}{15}$ **89.** $\frac{3}{5}$ **91.** $\frac{2m}{3}$ **93.** $\frac{y}{2}$ **95.** $\frac{2a}{3}$ **97.** c **99.** $6k$ **101.** $\frac{3}{8} < \frac{2}{5}$ **103.** $\frac{3}{4} < \frac{7}{9}$ **105.** $\frac{2}{3} > \frac{7}{11}$
107. $\frac{17}{24} > \frac{11}{16}$ **109.** $\frac{7}{15} > \frac{5}{12}$ **111.** $\frac{5}{9} > \frac{11}{21}$ **113.** $\frac{7}{12} < \frac{13}{18}$ **115.** $\frac{4}{5} > \frac{7}{9}$ **117.** $\frac{9}{16} > \frac{5}{9}$ **119.** $\frac{5}{8} < \frac{13}{20}$ **121.** $\frac{4}{5}$
123. $\frac{5}{6}$ **125. a.** Location **b.** Location **127.** $\frac{2}{25}$

SECTION 2.3

1. $\frac{9}{11}$ **3.** 1 **5.** $1\frac{2}{3}$ **7.** $1\frac{1}{6}$ **9.** $\frac{16}{b}$ **11.** $\frac{9}{c}$ **13.** $\frac{11}{x}$ **15.** $\frac{11}{12}$ **17.** $\frac{11}{12}$ **19.** $1\frac{7}{12}$ **21.** $2\frac{2}{15}$
23. $15\frac{2}{3}$ **25.** $5\frac{2}{3}$ **27.** $15\frac{1}{20}$ **29.** $10\frac{7}{36}$ **31.** $7\frac{5}{12}$ **33.** A whole number other than 1 **35.** The number 1
37. $\frac{3}{4}$ **39.** $6\frac{5}{24}$ **41.** $2\frac{5}{24}$ **43.** $1\frac{2}{5}$ **45.** $1\frac{13}{18}$ **47.** $1\frac{5}{24}$ **49.** $11\frac{2}{3}$ **51.** $14\frac{3}{4}$ **53.** Yes **55.** $\frac{1}{6}$ **57.** $\frac{1}{6}$
59. $\frac{5}{d}$ **61.** $\frac{5}{n}$ **63.** $\frac{1}{14}$ **65.** $\frac{1}{2}$ **67.** $\frac{1}{4}$ **69.** $2\frac{1}{3}$ **71.** $6\frac{3}{4}$ **73.** $1\frac{1}{12}$ **75.** $3\frac{3}{8}$ **77.** $5\frac{1}{9}$ **79.** $2\frac{3}{4}$ **81.** $1\frac{17}{24}$
83. $4\frac{19}{24}$ **85.** $1\frac{7}{10}$ **87.** Yes **89.** $\frac{5}{24}$ **91.** $6\frac{5}{12}$ **93.** $\frac{1}{3}$ **95.** $\frac{1}{6}$ **97.** $1\frac{1}{9}$ **99.** $2\frac{2}{9}$ **101.** Yes
103. The length is $3\frac{1}{4}$ ft. **105.** The horses run $\frac{1}{16}$ mi farther in the Kentucky Derby. They run $\frac{5}{16}$ mi farther in the Belmont Stakes. **107.** The difference is $\frac{3}{32}$ in. **109.** You need $29\frac{1}{2}$ ft of fencing. **111.** The difference represents the distance that will remain to be traveled after the first day. **113.** The difference between the heights of the jumps is $1\frac{1}{2}$ ft. **115.** $\frac{17}{20}$ of the men in the United States are not left-handed. **117.** The wall is $6\frac{5}{8}$ in. thick.

SECTION 2.4

3. $\frac{3}{5}$ **5.** $\frac{4}{5}$ **7.** 0 **9.** $\frac{63}{xy}$ **11.** $\frac{1}{9}$ **13.** 1 **15.** 6 **17.** 0 **19.** $\frac{1}{2}$ **21.** 19 **23.** 42 **25.** $5\frac{1}{2}$
27. For example, $\frac{3}{4}$ and $\frac{4}{3}$ **29.** $\frac{7}{10}$ **31.** $1\frac{4}{5}$ **33.** A typical household spends \$13,000 on housing per year. **35.** $\frac{7}{48}$
37. $3\frac{1}{2}$ **39.** $\frac{1}{5}$ **41.** $\frac{1}{6}$ **43.** $3\frac{2}{3}$ **45.** No **47.** $\frac{9}{16}$ **49.** $\frac{5}{128}$ **51.** $\frac{4}{45}$ **53.** $1\frac{1}{7}$ **55.** $\frac{16}{81}$ **57.** $\frac{2}{3}$
59. $1\frac{11}{14}$ **61.** 0 **63.** 8 **65.** $\frac{1}{8}$ **67.** Undefined **69.** $\frac{bd}{30}$ **71.** $5\frac{1}{3}$ **73.** $\frac{1}{2}$ **75.** $5\frac{2}{7}$ **77.** $1\frac{29}{31}$ **79.** True
81. $1\frac{1}{5}$ **83.** $\frac{7}{26}$ **85.** $\frac{1}{12}$ **87.** 48 **89.** There are 32 servings in the box. **91.** $\frac{3}{4}$ **93.** $\frac{1}{6}$ **95.** 6 **97.** $\frac{18}{35}$
99. $1\frac{7}{25}$ **101.** $3\frac{3}{11}$ **103.** True **105.** $\frac{8}{19}$ **107.** $1\frac{7}{16}$ **109.** No. **111.** There are 354 days in 1 year in the Assyrian calendar. **113.** Less than **115.** The cost is \$11. **117.** The worker can assemble 8 products in 1 h. **119.** You should buy $12\frac{1}{2}$ lb of hamburger meat. **121.** The employee earned \$318 this week. **123.** $5\frac{1}{2}$ billion bushels of corn are turned into ethanol each year. **125.** Two bags of seed should be purchased. **127.** The rate of the hiker is $3\frac{1}{2}$ mph. **129.** The distance between the two cities is 1250 mi.

SECTION 2.5

1. Thousandths **3.** Ten-thousandths **5.** Hundredths **7.** 0.3 **9.** 0.21 **11.** 0.461 **13.** 0.093 **15.** $\frac{1}{10}$
17. $\frac{47}{100}$ **19.** $\frac{289}{1000}$ **21.** $\frac{9}{100}$ **23.** Thirty-seven hundredths **25.** Nine and four tenths **27.** Fifty-three ten-thousandths
29. Forty-five thousandths **31.** Twenty-six and four hundredths **33.** 3.0806 **35.** 407.03 **37.** 246.024
39. 73.02684 **41.** $0.16 < 0.6$ **43.** $5.54 > 5.45$ **45.** $0.047 < 0.407$ **47.** $1.0008 < 1.008$ **49.** $7.6005 < 7.605$
51. $0.31502 < 0.3152$ **53.** $0.172 < 17.2$ **55.** 0.309, 0.39, 0.399 **57.** 0.0024, 0.024, 0.204, 0.24
59. 0.0061, 0.059, 0.06, 0.061 **61.** 6.2 **63.** 21.0 **65.** 18.41 **67.** 72.50 **69.** 936.291 **71.** 47 **73.** 7015
75. 2.97527 **77.** For example, 0.2701 **79.** Tony Dorsett had the greater average number of yards per carry.
81. The average life expectancy is higher in Italy. **83.** The length of the race is 42.2 km. **85.** **a.** \$2.40 **b.** \$3.60 **c.** \$6.00 **d.** \$7.00 **e.** \$4.70 **f.** \$2.40 **g.** \$2.40 **87.** For example: **a.** 0.15 **b.** 1.05 **c.** 0.001

SECTION 2.6

1. 65.9421 **3.** 190.857 **5.** 21.26 **7.** 21.26 **9.** 2.768 **11.** 56.361 **13.** 53.67 **15.** 12; 12.325
17. 40; 33.63 **19.** 0.3; 0.303 **21.** 40; 38.618 **23.** 137.505 **25.** 24.53 **27.** 11.789 **29.** $8.07 - 5.392$
31. 0.27 **33.** 2.664 **35.** 592 **37.** 82 **39.** 10^3 **41.** 8.0; 7.5537 **43.** 70; 68.5936 **45.** 30; 32.1485
47. 50.16 **49.** 48 **51.** 32.3 **53.** 67.7 **55.** 6.3 **57.** 5.8 **59.** 0.81 **61.** 0.08 **63.** 5.278 **65.** 0.4805
67. Greater than 1 **69.** 1000; 954.93 **71.** 2; 2.18 **73.** 100; 103.03 **75.** 25; 28.94 **77.** 5.06 **79.** 0.24

81. 2.06 **83.** $0.4\overline{6}$ **85.** 0.5625 **87.** $1.\overline{6}$ **89.** 2.75 **91.** $3.\overline{2}$ **93.** 0.12 **95.** 6.6 **97.** Less than 1 **99.** Greater than 1 **101.** $\frac{3}{5}$ **103.** $\frac{1}{4}$ **105.** $\frac{12}{25}$ **107.** $\frac{13}{40}$ **109.** $3\frac{2}{5}$ **111.** $9\frac{19}{20}$ **113.** $5\frac{17}{25}$ **115.** $\frac{17}{200}$ **117.** $\frac{7}{20} > 0.34$ **119.** $\frac{3}{4} > 0.706$ **121.** $0.72 > \frac{5}{7}$ **123.** $0.25 < \frac{13}{50}$ **125.** $\frac{7}{18} < 0.39$ **127.** $\frac{8}{15} < 0.543$ **129.** Your monthly salary is \$3968.25. **131.** The new balance is \$473.72. **133.** On average, a second-grade student uses a computer 36.4 h more per year than a fifth-grade student. **135.** The perimeter is 14 ft. **137.** The perimeter is 18.4 m. **139.** The area is 35.88 cm^2. **141.** The perimeter is 18.5 m. **143.** The federal earnings for the employee are \$562.20. **145.** The force is 41.65 newtons. **147.** The equity is \$57,146.75. **149.** The cost is \$11.30. **151. a.** \$79.90 **b.** \$125.60 **c.** \$108.05

153. a.

Quantity	*Item Number*	*Description*	*Unit Price*	*Total*
1	45837	Gasket set	\$174.90	\$174.90
1	29753	Ring set	\$169.99	\$169.99
8	54678	Valve	\$ 16.99	\$135.92
8	28632	Wrist pin	\$ 23.55	\$188.40
16	27345	Valve spring	\$ 9.25	\$148.00
8	34922	Rod bearing	\$ 13.69	\$109.52
5	41257	Main bearing	\$ 17.49	\$ 87.45
16	2871	Valve seal	\$ 1.69	\$ 27.04
1	23751	Timing chain	\$ 50.49	\$ 50.49

b. \$1091.71 **c.** \$1799.88 **d.** \$2891.59

SECTION 2.7

1. $1\frac{1}{5}$ **3.** $\frac{5}{36}$ **5.** $\frac{11}{32}$ **7.** 1 **9.** 4 **11.** $\frac{8}{9}$ **13.** 1.72 **15.** 1.84 **17.** 2.04 **19.** $1\frac{3}{10}$ **21.** $1\frac{1}{9}$ **23.** $1\frac{15}{16}$ **25.** $\frac{1}{2}$ **27.** 1 **29.** 18.09 **31.** 30.5 **33.** $\frac{6}{x}$

CHAPTER 2 CONCEPT REVIEW*

1. To find the LCM of 75, 30, and 50, find the prime factorization of each number and write the factorization of each number in a table. Circle the greatest product in each column. The LCM is the product of the circled numbers.

	2	3	5
75 =		(3)	(5 · 5)
30 =	(2)	3	5
50 =	2		5 · 5

LCM = 2 · 3 · 5 · 5 = 150 [2.1A]

2. To find the GCF of 42, 14, and 21, find the prime factorization of each number and write the factorization of each number in a table. Circle the least product in each column that does not have a blank. The GCF is the product of the circled numbers.

	2	3	7
42 =	2	3	(7)
14 =	2		7
21 =		3	7

GCF = 7 [2.1B]

3. To write an improper fraction as a mixed number, divide the numerator by the denominator. The quotient without the remainder is the whole number part of the mixed number. To write the fractional part of the mixed number, write the remainder over the divisor. [2.2A]

**Note:* The numbers in brackets following the answers in the Concept Review are a reference to the objective that corresponds to that problem. For example, the reference [1.2A] stands for Section 1.2, Objective A. This notation will be used for all Prep Tests, Concept Reviews, Chapter Reviews, Chapter Tests, and Cumulative Reviews throughout the text.

4. A fraction is in simplest form when the numerator and denominator have no common factors other than 1. For example, $\frac{8}{12}$ is not in simplest form because 8 and 12 have a common factor of 4. $\frac{5}{7}$ is in simplest form because 5 and 7 have no common factors other than 1. [2.2B]

5. When adding fractions, you have to convert to equivalent fractions with a common denominator. One way to explain this is that you can combine like things, but you cannot combine unlike things. You can combine 3 apples and 4 apples and get 7 apples. You cannot combine 4 apples and 3 oranges and get a sum consisting of just one item. In adding whole numbers, you add like things: ones, tens, hundreds, and so on. In adding fractions, you can combine 2 *ninths* and 5 *ninths* and get 7 *ninths*, but you cannot add 2 *ninths* and 3 *fifths*. [2.3A]

6. To add mixed numbers, add the fractional parts and then add the whole number parts. Then reduce the sum to simplest form. [2.3A]

7. To subtract mixed numbers, the first step is to subtract the fractional parts. If we are subtracting a mixed number from a whole number, there is no fractional part in the whole number from which to subtract the fractional part of the mixed number. Therefore, we must borrow a 1 from the whole number and write 1 as an equivalent fraction with a denominator equal to the denominator of the fraction in the mixed number. Then we can subtract the fractional parts and then subtract the whole numbers. [2.3B]

8. Write the decimal point in the product of two decimals so that the number of decimal places in the product is the sum of the numbers of decimal places in the factors. [2.6B]

9. To estimate the product of two decimals, round each number so that it contains one nonzero digit. Then multiply. For example, to estimate the product of 0.068 and 0.0052, round the two numbers to 0.07 and 0.005; then multiply $0.07 \times 0.005 = 0.00035$. [2.6B]

10. When dividing decimals, move the decimal point in the divisor to the right to make the divisor a whole number. Move the decimal point in the dividend the same number of places to the right. Place the decimal point in the quotient directly over the decimal point in the dividend, and then divide as with whole numbers. [2.6C]

11. When dividing 0.763 by 0.6, the decimal points will be moved one place to the right: $7.63 \div 6$. The decimal 7.63 has digits in the tenths and hundredths places. We need to write a zero in the thousandths place in order to determine the digit in the thousandths place of the quotient so that we can then round the quotient to the nearest hundredth. [2.6C]

12. To subtract a decimal from a whole number that has no decimal point, write a decimal point in the whole number to the right of the ones place. Then write as many zeros to the right of that decimal point as there are places in the decimal being subtracted from the whole number. For example, the subtraction $5 - 3.578$ would be written $5.000 - 3.578$. [2.6A]

CHAPTER 2 REVIEW EXERCISES

1. $9\frac{1}{2}$ [2.2A] **2.** $2\frac{5}{6}$ [2.3B] **3.** $1\frac{1}{2}$ [2.4B] **4.** 5.034 [2.5A] **5.** $\frac{7}{25}$ [2.6D] **6.** $2\frac{2}{3}$ [2.4A] **7.** $8.039 < 8.31$ [2.5B] **8.** $\frac{3}{5} > \frac{7}{15}$ [2.2C] **9.** 150 [2.1A] **10.** 91,800 [2.6B] **11.** $3\frac{1}{3}$ [2.4A] **12.** $\frac{10}{7}$; $1\frac{3}{7}$ [2.2A] **13.** $\frac{3}{7} < 0.429$ [2.6D] **14.** $\frac{3}{5}$ [2.4C] **15.** $\frac{32}{72}$ [2.2B] **16.** $\frac{1}{3}$ [2.4A] **17.** $\frac{2}{7}$ [2.7A] **18.** 21 [2.1B] **19.** 0.0142 [2.6C] **20.** 0.1 [2.6C] **21.** $\frac{5}{6}$ [2.4B] **22.** 0.11 [2.6C] **23.** 440 [2.6A] **24.** 2.4622 [2.6B] **25.** 50.743 [2.6A] **26.** $2\frac{1}{4}$ [2.4A] **27.** $9\frac{1}{12}$ [2.3A] **28.** $\frac{2}{7}$ [2.2B] **29.** $4\frac{7}{10}$ [2.3B] **30.** The average amount spent by each visitor was $2146.85. [2.6C] **31.** The wrestler must gain $6\frac{1}{4}$ lb during the third and fourth weeks. [2.3C] **32.** The employee can assemble 192 units during an 8-hour day. [2.4D] **33.** The employee is due $150 in overtime pay. [2.4D] **34.** The final velocity is 496 ft/s. [2.4D]

CHAPTER 2 TEST

1. $2\frac{4}{7}$ [2.2A, Example 2] **2.** $3\frac{11}{12}$ [2.3B, How To 12] **3.** $22\frac{1}{2}$ [2.4A, Example 5] **4.** $\frac{7}{12}$ [2.4A, How To 5] **5.** 90 [2.1A, Example 1] **6.** 9.033 [2.5A, You Try It 5] **7.** $2\frac{11}{32}$ [2.4A, Example 5] **8.** $\frac{19}{5}$ [2.2A, How To 1] **9.** $\frac{7}{9}$ [2.4B, Example 6] **10.** $4.003 < 4.009$ [2.5B, How To 2] **11.** 7 [2.4C, Example 11] **12.** 18 [2.1B, Example 2] **13.** $\frac{1}{6}$ [2.3B, Example 7] **14.** $\frac{4}{5}$ [2.2B, How To 4] **15.** $2\frac{17}{24}$ [2.3A, Example 5] **16.** $\frac{5}{6} > \frac{11}{15}$ [2.2C, Example 12] **17.** $3\frac{16}{25}$ [2.7A, How To 3] **18.** $0.22 < \frac{2}{9}$ [2.6D, Example 17] **19.** 6.051 [2.5C, You Try It 8] **20.** $1\frac{1}{2}$ [2.4B, Example 9] **21.** 22.753 [2.6A, Example 2] **22.** 70 [2.6A, How To 6] **23.** 14.497 [2.6A, How To 2] **24.** 64 [2.6B, Example 8] **25.** 0.8496 [2.6C, How To 12] **26.** $\frac{12}{28}$ [2.2B, How To 2] **27.** The gross was $40.8 million greater. [2.6E, Example 18] **28.** Ten more hours of community service are still required of you. [2.3C, Example 9] **29.** The employee can assemble 80 units in 6 h. [2.4D, Example 13] **30.** The stockholders' equity is \$20.6 million. [2.6E, Example 20] **31.** The perimeter is 18.5 m. [2.6E, Example 20]

CUMULATIVE REVIEW EXERCISES

1. 0.03879 [2.6C] **2.** 15 [1.4A] **3.** $2^2 \cdot 5 \cdot 7$ [1.3D] **4.** 8,072,092 [1.1B] **5.** $\frac{7}{11} < \frac{4}{5}$ [2.2C] **6.** 36 [2.1B] **7.** $\frac{1}{7}$ [2.3B] **8.** 1900 [1.2A] **9.** 11,272 [1.2A] **10.** $1\frac{1}{2}$ [2.4B] **11.** 55.42 [2.6A] **12.** $\frac{1}{7}$ [2.4A] **13.** 1600 [1.3B] **14.** $2^2 \cdot 5 \cdot 13$ [1.3D] **15.** 0.76 [2.6D] **16.** 20,000 [1.2B] **17. a.** Sweden mandates more vacation days. **b.** Austria mandates 1.5 times more vacation days than Switzerland. [2.6E] **18.** Undefined [1.3C] **19.** $\frac{19}{21}$ [2.3A] **20.** $7\frac{1}{28}$ [2.3B] **21.** 39 [1.4A] **22.** 17 [1.4A] **23.** $\frac{3}{10}$ [2.4C] **24.** $\frac{3}{28}$ [2.4A] **25.** 2.8 [2.6C] **26.** You would burn 40 more calories. [1.3E] **27.** The projected increase in the population is 1,740,000 people. [1.2C] **28. a.** On average, a salesperson works 46.5 h per week. **b.** The average salesperson spends more time face-to-face selling. [2.6E] **29.** The distance traveled by the bicyclist was $4\frac{1}{8}$ mi. [2.4D] **30.** The cost is \$3.12 per visit. [2.6E]

ANSWERS TO CHAPTER 3 SELECTED EXERCISES

PREP TEST

1. $54 > 45$ [1.1A] **2.** 4 units [1.1A] **3.** 15,847 [1.2A] **4.** 3779 [1.2B] **5.** 26,432 [1.3A] **6.** 6 [1.3C] **7.** $1\frac{4}{15}$ [2.3A] **8.** $\frac{7}{16}$ [2.3B] **9.** 11.058 [2.6A] **10.** 3.781 [2.6A] **11.** $\frac{2}{5}$ [2.4A] **12.** $\frac{5}{9}$ [2.4B] **13.** 9.4 [2.6B] **14.** 0.4 [2.6C] **15.** 31 [1.4A]

SECTION 3.1

1. −6 −5 −4 −3 −2 −1 0 1 2 3 4 5 6 **3.** −6 −5 −4 −3 −2 −1 0 1 2 3 4 5 6

5. −6 −5 −4 −3 −2 −1 0 1 2 3 4 5 6 **7.** −6 −5 −4 −3 −2 −1 0 1 2 3 4 5 6 **9.** 1 **11.** −1 **13.** 3

15. *A* is −4. *C* is −2. **17.** *A* is −7. *D* is −4. **19.** $-2 > -5$ **21.** $3 > -7$ **23.** $-42 < -7$ **25.** $53 > -46$ **27.** $-51 < -20$ **29.** $-131 < 101$ **31.** −7, −2, 0, 3 **33.** −5, −3, 1, 4 **35.** −4, 0, 5, 9 **37.** −10, −7, −5, 4, 12 **39.** −11, −7, −2, 5, 10 **41.** Always true **43.** Sometimes true **45.** −45 **47.** 88 **49.** $-n$ **51.** d **53.** The opposite of negative thirteen **55.** The opposite of negative p **57.** Five plus negative ten **59.** Negative fourteen minus negative three **61.** Negative thirteen minus eight **63.** m plus negative n **65.** 7 **67.** 61 **69.** −46 **71.** 73 **73.** z **75.** $-p$ **77.** Positive integers **79.** 4 **81.** 9 **83.** 11 **85.** 12 **87.** 23 **89.** −27 **91.** 25 **93.** −41 **95.** −93 **97.** 10 **99.** 8 **101.** 6 **103.** $|-12| > |8|$ **105.** $|6| < |13|$ **107.** $|-1| < |-17|$ **109.** $|x| = |-x|$ **111.** $-|6|, -(4), |-7|, -(-9)$ **113.** $-9, -|-7|, -(5), |4|$ **115.** $-|10|, -|-8|, -(-2), -(-3), |5|$ **117.** Negative integers **119.** The wind chill factor is −9°F. **121.** The cooling power is −35°F. **123.** A temperature of −30°F with a 5-mph wind would feel colder. **125.** Stock B showed the least net change. **127.** −1 **129.** 11 and −11

SECTION 3.2

1. −11 **3.** −5 **5.** 8 **7.** −4 **9.** −2 **11.** −9 **13.** 1 **15.** −15 **17.** 0 **19.** −21 **21.** −14 **23.** 19 **25.** −5 **27.** −30 **29.** 9 **31.** −12 **33.** −28 **35.** −13 **37.** −18 **39.** 11 **41.** 1 **43.** $x + (-7)$ **45.** −12 **47.** −11 **49.** −7 **51.** 20 **53.** The Commutative Property of Addition **55.** The Inverse Property of Addition **57.** −11 **59.** −4 **61.** Yes **63.** Yes **65.** Always true **67.** Sometimes true **71.** −3 **73.** −13 **75.** 7 **77.** 0 **79.** −17 **81.** −3 **83.** 12 **85.** 27 **87.** −106 **89.** −67 **91.** −6 **93.** −15 **95.** $-t - r$ **97.** 82°C **99.** −9 **101.** 11 **103.** 0 **105.** −138 **107.** 26 **109.** 13 **111.** −8 **113.** 5 **115.** 2 **117.** −6 **119.** 12 **121.** −3 **123.** 18 **125.** Yes **127.** Yes **129.** Never true

131. Sometimes true **133. a.** The difference in elevation is 7046 m. **b.** The difference in elevation is 6051 m. **135.** The difference between the highest and lowest elevations is smallest in Africa. **137.** The temperature is 3°C. **139.** The difference is 0°F. **141.** The new temperature is below 0°C. **143.** The golfer's score was −3. **145.** 19 **147. a.** Sometimes true **b.** Always true

SECTION 3.3

3. −21 **5.** 5 **7.** 24 **9.** −28 **11.** −27 **13.** 18 **15.** −11 **17.** 558 **19.** −72 **21.** 140 **23.** −320 **25.** −228 **27.** 204 **29.** −70 **31.** 162 **33.** 36 **35.** −700 **37.** 360 **39.** $-fgh$ **41.** Positive **43.** Negative **45.** The Multiplication Property of One **47.** The Associative Property of Multiplication **49.** −6 **51.** 1 **53.** −24 **55.** −60 **57.** 357 **59.** −56 **61.** −1600 **63.** No **65.** Yes **67.** −6 **69.** 8 **71.** −49 **73.** 8 **75.** −11 **77.** 14 **79.** 13 **81.** 1 **83.** 26 **85.** 23 **87.** Never true **89.** Always true **91.** −110 **93.** 111 **95.** $\frac{-9}{x}$ **97.** −9 **99.** 9 **101.** −6 **103.** 6 **105.** No **107.** Yes **109.** The average score was −3. **111.** The average record low temperature is −62°F. **113.** The average daily high temperature was −26°. **115.** The low temperature on the sixth day was lower than −12°C. **117.** The student's score is 93 points. **119.** The wind chill factor is −45°F. **121.** −16, 32, −64 **123.** −125, −625, −3125 **125. a.** 81 **b.** −17

SECTION 3.4

1. $-\frac{5}{24}$ **3.** $-\frac{19}{24}$ **5.** $\frac{5}{26}$ **7.** $\frac{7}{24}$ **9.** $-\frac{19}{60}$ **11.** $-1\frac{3}{8}$ **13.** $\frac{3}{4}$ **15.** $-\frac{47}{48}$ **17.** $\frac{3}{8}$ **19.** $-\frac{7}{60}$ **21.** $\frac{13}{24}$ **23.** −3.4 **25.** −8.89 **27.** −8.0 **29.** −0.68 **31.** −181.51 **33.** 2.7 **35.** −20.7 **37.** −37.19 **39.** −34.99 **41.** $-\frac{5}{48}$ **43.** $-1\frac{5}{36}$ **45.** $1\frac{3}{10}$ **47.** −649.36 **49.** $-\frac{1}{6}$ **51.** $-1\frac{5}{24}$ **53.** −25.665 **55.** $-1\frac{1}{4}$ **57.** $\frac{5}{36}$ **59.** −2.163 **61.** Yes **63.** No **65.** 2 **67.** 1 **69.** $-\frac{3}{8}$ **71.** $\frac{1}{10}$ **73.** $-\frac{4}{9}$ **75.** $-\frac{7}{26}$ **77.** 10 **79.** $-\frac{7}{30}$ **81.** $-\frac{2}{3}$ **83.** $4\frac{2}{7}$ **85.** $-\frac{8}{9}$ **87.** $\frac{9}{10}$ **89.** $-\frac{1}{6}$ **91.** 28.14 **93.** −7.84 **95.** 0.117 **97.** 9.91 **99.** −84.3 **101.** −49.8 **103.** −1.7 **105.** $-\frac{1}{12}$ **107.** $-\frac{1}{21}$ **109.** $-1\frac{1}{24}$ **111.** −131.328 **113.** −25.4 **115.** $-17\frac{1}{2}$ **117.** $\frac{1}{6}$ **119.** −48 **121.** $\frac{1}{12}$ **123.** −48 **125.** 10.5 **127.** −1.7 **129.** Yes **131.** No **133.** Greater than 1 **135.** The temperature fell 32.22°C. **137.** The difference is greater than 4.8. **139.** The difference is 14.06°C. **141. a.** True **b.** True **c.** False **d.** False

SECTION 3.5

1. −3 **3.** −6 **5.** −5 **7.** −12 **9.** −3 **11.** 19 **13.** 2 **15.** 1 **17.** 14 **19.** 42 **21.** −13 **23.** −12 **25.** −6 **27.** 0.21 **29.** −0.96 **31.** −0.29 **33.** −1 **35.** 0 **37.** $-\frac{5}{8}$ **39.** (i) **41.** 2 **43.** 1 **45.** 15 **47.** 32 **49.** 1 **51.** 1 **53.** 5 **55.** 28 **57.** $\frac{13}{18}$ **59.** −4

CHAPTER 3 CONCEPT REVIEW*

1. There are two numbers that are 6 units from 4 on the number line. The numbers are −2 and 10. [3.1A]
2. Both 6 and −6 have an absolute value of 6. [3.1C]
3. Rule for adding two integers:

   To add numbers with the same sign, add the absolute values of the numbers. Then attach the sign of the addends.

   To add numbers with different signs, find the difference between the absolute values of the numbers. Then attach the sign of the addend with the greater absolute value. [3.2A]
4. The rule for subtracting two integers is to add the opposite of the second number to the first number. [3.2B]
5. To find the change in temperature from −5°C to −14°C, use subtraction. [3.2C]

Note: The numbers in brackets following the answers in the Concept Review are a reference to the objective that corresponds to that problem. For example, the reference [1.2A] stands for Section 1.2, Objective A. This notation will be used for all Prep Tests, Concept Reviews, Chapter Reviews, Chapter Tests, and Cumulative Reviews throughout the text.

6. $4 - 9 = 4 + (-9)$. To show this addition on the number line, start at 4 on the number line. Draw an arrow 9 units long pointing to the left, with its tail at 4. The head of the arrow is at -5. $4 + (-9) = -5$. [3.2B]

7. If you multiply two numbers with different signs, the product is negative. [3.4B]

8. If you divide two numbers with the same sign, the quotient is positive. [3.4B]

9. When a number is divided by zero, the result is undefined because division by zero is undefined. [3.3B]

10. A terminating decimal is one that ends. For example, 0.75 is a terminating decimal. [3.4A]

11. The steps in the Order of Operations Agreement are:

1. Do all operations inside parentheses.
2. Simplify any numerical expressions containing exponents.
3. Do multiplication and division as they occur from left to right.
4. Rewrite subtraction as addition of the opposite. Then do additions as they occur from left to right. [3.5A]

CHAPTER 3 REVIEW EXERCISES

1. Eight minus negative one [3.1B] **2.** -36 [3.1C] **3.** 200 [3.3A] **4.** -9 [3.3B] **5.** -14 [3.2A]
6. 13 [3.1B] **7.** (number line from −6 to 6 with a dot at −2) −6 −5 −4 −3 −2 −1 0 1 2 3 4 5 6 [3.1A] **8.** -98.38 [3.4A] **9.** 17 [3.3B]
10. -210 [3.3B] **11.** -2 [3.2B] **12.** -18 [3.3A] **13.** -1 [3.2A] **14.** -72 [3.3A] **15.** -4 [3.5A]
16. -2 [3.2B] **17.** 13 [3.2B] **18.** $\frac{2}{7}$ [3.4B] **19.** Yes [3.2B] **20.** 14 [3.2B] **21.** 0 [3.3B]
22. -60 [3.3A] **23.** -12 [3.2A] **24.** 5 [3.5A] **25.** $-8 > -10$ [3.1A] **26.** 9 [3.5A] **27.** 27 [3.1C]
28. -2.8 [3.4B] **29.** $-\frac{5}{48}$ [3.4A] **30.** A temperature of -12°C is colder. [3.1D] **31.** The boiling point of neon is -238°C. [3.3C] **32.** The temperature is -3°C after the increase. [3.4C] **33.** 12 [3.2C]

CHAPTER 3 TEST

1. Negative three plus negative five [3.1B, You Try It 6] **2.** -34 [3.1C, Example 9] **3.** 18 [3.2B, Example 10]
4. -20 [3.2A, How To 8] **5.** 24 [3.3A, How To 5] **6.** $-\frac{7}{18}$ [3.4A, How To 1] **7.** 12 [3.3B, You Try It 5]
8. 2 [3.2A, Example 3] **9.** $16 > -19$ [3.1A, Example 3] **10.** -2 [3.2B, How To 13] **11.** -3 [3.2B, Example 13]
12. 49 [3.1B, Example 7] **13.** -250 [3.3A, How To 1] **14.** $-|5|, -(3), |-9|, -(-11)$ [3.1C, Example 11]
15. No [3.2B, Example 14] **16.** -3 [3.1A, Example 1] **17.** 0 [3.3B, You Try It 5] **18.** 19 [3.5A, How To 3]
19. -25 [3.1B, Example 5] **20.** -11.613 [3.4A, How To 8] **21.** -11 [3.2B, How To 12] **22.** 24 [3.3B, How To 8]
23. 10 [3.5A, How To 1] **24.** -7 [3.3B, Example 8] **25.** 60 [3.3A, How To 5] **26.** -107 [3.5A, Example 2]
27. -27 [3.4B, How To 12] **28.** -1.53 [3.4B, You Try It 13] **29.** -10 [3.2B, You Try It 10] **30.** The temperature is 4.5°C after the increase. [3.4C, Example 15] **31.** The wind chill factor is -64°F. [3.3C, Example 10] **32.** The high temperature was -5°C. [3.2C, Example 17] **33.** 16 units [3.2C, Example 18]

CUMULATIVE REVIEW EXERCISES

1. 5 [3.2B] **2.** 12,000 [1.3A] **3.** 32.3 [2.6C] **4.** 2 [1.4A] **5.** -82 [3.1C] **6.** 309,480 [1.1B]
7. 2400 [1.3A] **8.** 21 [3.3B] **9.** -11 [3.2B] **10.** -40 [3.2A] **11.** 1, 2, 4, 11, 22, 44 [1.3D]
12. 1 [2.4A] **13.** 630,000 [1.1C] **14.** 1300 [1.2A] **15.** 9 [3.2B] **16.** -2500 [3.3A] **17.** 8.77 [2.6A]
18. $5\frac{2}{3}$ [2.3A] **19.** -32 [3.5A] **20.** -4 [3.3B] **21.** $1\frac{1}{5}$ [2.4B] **22.** $-62 < 26$ [3.1A] **23.** 126 [3.3A]
24. 4.14 [3.4B] **25.** $2^5 \cdot 7^2$ [1.3B] **26.** 47 [1.4A] **27.** 10,062 [1.2A] **28.** -26 [3.2B] **29.** 5000 [1.2B]
30. 2025 [1.3B] **31.** The land area was 1,722,685 mi^2 after the purchase. [1.2C] **32.** Albert Einstein was 76 years old when he died. [1.2C] **33.** The amount that remains to be paid is $14,200. [1.2C] **34.** The total cost of the land is $92,250. [1.3E] **35.** The temperature is -5°C after the increase. [3.2C] **36. a.** The difference is 168°F.
b. The difference is greatest for Alaska. [3.2C] **37.** Your sales for the fourth quarter must be $24,900. [1.2C]
38. The golfer's score is -8. [3.2C]

ANSWERS TO CHAPTER 4 SELECTED EXERCISES

PREP TEST

1. 3 [3.2B] **2.** 4 [3.3B] **3.** $\frac{1}{12}$ [3.4A] **4.** $-\frac{4}{9}$ [2.4A] **5.** $\frac{3}{10}$ [3.4B] **6.** -16 [3.5A] **7.** $\frac{8}{27}$ [2.4A]
8. 48 [1.4A] **9.** 1 [1.4A] **10.** 12 [3.5A]

SECTION 4.1

1. $2x^2, 5x, \underline{-8}$ **3.** $-a^4, \underline{6}$ **5.** $\underline{7x^2y}, \underline{6xy^2}$ **7.** $1, -9$ **9.** $1, -4, -1$ **13.** 10 **15.** 32 **17.** 21 **19.** 16 **21.** -9
23. 41 **25.** -7 **27.** 13 **29.** -15 **31.** 41 **33.** 1 **35.** 5 **37.** 1 **39.** 57 **41.** 5 **43.** 8 **45.** -3
47. -2 **49.** -4 **51.** Positive **53.** Negative **55.** 41 **57.** 1 **59.** -23 **61. a.** 2 **b.** 5 **c.** 6 **d.** 7

SECTION 4.2

3. $14x$ **5.** $5a$ **7.** $-6y$ **9.** $7 - 3b$ **11.** $5a$ **13.** $-2ab$ **15.** $5xy$ **17.** 0 **19.** $-\frac{5}{6}x$ **21.** $6.5x$
23. $0.45x$ **25.** $7a$ **27.** $-14x^2$ **29.** $-\frac{11}{24}x$ **31.** $17x - 3y$ **33.** $-2a - 6b$ **35.** $-3x - 8y$
37. $-4x^2 - 2x$ **39.** (iv) and (v) **41.** $60x$ **43.** $-10a$ **45.** $30y$ **47.** $72x$ **49.** $-28a$ **51.** $108b$
53. $-56x^2$ **55.** x^2 **57.** x **59.** a **61.** b **63.** x **65.** n **67.** $2x$ **69.** $-2x$ **71.** $-15a^2$ **73.** $6y$
75. $3y$ **77.** $-2x$ **79.** $-9y$ **81.** $8x - 6$ **83.** $-2a - 14$ **85.** $-6y + 24$ **87.** $-x - 2$ **89.** $35 - 21b$
91. $2 - 5y$ **93.** $15x^2 + 6x$ **95.** $2y - 18$ **97.** $-15x - 30$ **99.** $-6x^2 - 28$ **101.** $-6y^2 + 21$
103. $3x^2 - 3y^2$ **105.** $-4x + 12y$ **107.** $-6a^2 + 7b^2$ **109.** $4x^2 - 12x + 20$ **111.** $\frac{3}{2}x - \frac{9}{2}y + 6$
113. $-12a^2 - 20a + 28$ **115.** $12x^2 - 9x + 12$ **117.** $10x^2 - 20xy - 5y^2$ **119.** $-8b^2 + 6b - 9$ **121.** (iii)
123. $a - 7$ **125.** $-11x + 13$ **127.** $-4y - 4$ **129.** $-2x - 16$ **131.** $14y - 45$ **133.** $a + 7b$
135. $6x + 28$ **137.** $5x - 75$ **139.** $4x - 4$ **141.** $2x - 9$ **143.** $1.24x + 0.36$ **145.** $-0.01x + 40$

SECTION 4.3

1. $8 + y$ **3.** $t + 10$ **5.** $z + 14$ **7.** $x^2 - 20$ **9.** $\frac{3}{4}n + 12$ **11.** $8 + \frac{n}{4}$ **13.** $3(y + 7)$ **15.** $t(t + 16)$
17. $\frac{1}{2}x^2 + 15$ **19.** $5n^3 + n^2$ **21.** $r - \frac{r}{3}$ **23.** $x^2 - (x + 17)$ **25.** $9(z + 4)$ **27.** Answers may vary. For example: The product of 5 and 1 more than the square of n; 5 times the sum of 1 plus the square of n **29.** $\frac{x}{18}$ **31.** $x + 20$
33. $11x - 8$ **35.** $\frac{7}{5 + x}$ **37.** $40 - \frac{x}{20}$ **39.** $x^2 + 2x$ **41.** $10(x - 50)$; $10x - 500$ **43.** $x - (x + 3)$; -3
45. $(2x - 4) + x$; $3x - 4$ **47.** $x - (3x - 8)$; $-2x + 8$ **49.** $x + 3x$; $4x$ **51.** $(x + 6) + 5$; $x + 11$
53. $x - (x + 10)$; -10 **55.** $\frac{1}{6}x + \frac{4}{9}x$; $\frac{11}{18}x$ **57.** s represents the number of students enrolled in fall-term science classes.
59. Number of visitors to the Metropolitan Museum of Art: M; number of visitors to the Louvre: $M + 3{,}800{,}000$
61. Number of visitors to Google websites: G; number of visitors to Microsoft websites: $G - 63{,}000{,}000$
63. Length of one piece: S; length of second piece: $12 - S$ **65.** Distance traveled by the faster car: x; distance traveled by the slower car: $200 - x$ **67.** Number of bones in your body: N; number of bones in your foot: $\frac{1}{4}N$
69. Number of people surveyed: N; number of people who would pay down their debt: $0.43N$ **71.** $\frac{1}{4}x$

CHAPTER 4 CONCEPT REVIEW*

1. In a term, the numerical coefficient is the number. The variable part consists of the variables and their exponents. [4.1A]
2. When evaluating a variable expression, the Order of Operations Agreement must be used to simplify the resulting numerical expression. [4.1A]
3. For two terms to be like terms, the variable parts of the two terms must be the same. [4.2A]

**Note:* The numbers in brackets following the answers in the Concept Review are a reference to the objective that corresponds to that problem. For example, the reference [1.2A] stands for Section 1.2, Objective A. This notation will be used for all Prep Tests, Concept Reviews, Chapter Reviews, Chapter Tests, and Cumulative Reviews throughout the text.

4. Like terms of a variable expression are terms with the same variable part. Constant terms are also considered like terms. [4.2A]

5. The Commutative Property of Multiplication states that two numbers can be multiplied in either order; the product is the same. The Associative Property of Multiplication states that changing the grouping of three or more factors does not change their product. In the Commutative Property of Multiplication, the order in which the numbers appear changes, while in the Associative Property of Multiplication, the order in which the numbers appear does not change. [4.2B]

6. By the Inverse Property of Addition, the result of adding a number and its opposite is zero. [4.2A]

7. To evaluate $6 \cdot \frac{1}{6}$, use the Inverse Property of Multiplication. [4.2B]

8. The reciprocal of a number is the number with the numerator and denominator interchanged. The reciprocal of a number is also called the multiplicative inverse of the number. [4.2B]

9. Some mathematical terms that translate into multiplication are *times, twice, of, the product of,* and *multiplied by.* [4.3A]

10. Some mathematical terms that translate into subtraction are *minus, less than, decreased by, subtract . . . from . . . ,* and *the difference between.* [4.3A]

CHAPTER 4 REVIEW EXERCISES

1. $3x^2 - 24x - 21$ [4.2C] **2.** $11x$ [4.2A] **3.** $8a - 4b$ [4.2A] **4.** $-5n$ [4.2B] **5.** 79 [4.1A] **6.** $10x - 35$ [4.2C] **7.** $12y^2 + 8y - 10$ [4.2C] **8.** $-6a$ [4.2B] **9.** $-42x^2$ [4.2B] **10.** $-63 - 36x$ [4.2C] **11.** $-5y$ [4.2A] **12.** -4 [4.1A] **13.** $-6x - 1$ [4.2D] **14.** $-40a + 40$ [4.2D] **15.** $24y + 30$ [4.2D] **16.** $9c - 5d$ [4.2A] **17.** $20x$ [4.2B] **18.** $7x + 46$ [4.2D] **19.** 29 [4.1A] **20.** $-9r + 8s$ [4.2A] **21.** 50 [4.1A] **22.** 28 [4.1A] **23.** $-4x^2 + 6x$ [4.2A] **24.** $-90x + 25$ [4.2D] **25.** $-0.2x + 150$ [4.2D] **26.** $-\frac{1}{12}x$ [4.2A] **27.** $28a^2 - 8a + 12$ [4.2C] **28.** $-4x + 20$ [4.2D] **29.** -7 [4.1A] **30.** $36y$ [4.2B] **31.** $\frac{2}{3}(x + 10)$ [4.3A] **32.** $4x$ [4.3A] **33.** $x - 6$ [4.3A] **34.** $x + 2x$; $3x$ [4.3B] **35.** $2x - \frac{1}{2}x$; $\frac{3}{2}x$ [4.3B] **36.** $3x + 5(x - 1)$; $8x - 5$ [4.3B] **37.** Number of American League cards: A; number of National League cards: $5A$ [4.3C] **38.** Number of ten-dollar bills: T; number of five-dollar bills: $35 - T$ [4.3C] **39.** Number of calories in an apple: a; number of calories in the candy bar: $2a + 8$ [4.3C] **40.** Width of Parthenon: w; length of Parthenon: $1.6w$ [4.3C] **41.** Kneeling height: h; standing height: $1.3h$ [4.3C]

CHAPTER 4 TEST

1. $5x$ [4.2A, How To 1] **2.** $-6x^2 + 21y^2$ [4.2C, How To 9] **3.** $-x + 6$ [4.2D, Example 11] **4.** $-7x + 33$ [4.2D, Example 13] **5.** $-9x - 7y$ [4.2A, Example 1] **6.** 22 [4.1A, How To 1] **7.** $2x$ [4.2B, Example 5] **8.** $7x + 38$ [4.2D, Example 12] **9.** $-10x^2 + 15x - 30$ [4.2C, Example 9] **10.** $-2x - 5y$ [4.2A, Example 1] **11.** 3 [4.1A, Example 3] **12.** $3x$ [4.2B, How To 7] **13.** y^2 [4.2A, Example 2] **14.** $-4x + 8$ [4.2C, Example 8] **15.** $-10a$ [4.2B, Example 5] **16.** $2x + y$ [4.2D, Example 13] **17.** $36y$ [4.2B, Example 4] **18.** $15 - 35b$ [4.2C, Example 6] **19.** $a^2 - b^2$ [4.3A, How To 2] **20.** $10(x - 3) = 10x - 30$ [4.3B, How To 3] **21.** $x + 2x^2$ [4.3B, Example 3] **22.** $\frac{6}{x} - 3$ [4.3B, How To 3] **23.** $b - 7b$ [4.3A, Example 2] **24.** Speed of return throw: s; speed of fastball: $2s$ [4.3C, How To 4] **25.** Shorter piece: x; longer piece: $4x - 3$ [4.3C, Example 5]

CUMULATIVE REVIEW EXERCISES

1. -7 [3.2A] **2.** 5 [3.2B] **3.** 24 [3.3A] **4.** -5 [3.3B] **5.** $2 \cdot 5 \cdot 11$ [1.3D] **6.** $\frac{11}{48}$ [3.4A] **7.** $-\frac{1}{6}$ [3.4B] **8.** $\frac{1}{4}$ [3.4B] **9.** 1300 [1.2A] **10.** -5 [3.5A] **11.** $-\frac{27}{26}$ [3.5A] **12.** 16 [4.1A] **13.** $5x^2$ [4.2A] **14.** $-7a - 10b$ [4.2A] **15.** 8.357 [2.5A] **16.** $5.101 > 5.013$ [2.5B] **17.** $24 - 6x$ [4.2C] **18.** $6y - 18$ [4.2C] **19.** 10 [2.6A] **20.** 8.7 [2.5C] **21.** $-8x^2 + 12y^2$ [4.2C] **22.** $-9y^2 + 9y + 21$ [4.2C] **23.** $-7x + 14$ [4.2D] **24.** $5x - 43$ [4.2D] **25.** $17x - 24$ [4.2D] **26.** $-3x + 21y$ [4.2D] **27.** $\frac{1}{2}b + b$ [4.3A] **28.** $\frac{10}{y - 2}$ [4.3A] **29.** $8 - \frac{x}{12}$ [4.3B] **30.** $x + (x + 2)$; $2x + 2$ [4.3B] **31.** The perimeter is 9 in. [2.6E] **32.** Speed of dial-up connection: s; speed of DSL connection: $10s$ [4.3C]

ANSWERS TO CHAPTER 5 SELECTED EXERCISES

PREP TEST

1. −4 [3.2B] **2.** 1 [3.4B] **3.** −10 [3.4B] **4.** 1 [3.4B] **5.** $7y$ [4.2A] **6.** −9 [4.2A] **7.** −5 [4.1A]

SECTION 5.1

1. Yes **3.** No **5.** No **7.** Yes **9.** No **11.** Yes **13.** No **15.** Yes **17.** No **19.** Negative **23.** 2 **25.** 15 **27.** 6 **29.** 3 **31.** 0 **33.** −7 **35.** −7 **37.** −12 **39.** −5 **41.** 15 **43.** 9 **45.** 14 **47.** −1 **49.** 1 **51.** $-\frac{1}{2}$ **53.** $-\frac{3}{4}$ **55.** $\frac{1}{12}$ **57.** $-\frac{7}{12}$ **59.** 0.6529 **61.** −0.283 **63.** 9.257 **65.** −3 **67.** 0 **69.** −2 **71.** 9 **73.** 80 **75.** 0 **77.** −7 **79.** 12 **81.** −18 **83.** 15 **85.** −20 **87.** 0 **89.** $\frac{8}{3}$ **91.** $\frac{1}{3}$ **93.** $-\frac{1}{2}$ **95.** $-\frac{3}{2}$ **97.** $\frac{15}{7}$ **99.** 4 **101.** 3 **103.** 4.745 **105.** 2.06 **107.** −2.13 **109.** Positive **111.** Negative **113. a.** The distance walked by Joe is greater than the distance walked by John. **b.** Joe's time spent walking is equal to John's time spent walking. **c.** The total distance traveled is 2 mi. **115.** The runner will travel 3 mi. **117.** Marcella's average rate of speed is 36 mph. **119.** It would take Palmer 2.5 h to walk the course. **121.** The two joggers will meet 40 min after they start. **123.** It will take them 0.5 h. **125. a.** Answers will vary. **b.** Answers will vary.

SECTION 5.2

1. 3 **3.** 6 **5.** −1 **7.** −3 **9.** 2 **11.** 2 **13.** 5 **15.** −3 **17.** 6 **19.** 3 **21.** 1 **23.** 6 **25.** −7 **27.** 0 **29.** $\frac{3}{4}$ **31.** $\frac{4}{9}$ **33.** $\frac{1}{3}$ **35.** $-\frac{1}{2}$ **37.** $-\frac{3}{4}$ **39.** $\frac{1}{3}$ **41.** $-\frac{1}{6}$ **43.** 1 **45.** 1 **47.** 0 **49.** 0.15 **51.** $\frac{2}{5}$ **53.** $-\frac{4}{3}$ **55.** $-\frac{3}{2}$ **57.** 18 **59.** 8 **61.** −16 **63.** 25 **65.** $\frac{3}{4}$ **67.** $\frac{3}{8}$ **69.** $\frac{16}{9}$ **71.** $\frac{1}{18}$ **73.** $\frac{15}{2}$ **75.** $-\frac{18}{5}$ **77.** 2 **79.** 3 **81.** Negative **83.** Negative **85.** $x = 7$ **87.** $y = 3$ **89.** 19 **91.** −1 **93.** The average crown spread of the baldcypress is 57 ft. **95.** There are 9 g of protein in an 8-ounce serving of the yogurt. **97.** The initial velocity is 8 ft/s. **99.** The depreciated value will be $38,000 after 2 years. **101.** The approximate length is 31.8 in. **103.** The distance the car will skid is 168 ft. **105.** Negative **107.** $a = 7$ **109.** 385

SECTION 5.3

1. Subtract $2x$ from each side. **3.** 3 **5.** −2 **7.** −3 **9.** 2 **11.** −2 **13.** −0.2 **15.** 0 **17.** −2 **19.** −2 **21.** −2 **23.** 4 **25.** $\frac{3}{4}$ **27.** $\frac{3}{2}$ **29.** −14 **31.** 7 **33.** (ii) **35.** 1 **37.** 4 **39.** −1 **41.** −1 **43.** 24 **45.** 495 **47.** $\frac{1}{2}$ **49.** $-\frac{1}{3}$ **51.** $\frac{10}{3}$ **53.** $-\frac{1}{4}$ **55.** 0 **57.** The score given by the third judge was 8.5. **59.** The score given by the third judge was 8. **61. a.** The fulcrum is 5 ft from the other person. **b.** The person who is 3 ft from the fulcrum is heavier. **c.** No, the seesaw will not balance. **63.** The fulcrum is 10 ft from the child. **65.** The fulcrum must be placed 4.8 ft from the 90-pound child. **67.** The force on the lip of the can is 1770 lb. **69.** The break-even point is 260 barbecues. **71.** The break-even point is 520 recorders. **73.** The oxygen consumption is 54.8 ml/min.

SECTION 5.4

1. $x - 15 = 7$; 22 **3.** $9 - x = 7$; 2 **5.** $5 - 2x = 1$; 2 **7.** $2x + 5 = 15$; 5 **9.** $4x - 6 = 22$; 7 **11.** $3(4x - 7) = 15$; 3 **13.** $3x = 2(20 - x)$; 8, 12 **15.** $2x - (14 - x) = 1$; 5, 9 **17.** 15, 17, 19 **19.** −1, 1, 3 **21.** 4, 6 **23.** (iii) **25.** The yearly costs for a robot are $600. **27.** The lengths of the sides are 20 m, 20 m, and 6 m. **29.** The customer used the service for 11 min. **31.** The executive used the phone for 162 min. **33.** The customer pays $.15 per text message over 300 messages. **35.** 3 h of labor were required for the job. **37.** The shorter piece is 3 ft; the longer piece is 9 ft. **39.** The larger scholarship is $5000.

SECTION 5.5

1. (iii) and (v) **3.** The amount of \$.80 meal used is 300 lb. **5.** The cost per ounce is \$3. **7.** 25 lb of sunflower seeds are needed. **9.** The amount of diet supplement is 2 lb; the amount of vitamin supplement is 3 lb. **11.** The cost per pound of the trail mix is \$2.90. **13.** The goldsmith used 56 oz of the \$4.30 alloy and 144 oz of the \$1.80 alloy. **15.** The cost per pound of the coffee mixture is \$6.98. **17.** 1016 adult tickets were sold. **19.** (i) **21.** The first plane is traveling at a rate of 105 mph; the second plane is traveling at a rate of 130 mph. **23.** The planes will be 3000 km apart at 11 A.M. **25.** After 2 h, the cabin cruiser will be alongside the motorboat. **27.** The corporate offices are 120 mi from the airport. **29.** The rate of the car is 68 mph. **31.** The distance between the airports is 300 mi. **33.** The planes will pass each other 2.5 h after the plane leaves Seattle. **35.** The cyclists will meet after 1.5 h. **37.** The car overtakes the cyclist 48 mi from the starting point. **39.** The cyclist's average speed is $13\frac{1}{3}$ mph.

CHAPTER 5 CONCEPT REVIEW*

1. An equation contains an equals sign; an expression does not. [5.1A]
2. A number is not a solution of an equation if, when you substitute the number for the variable in the equation, the result is a false equation. [5.1A]
3. The Addition Property of Equations is used to remove a term from one side of the equation by adding the opposite of that term to each side of the equation. [5.1B]
4. The Multiplication Property of Equations is used to remove a coefficient by multiplying each side of the equation by the reciprocal of the coefficient. [5.1C]
5. To check the solution of an equation, substitute the proposed solution for the variable in the original equation. Simplify the resulting numerical expressions. If the left and right sides are equal, the proposed solution is the solution of the equation. If the left and right sides are not equal, the proposed solution is not a solution of the equation. [5.1A]
6. To solve the equation $-14x = 28$, divide both sides of the equation by the coefficient of the variable term, -14. [5.1C]
7. To solve the equation $\frac{1}{3}x - \frac{2}{9} = \frac{1}{3}$, first multiply both sides of the equation by the LCM of the denominators, 9. The result is $3x - 2 = 3$. Then add 2 to both sides of the equation. The result is $3x = 5$. Next divide each side of the equation by 3. The result is $x = \frac{5}{3}$. The solution of the equation is $\frac{5}{3}$. [5.2A]
8. To solve an equation containing parentheses, first use the Distributive Property to remove the parentheses. Combine like terms on each side of the equation. Use the Addition Property of Equations to rewrite the equation with only one variable term. Then use the Addition Property of Equations to rewrite the equation with only one constant term. Use the Multiplication Property of Equations to rewrite the equation in the form *variable* = *constant*. The constant is the solution. [5.3B]
9. The formula $d = rt$ is used to solve a uniform motion problem. [5.1D]
10. The solution of a value mixture problem is based on the equation $AC = V$, where A is the amount of the ingredient, C is the cost per unit of the ingredient, and V is the value of the ingredient. [5.5A]
11. The formula $F_1x = F_2(d - x)$ is used to solve a lever system problem. [5.3C]
12. Consecutive integers differ by 1. For example, 6 and 7 are consecutive integers. Consecutive even integers are even integers that differ by 2. For example, 6 and 8 are consecutive even integers. [5.4A]

CHAPTER 5 REVIEW EXERCISES

1. 21 [5.1B] **2.** 10 [5.3B] **3.** 7 [5.2A] **4.** No [5.1A] **5.** 20 [5.1C] **6.** -2 [5.3B] **7.** -4 [5.1B] **8.** 4 [5.3A] **9.** -1 [5.3B] **10.** 4 [5.3A] **11.** $-\frac{5}{2}$ [5.3A] **12.** $\frac{5}{4}$ [5.1C] **13.** 10 [5.2A] **14.** $\frac{4}{3}$ [5.3B] **15.** The force is 24 lb. [5.3C] **16.** The average speed on the winding road was 32 mph. [5.5B] **17.** The number is 2. [5.4B] **18.** The number is 10. [5.4B] **19.** The amount of cranberry juice is 7 qt; the amount of apple juice is 3 qt. [5.5A] **20.** The three integers are -1, 0, 1. [5.4A] **21.** $5n - 4 = 16$; 4 [5.4A] **22.** The height of the Eiffel Tower is 1063 ft. [5.4B] **23.** The temperature is 37.8°C. [5.2B] **24.** The jet overtakes the propeller-driven plane 600 mi from the starting point. [5.5B] **25.** The numbers are 8 and 13. [5.4A] **26.** The farmer harvested 25,300 bushels of corn last year. [5.4B]

**Note:* The numbers in brackets following the answers in the Concept Review are a reference to the objective that corresponds to that problem. For example, the reference [1.2A] stands for Section 1.2, Objective A. This notation will be used for all Prep Tests, Concept Reviews, Chapter Reviews, Chapter Tests, and Cumulative Reviews throughout the text.

CHAPTER 5 TEST

1. -5 [5.3A, Example 1] **2.** -5 [5.1B, How To 2] **3.** -3 [5.2A, Example 1] **4.** 2 [5.3B, How To 2] **5.** No [5.1A, How To 1] **6.** 5 [5.2A, Example 2] **7.** $-\frac{1}{2}$ [5.2A, Example 2] **8.** $-\frac{1}{3}$ [5.3B, How To 2] **9.** 2 [5.3A, Example 2] **10.** -12 [5.1C, How To 4] **11.** 95 [5.2A, Example 3] **12.** $\frac{16}{5}$ [5.3A, How To 1] **13.** -3 [5.2A, Example 2] **14.** 11 [5.3B, How To 2] **15.** The amount of rye flour is 10 lb; the amount of wheat flour is 5 lb. [5.5A, How To 1] **16.** 200 calculators were produced. [5.2B, Example 6] **17.** The numbers are 10, 12, and 14. [5.4A, How To 2] **18.** 4000 clocks were made during the month. [5.2B, Example 6] **19.** $3x - 15 = 27$; 14 [5.4A, How To 1] **20.** The rate of the snowmobile was 6 mph. [5.5B, How To 2] **21.** The company makes 110 25-inch TVs each day. [5.4B, Example 4] **22.** The smaller number is 8; the larger number is 10. [5.4A, Example 1] **23.** The distance between the airports is 360 mi. [5.5B, You Try It 3] **24.** The time required is 11.5 s. [5.2B, You Try It 6] **25.** The final temperature is 60°C. [5.3C, Example 5]

CUMULATIVE REVIEW EXERCISES

1. 6 [3.2B] **2.** -48 [3.3A] **3.** $-\frac{19}{48}$ [3.4A] **4.** -2 [3.4B] **5.** 54 [3.5A] **6.** 24 [3.5A] **7.** 6 [4.1A] **8.** $-17x$ [4.2A] **9.** $-5a - 2b$ [4.2A] **10.** $2x$ [4.2B] **11.** $36y$ [4.2B] **12.** $2x^2 + 6x - 4$ [4.2C] **13.** $-4x + 14$ [4.2D] **14.** $6x - 34$ [4.2D] **15.** Yes [5.1A] **16.** No [5.1A] **17.** $\frac{11}{18}$ [3.5A] **18.** -25 [5.1C] **19.** -3 [5.2A] **20.** 3 [5.2A] **21.** 0.047383 [2.6B] **22.** 7 [1.2B] **23.** 13 [5.3B] **24.** 2 [5.3B] **25.** -3 [5.3A] **26.** $\frac{1}{2}$ [5.3A] **27.** The final temperature is 60°C. [5.3C] **28.** $12 - 5x = -18$; 6 [5.4A] **29.** The area of the garage is 600 ft^2. [5.4B] **30.** 20 lb of oat flour are needed for the mixture. [5.5A] **31.** $3n + 4$ [4.3B] **32.** The number is 2. [5.4B] **33.** The length of the track is 120 m. [5.5B]

ANSWERS TO CHAPTER 6 SELECTED EXERCISES

PREP TEST

1. $\frac{4}{5}$ [2.2B] **2.** 24.8 [2.6D] **3.** 4×33 [1.1A/1.3A] **4.** $\frac{19}{100}$ [2.4A] **5.** 0.23 [2.6B] **6.** 47 [2.6B] **7.** 2850 [2.6B] **8.** 4000 [2.6C] **9.** 8000 [2.6C] **10.** 62.5 [3.4B] **11.** $66\frac{2}{3}$ [2.2A] **12.** 1.75 [2.6C]

SECTION 6.1

1. $\frac{2}{3}$, 2:3, 2 to 3 **3.** $\frac{3}{8}$, 3:8, 3 to 8 **5.** $\frac{9}{2}$, 9:2, 9 to 2 **7.** $\frac{1}{2}$, 1:2, 1 to 2 **9.** The ratio is $\frac{9}{13}$. **11.** $\frac{\$85}{3 \text{ shirts}}$ **13.** $\frac{\$19}{2 \text{ h}}$ **15.** $\frac{42 \text{ trees}}{1 \text{ acre}}$ **17.** \$11.50/h **19.** 55.4 mph **21.** \$4.24/lb **23.** The ratio is $\frac{1}{56}$. **25.** The rate of travel is 462 mph. **27.** The investor's profit per share was \$7. **29.** 15 feet in 1 second **31.** The highest wage would be \$18/h.

SECTION 6.2

1. Not true **3.** True **5.** True **7.** True **9.** True **11.** True **13.** 10 **15.** 2.4 **17.** 4.5 **19.** 17.14 **21.** 25.6 **23.** 20.83 **25.** 4.35 **27.** 10.97 **29.** 1.15 **31.** 38.73 **33.** 0.5 **35.** 10 **37.** 0.43 **39.** -1.6 **41.** 6.25 **43.** 32 **45.** 6.2 **47.** 5.8 **49.** Yes **51.** Yes **53.** The number of morbidly obese Americans is 6,000,000. **55.** He should use 70 lb of fertilizer. **57.** It would take 2496 bricks to build a wall 48 ft in length. **59.** The length is 24 ft. The width is 15 ft. **61.** The monthly payment is \$133.80. **63.** 3.5 gal of paint would be required. **65.** 750 defective boards can be expected in a run of 25,000 circuit boards. **67.** Carlos would receive a dividend of \$1071. **69.** It would weigh 2.67 lb on the moon. **71.** The candle will burn 9 in. in 4 h. **73.** The yearly midmanagement salary would be \$126,000.

SECTION 6.3

1. $\frac{1}{4}$, 0.25 **3.** $1\frac{3}{10}$, 1.30 **5.** 1, 1.00 **7.** $\frac{73}{100}$, 0.73 **9.** $3\frac{83}{100}$, 3.83 **11.** $\frac{7}{10}$, 0.70 **13.** $\frac{22}{25}$, 0.88 **15.** $\frac{8}{25}$, 0.32 **17.** $\frac{2}{3}$ **19.** $\frac{5}{6}$ **21.** $\frac{1}{9}$ **23.** $\frac{5}{11}$ **25.** $\frac{3}{70}$ **27.** $\frac{1}{15}$ **29.** 0.065 **31.** 0.123 **33.** 0.0055 **35.** 0.0825 **37.** 0.0505 **39.** 0.02 **41.** Greater than **43.** 73% **45.** 1% **47.** 294% **49.** 0.6% **51.** 310.6% **53.** 70% **55.** 37% **57.** 40% **59.** 12.5% **61.** 150% **63.** 166.7% **65.** 87.5% **67.** 48% **69.** $33\frac{1}{3}$% **71.** $166\frac{2}{3}$% **73.** $87\frac{1}{2}$% **75.** Less than **77.** 6% of those surveyed named something other than corn on the cob, cole slaw, corn bread, or fries. **79.** This represents $\frac{1}{2}$ off the regular price.

SECTION 6.4

1. 8 **3.** 0.075 **5.** $16\frac{2}{3}$% **7.** 37.5% **9.** 100 **11.** 1200 **13.** 51.895 **15.** 13 **17.** 2.7% **19.** 400% **21.** 7.5 **23.** False **25.** 65 **27.** 25% **29.** 75 **31.** 12.5% **33.** 400 **35.** 19.5 **37.** 14.8% **39.** 62.62 **41.** 5 **43.** 45 **45.** 300% **47.** Less than **49.** There are 58,747 new student pilots flying single-engine planes this year. **51.** 77 million of the returns were filed electronically. **53.** 49.59 billion email messages sent per day are not spam. **55.** 0.4% of the total energy production was generated by wind machines. **57.** 0.18% of the millionaires were audited by the IRS. **59.** About 15.8 million travelers allowed their children to miss school to go along on a trip. **61.** 22,366 runners started the Boston Marathon. **63.** The cargo ship's daily fuel bill without the kite is $8000. **65.** 27.4% of the total amount is spent on food. **67.** 91.8% of the total amount is spent on all categories except training.

SECTION 6.5

3. a. $25, $50, $75, $100, $125 **b.** $150 **c.** $175 **d.** $200 **e.** $225 **5.** The simple interest due on the loan is $273.70. **7.** The simple interest due on the loan is $6750. **9.** The interest owed is $20. **11.** The simple interest owed is $1440. **13.** The maturity value of the loan is $164,250. **15.** The maturity value of the loan is $15,061.51. **17.** The annual simple interest rate is 7.7%. **19.** The annual simple interest rate earned on the investment is 9.125%. **21. a.** Equal to **b.** Greater than **c.** Less than

CHAPTER 6 CONCEPT REVIEW*

1. If the units in a comparison are different, it is a rate. [6.1A]
2. To find a unit rate, divide the number in the numerator of the rate by the number in the denominator of the rate. [6.1A]
3. To write the ratio 12:15 in simplest form, divide both numbers by the GCF of 3: $\frac{12}{3}:\frac{15}{3} = 4:5$. [6.1A]
4. To write the ratio 19:6 as a fraction, write the first number as the numerator of the fraction and the second number as the denominator: $\frac{19}{6}$. [6.1A]
5. A proportion is true if the fractions are equal when written in lowest terms. Another way to describe a true proportion is to say that in a true proportion, the cross products are equal. [6.2A]
6. When one of the numbers in a proportion is unknown, we can solve the proportion to find the unknown number. We do this by setting the cross products equal to each other and then solving for the unknown number. [6.2A]
7. When setting up a proportion, keep the same units in the numerators and the same units in the denominators. [6.2B]
8. To check the solution of a proportion, replace the unknown number in the proportion with the solution. Then find the cross products. If the cross products are equal, the solution is correct. If the cross products are not equal, the solution is not correct. [6.2A]
9. The basic percent equation is Percent × base = amount. [6.4A]
10. To find what percent of 40 is 30, use the basic percent equation: $n \times 40 = 30$. To solve for n, we *divide* 30 by 40: $30 \div 40 = 0.75 = 75\%$. [6.4A]

**Note:* The numbers in brackets following the answers in the Concept Review are a reference to the objective that corresponds to that problem. For example, the reference [1.2A] stands for Section 1.2, Objective A. This notation will be used for all Prep Tests, Concept Reviews, Chapter Reviews, Chapter Tests, and Cumulative Reviews throughout the text.

11. To find 11.7% of 532, use the basic percent equation and write the percent as a decimal: $0.117 \times 532 = n$. To solve for n, we *multiply* 0.117 by 532: $0.117 \times 532 = 62.244$. [6.4A]

12. To answer the question "36 is 240% of what number?", use the basic percent equation and write the percent as a decimal: $2.4 \times n = 36$. To solve for n, we *divide* 36 by 2.4: $36 \div 2.4 = 15$. [6.4A]

13. To use the proportion method to solve a percent problem, identify the percent, the amount, and the base. Then use the proportion $\frac{\text{percent}}{100} = \frac{\text{amount}}{\text{base}}$. Substitute the known values into this proportion and solve for the unknown. [6.4B]

CHAPTER 6 REVIEW EXERCISES

1. $\frac{1}{1}$, 1:1, 1 to 1 [6.1A] **2.** $\frac{\text{2 roof supports}}{\text{1 ft}}$ [6.1A] **3.** \$15.70/h [6.1A] **4.** $\frac{8}{15}$ [6.1A] **5.** 1.6 [6.2A] **6.** $\frac{\text{5 lb}}{\text{4 trees}}$ [6.1A] **7.** 57 mph [6.1A] **8.** 6.86 [6.2A] **9.** $\frac{8}{25}$ [6.3A] **10.** 0.22 [6.3A] **11.** $\frac{1}{4}$, 0.25 [6.3A] **12.** $\frac{17}{500}$ [6.3A] **13.** 17.5% [6.3B] **14.** 128.6% [6.3B] **15.** 280% [6.3B] **16.** 21 [6.4A/6.4B] **17.** 500% [6.4A/6.4B] **18.** 66.7% [6.4A/6.4B] **19.** 30 [6.4A/6.4B] **20.** 15.3 [6.4A/6.4B] **21.** 562.5 [6.4A/6.4B] **22.** 5.625% [6.4A/6.4B] **23.** The ratio is $\frac{2}{5}$. [6.1A] **24.** \$12,000 must be invested to earn \$780 in dividends. [6.2B] **25.** 2.75 lb of plant food should be used. [6.2B] **26.** The other attorney receives \$64,000. [6.2B] **27.** 34.0% of the tourists will be visiting China. [6.4C] **28.** The company spent \$8400 for advertising. [6.4C] **29.** 3952 of the phones were not defective. [6.4C] **30.** Cable households spend 36.5% of the week watching TV. [6.4C] **31.** The simple interest is \$31.81. [6.5A] **32.** The maturity value of the loan is \$10,630. [6.5A] **33.** The airline would sell 196 tickets for an airplane that has 175 seats. [6.4C]

CHAPTER 6 TEST

1. $\frac{1}{8}$, 1:8, 1 to 8 [6.1A, Example 1] **2.** $\frac{\text{1 oz}}{\text{4 cookies}}$ [6.1A, Example 2] **3.** 0.6 mi/min [6.1A, How To 1] **4.** $\frac{2}{1}$ [6.1A, Example 1] **5.** 0.75 [6.2A, Example 2] **6.** 2 ft/s [6.1A, Example 3] **7.** 476.67 ft^2/h [6.1A, Example 3] **8.** 3.56 [6.2A, Example 3] **9.** 0.864 [6.3A, Example 3] **10.** 40% [6.3B, How To 2] **11.** 125% [6.3B, How To 1] **12.** $\frac{5}{6}$ [6.3A, Example 2] **13.** $\frac{11}{25}$ [6.3A, Example 1] **14.** 118% [6.3B, How To 2] **15.** 90 [6.4A/6.4B, How To 3] **16.** 49.64 [6.4A/6.4B, How To 1] **17.** 56.25% [6.4A/6.4B, How To 2] **18.** 200 [6.4A/6.4B, How To 3] **19.** The ratio is $\frac{33}{38}$. [6.1A, How To 1] **20.** The sales tax is \$3136. [6.2B, You Try It 5] **21.** 243,750 voters would vote. [6.2B, Example 5] **22.** The room is 50 ft long. [6.2B, Example 5] **23. a.** The cash generated from sales of Thin Mints is \$175 million. **b.** The cash generated from sales of Trefoil cookies is \$63 million. [6.4C, Example 8] **24.** The American Red Cross's total revenue is \$5,900,000,000. [6.4C, Example 9] **25.** The U.S. total turkey production was 7 billion pounds. [6.4C, Example 9] **26.** The number of fat grams in the beef burger is 600% of the number in the soy burger. [6.4C, Example 7] **27.** The maturity value of the loan is \$41,520.55. [6.5A, Example 1]

CUMULATIVE REVIEW EXERCISES

1. 57 [3.5A] **2.** 625 [1.3B] **3.** $3\frac{31}{36}$ [2.3B] **4.** $1\frac{2}{3}$ [2.7A] **5.** -114 [3.3B] **6.** 22 [3.5A] **7.** 4 [5.2A] **8.** 3 [5.3B] **9.** $\frac{23}{24}$ [3.4A] **10.** 100,500 [2.6B] **11.** 21 [3.5A] **12.** $-\frac{3}{8}$ [3.5A] **13.** 4 [3.2B] **14.** $8a - 3$ [4.2D] **15.** -12 [5.1C] **16.** $-4y^2 - 3y$ [4.2A] **17.** $1\frac{25}{62}$ [2.4B] **18.** $\frac{3}{10}$ [6.1A] **19.** \$3885/month [6.1A] **20.** 24.954 [2.6A] **21.** 32 [6.2A] **22.** $\frac{10}{11}$ [2.4C] **23.** 8.3% [6.4A/6.4B] **24.** 25.2 [6.4A/6.4B] **25.** The difference is \$28.67. [2.6E] **26.** The number is 12. [5.4A] **27.** $4x - 3(x + 2)$; $x - 6$ [4.3B] **28.** There are 64 mi left to drive. [1.2C] **29.** The new balance is \$265.48. [2.6E]

30. $\frac{4}{15}$ of the job remains to be finished on the third day. [2.3C] **31.** $\frac{1}{10}$ of the population aged 75–84 is affected by Alzheimer's disease. [6.3A] **32.** The car traveled 35 mi per gallon of gas. [6.1A] **33.** The rpm of the engine in third gear is 3750. [5.4B]

ANSWERS TO CHAPTER 7 SELECTED EXERCISES

PREP TEST

1. 56 [1.4A] **2.** 56.52 [4.1A] **3.** 113.04 [4.1A] **4.** 43 [5.1B] **5.** 51 [5.1B] **6.** 14.4 [6.2A]

SECTION 7.1

1. 40°; acute **3.** 115°; obtuse **5.** 90°; right **7.** The complement is 28°. **9.** The supplement is 18°. **11.** The length of BC is 14 cm. **13.** The length of QS is 28 ft. **15.** The length of EG is 30 m. **17.** The measure of $\angle MON$ is 86°. **19.** 71° **21.** 30° **23.** 36° **25.** 127° **27.** 116° **29.** 20° **31.** 20° **33.** 20° **35.** 141° **37.** An acute angle **39.** An obtuse angle **41.** 106° **43.** 11° **45.** $\angle a = 38°$, $\angle b = 142°$ **47.** $\angle a = 47°$, $\angle b = 133°$ **49.** 20° **51.** 47° **53.** True **55.** $\angle x = 155°$, $\angle y = 70°$ **57.** $\angle a = 45°$, $\angle b = 135°$ **59.** $90° - x$ **61.** The measure of the third angle is 60°. **63.** The measure of the third angle is 35°. **65.** The measure of the third angle is 102°. **67.** False

SECTION 7.2

1. Hexagon **3.** Pentagon **5.** Scalene **7.** Equilateral **9.** Obtuse **11.** Acute **13.** 56 in. **15.** 14 ft **17.** 47 mi **19.** 8π cm or approximately 25.13 cm **21.** 11π mi or approximately 34.56 mi **23.** 17π ft or approximately 53.41 ft **25.** The perimeter is 17.4 cm. **27.** The perimeter is 8 cm. **29.** The perimeter is 24 m. **31.** The perimeter is 48.8 cm. **33.** The perimeter is 17.5 in. **35.** The length of a diameter is 8.4 cm. **37.** The circumference is 1.5π in. **39.** The circumference is 226.19 cm. **41.** 60 ft of fencing should be purchased. **43.** The carpet must be nailed down along 44 ft. **45.** The length is 120 ft. **47.** The length of the third side is 10 in. **49.** The length of each side is 12 in. **51.** The length of a diameter is 2.55 cm. **53.** The length is 13.19 ft. **55.** The bicycle travels 50.27 ft. **57.** The circumference is 39,935.93 km. **59.** The perimeter of the square is greater. **61.** 60 ft^2 **63.** 20.25 in^2 **65.** 546 ft^2 **67.** 16π cm^2 or approximately 50.27 cm^2 **69.** 30.25π mi^2 or approximately 95.03 mi^2 **71.** 72.25π ft^2 or approximately 226.98 ft^2 **73.** The area is 156.25 cm^2. **75.** The area is 570 in^2. **77.** The area is 192 in^2. **79.** The area is 13.5 ft^2. **81.** The area is 330 cm^2. **83.** The area is 25π in. **85.** The area is 9.08 ft^2. **87.** The area is $10{,}000\pi$ in^2. **89.** The area is 126 ft^2. **91.** 7500 yd^2 must be purchased. **93.** The width is 10 in. **95.** The length of the base is 20 m. **97.** You should buy 2 qt. **99.** The cost is $98. **101.** The increase in area is 113.10 in^2. **103.** The area is 216 m^2. **105.** The cost will be $878. **107.** The area of the first figure is equal to the area of the second figure. **109.** The area of the resulting triangle is 4 times larger.

SECTION 7.3

1. 5 in. **3.** 8.6 cm **5.** 11.2 ft **7.** 4.5 cm **9.** 12.7 yd **11.** 8.5 cm **13.** 24.3 cm **15.** (i) **17.** $\frac{1}{2}$ **19.** 7.2 cm **21.** 3.3 m **23.** 12 m **25.** 12 in. **27.** 56.3 cm^2 **29.** 18 ft **31.** 16 m **33. a.** Always true **b.** Sometimes true **c.** Always true **35.** Yes, SAS Rule **37.** Yes, SSS Rule **39.** Yes, ASA Rule **41.** Yes, SAS Rule **43.** No **45.** No

SECTION 7.4

1. 840 in^3 **3.** 15 ft^3 **5.** 4.5π cm^3 or approximately 14.14 cm^3 **7.** The volume is 34 m^3. **9.** The volume is 42.875 in^3. **11.** The volume is 36π ft^3. **13.** The volume is 8143.01 cm^3. **15.** The volume is 75π in^3. **17.** The volume is 120 in^3. **19.** The width is 2.5 ft. **21.** The height is 6.40 in. **23.** The volume is 172,800 ft^3. **25.** The sphere has the greater volume. **27.** 94 m^2 **29.** 56 m^2 **31.** 96π in^2 or approximately 301.59 in^2 **33.** The surface area is 184 ft^2. **35.** The surface area is 69.36 m^2. **37.** The surface area is 225π cm^2. **39.** The surface area is 402.12 in^2. **41.** The surface area is 6π ft^2. **43.** The surface area is 297 in^2. **45.** The width is 3 cm. **47.** 11 cans of paint should be purchased. **49.** 456 in^2 of glass are needed. **51.** The surface area of the pyramid is 22.53 cm^2 larger. **53. a.** Always true **b.** Never true **c.** Sometimes true

CHAPTER 7 CONCEPT REVIEW*

1. Perpendicular lines are intersecting lines that form right angles. [7.1A]

2. Two angles are complementary when the sum of their measures is 90°. [7.1A]

3. The sum of the measures of the three angles in a triangle is 180°. [7.1C]

4. To find the radius of a circle when you know the diameter, multiply the diameter by $\frac{1}{2}$. [7.2A]

5. The formula for the perimeter of a rectangle is $P = 2L + 2W$, where P is the perimeter, L is the length, and W is the width. [7.2A]

6. To find the circumference of a circle, multiply π times the diameter or multiply 2π times the radius. [7.2A]

7. The formula for the area of a triangle is $A = \frac{1}{2}bh$, where A is the area, b is the base, and h is the height of the triangle. [7.2B]

8. To find the volume of a rectangular solid, you need to know the length, the width, and the height. [7.4A]

9. To calculate the surface area of a right circular cylinder, you are calculating the area of two circles and a rectangle. [7.4B]

10. The hypotenuse of a right triangle is the side of the triangle that is opposite the right angle. [7.3A]

11. In similar triangles, the ratios of corresponding sides are equal. Therefore, we can write proportions setting the ratios of corresponding sides equal to each other. If one side of a triangle is unknown, we can write a proportion using two ratios and then solve for the unknown side. [7.3B]

12. The side-angle-side rule states that two triangles are congruent if two sides and the included angle of one triangle equal the corresponding sides and included angle of the second triangle. [7.3C]

CHAPTER 7 REVIEW EXERCISES

1. $\angle x = 22°$, $\angle y = 158°$ [7.1C] **2.** 24 in. [7.3B] **3.** No [7.3C] **4.** 68° [7.1B] **5.** Yes, by the SAS Rule [7.3C] **6.** 7.21 in. [7.3A] **7.** 44 cm [7.1A] **8.** 19° [7.1A] **9.** 32 in^2 [7.2B] **10.** 96 cm^3 [7.4A] **11.** 42 in. [7.2A] **12.** $\angle a = 138°$, $\angle b = 42°$ [7.1B] **13.** 220 ft^2 [7.4B] **14.** 9.75 ft [7.3A] **15.** 42.875 in^3 [7.4A] **16.** 148° [7.1A] **17.** 39 ft^3 [7.4A] **18.** 95° [7.1C] **19.** 8 cm [7.2B] **20.** 288π mm^3 [7.4A] **21.** 21.5 cm [7.2A] **22.** 4 cans [7.4B] **23.** 208 yd [7.2A] **24.** 90.25 m^2 [7.2B] **25.** 276 m^2 [7.2B]

CHAPTER 7 TEST

1. 7.55 cm [7.3A, How To 1] **2.** Congruent, SAS [7.3C, You Try It 3] **3.** 111 m^2 [7.2B, How To 5] **4.** 42 ft^2 [7.2B, How To 8] **5.** $\frac{784\pi}{3}$ cm^3 [7.4A, Example 2] **6.** 75 m^2 [7.4B, Example 3] **7.** 4618.14 cm^3 [7.4A, You Try It 2] **8.** 159 in^2 [7.2B, How To 9] **9.** 34 ft [7.2A, How To 2] **10.** 100° [7.1C, Example 8] **11.** 34° [7.1B, Example 5] **12.** Octagon [7.2A] **13.** Not necessarily congruent [7.3C, Example 3] **14.** 168 ft^3 [7.4A, Example 1] **15.** 8.06 m [7.3A, Example 1] **16.** 143° [7.1B, How To 4] **17.** 500π cm^2 [7.4B, You Try It 3] **18.** 61° [7.1C, You Try It 7] **19.** 6.67 ft [7.3B, Example 2] **20.** 4.27 ft [7.3B, How To 2] **21.** 20 m [7.2A, How To 3] **22.** 26 cm [7.2A, How To 2] **23.** 51.6 ft [7.3A, Example 1] **24.** 102° [7.1C, Example 8] **25.** 113° [7.1A, You Try It 3]

CUMULATIVE REVIEW EXERCISES

1. 204 [6.4A/6.4B] **2.** 1, 2, 3, 6, 13, 26, 39, 78 [1.3D] **3.** $\frac{5}{6}$ [2.4B] **4.** 18 [3.1C] **5.** 12.8 [2.6C] **6.** -2 [3.2B] **7.** 131° [7.1B] **8.** 26 cm [7.3A] **9.** $1\frac{11}{30}$ [2.3A] **10.** $-7b$ [4.2B] **11.** -28 [3.5A] **12.** $\frac{1}{2}$ [5.3B] **13.** $2a$ [4.2A] **14.** 75 [3.3A] **15.** $7x + 18$ [4.2D] **16.** -2 [4.1A] **17.** $\frac{4}{5}$ [4.1A]

**Note:* The numbers in brackets following the answers in the Concept Review are a reference to the objective that corresponds to that problem. For example, the reference [1.2A] stands for Section 1.2, Objective A. This notation will be used for all Prep Tests, Concept Reviews, Chapter Reviews, Chapter Tests, and Cumulative Reviews throughout the text.

18. $2 \cdot 3 \cdot 13$ [1.3D] **19.** 5 [5.3A] **20.** 37.5% [6.3B] **21.** $8(2n)$; $16n$ [4.3B] **22.** The two cars are traveling 50 mph and 55 mph. [5.5B] **23.** The simple interest is $1313.01. [6.5A] **24.** 24 oz of the silver alloy that costs $3.50 per ounce must be used. [5.5A] **25.** The executive used the cell phone for 87 min. [5.4B] **26.** The sales tax on a $75 purchase is $4.50. [6.2B] **27.** The increase in the value of the imports is $.08 trillion. [2.6E] **28.** The height of the box is 3 ft. [7.4A] **29.** The depth when the pressure is 35 lb/in^2 is 40 ft. [5.2B] **30.** It takes these elevators 8.8 s to travel from the fifth to the 25th floor. [6.2B]

ANSWERS TO CHAPTER 8 SELECTED EXERCISES

PREP TEST

1. 48.0% was bill-related mail. [6.4A/6.4B] **2. a.** The greatest cost increase is between 2009 and 2010. **b.** Between those years, there was an increase of $5318. [1.2C] **3.** The ratio is $\frac{5}{3}$. [6.1A] **4.** $\frac{3}{50}$ of the women in the military are in the Marine Corps. [6.3A]

SECTION 8.1

1. 128 units are required to graduate with a degree in accounting. **3.** 35.2% of the units required to graduate are taken in accounting. **5.** $1,085,000,000 is spent on TV game machines. **7.** $\frac{2}{25}$ of the total money spent was spent on accessories. **9.** 39 million passenger cars were produced worldwide. **11.** 28% of the passenger cars were produced in Asia. **13.** Women ages 75 and up have the lowest recommended number of Calories. **15.** True

SECTION 8.2

1. a. Median **b.** Mean **c.** Mode **d.** Median **e.** Mode **f.** Mean **3.** The mean number of seats filled is 381.5625 seats. The median number of seats filled is 394.5 seats. Since each number occurs only once, there is no mode. **5.** The mean cost is $85.615. The median cost is $85.855. **7.** The mean monthly rate is $403.625. The median monthly rate is $404.50. **9.** The mean life expectancy is 73.4 years. The median life expectancy is 74 years. **13. a.** 25% **b.** 75% **c.** 75% **d.** 25% **15.** Lowest is $46,596. Highest is $82,879. Q_1 = $56,067. Q_3 = $66,507. Median = $61,036. Range = $36,283. Interquartile range = $10,440. **17. a.** There were 40 adults who had cholesterol levels above 217. **b.** There were 60 adults who had cholesterol levels below 254. **c.** There are 20 cholesterol levels in each quartile. **d.** 25% of the adults had cholesterol levels of not more than 198. **19. a.** Range = 5.6 million metric tons. Q_1 = 0.59 million metric tons. Q_3 = 1.52 million metric tons. Interquartile range = 0.93 million metric tons. **b.** 0.45 0.59 1.035 1.52 6.05 **c.** 6.05

21. a. No, the difference in the means is not greater than 1 inch. **b.** The difference in medians is 0.3 inch.

c.

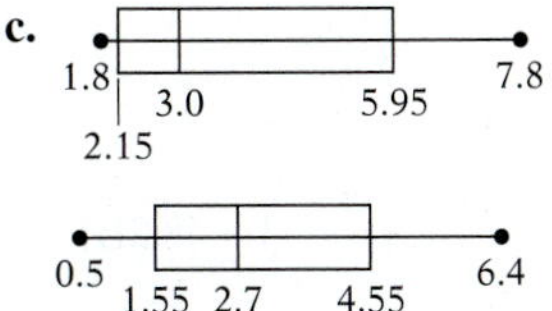

23. Answers will vary. For example, 55, 55, 55, 55, 55, or 50, 55, 55, 55, 60 **27.** Answers will vary. For example, 20, 21, 22, 24, 26, 27, 29, 31, 31, 32, 32, 33, 33, 36, 37, 37, 39, 40, 41, 43, 45, 46, 50, 54, 57

SECTION 8.3

1. {(HHHH), (HHHT), (HHTT), (HHTH), (HTTT), (HTHH), (HTTH), (HTHT), (TTTT), (TTTH), (TTHH), (THHH), (TTHT), (THHT), (THTT), (THTH)} **3.** {(1, 1), (1, 2), (1, 3), (1, 4), (2, 1), (2, 2), (2, 3), (2, 4), (3, 1), (3, 2), (3, 3), (3, 4), (4, 1), (4, 2), (4, 3), (4, 4)} **5. a.** {1, 2, 3, 4, 5, 6, 7, 8} **b.** {1, 2, 3} **7. a.** The probability that the sum is 5 is $\frac{1}{9}$. **b.** The probability that the sum is 15 is 0. **c.** The probability that the sum is less than 15 is 1. **d.** The probability that the sum is 2 is $\frac{1}{36}$.

9. a. The probability that the number is divisible by 4 is $\frac{1}{4}$. **b.** The probability that the number is a multiple of 3 is $\frac{1}{3}$.

11. The probability of throwing a sum of 5 is greater. **13. a.** The probability is $\frac{4}{11}$ that the letter *I* is drawn. **b.** The probability of choosing an *S* is greater. **15.** The fraction $\frac{1}{11}$ represents the probability that the card has the letter *M* on it. **17.** Drawing a spade has the greater probability. **19.** The empirical probability that a person prefers a cash discount is 0.39. **21.** The probability is $\frac{185}{377}$ that a customer rated the service as satisfactory or excellent.

CHAPTER 8 CONCEPT REVIEW*

1. A sector of a circle is one of the "pieces of the pie" into which a circle graph is divided. [8.1A]
2. A jagged portion of the vertical axis on a bar graph is used to indicate that the vertical scale is missing numbers from 0 to the lowest number shown on the vertical axis. [8.1A]
3. In a broken-line graph, points are connected by line segments to show data trends. If a line segment goes up from left to right, it indicates an increase in the quantity represented on the vertical axis during that time period. If a line segment goes down from left to right, it indicates a decrease in the quantity represented on the vertical axis during that time period. [8.1A]
4. The formula for the mean is $\bar{x} = \frac{\text{sum of the data values}}{\text{number of data values}}$. [8.2A]
5. To find the median, the data must be arranged in order from least to greatest in order to determine the "middle" number, or the number which separates the data so that half of the numbers are less than the median and half of the numbers are greater than the median. [8.2A]
6. If a set of numbers has no number occurring more than once, then the data have no mode. [8.2A]
7. A box-and-whiskers plot shows the least number in the data; the first quartile, Q_1; the median; the third quartile, Q_3; and the greatest number in the data. [8.2B]
8. To find the first quartile, find the median of the data values that lie below the median. [8.2B]
9. The empirical probability formula states that the empirical probability of an event is the ratio of the number of observations of the event to the total number of observations. [8.3A]
10. The theoretical probability formula states that the theoretical probability of an event is a fraction with the number of favorable outcomes of the experiment in the numerator and the total number of possible outcomes in the denominator. [8.3A]

CHAPTER 8 REVIEW EXERCISES

1. The agencies spent $349 million on maintaining websites. [8.1A] **2.** The ratio is $\frac{9}{8}$. [8.1A] **3.** 8.9% of the total amount of money was spent by NASA. [8.1A] **4.** The difference was 50 days. [8.1A] **5.** The percent is 50%. [8.1A]
6. a. The Southeast had the lowest number of days of full operation. **b.** This region had 30 days of full operation. [8.1A]
7. The mean is $214.\overline{54}$. The median is 210. [8.2A] **8.** The mean is 7.17 lb. The median is 7.05 lb. [8.2A]
9. The modal response was "good." [8.2A] **10.** [8.2B]

Q_1 median Q_2

89 105 111 124 134

11. There are 12 elements in the sample space. [8.3A] **12.** The probability is $\frac{1}{500}$. [8.3A] **13.** $Q_1 = 1$; $Q_3 = 3$ [8.2B]
14. The probability is $\frac{1}{6}$. [8.3A] **15.** The probability is $\frac{5}{14}$. [8.3A] **16. a.** The mean is 91.6 heartbeats per minute. The median is 93.5 heartbeats per minute. The mode is 96 heartbeats per minute. [8.2A] **b.** The range is 36 heartbeats per minute. The interquartile range is 15 heartbeats per minute. [8.2B]

CHAPTER 8 TEST

1. There were 355 more films rated R. [8.1A, Example 1] **2.** The number of PG-13 films was 16 times the number of NC-17 films. [8.1A, Example 1] **3.** The percent of films rated G was 5.6%. [8.1A, Example 1] **4.** The student enrollment increased the least during the 1990s. [8.1A, Example 1] **5.** The increase in the enrollment was 11 million students.

**Note:* The numbers in brackets following the answers in the Concept Review are a reference to the objective that corresponds to that problem. For example, the reference [1.2A] stands for Section 1.2, Objective A. This notation will be used for all Prep Tests, Concept Reviews, Chapter Reviews, Chapter Tests, and Cumulative Reviews throughout the text.

[8.1A, Example 1] **6.** The mean score was 152.875. [8.2A, How To 1] **7.** The median response time was 14 min. [8.2A, How To 2] **8.** The modal response was "very good." [8.2A] **9.** The first quartile is 9.8. [8.2B, Example 3] **10. a.** The range is 22 days. **b.** The median is 14 vacation days. [8.2B, How To 3]

11. [8.2B, You Try It 3]

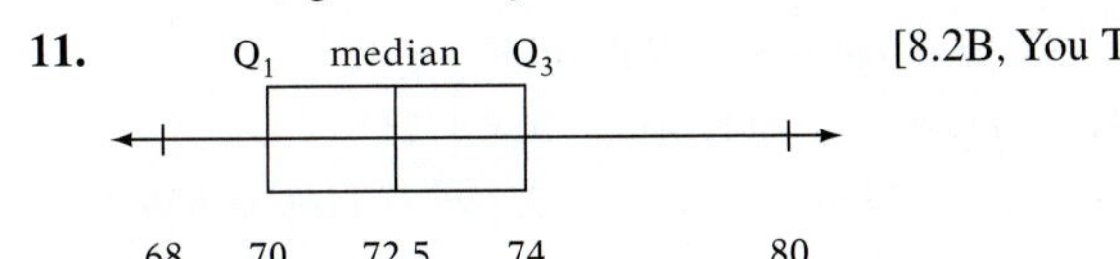

12. The sample space is {(N, D, Q), (N, Q, D), (D, N, Q), (D, Q, N), (Q, N, D), (Q, D, N)}. [8.3A, How To 1]

13. The probability is $\frac{4}{31}$. [8.3A, How To 2] **14.** The probability is $\frac{1}{3}$. [8.3A, How To 3]

15. The probability is $\frac{1}{8}$. [8.3A, Example 1] **16.** The probability is $\frac{2}{3}$. [8.3A, How To 2]

17. The probability is $\frac{5}{12}$. [8.3A, How To 3] **18. a.** The mean is 2.53 days. **b.** The median is 2.55 days. [8.2A, Example 1]

c. [8.2B, How To 3]

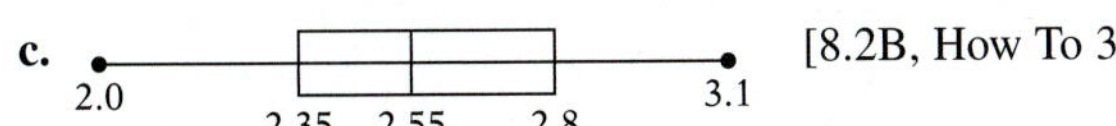

CUMULATIVE REVIEW EXERCISES

1. 540 [1.3B] **2.** 14 [1.4A] **3.** 120 [2.1A] **4.** $\frac{5}{12}$ [2.2B] **5.** $12\frac{3}{40}$ [2.3A] **6.** $4\frac{17}{24}$ [2.3B] **7.** 2 [2.4A] **8.** $\frac{64}{85}$ [2.4B] **9.** $8\frac{1}{4}$ [2.7A] **10.** 209.305 [2.5A] **11.** 2.82348 [2.6B] **12.** 16.67 [2.6D] **13.** 26.4 mpg [6.1A] **14.** 3.2 [6.2A] **15.** 80% [6.3B] **16.** 80 [6.4A/6.4B] **17.** 16.34 [6.4A/6.4B] **18.** 40% [6.4A/6.4B] **19.** The salesperson's income for the week was \$650. [6.4C] **20.** The cost is \$407.50. [6.2B] **21.** The interest due is \$3750. [6.5A] **22.** The price is \$279. [6.4C] **23.** The amount budgeted for food is \$570. [8.1A] **24.** The difference is 12 points. [8.1A] **25.** The mean high temperature is 69.6°F. [8.2A] **26.** The probability is $\frac{5}{36}$. [8.3A]

ANSWERS TO CHAPTER 9 SELECTED EXERCISES

PREP TEST

1. 1 [3.2B] **2.** −18 [3.3A] **3.** $\frac{2}{3}$ [3.3B] **4.** 48 [4.1A] **5.** 0 [1.3C] **6.** No [4.1A] **7.** $5x^2 - 9x - 6$ [4.2A] **8.** 0 [4.2A] **9.** $-6x + 24$ [4.2C] **10.** $-7xy + 10y$ [4.2D]

SECTION 9.1

1. Yes **3.** No **5.** Yes **7.** Yes **9.** Binomial **11.** Trinomial **13.** None of these **15.** Binomial **17.** $5x^2 + 8x$ **19.** $7x^2 + xy - 4y^2$ **21.** $3a^2 - 3a + 17$ **23.** $5x^3 + 10x^2 - x - 4$ **25.** $3r^3 + 2r^2 - 11r + 7$ **27.** $-2x^2 + 3x$ **29.** $y^2 - 8$ **31.** $5x^2 + 7x + 20$ **33.** $x^3 + 2x^2 - 6x - 6$ **35.** $2a^3 - 3a^2 - 11a + 2$ **37.** (iv) **39.** $-y^2 - 13xy$ **41.** $2x^2 - 3x - 1$ **43.** $-2x^3 + x^2 + 2$ **45.** $3a^3 - 2$ **47.** $4y^3 + 2y^2 + 2y - 4$ **49.** $4x$ **51.** $3y^2 - 4y - 2$ **53.** $-7x - 7$ **55.** $4x^3 + 3x^2 + 3x + 1$ **57.** $y^3 + 5y^2 - 2y - 4$ **59.** $x^2 + 9x - 11$

SECTION 9.2

1. a. No **b.** Yes **3.** $30x^3$ **5.** $-42c^6$ **7.** $9a^7$ **9.** x^3y^4 **11.** $-10x^9y$ **13.** $12x^7y^8$ **15.** $-6x^3y^5$ **17.** x^4y^5z **19.** $a^3b^5c^4$ **21.** $-30a^5b^8$ **23.** $6a^5b$ **25.** $40y^{10}z^6$ **27.** $x^3y^3z^2$ **29.** $-24a^3b^3c^3$ **31.** $8x^7yz^6$ **33.** $30x^6y^8$ **35.** $-36a^3b^2c^3$ **37. a.** No **b.** Yes **39.** x^{15} **41.** x^{14} **43.** x^8 **45.** y^{12} **47.** $-8x^6$ **49.** x^4y^6 **51.** $9x^4y^2$ **53.** $-243x^{15}y^{10}$ **55.** $-8x^7$ **57.** $24x^8y^7$ **59.** a^4b^6 **61.** $64x^{12}y^3$ **63.** $-18x^3y^4$ **65.** $-8a^7b^5$ **67.** $-54a^9b^3$ **69.** $12x^2$ **71.** $2x^6y^2 + 9x^4y^2$ **73.** 0 **75.** $17x^4y^8$ **77.** No. $2^{(3^2)}$ is larger.

SECTION 9.3

1. $x^2 - 2x$ **3.** $-x^2 - 7x$ **5.** $3a^3 - 6a^2$ **7.** $-5x^4 + 5x^3$ **9.** $-3x^5 + 7x^3$ **11.** $12x^3 - 6x^2$ **13.** $6x^2 - 12x$ **15.** $3x^2 + 4x$ **17.** $-x^3y + xy^3$ **19.** $2x^4 - 3x^2 + 2x$ **21.** $2a^3 + 3a^2 + 2a$ **23.** $3x^6 - 3x^4 - 2x^2$

25. $-6y^4 - 12y^3 + 14y^2$ **27.** $-2a^3 - 6a^2 + 8a$ **29.** $6y^4 - 3y^3 + 6y^2$ **31.** $x^3y - 3x^2y^2 + xy^3$ **33.** (ii) and (iii) **35.** $x^3 - 4x^2 + 11x - 14$ **37.** $2x^3 - 9x^2 + 19x - 15$ **39.** $-2a^3 + 7a^2 - 7a + 2$ **41.** $-2a^3 - 3a^2 + 8a - 3$ **43.** $2y^3 + y^2 - 10y$ **45.** $2y^4 + 7y^3 - 4y^2 - 16y + 8$ **47.** $12y^3 + 3y^2 - 29y + 15$ **49.** $18b^4 - 33b^3 + 5b^2 + 42b - 7$ **51.** $4a^4 - 12a^3 + 13a^2 - 8a + 3$ **53.** $x^2 + 4x + 3$ **55.** $a^2 + a - 12$ **57.** $y^2 - 5y - 24$ **59.** $y^2 - 10y + 21$ **61.** $2x^2 + 15x + 7$ **63.** $3x^2 + 11x - 4$ **65.** $4x^2 - 31x + 21$ **67.** $3y^2 - 2y - 16$ **69.** $9x^2 + 54x + 77$ **71.** $21a^2 - 83a + 80$ **73.** $6a^2 - 25ab + 14b^2$ **75.** $2a^2 - 11ab - 63b^2$ **77.** $100a^2 - 100ab + 21b^2$ **79.** $15x^2 + 56xy + 48y^2$ **81.** $14x^2 - 97xy - 60y^2$ **83.** $56x^2 - 61xy + 15y^2$ **85.** $12x^2 - x - 20$ **87.** $y^2 - 36$ **89.** $16x^2 - 49$ **91.** $81x^2 - 4$ **93.** $16x^2 - 81y^2$ **95.** $y^2 - 6y + 9$ **97.** $36x^2 - 60x + 25$ **99.** $x^2 - 4xy + 4y^2$ **101.** $4a^2 - 36ab + 81b^2$ **103.** $a^3 + 9a^2 + 27a + 27$ **105.** The area of the rectangle is $(18x^2 + 12x + 2)$ in^2. **107.** The area of the circle is $(\pi x^2 + 8\pi x + 16\pi)$ cm^2. **109.** The total area of the softball diamond and the base paths is $(90x + 2025)$ ft^2. **111. a.** The length of the Water Cube is $(5h + 22)$ ft. **b.** The area of an exterior wall of the Water Cube is $(5h^2 + 22h)$ ft^2. **113.** $7x^2 - 11x - 8$

SECTION 9.4

1. y^4 **3.** a^3 **5.** p^4 **7.** $2x^3$ **9.** $2k$ **11.** m^5n^2 **13.** $\frac{3r^2}{2}$ **15.** $-\frac{2a}{3}$ **17.** $\frac{1}{y^5}$ **19.** $\frac{1}{a^6}$ **21.** $\frac{1}{3x^3}$ **23.** $\frac{2}{3x^5}$ **25.** $\frac{y^4}{x^2}$ **27.** $\frac{2}{5m^3n^8}$ **29.** $\frac{1}{p^3q}$ **31.** $\frac{1}{2y^3}$ **33.** $\frac{7xz}{8y^3}$ **35.** $\frac{p^2}{2m^3}$ **37.** $\frac{1}{25}$ **39.** 64 **41.** $\frac{1}{27}$ **43.** 2 **45.** $\frac{1}{x^2}$ **47.** a^6 **49.** $\frac{4}{x^7}$ **51.** $\frac{2}{3z^2}$ **53.** $5b^8$ **55.** $\frac{x^2}{3}$ **57.** 1 **59.** -1 **61.** $-\frac{8x^3}{y^6}$ **63.** $\frac{9}{x^2y^4}$ **65.** $\frac{2}{x^4}$ **67.** $-\frac{5}{a^8}$ **69.** $-\frac{a^5}{8b^4}$ **71.** $\frac{10y^3}{x^4}$ **73.** $\frac{1}{2x^3}$ **75.** $\frac{3}{x^3}$ **77.** $\frac{1}{2x^3y^6}$ **79.** $\frac{1}{x^6y}$ **81.** $\frac{a^4}{y^{10}}$ **83.** $-\frac{1}{6x^3}$ **85.** $-\frac{a^2b}{6c^2}$ **87.** $-\frac{7b^6}{a^2}$ **89.** $\frac{s^8t^4}{4r^{12}}$ **91.** $\frac{125p^3}{27m^{15}n^6}$ **93.** False **95.** True **97.** 3.24×10^{-9} **99.** 3×10^{-18} **101.** 3.2×10^{16} **103.** 1.22×10^{-19} **105.** 5.47×10^8 **107.** 0.000167 **109.** 68,000,000 **111.** 0.0000305 **113.** 0.00000000102 **115.** $-n - 1$ **117.** 1.5×10^{-8} m **119.** 3.7×10^{-6} **121.** 1.6×10^{10} **123.** 5,115,600,000,000,000 **125.** $\frac{1}{4}, \frac{1}{2}$, 1, 2, 4

SECTION 9.5

1. $15x^2 + 12x = 3x(5x + 4)$ **3.** $2b - 5$ **5.** $4b^2 - 3$ **7.** $5y - 3$ **9.** $-y + 9$ **11.** $a^2 - 5a + 7$ **13.** $a^6 - 5a^3 - 3a$ **15.** $xy - 3$ **17.** $-2x^2 + 3$ **19.** $8y + 2 - \frac{3}{y}$ **21.** $2a - 1 - 3b$ **23.** $b - 7$ **25.** $y - 5$ **27.** $2y - 7$ **29.** $2y + 6 + \frac{25}{y - 3}$ **31.** $x - 2 + \frac{8}{x + 2}$ **33.** $3y - 5 + \frac{20}{2y + 4}$ **35.** $6x - 12 + \frac{19}{x + 2}$ **37.** $b - 5 - \frac{24}{b - 3}$ **39.** $3x + 17 + \frac{64}{x - 4}$ **41.** $5y + 3 + \frac{1}{2y + 3}$ **43.** $4a + 1$ **45.** $2a + 9 + \frac{33}{3a - 1}$ **47.** $x^2 - 5x + 2$ **49.** $x^2 + 5$ **51.** $3ab$

CHAPTER 9 CONCEPT REVIEW*

1. Writing the terms in descending order before adding helps us to add the like terms of the polynomials. [9.1A]
2. The opposite of $-7x^3 + 3x^2 - 4x - 2$ is the polynomial with the sign of every term changed: $7x^3 - 3x^2 + 4x + 2$. [9.1B]
3. When multiplying the terms $4p^3$ and $7p^6$, we add the exponents 3 and 6 to get $28p^9$. [9.2A]
4. The simplification of $-4b(2b^2 - 3b - 5)$ is incorrect because $-4b$ times $-3b$ is $12b^2$, and $-4b$ times -5 is $20b$. $-4b(2b^2 - 3b - 5) = -8b^3 + 12b^2 + 20b$. [9.3A]
5. To multiply two binomials, use the FOIL method: Add the products of the First terms, the Outer terms, the Inner terms, and the Last terms. [9.3C]
6. To simplify $\frac{w^2x^4yz^6}{w^3xy^4z^0}$, subtract the exponents of the like bases. Then use the Definition of a Negative Exponent.

$$\frac{w^2x^4yz^6}{w^3xy^4z^0} = w^{2-3}x^{4-1}y^{1-4}z^{6-0} = w^{-1}x^3y^{-3}z^6 = \frac{x^3z^6}{wy^3}. \quad [9.4A]$$

*Note: The numbers in brackets following the answers in the Concept Review are a reference to the objective that corresponds to that problem. For example, the reference [1.2A] stands for Section 1.2, Objective A. This notation will be used for all Prep Tests, Concept Reviews, Chapter Reviews, Chapter Tests, and Cumulative Reviews throughout the text.

7. To simplify $\left(\frac{a^0}{b^{-2}}\right)^{-2}$, first use the Rule for Simplifying the Power of a Quotient: $\left(\frac{a^0}{b^{-2}}\right)^{-2} = \frac{a^0}{b^4}$.

Then simplify a^0. $\frac{a^0}{b^4} = \frac{1}{b^4}$. [9.4A]

8. To write a large number in scientific notation, move the decimal point to the right of the first digit. The exponent on 10 is positive and equal to the number of places the decimal point has been moved. For example, $35{,}000{,}000{,}000{,}000 = 3.5 \times 10^{13}$. [9.4B]

9. The simplification $\frac{14x^3 - 8x^2 - 6x}{2x} = 7x^2 - 8x^2 - 6x$ is incorrect because each term in the numerator must be divided by the term in the denominator. The correct simplification is $\frac{14x^3 - 8x^2 - 6x}{2x} = 7x^2 - 4x - 3$. [9.5A]

10. The equation used to check polynomial division is (Quotient × divisor) + remainder = dividend. [9.5B]

CHAPTER 9 REVIEW EXERCISES

1. $8b^2 - 2b - 15$ [9.3C] **2.** $21y^2 + 4y - 1$ [9.1A] **3.** $x^4y^8z^4$ [9.2A] **4.** $\frac{2x^3}{3}$ [9.4A] **5.** $-8x^3 - 14x^2 + 18x$ [9.3A] **6.** $-\frac{1}{2a}$ [9.4A] **7.** $16u^{12}v^{16}$ [9.2B] **8.** 64 [9.2B] **9.** $2x^2 + 3x - 8$ [9.1B] **10.** $\frac{b^6}{a^4}$ [9.4A] **11.** $-108x^{18}$ [9.2B] **12.** $25y^2 - 70y + 49$ [9.3D] **13.** $100a^{15}b^{13}$ [9.2B] **14.** $4b^4 + 12b^2 - 1$ [9.5A] **15.** $-\frac{1}{16}$ [9.4A] **16.** $13y^3 - 12y^2 - 5y - 1$ [9.1B] **17.** $-x + 2 + \frac{1}{x+3}$ [9.5B] **18.** $2ax - 4ay - bx + 2by$ [9.3C] **19.** $6y^3 + 17y^2 - 2y - 21$ [9.3B] **20.** $b^2 + 5b + 2 + \frac{7}{b-7}$ [9.5B] **21.** $8a^3b^3 - 4a^2b^4 + 6ab^5$ [9.3A] **22.** $4a^2 - 25b^2$ [9.3D] **23.** $12b^5 - 4b^4 - 6b^3 - 8b^2 + 5$ [9.3B] **24.** $2x^3 + 9x^2 - 3x - 12$ [9.1A] **25.** $-4y + 8$ [9.5A] **26.** $a^2 - 49$ [9.3D] **27.** 3.756×10^{10} [9.4B] **28.** 14,600,000 [9.4B] **29.** $-54a^{13}b^5c^7$ [9.2A] **30.** $2y - 9$ [9.5B] **31.** $\frac{x^4y^6}{9}$ [9.4A] **32.** $10a^2 + 31a - 63$ [9.3C] **33.** 1.27×10^{-7} [9.4B] **34.** 0.0000000000032 [9.4B] **35.** The area is $(2w^2 - w)$ ft^2. [9.3E] **36.** The area is $(9x^2 - 12x + 4)$ in^2. [9.3E]

CHAPTER 9 TEST

1. $4x^3 - 6x^2$ [9.3A, How To 1] **2.** $4x - 1 + \frac{3}{x^2}$ [9.5A, How To 1] **3.** $-\frac{4}{x^6}$ [9.4A, How To 14] **4.** $-6x^3y^6$ [9.2A, How To 2] **5.** $x - 1 + \frac{2}{x+1}$ [9.5B, Example 3] **6.** $x^3 - 7x^2 + 17x - 15$ [9.3B, How To 2] **7.** $-8a^6b^3$ [9.2B, Example 3] **8.** $\frac{9y^{10}}{x^{10}}$ [9.4A, Example 2] **9.** $a^2 + 3ab - 10b^2$ [9.3C, Example 4] **10.** $4x^4 - 2x^2 + 5$ [9.5A, How To 1] **11.** $x + 7$ [9.5B, How To 2] **12.** $6y^4 - 9y^3 + 18y^2$ [9.3A, How To 1] **13.** $-4x^4 + 8x^3 - 3x^2 - 14x + 21$ [9.3B, You Try It 3] **14.** $16y^2 - 9$ [9.3D, Example 6] **15.** a^4b^7 [9.2A, Example 2] **16.** $8ab^4$ [9.4A, How To 14] **17.** $4a - 7$ [9.5A, How To 1] **18.** $-5a^3 + 3a^2 - 4a + 3$ [9.1B, You Try It 4] **19.** $4x^2 - 20x + 25$ [9.3D, How To 7] **20.** $2x + 3 + \frac{2}{2x-3}$ [9.5B, How To 3] **21.** $-2x^3$ [9.4A, Example 2] **22.** $10x^2 - 43xy + 28y^2$ [9.3C, How To 5] **23.** $3x^3 + 6x^2 - 8x + 3$ [9.1A, How To 1] **24.** 3.02×10^{-9} [9.4B, You Try It 4] **25.** The area of the circle is $(\pi x^2 - 10\pi x + 25\pi)$ m^2. [9.3E, You Try It 8]

CUMULATIVE REVIEW EXERCISES

1. $\frac{5}{144}$ [3.4A] **2.** $\frac{5}{3}$ [3.5A] **3.** $\frac{25}{11}$ [3.5A] **4.** $-\frac{22}{9}$ [4.1A] **5.** $5x - 3xy$ [4.2A] **6.** $-9x$ [4.2B] **7.** $-18x + 12$ [4.2D] **8.** -16 [5.1C] **9.** -16 [5.3A] **10.** 15 [5.3B] **11.** 22% [6.4A/6.4B] **12.** $4b^3 - 4b^2 - 8b - 4$ [9.1A] **13.** $3y^3 + 2y^2 - 10y$ [9.1B] **14.** a^9b^{15} [9.2B] **15.** $-8x^3y^6$ [9.2A] **16.** $6y^4 + 8y^3 - 16y^2$ [9.3A] **17.** $10a^3 - 39a^2 + 20a - 21$ [9.3B] **18.** $15b^2 - 31b + 14$ [9.3C]

19. $\frac{1}{2b^2}$ [9.4A] **20.** $a - 7$ [9.5B] **21.** 0.0000609 [9.4B] **22.** $8x - 2x = 18; 3$ [5.4B]
23. The cost is $.13 per ounce. [5.5A] **24.** The car overtakes the cyclist 25 mi from the starting point. [5.5B]
25. The length is 15 m and the width is 6 m. [7.2A]

ANSWERS TO CHAPTER 10 SELECTED EXERCISES

PREP TEST

1. $2 \cdot 3 \cdot 5$ [1.3D] **2.** $-12y + 15$ [4.2C] **3.** $-a + b$ [4.2C] **4.** $-3a + 3b$ [4.2D] **5.** 0 [5.1C]
6. $-\frac{1}{2}$ [5.2A] **7.** $x^2 - 2x - 24$ [9.3C] **8.** $6x^2 - 11x - 10$ [9.3C] **9.** x^3 [9.4A] **10.** $3x^3y$ [9.4A]

SECTION 10.1

3. $5(a + 1)$ **5.** $8(2 - a^2)$ **7.** $4(2x + 3)$ **9.** $6(5a - 1)$ **11.** $x(7x - 3)$ **13.** $a^2(3 + 5a^3)$ **15.** $y(14y + 11)$
17. $2x(x^3 - 2)$ **19.** $2x^2(5x^2 - 6)$ **21.** $4a^5(2a^3 - 1)$ **23.** $xy(xy - 1)$ **25.** $3xy(xy^3 - 2)$ **27.** $xy(x - y^2)$
29. $5y(y^2 - 4y + 1)$ **31.** $3y^2(y^2 - 3y - 2)$ **33.** $3y(y^2 - 3y + 8)$ **35.** $a^2(6a^3 - 3a - 2)$ **37.** $ab(2a - 5ab + 7b)$
39. $2b(2b^4 + 3b^2 - 6)$ **41.** $x^2(8y^2 - 4y + 1)$ **43. a.** (i), (ii), and (iii) **b.** (iii) **45.** $(a + z)(y + 7)$
47. $(a - b)(3r + s)$ **49.** $(m - 7)(t - 7)$ **51.** $(4a + b)(2y - 1)$ **53.** $(x + 2)(x + 2y)$ **55.** $(p - 2)(p - 3r)$
57. $(a + 6)(b - 4)$ **59.** $(2z - 1)(z + y)$ **61.** $(2v - 3y)(4v + 7)$ **63.** $(2x - 5)(x - 3y)$ **65.** $(y - 2)(3y - a)$
67. $(3x - y)(y + 1)$ **69.** $(3s + t)(t - 2)$ **71.** P doubles.

SECTION 10.2

1. $(x + 1)(x + 2)$ **3.** $(x + 1)(x - 2)$ **5.** $(a + 4)(a - 3)$ **7.** $(a - 1)(a - 2)$ **9.** $(a + 2)(a - 1)$
11. $(b - 3)(b - 3)$ **13.** $(b + 8)(b - 1)$ **15.** $(y + 11)(y - 5)$ **17.** $(y - 2)(y - 3)$ **19.** $(z - 5)(z - 9)$
21. $(z + 8)(z - 20)$ **23.** $(p + 3)(p + 9)$ **25.** $(x + 10)(x + 10)$ **27.** $(b + 4)(b + 5)$ **29.** $(x + 3)(x - 14)$
31. $(b + 4)(b - 5)$ **33.** $(y + 3)(y - 17)$ **35.** $(p + 3)(p - 7)$ **37.** Nonfactorable over the integers
39. $(x - 5)(x - 15)$ **41.** $(p + 3)(p + 21)$ **43.** $(x + 2)(x + 19)$ **45.** $(x + 9)(x - 4)$ **47.** $(a + 4)(a - 11)$
49. $(a - 3)(a - 18)$ **51.** $(z + 21)(z - 7)$ **53.** $(c + 12)(c - 15)$ **55.** $(p + 9)(p + 15)$ **57.** $(c + 2)(c + 9)$
59. $(x + 15)(x - 5)$ **61.** $(x + 25)(x - 4)$ **63.** $(b - 4)(b - 18)$ **65.** $(a + 45)(a - 3)$ **67.** $(b - 7)(b - 18)$
69. $(z + 12)(z + 12)$ **71.** $(x - 4)(x - 25)$ **73.** $(x + 16)(x - 7)$ **75.** Positive **77.** $3(x + 2)(x + 3)$
79. $-(x - 2)(x + 6)$ **81.** $a(b + 8)(b - 1)$ **83.** $x(y + 3)(y + 5)$ **85.** $-2a(a + 1)(a + 2)$
87. $4y(y + 6)(y - 3)$ **89.** $2x(x^2 - x + 2)$ **91.** $6(z + 5)(z - 3)$ **93.** $3a(a + 3)(a - 6)$ **95.** $(x + 7y)(x - 3y)$
97. $(a - 5b)(a - 10b)$ **99.** $(s + 8t)(s - 6t)$ **101.** Nonfactorable over the integers **103.** $z^2(z + 10)(z - 8)$
105. $b^2(b + 2)(b - 5)$ **107.** $3y^2(y + 3)(y + 15)$ **109.** $-x^2(x + 1)(x - 12)$ **111.** $3y(x + 3)(x - 5)$
113. $-3x(x - 3)(x - 9)$ **115.** $(x - 3y)(x - 5y)$ **117.** $(a - 6b)(a - 7b)$ **119.** $(y + z)(y + 7z)$
121. $3y(x + 21)(x - 1)$ **123.** $3x(x + 4)(x - 3)$ **125.** $2(t - 5s)(t - 7s)$ **127.** $3(a + 3b)(a - 11b)$
129. $5x(x + 2y)(x + 4y)$ **131. a.** Yes **b.** No **133.** $-19, -11, -9, 9, 11, 19$ **135.** 3, 4 **137.** 5, 8, 9
139. 2 **141.** An infinite number

SECTION 10.3

1. $(x + 1)(2x + 1)$ **3.** $(y + 3)(2y + 1)$ **5.** $(a - 1)(2a - 1)$ **7.** $(b - 5)(2b - 1)$ **9.** $(x + 1)(2x - 1)$
11. $(x - 3)(2x + 1)$ **13.** $(t + 2)(2t - 5)$ **15.** $(p - 5)(3p - 1)$ **17.** $(3y - 1)(4y - 1)$ **19.** Nonfactorable over the integers **21.** $(2t - 1)(3t - 4)$ **23.** $(x + 4)(8x + 1)$ **25.** Nonfactorable over the integers
27. $(3y + 1)(4y + 5)$ **29.** $(a + 7)(7a - 2)$ **31.** $(b - 4)(3b - 4)$ **33.** $(z - 14)(2z + 1)$ **35.** $(p + 8)(3p - 2)$
37. $2(x + 1)(2x + 1)$ **39.** $5(y - 1)(3y - 7)$ **41.** $x(x - 5)(2x - 1)$ **43.** $b(a - 4)(3a - 4)$ **45.** Nonfactorable over the integers **47.** $-3x(x + 4)(x - 3)$ **49.** $4(4y - 1)(5y - 1)$ **51.** $z(2z + 3)(4z + 1)$ **53.** $y(2x - 5)(3x + 2)$
55. $5(t + 2)(2t - 5)$ **57.** $p(p - 5)(3p - 1)$ **59.** $2(z + 4)(13z - 3)$ **61.** $2y(y - 4)(5y - 2)$
63. $yz(z + 2)(4z - 3)$ **65.** $3a(2a + 3)(7a - 3)$ **67.** $y(3x - 5y)(3x - 5y)$ **69.** $xy(3x - 4y)(3x - 4y)$ **71.** Odd
73. $(2x - 3)(3x - 4)$ **75.** $(b + 7)(5b - 2)$ **77.** $(3a + 8)(2a - 3)$ **79.** $(z + 2)(4z + 3)$ **81.** $(2p + 5)(11p - 2)$

83. $(y + 1)(8y + 9)$ **85.** $(6t - 5)(3t + 1)$ **87.** $(b + 12)(6b - 1)$ **89.** $(3x + 2)(3x + 2)$ **91.** $(2b - 3)(3b - 2)$ **93.** $(3b + 5)(11b - 7)$ **95.** $(3y - 4)(6y - 5)$ **97.** $(3a + 7)(5a - 3)$ **99.** $(2y - 5)(4y - 3)$ **101.** $(2z + 3)(4z - 5)$ **103.** Nonfactorable over the integers **105.** $(2z - 5)(5z - 2)$ **107.** $(6z + 5)(6z + 7)$ **109.** $(x + y)(3x - 2y)$ **111.** $(a + 2b)(3a - b)$ **113.** $(y - 2z)(4y - 3z)$ **115.** $-(z - 7)(z + 4)$ **117.** $-(x - 1)(x + 8)$ **119.** $3(x + 5)(3x - 4)$ **121.** $4(2x - 3)(3x - 2)$ **123.** $a^2(5a + 2)(7a - 1)$ **125.** $5(b - 7)(3b - 2)$ **127.** $(x - 7y)(3x - 5y)$ **129.** $3(8y - 1)(9y + 1)$ **131.** $-(x - 1)(x + 21)$ **133.** Two positive **135.** Two negative **139.** $x(x - 1)$ **141.** $(2y + 1)(y + 3)$ **143.** $(4y - 3)(y - 3)$

SECTION 10.4

1. a. Answers will vary. For instance, $x^2 - 25$. **b.** Answers will vary. For instance, $x^2 + 6x + 9$. **3.** $(x + 2)(x - 2)$ **5.** $(a + 9)(a - 9)$ **7.** $(y + 1)^2$ **9.** $(a - 1)^2$ **11.** $(2x + 1)(2x - 1)$ **13.** $(x^3 + 3)(x^3 - 3)$ **15.** Nonfactorable over the integers **17.** $(x + y)^2$ **19.** $(2a + 1)^2$ **21.** $(3x + 1)(3x - 1)$ **23.** $(1 + 8x)(1 - 8x)$ **25.** Nonfactorable over the integers **27.** $(3a + 1)^2$ **29.** $(b^2 + 4a)(b^2 - 4a)$ **31.** $(2a - 5)^2$ **33.** $(3a - 7)^2$ **35.** $(5z + y)(5z - y)$ **37.** $(ab + 5)(ab - 5)$ **39.** $(5x + 1)(5x - 1)$ **41.** $(2a - 3b)^2$ **43.** $(2y - 9z)^2$ **45.** (i) and (iii) **47.** $12(n + 2)(n - 2)$ **49.** $r(2s - 1)^2$ **51.** $(9 + t^2)(3 + t)(3 - t)$ **53.** $(x + 2)(x + 8)$ **55.** $4c^2(3c - 2)^2$ **57.** $2(4s^2 + 1)(2s + 1)(2s - 1)$ **59.** $(3 + 4a)^2$ **61.** $4x(x + 5)$ **63.** $(2x + 3 + 2y)(2x + 3 - 2y)$ **65.** $2(x + 3)(x - 3)$ **67.** $y(y - 5)^2$ **69.** $a^2(a - 3)(a - 8)$ **71.** $6(y - 2)(y - 6)$ **73.** Nonfactorable over the integers **75.** $3b(a + 9)(a - 2)$ **77.** $b(b^2 - 8b - 7)$ **79.** $3(y + 7)(y - 7)$ **81.** $3(2a - 3)^2$ **83.** $a^2(b + 11)(b - 8)$ **85.** $4(2x - y)(2x - 3y)$ **87.** $-2(x + 6)(x - 6)$ **89.** $b^2(a + 3)^2$ **91.** $xy(x - 2y)(2x - 3y)$ **93.** $2a(3a + 2)^2$ **95.** $3(25 + 9y^2)$ **97.** $3x(2x - 5)(4x - 1)$ **99.** $2a(a - 2b)^2$ **101.** $b^2(a - 3)^2$ **103.** $-x^2(2x - 3)(x + 7)$ **105.** $b^2(b + a)(b - a)$ **107.** $2x(4y - 3)^2$ **109.** $-y^2(x - 3)(x + 5)$ **111.** $3(x + 3y)(x - 3y)$ **113.** $y(y + 3)(y - 3)$ **115.** $x^2y^2(5x + 4y)(3x - 5y)$ **117.** $2(x - 1)(a + b)$ **119.** $(x - 2)(x + 1)(x - 1)$ **121.** $(x + 2)(x - 2)(a + b)$ **123.** $(x - 5)(2 + x)(2 - x)$ **125.** $-12, 12$ **127.** $-16, 16$ **129.** $-10, 10$

SECTION 10.5

3. $-3, -2$ **5.** $7, 3$ **7.** $0, 5$ **9.** $0, 9$ **11.** $0, -\frac{3}{2}$ **13.** $0, \frac{2}{3}$ **15.** $-2, 5$ **17.** $-9, 9$ **19.** $-\frac{7}{2}, \frac{7}{2}$ **21.** $-\frac{1}{3}, \frac{1}{3}$ **23.** $-2, -4$ **25.** $-7, 2$ **27.** $-\frac{1}{2}, 5$ **29.** $-\frac{1}{3}, -\frac{1}{2}$ **31.** $0, 3$ **33.** $0, 7$ **35.** $-1, -4$ **37.** $2, 3$ **39.** $\frac{1}{2}, -4$ **41.** $\frac{1}{3}, 4$ **43.** $3, 9$ **45.** $-2, 9$ **47.** $-1, -2$ **49.** $-9, 5$ **51.** $-7, 4$ **53.** $-2, -3$ **55.** $-8, 9$ **57.** $1, 4$ **59.** $-5, 2$ **61.** Less than **63.** The number is 6. **65.** The numbers are 2 and 4. **67.** (ii) **69.** The numbers are 4 and 5. **71.** The numbers are 3 and 7. **73.** There will be 12 consecutive numbers. **75.** There are 6 teams in the league. **77.** The object will hit the ground 3 s later. **79.** The golf ball will return to the ground 3.75 s later. **81.** The length is 15 in. The width is 5 in. **83.** The height of the triangle is 14 m. **85.** The width of the lane is 16 ft. **87.** The dimensions of the type area are 4 in. by 7 in. **89.** The width of the iris is 3 mm. **91.** 2, 32 **93.** $-9, -1$ **95.** 1, 18

CHAPTER 10 CONCEPT REVIEW*

1. In factoring a polynomial, we always first check to see if the terms of the polynomial have a common factor. If they do, we factor out the GCF of the terms. [10.4A]
2. When factoring, the terms of a polynomial do not have to be like terms. If they were like terms, we would combine them, and the result would be a monomial. [10.1A]
3. To check the answer to a factorization, multiply the factors. The product must be the original polynomial. [10.2A]
4. When factoring by grouping, after we group the first two terms and group the last two terms, we factor the GCF from each group. [10.1B]

*Note: The numbers in brackets following the answers in the Concept Review are a reference to the objective that corresponds to that problem. For example, the reference [1.2A] stands for Section 1.2, Objective A. This notation will be used for all Prep Tests, Concept Reviews, Chapter Reviews, Chapter Tests, and Cumulative Reviews throughout the text.

5. A polynomial of the form $x^2 + bx + c$ or $ax^2 + bx + c$ is nonfactorable over the integers when it does not factor into the product of two binomials that have integer coefficients and constants. [10.2A]

6. To factor a polynomial completely means to write the polynomial as a product of factors that are nonfactorable over the integers. [10.2B]

7. When factoring a polynomial of the form $x^2 + bx + c$, we begin by finding the possible factors of c because we are looking for two numbers whose product is c and whose sum is b. [10.2A]

8. Trial factors can be used when factoring a trinomial of the form $ax^2 + bx + c$. We use the factors of a and the factors of c to write all the possible binomial pairs that, when multiplied, have ax^2 and c in their product. We test each pair of trial factors to find which one has bx as the middle term of the product when the factors are multiplied. [10.3A]

9. The middle term of a trinomial of the form $x^2 + bx + c$ or $ax^2 + bx + c$ is bx. [10.3A]

10. The binomial factors of the difference of two squares $a^2 - b^2$ are $a + b$ and $a - b$. [10.4A]

11. The square of a binomial is a perfect-square trinomial. For example, $(2x - 5)^2$ is the square of a binomial. $(2x - 5)(2x - 5) = 4x^2 - 20x + 25$, so $4x^2 - 20x + 25$ is a perfect-square trinomial. [10.4A]

12. To solve an equation by factoring, the equation must be set equal to zero in order to use the Principle of Zero Products, which states that if the product of two factors is zero, then at least one of the factors must be zero. [10.5A]

CHAPTER 10 REVIEW EXERCISES

1. $(b - 10)(b - 3)$ [10.2A] **2.** $(x - 3)(4x + 5)$ [10.1B] **3.** Nonfactorable over the integers [10.3A] **4.** $5x(x^2 + 2x + 7)$ [10.1A] **5.** $7y^3(2y^6 - 7y^3 + 1)$ [10.1A] **6.** $(y + 9)(y - 4)$ [10.2A] **7.** $(2x - 7)(3x - 4)$ [10.3A] **8.** $3ab(4a + b)$ [10.1A] **9.** $(a^3 + 10)(a^3 - 10)$ [10.4A] **10.** $n^2(n - 3)(n + 1)$ [10.2B] **11.** $(6y - 1)(2y + 3)$ [10.3A] **12.** $2b(2b - 7)(3b - 4)$ [10.4B] **13.** $(3y^2 + 5z)(3y^2 - 5z)$ [10.4A] **14.** $(c + 6)(c + 2)$ [10.2A] **15.** $(6a - 5)(3a + 2)$ [10.3B] **16.** $\frac{1}{4}, -7$ [10.5A] **17.** $4x(x - 6)(x + 1)$ [10.2B] **18.** $3(a - 7)(a + 2)$ [10.2B] **19.** $(a - 12)(2a + 5)$ [10.3B] **20.** $7, -3$ [10.5A] **21.** $(3a - 5b)(7x + 2y)$ [10.1B] **22.** $(ab + 1)(ab - 1)$ [10.4A] **23.** $(2x + 5)(5x + 2y)$ [10.1B] **24.** $5(x - 3)(x + 2)$ [10.2B] **25.** $3(x + 6)^2$ [10.4B] **26.** $(x - 5)(3x - 2)$ [10.3B] **27.** The length is 100 yd. The width is 60 yd. [10.5B] **28.** The distance is 20 ft. [10.5B] **29.** The width of the frame is 1.5 in. [10.5B] **30.** A side of the original garden plot was 20 ft. [10.5B]

CHAPTER 10 TEST

1. $(b + 6)(a - 3)$ [10.1B, Example 6] **2.** $2y^2(y - 8)(y + 1)$ [10.2B, Example 3] **3.** $4(x + 4)(2x - 3)$ [10.3B, Example 4] **4.** $(2x + 1)(3x + 8)$ [10.3A, How To 2] **5.** $(a - 16)(a - 3)$ [10.2A, Example 1] **6.** $2x(3x^2 - 4x + 5)$ [10.1A, How To 2] **7.** $(x + 5)(x - 3)$ [10.2A, Example 2] **8.** $-\frac{1}{2}, \frac{1}{2}$ [10.5A, Example 2] **9.** $5(x^2 - 9x - 3)$ [10.1A, How To 2] **10.** $(p + 6)^2$ [10.4A, Example 3] **11.** $3, 5$ [10.5A, Example 3] **12.** $3(x + 2y)^2$ [10.4B, Example 7] **13.** $(b + 4)(b - 4)$ [10.4A, How To 1] **14.** $3y^2(2x + 1)(x + 1)$ [10.3B, Example 4] **15.** $(p + 3)(p + 2)$ [10.2A, You Try It 1] **16.** $(x - 2)(a + b)$ [10.1B, Example 4] **17.** $(p + 1)(x - 1)$ [10.1B, Example 4] **18.** $3(a + 5)(a - 5)$ [10.4B, Example 5] **19.** Nonfactorable over the integers [10.3B, How To 6] **20.** $(x - 12)(x + 3)$ [10.2A, How To 2] **21.** $(2a - 3b)^2$ [10.4A, How To 4] **22.** $(2x + 7y)(2x - 7y)$ [10.4A, Example 1] **23.** $\frac{3}{2}, -7$ [10.5A, How To 1] **24.** The two numbers are 7 and 3. [10.5B, Example 4] **25.** The length is 15 cm. The width is 6 cm. [10.5B, Example 4]

CUMULATIVE REVIEW EXERCISES

1. 7 [3.2B] **2.** 4 [3.5A] **3.** -7 [4.1A] **4.** $15x^2$ [4.2B] **5.** 12 [4.2D] **6.** $\frac{2}{3}$ [5.1C] **7.** $\frac{7}{4}$ [5.3A] **8.** 3 [5.3B] **9.** 45 [6.4A/6.4B] **10.** $9a^6b^4$ [9.2B] **11.** $x^3 - 3x^2 - 6x + 8$ [9.3B] **12.** $4x + 8 + \frac{21}{2x - 3}$ [9.5B] **13.** $\frac{y^6}{x^8}$ [9.4A] **14.** $(a - b)(3 - x)$ [10.1B] **15.** $5xy^2(3 - 4y^2)$ [10.1A] **16.** $(x - 7y)(x + 2y)$ [10.2A] **17.** $(p - 10)(p + 1)$ [10.2A] **18.** $3a(3a + 2)(2a + 5)$ [10.4B] **19.** $(6a + 7b)(6a - 7b)$ [10.4A] **20.** $(2x + 7y)^2$ [10.4A] **21.** $(3x + 7)(3x - 2)$ [10.3A] **22.** $2(3x - 4y)^2$ [10.4B] **23.** $(x - 3)(3y - 2)$ [10.1B]

24. $\frac{2}{3}, -7$ [10.5A] **25.** The shorter piece is 4 ft long. The longer piece is 6 ft long. [5.4B] **26.** The discount rate is 40%. [5.2B/6.3B] **27.** $m\angle a = 72°$; $m\angle b = 108°$ [7.1B] **28.** The distance to the resort is 168 mi. [5.5B] **29.** The integers are 10, 12, and 14. [5.4A] **30.** The length of the base of the triangle is 12 in. [10.5B]

ANSWERS TO CHAPTER 11 SELECTED EXERCISES

PREP TEST

1. 36 [2.1A] **2.** $\frac{3x}{y^3}$ [9.4A] **3.** $-\frac{5}{36}$ [3.4A] **4.** $-\frac{10}{11}$ [3.4B] **5.** No [1.3C] **6.** $\frac{19}{8}$ [5.2A] **7.** $(x-6)(x+2)$ [10.2A] **8.** $(2x-3)(x+1)$ [10.3A] **9.** 9:40 A.M. [5.5B]

SECTION 11.1

3. $\frac{3}{4x}$ **5.** $\frac{1}{x+3}$ **7.** -1 **9.** $\frac{2}{3y}$ **11.** $-\frac{3}{4x}$ **13.** $\frac{a}{b}$ **15.** $-\frac{2}{x}$ **17.** $\frac{y-2}{y-3}$ **19.** $\frac{x+5}{x+4}$ **21.** $\frac{x+4}{x-3}$ **23.** $-\frac{x+2}{x+5}$ **25.** $\frac{2(x+2)}{x+3}$ **27.** $\frac{2x-1}{2x+3}$ **29.** $-\frac{x+7}{x+6}$ **31.** $\frac{2}{3xy}$ **33.** $\frac{8xy^2ab}{3}$ **35.** $\frac{2}{9}$ **37.** $\frac{y^2}{x}$ **39.** $\frac{y(x+4)}{x(x+1)}$ **41.** $\frac{x^3(x-7)}{y^2(x-4)}$ **43.** $-\frac{y}{x}$ **45.** $\frac{x+3}{x+1}$ **47.** $\frac{x-5}{x+3}$ **49.** $-\frac{x+3}{x+5}$ **51.** $-\frac{x+3}{x-12}$ **53.** $\frac{x+2}{x+4}$ **55.** 1 **57.** 1 **61.** $\frac{7a^3y^2}{40bx}$ **63.** $\frac{4}{3}$ **65.** $\frac{3a}{2}$ **67.** $\frac{x^2(x+4)}{y^2(x+2)}$ **69.** $\frac{x(x-2)}{y(x-6)}$ **71.** $-\frac{3by}{ax}$ **73.** $\frac{(x+6)(x-3)}{(x+7)(x-6)}$ **75.** 1 **77.** $-\frac{x+8}{x-4}$ **79.** $\frac{2n+1}{2n-3}$ **81.** Yes **83.** No **85.** $\frac{4}{25}$

SECTION 11.2

1. $24x^3y^2$ **3.** $30x^4y^2$ **5.** $8x^2(x+2)$ **7.** $6x^2y(x+4)$ **9.** $36x(x+2)^2$ **11.** $6(x+1)^2$ **13.** $(x-1)(x+2)(x+3)$ **15.** $(2x+3)^2(x-5)$ **17.** $(x-1)(x-2)$ **19.** $(x-3)(x+2)(x+4)$ **21.** $(x+4)(x+1)(x-7)$ **23.** $(x-6)(x+6)(x+4)$ **25.** $(2x-1)(x-3)(x+1)$ **27.** $(x+2)(x-3)$ **29.** $(x+6)(x-3)$ **31. a.** One **b.** Zero **c.** Two **33.** $\frac{4x}{x^2}, \frac{3}{x^2}$ **35.** $\frac{4x}{12y^2}, \frac{3yz}{12y^2}$ **37.** $\frac{xy}{x^2(x-3)}, \frac{6x-18}{x^2(x-3)}$ **39.** $\frac{9x}{x(x-1)^2}, \frac{6x-6}{x(x-1)^2}$ **41.** $\frac{3x}{x(x-3)}, -\frac{5}{x(x-3)}$ **43.** $\frac{3}{(x-5)^2}, -\frac{2x-10}{(x-5)^2}$ **45.** $\frac{3x}{x^2(x+2)}, \frac{4x+8}{x^2(x+2)}$ **47.** $\frac{x^2-6x+8}{(x+3)(x-4)}, \frac{x^2+3x}{(x+3)(x-4)}$ **49.** $\frac{3}{(x+2)(x-1)}, \frac{x^2-x}{(x+2)(x-1)}$ **51.** $\frac{x^2-3x}{(x+3)(x-3)(x-2)}, \frac{2x^2-4x}{(x+3)(x-3)(x-2)}$

SECTION 11.3

1. $\frac{11}{y^2}$ **3.** $-\frac{7}{x+4}$ **5.** $\frac{8x}{2x+3}$ **7.** $\frac{5x+7}{x-3}$ **9.** $\frac{2x-5}{x+9}$ **11.** $\frac{-3x-4}{2x+7}$ **13.** $\frac{1}{x+5}$ **15.** $\frac{1}{x-6}$ **17.** $\frac{3}{2y-1}$ **19.** $\frac{1}{x-5}$ **21.** (i) and (iv) **23.** $\frac{4y+5x}{xy}$ **25.** $\frac{19}{2x}$ **27.** $\frac{5}{12x}$ **29.** $\frac{19x-12}{6x^2}$ **31.** $\frac{52y-35x}{20xy}$ **33.** $\frac{13x+2}{15x}$ **35.** $\frac{7}{24}$ **37.** $\frac{x+90}{45x}$ **39.** $\frac{x^2+2x+2}{2x^2}$ **41.** $\frac{2x^2+3x-10}{4x^2}$ **43.** $\frac{-x^2-4x+4}{x+4}$ **45.** $\frac{4x+7}{x+1}$ **47.** $\frac{4x^2+9x+9}{24x^2}$ **49.** $\frac{3x-1-2xy-3y}{xy^2}$ **51.** $\frac{20x^2+28x-12xy+9y}{24x^2y^2}$ **53.** $\frac{9x^2-3x-2xy-10y}{18xy^2}$ **55.** $\frac{7x-23}{(x-3)(x-4)}$ **57.** $\frac{-y-33}{(y+6)(y-3)}$ **59.** $\frac{3x^2+20x-8}{(x-4)(x+6)}$ **61.** $\frac{3(4x^2+5x-5)}{(x+5)(2x+3)}$ **63.** $\frac{-4x+5}{x-6}$ **65.** $\frac{2(y+2)}{(y+4)(y-4)}$ **67.** $-\frac{4x}{(x+1)^2}$ **69.** $\frac{2x-1}{(1+x)(1-x)}$ **71.** $\frac{14}{(x-5)^2}$ **73.** $\frac{-2(x+7)}{(x+6)(x-7)}$ **75.** $\frac{x-4}{x-6}$ **77.** $\frac{2x+1}{x-1}$ **79.** $\frac{-3(x^2+8x+25)}{(x-3)(x+7)}$ **81. a.** $\frac{44{,}400}{x}$ dollars **b.** $\frac{222{,}000}{x(x+5)}$ dollars **c.** $296

SECTION 11.4

1. $\frac{x}{x-3}$ **3.** $\frac{2}{3}$ **5.** $\frac{y+3}{y-4}$ **7.** $\frac{2(2x+13)}{5x+36}$ **9.** $\frac{x+2}{x+3}$ **11.** $\frac{x-6}{x+5}$ **13.** $\frac{-x+2}{x+1}$ **15.** $x-1$ **17.** $\frac{1}{2x-1}$ **19.** $\frac{x-3}{x+5}$ **21.** $\frac{x-7}{x-8}$ **23.** $\frac{2y-1}{2y+1}$ **25.** $\frac{x-2}{2x-5}$ **27.** $\frac{-x-1}{4x-3}$ **29.** $\frac{x+1}{2(5x-2)}$ **31.** True **33.** $\frac{8}{5}$ **35.** $\frac{x+1}{x-1}$ **37.** $\frac{y^2+x^2}{xy}$

SECTION 11.5

1. −1, 2 **3.** 0, 9 **5.** 1 **7.** 3 **9.** 2 **11.** 2 **13.** $\frac{2}{3}$ **15.** 4 **17.** −3 **19.** $\frac{3}{4}$ **21.** 7 **23.** −7

25. −1 **27.** −1 **29.** No solution **31.** 2, −6 **33.** $-\frac{2}{3}$, 5 **35.** −1, 6 **39.** 1 **41.** 0, $-\frac{5}{2}$ **43.** −3

SECTION 11.6

1. $y = -3x + 10$ **3.** $y = 4x - 3$ **5.** $y = -\frac{3}{2}x + 3$ **7.** $y = \frac{2}{5}x - 2$ **9.** $y = -\frac{2}{7}x + 2$ **11.** $y = -\frac{1}{3}x + 2$

13. $y = 3x + 8$ **15.** $y = -\frac{2}{3}x - 3$ **17.** $x = -6y + 10$ **19.** $x = \frac{1}{2}y + 3$ **21.** $x = -\frac{3}{4}y + 3$

23. $x = 4y + 3$ **25.** $t = \frac{d}{r}$ **27.** $T = \frac{PV}{nR}$ **29.** $l = \frac{P - 2w}{2}$ **31.** $b_1 = \frac{2A - hb_2}{h}$ **33.** $h = \frac{3V}{A}$

35. $S = C - Rt$ **37.** $P = \frac{A}{1 + rt}$ **39.** $w = \frac{A}{S + 1}$ **41. a.** $S = \frac{F + BV}{B}$ **b.** The required selling price is $180.

c. The required selling price is $75.

SECTION 11.7

3. t is less than k. **5.** With both skiploaders working together, it would take 3 h to remove the earth. **7.** It would take the new machine 12 h to complete the task. **9.** It would take 30 min to print the first edition with both presses operating. **11.** Working alone, the apprentice could construct the wall in 15 h. **13.** It will take the second technician 3 h to complete the wiring. **15.** Working alone, it would have taken one of the welders 40 h to complete the welds. **17.** It would have taken one machine 28 h to fill the boxes. **19.** Zachary picked $\frac{m}{n}$ of the row of peas. Eli picked $\frac{n - m}{n}$ of the row of peas.

21. Time $= \frac{\text{distance}}{\text{rate}}$; Rate $= \frac{\text{distance}}{\text{time}}$ **23.** The rate of the jet plane was 600 mph. **25.** The rate of the freight train is 30 mph. The rate of the express train is 50 mph. **27.** Camille's walking speed is 4 mph. **29.** The rate of the car is 48 mph. **31.** The hiker walked the first 9 mi at a rate of 3 mph. **33.** The rate of the jet stream is 50 mph. **35.** The rate of the current is 5 mph. **39.** The bus usually travels 60 mph.

CHAPTER 11 CONCEPT REVIEW*

1. A rational expression is in simplest form when the numerator and denominator have no common factors other than 1. [11.1A]

2. To divide two rational expressions, change the division to a multiplication and change the divisor to its reciprocal. Then multiply. [11.1C]

3. To find the LCM of two polynomials, first factor each polynomial completely. The LCM is the product of each factor the greatest number of times it occurs in any one factorization. [11.2A]

4. When subtracting two rational expressions, both expressions must have the same denominator before subtraction can take place. [11.3B]

5. To add rational expressions:

1. Find the LCM of the denominators.

2. Write each fraction as an equivalent fraction using the LCM as the denominator.

3. Add the numerators and place the result over the common denominator.

4. Write the answer in simplest form. [11.3B]

6. To simplify a complex fraction by Method 1:

1. Determine the LCM of the denominators of the fractions in the numerator and denominator of the complex fraction.

2. Multiply the numerator and denominator of the complex fraction by the LCM.

3. Simplify.

To simplify a complex fraction by Method 2:

1. Simplify the numerator to a single fraction and simplify the denominator to a single fraction.

2. Using the definition for dividing fractions, multiply the numerator by the reciprocal of the denominator.

3. Simplify. [11.4A]

**Note:* The numbers in brackets following the answers in the Concept Review are a reference to the objective that corresponds to that problem. For example, the reference [1.2A] stands for Section 1.2, Objective A. This notation will be used for all Prep Tests, Concept Reviews, Chapter Reviews, Chapter Tests, and Cumulative Reviews throughout the text.

7. When solving an equation that contains fractions, we first clear the denominators in order to rewrite the equation without any fractions. [11.5A]

8. When solving a literal equation for a particular variable, the goal is to rewrite the equation so that the variable being solved for is on one side of the equation and all numbers and other variables are on the other side. [11.6A]

9. If a job is completed in x hours, the rate of work is $\frac{1}{x}$ of the job each hour. [11.7A]

10. The rate of the boat when traveling down this river can be expressed as $r + c$, and the rate of the boat when traveling up this river can be expressed as $r - c$. [11.7B]

CHAPTER 11 REVIEW EXERCISES

1. $\frac{b^3y}{10ax}$ [11.1C] **2.** $\frac{7x + 22}{60x}$ [11.3B] **3.** $\frac{2xy}{5}$ [11.1B] **4.** $\frac{2xy}{3(x + y)}$ [11.1C] **5.** $\frac{x - 2}{3x - 10}$ [11.4A] **6.** $-\frac{x + 6}{x + 3}$ [11.1A] **7.** $\frac{2x^4}{3y^7}$ [11.1A] **8.** 62 [11.5A] **9.** $\frac{(3y - 2)^2}{(y - 1)(y - 2)}$ [11.1C] **10.** $x = \frac{5}{3a - 1}$ [11.6A] **11.** 8 [11.5A] **12.** $\frac{x^2 + 3y}{xy}$ [11.3B] **13.** $y = -\frac{5}{4}x + 5$ [11.6A] **14.** $\frac{by^3}{6ax^2}$ [11.1B] **15.** $\frac{x}{x - 7}$ [11.4A] **16.** $\frac{3x^2 - x}{(6x - 1)(2x + 3)(3x - 1)}, \frac{24x^3 - 4x^2}{(6x - 1)(2x + 3)(3x - 1)}$ [11.2B] **17.** $a = \frac{T - 2bc}{2b + 2c}$ [11.6A] **18.** 2 [11.5A] **19.** $\frac{2x + 1}{3x - 2}$ [11.4A] **20.** $\frac{x^2 + 5}{(x - 5)(x - 2)}$ [11.3B] **21.** $c = \frac{100m}{i}$ [11.6A] **22.** No solution [11.5A] **23.** $\frac{1}{x^2}$ [11.1C] **24.** $\frac{2y - 3}{5y - 7}$ [11.3B] **25.** $\frac{1}{x + 3}$ [11.3A] **26.** $(5x - 3)(2x - 1)(4x - 1)$ [11.2A] **27.** $y = -\frac{4}{9}x + 2$ [11.6A] **28.** $\frac{2x + 1}{x + 2}$ [11.1B] **29.** 5 [11.5A] **30.** $\frac{3x - 1}{x - 5}$ [11.3B] **31.** 10 [11.5A] **32.** 12 [11.5A] **33.** It would take 6 h to fill the pool. [11.7A] **34.** The rate of the car is 45 mph. [11.7B] **35.** The rate of the wind is 20 mph. [11.7B]

CHAPTER 11 TEST

1. $\frac{x^2 - 4x + 5}{(x + 3)(x - 2)}$ [11.3B, How To 1] **2.** -1 [11.5A, You Try It 1] **3.** $\frac{(2x - 1)(x - 5)}{(x + 3)(2x + 5)}$ [11.1B, Example 5] **4.** $\frac{2x^3}{3y^3}$ [11.1A, Example 1] **5.** $t = \frac{d - s}{r}$ [11.6A, Example 3] **6.** 2 [11.5A, Example 2] **7.** $-\frac{x + 5}{x + 1}$ [11.1A, Example 2] **8.** $3(2x - 1)(x + 1)$ [11.2A, How To 1] **9.** $\frac{5}{(2x - 1)(3x + 1)}$ [11.3B, Example 5] **10.** $\frac{x + 5}{x + 4}$ [11.1C, Example 7] **11.** $\frac{x - 3}{x - 2}$ [11.4A, Example 2] **12.** $\frac{3x + 6}{x(x - 2)(x + 2)}, \frac{x^2}{x(x - 2)(x + 2)}$ [11.2B, Example 4] **13.** $\frac{2}{x + 5}$ [11.3A, Example 1] **14.** $y = \frac{3}{8}x - 2$ [11.6A, Example 1] **15.** No solution [11.5A, How To 2] **16.** $\frac{x + 1}{x^3(x - 2)}$ [11.1B, Example 4] **17.** $\frac{6b^4x}{ay^3}$ [11.1C, You Try It 6] **18.** $\frac{1}{x^2y}$ [11.3A, Example 1] **19.** It will take 3 h until the run can be opened. [11.7A, How To 1] **20.** It would take 4 h to fill the pool. [11.7A, How To 1] **21.** The rate of the wind is 20 mph. [11.7B, How To 2] **22.** The rate of the current is 5 mph. [11.7B, How To 2]

CUMULATIVE REVIEW EXERCISES

1. $\frac{31}{30}$ [2.7A] **2.** 21 [4.1A] **3.** $5x - 2y$ [4.2A] **4.** $-8x + 26$ [4.2D] **5.** $-\frac{9}{2}$ [5.2A] **6.** -12 [5.3B] **7.** 10 [6.4A/6.4B] **8.** a^3b^7 [9.2A] **9.** $a^2 + ab - 12b^2$ [9.3C] **10.** $3b^3 - b + 2$ [9.5A] **11.** $x^2 + 2x + 4$ [9.5B] **12.** $(3x - 1)(4x + 1)$ [10.3A] **13.** $(y - 6)(y - 1)$ [10.2A] **14.** $a(a + 5)(2a - 3)$ [10.3A] **15.** $4(b + 5)(b - 5)$ [10.4B] **16.** $-3, \frac{5}{2}$ [10.5A] **17.** $\frac{2x^3}{3y^5}$ [11.1A] **18.** $-\frac{x - 2}{x + 5}$ [11.1A] **19.** 1 [11.1C] **20.** $\frac{3}{(2x - 1)(x + 1)}$ [11.3B] **21.** $\frac{x + 3}{x + 5}$ [11.4A] **22.** 4 [11.5A] **23.** 3 [11.5A] **24.** $t = \frac{f - v}{a}$ [11.6A] **25.** $5x - 13 = -8; x = 1$ [5.4A] **26.** The school-age population is 50 million people. [6.4C] **27.** The base is 10 in. The height is 6 in. [10.5B] **28.** The cost of a \$5000 policy is \$80. [6.2B] **29.** It would take both pipes 6 min to fill the tank. [11.7A] **30.** The rate of the current is 2 mph. [11.7B]

ANSWERS TO CHAPTER 12 SELECTED EXERCISES

PREP TEST

1. 3 [3.5A] **2.** -1 [4.1A] **3.** $-3x + 12$ [4.2C] **4.** -2 [5.2A] **5.** $x = 5$ [5.2A] **6.** $y = -2$ [5.2A] **7.** $-4x + 5$ [9.5A] **8.** 4 [6.2A] **9.** $y = \frac{3}{5}x - 3$ [11.6A] **10.** $y = -\frac{1}{2}x - 5$ [11.6A]

SECTION 12.1

1.

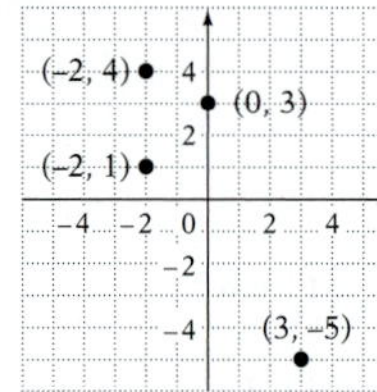

3.

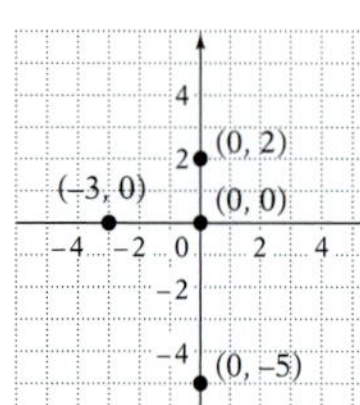

5.

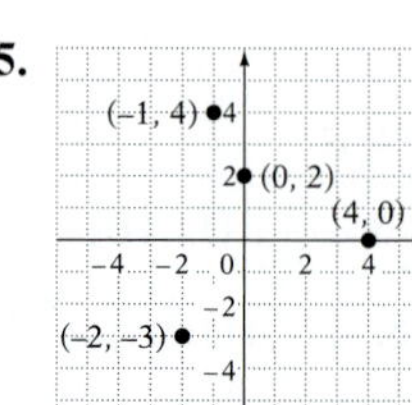

7. $A(2, 3)$, $B(4, 0)$, $C(-4, 1)$, $D(-2, -2)$

9. $A(-2, 5)$, $B(3, 4)$, $C(0, 0)$, $D(-3, -2)$ **11. a.** 2, -4 **b.** 1, -3 **13. a.** y-axis **b.** x-axis **15.** Yes **17.** No **19.** No **21.** Negative

23.

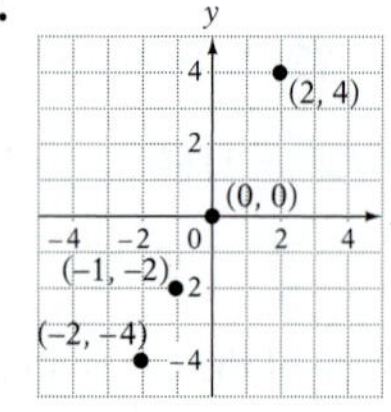

25.

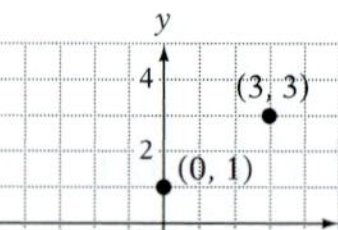

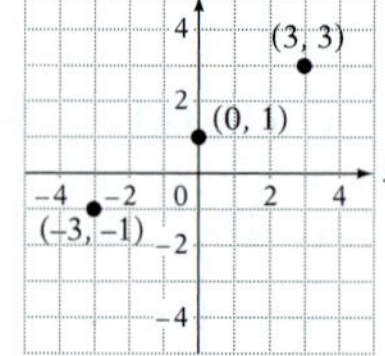

27.

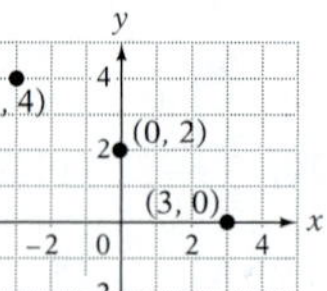

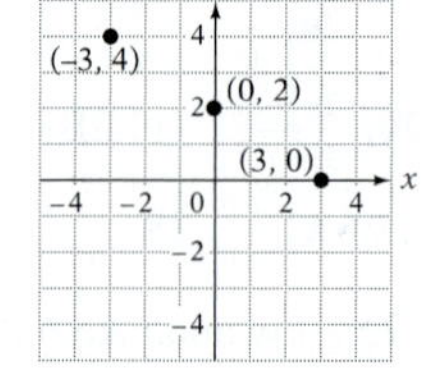

29. No

31. {(24, 600), (32, 750), (22, 430), (15, 300), (4.4, 68), (17, 370), (15, 310), (4.4, 55)}; No

33. {(352, 15.4), (353, 15.4), (354, 15.1), (355, 15.1), (356, 15.2), (358, 15.4), (360, 15.3), (361, 15.5)}; Yes

35. {(35, 7.50), (45, 7.58), (38, 7.63), (24, 7.78), (47, 7.80), (51, 7.86), (35, 7.89), (48, 7.92)}; No **37.** Yes **39.** Yes

41. 11 **43.** 0 **45.** 6 **47.** $-\frac{4}{3}$ **49.** 11 **51.** $\frac{6}{7}$ **53.** Either **55.** Positive

SECTION 12.2

1.

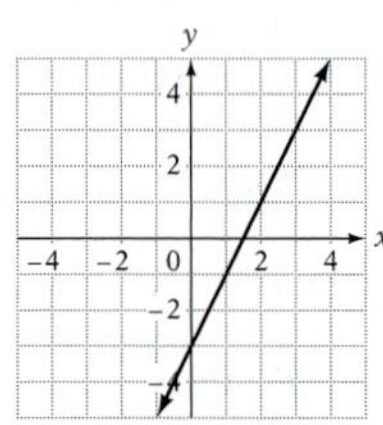

3.

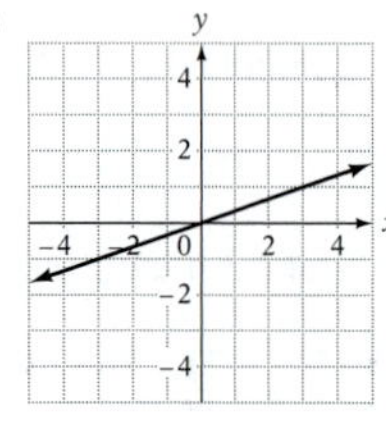

5.

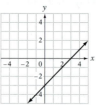

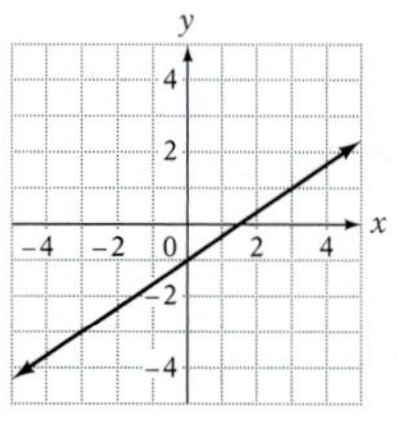

7.

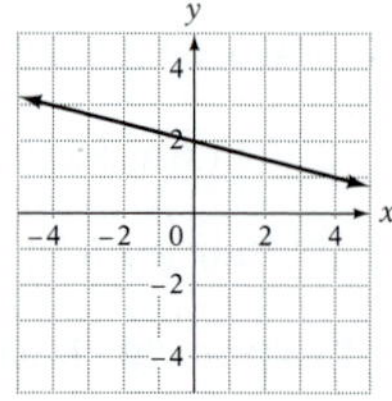

9.

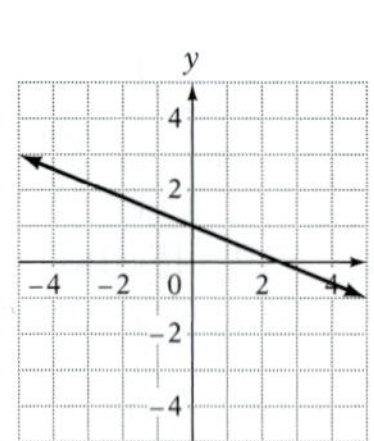

11.

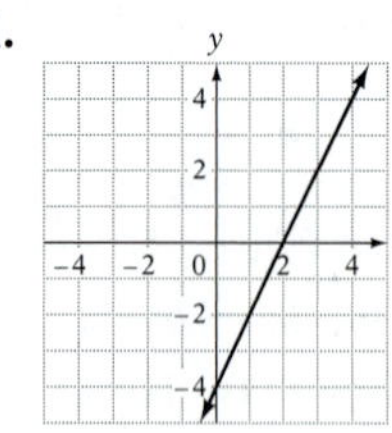

13.

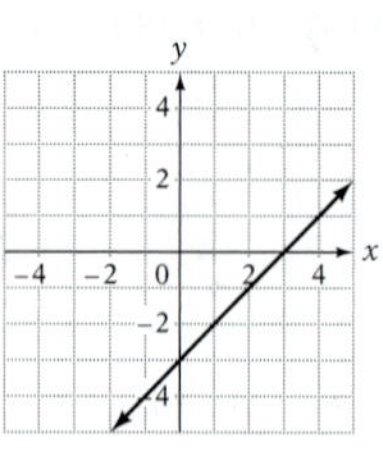

15.

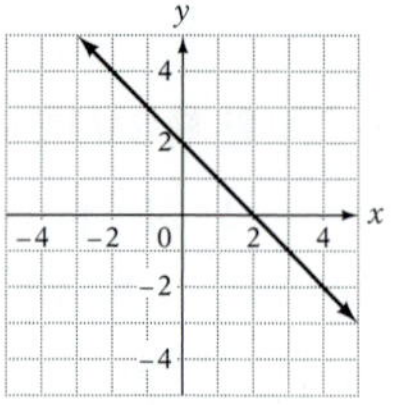

17.

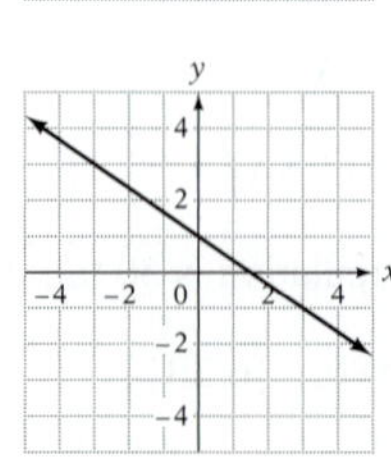

19. 0

21.

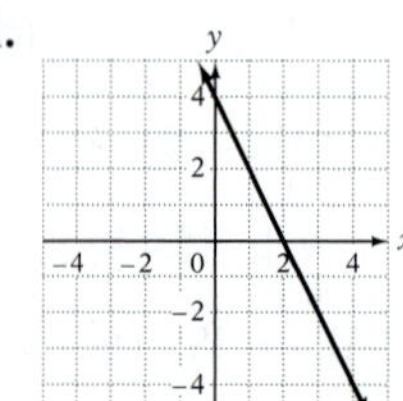

23.

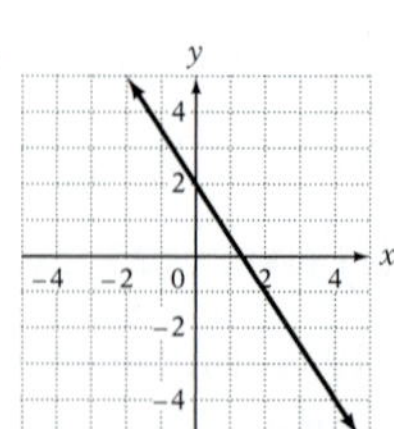

25.
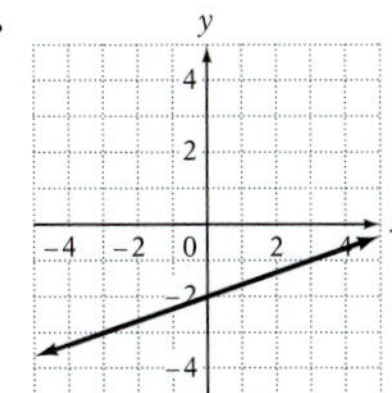

27.
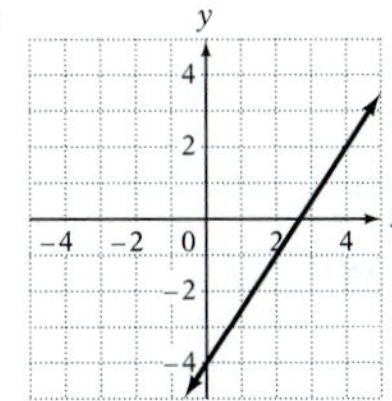

29.
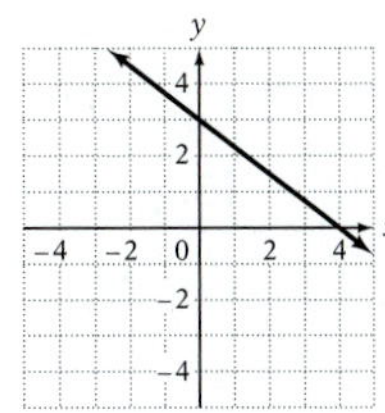

31.
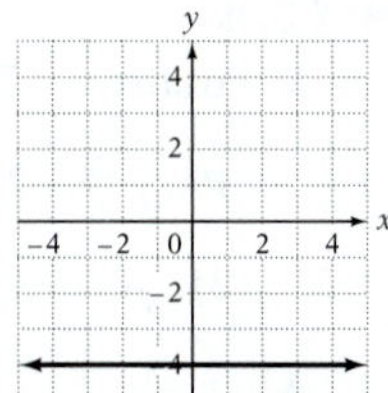

33.
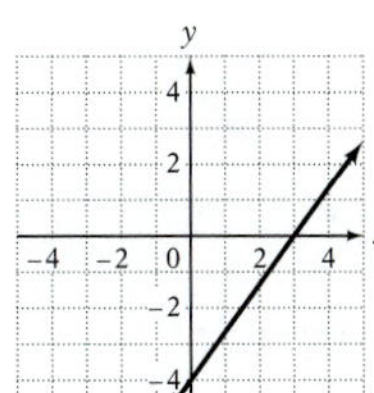

35.
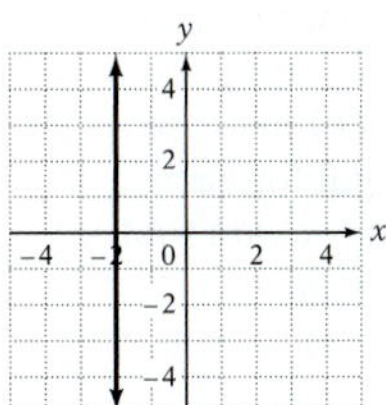

37.
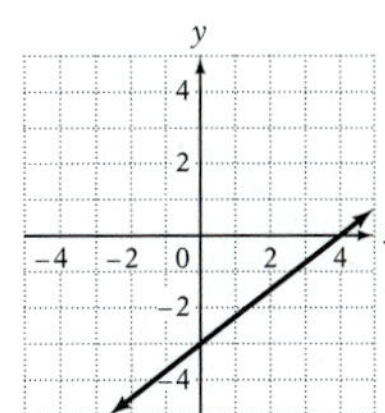
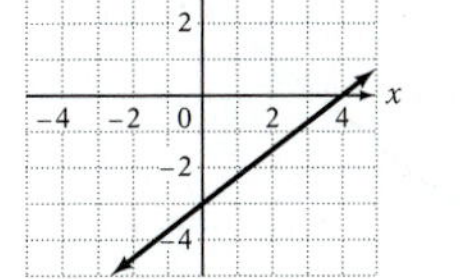

39. The statement is false.

41. After flying for 3 min, the helicopter is 3.5 mi away from the victims.

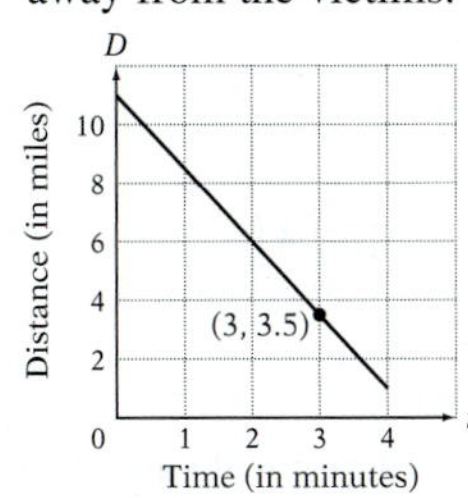

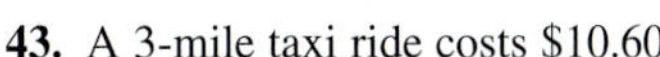
43. A 3-mile taxi ride costs $10.60.

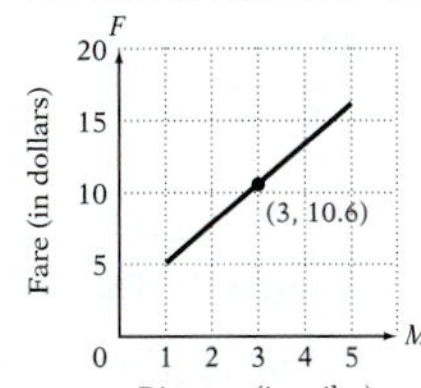

SECTION 12.3

1. (3, 0), (0, −3) **3.** (2, 0), (0, −6) **5.** (10, 0), (0, −2) **7.** (−4, 0), (0, 12) **9.** (0, 0), (0, 0) **11.** (6, 0), (0, 3)

13.
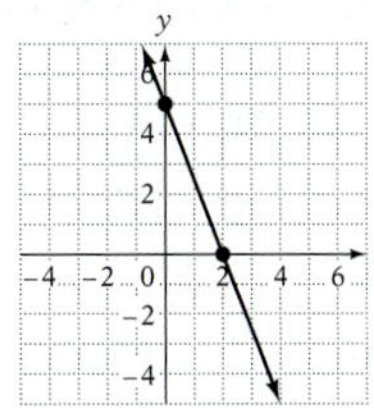

15.
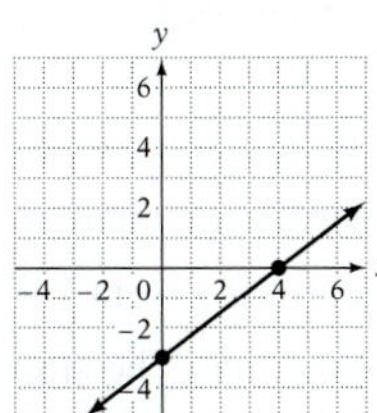

17.
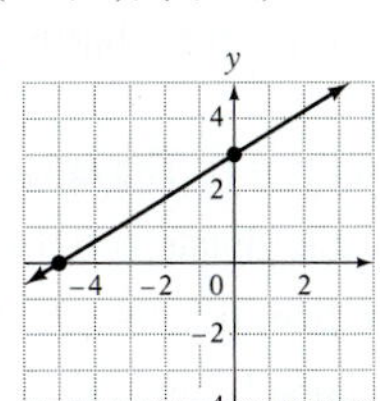

19. Above the x-axis **23.** 3

25. $\frac{1}{4}$ **27.** $-\frac{3}{2}$ **29.** Undefined **31.** Zero **33.** 1 **35.** $b = d, a \neq c$ **37.** Parallel **39.** Perpendicular

41. Neither **43.** Perpendicular **45.** $m = -\frac{19}{30}$. The water in the lock decreases by $0.6\overline{3}$ million gallons each minute.

47. $m = -3$. The number of deaths per 10 billion miles traveled has decreased by 3 deaths per year. **49.** $m = -1$, (0, 7)

51. $m = -\frac{4}{3}$, (0, 4) **53.** $m = -2$, (0, 0) **55.** $m = -\frac{2}{3}$, $\left(0, \frac{8}{3}\right)$

57.
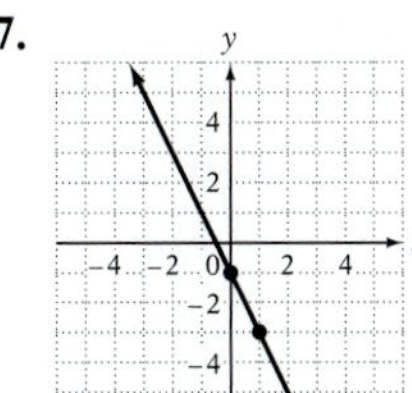

59.
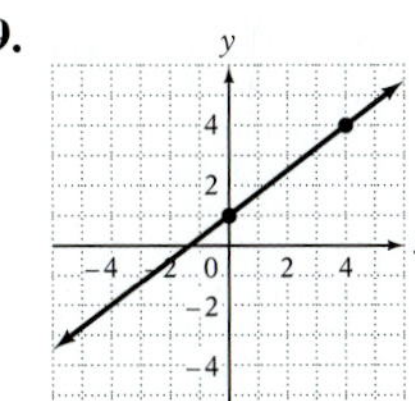

61.
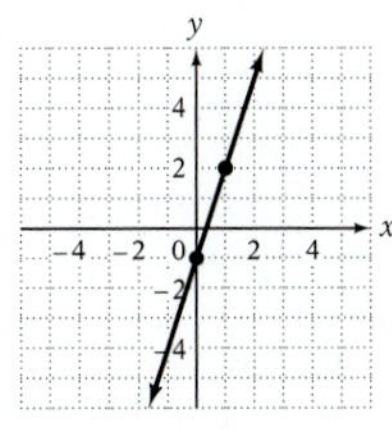

63.
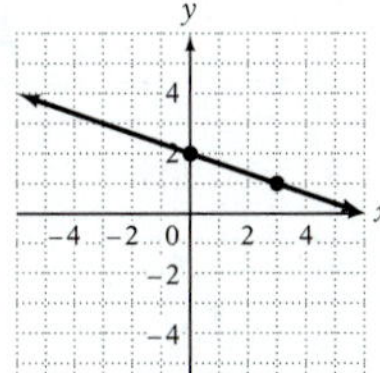

65.
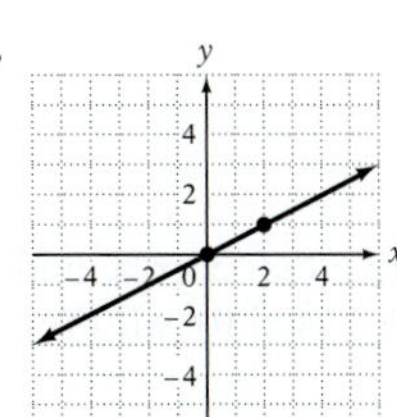

67.
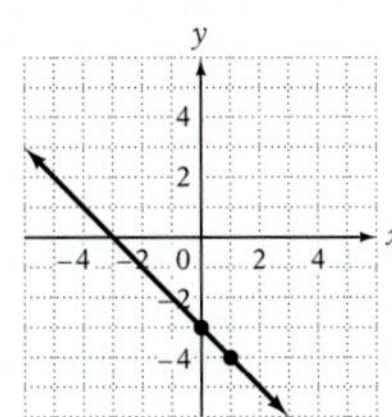

69.
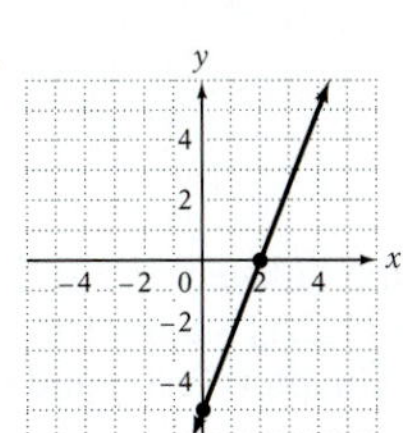

71. Downward to the right

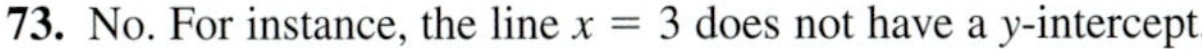
73. No. For instance, the line $x = 3$ does not have a y-intercept.

SECTION 12.4

3. Negative **5.** Positive **7.** $y = 2x + 2$ **9.** $y = -3x - 1$ **11.** $y = \frac{1}{3}x$ **13.** $y = \frac{3}{4}x - 5$ **15.** $y = -\frac{3}{5}x$

17. $y = \frac{1}{4}x + \frac{5}{2}$ **19.** $y = -2$ **21.** $x = -3$ **23.** $y - b = m(x - 0)$; Yes **25.** Negative **27.** Positive

29. $y = 2x - 3$ **31.** $y = -2x - 3$ **33.** $y = \frac{2}{3}x$ **35.** $y = \frac{1}{2}x + 2$ **37.** $y = -\frac{3}{4}x - 2$ **39.** $y = \frac{3}{4}x + \frac{5}{2}$

41. $y = -3$ **43.** $x = -2$ **45.** 2

47. The tennis player is using 1.55 g of carbohydrates per minute.

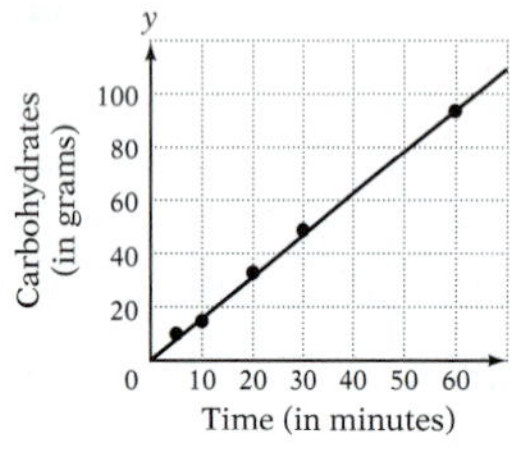

49. The amount of water that evaporates per day from a pool increases by 0.17 gal for each additional square foot of surface area.

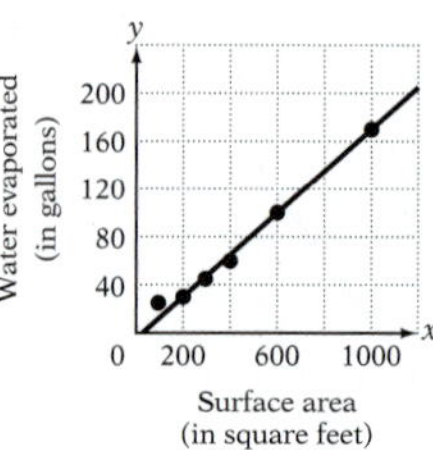

51. Increases; 3; 3 **53.** No **55.** Yes **57.** $-\frac{3}{2}$ **59.** -5

CHAPTER 12 CONCEPT REVIEW*

1. The ordinate is the second number in an ordered pair. The abscissa is the first number in an ordered pair. [12.1A]

2. A linear equation in two variables has an infinite number of solutions. [12.1B]

3. A relation is a function when no two ordered pairs of the relation have the same first coordinate. [12.1C]

4. The value of a dependent variable y depends on the value of the independent variable x. The value of the independent variable x is not dependent on the value of any other variable. [12.1D]

5. In the general equation $y = mx + b$, m represents the slope and b is the ordinate of the y-intercept. [12.3C]

6. A straight line is determined by two points. However, to ensure the accuracy of a graph, find three ordered-pair solutions. If the three solutions do not lie on a straight line, there has been an error in calculating an ordered-pair solution or in plotting the points. [12.2A]

7. The equation of a vertical line is of the form $x = a$. The equation of a horizontal line is of the form $y = b$. [12.2B]

8. In the ordered pair for an x-intercept, the y coordinate is 0. In the ordered pair for a y-intercept, the x coordinate is 0. [12.3A]

9. A line that has undefined slope is a vertical line. It is the graph of an equation of the form $x = a$. [12.3B]

10. Given two ordered pairs on a line, use the slope formula to find the slope of the line. The slope formula is $m = \frac{y_2 - y_1}{x_2 - x_1}$, where m is the slope, and (x_1, y_1) and (x_2, y_2) are the two points on the line. [12.3B]

11. Parallel lines never meet; the distance between them is always the same. In a rectangular coordinate system, parallel lines have the same slope. Perpendicular lines are two lines that intersect at right angles. In a rectangular coordinate system, two lines are perpendicular if the product of their slopes is -1. [12.3B]

12. The point-slope formula states that if (x_1, y_1) is a point on a line with slope m, then $y - y_1 = m(x - x_1)$. [12.4A]

CHAPTER 12 REVIEW EXERCISES

1. a.

(−2, 4)
(3, −2)

b. -2

c. -4 [12.1A]

2.

(−4, 0)
(−2, −1)
(0, −2)
(2, −3)

[12.1B]

3. $y = -\frac{8}{3}x + \frac{1}{3}$ [12.4B] **4.** $y = -\frac{5}{2}x + 16$ [12.4A]

**Note:* The numbers in brackets following the answers in the Concept Review are a reference to the objective that corresponds to that problem. For example, the reference [1.2A] stands for Section 1.2, Objective A. This notation will be used for all Prep Tests, Concept Reviews, Chapter Reviews, Chapter Tests, and Cumulative Reviews throughout the text.

5. 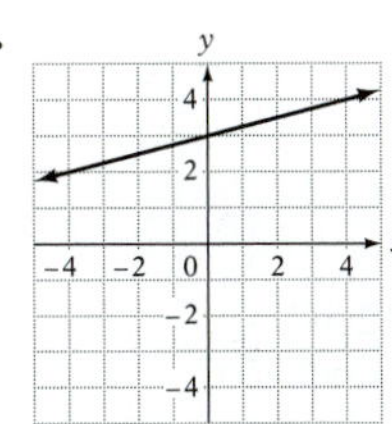 [12.2A]

6. 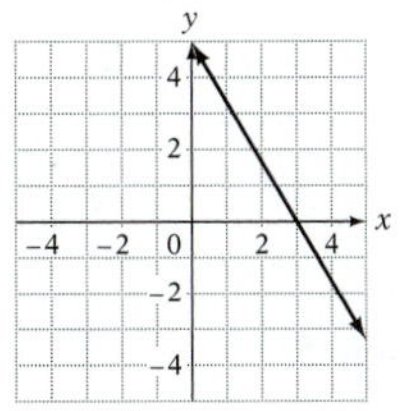 [12.2B]

7. Neither [12.3B] **8.** −1 [12.1D]

9. Yes [12.1C] **10.** $\frac{7}{11}$ [12.3B] **11.** (8, 0), (0, −12) [12.3A] **12.** 0 [12.3B]

13. 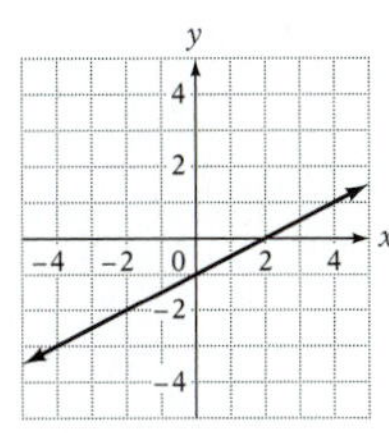 [12.3C]

14. 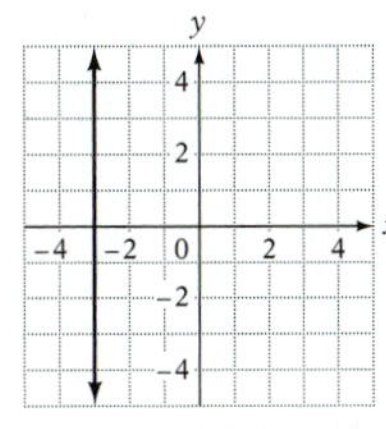 [12.2B]

15. 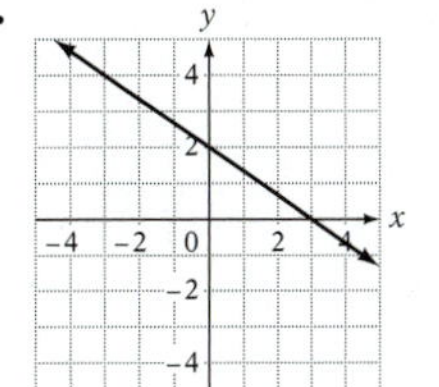 [12.3C]

16. 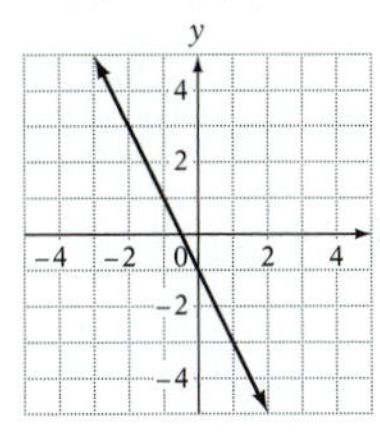 [12.2A]

17.  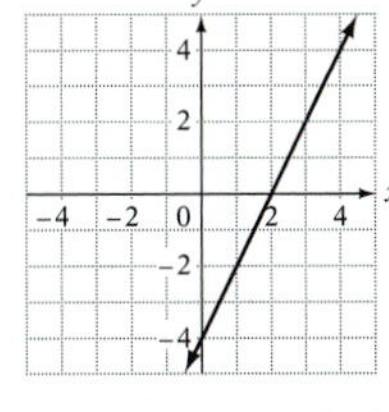 [12.3C]

18. 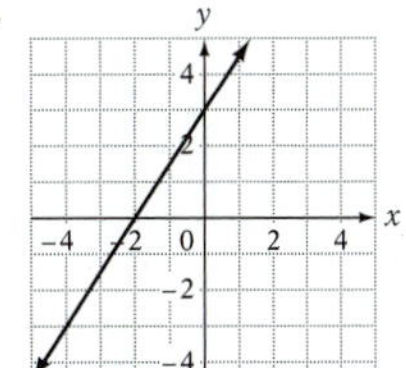 [12.2B]

19. {(55, 95), (57, 101), (53, 94), (57, 98), (60, 100), (61, 105), (58, 97), (54, 95)}; No [12.1C]

20. The cost of 50 min of access time for 1 month is \$97.50.

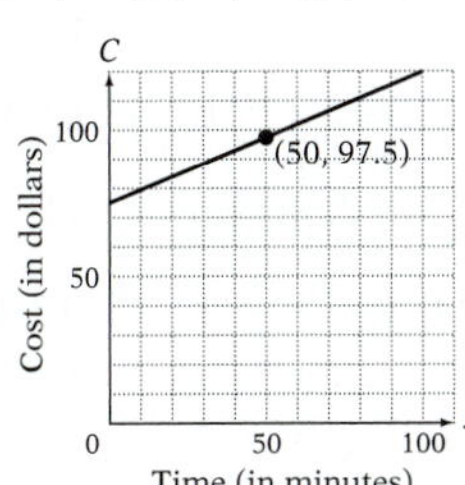

[12.2C]

21. The annual cost of telephone bills for the family increased by \$34 per year.

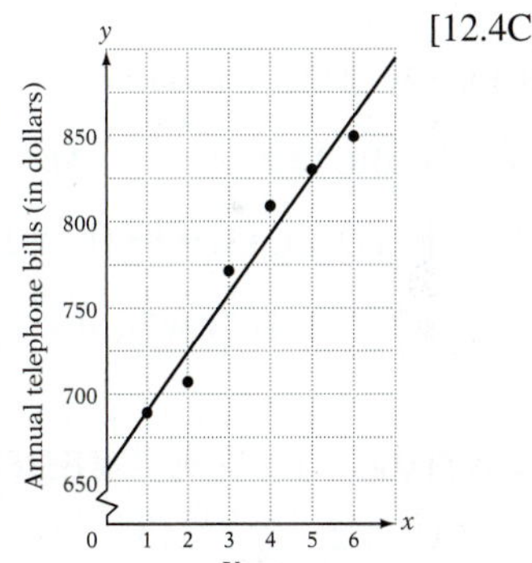

[12.4C]

CHAPTER 12 TEST

1. (3, −3) [12.1B, How To 2]

2. 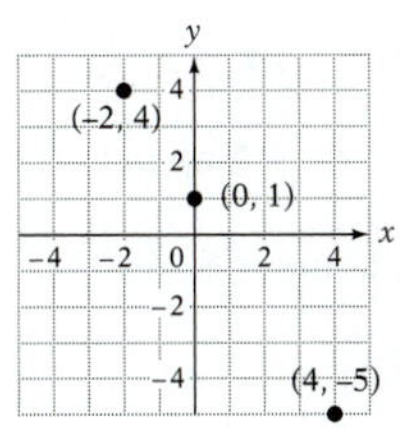

[12.1B, How To 3]

3. Yes [12.1C, Example 6] **4.** 6 [12.1D, How To 4]

5. 3 [12.1D, How To 4] **6.** {(3.5, 25), (4.0, 30), (5.2, 45), (5.0, 38), (4.0, 42), (6.3, 12), (5.4, 34)}; No [12.1C, Example 5]

7. 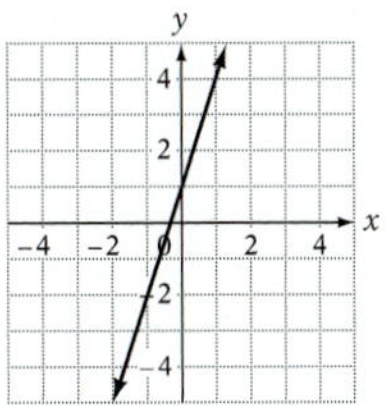 [12.2A, Example 1]

8. 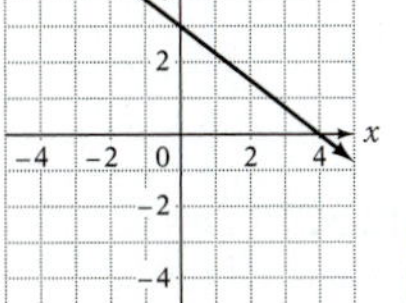 [12.2A, Example 3]

9. 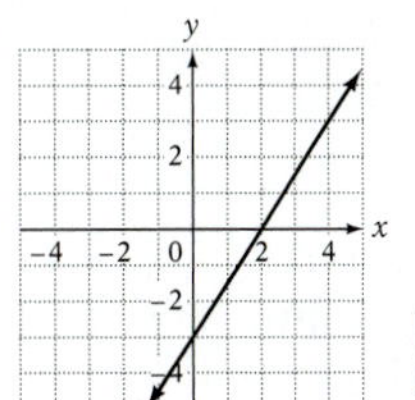 [12.2B, Example 4]

10. 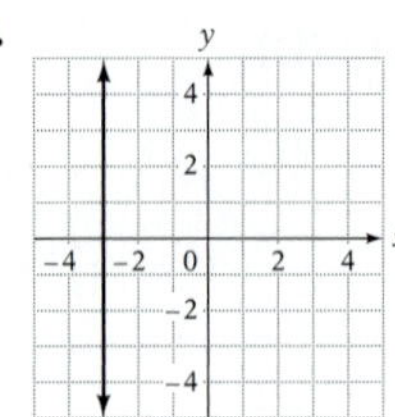 [12.2B, Example 7]

11. 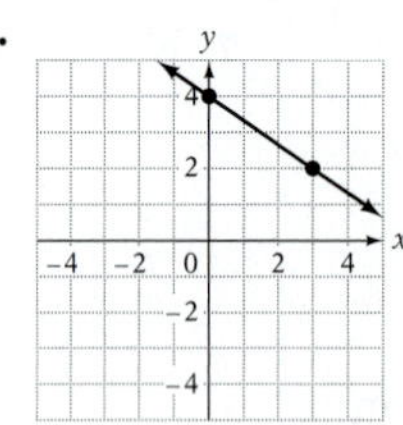 [12.3C, Example 5]

12. 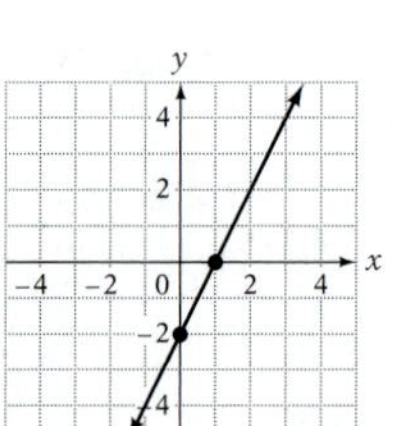 [12.3C, Example 5]

13. After 1 s, the speed of the ball is 96 ft/s. [12.2C, Example 8]

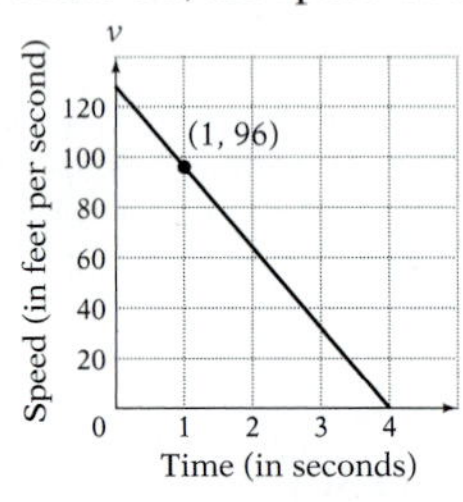

14. $m = 70$. Walking burns 70 calories per mile. [12.3B, Example 4]

15.

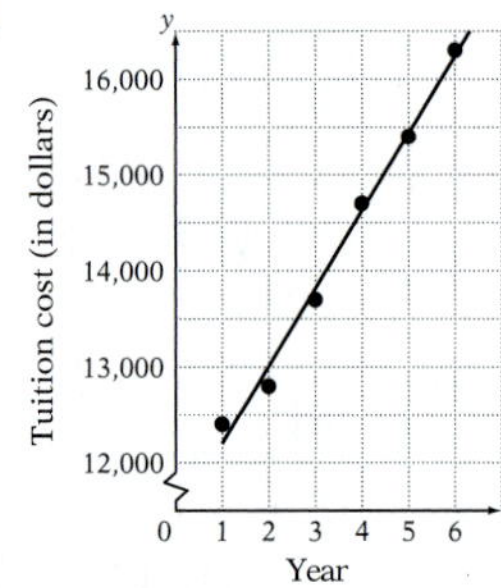

The annual tuition costs increased by $809 per year. [12.4C, Example 4]

16. (2, 0), (0, −3) [12.3A, Example 1] **17.** (−2, 0), (0, 1) [12.3A, Example 1] **18.** 2 [12.3B, Example 2]

19. Parallel [12.3B, Example 2] **20.** Undefined [12.3B, Example 3] **21.** $-\frac{2}{3}$ [12.3B, How To 2]

22. $y = 3x - 1$ [12.4A, Example 1] **23.** $y = \frac{2}{3}x + 3$ [12.4A, Example 2] **24.** $y = -\frac{5}{8}x - \frac{7}{8}$ [12.4B, How To 3]

25. $y = -\frac{2}{7}x - \frac{4}{7}$ [12.4B, How To 3]

CUMULATIVE REVIEW EXERCISES

1. −12 [3.5A] **2.** $-\frac{5}{8}$ [4.1A] **3.** $f(-2) = -\frac{2}{3}$ [12.1D] **4.** $\frac{3}{2}$ [5.2A] **5.** $\frac{19}{18}$ [5.3B] **6.** $\frac{1}{15}$ [6.3A]

7. $-32x^8y^7$ [9.2B] **8.** $-3x^2$ [9.4A] **9.** $x + 3$ [9.5B] **10.** $5(x + 2)(x + 1)$ [10.2B]

11. $(a + 2)(x + y)$ [10.1A] **12.** 4 and −2 [10.5A] **13.** $\frac{x^3(x + 3)}{y(x + 2)}$ [11.1B] **14.** $\frac{3}{x + 8}$ [11.3A] **15.** 2 [11.5A]

16. $y = \frac{4}{5}x - 3$ [11.6A] **17.** (−2, −5) [12.1B] **18.** Zero [12.3B] **19.** $y = \frac{1}{2}x - 2$ [12.4A]

20. $y = -3x + 2$ [12.4A] **21.** $y = 2x + 2$ [12.4A] **22.** $y = \frac{2}{3}x - 3$ [12.4A] **23.** The probability is $\frac{2}{3}$. [8.3A]

24. The angles measure 46°, 43°, and 91°. [7.1C] **25.** The value of the home is $1,100,000. [6.2B]

26. It would take $3\frac{3}{4}$ h for both, working together, to wire the garage. [11.7A]

27. 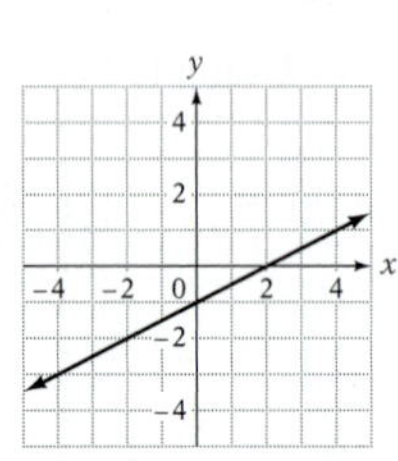 [12.2A]

28. 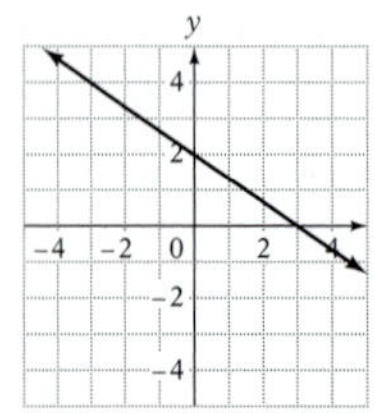 [12.3C]

ANSWERS TO CHAPTER 13 SELECTED EXERCISES

PREP TEST

1. $y = \frac{3}{4}x - 6$ [11.6A] **2.** 1000 [5.3B] **3.** $33y$ [4.2D] **4.** $10x - 10$ [4.2D] **5.** Yes [12.1B]

6. $(4, 0), (0, -3)$ [12.3A] **7.** Yes [12.3B] **8.** 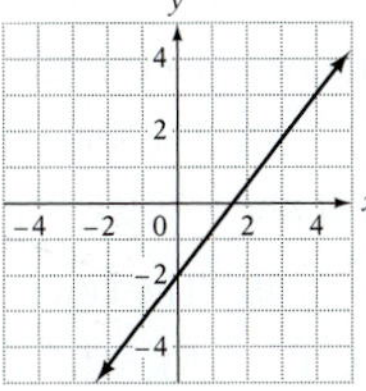 [12.3C]

9. The hikers will be side-by-side 1.5 h after the second hiker starts. [5.5B]

SECTION 13.1

1. Yes **3.** No **5.** Yes **7.** Yes **9.** I: c; II: a; III: b **11.** $(2, -1)$

13. The ordered-pair solutions of $y = -\frac{3}{2}x + 1$ **15.** No solution **17.** $(-2, 4)$

19. No solution **21.**  **23.**  **25.** 

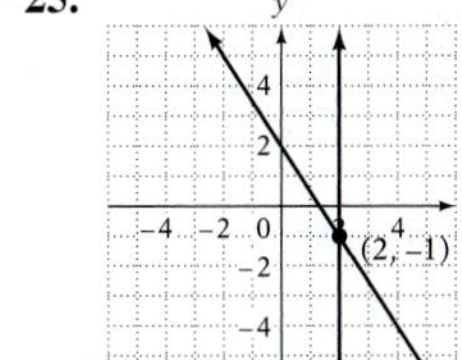

27. 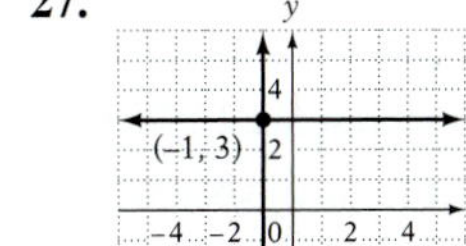 **29.**  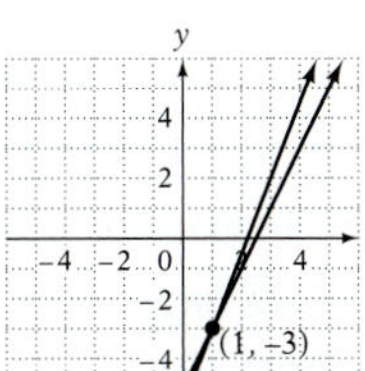 **31.** 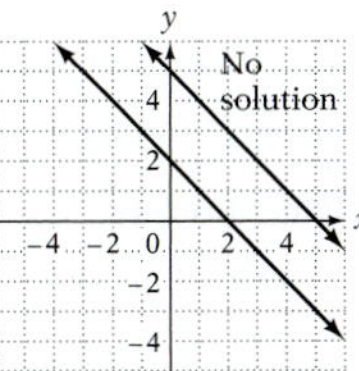

33. 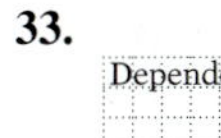 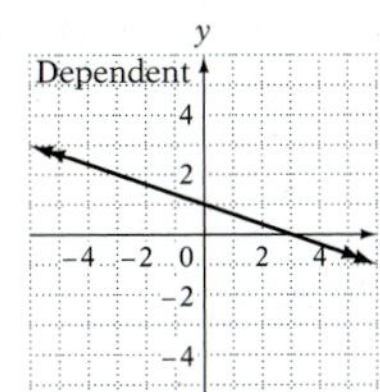 The ordered-pair solutions of $y = -\frac{1}{3}x + 1$ **35.**  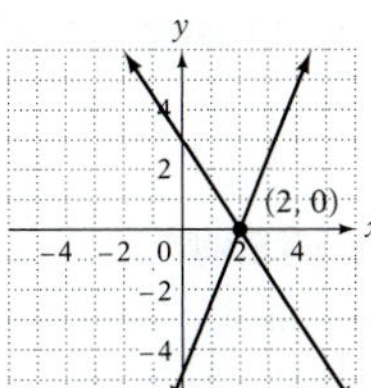 **37.** 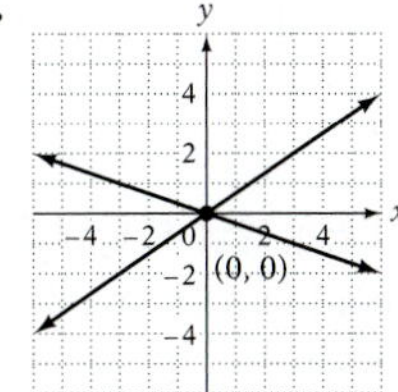

39. 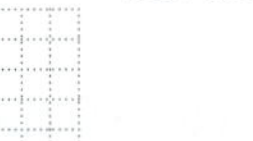 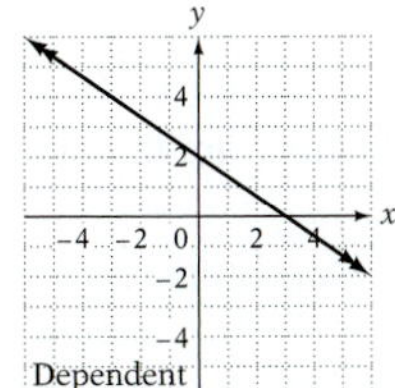 The ordered-pair solutions of $4x + 6y = 12$  **41.** Independent **43.** Answers will vary.

45. Answers will vary.

SECTION 13.2

3. (2, 1) **5.** (4, 1) **7.** (−1, 1) **9.** No solution **11.** No solution **13.** $\left(-\frac{3}{4}, -\frac{3}{4}\right)$ **15.** (1, 1) **17.** (2, 0) **19.** (1, −2) **21.** (0, 0) **23.** Dependent. The solutions satisfy the equation $2x - y = 2$. **25.** (−4, −2) **27.** (10, 31) **29.** (3, −10) **31.** (−22, −5) **33.** Dependent **35.** x = amount invested at 8%, y = amount invested at 6.5%; $x + y = 10{,}000$ represents the fact that the sum of the two investments is \$10,000; $0.08x + 0.065y = 710$ represents the fact that the total interest earned by the two investments is \$710. **37.** The amounts invested should be \$1900 at 5% and \$1600 at 7.5%. **39.** The amounts invested should be \$2400 at 9% and \$3600 at 6%. **41.** The amounts invested should be \$4400 at 8% and \$1600 at 11%. **43.** The amount invested at 6.5% was \$21,000. **45.** The amounts invested were \$12,000 at 8% and \$8000 at 7%. **47.** The amount invested in the trust deed was \$3750. **49.** 1 **51.** The assertion is not correct. The system of equations is independent. The solution is (0, 2). **53.** Simple interest: \$400; compounded monthly: \$415.00; compounded daily: \$416.39

SECTION 13.3

1. (5, −1) **3.** (1, 3) **5.** (1, 1) **7.** (3, −2) **9.** Dependent. The solutions satisfy the equation $2x - y = 1$. **11.** (3, 1) **13.** Dependent. The solutions satisfy the equation $2x - 3y = 1$. **15.** $\left(\frac{2}{3}, \frac{1}{2}\right)$ **17.** (2, 0) **19.** (0, 0) **21.** (5, −2) **23.** $\left(\frac{32}{19}, -\frac{9}{19}\right)$ **25.** $\left(\frac{7}{4}, -\frac{5}{16}\right)$ **27.** (1, −1) **29.** No solution **31.** (3, 1) **33.** (−1, 2) **35.** (1, 1) **37.** Dependent **39.** Independent **41.** $A = 3$; $B = -1$ **43. a.** 1 **b.** $\frac{3}{2}$ **c.** 4

SECTION 13.4

1. m is greater than n. **3.** The rate rowing in calm water was 14 km/h. The rate of the current was 6 km/h. **5.** The rate of the whale in calm water was 35 mph. The rate of the current was 5 mph. **7.** The rate of the Learjet was 525 mph. The rate of the wind was 35 mph. **9.** The rate of the helicopter in calm air was 225 mph. The rate of the wind was 45 mph. **11.** The rate of the boat in calm water was 7 mph. The rate of the current was 3 mph. **13.** x = cost of an adult ticket, y = cost of a child ticket; $4x + 2y = 320$ represents the fact that you spent \$320 on four adult tickets and two child tickets; $2x + 3y = 240$ represents the fact that your neighbor spent \$240 on two adult tickets and three child tickets. **15.** The cost per pound of the wheat flour was \$.65. The cost per pound of the rye flour was \$.70. **17.** The delicatessen charges \$6.25 for a turkey sandwich and \$1.90 for an order of fries. **19.** The pastry chef used 15 oz of the 20% solution and 35 oz of the 40% solution. **21.** Both formulas give the same ideal body weight at 72 in. **23. a.** The original postage value of the Lincoln stamp was \$.90. The original postage value of the Jefferson stamp was \$.10. **b.** The original postage value of the Henry Clay stamp was \$.12. **25.** It is impossible to earn \$600 in interest.

CHAPTER 13 CONCEPT REVIEW*

1. To determine the solution after graphing a system of linear equations, find the point of intersection of the lines. The point of intersection is the ordered-pair solution. [13.1A]
2. An independent system of equations has exactly one solution; the graphs of the equations intersect at one point. A dependent system has an infinite number of solutions; the graphs of the equations are the same line. [13.1A]
3. The graph of an inconsistent system of equations is the graph of two parallel lines. [13.1A]
4. The graph of a dependent system of equations looks like the graph of one line; because the graphs of the equations are the same line, one graph lies directly on top of the other. [13.1A]
5. To solve a system of linear equations by the substitution method, solve one of the equations in the system for one of its variables. Suppose you have solved one equation for y. Substitute the expression for y into the other equation. You now have an equation with only one variable in it. Let's suppose the variable is x. Solve this equation for x. This is the first coordinate of the ordered-pair solution of the system of equations. Substitute the value of x into the equation that has been solved for y. Evaluate the numerical expression. This is the second coordinate of the ordered-pair solution of the system of equations. [13.2A]
6. To solve a simple interest problem, use the formula $Pr = I$, where P is the principal, r is the simple interest rate, and I is the simple interest. [13.2B]

**Note:* The numbers in brackets following the answers in the Concept Review are a reference to the objective that corresponds to that problem. For example, the reference [1.2A] stands for Section 1.2, Objective A. This notation will be used for all Prep Tests, Concept Reviews, Chapter Reviews, Chapter Tests, and Cumulative Reviews throughout the text.

7. To solve a system of linear equations by the addition method, rewrite one or both of the equations in the system so that the coefficients of one of the variables are opposites. Suppose that the coefficients of the y terms are opposites. Add the two equations. The result is an equation with no y term; solve it for x. The solution of this equation is the first coordinate of the ordered-pair solution of the system of equations. Substitute the value of x into either of the original equations in the system, and solve the resulting equation for y. This is the second coordinate of the ordered-pair solution of the system of equations. [13.3A]

8. When using the addition method, after adding the two equations in the system of equations, a true equation that contains no variable, such as $0 = 0$, tells you that the system of equations is dependent. [13.3A]

9. In a rate-of-wind problem, the expression $p + w$ represents the rate of the plane flying with the wind. The expression $p - w$ represents the rate of the plane flying against the wind. [13.4A]

10. If we have only one equation in two variables, we cannot solve that equation for the values of both variables. In application problems with two variables, we need to write two equations. When we have two equations—a system of equations—we can solve the system of equations for the value of each of the two variables. [13.4A]

CHAPTER 13 REVIEW EXERCISES

1. Yes [13.1A] **2.** No [13.1A] **3.**

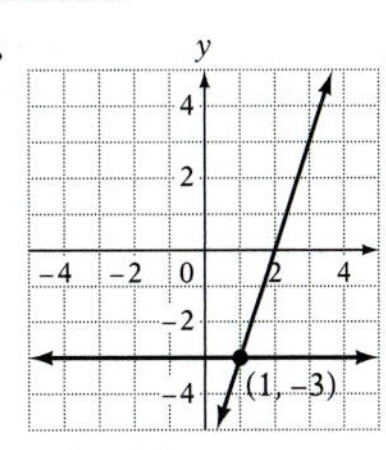

[13.1A]

4.

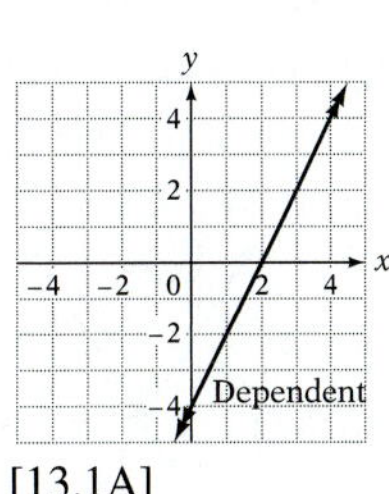

[13.1A] The solutions are the ordered-pair solutions of $y = 2x - 4$.

5. No solution [13.1A] **6.** $(-1, 1)$ [13.2A] **7.** $(1, 6)$ [13.2A] **8.** $(-3, 1)$ [13.3A] **9.** $\left(-\frac{5}{6}, \frac{1}{2}\right)$ [13.3A]

10. No solution [13.2A] **11.** $(1, 6)$ [13.2A] **12.** $(1, -5)$ [13.3A] **13.** No solution [13.3A] **14.** Dependent. The solutions satisfy the equation $y = -\frac{4}{3}x + 4$. [13.2A] **15.** $(-1, -3)$ [13.2A] **16.** Dependent. The solutions satisfy the equation $3x + y = -2$. [13.3A] **17.** $\left(\frac{2}{3}, -\frac{1}{6}\right)$ [13.3A] **18.** The rate of the sculling team in calm water was 9 mph. The rate of the current was 3 mph. [13.4A] **19.** 1300 \$6 shares were purchased, and 200 \$25 shares were purchased. [13.4B] **20.** The rate of the flight crew in calm air was 125 km/h. The rate of the wind was 15 km/h. [13.4A] **21.** The rate of the plane in calm air was 105 mph. The rate of the wind was 15 mph. [13.4A] **22.** The service charge per hour for regular service is \$1.00. The service charge per hour for premium service is \$2.50. [13.4B] **23.** The amounts invested are \$7000 at 7% and \$5000 at 8.5%. [13.2B] **24.** There were originally 350 bushels of lentils and 200 bushels of corn in the silo. [13.4B] **25.** The amounts invested were \$165,000 at 5.4% and \$135,000 at 6.6%. [13.2B]

CHAPTER 13 TEST

1. Yes [13.1A, Example 1] **2.** Yes [13.1A, Example 1] **3.**

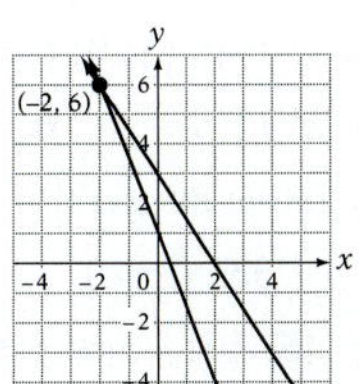

[13.1A, Example 2]

4. $(3, 1)$ [13.2A, Example 2]

5. $(1, -1)$ [13.2A, Example 2] **6.** $(2, -1)$ [13.2A, Example 1] **7.** $\left(\frac{22}{7}, -\frac{5}{7}\right)$ [13.2A, Example 1]

8. No solution [13.2A, Example 2] **9.** $(2, 1)$ [13.3A, Example 1] **10.** $\left(\frac{1}{2}, -1\right)$ [13.3A, Example 1]

11. Dependent. The solutions satisfy the equation $x + 2y = 8$. [13.3A, Example 2] **12.** $(2, -1)$ [13.3A, Example 3]

13. (1, −2) [13.3A, Example 3] **14.** The rate of the plane in calm air is 100 mph. The rate of the wind is 20 mph. [13.4A, Example 1] **15.** The price of a reserved-seat ticket was \$10. The price of a general-admission ticket was \$6. [13.4B, Example 2] **16.** The amounts invested were \$15,200 at 6.4% and \$12,800 at 7.6%. [13.2B, You Try It 4]

CUMULATIVE REVIEW EXERCISES

1. $\frac{3}{2}$ [4.1A] **2.** $-\frac{3}{2}$ [5.1C] **3.** 7 [12.1D] **4.** $-6a^3 + 13a^2 - 9a + 2$ [9.3B] **5.** $-2x^5y^2$ [9.4A] **6.** $2b - 1 + \frac{1}{2b - 3}$ [9.5B] **7.** $-\frac{4y}{x^3}$ [9.4A] **8.** $4y^2(xy - 4)(xy + 4)$ [10.4B] **9.** 4, −1 [10.5A] **10.** $x - 2$ [11.1C] **11.** $\frac{x^2 + 2}{(x + 2)(x - 1)}$ [11.3B] **12.** $\frac{x - 3}{x + 1}$ [11.4A] **13.** $-\frac{1}{5}$ [11.5A] **14.** $r = \frac{A - P}{Pt}$ [11.6A] **15.** x-intercept: (6, 0); y-intercept: (0, −4) [12.3A] **16.** $-\frac{7}{5}$ [12.3B] **17.** $y = -\frac{3}{2}x$ [12.4A] **18.** Yes [13.1A] **19.** (−6, 1) [13.2A] **20.** (4, −3) [13.3A] **21.** The amounts invested should be \$3750 at 9.6% and \$5000 at 7.2%. [13.2B] **22.** The rate of the passenger train is 56 mph. The rate of the freight train is 48 mph. [5.5B] **23.** The length of a side of the original square is 8 in. [10.5B] **24.** The rate of the wind is 30 mph. [13.4A]

25.

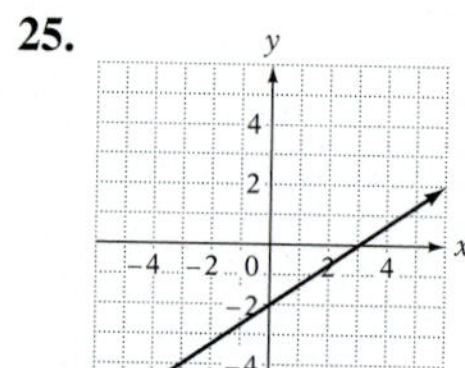

[12.2B]

26.

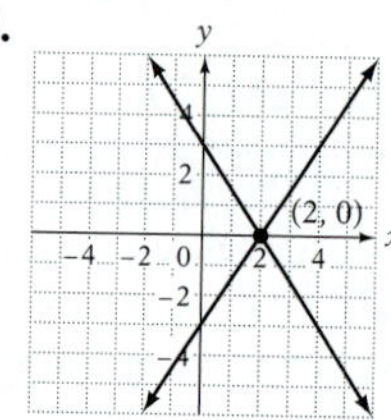

[13.1A]

27. The rate of the motorboat in calm water is 14 mph. [13.4A] **28.** 76% of the voting-age population was registered to vote. [6.4C]

ANSWERS TO CHAPTER 14 SELECTED EXERCISES

PREP TEST

1. < [3.1A] **2.** $-7x + 15$ [4.2D] **3.** The same number can be added to each side of an equation without changing the solution of the equation. [5.1B] **4.** Each side of an equation can be multiplied by the same nonzero number without changing the solution of the equation. [5.1C] **5.** There are 0.45 lb of fat in 3 lb of this grade of hamburger. [6.4C] **6.** $-\frac{1}{2}$ [5.2A] 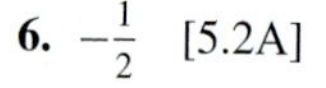 **7.** $-\frac{8}{3}$ [5.2A] **8.** 2 [5.3B] **9.**

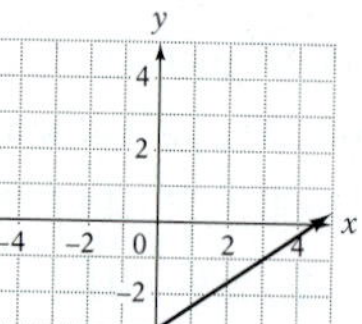

[12.2A]

10.

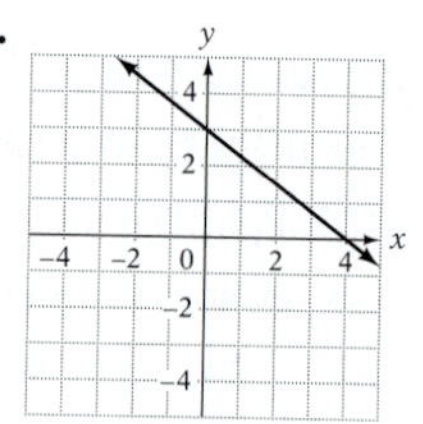

[12.2B]

SECTION 14.1

3. $A = \{16, 17, 18, 19, 20, 21\}$ **5.** $A = \{9, 11, 13, 15, 17\}$ **7.** $A = \{b, c\}$ **9.** $A \cup B = \{3, 4, 5, 6\}$ **11.** $A \cup B = \{-10, -9, -8, 8, 9, 10\}$ **13.** $A \cup B = \{a, b, c, d, e, f\}$ **15.** $A \cup B = \{1, 3, 7, 9, 11, 13\}$ **17.** $A \cap B = \{4, 5\}$ **19.** $A \cap B = \varnothing$ **21.** $A \cap B = \{c, d, e\}$ **23.** Answers may vary. For example, $A = \{1, 2, 3\}$ and $B = \{1, 2, 4, 5\}$. **25.** $\{x \mid x > -5, x \in \text{negative integers}\}$ **27.** $\{x \mid x > 30, x \in \text{integers}\}$ **29.** $\{x \mid x > 8, x \in \text{real numbers}\}$ **31.** (1, 2) **33.** (3, ∞) **35.** [−4, 5) **37.** (−∞, 2] **39.** [−3, 1] **41.** $\{x \mid -5 < x < -3\}$ **43.** $\{x \mid x \le -2\}$ **45.** $\{x \mid -3 \le x \le -2\}$ **47.** $\{x \mid x \le 6\}$

49. −5 −4 −3 −2 −1 0 1 2 3 4 5 **51.** −5 −4 −3 −2 −1 0 1 2 3 4 5 **53.** −5 −4 −3 −2 −1 0 1 2 3 4 5

55. −5 −4 −3 −2 −1 0 1 2 3 4 5 **57.** −5 −4 −3 −2 −1 0 1 2 3 4 5 **59.** −5 −4 −3 −2 −1 0 1 2 3 4 5

61. −5 −4 −3 −2 −1 0 1 2 3 4 5 **63.** −5 −4 −3 −2 −1 0 1 2 3 4 5 **65.** None **67.** $m \ge 250$

SECTION 14.2

1. $\{x \mid x < 2\}$ **3.** $\{x \mid x > 3\}$

5. $\{n \mid n \geq 3\}$ **7.** $\{x \mid x \leq -4\}$ **9.** $[-9, \infty)$

11. $(-\infty, 12)$ **13.** $[5, \infty)$ **15.** $(-\infty, -11)$ **17.** $(-\infty, 10]$ **19.** $[-6, \infty)$ **21.** $\{x \mid x > 2\}$ **23.** $\left\{d \mid d < -\frac{1}{6}\right\}$

25. $\left\{x \mid x \geq -\frac{31}{24}\right\}$ **27.** $\left\{x \mid x < \frac{5}{8}\right\}$ **29.** $\left\{x \mid x < \frac{5}{4}\right\}$ **31.** $\left\{x \mid x > \frac{5}{24}\right\}$ **33.** $\{x \mid x < -3.8\}$ **35.** $\{x \mid x \leq -1.2\}$

37. $\{x \mid x < 5.6\}$ **39.** Negative **41.** Negative and positive **43.** $\{x \mid x < 4\}$

45. $\{y \mid y \geq 3\}$ **47.** $\{x \mid x \leq 1\}$

49. $\{x \mid x < -1\}$ **51.** $\{b \mid b < -4\}$

53. $(-\infty, 0]$ **55.** $\left(\frac{2}{7}, \infty\right)$ **57.** $\left(-\infty, -\frac{5}{2}\right]$ **59.** $(-4, \infty)$ **61.** $(-2, \infty)$ **63.** $(-\infty, 18)$ **65.** $[6, \infty)$

67. $(-\infty, 6]$ **69.** $\{b \mid b \leq 33\}$ **71.** $\left\{n \mid n < \frac{3}{4}\right\}$ **73.** $\{x \mid x \leq 0\}$ **75.** $\left\{x \mid x > \frac{10}{7}\right\}$ **77.** $\left\{y \mid y \leq \frac{5}{6}\right\}$

79. $\{x \mid x \leq 4.2\}$ **81.** $\{d \mid d < -2.1\}$ **83.** $\{m \mid m > -8\}$ **85.** (ii) **87.** (iii) **89.** The couple's monthly household income is \$5394 or less. **91.** The organization must collect more than 440 lb of cans on the fourth drive.
93. The student must receive a grade of 78 or higher. **95.** The height of the tallest wind turbine you can install is 45 ft.
97. $\{c \mid c > 0\}$ **99.** $\{c \mid c \in \text{real numbers}\}$ **101.** $\{c \mid c > 0\}$

SECTION 14.3

1. $(-\infty, 4)$ **3.** $(-\infty, -4)$ **5.** $[1, \infty)$ **7.** $(-\infty, 5)$ **9.** $(-\infty, 0)$ **11.** $\{x \mid x < 20\}$ **13.** $\left\{y \mid y \leq \frac{5}{2}\right\}$

15. $\left\{x \mid x < \frac{25}{11}\right\}$ **17.** $\left\{n \mid n \leq \frac{11}{18}\right\}$ **19.** $\{x \mid x \geq 6\}$ **21.** (iii) and (iv) **23.** In 1 month, the agent expects to make sales totaling \$20,000 or less. **25.** The company must have 8 or more individual computers for the site license to be more economical.
27. The amount of artificial flavors that can be added is 8 oz or less. **29.** The *Sentry* can descend for 65 min or less before stopping again.

SECTION 14.4

1.
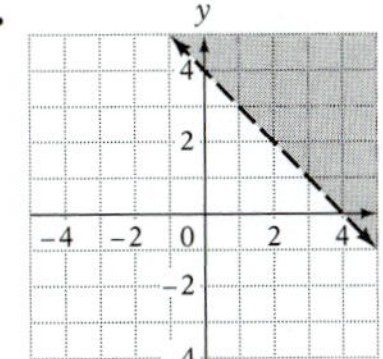

3.
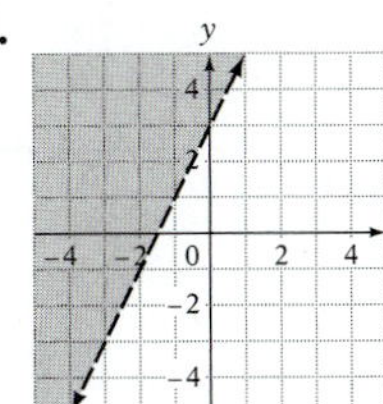

5.
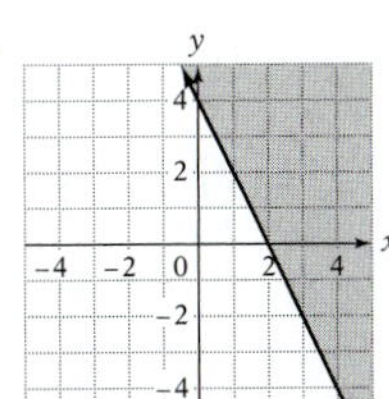

7.
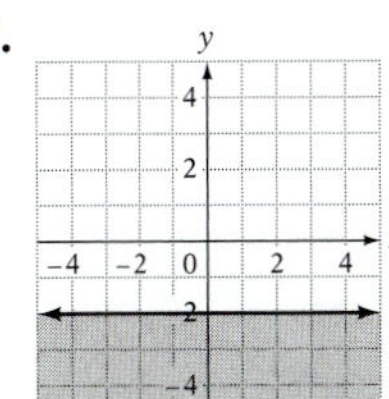

9.
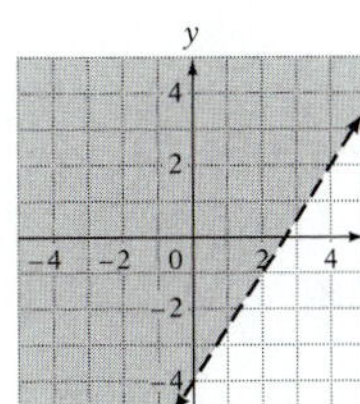

11.
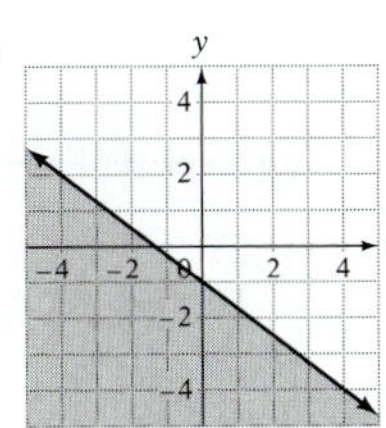

13. Negative

15.
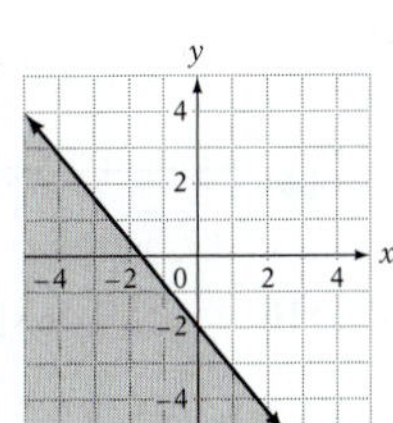

17.
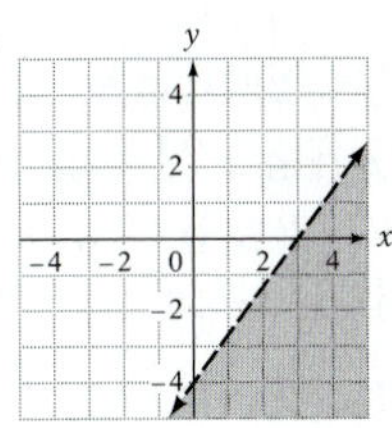

19.
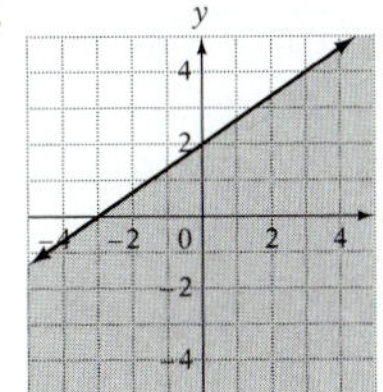

21.
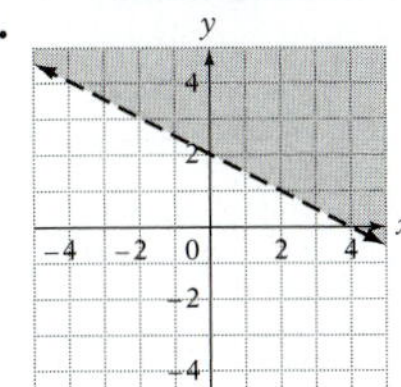

23.
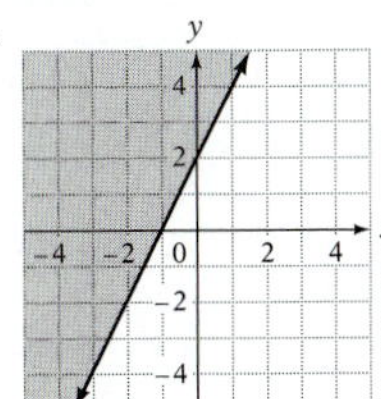

25. $x \leq 3$

CHAPTER 14 CONCEPT REVIEW*

1. The empty set or null set is represented by ∅ or { }. [14.1A]

2. $A \cup B$ is the union of two sets A and B. $A \cap B$ is the intersection of two sets A and B. [14.1A]

3. The roster method encloses a list of the elements of a set in braces; for example, {1, 2, 3, 4, 5}. Set-builder notation uses a rule to describe the elements of the set; for example, $\{x \mid x > 4, x \in \text{real numbers}\}$. [14.1A, 14.1B]

4. The solution set of an inequality is represented on a number line by shading the line over all the numbers in the solution set. At the start or end of the shading, a parenthesis indicates that a number is not included in the solution set, and a bracket indicates that a number is included in the solution set. An arrowhead indicates that the shading goes on forever in the direction of the arrow. [14.1B]

5. Yes, the same term can be added to each side of an inequality without changing the solution set of the inequality. [14.2A]

6. Each side of an inequality can be multiplied by the same *positive* number without changing the solution set of the inequality. [14.2B]

7. The Multiplication Property of Equations states that both sides of an equation can be multiplied by the same nonzero number without changing the solution of the equation. The Multiplication Property of Inequalities consists of two rules: (1) Each side of an inequality can be multiplied by the same positive number without changing the solution set; (2) If each side of an inequality is multiplied by the same negative number, the inequality symbol must be reversed in order to keep the solution set of the inequality unchanged. [14.2B]

8. Solving a general first-degree inequality is the same as solving a general first-degree equation except that at the last step, if you multiply or divide by a negative number, the inequality symbol must be reversed. [14.3A]

9. Graphing a linear inequality in two variables differs from graphing a linear equation in two variables because (1) the line being graphed in a linear inequality can be either solid or dashed, and (2) one side of the graphed line must be shaded when graphing a linear inequality. [14.4A]

10. When graphing an inequality in two variables, a dashed line is used when the inequality is > or <. [14.4A]

CHAPTER 14 REVIEW EXERCISES

1. $\{x \mid x > 18\}$ [14.2A] **2.** $A \cap B = \varnothing$ [14.1A] **3.** $\{x \mid x > -8, x \in \text{real numbers}\}$ [14.1B]

4. $A \cup B = \{2, 4, 6, 8, 10\}$ [14.1A] **5.** $A = \{1, 3, 5, 7\}$ [14.1A] **6.** $[4, \infty)$ [14.3A]

7. −5 −4 −3 −2 −1 0 1 2 3 4 5 [14.1B] **8.** $\{x \mid x \geq -4\}$ [14.3A] **9.**

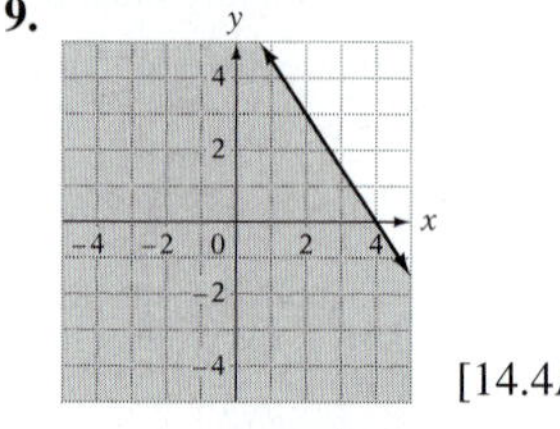

[14.4A]

10.

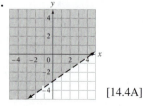

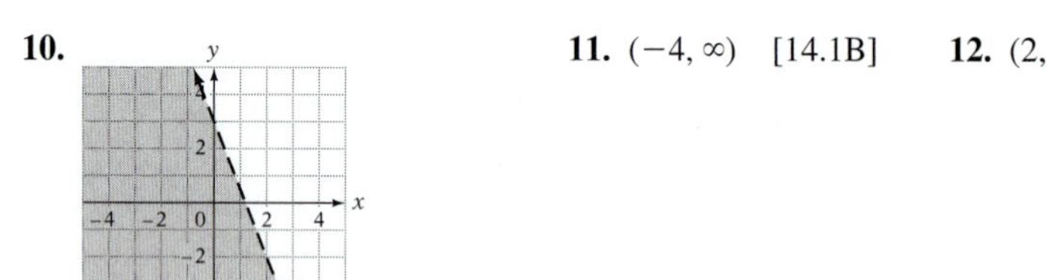

[14.4A]

11. $(-4, \infty)$ [14.1B] **12.** $(2, \infty)$ −5 −4 −3 −2 −1 0 1 2 3 4 5 [14.2A]

13. $A \cap B = \{1, 5, 9\}$ [14.1A] **14.** −5 −4 −3 −2 −1 0 1 2 3 4 5 [14.1B] **15.** −5 −4 −3 −2 −1 0 1 2 3 4 5 [14.1B]

16. $\{x \mid x \geq -3\}$ [14.2B] **17.** $(-18, \infty)$ [14.3A] **18.** $\left\{x \mid x < \frac{1}{2}\right\}$ [14.3A] **19.** $\left\{x \mid x < -\frac{8}{9}\right\}$ [14.2B]

20. $[4, \infty)$ [14.3A] **21.** y 4 2 x −4 −2 0 2 4 −2 −4 [14.4A]

22. For florist B to be more economical, there must be 2 or more residents in the nursing home. [14.3B]

**Note:* The numbers in brackets following the answers in the Concept Review are a reference to the objective that corresponds to that problem. For example, the reference [1.2A] stands for Section 1.2, Objective A. This notation will be used for all Prep Tests, Concept Reviews, Chapter Reviews, Chapter Tests, and Cumulative Reviews throughout the text.

23. The minimum length is 24 ft. [14.3B] **24.** The smallest integer that satisfies the inequality is 32. [14.2C] **25.** 72 is the lowest score that the student can receive and still attain a minimum of 480 points. [14.2C]

CHAPTER 14 TEST

1. –5 –4 –3 –2 –1 0 1 2 3 4 5 [14.1B, Example 13] **2.** $\{x \mid x < 50, x \in \text{positive integers}\}$ [14.1B, How To 5]

3. $A = \{4, 6, 8\}$ [14.1A, Example 1] **4.** $(-\infty, -3]$ [14.3A, Example 2] **5.** $\left\{x \mid x > \frac{1}{8}\right\}$ [14.2A, You Try It 1]

6. –5 –4 –3 –2 –1 0 1 2 3 4 5 [14.1B, Example 11] **7.** $(-\infty, -1)$ [14.3A, Example 1]

8. $\{x \mid x > -23, x \in \text{real numbers}\}$ [14.1B, How To 6] **9.**

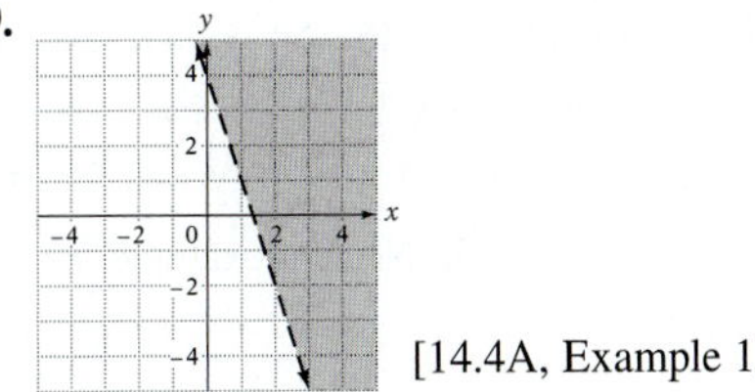

[14.4A, Example 1]

10.

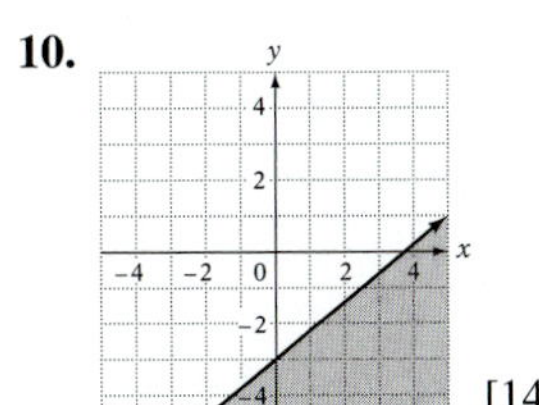

[14.4A, You Try It 2]

11. $A \cap B = \{12\}$ [14.1A, Example 4]

12. $\{x \mid x < -3\}$ –5 –4 –3 –2 –1 0 1 2 3 4 5 [14.2A, You Try It 1] **13.** $\left\{x \mid x \geq -\frac{40}{3}\right\}$ [14.2B, Example 4]

14. $\left(-\infty, -\frac{22}{7}\right)$ [14.3A, Example 2] **15.** $[3, \infty)$ –5 –4 –3 –2 –1 0 1 2 3 4 5 [14.2B, Example 4]

16. $\{x \mid x \geq -4\}$ [14.3A, Example 1] **17.** The child must grow 5 in. or more. [14.2C, Example 5] **18.** The width must be less than or equal to 11 ft. [14.3B, Example 3] **19.** The diameter must be between 0.0389 in. and 0.0395 in. [14.2C, Example 5] **20.** The total value of the stock processed by the broker was $75,000 or less. [14.3B, Example 3]

CUMULATIVE REVIEW EXERCISES

1. $40a - 28$ [4.2D] **2.** $\frac{1}{8}$ [5.2A] **3.** 4 [5.3B] **4.** $-12a^7b^4$ [9.2B] **5.** $-\frac{1}{b^4}$ [9.4A]

6. $4x - 2 - \frac{4}{4x - 1}$ [9.5B] **7.** 0 [12.1D] **8.** $3a^2(3x + 1)(3x - 1)$ [10.4B] **9.** $\frac{1}{x + 2}$ [11.1C]

10. $\frac{18a}{(2a - 3)(a + 3)}$ [11.3B] **11.** $-\frac{5}{9}$ [11.5A] **12.** $C = S + Rt$ [11.6A] **13.** $-\frac{7}{3}$ [12.3B]

14. $y = -\frac{3}{2}x - \frac{3}{2}$ [12.4A] **15.** (4, 1) [13.2A] **16.** (1, –4) [13.3A] **17.** $A \cup B = \{-10, -2, 0, 1, 2\}$ [14.1A]

18. $\{x \mid x < 48, x \in \text{real numbers}\}$ [14.1B] **19.** $(-\infty, 4)$ [14.1B] **20.** –5 –4 –3 –2 –1 0 1 2 3 4 5 [14.2B]

21. $x < -15$ [14.2B] **22.** $x > 2$ [14.3A] **23.** $\{x \mid x \leq -26, x \in \text{integers}\}$ [14.2C]

24. The maximum number of miles is 359 mi. [14.3B] **25.** There are an estimated 5000 fish in the lake. [6.2B]

26. The angle measures are 65°, 35°, and 80°. [7.1C] **27.**

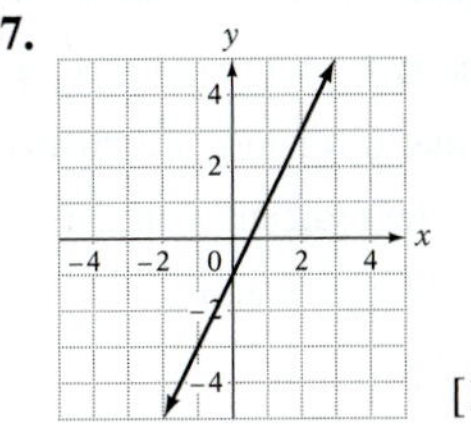

[12.2A]

28. [14.4A]

ANSWERS TO CHAPTER 15 SELECTED EXERCISES

PREP TEST

1. −14 [3.1C] **2.** $-2x^2y - 4xy^2$ [4.2A] **3.** 14 [5.1C] **4.** $\frac{7}{5}$ [5.3A] **5.** x^6 [9.2A]
6. $x^2 + 2xy + y^2$ [9.3D] **7.** $4x^2 - 12x + 9$ [9.3D] **8.** $4 - 9v^2$ [9.3D] **9.** $a^2 - 25$ [9.3D] **10.** $\frac{x^2y^2}{9}$ [9.4A]

SECTION 15.1

3. 4 **5.** 7 **7.** $4\sqrt{2}$ **9.** $2\sqrt{2}$ **11.** $-18\sqrt{2}$ **13.** $10\sqrt{10}$ **15.** $\sqrt{15}$ **17.** $\sqrt{29}$ **19.** $-54\sqrt{2}$ **21.** $3\sqrt{5}$ **23.** 0 **25.** $48\sqrt{2}$ **27.** −11 and −10 **29.** 2 and 3 **31.** 15.492 **33.** 16.971 **35.** 18.708 **37.** x^7 **39.** $y^7\sqrt{y}$ **41.** a^{10} **43.** x^2y^2 **45.** $2x^2$ **47.** $2x\sqrt{6}$ **49.** $2x^2\sqrt{15x}$ **51.** $7a^2b^4$ **53.** $3x^2y^3\sqrt{2xy}$ **55.** $2x^5y^3\sqrt{10xy}$ **57.** $4a^4b^5\sqrt{5a}$ **59.** $8ab\sqrt{b}$ **61.** x^3y **63.** $8a^2b^3\sqrt{5b}$ **65.** $6x^2y^3\sqrt{3y}$ **67.** $4x^3y\sqrt{2y}$ **69.** $5a + 20$ **71.** $2x^2 + 8x + 8$ **73.** $x + 2$ **75.** $y + 1$ **77.** Rational **79.** Irrational **81. a.** The speed of the car was $12\sqrt{15}$ mph. **b.** 46 mph **83.** No. For example, let $a = 16$ and $b = 9$. $\sqrt{a+b} = \sqrt{16+9} = \sqrt{25} = 5$. $\sqrt{a} + \sqrt{b} = \sqrt{16} + \sqrt{9} = 4 + 3 = 7$.

SECTION 15.2

1. 2, 20, and 50 **3.** $3\sqrt{2}$ **5.** $-\sqrt{7}$ **7.** $-11\sqrt{11}$ **9.** $10\sqrt{x}$ **11.** $-2\sqrt{y}$ **13.** $-11\sqrt{3b}$ **15.** $2x\sqrt{2}$ **17.** $-3a\sqrt{3a}$ **19.** $-5\sqrt{xy}$ **21.** $8\sqrt{5}$ **23.** $8\sqrt{2}$ **25.** $15\sqrt{2} - 10\sqrt{3}$ **27.** $\sqrt{x}$ **29.** $-12x\sqrt{3}$ **31.** $2xy\sqrt{x} - 3xy\sqrt{y}$ **33.** $-9x\sqrt{3x}$ **35.** $-13y^2\sqrt{2y}$ **37.** $4a^2b^2\sqrt{ab}$ **39.** $7\sqrt{2}$ **41.** $6\sqrt{x}$ **43.** $-3\sqrt{y}$ **45.** $-45\sqrt{2}$ **47.** $13\sqrt{3} - 12\sqrt{5}$ **49.** $32\sqrt{3} - 3\sqrt{11}$ **51.** $6\sqrt{x}$ **53.** $-34\sqrt{3x}$ **55.** $10a\sqrt{3b} + 10a\sqrt{5b}$ **57.** $-2xy\sqrt{3}$ **59. a.** False **b.** False **61.** $5\sqrt{2}$

SECTION 15.3

1. 5 **3.** 6 **5.** x **7.** x^3y^2 **9.** $3ab^6\sqrt{2a}$ **11.** $12a^4b\sqrt{b}$ **13.** $2 - \sqrt{6}$ **15.** $x - \sqrt{xy}$ **17.** $5\sqrt{2} - \sqrt{5x}$ **19.** $3a - 3\sqrt{ab}$ **21.** $10abc$ **23.** $x - 6\sqrt{x} + 9$ **25.** $-2 + 2\sqrt{5}$ **27.** $16 + 10\sqrt{2}$ **29.** $6x + 10\sqrt{x} - 4$ **31.** $15x - 22y\sqrt{x} + 8y^2$ **33.** 4 **35.** $9x - 16$ **37.** Less than **41.** 4 **43.** 7 **45.** 3 **47.** $x\sqrt{5}$ **49.** $\frac{a^2}{7}$ **51.** $\frac{y\sqrt{3}}{3}$ **53.** $\sqrt{2y}$ **55.** $-\frac{\sqrt{2}+3}{7}$ **57.** $\frac{15 - 3\sqrt{5}}{20}$ **59.** $-\frac{7\sqrt{2}+49}{47}$ **61.** $-\frac{42 - 26\sqrt{3}}{11}$ **63.** $\sqrt{15} + 2\sqrt{5}$ **65.** $-\frac{20 + 7\sqrt{3}}{23}$ **67.** $\frac{6 + 5\sqrt{x} + x}{4 - x}$ **69.** $\frac{x\sqrt{y} + y\sqrt{x}}{x - y}$ **71.** Equal to **73.** True **75.** False, $x + 2\sqrt{x} + 1$

SECTION 15.4

1. 25 **3.** 144 **5.** No solution **7.** 6 **9.** 24 **11.** −1 **13.** $-\frac{2}{5}$ **15.** $\frac{4}{3}$ **17.** 3 **19.** No solution **21.** 1 **23. a.** Yes **b.** Yes **c.** Yes **d.** Yes **25.** The hay was dropped from a height of 576 ft. **27.** The child is 18.75 ft from the center of the merry-go-round. **29.** The pitcher's mound is less than halfway between home plate and second base. **31. a.** A first-year student has an average of 2.3 credit cards. **b.** A sophomore has an average of 3.3 credit cards. **c.** A junior has an average of 4.0 credit cards. **d.** A senior has an average of 4.6 credit cards. **33. a.** The perimeter is $12\sqrt{2}$ cm. **b.** The area is 12 cm^2.

CHAPTER 15 CONCEPT REVIEW*

1. The principal square root of a number is the positive square root of the number. [15.1A]
2. A square root is an irrational number when the radicand is not a perfect square. [15.1A]
3. The Product Property of Square Roots states that if a and b are positive real numbers, then $\sqrt{ab} = \sqrt{a} \cdot \sqrt{b}$. [15.1A]
4. A radical expression is in simplest form when the radicand contains no factor greater than 1 that is a perfect square. [15.1B]
5. You can add two radical expressions that have like radicands. [15.2A]
6. The conjugate of $3 + \sqrt{8}$ is $3 - \sqrt{8}$. [15.3A]

**Note:* The numbers in brackets following the answers in the Concept Review are a reference to the objective that corresponds to that problem. For example, the reference [1.2A] stands for Section 1.2, Objective A. This notation will be used for all Prep Tests, Concept Reviews, Chapter Reviews, Chapter Tests, and Cumulative Reviews throughout the text.

7. To rationalize the denominator of $\frac{5}{\sqrt{2x}}$, multiply the expression by 1 in the form of $\frac{\sqrt{2x}}{\sqrt{2x}}$. [15.3B]

8. Conjugates are used to rationalize a denominator when the denominator contains a radical expression with two terms. The numerator and denominator are multiplied by the conjugate of the denominator. [15.3B]

9. The Property of Squaring Both Sides of an Equation states that if a and b are real numbers and $a = b$, then $a^2 = b^2$. [15.4A]

10. It is important to check your solution to a radical equation because when both sides of the equation are squared, the resulting equation may have a solution that is not a solution of the original equation. [15.4A]

11. A right triangle is a triangle with a right angle, or 90° angle. [15.4B]

12. The Pythagorean Theorem states that if a and b are the lengths of the legs of a right triangle and c is the length of the hypotenuse, then $c^2 = a^2 + b^2$. [15.4B]

CHAPTER 15 REVIEW EXERCISES

1. 3 [15.3A] **2.** $9a^2\sqrt{2ab}$ [15.1B] **3.** 12 [15.1A] **4.** $3a\sqrt{2} + 2a\sqrt{3}$ [15.3A] **5.** $2\sqrt{6}$ [15.3B] **6.** $-8\sqrt{2}$ [15.2A] **7.** 2 [15.3A] **8.** 1 [15.4A] **9.** $-x\sqrt{3} - x\sqrt{5}$ [15.3B] **10.** $-6\sqrt{30}$ [15.1A] **11.** 20 [15.4A] **12.** $20\sqrt{3}$ [15.1A] **13.** $7x^2y^4$ [15.3B] **14.** No solution [15.4A] **15.** $18a\sqrt{5b} + 5a\sqrt{b}$ [15.2A] **16.** $20\sqrt{10}$ [15.1A] **17.** $7x^2y\sqrt{15xy}$ [15.2A] **18.** $8y + 10\sqrt{5y} - 15$ [15.3A] **19.** $26\sqrt{3x}$ [15.2A] **20.** No solution [15.4A] **21.** $\frac{8\sqrt{x} + 24}{x - 9}$ [15.3B] **22.** $36x^8y^5\sqrt{3xy}$ [15.1B] **23.** $2y^4\sqrt{6}$ [15.1B] **24.** -1 [15.4A] **25.** $-6x^3y^2\sqrt{2y}$ [15.2A] **26.** $\frac{16\sqrt{a}}{a}$ [15.3B] **27.** The distance across the pond is approximately 43 ft. [15.4B] **28.** The explorer weighs 144 lb on the surface of Earth. [15.4B] **29.** The depth of the water is 100 ft. [15.4B] **30.** The radius of the corner is 25 ft. [15.4B]

CHAPTER 15 TEST

1. $11x^4y$ [15.1B, Example 5] **2.** $6x^2y\sqrt{y}$ [15.3A, Example 1] **3.** $-5\sqrt{2}$ [15.2A, Example 2] **4.** $3\sqrt{5}$ [15.1A, How To 1] **5.** 9 [15.3B, How To 8] **6.** 25 [15.4A, Example 1] **7.** $4a^2b^5\sqrt{2ab}$ [15.1B, Example 5] **8.** $7ab\sqrt{a}$ [15.3B, How To 8] **9.** $\sqrt{3} + 1$ [15.3B, How To 10] **10.** $4x^2y^2\sqrt{5y}$ [15.3A, How To 1] **11.** 9 [15.4A, Example 2] **12.** $21\sqrt{2y} - 12\sqrt{2x}$ [15.2A, Example 4] **13.** $6x^3y\sqrt{2x}$ [15.1B, Example 5] **14.** $y + 2\sqrt{y} - 15$ [15.3A, How To 3] **15.** $-2xy\sqrt{3xy} - 3xy\sqrt{xy}$ [15.2A, Example 4] **16.** $\frac{17 - 8\sqrt{5}}{31}$ [15.3B, Example 7] **17.** $a - \sqrt{ab}$ [15.3A, Example 2] **18.** $5\sqrt{3}$ [15.1A, How To 1] **19.** The length of the pendulum is 7.30 ft. [15.4B, You Try It 4] **20.** The rope should be secured about 7 ft from the base of the pole. [15.4B, Example 3]

CUMULATIVE REVIEW EXERCISES

1. $-\frac{1}{12}$ [3.5A] **2.** $2x + 18$ [4.2D] **3.** $\frac{1}{13}$ [5.3B] **4.** $6x^5y^5$ [9.2A] **5.** $-2b^2 + 1 - \frac{1}{3b^2}$ [9.5A] **6.** 1 [12.1D] **7.** $2a(a - 5)(a - 3)$ [10.2B] **8.** $\frac{1}{4(x + 1)}$ [11.1B] **9.** $\frac{x + 3}{x - 3}$ [11.3B] **10.** $\frac{5}{3}$ [11.5A] **11.** $y = \frac{1}{2}x - 2$ [12.4A] **12.** (1, 1) [13.2A] **13.** (3, −2) [13.3A] **14.** $x \le -\frac{9}{2}$ [14.3A] **15.** $6\sqrt{3}$ [15.1A] **16.** $-4\sqrt{2}$ [15.2A] **17.** $4ab\sqrt{2ab} - 5ab\sqrt{ab}$ [15.2A] **18.** $14a^5b^2\sqrt{2a}$ [15.3A] **19.** $3\sqrt{2} - x\sqrt{3}$ [15.3A] **20.** 8 [15.3B] **21.** $-6 - 3\sqrt{5}$ [15.3B] **22.** 6 [15.4A] **23.** The cost of the book is \$49.50. [5.2B] **24.** The rate of the faster cyclist is 15 mph. [5.5B] **25.** The numbers are 8 and 13. [10.5B] **26.** It would take the small pipe, working alone, 48 h. [11.7A]

27.

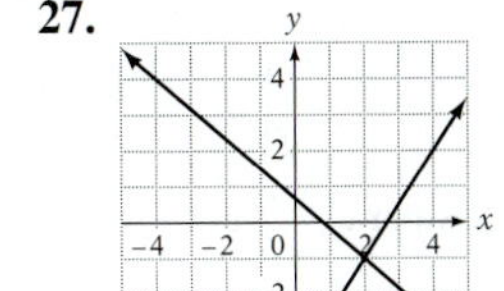

[13.1A]

28.

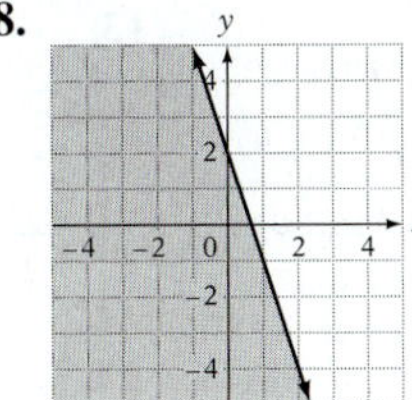

[14.4A]

29. The smaller integer is 40. [15.4B]

30. The height of the building is 400 ft. [15.4B]

ANSWERS TO CHAPTER 16 SELECTED EXERCISES

PREP TEST

1. 41 [4.1A] **2.** $-\frac{1}{5}$ [5.2A] **3.** $(x + 4)(x - 3)$ [10.2A] **4.** $(2x - 3)^2$ [10.4A] **5.** Yes [10.4A] **6.** 3 [11.5A]

7. 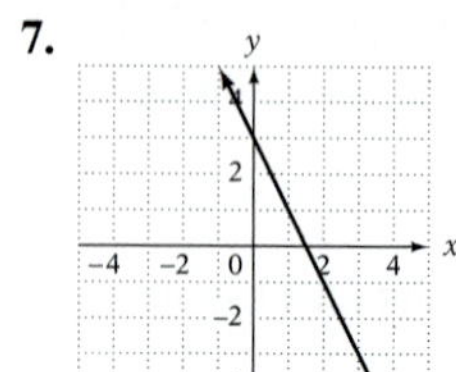

[12.2A] **8.** $2\sqrt{7}$ [15.1A] **9.** $|a|$ [15.1B] **10.** The length of the hiking trail is 3.6 mi. [5.5B]

SECTION 16.1

1. $-3, 5$ **3.** $-\frac{5}{2}, \frac{1}{3}$ **5.** $-5, 3$ **7.** $1, 3$ **9.** $-2, -1$ **11.** 3 **13.** $-\frac{2}{3}, 0$ **15.** $-2, 5$ **17.** $\frac{2}{3}, 1$
19. $-3, \frac{1}{3}$ **21.** $\frac{2}{3}$ **23.** $-\frac{1}{2}, \frac{3}{2}$ **25.** $\frac{1}{2}$ **27.** $-3, 3$ **29.** $-\frac{1}{2}, \frac{1}{2}$ **31.** $-3, 5$ **33.** $1, 5$ **35.** (iii)
37. $-7, 7$ **39.** $-8, 8$ **41.** $-\frac{8}{3}, \frac{8}{3}$ **43.** $-\frac{5}{2}, \frac{5}{2}$ **45.** $-\frac{8}{5}, \frac{8}{5}$ **47.** No real number solution **49.** $-4\sqrt{3}, 4\sqrt{3}$
51. $-9, 5$ **53.** $-2, 8$ **55.** $-\frac{15}{2}, \frac{3}{2}$ **57.** $\frac{10}{9}, \frac{26}{9}$ **59.** $-5 - 5\sqrt{2}, -5 + 5\sqrt{2}$ **61.** $\frac{1}{2} - \sqrt{6}, \frac{1}{2} + \sqrt{6}$
63. One **65.** Two **67.** The annual interest rate was 9%.

SECTION 16.2

1. $x^2 - 8x + 16, (x - 4)^2$ **3.** $x^2 + 5x + \frac{25}{4}, \left(x + \frac{5}{2}\right)^2$ **5.** $-3, 1$ **7.** $-2, 8$ **9.** 2 **11.** No real number solution
13. $-4, -1$ **15.** $-8, 1$ **17.** $-2 - \sqrt{3}, -2 + \sqrt{3}$ **19.** $-3 - \sqrt{14}, -3 + \sqrt{14}$ **21.** $1 - \sqrt{2}, 1 + \sqrt{2}$
23. $\frac{-3 - \sqrt{13}}{2}, \frac{-3 + \sqrt{13}}{2}$ **25.** $1, 2$ **27.** $\frac{-1 - \sqrt{13}}{2}, \frac{-1 + \sqrt{13}}{2}$ **29.** $-5 - 4\sqrt{2}, -5 + 4\sqrt{2}$ **31.** $\frac{3 - \sqrt{29}}{2}, \frac{3 + \sqrt{29}}{2}$
33. $\frac{1 - \sqrt{17}}{2}, \frac{1 + \sqrt{17}}{2}$ **35.** No real number solution **37.** $\frac{1}{2}, 1$ **39.** $-3, \frac{1}{2}$ **41.** $\frac{7 - \sqrt{41}}{4}, \frac{7 + \sqrt{41}}{4}$
43. $\frac{-1 - \sqrt{73}}{12}, \frac{-1 + \sqrt{73}}{12}$ **45.** $-\frac{1}{2}, \frac{4}{3}$ **47.** There is one negative and one positive solution. **49.** $-5.372, 0.372$
51. $-3.212, 1.712$ **53.** 7 **55.** $4 - \sqrt{3}, 4 + \sqrt{3}$ **57.** $0, 8$ **59.** The ball hits the basket 1.88 s after it is released.

SECTION 16.3

1. $-1, 5$ **3.** $-1, 3$ **5.** $-\frac{1}{2}, 1$ **7.** No real number solution **9.** $-\frac{5}{2}, \frac{3}{2}$ **11.** No real number solution
13. $1 - \sqrt{6}, 1 + \sqrt{6}$ **15.** $-3 - \sqrt{10}, -3 + \sqrt{10}$ **17.** $2 - \sqrt{13}, 2 + \sqrt{13}$ **19.** $\frac{-1 - \sqrt{2}}{3}, \frac{-1 + \sqrt{2}}{3}$
21. $-\frac{1}{2}$ **23.** No real number solution **25.** $\frac{-4 - \sqrt{5}}{2}, \frac{-4 + \sqrt{5}}{2}$ **27.** $\frac{1 - 2\sqrt{3}}{2}, \frac{1 + 2\sqrt{3}}{2}$ **29.** $\frac{-5 - \sqrt{2}}{3}, \frac{-5 + \sqrt{2}}{3}$
31. True **33.** $-3.690, 5.690$ **35.** $-1.690, 7.690$ **37.** $-1.089, 4.589$ **39.** $-2.118, 0.118$ **41.** $-0.905, 1.105$
43. 6 **45.** 3 **47.** $5 - \sqrt{13}, 5 + \sqrt{13}$ **49. a.** False **b.** False **c.** False **d.** True **51.** The planes are 1000 mi apart after 2 h.

SECTION 16.4

1. Down **3.** Up **5.** 4 **7.** -5 **9.** -42 **11.**

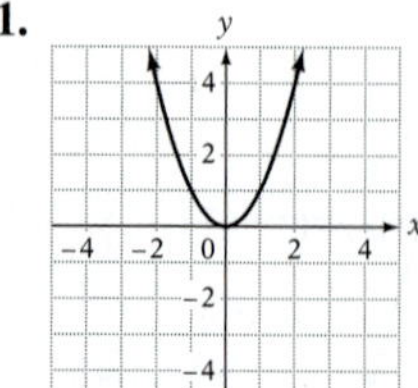

13.

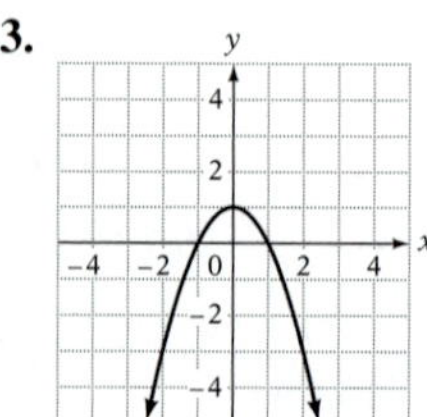

15. 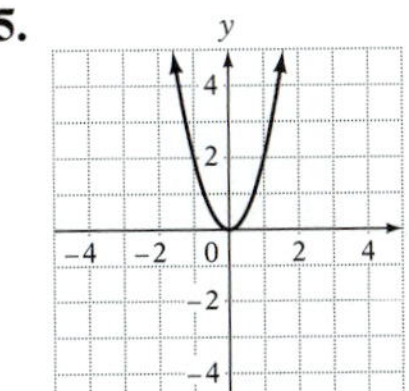

17. 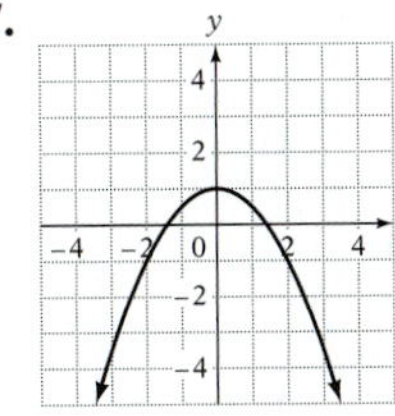

19. 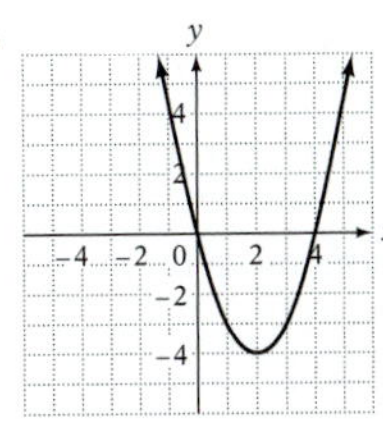

21. 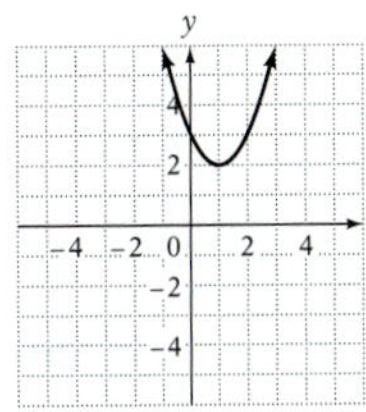

23. 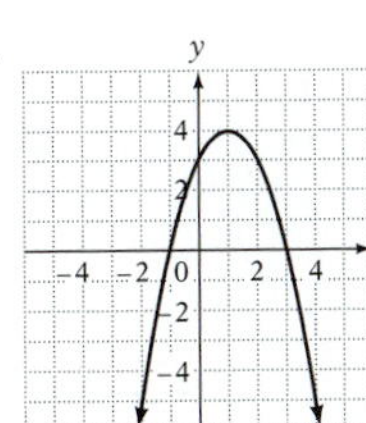

25. 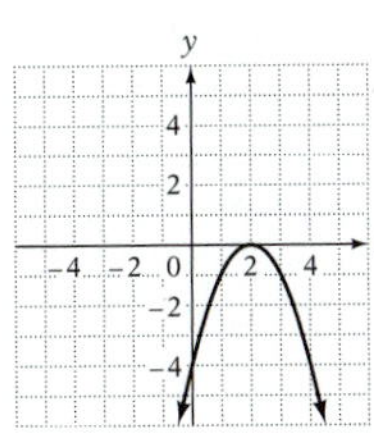

27. $(-6, 0), (1, 0); (0, -6)$ **29.** $(-6, 0); (0, 36)$

31. $(-2 - \sqrt{6}, 0), (-2 + \sqrt{6}, 0); (0, -2)$ **33.** No x-intercepts; $(0, 1)$ **35.** $\left(\frac{3}{2}, 0\right), (5, 0); (0, 15)$

37. $\left(\frac{-3 - \sqrt{33}}{6}, 0\right), \left(\frac{-3 + \sqrt{33}}{6}, 0\right); (0, 2)$ **39.** $y = x^2 - 8x + 15$ **41.** $y = 2x^2 - 12x + 22$

SECTION 16.5

1. The height is 10 m. The length is 4 m. **3.** The length was 70 ft. The width was 40 ft. **5.** The length is 5 ft. The width is 3 ft. **7.** The length is 78 ft. The width is 27 ft. **9.** The dimensions of the cardboard are 20 in. by 20 in. **11.** The radius of the large pizza is 7 in. **13.** The legs of the right triangle are 2 cm and 3 cm. **15.** True **17.** It would take the first drain 24 min. It would take the second drain 8 min. **19.** Working alone, it would take the apprentice mason 12 h and the experienced mason 4 h. **21.** The rate of the boat in calm water is 15 mph. **23.** There will be 50 million people aged 65 and older in the United States in 2017. **25.** The maximum velocity is 78 ft/s. **27.** The cost of a pizza with a diameter of 16 in. should be $40.

CHAPTER 16 CONCEPT REVIEW*

1. A second-degree equation is a quadratic equation. [16.1A]
2. When solving a quadratic equation by factoring, you know your solution is a double root when the solution of both equations is the same number. [16.1A]
3. The symbol $\pm$ means plus or minus. For example, ± 4 means plus 4 or minus 4. [16.1B]
4. To complete the square on a binomial of the form $x^2 + bx$, square the product of $\frac{1}{2}$ and b. Add this constant to the binomial. [16.2A]
5. The quadratic formula is used to solve a quadratic equation that cannot be solved by factoring. [16.3A]
6. In solving a quadratic equation using the quadratic formula, the value of a is the coefficient of x^2, the value of b is the coefficient of x, and the value of c is the constant. [16.3A]
7. The graph of a quadratic function is a parabola, which is U-shaped and opens either up or down. [16.4A]
8. To find the x-intercepts of a quadratic function, let $y = 0$ and solve for x. [16.4A]
9. To find the y-intercept of a quadratic function, let $x = 0$ and solve for y. [16.4A]
10. The graph of a quadratic function opens down when the coefficient of the x^2 term is negative. [16.4A]

CHAPTER 16 REVIEW EXERCISES

1. $-\frac{7}{2}, \frac{4}{3}$ [16.1A] **2.** $-\frac{5}{7}, \frac{5}{7}$ [16.1B] **3.** $-6, 4$ [16.2A] **4.** $-6, 1$ [16.3A] **5.** $-4, \frac{3}{2}$ [16.2A]
6. $2, \frac{5}{12}$ [16.1A] **7.** $-2 - 2\sqrt{6}, -2 + 2\sqrt{6}$ [16.1B] **8.** $1, \frac{3}{2}$ [16.3A] **9.** $-\frac{1}{2}, -\frac{1}{3}$ [16.1A]

*Note: The numbers in brackets following the answers in the Concept Review are a reference to the objective that corresponds to that problem. For example, the reference [1.2A] stands for Section 1.2, Objective A. This notation will be used for all Prep Tests, Concept Reviews, Chapter Reviews, Chapter Tests, and Cumulative Reviews throughout the text.

10. No real number solution [16.1B] **11.** $2 - \sqrt{3}, 2 + \sqrt{3}$ [16.2A] **12.** $\frac{3 - \sqrt{29}}{2}, \frac{3 + \sqrt{29}}{2}$ [16.3A]

13. No real number solution [16.2A] **14.** $-10, -7$ [16.1A] **15.** $-1, 2$ [16.1B]

16. $\frac{-4 - \sqrt{23}}{2}, \frac{-4 + \sqrt{23}}{2}$ [16.2A] **17.** No real number solution [16.3A] **18.** $-2, -\frac{1}{2}$ [16.3A]

19. [16.4A] **20.** [16.4A] **21.** [16.4A]

22.

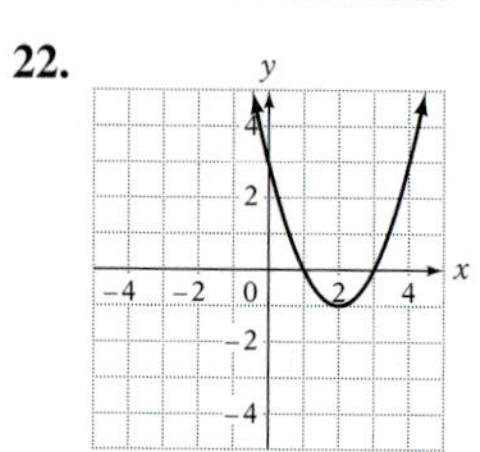

[16.4A]

23.

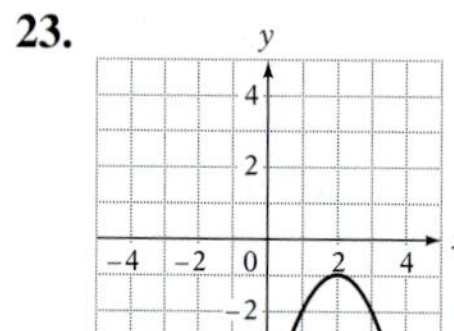

[16.4A]

24. x-intercepts: $(-3, 0), (5, 0)$; y-intercept: $(0, -15)$ [16.4A] **25.** The rate of the hawk in calm air is 75 mph. [16.5A]

CHAPTER 16 TEST

1. $-1, 6$ [16.1A, How To 1] **2.** $-4, \frac{5}{3}$ [16.1A, How To 1] **3.** $-\frac{1}{2}, 0$ [16.1A, How To 1] **4.** $-\frac{3}{2}, \frac{3}{2}$ [16.1B, How To 6] **5.** $0, 10$ [16.1B, How To 7] **6.** $-4 - 2\sqrt{5}, -4 + 2\sqrt{5}$ [16.1B, How To 7] **7.** $-2 - 2\sqrt{5}, -2 + 2\sqrt{5}$ [16.2A, How To 3] **8.** $\frac{-3 - \sqrt{41}}{2}, \frac{-3 + \sqrt{41}}{2}$ [16.2A, How To 3] **9.** $\frac{3 - \sqrt{7}}{2}, \frac{3 + \sqrt{7}}{2}$ [16.2A, Example 1] **10.** $\frac{-4 - \sqrt{22}}{2}, \frac{-4 + \sqrt{22}}{2}$ [16.2A, Example 1] **11.** $-2 - \sqrt{2}, -2 + \sqrt{2}$ [16.3A, How To 1] **12.** $\frac{3 - \sqrt{33}}{2}, \frac{3 + \sqrt{33}}{2}$ [16.3A, How To 1] **13.** $-\frac{1}{2}, 3$ [16.3A, Example 1] **14.** $\frac{1 - \sqrt{13}}{6}, \frac{1 + \sqrt{13}}{6}$ [16.3A, How To 1] **15.** $-1.651, 0.151$ [16.3A, Example 1] **16.** $-1.387, 0.721$ [16.3A, Example 1] **17.**

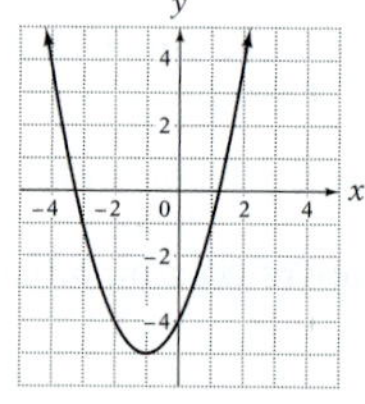

[16.4A, How To 2]

18. x-intercepts: $(-4, 0), (3, 0)$; y-intercept: $(0, -12)$ [16.4A, Example 2] **19.** The length is 8 ft. The width is 5 ft. [16.5A, You Try It 1] **20.** The rate of the boat in calm water is 11 mph. [16.5A, How To 1]

CUMULATIVE REVIEW EXERCISES

1. $-28x + 27$ [4.2D] **2.** $\frac{3}{2}$ [5.1C] **3.** 3 [5.3B] **4.** $-12a^8b^4$ [9.2B] **5.** $x + 2 - \frac{4}{x - 2}$ [9.5B] **6.** $x(3x - 4)(x + 2)$ [10.3A/10.3B] **7.** $\frac{9x^2(x - 2)^2}{(2x - 3)^2}$ [11.1C] **8.** $\frac{x + 2}{2(x + 1)}$ [11.3B] **9.** $\frac{x - 4}{2x + 5}$ [11.4A] **10.** x-intercept: $(3, 0)$; y-intercept: $(0, -4)$ [12.3A] **11.** $y = -\frac{4}{3}x - 2$ [12.4A] **12.** $(2, 1)$ [13.2A] **13.** $(2, -2)$ [13.3A] **14.** $x > \frac{1}{9}$ [14.3A] **15.** $a - 2$ [15.3A] **16.** $6ab\sqrt{a}$ [15.3B] **17.** $\frac{-6 + 5\sqrt{3}}{13}$ [15.3B] **18.** 5 [15.4A] **19.** $\frac{1}{3}, \frac{5}{2}$ [16.1A] **20.** $5 - 3\sqrt{2}, 5 + 3\sqrt{2}$ [16.1B] **21.** $\frac{-7 - \sqrt{13}}{6}, \frac{-7 + \sqrt{13}}{6}$ [16.2A] **22.** $-\frac{1}{2}, 2$ [16.3A] **23.** The cost of the mixture is \$2.90 per pound. [5.5A] **24.** 250 additional shares are required. [6.2B] **25.** The rate of the plane in still air is 200 mph. The rate of the wind is 40 mph. [13.4A] **26.** The score on the last test must be 77 or better. [14.2C] **27.** The middle integer can be -5 or 5. [16.5A] **28.** The rate for the last 8 mi is 4 mph. [16.5A]

29. 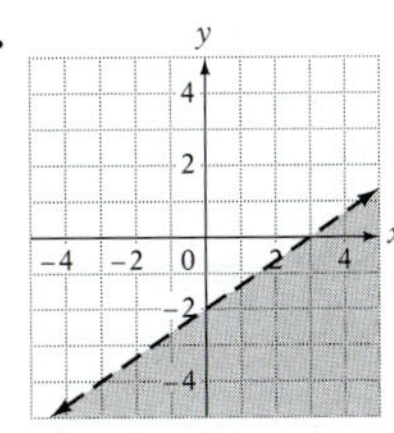 [14.4A]

30. 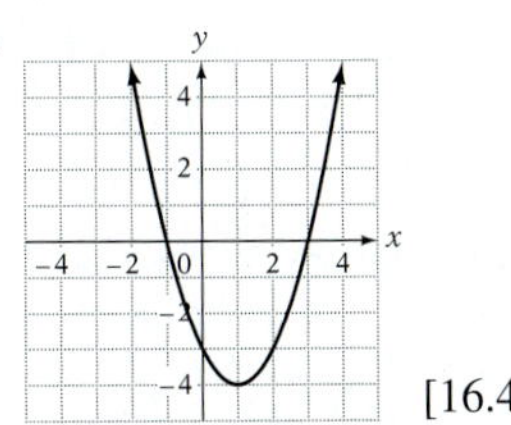 [16.4A]

FINAL EXAM

1. -3 [3.1C] **2.** -6 [3.2B] **3.** $-\frac{1}{2}$ [3.4A] **4.** -11 [3.5A] **5.** $-\frac{15}{2}$ [4.1A] **6.** $9x + 6y$ [4.2A] **7.** $6z$ [4.2B] **8.** $16x - 52$ [4.2D] **9.** -50 [5.1C] **10.** -3 [5.3B] **11.** 12.5% [6.3B] **12.** 15.2 [6.4A/6.4B] **13.** $-3x^2 - 3x + 8$ [9.1B] **14.** $81x^4y^{12}$ [9.2B] **15.** $6x^3 + 7x^2 - 7x - 6$ [9.3B] **16.** $-\frac{x^4y}{2}$ [9.4A] **17.** $\frac{3x}{y} - 4x^2 - \frac{5}{x}$ [9.5A] **18.** $5x - 12 + \frac{23}{x + 2}$ [9.5B] **19.** $\frac{4y^6}{x^6}$ [9.4A] **20.** $\frac{3}{4}$ [12.1D] **21.** $(x - 6)(x + 1)$ [10.2A] **22.** $(3x + 2)(2x - 3)$ [10.3A/10.3B] **23.** $4x(2x - 1)(x - 3)$ [10.4B] **24.** $(5x + 4)(5x - 4)$ [10.4A] **25.** $2(a + 3)(4 - x)$ [10.1B] **26.** $3y(5 + 2x)(5 - 2x)$ [10.4B] **27.** $\frac{1}{2}, 3$ [10.5A] **28.** $\frac{2(x + 1)}{x - 1}$ [11.1B] **29.** $\frac{-3x^2 + x - 25}{(x + 3)(2x - 5)}$ [11.3B] **30.** $\frac{x^2 - 2x}{x - 1}$ [11.4A] **31.** 2 [11.5A] **32.** $a = b$ [11.6A] **33.** $\frac{2}{3}$ [12.3B] **34.** $y = -\frac{2}{3}x - 2$ [12.4A] **35.** (6, 17) [13.2A] **36.** (2, −1) [13.3A] **37.** $x \le -3$ [14.2B] **38.** $y \ge \frac{5}{2}$ [14.3A] **39.** $7x^3$ [15.1B] **40.** $38\sqrt{3a}$ [15.2A] **41.** $\sqrt{15} + 2\sqrt{3}$ [15.3B] **42.** 2 [15.4A] **43.** $-1, \frac{4}{3}$ [16.1A] **44.** $\frac{1 - \sqrt{5}}{4}, \frac{1 + \sqrt{5}}{4}$ [16.3A] **45.** $2x + 3(x - 2)$; $5x - 6$ [4.3B] **46.** The original value was \$3000. [6.4C] **47.** The mean is 2.025 in., the median is 1 in., and the mode is 0 in. [8.2A] **48.** The amount invested in the 4.8% account was \$4000. [13.2B] **49.** The cost for the mixture is \$4 per pound. [5.5A] **50.** The probability is $\frac{2}{3}$. [8.3A] **51.** The distance traveled in the first hour was 215 km. [5.5B] **52.** The angles are 50°, 60°, and 70°. [7.1C] **53.** The middle integer can be −4 or 4. [16.5A] **54.** The width is 5 m. The length is 10 m. [10.5B] **55.** 16 oz of dye are required. [6.2B] **56.** Working together, it would take them 36 min or 0.6 h. [11.7A] **57.** The rate of the boat in calm water is 15 mph. The rate of the current is 5 mph. [13.4A] **58.** The rate of the wind is 25 mph. [16.5A]

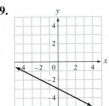

59. [12.3C]

60. 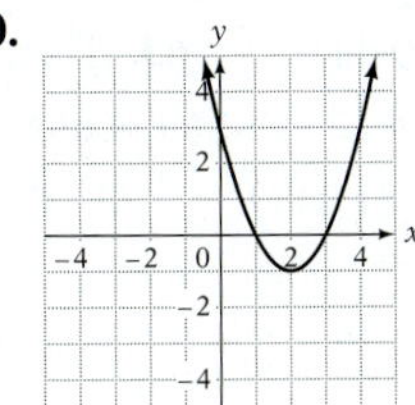 [16.4A]

ANSWERS TO CHAPTER R SELECTED EXERCISES

SECTION R.1

1. $\frac{3}{7}$ **3.** 1 **5.** $1\frac{3}{8}$ **7.** $1\frac{1}{6}$ **9.** $\frac{13}{14}$ **11.** $\frac{53}{60}$ **13.** $\frac{23}{60}$ **15.** $1\frac{17}{18}$ **17.** $1\frac{9}{20}$ **19.** $\frac{1}{3}$ **21.** $\frac{5}{8}$ **23.** $\frac{13}{45}$ **25.** $\frac{1}{10}$ **27.** $\frac{1}{4}$ **29.** $\frac{1}{2}$ **31.** $\frac{11}{60}$ **33.** $\frac{1}{12}$ **35.** $\frac{5}{24}$ **37.** $\frac{7}{12}$ **39.** $\frac{7}{16}$ **41.** $\frac{1}{3}$ **43.** $\frac{6}{11}$ **45.** 6 **47.** $\frac{5}{32}$ **49.** $\frac{3}{16}$ **51.** $\frac{2}{9}$ **53.** $\frac{5}{54}$ **55.** 4 **57.** $\frac{2}{7}$ **59.** 0 **61.** $\frac{2}{9}$ **63.** $\frac{7}{10}$ **65.** 2 **67.** $\frac{2}{3}$ **69.** 6 **71.** $\frac{1}{2}$

SECTION R.2

1. −5 **3.** −17 **5.** 3 **7.** −20 **9.** 0 **11.** 21 **13.** −7 **15.** −11 **17.** 8 **19.** 0 **21.** −25 **23.** −9 **25.** 22 **27.** −6 **29.** 16 **31.** −32 **33.** −15 **35.** −14 **37.** 11 **39.** 16 **41.** 13 **43.** 5

45. -9 **47.** -21 **49.** 35 **51.** -36 **53.** -50 **55.** 49 **57.** 0 **59.** 60 **61.** -126 **63.** 81 **65.** -192 **67.** -85 **69.** 98 **71.** -80 **73.** -144 **75.** -3 **77.** 9 **79.** 0 **81.** -11 **83.** -9 **85.** 22 **87.** 14 **89.** Undefined **91.** -21 **93.** -40

SECTION R.3

1. $-\frac{1}{12}$ **3.** $-\frac{1}{3}$ **5.** $-\frac{7}{8}$ **7.** $-1\frac{1}{10}$ **9.** $-\frac{1}{24}$ **11.** $-\frac{3}{4}$ **13.** $\frac{11}{24}$ **15.** $\frac{1}{12}$ **17.** -50.7 **19.** -3.312 **21.** 9.39 **23.** -16.35 **25.** -9.55 **27.** -19.189 **29.** $\frac{11}{12} - \left(-\frac{1}{4}\right)$, because $\frac{11}{12} - \left(-\frac{1}{4}\right) = \frac{11}{12} + \frac{1}{4}$, so the difference is positive, whereas the difference $-\frac{1}{8} - \frac{3}{4}$ is negative. **31.** $-\frac{1}{4}$ **33.** $\frac{1}{10}$ **35.** $-\frac{3}{8}$ **37.** $-\frac{7}{30}$ **39.** $\frac{3}{16}$ **41.** $-7\frac{1}{2}$ **43.** $-4\frac{1}{5}$ **45.** -10 **47.** $-\frac{3}{7}$ **49.** $\frac{9}{10}$ **51.** $-\frac{1}{12}$ **53.** $-\frac{9}{10}$ **55.** 1.47 **57.** -7.03 **59.** 60.75 **61.** -5.9 **63.** -67.7 **65.** 4.14 **67.** -6.1 **69.** -128.8 **71.** 37.0 **73.** $-\frac{5}{6} \div (-5)$, because the quotient is positive, whereas the product $-\frac{3}{4}\left(\frac{2}{9}\right)$ is negative. **75.** $-\frac{8}{343}$ **77.** 12.25 **79.** $\frac{1}{121}$ **81.** $-\frac{1}{75}$ **83.** 0.51 **85.** 1.7 **87.** $\frac{9}{32}$ **89.** $\frac{8}{9}$ **91.** 0 **93.** $\frac{1}{8}$

95. $16 + 15 \div (-5) - (-4) > 18 \div |2^3 - 9| + (-3) > 20 \div (6 - 2^4) + (-5)$; $[17 > 15 > -7]$

SECTION R.4

1. -12 **3.** -1 **5.** -8 **7.** -20 **9.** 8 **11.** 3 **13.** 5 **15.** -3 **17.** 2 **19.** -1 **21.** 1 **23.** 3 **25.** 3 **27.** 2 **29.** -1 **31.** $\frac{1}{2}$ **33.** 5 **35.** $-\frac{1}{9}$

ANSWERS TO APPENDIX SELECTED EXERCISES

1. Meter, liter, gram **3.** Kilometer **5.** Centimeter **7.** Gram **9.** Meter **11.** Gram **13.** Milliliter **15.** Milligram **17.** Millimeter **19.** Milligram **21.** Gram **23.** Kiloliter **25.** Milliliter **27.** 420 mm **29.** 0.360 kg **31.** 5.194 L **33.** 2 000 mm **35.** 0.217 g **37.** 4 520 ml **39.** 8.406 km **41.** 2 400 g **43.** 6 180 L **45.** 9 612 m **47.** 240 mg **49.** 2.98 m **51.** 2.431 kl **53.** 66 cm **55.** 24.3 cm

57. a. Column 2: k, c, m; column 3: 10^9, 10^3, 10^2, $\frac{1}{10^3}$, $\frac{1}{10^{12}}$; column 4: 1 000 000, 10, 0.1, 0.01, 0.000 001, 0.000 000 001

59. The weight is 2 g. **61.** 24 L of chlorine are used in a month of 30 days. **63.** There are 16 servings in the container. **65.** The patient should take 4 tablets per day. **67.** The case containing 12 one-liter bottles costs less per milliliter. **69.** It takes light 500 s to reach Earth from the sun. **71.** The profit is $17,430. **73.** The profit is $66.50.

Glossary

abscissa The first number in an ordered pair. It measures a horizontal distance and is also called the first coordinate. [12.1]

absolute value of a number The distance of a number from zero on the number line. [3.1]

acute angle An angle whose measure is between 0° and 90°. [7.1]

acute triangle A triangle that has three acute angles. [7.2]

addend In addition, one of the numbers added. [1.2]

addition The process of finding the total of two numbers. [1.2]

addition method An algebraic method of finding an exact solution of a system of linear equations, in which the equations in the system are added. [13.3]

additive inverses Numbers that are the same distance from zero on the number line, but on opposite sides; also called opposites. [3.2]

adjacent angles Two angles that share a common side. [7.1]

alternate exterior angles Two nonadjacent angles that are on opposite sides of the transversal and outside the parallel lines. [7.1]

alternate interior angles Two non-adjacent angles that are on opposite sides of the transversal and between the parallel lines. [7.1]

analytic geometry Geometry in which a coordinate system is used to study the relationships between variables. [12.1]

angle The figure formed when two rays start at the same point; it is measured in degrees. [1.2]

area A measure of the amount of surface in a region. [7.2]

axes The two number lines that form a rectangular coordinate system; also called coordinate axes. [12.1]

bar graph A graph that represents data by the height of the bars. [1.1]

base In exponential notation, the factor that is multiplied the number of times shown by the exponent. [1.3]

base of a triangle The side of a triangle that the triangle rests on. [2.4]

basic percent equation The equation that states that percent times base equals amount. [6.4]

binomial A polynomial of two terms. [9.1]

binomial factor A factor that has two terms. [10.1]

borrowing In subtraction, taking a unit from the next larger place value in the minuend and adding it to the number in the given place value in order to make that number larger than the number to be subtracted from it. [1.2]

box-and-whiskers plot A graph that shows the smallest value in a set of numbers, the first quartile, the median, the third quartile, and the greatest value. [8.2]

broken-line graph A graph that represents data by the positions of the lines and shows trends and comparisons. [1.1]

center of a circle The point from which all points on the circle are equidistant. [7.2]

center of a sphere The point from which all points on the surface of the sphere are equidistant. [7.4]

circle A plane figure in which all points are the same distance from point O, which is the figure's center. [7.2]

circle graph A graph that represents data by the size of the sectors. [1.1]

circumference The distance around a circle. [7.2]

clearing denominators Removing denominators from an equation that contains fractions by multiplying each side of the equation by the LCM of the denominators. [5.2]

coefficient The number part of a variable term; for example, the 2 in the variable expression $2x$. [4.1]

combining like terms Adding like terms of a variable expression. [4.2]

common factor A number that is a factor of two or more numbers. [2.1]

common multiple A number that is a multiple of two or more numbers. [2.1]

complementary angles Two angles whose sum is 90°. [7.1]

completing the square Adding to a binomial the constant term that makes it a perfect-square trinomial. [16.2]

complex fraction A fraction whose numerator or denominator contains one or more fractions. [2.4]

composite number A number that has whole number factors besides 1 and itself. For instance, 10 has whole number factors of 2 and 5. [1.3]

congruent objects Objects that have the same shape and the same size. [7.3]

congruent triangles Triangles that have the same shape and the same size. [7.3]

conjugates Binomial expressions that differ only in the sign of a term; for example, the expressions $a + b$ and $a - b$. [15.3]

consecutive even integers Even integers that follow one another in order. [5.4]

consecutive integers Integers that follow one another in order. [5.4]

consecutive odd integers Odd integers that follow one another in order. [5.4]

constant term A term that includes no variable part. [4.1]

coordinate axes The two number lines that form a rectangular coordinate system. [12.1]

coordinates of a point The numbers in an ordered pair that is associated with a point. [12.1]

corresponding angles Two angles that are on the same side of the transversal and are both acute angles or are both obtuse angles. [7.1]

cross product In a proportion, either the product of the numerator on the left side of the proportion times the denominator on the right, or the product of the denominator on the left side of the proportion times the numerator on the right. [6.2]

cube A rectangular solid in which all six faces are squares. [7.4]

data Numerical information. [8.1]

decimal A number written in decimal notation. [2.5]

decimal notation Notation in which a number is written with a whole number part, a decimal point, and a decimal part. [2.5]

decimal part In decimal notation, that part of the number that appears to the right of the decimal point. [2.5]

decimal point In decimal notation, the point that separates the whole number part of a number from the decimal part. [2.5]

degree A unit used to measure angles. [1.2]

degree of a polynomial in one variable For a polynomial in one variable, the largest exponent that appears on a variable in the expression. [9.1]

denominator The part of a fraction that appears below the fraction bar. [2.2]

dependent system A system of equations that has an infinite number of solutions. [13.1]

dependent variable In a function, the variable whose value depends on the value of another variable, known as the independent variable. [12.1]

descending order A polynomial in one variable arranged so that the exponents on the variable terms decrease from left to right. For example, the polynomial $9x^5 - 2x^4 + 7x^3 + x^2 - 8x + 1$. [9.1]

diameter of a circle A line segment with endpoints on the circle and passing through the center. [7.2]

diameter of a sphere A line segment with endpoints on the sphere and passing through the center. [7.4]

difference In subtraction, the result of subtracting two numbers. [1.2]

difference of two squares A polynomial of the form $a^2 - b^2$. [10.4]

dividend In division, the number into which the divisor is divided to yield the quotient. [1.3]

division The process of finding the quotient of two numbers. [1.3]

divisor In division, the number that is divided into the dividend to yield the quotient. [1.3]

domain The set of first coordinates of the ordered pairs in a relation. [12.1]

double-bar graph A graph used to display data for purposes of comparison. [1.1]

double root Two equal roots of a quadratic equation. [16.1]

element of a set One of the objects in a set. [14.1]

empirical probability Probability expressed as the ratio of the number of observations of an event to the total number of observations. [8.3]

empty set The set that contains no elements; also called the null set. [14.1]

equation A statement of the equality of two mathematical expressions. [1.2]

equilateral triangle A triangle that has three sides of equal length; the three angles are also of equal measure. [7.2]

equivalent equations Equations that have the same solution. [5.1]

equivalent fractions Equal fractions with different denominators; for example, $\frac{2}{3}$ and $\frac{4}{6}$. [2.2]

evaluating a function Replacing x in $f(x)$ with some value and then simplifying the numerical expression that results. [12.1]

evaluating a variable expression Replacing the variable or variables in an expression with numbers and then simplifying the resulting numerical expression. [1.2]

even integer An integer that is divisible by 2. [5.4]

event One or more outcomes of an experiment. [8.3]

expanded form The form of the number 46,208 when written as $40{,}000 + 6000 + 200 + 0 + 8$. [1.1]

experiment Any activity that has an observable outcome. [8.3]

exponent In exponential notation, the raised number that indicates how many times the base is taken as a factor. [1.3]

exponential form The expression of a number to a power, indicated by an exponent. [1.3]

exterior angle of a triangle The angle adjacent to an interior angle of a triangle. [7.1]

factor by grouping The process of grouping and factoring terms of a polynomial in such a way that a common binomial factor is found. [10.1]

factor completely The process of writing a polynomial as a product of factors that are nonfactorable over the integers. [10.2]

factor of a number In multiplication, a number being multiplied. [1.3]

factor a polynomial The process of writing the polynomial as a product of other polynomials. [10.1]

factor a trinomial of the form $x^2 + bx + c$ To express the trinomial as the product of two binomials. [10.2]

factored form The expression of a number as a product of its factors; for example, $2 \cdot 2 \cdot 2 \cdot 2 \cdot 2$. [1.3]

favorable outcomes The outcomes of an experiment that satisfy the requirements of a particular event. [8.3]

first coordinate The first number in an ordered pair. It measures a horizontal distance and is also called the abscissa. [12.1]

first-degree equation in two variables An equation of the form $y = mx + b$, where m is the coefficient and b is a constant; also called a linear equation in two variables or a linear function. [12.2]

first quartile In a set of numbers, the number below which one-quarter of the data lie. [8.2]

FOIL A method of finding the product of two binomials; it ensures that each term of one binomial is multiplied by each term of the other binomial. [9.3]

formula An equation that expresses a relationship among variables; for example, $A = LW$. [11.6]

fraction The notation used to represent the number of equal parts of a whole. [2.2]

fraction bar The horizontal line that separates the numerator of a fraction from the denominator. [2.2]

function A relation in which no two ordered pairs that have the same first coordinate have different second coordinates. [12.1]

function notation The notation $f(x)$, which is used to designate a function and represents the value of the function at x. [12.1]

geometric solid A figure in space. [7.4]

graph of an equation in two variables A graph of the ordered-pair solutions of an equation in two variables. [12.2]

graph of an ordered pair The dot drawn at the coordinates of the point in the plane. [12.1]

graph a point in the plane To place a dot at the location given by the ordered pair; also called plotting a point. [12.1]

graph of a relation The graph of the ordered pairs that belong to the relation. [12.1]

graph of a whole number A heavy dot placed directly above a number on the number line. [1.1]

greater than The meaning of the symbol $>$. [1.1]

greatest common factor (GCF) The largest common factor of two or more numbers. [2.1]

greatest common factor (GCF) of two or more monomials The product of the GCF of the coefficients and the common variable factors of the monomials. [10.1]

half-plane The solution set of an inequality in two variables. [14.4]

height of a parallelogram The distance between parallel sides of a parallelogram. [7.2]

height of a triangle In a triangle, a line segment perpendicular to the base from the opposite vertex. [2.4]

hypotenuse The side opposite the right angle in a right triangle. [7.3]

improper fraction A fraction in which the numerator is greater than or equal to the denominator. [2.2]

inconsistent system A system of equations that has no solution. [13.1]

independent system A system of equations that has one solution. [13.1]

independent variable In a function, the variable that varies independently and whose value determines the value of the dependent variable. [12.1]

inequality An expression that contains the symbol $>$, $<$, $\geq$ (is greater than or equal to), or $\leq$ (is less than or equal to). [1.1]

integers The numbers . . . , -3, -2, -1, 0, 1, 2, 3, ... [3.1]

interest Money paid for the privilege of using someone else's money. [6.5]

interest rate The percent used to determine the amount of interest to be paid. [6.5]

interior angle of a triangle An angle within the region enclosed by a triangle. [7.1]

interquartile range The difference between the third quartile and the first quartile. [8.2]

intersecting lines Lines that cross at a point in the plane. [1.2]

intersection of sets *A* and *B* The set that contains the elements that are common to both A and B. [14.1]

interval notation A type of set notation in which the property that distinguishes the elements of the set is their location within a specified interval. [1.2]

inverting a fraction The process of interchanging the numerator and denominator of a fraction. [2.4]

irrational number A number whose decimal representation never repeats or terminates. [3.4]

isosceles triangle A triangle that has two sides of equal length; the angles opposite the equal sides are of equal measure. [7.2]

least common denominator (LCD) The least common multiple of the denominators of two or more fractions. [2.2]

least common multiple (LCM) The smallest common multiple of two or more numbers. [2.1]

least common multiple (LCM) of two or more polynomials The polynomial of least degree that contains all the factors of each polynomial. [11.2]

legs of a right triangle The two shortest sides of a right triangle. [7.3]

less than The meaning of the symbol $<$. [1.1]

like terms Terms of a variable expression that have the same variable part. [4.2]

line A geometric figure that extends indefinitely in two directions in a plane; it has no width. [1.2]

line of best fit A line drawn to approximate data that are graphed as points in a coordinate system. [12.4]

line segment Part of a line; it has two endpoints. [1.2]

linear equation in two variables An equation of the form $y = mx + b$, where m and b are constants; also called a linear function or a first-degree equation in two variables. [12.2]

linear function A function that can be expressed in the form $f(x) = mx + b$. [12.2]

linear model A first-degree equation that is used to describe a relationship between quantities. [12.4]

literal equation An equation that contains more than one variable. [11.6]

maturity value of a loan The principal of a loan plus the interest owed on it. [6.5]

mean The sum of a set of values divided by the number of those values; also known as the average value. [8.2]

median The average that separates a list of values in such a way that the number of values below it is the same as the number of values above it. [8.2]

minuend In subtraction, the number from which another number (the subtrahend) is subtracted. [1.2]

mixed number A number greater than 1 that has a whole number part and a fractional part. [2.2]

mode In a set of numbers, the value that occurs most frequently. [8.2]

monomial A number, a variable, or a product of a number and variables; a polynomial of one term. [9.1]

multiples of a number The products of a number and the numbers 1, 2, 3, 4, ... [2.1]

multiplication The process of finding the product of two numbers. [1.3]

multiplicative inverse The reciprocal of a nonzero number. [2.4]

natural numbers The numbers 1, 2, 3, . . . ; also called the positive integers. [1.1]

negative integers The numbers . . . , -4, -3, -2, -1. [3.1]

negative numbers The numbers less than zero. [3.1]

negative slope A property of a line that slants downward to the right. [12.3]

nonfactorable over the integers A polynomial that does not factor using only integers. [10.2]

null set The set that contains no elements; also called the empty set. [14.1]

number line A line on which points are marked off at regular, evenly spaced intervals and are labeled with ordered numbers. [1.1]

numerator The part of a fraction that appears above the fraction bar. [2.2]

numerical coefficient The number part of a variable term; for example, the 2 in the variable expression $2x$. [4.1]

obtuse angle An angle whose measure is between 90° and 180°. [7.1]

obtuse triangle A triangle in which one angle measures more than 90°. [7.2]

odd integer An integer that is not divisible by 2. [5.4]

opposite numbers Two numbers that are the same distance from zero on the number line, but on opposite sides. [3.1]

opposite of a polynomial The polynomial created when the sign of each term of the original polynomial is changed. [9.1]

Order of Operations Agreement A set of rules that tells us in what order to perform the operations that occur in a numerical expression. [1.4]

ordered pair A pair of numbers, such as (a, b), that can be used to identify a point in the plane determined by the axes of a rectangular coordinate system. [12.1]

ordinate The second number in an ordered pair. It measures a vertical distance and is also called the second coordinate. [12.1]

origin The point of intersection of the two coordinate axes that form a rectangular coordinate system. [12.1]

parabola The graph of a quadratic equation in two variables. [16.4]

parallel lines Lines that never meet; the distance between them is always the same. [1.2/7.3]

parallelogram A quadrilateral that has equal and parallel opposite sides. [7.2]

percent The word used to mean "parts per hundred." [6.3]

perfect square The square of an integer. [15.1]

perfect-square trinomial A trinomial that is a product of a binomial and itself. [10.4]

perimeter The distance around a plane figure. [1.2]

period In a number written in standard form, each group of digits separated from other digits by a comma or commas. [1.1]

perpendicular lines Intersecting lines that form right angles. [7.1]

pictograph A graph in which the data are displayed using pictures or symbols. [1.1]

place value The value associated with the position of a digit in a number; it indicates the value of the digit. [1.1]

place-value chart A chart that indicates the place value of every digit in a number. [1.1]

plane A flat surface that extends forever in all directions. [1.2]

plane figure A figure that lies entirely in a plane. [1.2]

plot a point in the plane To place a dot at the location given by an ordered pair; also called graphing a point. [12.1]

point-slope formula The formula that states that if (x_1, y_1) is a point on a line with slope m, then $y - y_1 = m(x - x_1)$. [12.4]

polygon A closed figure determined by three or more line segments that lie in a plane. [1.2]

polynomial A variable expression in which the terms are monomials. [9.1]

positive integers The numbers 1, 2, 3, 4, . . . ; also called the natural numbers. [3.1]

positive numbers The numbers greater than zero. [3.1]

positive slope A property of a line that slants upward to the right. [12.3]

prime factorization The expression of a number as the product of numbers whose only whole number factors are 1 and themselves. [1.3]

prime number A positive number other than 1, such as 5 or 13, whose only whole number factors are 1 and itself. [1.3]

prime polynomial A polynomial that is nonfactorable over the integers. [10.2]

principal The amount of money originally deposited or borrowed. [6.5]

principal square root The positive square root of a number. [15.1]

probability A number from 0 to 1 that tells us how likely it is that a certain outcome of an experiment will happen. [8.3]

product In multiplication, the result of multiplying two numbers. [1.3]

proper fraction A fraction in which the numerator is less than the denominator. [2.2]

proportion An equation that states the equality of two ratios or rates. [6.2]

Pythagorean Theorem The theorem that states that the square of the hypotenuse of a right triangle is equal to the sum of the squares of the two legs. [15.4]

quadrant One of the four regions into which the two axes of a rectangular coordinate system divide a plane. [12.1]

quadratic equation An equation of the form $ax^2 + bx + c = 0$, where a, b, and c are constants and a is not equal to zero; also called a second-degree equation. [10.5/11.1]

quadratic equation in two variables An equation of the form $y = ax^2 + bx + c$, where a is not equal to zero. [16.4]

quadratic function A function of the form $f(x) = ax^2 + bx + c$, where a is not equal to zero. [16.4]

quadrilateral A four-sided closed figure. [1.2]

quotient In division, the result of dividing the divisor into the dividend. [1.3]

radical equation An equation that contains a variable expression in a radicand. [15.4]

radical sign The symbol $\sqrt{}$, which is used to indicate the positive, or principal, square root of a number. [15.1]

radicand In a radical expression, the expression under the radical sign. [15.1]

radius of a circle A line segment going from the center of a circle to a point on the circle. [7.2]

radius of a sphere A line segment going from the center of a sphere to a point on the sphere. [7.4]

range of a relation The set of second coordinates of the ordered pairs in a relation. [12.1]

range of a set of data In a set of numbers, the difference between the largest and smallest values. [8.2]

rate The quotient of two quantities that have different units. [6.1]

rate of work That part of a task that is completed in one unit of time. [11.7]

ratio The quotient of two quantities that have the same unit. [6.1]

rational expression A fraction in which the numerator and denominator are polynomials. [11.1]

rational number A number that can be written in the form $\frac{a}{b}$, where a and b are integers and b is not equal to zero. [3.4]

rationalizing the denominator The procedure used to remove a radical from the denominator of a fraction. [15.3]

ray A geometric figure that starts at a point and extends indefinitely in one direction. [1.2]

real numbers The rational numbers and the irrational numbers taken together. [3.4]

reciprocal of a fraction The word used to describe a fraction with the numerator and denominator interchanged. [2.4]

reciprocal of a rational expression A rational expression in which the numerator and denominator have been interchanged. [11.1]

rectangle A parallelogram that has four right angles. [1.2]

rectangular coordinate system A system formed by two number lines, one horizontal and one vertical, that intersect at the zero point of each line. [12.1]

rectangular solid A solid in which all six faces are rectangles. [7.4]

regular polygon A polygon in which each side has the same length and each angle has the same measure. [7.2]

relation Any set of ordered pairs. [12.1]

remainder In division, the quantity left over when it is not possible to separate objects or numbers into a whole number of equal groups. [1.3]

repeating decimal A decimal in which a block of one or more digits repeats forever. [2.6]

right angle A 90° angle. [1.2]

right triangle A triangle that contains one right angle. [7.2]

roster method The method of writing a set by enclosing a list of the elements of the set in braces. [14.1]

rounding Giving an approximate value of an exact number. [1.1]

sample space All possible outcomes of an experiment. [8.3]

scalene triangle A triangle that has no sides of equal length; no two of its angles are of equal measure. [7.2]

scatter diagram A graph of collected data as points in a coordinate system. [12.4]

scientific notation A notation in which a number is expressed as the product of two factors, one a number between 1 and 10 and the other a power of 10. [9.4]

second coordinate The second number in an ordered pair. It measures a vertical distance and is also called the ordinate. [12.1]

second-degree equation An equation of the form $ax^2 + bx + c = 0$, where a, b, and c are constants and a is not equal to zero; also called a quadratic equation. [16.1]

set A collection of objects. [14.1]

set-builder notation A method of designating a set that makes use of a variable and a certain property that only elements of that set possess. [14.1]

sides of a polygon The line segments that form the polygon. [1.2]

similar objects Objects that have the same shape but not necessarily the same size. [7.3]

similar triangles Triangles that have the same shape but not necessarily the same size. [7.3]

simple interest Interest computed on the original principal. [6.5]

simplest form of a fraction The form of a fraction in which the numerator and denominator contain no common factors other than 1. [2.2]

simplest form of a rate A rate is in simplest form when the numbers that make up the rate have no common factor. [6.1]

simplest form of a ratio A ratio is in simplest form when the two numbers do not have a common factor. [6.1]

simplest form of a rational expression The form of a rational expression in which the numerator and denominator have no common factors other than 1. [11.1]

simplifying a variable expression Combining like terms of an expression by adding their numerical coefficients. [4.2]

slope The measure of the slant of a line, symbolized by m. [12.3]

slope-intercept form The form of an equation of a straight line written as $y = mx + b$. [12.3]

solid An object that exists in space. [7.4]

solution of an equation A number that, when substituted for the variable in an equation, results in a true equation. [1.2]

solution of an equation in two variables An ordered pair whose coordinates make an equation in two variables a true statement. [12.1]

solution set of an inequality A set of numbers each element of which, when substituted for the variable in a variable inequality, results in a true inequality. [14.2]

solution of a system of equations in two variables An ordered pair that is a solution of each equation in a system of equations. [13.1]

solving an equation Finding a solution of the equation. [5.1]

sphere A solid in which all points are the same distance from point O, which is the sphere's center. [7.4]

square A rectangle that has four equal sides. [7.2]

square root One of two identical factors of a number; for example, 3 is one of two identical factors of 9. [15.1]

standard form of a linear equation in two variables The form of an equation in two variables when it is written as $Ax + By = C$, where A and B are coefficients and C is a constant. [12.2]

standard form of a number The form of a number when it is written using the digits 0, 1, 2, . . . , 9. An example is 46,208. [1.1]

standard form of a quadratic equation The form of a quadratic equation when it is written with the polynomial in descending order and equal to zero. [10.5/11.1]

statistics The branch of mathematics concerned with data, or numerical information. [8.1]

straight angle An angle whose measure is 180°. [7.1]

substitution method An algebraic method of finding an exact solution of a system of equations, in which one variable is expressed in terms of another variable. [13.2]

subtraction The process of finding the difference between two numbers. [1.2]

subtrahend In subtraction, the number that is subtracted from another number (the minuend). [1.2]

sum In addition, the total of the numbers being added. [1.2]

supplementary angles Two angles whose sum is 180°. [7.1]

system of equations Two or more equations considered together. [13.1]

terminating decimal A decimal that has a finite number of digits after the decimal point, which means that it comes to an end and does not go on forever. [2.6]

terms of a variable expression The addends of a variable expression. [4.1]

theoretical probability A fraction consisting of the number of favorable outcomes of an experiment in the numerator and the total number of possible outcomes of the experiment in the denominator. [8.3]

third quartile In a set of numbers, the number above which one-quarter of the data lie. [8.2]

transversal A line intersecting two other lines at two different points. [7.1]

triangle A three-sided closed figure. [1.2]

trinomial A polynomial of three terms. [9.1]

undefined slope The slope of a vertical line. [12.3]

uniform motion The motion of a moving object whose speed and direction do not change. [5.1]

union of sets A and B The set that contains all the elements of A and all the elements of B. [14.1]

unit rate A rate in which the number in the denominator is 1. [6.1]

value of a function at x The result of evaluating a variable expression, represented by the symbol $f(x)$. [12.1]

value of a variable The number assigned to a variable. [4.1]

variable A letter of the alphabet used to represent a number that is unknown or that can change. [1.2]

variable expression An expression that contains one or more variables. [1.2]

variable part In a variable term, the variable or variables and their exponents. [4.1]

variable term A term composed of a numerical coefficient and a variable part. [4.1]

vertex The point at which the rays of an angle meet. [7.1]

vertical angles Two angles that are on opposite sides of the intersection of two lines. [7.1]

volume A measure of the amount of space inside a closed surface. [7.4]

whole numbers The numbers 0, 1, 2, 3, 4, ... [1.1]

whole number part In decimal notation, that part of the number that appears to the left of the decimal point. [2.5]

x-coordinate The abscissa, or first coordinate, of an ordered pair in an xy-coordinate system. [12.1]

x-intercept The point at which a graph crosses the x-axis. [12.3]

xy-coordinate system A rectangular coordinate system in which the horizontal axis is labeled x and the vertical axis is labeled y. [12.1]

y-coordinate The ordinate, or second coordinate, of an ordered pair in an xy-coordinate system. [12.1]

y-intercept The point at which a graph crosses the y-axis. [12.3]

zero slope The slope of a horizontal line. [12.3]

Index

A

B

Q

R

S

T

Index of Applications

TI-30X IIS

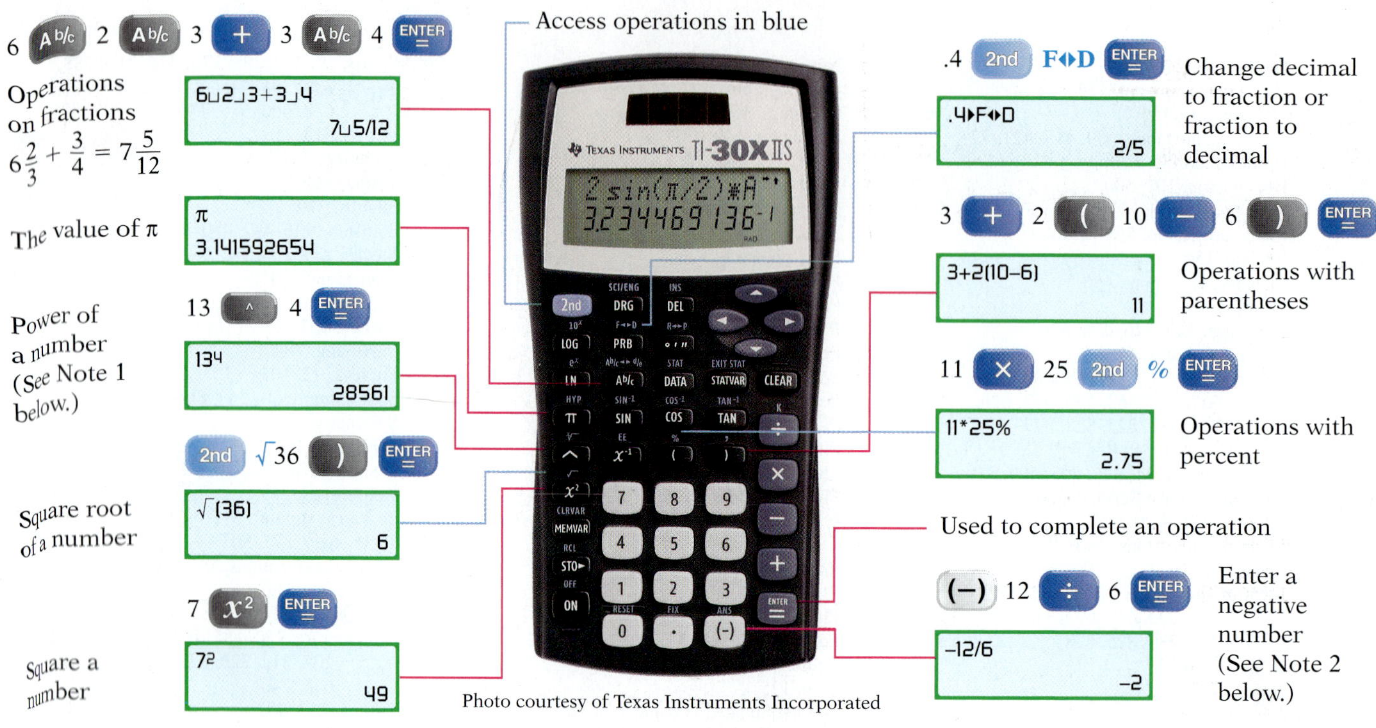

Photo courtesy of Texas Instruments Incorporated

fx-300MS

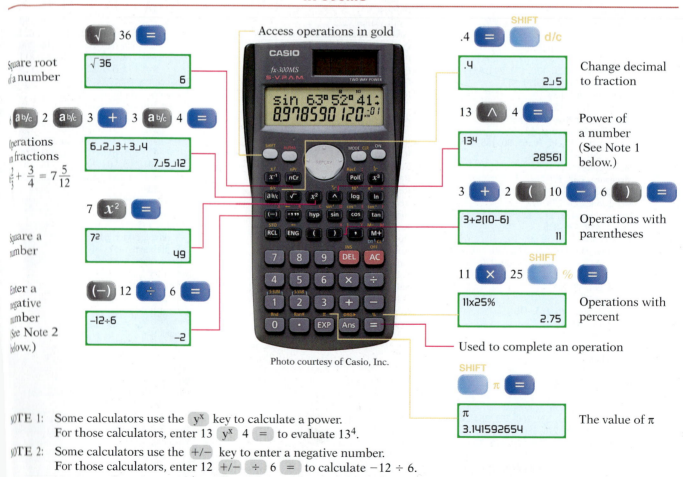

√ 36 =

Square root of a number

√36
6

a b/c 2 a b/c 3 + 3 a b/c 4 =

Operations on fractions $6\frac{2}{3} + \frac{3}{4} = 7\frac{5}{12}$

6⌟2⌟3+3⌟4
7⌟5⌟12

7 x^2 =

Square a number

7²
49

(−) 12 ÷ 6 =

Enter a negative number (See Note 2 below.)

−12÷6
−2

Access operations in gold

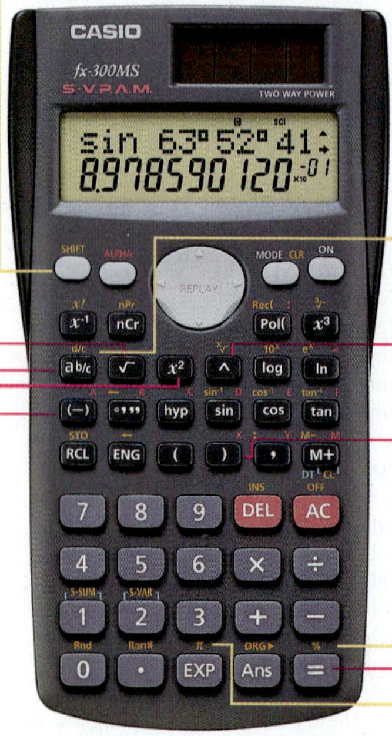

.4 = SHIFT d/c

.4
2⌟5

Change decimal to fraction

13 ∧ 4 =

13^4
28561

Power of a number (See Note 1 below.)

3 + 2 (10 − 6) =

3+2(10−6)
11

Operations with parentheses

11 × 25 SHIFT % =

11x25%
2.75

Operations with percent

Used to complete an operation

SHIFT π =

π
3.141592654

The value of π

Photo courtesy of Casio, Inc.

NOTE 1: Some calculators use the y^x key to calculate a power. For those calculators, enter 13 y^x 4 = to evaluate 13^4.

NOTE 2: Some calculators use the +/− key to enter a negative number. For those calculators, enter 12 +/− ÷ 6 = to calculate −12 ÷ 6.